SATURN IN THE 21ST CENTURY

The Cassini Orbiter mission, launched in 1997, has transformed our understanding of the origins and workings of Saturn. Drawing from new discoveries and scientific insights from the mission, this book provides a detailed overview of the planet as revealed by Cassini. Chapters by eminent planetary scientists and researchers from across the world comprehensively review the current state of knowledge regarding Saturn's formation, interior, atmosphere, ionosphere, thermosphere, and magnetosphere. Specialized chapters discuss the planet's seasonal variability; the circulation of strong zonal winds; the constantly changing polar aurorae; and the Great Storm of 2010–2011, the most powerful convective storm ever witnessed by humankind. Documenting the latest research on the planet, from its formation to how it operates today, this is an essential reference for graduate students, researchers, and planetary scientists.

KEVIN H. BAINES is Senior Scientist at the Space Science and Engineering Center at the University of Wisconsin–Madison, and Principal Scientist at Caltech/Jet Propulsion Laboratory, Pasadena. He has over 35 years of experience in the development, planning, data analysis, and publication of science results from NASA and ESA planetary orbital missions. Specializing in the 3D nature of planetary atmospheres as gleaned from spacecraft-borne visual-to-near-infrared spectral imagers, he has been a NASA-selected scientist on the Cassini–Huygens and Galileo orbiter missions to Jupiter and Saturn and was the leader of the NASA science team on ESA's Venus Express orbiter mission.

F. MICHAEL FLASAR is Space Scientist at the Planetary Systems Laboratory at the NASA Goddard Space Flight Center. He has devoted 45 years to the study of solar system planets and their atmospheres, particularly from thermal-infrared spectroscopy and radio-occultation data. He has been an investigator on the Voyager mission to the giant planets, the Galileo mission to Jupiter, the Mars Global Surveyor mission, and the Cassini–Huygens mission to Saturn. He is a recipient of NASA Goddard Space Flight Center's John C. Lindsay Memorial Award for Space Science and he is a fellow of the American Geophysical Union.

NORBERT KRUPP is Scientist at the Max Planck Institute for Solar System Research, Göttingen, Germany. He has 25 years of experience in data analysis and the development of space instrumentation. His main interest is the understanding of processes driving the global configuration and dynamics of particles around planets, including the interaction with moons, rings, and neutral clouds. He has been involved in several space missions, including Mars Express, Venus Express, Ulysses, Bepi Colombo, Juice, Galileo, Cassini–Huygens, and Europa Clipper. On Cassini, he co-led the magnetosphere and plasma science working group MAPS, and is Co-Investigator of the MIMI instrument.

TOM STALLARD is Associate Professor in Planetary Astronomy at the University of Leicester. He is a world-leading planetary astronomer who has observed the gas giants of our solar system from many of the largest telescopes around the world. Focusing on the investigation of aurorae of these planets, he has also been extensively involved in analyzing spacecraft data, including images of Saturn's aurora taken by the Cassini spacecraft. He has also appeared on numerous television and radio programs to discuss recent science advances. His public outreach has included involvement in BBC Stargazing live events and he was awarded the honorary title of "Hoku Kolea" for his extensive work with the Mauna Kea visitors' center.

Cambridge Planetary Science

Series Editors:
Fran Bagenal, David Jewitt, Carl Murray, Jim Bell, Ralph Lorenz, Francis Nimmo, Sara Russell

Books in the Series

1. Jupiter: The Planet, Satellites and Magnetosphere†
   Edited by Bagenal, Dowling and McKinnon
   978-0-521-03545-3
2. Meteorites: A Petrologic, Chemical and Isotopic Synthesis†
   Hutchison
   978-0-521-03539-2
3. The Origin of Chondrules and Chondrites†
   Sears
   978-1-107-40285-0
4. Planetary Rings†
   Esposito
   978-1-107-40247-8
5. The Geology of Mars: Evidence from Earth-Based Analogs†
   Edited by Chapman
   978-0-521-20659-4
6. The Surface of Mars†
   Carr
   978-0-521-87201-0
7. Volcanism on Io: A Comparison with Earth†
   Davies
   978-0-521-85003-2
8. Mars: An Introduction to its Interior, Surface and Atmosphere†
   Barlow
   978-0-521-85226-5
9. The Martian Surface: Composition, Mineralogy and Physical Properties
   Edited by Bell
   978-0-521-86698-9
10. Planetary Crusts: Their Composition, Origin and Evolution†
    Taylor and McLennan
    978-0-521-14201-4
11. Planetary Tectonics†
    Edited by Watters and Schultz
    978-0-521-74992-3
12. Protoplanetary Dust: Astrophysical and Cosmochemical Perspectives†
    Edited by Apai and Lauretta
    978-0-521-51772-0
13. Planetary Surface Processes
    Melosh
    978-0-521-51418-7

14. Titan: Interior, Surface, Atmosphere and Space Environment
    Edited by Müller-Wodarg, Griffith, Lellouch and Cravens
    978-0-521-19992-6
15. Planetary Rings: A Post-Equinox View (Second Edition)
    Esposito
    978-1-107-02882-1
16. Planetesimals: Early Differentiation and Consequences for Planets
    Edited by Elkins-Tanton and Weiss
    978-1-107-11848-5
17. Asteroids: Astronomical and Geological Bodies
    Burbine
    978-1-107-09684-4
18. The Atmosphere and Climate of Mars
    Edited by Haberle, Clancy, Forget, Smith and Zurek
    978-1-107-01618-7
19. Planetary Ring Systems: Properties, Structure and Evolution
    Edited by Tiscareno and Murray
    978-1-107-11382-4
20. Saturn in the 21st Century
    Edited by Baines, Flasar, Krupp and Stallard
    978-1-107-10677-2

†Reissued as a paperback

SATURN IN THE 21ST CENTURY

Edited by

KEVIN H. BAINES
University of Wisconsin–Madison

F. MICHAEL FLASAR
NASA Goddard Space Flight Center

NORBERT KRUPP
Max Planck Institute for Solar System Research

TOM STALLARD
University of Leicester

CAMBRIDGE
UNIVERSITY PRESS

CAMBRIDGE
UNIVERSITY PRESS

University Printing House, Cambridge CB2 8BS, United Kingdom

One Liberty Plaza, 20th Floor, New York, NY 10006, USA

477 Williamstown Road, Port Melbourne, VIC 3207, Australia

314–321, 3rd Floor, Plot 3, Splendor Forum, Jasola District Centre, New Delhi – 110025, India

79 Anson Road, #06–04/06, Singapore 079906

Cambridge University Press is part of the University of Cambridge.

It furthers the University's mission by disseminating knowledge in the pursuit of education, learning, and research at the highest international levels of excellence.

www.cambridge.org
Information on this title: www.cambridge.org/9781107106772
DOI: 10.1017/9781316227220

First published 2019

Printed in the United Kingdom by TJ International Ltd. Padstow Cornwall

A catalogue record for this publication is available from the British Library.

Library of Congress Cataloging-in-Publication Data
Names: Baines, Kevin Hays, editor. | Flasar, F. Michael, editor. | Krupp, Norbert, editor. | Stallard, Tom, editor.
Title: Saturn in the 21st century / edited by Kevin H. Baines (University of Wisconsin, Madison), F. Michael Flasar (NASA-Goddard Space Flight Center), Norbert Krupp (Max Planck Institute for the Study of Societies, Cologne), Tom Stallard (University of Leicester).
Description: Cambridge : Cambridge University Press, 2019. | Series: Cambridge planetary science series ; 20 | Includes bibliographical references and index.
Identifiers: LCCN 2017054700 | ISBN 9781107106772
Subjects: LCSH: Saturn (Planet) | Saturn (Planet) – Geology.
Classification: LCC QB671 .S2445 2018 | DDC 523.46–dc23
LC record available at https://lccn.loc.gov/2017054700

ISBN 978-1-107-10677-2 Hardback

Contents

List of Contributors *pages* ix
List of Reviewers xiii

1 Introduction to *Saturn in the 21st Century* 1
KEVIN H. BAINES, F. MICHAEL FLASAR, NORBERT KRUPP AND TOM STALLARD

2 The Origin and Evolution of Saturn, with Exoplanet Perspective 5
SUSHIL K. ATREYA, AURÉLIEN CRIDA, TRISTAN GUILLOT, JONATHAN I. LUNINE, NIKKU MADHUSUDHAN AND OLIVIER MOUSIS

3 The Interior of Saturn 44
JONATHAN J. FORTNEY, RAVIT HELLED, NADINE NETTELMANN, DAVID J. STEVENSON, MARK S. MARLEY, WILLIAM B. HUBBARD AND LUCIANO IESS

4 Saturn's Magnetic Field and Dynamo 69
ULRICH R. CHRISTENSEN, HAO CAO, MICHELE K. DOUGHERTY AND KRISHAN KHURANA

5 The Mysterious Periodicities of Saturn *Clues to the Rotation Rate of the Planet* 97
JAMES F. CARBARY, MATTHEW M. HEDMAN, THOMAS W. HILL, XIANZHE JIA, WILLIAM KURTH, LAURENT LAMY AND GABRIELLE PROVAN

6 Global Configuration and Seasonal Variations of Saturn's Magnetosphere 126
NORBERT KRUPP, PETER KOLLMANN, DONALD G. MITCHELL, MICHELLE THOMSEN, XIANZHE JIA, ADAM MASTERS AND PHILIPPE ZARKA

7 Saturn's Aurorae 166
TOM STALLARD, SARAH V. BADMAN, ULYANA DYUDINA, DENIS GRODENT AND LAURENT LAMY

8 Saturn's Ionosphere 196
LUKE MOORE, MARINA GALAND, ARVYDAS J. KLIORE, ANDREW F. NAGY AND JAMES O'DONOGHUE

9 Saturn's Variable Thermosphere 224
DARRELL F. STROBEL, TOMMI T. KOSKINEN AND INGO MÜLLER-WODARG

10 Saturn's Seasonally Changing Atmosphere: *Thermal Structure, Composition and Aerosols* 251
LEIGH N. FLETCHER, THOMAS K. GREATHOUSE, SANDRINE GUERLET, JULIANNE I. MOSES AND ROBERT A. WEST

11 The Global Atmospheric Circulation of Saturn 295
ADAM P. SHOWMAN, ANDREW P. INGERSOLL, RICHARD ACHTERBERG AND YOHAI KASPI

12 Saturn's Polar Atmosphere 337
KUNIO M. SAYANAGI, KEVIN H. BAINES, ULYANA DYUDINA, LEIGH N. FLETCHER, AGUSTÍN SÁNCHEZ-LAVEGA AND ROBERT A. WEST

13 The Great Saturn Storm of 2010–2011 377
AGUSTÍN SÁNCHEZ-LAVEGA, GEORG FISCHER, LEIGH N. FLETCHER, ENRIQUE GARCÍA-MELENDO, BRIGETTE HESMAN, SANTIAGO PÉREZ-HOYOS, KUNIO M. SAYANAGI AND LAWRENCE A. SROMOVSKY

14 The Future Exploration of Saturn 417
KEVIN H. BAINES, SUSHIL K. ATREYA, FRANK CRARY, SCOTT G. EDGINGTON, THOMAS K. GREATHOUSE, HENRIK MELIN, OLIVIER MOUSIS, GLENN S. ORTON, THOMAS R. SPILKER AND ANTHONY WESLEY

Index 442

Color plates can be found between pages 210 and 211

Contributors

RICHARD ACHTERBERG
University of Maryland

SUSHIL K. ATREYA
University of Michigan

SARAH V. BADMAN
Lancaster University

KEVIN H. BAINES
University of Wisconsin–Madison, Space Science and Engineering Center, and Jet Propulsion Laboratory,
California Institute of Technology

HAO CAO
California Institute of Technology

JAMES F. CARBARY
Johns Hopkins University,
Applied Physics Laboratory

ULRICH R. CHRISTENSEN
Max Planck Institute for Solar System Research

FRANK CRARY
University of Colorado,
Laboratory for Atmospheric and Space Physics

AURÉLIEN CRIDA
Observatoire de la Côte d'Azur
Institut Universitaire de France

MICHELE K. DOUGHERTY
Imperial College London

ULYANA DYUDINA
California Institute of Technology

SCOTT G. EDGINGTON
Jet Propulsion Laboratory,
California Institute of Technology

GEORG FISCHER
Space Research Institute–Graz

F. MICHAEL FLASAR
NASA Goddard Space Flight Center

LEIGH N. FLETCHER
University of Leicester

JONATHAN J. FORTNEY
University of California–Santa Cruz

MARINA GALAND
Imperial College London

ENRIQUE GARCÍA-MELENDO
University of the Basque Country UPV/EHU

THOMAS K. GREATHOUSE
Southwest Research Institute

DENIS GRODENT
University of Liège

SANDRINE GUERLET
LMD, CNRS, Sorbonne Université

TRISTAN GUILLOT
Observatoire de la Côte d'Azur

MATTHEW M. HEDMAN
University of Idaho

RAVIT HELLED
University of Zürich

BRIGETTE HESMAN
University of Maryland

THOMAS W. HILL
Rice University

WILLIAM B. HUBBARD
University of Arizona, Lunar and Planetary Laboratory

LUCIANO IESS
Sapienza University of Rome

ANDREW P. INGERSOLL
California Institute of Technology

XIANZHE JIA
University of Michigan

YOHAI KASPI
Weizmann Institute of Science

KRISHAN KHURANA
University of California–Los Angeles

ARVYDAS J. KLIORE
Jet Propulsion Laboratory, California Institute of Technology

PETER KOLLMANN
Johns Hopkins University, Applied Physics Laboratory

TOMMI T. KOSKINEN
University of Arizona, Lunar and Planetary Laboratory

NORBERT KRUPP
Max Planck Institute for Solar System Research

WILLIAM KURTH
University of Iowa

LAURENT LAMY
LESIA, Observatoire de Paris, Université PSL, CNRS

JONATHAN I. LUNINE
Cornell University

NIKKU MADHUSUDHAN
University of Cambridge

MARK S. MARLEY
NASA Ames Research Center

ADAM MASTERS
Imperial College London

HENRIK MELIN
University of Leicester

DONALD G. MITCHELL
Johns Hopkins University, Applied Physics Laboratory

LUKE MOORE
Boston University, Center for Space Physics

JULIANNE I. MOSES
Space Sciences Institute

OLIVIER MOUSIS
Aix-Marseille University

INGO MÜLLER-WODARG
Imperial College London

ANDREW F. NAGY
University of Michigan

NADINE NETTELMANN
University of Rostock

JAMES O'DONOGHUE
NASA Goddard Space Flight Center

GLENN S. ORTON
Jet Propulsion Laboratory,
California Institute of Technology

SANTIAGO PÉREZ-HOYOS
University of the Basque Country UPV/EHU

GABRIELLE PROVAN
University of Leicester

AGUSTÍN SÁNCHEZ-LAVEGA
University of the Basque Country UPV/EHU

KUNIO M. SAYANAGI
Hampton University

ADAM P. SHOWMAN
University of Arizona

THOMAS R. SPILKER
Solar System Science and Exploration

LAWRENCE A. SROMOVSKY
University of Wisconsin–Madison,
Space Science and
Engineering Center

TOM STALLARD
University of Leicester

DAVID J. STEVENSON
California Institute of Technology

DARRELL F. STROBEL
Johns Hopkins University

MICHELLE THOMSEN
Planetary Science Institute

ANTHONY WESLEY
Astronomical Society of Australia

ROBERT A. WEST
Jet Propulsion Laboratory,
California Institute of Technnology

PHILIPPE ZARKA
LESIA, Observatoire de Paris,
Université PSL, CNRS

Reviewers

Nicholas Achilleos
Richard Achterberg
David H. Atkinson
Fran Bagenal
Gordon L. Bjoraker
Scott G. Edgington
Thérèse Encrenaz
Leigh N. Fletcher
Thierry Fouchet
A. James Friedson
Jean-Claude Gérard
Peter J. Gierasch
Tristan Guillot
Andrew P. Ingersoll
Wing Ip
Margaret G. Kivelson
Katia I. Matcheva
Julianne I. Moses
Morris Podolak
Wayne R. Pryor
Peter L. Read
Christophe Sotin
Sabine Stanley
Peter Storer, Jr.
Vytenis M. Vasyliunas
Ronald J. Vervack Jr.

1

Introduction to *Saturn in the 21st Century*

KEVIN H. BAINES, F. MICHAEL FLASAR, NORBERT KRUPP AND TOM STALLARD

Saturn, with its exquisite rings, is perhaps the most photogenic planet in the solar system, a favorite for amateur and professional astronomers alike, as well as for atmospheric scientists, magnetospheric researchers, and auroral specialists. Together with the other giant planets, Saturn serves as a prototype of giant, rapidly rotating fluid planets, which are increasingly used to interpret spatially unresolved observations of large exoplanets.

During the past decade, there has been a sea change in our perception and understanding of Saturn, in large part stemming from its intense scrutiny by the Cassini spacecraft since it entered Saturn's orbit in 2004. The Cassini orbiter's 252 (as of late 2016) revolutions of Saturn during its 12-year reconnaissance have not only provided detailed maps of the three-dimensional spatial structure of the atmosphere and surrounding magnetosphere, but also an extended record of the temporal behavior of these regions, extending over parts of three seasons for each hemisphere. Such remarkable datasets are a result of the comprehensive nature of the Cassini mission. First, its broad complement of remote sensing instruments thoroughly covers the spectrum from the ultraviolet through the thermal infrared, while a variety of in situ experiments directly sample magnetospheric plasmas, ions, magnetic fields, and – on close flybys (<3,000 km above the 1-bar level) – atmospheric molecules. Second, its prolonged, on-orbit observing phase exceeding thirteen terrestrial years has spanned nearly half of a Saturnian year, thus providing an extensive record of seasonal responses under both winter and summer conditions. Third, the unusually large range of viewing geometries provided by the orbiter's repeated multi-year stints in both polar and equatorial orbits has enabled unprecedentedly sharp and direct views of both high- and low-latitude features over the full range of lighting conditions, including nighttime views unattainable from Earth. Such comprehensive scrutiny has revealed the structure and dynamics of both polar regions, the nature of powerfully eruptive, global-scale storms, the temporally varying thermosphere, the effervescent aurorae, and the structure and energetics of the magnetosphere.

With the contributions of 66 experts from ten countries (Australia, Austria, Belgium, France, Germany, Israel, Italy, Spain, the United Kingdom, and the United States), the thirteen subsequent chapters provide a thorough review of Saturn's atmosphere and its magnetosphere, including the planet's origin and evolution, its interior, the dynamic meteorology of its troposphere and middle atmosphere, its thermosphere, ionosphere, and aurorae. While this book builds upon the earlier review volume, *Saturn from Cassini–Huygens* (Dougherty et al. 2009), it focuses on new revelations, perspectives, and theoretical/modeling advances achieved over the near-decade since the previous work, highlighting new results from some 550 peer-reviewed papers on Saturn and its magnetosphere, as obtained from the ~400 Gbits of data transmitted to Earth over 160 Saturn orbits since 2008.

Each of the thirteen subsequent chapters of this book reviews the current knowledge associated with its subject, in light of recent data and interpretation, and lists open questions to be addressed by future observations, usually from spacecraft missions. The questions and issues raised by Cassini observations have led to some departures from the usual pattern in chapter organization. For instance, a key question important for interior modeling is the value of the internal rotation rate – which, unlike any other planet in our solar system, is unknown to better precision than roughly a quarter of an hour. Here, a separate section (Chapter 5) describes the challenges faced in measuring the length of a day on Saturn, reviewing the surprisingly complicated periodicities exhibited by radio emission, the rotating magnetic field structure, variabilities in charged particle flux, the ever-changing aurorae, and discrete atmospheric features. In general, these ~10.7-hour periodicities differ from one another, and sometimes vary with hemisphere and time. Saturn's surprisingly cryovolcanically active moon, Enceladus, appears to be an unexpected culprit in the confusion, injecting watery jets of materials into the Saturn system that perhaps affects the expected "heart beat" signal of the planet's rotation. Cassini scientists hope that the unique close-up measurements provided by Grand Finale observations in 2017 – as well as a new technique using Saturn-generated wave-like signatures in the rings that result from slight internal longitudinal density variations – will settle the issue.

Chapter 2 by Atreya et al. covers models of Saturn's origin and evolution and lays out the case for favoring core accretion over the disk instability model of formation. The chapter emphasizes the importance of atmospheric elemental and isotopic composition in constraining models of Saturn and reviews the limitations of the existing data sets. The enrichment of carbon and other heavy elements in Saturn (and Jupiter) requires special delivery mechanisms for volatiles to these planets, which are not well understood. Current data on extrasolar giant planets and hydrodynamic and chemical models offer an avenue for further insights. Chapter 3 by Fortney et al. describes recent progress in understanding the physical inputs for models of Saturn's interior. These include the equations of state of hydrogen–helium–heavy-element mixtures at high pressures, uncertainties in physical parameters such as the gravity field and rotation rate, and constraints on Saturnian free oscillations. Nonetheless, interior models are still not well constrained, in part because of uncertainties in compositional gradients and in the rotational structure of Saturn's interior. The improvements in the gravity field expected from the close-in Cassini Grand Finale orbits should help reduce these uncertainties.

Chapter 4 by Christensen et al. summarizes Cassini's findings of Saturn's internal magnetic field and dynamo, from which important details about the interior structure, the depth of the dynamo region, and the extremely small degree of tilt between the spin and magnetic pole axes are obtained. As well, Cassini has enabled the exploration of the magnetosphere of Saturn in three dimensions and over a large range of temporal scales, spanning seconds to seasons, all of which have provided new insights into its dynamics. Chapter 6 by Krupp et al. reviews the current global picture of the second-largest magnetosphere in our solar system and describes Cassini's measurements of charged and neutral particles, local magnetic fields, and plasma waves obtained over more than thirteen years of on-orbit scrutiny. The chapter explains the structure and the source of the trapped particles in the radiation belts of Saturn as well as the losses caused by the icy moons in the inner magnetosphere. The chapter also reviews the findings of Cassini about the structure, composition, and dynamics of the equatorial plasma disk in the middle magnetosphere, and describes new measurements of processes in the magnetotail. This contribution also describes how the cryovolcanically active moon Enceladus has been revealed as the major source of plasma in the Saturnian magnetosphere and how such plasmic particles are accelerated, transported, and lost in the magnetosphere.

As reviewed in Chapter 7 by Stallard et al., the Cassini mission has revolutionized our understanding of the current systems that generate Saturn's aurorae, including how these currents influence the shape and size of the aurorae. Beyond revealing the faint pinkish purple visible aurorae hidden on Saturn's nightside, the Cassini mission has enabled the comparison of the bright auroral emission from atomic and molecular hydrogen and the molecular ion H_3^+ in, respectively, ultraviolet and infrared light. The Cassini spacecraft's periodic sojourns into highly inclined orbits have passed it directly through magnetic field lines that close in the auroral region, enabling direct in situ measurements of the auroral currents and revealing the source region for Saturn's radio aurorae. This has provided the first direct evidence that Saturn has aurorae that are generated, in effect, by weather systems within the planet's upper atmosphere, the first such aurorae found in the Solar System.

As detailed in Chapter 8 by Moore et al., Cassini's observations over the past decade have resulted in

Saturn's upper atmosphere and ionosphere becoming the best understood of any gas giant. Measurements of the ionosphere have been greatly enhanced by ~60 radio occultation soundings, from which vertical profiles of the ionospheric electron density can be determined. In addition, Cassini has conclusively proven that short-lived, impulsive radio emissions known as Saturn electrostatic discharges (SEDs) are caused by lightning storms in the lower atmosphere, tying them to known giant convective storms in Saturn's atmosphere. Measurements of the H_3^+ emission near Saturn's equator have revealed significant latitudinal structure, suggesting that the ionosphere is being modified by the infall of charged water from Saturn's rings along magnetic field lines into the planet.

As discussed in Chapter 9 by Strobel et al., Cassini has obtained a large collection of stellar and solar occultations of Saturn's thermosphere, providing a set of altitudinal profiles an order of magnitude larger than what we currently have for Jupiter. These have revealed not only the broad structure of the upper atmosphere, but unexpected changes in both latitude and time. This has shown that the thermosphere is much deeper and warmer at the poles than at the equator. Interestingly, the thermosphere has also significantly varied during Cassini's time in orbit, with the top of the atmosphere expanding by 500 km between 2006 and 2011 (and more recently contracting), likely in response to the 100–200 K thermospheric warming observed over the same period.

The longevity of Cassini observations, close to half a Saturnian year, together with the even longer interval afforded by Earth-based observations, have facilitated a detailed mapping of seasonal behavior, as described in Chapter 10 by Fletcher et al. Hemispherical asymmetries in (1) seasonal variations in temperatures, (2) abundances of photochemically generated and destroyed species, and (3) aerosol hazes are noted. All of these are perhaps predominantly driven by the ~20 percent difference in insolation of the northern and southern hemispheres during the heights of their respective summers, due to the large planet's orbital eccentricity of 0.054 (vs. 0. 017 for the Earth).

Chapter 11 by Showman et al. provides a review of Saturn's global circulation, including the progress being made in three-dimensional convectively forced models, both in the troposphere and in Saturn's interior. Such models are starting to identify which mechanisms power and sustain the zonal jets and other salient wind systems on both Saturn and Jupiter, including the fast (nearly supersonic) equatorial super-rotation, multiple jets at higher latitudes, vortices, and storms.

The high-inclination orbits achieved during the Cassini mission have provided the first extensive measurements of Saturn's poles. Chapter 12 by Sayanagi et al. describes the structure of these regions and the physical processes that constrain them, including the ~10,000-km-radius vortices that are centered on both poles and the enigmatic north polar hexagon that effectively encircles the north polar vortex.

Saturn is capable of surprise, as amply demonstrated by the unexpected eruption of a giant, latitude-girdling northern storm in late 2010. Its eruption and decay in the troposphere and stratosphere are described in Chapter 13 by Sánchez-Lavega et al. This storm currently holds the record for the most powerful single convective storm ever recorded, lofting water ice more than 200 km in altitude above the 10-bar freezing level. The subsequent long-term (> 5 years) drying-out of all condensables – including both vapors and condensates – within the entire latitudinal storm band was also a surprise. Downward rebound of convective overshooting appears to be an explanation, and may also explain the episodic (every 2–3 decades) eruptive nature of such "super storms."

Chapter 14 by Baines et al. covers the prospects for the exploration of Saturn after Cassini and reviews several mission concepts proposed by both NASA and ESA researchers for a variety of mission opportunities. A common theme is to probe, in situ, the depths of the planet in order to reveal its dynamical structure and chemical makeup in the well-mixed portion of the atmosphere, as well as to determine the abundances of telltale noble gases and their isotopes that provide unique and fundamental information on how the planet formed and evolved. Over the next decade or so (2018–2030), the advent of the James Webb Space Telescope and a new class of ground-based telescopes with apertures exceeding 10 meters in diameter promise new views of Saturn with remarkable clarity. Finally, a growing worldwide army of amateur ground-based observers using state-of-the-art commercially available cameras and techniques are today regularly obtaining sharp views of the planet rivaling those achievable at the end of the previous century only by the largest ground-based observatories. Their continued scrutiny of the ringed world in the decades to come should prove invaluable.

The end of Cassini. At the time of the writing of this Introduction in Spring, 2017, Cassini has entered the final 5-month phase of its ambitious mission, known as the Grand Finale, which in essence is an entirely new science mission akin to the Juno mission at Jupiter. During 22 orbital passes, the orbiter threads inside of Saturn's rings to 1,500–2,000 km above the cloud tops – up to 100 times closer than ever before. Cassini will directly sample gases within Saturn's exosphere and achieve unprecedented views of Saturn's cloud structure at sub-km scales. As well, Cassini will obtain uniquely detailed measurements of the planet's gravity field and magnetosphere. These latter two are expected to provide fundamental new insights into the nature of Saturn's interior, including the size of its core and perhaps a definitive answer to the planet's bulk rotation rate. All of these topics will be thoroughly covered in a subsequent volume of *Saturn in the 21st Century*, currently slated for publication in 2022.

References

Dougherty, M. K., Esposito, L. W., and Krimigis, S. M., Eds. (2009). *Saturn from Cassini-Huygens*. Springer.

2

The Origin and Evolution of Saturn, with Exoplanet Perspective

SUSHIL K. ATREYA, AURÉLIEN CRIDA, TRISTAN GUILLOT, JONATHAN I. LUNINE, NIKKU MADHUSUDHAN AND OLIVIER MOUSIS

Abstract

Saturn formed beyond the snow line in the primordial solar nebula, and that made it possible for it to accrete a large mass. Disk instability and core accretion models have been proposed for Saturn's formation, but core accretion is favored on the basis of its volatile abundances, internal structure, hydrodynamic models, chemical characteristics of protoplanetary disk, etc. The observed frequency, properties, and models of exoplanets provide additional supporting evidence for core accretion. The heavy elements with mass greater than ^{4}He make up the core of Saturn, but are presently poorly constrained, except for carbon. The C/H ratio is super-solar, and twice that in Jupiter. The enrichment of carbon and other heavy elements in Saturn and Jupiter requires special delivery mechanisms for volatiles to these planets. In this chapter we will review our current understanding of the origin and evolution of Saturn and its atmosphere, using a multifaceted approach that combines diverse sets of observations on volatile composition and abundances, relevant properties of the moons and rings, comparison with the other gas giant planet, Jupiter, and analogies to the extrasolar giant planets, as well as pertinent theoretical models.

2.1 Introduction

Saturn, though about one-third the mass of Jupiter, is the largest planetary system in the solar system, considering the vast reach of its rings and dozens of known moons. Thus, Saturn is key to understanding the origin and evolution of the solar system itself. Models, observations, comparison with Jupiter, the other gas giant planet, and analogies with extrasolar giant planets have begun to give a sense of how Saturn, in particular, and the giant planets in general, originated and evolved.

Two distinct mechanisms of giant planet formation have been proposed in the literature: (1) disk instability and (2) nucleated instability (or core accretion). The latter goes back to papers by Hayashi (1981) and his colleagues (e.g. Mizuno 1980), and requires the accretion of a solid body (rock/metal, ice, and, possibly, refractory organics) up to a critical mass threshold at which rapid accretion of gas becomes inevitable – typically 10 times the mass of the Earth (see Armitage 2010, for a discussion). The former theory had its origin in the 1970s (see Cameron 1979) for hot, massive disks, but it was determined later (Boss 2000; Mayer et al. 2002) that the instabilities required to break up a portion of a gaseous disk into clumps are a feature of cold, massive disks. We will focus on each of these contrasting models in turn, and then discuss the observational indicators in our own and extrasolar planetary systems that might distinguish between the two models.

The disk instability model is based on numerical simulations showing that massive, relatively cold disks will spontaneously fragment due to a gravitational instability, leading to multiple discrete, self-gravitating masses. In computer simulations of the process these features seem somewhat ill defined, and it is not possible to track the subsequent condensation of these features in the same hydrodynamical simulation that tracks the onset of the instability itself.

Nonetheless, basic disk physics dictates that such fragmentation will occur for a sufficiently massive or cold disk (Armitage 2010) and that the timescale for the fragmentation, once the instability does occur, is extremely short – hundreds to thousands of years.

Once formed, the fragments (assuming they continue to contract to form giant planets) are usually sufficiently numerous that the aggregate planetary system is dynamically unstable. The planets will gravitationally interact, scattering some out of the system and leaving the others in a variety of possible orbits. The evidence from microlensing of a substantial population of free-floating Jupiter-mass objects (Sumi et al. 2011) not associated with a parent star constitutes one argument in favor of the importance of this formation mechanism.

On the other hand, it is not evident how giant planets formed by the disk instability mechanism acquire significant amounts of heavy elements over and above their parent star's abundances. It has been argued that subsequent accretion of planetesimals would generate the increased metallicity, but the disruption of the disk associated with the gravitational instability might have removed the raw material for large amounts of planetesimals – the materials going into numerous giant planets that are then kicked out of the system. A subsequent phase of disk building or direct accretion of planetesimals from the surrounding molecular cloud may have to be invoked. And this begs the question of core formation – giant planets formed in this way may have super-solar metallicities but lack a heavy element core unless (as seems unlikely) very large (Earth-sized) planets are consumed by these objects.

The core accretion model, in contrast, begins by building a heavy element core through planetesimal and embryo accretion in the gaseous disk (embryo is usually reserved for lunar-sized bodies and upward). At some point, the gravitational attraction of the large core leads to an enhanced accretion of gas, so much so that gas accretion quickly dominates in a runaway process and the object gains largely nebular-composition gas until its mass is large enough to create a gap in the disk and slow accretion. Such a model produces, by definition, a heavy element core, and through co-accretion of gas and planetesimals, an envelope enrichment of heavy elements as well. The model's Achilles heel is the time required to build the heavy element core to the point where rapid gas accretion occurs – millions of years or more. The onset of rapid gaseous accretion, by which point further growth may be rapid, depends not only on the core accretion rate but also, through the critical core mass (roughly 10 M_E, where M_E is an Earth mass) needed to trigger rapid gas accretion, the envelope opacity, and hence metallicity. Furthermore, the core accretion rate itself is a sensitive function of what one assumes about the planetesimal size distribution and surface density in the disk.

A plausible timescale for the formation of Saturn must be consistent with the lifetime of gas in disks, but may also be constrained by the 3–5 million-year (Myr) estimate of the formation duration of Iapetus, from its geophysical shape and thermal history (Castillo-Rogez et al. 2009). Earliest models had lengthy formation times (e.g. 8 Myr; Pollack et al. 1996), but more recent models can make Saturn in a few million years by appropriate selection of nebular parameters such as grain distribution and opacity (Dodson-Robinson et al. 2008).

The overall history of the solar system and presence of a substantial terrestrial planet system inward of Jupiter and Saturn suggests that the extreme dynamical scattering suffered after disk instability protoplanets are formed did not happen in our solar system. Furthermore, if the 3–5-million-year estimate of the interval between the formation of the first solids and the formation of Iapetus (Castillo-Rogez et al. 2009) is correct, the disk instability – if it occurred – would have produced Saturn much too soon after (or even before) the first solids in the solar system condensed out. There is sufficient evidence that the first solids, millimeter-sized chondrules and calcium aluminum inclusions (CAIs) in chondrites, date back to 4.5682 Gyr (billion years) (Amelin et al. 2010), which provides clear evidence that submicron-sized interstellar grains were sticking and accumulating to form solids at the very beginning of the solar system.

Measurement by the Juno mission of the water abundance below the meteorology layer in Jupiter, tied to the abundances of other major elements measured by the Galileo probe, will also provide an indication of how much planetesimal material was accreted (Helled and Lunine 2014), and to some extent, the nature of the carrier species (e.g. Mousis 2012). Although it is possible to enrich the envelopes of the giant planets even in the disk instability model by adding planetesimals much later, the presence of both a substantial (10 M_E) core and envelope enrichment of heavy elements would strongly militate in favor of the core accretion model. Saturn's core mass may be measured by Cassini, but an inventory of the envelope enrichment of heavy elements and measurement of the deep water abundance will have to await a future Saturn probe.

The core accretion model gets a boost also from observational surveys of exoplanets. An analysis of the frequency of planets with different masses, sizes, orbits, and host characteristics reveals that a greater percentage of giant planets are found around higher-metallicity stars, and smaller planets between Earth's and Neptune's mass far exceed Jupiter-sized planets (Howard 2013; Johnson et al. 2010). This is what one would expect if core accretion were prerequisite for planetary formation. Thus, for our planetary system, at least, core accretion seems to make more sense. Trying to constrain detailed formation mechanisms by matching orbital properties is much more difficult because of the profound effects of migration (Mordasini, et al. 2009; Ida et al. 2013 and references therein).

In addition to their occurrence rates and orbital characteristics, the masses, radii, and atmospheric volatile gas compositions of giant exoplanets may also provide important clues regarding their formation processes, and in turn, formation of Saturn and Jupiter in the solar system. With rapid advances in spectroscopic observations of exoplanets, a number of gases relevant to formation models, including water vapor, methane and carbon monoxide have been detected in several giant exoplanets (Section 2.5), revealing diversity in chemical abundances. For example, there are some planets (e.g. HD 209458b) with seemingly lower H_2O abundances than expected from solar elemental composition (e.g. Deming et al. 2013; Madhusudhan et al. 2011a, 2014a), while others (e.g. WASP-43b) appear consistent with super-solar H_2O (e.g. Kreidberg et al. 2014). The latter is consistent with super-solar abundance of measured heavy elements in Saturn and Jupiter (Section 2.2.1), with a good likelihood that their original cores were rich in water ice. On the other hand, WASP-12b – which indicates a C/O ratio (≥1) twice solar (~0.5) – argues for a core made up of largely carbon-bearing constituents. If this result is confirmed for a multitude of similar exoplanets, it would have important implications for their formation and the formation of the gas giant planets of the solar system. More generally, new theoretical studies are suggesting that the observable O/H, C/H, and, hence, C/O ratios in giant exoplanetary atmospheres can place powerful constraints on their formation and migration mechanisms, as discussed in Section 2.5.3.

2.2 Observational Constraints

The models of Saturn's formation and evolution are constrained by data presently available on the planet's chemical composition and its interior. This section elaborates on each of these aspects and forms the basis for the discussions in subsequent sections.

2.2.1 Elemental Composition of Saturn's Atmosphere and Comparison to Jupiter

The composition of Saturn's atmosphere has been measured by remote sensing from ground-based and Earth-orbiting telescopes and flyby and orbiting spacecraft for over half a century. These observations have been instrumental in revealing the chemical makeup of Saturn's stratosphere and upper troposphere. As a result, mole fractions of helium (He), methane (CH_4), and a number of its photochemical products including methyl radical (CH_3), ethane (C_2H_6), acetylene (C_2H_2), methyl acetylene (C_3H_4), and benzene (C_6H_6), ammonia (NH_3), hydrogen sulfide (H_2S), and those species that are in thermochemical disequilibrium in Saturn's upper troposphere and stratosphere such as phosphine (PH_3), carbon monoxide (CO), germane (GeH_4), and Arsine (AsH_3) have been measured to varying degrees of precision. Some of the most precise data have come from observations made by the Cassini spacecraft (Fletcher et al., this book) that attained orbit around Saturn in 2004 and will embark on proximal orbits toward the end of the mission in 2017 (Baines et al., this book).

The abundances of certain heavy elements (m/z >^{4}He) and their isotopes can be derived from their principal chemical reservoirs in the atmosphere. As discussed earlier, heavy elements are key to constraining the models of the formation of Saturn and its atmosphere. Current best data on the abundances of elements relative to hydrogen in Saturn are listed in

Table 2.1. As Jupiter, the other gas giant planet in the solar system, is a good analog for Saturn, we list for comparison also the elemental abundances in Jupiter's atmosphere.

Many more heavy elements have been determined at Jupiter, in contrast to Saturn, because of in situ Galileo Jupiter entry probe measurements from 1995. Enrichment factors of the elements relative to protosolar values are also listed in Table 2.1, using currently available solar elemental abundances from two different sources (Asplund et al. 2009; Lodders et al. 2009). Further insight into key elemental abundances is given below, and the reader is referred also to the table footnotes.

After hydrogen, helium is the most abundant element in the universe, the sun, and the giant planets. Conventional thinking has been that the current abundance of helium ratioed to hydrogen in the giant planets should be the same as in the primordial solar nebula from which these planets formed, and originally the Big Bang, in which helium was created. Thus, precise determination of the helium abundance is essential to understanding the formation of the giant planets, in particular, and to shedding light on the solar nebula and the universe in general. Whereas helium has been measured very accurately at Jupiter by two independent techniques on the Galileo probe (Table 2.1), such is not the case for Saturn. In the absence of an entry probe at Saturn, helium abundance at Saturn was derived from atmospheric mean molecular weight (μ), using a combination of the Voyager infrared spectrometer (IRIS) and the radio science (RSS) investigations. RSS measured radio refractivity that provides the information on T/μ, where T is the temperature measured by both instruments.

Initial analysis using the IRIS-RSS data (Conrath et al. 1984) yielded a greatly sub-solar He/H=0.017 ±0.012 (He/H_2=2×He/H). Subsequent reanalysis of the data employing IRIS alone gave He/H between 0.055 and 0.08 (Conrath and Gautier 2000). The authors emphasize, however, the retrieval of He/H is non-unique, but strongly suggests a value significantly greater than the earlier result that was based on the combined IRIS-RSS approach. For the purpose of this chapter, we take an average of the range of Saturn's He/H of 0.055–0.08, and express it as 0.0675±0.0125 (Table 2.1), but with the caveat that the value could well change following detailed analysis of the Cassini CIRS data and, especially, future in situ measurements at Saturn, as did Jupiter's He/H_2 following in situ measurements by the Galileo probe compared to the value derived from Voyager's remote sensing observations. The current estimate of He/H in Saturn's upper troposphere is about 0.7× solar compared to Jupiter's 0.8× solar. The sub-solar He/H_2 in the tropospheres of Jupiter and Saturn presumably results from the removal of some fraction of helium vapor through condensation as liquid at 1–2 megabar pressure in the interiors of these planets, followed by separation of helium droplets from metallic hydrogen. The severe depletion of Ne observed by the Galileo probe (Table 2.1) in Jupiter is excellent evidence of the helium-hydrogen immiscibility layer, as helium droplets absorb neon vapor, separate from hydrogen, rain toward the core, and this results in the depletion of helium and neon in the upper troposphere (Roulston and Stevenson 1995; Wilson and Militzer 2010). Models predict that the cooler interior of Saturn is expected to result in a greater degree of helium condensation and therefore a tropospheric He/H_2 ratio lower for Saturn than for Jupiter. Although the central value for Saturn is smaller than Jupiter's, the large uncertainty of Saturn's result does not provide a definite answer. Helium differentiation in Saturn's interior is invoked also as a way to explain the planet's large energy balance (Conrath et al. 1989). Without such chemical differentiation, models predict the heat flux excess at Saturn to be about three times lower than observed (Grossman et al. 1980), but the equation of state for the high-pressure, high-temperature interior is uncertain, so the modeled excess is not that well constrained (see Chapter 3 by Fortney et al. for additional details). Saturn and Jupiter both emit nearly twice the thermal radiation compared to the radiation the absorb from the sun. Whereas the release of heat of accretion from conversion of the gravitational potential energy as these planets cool and contract over time accounts for a good fraction of the energy balance of Jupiter, helium differentiation may play a significant role at Saturn. Since helium is denser than hydrogen, gravitational potential energy available for conversion to heat increases as helium raindrops begin to separate from hydrogen and precipitate upon reaching centimeter size. In summary, there are indications that helium is depleted relative to solar in Saturn's

Table 2.1 *Elemental Abundances in Jupiter and Saturn and Ratios to Protosolar Values*

Elements	Jupiter	Saturn	Sun-Protosolar (Asplund et al. 2009) [a,b]	Jupiter/ Protosolar (using Asplund et al. 2009) [a,b]	Saturn/Protosolar (using Asplund et al. 2009) [a,b]	Sun-Protosolar (Lodders et al. 2009) [m]	Jupiter/ Protosolar (using Lodders et al. 2009) [m]	Saturn/ Protosolar (using Lodders et al. 2009) [m]
He/H	$7.85 \pm 0.16 \times 10^{-2}$ [c]	$5.5–8.0 \times 10^{-2}$ [i], taken as $6.75 \pm 1.25 \times 10^{-2}$	9.55×10^{-2}	0.82 ± 0.02	0.71 ± 0.13 (?)	9.68×10^{-2}	0.81 ± 0.02	0.70 ± 0.13 (?)
Ne/H	$1.24 \pm 0.014 \times 10^{-5}$ [d]		9.33×10^{-5}	0.13 ± 0.001		1.27×10^{-4}	0.098 ± 0.001	
Ar/H	$9.10 \pm 1.80 \times 10^{-6}$ [d]		2.75×10^{-6}	3.31 ± 0.66		3.57×10^{-6}	2.55 ± 0.50	
Kr/H	$4.65 \pm 0.85 \times 10^{-9}$ [d]		1.95×10^{-9}	2.38 ± 0.44		2.15×10^{-9}	2.16 ± 0.39	
Xe/H	$4.45 \pm 0.85 \times 10^{-10}$ [d]		1.91×10^{-10}	2.34 ± 0.45		2.1×10^{-10}	2.11 ± 0.40	
C/H	$1.19 \pm 0.29 \times 10^{-3}$ [e]	$2.65 \pm 0.10 \times 10^{-3}$ [j]	2.95×10^{-4}	4.02 ± 0.98	8.98 ± 0.34	2.77×10^{-4}	4.29 ± 1.05	9.56 ± 0.36
N/H	$3.32 \pm 1.27 \times 10^{-4}$ [e]	$0.80–2.85 \times 10^{-4}$ [k];	7.41×10^{-5}	4.48 ± 1.71[e]	1.08–3.84;	8.19×10^{-5}	4.06 ± 1.55[e]	0.98–3.48;
	$4.00 \pm 0.50 \times 10^{-4}$ [f]	$2.27 \pm 0.57 \times 10^{-4}$ with		5.40 ± 0.68[f]	3.06 ± 0.77 with		4.89 ± 0.62[f]	2.78 ± 0.73 with
	$2.03 \pm 0.46 \times 10^{-4}$ [g]	$f NH_3 = 4 \pm 1 \times 10^{-4}$		2.70 ± 0.60[g]	$f NH_3 = 4 \pm 1 \times 10^{-4}$		2.50 ± 0.55[g]	$f NH_3 = 4 \pm 1 \times 10^{-4}$
O/H	$2.45 \pm 0.80 \times 10^{-4}$ [e]		5.37×10^{-4}	0.46 ± 0.15 (hotspot)		6.07×10^{-4}	0.40 ± 0.13 (hotspot)	
S/H	$4.45 \pm 1.05 \times 10^{-5}$ [e]	1.88×10^{-4} [l]	1.45×10^{-5}	3.08 ± 0.73	13.01	1.56×10^{-5}	2.85 ± 0.67	12.05
P/H	$1.08 \pm 0.06 \times 10^{-6}$ [h]	$3.64 \pm 0.24 \times 10^{-6}$ [h]	2.82×10^{-7}	3.83 ± 0.21	12.91 ± 0.85	3.26×10^{-7}	3.30 ± 0.18	11.17 ± 0.74

(a) Protosolar values calculated from the solar photospheric values of Asplund et al. (2009, table 1).

(b) According to Asplund et al. (2009), the protosolar metal abundances relative to hydrogen can be obtained from the present-day photospheric values (table 1 of Asplund et al. 2009) increased by +0.04 dex, i.e. ~11%, with an uncertainty of ±0.01 dex; the effect of diffusion on He is very slightly larger: +0.05 dex (±0.01). Note that Grevesse et al. (2005, 2007) used the same correction of +0.05 dex for all elements (dex stands for "decimal exponent," so that 1 dex=10).

(c) von Zahn et al. (1998), using helium detector on Galileo Probe; independently confirmed by the Galileo Probe Mass Spectrometer (GPMS, Niemann et al. 1998).

(d) Mahaffy et al. (2000); Kr and Xe represent the sum of all isotopes except for ^{126}Xe and ^{124}Xe that could not be measured by the GPMS but are probably negligible, as together they make up 0.2% of the total xenon in the sun.

(e) Wong et al. (2004), based on re-calibration of the GPMS data on CH_4, NH_3, H_2O, and H_2S down to 21 bars, using an experiment unit; represents an update of the values reported in Niemann et al. (1998) and Atreya et al. (1999, 2003).

(f) Folkner et al. (1998), by analyzing the attenuation of the Galileo probe-to-orbiter radio communication signal (L-band at 1387 MHz or 21.6 cm) by ammonia in Jupiter's atmosphere.

(g) Juno microwave radiometer (MWR) preliminary result in the equatorial region and two different longitudes (Bolton et al. 2017).

(h) Fletcher et al. (2009a) derived global PH_3 mole fractions of 1.86±0.1 ppm and 6.41±0.42 ppm, respectively, in the upper tropospheres of Jupiter and Saturn from an analysis of the mid-IR emission measured by the Cassini Composite Infrared Spectrometer (CIRS).

(i) Conrath and Gautier (2000) give a range of 0.11–0.16 for the He/H_2 mole fraction from re-analysis of the Voyager IRIS data at Saturn, but the result is tentative. We use an average He/H=0.0675 for the purpose of calculating the ratios of other elements relative to hydrogen in Saturn.

(j) Fletcher et al. (2009b) report mole fraction of $CH_4=4.7\pm0.2\times10^{-3}$ from an analysis of the CIRS data.

(k) Fletcher et al. (2011), using VIMS data giving an ammonia mole fraction, fNH_3, in the 1–3 bar region that is 140±50 ppm (scattering), 200±80 ppm (non-scattering) and rises to 300–500 ppm at the equator. If the maximum in ammonia measured at the equator (300–500 ppm, or 400±100 ppm) represents deep atmospheric NH_3, the corresponding $NH_3/H = 2.27\pm0.6\times10^{-4}$.

(l) Briggs and Sackett (1989), using the VLA and the Arecibo microwave and radio data. The authors reported 10× solar H_2S, using solar $S/H = 1.88\times10^{-5}$ from then current listing (Cameron 1982). The S/H result is questionable (see text).

(m) Protosolar values based on present-day solar photospheric values of Lodders et al. (2009, table 4). The proto-solar abundances are calculated from the present-day values using the following corrections: +0.061 dex for He and +0.053 dex for all other elements.

troposphere, but the extent of such depletion will continue to be a subject of debate until precise in situ measurements can be made. In this regard, the final proximal orbits of Cassini in September 2017 are promising for the measurement of helium by the Ion and Neutral Mass Spectrometer down to ~1700 km or ≤0.1 nanobar (S. Edgington, personal comm., 2015), which is above Saturn's homopause level (1000–1100 km, or ~10–100 nanobar; Atreya 1986; Strobel et al., this book), and perhaps deeper in the final trajectory when the spacecraft plunges into Saturn. Extrapolation to a well-mixed troposphere would be model dependent even if the homopause level could be derived independently from the Cassini occultation data in the proximal orbits. Hence, precise helium abundance measurement directly in the well-mixed troposphere will still be essential, and that can only be done from an entry probe.

The nitrogen elemental abundance in Saturn is obtained from Saturn's principal nitrogen-bearing molecule, NH_3. From an analysis of the Cassini Visual and Infrared Mapping Spectrometer (VIMS) data, Fletcher et al. (2011) derive an ammonia mole fraction, fNH_3, in the 1–3 bar region that is 140±50 ppm (scattering), 200±80 ppm (non-scattering), and rising to 300–500 ppm at the equator. If we assume that maximum in ammonia measured at the equator (300–500 ppm, taken as $4\pm1\times10^{-4}$ here) represents also the NH_3 mole fraction in Saturn's deep well-mixed troposphere, then the corresponding $NH_3/H = 2.27\pm0.6\times10^{-4}$. That would imply an N/H enrichment of about 3× solar at Saturn, in contrast to Jupiter's roughly 3–5× solar. Previously, de Pater and Massie (1985) also found a 3× solar enhancement in Saturn's N/H in the 3-bar region, based on the VLA observations. The VLA and the Cassini RADAR 2.2 cm data (Laraia et al. 2013) also show that ammonia is subsaturated down to several bars, which most likely results from the loss of NH_3 in the lower clouds of NH_4SH (or another form such as $(NH_4)_2S$) at ≥5 bars and the NH_3-H_2O (aqueous-ammonia) solution cloud between approximately 10 and 20 bars, depending on the enhancement of O/H (H_2O) above solar (Atreya et al. 1999; Atreya and Wong 2005; see also Section 2.6 and Figure 2.9 therein). Whether the above 3× solar N/H in the 3-bar region is representative of the true nitrogen elemental ratio in Saturn's deep well-mixed troposphere is presently an open question, as the infrared or radio data can neither confirm it nor rule it out. Unlike Saturn, there is no such ambiguity in the determination of Jupiter's N/H, since direct in situ measurements of NH_3 could be made by the Galileo probe mass spectrometer (GPMS; Niemann et al. 1998) down to 21 bars, which is well below the expected NH_3 condensation level of 0.5–1 bar. Independently, NH_3 was derived also by analyzing the attenuation of the Galileo probe-to-orbiter radio communication signal (L-band at 1387 MHz or 21.6 cm) by ammonia in Jupiter's atmosphere (Folkner et al. 1998). NH_3 from the two sets of data agree to within 20%, with tighter constraints coming from the radio attenuation data, which yields N/H = 5.40±0.68× solar (Table 2.1). It is generally assumed the Galileo probe value is likely representative of the global N/H in Jupiter, as the measurements were done well below any possible traps of ammonia, including condensation clouds of NH_3, NH_4SH, and NH_3-H_2O. Preliminary deep NH_3 values from the Juno microwave radiometer (Bolton et al. 2017) overlap the Galileo mass spectrometer value within the range of uncertainty of the two datasets, but not the Galileo radio attenuation data (Table 2.1). At Saturn, NH_3 from remote sensing extends to ~3 bars; however, an entry probe to deeper levels can answer whether that value is representative of the global well-mixed N/H or similar to C/H.

Sulfur is sequestered largely in the H_2S gas in the atmospheres of Jupiter and Saturn. Whereas Jupiter's H_2S could be measured directly and precisely in situ by the Galileo probe (Table 2.1), it was derived indirectly at Saturn by fitting the VLA and Arecibo microwave and radio data to assumed NH_3 abundances (Briggs and Sackett 1989). Although direct microwave absorption by H_2S could not be measured in these observations, they deduced H_2S by analyzing NH_3, whose abundance is controlled to some extent by H_2S, since models predict it would remove a portion of the NH_3 vapor via the formation of an NH_4SH cloud below. Using the then-available solar $S/H=1.88\times10^{-5}$ (Cameron 1982), they derived a 10× solar enrichment of S/H in Saturn's atmosphere, which translates into 12–13× solar S/H using current solar S/H values, or about four times the value determined by the Galileo probe in Jupiter (Table 2.1). It is important to add a caveat, however. Whereas the Jupiter result comes from direct, in situ measurement of H_2S, the above result for Saturn is highly model-dependent, as it depends on the assumption of the formation of purported NH_4SH cloud whose thermochemical properties are poorly constrained. Since sulfur is a key heavy element in the models of Saturn's formation, a fresh set of data on Saturn's H_2S is warranted.

We list P/H in Table 2.1, but add a caveat that it may not represent the true P/H value in the deep well-mixed atmospheres of Jupiter or Saturn. This is because PH_3, the principal reservoir of phosphorus in the atmospheres of Jupiter and Saturn, is a disequilibrium species that is thermochemically stable in the deep atmosphere at pressures of about one thousand bars where the temperature is ~1000 K or greater (Fegley and Prinn 1985; Visscher and Fegley 2005), but it could only be measured in the upper troposphere/lower stratosphere. As PH_3 is dredged up from deep in the atmosphere to the upper atmosphere, it may potentially undergo loss due to oxidation to P_4O_6 by water vapor and solution in any water clouds along the way, or by other chemical reactions. Thus, the P/H ratio deduced from observations of PH_3 for Saturn and Jupiter in the upper atmosphere may represent a lower limit to the P/H ratio in their deep well-mixed atmosphere. Hence, the P/H values listed in Table 2.1 should not automatically be taken as a good proxy for the enrichment of other heavy elements not yet measured in Jupiter or Saturn. On the other hand, disequilibrium species such as PH_3, GeH_4, AsH_3, and CO are excellent tracers of the strength of convective mixing in the deep atmospheres of Saturn and Jupiter, and some could potentially be exploited to yield also a rough estimate of the deep water abundance.

Oxygen is arguably the most crucial of all heavy elements for constraining the formation models of Jupiter and Saturn. This is because in the reducing environments of the giant planets, oxygen is predominantly sequestered in water, which was presumably the original carrier of the heavy elements that formed the core and made it possible to accrete gas and complete the planet formation. (CO is another oxygen bearing species, but is a million times less abundant than water.) Yet the deep well-mixed abundance of water, and hence of O/H, remains a mystery. In the case of Jupiter, the Galileo probe entered an anomalously dry region known as a 5-micron hot spot. In this "Sahara Desert of Jupiter," water was found to be severely depleted (Niemann et al. 1998; Atreya et al. 1999, 2003). Although the probe mass spectrometer measured water vapor down to 21 bars, i.e. well below the expected condensation level of H_2O between 5 and 10 bars, it was still sub-solar at that level (Table 2.1), but rising. The determination of Jupiter's water abundance must await the analysis of Juno microwave radiometer observations in 2016–2017. No measurements of water vapor are available for Saturn's troposphere, however. The presence of water in Saturn's atmosphere is inferred indirectly from observations of visible lightning by Cassini's imaging spectrometer where lightning storms were predicted by Cassini's radio observations (Dyudina et al. 2010). Broadband clear filter observations showed visible lightning at ~35°S on the nightside in 2009 (Dyudina et al. 2010) and in blue wavelengths only on the dayside in the 2010–2011 giant lightning storm at ~35°N (Dyudina et al. 2013). These authors conjecture that a 5- to 10-times enhancement of water over solar can explain Saturn's lower occurrence rate for moist convection, an indicator of lightning, compared to Jupiter's (Dyudina et al. 2010). Similarly, using thermodynamic arguments Li and Ingersoll (2015) conclude that Saturn's quasi-periodic giant storms, which recur every few decades, result from interaction between moist convection and radiative cooling above the water cloud base, provided that the tropospheric water vapor abundance is 1 or greater, i.e. O/H $\geq$10× solar. Such an enrichment in O/H would result in a droplet cloud of NH_3-H_2O at ~20-bar level at Saturn (Atreya and Wong 2005; see also section 2.6 and figure 2.9 therein). Although direct measurements of Saturn's well-mixed water may have to wait for future missions, as discussed in Section 2.5, the recent discoveries of hot giant exoplanets and a Saturn-analog exoplanet are making it possible to measure H_2O abundances in their atmospheres, and in turn informing possible H_2O abundances in solar system giant planets.

Highly precise measurements of methane in the atmosphere of Saturn have been carried out with Cassini's composite infrared spectrometer (CIRS) instrument (Flasar et al. 2005), which yield a mole fraction of $CH_4 = 4.7\pm0.2\times10^{-3}$ (Fletcher et al. 2009b). This results in a robust determination of the C/H ratio in Saturn (about twice the Jupiter value) that can be compared with rather imprecise but definitely

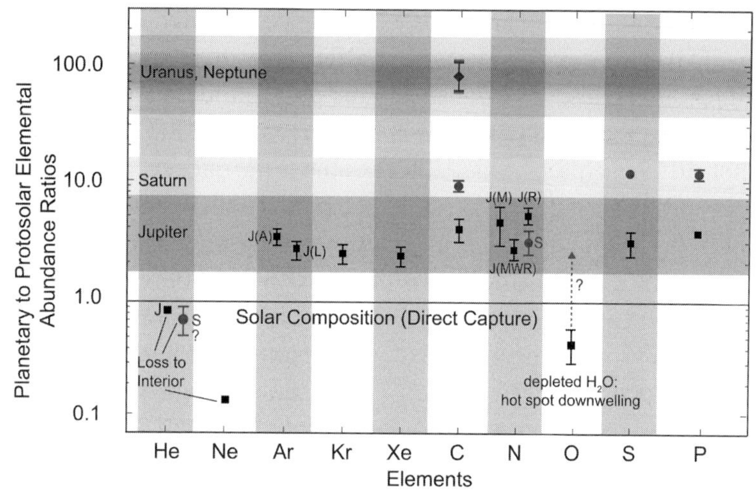

Figure 2.1 Abundances of key elements in the atmospheres of Saturn (brown dots, and label S) and Jupiter (black squares) relative to protosolar values derived from the present-day photospheric values of Asplund et al. (2009). Only C/H is presently determined for Uranus and Neptune, though poorly; its best estimate from Earth-based observations is shown. The values are listed in Table 2.1. All values are ratioed to H (multiply by 2 for ratio to H_2). Direct gravitational capture would result in solar composition, i.e. no volatile enrichment, hence they would all fall on the horizontal line (normalized to solar) in the middle of the figure. Only He, C, N, S, and P have been determined for Saturn, but only C/H is robust for the well-mixed atmosphere (see text). The Jupiter values are from the Galileo probe mass spectrometer (GPMS), except for N/H from NH_3 that was measured on the Galileo probe by the GPMS [J(M)] and from attenuation of the probe radio signal through the atmosphere [J(R)] as well as Juno microwave radiometer [J(MWR)], whose preliminary result is shown. For Ar, enrichments using both Asplund et al. [J(A)] and Lodders et al. [J(L)] solar values are shown. O/H is sub-solar in the very dry entry site of the Galileo Probe at Jupiter, but was still on the rise at the deepest level probed. Helium is depleted in the shallow troposphere due to condensation and differentiation in the planetary interior. Ne was also depleted in Jupiter as neon vapor dissolves in helium droplets. (A black-and-white version of this figure appears in some formats. For the color version, please refer to the plate section.)

higher estimates of C/H in Uranus and Neptune, as a way of constraining the giant planet formation scenarios.

Heavy noble gases, Ne, Ar, Kr, and Xe, have been measured only in Jupiter's atmosphere (Table 2.1), since they can only be detected in situ by an entry probe, not by remote sensing. As noble gases are chemically inert, their abundance is unaffected by chemistry and condensation processes that control NH_3, H_2S, H_2O, and PH_3. Thus, the heavy noble gas enrichments are expected to be the same everywhere in the atmosphere. At Jupiter, with the exception of neon, they range from a factor of 2 to 3× solar within the range of uncertainty of their planetary measurements and the solar values (Table 2.1). As neon dissolves in liquid helium, it is removed along with helium, which condenses in the 3 megabar region in Jupiter's interior, and is thus found depleted at observable shallow tropospheric levels (Wilson and Militzer 2010). At Saturn, neon is expected to meet the same fate.

Figure 2.1 shows the enrichment factors of the heavy elements and He in the atmospheres of Saturn and Jupiter relative to their protosolar values (all ratioed to H). Here we use the Asplund et al. (2009) compilation of photospheric elemental abundances (their table 1), as they represent an improvement over previous conventional standards (e.g. Anders and Grevesse 1989; Grevesse et al. 2005, 2007) and result from the use of a 3D hydrodynamic model of the solar atmosphere, nonlocal thermodynamic equilibrium effects, and improved atomic and molecular data. The photospheric values are then converted to protosolar elemental abundance (see table footnote). The latter account for the effects of diffusion at the bottom of the convective zone on the chemical composition of the photosphere, together with the effects of gravitational settling and radiative accelerations. According to Asplund et al. (2009), the protosolar metal abundances relative to hydrogen can be obtained from the present-day values increased by +0.04 dex, i.e. ~11%, with an uncertainty of ±0.01 dex; the effect of diffusion on He is very slightly larger: +0.05 dex (±0.01) (dex stands for "decimal exponent," so that 1 dex=10; it is a commonly used unit in astrophysics). Lodders et al. (2009) suggest a slightly larger correction of +0.061 dex for He and +0.053 dex for all other elements. Previously, Grevesse et al. (2005, 2007) used the same protosolar correction of +0.05 dex for all elements.

Figure 2.1 is based on protosolar correction to Asplund et al. (2009) photospheric abundances, while Table 2.1 lists planetary elemental enrichment factors also for Lodders et al. (2009) protosolar values. Whereas the difference between the enrichment factors based on Asplund et al. and Lodders et al. values is at most 10 to 15% for most elements, Asplund et al. estimate nearly 30% greater enrichment for Ar/H, compared to Lodders et al. (Table 2.1).

The difference in Jupiter's Ar enrichment factors based on Asplund et al. (2009) and Lodders et al. (2009) can be traced back largely to the choice of O/H employed by the two sets of authors. Because of their high excitation potentials, noble gases do not have photospheric spectral features; hence their solar abundances are derived indirectly. Asplund et al. (2009) infer solar Ar/H following the same procedure as Lodders (2008), i.e. by using, amongst other things, the Ar/O data from the solar wind, solar flares, and solar energetic particles, but employing their own photospheric abundances of O/H that have a somewhat lower uncertainty than Lodders et al. (2009). This accounts for much of the abovementioned 30% difference in Jupiter's Ar/H enrichment factor. Nevertheless, within the range of uncertainty of Jupiter's Ar abundance and the dispersion in the solar values, the Ar/H enrichment in Jupiter relative to the solar Ar/H is nearly the same whether one uses Asplund et al. (2009) or Lodders et al. (2009) solar Ar/H. We show both results in Figure 2.1. A word of caution about oxygen, which is used by the above authors as a proxy for deriving the solar Ar/H, is in order, however, as explained below.

Ever since concerted efforts were made to determine the solar elemental abundances, particular attention has been paid to oxygen, as oxygen is the most abundant element that was not created in the Big Bang, and third only to H and He, which were created in the Big Bang. Furthermore, the principal reservoir of oxygen in Saturn and Jupiter, H_2O, was presumably the original carrier of the heavy elements to these planets. Thus, oxygen is centrally important to the question of origin of all things. Yet, its abundance in the sun has been revised constantly. As illustrated in Figure 2.2, the solar O/H values have gyrated up and down several times in the past four decades, starting with the classic work of Cameron (1973) to the present. The highest solar O/H value is the one recommended by Anders and Grevesse (1989), which remained the standard for a good fifteen years, only to be revised downward by nearly a factor of two in 2005 (Grevesse et al. 2005), and having crept up a bit since then. Not surprisingly, the solar Ar/H, also plotted in Figure 2.2, shows the same trend as O/H over time, though they are not completely proportional to each other, nor are they expected to be. Thus, one needs to be vigilant about changes in the photospheric abundance of oxygen and other elements such as argon that use oxygen as a reference.

In summary, the most robust elemental abundance determined to date in Saturn is that of carbon. At 9× solar, Saturn's C/H is a little over twice the C/H ratio in Jupiter. This is consistent with the core accretion model of giant planet formation, according to which progressively increasing elemental abundance ratios are expected from Jupiter to Neptune. Carbon is the only heavy element ever determined for all four giant planets (Figure 2.1), and indeed it is found to increase from 4× solar in Jupiter to 9× solar in Saturn, rising to 80(±20)× solar or greater in both Uranus (Sromovsky et al. 2011; E. Karkoschka and K. Baines personal communication, 2015) and Neptune (Karkoschka and Tomasko 2011), using the current solar C/H from Table 2.1. The same trend is also seen in the S/H ratio of Saturn compared to Jupiter, except for a fourfold increase from Jupiter to Saturn, but Saturn's S/H is less secure, as discussed above. The difference in the relative changes of C/H and S/H is worth noting, but caution should be exercised to not overinterpret it. This is because H_2S is a thermochemically condensible volatile in the gas giants, unlike CH_4. Saturn's S/H would benefit greatly from a fresh set of modern data. A similar fourfold increase is also seen in the P/H ratio in Saturn compared to that in Jupiter, and the relative change may be valid if the disequilibrium species PH_3 meets a similar fate in the tropospheres of Saturn and Jupiter. On the other hand, the observed 3× solar N/H ratio in Saturn seems puzzling, as it is about a factor of two less, not more, than Jupiter's N/H, contrary to the predictions of conventional formation models. However, the

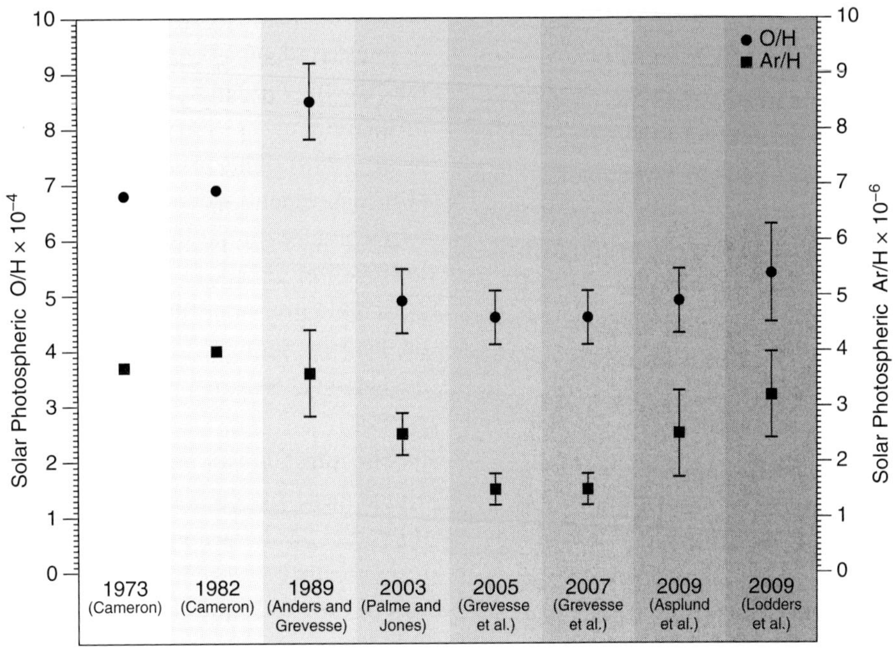

Figure 2.2 Time history of the solar photospheric O/H and Ar/H, showing only the major milestones. Although Ar/H shows the same trend as O/H, they do not track each other exactly. The solar photospheric values for O/H ($\times 10^{-4}$) and Ar/H ($\times 10^{-6}$) plotted here are, respectively, 6.8 and 3.7 (Cameron 1973), 6.9 and 4 (Cameron 1982), 8.5±0.7 and 3.6±0.8 (Anders and Grevesse 1989), 4.9±0.6 and 2.5±0.4 (Palme and Jones 2003), 4.6±0.5 and 1.5±0.3 (Grevesse et al. 2005, 2007), 4.9±0.6 and 2.5±0.8 (Asplund et al. 2009), and 5.4±0.9 and 3.2±0.8 (Lodders et al. 2009).

present data on Saturn's NH_3 in the 3 bar region do not rule out much greater ammonia abundance in the deep well-mixed atmosphere of Saturn, as discussed earlier. Presence of water is inferred in Saturn's troposphere indirectly from localized lightning observations, but no firm conclusions can be drawn from it on the global O/H ratio in Saturn. The Juno spacecraft is designed to measure and map water to several hundred bars in Jupiter's troposphere, which will provide a definitive answer on Jupiter's O/H ratio. In Jupiter at least, for which data are available for most of the heavy elements, except for O/H, it is striking that the heavy noble gases Ar, Kr, and Xe all display similar enrichment over solar by a factor of 2 to 3 (or, 2 to 2.5 with Lodders' solar values, Table 2.1), whereas enrichment of non-noble gas elements, carbon, nitrogen, and sulfur, is greater, ranging from 4 to 6. (Regarding S/H, from their clathrate hydrate model, Gautier et al. (2001) calculate an S/H enrichment in Jupiter that is twice the value measured by the Galileo probe (Table 2.1), or ~6× solar, and attribute the lower measured value to the loss of H_2S in troilite (FeS) in the inner solar nebula.) Though it may seem tempting and convenient to lump them all together and suggest that the heavy elements in Jupiter are enriched uniformly by a factor of 4±2 relative to their solar abundances, we advise caution.

The differences between the enrichments of the heavy noble gases and those of the non-noble gas heavy elements are apparently real, and may indicate two distinct populations arising from differences in the way noble gases were delivered (see also Section 2.4.3). Robust measurements of the same set of heavy elements at Saturn as at Jupiter are crucial to determining whether they have solar composition, as proposed by Owen and Encrenaz (2006), which will in turn have a bearing on the models of the origin, nature, and delivery of the Saturn-forming planetesimals. Similar efforts are now underway to measure key elemental abundances, particularly of O and C, in the atmospheres of giant exoplanets, and in using them to constrain formation conditions of exoplanetary systems (see Section 2.5).

2.2.2 *Isotopic Composition of Saturn's Atmosphere and Comparison to Jupiter*

Isotope ratios provide an insight into the conditions prevailing at the time of formation of the solar system and even early in the beginning of the universe. The giant planets and the terrestrial planets formed from much of the same initial inventory of material in the primordial solar nebula. Thus, the stable gas isotope ratios originally were the same in all planets. Abiotic fractionation of isotopes can occur due to escape of gases to space, loss to surface, phase change, or photochemistry. Indeed, fractionation of various stable gas isotopes has been found in the atmospheres of comparatively small solar system objects including Venus, Earth, Mars, and Titan (e.g. von Zahn et al. 1983; Niemann et al. 2010; Atreya et al. 2013; Webster et al. 2013; Mahaffy et al. 2014), and has been attributed mainly to the loss of their volatiles to space over geologic time. On the other hand, the sheer mass of the giant planets, in particular Jupiter and Saturn, does not permit loss of volatiles either by thermal, charged particle or other processes, hence their original isotopic ratios of elements are expected to be preserved, for all practical purposes. Thus, their present atmospheric isotope ratios should, in theory, also represent protosolar values.

Only a handful of the isotopes have been measured in Saturn's atmosphere: $^{13}C/^{12}C$, D/H, and an upper limit on $^{15}N/^{14}N$. In the atmosphere of Jupiter, $^{3}He/^{4}He$, $^{36}Ar/^{38}Ar$, and all isotopes of Xe except for ^{124}Xe and ^{126}Xe that together comprise 0.2% of total xenon in the sun have been measured, in addition to $^{13}C/^{12}C$, D/H, $^{15}N/^{14}N$. The measurement of noble gas isotopes in Jupiter was facilitated by in situ measurements with a mass spectrometer on the Galileo probe (GPMS). The isotope ratios for the atmosphere of Saturn and Jupiter are listed in Table 2.2. The helium, carbon, and xenon isotope ratios of Jupiter are nearly identical to the solar values, as expected.

The hydrogen isotope ratio, D/H, in Jupiter and Saturn is important for understanding the very beginnings of the universe and galactic evolution. Deuterium was formed following the Big Bang, but has been declining ever since because its destruction in the stars far outweighs any new creation. Thus, the D/H ratio in Jupiter and Saturn represents the protosolar value of D/H in the sun, in which it cannot be measured directly today. The value derived by the GPMS in Jupiter's atmosphere was thus the first measurement of the protosolar D/H ratio (Mahaffy et al. 1998). The result is in agreement with the D/H measurements done later with the short-wavelength spectrometer on the Infrared Space Observatory (ISO, Lellouch et al. 2001) and theoretical estimates (Table 2.2). This gives confidence in the D/H ratio measured by ISO in Saturn's atmosphere. Within the range of uncertainty, Saturn's D/H ratio is similar to that in Jupiter.

The nitrogen isotope ratio was measured in Jupiter's atmosphere by the Galileo probe mass spectrometer (Owen et al. 2001), and represented the first measurement of the protosolar $^{15}N/^{14}N$ ratio. The value in the sun is now available from the Genesis measurements (Marty et al. 2011) and is identical to the GPMS result for Jupiter. The ISO data give a slightly lower $^{15}N/^{14}N$, probably resulting from isotope fractionation below the ammonia clouds to which the ISO data apply. Note, however, that $^{15}N/^{14}N$ from ISO has large uncertainties that can easily envelop the GPMS result. Unlike Jupiter, only an upper limit on the $^{15}N/^{14}N$ ratio in Saturn's atmosphere is available. Using the Texas Echelon Cross Echelle Spectrograph (TEXES) on NASA's Infrared Telescope Facility (IRTF), Fletcher et al. (2014) observed spectral features of $^{14}NH_3$ and $^{15}NH_3$ in 900 cm^{-1} and 960 cm^{-1}, and derived an upper limit on the $^{15}N/^{14}N$ ratio of 2×10^{-3} for the 900 cm^{-1} channel and 2.8×10^{-3} for the 960 cm^{-1} channel. Though these values fall in the range of Jupiter's $^{15}N/^{14}N$ ratio, in the absence of actual measurement they represent only upper limits of $^{15}N/^{14}N$ in Saturn's atmosphere. In Figure 2.3, we show the best available data on this important ratio in the sun, interstellar medium, Jupiter, Saturn, and comets (from CN, HCN, and NH_2), which represent the original reservoirs of nitrogen (left panel, labeled "Primordial"), and in N_2 of the terrestrial planets and Titan, where nitrogen is secondary (right panel, labeled "Secondary"). The corresponding nitrogen isotope ratios are listed in Table 2.3. Nitrogen isotope fractionation is clearly evident in the terrestrial bodies. The lighter isotope floats up to the top of the atmosphere and escapes preferentially, leading to the build-up of the heavier isotope.

Table 2.2 *Elemental Isotopic Ratios in the Sun, Jupiter, and Saturn*

Elements	Sun	Jupiter	Saturn
$^{13}C/^{12}C$	0.0112[a]	0.0108 ± 0.0005[i]	0.0109 ± 0.001[o]
$^{15}N/^{14}N$	2.27 ± 0.0810^{-3} [b]	$(2.3\pm0.3)\times10^{-3}$ (0.8–2.8 bar)[j] $1.9(+0.9, -1.0)\times10^{-3}$ (0.2–1.0 bar)[k]	$<2.0\times10^{-3}$ [p] (900 cm^{-1} channel) $<2.8\times10^{-3}$ [p] (960 cm^{-1} channel)
$^{36}Ar/^{38}Ar$	5.5 ± 0.0[c]	5.6 ± 0.25[l]	
$^{136}Xe/Xe$	0.0795[a]	0.076 ± 0.009[l]	
$^{134}Xe/Xe$	0.0979[a]	0.091 ± 0.007[l]	
$^{132}Xe/Xe$	0.2651[a]	0.290 ± 0.020[l]	
$^{131}Xe/Xe$	0.2169[a]	0.203 ± 0.018[l]	
$^{130}Xe/Xe$	0.0438[a]	0.038 ± 0.005[l]	
$^{129}Xe/Xe$	0.2725[a]	0.285 ± 0.021[l]	
$^{128}Xe/Xe$	0.0220[a]	0.018 ± 0.002[l]	
$^{20}Ne/^{22}Ne$	13.6[a]	13 ± 2[l]	
$^{3}He/^{4}He$	1.66×10^{-4} [a] $(1.5\pm0.3)\times10^{-4}$ (meteoritic)[d,e,f,g]	$(1.66\pm0.05)\times10^{-4}$ [m]	
D/H	$(2.0\pm0.5)\times10^{-5}$ [a] $(2.1\pm0.5)\times10^{-5}$ [h] *protosolar* values	$(2.6\pm0.7)\times10^{-5}$ [m] $(2.25\pm0.35)\times10^{-5}$ [n]	$1.7(+0.75, -0.45)\times10^{-5}$ [n]

[a] Asplund et al. (2009), updated from Rosman and Taylor (1998);
[b] Marty et al. (2011), from Genesis;
[c] Vogel et al. (2011);
[d] Black (1972);
[e] Eberhardt (1974);
[f] Geiss and Reeves (1972);
[g] Geiss (1993);
[h] Geiss and Gloeckler (1998);
[i] Niemann et al. (1998);
[j] Owen et al. (2001), from Galileo probe mass spectrometer (GPMS) in situ measurements, largely below the NH_3 condensation level;
[k] Fouchet et al., (2000), from ISO infrared remote sensing measurements, largely above the NH_3 condensation level;
[l] Mahaffy et al. (2000), normalized to 1.0 for xenon isotopes measured, only ^{126}Xe and ^{124}Xe, which together make up 0.2% of the total xenon in the sun, could not be measured by the GPMS in Jupiter, and the xenon error bars are with respect to the ratio of each isotope to its non-radiogenic terrestrial value;
[m] Mahaffy et al. (1998), from GPMS;
[n] Lellouch et al. (2001), from ISO;
[o] Fletcher et al. (2009b);
[p] Fletcher et al. (2014).

2.2.3 Saturn's Interior

Saturn's interior may be probed indirectly through models and the measurement of the planet's mean density and gravitational moments (Fortney et al., this book) and measurement of the planet's dissipation factor (e.g. Remus et al. 2012). It has long been known that it is mostly made of hydrogen and helium, except for the presence of a central dense core. Detailed models show that in spite of its low global density of 0.688 g/cm^3, Saturn must contain a significant fraction

Table 2.3 *Nitrogen Isotope Ratios in the Solar System*

Objects	$^{14}N/^{15}N$	$^{15}N/^{14}N$ ($\times 10^{-3}$)
Sun (protosolar)[(a)]	441±5	2.27±0.03
Jupiter[(b)]	442±58	2.30±0.3
Saturn[(c)]	<357	<2.8
Interstellar medium (ISM)[(d)]	450±98	2.2±0.5
Comet Hale-Bopp (CN)[(e)]	140±30	7.1(+2.0, −1.3)
Comet Hale-Bopp (CN)[(f)]	140±35	7.1(+2.4, −1.4)
Comet Hale-Bopp (HCN)[(g)]	323±46	3.1(+0.5, −0.4)
Comet Hale-Bopp (HCN)[(f)]	205±70	4.9(+2.5, −1.3)
Comet Holmes (CN)[(f)]	139±26	7.2(+1.7, −1.1)
Comet Holmes (HCN)[(f)]	165±40	6.1(+1.9, −1.2)
Comets (NH_2)[(h)]	80–190	5.26–12.5
Earth	272	3.68
Venus[(i)]	272±54	3.7(+0.9, −0.6)
Mars (atmosphere)[(j)]	173±11	5.8±0.4
Mars (solid body)[(k)]	276.5±0.25	3.62
Titan[(l)]	167.7±0.6	6.0±0.02

(a) Marty et al. (2011), from Genesis sample analysis;
(b) Owen et al. (2001), Galileo probe mass spectrometer;
(c) Fletcher et al. (2014), IRTF;
(d) Dahmen et al. (1995);
(e) Arpigny et al. (2003);
(f) Bockelée-Morvan et al. (2008);
(g) Jewitt et al. (1997);
(h) Rousselot et al. (2014), derived from emission lines of NH_2 in twelve comets between 2002 and 2013, and the authors state that the range in $^{14}N/^{15}N$ is probably more appropriate to use than the average value of 127, which does not account for uncertainties because of the difficulty in accurately subtracting the solar continuum for each region of interest;
(i) Hoffman et al. (1979), Pioneer Venus;
(j) Wong et al. (2013), from MSL;
(k) Matthew and Marti (2001), from the oldest known Martian meteorite, ALH84001 (4.1 Gyr old);
(l) Niemann et al. (2010), Huygens-GCMS.

of its mass as heavy elements: between about 12 and 28 M_E (Nettelmann et al. 2013; Helled and Guillot 2013), corresponding to a mass fraction Z=0.13 to 0.29 or a global enrichment in heavy elements of 8.9 to 20 times the solar value.

In classical three-layer models, most of the heavy elements are embedded in a central core. The solutions of Helled and Guillot (2013), assuming a well-defined central core and a homogeneous abundance of heavy elements in the envelope, indicate a core with a mass between 10 and 20 M_E and an envelope with 4 to 8 M_E of heavy elements, corresponding to an enrichment of 4 to 8 times the solar value. The abundances of C-, N-, and S-bearing species in the atmosphere account for the lower limit of this range, meaning that the elusive O can be enriched only as much as C. As the "total" enrichment in heavy elements is constrained by the interior models, the addition of other species (e.g. silicates) into the envelope would mean less enrichment for others, which would make O even less enriched, implying a C/O ratio that is very likely to be supersolar in Saturn's atmosphere. Solutions by Nettelmann et al. (2013) add one degree of freedom, the possibility for the abundance of heavy elements to vary in the envelope. That leads to the possibility of even smaller core masses, but with a deep envelope that is enriched in heavy

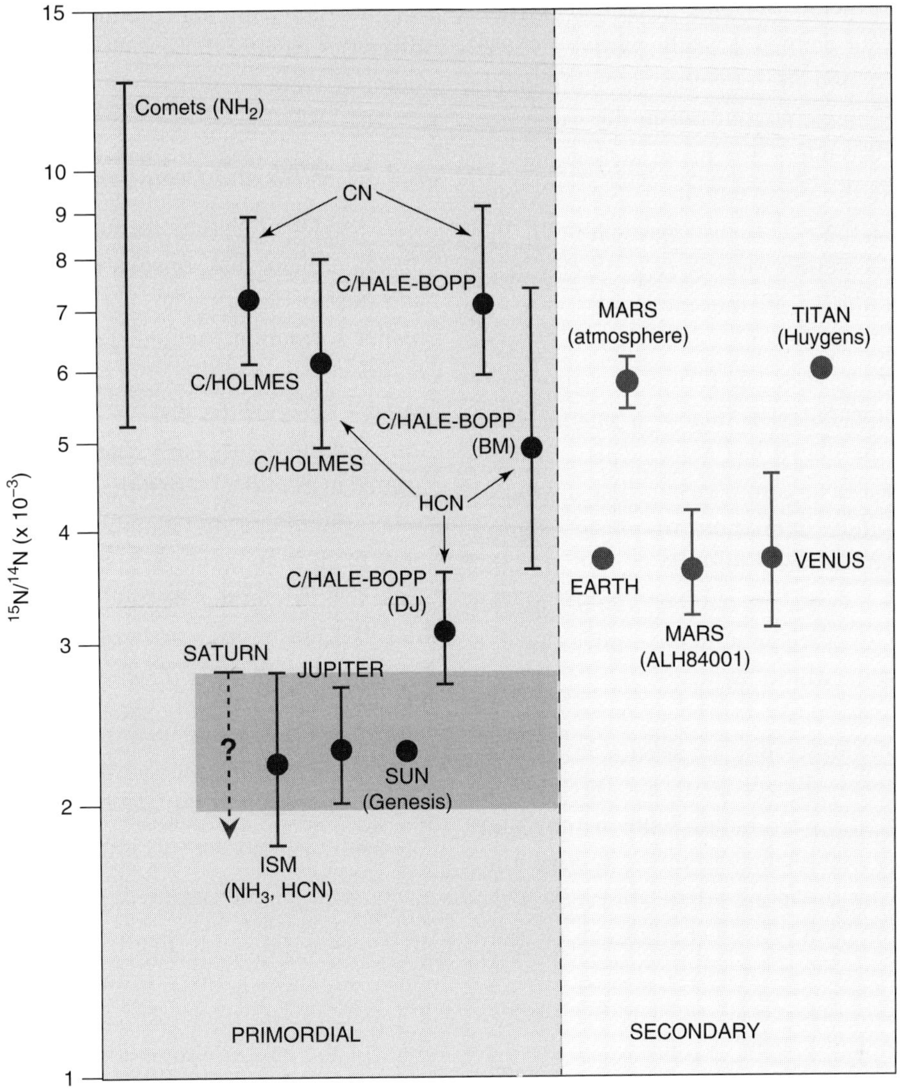

Figure 2.3 A comparison of the current nitrogen isotope ratios in primordial (sun, ISM, Jupiter, Saturn, and comets; left panel) and secondary (Venus, Earth, Mars, and Titan; right panel) reservoirs. The values for the secondary reservoirs illustrate loss of nitrogen from these objects over geologic time. A subset of the $^{15}N/^{14}N$ ratios based on then available data can be found in Owen et al. (2001) and Atreya et al. (2009). References for current ratios are listed in Table 2.3. For Mars, the value in ALH84001 is also shown, and represents $^{15}N/^{14}N$ for solid Mars, which is nearly the same as for the atmosphere of Earth, as expected, considering that these planets presumably acquired their original building material from the same source. The substantially higher $^{15}N/^{14}N$ in Mars atmosphere today compared to Earth's atmosphere is the consequence of thermal- and solar–wind-induced escape of nitrogen from (lighter and non-magnetic) Mars over geologic time. Higher $^{15}N/^{14}N$ in Titan's present atmosphere compared to Earth's atmosphere may reflect the value in its building blocks, coupled with atmospheric loss over time. In comets, a similar value of $^{15}N/^{14}N$ has been measured in CN of a dozen comets, and in CN and HCN of comets Hale Bopp and Holmes. However, the $^{15}N/^{14}N$ in HCN in the latter two comets is substantially different from that in CN, and different observers (BM: Bockelée-Morvan et al. 2008, DJ: Jewitt et al. 1997) report different values, though all have large error bars. $^{15}N/^{14}N$ from NH_2 in a number of comets (Rousselot et al. 2014) is shown as a range, which is more appropriate than an average value because of the difficulty of accounting for solar continuum for each emission feature of NH_2, according to the authors. The reader is referred to Atreya et al. (2009) for additional discussion on the cometary $^{15}N/^{14}N$ and implications for Titan. For Saturn only upper limits from 900 cm^{-1} and 960 cm^{-1} spectral channels are available. Parts of this figure are taken from Figure 7.2 of Atreya et al. (2009), with permission from Springer. (A black-and-white version of this figure appears in some formats. For the color version, please refer to the plate section.)

elements. Thus, the picture that emerges is one in which a core is either well-defined or only partially mixed with the envelope, and the envelope is significantly enriched in heavy elements, but not in the same way for all species. Accounting for material that may be partially mixed in the deep envelope, Saturn's core appears to have a mass that is consistent with that required by core-accretion models (e.g. Ikoma et al. 2001). The enrichment of the envelope is to be explained either by planetesimal impacts or by upward mixing and/or core erosion. The former is traditionally difficult because the cross-section of a mature giant planet (i.e. when the planet has accreted its full mass and does not possess a circumplanetary disk anymore) is small. For example, simulations of impacts during the great heavy bombardment indicate that of an initial disk mass of 35 M_E, only between 0.05 and 0.1 M_E hit Saturn (the values are about double for Jupiter, due to a larger focusing factor; Matter et al. 2009).

Core erosion is made possible from a physical point of view because of the miscibility of species in metallic hydrogen (Wilson and Militzer 2010, 2012). However, while it is effective at Jupiter, Saturn's smaller envelope implies that only about 2 M_E may be mixed upward from a massive central core, assuming a 10% efficiency of the process (Guillot et al. 2004). A higher efficiency, or, more likely, the upward mixing of an initially heavy-element-rich primordial envelope, could explain the planet's heavy-element-rich atmosphere. Variations in the elemental composition (such as those leading to a supersolar C/O ratio) could be explained by a selective retention of species (e.g. silicates, water) in the deeper regions.

2.3 Saturn's Formation: Hydrodynamical Point of View

Standard models of Saturn's interior with a core surrounded by a hydrogen-helium envelope that is enriched in heavy elements fits well with the picture of its formation by core accretion followed by the capture of the gas envelope from the protoplanetary disk. However, considerable uncertainties remain, both on the internal structure itself and on formation models. To understand the end-to-end origin and evolution of Saturn, it is important to consider, then, the starting protosolar disk, the manner of formation and growth of the core, Saturn's circumplanetary disk, and any insight from the moons and rings. This section discusses each of these aspects from a hydrodynamical point of view.

2.3.1 Birth and Evolution of the Protosolar Disk

Any model of Saturn formation must begin with the protoplanetary disk, or solar nebula, from which the gas and dust of Saturn were derived. Constraints on giant planet formation include the disk lifetime, elemental composition (specifically, C/H, O/H, etc.,), and the overall mass of the disk. A low-opacity massive disk may fragment very early due to disk instability, but we argued in Section 2.1 that the overall architecture of our solar system does not match such an event. Core accretion, then, is constrained to build Saturn within a plausible disk lifetime. The model of Dodson-Robinson et al. (2008) provides a particular example of the detailed specification of a solar nebula model needed to build Saturn in an acceptably short length of time.

2.3.2 Formation and Growth of Giant Planet Cores

Core Formation

In the framework of the core accretion model, the first step is obviously to form a 10-Earth-masses core. In the classical view, gravity is the dominant process, and kilometer-sized planetesimals merge when they collide. In the end, a population of so-called "oligarchs" is produced, which accrete all the planetesimals within reach of their orbit (Kokubo and Ida 1998). Their mass is then typically $0.05(r/1AU)^{0.75}$ M_E.

Another model suggests that centimeter-sized dust aggregates are concentrated by vortices in the gas up to the point where the concentration of solids becomes gravitationally unstable, leading possibly directly to the formation of solid bodies of hundreds of kilometers (Johansen et al. 2007; see Turner et al. 2014a for a review of turbulent processes).

It has been shown recently that such embryos are very efficient at accreting pebbles, i.e. cm-sized aggregates moderately coupled to the gas (Lambrechts and Johansen 2012; Morbidelli and Nesvorny 2012). As such pebbles drift radially, nothing stops this growth, whose rate is exponential. Pebble accretion is to date the most promising mechanism to form a few-Earth-masses core within the lifetime of a protoplanetary disk. Furthermore, Lambrechts et al. (2014) show that pebble accretion naturally stops when the core becomes massive enough to carve a dip in the gas that stops the radial drift of pebbles (~20 M_E). The end of the accretion of solids by the core then triggers the onset of the runaway accretion of gas.

Planet Migration

Cores and planets interact gravitationally with the gas disk. This leads to exchanges of energy and angular momentum, hence to a variation of the orbit of the planet. This is called planetary migration. Bodies of less than roughly 50 M_E do not perturb the gas profile much and are in the type I migration regime (Ward 1997). It has been shown in the last decade that type I migration can be directed inwards or outwards, depending on the thermodynamics of the gas disk (Paardekooper and Mellama 2006; Paardekooper et al. 2010, 2011). Typically, migration would be directed inwards in the outer, optically thin regions of the disk, while it can be directed outwards in the inner, optically thick regions. This opens the possibility of convergent migration towards a zero-torque migration radius where bodies of few Earth masses should gather, and hopefully merge (Lyra et al. 2010; Cossou et al. 2013). In general, there are two such radii, and their locations depend on the disk structure (Bitsch et al. 2013). One is inside the snowline and vanishes when the accretion rate in the disk decreases, and one is beyond the snowline, moving from roughly 10 AU to 4 AU as the disk ages (Bitsch et al. 2014). In any case, this new vision of planet migration (see Baruteau et al. 2014 for a recent and complete review) opens the possibility to keep the core of a giant planet safe at the zero-torque migration radius, instead of losing it into the star. It can then grow by slowly accreting its gas envelope.

When it is massive enough, the planet will open a gap in the gas disk (Papaloizou and Lin 1984; Crida et al. 2006), and thereby leave the type I migration regime. Planets opening gaps are in the type II, slower mode of migration, in which they follow roughly the viscous evolution of the disk (Lin and Papaloizou 1986; Crida and Morbidelli 2007; Dürmann and Kley 2015).

It should be noted that the migration of the Jupiter-Saturn pair is, however, more complex than that of a single giant planet. Jupiter and Saturn most likely enter in mean motion resonance, which can reverse their migration (Masset and Snellgrove 2001). A fine tuning of the disk parameters allows the Jupiter and Saturn pair to avoid any significant migration in the protosolar nebula (Morbidelli and Crida 2007). Another possibility is that Jupiter grew and migrated inwards first, then was caught up by Saturn, which made the pair migrate back outwards (Walsh et al. 2011). In this so-called "Grand Tack" scenario, the main asteroid belt is satisfactorily reproduced, and Jupiter's excursion sculpts the inner disk of embryos and planetesimals in a very favorable way for the formation of the terrestrial planets. It implies that Saturn came as close as about 2 AU from the Sun. Little room is left for gas accretion in this scenario, as the final masses of Jupiter and Saturn are ideal for such a tack, but Saturn could have been half its present mass, gaining the rest on the way out.

In any case, an unavoidable consequence of migration is that Saturn most likely was in resonance with Jupiter, on a circular orbit ~8 AU from the Sun, when the protosolar nebula dissipated. It reached its final orbit ~650 million years later, during a global dynamical instability among the giant planets, often referred to as the "Nice model" (Tsiganis et al. 2005).

2.3.3 Formation of Saturn and Its Circumplanetary Disk

As discussed above and illustrated in Figure 2.4, giant planets open gaps in the protoplanetary disk. The neighborhood of their orbit is depleted, splitting the disk into an inner and an outer disk. While the width of the gap is set solely by the Hill radius of the planet, the depth of the gap increases with the planet mass and is also a smooth function of the viscosity and aspect ratio of the disk (Crida et al. 2006). Even for massive planets like Saturn or Jupiter, the opening of the gap does not terminate gas accretion. Indeed, as can be seen in Figure 2.4, gas still flows towards the planet through the spiral wake. As a consequence, the final phase of runaway gas accretion corresponding to the collapse of the gas envelope (see Section 2.1) has no reason to end until a few Jupiter masses are reached. However, numerical simulations show that massive planets are capable of creating their own gas disk around them, inside the gap (Bate et al. 2003; Ayliffe and Bate 2012).

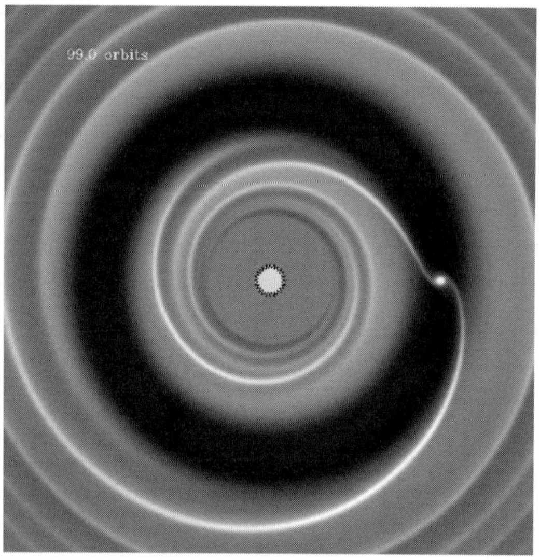

Figure 2.4 Gas density map from a 2D hydrodynamical simulation. Light color corresponds to high density, black to low density. The star is in the center of the image; the giant planet is on the right. The black annulus is the gap around the planetary orbit, and the white spot around the planet (not shown) is the CPD. (A black-and-white version of this figure appears in some formats. For the color version, please refer to the plate section.)

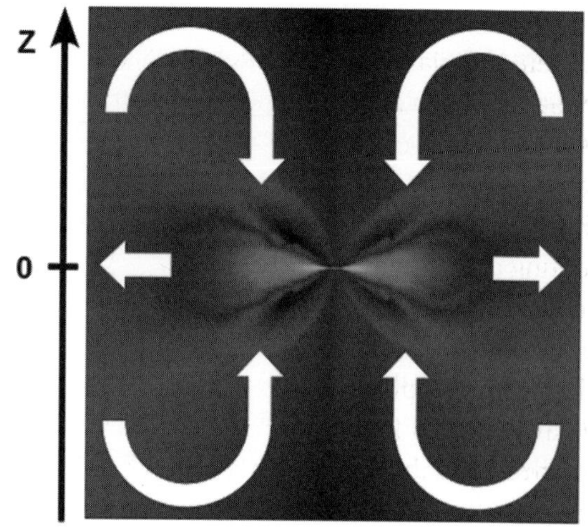

Figure 2.5 Vertical cut through the CPD of a Jupiter mass planet in a 3D simulation. The horizontal axis represents the distance to the star and the vertical axis is perpendicular to the orbital plane of the planet, with the planet in the center. The color corresponds to the gas density, and the arrows sketch the meridional circulation of the gas. (A black-and-white version of this figure appears in some formats. For the color version, please refer to the plate section.)

This circumplanetary disk (hereafter CPD) is poorly resolved in Figure 2.4, but has been studied in greater detail in other works.

The simulations reveal that the gas flow around a giant planet is 3D, and cannot be accurately modeled by a 2D simulation. Actually, most of the gas that reaches the CPD comes from a vertical direction, which is perpendicular to the orbital plane (Bate et al. 2003; Machida et al. 2008; Ayliffe and Bate 2009a,b; Tanigawa et al. 2012; Szulagyi et al. 2014). An explanation for this unexpected flow pattern is given by Morbidelli et al. (2014): in the upper layers of the disk the gravitational force from the planet is weaker, and therefore the gap tends to be narrower than in the midplane. Gas comes in, and being not supported by pressure, falls towards the midplane, where the planet ejects it back out of the gap, still in the midplane. Therefore, a meridional circulation pattern is created, as sketched by white arrows in Figure 2.5: gas ejected from the midplane by the planet expands vertically further away, and then slowly penetrates inside the gap from the upper layers, before falling back on the midplane. Note that such a full loop is much longer than an orbital period. Part of this vertical inflow falls on the CPD and the planet, contributing to the planet's growth.

This may have strong implications on the nature and rate of the solids in this flow. By the time a giant planet forms, dust is supposed to have sedimented in the midplane of the disk, being only stirred by turbulence. Micrometric grains, well coupled to the gas, should follow the gas flow, but larger aggregates may well be unable to reach the planet and its CPD.

In isothermal simulations (the only ones available so far in the literature), gas falls at a supersonic speed onto the CPD and shocks at the surface of the latter. In the absence of viscosity in the CPD, gas should in principle reach its centrifugal radius and orbit around the planet forever. This opens the possibility that the CPD acts as a bottleneck for planet growth, and limits the final mass of the giant planets (Rivier et al. 2012). There are good reasons to think that the CPD is really inviscid (Turner et al. 2010, 2014b; Fujii et al. 2011, 2014). Hence, Szulagyi et al. (2014) have measured the different sources of angular momentum loss in the CPD (torque from the central star, contact with the infalling gas), using mesh refinement around the planet in a 3D global simulation, allowing for the gap to form accurately. They deduce a mass-doubling time on the order of half a million years for a Jupiter-mass planet. This is much slower than the standard, 1D model of Pollack et al. (1996), and could be the reason why Saturn, like most giant exoplanets (see section 2.5), did not grow more massive than Jupiter: depending when gas accretion starts, there is not enough time to grow super giant planets.

This type of research is demanding, both in terms of computing capability and time, hence it is still open. In particular, non-isothermal simulations are necessary to better determine the structure of the CPD. Its temperature has strong implications for the composition of the solids available to form the satellites and for the chemical species in the gas. Novel and promising results on gas accretion by a giant planet are expected in the future.

2.3.4 Formation of Saturn, from Consideration of the Formation of Moons and Rings

For a comprehensive understanding of the formation and evolution of Saturn, it is necessary also to gain an insight into the formation of Saturn's moons. Here we provide a brief discussion of this aspect of the Saturn system. About 60 satellites with confirmed orbits have been detected so far. Among them, 23 have quasi-circular orbits of radius smaller than 2 million kilometers in the plane of Saturn's equator: the so-called regular satellites. The others have eccentric, inclined (sometimes even retrograde), and larger orbits. They are called irregular satellites, and are supposedly captured. Hence, the irregular satellites do not inform much about Saturn's formation. In contrast, the rings and the regular satellites most likely formed together with Saturn in some way, and therefore provide constrains.

Titan, the largest moon of Saturn, dominates the population of satellites, being 60 times more massive than the second largest moon, Rhea. Titan's composition can provide further insight into the physico-chemical conditions prevailing in Saturn's CPD that must have played a crucial role in the make-up of Titan's building blocks. In situ measurements with the Huygens gas chromatograph mass spectrometer (Niemann et al. 2005, 2010) revealed for the first time that the bulk atmosphere of Titan is approximately 94% by volume nitrogen (N_2) and ~6% methane (CH_4). Methane may have originated on Titan, but direct external contribution is also possible. N_2, on the other hand, is almost certainly "secondary," i.e. instead of being delivered directly as N_2, it resulted from other nitrogen-bearing molecules originally captured in Titan's building blocks. Before nitrogen was actually detected on Titan by Voyager in 1980, Atreya et al. (1978) showed that the solar UV photolysis of ammonia (NH_3) could produce a substantial atmosphere of nitrogen on Titan in the past, which was eventually confirmed by Huygens in 2005. The dissociation of ammonia by impact shock heating has also been proposed (Jones and Lewis 1987; McKay et al. 1988; Sekine et al. 2011; Ishimaru et al. 2011). While it seems like an attractive hypothesis, it faces many hurdles, including the removal of accompanying copious oxygen-bearing species and hydrogen, not found on Titan (see, e.g. Atreya et al. 2009). The fact that Titan's N_2 is not primordial but formed from ammonia has important implications for Saturn's CPD, considering possible scenarios of Titan's formation. Similarly, the origin of Titan's methane has a bearing on Saturn's CPD, so it is also discussed here briefly.

Two possibilities for the origin of Titan's methane have been proposed: production on Titan, or delivery to Titan. In the former case, methane was produced by hydro-geochemistry, i.e. water-rock reactions or serpentinization in the interior of Titan during its accretionary heating phase, when water was presumably in contact with the rocky core (Atreya et al. 2006; 2009). In this scenario, H_2 liberated in serpentinization reacts with primordial carbon in the form of CO, CO_2, or carbon grains in a metal-catalyzed Fischer–Tropsch process to produce methane. Mousis et al. (2009a) surmised that if water-rock reactions were responsible for Titan's methane, the D/H ratio in Titan's CH_4 ($\sim 1.3 \times 10^{-4}$) should then reflect the value in Titan's water ice. As no measurements are available for D/H in Titan's water ice, they assumed that the D/H ratio measured in the water vapor plumes of Enceladus could serve as a proxy for the D/H in Titan's water. The Enceladus D/H value is 3×10^{-4}, which is more than twice the value in Titan's CH_4. This discrepancy led Mousis et al. (2009a) to propose that Titan's methane was trapped in its building blocks, which agglomerated from icy grains condensed in the protosolar nebula. In this scenario, Titan's methane would originate from ISM and its inferred D/H value would have resulted from the isotopic exchange with the nebula's hydrogen that occurred until it condensed and was agglomerated by the building blocks of Titan. This conclusion was supported by the measurements of D/H in water in six Oort cloud comets available at that time, all of which have a value that is nearly identical to that measured in Enceladus' H_2O plumes by the Cassini ion and neutral mass spectrometer, and corresponded to more than two times the value of D/H in Titan's methane. However, later observations of Jupiter family comet Hartley 2 yield a D/H ratio of 1.56×10^{-4} in water (Hartogh et al. 2011), which is similar to the value in Titan's CH_4 within the range of uncertainty for both objects. Another Jupiter family comet, 67P/Churyumov–Gerasimenko, on the other hand, yields D/H=5.3×10^{-4} in water (Altwegg et al. 2015), four times higher than in Titan's CH_4. Neither of these two comets has a D/H ratio in water similar to that assumed for Titan's water ice, the Enceladus value. Though the argument of methane trapping from the protosolar nebula appears to be weakened in view of these findings, it remains a plausible mechanism that needs to be validated by future observations, including, for example, direct D/H measurement in Titan's water, D/H in a large number of comets, as well as laboratory studies and modeling.

Two main models have been proposed for the formation of large satellites of the gas giant planets, including Titan. In the first model, the satellite formation takes place in a dense and hot disk at the early stages of the gas giant planet formation (Prinn and Fegley 1981, 1989; Lunine et al. 1989), while the other model uses a thin and cold disk to depict satellite formation (Canup and Ward 2002). In the former model, the chemical composition of Titan's proto-atmosphere would have been primarily CH_4 and NH_3. These gases would have been produced, prior to planetesimal condensation, from CO and N_2 initially present in the dense, hot, and chemically active CPD (Prinn and Fegley 1989; Sekine et al. 2005). This scenario may be ruled out for the bulk of Titan's nitrogen on the basis of the nitrogen isotope ratio, with the caveat about atmospheric escape, as discussed below. In Titan's atmosphere, $^{14}N/^{15}N = 167.7$ (Niemann et al. 2010), which is much less than, not similar to, the value in Saturn (>357; Table 2.3, Section 2.2.1), which implies that the ammonia accreted by Titan did not originate from the protosolar nebula. On the other hand, the present nitrogen isotope ratio depends on the evolutionary history and the processes of escape of nitrogen from Titan early on and in the past 4.5 Gyr, which are very poorly constrained. Even on Earth and Venus, which have atmospheres as dense as or even denser than Titan, escape processes have shaped their present atmospheric isotope ratios. Additional modeling and observations are needed to resolve the issue of evolution of Titan's nitrogen isotope ratio over time.

The other model suggests that icy planetesimals were actively supplied into the CPD from Saturn's feeding zone in the solar nebula (Canup and Ward 2002; Alibert and Mousis 2007). In this scenario, the chemical composition of Titan's proto-atmosphere would derive from that of CO- and N_2-rich icy planetesimals formed at low temperature (~20 K) in the protosolar nebula (see Section 2.4.1 for details concerning the composition of the protosolar nebula). However, similar to the previous case, this scenario is found to be inconsistent with the low $^{14}N/^{15}N$ ratio measured in Titan's nitrogen, but with the caveat of nitrogen escape mentioned above. In order to solve these discrepancies, Mousis et al. (2009b) proposed that Titan was formed from icy planetesimals initially produced in the solar nebula that were partially devolatilized during their migration within Saturn's CPD. By doing so, Titan's building blocks preserved the ammonia and methane they acquired from the protosolar nebula and released most of the carbon monoxide and nitrogen prior to satellite formation. However, as discussed above, production of methane on Titan by serpentinization, rather than direct delivery of CH_4, is also an attractive possibility.

Considering the lack of full complement of relevant observational constraints for Titan, Saturn, and the comets, clear discrimination between the two scenarios is not possible at this time. It is also plausible that both mechanisms could have played a role to varying degrees. Nevertheless, above considerations about Titan's composition still allow us to place important constraints on the thermodynamic state of Saturn's CPD at the time of formation of its largest satellites. In view of the low $^{14}N/^{15}N$ ratio measured in Titan's atmosphere, it seems that the CPD may not have been warm and dense enough to allow in situ condensation of its building blocks, but available observational constraints are insufficient to make a firm statement, as discussed above. On the other hand, a temperature-density gradient probably did exist throughout Saturn's CPD, not important enough to allow the vaporization of water ice at Titan's formation zone, but probably sufficient to explain why Titan's primordial nitrogen reservoir is NH_3 and not N_2, as is the case for Saturn (Mandt et al. 2014).

The above two models focus on the dominant body only (Titan) and somehow disregard the system of regular satellites of Saturn as a whole. However, a recent scenario for the formation of the regular satellites interior to Titan provides constraints on Saturn's history and internal structure. The rings spread, and spread faster when they are more massive, so that after about 4 Gyr of evolution, they should have roughly the present mass, whatever their initial mass (Salmon et al. 2010). Hence, it is possible that they originally were thousands of times more massive than now (e.g. Canup 2010). As the rings spread beyond the Roche radius, they agglomerate into moonlets, which migrate outwards due to their interaction with the rings (Charnoz et al. 2010). Numerical simulations show that this process can generate the 10 regular satellites inside Titan, and even explain the irregular silicate cores of the 5 largest ones (Charnoz et al. 2011). Crida and Charnoz (2012) solved analytically the equations governing the formation and migration of satellites from the spreading of rings beyond the Roche limit and found that the mass-distance relation in a system of satellites formed in this manner must follow a particular power law, which is represented in Figure 2.6. The agreement with the observed distribution supports this model. Even Titan lies on the theoretical line, which could be a coincidence, or not. Iapetus doesn't fit in this model and is not shown in the figure (it would be further on the middle right), but Iapetus is thought to have formed concurrently with Saturn in the circumplanetary disk (Castillo-Rogez et al. 2009).

It is possible that the regular satellites inside Titan formed after Titan and Saturn, from the spreading of initially massive rings. But for this to happen within the age of the solar system, strong tidal dissipation is needed inside Saturn. Tidal dissipation is characterized by a dimensionless factor, generally denoted as Q; the lower the value of Q, the stronger the dissipation. Charnoz et al. (2011) found that with Q on the order of 1700, as argued by Lainey et al. (2012, 2015), the formation of Saturn's satellite system takes about 3.5 Gyr (present satellite crater density record does not provide unambiguous evidence either in favor of or against this timescale). The standard value of Saturn's Q used to be an order of magnitude larger (corresponding to ten times less dissipation), but one should consider that this high value was supported by the constraint that the satellites were supposed to have hardly moved since the formation of the solar system, which may be incorrect. In contrast, the Lainey et al. value is based on observations and is consistent with this new model for satellite formation. In the end, models of the formation of the satellites allow us to place constraints on the efficiency of the dissipation inside Saturn, and hence on its interior. Remus et al. (2012) show that low values of Q (high dissipation rates) are possible in the framework of a model in which tidal dissipation occurs at the interface between the core and the envelope as a result of different anelastic deformations. Values as low as 10^3 require a specific range of values of the shear modulus and viscous modulus in the core. Unfortunately, these two quantities are almost unknown, given the uncertainties on the size, composition, and physical state of the core.

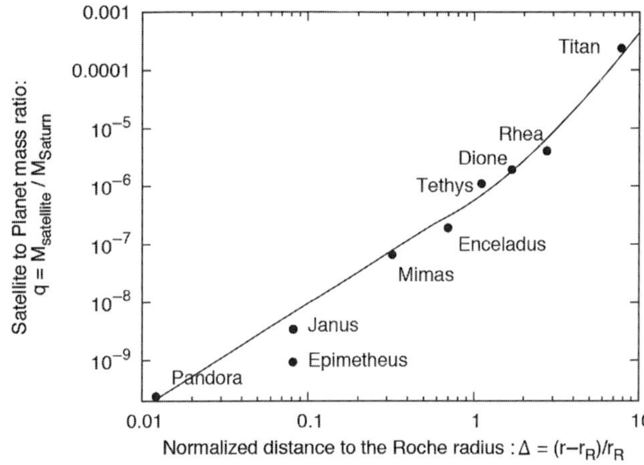

Figure 2.6 Satellite-to-planet mass ratio q as a function of the normalized distance to the Roche radius, located at the outer edge of the rings, 140,000 km away from Saturn's center. The line corresponds to the model described in Section 2.3.4. Adapted from Figure 1 of Crida and Charnoz (2012). Reprinted with permission from AAAS.

2.4 Saturn's Formation: Chemical Point of View

Just as the hydrodynamical scenario discussed in the previous section provides insight into Saturn's origin, chemical evolution of the protosolar disk and the manner in which volatiles are sequestered in grains or planetesimals, together with their nature and delivery to Saturn, provide valuable constraints to the models of Saturn's formation and evolution. This section elaborates on these processes.

2.4.1 Chemical Evolution of the Protosolar Disk

Formation scenarios of the protosolar nebula invoke two reservoirs of ices, namely an inner and an outer reservoir, which took part in the production of icy planetesimals. The first reservoir contains ices (mostly water ice) originating from the ISM, which were initially vaporized due to their proximity to the Sun. With time, the decrease of temperature and pressure conditions allowed the water in this reservoir to condense at ~150 K in the form of microscopic crystalline ice on the surface of pre-existing refractory grains (Kouchi et al. 1994). The other reservoir, located at larger heliocentric distances, is composed of ices originating from the ISM that were preserved when entering into the disk. In this reservoir, water ice was essentially in the amorphous form and the other volatiles remained trapped in the amorphous matrix (Notesco and Bar-Nun 2005). The exact localization of the boundary between these two reservoirs, corresponding to the initial location of the so-called "iceline," depends on the assumed thermal structure of the disk, which is still poorly constrained. Optically thin disks such as the steady nebula model of Hayashi (1981) predict that the water iceline is located just beyond the Main Belt (also known as the Asteroid Belt, which is located between the orbits of Mars and Jupiter, or 2.2–3.2 AU). On the other hand, optically thick models of the protosolar nebula suggest that the water iceline might have been initially up to ~30 AU from the Sun (Chick and Cassen 1997).

The O-, C-, and N-bearing ices delivered from ISM to the protosolar nebula are expected to be essentially constituted from H_2O, CO, CO_2, CH_3OH, CH_4, N_2 and NH_3, with H_2O, CO, CO_2 and N_2 being the most abundant molecules in decreasing order (Öberg et al. 2011a; Gibb et al. 2004). H_2O ice is expected to be dominant because (i) of its high abundance (due to the cosmic abundance of H and O) and (ii) it is by far the first volatile to condense as the temperature decreases in the nebula.

Regardless of the considered volatile reservoir in the protosolar nebula, dust and ice particles coagulated until they reached cm-sized pebbles. Once formed, these pebbles agglomerated into large planetesimals (10–1000 km) by streaming instabilities (Youdin and Goodman 2005; Johansen and Youdin 2007; Johansen et al. 2009) and formed the cores of the giants on timescales that were sufficiently short to allow in situ formation of these planets prior to their migration in the protosolar nebula (Lambrechts et al. 2014). Pebbles and planetesimals formed in the outer reservoir should have coagulated from pristine amorphous ice originating from ISM. In contrast, pebbles and planetesimals formed during the cooling of the inner reservoir coagulated from a mixture of microscopic icy grains made of pure condensates, stoichiometric hydrates (such as NH_3-H_2O) and clathrates, whose proportions depended on the availability of water ice and the temperature to which the disk had cooled down.

2.4.2 Delivery of Volatiles to Saturn via the Accretion of Planetesimals

Several hypotheses relating the thermodynamic evolution of the protosolar nebula to the formation conditions of the giant planets have been developed in order to interpret their observed volatile enrichments. In particular, the volatile enrichments observed in the giant planets can be explained by the accretion of icy planetesimals and their vaporization in the envelopes at the time of their growth from nebular gas. The two main scenarios proposed in the literature, each based on the hypothesis that the giant planets accreted planetesimals originating from one of the two abovementioned reservoirs of ices, are discussed below.

Delivery of Amorphous Ices to Saturn

Owen et al. (1999) proposed a cold icy planetesimal model, according to which the volatile enrichments observed by the Galileo probe in Jupiter result from the accretion of planetesimals agglomerated from amorphous ice at temperatures below approximately 30 K (such low temperatures are needed to trap N_2 and Ar; Owen et al. 1999). Owen et al. postulated that either Jupiter was formed at large heliocentric distances of 40–50 AU, where the cold temperature favored the preservation of amorphous ice in the disk, and then migrated to its current location, or the protosolar nebula was much cooler at the current location of Jupiter (~5 AU) than what is predicted by current turbulent accretion disk models. In either case, the icy material originated from the protosolar cloud and survived the formation of the protosolar nebula. If correct, this scenario predicts that the volatile enrichments should be uniform (also, Owen and Encrenaz 2006), since volatiles are not fractionated when trapped in amorphous ice. However, as discussed in Section 2.2.1, current analysis of the Galileo Probe data shows that the enrichment of the observed heavy elements spans a range of 2 to 6 times the current solar elemental abundances. For Saturn, key data needed to assess the validity of the icy planetesimal model or another model are presently lacking. Noble gases are not measured. Amongst non-noble gases, the only heavy element with a robust value is carbon, with C/H approximately 9× solar (Table 2.1). Sulfur enrichment is similar to carbon, but the result is tentative (Section 2.2.1). NH_3 is a good measurement, but currently provides N/H only in the 1- to 3-bar region. As discussed earlier (Section 2.2.1), it is far from certain that the N/H value in the deep well-mixed atmosphere of Saturn is going to be similar; it could be greater. If one assumes that the N/H in Saturn's deep atmosphere is unchanged from the value at 3 bars, then the C-enrichment is greater than N-enrichment by a factor of 3, not the same, which would argue against the cold icy planetesimal model as presented in Owen et al. (1999). On the other hand, C/S would favor it if the H_2S result were confirmed by future observations.

Delivery of Crystalline Ices to Saturn

An alternative interpretation of the volatile enrichments measured in Jupiter is based on the hypothesis that most of the volatiles were trapped in clathrates in the giant planet's feeding zone (Gautier et al. 2001; Alibert et al. 2005a, 2005b). These authors assumed that Jupiter's building blocks formed in the inner zone of the protosolar nebula, in which the gas phase has been enriched at early epochs by the vaporization of amorphous ice entering from the Interstellar Medium (ISM). During the cooling of this region of the disk, water vapor crystallized and trapped the volatiles in the form of clathrates or stoichiometric hydrates in the 40–90 K range instead of condensing at lower temperatures. These ices then agglomerated and formed the solids that were ultimately accreted in the envelope of the growing Jupiter. These scenarios, which assume the full (100%) clathration of volatiles, are based on the hypothesis that the amount of available crystalline water ice was large enough ($H_2O/H_2 > 2\times(O/H)_{protosolar}$) to trap the other volatiles in the feeding zone of Jupiter. Later studies have shown that it is also possible to explain the volatile enrichments in Jupiter in terms of the accretion and the vaporization in its envelope of icy planetesimals made from a mixture of clathrates and pure condensates (Mousis et al. 2009c, 2012), assuming a full protosolar composition for the gas phase of the disk and provided that the disk's temperature decreased down to ~20 K at their formation location. Figure 2.7 represents a clathration/condensation sequence of volatiles that has been used by Mousis et al. (2012) to interpret the volatile enrichments in Jupiter.

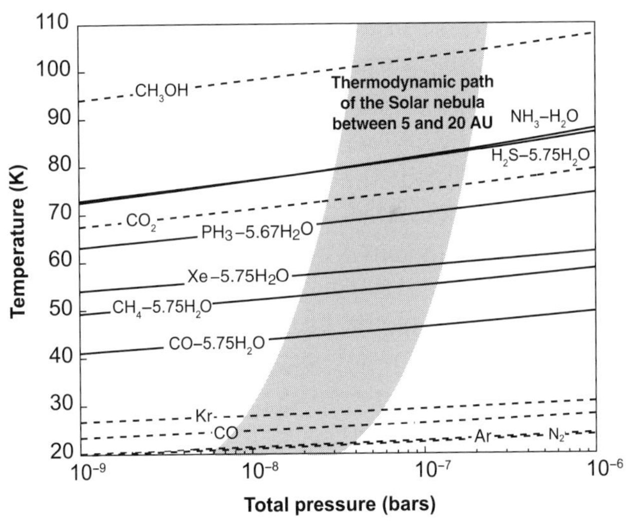

Figure 2.7 Formation conditions of icy planetesimals in the solar nebula. Equilibrium curves of hydrate (NH_3–H_2O; solid line), clathrates (X-5.75H_2O or X-5.67H_2O; solid lines), and pure condensates (dashed lines) are superimposed with the thermodynamic path of the solar nebula in the 5–20 AU range, assuming full clathration efficiency. Elemental abundances are solar, with molecular ratios specified in Mousis et al. (2012). Species remain in the gas phase above the equilibrium curves. Below, they are trapped as clathrates or simply condense. The clathration process stops when no more crystalline water ice is available to trap the volatile species.

In the case of Saturn, if only C enrichment is considered (Section 2.2.1), it is easily explained via the delivery of planetesimals formed at low temperatures, similarly to those accreted by Jupiter. When considering both C and N enrichments measured in Saturn with C/N=3, provided that N/H in the bulk atmosphere is the same as in 1- to 3-bar region, which is far from certain (Section 2.2.1), the scenario of full volatile clathration may not hold anymore because it would result in a uniform enrichment of these two species (Mousis et al. 2006). It has thus been argued that Saturn might have formed at a higher temperature than those required for the formation of CO and N_2 clathrates in the protosolar nebula (Hersant et al. 2008). However, this scenario does not match the high $^{14}N/^{15}N$ ratio recently estimated for Saturn (>357, Table 2.3; Section 2.2.2), since it predicts a value intermediate between the values for Jupiter (434) and the Earth (272). As discussed above, much of the critical heavy element abundance data for Saturn needed to fully evaluate the validity of the clathrate model are missing, and to some extent they are also missing for Jupiter. With 100% efficiency of clathration, models predict approximately 15× solar O/H at Jupiter (e.g. Gautier et al. 2001, using current solar O/H of Asplund et al. 2009), whereas the icy planetesimal model predicts it to be four times less (Owen et al. 1999). Water is critical for discriminating between various formation scenarios. Little laboratory data are presently available for the relatively low pressure conditions of the solar/protoplanetary nebula. In summary, both the cold icy planetesimal model and the clathrate model have their strengths and weaknesses, and discrimination between them requires new sets of data, particularly for Saturn (see Chapter 14 by Baines et al. for additional details).

2.4.3 Role of Photoevaporation of the Protosolar Disk in Determining Present-Day Composition

The atmospheres of the giant planets result from the accretion of both gaseous and solid material by planetary cores. The clathrate scenario implicitly assumes that all species other than hydrogen and helium were delivered with the solids. However, processes affecting the protosolar disk itself may also play an important role in determining the final atmospheric compositions. A plausible scenario proposed by Guillot and Hueso (2006) to explain the homogeneous enrichment of noble gases in Jupiter is illustrated in Figure 2.8. It is based on the fact that protoplanetary disks can extend to hundreds of AU and that their outer parts are generally very cold (e.g. Dartois et al. 2003). Temperatures of 10–30 K in the outer disks allow the direct condensation of most noble gases onto small grains (e.g. Owen et al. 1999). These grains will grow, settle towards the disk mid-plane and migrate inward (e.g. Adachi et al. 1976; Weidenschilling 1984; Dubrulle et al. 1995). In parallel, the gas disk is being accreted by the central star and photoevaporated both due to direct irradiation from the central star (Gorti et al. 2009) and by ambient FUV irradiation from other stars in the cluster (Adams et al. 2004). This evaporation takes place in the disk atmosphere, a region in which the temperature gradient is strongly negative (Chiang and Goldreich 1997). This would prevent a convective transport of species in the mid-plane regions and therefore, Guillot and Hueso (2006) conjecture, hydrogen and helium would preferentially evaporate. This would lead to a progressive homogeneous enrichment of the disk in condensing species. If formed late, giant planets would incorporate gas that is heavy-element rich, and in particular it would be rich in species such as Ar, Kr, and Xe. This theory explains the enrichment in noble gases in Jupiter's atmosphere measured by the Galileo probe and predicts a similar, homogeneous enrichment in Saturn (i.e. with solar Kr/Ar and Xe/Ar ratios). It cannot make predictions on elements that are delivered into giant planets with the solids and for which the story may be more complicated, as illustrated in the previous section.

2.5 Extrasolar Giant Planets Context

The discovery of numerous extrasolar planets in recent years is now allowing us to place the solar system planets in a cosmic context. Over 3600 confirmed exoplanets are known as of July 2017, of which over 500 are giant planets larger (in mass and/or size) than Saturn.[1] The majority of these planets have been detected either through Doppler spectroscopy of their host stars, i.e. the "radial velocity" (RV) method, or by observing transits of the planets in front of their host

[1] Extrasolar Planets Encyclopedia (http://exoplanet.eu) list of currently known exoplanets and their properties

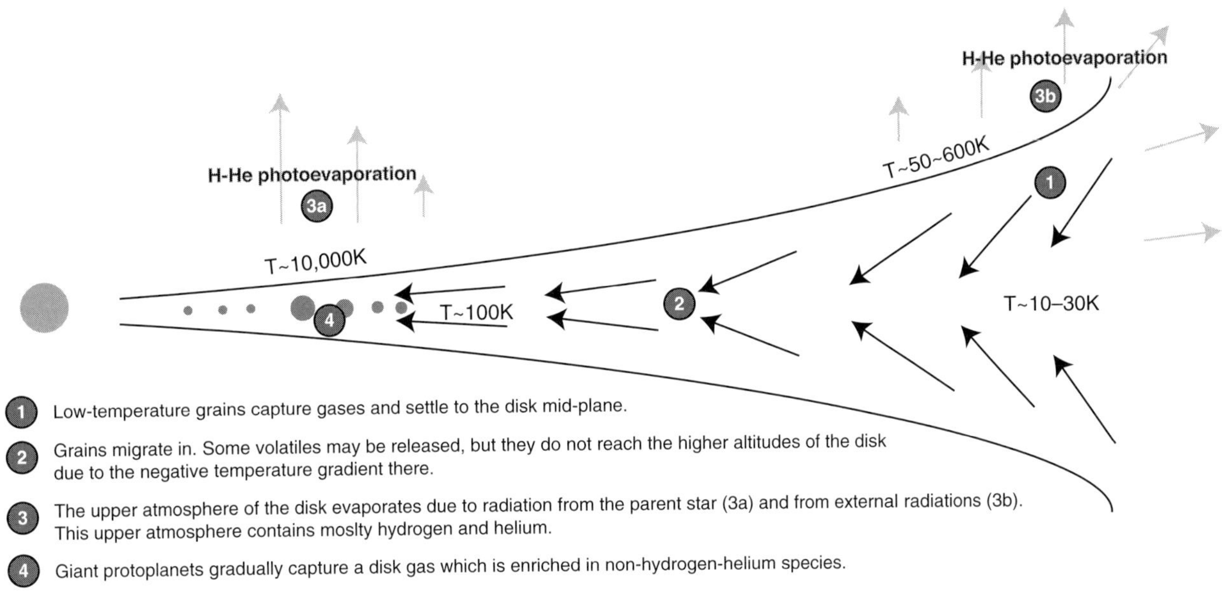

Figure 2.8 Sketch illustrating the scenario proposed by Guillot and Hueso (2006) to explain a homogeneous enrichment of noble gases in the envelopes of giant planets. A gaseous protosolar disk is shown edge-on. Forming protoplanets are shown by grey circles. Black arrows represent the dynamic evolution of grains and noble gases. Yellow arrows correspond to photoevaporation of gas from the disk due to both internal and external UV irradiation. The four circles correspond to the important evolution steps from the condensation of noble gases into cold grains in the outer disk to their incorporation in the envelopes of growing giant planets. (A black-and-white version of this figure appears in some formats. For the color version, please refer to the plate section.)

stars, i.e. the "transit" method, while a few tens of the planets have been detected via direct imaging. The observational sensitivities of the various exoplanet detection methods have precluded conclusive detections of exact analogues of Saturn and Jupiter in exoplanetary systems. Currently, the RV and transit methods, which together allow measurements of the mass and radius of exoplanets, are preferentially sensitive to planets at short orbital separations. Giant exoplanets, with both mass and radius measured, are known at orbital separations of ~0.01 to 0.5 AU, implying equilibrium temperatures over ~1000 K. On the other hand, while the direct imaging method is more sensitive to planets with large orbital separations ($\gtrsim$10 AU), current instruments are only sensitive to young, and hence also hot, giant planets, whose large fluxes make them detectable.

These diverse giant exoplanets form a starting point for placing the properties of solar system giant planets in perspective. In this section, we review our current understanding of the interior, atmosphere, and formation conditions of extrasolar giant planets and their analogies with Saturn and Jupiter in the solar system.

2.5.1 Interiors of Giant Exoplanets

Constraints on the interior composition of giant exoplanets are based primarily on their masses and radii, which are both known for about 200 transiting exoplanets with a mass and radius greater than those of Saturn (0.30 M_J and 0.84 R_J), generally referred to as "hot Jupiters." These planets have revealed an extreme diversity in their bulk parameters, with masses of 0.3 to 20 M_J, radii of 0.84 to 2 R_J, and temperatures of ~1000 to 3000 K. The masses and radii of these planets are consistent with a gaseous interior dominated by H_2 and He, similar to the interior of Saturn and Jupiter in the solar system. However, the diversity in mass and radius also implies a wide range of possible core masses, ranging from no core to ~200 M_E for the heaviest planets (e.g. Guillot et al. 2006; Baraffe et al. 2008), while upper limits on the core mass in Saturn and Jupiter are estimated to be ~10 to 20 M_E, as discussed in Section 2.2.3. Some studies have also suggested a possible positive correlation between the planetary core mass in hot Jupiters and the metallicity of the host star (e.g. Guillot et al. 2006). However, constraints on the core mass in giant exoplanets and on their internal structure in general are confounded by several complexities (see e.g. review by Baraffe et al. 2014).

Unlike the solar system giants, a large fraction of hot Jupiters have a radius that is significantly larger than would be predicted by internal structure models of giant planets, even assuming no solid core (the presence of a core contracts the planet). The largest planet known, WASP-17b, has a radius of ~2 R_J but a mass of only 0.5 M_J (Triaud et al. 2010). Such bloated giant planets have no analogy in the solar system and cannot be explained by canonical interior models of Saturn and Jupiter. Several mechanisms have been proposed to address the problem that bear on the key factors in which hot Jupiters differ from solar system giant planets. Hot Jupiters receive stellar irradiations that are 3 to 4 orders of magnitude higher than the insolation at Jupiter, and because of their close proximity to their host stars they are also subject to strong tidal and magnetic interactions. The various proposed mechanisms to explain inflated giants, though still actively debated, broadly include (a) deposition of incident energy deep in the planetary atmosphere, causing an extra energy source in the planetary interior and slowing down the thermal evolution, i.e. cooling, of the planet (Guillot and Showman 2002; Batygin and Stevenson 2010; Youdin and Mitchell 2010); (b) intrinsic heating caused by tidal dissipation in the planetary interior due to an eccentric close-in orbit that is being tidally circularized (e.g. Bodenheimer et al. 2001; Leconte et al. 2010); and (c) strong atmospheric opacity that inhibits the emergent flux, thereby delaying the cooling, and hence contraction, of the planet during its evolution (Burrows et al. 2007). However, none of these mechanisms conclusively explains the radius distributions in all the hot Jupiters currently known (see, e.g. Spiegel and Burrows 2013; Baraffe et al. 2014). Consequently, even though hundreds of giant exoplanets are known with similar masses and sizes as solar system giants, their interior structures and compositions are likely extremely diverse, albeit currently underconstrained.

2.5.2 Atmospheres of Giant Exoplanets

Remarkable progress has been made in the past decade in spectroscopic observations of exoplanetary atmospheres, primarily of hot gas giants that are most accessible to current instruments (see, e.g. review by Madhusudhan et al. 2014b). Currently observable gas giant atmospheres fall into two distinct categories: (a) highly irradiated giant planets ("hot Jupiters") in very close orbits (as close as 0.01 AU) and (b) young and self-luminous directly imaged planets at wide orbital separations (beyond ~10 AU). The effective temperatures of either class of planets are in the range of ~1000 to 3000 K, which are an order of magnitude hotter than those of solar system giant planets (~100 to 200 K). Since the radiation field is intricately linked to the physicochemical characteristics of the atmospheres, the atmospheric temperature structure, chemistry and dynamics in these giant exoplanets can be markedly different from those of Saturn and Jupiter in the solar system, even if the masses, radii, and bulk elemental abundances turn out to be similar. Here, we review current understanding of giant exoplanetary atmospheres vis-a-vis our understanding about the atmospheres of Saturn and Jupiter.

Atmospheric Observations

Spectra of exoplanetary atmospheres are inherently disk-integrated, unlike spectra of solar system giant planets, which can be spatially resolved over the planetary disk. Observations of exoplanetary spectra have been obtained using three key methods. Firstly, the atmospheres of close-in hot Jupiters have been observed primarily through transit spectroscopy, obtained during the planet's "transit" in front of the host star or "occultation" behind the star. While a transit (or transmission) spectrum probes the atmosphere of the day-night terminator region of the planet, the occultation (or emission) spectrum probes the dayside atmosphere of the planet. Spectra of transiting hot Jupiters have been observed both from space, using the Hubble and Spitzer space telescopes, as well as from ground-based facilities. While Spitzer and ground-based facilities have typically provided photometric observations of transiting exoplanets in the near-infrared (e.g. Charbonneau et al. 2008; Croll et al. 2011), the Hubble telescope has been instrumental in obtaining spectra across multiple spectral regimes, from the ultraviolet to near-infrared, for a few planets (Vidal-Madjar et al. 2003; Sing et al. 2011; Deming et al. 2013). These state-of-the-art observations have provided both the high precision and a long spectral baseline required to constrain the atmospheric properties of several transiting hot Jupiters. Secondly, it has also been possible to detect molecules in the atmospheres of a few transiting and non-transiting close-in hot Jupiters using very high resolution (R ~10^5) infrared Doppler spectroscopy using large ground-based telescopes (Snellen et al. 2010). Thirdly, ground-based spectroscopy of directly imaged planets has led to both photometry and high-resolution spectra of thermal emission from several young self-luminous planets in the near infrared (e.g. Marois et al. 2010; Konopacky et al. 2013; Janson et al. 2013).

Atmospheric Chemistry

The chemical compositions of hot giant exoplanets are expected to be markedly different from those of solar system giant planets, even if the bulk elemental abundances may be identical. The bulk molecular composition of the atmospheres of Saturn and Jupiter is generally consistent with expectations for low-temperature (~100 to 200 K) H-rich atmospheres, i.e. those dominated by methane, ammonia and higher-order hydrocarbons (Section 2.2.1). H_2O is expected to be the dominant O carrier but its abundance is presently undetermined in both Saturn and Jupiter because of their low temperatures, as discussed in Section 2.2.1. On the other hand, H_2O is more observable in the high-temperature atmospheres of giant exoplanets. However, the expected molecular composition depends strongly not only on the atmospheric temperatures but also on the elemental abundance ratios, particularly the overall metallicity and the C/O ratio (Madhusudhan 2012; Moses et al. 2013). Assuming solar abundances (i.e. C/O = 0.5), in the 1000 to 3000 K temperature range of hot giant exoplanets H_2O is expected to be the dominant carrier of O in the observable atmosphere and CO is expected to be the dominant C carrier above ~1300 K, while at lower temperatures CH_4 and NH_3 are expected to be abundant, along with trace quantities of CO_2 (Lodders and Fegley 2002; Madhusudhan 2012). Other species expected in hot Jupiters include Na, K, TiO, and VO (Seager et al. 2000; Hubeny et al. 2003; Madhusudhan 2012), which are not found in solar system gas giants because of their low temperatures. The chemistry can be even more drastic for super-solar abundance ratios, e.g. C/O = 1, in which case H_2O can be substantially underabundant and carbon-rich species overabundant, even in very high-temperature atmospheres (Madhusudhan 2012; Moses et al. 2013). Therefore, molecular abundances in hot Jupiters serve as key indicators of their elemental abundance ratios, such as the C/O ratio.

Chemical species have been detected in several giant exoplanetary atmospheres using all three observational methods discussed above. Recently, H_2O has been detected at high statistical significance in the atmospheres of several transiting hot Jupiters using the HST WFC3 spectrograph in the near-infrared (1.1–1.7 μm), e.g. in HD 209458b, HD 189733b, WASP-43b, and WASP-17b (Deming et al. 2013; Mandell et al. 2013; Kreidberg et al. 2014; McCullough et al. 2014). Additionally, transmission spectroscopy in the visible has been used to detect several atomic species in hot Jupiter atmospheres, e.g. Na and K (Redfield et al. 2008; Sing et al. 2011).

More recently, CO and H_2O have been detected in some transiting as well as non-transiting hot Jupiters using ground-based high-resolution infrared Doppler spectroscopy (e.g. Brogi et al. 2012; Birkby et al. 2013). On the other hand, H_2O, CO, and CH_4 have also been detected robustly in the atmospheres of directly imaged young giant exoplanets using high-resolution ground-based spectroscopy (e.g. Janson et al. 2013; Konopacky et al. 2013). The recent detection of a methane-rich giant exoplanet 51 Eri b, roughly twice the mass of Jupiter (Macintosh et al. 2015), represents the closest, albeit young (~20 Myr), analogue to solar-system giant planets. 51 Eri b orbits a Sun-like star, 51 Eridiani, at a Saturn-like orbital separation (13 AU), and like Saturn and Jupiter contains CH_4 as the dominant C-bearing molecule in its atmosphere.

In addition to molecular detections, recent observations are beginning to place notable statistical constraints on the molecular abundances in giant exoplanetary atmospheres, suggesting likely diverse elemental compositions. On one hand, some of the highest-precision HST WFC3 near-infrared spectra of transiting hot Jupiters are revealing significantly weaker H_2O features than expected for solar-composition atmospheres. For example, the thermal emission spectrum of the hot Jupiter WASP-12b suggest 100× sub-solar H_2O and a C/O ≥ 1, in its dayside atmosphere (Madhusudhan 2012; Stevenson et al. 2014), and that of WASP-33b suggests ~5 to 10× sub-solar H_2O, but with C/O < 1 (Haynes et al. 2015). Similarly, high-precision transmission spectra of the day-night terminator regions of hot Jupiters HD 189733b and HD 209458b suggest H_2O abundances as low as 100× sub-solar, assuming cloud-free models (Deming et al. 2013; Madhusudhan et al. 2014b). It is possible that the presence of high-temperature silicate clouds/hazes as discussed below could be masking the spectral features in some transmission spectra. On the other hand, thermal emission and transmission spectra of the hot Jupiter WASP-43b (semimajor axis 0.01526 AU, orbital period 0.81 days, planetary mass 2 M_J, host star mass 0.717 M⊙ and T⊙ 4520 K) reveal H_2O abundances in the range of 0.4 to 3.5× solar at 1σ confidence level and an upper limit of 20× solar at 3σ confidence level (Kreidberg et al. 2014). Thus, there is a real possibility of super-solar O/H in at least some extrasolar giant planets. Only when H_2O is measured in well-mixed atmospheres of Saturn and Jupiter, direct comparison with O/H in exoplanets will be possible. Meanwhile, consistency between the super-solar O/H in WASP-43b and super-solar C/H in all solar-system giant planets and super-solar Ar, Kr, Xe, N, and S in Jupiter seems to indicate similar formation processes of at least some hot Jupiters and Jupiter and Saturn in the solar system, but much further work is needed to be confident.

Clouds and Hazes

Clouds are ubiquitous in Saturn and Jupiter, but with quite different chemical compositions (Sections 2.2.1 and 2.6) from those expected in giant exoplanets. While clouds in Saturn and Jupiter are presumably made of low-temperature (150 to 300 K) condensates of ammonia, hydrogen sulfide (combined with ammonia), and water, as discussed in Section 2.2.1 and illustrated in Figure 2.9, those in hot giant exoplanets (at 1000 to 3000 K) are expected to be composed of refractory compounds such as silicates, alkali chlorides, Fe, etc. (Sudarsky et al. 2003). To date there is no spectral signature of a cloud-forming condensate in an exoplanetary atmosphere. Instead, the inferences of clouds in these atmospheres are derived from non-detections of expected atomic or molecular features (i.e. due to possible obscuration from clouds) or from modulations in the planetary spectrum indicative of particulate scattering (e.g. a steeper slope rising blue-ward than would be expected from pure gaseous Rayleigh scattering). For example, non-detections of strong Na and K absorption in the visible wavelengths (at 589 nm and 770 nm, respectively) along with a steep power-law spectrum have been suggested as indicative of haze in the hot Jupiter HD 189733b (Sing et al. 2011; Pont et al. 2013). The presence of clouds has also been inferred from observations of visible reflected light and phase

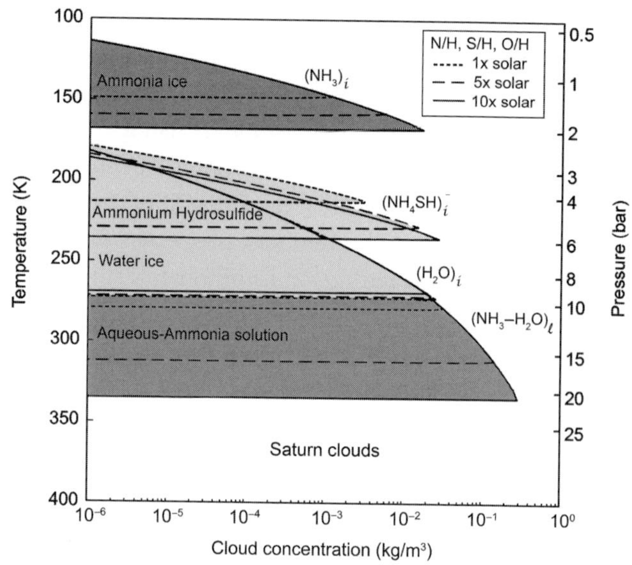

Figure 2.9 Equilibrium cloud condensation model of Saturn, assuming uniform enrichment of 1× solar, 5× solar, and 10× solar abundances for each of the condensible volatiles, NH_3, H_2S, and H_2O, and hence the elemental ratios N/H, S/H, and O/H, respectively. Calculations are based on current solar elemental abundances (Asplund et al., 2009) from Table 2.1. (Note that H_2S does not directly condense, but NH_4SH produced by the vapor-phase reaction between H_2S and NH_3 does in solid form [$(NH_4SH)_i$].) Water can condense as ice $(H_2O)_i$, and liquid of water-ammonia solution ($(NH_3$-$H_2O)_l$, colloquially, windex cloud) for large enrichment of H_2O. The cloud concentrations represent *upper limits*. Precipitation and dynamics would almost certainly deplete cloud densities by up to several orders of magnitude, as in the water clouds in the Earth's troposphere. The cloud bases (or, lifting condensation levels) are robust, however. More realistic cloud densities are formulated in Wong et al. (2015). Figure format adapted from Figure 23.2 in "Saturn Exploration Beyond Cassini–Huygens," by T. Guillot, S. K. Atreya, S. Charnoz, M. Dougherty, P. Read, in Saturn from Cassini–Huygens (M. K. Dougherty et al., eds.), Chapter 23, pp. 745–761, 2009, with permission from the publisher. The present figure is based on entirely new calculations, however. (A black-and-white version of this figure appears in some formats. For the color version, please refer to the plate section.)

curves of hot Jupiters (Demory et al. 2013; Evans et al. 2013; Barstow et al. 2014). On the other hand, several studies have used near-infrared observations of thermal emission from directly imaged giant planets, such as HR 8799b,c,d,e to suggest the presence of thick clouds in their atmospheres (e.g. Marois et al. 2010; Currie et al. 2011; Marley et al. 2012). Although hazes and clouds are not expected to be made up of H_2O in hot Jupiters due to their atmospheric temperatures that are high enough to vaporize water, they could still remove O from gas phase in the form of condensed silicates, thereby decreasing the amount of O available for H_2O. They could also provide adsorption/sequestration sites for water vapor (and other volatiles) that may result in the removal of H_2O by heterogeneous chemistry or surface processes, depending on the nature of hazes and temperatures.

Temperature Profiles and Stratospheres

Accurate determination of the atmospheric temperature profiles is important to constrain various thermal processes in exoplanetary atmospheres, and also because the temperature gradient is degenerate with chemical composition in their contributions to an emission spectrum (Madhusudhan and Seager 2010). One of the long-standing conundrums in the field concerns the possibility of temperature inversions (or "stratospheres") in exoplanetary atmospheres, i.e. temperatures increasing with altitude in the atmosphere, as opposed to a monotonically decreasing temperature profile, which would be expected for an isolated body. The Earth and larger planets in the solar system all have thermal inversions; on Earth it is due to ozone, in giant planets it is due to hydrocarbon haze. Early theoretical studies (Hubeny et al. 2003; Fortney et al. 2008) predicted that atmospheres of hot Jupiters could also host thermal inversions, but due to very different sources than those in Jupiter or Saturn, namely from gaseous TiO and VO, which can survive at high temperatures. Current observations suggest that some hot Jupiters show "tentative" evidence for thermal inversions whereas others do not (e.g. Stevenson et al. 2014; Haynes et al. 2015). Various processes have been proposed to explain possible trends, e.g. correlations with stellar irradiation (Fortney et al. 2008), TiO condensation (Spiegel et al. 2009), stellar activity (Knutson et al. 2010), C/O ratios (Madhusudhan 2012), and thermoresistive instability (Menou 2012).

Overall, there are presently no conclusive constraints on the presence of thermal inversions in exoplanetary atmospheres or on any inversion-causing absorbers. High-resolution spectra from future facilities would be required to make robust detections of thermal inversions. Although detailed data on thermal structure of the atmospheres of Saturn and Jupiter exists as a result of spacecraft remote sensing and entry probe (at Jupiter) measurements, thermal structure of cold gas planets is not a suitable guide for what to expect in hot Jupiters whose structure is controlled by extreme stellar forcing. However, with appropriate modifications, radiative transfer models used for interpreting temperature observations of cold, clear, cloudy, or hazy gas planets are to some degree applicable to hot Jupiters (Lee et al. 2012).

2.5.3 Formation of Giant Exoplanets

The large population of giant exoplanets provides a diverse sample to test theories of formation of giant planets in the solar system. As discussed in Section 2.1, two main formation mechanisms have been proposed to explain the formation of Saturn and Jupiter: core accretion (CA) and gravitational instability (GI), with clear preference for CA. Various efforts have been made to identify if either of these formation mechanisms could explain the formation of giant exoplanets based on their observed orbital parameters and chemical compositions.

Dynamical Constraints

The diverse orbital parameters of different classes of giant exoplanets (close-in versus distant) constrain the different possible formation mechanisms. In the CA model (Pollack et al. 1996), the planetary embryos start out as ~10 M_E cores in the protoplanetary disk that subsequently undergo runaway accretion of a large volume of gas and planetesimals to form a massive gaseous envelope. On the other hand, a GI in a young disk can cause rapid collapse of a large volume of ambient gas and solids to form a giant planet (Boss et al. 2000). Both scenarios occur in planet-forming disks, but at different orbital separations. While CA is favored closer to the snowline (within ~2 to 10 AU) because cores take too long to form at larger distances and only reach large masses after the disk has dispersed, GI is favored at larger distances ($\gtrsim$10 AU) where the disk can cool sufficiently on orbital timescales to fragment. In this regard, GI may be the favored mechanism for the formation of distant gas giant exoplanets detected via direct imaging. However, neither GI nor CA is thought to operate in such a way that allows hot Jupiters to form in situ at their current locations close to the host stars. The disk cannot fragment at those distances, and cores with sufficient mass to attract significant envelopes cannot form. Therefore, the existence of hot Jupiters requires some form of "migration" from their original formation locations to their present orbits (see Section 2.3.2).

Migration may occur relatively early in the planet's history via the planet's interaction with, and transport through, the protoplanetary disk while the gas in the disk is still present (Lin et al. 1996). Alternately, migration may also occur at any time via scattering (Rasio and Ford 1996) or secular interactions, such as Kozai resonances (Fabrycky and Tremaine 2007), of the planet with other massive planetary or stellar components in the system. Measurements of orbital obliquities, i.e. the degree of alignment between the stellar equatorial plane and the planetary orbital plane, have been proposed to distinguish between the two migration scenarios (Gaudi and Winn 2007). Whereas migration of a planet through a viscous disk would be expected to damp any initial misalignment, migration by scattering or Kozai resonances could lead to very high spin-orbit misalignments. The observations of a significant number of large spin-orbit misalignments in hot Jupiter systems in recent years initially supported the role of migration by scattering phenomena (Winn et al. 2010; Triaud et al. 2010). However, more recent studies have shown that spin-orbit misalignments can also be caused by planet migration through disks, which are themselves misaligned due to torques induced by a distant stellar companion (Crida and Batygin 2014). Consequently, dynamical measurements alone have not been able to conclusively constrain the formation of hot Jupiters, though directly imaged planets at wide separations ($\gtrsim$10 AU) seem more likely to be formed via GI.

Chemical Constraints

Atmospheric elemental abundances of solar-system giant planets have led to important constraints on the origins of the solar system. For example, the observed super-solar enrichments of C, S, N and the heavy noble gases (Section 2.2.1) support the formation of Jupiter and Saturn by core-accretion (see Sections 2.1 and 2.2.1). However, the oxygen abundance, which is a critical parameter in formation models, is not known in Saturn and Jupiter (see Sections 2.2.1 and 2.6). On the other hand, as discussed in Section 2.5.2, given the high temperatures of currently known giant exoplanets (T ~1000 to 3000 K), several key molecules are expected to be observable in their atmospheres and allow estimations of elemental abundance ratios involving H, C, O, and N. Nominal constraints on atmospheric C/H, O/H, and C/O ratios have already been reported for a few exoplanets and reveal both oxygen-rich (C/O $<$ 1) as well as carbon-rich (C/O $\geq$ 1) compositions; the solar composition is oxygen-rich with C/O = 0.5.

Findings of super-solar C/O ratios in giant exoplanets are beginning to motivate new ideas on their formation mechanisms. The C/O ratios of most planet-hosting stars in the solar neighborhood are solar-like, i.e. oxygen-rich (e.g. Delgado-Mena et al. 2010). Thus, in the standard core-accretion model of planet formation, it is expected that oxygen-rich planetesimals with abundant H_2O ice would dominate the planetesimal composition. Thus, the possibility of C-rich giant planet atmospheres orbiting O-rich stars poses a challenge to standard formation models of Jupiter and Saturn. An early investigation into this question was pursued in the context of Jupiter in the solar system for which, as discussed above, only a lower limit on the O/H is known, which may allow for the possibility of C/O $>$ 1. Lodders (2004) suggested the possibility of Jupiter forming by accreting tar-dominated planetesimals instead of those dominant in water ice, as expected in the solar system based on the composition of minor bodies in the solar system. Following the inference of C/O $\geq$ 1 in the hot Jupiter WASP-12b (Madhusudhan et al. 2011a), Öberg et al. (2011b) suggested that C/O ratios in giant exoplanetary envelopes depend on the formation location of the planets in the disk relative to the icelines of major C- and O-bearing volatile species, such as H_2O, CO, and CO_2. The C/O ratio of the gas in the nebula approaches 1 outside the CO and CO_2 icelines. By predominantly accreting such C-rich gas, more so than O-rich planetesimals, gas giants could host C-rich atmospheres even when orbiting O-rich stars. It may also be possible that inherent inhomogeneities in the C/O ratios of the disk itself may contribute to higher C/O ratios of the planets relative to the host stars (Madhusudhan et al. 2011b). Additionally, the composition of the planet is also influenced by the temporal evolution of the chemical and thermodynamic properties of the disk at the formation location of the planet Saturn (Ali-Dib et al. 2014; Helling et al. 2014; Marboeuf et al. 2014). More recently, Madhusudhan et al. (2014c) suggested that O and C abundances of hot Jupiters could also provide constraints on their migration mechanisms. In particular, hot Jupiters with sub-solar elemental abundances are more likely to have migrated to their close-in orbits by disk-free mechanisms (e.g. scattering) rather than through the disk, regardless of their formation by core accretion or gravitational instability process.

Thus, various scenarios of giant planet formation and migration predict different limits on the metallicites and C/O ratios of giant exoplanets, which are testable with future high-precision and high-resolution observations of their atmospheres as will be possible with facilities like the *James Webb Space Telescope*, large ground-based telescopes of the future and dedicated space missions. As tighter constraints on the elemental abundances in exoplanets become available, investigating them together with elemental abundances in Saturn and Jupiter will allow development of convincing scenarios of the formation of gas giant planets in the solar system and extrasolar systems.

2.6 Outstanding Issues and Looking to the Future

Existing observations of Saturn, its atmosphere, rings and the moons have provided tantalizing clues into the formation and evolution scenarios of the Saturnian system. Additional insight has come from volatile composition and abundance data of giant exoplanets. Yet, the current observational constraints for developing robust models are either inadequate, poor, or simply non-existent, including those needed to address such fundamental questions as "does Saturn have a core today," "how does the size of Saturn's core compare to Jupiter's core," "what's Saturn's true intrinsic rotation rate," "what's Saturn's bulk composition – in particular, the abundance of heavy elements – and how does it compare with Jupiter's bulk composition," "what's the helium abundance in the troposphere of Saturn," "is the history of heavy noble gases different from that of other heavy elements," and "what are the isotope ratios of H, He, N, S, Ar, Ne, Kr and Xe, and what are their implications." New types of observations are required to address these issues. In the near future, the Cassini Grand Finale Mission appears promising for answering some of these questions.

Following a spectacular tour of the Saturnian system since reaching Saturn in 2004, the Cassini orbiter will enter its final phase of the mission in 2016, aptly named the Cassini Grand Finale, before the spacecraft crashes and burns in Saturn's atmosphere mid-2017. In the final 22 proximal orbits, Cassini's trajectory will take it high above the north pole, flying outside the F-ring and then plunging between Saturn and its innermost ring, skimming as close as ~1700 km above Saturn's cloud tops. These proximal orbits will give an unprecedented opportunity to carry out high-precision measurements of higher-order moments of gravity and magnetic fields and the ring mass and particle distribution. These observations will provide useful constraints on the internal structure, rotation rate and the age of Saturn's rings. As the orbits of the Juno spacecraft at Jupiter will be very similar to Cassini proximal orbits, a comparison between Jupiter and Saturn results in terms of the gravitational and magnetic fields will be possible. This extraordinary opportunity to gather comparable data on Jupiter and Saturn will help us not only to understand the intrinsic differences between these bodies but also to get a sense of the variation we might expect among extrasolar giant planets within the same stellar system. The atmospheric composition relevant to Saturn's formation models requires in situ measurements, however.

Bulk composition and the atmospheric isotope determination of the giant planets cannot be carried out by remote sensing, for the most part. The abundances of He and the heavy elements C, N, S, O, Ne, Ar, Kr, and Xe and isotope ratios D/H in H_2, $^3He/^4He$,

$^{13}C/^{12}C$, $^{15}N/^{14}N$, and $^{34}S/^{32}S$ and the isotope ratios of the heavy noble gases are crucial constraints on the formation models. With the exception of carbon, their determination requires an entry probe at Saturn, as was done at Jupiter with the Galileo probe in 1995 (note the remote sensing result on D/H at Saturn is imprecise). A shallow-entry probe to 10 bars at Saturn is expected to deliver meaningful data on all of the above elements and isotopes, except perhaps oxygen, unless O/H is substantially sub-solar in Saturn. This is evident from Figure 2.9, which shows the equilibrium cloud condensation levels of the condensible volatiles in Saturn's troposphere. For solar O/H, the base of the water cloud is found to be at 10 bars (cloud densities in the figure are *upper limits*; cloud bases are robust, however). As discussed earlier, water may be enriched similarly to carbon, i.e. roughly 10× solar. In that case, the base of the water cloud would be at ~20 bars. Because of convective and dynamical processes, well-mixed water may not be reached above two to three times these pressure levels, however. Thus, even in the unlikely scenario of solar water, only probe measurements to at least 20 to 30 bars can ensure reliable O/H determination in Saturn. If water is 10× solar, measurements down to at least 50 bars, preferably 100 bars, will be required for the O/H determination. If water in Saturn is greatly sub-solar, probes to 10 bars will be able to determine the O/H directly in Saturn. Deep probes to such extreme environments of high pressures and temperatures and large radio opacity are presently unfeasible. However, Juno-like microwave radiometry from orbit at Saturn could potentially map the deep water abundance over the planet, thus allowing the determination of the O/H ratio. Although O/H ratio in Saturn is desirable, its absence due to technical hurdles or cost constraints would not be a disaster. Comparison of all other elements and isotopes in Saturn, particularly the noble gases, with those in Jupiter measured by the Galileo probe and Juno's O/H would establish a trend or pattern from one gas giant planet to the other, which may still provide meaningful constraints on Saturn's O/H. Other reservoirs of oxygen, such as CO, though much less abundant than H_2O, could also be exploited to obtain clues to the limits of O/H in Saturn. Future ground-based microwave measurements with improved capability are also promising for the deep water abundance. Refer to Chapter 14 by Baines et al. for additional details on future exploration of Saturn.

Finally, composition data including especially the profiles of H_2O, CO, and CH_4 in the atmospheres of giant exoplanets can provide a useful guide for Saturn. Similarly, in many respects, Saturn and Jupiter are ideal analogs for similar-sized exoplanets around sun-like stars, despite the differences in their current orbital distances and resulting temperatures. Spectroscopic characterization of exoplanet atmospheres is proceeding rapidly, and there is a good prospect of addressing many of the outstanding issues including temperature structure and aerosol distribution. A comparison between atmospheric properties of a multitude of giant exoplanets is also essential. This chapter demonstrates that cross-fertilization between the giant planet research and the giant exoplanet research is beneficial both fields, and leads to a deeper understanding of the origin and evolution of this solar system and the extra-solar systems.

Acknowledgments

Discussions with several colleagues on the Juno, Cassini and Galileo science teams and ground-based planetary astronomers were beneficial in preparing this chapter. Joong Hyun In, Gloria Kim, and Carmen Lee assisted with references and formatting.

References

Adachi, I., C. Hayashi and K. Nakazawa (1976), The gas drag effect on the elliptical motion of a solid body in the primordial solar nebula, *Prog. Theor. Phys.*, **56**, 1756–1771.

Adams, F. C., D. Hollenbach, G. Laughlin et al. (2004), Photoevaporation of circumstellar disks due to external far-ultraviolet radiation in stellar aggregates, *Astrophys. J.*, **611**, 360–379.

Alibert, Y. and O. Mousis (2007), Formation of Titan in Saturn's subnebula: Constraints from Huygens probe measurements, *Astron. Astrophys.*, **465**, 1051–1060.

Alibert, Y., Mousis, O. and Benz, W. (2005a), On the Volatile enrichments and composition of Jupiter, *Astrophys. J.*, **622**, L145–L148.

Alibert, Y., O. Mousis, C. Mordasini et al. (2005b), New Jupiter and Saturn formation models meet observations, *Astrophys. J.*, **626**, L57–L60.

Ali-Dib, M., O. Mousis, J. M. Petit et al. (2014), Carbon-rich planet formation in a solar composition disk, *Astrophys. J.*, **785**, 125–131.

Altwegg, K., H. Balsiger, A. Bar-Nun et al. (2015), 67P/Churyumov-Gerasimenko: A Jupiter family comet with a high D/H ratio, *Science*, **347**(6220). 1261952–1 – 1261952–3.

Amelin, Y. A., A. Kaltenbach, T. Iizuka et al. (2010), U-Pb chronology of the solar system's oldest solids with variable $^{238}U/^{235}U$, *Earth Planet. Sci. Lett.*, **300**, 343–350.

Anders, E. and N. Grevesse (1989), Abundances of the elements: meteoritic and solar, *Geochim. Cosmochim. Acta*, **53**, 197–214.

Armitage, P. (2010), *Astrophysics of Planet Formation*, Cambridge University Press, Cambridge.

Arpigny, C., E. Jehin, J. Manfroid et al. (2003), Anomalous nitrogen isotope ratio in comets, *Science*, **301**, 1522–1524.

Asplund, M., N. Grevesse, J. Sauval et al. (2009), The chemical composition of the Sun, *Annu. Rev. Astron. Astrophys*, **47**, 481–522.

Atreya, S. K. (1986), *Atmospheres and Ionospheres of the Outer Planets and their Satellites*, pp. 66–79. Springer-Verlag, New York-Heidelberg.

Atreya, S. K., E. Y. Adams, H. B. Niemann et al. (2006), Titan's methane cycle, *Planet. Space Sci.*, **54**, 1177–1187.

Atreya, S. K., T. M. Donahue and W. R. Kuhn (1978), Evolution of a nitrogen atmosphere on Titan, *Science*, **201**, 611–613.

Atreya, S. K., R. D. Lorenz and J. H. Waite (2009), Volatile origin and cycles: Nitrogen and methane, in *Titan from Cassini-Huygens*, edited by R. H. Brown, J. P. Lebreton, and J. Waite, pp. 177–199, Springer Dordrecht, Heidelberg-London-New York.

Atreya, S. K., P. R. Mahaffy, H. B. Niemann et al. (2003), Composition and origin of the atmosphere of Jupiter: An update, and implications for the extrasolar giant planets, *Planet. Space Sci.*, **51**, 105–112.

Atreya, S. K., M. G. Trainer, H B. Franz et al. (2013), Primordial argon isotope fractionation in the atmosphere of Mars measured by the SAM instrument on Curiosity, and implications for atmospheric loss, *Geophys. Res. Lett.*, **40**, 5605–5609.

Atreya, S. K. and A. S. Wong (2005), Coupled chemistry and clouds of the giant planets: A case for multiprobes, *Space Sci. Rev.*, **116**, 121–136.

Atreya, S. K., M. H. Wong, T. C. Owen et al. (1999), A Comparison of the atmospheres of Jupiter and Saturn: Deep atmospheric composition, cloud structure, vertical mixing, and origin, *Planet. Space Sci.*, **47**, 1243–1262.

Ayliffe, B. and M. Bate (2009a), Circumplanetary disc properties obtained from radiation hydrodynamical simulations of gas accretion by protoplanets, *MNRAS*, **397**, 657–665.

(2009b), Gas accretion on to planetary cores: Three-dimensional self-gravitating radiation hydrodynamical calculations, *MNRAS*, **393**, 49–64.

(2012), The growth and hydrodynamic collapse of a protoplanet envelope, *MNRAS*, **427**, 2597–2612.

Baraffe, I., G. Chabrier and T. Barman (2008), Structure evolution of super-Earth to super-Jupiter exoplanets. I. Heavy element enrichment in the interior, *Astron. Astrophys.*, **482**, 315–332.

Baraffe, I., G. Chabrier, J. Fortney et al. (2014), Planetary internal structures, in *Protostars and Planets VI*, edited by H. Beuther, R. Klessen, C. Dullemond, and Th. Henning, University of Arizona Press, Tucson, AZ.

Barstow, J. K., S. Aigrain, P. G. J. Irwin et al. (2014), Clouds on the Hot Jupiter HD189733b: Constraints from the Reflection Spectrum, *Astrophys. J.* **786**, 154.

Baruteau, C., A. Crida, S. J. Paardekooper et al. (2014), Planet-disc interactions and early evolution of planetary systems, in *Protostars and Planets VI*, edited by H. Beuter, R. Klessen, C. Dullemond, and T. Henning. University of Arizona Press, Tucson, AZ.

Bate, M. R., S. H. Lubow, G. I. Ogilvie et al. (2003), Three-dimensional calculations of high- and low-mass planets embedded in protoplanetary discs, *MNRAS*, **341**, 213–229.

Batygin, K. and D. J. Stevenson (2010), Inflating hot Jupiters with ohmic dissipation, *Astrophys. J.*, **714**, L238–L243.

Birkby, J. L., R. J. de Kok, M. Brogi et al. (2013), Detection of water absorption in the day side atmosphere of HD 189733 b using ground-based high-resolution spectroscopy at 3.2 μm, *MNRAS*, **436**, 35–39.

Bitsch, B., A. Crida, A. Morbidelli et al. (2013), Stellar irradiated discs and implications on migration of embedded planets. I. Equilibrium discs, *Astron. Astrophys.*, **549**, A124–137.

Bitsch, B., A. Morbidelli, E. Lega et al. (2014), Stellar irradiated discs and implications on migration of embedded planets. II. Accreting-discs, *Astron. Astrophys.*, **564**, A135–146.

Black, D. C. (1972), The origins of trapped helium, neon, and argon isotopic variations in meteorites-I, *Geochim. Cosmochim. Acta*, **36**, 347–375.

Bockelée-Morvan, D., N. Biver, E. Jehin et al. (2008), Large excess of heavy nitrogen in both hydrogen cyanide and cyanogen from comet 17P/Holmes, *Astrophys. J.*, **679**, L49–L52.

Bolton, S. J., S. Adriani, V. Adumitroaie, M. Allison, J. Anderson, S. K. Atreya et al. (2017), Jupiter's interior and deep atmosphere: the initial pole-to-pole passes with the Juno spacecraft, *Science*, **356**, 821–825.

Bodenheimer, P., D. N. C. Lin and R. A. Mardling (2001), On the tidal inflation of short-period extrasolar planets, *Astrophys. J.*, **548**, 466–472.

Boss, A. P. (2000), Possible rapid gas giant planet formation in the solar nebula and other protoplanetary disks, *Astrophys. J.*, **536**, L101–L104.

Briggs, F. H. and P. D. Sackett (1989), Radio observations of Saturn as a probe of its atmosphere and cloud structure, *Icarus*, **80**, 77–103.

Brogi, M., L. A. G. Snellen, R. J. de Kok et al. (2012), The signature of orbital motion from the dayside of the planet τ Boötis b, *Nature*, **486**, 502–504.

Burrows, A., I. Hubeny, J. Budaj et al. (2007), Possible solutions to the radius anomalies of transiting giant planets, *Astrophys. J.*, **661**, 502–514.

Cameron, A. G. W. (1973), Abundances of the elements in the solar system, *Space Sci. Rev.* **15**, 121–146.

(1979), The interaction between giant gaseous protoplanets and the primitive solar nebula, *The Moon and the Planets*, **21**, 173–183.

(1982), Elemental and nuclidic abundances in the solar system, in *Essays in Nuclear Astrophysics*, edited by W. A. Fowler, pp. 23–43, Cambridge University Press.

Canup, R. M. (2010), Origin of Saturn's rings and inner moons by mass removal from a lost Titan-sized satellite, *Nature*, **468**, 943–946.

Canup, R. M. and W. R. Ward (2002), Formation of the Galilean satellites: Conditions of accretion, *Astron. J.*, **124**, 3404–3423.

Castillo-Rogez, J., T. V. Johnson, M. H. Lee et al. (2009), ^{26}Al decay: Heat production and a revised age for Iapetus, *Icarus*, **204**, 658–662.

Charbonneau, D., H. A. Knutson, T. Barman et al. (2008), The broadband infrared emission spectrum of the exoplanet HD 189733b, *Astrophys. J.*, **686**, 2, 1341–1348.

Charnoz, S., A. Crida, J. C. Castillo-Rogez et al. (2011), Accretion of Saturn's mid-sized moons during the viscous spreading of young massive rings: Solving the paradox of silicate-poor rings versus silicate-rich moons, *Icarus*, **216**, 535–550.

Charnoz, S., J. Salmon and A. Crida (2010), The recent formation of Saturn's moonlets from viscous spreading of the main rings, *Nature*, **465**, 752–754.

Chiang, E. and P. Goldreich (1997), Spectral energy distributions of T Tauri stars with passive circumstellar disks, *Astrophys. J.*, **490**, 368–376.

Chick, K. M. and P. Cassen (1997), Thermal processing of interstellar dust grains in the primitive solar environment, *Astrophys. J.*, **477**, 398–409.

Conrath, B. J. and D. Gautier (2000), Saturn helium abundance: A reanalysis of Voyager measurements, *Icarus*, **144**, 124–134.

Conrath, B. J., D. Gautier, R. A. Hanel et al. (1984), The helium abundance of Saturn from Voyager measurements, *Astrophys. J.*, **282**, 807–815.

Conrath, B. J., R. A. Hanel and R. E. Samuelson (1989), Thermal structure and heat balance of the outer planets, in *Origin and Evolution of Planetary and Satellite Atmospheres*, edited by. S. K. Atreya, J. B. Pollack, and M. S. Matthews, pp. 513–538, University of Arizona Press, Tucson, AZ.

Cossou, C., S. Raymond and A. Pierens (2013), Convergence zones for Type I migration: An inward shift for multiple planet systems, *Astron. Astrophys.*, **553**, L2 (5 pp.).

Crida, A. and K. Batygin (2014), Spin-Orbit angle distribution and the origin of (mis)aligned hot Jupiters, *Astron. Astrophys.*, **567**, A42 (8 pp.).

Crida, A. and S. Charnoz (2012), Formation of regular satellites from ancient massive rings in the solar system, *Science*, **338**, 1196–1199.

Crida, A. and A. Morbidelli (2007), Cavity opening by a giant planet in a protoplanetary disc and effects on planetary migration, *MNRAS*, **377**, 1324–1336.

Crida, A., A. Morbidelli and F. Masset (2006), On the width and shape of gaps in protoplanetary disks, *Icarus*, **181**, 587–604.

Croll, B., D. Lafreniere, L. Albert et al. (2011), Near-infrared thermal emission from WASP-12b: Detections of the secondary eclipse in Ks, H, and J, *Astron. J.*, **141**, 30–42.

Currie, T., A. Burrows, Y. Itoh et al. (2011), A combined Subaru/VLT/MMT 1–5 μm study of planets orbiting HR 8799: Implications for atmospheric properties, masses, and formation, *Astrophys. J.*, **729**, 128–147.

Dahmen, G., T. L. Wilson and F. Matteucci (1995), The nitrogen isotope abundance in the galaxy 1: The galactic disk gradient, *Astron. Astrophys.*, **295**, 194–198.

Dartois, E., A. Dutrey and S. Guilloteau (2003), Structure of the DM Tau outer disk: probing the vertical kinetic temperature gradient, *Astron. Astrophys.*, **399**, 773–787.

Delgado-Mena, E., G. Israelian, J. I. González Hernández et al. (2010), Chemical clues on the formation of planetary systems: C/O versus Mg/Si for HARPS GTO sample, *Astrophys. J.*, **725**, 2349–2358.

Deming, D., A. Wilkins, P. McCullough et al. (2013), Infrared transmission spectroscopy of the exoplanets HD 209458b and XO-1b using the wide field camera-3 on the Hubble Space Telescope, *Astrophys. J.*, **774**, 95–112.

Demory, B. J., J. de Wit, N. Lewis et al. (2013), Inference of inhomogeneous clouds in an exoplanet atmosphere, *Astrophys. J. Lett.*, **776**, L25 (7 pp.).

de Pater, I. and S. T. Massie (1985), Models of the millimeter-centimeter spectra of the giant planets, *Icarus*, **62**, 143–171.

Dodson-Robinson, S. E., P. Bodenheimer, G. Laughlin et al. (2008), Saturn forms by core accretion in 3.4 Myr, *Astrophys. J.*, **688**, L99–L102.

Dubrulle, B., G. Morfill and M. Sterzik (1995), The dust subdisk in the protoplanetary nebula, *Icarus*, **114**, 237–246.

Dürmann, C. and W. Kley (2015), Migration of massive planets in accreting disks, *Astron. Astrophys.*, **574**, A52.

Dyudina, U. A., A. P. Ingersoll, S. P. Ewald et al. (2010), Detection of visible lightning on Saturn, *Geophys. Res. Lett.*, **37**, L09205.

(2013), Saturn's visible lightning, its radio emissions, and the structure of the 2009–2011 lightning storms, *Icarus*, **226**, 1020–1037.

Eberhardt, P. (1974), A Neon-E-rich phase in the orgueil carbonaceous chondrite, *Earth Planet. Sci. Lett.*, **24**, 182–187.

Estrada, P. R., I. Mosqueira and S. Charnoz (2009), The Ganymede, Titan, Callisto, Iapetus trend: Interpretation of Iapetus' composition, *AGU Fall Meeting Abstracts*, C1148.

Evans, T. M., F. Pont, D. K. Sing et al. (2013), The deep blue color of HD 189733b: albedo measurements with Hubble Space Telescope/space telescope imaging spectrograph at visible wavelengths, *Astrophys. J. Lett.* **772**, L16 (5 pp.).

Fabrycky, D. and S. Tremaine (2007), Shrinking binary and planetary orbits by Kozai cycles with Tidal friction, *Astrophys. J.*, **669**, 1298–1315.

Fegley, B. and R. G. Prinn (1985), Equilibrium and nonequilibrium chemistry of Saturn's atmosphere: Implications for the observability of PH_3, N_2, CO, and GeH_4, *Astrophys. J.*, **299**, 1067–1078.

Flasar, F. M., R. K. Achterberg, B. J. Conrath et al. (2005), Temperatures, winds, and composition in the Saturnian system, *Science*, **307**, 1247–1251.

Fletcher, L. N., K. H. Baines, T. W. Momary et al. (2011), Saturn's tropospheric composition and clouds from Cassini/VIMS 4.6–5.1 μm nightside spectroscopy, *Icarus*, **214**, 510–533.

Fletcher, L. N., T. K. Greathouse, G. S. Orton et al. (2014), The origin of nitrogen on Jupiter and Saturn from the $^{15}N/^{14}N$ ratio, *Icarus*, **238**, 170–190.

Fletcher, L. N., G. S. Orton, N. A. Teanby et al. (2009a), Phosphine on Jupiter and Saturn from Cassini/CIRS, *Icarus*, **202**, 543–564.

(2009b), Methane and its isotopologues on Saturn from Cassini/CIRS Observations, *Icarus*, **199**, 351–367.

Folkner, W. M., R. Woo and S. Nandi (1998), Ammonia abundance in Jupiter's atmosphere derived from attenuation of the Galileo Probe's radio signal, *J. Geophys. Res.*, **103**, 22847–22856.

Fortney, J. J., K. Lodders, M. S. Marley et al. (2008), A unified theory for the atmospheres of the hot and very hot Jupiters: Two classes of irradiated atmospheres, *Astrophys. J.*, **678**, 1419–1435.

Fouchet, T., E. Lellouch, B. Bezard et al. (2000), ISO-SWS observations of Jupiter: measurement of the ammonia tropospheric profile and of the $^{15}N/^{14}N$ isotopic ratio, *Icarus*, **143**, 223–243.

Fujii, Y. I., S. Okuzumi and S. Inutsuka (2011), A fast and accurate calculation scheme for ionization degrees in protoplanetary and circumplanetary disks with charged dust grains, *Astrophys. J.*, **743**, 53–61.

Fujii, Y. I., S. Okuzumi, T. Tanigawa et al. (2014), On the viability of the magnetorotational instability in circumplanetary disks, *Astrophys. J.*, **785**, 101–108.

Gaudi, B. S. and J. N. Winn (2007), Prospects for the characterization and confirmation of transiting exoplanets via the Rossiter–McLaughlin effect, *Astrophys. J.*, **655**, 550–563.

Gaulme, P., F. X. Schmider, J. Gay et al. (2011), Detection of Jovian seismic waves: A new probe of its interior structure, *Astron. Astrophys.*, **531**, A104 (7 pp.).

Gautier, D., F. Hersant, O. Mousis et al. (2001), Enrichments in volatiles in Jupiter: A new interpretation of the Galileo measurements. *Astrophys. J.*, **550**, L227–L230.

Geiss, J. (1993), Primordial abundance of hydrogen and helium isotopes, in *Origin and Evolution of the Elements*, edited by N. Prantzos, E. Vangioni-Flam, and M. Cassé, pp. 89–106, Cambridge University Press, Cambridge.

Geiss, J. and G. Gloeckler (1998), Abundances of deuterium and helium-3 in the protosolar cloud, *Space Sci. Rev.*, **84**, 239–250.

Geiss, J. and H. Reeves (1972), Cosmic and solar system abundances of deuterium and helium-3, *Astron. Astrophys.*, **18**, 126–132.

Gibb, E. L., D. C. B. Whittet, A. C. A. Boogert et al. (2004), Interstellar ice: The infrared space observatory legacy, *Astrophys. J. Supp. Ser.*, **151**, 35–73.

Gorti, U., C. P. Dullemond and D. Hollenbach (2009), Time evolution of viscous circumstellar disks due to photoevaporation by far-ultraviolet, extreme-ultraviolet, and X-ray radiation from the central star, *Astrophys. J.*, **705**, 1237–1251.

Grevesse, N., M. Asplund and A. J. Sauval (2005), The new solar chemical composition, in Elements Stratification in Stars, 40 Years of Atomic Diffusion, edited by G. Alecian, O. Richard, and S. Vauclair, **17**, pp. 21–30, *EAS Publications Series*.

(2007), The solar chemical composition, *Space Sci. Rev.*, **130**, 105–114.

Grossman, A. S., J. B. Pollack, R. T. Reynolds et al. (1980), The effect of dense cores on the structure and evolution of Jupiter and Saturn, *Icarus*, **42**, 358–379.

Guillot, T. and R. Hueso (2006), The composition of Jupiter: Sign of a (relatively) late formation in a chemically evolved protosolar disc, *MNRAS*, **367**, L47–L51.

Guillot, T., N. C. Santos, F. Pont et al. (2006), A correlation between the heavy element content of transiting extrasolar planets and the metallicity of their parent stars, *Astron. Astrophys*, **453**, L21–L24.

Guillot, T. and A. P. Showman (2002), Evolution of "51 Pegasus b-like" planets, *Astron. Astrophys.*, **385**, 156–165.

Guillot, T., D. J. Stevenson, W. B. Hubbard et al. (2004), The interior of Jupiter, in *Jupiter. The Planet, Satellites and Magnetosphere*, edited by F. Bagenal, T. E. Dowling, and W. B. McKinnon, pp. 35–57, Cambridge University Press, Cambridge.

Hartogh, P., D. C. Lis, D. Bockelée-Morvan et al. (2011), Ocean-like water in the Jupiter-family comet 103P/Hartley 2, *Nature*, **478**, 218–220. doi:10.1038/nature10519; pmid: 21976024.

Hayashi, C. (1981), Structure of the solar nebula, growth and decay of magnetic fields and effects of magnetic and turbulent viscosities on the nebula, *Prog. Theor. Phys. Supp.*, **70**, 35–53.

Haynes, K., A. M. Mandell, N. Madhusudhan et al. (2015), Spectroscopic evidence for a temperature inversion in the dayside atmosphere of hot Jupiter WASP-33b, *Astrophys. J.*, **806**, 146, 12 pp.

Helled, R. and T. Guillot (2013), Interior models of Saturn: Including the uncertainties in shape and rotation, *Astrophys. J.*, **767**, 113–119.

Helled, R. and J. I. Lunine (2014), Measuring Jupiter's water abundance by Juno: The link between interior and formation models, *MNRAS*, **441**, 2273–2279.

Helling, C., P. Woitke, P. B. Rimmer et al. (2014), Disk evolution, element abundances and cloud properties of young gas giant planets, *Life*, **4**, 142–173.

Hersant, F., D. Gautier, G. Tobie et al. (2008), Interpretation of the carbon abundance in Saturn measured by Cassini, *Planet. Space Sci.*, **56**, 1103–1111.

Hoffman, J. H., R. R. Hodges, M. B. McElroy et al. (1979), Composition and structure of the Venus atmosphere: Results from Pioneer Venus, *Science*, **205**, 49–52.

Howard, A. W. (2013), Observed properties of extrasolar planets, *Science*, **340**, 572–576.

Hubeny, I., A. Burrows and D. Sudarsky (2003), A possible bifurcation in atmospheres of strongly irradiated stars and planets, *Astrophys. J.*, **594**, 1011–1018.

Ida, S., D. N. C. Lin and M. Nagasawa (2013), Toward a deterministic model of planetary formation. VII. Eccentricity distribution of gas giants, *Astrophys. J.*, **775**, 42–66.

Ikoma, M., H. Emori and K. Nakazawa (2001), Formation of giant planets in dense nebulae: Critical core mass revisited, *Astrophys. J.*, **553**, 99–1005.

Ishimaru, R., Y. Sekine, T. Matsui et al. (2011), Oxidizing proto-atmosphere on Titan: Constraint from N_2 formation by impact shock, *Astrophys. J. Lett.*, **741**, L10 (6 pp.).

Janson, M., T. D. Brandt, M. Kuzuhara et al. (2013), Direct imaging detection of methane in the atmosphere of GJ 504 b, *Astrophys. J. Lett.*, **778**, L4 (6 pp.).

Jewitt, D., H. E. Matthews, T. Owen et al. (1997), The $^{12}C/^{13}C$, $^{14}N/^{15}N$ and $^{32}S/\ ^{34}S$ isotope ratios in comet Hale–Bopp (C/1995 O1), *Science*, **278**, 90–93.

Johansen, A., J. S. Oishi, M. Mac Low et al. (2007), Rapid planetesimal formation in turbulent circumstellar disks, *Nature*, **448**, 1022–1025.

Johansen, A. and A. Youdin (2007), Protoplanetary disk turbulence driven by the streaming instability: Nonlinear saturation and particle concentration, *Astrophys. J.*, **662**, 627–641.

Johansen, A., A. Youdin and M. M. Mac Low (2009), Particle clumping and planetesimal formation depend strongly on metallicity, *Astrophys. J.*, **704**, L75–L79.

Johnson, J. A., K. M. Aller, A. W. Howard et al. (2010), Giant planet occurrence in the stellar mass-metallicity plane, *PASP*, **122**, 905–915.

Jones, T. D. and J. S. Lewis (1987), Estimated impact shock production of N_2 and organic compounds on early Titan, *Icarus*, **72**, 381–393.

Karkoschka, E. and M. G. Tomasko (2011), The haze and methane distributions on Neptune from HST-STIS spectroscopy, *Icarus*, **211**, 328–340.

Knutson, H. A., A. W. Howard and H. Isaacson (2010), A correlation between stellar activity and hot Jupiter emission spectra, *Astrophys. J.*, **720**, 1569–1576.

Kokubo, E. and S. Ida (1998), Oligarchic growth of protoplanets, *Icarus*, **131**, 171–178.

Konopacky, Q. M., T. S. Barman, B. A. Macintosh et al. (2013), Detection of carbon monoxide and water absorption lines in an exoplanet atmosphere, *Science*, **339**, 1398–1401.

Kouchi, A., T. Yamamoto, T. Kozasa et al. (1994), Conditions for condensation and preservation of amorphous ice and crystallinity of astrophysical ices, *Astron. Astrophys.*, **290**, 1009–1018.

Kreidberg, L., J. L. Bean, J. M. Désert et al. (2014), A precise water abundance measurement for the hot Jupiter WASP-43b, *Astrophys J. Lett.*, **793**, L27 (6 pp.).

Lainey, V., R. A. Jacobson, R. Tajeddine et al. (2015), New constraints on Saturn's interior from Cassini astrometric data. Submitted.

Lainey, V., Ö. Karatakin, J. Desmars et al. (2012), Strong tidal dissipation in Saturn and constraints on Enceladus' thermal state from astrometry, *Astrophys. J.*, **752**, 14–32.

Lambrechts, M. and A. Johansen (2012), Rapid growth of gas-giant cores by pebble accretion, *Astron. Astrophys.*, **544**, A32.

Lambrechts, M., A. Johansen and A. Morbidelli (2014), Separating gas-giant and ice-giant planets by halting pebble accretion, *Astron. Astrophys.*, **572**, A35.

Laraia, A. L., A. P. Ingersoll, M. A. Janssen et al. (2013), Analysis of Saturn's thermal emission at 2.2-cm wavelength: Spatial distribution of ammonia vapor, *Icarus*, **226**, 641–654.

Leconte, J., G. Chabrier, I. Baraffe et al. (2010), Is tidal heating sufficient to explain bloated exoplanets? Consistent calculations accounting for finite initial eccentricity, *Astron. Astrophys.*, **516**, A64 (13 pp.).

Lee, J. M., L. N. Fletcher and P. G. J. Irwin (2012), Optimal estimation retrievals of the atmospheric structure and composition of HD 189733b from secondary eclipse spectroscopy, *MNRAS*, **420**, 170–182.

Lellouch, E., B. Bézard, T. Fouchet et al. (2001), The deuterium abundance in Jupiter and Saturn from ISO-SWS observations, *Astron. Astrophys.*, **370**, 610–622.

Li, C. and A. P. Ingersoll (2015), Moist convection in hydrogen atmospheres and the frequency of Saturn's giant storms, *Nat. Geosci*. **8**, 398–403.

Lin, D. N. C., P. Bodenheimer and D. C. Richardson (1996), Orbital migration of the planetary companion of 51 Pegasi to its present location, *Nature*, **380**, 606–607.

Lin, D. N. C. and J. Papaloizou (1986), On the tidal interaction between protoplanets and the protoplanetary disk. III. Orbital migration of protoplanets, *Astrophys. J.*, **309**, 846–857.

Lodders, K. (2004), Jupiter formed with more tar than ice, *Astrophys. J.*, **611**, 587–597.

(2008), The solar argon abundance, *Astrophys. J.*, **674**, 607–611.

Lodders, K. and B. Fegley (2002), Atmospheric chemistry in giant planets, brown dwarfs, and low-mass dwarf stars. I. Carbon, nitrogen, and oxygen, *Icarus*, **155**, 393–424.

Lodders, K., H. Palme and H. P. Gail (2009), Abundances of the elements in the solar system, in *Landolt-Börnstein New Series, Astron. and Astrophys.*, edited by J. E. Trümper, vol. VI/4B, pp. 560–630, Springer-Verlag, Berlin.

Lunine, J. I., S. K. Atreya and J. B. Pollack (1989), Present state and chemical evolution of the atmospheres of

Titan, Triton, and Pluto, in *Origin and Evolution of Planetary and Satellite Atmospheres*, edited by S. K. Atreya, J. B. Pollack, and M. S. Matthews, pp. 605–665, University of Arizona Press, Tucson, AZ.

Lyra, W., S. Paardekooper and M. Mac Low (2010), Orbital migration of low-mass planets in evolutionary radiative models: Avoiding catastrophic infall, *Astrophys. J. Lett.*, **715**, L68–L73.

Machida, M. N., E. Kokubo, S. I. Inutsuka et al. (2008), Angular momentum accretion onto a gas giant planet, *Astrophys. J.*, **685**, 1220–1236.

Macintosh, B., J. R. Graham, T. Barman et al. (2015), Discovery and spectroscopy of the young Jovian planet 51 Eri b with the Gemini Planet Imager, *Science*, **350**, 64–67.

Madhusudhan, N. (2012), C/O ratio as a dimension for characterizing exoplanetary atmospheres, *Astrophys. J.*, **758**, 36–57.

Madhusudhan, N., M. A. Amin and G. M. Kennedy (2014c), Toward chemical constraints on hot Jupiter migration, *Astrophys. J. Lett.*, **794**, 12.

Madhusudhan, N., N. Crouzet, P. R. McCullough et al. (2014a), H_2O abundances in the atmospheres of three hot Jupiters, *Astrophys. J. Lett.*, **791**, L9 (5 pp.).

Madhusudhan, N., J. Harrington, K. B. Stevenson et al. (2011a), A high C/O ratio and weak thermal inversion in the atmosphere of exoplanet WASP-12b, *Nature*, **469**, 64–67.

Madhusudhan, N., H. Knutson, J. Fortney et al. (2014b), Exoplanetary atmospheres, in *Protostars and Planets VI*, edited by H. Beuther, R. Klessen, C. Dullemond, and Th. Henning, University of Arizona Press, Tucson, AZ.

Madhusudhan, N., O. Mousis, T. V. Johnson et al. (2011b), Carbon-rich giant planets: atmospheric chemistry, thermal inversions, spectra, and formation conditions, *Astrophys. J.*, **743**, 191–202.

Madhusudhan N. and S. Seager (2010), On the inference of thermal inversions in hot Jupiter atmospheres, *Astrophys. J.*, **725**, 261–274.

Mahaffy, P. R., T. M. Donahue, S. K. Atreya et al. (1998), Galileo Probe measurements of D/H and $^3He/^4He$ in Jupiter's atmosphere, *Space Sci. Rev.*, **84**, 251–263.

Mahaffy, P. R., H. B. Niemann, A. Alpert et al. (2000), Noble gas abundances and isotope ratios in the atmosphere of Jupiter from the Galileo probe mass spectrometer, *J. Geophys. Res.* (*Planets*), **105 (E6)**, 15061–15071.

Mahaffy, P. R., C. R. Webster, J. C. Stern et al. (2014), The imprint of atmospheric evolution in the D/H of Hesperian clay minerals on Mars, *Science*, **347**, 415–417.

Mandell, A., K. Haynes, E. Sinukoff et al. (2013), Exoplanet transit spectroscopy using WFC3: WASP-12 b, WASP-17 b, and WASP-19 b, *Astrophys. J.*, **779**, 128.

Mandt, K. E., O. Mousis, J. I. Lunine et al. (2014), Protosolar ammonia as the unique source of Titan's nitrogen, *Astrophys. J. Lett.*, **788**, L24 (5 pp.).

Marboeuf, U., A. Thiabaud, A. Alibert et al. (2014), From stellar nebula to planetesimals, *Astron. Astrophys.*, **570**, 35.

Marley, M. S., D. Saumon, M. Cushing et al. (2012), Masses, radii, and cloud properties of the HR 8799 planets, *Astrophys. J.*, **753**, 135–152.

Marois, C., B. Zuckerman, Q. M. Konopacky et al. (2010), Images of a fourth planet orbiting HR 8799, *Science*, **468**, 1080–1083.

Marty, B., M. Chaussidon, R. C. Wiens et al. (2011), A ^{15}N-poor isotopic composition for the solar system as shown by genesis solar wind samples, *Science*, **332**, 1533–1536.

Masset, F. and M. Snellgrove (2001), Reversing type II migration: Resonance trapping of a lighter giant protoplanet, *MNRAS*, **320**, L55–L59.

Mathew, K. J. and K. Marti (2001), Early evolution of Martian volatiles: Nitrogen and noble gas components in ALH84001 and Chassigny, *J. Geophys. Res.*, **106**, 1401–1422.

Matter, A., T. Guillot and A. Morbidelli (2009), Calculation of the enrichment of the giant planet envelopes during the "late heavy bombardment," *Planet. Space Sci.*, **57**, 816–821.

Mayer, L., T. Quinn, K. Wadsley et al. (2002), Formation of giant planets by fragmentation of protoplanetary disks, *Science*, **298**, 1756–1759.

McCullough, P. R., N. Crouzet, D. Deming et al. (2014), Water vapor in the spectrum of the extrasolar planet HD 189733b: 1. The transit, *Astrophys. J.*, **791**, 55.

McKay, C. P., T. W. Scattergood, J. B. Pollack et al. (1988), High-temperature shock formation of N_2 and organics on primordial Titan, *Nature*, **332**, 520–522.

Menou, K. (2012), Thermo-resistive instability of hot planetary atmospheres, *Astrophys. J.*, **754**, L9–L14.

Mizuno, H. (1980), Formation of the giant planets, *Progr. Theor. Phys.* **64**, 544–557.

Morbidelli, A. and A. Crida (2007), The dynamics of Jupiter and Saturn in the gaseous protoplanetary disk, *Icarus*, **191**, 158–171.

Morbidelli, A. and D. Nesvorny (2012), Dynamics of pebbles in the vicinity of a growing planetary embryo: Hydro-dynamical simulations, *Astron. Astrophys.*, **546**, A18–25.

Morbidelli, A., J. Szulágyi, A. Crida et al. (2014), Meridional circulation of gas into gaps opened by giant planets in three-dimensional low-viscosity disks, *Icarus*, **232**, 266–270.

Mordasini, C., Y. Alibert and W. Benz (2009), Extrasolar planet population synthesis. I. Method, formation, tracks, and mass-distance distribution, *Astron. Astrophys.*, **501**, 1139–1160.

Moses, J. I., N. Madhusudhan, C. Visscher et al. (2013), Chemical consequences of the C/O ratio on hot Jupiters: Examples from WASP-12b, CoRoT-2b, XO-1b, and HD 189733b, *Astrophys. J.*, **763**, 25–51.

Mousis, O., Y. Alibert and W. Benz (2006), Saturn's internal structure and carbon enrichment, *Astron. Astrophys.*, **449**, 411–415.

Mousis, O., D. Gautier and D. Bockelée-Morvan (2002), An evolutionary turbulent model of Saturn's subnebula:

Implications for the origin of the atmosphere of Titan, *Icarus*, **156**, 162–175.

Mousis, O., J. I. Lunine, N. Madhusudhan et al. (2012), Nebular water depletion as the cause of Jupiter's low oxygen abundance, *Astrophys. J.* **751**, L7 (5 pp.).

Mousis, O., J. I. Lunine, M. Pasek et al. (2009a), A primordial origin for the atmospheric methane of Saturn's moon Titan, *Icarus*, **204**, 749–751.

Mousis, O., J. I. Lunine, C. Thomas et al. (2009b), Clathration of volatiles in the solar nebula and implications for the origin of Titan's atmosphere, *Astrophys. J.*, **691**, 1780–1786.

Mousis, O., U. Marboeuf, J. I. Lunine et al. (2009c), Determination of the minimum masses of heavy elements in the envelopes of Jupiter and Saturn, *Astrophys. J.*, **696**, 1348–1354.

Nettelmann, N., R. Püstow and R. Redmer (2013), Saturn layered structure and homogeneous evolution models with different EOSs, *Icarus*, **225**, 548–557.

Niemann, H. B., S. K. Atreya, S. J. Bauer et al. (2005), The abundances of constituents of Titan's atmosphere from the GCMS instrument on the Huygens probe, *Nature*, **438**, 779–784.

Niemann, H. B., S. K. Atreya, G. R. Carignan et al. (1998), The composition of the jovian atmosphere as determined by the Galileo probe mass spectrometer, *J. Geophys. Res.*, **103**, 22831–22845.

Niemann, H. B., S. K. Atreya, J. E. Demick et al. (2010), The composition of Titan's lower atmosphere and simple surface volatiles as measured by the Cassini–Huygens probe gas chromatograph mass spectrometer experiment, *J. Geophys. Res. (Planets)*, **115**, E12006.

Notesco, G. and A. Bar-Nun (2005), A ~25 K temperature of formation for the submicron ice grains which formed comets, *Icarus*, **175**, 546–550.

Öberg, K. I., A. C. A. Boogert, K. M. Pontoppidan et al. (2011a), II, The Spitzer ice legacy: Ice evolution from cores to protostars, *Astrophys. J.*, **740**, 109–124.

Öberg, K. I., R. Murray-Clay and E. A. Bergin (2011b), The effects of snowlines on C/O in planetary atmospheres, *Astrophys. J. Lett.*, **743**, L16 (5 pp.).

Owen, T. and T. Encrenaz (2006), Compositional constraints on the giant planet formation, *Planet. Space Sci.* **54**, 1188–1196.

Owen, T., P. Mahaffy, H. Niemann et al. (1999), A low-temperature origin for the planetesimals that formed Jupiter, *Nature*, **402**, 269–270.

(2001), Protosolar nitrogen, *Astophys. J.*, **553**, L77–L79.

Paardekooper, S. J., C. Baruteau, A. Crida et al. (2010), A torque formula for non-isothermal type I planetary migration: I. Unsaturated horseshoe drag, *MNRAS*, **401**, 1950–1964.

Paardekooper, S. J., C. Baruteau and W. Kley (2011), A torque formula for non-isothermal Type I planetary migration: II. Effects of diffusion, *MNRAS*, **410**, 293–303.

Paardekooper, S. J. and G. Mellama (2006), Halting type I planet migration in non-isothermal disks, *Astron. Astrophys.*, **459**, L17–L20.

Palme, H. and A. Jones (2003), Solar system abundances of the elements, in *Treatise on Geochemistry*, edited by H. D. Holland and K. Turekian, Elsevier, pp. 41–61.

Papaloizou, J. and D. N. C. Lin (1984), On the tidal interaction between protoplanets and the primordial solar nebula: I. Linear calculation of the role of angular momentum exchange, *Astrophys. J.*, **285**, 818–834.

Pollack, J. B., O. Hubickyj, P. Bodenheimer et al. (1996), Formation of the giant planets by concurrent accretion of solids and gas, *Icarus*, **124**, 62–85.

Pont, F., D. K. Sing, N. P. Gibson et al. (2013), The prevalence of dust on the exoplanet HD 189733b from Hubble and Spitzer observations, *Mon. Not. R. Astron. Soc.*, **432**, 2917–2944.

Prinn, R. G. and B. Fegley (1981), Kinetic inhibition of CO and N_2 reduction in circumplanetary nebulae: Implications for satellite composition, *Astrophys. J.*, **249**, 308–317.

(1989), Solar nebula chemistry: Origins of planetary, satellite and cometary volatiles, in *Origin and Evolution of Planetary and Satellite Atmospheres*, edited by S. K. Atreya, J. B. Pollack, and M. S. Matthews, pp. 78–136, University of Arizona Press, Tucson, AZ.

Rasio, F. A. and E. B. Ford (1996), Dynamical instabilities and the formation of extrasolar planetary systems, *Science*, **274**, 954–956.

Redfield, S., M. Endl, W. D. Cochran et al. (2008), Sodium absorption from the exoplanetary atmosphere of HD 189733b detected in the optical transmission spectrum, *Astrophys. J.*, **673**, L87–L90.

Remus, F., S. Mathis, J. P. Zahn et al. (2012), Anelastic tidal dissipation in multi-layer planets, *Astron. Astrophys.*, **541**, A165 (17 pp.).

Rivier, G., A. Crida, A. Morbidelli et al. (2012), Circumplanetary discs as bottlenecks for gas accretion onto giant planets, *Astron. Astrophys.*, **548**, A116 (7 pp.).

Rosman, K. J. R. and P. D. P. Taylor (1998), Isotopic compositions of the elements, *J. Phys. Chem. Ref. Data*, **27**, 1275–1287.

Roulston, M. S. and D. J. Stevenson (1995), Prediction of neon depletion in Jupiter's atmosphere (abstract), *EOS Abstr., AGU Fall Mtg.*, **76**, 343.

Rousselot, P., O. Pirali, E. Jehin et al. (2014), Toward a unique nitrogen isotopic ratio in cometary ices, *Astrophys. J. Lett*, **780**, L17 (5 pp.).

Salmon, J., S. Charnoz, A. Crida et al. (2010), Long-term and large-scale viscous evolution of dense planetary rings, *Icarus*, **209**, 771–785.

Seager, S. and D. Sasselov (2000), Theoretical transmission spectra during extrasolar giant planet transits, *Astrophys. J.*, **537**, 916–921.

Sekine, Y., H. Genda, S. Sugita et al. (2011), Replacement and late formation of atmospheric N_2 on undifferentiated Titan by impacts, *Nature Geoscience*, **4**, 359–362.

Sekine, Y., S. Sugita, T. Shido et al. (2005), The role of Fischer Tropsch catalysis in the origin of methane-rich Titan, *Icarus*, **178**, 154–164.

Sing, D. K., F. Pont, S. Aigrain et al. (2011), Hubble Space Telescope transmission spectroscopy of the exoplanet

HD 189733b: High-altitude atmospheric haze in the optical and near-ultraviolet with STIS, *Mon. Not. R. Astron. Soc.*, **416**, 1443–1455.

Snellen, I. A. G., R. J. de Kok, E. J. W. de Mooij et al. (2010), The orbital motion, absolute mass and high-altitude winds of exoplanet HD209458b, *Nature*, **465**, 1049–1053.

Spiegel, D. S. and A. Burrows (2013), Thermal processes governing hot-Jupiter radii, *Astrophys. J.*, **772**, 76–89.

Spiegel, D. S., K. Silverio and A. Burrows (2009), Can TiO explain thermal inversions in the upper atmospheres of irradiated giant planets?, *Astrophys. J.*, **699**, 1487–1500.

Sromovsky, L. A., P. M. Fry and J. H. Kim (2011), Methane on Uranus: The case for a compact CH_4 cloud layer at low latitudes and a severe CH_4 depletion at high-latitudes based on re-analysis of Voyager occultation measurements and STIS spectroscopy, *Icarus*, **215**, 292–312.

Stevenson, K. B., J.-M. Desert, M. R. Line et al. (2014), Thermal structure of an exoplanet atmosphere from phase-resolved emission spectroscopy, *Science*, **346**, 838–841.

Sudarsky, D., A. Burrows and I. Hubeny (2003), Theoretical spectra and atmospheres of extrasolar giant planets, *Astrophys. J.*, **588**, 1121–1148.

Sumi, T., K. Kamiya, A. Udalski et al. (2011), Unbound or distant planetary mass population detected by gravitational microlensing, *Nature*, **473**, 349–352.

Szula×i, J., A. Morbidelli, A. Crida et al. (2014), Accretion of Jupiter-mass planets in the limit of vanishing viscosity, *Astrophys. J.*, **782**, 65–75.

Tanigawa, T., K. Ohtsuki and M. N. Machida (2012), Distribution of accreting gas and angular momentum onto circumplanetary disks, *Astrophys. J.*, **747**, 47–62.

Triaud, A. H. M. J., A. Collier Cameron, D. Queloz et al. (2010), Spin-orbit angle measurements for six southern transiting planets: New insights into the dynamical origins of hot Jupiters, *Astron. and Astrophys*, **524**, A25.

Tsiganis, K., R. Gomes, A. Morbidelli et al. (2005), Origin of the orbital architecture of the giant planets of the Solar System, *Nature*, **435**, 459–461.

Turner N. J., S. Fromang, C. Gammie et al. (2014), Transport and accretion in planet-forming disks, in *Protostars and Planets VI*, edited by H. Beuter, R. Klessen, C. Dullemond, and T. Henning. University of Arizona Press, Tucson, AZ.

Turner, N. J., M. H. Lee and T. Sano (2010), The formation environment of the Galilean moons. American Astronomical Society, DPS Meeting No. 42, Abst. 24.08.

(2014), Magnetic coupling in the disks around young gas giant planets, *Astrophys. J.*, **783**, 14–28.

Vidal-Madjar, A., A. Lecavelier des Etangs, J. M. Désert et al. (2003), An extended upper atmosphere around the extrasolar planet HD209458b, *Nature*, **422**, 143–146.

Visscher, C. and B. Fegley (2005), Chemical constraints on the water and total oxygen abundances in the deep atmosphere of Saturn, *Astrophys. J.*, **623**, 1221–1227.

Vogel, N., V. S. Heber, H. Baur et al. (2011), Argon, krypton, and xenon in the bulk solar wind as collected by the genesis mission, *Geochim. Cosmochim. Acta*, **75**, 3057–3071.

von Zahn, U., D. M. Hunten and G. Lehmacher (1998), Helium in Jupiter's atmosphere: results from the Galileo probe helium interferometer experiment, *J. Geophys. Res.*, **103**, 22815–22830.

von Zahn, U., S. Komer, H. Wieman et al. (1983), Composition of the Venus atmosphere, in *Venus*, edited by D. M. Hunten, L. Colin, T. M. Donahue, and V. I. Moroz, pp. 297–430, University of Arizona Press, Tucson, AZ.

Walsh, K., A. Morbidelli, S. N. Raymond et al. (2011), A low mass for Mars from Jupiter's early gas-driven migration, *Nature*, **475**, 206–209.

Ward, W. R. (1997), Protoplanet migration by nebula tides, *Icarus*, **126**, 261–281.

Webster, C. R., P. R. Mahaffy, G. J. Flesch et al. (2013), Isotope ratios of H, C, and O in CO_2 and H_2O of the martian atmosphere, *Science*, **341**, 260–263.

Weidenschilling, S. J. (1984), Evolution of grains in a turbulent solar nebula, *Icarus*, **60**, 553–568.

Wilson, H. F. and B. Militzer (2010), Sequestration of noble gases in giant planet interiors, *Phys. Rev. Lett.*, **104**, 121101 (4 pp.).

(2012), Solubility of water ice in metallic hydrogen: Consequences for core erosion in gas giant planets, *Astrophys. J.*, **745**, 54–58.

Winn, J. N., D. Fabrycky, S. Albrecht et al. (2010), Hot stars with hot Jupiters have high obliquities, *Astrophys. J. Lett.*, **718**, L145–L149.

Wong, M. H., S. K. Atreya, W. R. Kuhn et al. (2015), Fresh clouds: A parameterized updraft method for calculating cloud densities in one-dimensional models, *Icarus*, **245**, 273–281.

Wong, M. H., S. K. Atreya, P. N. Mahaffy et al. (2013), Isotopes of nitrogen on Mars: atmospheric measurements by Curiosity's mass spectrometer, *Geophys. Res. Lett.*, **40**, 6033–6037.

Wong, M. H., P. R. Mahaffy, S. K. Atreya et al. (2004), Updated Galileo probe mass spectrometer measurements of carbon, oxygen, nitrogen, and sulfur on Jupiter, *Icarus*, **171**, 153–170.

Youdin, A. N. and J. Goodman (2005), Streaming instabilities in protoplanetary disks, *Astrophys. J.*, **620**, 459–469.

Youdin, A. N. and J. L. Mitchell (2010), The mechanical greenhouse: Burial of heat by turbulence in hot Jupiter atmospheres, *Astrophys. J.*, **721**, 1113–1126.

3

The Interior of Saturn

JONATHAN J. FORTNEY, RAVIT HELLED, NADINE NETTELMANN, DAVID J. STEVENSON, MARK S. MARLEY, WILLIAM B. HUBBARD AND LUCIANO IESS

Abstract

We review our current understanding of the interior structure and thermal evolution of Saturn, with a focus on recent results in the Cassini era. There has been important progress in understanding physical inputs, including equations of state of planetary materials and their mixtures, physical parameters like the gravity field and rotation rate, and constraints on Saturnian free oscillations. At the same time, new methods of calculation, including work on the gravity field of rotating fluid bodies, and the role of interior composition gradients, should help to better constrain the state of Saturn's interior, now and earlier in its history. However, a better appreciation of modeling uncertainties and degeneracies, along with a greater exploration of modeling phase space, still leave great uncertainties in our understanding of Saturn's interior. Further analysis of Cassini data sets, as well as precise gravity field measurements from the Cassini Grand Finale orbits, will further revolutionize our understanding of Saturn's interior over the next few years.

3.1 Introduction

In investigations into the interior structure, composition, and thermal evolution of giant planets, Saturn can sometimes receive "Second City" status compared to the bright lights of Jupiter. Both planets are natural laboratories for understanding the physics of hydrogen, helium and their mixtures under high pressure. Since both planets are pre-dominantly composed of H/He, an understanding of their compositions sheds important and unique light on the composition of the solar nebula during the era of planet formation.

While Jupiter is often thought of as the benchmark giant planet for this class of astrophysical object, now known to be abundant in the universe, Saturn provides an interesting point of comparison and departure for understanding giant planet structure and evolution (Figure 3.1). For instance, Jupiter models are highly sensitive to the equation of state (*EOS*, the relation between important quantities such as temperature, pressure, and density) of hydrogen, the most abundant element in the universe, and thus can help to probe the phase space region around a few megabars and ten thousand Kelvin, for which accurate lab experimental data are not available yet. Saturn on the other hand, with 30% of Jupiter's mass, probes less of hydrogen's phase space but has its own host of complex issues. With its peculiar magnetic field and high intrinsic luminosity, Saturn provides challenges to our understanding of the first-order properties that define a gas giant planet. For

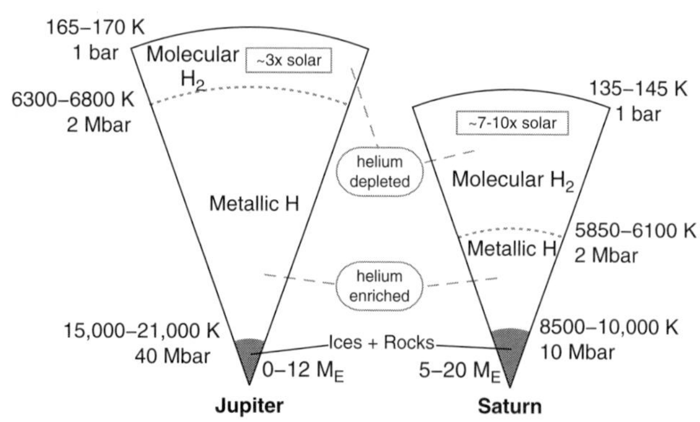

Figure 3.1 Highly idealized comparative view on the interiors of Jupiter and Saturn.

both planets, an understanding of their bulk composition can come from interior modeling, which is an important constraint on formation scenarios.

In looking back at the post-Voyager Saturn review chapter of Hubbard and Stevenson (1984), it is apparent that a number of the important issues of the day are still unsolved. What is the enrichment of heavy elements compared to the Sun and Jupiter, and what is their distribution within the planet? What is the mass of any heavy-element core? To what degree has the phase separation of helium in the planet's deep interior altered the evolutionary history of the planet? Are there deviations from adiabaticity?

Understanding the interior of Saturn crosses diverse fields, from condensed-matter physics to planet formation, but progress is challenging due to uncertainties in input physics as well as in the indirect nature of our constraints on the planet. This era near the end of the Cassini Mission is an excellent time to review our understanding of Saturn's interior. We are at a time where new observational constraints, such as a refined measurement of the gravity field as well as ring seismology, new theoretical and experimental constraints on input physics like the hydrogen-helium phase diagram, and new methods of calculating interior models, are all coming together to allow for a new understanding of Saturn's interior, and by extension the interiors of giant planets as a class of astrophysical object.

3.2 Available Data and Its Applications

3.2.1 Energy Balance

Like Jupiter, the power incident upon Saturn due to solar radiation is on the same order as the intrinsic power from the planet. The thermal flux detected from the planet today is a combination of this intrinsic flux, which is a remnant of formation, as well as thermalized solar energy. To distinguish between these components, the Bond albedo of Saturn must be determined, as this quantity yields the total flux absorbed by the planet, which is then re-radiated. Models aim to understand the intrinsic flux from the planet's interior, and how the planet cooled to this flux level at 4.56 Gyr (e.g. Fortney et al. 2011). In addition, the 1-bar temperature dictates the specific entropy of an isentropic deep interior (see Section 3.5.2), setting the upper boundary for the thermal structure of the interior. A self-consistent model should fit the intrinsic flux as well as the 1-bar temperature, in addition to other constraints detailed below. Table 3.1 shows the current energy budget of the planet.

3.2.2 Atmospheric Composition

Saturn's atmospheric composition is an important constraint on its interior structure and formation history (see Chapter 2 by Atreya et al.). In particular, if the H-He envelope is fully convective and well-mixed, atmospheric abundances that can be measured either spectroscopically or in situ should be representative of the entire H-He envelope. This could yield complementary information on the heavy-element enrichment of the H-He envelope that would be distinct from that of the gravity field. While Jupiter's atmosphere above 22 bars was directly sampled by the Galileo Entry Probe, no similar probe for Saturn is currently scheduled. Jupiter's atmosphere is enhanced (by number) by factors of ~2–5 relative to abundances in the Sun in most elements, with depletions in helium and neon attributed to interior processes (discussed below). The well-known depletion

Table 3.1 *Classic values for Saturn heat balance: All data after Pearl and Conrath (1991) except* $T_{1\ bar}$, *from Lindal et al. (1985); Lindal (1992). It should be noted that the radio occultation retrieved profile depends on the atmospheric composition which was assumed to be 94% (by number) of molecular hydrogen with the rest being helium. Furthermore, analysis of Cassini data by Li et al. (2015) yields revised, appreciably higher, values for the intrinsic flux and for the* T_{eff}, *found to be 96.67 ± 0.17 K.*

Absorbed power	Emitted power	Intrinsic power	Intrinsic flux	Bond albedo	T_{eff}	$T_{1\ bar}$
10^{23} erg s^{-1}	10^{23} erg s^{-1}	10^{23} erg s^{-1}	erg s^{-1} cm^{-2}		K	K
11.14(50)	19.77(32)	8.63(60)	2010(140)	0.342(30)	95.0(4)	135(5)

in water is still an area of active discussion concerning whether the depletion reflects the true abundance of Jovian water (Showman and Ingersoll 1998; Lodders 2004; Mousis et al. 2012; Helled and Lunine 2014), which will hopefully be settled by the Juno mission.

In terms of comparative planetary science, the only elemental abundance that has been accurately determined for each of the solar system's four giant planets is that of carbon, found in methane, where the supersolar enhancement grows with decreasing planet mass, from a factor of ~4 for Jupiter (Wong et al. 2004), ~10 for Saturn (Fletcher et al. 2009), and ~80 for Uranus and Neptune (Sromovsky et al. 2011; Karkoschka and Tomasko 2011). For Saturn, this suggests an H-He envelope that on the whole may be strongly enhanced in heavy elements. It has been problematic to directly include an implementation of an enriched envelope as a constraint in interior modeling, however. One would also like to know the water and ammonia abundances, as these, along with methane, would likely correspond to ~60–80% of the heavy element mass. Given the Galileo results, it may be very unwise to assume that O and N scale with C in giant planets. If one knew the "metallicity" of the envelope of Saturn, that would place important constraints on the core mass, as a wide range of solutions for the bulk abundance of heavy elements allowed by the gravity field find a wide diversity of the amount of heavy elements in the envelope. The abundance of helium is an essential constraint on interior models, as it affects the planet's density and temperature distribution with radius, as well as the planet's thermal evolution, and is discussed in detail later in the chapter.

3.2.3 Gravity Field

The mass of Saturn is obtained from the observation of the motions of natural satellites: 95.161 $M_\oplus$, where 1 $M_\oplus$= 5.97369 × 10^{24} kg. More precise measurements of the planet's gravity field can be obtained through the analysis of the trajectories of spacecraft (e.g. Voyager, Cassini) during flyby (obtained via Doppler shift of radio emission). The most precise constraints come from close-in passes to the planet, in a near-polar orbit. Because of the rapid rotation of Saturn, its gravitational field departs from that of a point mass (a purely spherical field).

Of particular interest for using the gravity field is the need for a suitable theory to invert the gravity information to provide constraints on the planet's density as a function of radius. This "Theory of Figures" is a classical problem (Zharkov and Trubitsyn 1978), which is discussed in Section 3.5.1. For Saturn, a limitation in the application of this theory is our uncertainty in the rotation rate of the planet, which is discussed in detail in Section 3.6.

3.2.4 Magnetic Field

All of the solar system planets and some moons have or have had large-scale magnetic fields during their history, with Venus as the only possible large exception. Six out of eight planets in our Solar system have present-day planetary-scale magnetic fields of internal origin, and all of the giant planets have large fields (e.g. Stevenson 2003). The planetary magnetic fields are as diverse as the host planets, yet no simple correlations have been found between the basic features of the magnetic fields (e.g. field strength, field morphology) and the basic features of the host planets (e.g. composition, mass, radius, rotation, heat flux). It is striking that Jupiter and Saturn, planets of a similar kind, should have such different fields, and this remains one of the biggest challenges to our understanding (see Chapter 4 by Christensen et al.).

In situ magnetic field measurements made by the Pioneer 11 Saturn flyby in 1979 showed for the first time the existence of a dipole-dominant, global-scale magnetic field at Saturn with surface field strength around 30,000 nT (Smith et al. 1980). Subsequent MAG measurements made during the Voyager 1 and Voyager 2 Saturn flybys (Ness et al. 1981, 1982) and those made with the ongoing Cassini orbital mission (Dougherty et al. 2005; Burton et al. 2009; Sterenborg and Bloxham 2010; Cao et al. 2011, 2012) have established the low-degree structures of Saturn's magnetic field. Cao et al. (2011, 2012) employed spherical harmonic analysis based on the close-in part of the Cassini MAG measurements from Saturn orbital insertion (SOI) to early 2010 and showed that Saturn's magnetic field is extremely axisymmetric, with an upper bound on its dipole tilt of 0.06°. This is in striking contrast to the Earthlike dipole tilt of 10° exhibited by Jupiter. It also explains the persistent uncertainty in the spin rate of Saturn (discussed in Section 3.6), whose value for a fluid planet can probably only be meaningfully defined by consideration of the magnetic field non-spin axisymmetry.

One cannot rule out for certain the presence of a non-axisymmetric gravity field component, the value of which might need only be one part in ~10^8 or even smaller to be detectable. However, there is no assurance that this would represent rotation of the deep interior, whereas the magnetic field deep down is prevented from having a significant differential rotation because of the large toroidal field and resulting torques that would otherwise result. Saturn also has a modest north-south asymmetry with an axial quadrupole moment that amounts to 7.5% of the axial dipole moment on the surface, and has extremely slow time evolution between the Cassini era and the Pioneer-Voyager era, with an upper bound one order of magnitude smaller than that of the geomagnetic field.

The external magnetic field B is often expressed as an expansion in spherical harmonics of the scalar potential W, with $B = -\nabla W$:

$$W = R_{\text{eq}} \sum_{n-1}^{\infty} \left(\frac{R_{eq}}{r} \right)^{n+1} \sum_{m=0}^{n} \tag{3.1}$$

$$(g_n^m \cos(m\Phi) + h_n^m \sin(m\Phi))\, P_n^m \cos(\theta),$$

where Φ is the longitude and the P_n^m terms are the associated Legendre polynomials. The coefficients g_n^m and h_n^m are the magnetic moments that characterize the field, and are in units of Tesla.

Since the field is indistinguishable from axisymmetry in the current data, it suffices to list the axial terms in the usual expansion of the potential whose gradient provides the field. Given here in Table 3.2 is the Cassini 5 model of Cao et al. (2012). As is usual in field models, one cannot exclude higher harmonics at level of tens of nT, and accordingly, the uncertainties in the listed values are of that order.

As with Jupiter and Earth, the quadrupole is suppressed (that is, the octupole is larger) but since the field is so nearly axisymmetric, it is also evident that the field has very little radial flux near the equator (recall that the dipole has none and the octupole has none). This may be of significance for interpreting the behavior of the dynamo that produces the field.

Table 3.2 *Axial terms of Cassini 5 magnetic field model of Cao et al. (2012)*

Spherical harmonic coefficient	Amplitude (nT)
g° 1	21191
g° 2	1586
g° 3	2374
g° 4	65
g° 5	185

It is common (though perhaps dangerous) practice to infer the size of the dynamo region by asking for the radius at which the field at higher harmonics approaches the dipole in magnitude. By this measure (and using the octupole as a better guide than the suppressed quadrupole), we can infer a "core radius" of $(2374 \times 4/21191 \times 2)^{1/2} \sim 0.47$ in units of Saturn's radius, not greatly different from the radius at which hydrogen becomes highly conducting within Saturn. However, this should not be over-interpreted.

3.3 Input Physics

3.3.1 Hydrogen

Within giant planets, hydrogen is found in the fluid, not solid, phase. Most of the mass of Saturn is beyond the realm of current experiments on hydrogen, so a mix of constraints from lower-pressure experiments, together with simulations of hydrogen under high pressure, are used to understand its EOS. The EOS of hydrogen is the most important physical input into Saturn interior models. However, uncertainties in the EOS for Saturn models are not as important as for Jupiter (Saumon and Guillot 2004), since Saturn is lower in mass, and therefore less of its interior is found in the higher-pressure regions above several Mbar that is not yet accessible to experiment.

For a given interior isentrope, the Grüneisen parameter indicates the change of temperature with density within the interior, and hence the bulk energy reservoir of a given model. The compressibility of hydrogen, as a function of pressure (and hence, radius) directly affects inferences for the amount of heavy elements within the planet needed to explain its radius and gravity field, as well as *where* these heavy elements are found within the planet – meaning, perhaps, within a core, which may not be fully distinct from the overlying envelope, or distributed within the H/He-dominated envelope.

The past 15 years has seen dramatic advances in our understanding of hydrogen, both in the realm of experiment and simulation. On the experimental side, reverberation shock measurements of the conductivity of hydrogen at the fluid insulator (H_2) to fluid metal (H^+) transition indicate a continuous transition from the molecular to the metallic phase (Nellis et al. 1999). First-principles simulations of this transition also find a continuous transition over the temperatures of interest for giant planets (see McMahon et al. 2012, for a comprehensive review).

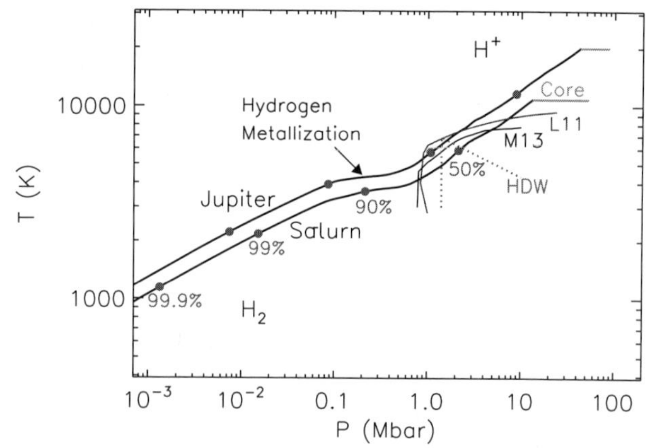

Figure 3.2 Interior $P - T$ profiles of Jupiter and Saturn, following the methods of Nettelmann et al. (2013). The H/He envelopes are shown, including the assumed isothermal cores at the highest pressures. For both planets, dots indicate where the enclosed mass of the model planet is 50%, 90%, 99%, and 99.9%. Note the shift outward to lower pressures at a given mass shell for Saturn, compared to Jupiter. The gradual transition from fluid H_2 to liquid metallic hydrogen (H^+) is shown with a black arrow. Three predicted regions of He immiscibility (at Y=0.27, the protosolar abundances) in hydrogen are depicted. The theory of Hubbard and Dewitt (1985), analogous to Stevenson (1975), is labeled "HDW." The theories of Lorenzen et al. (2011) and Morales et al. (2013) are labelled "L11" and "M13" respectively. Both of these recent *ab initio* simulations predict that large regions of Saturn's interior mass are within the He immiscibility region.

3.3.2 H/He Mixtures

How the physics of mixtures of hydrogen and helium may differ from a simple linear mixture of the two components has been an area of active study for decades. A number of earlier investigations suggested the helium may phase separate from liquid metallic hydrogen under giant planet conditions and "rain down" to deeper layers within the planet (Salpeter 1973; Stevenson 1975; Stevenson and Salpeter 1977b,a). Phase separation occurs when the Gibbs free energy of a mixture can be minimized when the mixture separates into two distinct phases – here, where one is helium poor, and the other, helium rich.

Early work on trying to understand the phase diagram of H-He mixtures focused on systems that were readily amenable to calculation, for instance mixtures of fully ionized H and He (Stevenson 1975; Hubbard and Dewitt 1985).

These calculations suggested that Saturn's current isentrope, and perhaps Jupiter's, intercepted regions of $P - T$ where phase separation would occur. As *ab initio* methods became possible, the phase diagram was investigated with these tools (Klepeis et al. 1991; Pfaffenzeller et al. 1995), but the results of these early 90s works were significantly inconsistent with each other. With the rise of more modern *ab initio* tools that were able to make fewer approximations, H-He mixtures have again been investigated. Deviations from linear mixing have been found, and new calculations of the phase diagram have been published by two groups, Lorenzen et al. (2009, 2011) and Morales et al. (2009, 2013). Isentropic interior models of Jupiter and Saturn, compared to these various H/He phase diagrams, are shown in Figure 3.2. All recent work finds that Saturn's estimated interior $P - T$ profile intersects regions of He phase separation. He-rich droplets, being denser than their surroundings, may rain down to deeper layers of the planet, redistributing significant mass and altering the cooling history of the planet (Stevenson and Salpeter 1977b; Fortney and Hubbard 2003).

Beside the onset temperature for immiscibility of the mixture, the shape of the phase diagram is important, as that directly controls the fraction of the planet's mass that falls within the phase separation region. Earlier work (Stevenson 1975; Hubbard and Dewitt 1985) suggested He immiscibility in a relatively narrow pressure range, which grew slowly as the planet's interior cooled. However, Fortney and Hubbard (2003) suggested that Saturn's current luminosity can be explained by settling of helium droplets throughout most of the planetary interior. Modern phase diagrams (Lorenzen et al. 2011; Morales et al. 2013) basically agree that this immiscibility region includes the bulk of Saturn's interior.

3.3.3 Water and Rock

The EOS for the heavier elements have generally received somewhat less attention than those for hydrogen and helium. However, the past five years have seen substantial advances in the *ab initio* calculations of the EOS, as well as miscibility properties, for water, ammonia, rock and iron.

Perhaps most importantly, an accurate *ab initio* EOS for water has been published by French et al. (2009), which fares extremely well against data from single- and double-shock experiments up to an impressive 7 Mbar (Knudson et al. 2012). The phase diagram of water has been explored, and it appears rather conclusive that any water in Saturn's core is found in the fluid, not solid, state. Wilson and Militzer (2012b) have also looked at whether water, at Saturn and Jupiter's core conditions, is miscible in liquid metallic hydrogen. They find that it is, such that for both planets diffusion of core material into the overlying H/He envelope is probable, although the efficiency of this process is still unknown.

The details of the EOS of rock and iron are less essential, as the temperature dependence on the density of these components is relatively weak at giant-planet interior conditions. Approximate EOS values for rock-iron mixtures can be found in Hubbard and Marley (1989), and Saumon and Guillot (2004) use a "dry sand" EOS from the Sesame database. Wilson and Militzer (2012a) and Wahl et al. (2013) have looked at miscibility of silicate rock MgO, and iron, respectively and, like for water, find that these components are also miscible in liquid metallic hydrogen. However, MgO was found to be solid under Saturnian conditions, while iron is likely solid today, but perhaps liquid at earlier times when the core was hotter. The miscibility behavior of core material is an important, but by itself incomplete, indicator for the efficiency of core erosion. Thus, at present the amount of possibly redistributed core material in Jupiter and Saturn is not known.

3.4 First-Order Deductions from Simple Models

3.4.1 Heavy-Element Enrichment from M–R

While there are many open questions regarding Saturn's bulk composition and internal structure, first-order knowledge of its composition can be inferred from the relation between its mass and radius (the Mass-Radius relation). The Mass-Radius relation of planetary (or astrophysical) objects is often used to infer their bulk composition simply by calculating the mean density from $\bar{\rho} = 3M/4\pi\bar{R}^3$, where M is the planet's mass and $\bar{R}$ is its mean radius. As the mass of the object increases, the increase of density due to high pressures becomes important. Eventually, the density becomes so high with increasing planet mass that the radius will start to decrease. Such behavior is clearly seen in M–R relations for exoplanets, which exhibit a wide range of masses, as shown in Figure 3.3. However, inferring the bulk heavy-element abundance of giant exoplanets from their M–R relation becomes challenging when their natural shrinkage in radius over time is retarded by strong stellar irradiation. This can be the case for

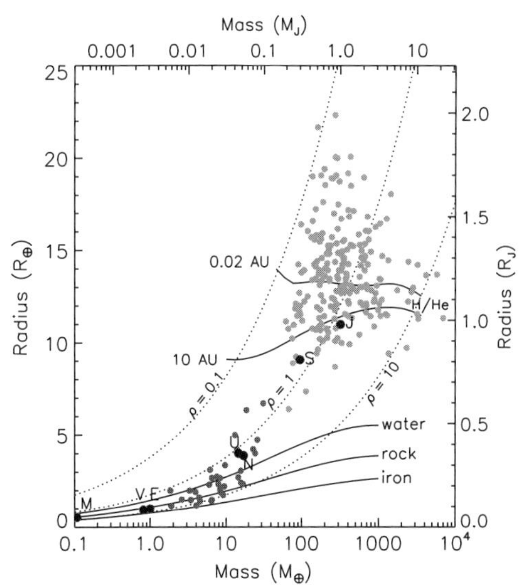

Figure 3.3 The mass-radius relation of planets with well-determined masses, radii and orbits. Curves of constant bulk density (ρ = 0.1, 1, 10) are shown as dotted lines. Models for solar-composition planets at 4.5 Gyr, at 10 and 0.02 AU from the Sun, are shown as thick black curves (Fortney et al. 2007). Jupiter, Saturn, Uranus, Neptune, Earth and Venus are labeled by their first letter. Planets more massive than 0.1 M_J are shown in red. Saturn is appreciably lower that the 10 AU curve, indicating it is enriched in heavy elements.

close-in exoplanets, e.g. those with orbital distances of 0.02 AU. On the other hand, the radius of evolved planets at large orbital distances should be mainly affected by the bulk composition.

From Figure 3.3 we can conclude that Saturn is enriched with heavy elements compared to proto-solar composition, as its radius is smaller than the one expected for a solar-composition planet at 10 AU.

3.4.2 Dimensionless Parameters

There are several dimensionless parameters that are associated with the characterization of giant planets. In particular, the rotation parameter m or q, the nondimensional moment of inertia (NMOI), and the flattening f are linked by the Radau–Darwin approximation (Jeffreys 1924) and will be described in the following.

Flattening (Oblateness)

The flattening of a planet is defined by

$$f \equiv \frac{R_{\mathrm{eq}} - R_{\mathrm{p}}}{R_{\mathrm{eq}}} \tag{3.2}$$

where R_{eq} and R_{p} are the equatorial and polar radius, respectively. While knowledge of the continuous shape of a planet (i.e. radius vs. latitude) is also desirable and available, typically, interior models use only the oblateness, or mean radius $\overline{R}$ that can be estimated from $\overline{R} \tilde{} (R_{\mathrm{eq}}^2 R_{\mathrm{p}})^{1/3}$. The oblateness provides information on the planetary rotation rate: planets that rotate rapidly are more oblate, as we will see below by using the Darwin–Radau relation.

The Rotation Parameter

Typically, the density profile of giant planets is derived by using the theory of figures that is expanded in powers of a small parameter, the rotation parameter (e.g. Zharkov and Trubitsyn 1978), and discussed in Section 3.5.1. It is defined as the ratio of the centrifugal to gravitational force at the equator, $q \equiv \omega^2 R_{\mathrm{eq}}^3 / GM$, or alternatively, with respect to the mean radius, $m \equiv \omega^2 \overline{R}^3 / GM$. The values for Saturn are $q \sim 0.155$ and $m \sim 0.139$ when using the Voyager radio period. Since the rotation parameter depends on ω (i.e. the rotation rate) which is unknown for Saturn to within several minutes (see Section 3.6) there is an uncertainty associated with the rotation parameter. For a rotation period that is ten minutes shorter than the Voyager period the rotation parameters for Saturn are $q \sim 0.160$ and $m \sim 0.144$. The smaller the values of q and m are, the better is the approximation of the theory of figures (Hubbard 2012).

Moment of Inertia

The axial moment of inertia of a planetary body provides information on its density profile. Since giant planets are in hydrostatic equilibrium and therefore symmetric around the axis of rotation, their moment of inertia is derived from

$$I = 2\pi \int_0^{\Pi} \int_0^{R} d\vartheta \; dr \; \rho(\vartheta, r) r^4 \; \sin^3\vartheta. \tag{3.3}$$

It is common to define the nondimensional moment of inertia factor (hereafter, NMOI) as I/MR^2. Then, the NMOI can be directly linked to the density (radial) distribution. An NMOI of a constant-density object is 0.4; lower NMOI values correspond to objects that are more centrally condensed, i.e. increase of density toward the center. Therefore, just like J_{2n}, which is defined below in 3.5.1, the MOI can be used as an independent constraint on the internal density distribution.

The Radau–Darwin Approximation

Finally, one can link the dimensionless parameters by using the Radau–Darwin approximation (e.g. Zharkov and Trubitsyn 1978). There are several forms for this approximation. One of them relates the planetary NMOI and $\Lambda_2 \equiv J_2/q$. The Radau–Darwin formula suggests that there is a one-to-one correspondence between MOI and Λ_2 by

$$\mathrm{MOI} = \frac{I}{MR^2} = \frac{2}{3}\left[1 - \frac{2}{5}\left(\frac{5}{(\Lambda_2 + 1) - 1}\right)^{1/2}\right]. \tag{3.4}$$

Another form of the Radau–Darwin relates the MOI with the flattening f and rotation parameter q via

$$\frac{I}{MR^2} = \frac{2}{3}\left[1 - \frac{2}{5}\left(\frac{5q}{2f} - 1\right)^{1/2}\right]. \tag{3.5}$$

For Saturn, the Radau–Darwin relation suggests an NMOI of 0.220 (e.g. Guillot and Gautier 2014; Helled 2011). For comparison, Jupiter's NMOI is estimated to be ~0.265 (Jeffreys 1924; Helled et al. 2011), therefore indicating that Saturn is more centrally condensed than Jupiter, and potentially indicating that Saturn has a larger core mass.

The Radau–Darwin approximation is quite good. Helled (2011) investigated the possible range of Saturn's MOI values accounting for the uncertainty in rotation period and internal structure. It was found that the MOI value can differ by up to 10% from the value derived by the Radau–Darwin relation. A similar analysis was also done for Jupiter (Helled et al. 2011), for which the MOI range was found to be 0.264 with an uncertainty of up to 6%.

3.5 Current Modeling Methods and Assumptions

3.5.1 *Theory of the Gravity Field*

In the theory of figures (TOF), one uses potential theory to solve for the structure of rotationally distorted Saturn, assuming hydrostatic equilibrium for a given pressure-density relation $P(\rho)$. The external gravitational potential of a rotating planet in hydrostatic equilibrium is given by

$$V = \frac{GM}{r}\left(1 - \sum_{n=1}^{\infty}\left(\frac{R_{eq}}{r}\right)^{2n} J_{2n}P_{2n}(\cos\theta)\right) + \frac{1}{2}\omega^2 r^2 \sin^2\theta, \tag{3.6}$$

where (r, θ, φ) are spherical polar coordinates and R_{eq} is the equatorial radius. It can be shown (Zharkov and Trubitsyn 1978) that the potential can be expressed as a double power-series expansion in the dimensionless small parameter m defined above. Each term J_{2n} in the multipole expansion for Saturn's external gravitational potential V, then, has a coefficient whose further expansion takes the form

$$J_{2n} = m^n \sum_{t=0}^{\infty} \Lambda_{2n}^{(t)} m^t, \tag{3.7}$$

where the dimensionless response coefficients $\Lambda_{2n}^{(t)}$ must be obtained from the solution of a hierarchy of integrodifferential equations. Since Saturn's m ~0.14, the expansion does not converge rapidly. In principle, for comparison with expected high-precision Cassini measurements of Saturn's J_{2n}, using the expansion method one would need to derive all of the response coefficients for a test $P(\rho)$ out to terms ~m^9! This situation thus indicates the need for a nonperturbative approach.

The multipole coefficients J_{2n} are measurable by fitting a multiparameter model to spacecraft Doppler residuals. However, the corresponding model values of J_{2n} for a specified $P(\rho)$ are not obtained by expanding in powers of m, but rather are calculated directly using an iterative self-consistent solution to a prescribed precision, usually ~10^{-12}. Two algorithms for nonperturbative calculations are available. One method (J. Wisdom, 1996, unpublished, available at http://web.mit.edu/wisdom/www/interior.pdf) assumes that the interior density distribution is a continuous function of position and can be expanded on a set of polynomials. The other method (called the CMS, or concentric Maclaurin spheroid, method) represents the interior density distribution by a nested set of spheroids, each of constant density (Hubbard 2012, 2013). A CMS model can be made to approach a continuous-density model by increasing the number of spheroids, at the price of lengthier computations.

It is not widely appreciated that traditional TOF methods employ a formally nonconvergent expansion attributed to Laplace. The suspect expansion is in fact intimately related to the standard J_{2n} expansion of the external gravity potential. Although criticisms of the expansion have been published over the years, e.g. Kong et al. (2013), it can be shown (Hubbard et al. 2014) that both Jupiter and Saturn are in the domain of m where Laplace's "swindle" works exactly, or at least to a precision ~10^{-12}, more than adequate for quantitative comparison within the expected precision

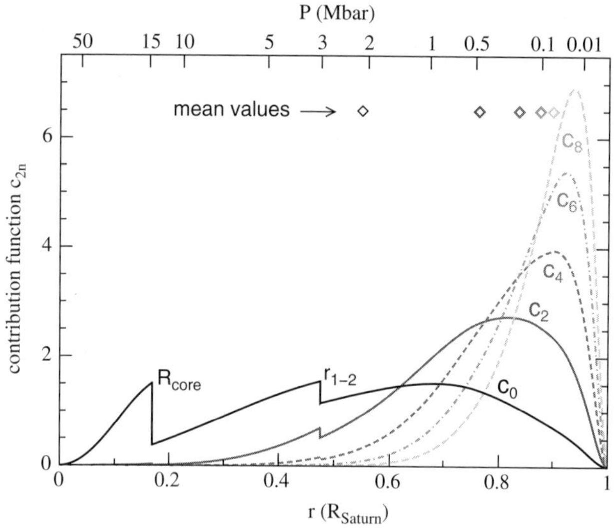

Figure 3.4 Contribution functions of the gravitational harmonics J_2 (solid, labelled C_2), J_4 (short-dashed), J_6 (dot-dashed) and J_8 (long-dashed) for the Saturn model "S12-3a" from Nettelmann et al. (2013). The bottom x-axis is the fraction of Saturn's radius; the top x-axis shows the radii where the pressures of 0.01–50 Mbar occur. Layer boundaries due to the heavy element core, and between the helium-enriched interior and helium-depleted exterior are clearly seen in the function c_0 which tracks the contribution of radius shells to the planet's total mass (solid, labelled C_0). Diamonds show the radius where half of the final $J_{2\,n}$ value is reached.

of Cassini measurements. An earlier proof of the validity of the traditional expansion is given by Wavre (1930).

In summary, assuming that the Juno and Cassini gravity experiments can successfully measure higher moments out to $\sim J_{10}$ to a precision $\sim 10^{-8}$ or even $\sim 10^{-9}$, and assuming that the value of Saturn's m can be accurately determined, one can use TOF methods with adequate accuracy to provide strong constraints on acceptable barotropes, particularly at radius levels relatively close to the surface, as shown in Figure 3.4. This in turn feeds back into the inferred core mass, even though the core region does not contribute directly to the multipole weighting functions.

3.5.2 Adiabatic Assumption

The thermal structure of a planet interior is of importance in several ways. It will determine the phase of the material (liquid or solid) and it will likely play a role in its electronic properties (important for maintaining a magnetic field). It also contributes significantly to the pressure at a given density (of order 10% is typical in the deep interior of Saturn). It plays a central role in the thermal evolution of the planet; by the Virial theorem, the decrease of total heat content with time can be a large source of luminosity. It is also evident from the First Law that the planet is likely to be hot immediately after formation. One clearly needs a prescription for the thermal structure and how it evolves with time.

It is often said that the giant planets are "adiabatic." There are good reasons for thinking that this is a good starting approximation, but also good reasons for doubting that it is correct for the planets as a whole; here we review both perspectives. But first, a definition.

The adiabatic assumption is more precisely stated as follows. Throughout most of the planet, the specific entropy within well-mixed layers is nearly constant. If there are several well-mixed layers (but with each layer having a different composition), then the entropy within each layer may also be nearly constant and the "jump condition" across layer interfaces is nearly isothermal. The assumption must of course break down in the outermost region of the atmosphere (optical depth unity or less), where outgoing IR photons are free to escape to space. But our concern here is whether and to what extent it breaks down deeper in the planet. Note that the preferred word is "isentropic," not "adiabatic," since the latter word describes a process while the former is a thermodynamic statement, and that is what we need to define the thermal state. Nevertheless, we here apply the latter word in line with common usage.

It is instructive to first consider a completely homogeneous, fluid planet that is emitting energy from its interior (as Jupiter and Saturn are observed to do). As one proceeds deeper than optical depth unity, the opacity increases, primarily because of pressure-induced molecular hydrogen opacity (Guillot et al. 1994). As a consequence, the temperature gradient that would be needed to carry out the heat by radiation alone exceeds the adiabatic temperature gradient, at least for most of the interior. As shown by Guillot et al. (1994), a small radiative window may be possible at 2000–3000 K in Saturn. However, application of improved opacities for the alkali metals sodium and potassium leads to closure of such a window for Jupiter (Guillot et al. 2004). The same is expected to hold for Saturn, so that we safely assume that the opacities are too high for radiation to carry the heat efficiently throughout Saturn's interior. The heat carried by conduction throughout the interior, once the hydrogen becomes an electrical conductor, is also too small to be important. This means the interior is convectively unstable. Should such a planet be convecting heat outwards, then there is no doubt that the temperature gradient is extremely close to being adiabatic.

The conclusion is reached in two steps. First, one asks what the state of the material would be if it were in fact isentropic. The answer is that it is everywhere fluid and of low viscosity. (In this context, even a viscosity six orders of magnitude greater than everyday water would qualify as "low." In fact, the viscosity is comparable to that of everyday water.) The second step is to recognize that convection in such a medium is extremely efficient in carrying heat over large distances.

Consider, for example, the flows that could carry a few Watts m^{-2} (typical of Saturn's interior). We can write the heat flux as $F \sim \rho C_p v \delta T$, where ρ is fluid density, C_p is specific heat, v is convective velocity and δT is the temperature anomaly whose resulting density anomaly leads to the buoyancy that is responsible for v. We further expect $v \sim (g\alpha\delta T\, L)^{1/2}$ where α is the coefficient of thermal expansion and L is the characteristic length scale of the motions, because viscosity is too small to be relevant. For the choices $\rho = 1000$ kg/m^{-3}, $C_p = 2 \times 10^4$ J kg^{-1} K^{-1}, $\delta T = 10^{-4}$ K, $\alpha = 10^{-5}$ K^{-1}, $L = 10^6$ m, one finds $v \sim 0.1$ m s^{-1} and $F \sim 10^2$ W/m^{-2}. This crude order of magnitude argument (mixing length theory) may well be wrong by an order of magnitude or even several (perhaps the value of L is smaller because of the Coriolis effect), but the conclusion is inescapable: temperature deviations from an isentrope are expected to be extremely small. This implies something remarkable: given the outer boundary condition (the specific entropy at the top of the convective zone), one can determine the temperature at the center of the planet (e.g. Hubbard 1973). Adding an isolated core (i.e. a core that does not dissolve in the overlying material) to an otherwise homogeneous planet does not significantly change this story. However, there are several respects in which this picture could be in error:

1 Although the planet was heated by accretion, it may have formed in such a way that the deep interior had lower entropy than the outer regions. Since gravitational energy release per unit mass increases as the planet grows, this could even be a likely outcome.
2 Although the planet may be fluid almost everywhere, that does not preclude first-order phase transitions. The temperature structure may be altered by these transitions.
3 The core, if any, may be soluble in the overlying hydrogen. This creates compositional gradients (even if no gradient were present at the end of accretion).
4 The accretion process might create compositional gradients because of the imperfect mixing that arises when incoming planetesimals break up in the envelope.
5 The presence of H/He phase separation may act as a barrier to convection if the growing He droplets do not rain down instantaneously.
6 Condensation of ices and latent heat release in the weather layer can lead to molecular weight gradients and to deviation from adiabaticity due to the different temperature gradients along moist and dry adiabats.

The first of these concerns is probably less important than the others. While it is indeed true that the specific entropy distribution may be initially stable (low entropy towards the center of the planet), the outer regions undergo entropy decrease with time, and this is rapid when the planet has a high effective temperature, with the result that the planet can naturally evolve towards an isentropic state.

The second concern has two important cases to consider. First there might be a first-order phase transition from molecular to metallic hydrogen. In the fluid phase, this is referred to as the Plasma Phase Transition (PPT). Many aspects of this transition are still uncertain, but there is currently no evidence that it persists to the high temperatures typical of interior models at the relevant pressure (McMahon et al. 2012), suggesting a continuous transition in planets. Were such a transition to exist, it could lead to a stable interface between the two phases, with an entropy discontinuity across the interface (Stevenson and Salpeter 1977a). The more important case of relevance to Saturn (and apparently to some extent Jupiter as well) is the limited solubility of helium in hydrogen, which can impose a helium composition gradient, which is discussed in Section 3.8.2.

3.6 Rotation Rate Uncertainty

The rotation period of a giant planet is a fundamental physical property that is used for constraining the internal structure and has implications for the dynamics of the planetary atmosphere. Saturn's rotation period is still not well constrained. Cassini has confirmed a time-dependence in Saturn's auroral radio emission found from a comparison of Ulysses data to the earlier Voyager 1 and 2 spacecraft observations (Galopeau and Lecacheux 2000). Because Saturn's magnetic pole is aligned with Saturn's rotation axis, Saturn's standard rotation period is set to the Voyager 2 radio period, 10h 39m 22.4s. This rotation period was derived from the periodicity in Saturn's kilometric radiation SKR (e.g. Dessler 1983). Surprisingly, Cassini's SKR measured a rotation period of 10h 47m 6s (e.g. Gurnett et al. 2007), about eight minutes longer, using the exact same method. It is now accepted that Saturn's exact rotation period is unknown to within several minutes and cannot be inferred from SKR measurements. In addition, atmospheric features such as clouds cannot directly be used to derive Saturn's rotation period because it is unclear how they are linked to the rotation of the deep interior, and in fact Saturn's observed wind velocities are always given relative to an *assumed* solid-body rotation period.

Recently, several theoretical approaches to determine Saturn's rotation period have been presented. The first approach was based on minimizing the dynamical heights at the 100-mbar pressure level above the geopotential surface caused by the atmospheric winds (Anderson and Schubert 2007). The dynamical heights (as well as the wind speeds) were found to be minimized for a rotation period of 10h 32m 35s ± 13s, about 7 minutes shorter than the Voyager 2 value. In a second study, an analysis of the potential vorticity based on considerations of their dynamic meteorology was presented. Saturn's derived rotation period was found to be 10h 34m 13s ± 20s (Read et al. 2009). Both of these studies relied on the measured wind velocities obtained from cloud tracking at the observed cloud level.

In a third study, Saturn's rotation period was derived from its observed gravitational moments and its observed shape, including uncertainties in these measurements by taking an optimization approach (Helled et al. 2015). The gravitational data are insufficient to uniquely determine the rotation period, and therefore the problem is underdetermined (there are more unknowns than constraints). Accordingly, a statistical optimization approach was taken, and by using the constraints on the radius and the gravitational field, the rotation period of Saturn was determined (statistically) with a fairly small uncertainty. When only the gravitational field is used as a constraint, the rotation period was found to be 10h 43m 10s ± 4m. With the constraints on Saturn's shape and internal density structure, the rotation period was found to be 10h 32m 45s ± 46s, in excellent agreement with Anderson and Schubert (2007). This is because Saturn's mean radius is more consistent with shorter rotation periods, *if* dynamical distortions on the shape are not included. Interestingly, all of these studies infer a shorter rotation period for Saturn than the Voyager 2 rotation period, leading to smaller wind velocities and atmosphere dynamics more similar to those of Jupiter. The validity of these theoretical approaches, however, is yet to be proven and a more complete understanding of the shape-dynamics-internal-rotation feedback is required. However, all of the three methods described above, when applied to Jupiter, yield a rotation period that is consistent with Jupiter's generally accepted value, based on the rotation of its non-axisymmetric magnetic field.

Besides implications of the rotation period of Saturn to its inferred internal structure (see below), it also directly affects the atmospheric wind velocities. Saturn's wind profile with respect to three different rotation periods is shown in Figure 3.5. Shown are the wind velocities vs. latitude (degrees) for rotation

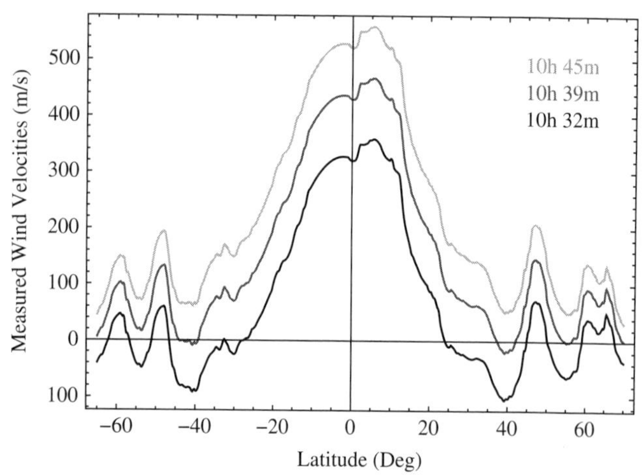

Figure 3.5 Saturn's wind velocities for three different underlying rotation periods: 10h 32m (black), 10h 39m (medium gray), 10h 45m (light gray). The measured wind velocities at the cloud-layer are obtained from Sanchez-Lavega et al. (2000).

periods of 10h 32m (black), 10h 39m (medium gray), and 10h 45m (light gray). A rotation period of about 10h 32m implies that the latitudinal wind structure is more symmetric, containing both easterly and westerly jets, as observed on Jupiter.

Finally, an additional complication regarding Saturn's internal rotation period arises from the fact that Saturn could rotate differentially on cylinders and/or that its atmospheric winds penetrate deep into the interior. This can also affect interior models (e.g., Hubbard 1982, 1999; Helled and Guillot 2013). The realization that Saturn's rotation period is not constrained within a few percent, and the possibility of differential rotation, introduce an uncertainty for interior models, as discussed below.

3.7 Current Structure from Isentropic Models

Saturn's internal structure has been studied for decades. In this section, we concentrate on results from recent interior models of Saturn and on the physical parameters that are used to constrain the planetary interior.

3.7.1 Observational Constraints and Model Assumptions

Constraints on Saturn's internal structure have been provided through several spacecraft missions and ground-based observations. For brevity, we call such constraints "observational," although most of them are not directly obtained from measurements, but through a variety of theoretical models that fit the measured data. Prior to the Cassini era, constraints were derived for Saturn's total mass, shape, gravitational harmonics (J_{2n}), periodicities in its surrounding plasma disk, magnetic field, kilometric radio-emission, the atmospheric helium abundance and temperature profile and the luminosity. Cassini has improved our precision on most of these quantities.

The uncertainty in Saturn's internal structure is not only linked to the EOS and the uncertainties in the observational constraints, but is in fact in the philosophy adopted when modeling the interior. Even under the assumption of an isentropic interior, model properties such as the existence of differential rotation, the number of layers, and the distribution of heavy elements can lead to rather different inferred structures and compositions. For example, the consideration of the expected correction of differential rotation to the gravitational field calculated by static models relaxes the otherwise rather stringent constraints of the measured gravitational field, assuming solid-body rotation. In fact, there are various estimates for the magnitude of this effect (e.g. Hubbard 1982; Kaspi 2013; and Chapter 11 on Saturn's atmosphere dynamics by Showman et al.) and they are crucial for the interpretation of Cassini gravity data. It should be noted, however, that even for a perfectly known gravitational field and the contribution of dynamics (differential rotation), there is still ambiguity regarding the internal density distribution. This is reflected in the various assumptions adopted by different authors, such as inhomogeneities in heavy elements and the location of the transition from helium-poor to helium-rich envelopes. The internal density distribution can be affected by various physical processes such as helium rain and core erosion that are not well understood. As a consequence, there is some freedom in constructing an interior model – some authors include corrections due to differential rotation, while others add an inhomogeneity in heavy elements between the inner and outer H/He envelope (e.g. $Z_{\rm out}$ 6 $\neq Z_{\rm in}$ – see below). Standard interior models assume a three-layer structure, since the goal is to minimize the number of free parameters in the models.

The uncertainties in Saturn's J_2, J_4, J_6 have been significantly reduced compared to the pre-Cassini era through the combined analysis of Pioneer 11, Voyager, Cassini, and long-term ground-based and *HST* astrometry data (Jacobson et al. 2006). Smaller error bars provide an opportunity to narrow down the possible internal density distributions and thus Saturn's internal structure. Table 3.3 lists some of the values applied to Saturn interior models. As discussed above, uncertainties in Saturn's rotation rate and the fact that atmospheric winds and/or differential rotation affect the planetary shape cause an additional uncertainty in values of the gravitational harmonics used by hydrostatic interior models. Several models have accounted for this uncertainty, as presented below.

We first describe the different model assumptions and imposed constraints of recent isentropic, quasi-

Table 3.3 *Values for the gravitational coefficients applied to models of Saturn with a nominal rotation period of 10h 39m 24s. Saturn's measured gravitational coefficients are determined to be* $J_2 \times 10^2 = 1.62907(3)$; $J_4 \times 10^4 = -9.358(28)$ $J_6 \times 10^4 = 86.1(9.6)$ *for a reference radius of 60,330 km (Jacobson et al. 2006).* R_{eq} *corresponds to the equatorial radius at the 1-bar pressure level.* T_{1-bar} *is the assumed temperature at 1 bar,* Y_{atm} *is the atmospheric helium mass fraction and* Z_{env} *being the envelope metallicity which is either assumed to be homogeneous (hom) or inhomogeneous (inhom), and Diff. Rot. corresponds to whether corrections linked to differential rotation were considered (Yes/No). Values marked by (*) refer to the 1-bar level. G99: Guillot (1999); HG13: Helled and Guillot (2013); H11: Helled (2011); N13: Nettelmann et al. (2013); SG04: Saumon and Guillot (2004)*

Ref.	Req (km)	$J2 \times 10^2$	$J4 \times 10^4$	$J6 \times 10^4$	$T_{1\text{-bar}}$ (K)	Yatm	Zenv	Diff. Rot.
Voyager constraints								
G99	60268(4)	1.63320(100)*	−9.190(400)*	1.04(50)*	130 – 140	0.11 – 0.21	inhom	Yes
SG04	60268(4)	1.63320(100)*	−9.190(400)*	1.04(50)*	135 – 145	0.11 – 0.21	hom	Yes
HG13	60269	1.62580(410)	−9.050(410)	98(51)	130 – 145	0.11 – 0.25	hom	Yes
Cassini constraints								
H11	60141.4	1.63931*	−9.476*	87.8*	≈135	–	inhom	No
N13	60268	1.63242(3)*	−9.396(28)*	86.6(9.6)*	140	0.18	inhom	No
HG13	60269	1.62510(400)	−9.260(110)	81(11)	130 – 145	0.11 – 0.25	hom	Yes

homogeneous Saturn models (Guillot 1999; Saumon and Guillot 2004; Helled and Guillot 2013; Nettelmann et al. 2013).

Saturn models by Guillot (1999) (enclosed by *thin black dashed lines* in Figure 3.6) were designed for consistency with the Voyager constraints. A surface temperature of $T_{1\ \mathrm{bar}} = 135 - 145$ K, a rotation period of 10h 39m, a bulk helium mass fraction of $Y = 0.265 - 0.285$ with an atmospheric helium mass fraction of $Y_{\mathrm{atm}} = 0.11 - 0.21$ were used. The interior models were derived using the SCvH EOS for helium and the SCvH-i EOS for hydrogen (Saumon et al. 1995), which interpolates between the EOSs for the molecular and the metallic phases of hydrogen. Furthermore, they assume a sharp layer boundary between the molecular and metallic hydrogen envelope, where the abundances of helium and of heavy elements (Z_{out} for heavy elements in the outer envelope and Z_{in} for heavy elements in the inner envelope, respectively) change discontinuously. Heavy elements were assumed to have a water-like mean molecular weight and specific heat. Their mass fraction is derived from any "excess" helium abundance needed to fit the Voyager constraints.

Helled and Guillot (2013) modeled Saturn's interior, assuming a three-layer structure that consists of a central ice/rock core and an envelope that is split into a helium-rich metallic hydrogen region and a helium-poor molecular region. The main differences in underlying model assumptions from those of Guillot (1999) consist in $Y_{\mathrm{atm}} = 0.11 - 0.25$, a global $Y = 0.265 - 0.275$ consistent with the protosolar value (e.g. Bahcall et al. 1995), $T_{1\ \mathrm{bar}} = 130 - 145$ K, and, perhaps most importantly, in imposing $Z_{\mathrm{in}} = Z_{\mathrm{out}}$ and using a physical EOS of water and sand for heavy elements in the envelopes. The transition pressure between the helium-rich to helium-poor (hereafter P_{trans}) was assumed to be between 1 and 4 Mbars. The envelope heavy elements were assumed to be homogeneously mixed within the planetary envelope, as may well be expected in models that lack a first-order phase transition for hydrogen. In order to account for the uncertainty associated with differential rotation, the uncertainty in the gravitational harmonics as expected from differential rotation was also included (e.g. Hubbard 1982). In addition, two sets of gravitational data were used: the gravitational moments as measured by Voyager and those measured by Cassini (see Table 3.3). Two solid-body rotation periods were considered: the Voyager 2 radio period, and a shorter period that was set to the 10h 32m 35s as derived by Anderson and Schubert (2007). Finally, to account for the uncertainty in Saturn's shape, three different cases in terms of mean radius were considered: c_0, the "standard" equatorial radius of 60,269 km as previously assumed by interior models; c_1, a fixed polar radius with a corresponding equatorial radius of 60,148 km;

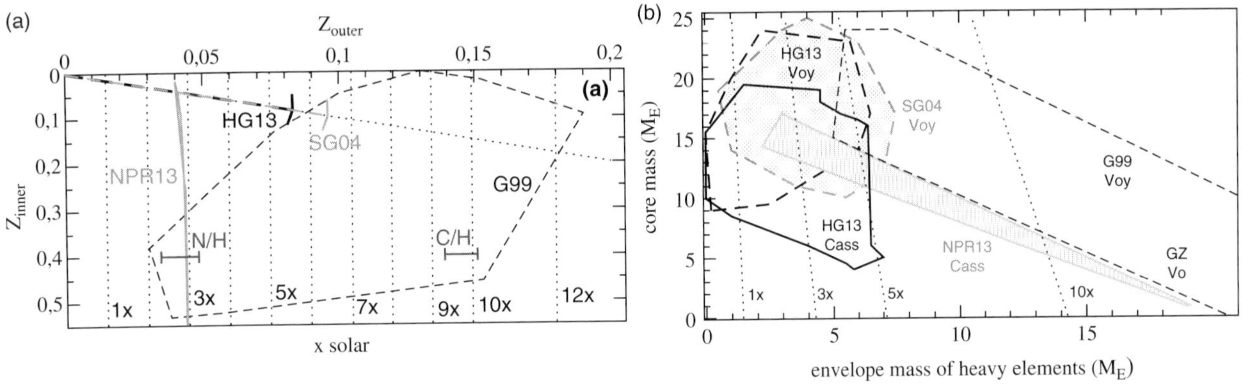

Figure 3.6 Saturn structure model results under various assumptions made by different authors (see text for details). (a) Heavy element enrichment by mass in the outer and inner envelope; enrichment factors over the bulk protosolar abundance of $Z_0 = 0.0149$ refer to Z_{outer} and are compared to the measured C/H and N/H ratios in Saturn's atmosphere. (b) Core mass and total mass of heavy elements in the homogeneous, isentropic envelopes.

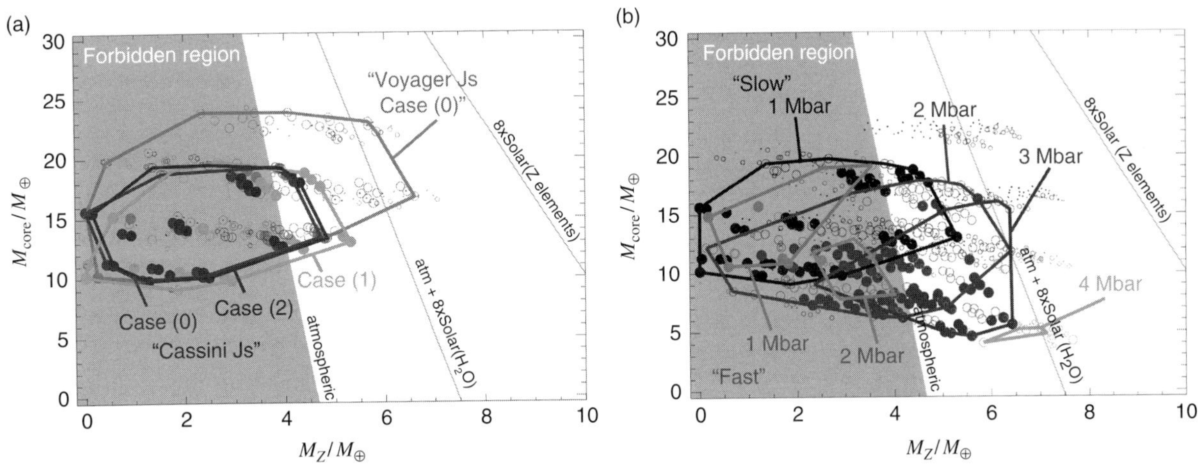

Figure 3.7 Saturn's core mass (M_{core}) vs. the mass of heavy elements in the envelope (M_Z) for interior models matching the available observational constraints. Left: Solutions for $P_{trans} = 1$ Mbar using the Voyager rotation period with Voyager's *J*s, and model c_0 (red), and for Cassini *J*s and models c_0 (purple), c_1 (blue), c_2 (light blue). **Right:** Solutions when using the Cassini *J*s, combining three different cases for the planetary shape (c_0, c_1, c_2): (i) Voyager rotation period and *P*trans = 1 Mbar (black), 2 Mbar (purple), 3 Mbar (blue), 4 Mbar (light blue). (ii) A rotation period of 10h 32 m 35s and $P_{trans} = 1$ Mbar (red), 2 Mbar (orange). From Helled and Guillot (2013). (A black-and-white version of this figure appears in some formats. For the color version, please refer to the plate section.)

and c_2, an intermediate case – with an equatorial radius set to 60,238 km. Figures 3.7a,b show the resulting models, while Figure 3.6 shows subsets for the Voyager rotation rate and, respectively, the Voyager constraints (within the *thick black dashed lined*) and the Cassini constraints (within the *black solid lined area*).

Models by Saumon and Guillot (2004) (*grey dashed lined* in Figure 3.6) use the same observational constraints as in Guillot (1999), but assume $Z_{in} = Z_{out}$ as Helled and Guillot (2013); P_{trans} between helium-poor and helium-rich varies from 1 to 3 Mbar. The *grey shaded areas* in Figure 3.6 shows interior models derived by Nettelmann et al. (2013), based on *ab initio* EOS for hydrogen, helium and H_2O, with $T_{1\ bar} = 140$ K, $Y_{atm} = 0.18$ and $Y = 0.275$, an allowance for Z_{in} 6 = Z_{out}, cores made of pure rock or pure water, and no imposed limit on P_{trans}. Finally, the models by Gudkova and Zharkov (1999) resemble those of Guillot (1999), but assume a five-layer structure with a helium layer on top of the core to formally account for hydrogen-helium demixing and helium sedimentation in the entire inner envelope.

3.7.2 Results for Isentropic Models

Figure 3.6b presents the derived masses of heavy elements in the core and in the envelope, $M_{Z,env}$. Figure 3.6a presents the envelope heavy element mass fractions for isentropic Saturn models obtained under various model assumptions and by different authors as described in Section 3.7.1. The sensitivity of the derived internal structure to the assumed shape and rotation rate is demonstrated in Figure 3.7a,b, where we show the results by Helled and Guillot (2013) when assuming various rotation periods and shapes.

From Figure 3.6, several striking properties can be seen: (i) the Cassini gravitational data reduces Saturn's core mass by $\approx 5\ M_\oplus$, i.e. from $\sim 10 - 25\ M_\oplus$ to $\sim 5 - 20\ M_\oplus$; (ii) models that allow for a larger enrichment in the deep envelope than in the atmosphere ($Z_{out} < Z_{in}$) allow for no-core solutions if $P_{\rm trans} \approx 5$ Mbar; (iii) the models differ largely in their predicted heavy element mass fraction and Z_{out} values, which for the Cassini data are found to range from 0.1× to 5× solar (Helled and Guillot 2013), or from 2× to 13× solar (Nettelmann et al. 2013). The difference is mostly due to the EOS of heavy elements in the envelopes (water + sand vs. pure water) and to the J_2 and J_4 values used.

Since the heavy element mass fraction is unlikely to decrease with depth, as it would trigger an instability that would tend to equilibrate the abundances, it is reasonable to assume $Z_{atm} \leq Z_{out}$ (and potentially that $Z_{atm} \sim Z_{out}$) and $Z_{out} \leq Z_{in}$. Therefore, an observational determination of the bulk atmospheric heavy element abundance Z_{atm} through measured O/H, C/H and N/H ratios below the respective cloud decks can be used to rule out a vast amount of Saturn models. This idea is highlighted in Figure 3.6a, where measured C and N enrichments (Guillot and Gautier 2014) are plotted in comparison to Z_{out} values from the structure models. Models based on the Cassini constraints (Nettelmann et al. 2013; Helled and Guillot 2013) are consistent with the measured ~3× solar enrichment of N/H, which, however, may only be a lower limit to the abundance at greater depths, whereas only the early models of Guillot (1999) seem to allow for bulk 9× solar enrichments as indicated by the C/H ratio. We conclude that modern Saturn models based on tighter constraints for the gravitational field and from updated EOS calculations predict Z_{atm} less than ~5× solar, and thus O/H less than ~7× solar (Nettelmann et al. 2013).

The results for Saturn's internal structure as derived by Helled and Guillot (2013) are shown in detail in Figure 3.7. The left panel presents a comparison of the derived interior model solutions for Saturn with $P_{\rm trans} = 1$ Mbar, for the Voyager rotation period and with Voyager *J*s, and model c_0 (red), and for Cassini *J*s and models c_0 (purple), c_1(blue), c_2 (light blue). The contours correspond to interior models that fit within 2 sigma in equatorial radius, J_2, and J_4 (J_6 fits within 1 sigma). The grey area represents a "forbidden zone" corresponding to a region that its atmospheric abundances are inconsistent with the atmospheric abundances derived from measurements (e.g. Guillot and Gautier 2014). The first grey line adds 8 times the solar abundance (Asplund et al. 2009) of water, while the second grey line assumes that all the heavy elements are enriched by a factor of 8 compared to solar. It is found that the possible parameter-space of solutions is smaller when Cassini's *J*s are used. This is not surprising, given that the uncertainties of the gravitational harmonics are smaller. With the Voyager rotation period, the inferred heavy element mass in Saturn's envelope is $0 - 7\ {\rm M}_\oplus$ and the core mass is $10 - 20\ {\rm M}_\oplus$. The right panel of Figure 3.7 shows how the assumed rotation period affects the inferred composition of Saturn. For the Voyager period, Saturn's core mass ranges between 5 and 20 $\rm M_\oplus$, with the lower values corresponding to higher P_{trans}. Saturn's core mass strongly depends on the P_{trans} where the derived core mass decreases significantly for higher transition pressures. The heavy element mass in the envelope remains $0 - 7\ {\rm M}_\oplus$. Interior models with the shorter rotation period with $P_{\rm trans} = 1$ and 2 Mbar were found to have heavy element mass in the envelope less than 4 $\rm M_\oplus$, below the values derived from atmospheric spectroscopic measurements. Solutions with this rotation period can be found when a discontinuity in the heavy elements distribution is considered (see e.g. Nettelmann et al. 2013).

It should also be noted that *interior models of Saturn with no ice/rock core are possible*. The lack of knowledge on the depth of differential rotation in Saturn, its rotation period, and whether the heavy elements are homogeneously distributed within the planet are major sources of uncertainty on the internal structure and global composition of the planet.

Despite the uncertainties in Saturn's derived internal structure, an important conclusion from these results can be drawn: none of the Saturn models has a P_{trans} value near the core-mantle boundary of 10 – 15 Mbar, although a wide range of uncertainties has been considered. Hence, an interior where helium rains down all the way to the core and leaves the envelope above homogeneous and isentropic, is *excluded*. In other words, if helium in Saturn rains down to the core, the mantle above should have an inhomogeneity in helium (or heavy elements). This inhomogeneous region could be non-isentropic. This conclusion holds unless the helium layer on top of the core reaches upward to ~4 Mbar, in which case Saturn's atmosphere should be highly depleted in helium, or unless the gravitational harmonics are severely altered by deep winds. A measurement of high-order gravitational harmonics and of the atmospheric helium abundance are thus important for establishing a better understanding of Saturn's internal structure.

3.7.3 Comparison with Jupiter

It is useful to compare and contrast Jupiter and Saturn. While both Jupiter and Saturn are massive giant planets that consist mainly of hydrogen and helium, their relative enrichment compared to protosolar composition is somewhat different, and according to most of the available interior models Saturn is predicted to be more enriched with heavy elements. In addition, while interior models for both planets suggest that solutions with no cores are valid, typically Saturn's interior models include a core, and indeed the existence of a dense inner region in Saturn is supported by its lower MOI value. Both planets are fast rotators, having a large equatorial jet, but without constraining the rotation period of Saturn exactly we cannot say whether both planets have eastern and western jets at high latitudes or whether this feature exists only for Jupiter, nor can we say how the two planets compare in terms of wind speeds. In addition, while the atmospheres of both planets are depleted in helium compared to protosolar composition, the depletion appears to be more significant for Saturn. As discussed below, this leads to the "slower" cooling of Saturn and its luminosity excess. Finally, other important differences are the heat flux, tilt, strength of the magnetic fields, and the rings and satellites systems.

Is Saturn simply a smaller version of Jupiter? Not necessarily. Given our current understanding of planet formation, the cause of the difference in terms of total mass and composition seems to be the slower formation of Saturn compared to Jupiter. In the standard picture of giant planet formation, also known as core accretion (see e.g. Helled et al. 2014 and Chapter 2 by Atreya et al. on Saturn's origin), the growth rate of a planetary embryo is larger at small radial distances, which explains why Jupiter could have reached a critical mass for runaway gas accretion before Saturn. Thus, both planets have reached critical masses and accreted significant amounts of hydrogen and helium. Why is Saturn's gaseous envelope smaller? This is not yet fully understood, but is most likely related to Saturn's relatively slower growth rate and the interaction with Jupiter and, possibility, other growing planets (i.e. Uranus and Neptune). This suggestion has been investigated in several versions of the Nice model (e.g. Thommes et al. 1999; Tsiganis et al. 2005; Walsh et al. 2011 and references therein).

There are still many open questions regarding the origin and internal structure of Jupiter and Saturn and it is certainly beneficial to study the two planets together. An opportunity to investigate and compare Jupiter and Saturn will be available in the upcoming years. By 2017, accurate measurements of the gravitational fields of the planets will be available from the Juno and Cassini Solstice missions. A detailed comparison of Jupiter and Saturn will be very inspiring and will improve our understanding of the origin of the solar system and provide insights on the characteristics of gas giant planets in general.

3.8 Results from Thermal Evolution Modeling

3.8.1 Simple Models with Helium Phase Separation

The first thermal evolution calculations for warm, fluid, adiabatic models of Jupiter and Saturn were performed by a number of authors in the mid 1970s (Graboske et al. 1975; Bodenheimer 1976; Hubbard 1977; Pollack et al. 1977), and most of their findings remain relevant today. Models for Jupiter, starting from a hot, post-

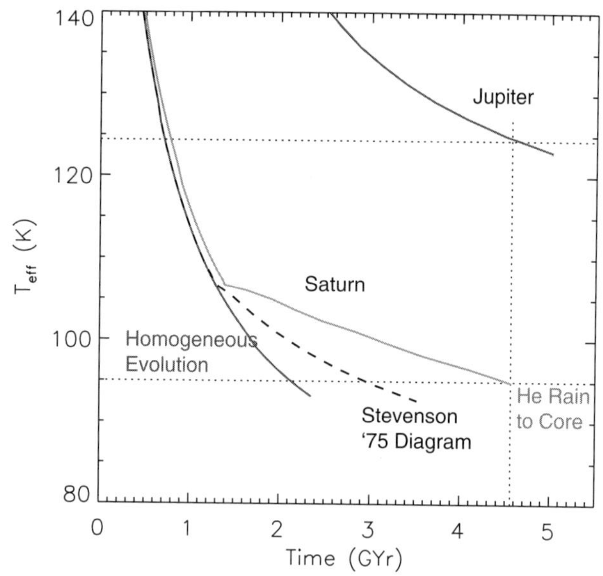

Figure 3.8 Fully isentropic, homogeneous models of the thermal evolution of Jupiter and Saturn, after Fortney and Hubbard (2003). The current T_{eff} of each planet is shown with a dotted line. For Saturn, the real planet (age 4.55 Gyr) has a much higher T_{eff} than the model, indicating the model is missing significant physics. A model including helium-rain using the Stevenson (1975) phase diagram is shown in dashed black. A model that uses an ad-hoc phase diagram, designed to rain helium down to the top of the planet's core, liberating more gravitational energy, is shown in lightest gray.

formation state with a hydrogen-helium envelope that was assumed to be homogeneous, isentropic and well-mixed, cooled to Jupiter's known T_{eff} of 124 K in ~4.5 Gyr. However, these calculations failed to reproduce Saturn's cooling history. A Saturnian cooling age of 2 – 2.5 Gyr was found, implying that Saturn today (T_{eff} = 95 K) is much too hot, by a *factor of 50% in luminosity* (Pollack et al. 1977; Stevenson and Salpeter 1977a) (see Figure 3.8). If giant planets are fully or mostly isentropic below their radiative atmospheres, then it is the atmosphere that serves as the bottleneck for interior cooling. However, advances in atmosphere modeling used in several generations of thermal evolution models (Graboske et al. 1975; Hubbard et al. 1999; Fortney et al. 2011) did not alter this dichotomy in the cooling history between Jupiter and Saturn.

The leading explanation to remedy the cooling shortfall has long focused on the separation of helium from hydrogen (or "demixing") (e.g. Stevenson and Salpeter 1977a,b; Fortney and Hubbard 2003). Stevenson and Salpeter found that a "rain" of helium was likely within Saturn, and perhaps Jupiter, and that this differentiation (a conversion of gravitational potential energy to thermal energy) could prolong Saturn's evolution, keeping it warmer longer (see Figure 3.8). The evidence for this process is strong if Jupiter and Saturn are considered together. The atmospheres of both Jupiter and Saturn are depleted in helium, relative to the protosolar abundance (mass fraction Y_{proto} = 0.270 ± 0.005, derived from helioseismology (Asplund et al. 2009)). For Jupiter, Y_{atmos} = 0.234 ± 0.008, from the Galileo Entry Probe (von Zahn et al. 1998). Furthermore, Ne is strongly depleted in the atmosphere as well, and calculations suggest it is lost into He-rich droplets (Roulston and Stevenson 1995; Wilson and Militzer 2010). Taken together, this is convincing evidence for phase separation. One would expect that Saturn, being cooler than Jupiter, should be more depleted in He. However, Voyager 2 estimates from spectroscopy run from Y_{atmos} = 0.01 – 0.11 (Conrath et al. 1984) to Y_{atmos} = 0.18 – 0.25 (see Conrath and Gautier 2000 for details). Revised estimates from Cassini have not yet been published, but preliminary values cluster around Y_{atmos} ~ 0.14 (P. Gierasch, pers. communication).

These depletions imply that helium phase separation has begun in Jupiter, perhaps relatively recently, and that it has been ongoing in Saturn for a longer time. There is no other published explanation for the planets' He depletions. Hubbard et al. (1999) and Fortney and Hubbard (2003) found that by raining He down all the way to the core at a late evolution state, the maximum allowed Y_{atmos} must be < 0.20 for Saturn today to explain its current luminosity. However, our understanding of Saturn is clearly not complete, as the correct amount of He depletion, along with its distribution within the interior, must be accounted for in models.

Barriers to a better quantitative understanding of the phase separation process include: (1) the H/He phase diagram is still not precisely known, which dramatically impacts the amount of Saturn's mass within the He-immiscibility region, as well as the extent of any He-gradient region; and (2) the issue of how He composition gradients would affect the temperature gradient in Saturn's deep interior is poorly understood, but new work in this area is described below.

3.8.2 *Variants for Inhomogeneous Structure and Evolution Models*

Homogeneous models of Saturn are successful in providing a good match to many observed properties, such as the low-order gravitational harmonics. However, they fail to explain Saturn's high luminosity, its dipolar magnetic field (see Section 3.2.4) and the fine-spitting of density waves in its rings (see Section 3.9). Saturn has therefore been suggested to contain at least one inhomogeneous zone in its interior. We review recent attempts of inhomogeneous model developments for Saturn and discuss their physical justifications as well as their abilities in explaining the observational constraints.

As discussed in Section 3.5.2, an inhomogeneous zone in a giant planet can have different possible origins, due to the formation process itself, subsequent erosion of an initially massive core or phase separation and sedimentation of abundant constituents. While details of each of these processes are not well understood yet, some basic properties can naturally lead to deviations from homogeneity. For instance, the core accretion scenario for giant planets suggests that during the period of massive core formation, before rapid runaway gas accretion sets in, both gaseous material and planetesimals of various sizes are accreted. While small planetesimals may easily dissolve in the gaseous component, and heavy ones sink down to the initial core, and medium-sized bodies may dissolve at different altitudes and cause a compositional gradient. Moreover, during the subsequent long-term evolution, the initial massive core may then erode, as the heavy elements (O, Si, Fe) are miscible in the metallic hydrogen envelope above (Wilson and Militzer 2012b,a; Wahl et al. 2013). The efficiency of upward mixing by thermal convection may be low and may take billions of years, so that today an inhomogeneous region may still exist atop the core (Stevenson 1985). For Saturn in particular, the H/He phase separation and helium-rain are likely to occur and could result in an extended inhomogeneous zone at several Mbars, and/or in an He-layer atop the core. H/He phase separation in Saturn's entire interior below about 1 Mbar (~ two-thirds of its radius) is supported by modern H/He phase diagrams based on *ab initio* simulations (Lorenzen et al. 2011; Morales et al. 2013). Application of those new predictions to Saturn's inhomogeneous evolution remains to be done.

A different inhomogeneous model has recently been shown to provide a possible alternative explanation for Saturn's high luminosity (Leconte and Chabrier 2013). In that case, the internal structure is assumed to have an inhomogeneous zone where the abundance of heavy elements increases with depth, and through which heat is transported by layered semi-convection. While details of the dynamical behavior like layer formation or merging in such a semi-convective zone are poorly understood, such a scenario can in principle explain an enhanced luminosity (the Saturn case) or a reduced luminosity (the Uranus case). In fact, by adjusting the zone's extent and the *a priori* unknown height of the convective layers, Saturn's observed luminosity can be reproduced, without requiring – albeit not excluding – an additional energy source like helium-rain (Leconte and Chabrier 2013). A semi-convective zone may lead to significantly higher deep internal temperatures, so that the He rain region may terminate before the core is reached. In the region between the semi-convective zone and the core, convection could be maintained, and the magnetic field generated. Higher internal temperatures in the deep interior lead to a lower density for the H/He mixture, which also necessitates a larger mass of heavy elements in the planet's interior (Leconte and Chabrier 2012). These authors find Saturn and Jupiter models with total masses of heavy elements *nearly double those of adiabatic models*, including up to ~30 $M_{\oplus}$ in the H/He envelope. Therefore, isentropic models should be thought of as providing a lower limit on heavy element content of Saturn.

New work on the influence of a double-diffusive region created by He phase separation on the temperature gradient and cooling history of giant planets was recently published by Nettelmann et al. (2015). These authors couple the interior composition gradient and temperature gradient via an iterative procedure, since there is feedback between the two gradients. They allow the He gradient region to evolve with time, given an H/He phase diagram and a prescription for energy transport in the gradient region (Mirouh et al. 2012; Wood et al. 2013). In application to Jupiter, they find that He rain can either prolong, or even *shorten*,

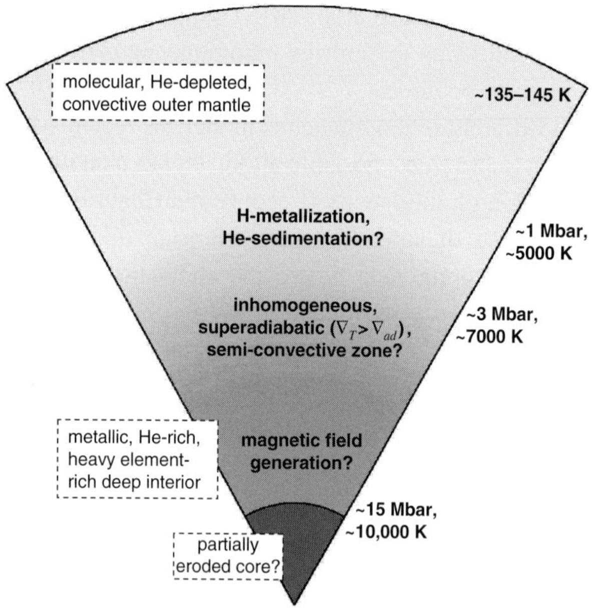

Figure 3.9 Illustration of Saturn's possible inhomogeneous internal structure. The outer third of the planet is shown to be He-depleted, convective and homogeneous. A semi-convective region with compositional gradient (He, possibly heavy elements) separates the outer envelope from the convective, homogeneous and metallic deep interior in which the magnetic field may be generated. Some of the initial core material may today be mixed into the deep envelope.

the cooling time for Jupiter to its measured T_{eff}, depending on the efficiency of energy transport through the He-gradient region.

Figure 3.9 illustrates a possible inhomogeneous model for Saturn that could be consistent with its high luminosity, helium-depleted atmosphere, the dipolar magnetic field, and some of the observed waves in the rings (see Section 3.9). The abundance of helium (or heavy elements) is shown to increase between 1 Mbar and 5 Mbar as a result of H/He phase separation and He rain in a semi-convective, superadiabatic zone. In that thick zone, non-dipolar moments of the magnetic field may be filtered out (see the chapter on Saturn's magnetic field), leading to a dipolar field on top of it. Furthermore, the non-zero Brunt–Väisälä frequency could allow for the generation of gravity waves that then, through mode-mixing with f-modes, may explain the observed fine-splitting between the $m = -2$ modes (Fuller 2014; see also Section 3.9).

These recent developments suggest the consideration of interior models for gaseous giant planets beyond the standard assumption of largely isentropic, homogeneous envelopes atop a well-defined core, in order to be explain the observed properties.

3.9 Seismology

Perhaps the most straightforward method for constraining the interior properties of a planet is to study waves that propagate directly through its interior. This is the realm of seismology, and on Earth both waves launched by discrete events (earthquakes) and resonant normal mode oscillations are studied. The detection and measurement of the frequencies of individual resonant modes trapped in the solar interior revolutionized the study of the Sun (Christensen-Dalsgaard 2002) and, later, stars (Chaplin and Miglio 2013), and led to suggestions that the study of similar trapped modes would open a new window into the internal structure of the giant planets (Vorontsov et al. 1989).

Several types of oscillations can be found in a fluid sphere, and they are denoted by their primary restoring force. The most commonly discussed include pressure modes (or p-modes) that are essentially trapped acoustic waves and gravity modes (or g-modes) that are resonant waves found in regions of varying static stability. In a purely isentropic sphere, g-modes would not be present. The f-modes are the equivalent of surface waves in a lake and are modes with no radial nodes in displacement. An individual mode is denoted by the three integers, (l, m, n) that count the total number of nodal lines on the surface, in azimuth, and in radius. Modes with $l = m$ are sectoral, like segments of an orange, and the f-modes have $n = 0$ by definition.

By comparing observed frequencies of these with those expected from theory, the density profile throughout the interior can be tightly constrained (for a discussion of inversion methods, see e.g. Vorontsov et al. (1989); Jackiewicz et al. (2012)). Since lower-order modes probe more deeply into the planet, such modes are of the greatest interest. Gudkova et al. (1995) showed that the observation of oscillation modes up to degree $l = 25$ would constrain both the core size and the nature of the metallic hydrogen phase transition. Several attempts to observe these oscillations on Jupiter were made, and Schmider et al. (1991) and Mosser et al. (2000) reported excess oscillatory power in the expected frequency range for Jovian p-modes, although they were not able to identify specific modes or frequencies. More recently, Gaulme et al. (2011) detected Jupiter's oscillations, a promising first result, although the detection was not able to strongly constrain Jupiter interior models.

Telescopic searches for modes on Saturn are complicated by the rings and there have been no systematic surveys. However, following a suggestion by Stevenson (1982), Marley (1990) and Marley and Porco (1993) explored whether Saturn's rings might serve as a seismograph, recording slight perturbations to the gravity field produced by periodic density perturbations inside the planet. They found that even 1-m amplitude f-mode oscillations could indeed induce perturbations to the external gravity field that in the nearby C-ring were comparable to those induced by distant external satellites. The f-modes are favorable for detection because they have no radial nodes as they perturb the density within the planet. Consequently, the integrated density perturbation from surface to deep interior is always in phase and the effect on the external gravitation field is greater than for any p- or g-mode, which always have at least one radial node. Furthermore, the frequencies of low order (small l) modes serendipitously produce first-order resonances in the C-ring, which lies near the planet.

Those f-modes, which propagate in the same direction as Saturn rotates, appear to a fixed external observer to be of even higher frequency, as their gravitational perturbation is swept around the planet by rotation. This slight, regular, perturbation tugs on those ring particles that happen to orbit at an orbital radius where the apparent frequency of the mode is resonant with their orbit. If the perturbation is sufficiently large the collective response of the ring particles produces a wave feature or even a gap in the rings. By measuring the precise location of such ring features it would in principle be possible to infer the resonant frequency (from the orbital dynamics) and thereby the mode frequency.

After Marley (1991) computed new Saturn mode oscillation frequencies, Marley and Porco (1993) proposed that certain wave features in Saturn's C-ring discovered by Rosen et al. (1991), as well as the Maxwell gap, were created by resonant interaction with low-order internal f-mode oscillations of Saturn. While Marley and Porco (1993) argued that the Rosen waves could be associated with Saturnian oscillation modes, their detailed predictions for the characteristics of the waves expected to be excited in the rings could not be tested by the Voyager data available at the time. Almost 25 years later, however, Cassini stellar occultation data, obtained as stars pass behind the rings, finally allowed a test of the ring seismology hypothesis.

After an exhaustive analysis of the C-ring occultation data, Hedman and Nicholson (2013, 2014) confirmed that indeed at least 8 unexplained C-ring wave features have the appropriate characteristics to be excited by f-mode oscillations of Saturn. Since the precise orbital frequency is known from the location of the wave feature, this essentially provides a very precise measurement of several specific Saturn oscillation frequencies, fulfilling the promise of ring seismology. However instead of a single f-mode (with a specific (l, m)) being associated with the expected single C-ring feature, Hedman and Nicholson (2013, 2014) found that two f-modes were associated, with three features each. This "fine-splitting" of mode frequencies was unexpected and is not the result of rotation alone, as the usual rotational splitting of modes has already been accounted for in the seismological predictions. The confusion over the multiple mode frequencies has rendered the modes' value for constraining Saturn's interior structure somewhat problematic, at least until an appropriate theory to explain the splitting is developed.

Fuller (2014) has attempted to develop such a theory. He investigated mode-mixing, where distinct oscillation modes can interact with one another if they have similar frequencies. He found that the $l = 2$ f-mode, for example, can interact with a gravity mode of Saturn if there is a convectively stable region above Saturn's core. In essence, the f-mode and gravity mode interact, and the result is a mode of mixed character that splits the $l = 2$ f-mode oscillation frequency. If this is indeed the cause of the observed splitting, then this may be offering a precise measure of core erosion (or a deep He-gradient) in the deep interior of Saturn. Combined with the other measured mode frequencies, seismology may hold promise for constraining not only the size of Saturn's core, but also the deep composition of the planet. More theoretical development is required, however, to fully exploit this opportunity.

Nevertheless, ring seismology likely now has great promise for opening a unique window into Saturn's interior structure.

3.10 Future Prospects

3.10.1 Cassini Grand Finale

Before the planned termination and plunge into Saturn's atmosphere in September 2017, the Cassini mission will execute 20 orbits with a 7.2-day period and pericenter at about 2.5 Saturn radii, and 22 highly inclined (63.4 degree) orbits with a period of 6.2 day and a periapsis altitude about 1700 km above the 1-bar pressure level (see Figure 3.10). This set of orbits, named the Cassini Grand Finale, has been tailored to carry out close observations of Saturn and to probe its interior structure by means of gravity and magnetic field measurements. Although neither the spacecraft nor its instruments were designed for this kind of observations, the scientific return is expected to be high. Thanks to the proximity of the spacecraft to Saturn in the final 22 orbits (just inside the inner edge of the D ring), Cassini will return the harmonic coefficients of the magnetic and gravity field to about degree 10 or larger.

While the onboard magnetometer will carry out continuous measurements throughout the last 22 orbits, starting from April 2017 there are fewer opportunities for gravity measurements, which are only possible when the high-gain antenna is Earth-pointed. Operational constraints (such as the elevation angle at the ground station and protection of the spacecraft from dust hazard using the antenna as a shield), and the need to share the observation time between the onboard instruments, drastically reduce to six the number of orbits devoted to gravity science.

The determination of Saturn's gravity field will be carried out by means of range rate measurements and sophisticated orbit determination codes. Range rate is routinely measured at a ground antenna by transmitting a highly stable monochromatic microwave signal to the spacecraft. An onboard transponder receives the signal and coherently re-transmits it back to ground, where the Doppler shift between the outgoing and incoming signal is measured. The antennas of NASA's Deep Space Network (DSN), operating at X-band (7.2–8.4 GHz), provide range rate accuracies of about 12 micron/s at 1000 s integration times, and about a factor of four larger at a time scale of 60 s. Thanks to the use of higher

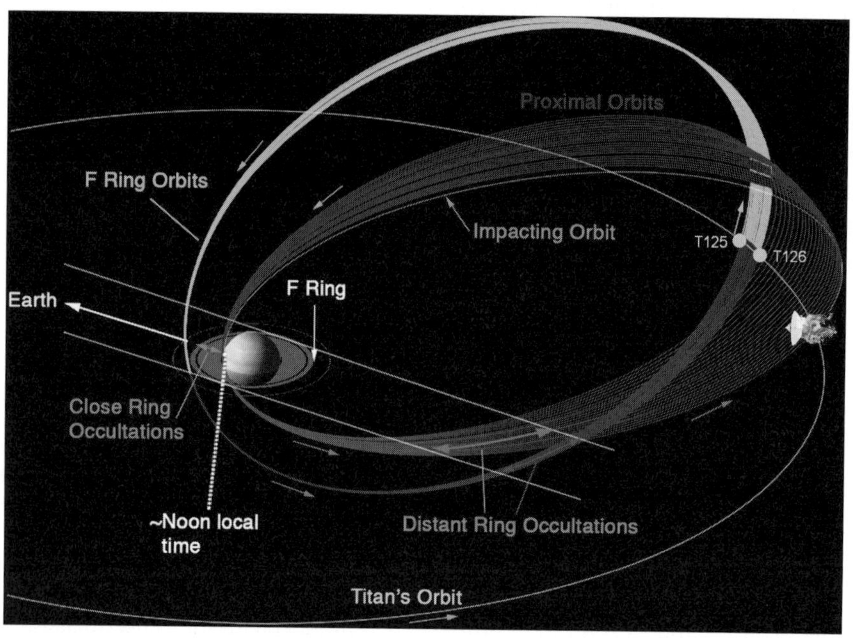

Figure 3.10 The Cassini Grand Finale orbits. After completion of the first 20 F ring orbits, Cassini will pass 22 times between the inner edge of the D ring and Saturn's atmosphere, with an orbital inclination of 63.4 degrees and a pericenter latitude between 5.5 and 7.5 degrees south. The final plunge into Saturn's atmosphere, required by planetary protection rules, is currently scheduled for September 15, 2017. (A black-and-white version of this figure appears in some formats. For the color version, please refer to the plate section.)

frequencies (32 – 34 GHz) and the consequent immunity to propagation noise from interplanetary plasma, a similar experiment on the Juno mission will exploit observable quantities that are less noisy by about a factor of four.

In spite of the limited number of orbits, Cassini's gravity measurements will take advantage of a remarkably favorable orbital geometry, which always ensures a large projection of the spacecraft velocity along the line of sight when the spacecraft is close to the planet. Numerical simulations of the gravity experiment in a realistic scenario indicate that Cassini will be able to estimate the (unnormalized) zonal harmonics with accuracies ranging from 2×10^{-9} for J_2 and 1.5×10^{-7} for J_{12}. This precision will provide unprecedented constraints on the density structure of the outer H/He envelope, and allow for the detection of the depth of zonal flows seen in the visible atmosphere (e.g. Kaspi et al. 2010, 2013). It is expected that the Cassini data will also provide an estimate of Saturn's k_2 and k_3 Love numbers to accuracies respectively of 0.015 and 0.12, and the gravitational parameter GM of the B ring to 0.15 km^3 s^{-2}.

3.10.2 What Is Needed for Future Progress

What has emerged from the still-unfolding Cassini era is a picture of Saturn's interior that is full of complexity. The completion of the Cassini revolution will utilize the tremendously more precise data on the planet's gravity field and magnetic field that are essentially assured from the Cassini Grand Finale orbits. To maximize the science from these unique data sets will require a concomitant significant push in the analysis of existing Cassini data sets, laboratory studies and theoretical work. Below we suggest areas for future work.

- Better knowledge of the phase diagram for helium immiscibility. The most recent *ab initio* phase diagrams (Lorenzen et al. 2011; Morales et al. 2013) yield similar, but clearly discrepant predictions for the temperatures of the onset of phase separation. While it appears that most of Saturn's interior is in a region with a gradient in helium abundance, additional theoretical and experimental work are needed to bring confidence to our understanding of the phase diagram.
- A determination of Saturn's atmospheric He abundance from Cassini spectroscopic and occultation observations. This is a complicated issue (e.g. Conrath and Gautier 2000), but uncertainty in this value will long dominate our uncertain knowledge of Saturn's cooling history. In tandem, a derivation of atmospheric $P - T$ profile, including the 1-bar temperature, would put our knowledge of the planet's thermal profile on more solid footing.
- Three-dimensional simulations of the transport of helium and energy within the helium immiscibility region. Recent work (Wood et al. 2013) on double-diffusive convection is an important step in this direction, but the coupled nature of the helium and temperature gradients within giant planets warrants additional focused work.
- If the first three items are addressed, a new generation of Saturn thermal evolution models is certainly warranted, which could simultaneously match the planet's intrinsic flux, 1-bar temperature and current atmospheric helium abundance.
- The most dramatic advance in our understanding of the planet's interior would surely come from a wider exploration of the range of interior structures that are consistent with both the gravity field and seismology data. Fuller (2014) has begun this work with a limited range of Saturn interior models. However, a wider exploration of the utility of the seismology data sets, in tandem with the Grand Finale gravity field constraints, is clearly needed.
- Measurements from Cassini's Solstice mission and contemporary model improvements will not close our current knowledge gaps on fundamental properties of Saturn and our knowledge of giant planet formation in the solar system. In line with Mousis et al. (2014), we suggest future space exploration of Saturn by means of an entry probe into its atmosphere in order to determine accurately the abundances and isotopic ratios of noble gases. Such values contain unique information not only on internal properties such as helium rain and bulk composition but also on the early history of the solar system.

References

Anderson, J. D. and G. Schubert (2007), Saturn's Gravitational Field, Internal Rotation, and Interior Structure. *Science* **317**, 1384–1387.

Asplund, M., N. Grevesse, A. J. Sauval and P. Scott (2009), The Chemical Composition of the Sun. *Annu. Rev. Astron. Astrophys*. **47**, 481–522.

Bahcall, J. N., M. H. Pinsonneault and G. J. Wasserburg (1995), Solar models with helium and heavy-element diffusion. *Reviews of Modern Physics* **67**, 781–808.

Bodenheimer, P. (1976), Contraction models for the evolution of Jupiter. *Icarus* **29**, 165–171.

Burton, M. E., M. K. Dougherty and C. T. Russell (2009), Model of Saturn's internal planetary magnetic field based on Cassini observations. *Planet. Space Sci*. **57**, 1706–1713.

Cao, H., C. T. Russell, U. R. Christensen, M. K. Dougherty and M. E. Burton (2011), Saturn's very axisymmetric magnetic field: No detectable secular variation or tilt. *Earth and Planetary Science Letters* **304**, 22–28.

Cao, H., C. T. Russell, J. Wicht, U. R. Christensen and M. K. Dougherty (2012), Saturn's high degree magnetic moments: Evidence for a unique planetary dynamo. *Icarus* **221**, 388–394.

Chaplin, W. J. and A. Miglio (2013), Asteroseismology of Solar-Type and Red-Giant Stars. *Annu. Rev. Astron. Astrophys.* **51**, 353–392.

Christensen-Dalsgaard, J. (2002), Helioseismology. *Reviews of Modern Physics* **74**, 1073–1129.

Conrath, B. J. and D. Gautier (2000), Saturn Helium Abundance: A Reanalysis of Voyager Measurements. *Icarus* **144**, 124–134.

Conrath, B. J., D. Gautier, R. A. Hanel and J. S. Hornstein (1984), The helium abundance of Saturn from Voyager measurements. *Astrophys. J.* **282**, 807–815.

Dessler, A. J. (1983), *Physics of the Jovian magnetosphere*.

Dougherty, M. K., N. Achilleos, N. Andre et al. (2005), Cassini Magnetometer Observations During Saturn Orbit Insertion. *Science* **307**, 1266–1270.

Fletcher, L. N., G. S. Orton, N. A. Teanby, P. G. J. Irwin and G. L. Bjoraker (2009), Methane and its isotopologues on Saturn from Cassini/CIRS observations. *Icarus* **199**, 351–367.

Fortney, J. J. and W. B. Hubbard (2003), Phase separation in giant planets: inhomogeneous evolution of Saturn. *Icarus* **164**, 228–243.

Fortney, J. J., M. Ikoma, N. Nettelmann et al. (2011), Self-consistent Model Atmospheres and the Cooling of the Solar System's Giant Planets. *Astrophys. J.* **729**, 32.

Fortney, J. J., M. S. Marley and J. W. Barnes (2007), Planetary Radii across Five Orders of Magnitude in Mass and Stellar Insolation: Application to Transits. *Astrophys. J.* **659**, 1661–1672.

French, M., T. R. Mattsson, N. Nettelmann and R. Redmer (2009), Equation of state and phase diagram of water at ultrahigh pressures as in planetary interiors. *Phys. Rev. B 79*(5), 054107.

Fuller, J. (2014), Saturn ring seismology: Evidence for stable stratification in the deep interior of Saturn. *Icarus* **242**, 283–296.

Galopeau, P. H. M. and A. Lecacheux (2000), Variations of Saturn's radio rotation period measured at kilometer wavelengths. *J. Geophys. Res.* **105**, 13089–13102.

Gaulme, P., F.-X. Schmider, J. Gay, T. Guillot and C. Jacob (2011), Detection of Jovian seismic waves: a new probe of its interior structure. *Astron. Astrophys.* **531**, A104.

Graboske, H. C., R. J. Olness, J. B. Pollack and A. S. Grossman (1975), The structure and evolution of Jupiter – The fluid contraction stage. *Astrophys. J.* **199**, 265–281.

Gudkova, T., B. Mosser, J. Provost et al. (1995), Seismological comparison of giant planet interior models. *Astron. Astrophys.* **303**, 594.

Gudkova, T. V. and V. N. Zharkov (1999), Models of Jupiter and Saturn after Galileo mission. *Planet. Space Sci.* **47**, 1201–1210.

Guillot, T. (1999), A comparison of the interiors of Jupiter and Saturn. *Planet. Space Sci.* **47**, 1183–1200.

Guillot, T. and D. Gautier (2014), Giant Planets. *ArXiv:1405.3752, second edition of the Treatise on Geophysics, in press*.

Guillot, T., D. Gautier, G. Chabrier and B. Mosser (1994), Are the giant planets fully convective? *Icarus* **112**, 337–353.

Guillot, T., D. J. Stevenson, W. B. Hubbard and D. Saumon (2004), The interior of Jupiter. In *Jupiter – The Planet, Satellites and Magnetosphere*, pp. in press.

Gurnett, D. A., A. M. Persoon, W. S. Kurth et al. (2007), The Variable Rotation Period of the Inner Region of Saturns Plasma Disk. *Science* **316**, 442–445.

Hedman, M. M. and P. D. Nicholson (2013), Kronoseismology: Using Density Waves in Saturn's C Ring to Probe the Planet's Interior. *Astron. J.* **146**, 12.

Hedman, M. M. and P. D. Nicholson (2014), More Kronoseismology with Saturn's rings. *Mon. Not. R. Astron. Soc.* **444**, 1369–1388.

Helled, R. (2011), Constraining Saturn's Core Properties by a Measurement of Its Moment of InertiaImplications to the Cassini Solstice Mission. *Astrophys. J. Lett.* **735**, L16.

Helled, R., J. D. Anderson, G. Schubert and D. J. Stevenson (2011), Jupiter's moment of inertia: A possible determination by Juno. *Icarus* **216**, 440–448.

Helled, R., P. Bodenheimer, M. Podolak et al. (2014), Giant Planet Formation, Evolution, and Internal Structure. *Protostars and Planets VI*, 643–665.

Helled, R., E. Galanti and Y. Kaspi (2015), Saturn's fast spin determined from its gravitational field and oblateness. *Nature* **520**, 202–204.

Helled, R. and T. Guillot (2013), Interior Models of Saturn: Including the Uncertainties in Shape and Rotation. *Astrophys. J.* **767**, 113.

Helled, R. and J. Lunine (2014), Measuring Jupiter's water abundance by Juno: the link between interior and formation models. *Mon. Not. R. Astron. Soc.* **441**, 2273–2279.

Hubbard, W. (2012), High-precision maclaurin-based models of rotating liquid planets. *Astrophys. J.* **756**, L15.

Hubbard, W.(2013), Concentric maclaurin spheroid models of rotating liquid planets. *Astrophys. J.* **768**, 43.

Hubbard, W. B. (1973), The Significance of Atmospheric Measurements for Interior Models of the Major Planets. *Space Science Reviews* **14**, 424–432.

Hubbard, W. B. (1977), The Jovian surface condition and cooling rate. *Icarus* **30**, 305–310.

Hubbard, W. B. (1982), *Icarus* **52**, 509.

Hubbard, W. B.(1982), Effects of differential rotation on the gravitational figures of Jupiter and Saturn. *Icarus* **52**, 509–515.

Hubbard, W. B.(1999), NOTE: Gravitational Signature of Jupiter's Deep Zonal Flows. *Icarus* **137**, 357–359.

Hubbard, W. B. and H. E. Dewitt (1985), Statistical mechanics of light elements at high pressure. VII – A perturbative free energy for arbitrary mixtures of H and He. *Astrophys. J.* **290**, 388–393.

Hubbard, W. B., T. Guillot, M. S. Marley et al. (1999), Comparative evolution of Jupiter and Saturn. *Planet. Space Sci.* **47**, 1175–1182.

Hubbard, W. B. and M. S. Marley (1989), Optimized Jupiter, Saturn, and Uranus interior models. *Icarus* **78**, 102–118.

Hubbard, W. B., G. Schubert, D. Kong and K. Zhang (2014), On the convergence of the theory of figures. *Icarus* **242**, 138–141.

Hubbard, W. B. and D. J. Stevenson (1984), Interior structure of Saturn. In *Saturn*, pp. 47–87.

Jackiewicz, J., N. Nettelmann, M. Marley and J. Fortney (2012), Forward and inverse modeling for jovian seismology. *Icarus* **220**, 844–854.

Jacobson, R. A., P. G. Antresian, J. J. Bordi et al. (2006), The gravity field of the Saturnian system from satellite observations and spacecraft tracking data132. *Astronom. J.* **132**, 2520.

Jeffreys, H. (1924), On the inernal constitution of jupiter and saturn.*MNRAS* **84**, 534.

Karkoschka, E. and M. G. Tomasko (2011), The haze and methane distributions on Neptune from HST-STIS spectroscopy. *Icarus* **211**, 780–797.

Kaspi, Y. (2013), Inferring the depth of the zonal jets on Jupiter and Saturn from odd gravity harmonics. *Geophys. Res. Letbt.* **40**, 676–680.

Kaspi, Y., E. Galanti, W. B. Hubbard and J. E. Davighi (2013), Estimating the depth of the zonal jet streams on Jupiter and Saturn through inversion of gravity measurements by Juno and Cassini. *AGU Fall Meeting Abstracts*, C1740.

Kaspi, Y., W. B. Hubbard, A. P. Showman and G. R. Flierl (2010), Gravitational signature of Jupiter's internal dynamics. *Geophys. Res. Lett.* **37**, 1204.

Klepeis, J. E., K. J. Schafer, T. W. Barbee and M. Ross (1991), Hydrogen-helium mixtures at megabar pressures – Implications for Jupiter and Saturn. *Science* **254**, 986–989.

Knudson, M. D., M. P. Desjarlais, R. W. Lemke et al. (2012), *Phys.Rev. Lett.* **108**, 091102.

Kong, D., K. Zhang and G. Schubert (2013), On the Gravitational Fields of Maclaurin Spheroid Models of Rotating Fluid Planets. *Astrophys. J.* **764**, 67.

Leconte, J. and G. Chabrier (2012), A new vision of giant planet interiors: Impact of double diffusive convection. *Astron. Astrophys*. **540**, A20.

Leconte, J. and G. Chabrier (2013), Layered convection as the origin of Saturn's luminosity anomaly. *Nature Geoscience* **6**, 347–350.

Li, L., X. Jiang, H. J. Trammell et al. (2015), Saturn's giant storm and global radiant energy. *Geophys. Res. Lett.* **42**, 2144–2148.

Lindal, G. F. (1992), The atmosphere of Neptune – an analysis of radio occultation data acquired with Voyager 2. *Astron. J.* **103**, 967–982.

Lindal, G. F., D. N. Sweetnam and V. R. Eshleman (1985), The atmosphere of Saturn – an analysis of the Voyager radio occultation measurements. *Astron. J.* **90**, 1136–1146.

Lodders, K. (2004), Jupiter Formed with More Tar than Ice. *Astrophys. J.* **611**, 587–597.

Lorenzen, W., B. Holst and R. Redmer (2009), Demixing of Hydrogen and Helium at Megabar Pressures. *Physical Review Letters* **102**(11), 115701.

Lorenzen, W., B. Holst and R. Redmer (2011), Metallization in hydrogen-helium mixtures. *Phys. Rev. B 84*(23), 235109.

Marley, M. S. (1990), *Nonradial oscillations of Saturn: Implications for ring system structure*. Ph. D. thesis, Arizona Univ., Tucson.

Marley, M. S. (1991), Nonradial oscillations of Saturn. *Icarus* **94**, 420–435.

Marley, M. S. and C. C. Porco (1993), Planetary acoustic mode seismology – Saturn's rings. *Icarus* **106**, 508.

McMahon, J. M., M. A. Morales, C. Pierleoni and D. M. Ceperley (2012), The properties of hydrogen and helium under extreme conditions. *Reviews of Modern Physics* **84**, 1607–1653.

Mirouh, G. M., P. Garaud, S. Stellmach et al. (2012), A New Model for Mixing by Double-diffusive Convection (Semi-convection).I. The Conditions for Layer Formation. *Astrophys. J.* **750**, 61.

Morales, M. A., S. Hamel, K. Caspersen and E. Schwegler (2013), Hydrogen-helium demixing from first principles: From diamond anvil cells to planetary interiors. *Phys. Rev. B 87*(17), 174105.

Morales, M. A., E. Schwegler, D. Ceperley et al. (2009), Phase separation in hydrogen-helium mixtures at Mbar pressures. *Proceedings of the National Academy of Science* **106**, 1324–1329.

Mosser, B., J. P. Maillard and D. Mékarnia (2000), New Attempt at Detecting the Jovian Oscillations. *Icarus* **144**, 104–113.

Mousis, O., L. N. Fletcher, J.-P. Lebreton et al. (2014), Scientific rationale for Saturn's in situ exploration. *Planet. Space Sci*. **104**, 29–47.

Mousis, O., J. I. Lunine, N. Madhusudhan and T. V. Johnson (2012), Nebular Water Depletion as the Cause of Jupiter's Low Oxygen Abundance. *Astrophys. J. Lett.* **751**, L7.

Nellis, W. J., S. T. Weir and A. C. Mitchell (1999), Minimum metallic conductivity of fluid hydrogen at 140 GPa (1.4 Mbar). *Phys. Rev. B* **59**, 3434–3449.

Ness, N. F., M. H. Acuna, K. W. Behannon et al. (1982), Magnetic field studies by Voyager 2 – Preliminary results at Saturn. *Science* **215**, 558–563.

Ness, N. F., M. H. Acuna, R. P. Lepping et al. (1981), Magnetic field studies by Voyager 1 – Preliminary results at Saturn. *Science* **212**, 211–217.

Nettelmann, N., J. J. Fortney, K. Moore and C. Mankovich (2015), An exploration of double diffusive convection in Jupiter as a result of hydrogen-helium phase separation. *Mon. Not. R. Astron. Soc.* **447**, 3422–3441.

Nettelmann, N., R. Püstow and R. Redmer (2013), Saturn layered structure and homogeneous evolution models with different EOSs. *Icarus* **225**, 548–557.

Pearl, J. C. and B. J. Conrath (1991), The albedo, effective temperature, and energy balance of Neptune, as determined from Voyager data. *J. Geophys. Res.* **96**, 18921.

Pfaffenzeller, O., D. Hohl and P. Ballone (1995), Miscibility of Hydrogen and Helium under Astrophysical Conditions. *Physical Review Letters* **74**, 2599–2602.

Pollack, J. B., A. S. Grossman, R. Moore and H. C. Graboske (1977), A calculation of Saturn's gravitational contraction history. *Icarus* **30**, 111–128.

Read, P. L., T. E. Dowling and G. Schubert (2009), Saturn's rotation period from its atmospheric planetary-wave configuration. *Nature* **460**, 608–610.

Rosen, P. A., G. L. Tyler, E. A. Marouf and J. J. Lissauer (1991), Resonance structures in Saturn's rings probed by radio occultation. II – Results and interpretation. *Icarus* **93**, 25–44.

Roulston, M. S. and D. J. Stevenson (1995), Prediction of neon depletion in Jupiter's atmosphere. *EOS* **76**, 343 (abstract).

Salpeter, E. E. (1973), On Convection and Gravitational Layering in Jupiter and in Stars of Low Mass. *Astrophys. J. Lett.* **181**, L83.

Sanchez-Lavega, A., J. F. Rojas and P. V. Sada (2000), Saturn's Zonal Winds at Cloud Level. *Icarus* **147**, 405–420.

Saumon, D., G. Chabrier and H. M. van Horn (1995), An Equation of State for Low-Mass Stars and Giant Planets. *Astrophys. J. Suppl.* **99**, 713.

Saumon, D. and T. Guillot (2004), Shock Compression of Deuterium and the Interiors of Jupiter and Saturn. *Astrophys. J.* **609**, 1170–1180.

Schmider, F.-X., E. Fossat and B. Mosser (1991), Possible detection of Jovian global oscillations. *Astron. Astrophys.* **248**, 281–291.

Showman, A. P. and A. P. Ingersoll (1998), Interpretation of Galileo Probe Data and Implications for Jupiter's Dry Down-drafts. *Icarus* **132**, 205–220.

Smith, E. J., L. Davis, D. E. Jones, P. J. Coleman, D. S. Colburn, P. Dyal and C. P. Sonett (1980), Saturn's magnetic field and magnetosphere. *Science* **207**, 407–410.

Sromovsky, L. A., P. M. Fry and J. H. Kim (2011), Methane on Uranus: The case for a compact CH 4 cloud layer at low latitudes and a severe CH 4 depletion at high-latitudes based on re-analysis of Voyager occultation measurements and STIS spectroscopy. *Icarus* **215**, 292–312.

Sterenborg, M. G. and J. Bloxham (2010), Can Cassini magnetic field measurements be used to find the rotation period of Saturn's interior? *Geophys. Res. Lett.* **37**, 11201.

Stevenson, D. J. (1975), Thermodynamics and phase separation of dense fully ionized hydrogen-helium fluid mixtures. *Phys. Rev. B* **12**, 3999–4007.

Stevenson, D. J. (1982), Are Saturn's Rings a Seismograph for Planetary Inertial Oscillations? *EOS* **62**, 1020.

Stevenson, D. J. (1985), Cosmochemistry and structure of the giant planets and their satellites. *Icarus* **62**, 4–15.

Stevenson, D. J. (2003), Planetary magnetic fields. *Earth and Planetary Science Letters* **208**, 1–11.

Stevenson, D. J. and E. E. Salpeter (1977a), The dynamics and helium distribution in hydrogen-helium fluid planets. *Astrophys. J. Suppl.* **35**, 239–261.

Stevenson, D. J. and E. E. Salpeter (1977b), The phase diagram and transport properties for hydrogen-helium fluid planets. *Astrophys. J. Suppl.* **35**, 221–237.

Thommes, E. W., M. J. Duncan and H. F. Levison (1999), The formation of Uranus and Neptune in the Jupiter-Saturn region of the Solar System. *Nature* **402**, 635–638.

Tsiganis, K., R. Gomes, A. Morbidelli and H. F. Levison (2005), Origin of the orbital architecture of the giant planets of the Solar System. *Nature* **435**, 459–461.

von Zahn, U., D. M. Hunten and G. Lehmacher (1998), Helium in Jupiter's atmosphere: Results from the Galileo probe helium interferometer experiment. *J. Geophys. Res.* **103**, 22815–22830.

Vorontsov, S. V., T. V. Gudkova and V. N. Zharkov (1989), Jovian Seismology. *Soviet Astronomy Letters* **15**, 278.

Wahl, S. M., H. F. Wilson and B. Militzer (2013), Solubility of Iron in Metallic Hydrogen and Stability of Dense Cores in Giant Planets. *Astrophys. J.* **773**, 95.

Walsh, K. J., A. Morbidelli, S. N. Raymond, D. P. O'Brien and A. M. Mandell (2011), A low mass for Mars from Jupiter's early gas-driven migration. *Nature* **475**, 206–209.

Wavre, R. (1930), Sur les figures d'équilibre et la géodésie. *Publications of the Observatoire Geneve Series A* **10**, 9–25.

Wilson, H. F. and B. Militzer (2010), Sequestration of Noble Gases in Giant Planet Interiors. *Physical Review Letters* **104**(12), 121101.

Wilson, H. F. and B. Militzer (2012a), Rocky Core Solubility in Jupiter and Giant Exoplanets. *Physical Review Letters* ***108***(11), 111101.

Wilson, H. F. and B. Militzer (2012b), Solubility of Water Ice in Metallic Hydrogen: Consequences for Core Erosion in Gas Giant Planets. *Astrophys. J.* **745**, 54.

Wong, M. H., P. R. Mahaffy, S. K. Atreya, H. B. Niemann and T. C. Owen (2004), Updated Galileo probe mass spectrometer measurements of carbon, oxygen, nitrogen, and sulfur on Jupiter. *Icarus* **171**, 153–170.

Wood, T. S., P. Garaud and S. Stellmach (2013), A New Model for Mixing by Double-diffusive Convection (Semi-convection). II. The Transport of Heat and Composition through Layers. *Astrophys. J.* **768**, 157.

Zharkov, V. N. and V. P. Trubitsyn (1978), *Physics of planetary interiors*. Astronomy and Astrophysics Series, Tucson: Pachart, 1978.

4

Saturn's Magnetic Field and Dynamo

ULRICH R. CHRISTENSEN, HAO CAO, MICHELE K. DOUGHERTY AND KRISHAN KHURANA

Abstract

The magnetometer measurements taken by Cassini have confirmed the unusual character of Saturn's internal magnetic field known from previous flybys and have revealed additional properties that suggest a rather unique dynamo in this planet. Within measurement uncertainty, the internal magnetic field is completely symmetric with respect to Saturn's spin axis. The upper limit on the tilt of the magnetic dipole could be reduced from 1 to 0.06 degree. Moreover, only axisymmetric quadrupole and octupole moments are needed to fit the data. The lack of non-axisymmetric field components prevents a reliable determination of the bulk rotation rate of Saturn's deep interior. Using data from Cassini's closest approach to Saturn during orbit insertion, the magnetic moments of degrees four and five have been determined. The spatial power spectrum shows a zig-zag pattern with high power in odd spherical harmonic degrees and low power in even degrees. Compared to a simple dipole field, this corresponds to a concentration of magnetic flux towards the rotation poles. The flux concentration becomes progressively more pronounced when the field is continued into the interior. Comparison of the Cassini field model with that based on the Pioneer 11 and Voyager 1 and 2 measurements taken roughly 30 years earlier suggests that the secular variation of Saturn's field is at least one order of magnitude slower than that of the Earth. A viable explanation for most of the unusual field properties is that a stably stratified and electrically conducting layer, formed by a partial demixing of helium from metallic hydrogen, exists on top of a "standard" dynamo in Saturn's deep interior. This dynamo, driven by thermal and compositional convection, generates a magnetic field that is moderately asymmetric and time dependent. Rapid time variations and non-axisymmetric field components are filtered out in the stable layer by a skin effect. This model also implies that the top of the active dynamo may be located rather deep in Saturn's interior and the geometric drop-off of the dipole strength with the radius cubed could explain the unexpectedly low field strength at Saturn's surface. The stable layer model does not provide an explanation for the magnetic flux concentration towards the poles. Strong differential rotation in the dynamo region can have this effect, but a physical mechanism for such a flow state remains to be explored. From magnetic measurements to be taken during the very close approaches in the Grand Finale of the Cassini mission, we can expect to characterize Saturn's magnetic field up to at least spherical harmonic degree nine and possibly to detect weak non-axisymmetric field components, which would enable an accurate determination of Saturn's rotation period.

4.1 Introduction

The internal magnetic field of a planetary body is one of the fundamental properties and yields important information on the interior structure and evolution of the planet. Among the bodies of the solar system, examples for global-scale fields, local-scale fields that could be relics of now-extinct dynamos, and the absence of a magnetic field are found (e.g. Stevenson 2003; Schubert and Soderlund 2011; Jones 2011). A global magnetic field requires a dynamo process for its maintenance (e.g. Wicht and Tilgner, 2010). The strength of the field, its spatial structure and power spectrum and its time variation can be used to derive constraints on the depth of the dynamo region and on structural or dynamical conditions in the planet.

Before Pioneer 11 passed on September 1, 1979 within 0.4 planetary radii above the surface of Saturn (defined as the 1-bar pressure level) and provided the first magnetometer observations, it was expected that Saturn's magnetic field would resemble that of Jupiter or Earth. In particular, the expectation was that Saturn's field would be dominated by a dipole, moderately tilted with respect to the rotation axis, and would have a surface field strength that is intermediate between those of Earth and Jupiter. However, while Pioneer 11 found a dipolar field, the surface field strength reaches only two-thirds that of the geomagnetic field and, most surprisingly, the tilt angle of the dipole turned out to be limited to less than a degree (Smith et al. 1980). In addition, the quadrupole was found to be weaker relative to the dipole than it is at Earth and at Jupiter. Analysis of the magnetometer data, including data from the later flybys of Voyager 1 and Voyager 2 and, in particular, from Cassini during its numerous orbits around Saturn and from its close approach during the Saturn orbit insertion, reinforced the unique nature of Saturn's magnetic field. The upper limit for the deviation of the magnetic dipole axis from the rotation axis could be lowered by more than an order of magnitude. In addition, only zonal terms are needed in the higher harmonics for an accurate description of the observed field. This situation contrasts with Earth and Jupiter, where non-zonal terms contribute more to the higher-degree magnetic moments than the zonal ones. The lack of detectable non-axisymmetric components in Saturn is a puzzle to dynamo theory, because Cowling's theorem (e.g. Roberts and Gubbins 1987), which is based on first principles, states that a dynamo cannot generate a perfectly axisymmetric field. Furthermore, it also deprives us of a convenient means to determine the internal rotation rate of Saturn (which, because of magnetic coupling, must be almost uniform in electrically conducting regions). As a consequence, the uncertainty in the bulk rotation rate based on measurements of Saturn's kilometric radiation is up to 3% (see Chapter 5). A comparison of field models derived from the Pioneer–Voyager data with those based on Cassini measurements taken approximately 25–30 years later suggests that the rate of secular change of Saturn's field is at least an order of magnitude less than it was for the geomagnetic field during the past century. Finally, to add to the quaintness of Saturn's magnetic field, a determination, from the Saturn orbital insertion (SOI) data, of the axial multipole components beyond the octupole shows that deeper inside Saturn the magnetic flux becomes strongly concentrated into the north and south polar cap regions.

While Saturn is usually considered as the "little brother" of Jupiter, at least with respect to their magnetic fields they are rather unlike siblings. The similar morphology of Jupiter's field and Earth's field suggests that many of the concepts and models for the geodynamo might, in principle, also be applicable to Jupiter's dynamo. Like Jupiter's dynamo, that of Saturn is thought to operate in a shell in the interior where high pressure turns hydrogen into a metallic fluid that is set into motion by thermo-chemical convection. However, to explain the striking differences in the magnetic field structure between both planets, some fundamental difference must exist in their internal structure or in the dynamics of fluid motion in their interiors. As discussed in Chapter 3, aside from a small rocky core, the interiors of Jupiter and Saturn are composed mainly of hydrogen and helium, in a molecular state in the outer layers and in metallic form at depth. The mixing ratio of hydrogen and helium may be non-uniform in the interior of Jupiter and Saturn, because in the metallic region the miscibility of helium and hydrogen is limited. Demixing of helium from hydrogen is probably more important in Saturn than in Jupiter because temperatures inside Jupiter are higher than they are in Saturn at the same pressure level, assuming an adiabatic temperature gradient. Stevenson (1980) suggested that the precipitation of helium rain and its remixing at greater depth could explain two enigmatic properties of Saturn at the same time. On the one hand, the gravitational energy released by this internal differentiation could explain the excess infrared luminosity of Saturn, which is larger than what homogeneous models of cooling and contraction predict. On the other hand, helium precipitation creates a layer at the top of the metallic region in Saturn with a compositional gradient that stabilizes it against convective overturn. Stevenson (1980) proposed that a "standard" dynamo operates in the deeper homogeneous metallic region and that differential rotation in the stably stratified layer filters out all non-zonal field components, which become invisible from the outside. As will be discussed in this chapter, subsequent work has shown that Stevenson's model still provides an attractive explanation for most (but not all) of the observed peculiar features of Saturn's magnetic field.

The chapter is structured as follows. In the next section the available magnetometer observations of Saturn's magnetic field will be reviewed. Section 4.3 is dedicated to the models of Saturn's internal magnetic field that have been derived from the measurements. In addition to models for the zonal main field, we also discuss attempts to identify non-zonal field components and to derive values for the secular variation of the field. Section 4.4 deals with conceptual and numerical models for Saturn's dynamo. We conclude the chapter by pointing out the expected substantial improvements in the characterization of Saturn's field from the magnetometer measurements to be taken during the "Grand Finale" part of the Cassini mission that will bring the spacecraft into unprecedented proximity of the planet.

4.2 Observation of Saturn's Magnetic Field

Our knowledge about Saturn's magnetic field is mainly derived from in situ magnetic field measurements made by magnetometers onboard three flyby missions and the ongoing Cassini mission. Pioneer 11 carried a vector helium magnetometer and a high-field fluxgate magnetometer. Both Voyager 1 and 2 carried dual low-field and high-field fluxgate magnetometers (FGM). Cassini carried a fluxgate magnetometer and a vector/scalar helium magnetometer (V/SHM, Dougherty et al. 2004). However, the helium magnetometer onboard Cassini stopped functioning in late 2005. The dual instrument suite was flown for reliability, science return and calibration reasons. The scalar/vector helium instrument was capable of determining the *absolute* magnitude of the field to an accuracy of 1 nanoTesla (nT) in a background field close to Saturn of order 10,000 nT. By calibrating the fluxgate measurements with the absolute scalar measurements, determination of the fields in the Saturnian environment would have been possible with hitherto-unachieved accuracy. The combination of V/SHM and FGM allowed very sensitive determination of wave fields. The VHM is more sensitive below approximately 1 Hz, with the FGM being more sensitive at high frequencies. Now, with the single fluxgate instrument, regular spacecraft rolls are required in order to calibrate the fluxgate magnetometer, which mitigates the issue of calibrating without the scalar instrument. Various magnetic field models have been developed for Saturn based on these magnetic field measurements (e.g. Connerney et al. 1982; Connerney et al. 1984; Davis and Smith 1985, 1986, 1990; Dougherty et al. 2005; Burton et al. 2009; Cao et al. 2011; Cao et al. 2012).

4.2.1 *Flyby Magnetometer Measurements: Pioneer 11, Voyagers 1 and 2*

Pioneer 11, Voyager 1 and Voyager 2 made respective close flybys of Saturn in 1979, 1980 and 1981, with corresponding closest distances to the center of Saturn at 1.34 R_S, 3.07 R_S and 2.69 R_S. Here one Saturnian radius is set to R_S = 60,268 km (equatorial radius), corresponding to a pressure level of 1 bar. Figure 4.1 shows projections of the flyby trajectories of the three spacecraft. In the vicinity of Saturn, Pioneer 11 remained close to Saturn's equatorial plane within ±6 degrees latitude, Voyager 1 was near 40 degrees south at its closest approach to Saturn, while Voyager 2 traveled from about 30 degrees north to about 30 degrees south.

The highest magnetic field strengths encountered during the Pioneer 11, Voyager 1 and Voyager 2 Saturn flybys were 8142 nT, 1092 nT and 1187 nT, respectively. Magnetometer measurements made during the Pioneer 11 flyby established that Saturn's magnetic field is dominated by the spin-axial dipole. This can be deduced by examining the spatial dependence of the field amplitude as well as the partitioning of the field among radial, meridional and azimuthal components (Smith et al. 1980). Pioneer 11 also revealed two surprising facts about Saturn's intrinsic magnetic field.

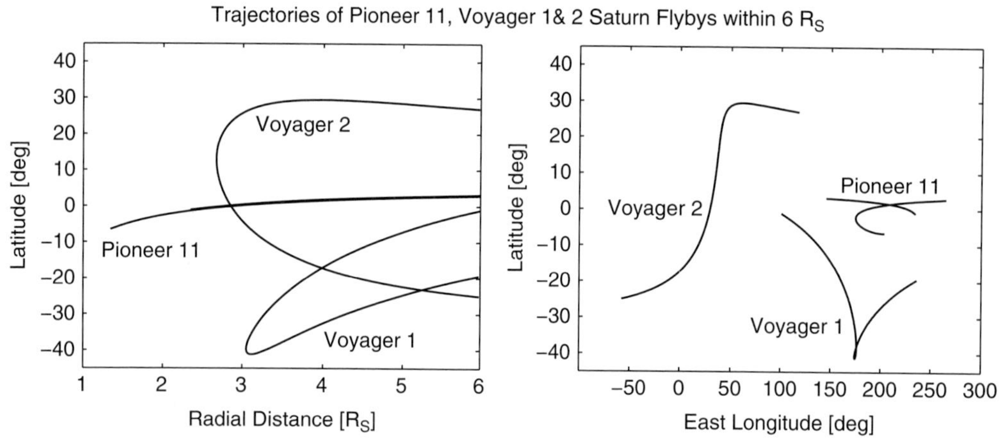

Figure 4.1 Trajectories of the Pioneer 11, Voyager 1 and 2 Saturn flybys. Only the portions inside r = 6 R_S with magnetic field measurements are shown. Trajectories are projected into the latitude-radial distance plane in the left panel and into the latitude-longitude plane in the right panel. The longitude is defined in the IAU System III Kronographic coordinate system with a fixed rotation period of 10h 39m 22.4s.

Firstly, although its dipole moment is much larger than that of Earth, in terms of field strength it is relatively weak. The surface field strength of Saturn is only two-thirds of the Earth's strength and is less than one-twentieth than at Jupiter. Secondly, Saturn's intrinsic magnetic field is very symmetric with respect to the spin axis. The Pioneer 11 measurements constrained the dipole tilt of Saturn to be within one degree, which is one order of magnitude smaller than that of the Earth and of Jupiter. Magnetometer measurements made during the subsequent Saturn flybys of Voyager 1 and Voyager 2, with mid-latitude coverage but further away from the planet, confirmed what Pioneer 11 has discovered about Saturn's intrinsic magnetic field and expanded on this knowledge by improving the latitudinal coverage of the observations.

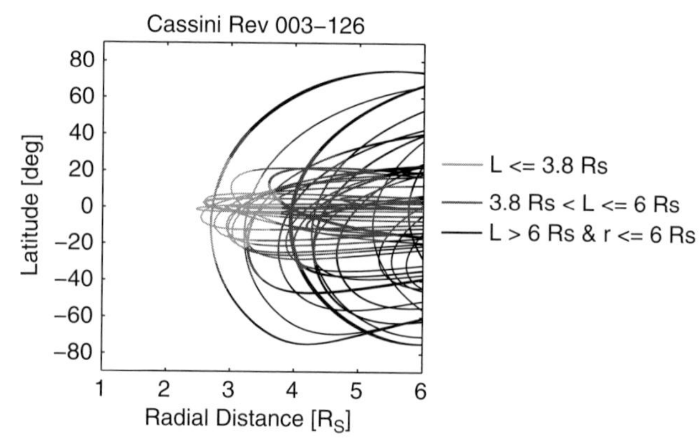

Figure 4.2 Trajectories of the Cassini spacecraft from Rev 003 to Rev 126 in the latitude-radial distance plane. Only the portions inside r = 6 R_S are shown, with red traces representing portions inside L = 3.8 R_S and blue traces represents portions between L = 3.8 R_S and L = 6 R_S. (A black-and-white version of this figure appears in some formats. For the color version, please refer to the plate section.)

4.2.2 Cassini Magnetometer Measurements

Since the arrival of the Cassini spacecraft at Saturn in mid-2004, the fluxgate magnetometer onboard has been making continuous measurements of Saturn's magnetic field over a wide range of latitudes and planetocentric distances. Figures 4.2 and 4.3 show the spatial coverage of Cassini orbits between 2005 and 2010 inside six Saturn radii. Figure 4.2 shows the trajectory in the latitude-radial distance space, which is unaffected by any uncertainty in our knowledge about the rotation rate of the planet. Figure 4.3 shows the trajectory in the latitude-longitude space for two different rotation periods. Reasonable lower and upper bounds on the rotation period are 10h 30m 00s and 10h 50m 00s, respectively (see Chapter 5). The rotation period derived from the gravity field is 10h 32m 35s (Anderson and Schubert 2007), while 10h 34m 13s is the rotation period estimated from the potential vorticity of the surface winds (Read et al. 2009). It can be seen from Figure 4.3 that Cassini provided good latitude-longitude coverage for the rotation periods of interest. The three magnetic field components from Cassini's magnetometer measurements along these

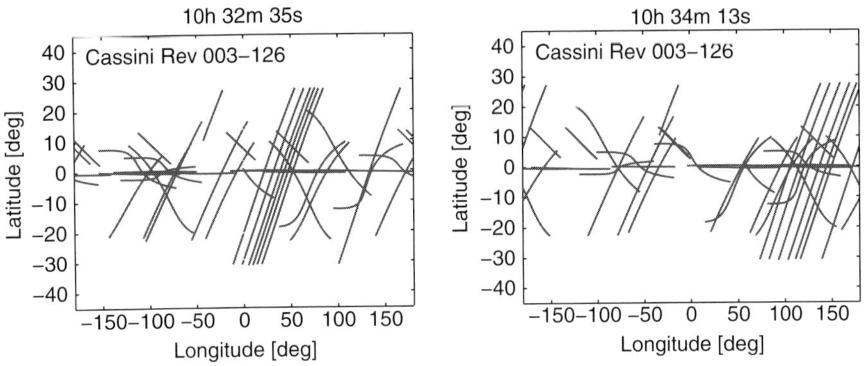

Figure 4.3 Latitude-longitude coverage of the Cassini spacecraft from Rev 003 to Rev 126 inside L = 3.8 R_S at two assumed rotation periods.

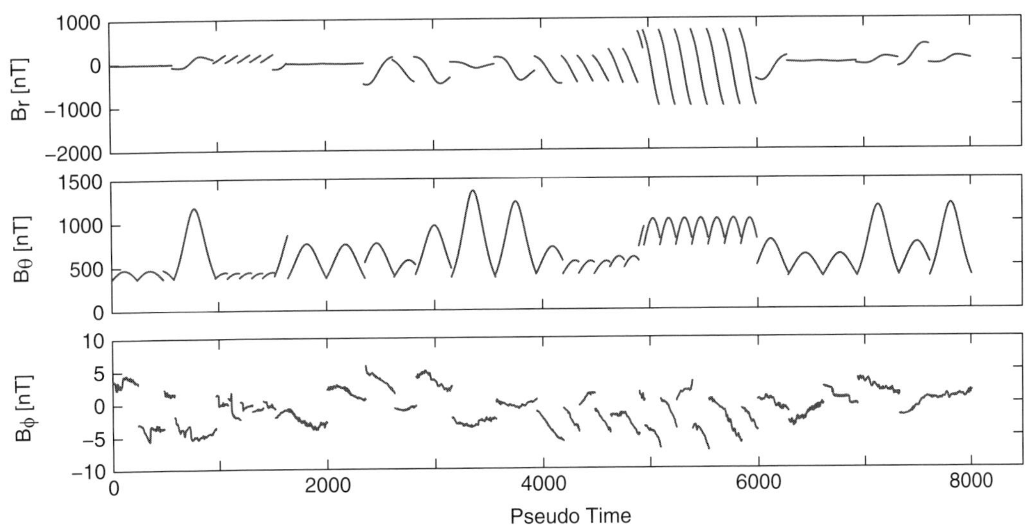

Figure 4.4 Cassini's magnetometer measurements of three magnetic field components inside $L = 3.8\ R_S$ between 2005 and 2010. The local maxima of the azimuthal component are almost two orders of magnitude smaller than the radial and meridional components. Pseudo time here is the sequential count of the measurements used in the analysis, a small fraction of the total data obtained by the magnetometer. Adapted from Cao et al. (2011).

orbits are shown in Figure 4.4. Only the measurements inside the dipole L-shell with L = 3.8 R_S are shown (the dipole L-shell is formed by the magnetic field lines of Saturn's dipole field that intersect the equatorial plane at distance L). The three components are defined in KRTP (Kronian Radial Theta Phi), spherical polar Saturn-centered coordinates, rotating at 10h 39m 22.4s, with r, θ and ϕ pointing along the radial, meridional and azimuthal directions, respectively. The chosen value of the rotation rate determines the longitude to which field vectors measured at different times are mapped. The maximum values of the azimuthal component are orders of magnitude smaller than the maximum radial and meridional components, which is direct evidence for Saturn's magnetic field being nearly axisymmetric.

The measurements inside L = 3.8 R_S can be regarded as almost "current-free," since they avoid several local currents in the magnetosphere. The bulk of the ring current in the magnetosphere extends between 5 R_S and 20 R_S (Bunce et al. 2007; Achilleos et al. 2010; Carbary et al. 2012). The current generated by the Enceladus interaction with the rotating magnetospheric plasma is confined to the vicinity of the dipole L-shell with L = 3.95R_S (Dougherty et al. 2006; Pryor et al. 2011) and the strong field-aligned currents originating from various sources are outside of L = 8R_S (Talboys et al. 2011).

4.3 Saturn Magnetic Field Models

Intrinsic planetary magnetic fields are generally represented by the gradient of a scalar potential V that is expanded in spherical harmonic functions multiplied with the internal Gauss coefficients g_l^m and h_l^m:

$$V(r,\theta,\phi) = \sum_{l=1}^{l_{int}} \sum_{m=0}^{l} \Big\{ R_P (\frac{R_P}{r})^{l+1} [g_l^m \cos(m\phi) + h_l^m \sin(m\phi)] S_l^m(\cos\theta) \Big\}, \quad (4.1)$$

$$B_r = -\frac{\partial V}{\partial r}, \; B_\theta = -\frac{1}{r}\frac{\partial V}{\partial \theta}, \; B_\phi = -\frac{1}{r\sin\theta}\frac{\theta V}{\partial \phi}. \quad (4.2)$$

Here, S_l^m are Schmidt quasi-normalized associated Legendre polynomials of degree l and order m, l_{int} is the maximum degree used for the expansion, r is the radial distance from the center of the planet and R_P is a reference radius, which is usually taken to be the radius of the planet. Note that for zonal (axisymmetric) field components, which are described by terms with $m = 0$, B_ϕ is zero. It should be emphasized here that the description of the field by a scalar potential is only valid in regions outside the source electrical currents.

Reconstructing the intrinsic magnetic field of a planet from in situ magnetometer measurements is a linear inverse problem for which a general theory and various techniques are well developed. However, in the planetary magnetic field inversion problem, spatial coverage is essential and is often far from ideal. Inadequate coverage, either in the radial direction or in the longitude-latitude extent, results in problems of instability and non-uniqueness of the inversion.

4.3.1 External Magnetic Field Components and Field Separation

The magnetic field measured near any magnetized planet inevitably contains two contributions: the intrinsic magnetic field of the planet and the external magnetic field created by the interaction of the internal field with the solar wind and/or interactions with additional plasma sources within the planetary system (e.g. Enceladus and Titan are two plasma sources for the Saturnian magnetosphere). For the purpose of modeling the intrinsic magnetic field of a planet, we can either adopt external Gauss coefficients in a current-free region and invert for them simultaneously with the internal coefficients or use specific external field models calculated from a vector potential to describe the low-frequency, large-scale, external magnetic field in the magnetosphere. For external Gauss coefficients, the scalar potential reads

$$V_{ext} = \sum_{l=1}^{l_{ext}} \sum_{m=0}^{l} \Big\{ R_P (\frac{r}{R_P})^{l} [G_l^m \cos(m\phi) + H_l^m \sin(m\phi)] S_l^m(\cos\theta) \Big\}, \quad (4.3)$$

where l_{ext} is the maximum spherical harmonic degree for the external potential and G_l^m and H_l^m are the external Gauss coefficients. It is worth pointing out that the degree-one external Gauss coefficients represent a uniform field with its orientation determined by the spherical harmonic order, e.g. for order zero it is aligned with the spin axis.

Ring currents, sometimes referred to as the magnetodisk currents, are the major contributions to the global external magnetic fields in the magnetospheres of Jupiter and Saturn. Close to the planet, these are azimuthal currents flowing near the plane of the planet's magnetic equator in the magnetosphere. We adopt a cylindrical coordinate system (ρ,z,ϕ), where the z-axis is parallel to the magnetic dipole axis and $z = 0$ is usually taken to be the equatorial plane of the magnetic dipole field. To first order, the ring current can be assumed to be purely azimuthal and axisymmetric, and its density is taken to be inversely proportional to the cylindrical radial distance from the z-axis ($J_\phi = I_0/\rho$). The magnetic field of such a simplified ring current model that is confined to the region $|z| \leq D$, $a \leq \rho < b$ can be expressed analytically (Connerney et al. 1981, 1983). Figure 4.5 shows the magnetic field for a model of the Saturnian ring current. Close to the planet, e.g. for $r < a/2$, the field can be well approximated in terms of degree-one external Gauss coefficients.

In Saturn's magnetosphere, an additional source of external field is a current system rotating nearly with the planetary rotation period. Before equinox in 2009, different periods of 10.6 and 10.8 hours were found in the northern and southern hemispheres, respectively

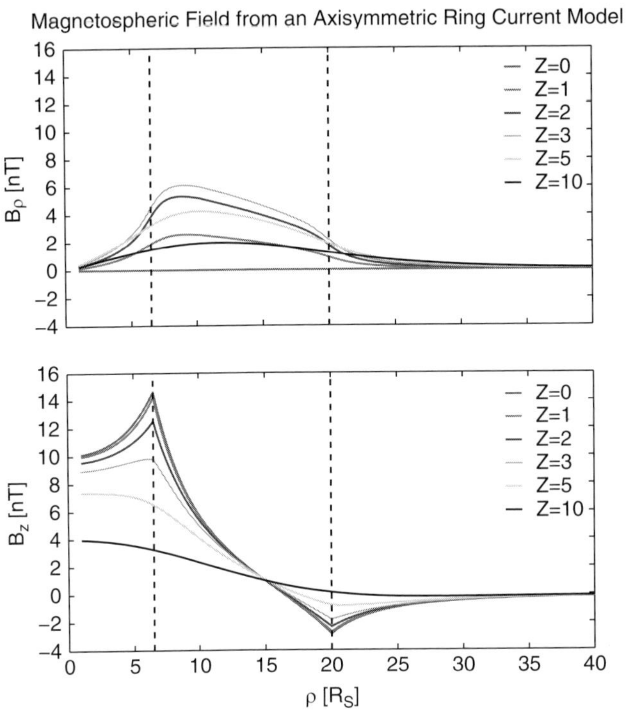

Figure 4.5 Magnetic field from an axisymmetric ring current model. Shown here are the cylindrical radial component (B_ρ) and axial component (B_Z) of the field as a function of cylindrical radial (ρ) distance at various distances (Z) from the equatorial plane. The parameters for this particular ring current model are inner edge $a = 6.5R_S$, outer edge $b = 20R_S$, half thickness $D = 2.5R_S$ and $\mu_0 I_0 = 40$ nT. (A black-and-white version of this figure appears in some formats. For the color version, please refer to the plate section.)

(Provan et al. 2009; Andrews et al. 2010). However, the two periods merged into a single one thereafter (Fischer et al. 2015). It appears to be in phase with the amplitude modulation period of Saturn's Kilometric Radiation (SKR) emitted from Saturn's high-latitude ionosphere (Gurnett et al. 2007). The source and form of this current system is not yet fully understood. Both magnetospheric driving, from a two-cell convection system (Gurnett et al. 2007; Khurana et al. 2009; Goldstein et al. 2014), and atmospheric driving, from a two-cell vortex moving with the SKR period in the ionosphere (Smith and Achilleos 2012; Jia et al. 2012; Jia and Kivelson 2012), have been suggested as possible source mechanisms. Regardless of the source of this time-varying field, in the inner magnetosphere ($r < 10R_S$, called the core region), it appears as a uniform rotating field parallel to Saturn's equator with a magnitude of 3–5 nT and with the perturbation in B_ϕ lagging that in B_r by 90 degrees (Southwood and Kivelson 2007; Andrews et al. 2008), while B_r and B_θ perturbations are in phase. In the region $10R_S < r < 14R_S$, the main manifestation of this current system is a rotating partial ring current with an azimuthal wave number of 1, a field magnitude of 6–10 nT and a period close to the SKR period. In the outer magnetosphere ($r > 14R_S$), the current systems produce a tilt of the current sheet, which is fixed in a frame rotating at the SKR period (Khurana et al. 2009; Arridge et al. 2011). There are two implications of these rotating current systems for modeling of the internal magnetic field of Saturn. First, at latitudes near ±70 degrees, strong field-aligned currents are observed in the ionosphere that must be modeled or excised from the data. Second, the external core rotating field must be modeled in the SKR phase and removed from the low- to mid-latitude observations when one seeks a signal in the internal field with an amplitude in the 3–5 nT range.

Two other sources of external magnetic fields are Chapman–Ferraro currents and crosstail currents. The magnetic field associated with these two current systems is generally one order of magnitude smaller than that associated with the ring currents and is quite uniform in the inner magnetosphere of the planet.

4.3.2 Internal Magnetic Field Models

4.3.2.1 Pioneer–Voyager Era

Several different intrinsic magnetic field models have been derived for Saturn based on the in situ magnetometer measurements from one or all of the three Saturn flybys made by Pioneer 11, Voyager 1 and Voyager 2. Only zonal harmonics are needed to give an adequate description of these measurements. Table 4.1 summarizes four of these models. The $P_{11}84$ model was solely based on the Pioneer 11 measurements (Davis and Smith 1985, 1986), the Z3 model was derived using Voyager 1 and 2 measurements only (Connerney et al. 1982), while the SPV model was derived using magnetometer measurements from all three flybys (Davis and Smith 1990). Although the exact values of the Gauss coefficients differ slightly between these models, all models give the same general picture of Saturn's intrinsic magnetic field. The non-zero axial quadrupole

Table 4.1 *Models for Saturn's intrinsic magnetic field based on different measurements. The Cassini 3 model up to degree 3 is based on Cassini in situ magnetometer measurements from Rev 3 to Rev 126 (Cao et al. 2011). The Cassini 5 model up to degree 5 is based on the Cassini 3 model and Cassini in situ magnetometer measurements made during Saturn Orbital Insertion (SOI) (Cao et al. 2012).* R_S *= 60,268 km in all models.*

Gauss coef. [nT]	$P_{11}84$	Z3	SPV	Cassini 3	Cassini 5
g_1^0	20,859	21,249	21,225	21,191 ± 24	21,191 ± 24
g_2^0	1,572	1,613	1,566	1,586 ± 7	1,586 ± 7
g_3^0	2,210	2,682	2,332	2,374 ± 47	2,374 ± 47
g_4^0					65 ± 70
g_5^0					185 ± 100

moment indicates a north-south asymmetry in the field and the positive octupole moment indicates that the field is weaker near the equator and stronger at the poles than that of a dipole with the same root-mean-square (RMS) surface field strength.

Although Pioneer 11 approached Saturn as close as 1.34 R_S, measurements made by the vector helium magnetometer only achieved an uncertainty level about 1%, according to Smith et al. (1980), which amounts to errors as high as 80 nT for the highest field encountered. The digitization step in the high field range of the vector helium magnetometer onboard Pioneer 11 was as large as 93.8 nT. For comparison, the digitization uncertainty in the highest field range of the Cassini fluxgate magnetometer, which can measure field stronger than 10,000 nT and weaker than 44,000 nT, is as low as 5 nT. The uncertainties in the magnetometer, as well as the limited latitude-longitude range covered in these three flybys, are limiting factors on the accuracy of the field model. As one consequence, only an upper bound of about 1 degree could be placed on the dipole tilt of Saturn's magnetic field based on these measurements. The uncertainties also prevented a meaningful derivation of higher-degree magnetic moments beyond the octupole.

4.3.2.2 Cassini Era

Various authors used Cassini magnetometer measurements to derive internal magnetic field models of Saturn (Dougherty et al. 2005; Burton et al. 2009; Cao et al. 2011; Cao et al. 2012). Two of them, the Cassini 3 model and the Cassini 5 model, are listed in Table 4.1 along with the three internal field models based on the Pioneer–Voyager measurements. The Cassini 3 model was derived using Cassini magnetometer measurements within L < 3.8 R_S taken between 2005 and 2010 (e.g. Figure 4.4). The closest approach during these measurements was at 2.46 R_S. In deriving this model, the weighted-least-square solution with Cauchy weighting functions is employed (Holland and Welsch 1977). The external field is taken as uniform and aligned with the spin axis. Attempts to derive the magnetic moments of higher degrees based on these measurements have also been made, but only resulted in large upper bounds on the Gauss coeffcients g_4^0 and g_5^0 (Cao et al. 2011). The error bars of the Cassini 3 model were estimated in a conservative fashion by inverting multiple subsets of the data and using the deviation in the results for the Gauss coefficients. The error bars quoted for g_1^0, g_2^0 and g_3^0 are 7 times, 4 times and 5 times the formal 1σ error bars in the weighted-least-square solution, respectively.

Cao et al. (2012) revisited the magnetometer measurements along the trajectory of the Cassini Saturn Orbital Insertion (SOI), which came as close as 1.33 R_S. After explicitly modeling the external ring current as a warped current sheet, the deviation of the SOI measurements from the predictions of the Cassini 3 model, which are shown in Figure 4.6 as the blue traces, can be best described as signals from high-degree internal magnetic moments. It is assumed that non-zonal field components are much smaller than the zonal ones at degrees 4 and 5, as they are for degrees 1 to 3. This assumption is supported by the magnetic field measurements at SOI, which show that the azimuthal component stays within ±7.5 nT even

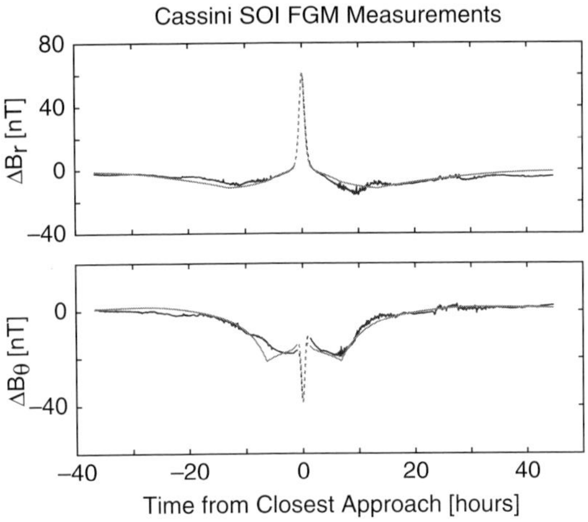

Figure 4.6 Deviation of the Cassini SOI measurements from the Cassini 3 model (blue) compared to the field corresponding to (g_4^0, g_5^0) = (65 nT, 185 nT) plus a modeled external field (magenta). The top panel shows the radial component, B_r, while the bottom panel shows the meridional component, B_θ. The broken parts of the traces are for measurements within 2.2 R_S from Saturn's center and the solid parts are outside 2.2 R_S. From Cao et al. (2012). (A black-and-white version of this figure appears in some formats. For the color version, please refer to the plate section.)

at such close distances. By searching the parameter space of the axial degree 4 and degree 5 internal components, a best solution of (g_4^0, g_g^0) = (65, 185) nT is obtained. The associated magnetic field components are shown as the magenta traces in Figure 4.6. The error bars on g_4^0 and g_5^0 are calculated by propagating the errors on the low-degree moments, as described in Cao et al. (2012). While the g_4^0 coefficient could be zero within the estimated error, the non-zero value of the degree-5 internal moment is significant.

The Cassini 5 model indicates that the Lowes–Mauersberger power spectrum of Saturn's intrinsic magnetic field takes a zig-zag pattern, with higher power in odd degrees than in even degrees up to degree 5 (Figure 4.7a). In the geomagnetic field at Earth's surface, the power decreases systematically with harmonic degree l. For a potential field of internal origin, the power varies with radius as $r^{-(2l+4)}$, hence the power in higher harmonics becomes relatively more enhanced as the field is downward-continued into the interior of the planet. In case of the Earth, the spectrum is known to be nearly flat at the top of Earth's core aside from elevated power in the dipole (Lowes 1974). Assuming that this property holds in general at the surface of a planetary dynamo, one can attempt to infer the depth to the dynamo also for other planets (e.g. Elphic and Russell 1978). Considering only the odd-degree terms in Saturn's spectrum, the power still decreases with l when calculated at $0.63R_S$ (the radius where hydrogen attains high metallic conductivity values; see Section 4.4.1). Only at about $0.4R_S$ does the odd-degree power spectrum become nearly flat, which also includes the dipole (Figure 4.7a).

The zig-zag pattern of the spectrum, together with the same sign for all odd Gauss coefficients of order zero, implies a trend of poleward concentration of the radial magnetic flux inside Saturn. The magnetic flux in Saturn appears to be expelled from the equatorial region and collected at high latitudes. The geometry of Saturn's magnetic field in the interior of the planet is in contrast to what is understood about the Earth's magnetic field at the core-mantle boundary (CMB). As shown in Figure 4.7b, the longitudinally averaged magnetic flux at the surface of the Earth's core has a local minimum at the poles, whilst at Saturn's poles it is at its maximum. At low to mid-latitudes there is significant flux at the surface of the Earth's core, in contrast to the radial field in Saturn's interior, which is very weak in this latitude band. The polar field minimum region at the Earth coincides with the cylindrical projection of the inner core on the outer core surface, and is believed to be controlled by the inner core (Olson and Aurnou 1999). Although both the fields of Earth and Saturn are generally classified as dipolar, it can be seen that in terms of the distribution of magnetic flux with latitude they share little similarity at the top of the dynamo region, except for the change in sign near the equator.

4.3.3 Search for Non-zonal Internal Field Components

The search for non-axisymmetric (non-zonal) components of Saturn's internal magnetic field is important for several reasons. A definitive identification of a non-axisymmetric component of Saturn's internal magnetic field would allow an accurate determination of the rotation rate of Saturn. The secular change of the non-zonal field in particular is used to map the flow at the surface of the Earth's core (e.g. Holme 2007) and could be used for the same purpose at other planets if it can be detected. Giampieri et al. (2006) tentatively associated

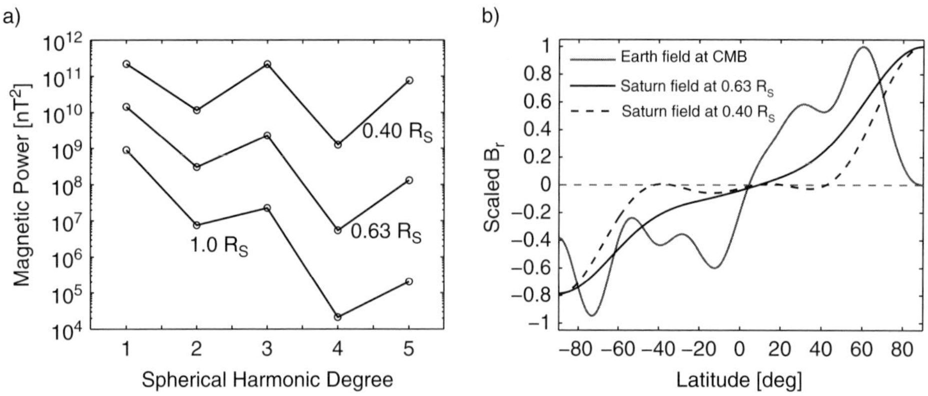

Figure 4.7 (a) Lowes–Mauersberger spatial power spectrum of Saturn's intrinsic magnetic field according to the Cassini 5 model at various radii. Adapted from Cao et al. (2012). (b) Zonally averaged radial magnetic field B_r, normalized by its absolute maximum value, vs. latitude at 0.63 R_S and 0.40 R_S according to the Cassini 5 model. For comparison, the zonally averaged B_r according to the geomagnetic IGRF10 model at the Earth's core-mantle boundary is shown.

an approximate 10h 47m periodicity observed in magnetic field measurements inside of 10 R_S with a non-zonal internal field component. However, their signal is most likely of external origin and as yet no unambiguous bulk rotation rate of Saturn could be derived (see Chapter 5).

Cowling's theorem requires that a non-axisymmetric field component must exist inside the dynamo region, although it does not quantify how large it must be. Constraining the amplitude of the weak non-zonal field component above the planet's surface will help either (i) to assess the processes that may axisymmetrize a field that is of complex geometry in the deep interior in a quantitative way (Section 4.4.3) or (ii) to better understand possible dynamos that may produce very little non-zonal field in the first place (Section 4.4.4).

Several attempts have been made to identify non-axisymmetric components of Saturn's internal magnetic field based on Cassini magnetometer measurements (Sterenborg and Bloxham 2010; Burton et al. 2010; Cao et al. 2011). A complication is that because of the uncertain rotation period it is not possible to assign a unique longitude to any one measurement. A variational principle, which asserts that any non-axisymmetric magnetic moments would be maximized in the inversion at the correct rotation rate of Saturn while yielding simultaneously a minimum of the misfit to the data, has been taken by all attempts. A test of this variational principle with synthetic data that assume a non-zonal component in Saturn's magnetic field was presented by Sterenborg and Bloxham (2010), who showed that it can indeed yield the correct rotation period of Saturn and the correct amplitude of the non-axisymmetric magnetic moments. However, applying this variational principle to the Cassini magnetometer measurements only tightened the upper bounds on the non-axisymmetry of Saturn's intrinsic magnetic field.

Figure 4.8 shows one example of applying the variational principle to the Cassini data. The amplitude of the non-axisymmetric power for the intrinsic dipole and quadrupole moments as well as the RMS error in the fit to the data are shown as a function of the assumed rotation period. Although maxima and minima can be found in the non-axisymmetric power and RMS error, respectively, no correlation or anti-correlation exists between any of these diagnostics. The maximum of the non-axisymmetric dipole moment corresponds to a dipole tilt of 0.06 degree, which should only be regarded as an upper limit to Saturn's true dipole tilt. This extreme alignment between the magnetic axis and the spin axis of Saturn places strong constraints on Saturn's dynamo. The Cassini "Grand Finale" will provide a better chance to detect any departure from axisymmetry in Saturn's intrinsic magnetic field (see Section 4.5).

4.3.4 Secular Variation

The dynamo process in planets is chaotic and turbulent and the generated magnetic field is therefore expected

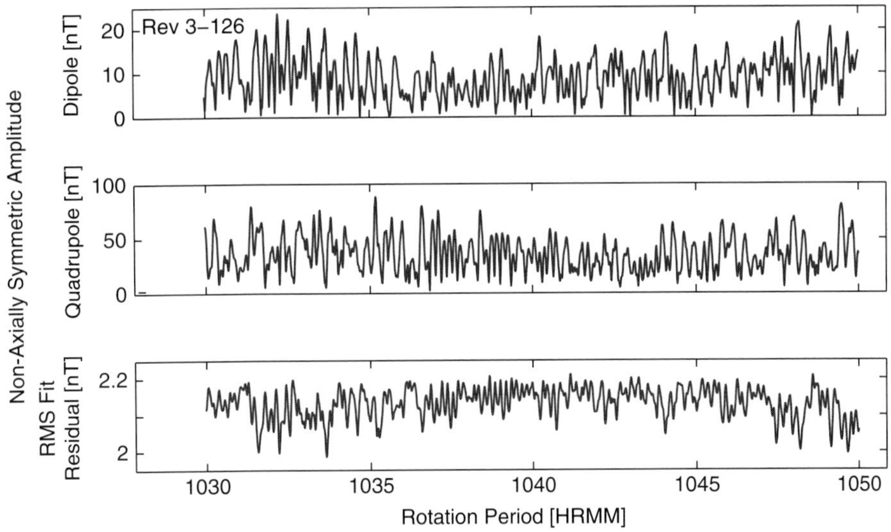

Figure 4.8 Amplitude of non-axisymmetric intrinsic dipole and quadrupole terms and root mean square (RMS) residuals for assumed rotation rates ranging from 10h 30m 00s to 10h 50m 00s. The maxima of non-zonal dipole and quadrupole power most likely result from the re-arrangement of the spatial coverage as the assumed rotation rate changes. The biggest improvement in the residuals is only ~0.2 nT, likely smaller than the overall uncertainties of the magnetometer measurements onboard Cassini in the second highest field range which measures 400 nT to 10,000 nT. Adapted from Cao et al. (2011).

to be complex both spatially and temporally. The internal geomagnetic field changes on time scales of a century, which is referred to as secular variation. We use the same term for changes of Saturn's dynamo-generated magnetic field. While the Earth's dipole field is currently changing at a rate about 0.1% per year (which includes changes of the tilt angle), more rapid changes occur in higher multipole components. These changes are at least partly attributed to the advection of magnetic field by core surface flows. Under the assumption that the magnetic flux is frozen into the moving fluid, flow velocities of around 15 km/year are inferred (Holme 2007).

A comparison between axisymmetric models derived from Pioneer 11 and Voyager 1 and 2 (the SPV model by Davis and Smith 1990) with the most recent Cassini model by Cao et al. (2011) shows little evidence for secular variation in Saturn's internal magnetic field. Estimates of secular variation are -1.2 ± 1.6 nT/yr, 0.7 ± 0.5 nT/yr and 1.5 ± 3.2 nT/yr for the axial dipole, quadrupole and octupole coefficients, respectively. Thus, upper bounds on Saturn's magnetic secular variation are estimated to be less than 0.013% per year, about one order of magnitude smaller than that of the geomagnetic field.

4.4 Saturn's Dynamo

4.4.1 Internal Structure of Gas Giants: Properties Relevant to the Dynamo

Saturn's interior structure is discussed in detail in Chapter 3. Here we summarize those points that are important for the dynamo and where Saturn may differ from Jupiter, in particular as regards the extent of helium-hydrogen demixing.

Jupiter and Saturn have basically solar composition, dominated by hydrogen and helium, although with a larger complement of heavy elements than the Sun. Most models assume a core of rock and ices making up as much as 25% of the planet's mass, which takes up to 15% of the planet's radius in the case of Saturn (e.g. Nettelmann et al. 2013). However, some or all of the heavy elements could also be mixed with hydrogen and helium deep inside the planet. The properties of hydrogen-helium mixtures at the relevant pressures and temperatures have been studied by various groups using ab-initio quantum mechanical simulations with increasing agreement of the results over the past years (e.g. Morales et al. 2009; Lorenzen et al. 2011). Of particular interest in the context of the dynamo is the metallization of hydrogen and the associated increase of electrical conductivity, the phase separation in H-He mixtures and the possible subdivision of the H-He region into compositionally distinct sublayers. For modeling the internal structure of the gas planet, it is usually assumed that the temperature follows an adiabat, starting from the observed surface temperature at the one-bar level. Because it is lower in the case of Saturn (134 K) compared to Jupiter (165 K), temperatures in Saturn are lower than in Jupiter throughout the interior at the same pressure level. At a pressure of 1 Mbar, where temperatures are of the order 5000 K, the difference has increased to more than 1000 K. However, while an adiabatic thermal structure is a perfect assumption when there is no barrier to radial convective mixing, it breaks down when compositional boundaries or gradients are present.

The metallization in hydrogen-helium mixtures is a gradual process, i.e. there is probably no first-order phase transition (Lorenzen et al. 2011). The electrical conductivity increases continuously with depth. Shock experiments (e.g. Nellis et al. 1999) found a steep increase of the electrical conductivity σ up to metallic values of 10^5 S/m between 0.9 and 1.4 Mbar. Temperatures are lower in these experiments than they are in the giant planets at the same pressures, and ab-initio simulations using the Jupiter adiabat (French et al. 2012) predict that metallic conductivity of $\sigma > 10^5$ S/m is reached in Jupiter at a pressure of 0.6 Mbar at about $0.9R_J$ (R_J is Jupiter's radius). Below $0.7R_J$, the conductivity is on the order of 10^6 S/m. Similar calculations for Saturn are not available, and here a somewhat higher pressure is probably required to reach a certain conductivity value because of lower temperatures. In Saturn the pressure range of 0.6–1.0 Mbar corresponds to 0.65–0.75 Saturn radii (R_S). However, lower values of electrical conductivity can be dynamically important. A relevant criterion is that a local magnetic Reynolds number, $Rm_\ell = Ud_\sigma\mu\sigma$, is of order one or larger. Here d_σ is a "conductivity scale height," i.e. the radial distance on which the conductivity changes by a factor of e, U is the characteristic flow velocity and μ the magnetic permeability (Liu et al. 2008). Using this criterion and setting $U \approx 1$ cm/s (a value that is expected in the dynamo region of a gas planet, see below), electrical conductivity becomes relevant when it exceeds about 100 S/m, which is reached at 0.94 R_J according to the conductivity model of French et al. (2012). If the jets observed at cloud level with speeds of more than 10 m/s were able to penetrate into the conducting region, 0.3 S/m (reached at 0.965 R_J) would be sufficient. The levels at which Rm_ℓ reaches unity are somewhat deeper in Saturn, at about $0.87R_S$ and $0.77R_S$ for the higher and lower velocity, respectively, when using the conductivity profile of Liu et al. (2008).

Jupiter's and Saturn's infrared luminosity exceeds the values that would result from the re-emission of absorbed sunlight. The excess flux at Saturn's surface is 2.4 W/m^2 (Ingersoll et al. 1980), about thirty times the geothermal flux from the Earth's interior. The internal heat flow is primarily a consequence of the cooling and contraction of the planets. An ongoing separation of a helium-rich phase from the hydrogen-helium mixture can lead to a "helium rain" in the gas planets as they cool (Stevenson and Salpeter 1977). In the gaseous envelope, hydrogen and helium can mix in any proportion, but at pressures where hydrogen starts to metallize, helium miscibility becomes limited (Morales et al. 2009; Lorenzen et al. 2011). The miscibility increases with temperature, and therefore phase separation should play a larger role in the cooler interior of Saturn than it does in Jupiter. Also, because of the temperature influence on miscibility, phase separation may be restricted to the top part of the metallic hydrogen region and the helium rain could dissolve and remix again at greater depth, where the temperature is higher. The simulations of Lorenzen et al. (2011) suggest that for a protosolar helium abundance the miscibility limit is exceeded in the pressure range of 1–2.5 Mbar in Jupiter for an adiabatic temperature structure, whereas the demixing region would reach to much higher pressures in Saturn. Therefore, some models consider that helium may rain down all the way to Saturn's rocky core (Fortney and Hubbard 2003; Nettelmann et al. 2013). However, because the stable density stratifation which is expected to develop in the immiscibility region suppresses convection in its standard form (see below), the temperature would rise more steeply with depth than along an adiabat. As a consequence, full miscibility could be reached again at some level above Saturn's core.

Figure 4.9 illustrates in a schematic way the formation of a stably stratified layer by helium rain in a gas giant planet. The curved dotted line is the miscibility

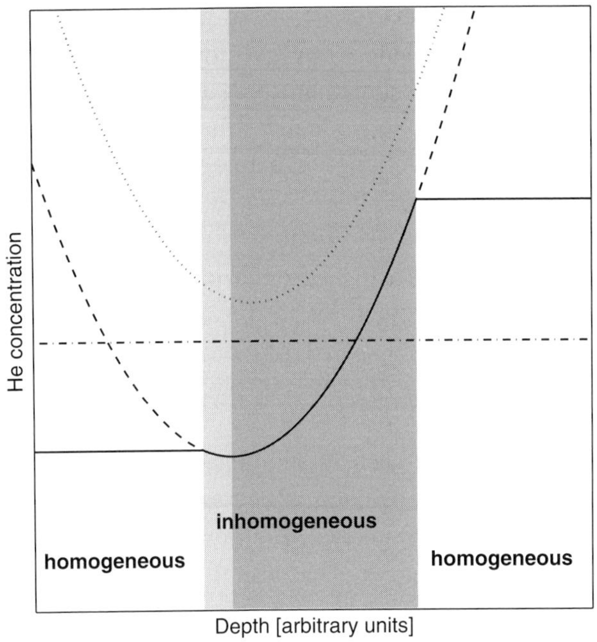

Figure 4.9 Schematic illustration for the formation of an inhomogeneous stably stratified layer by helium phase separation in a gas planet. See text for detailed explanation.

limit of helium at an early time after planet formation when the planet was hot. The miscibility first decreases with depth because of hydrogen metallization and increases again at greater depth because of the rising temperature. The dash-dotted line indicates the initially uniform helium concentration. The dashed curved line is the present miscibility limit after the planet has cooled. The solid line shows the present helium concentration vs. depth; white regions are homogeneous while inhomogeneous regions are shaded in grey. In the top part of the inhomogeneous region in light grey (whose width is strongly exaggerated), the helium rain is formed. Here the He-concentration decreases with depth; hence, the region is convectively unstable and exchanges material with the overlying layer, which becomes depleted in He. In the main part of the inhomogeneous region in dark grey, the He concentration increases with depth; hence, it is stably stratified. Here some more He rain is formed due to the secular cooling. Below the bottom of the inhomogeneous region, the precipitating He drops will dissolve again in the fluid that is here undersaturated. The helium will be convectively mixed into the deep layer and will raise its helium concentration.

While the homogeneous layers overturn by thermal convection, augmented by compositional convection due to the extraction of helium at the bottom of the upper layer and addition of helium at the top of the lower layer, in the stable inhomogeneous region only some form of double-diffusive convection is possible (e.g. Leconte and Chabrier 2013). It likely occurs in a large number of thin stacked sublayers. In each sublayer the He concentration is nearly uniform, with step-like changes at the boundaries between them. This form of convection involves no large-scale radial mass exchange. Although it is more efficient in transporting heat than simple conduction, it is less so compared to free convection and a net superadiabatic thermal gradient is set up across the stable region.

For Saturn, the helium rain scenario has been intensely discussed in the literature. The helium concentration in Saturn's atmosphere is less than the protosolar value (Conrath and Gautier 2000), in agreement with downward segregation of helium. Homogeneous evolution models significantly underpredict Saturn's present luminosity. The release of gravitational potential energy by helium segregation provides an additional energy source which could at least partly compensate for the missing heat (Stevenson and Salpeter 1977). To fully make up for the deficit, it may be necessary that helium segregates all the way down to the rocky core, rather than remix again below an inhomogeneous layer of limited depth extent (Fortney and Hubbard 2003). However, Leconte and Chabrier (2013) argue that the restricted heat transport capability in a stratified inhomogeneous layer of finite width retards the cooling of Saturn's interior sufficiently, compared to a homogeneous evolution history, to raise the present-day luminosity to its observed value. Finally, Stevenson (1980) proposed that the stable layer could explain many of the unexpected properties of Saturn's magnetic field. We will elaborate on this in Section 4.4.3. In summary, helium phase separation and the formation of a stably stratified layer inside Saturn is highly probable, although the actual thickness of this layer is rather unconstrained.

The higher temperatures along a Jupiter adiabat imply that in this planet helium phase separation should play a smaller role or perhaps not occur at all. Jupiter's internal heat flow can be explained with a homogeneous cooling model. The absence of a stably stratified layer might explain why Jupiter's magnetic field lacks the "abnormal" properties exhibited by Saturn's field. However, the He abundance in Jupiter's atmosphere is also less than the protosolar value (von Zahn et al. 1998), although it is perhaps higher than at Saturn. Furthermore, the observed strong depletion of neon in Jupiter's atmosphere hints at helium phase separation, because during this process neon strongly partitions into the helium phase (Wilson and Militzer 2010).

4.4.2 Fundamentals of Planetary Dynamos and Dynamo Models for Gas Giants

In order for a dynamo to operate inside a planetary body, some basic requirements must be fulfilled. The dynamo effect relies on Faraday's law of electromagnetic induction of electrical currents in a conductor that is moving in an existing magnetic field. Hence, we need a region inside the planet with a sufficiently high electrical conductivity that is in sufficiently rapid motion. The latter is only possible in a fluid or a plasma. The degree of "sufficiency" is quantified by the magnetic Reynolds number,

$$Rm = \frac{UL}{\lambda}. \tag{4.4}$$

In contrast to the local magnetic Reynolds number introduced in Section 4.4.1, here a global length scale L, such as the outer radius or the thickness of the conducting shell in a planet, is used, and the magnetic diffusivity $\lambda = 1/(\mu\sigma)$ is introduced. Rm can be understood as representing the ratio between creation of magnetic field by induction and destruction of field by Ohmic dissipation and must exceed a critical value Rm_c to allow for a dynamo. The value of Rm_c depends on the details of the dynamo, e.g. the flow structure. A self-sustained dynamo arises when the induced currents have the right strength and geometry to reproduce, through Ampère's law, the magnetic field that is needed for the induction process in the first place. In a technical dynamo, the right geometry of the current is ascertained by the complex arrangement of wires. In natural dynamos, which operate in a nearly homogeneous unstructured conductor, a certain complexity of the flow pattern is required. For example, simple differential rotation is insufficient for the dynamo effect. Particularly suitable for dynamo action is helical fluid motion, where fluid parcels move on corkscrew-type trajectories. It has been found that rapid bulk rotation, i.e. a strong influence of Coriolis forces, favors this kind of flow pattern, although it is not strictly a necessary requirement for a dynamo. In numerical dynamo simulations with rapid rotation, the critical magnetic Reynolds number is on the order of 50 (Christensen and Aubert 2006).

Can we expect that Jupiter and Saturn meet the requirements for a dynamo? They are fluid planets in rapid rotation. Their internal heat flow is two orders of magnitude larger than what can be transported by thermal conduction along an adiabatic gradient, even with the rather high thermal conductivity of a metallic fluid (French et al. 2012). Hence, some form of convection must occur, and in homogeneous parts of the planet this will involve global overturn. While the speed of the surface jets can exceed 100 m/s, in the deep interior the velocities are expected to be much smaller. Different rules have been suggested for the scaling of convective velocity U in a rotating dynamo (Christensen 2010). One that is decently consistent with the results of numerical dynamo simulations is the so-called CIA-rule (based on a balance of **C**oriolis, **I**nertial and **A**rchimedean forces) and predicts

$$U \propto \left(\frac{q}{\rho H_T}\right)^{2/5} \left(\frac{\mathrm{L}}{\Omega}\right)^{1/5}, \tag{4.5}$$

where q is the heat flow, ρ the density, H_T the temperature scale height, L a characteristic global length scale such as the outer radius or the thickness of the convecting layer and Ω the planetary rotation rate. Applied to Saturn, it predicts a typical flow velocity of 1 cm/s, but some uncertainty remains. The exponents in Equation 4.5 could be slightly different (Davidson 2013) and in the parameter regime of numerical dynamo simulations, viscosity, which is thought to be negligible in planets, can still have an influence (Stelzer and Jackson 2013; King and Buffett 2013). Alternative scaling rules are based on a MAC-balance (**M**agnetic, **A**rchimedean, **C**oriolis) (e.g. Starchenko and Jones 2002), predicting a velocity of the order 1 mm/s for Saturn, and the mixing length rule (Kippenhahn and Weigert 1990), which results in about 10 cm/s. For $\lambda \approx 2$ m^2/s, L $\approx 3 \times 10^4$ km and a velocity in the range 0.1–10 cm/s, the magnetic Reynolds number falls in the range 10^4–10^6, with a preferred value around 10^5, and in any case, far above the critical value for a dynamo in a rapidly rotating system. All requirements for a self-sustained dynamo are satisfied; hence, it does not come as a surprise that Jupiter and Saturn have intrinsic magnetic fields.

Much of our present understanding of planetary dynamos is based on the results of numerical simulations of the geodynamo during the past two decades (see Jones 2011 and Christensen 2011 for reviews). Generically, these models solve for buoyancy-driven flow and magnetic field generation in a rotating spherical shell filled with an electrically conducting fluid. Because the density variation across the Earth's fluid core is only on the order of 20%, incompressible flow is assumed in the so-called Boussinesq approximation, where density differences enter only in a buoyancy term. Usually the equations are cast into a non-dimensional form by introducing appropriate scales for length, time, magnetic field, etc. The physical properties of the system are here condensed into a set of non-dimensional control parameters. An example for the set of equations is (Christensen and Wicht 2015):

$$E(\frac{\partial \mathrm{u}}{\partial t}+\mathrm{u}.\nabla \mathrm{u})+2\hat{z}\times \mathrm{u}+\nabla\pi=E\nabla^2\mathrm{u}$$

$$+Ra\frac{\mathrm{r}}{r_o}T+\frac{1}{Pm}(\nabla\times \mathrm{B})\times \mathrm{B}, \tag{4.6}$$

$$\frac{\partial \mathrm{B}}{\partial t}-\nabla\times(\mathrm{u}\times \mathrm{B})=\frac{1}{Pm}\nabla^2\mathrm{B}, \tag{4.7}$$

$$\frac{\partial T}{\partial t}+\mathrm{u}.\nabla T=\frac{1}{Pr}\nabla^2 T+\epsilon, \tag{4.8}$$

$$\nabla.\mathrm{u}=0,\ \nabla.\mathrm{B}=0. \tag{4.9}$$

Here, u is the flow velocity, T is temperature and Π is the non-hydrostatic pressure. In the generalized Navier–Stokes Equation 4.6, the first term on the left-hand side is the inertial force and the second one the Coriolis force (the unit vector $\hat{z}$ indicates the direction of the rotation axis). On the right-hand side, the terms describe, in order, the viscous force, the buoyancy force and the electromagnetic Lorentz force. The induction Equation 4.7 is derived from Maxwell's equations and a generalized Ohm's law for a moving conductor. The second term on the left-hand side describes advection and induction of magnetic field and the term on the right-hand side the destruction of field by a diffusion process due to the finite conductivity. The temperature Equation 4.8 contains terms for advection and diffusion and ϵ is a volumetric heating rate.

The four non-dimensional control parameters in these equations are the Ekman number

$$E=\frac{\nu}{\Omega L^2}, \tag{4.10}$$

a (modified) Rayleigh number

$$Ra=\frac{\alpha g_0 \Delta T L}{\nu\,\Omega}, \tag{4.11}$$

the Prandtl number and the magnetic Prandtl number, respectively,

$$Pr=\frac{\nu}{\kappa},\quad Pm=\frac{\nu}{\lambda} \tag{4.12}$$

Here ν is the viscosity, Ω the rotation rate, α the thermal expansion coefficient, g_o gravity on the outer boundary, ΔT an imposed temperature contrast between the inner and outer shell boundaries, κ is thermal diffusivity and λ is magnetic diffusivity. It is not possible to use actual planetary values for the control parameters in the simulations. In particular, the Ekman number, which describes the ratio of viscous forces to the Coriolis force, is of order 10^{-15} in planets, whereas even the most advanced dynamo models barely reach values of the order 10^{-6}. Taking the correct planetary length scale and rotation rate, this can be interpreted as a value for the viscosity that is far too large. High viscosity is needed to suppress small eddies in the flow that would appear otherwise and require a numerical grid resolution that is unaffordable with present computational means. As a consequence of the high viscosity, the magnetic Prandtl number is also too large (of order unity in models, compared to order 10^{-6}

in planets), and the Rayleigh number, which describes the ratio of the driving buoyancy to forces that retard convection, is too small. Too much weight is given to viscous and inertial forces in the simulations compared to their relative contributions in the planets.

Despite the mismatch of control parameters, geodynamo models are able to reproduce properties of the geomagnetic field in remarkable detail. They not only generate magnetic fields dominated by the axial dipole of the right strength, but some also reproduce the (spatial) power spectrum of the Earth's field, show morphological structures that are thought to be generic for the geomagnetic field, and have a similar time-variability, ranging from secular variation to dipole reversals. One important reason for their success could be that geodynamo models can be run at the appropriate values of the magnetic Reynolds number, i.e. the proportion of magnetic advection and induction to diffusion in Equation 4.7 is correctly modeled. Values of Rm in gas planets are two to three orders of magnitude larger and cannot be reached (so far) in numerical dynamo models. Nonetheless, simulations may still be meaningful for Jupiter and Saturn when the results are properly scaled. To this end, scaling laws that are based on a large set of dynamo simulations which span the accessible range of control parameter space are very helpful. For example, to first approximation, the magnetic field strength is found to scale in these models with the cubic root of the power P (or energy flux) that is available for conversion into magnetic energy (Christensen and Aubert 2006; Aubert et al. 2009; Christensen et al. 2009)

$$B_{rms} \propto (\mu\rho)^{1/6} P^{1/3} \propto (\mu\rho)^{1/6} (Fq)^{1/3}. \tag{4.13}$$

For thermally driven convection, q is the heat flux at a representative level and F is a thermodynamic efficiency factor, which is of order one for gas planets (Christensen et al. 2009). Although there is some concern that the dynamo simulations may not be in the dynamical regime of planets (e.g. King and Buffett 2013), the field strength predicted by Equation 4.13 fits the observed strengths of Earth and Jupiter and, in addition, those of rapidly rotating stars, which suggests that it may be widely applicable. One complication in comparing the predictions of Equation 4.13 to observation is that the mean field strength inside the dynamo is predicted, whereas only the large-scale poloidal magnetic field at or above the planet's surface is observed. To link the two, an assumption must be made on the ratio of the field strength at the surface of the dynamo to that in the interior. Based on simulations, it can be estimated to lie in the range 2.5–5.5 in dynamos that generate a dominantly dipolar field. Furthermore, to relate the field strength at the planetary surface to that at the top of the dynamo, the outer radius of the dynamo must be known from structural models of the planetary interior. Figure 4.10 compares the predictions of Equation 4.13 with the observed fields of Earth, Jupiter and Saturn. While the fit is excellent for Jupiter and Earth, Saturn's observed field strength falls short of the prediction when we use $0.62R_S$ for the top of the dynamo region (labeled Saturn1 in Figure 4.10). This is a minimum estimate for the radius where hydrogen has metallized, and any shallower level would make the misfit worse.

The secular variation of the geomagnetic field is to a substantial degree caused by the advection of field by

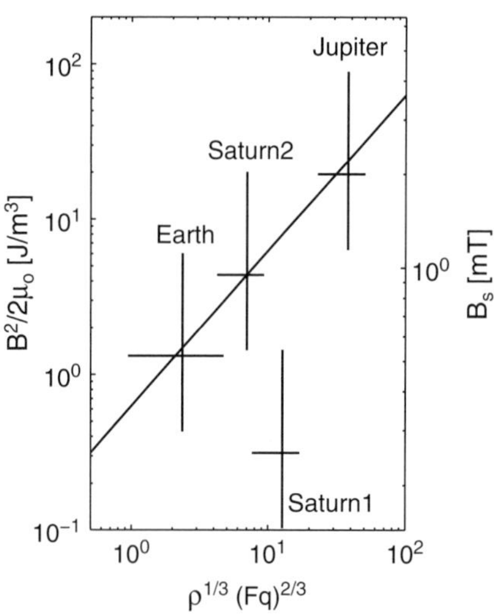

Figure 4.10 Comparison of the prediction of the scaling law Equation 4.13 with a constant of proportionality of 0.63 as deduced from dynamo simulations (solid line) with the observed field strength for various planets. The left axis refers to the magnetic energy density inside the dynamo and the right axis to the field strength at the surface of the dynamo. Error bars reflect uncertainties on the driving energy flux, the thermodynamic efficiency factor F and the relation between the observed large-scale field to the total field at the dynamo surface. Saturn1 is for the case where the top of the dynamo is located at 0.62 R_S and for Saturn2 it is at 0.4 R_S.

the flow near the core surface. The time constants of secular variation for a specific harmonic degree l can be defined as

$$\tau_l = \left(\frac{\sum_{m=0}^{l} \left(g_{lm}^2 + h_{lm}^2 \right)}{\sum_{m=0}^{l} \left(\dot{g}_{lm}^2 + \dot{h}_{lm}^2 \right)} \right)^{1/2}, \tag{4.14}$$

where the dot indicates the time derivative of the Gauss coefficients. The τ_l are linearly related to the characteristic advection time L/U, which is supported by numerical dynamo simulations (Christensen and Tilgner 2004). Comparing Saturn to Earth, both the length scale L and the velocity U are larger by an order of magnitude, hence their ratio is similar for both planets. Based on this, the expectation would be that the secular variation time scales of Saturn are similar to those of Earth. The finding of substantially weaker secular variation at Saturn therefore warrants a special explanation.

In addition to quantitative differences such as that in the magnetic Reynolds number, some qualitative differences exist between the dynamos in gas planets and the geodynamo. The density changes little across the Earth's fluid core, whereas the interior of gas planets is strongly density-stratified. Most of the density contrasts occur across the outer, poorly conducting, molecular hydrogen shell. Across the metallic region they are more moderate, but still cover a factor of 4–6, i.e. the dynamo region spans between one and two density scale heights. In terrestrial planets, the silicate mantle can be considered as a solid container to the fluid core. In the gas giants, the molecular hydrogen envelope is in vigorous motion, as testified by the strong zonal flow of alternating jets in east and west directions at the planetary surface. According to some models (Heimpel et al. 2005; Gastine and Wicht 2012), the jet flow is constant on cylinders coaxial with the rotation axis through the entire molecular hydrogen envelope (geostrophic flow, obeying the so-called Proudman–Taylor constraint), but its actual depth extent is an open question (see Chapter 11). There is no first-order phase change between the metallic and the molecular regions, and the electrical conductivity varies continuously, with significant values in a transition region (French et al. 2012). This dynamically couples the metallic and molecular hydrogen regions via Lorentz forces. Considerations based on maximum plausible values of ohmic dissipation suggest that the jets cannot penetrate to a depth where the electrical conductivity becomes non-negligible in the sense that Rm_ℓ reaches the order of one (Liu et al. 2008), i.e. a depth well above the point where the conductivity has reached metallic values. Strong ohmic dissipation could only be avoided if the flow and the magnetic field aligned each other in such a way that no shearing of field lines occurs (Glatzmaier 2008), for example if for geostrophic differential rotation the field were parallel to the rotation axis in regions with non-negligible conductivity. For Jupiter, such an arrangement seems unlikely, as non-zonal field components should be rapidly swept around by the zonal flow, but for Saturn, which lacks a non-zonal field, it cannot be excluded. Dynamo simulations with radially variable electrical conductivity (Heimpel and Gómez-Pérez 2011; Gastine et al. 2014; Jones 2014) showed that strong zonal circulation in the poorly conducting outer shell can coexist with slow convection in the dynamo region below.

In the past, many inferences on the dynamos in Jupiter and Saturn have been drawn based on Boussinesq models with constant density and conductivity. Recently, the consequences of deviating from these simplifications have been studied in numerical models. These simulations use a more complex set of equations in the so-called anelastic approximation, which allows for density stratification and variable conductivity (e.g. Glatzmaier 1984; Jones et al. 2011).

Gastine et al. (2012) found that, for constant conductivity, density stratification has an adverse influence on the generation of dipole-dominated magnetic fields. When the dynamo shell covers more than approximately 1.5 density scale heights, only multipolar solutions are found, which exhibit strong zonal circulation that is not found in dipole-dominated cases. Dipolar fields and strong zonal flow seem mutually exclusive. Dipolar magnetic fields in strongly density-stratified systems can be recovered when the electrical conductivity is variable and drops sharply in the outermost layer (Duarte et al. 2013). Strong zonal flow is here restricted to the poorly conducting outer shell and occurs mainly at equatorial latitudes. Jones (2014) simulated Jupiter's dynamo driven by secular cooling, using a detailed model of Jupiter's internal structure and conductivity distribution. He reports that

Jupiter-like (dipolar) fields are only found in a restricted region of parameter space and require a low value of the hydrodynamic Prandtl number *Pr*. In similar models Gastine et al. (2014) are able to match the first-order properties of Jupiter's magnetic field and predict that a secondary dynamo associated with the jet flow in the outer shell produces banded magnetic structures that would be detectable by NASA's Juno mission.

So far, dynamo models that took density and conductivity stratification into account have concentrated mainly on Jupiter – for example in the choice of the transition radius from low to high electrical conductivity. While some of the models match Jupiter's field decently, none of them exhibits characteristic traits of Saturn's field, such as the very strong axisymmetry or the relatively weak field strength. The strong differences between Jupiter's and Saturn's magnetic fields suggest that the dynamos in the two planets differ not only in terms of quantitative parameter values, but also on a more fundamental level. A successful dynamo model for Saturn must be able to explain, in addition to the dominance of the dipole, (i) the strong suppression of non-zonal field components, (ii) the peculiar spatial power spectrum and the concentration of flux in high latitudes, (iii) the lower-than-expected field strength compared to scaling law predictions and (iv) the weak secular variation.

4.4.3 Saturn Dynamo Models with Stably Stratified Layer

Stevenson (1980) suggested that both the unexpectedly low dipole moment and the high degree of axisymmetry would be explained if the dynamo operated below a stably stratified and electrically conducting layer created by helium precipitation. Because the dipole field drops with radius as r^{-3}, a deeper-seated dynamo will, for the same field strength on its outer boundary, create a weaker field at the planet's surface than a dynamo that extends out to the radius of hydrogen metallization. In addition, the heat flux driving a deep dynamo would probably be less than that in a larger dynamo and the thermodynamic efficiency factor F in Equation 4.13 would be smaller, although this might be compensated by the additional driving of convection by the influx of helium from above. The observed field strength at Saturn's surface can be brought in line with the prediction of the scaling law when the top of the dynamo region (the bottom of the inhomogeneous zone) is located at $0.4R_S$ (see Saturn2 in Figure 4.10).

A dynamo operating in the homogeneous region below a stable layer in Saturn can be assumed to generate a magnetic field with sizeable non-axisymmetric components. In the stably stratified layer, large-scale radial motion is suppressed, but horizontal flow is possible. Stevenson (1980) proposed that axisymmetric differential rotation occurs in this layer, driven as a thermal wind by an equator-to-pole temperature difference. Although this flow will involve horizontal shear, a very simple cartoon model (Figure 4.11) of a conducting hollow sphere in rigid-body rotation illustrates the mechanism for axisymmetrizing the observable field above the stable layer. Imagine that inside a sphere with outer radius R, a bar magnet is located with a fixed orientation. Its dipole axis is tilted against the sphere's rotation axis. As seen by an observer who rotates with the shell, the axial component of the dipole field is stationary, whereas the equatorial component oscillates in time with the rotation frequency. The equatorial field is therefore attenuated in the shell by the skin effect, whereas the axial field is unaffected. The degree of damping of the equatorial dipole field depends on the rotation rate Ω, the thickness D of the

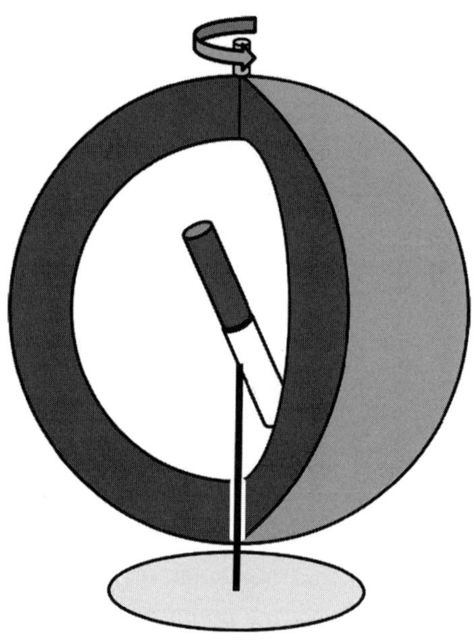

Figure 4.11 Cartoon model with conducting and rotating hollow sphere hosting a bar magnet whose dipole axis is tilted against the rotation axis.

shell and its magnetic diffusivity λ. Its amplitude decreases across the shell by the damping factor f_d (Stevenson 1982),

$$f_d \approx \exp(-[\alpha Rm_s/2]^{1/2}) \,, \tag{4.15}$$

where

$$Rm_s = UD/\lambda \tag{4.16}$$

is the stable shell magnetic Reynolds number with U = ΩR the velocity of differential rotation and

$$\alpha = mD/R \tag{4.17}$$

is a geometry factor. For the equatorial dipole, the harmonic order is $m = 1$. Strong damping occurs for $\alpha Rm_s \gg 1$. Applying the simple model to Saturn, assuming $U = 1$ cm/s (comparable to predicted convective velocities in the dynamo) and $\lambda = 2$ m^2/s, a stable layer of several hundred kilometers' thickness would be sufficient to cause a significant effect. For $D =$ 1000 km, $\alpha Rm_s \approx 170$ and the damping factor reaches 10^{-4}. As discussed in Section 4.4.1, the possible thickness of a stable layer in Saturn is poorly known and could be much larger. Putting its bottom at $0.4R_S$, which is suggested by the match achieved for Saturn's field strength when assuming this as the outer boundary of the dynamo (Figure 4.10), implies that $D \approx 15{,}000$ km, $\alpha Rm_s \approx 4 \times 10^4$ and $f_d \approx 10^{-60}$! Stevenson (1982) also studied more complex models with shear flow in the stable layer, but Equation 4.15 gives, in any case, a rough idea of the damping factor.

In the cartoon model in Figure 4.11, nothing changes when the sphere is fixed but the magnet rotates around the vertical axis. The rotation of the tilted magnet may serve as a crude analogue to the time-variation of the field generated by Saturn's dynamo below a stable layer. Because of the strong influence of Coriolis forces on the flow, the axial dipole and perhaps other axisymmetric field components have a special role in dipolar dynamos. In the geomagnetic field, all field components fluctuate on time scales that are roughly of the order of the advection time $R_c/U \approx 100$ yr, where R_c is the radius of Earth's core and $U \approx 1$ mm/s a typical velocity in it. On much longer time scales, non-axisymmetric field components have nearly a zero mean, whereas the axial dipole, even though it fluctuates in amplitude, keeps the same polarity on time scales of several hundred thousand years. In the case of the Earth, heterogeneous boundary conditions imposed by the solid mantle on the core probably lead to small non-zero mean values on long time scales for some non-zonal field components (e.g. Gubbins and Kelly 1993). However, for Saturn this appears very unlikely because, in its case, there is nothing that could impose a longitudinal preference on the dynamo. We can assume that the temporal change of the axial dipole (and other zonal field components) generated by Saturn's dynamo can be represented on some long time scale by a combination of a DC component and of AC components, whereas the equatorial dipole and other non-zonal terms in the field only have AC components. The DC part is unaffected by the stable layer, whereas the AC part is attenuated by the skin effect. This also works for a completely stagnant layer. Hence, differential rotation in the stable layer is not necessarily a prerequisite for the damping of non-zonal field components.

For a stagnant stable layer, the magnitude of the damping effect is more difficult to quantify. The characteristic frequency for secular variation of the non-axial dipole components is found in dynamo models and in the geomagnetic field to be related to the inverse advection time $\omega_o \approx U/R$, where U is now the mean convective velocity (Christensen and Tilgner 2004; Lhuillier et al. 2011). The damping factor for frequency ω_o in a stagnant shell is the same as obtained previously (Equation 4.15), but with the convective velocity instead of that of differential rotation. However, this overestimates the damping factor because fluctuations occur at various frequencies and those at low frequencies are less damped. Figure 4.12 shows the temporal spectra for fluctuations of the axial dipole and the equatorial dipole in a numerical dynamo simulation. For the equatorial dipole, the power spectral density is nearly flat for frequencies up to approximately ω_o (which is equal to one in this plot) and drops off strongly for higher frequencies. A rough estimate for the attenuation of the equatorial dipole component can be obtained by assuming that all the energy is contained in a white spectrum in the range $0 \leq \omega \leq \omega_o$. Replacing U by ωR in the frequency-dependent attenuation factor given by Equation 4.15 and folding

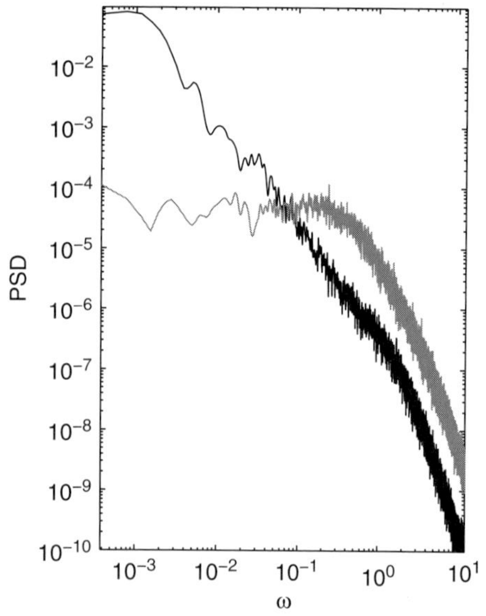

Figure 4.12 Temporal power spectral density of fluctuations of the axial dipole (black line) and the equatorial dipole (grey line) vs. angular frequency ω from a dynamo simulation, case c in Olson et al. (2012). Frequency has been scaled by the inverse advection time R/U_{rms} and both spectra have been normalized by the total power in the axial dipole.

the white spectrum up to ω_o with it results in the limit of $D^2\omega_o/\lambda \gg 1$, in a net damping factor

$$f_d \approx \left(\frac{\lambda}{2D^2\omega_o}\right)^{1/2} = (2aRm_s)^{-1/2}. \tag{4.18}$$

The damping effect is much more moderate than for the case of a differentially rotating stable layer. Nonetheless, for D = 10,000 km, R = 30,000 km, U = 1 cm/s and $\lambda = 2$ m^2/s, we obtain f_d = 1/173. Assuming that the average dipole tilt of the dynamo field below the stable layer is less than 10 degrees, it would be reduced to < 0.06 degree above the layer, in agreement with the upper limit from observation.

If Saturn's dynamo happened to undergo dipole reversals on a time scale similar to that of the geodynamo, a point of concern would be that a thick stable layer could also attenuate the axial dipole at this low frequency. For a duration of a polarity interval of 500,000 yr, a stable layer would start to show an effect if the thickness were larger than a few thousand kilometers – for D = 10,000 km, the damping reaches a factor of ten. However, it can be expected that a thick stable layer feeds back on the dynamo. It has been envisioned that the Earth's solid inner core stabilizes the dipole against too-frequent reversals, because a reversal can only be completed when a polarity fluctuation of the dipole in the fluid outer core persists for more than a magnetic diffusion time of the inner core (Hollerbach and Jones 1993; Dharmaraj and Stanley 2012). Because the magnetic diffusion time of a thick stable shell inside Saturn is much longer than that of the Earth's inner core, it might inhibit polarity reversals more efficiently and might prevent the field from reversing at all. However, for a thin stable shell, Stanley and Mohammadi (2008) observed an opposite effect in numerical simulations.

If U = 1 cm/s is a realistic estimate for the velocity in Saturn's dynamo, the characteristic time scale for the secular variation of the field R/U is of the order 100 yr. This is fortuitously similar to the secular variation time of the geodynamo, since both the velocity and the linear dimensions of the dynamo are larger by a factor of ten in Saturn. It is obvious that a stable conducting layer will prolong the time scales of secular variation that would be observable from the outside compared to those that are intrinsic to the dynamo, as the skin effect attenuates high frequencies more than low frequencies. Frequencies shorter than $2\lambda/D^2$ are effectively eliminated, which means that the secular variation time scale exceeds 10,000 yr for a stable layer thickness of only 1000 km. Hence, the concept of a stable layer above Saturn's dynamo is completely consistent with the small upper limit on field changes inferred from the comparison between the Pioneer/Voyager and the Cassini observations.

In the stable layer, large-scale horizontal flow that is not symmetric with respect to the planet's spin axis is possible in principle and would generate non-zonal field components from a zonal field. The importance of this flow component relative to the zonal flow is difficult to estimate on general principles. Furthermore, it is probably too simplistic to assume that the dynamo itself is unaffected by the stable layer and that this layer simply acts as a filter to the dynamo-generated field. It is quite probable that the existence of the stable conducting layer influences the working of the dynamo; a possible example for such feedback (stabilizing the axial dipole) has been discussed above. Here, some understanding can possibly be achieved through numerical simulations of dynamos below a stably stratified layer. While all planetary dynamo models suffer from the mismatch of control parameters such as the Ekman number and magnetic Prandtl number, the problem is aggravated in the case of the gas giant planets. In models for the Earth or other small solid planets, it is at least possible to match the magnetic Reynolds number as one of the key characteristic numbers of the dynamo, whereas, in Jupiter and Saturn models, Rm is lower than the planetary value by at least two orders of magnitude. Therefore, Saturn dynamo models must be considered even more so than geodynamo models as "tests of concept" instead of as representing the actual dynamo, and care must be exercised when giving a quantitative interpretation to their results.

So far, dynamo models with a stable layer that might be applicable for Saturn have only been conducted in the framework of the Boussinesq approximation. Christensen and Wicht (2008) studied dynamo models with density gradients that change gradually from unstable stratification at depth to increasingly stable stratification in the top part of a rotating spherical shell. Most of their models were conducted in a parameter regime (e.g. slow rotation relative to convection velocities) where the dynamo generates a multipolar field, and these cases have been applied to Mercury. Some of the models produced a dipole-dominated intrinsic dynamo field, and these have been interpreted in the context of Saturn. For one such case with a thick stable layer that extends from 0.4 R_S to 0.65 R_S, i.e. $\alpha = D/R \approx 0.4$, a snapshot of the radial magnetic field calculated for the surface of Saturn is shown in the top panel of Figure 4.13. The field is strongly dipolar and strongly axisymmetric. Averaged over the run time of the simulation, the tilt of the dipole axis is 1.5 degrees. For comparison, the bottom panel of Figure 4.13 shows the field of the same model truncated at the neutral stability radius at 0.4 RS, i.e. the stable layer has been replaced by a stagnant insulator. In this case, the field at Saturn's surface has significant non-zonal components and the time-average tilt of the dipole axis is 8.2 degrees. The value of αRm_S, based on the mean convective velocity, is approximately 40 in the model, and Equation 4.18 predicts a damping factor for the equatorial dipole of ≈ 9. Based on the zonal flow amplitude in the stable layer $\alpha Rm_s = 13$ and Equation 4.15 predicts a damping factor of 13. The factor 5.5 difference in dipole tilt angle between the models with and without stable layer is slightly less than these predictions. This could possibly be explained by the gradual change of the density gradient in the model, which is only weakly stabilizing in the bottom part of the stable layer. Its effective thickness may therefore be somewhat overestimated by the nominal value. At least qualitatively, the difference is consistent with the estimates for the damping factor obtained by the simple models.

Stanley and Mohammadi (2008) calculated dynamo models with a thin stable layer ($D/R = 0.1 - 0.2$), motivated by the early estimates by Stevenson (1980) of a stable layer thickness of a few thousand kilometers. In their models, the main effect of the stable

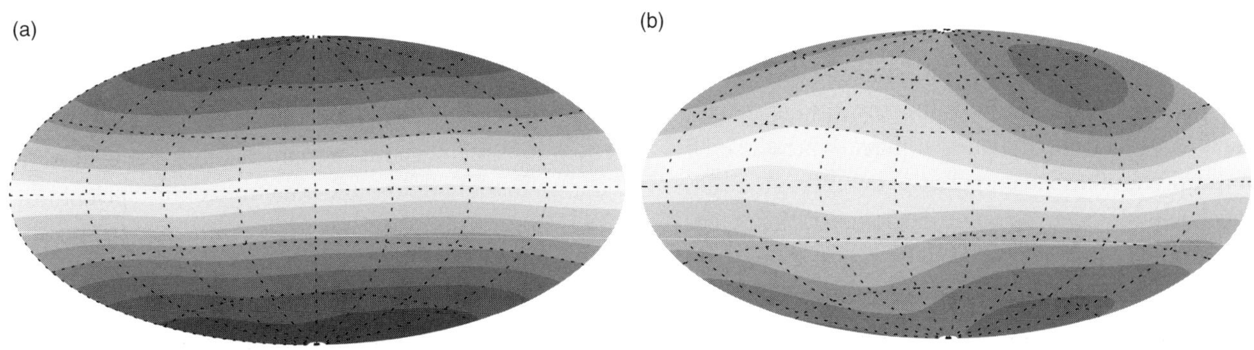

Figure 4.13 Radial magnetic field at Saturn's surface for dynamo models from Christensen and Wicht (2008); their cases 13 (left) and 13a (right). One contour step is 5000 nT when the non-dimensional model field is scaled by $\sqrt{\rho\mu\lambda\Omega}$ with $\rho = 1000$ kg/m^3, $\lambda = 2$ m^2/s and $\Omega = 1.6 \times 10^{-4}$ s^{-1}. In the model shown at the top, the dynamo operates below a thick stable layer. In the model at the bottom, the stable layer has been replaced by an insulating region. Figure 12 in Christensen and Wicht (2008) shows results from the same model calculations, but here they have been scaled differently to Saturn and different snapshots in time have been selected. (A black-and-white version of this figure appears in some formats. For the color version, please refer to the plate section.)

layer was to destabilize the strongly dipolar solution, which they found in the reference case without a stable layer, in favor of a multipolar field geometry. Only in one of their cases with $D/R = 0.1$ did the dynamo eventually settle into a dipolar state. A significant suppression of non-zonal field components above the stable layer was not found. The results underline that the role of a stable layer is not necessarily a passive one, but that it may control the dynamo mode. However, the lack of a strong axisymmetrizing effect in these models does not rule out that a thin stable layer inside Saturn could play this role. Because the magnetic Reynolds number is generally much smaller in the available dynamo models than it is in Saturn, this is also true for the parameter αRm_s at the same value of D/R. αRm_s is about 2–4 in the dynamo models by Stanley and Mohammadi (2008) with $D/R = 0.1$, which means that the skin effect is weak.

Stevenson (1980, 1982) had envisioned that the externally imposed equator-to-pole temperature difference due to the latitude-dependent insolation at the planet's surface drives a thermal wind circulation in the stable layer which axisymmetrizes the magnetic field. As the effective surface temperature (luminosity) of Jupiter and Saturn shows little variation with latitude (Hanel et al. 1983), the stronger heating by insolation at low latitudes must be compensated by a reduced internal heat flow. Models by Aurnou et al. (2008) suggest that convection in the molecular envelope of the gas planets may naturally lead to lower heat flow at the equator and higher heat flow at high latitude. Stanley (2010) studied the effect on the dynamo by imposing strong latitudinal heat flow variations on the outer boundary of her model with a thin stable layer. In models with a hot equator and cold poles (low equatorial heat flow), she finds that the dipolar dynamo mode is stabilized and the mean dipole tilt is reduced by a factor of about three compared to the reference cases without a stable layer, to values as low as 0.8 degree.

To sum up: Simple conceptual models suggest that a stably stratified layer above Saturn's dynamo can, through a skin effect, strongly damp non-zonal field components in the outside magnetic field. Differential rotation in the stable layer can enhance the damping effect, but is perhaps not a necessity when the stable layer is sufficiently thick. The skin effect also reduces the secular variation, in agreement with inferred rates that are substantially lower than those of the geomagnetic field. If the stable layer is thick, putting the top of the active dynamo region to about 0.4 R_S, this can also explain the relatively low strength of the surface magnetic field. However, the stable layer model does not offer an obvious explanation for the zig-zag pattern of the spatial power spectrum of Saturn and the related concentration of magnetic flux into polar regions at the top of Saturn's dynamo. Many open questions remain concerning the flow pattern and dynamics in the stable layer and the influence of this layer on the dynamo process. Available dynamo models do not provide an unambiguous answer, which is probably largely a consequence of our inability to run them at the appropriate planetary conditions, e.g. the correct value of the magnetic Reynolds number. A question that has not yet been explored is whether double-diffusive convection of the layered (staircase) type, as envisioned by Leconte and Chabrier (2013) for the stable layer, could support a dynamo in its own right. The magnetic Reynolds number based on the thickness of the individual layers may be small, but, in particular if the flow in the layers were organized in such a way that it had a coherent helicity across layers, then, collectively, the layers might still act as a dynamo.

4.4.4 Spherical-Couette Dynamo

The peculiar features of Saturn's intrinsic magnetic field – in particular, the extreme axisymmetry and poleward concentration of magnetic flux – have also motivated unconventional dynamo explanations. One interesting example is the spherical-Couette dynamo model presented by Cao et al. (2012). While it remains unclear how such a system can be physically realized inside Saturn, the flow structures and magnetic field generation mechanism within the spherical-Couette dynamo model could have counterparts in Saturn.

A spherical-Couette system is one where the flow is driven by differentially rotating inner and outer boundaries of a spherical shell. In a system with finite viscosity, four dimensionless parameters can be used to specify the problem: the dimensionless outer boundary rotation rate $\Omega = \Omega' L^2/\nu$, the dimensionless differential rotation $\Delta\Omega = \Delta\Omega' L^2/\nu$, the magnetic Prandtl number $Pm = \nu/\lambda$ and the radius ratio of the inner boundary to the outer boundary, r_i/r_o. Here Ω' and $\Delta\Omega'$ are the

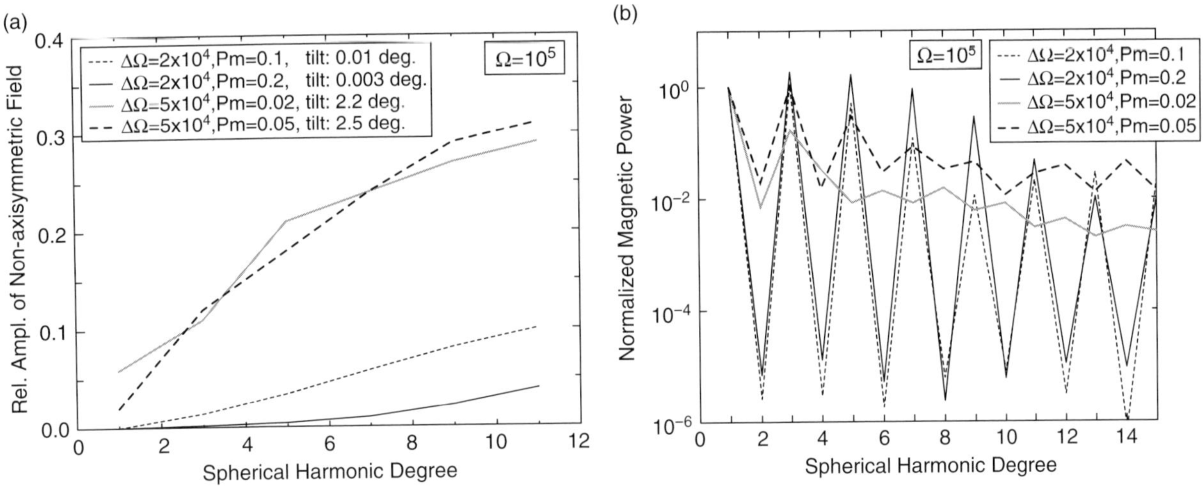

Figure 4.14 Results for spherical Couette dynamo model in Cao et al. (2012) with a superrotating inner boundary. (a) Amplitude of the non-axisymmetric components relative to the zonal component versus harmonic degree. (b) Normalized spatial power spectrum at the dynamo surface for spherical-Couette dynamo with superrotating inner boundary. Adapted from Cao et al. (2012).

dimensional rotation and differential rotation rates, $L = r_o - r_i$ is the shell thickness, ν is the kinematic viscosity and λ is the magnetic diffusivity. In the numerical spherical-Couette dynamo experiments presented in Cao et al. (2012), several Saturn-like solutions have been obtained. As shown in Figure 4.14, the resulting magnetic field is highly axisymmetric and the Lowes–Mauersberger power spectra of the spherical-Couette dynamo have a zig-zag pattern with enhanced power in the odd degrees. In addition, the magnetic field in the spherical-Couette dynamo models are indeed highly concentrated in the polar regions at the dynamo surface (Cao et al. 2012).

The extreme axisymmetry in the numerical spherical-Couette dynamos is intriguing and has a different origin from the electromagnetic shielding mechanism discussed in Section 4.4.3. It can be seen from Figure 4.14 that the relative amplitude of non-axisymmetry in the spherical-Couette dynamo model increases with spherical harmonic degree. The average dipole tilt in the spherical-Couette dynamo models can be as small as 0.003 degree. The extreme axisymmetry is thus an intrinsic property of this dynamo model. The flow in the spherical Couette dynamo shares one property with the flow in the stable layer envisioned for the electromagnetic shielding mechanism proposed by Stevenson (1980): the axisymmetric zonal flow dominates over the non-axisymmetric poloidal flows. In the spherical-Couette dynamo, the energy in the axisymmetric zonal flow is two orders of magnitude larger than that in the non-axisymmetric poloidal flow. The energy ratio between the axisymmetric toroidal magnetic field and the total poloidal magnetic field is also close to 100. In the convection-driven dipole-dominated dynamos, both ratios are of order unity or less. This raises the possibility that Saturn's very axisymmetric and relatively weak magnetic field could result from strong zonal flows within the dynamo region rather than from an electromagnetic shielding mechanism. In the spherical-Couette dynamo model, the strong zonal flow is due to boundary forcing, which may or may not be a good analogue to the mechanisms that could drive such flow in Saturn's deep interior. Strong zonal flow and an axial-dipole-dominated magnetic field are found to be mutually exclusive in present convection-driven dynamo simulations, but it remains to be explored if this also holds in a more realistic parameter regime.

The poleward magnetic flux concentration in the numerical spherical-Couette dynamo has a simple and straightforward origin: the non-axisymmetric radial flow and the helicity production are stronger inside the tangent cylinder, compared to those outside the tangent cylinder (the tangent cylinder is an imaginary spin-axis-aligned cylinder that encloses the inner core). Given the observational evidence that the intrinsic heat flow at Saturn is higher at the poles than at the equator, convective motion inside Saturn is probably stronger at high latitudes than at regions near the equator. If such a flow state is also maintained in Saturn's dynamo region, it may be natural to expect poleward magnetic flux concentration for Saturn.

4.5 Expections for the Cassini Proximal Orbits

The currently planned Cassini proximal orbits will take place between April 2017 and September 2017. During this time Cassini will pass through a broad range of latitudes close to Saturn. Periapse distances are as close as 1.03 R_S before the final descent into Saturn's atmosphere. Figure 4.15 shows the proximal orbit trajectories in red, compared with the orbits to date in black. Other than Saturn orbit insertion (SOI) with a periapse distance of 1.33 R_S, relatively few orbits have periapse distances within 2.5 R_S during the rest of the entire mission.

The expected field magnitudes along three typical proximal orbits are shown in Figure 4.16. The maximum is of order 20,000 nT, compared to about 1500 nT during the previous non-SOI orbits, well within the measurement capability of the instrument whose highest range reaches 44,000 nT (Dougherty et al. 2004).

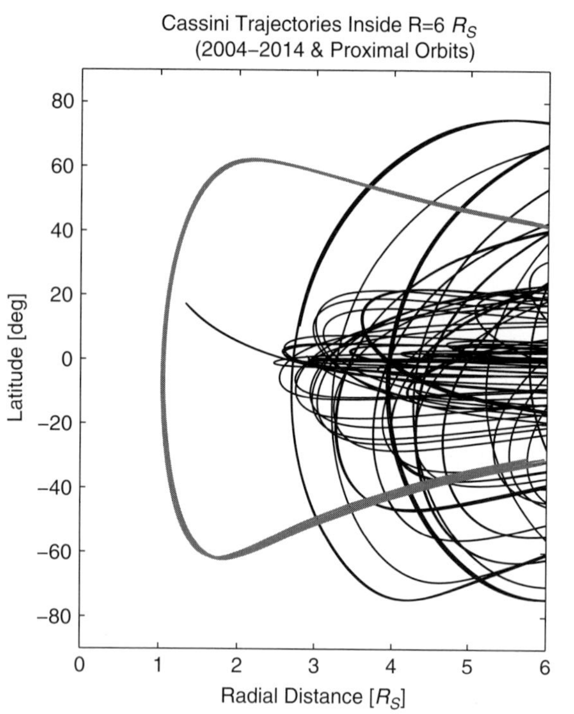

Figure 4.15 Cassini orbit trajectories from SOI to date (black traces) compared to those of the proximal orbits (grey). (A black-and-white version of this figure appears in some formats. For the color version, please refer to the plate section.)

The radial dependence of the intrinsic magnetic moments is $(1/r)^{(l+2)}$, where r is the radial distance to the center of the planet and l is the spherical harmonic degree. Thus, with measurements close to the planet we will be able to far better resolve higher-degree zonal and possible non-zonal components. The contribution of these components becomes much larger at smaller distance, both in absolute terms and also relative to the dominant dipole field. In addition, the highly inclined nature of the proximal orbits provides coverage of a wide range of latitudes close to the planet, which is another advantage when modelling the intrinsic field. Since Saturn's true rotation period is uncertain, the actual planetary longitude associated with each measurement is uncertain as well. However, for the planned proximal orbits, good longitudinal coverage is obtained for any plausible rotation period between 10h 30m and 10h 50m.

We expect the analysis of the MAG proximal orbit data to result in (1) determination of the high-degree magnetic moments of Saturn; (2) an estimate for the depth of the dynamo surface from the power spectrum; (3) the possible detection of non-zonal field components; and (4), in case of success with the previous point the determination of Saturn's rotation period.

Figure 4.17 shows two different estimates of Saturn's magnetic power spectrum (in blue and green). The estimates are based on the assumption

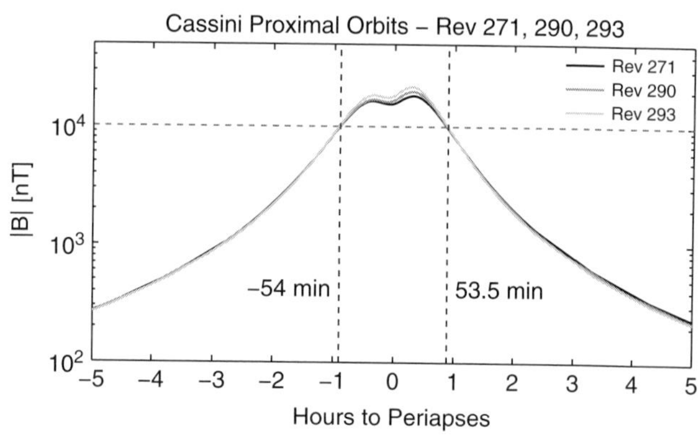

Figure 4.16 The expected field strength along three typical Cassini proximal orbits. The magnetometer will stay in the highest field range (10,000 nT to 44,000 nT) for about 2 hours centered near the time of the closest approach. (A black-and-white version of this figure appears in some formats. For the color version, please refer to the plate section.)

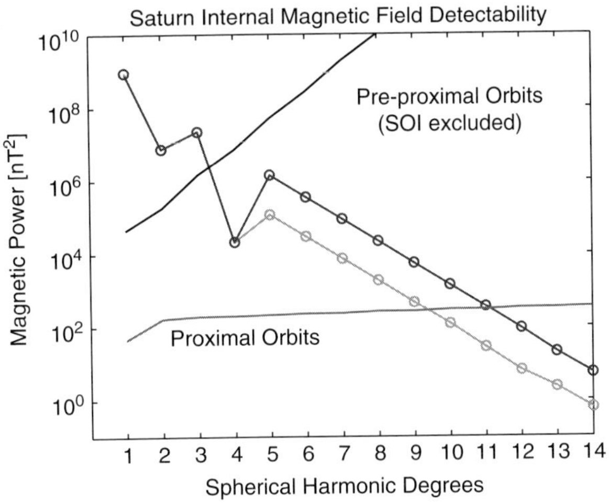

Figure 4.17 Magnetic power spectrum at the surface of Saturn extrapolated at degrees > 4 according to two different assumptions (blue and green lines). The red line is the detection limit for the proximal orbit phase and the black line is the limit for the pre-proximal orbits excluding SOI. (A black-and-white version of this figure appears in some formats. For the color version, please refer to the plate section.)

that the power spectrum becomes flat when taken at the top of the dynamo region, which is taken here to be at 0.5 R_S. The higher and lower estimates assume that the power associated with the high degree moments is equal to the power of the octupole or of the quadrupole, respectively, at 0.5 R_S. The black curve shows the limit of "detectability" for the pre-proximal orbits, whilst the red one is that for the proximal orbits. We expect magnetic moments that lie above each detectability curve to be measureable on the orbits which the curves represent. Based on this analysis, we hope to be able to resolve the field up to degree $l = 9$ or possibly even up to $l = 11$. It is clear that our knowledge of Saturn's high-degree magnetic moments will be significantly improved from MAG measurements during the Cassini proximal orbits.

The Cassini MAG measurements during the proximal orbit period will allow us the detection of a dipole tilt as small as 0.015 degree. Moreover, higher-degree non-axisymmetric magnetic field components can be better constrained. For example, if we set 10 nT as the detection threshold (which is mainly controlled by the digitization step and calibration limit in the strong-field range), the data will allow the determination of a non-zonal $|g_2^m, h_2^m|$ as small as 17 nT and $|g_3^m, h_3^m|$ as small as 20 nT. This is a dramatic improvement compared to the detection limits of the previous non-SOI orbits, which are approximately 80 nT for the degree-2 terms and 120 nT for degree-3 terms. During the one previous close approach at SOI, only a very limited range of longitudes was covered, which did not allow separation of possible non-zonal contributions from the zonal parts.

MAG data from Cassini's proximal orbits will provide the most accurate reference model of Saturn's intrinsic magnetic field for the coming decades. This will constitute an important point of comparison with other planetary magnetic fields and with dynamo simulations. At about the same time, the JUNO orbital mission at Jupiter will provide the magnetic field data for a much-improved magnetic field model of that planet. Comparing the detailed properties of the intrinsic magnetic fields of these two giant planets will assist our understanding of their internal structures and the dynamo processes that operate in their deep interior.

References

Achilleos, N., P. Guio and C. S. Arridge (2010), A model of force balance in Saturn's magnetodisc. *Mon. Not. R. astr. Soc.*, **401**, 2349–2371.

Anderson, J. D. and G. Schubert (2007), Saturn's gravitational field, internal rotation, and interior structure. *Science*, **317**, 1384–1387.

Andrews, D. J., E. J. Bunce, S. W. H. Cowley et al. (2008), Planetary period oscillations in Saturns magnetosphere: Phase relation of equatorial magnetic field oscillations and Saturn kilometric radiation modulation. *J. Geophys. Res.*, **113**, A09205.

Andrews, D. J., A. J. Coates, S. W. H. Cowley et al. (2010), Magnetospheric period oscillations at Saturn: Comparison of equatorial and high-latitude magnetic field periods with north and south Saturn kilometric radiation periods. *J. Geophys. Res.*, **115**, A12252.

Arridge, C. S., N. André, K. K. Khurana et al. (2011), Periodic motion of Saturns night-side plasma sheet. *J. Geophys. Res.*, **116**, A11205.

Aubert, J., S. Labrosse and C. Poitou (2009), Modelling the paleo-evolution of the geodynamo. *Geophys. J. Int.*, **179**, 1414–1429.

Aurnou, J. M., M. Heimpel, L. Allen, E. King and J. Wicht (2008), Convective heat transfer and the pattern of thermal emission on the gas planets. *Geophys. J. Int.*, **193**, 793–801.

Bunce, E. J., S. W. H. Cowley, I. I. Alexeev et al. (2007), Cassini observations of the variation of Saturn's ring current parameters with system size. *J. Geophys. Res.*, **112**, A10202.

Burton, M. E., M. K. Dougherty and C. T. Russell (2009), Model of Saturn's internal planetary magnetic field based on Cassini observations. *Planet. Space Sci.*, **57**, 1706–1713.

Burton, M. E., M. K. Dougherty and C. T. Russell (2010), Saturn's internal planetary magnetic field. *Geophys. Res. Lett.*, **37**, 24105.

Cao, H., C. T. Russell, U. R. Christensen, M. K. Dougherty and M. E. Burton (2011), Saturn's very axisymmetric magnetic field: no secular variation or tilt. *Earth Planet. Sci. Lett.*, **304**, 22–28.

Cao, H., C. T. Russell, J. Wicht, U. R. Christensen and M. K. Dougherty (2012), Saturn's high degree magnetic moments: Evidence for a unique planetary dynamo. *Icarus*, **221**, 388–394.

Carbary, J. F., N. Achilleos and C. S. Arridge (2012), Statistical ring current of Saturn. *J. Geophys. Res.*, **117**, 6223.

Christensen, U. R. (2010), Dynamo scaling laws: applications to the planets. *Space Sci. Rev.*, **152**, 565–590.

Christensen, U. R. (2011), Geodynamo models: tools for understanding properties of Earth's magnetic field. *Phys. Earth Planet. Inter.*, **187**, 157–169.

Christensen, U. R. and J. Aubert (2006), Scaling properties of convection-driven dynamos in rotating spherical shells and applications to planetary magnetic fields. *Geophys. J. Int.*, **166**, 97–114.

Christensen, U. R., V. Holzwarth and A. Reiners (2009), Energy flux determines magnetic field strength of planets and stars. *Nature*, **457**, 167–169.

Christensen, U. R. and A. Tilgner (2004), Power requirement of the geodynamo from ohmic losses in numerical and laboratory dynamos. *Nature*, **429**, 169–171.

Christensen, U. R. and J. Wicht (2008), Models of magnetic field generation in partly stable planetary cores: Applications to Mercury and Saturn. *Icarus*, **196**, 16–34.

Christensen, U. R. and J. Wicht (2015), Numerical dynamo simulations. Pages 245–277 of: G. Schubert (ed.), *Treatise on Geophysics, 2nd ed., Volume 8: Core Dynamics*. Amsterdam: Elsevier.

Connerney, J. E. P., M. H. Acûna and N. F. Ness (1981), Modeling the Jovian current sheet and inner magnetosphere. *J. Geophys. Res.*, **86**, 8370–8384.

Connerney, J. E. P., M. H. Acûna and N. F. Ness (1983), Currents in Saturn's magnetosphere. *J. Geophys. Res.*, **88**, 8779–8789.

Connerney, J. E. P., M. H. Acûna and N. F. Ness (1984), The Z3 model of Saturn's magnetic field and the Pioneer 11 vector helium magnetometer observations. *J. Geophys. Res.*, **89**, 7541–7544.

Connerney, J. E. P., N. F. Ness and M. H. Acûna (1982), Zonal harmonic models of Saturn's magnetic field from Voyager 1 and 2 observations. *Nature*, **298**, 44–46.

Conrath, B. and D. Gautier (2000), Saturn helium abundance: a reanalysis of Voyager measurements. *Icarus*, **144**, 124–134.

Davidson, P. A. (2013), Scaling laws for planetary dynamos. *Geophys. J. Int.*, **195**, 67–74.

Davis, L. and E. J. Smith (1985), Comments on "The Z3 model of Saturn's magnetic field and the Pioneer 11 vector helium magnetometer observations" by J. E. P. Connerney, M. H. Acuña and N. F. Ness. *J. Geophys. Res.*, **90**, 4461–4464.

Davis, L. and E. J. Smith (1986), New models of Saturn's magnetic field using Pioneer 11 Vector Helium Magnetometer data. *J.Geophys. Res.*, **91**, 1373–1380.

Davis Jr., L. and E. J. Smith (1990), A model of Saturn's magnetic field based on all available data. *J. Geophys. Res.*, **95**, 15257–15261.

Dharmaraj, G. and S. Stanley (2012), Effect of inner core conductivity on planetary dynamo models. *Phys. Earth Planet.Inter.*, **212–213**, 1–9.

Dougherty, M. K., N. Achilleos, N. André et al. (2005), Cassini magnetometer observations during Saturn orbit insertion. *Science*, **307**, 1266–1270.

Dougherty, M. K., S. Kellock, D. J. Southwood et al. (2004), The Cassini magnetic field investigation. *Space Sci. Rev.*, **114**, 331–383.

Dougherty, M. K., K. K. Khurana, F. M. Neubauer et al. (2006), Identification of a dynamic atmosphere at Enceladus with the Cassini magnetometer. *Science*, **311**, 1406–1409.

Duarte, L., T. Gastine and J. Wicht (2013), Anelastic dynamo models with variable electrical conductivity: an application to gas giants. *Phys. Earth Planet. Inter.*, **222**, 22–34.

Elphic, R. C. and C. T. Russell (1978), On the apparent source depth of planetary magnetic fields. *Geophys. Res. Lett.*, **5**, 211–214.

Fischer, G., D. A. Gurnett, W. S. Kurth, S.-Y., Ye and J. B. Groene (2015), Saturn kilometric radiation periodicity after equinox. *Icarus*, **254**, 72–91.

Fortney, J. J. and W. B. Hubbard (2003), Phase separation in giant planets: inhomogeneous evolution of Saturn. *Icarus*, **164**, 228–243.

French, M., A. Becker, W. Lorenzen et al. (2012), Ab initio simulations for material properties along the Jupiter adiabat. *Astrophys. J. Suppl.*, **202**, 5.

Gastine, T., L. Duarte and J. Wicht (2012), Dipolar versus multipolar dynamos: the influence of the background density stratification. *Astronom. Astrophys.*, **546**, A19.

Gastine, T. and J. Wicht (2012), Effects of compressibility on driving zonal flow in gas giants. *Icarus*, **219**, 428–442.

Gastine, T., J. Wicht, L. D. V. Duarte, M. Heimpel and A. Becker (2014), Explaining Jupiter's magnetic field and equatorial jet dynamics. *Geophys. Res. Lett.*, **41**, 5410–5419.

Giampieri, G., M. K. Dougherty, E. J. Smith and C. T. Russell (2006), A regular period for Saturn's magnetic field that may track its internal rotation. *Nature*, **441**, 62–64.

Glatzmaier, G. A. (1984), Numerical Simulation of Stellar Convective Dynamos. 1. The Model and Methods. *J. Comput. Phys.*, **55**, 461–484.

Glatzmaier, G. A. (2008), A note on constraints on deep-seated zonal winds inside Jupiter and Saturn. *Icarus*, **196**, 665–666.

Goldstein, J., T. W. Hill, H. Waite and J. L. Burch (2014), Analytical model of rotating two-cell convection at Saturn. *J. Geophys. Res.*, **119**, 1980–1993.

Gubbins, D. and P. Kelly (1993), Persistent patterns in the geomagnetic field over the past 2.5 Ma. *Nature*, **365**, 829–832.

Gurnett, D. A., A. M. Persoon, W. S. Kurth et al. (2007), The variable rotation period of the inner region of Saturn's plasma disk. *Science*, **316**, 442–445.

Hanel, R. A., B. J. Conrath, V. G. Kunde, J. C. Pearl and J. A. Pirraglia (1983), Albedo, internal heat flux, and energy balance of Saturn. *Icarus*, **53**, 262–285.

Heimpel, M. and N. Gómez-Pérez (2011), On the relationship between zonal jets and dynamo action in giant planets. *Geophys. Res. Lett.*, **38**, L1401.

Heimpel, M. H., J. M. Aurnou and J. Wicht (2005), Simulation of equatorial and high-latitude jets on Jupiter in a deep convection model. *Nature*, **438**(10), 193–196.

Holland, P. W. and R. E. Welsch (1977), Robust regression using iteratively reweighted least squares. *Commun. Stat. Theor. M.*, **6**, 813–827.

Hollerbach, R. and C. A. Jones (1993), Influence of the Earth's inner core on the geomagnetic fluctuations and reversals. *Nature*, **365**, 541–543.

Holme, R. (2007), Large-scale flow in the core. Pages 107–130 of: G. Schubert (ed.), *Treatise on Geophysics, Volume 8: Core Dynamics*. Amsterdam: Elsevier.

Ingersoll, A. P., G. S. Orton, G. Münch, G. Neugebauer and S. C. Chase (1980), Pioneer Saturn infrared radiometer experiment: preliminary results. *Science*, **207**, 439–443.

Jia, X. and M. G. Kivelson (2012), Driving Saturns magnetospheric periodicities from the upper atmosphere/ionosphere magnetotail response to dual sources. *J. Geophys. Res.*, **117**, A11219.

Jia, X., M. G. Kivelson and T. I. Gombosi (2012), Driving Saturns magnetospheric periodicities from the upper atmosphere/ionosphere. *J. Geophys. Res.*, **117**, A05215.

Jones, C. A. (2011), Planetary magnetic fields and fluid dynamos. *Ann. Rev. Fluid Mech.*, **43**, 583–614.

Jones, C. A. (2014), A dynamo model for Jupiter's magnetic field. *Icarus*, **241**, 148–159.

Jones, C. A., P. Boronski, A. S. Bruns et al. (2011), Anelastic convection-driven dynamo benchmarks. *Icarus*, **216**, 120–135.

Khurana, K. K., D. G. Mitchell, C. S. Arridge et al. (2009), Sources of rotational signals in Saturn's magnetosphere. *J. Geophys. Res.*, **114**, A02211.

King, E. M. and B. A. Buffett (2013), Flow speeds and length scales in geodynamo models: the role of viscosity. *Earth Planet. Sci. Lett.*, **371–372**, 156–162.

Kippenhahn, R. and A. Weigert (1990), *Stellar structure and evolution*. Berlin: Springer.

Leconte, J. and G. Chabrier (2013), Layered convection as the origin of Saturn's luminosity anomaly. *Nature Geosci.*, **6**, 347–350.

Lhuillier, F., A. Fournier, G. Hulot and J. Aubert (2011), The geomagnetic secular variation timescale in observations and numerical dynamo models. *Geophys. Res. Lett.*, **38**, L09306.

Liu, J., P. M. Goldreich and D. J. Stevenson (2008), Constraints on deep-seated zonal winds inside Jupiter and Saturn. *Icarus*, **196**, 653–664.

Lorenzen, W., B. Holst and R. Redmer (2011), Metallization in hydrogen-helium mixtures. *Phys. Rev. B*, **84**, 23519.

Lowes, F. J. (1974), Spatial power spectrum of the main geomagnetic field, and extrapolation to the core. *Geophys. J. Int.*, **36**, 717–730.

Morales, M. A., E. Schwelger, D. Ceperley et al. (2009), Phase separation in hydrogen-helium mixtures at Mbar pressures.*Proc. Natl. Acad. Sci.*, **106**, 1324–1329.

Nellis, W. J., S. T. Weir and A. C. Mitchell (1999), Minimum metallic conductivity of fluid hydrogen at 140 MPa (1.4 Mbar). *Phys. Rev. B*, **59**, 3434–3449.

Nettelmann, N., R. Püstow and R. Redmer (2013), Saturn layered structure and homogeneous evolution models with different EOSs. *Icarus*, **225**, 548–557.

Olson, P. and J. Aurnou (1999), A polar vortex in the Earth's core. *Nature*, **402**, 170–173.

Olson, P., U. R. Christensen and P. E. Driscoll (2012), From superchrons to secular variation: a broadband dynamo frequency spectrum for the geomagnetic dipole. *Earth Planet. Sci. Lett.*, **319–320**, 75–82.

Provan, G., S. W. H. Cowley and J. D. Nichols (2009), Phase relation of oscillations near the planetary period of Saturn's auroral oval and the equatorial magnetospheric magnetic field. *J. Geophys. Res.*, **114**, A04205.

Pryor, W. R., A. M. Rymer, D. G. Mitchell et al. (2011), The auroral footprint of Enceladus on Saturn. *Nature*, **472**, 331–333.

Read, P. L., T. E. Dowling and G. Schubert (2009), Saturn's rotation period from its atmospheric planetary-wave configuration. *Nature*, **460**, 608–610.

Roberts, P. H. and D. Gubbins (1987), Origin of the main field: Kinematics. Pages 185–249 of: J. A. Jacobs (ed.), *Geomagnetism, Vol. 2*. London: Academic Press.

Schubert, G. and K. M. Soderlund (2011), Planetary magnetic fields: observations and models. *Phys. Earth Planet. Inter.*, **187**, 92–108.

Smith, C. G. A. and N. Achilleos (2012), Axial symmetry breaking of Saturn's thermosphere. *Mon. Not. R. astr. Soc.*, **422**, 1460–1488.

Smith, E. J., L. Davis, D. E. Jones et al. (1980), Saturn's magnetic field and magnetosphere. *Science*, **207**, 407–410.

Southwood, D. J. and M. G. Kivelson (2007), Saturnian magnetospheric dynamics: Elucidation of a camshaft model. *J. Geophys. Res.*, **112**, A12222.

Stanley, S. (2010), A dynamo model for axisymmetrizing Saturn's magnetic field. *Geophys. Res. Lett.*, **37**, L05201.

Stanley, S. and A. Mohammadi (2008), Effects of an outer thin stably stratified layer on planetary dynamos. *Phys. Earth Planet. Inter.*, **168**, 179–190.

Starchenko, S. V. and C. A. Jones (2002), Typical velocities and magnetic field strengths in planetary interiors. *Icarus*, **157**, 426–435.

Stelzer, Z. and A. Jackson (2013), Extracting scaling laws from numerical dynamo models. *Geophys. J. Int.*, **193**, 1265–1276.

Sterenborg, M. G. and J. Bloxham (2010), Can Cassini magnetic field measurements be used to find the rotation period of Saturn's interior? *Geophys. Res. Lett.*, **37**, L11201.

Stevenson, D. J. (1980), Saturn's luminosity and magnetism. *Science*, **208**, 746–748.

Stevenson, D. J. (1982), Reducing the non-axisymmetry of a planetary dynamo and an application to Saturn. *Geophys. Astrophys. Fluid Dyn.*, **21**, 113–127.

Stevenson, D. J. (2003), Planetary magnetic fields. *Earth Planet. Sci. Lett.*, **208**, 1–11.

Stevenson, D. J. and E. E. Salpeter (1977), The dynamics and helium distribution in helium-hydrogen fluid planets. *Astrophys. J. Suppl.*, **35**, 239–261.

Talboys, D. L., E. J. Bunce, S. W. H. Cowley et al. (2011), Statistical characteristics of field-aligned currents in Saturn's nightside magnetosphere. *J. Geophys. Res.*, **116**, A04213.

von Zahn, U., D. M. Hunten and G. Lehmacher (1998), Helium in Jupiter's atmosphere: results from the Galileo probe helium interferometer experiment. *J. Geophys. Res.*, **103**, 22,815–22,829.

Wicht, J. and A. Tilgner (2010), Theory and modeling of planetary dynamos. *Space Sci. Rev.*, **152**, 501–542.

Wilson, H. F. and B. Militzer (2010), Sequestration of noble gases in giant planet interiors. *Phys. Rev. Lett.*, **104**, 121101.

5

The Mysterious Periodicities of Saturn

Clues to the Rotation Rate of the Planet

JAMES F. CARBARY, MATTHEW M. HEDMAN, THOMAS W. HILL, XIANZHE JIA, WILLIAM KURTH, LAURENT LAMY AND GABRIELLE PROVAN

Abstract

The rotation rate of a planet is a fundamental parameter, no less than its mass or composition, and planetary investigators require this rate to assess various other phenomena such as planetary wind speeds, internal and atmospheric models, ring dynamics and so forth. Saturn presents a conundrum, however, because none of its various planetary periods indicates the "true" rotation of the planet. Thus, although the planet displays an abundance of periodicities near 10.7 hours, the exact rotation period of Saturn is unknown. In the magnetosphere, "planetary-period oscillations" (PPOs) appear in charged particles, magnetic fields, energetic neutral atoms, radio emissions and motions of the plasma sheet and magnetopause. In Saturn's rings, the spoke phenomenon can exhibit periodicities near 10.7 hours, and ring phenomena themselves may be related to the interior rotation of the planet. In the high-latitude ionosphere, modulations near this period appear in auroral motions and intensities. In the upper atmosphere, some cloud features rotate near this period, although wind speeds are generally faster, and the well-known polar hexagon rotates with a period close to 10.7 hours. Some of the magnetospheric/ionospheric oscillations differ in the northern and southern hemispheres and their periods do not remain constant, sometimes varying on long time scales of a year or longer and sometimes on much shorter time scales. These variations in the period argue against a cause related to changes interior to Saturn, and because the magnetic and spin axes of Saturn are reported to be axisymmetric (unlike those of any other known planet), Saturn's periodicities cannot be explained as "wobble" caused by a geometric tilt or by a nondipolar magnetic anomaly. Several models have been proposed to account for the observed periodicities, including rotating atmospheric vortices, periodic plasma releases and a flapping magnetodisk, but none can satisfactorily explain all of Saturn's periodicities nor their common origin, and none can determine the exact rotation rate of the planet. This chapter reviews Saturn's periodicities, theories thereof, and how they might be used to determine the elusive rotation rate of the planet.

5.1 Introduction

The planet Saturn exhibits many periodicities near 10.7 hours. These so-called "planetary period oscillations," or PPOs, appear in the magnetosphere, the ionosphere and aurora, the upper atmosphere, and even in the rings. In many cases, these periodicities show differences of a few minutes between manifestations in the north and south. Also, the north and south periods vary slightly over timescales of weeks to years. The slow variation strongly implies that the periodicities do not indicate the true rotation rate of the planet. The origins of the planetary oscillations are uncertain because, unlike the Earth or Jupiter, the magnetic and spin axes of Saturn are coincident and Saturn's magnetic field is longitudinally uniform. Thus, the geometric tilt of the magnetic axis or a nondipolar magnetic anomaly cannot explain the periodicities. Rotating atmospheric vortices, periodic plasma releases in the magnetotail and a flapping magnetodisk have been proposed, but none of these can adequately explain all of Saturn's periodicities nor their supposedly common origin. This chapter surveys the observations of these periodicities made by the Cassini spacecraft and presents several possible explanations of their origins. In addition, Saturn displays "short" periodicities near ~1 hour and "long" periods near ~26 days, and these will also be mentioned herein.

Saturn displays a multitude of periodicities, but they are not consistent. Figure 5.1 summarizes where Saturn's periodic phenomena have been observed. Because periodicities in each region are usually examined independently, this separation will be maintained throughout our discussion. But these regions do overlap. For example, an ionospheric periodicity may generate periodic phenomena in the magnetosphere, or vice-versa. Indeed, all the periodicities should overlap if they are attributable to a common cause.

Saturn also has at least two periods different from 10.7 hours, one related to recurring solar wind structures (~26 days) and another that may be related to magnetospheric Alfvén-wave travel times (~1 hour). These periods are briefly discussed for completeness as an example of periodicities that are fairly well understood. The non-rotational periodicities, though, may be modulated by the rotational periodicities or the rotational periodicities may be modulated by the non-rotational periods, and this idea has yet to be fully explored.

5.2 Radio Periodicities

Of all the periodicities at Saturn, the magnetospheric periodicities have been the most thoroughly observed and investigated. They are observed by a multitude of instruments including radio receivers, magnetometers, particle detectors and even imagers. Figure 5.2 shows typical examples of the time profiles of magnetospheric observations from 11 days in 2005. The top panel exhibits a radio spectrogram in frequency, sampling essentially the Saturn kilometric radiation (SKR band, ~50–400 kHz), the second panel shows a spectrogram of energetic electrons, the third panel shows a spectrogram for thermal electrons and the bottom panel shows magnetic field oscillations from which the main (dipole) field has been subtracted. Each panel shows oscillations at close to the expected planetary period. Such oscillations are often referred to as

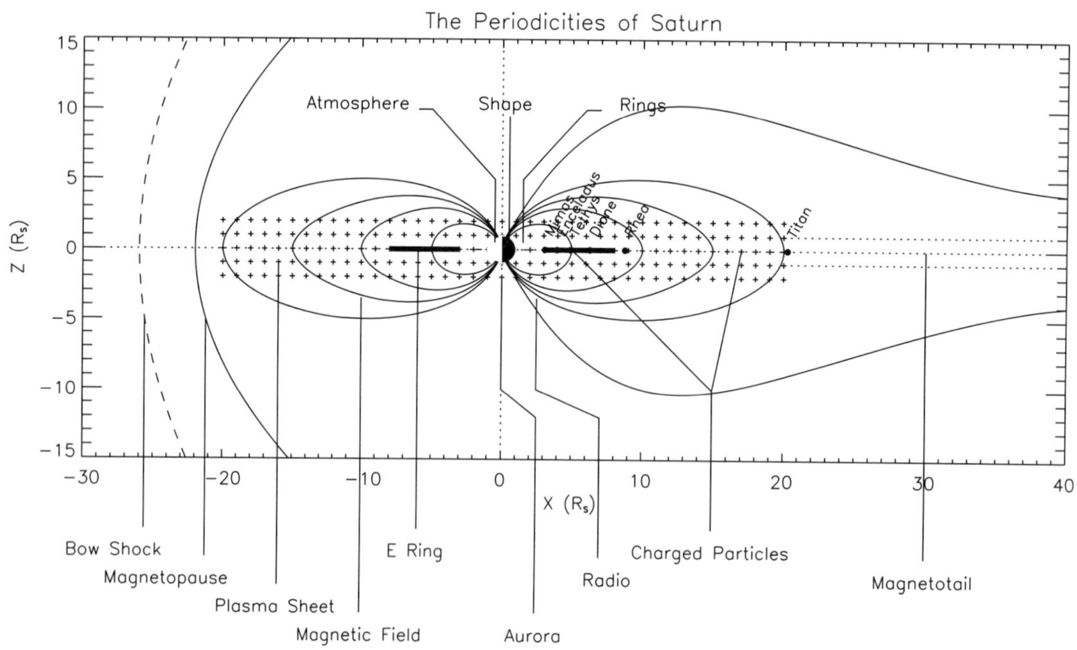

Figure 5.1 Overview of where the various periodicities of Saturn occur. Here, the observer is looking at the dusk flank of the magnetosphere from a local time of 18 h, with the Sun at the left and the planetary spin axis pointing up.

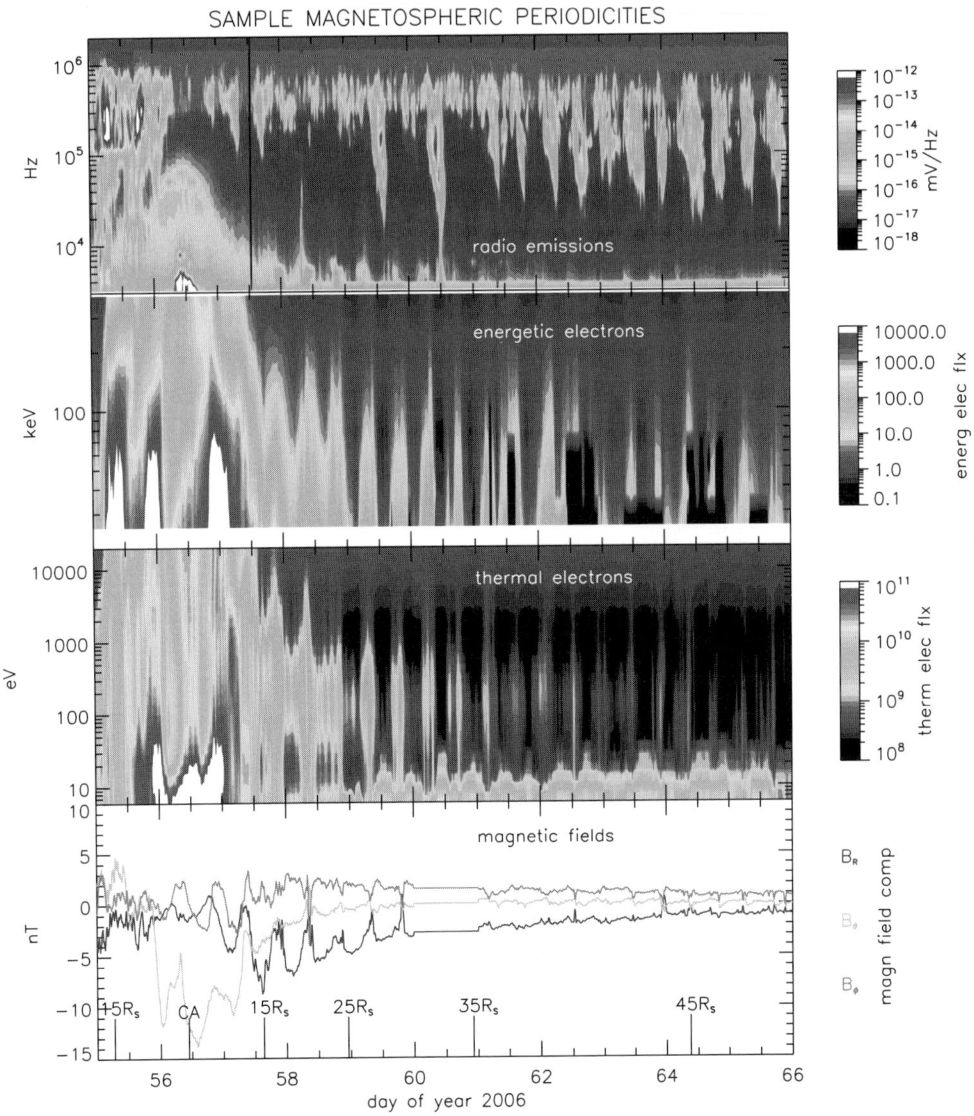

Figure 5.2 Sample periodicities from radio emissions (top), energetic electrons (second), thermal electrons (third) and magnetic fields (bottom). The radio spectrogram includes most of the SKR band (50–400 kHz), and the electron energy range from 7 eV to 400 keV. The day of the year 2006 is indicated at the bottom. The energetic electron fluxes are shown, while the CAP spectrogram shows differential energy flux. Conversion of a profile from the time domain to the frequency domain gives the period. These sample data were obtained during one pass of Cassini in the equatorial plane, with ranges indicated in the magnetometer panel. (A black-and-white version of this figure appears in some formats. For the color version, please refer to the plate section.)

PPOs. The periods of the oscillation are determined either by a Fourier or Lomb–Scargle spectral analysis, a phase-tracking method for radio emissions, or, as for the magnetic field oscillations, by correlating a model sinusoid of a known period with the observed oscillations.

Radio emissions are among the strongest of Saturn's magnetospheric periodicities. Periodicities in radio emissions have long been used to determine rotational rates of planets and other bodies. Jupiter is the prime example of this technique because its offset, tilted dipole (i.e. higher-order magnetic multipoles) coupled with the sharp beaming properties of decametric radiation ensure a strong rotational modulation of the received signal. Such modulations are the basis of the accepted rotational period of Jupiter (e.g. Davies et al. 1986; Higgins et al. 1997). Similarly, rotation periods for Uranus and Neptune as well as for pulsars are based on the rotational modulation of radio emissions (see Seidelman et al. (2007) for a definitive list of planetary rotation periods). Even Earth's rotation period may be determined by observing terrestrial auroral kilometric radiation (Panchenko et al. 2009; Lamy et al. 2010a; Morioka et al. 2013). In 1980–1981 when the Voyager 1 and 2 spacecraft flew by Saturn, it was thought that Saturn's rotation period was similarly well defined at 10.66h ± 7s (Desch and Kaiser 1981). Warwick et al. (1981) noted that SKR appeared to be modulated like a strobe light as opposed to a rotating beacon, as is the case for Jupiter. Lamy (2011) and Andrews et al. (2011) have demonstrated that the actual sources rotate in the co-rotational direction, but there is strong local time variation in intensity, with the most intense emission observed in the post-dawn sector, so Saturn is both a strobe and a rotating beacon. The primary mystery during the Voyager epoch was why the rotational modulation existed at all, given the nearly axisymmetric magnetic field determined by Pioneer 11 (Davis and Smith 1990) and the two Voyagers (Acuña et al. 1983; Connerney et al. 1983). In fact, the explanation of the rotational modulation of the Saturn kilometric radiation became an important objective of the Cassini mission during its formulation and actually remains a conundrum after 11 years of Cassini observations in orbit. More information about Saturn's radio sources can be found in Chapter 7 by Stallard et al. on the aurora.

Distant observations of SKR by the radio astronomy instrument on the Ulysses spacecraft after the Voyager flybys showed variations of the rotational modulation period of SKR of order 1% over several years (Lecacheux et al. 1997; Galopeau and Lecacheux 2000). Such variations cannot be due to actual variations of the rotational period of the planet (e.g. Read et al. 2009); hence, doubt was cast on the Voyager SLS period as being the true planetary rotation period. The initial SKR period observed by Cassini on approach (Gurnett et al. 2005) confirmed that the period had changed significantly from the Voyager period of 10.66h ± 7s to a new period of 10.76h ± 36s. Figure 5.3 compares the Voyager, Ulysses and Cassini SKR periods and demonstrates how the radio period has evolved over 30 years. In addition to the variation on time scales of months to years, Zarka et al. (2007) reported period variations on times scales of weeks or less that are similar to simultaneous solar wind speed variations.

Recognizing that the variable SKR period does not directly reflect the rotational period of Saturn, Kurth et al. (2007) devised a new SLS coordinate system (eventually referred to as SLS2) with the intent of providing a magnetospheric coordinate system within which observations of magnetic fields, charged particles and other phenomena could be compared. Kurth

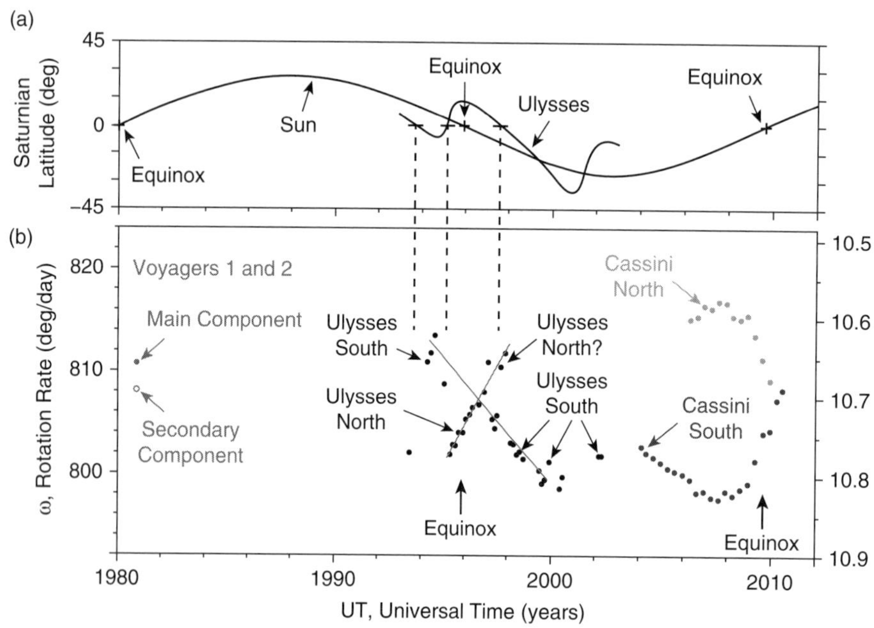

Figure 5.3 Comparison of the SKR periods from Voyager, Ulysses and Cassini. A clear shift in the periods had occurred between the initial 1981 measurement and successive ones. The faint lines suggest a possible "crossing" of northern and southern periods in the Ulysses era, although the Ulysses detector could not measure polarization to tell north from south (from Gurnett et al. 2010).

Table 5.1 *Longitude Systems and Periodicities of Saturn*

Name	Interval of applicability	Reference
IAU or SLS	Fixed (10.6562 h)	Desch and Kaiser (1981) Seidelman et al. (2002)
SKR S	mid-2003 to mid-2006	Zarka et al. (2007)
SLS2	2004–001 to 2006–240	Kurth et al. (2007)
SLS3	2004–001 to 2007–222	Kurth et al. (2008)
SLS4 – N	2006–305 to 2009–199	Gurnett et al. (2011)
SLS4 – S	2005–256 to 2010–044	
Meudon – N	2005–084 to 2010–193	Lamy (2011)
Meudon – S	2004–001 to 2010–193	
Leicester	2004–282 to 2010–085	Andrews et al. (2011)

et al. (2008) presented a revised system, SLS3, which extended the valid period of the SKR-based coordinate system to 10 August 2007. Based on polynomial fits to the rotation rate, the SLS3 system most accurately represents the periodicities within the time interval of the observations. In fact, several longitude systems have arisen in response to the various periodicities of Saturn, and these systems are given in Table 5.1.

An interesting side note in the Kurth et al. (2008) paper pointed out that a secondary, significantly faster period coexisted in the SKR data. Gurnett et al. (2009a) showed that the original period around 10.8 hours could be tied to observations of southern SKR sources and the shorter 10.6-hour period was associated with northern SKR sources. The northern component is separated from the southern component either using spacecraft latitude or on the basis of circular polarization of the extraordinary mode emission – because the resonant electrons spiral around the magnetic field lines, left-hand-polarized SKR originates in the southern hemisphere, while right-hand-polarized SKR comes from the north (Lamy 2008, 2011, and references therein). This duality was a new discovery by the Cassini mission. Subsequently, Gurnett et al. (2009b) showed that "auroral hiss" emissions displayed dual periods in the northern and southern hemispheres, similar to those of the SKR, and Ye et al. (2010) showed that narrowband radio emissions also exhibited north/south dual periods like SKR. As described in other sections of this chapter, other magnetospheric phenomena show periodicities, often segregated by northern and southern hemisphere and consistent with the SKR periods.

Both north and south periods are found to vary slowly over long time intervals of a year or more. When Cassini first arrived at Saturn during southern hemisphere summer the periods were well separated, ~10.8 h for the southern and ~10.6 h for the northern system. As Saturn approached its vernal equinox in 2009, the periods of SKR rotation modulation appeared to approach a common value near 10.7 h and the periods in the two hemispheres appeared to interchange (Gurnett et al. 2010). See also Lamy (2011) for the variation of the two hemispherical periodicities. However, subsequent studies of the SKR periodicity after equinox mainly concluded that the periodicities had become very complex and difficult to decipher. In fact, the reversal of the north and south periods is currently under debate. Using magnetic field data, Provan et al. (2013, 2014) reported that they did not reverse, while, more recently, Fischer et al. (2015) suggested, to the contrary, that SKR periods crossed for a short period of time just after equinox.

Figure 5.4 summarizes the SKR periodicities from 2004 to 2014. The north and south radio periods were separated by both polarization of the source and by latitude of the observer (Gurnett et al. 2009; Lamy 2011; Kimura et al. 2013). The "mixture" of north and south periods in Figure 5.4 is a real effect caused by oscillations of field-aligned currents on closed field lines (e.g. Provan et al. 2013, 2014; Hunt et al. 2014, 2015). Figure 5.4 also displays numerous secondary

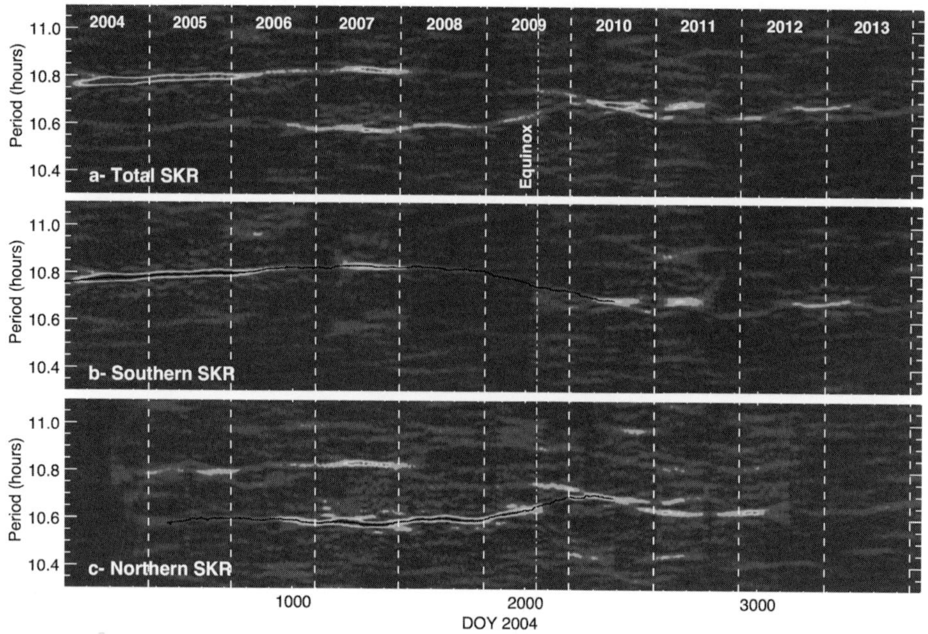

Figure 5.4 Lomb–Scargle 2-dimensional spectro-periodograms of (a) total, (b) southern X-mode and (c) northern X-mode SKR as a function of period and time. This is an update of the figure in Lamy (2011). Periodograms were generated within 200-day windows and show normalized peak power of Saturn kilometric radiation as a function of time from 2004 until late 2013. The ordinate shows the period in hours. (A black-and-white version of this figure appears in some formats. For the color version, please refer to the plate section.)

signals at strengths well below those of the primary north/south periodicities; these lesser signals are spurious, most likely an artifact of the processing, and may be ignored. The discrepancy between and temporal evolution of both the north and south SKR periods has been ascribed to various causes, such as seasonal variation of the ionosphere or solar cycle effects (Gurnett et al. 2009; Cowley and Provan 2013, and references therein).

5.3 Magnetic Field Periodicities

The magnetic field also displays planetary period oscillations. These are recognized in all components of the magnetic field, independent of coordinate system. To determine these periodicities, the analyst first subtracts the dipolar components of the total field, then applies a filter to smooth the data and then evaluates the remaining "perturbation" field components for periodic behavior. Both Voyager and Cassini observations of the magnetic field have revealed strong periodicities at or near the SKR period (Espinosa and Dougherty 2000; Espinosa et al. 2003a,b; Giampieri et al. 2006). Figure 5.5 exhibits power spectra from the Cassini magnetometer obtained for all three spherical components between July 2004 and August 2005. The main spectral peak in each spherical component lies close to but not at the IAU period measured by the Voyager radio experiment (black vertical line); the main peak at ~10.78 hours is essentially that of the SKR period measured by the Cassini radio instrument over the same interval. Notice that the secondary peaks evident in the Bθ and B_φ spectra can be explained as effects of the orbital period of the Cassini spacecraft, not as multiple periodicities (Giampieri et al. 2006).

Dual periodicities like those of the SKR are observed in the magnetometer data. In the northern polar region only the northern oscillations are observed and within the southern polar region only the southern oscillations are observed (Andrews et al. 2010a; Southwood 2011) (within a ~10% limit of experimental determination) (Andrews et al. 2012). In the "core" region of Saturn's magnetosphere (within dipole $L \approx 12$ in equatorial plane), both the northern and the southern oscillations are observed. The two oscillations constructively and destructively interfere in a manner described by the beat period of the two oscillations (e.g. Andrews et al. 2010a, 2011, 2012).

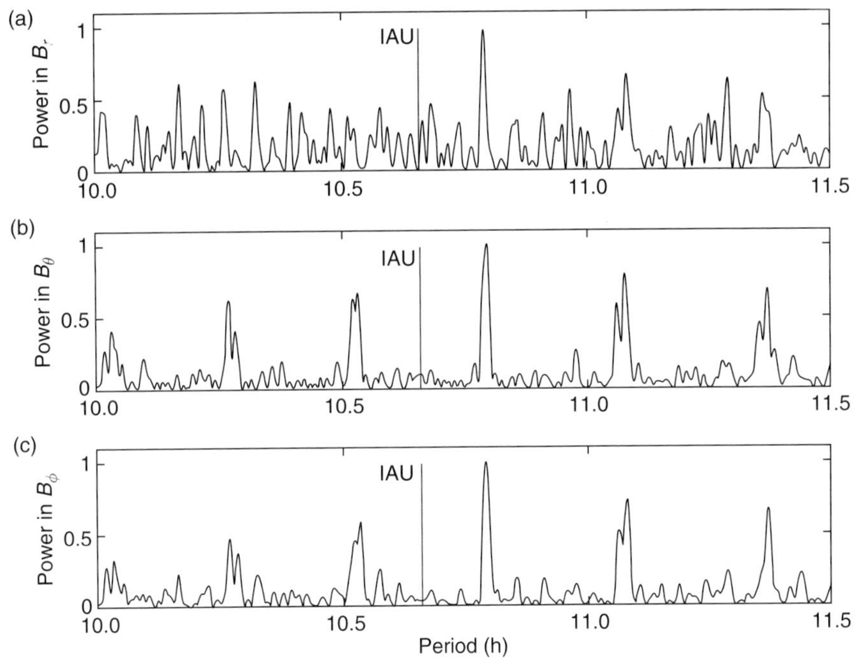

Figure 5.5 Lomb power spectra for the three spherical components of the magnetic field from observations made between July 2004 and August 2005. The vertical line indicates the IAU (Voyager) SKR period. The side lobes of the signal are associated with the orbital motion of the spacecraft and are artifacts of the spectral processing (from Giampieri et al. 2006).

The magnetic oscillations rotate at the PPO period, and the perturbation also propagates outward with a radial phase delay of ~3 degrees per Saturn radius (Andrews et al. 2010b). This oscillatory signal will be Doppler shifted because of the spacecraft motion, at least in the core region near periapsis, so it is not appropriate to utilize a Lomb–Scargle or Fourier analysis when determining the period of the oscillations (see Carbary (2015) for a discussion of Doppler effects). A signal processing technique has been developed wherein the oscillation amplitude and phase are obtained by cross-correlating the magnetic data for field component i (spherical polar r, θ or ϕ) with a wave described by the following formula,

$$B_i(\phi, r, t) = B_{oi} cos[\Phi_g(r, t) - \phi - \psi_i] \qquad (5.1)$$

where ϕ is azimuth measured from noon, t is time since 00:00 UT on 1 January 2004 and r is radial distance from the planet center. The function $\Phi_g(r,t)$ is a "guide" phase function that corresponds to the phase of an oscillation of guide period τ_g close to that of the oscillation of interest. For observations beyond 12 R_S, the radial phase delay of the signal is also taken into account. ψ_i is an angle that represents the phase of the oscillation in the magnetic field component concerned, relative to the guide phase. By determining the phase gradient of ψ_i with time it is possible to calculate the northern and southern PPO periods.

The northern and southern PPO periods were determined both within the equatorial "core" magnetosphere ($L < 12$) and separately in the northern and southern polar regions (Andrews et al. 2008, 2010a; Provan et al. 2013, 2014). The amplitude of the oscillations within the equatorial magnetosphere was also calculated. When Cassini first arrived at Saturn, only the southern oscillations were evident. However, as with the periods, the amplitude of the oscillations showed a clear drift with the changing seasons. By the time of the northern spring equinox of 2009 the oscillations had approximately equal amplitude within the core magnetosphere. It was expected the northern oscillations would dominate the southern oscillations as northern summer approached. But instead, a new and unusual behavior of the oscillations was reported, with sharp changes in the period, amplitude and relative amplitudes of the oscillations at a cadence of ~200 days. During some of the sharp changes, the southern oscillations resumed dominance, while during others the northern oscillations dominated. Towards the end of 2013, the amplitudes of the northern and southern oscillations were approximately equal within the "core" magnetosphere.

Intensive work has taken place on these periodicities since equinox, and there have been several attempts to find consistency between the MAG and SKR periods since equinox (Andrews et al. 2010, 2011, 2012; Provan et al. 2011, 2014; Fischer et al. 2014). One might expect the periods of the SKR/magnetic field to be the same, within experimental errors, if they are created by the same rotating current systems. As shown by Andrews et al. (2012) and Provan et al. (2014), there is in general close agreement between the magnetic and SKR PPO periods, and the differences often vary about zero through only a few tens of seconds (i.e. at the ~0.1% level). There is a clear discrepancy between the SKR and MAG periods in the immediate post-equinox interval, but overall there is very close agreement between the PPO periods as determined from the two datasets. This is despite the two datasets having different limitations. The periodicities in the magnetic field can only be observed when Cassini is within the magnetosphere and are local rather than global measurements. The radio observations are essentially global and not subject to local limitations; they must nevertheless take into account the strong anisotropy of SKR emission, which can cause significant visibility effects (Lamy et al. 2008, 2011, 2013; Andrews et al. 2011 and references therein). While there are differences in the details, there are some general conclusions for which there is consensus. Periods attributed to the two hemispheres converged during 2010 (within one year after equinox), but the southern magnetic period was consistently longer than the northern. The southern period returned to being longer than the northern period (as it was prior to equinox) around the end of 2010 (Provan et al. 2015). Rather abrupt changes in period (and hence in phase) were observed in both the magnetic field and SKR on several occasions in the interval after 2010. Figure 5.6 compares the post-equinox PPO rotation rates of the SKR and magnetic perturbation fields.

A recent controversy has arisen concerning the methods used to determine magnetic field periods and phases, as least in the post-equinox epoch. The fitting method used for magnetic field periodicities (e.g. Andrews et al. 2010; Provan et al. 2013) has been questioned by Yates et al. (2015). Furthermore, recent SKR analysis suggests that the north/south periods may have interchanged after equinox, which contradicts findings made by magnetometer analysis (Fischer et al. 2015). However, the analysis method of Andrews and Provan has been defended in a series of comments (Cowley et al. 2015a,b), to which the reader is referred for the current status of the controversy.

5.4 Particle Periodicities

Variations in charged particles constitute a third observation of magnetospheric oscillation. "Particle periodicities" always refers to the periodicities in particle fluxes or count rates, which are proportional to fluxes. Because Doppler effects may bias the periodicities of these fluxes in the inner magnetosphere (where spacecraft speeds are comparable to wave speeds), charged particle periodicities usually refer to those in the outer magnetosphere ($r > 15\ R_S$) where Doppler effects are assumed negligible (e.g. Carbary et al. 2007). See Carbary (2015) for a complete discussion of Doppler effects on Saturn's magnetospheric periodicities. Planetary period oscillations in energetic ions and electrons were discovered during the Voyager flybys of Saturn (Carbary and Krimigis 1982) and have been generally observed throughout the magnetosphere by Cassini instruments (Krimigis et al. 2005; Paranicas et al. 2005; Carbary et al. 2007). Charged particles also evidence dual periods (Carbary et al. 2009a), but they may also have displayed somewhat different periods than the SKR (Carbary et al. 2012, 2014). Figure 5.7 exhibits a power spectrum of typical charged particles. This example demonstrates not only the north and south SKR periods in energetic electrons, but also suggests that other planetary periodicities may be present at Saturn (see Chenette et al. 1976). The particle periodicities are strongest in energetic electrons ($E >$ 100 keV) and oxygen ions ($E > 5$ keV), but are virtually nonexistent in energetic protons (Carbary et al. 2007, 2014). The fact that particles of vastly different energies and species display common periods argues strongly against gradient and curvature drift causing the periodicities. The effects of gradient and curvature drifts in energetic particles have been demonstrated in magnetospheric dynamics, most notably in the

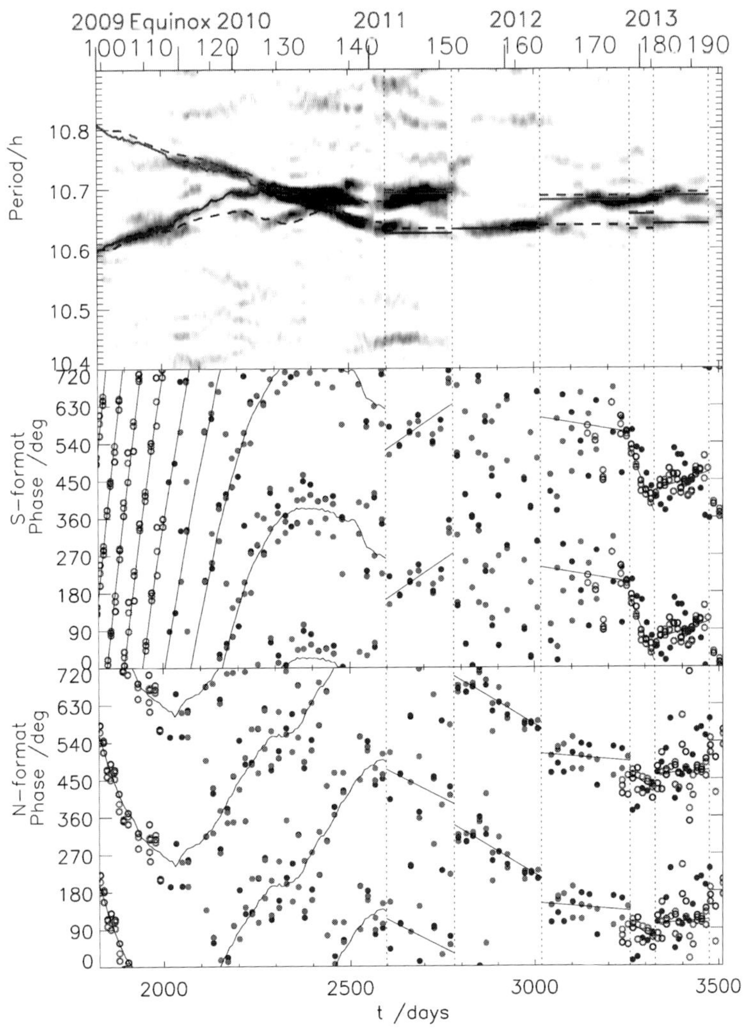

Figure 5.6 This figure presents the Lomb–Scargle periodograms for the total SKR emission (top panel). Overplotted are the SKR periods (solid lines) and MAG periods (dashed lines). The northern periods are shown in blue and the southern periods are shown in red. The second panel shows the phases of the northern perturbation magnetic field components with respect to a guide wave with a period of 10.64 h. The third panel shows the phases of the southern perturbation magnetic field components with respect to a guide wave of 10.69 h. For clarity, the phase panels show two repeating cycles of 360° each (from Provan et al. 2014). (A black-and-white version of this figure appears in some formats. For the color version, please refer to the plate section.)

hot-plasma injections occurring throughout the middle magnetosphere (Mauk et al. 2005; Paranicas et al. 2007; Mitchell et al. 2009) and in the inner magnetosphere (Burch et al. 2005; Hill et al. 2005).

Other magnetospheric oscillations may be considered to be different manifestations of the radio, magnetic or particle periodicities. For example, ions in Saturn's plasma sheet can be imaged by using energetic neutral atoms (ENA), which are formed by charge-exchange collisions between energetic ions and thermal neutral atoms. The ENA can be observed by the Ion Neutral Camera (INCA) on Cassini in much the same way as photons are observed by a conventional imager. Such imaging reveals that the plasma sheet can oscillate up and down with the SKR period (Paranicas et al. 2005; Carbary et al. 2008). Such oscillations can be recognized as "flapping" far down Saturn's magnetotail to distances exceeding ~25 R_S (Arridge et al. 2011; Provan et al. 2012). The phase of the ENA oscillations generally varies with respect to that of the SKR. During a 166-day period in 2004 (prior to orbit insertion), the neutral H signal led the SKR signal by ~1.5 h, while the neutral oxygen signal led SKR by ~12.2 h (Carbary et al. 2011). However, ENA "blobs" in the dawn-noon range are often seen in phase with SKR bursts and enhanced morning side aurora

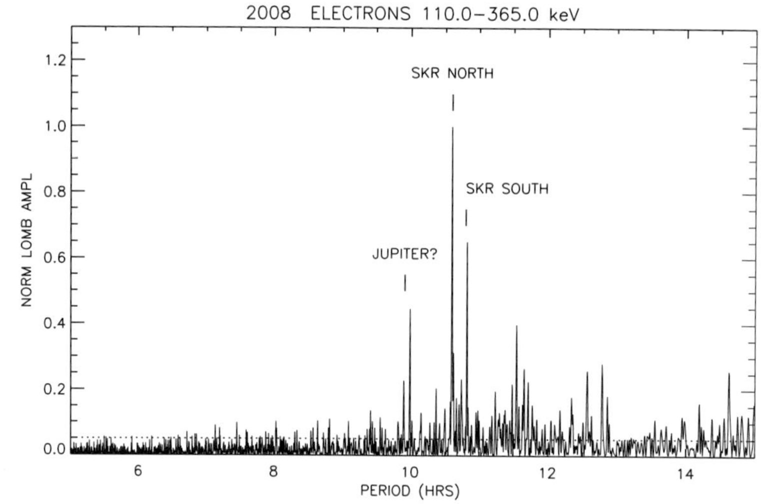

Figure 5.7 Periodogram of mildly relativistic electrons from data obtained over a full year (2008) using fluxes from the MIMI/LEMMS detector on Cassini. The "dual" SKR periods are not the only spectral feature in the spectrum. The feature labelled Jupiter may be relativistic electrons from Jupiter, which also exhibit the characteristic 9.92 h periodicity in interplanetary space (Chenette et al. 1974).

(Mitchell et al. 2009; Carbary et al. 2010, 2011; Lamy et al. 2013).

One very interesting aspect of measuring the ENA periodicities is the capability of the MIMI/INCA detector to globally image magnetospheric activity (e.g. Mitchell et al. 2005). The global coverage of INCA allows measurement of the simultaneous ENA intensities at multiple local times, which in turn allows estimates of the ENA (same as energetic ion) periodicities at different local times. Applying this technique to eight different local time sectors reveals that the periodicity actually varies with local time (Carbary et al. 2014). Generally, dual periodicities are evident on the nightside, but only single or no periodicities appear on the dayside. This is not to say that the ENA periods themselves vary with local time, but that the amplitudes of their signals do. Figure 5.8 shows periodograms from a 185-day interval spanning 2006 and 2007. If the strengths of magnetospheric periodicities all manifest a similar local time dependence, then one must be careful in interpreting in situ measurements such as those from the magnetometer and charged particle detectors that do not uniformly sample local times.

5.5 Auroral Periodicities

Saturn's aurora also exhibit periodicities. The aurora arises as a result of electron precipitation along magnetic field lines, and can be considered as a tracer of upward Birkeland currents (or downward-moving electrons). Several studies have noted rotational periodicities in both the auroral location and its intensity organized by SKR longitude (Nichols et al. 2008, 2010b; Carbary 2013a). Indeed, the centers of the northern and southern aurora ovals have been reported to oscillate with the northern and southern SKR periodicities (Nichols et al. 2008, 2010b). Interestingly, when organized in SLS4-south longitude, an enhancement of the auroral emissions has been detected in the southern aurora, possibly marking a rotational anomaly, but no such feature appears in the northern aurora when organized in the SLS4-northern longitude (Carbary 2013a). For more information on Saturn's aurora, consult Chapter 7 by Stallard et al. on the aurora.

5.6 Ring Periodicities

Rotating asymmetries in Saturn's magnetosphere can also appear to produce observable patterns in the distribution of small particles (1–100 μm) above and around Saturn's rings. Probably the best-known examples of this phenomenon are "spokes." The spokes are near-radial features in Saturn's B ring, ~10,000 km long (Grün et al. 1983, 1992). These structures consist of micron-sized particles that are most likely levitated

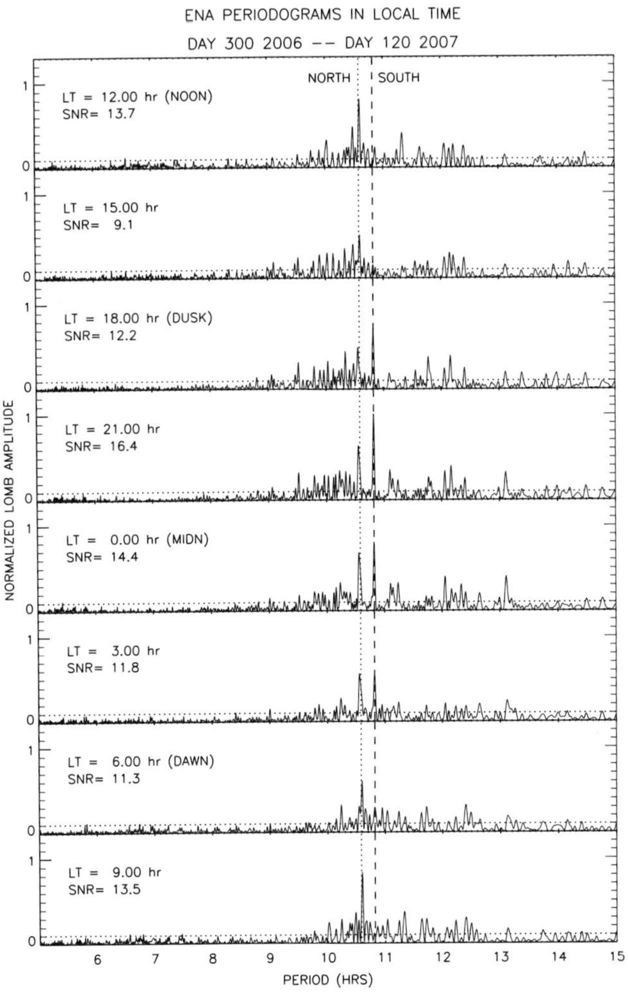

Figure 5.8 Periodograms from the time profiles of ENA intensities in eight different local time sectors for the interval from day 300, 2006 to day 120, 2007. The tics on the abscissa indicate the SKR periods during this interval. There is a strong single period on the dayside near noon, but dual periods on the nightside near midnight (from Carbary et al. 2014).

out of the ring plane by electromagnetic forces (Goertz and Morfill 1983; Farmer and Goldreich 2005; d'Aversa et al. 2010). Spokes appear to form primarily on the morning side of the rings around the synchronous radius (where the ring particle's Keplerian orbital period is close to the planet's spin period) and then drift around the planet. Analysis of the number and contrast of the spokes observed by the Voyager spacecraft reveals that these quantities are periodically modulated at a period close to that of the SKR (Porco and Danielson 1982). Recent Cassini observations have confirmed that these periodicities are close to the 10.7-hour standard and to the various magnetospheric periodicities (Mitchell et al. 2013). Figure 5.9 shows a spectrum of the ring-spoke periodicity obtained recently from Cassini's Imaging Science Subsystem and relates it to SKR and magnetic field periodicities. Only a few days' worth of spoke observations have been analyzed so far, so these data do not yet provide precise estimates of the associated periods.

Cassini images also reveal structures that may be linked to magnetospheric periodicities in the tenuous sheets of dust extending interior and exterior to the main rings, known as the D ring and Roche Division, respectively. Images of these regions taken in 2006 reveal patterns of alternating bright and dark bands that rotate around the planet with periods between 10.5 and 10.9 hours (Hedman et al. 2009). These patterns are most likely created by resonances with periodic perturbing forces, which generate structures by organizing the particles' eccentric orbits. Such patterns track the perturbation that produced them, so by comparing images taken at different times, one can obtain precise estimates of the perturbation periods. The physical process responsible for creating the periodic modulations in spoke activity and the patterns in the D ring and Roche Division are still not clear. However, in all these cases the patterns are created in populations of tiny (sub-millimeter) grains, which are especially sensitive to non-gravitational forces due to their high surface-area-to-volume ratios (Burns et al. 2001). This, together with the similar periods involved, strongly suggests that these ring phenomena are driven by asymmetries in Saturn's magnetosphere. If this is correct, then the magnetospheric oscillations at periods close to the planet's rotation rate must extend to low latitudes very close to the planet.

5.7 Wind-based Estimates of Saturn's Rotation Rate

The multiple, time-variable periods in Saturn's magnetosphere revealed by the Ulysses and Cassini measurements mean that magnetospheric data cannot yet provide an unambiguous, robust estimate of the rotation rate of Saturn's deep interior. Hence, there has been renewed interest in alternative methods of determining the planet's rotation rate from measurements of its winds, shape and gravitational field.

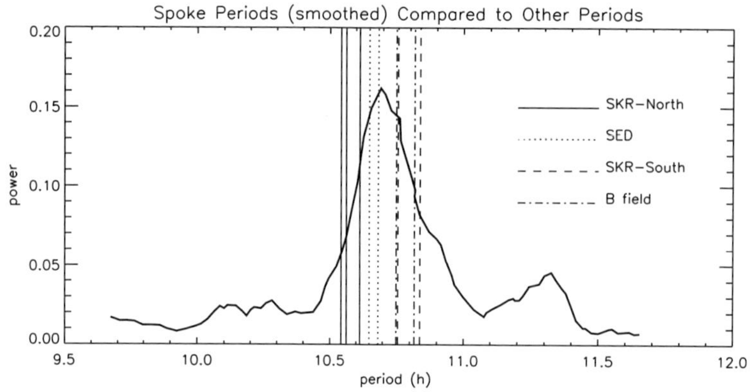

Figure 5.9 When the contrast or intensity of spokes in Saturn's B ring are subjected to a Fourier-type spectral analysis, this power spectrum emerges (adapted from Mitchell et al. 2013). The vertical lines indicate periodicities observed in various other phenomena. The spokes represent a manifestation of the elevation of dust particles from the ring plane by electrostatic forces.

The wind speeds at Saturn's cloud tops are measured by tracking clouds and various other atmospheric features in the zonal winds. The zonal winds well equatorward of the polar region have speeds that vary with latitude and have an alternating zonal pattern of "jets" and "troughs." Massive storms localized in latitude and associated with Saturn electrostatic discharges (SED) from lightning can also be observed (Fischer et al. 2011). While Saturn's wind speeds are usually given relative to the fixed IAU rotation period (e.g. Sánchez-Lavega et al. 2000; Vasaveda et al. 2006), one can readily convert such speeds to a frequency or period in inertial space. These "wind periods" partially overlap rotational periods found in the magnetosphere (Sánchez-Lavega 2005). This has led many to argue that winds in the upper atmosphere (i.e. to altitudes 500–1000 km above the visible clouds) may be a driving force behind the magnetospheric periodicities (Gurnett et al. 2009a; Jia et al. 2012; Smith 2011, 2014; Hunt et al. 2014). However, the detailed relationship between the winds and the magnetospheric periods is still obscure. For example, it is not clear why the strongest magnetospheric oscillations are at periods comparable to the planet's strongest westward jets, and thus significantly longer than the periods associated with the eastward jets (Cowley and Provan 2013).

Detailed studies of the winds can potentially constrain the rotation rate of planet's deeper interior. For instance, one can look for long-lived structures in the atmosphere that might have some persistent connection with the deep interior. At Saturn, probably the best example of such a feature is the "hexagon," a cloud pattern at ~75 degrees north latitude that encircles Saturn's pole and that seems to have persisted for many decades (Godfrey 1988, 1990; Sánchez-LaVega et al. 1997). Using both Cassini and ground-based images from 2008 through 2014, the vertices of the polar hexagon were found to have a constant speed that was consistent with a steady rotation period of 10.66h (Sánchez-LaVega et al. 2014). A similar analysis of Voyager-era images indicated a slightly faster rotation period (by only 3.5s), an effect ascribed to a large feature associated with the hexagon. The relative constancy of the hexagon's speed for much of a seasonal cycle could indicate that its rotation period (10.66h) is close to the actual rotation period of the planet. However, just as Jupiter's Great Red Spot rotates at a different rate from its deep interior, the rotation rate of Saturn's hexagon probably also deviates from the planet's bulk rotation rate.

Another way to estimate the rotation period of a planet from its wind patterns is by analyzing the zonally averaged potential vorticity, which is also obtained from cloud speeds (Read et al. 2009). This estimate is based on a nonlinear stability theorem of Arnol'd (McIntyre and Shepherd 1987). Indeed, the cloud-tracked winds on Jupiter approach marginal stability according to this theorem, and the application of such an analysis to Jupiter yields a reasonable estimate of its rotation period of 9.92 h. Conducting a similar analysis for Saturn yields a rotation period of 10.57 h (Read et al. 2009), which is comparable to the fastest SKR-north period. Interestingly, the rotational reference frame of this period produces a more symmetric pattern of alternating east-west zonal jets than the IAU period.

5.8 Shape-based Estimates of Saturn's Rotation Rate

Constraints on Saturn's rotation rate can also be derived from its overall shape. Saturn is an extremely oblate object; its equatorial radius exceeds its polar radius by nearly 10%. Presumably, the planet's rotation causes this oblate shape, so one expects that an analysis of the shape of Saturn might reveal something about its rotation. Indeed, Saturn's rotation period has been estimated based on its shape (Anderson and Schubert 2007; Helled et al. 2009). This approach finds the period that minimizes the dynamical height, i.e. the height of the 100 mb isobaric pressure above a reference geoid (a lower altitude than those discussed in Chapter 9 by Strobel et al.). Such a technique reveals a period of about 10.53 h. A new method for estimating the rotation period employs the planet's measured gravitational field and oblateness using statistical optimization (Helled et al. 2015). This approach yields a period of 10.54 h. Details of this method, as well as its uncertainties, can be found in Chapter 3 by Fortney et al. All these periods are shorter than the SKR periods.

The theory-based periods should not be considered definitive. One problem with this approach is that accurate models of the spin rate rely on accurate knowledge of the interior, which is not available. Furthermore, the depths of Saturn's winds, which are also unknown, also influence the shape and must be taken into account. Detailed measurements of Saturn's gravity field from the Cassini mission's Grand Finale may provide the information needed to constrain the depth of these zonal flows and thus the rotation state of the planet's interior.

5.9 New Insights from Ring Seismology

A new way to study Saturn's rotation has recently been provided by fine-scale structures in the planet's rings that can serve as seismological probes of the planet's interior. Figure 5.10 displays the wealth of wave-like structures available from a simple stellar occultation of the C ring. The basic idea behind this approach is that any axially asymmetric structure in Saturn's gravitational field, whether an external moon or an internal mass-concentration, will periodically perturb the orbits of ring particles at resonances (locations where the orbital period of the particles is close to a whole number ratio times the period of the perturbing force) and give rise to distinctive patterns in the main rings. (Note that, unlike the dust-sized grain populations described in Section 5.6, the particles in the main rings are too large to be perturbed significantly by electromagnetic forces.) The most dramatic and best-understood structures produced in this way are spiral density waves (Shu 1984); multi-armed, tightly wound spiral patterns

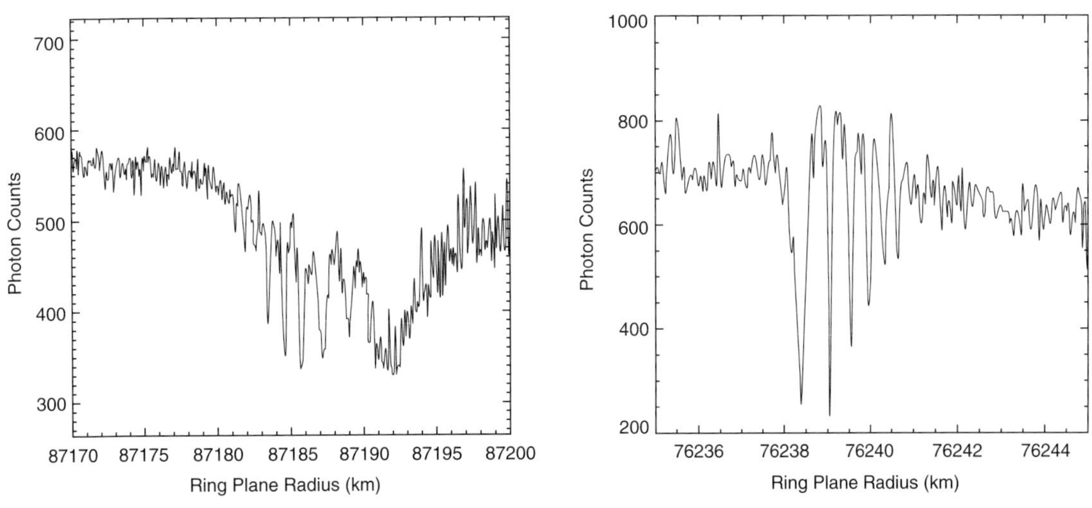

Figure 5.10 Radial profiles from stellar occultations of the C ring show radial wave patterns with dispersion of the wavelength. Any axially asymmetric structure in Saturn's gravitational field, whether an external moon or an internal mass-concentration, will periodically perturb the orbits of ring particles at resonances (locations where the orbital period of the particles is close to a whole number ratio times the period of the perturbing force) and give rise to distinctive patterns in the ring (from Baillé et al. 2011).

that appear as quasi-periodic opacity variations in the one-dimensional, high-resolution cuts through the ring provided by stellar occultations (see Figure 5.10, Colwell et al. 2009; Baillie et al. 2011; Hedman and Nicholson 2013, 2014). The Voyager missions revealed many examples of these sorts of structures, and while most could be attributed to resonances with Saturn's various moons, some appeared to require additional asymmetries in the planet's gravity field. Specifically, Marley and Porco (1993), building upon an earlier suggestion by Stevenson (1982), proposed that several features in Saturn's C ring might have been generated by fundamental normal-mode oscillations inside the planet. Since any resonantly generated patterns in the rings rotate at the same rate as the perturbation that produced them, detailed studies of the structures in the rings can not only confirm that some of these patterns are indeed due to structures inside the planet, but also yield very precise measurements of how fast the relevant structures inside the planet rotate. The ring features also encode information about the planet's internal structure, but that discussion is beyond the scope of this chapter.

In practice, the periods of the perturbing forces responsible for many of the C-ring waves were determined by comparing data from dozens of occultations that sample each wave at many different longitudes and times. The opacity maxima and minima in these different profiles appear at different locations, and these differences can be expressed in terms of an effective phase difference. The hundreds of observed phase differences can be compared to the expected phase differences for a pattern with a given number of arms rotating at a specific rate, enabling the pattern's azimuthal wavenumber m and its rotation rate to be determined (see Figure 5.11). This technique has revealed roughly a dozen waves with rotation rates that are too fast to be due to any moon, and thus are likely generated by structures inside the planet. Most of these waves appear to be generated by normal mode oscillations similar to those predicted by Marley and Porco (1993). However, at least five appear to be due to persistent gravitational anomalies with rotation periods between 10.365 h and 10.698 h (Hedman and Nicholson 2014). These C-ring periods span those of zonal wind speeds, as well as those of SKR (north), magnetic fields and energetic

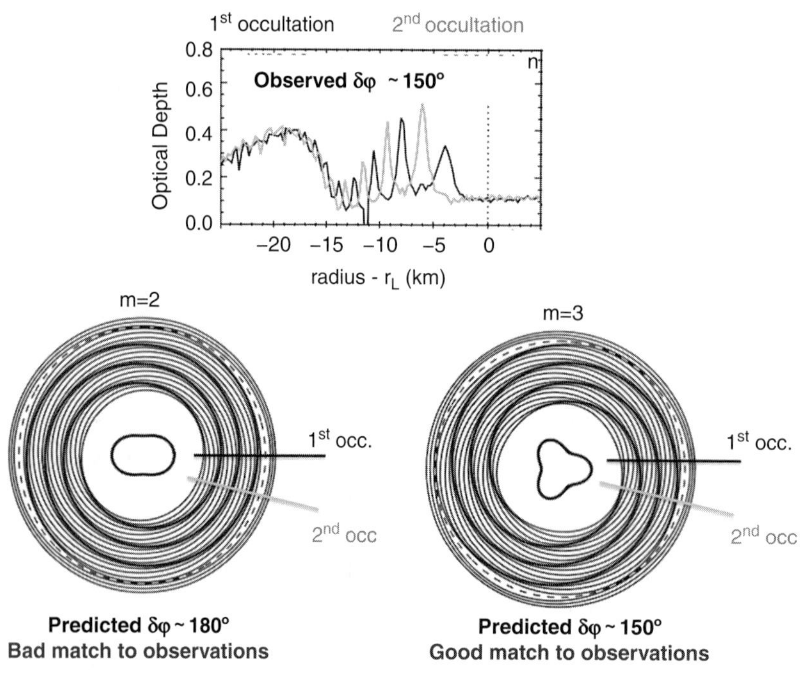

Figure 5.11 Two occultation profiles are compared (top). The expected density profiles for an $m = 2$ (bottom left) and $m = 3$ (bottom right) spiral are computed. A comparison of the expected profiles with the observed profiles allows a clean extraction of the wave number m from the observed phase difference $d\phi = 150°$.

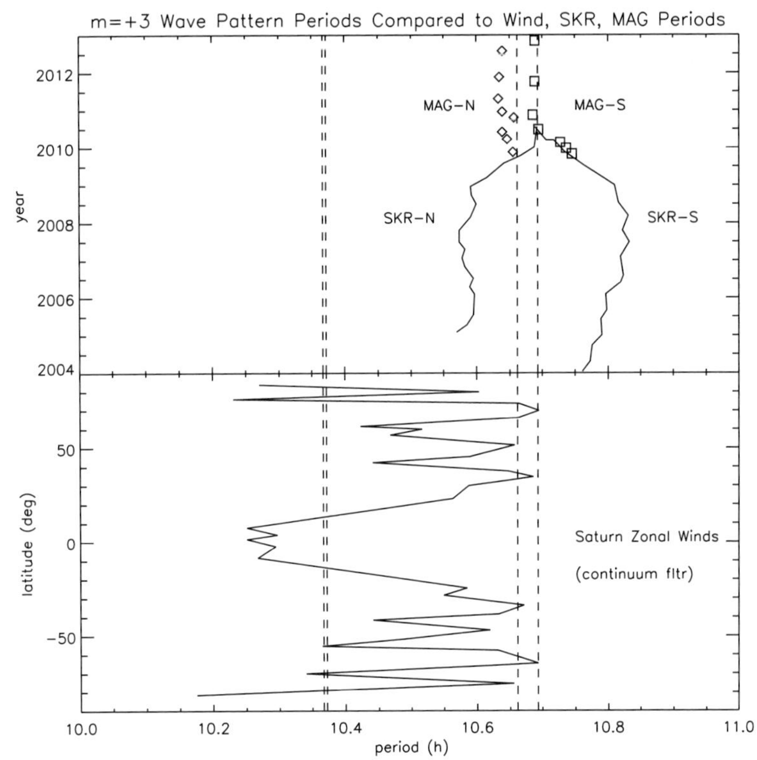

Figure 5.12 Wave patterns in the C ring exhibit periodicities between 10.365 h and 10.587 h, which are shown here as dotted vertical lines and compared to other periodicities. These wave periods are in the same general range as those of the Saturn winds (bottom) and the SKR (solid lines, top) and magnetic field (diamonds and squares, top) (from Hedman and Nicholson 2014).

particles (see Figure 5.12). As with the multiple magnetospheric periods, this range of periodic modulations in the planet's gravity field frustrates efforts to determine a unique rotation rate for the planet. However, it also hints that the dynamics of the planet's deep interior is richer than might have been expected.

5.10 Short and Long Periods

Non-rotational periodicities are those that are much different from the planetary rotation period. Saturn has two types of non-rotational periodicities: one has a period of about an hour and will be referred to as the "short period," and the other displays a period of ~26 days and harmonics thereof and will be called the "long period." Unlike the rotational periods, the short and long periods have plausible explanations.

The short period can generally be observed at high latitudes above ~50 degrees. One-hour variations can be seen in the intensities of certain auroral spots in the far ultraviolet, in low-frequency radio emissions (f < 100 Hz), in magnetic field components at high latitudes (e.g. Mitchell et al. 2009, 2014) and in relativistic electron fluxes (Roussos et al. 2015). Figure 5.13 displays an auroral keogram (top) for the intensities of an auroral feature between 10 and 15 degrees co-latitude. These intensities exhibit a ~1-hour period (bottom). The 1-hour oscillations have about the same period as the round-trip time of Alfvén waves travelling between the northern and southern ionosphere along the (non-dipolar) magnetic field lines. Visible aurorae observed near midnight brighten suddenly on the timescales of a few minutes and repeat with a period of ~1 hour ± 10 min (Dyudina et al. 2016). Such brightenings are seen in the brightest auroral arcs. Interestingly, at the parts of the auroral oval where the aurora is too faint to be spatially resolved by the Cassini camera, the auroral latitudes (72 to 78 degrees) are also brightening at the 1-hour period.

A detailed calculation of Alfvén times in the non-dipolar field of Saturn is beyond the scope of this chapter, but is basically the integration of the inverse Alfvén speed along magnetic field lines, which in turn

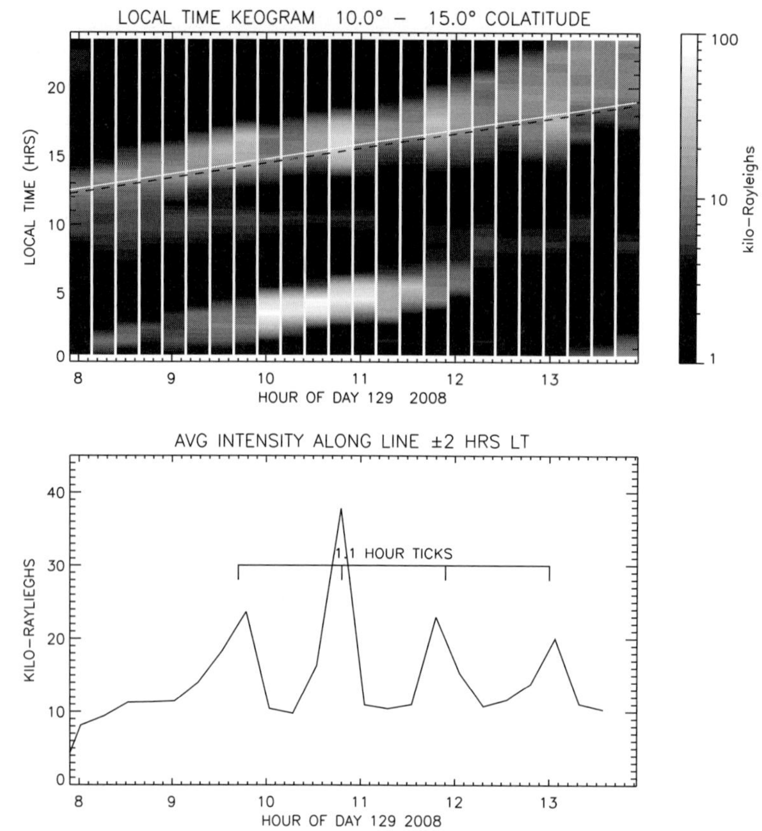

Figure 5.13 Keogram of auroral intensities between 10° and 15° co-latitude (top), and the intensity profile along the topmost feature (bottom). The solid line in the top panel indicates the track of the blob in local time; the intensity in the bottom panel is from integrating within ±2 h of local time relative to this line (from Mitchell et al. 2014).

requires a knowledge of the (non-dipolar) magnetic field and the plasma density. See Roussos et al. (2015) for a complete discussion of Alfvén times. Such waves may be generated, for example, by time-variable magnetic reconnection at the magnetopause, by Kelvin–Helmholtz instabilities near the magnetopause, by flux-tube interchange motion or by other processes that remain to be identified, and may be related to distributed field-aligned currents over the polar caps (Bunce et al. 2008; Badman et al. 2013; Mitchell et al. 2014). The ~1-hour oscillations in relativistic electrons are common in the outer magnetosphere and are much more common in the dusk sector than elsewhere (Roussos et al. 2015).

Long periods are close to the ~26-day recurrence period of the solar wind at Saturn. This solar wind period refers to the repetition of long-lived features known to exist in the solar wind, which have been measured to repeat at Earth and interplanetary space. The solar wind effect on SKR was clearly evident in the Voyager radio observations (Desch 1982) and can be easily perceived in periodograms of the SKR emissions in which the periodogram window stretches to ~50 days (Desch and Rucke 1983). Interestingly, the solar wind speeds have been shown to modulate changes in the SKR planetary period oscillations (Zarka et al. 2007). Solar wind periodicities can also be recognized in similar "long-window" spectra of charged particles (Carbary et al. 2009b, 2013b). No such solar wind periodicity has yet been reported in the magnetic field measurements. The solar wind effects on the radio emissions and charged particles are related to periodic compressions and expansions of the magnetosphere by the solar wind as co-rotating interaction regions sweep across the magnetosphere (e.g. Crary et al. 2005; Kurth et al. 2005; Badman et al. 2008). Figure 5.14 shows examples of solar wind periods in charged particle data. The energetic electrons shown here not only exhibit the solar wind 26-day period, but also its sub-harmonics. The energctic protons and oxygen ions also manifest

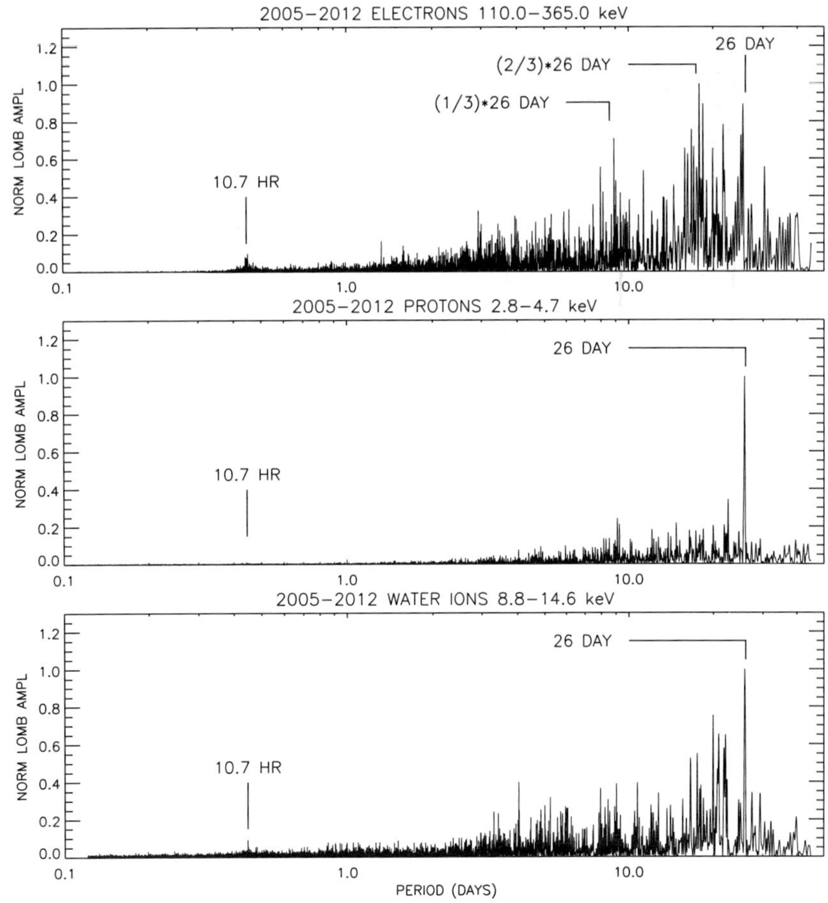

Figure 5.14 Solar wind periodicities in charged particle fluxes: very energetic electrons (top), energetic protons (middle) and energetic oxygen ions (bottom). Note that the solar wind periodicity is much stronger than the ~10.7 h period, which is not even observed in the proton spectrum.

a solar wind periodicity. In these cases, the solar wind signal is stronger than the planetary rotation signal at ~10.7 h.

5.11 General Dynamical Models of Saturn's Magnetosphere

Before proceeding to particular models of Saturn periodicities, first consider the general dynamical models of Saturn's magnetosphere. These can serve as a starting point for (possibly) understanding the periodicity models, which are described in more detail in Chapter 6 by Krupp et al.

First, consider an example of a periodicity driven by a magnetospheric phenomenon external to the atmosphere and rings. Saturn's magnetosphere rotates rapidly. The inner magnetosphere contains predominantly cool, dense plasma, while the outer magnetosphere contains predominantly hot, tenuous plasma. The interface between these two regimes is subject to an MHD dynamical instability called the "centrifugal interchange instability," wherein the hot tenuous plasma is injected inwards and exchanged with the outwardly pressing cold plasma. Such an injection is depicted schematically in Figure 6.9 of Chapter 6. Many examples of this type of interchange have been observed in both the thermal plasma and the energetic particles (Burch et al. 2005; Hill et al. 2005; Mauk et al. 2005; Paranicas et al. 2007). CAPS signatures of particle injections are shown in Figure 6.10 of Chapter 6. An $m = 1$ version of the centrifugal interchange instability, where m is the azimuthal wave number, was proposed as an explanation for the SKR periodicities by Goldreich and Farmer (2007). There is little evidence that "old" injections, characterized by well-developed energy-time dispersion signatures, are periodic (Chen and Hill 2008; Chen et al. 2010). "Young" injection events, less than ~1 hour old as revealed by the absence of dispersive signatures, are strongly ordered in SLS4-north longitude (rather than SLS4-south) when they occur near midnight (Kennelly et al. 2013).

Several models of the centrifugal interchange instability (CII) have been developed for both Jupiter and Saturn. The models generally solve the magnetohydrodynamic (MHD) equations for a rapidly rotating planet, and may include kinetic aspects and/or multiple fluids (e.g. Yang et al. 1994; Kidder et al. 2009; Liu et al. 2010; Liu and Hill 2012; Winglee et al. 2013). For Saturn, the models generally assume a longitudinally symmetric, radially distributed plasma source derived from ionization of water vapor from the Enceladus geysers, and allow the dynamics to unfold. After a few hours of simulation, the distribution develops numerous "fingers" of alternating inflow and outflow channels. Most of the CII models exhibit such fingers, although they differ in the widths and numbers of the fingers. If one finger or a group of neighboring fingers becomes dominant, an $m = 1$ convection mode might result and thus account for a spin periodicity in the surrounding magnetosphere. Figures in Chapter 6 show the $m = 1$ azimuthal asymmetry that results from a relatively small (10%) $m = 1$ asymmetry imposed on the assumed plasma source in a Rice Convection Model simulation (Hill et al. 2014).

Interactions between Titan and the magnetosphere have also been proposed as a means to force an $m = 1$ asymmetry to emerge from the multitude of CII fingers (Winglee et al. 2013), although it seems that such an asymmetry would be manifested in the Titan-fixed coordinate system, not in the longitude-fixed coordinate system.

The CII models can thus produce a planetary period oscillation driven by the magnetosphere if a modest (~10%) $m = 1$ asymmetry is imposed on the magnetospheric plasma source rate or, equivalently, on the Pedersen conductance of the ionosphere. A planetary period oscillation can also be driven by the ionosphere if the ionospheric flow pattern, and the resulting current systems are postulated to have a longitude asymmetry that co-rotates with the northern SKR rate in the northern hemisphere and the southern SKR rate in the southern hemisphere (Kivelson et al. 2011; Jia et al. 2012; Jia and Kivelson 2012). However, the current CII models do not address the problem of dual periodicities. Figure 5.15 depicts the ionospheric current systems that couple to the magnetosphere and generate dual periodicities in a global MHD simulation model (Jia et al. 2012).

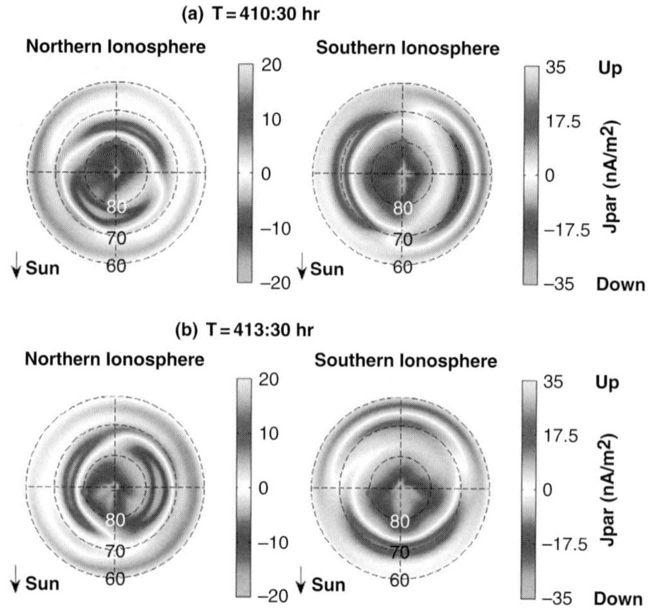

Figure 5.15 To drive magnetospheric and ionospheric periodicities, a dual-vortex neutral wind pattern and resulting current system are imposed on the ionosphere. The wind and current vortices are assumed to rotate at the northern SKR period in the north and the southern SKR period in the south. This figure represents the relation between the vortex and the plotted Birkeland (field-aligned) current density at a particular time. The entire polar cap is rotating and the vortex is embedded in it. To drive the dual vortex, a similar neutral wind is imposed on the ionosphere (from Jia and Kivelson 2012). (A black-and-white version of this figure appears in some formats. For the color version, please refer to the plate section.)

This so-called "dual vortex" system resembles the flow and current vortices observed in the polar ionospheres of the Earth, except that the vortex system at Saturn is assumed to rotate with the planet, while at Earth the vortices are fixed in local time. Apart from this difference, the current vortices could in principle be driven at Saturn by neutral winds, a process analogous to Earth's ionospheric neutral wind dynamo. If the vortices are assumed to rotate at the N and S radio/mag periods, then the magnetosphere faithfully reproduces these periodicities in a manner consistent with observations. Essentially all of the magnetospheric

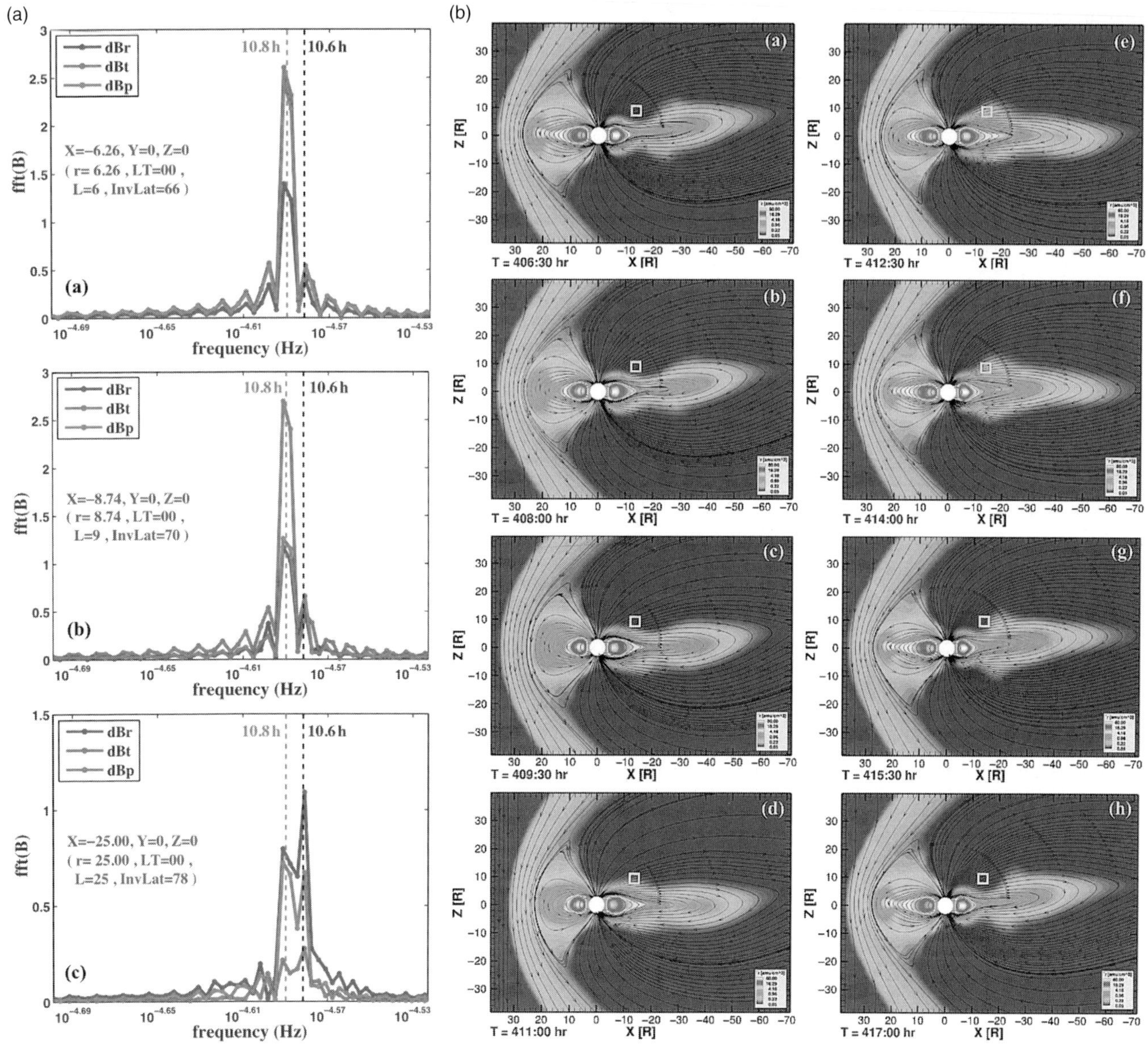

Figure 5.16 The dual vortices in the ionosphere can readily produce the dual periodicities in the magnetic field observed in the magnetosphere (left panels). The observations of these periods, however, would vary with location in the magnetosphere, as shown here for three different locations. The rotating vortex is also manifest in the magnetotail and can actually cause the tail to flap up and down with the SKR period (right panels) (from Jia and Kivelson 2012). (A black-and-white version of this figure appears in some formats. For the color version, please refer to the plate section.)

periodicities – radio, magnetic field, charged particles – can be reproduced if dual north and south ionospheric vortices are driving the magnetosphere. Figure 5.16 illustrates the periodograms that would be observed in magnetic field components at three different places in the magnetosphere as a result of a dual vortex current system in the ionosphere. This model, being global, even reproduces periodic oscillations in the bow shock, magnetopause and magnetotail (Jia et al. 2012; Jia and Kivelson 2012; Kivelson and Jia 2014).

5.12 Rotating Birkeland Current Models

If magnetospheric periodicities are driven by the ionosphere, the information must be communicated to the magnetosphere by Birkeland (field-aligned) currents (often abbreviated as FAC). Such currents are, in fact, observed by the magnetometer at high latitudes, and they appear to have the correct flow sense to provide such communication (e.g. Southwood and Kivelson 2007, 2009; Southwood and Cowley 2014). Birkeland currents are also responsible for the magnetosphere-ionosphere (MI) coupling whereby a rotating planetary ionosphere imposes (or partially imposes) its rotation on the magnetosphere. The MI coupling is not perfect (Wilson et al. 2008, 2009), so the local azimuthal flow is not responsible for the periodicities (Southwood and Cowley 2014). This result is in overall agreement with the modeling assumptions of Jia et al. (2012) and Jia and Kivelson (2012).

Both the SKR and the magnetic PPOs can be generated by two rotating systems of Birkeland currents. In each hemisphere, the rotating current system flows down the field lines from the magnetosphere to the ionosphere on one side of the polar cap, across the polar cap through the ionosphere, and back to the magnetosphere on the opposite side of the polar cap. These current systems rotate with the period of the northern or southern periods. The currents partially close in the opposite hemisphere, as first hypothesized by Southwood and Kivelson (2007), and partially in the equatorial plane of the magnetosphere, as first described by Southwood and Cowley (2014). Hunt et al. (2014) reported that the PPOs were driven outwards from the neutral atmosphere, from a study relating the co-latitude displacement of the Birkeland layer associated with the PPOs to the phase of the PPO oscillation, and considering the related plasma and atmospheric flows. They also provided clear evidence of such inter-hemispheric FAC current by showing that the southern FAC system partially closes in the northern hemisphere. The upward currents (downward-moving electrons) are associated with the generation of SKR, and the downward currents (upward-moving electrons) are believed to be associated with the generation of auroral hiss. The SKR is most intense when the rotating upward Birkeland current is in the post-dawn/pre-noon sector. Figure 5.17 shows a recent schematic of these rotating Birkeland current systems (Hunt et al. 2014, 2015).

Smith et al. (2011, 2014) have pointed out that such vortices must originate not in the ionosphere, but deeper in the atmosphere (~750 km above the 1 bar level) where a twin vortex system might be sustainable. Of course, the twin vortex idea begs the question: what causes the twin vortices in the first place?

Chapter 7 by Stallard et al. on Saturn's aurora describes how such currents close within the ionosphere and generate aurora there. That chapter also discusses how neutral winds might generate field-aligned currents associated with the periodicities.

5.13 Wavy Magnetodisk and Other Models

A different type of model describes Saturn's periodicities in terms of a wave in the magnetodisk of Saturn, where magnetodisk is understood to be the plasma sheet in a rotationally dominated magnetosphere such as Jupiter or Saturn. The wavy magnetodisk is similar to the heliospheric current sheet, except that the amplitude of the magnetodisk wave does not grow with radial distance, although the wavy magnetodisk does bear a strong resemblance to the well-known Parker spiral (Parker 1958). See Carbary (1980) for a concise discussion of the different types of magnetodisk. A wavy magnetodisk has been demonstrated to exist at Jupiter (e.g. Khurana and Kivelson 1989; Khurana 1992), and such a model has been used to explain the periodicities in the outer magnetosphere of Saturn (Arridge et al. 2011; Carbary 2013b). Because Saturn's magnetodisk also has a bowl shape (Arridge et al. 2008), the magnetodisk wave is envisioned as a wave on the warped surface. The equation for the center such a warped-wavy magnetodisk is, in cylindrical coordinates (ρ, ϕ, z), centered on Saturn,

$$z_C = \begin{cases} [\rho - r_H tanh(\rho/r_H)] tan\theta_{SUN} & \rho < \rho_0 \\ [\rho - r_H tanh(\rho/r_H)] tan\theta_{SUN} + & \\ (\rho - \rho_0) tan\theta_{TILT} cos\Psi_{PS}(t, \phi, \rho) & \rho > \rho_0 \end{cases} \tag{5.2}$$

where r_H is a "hinge" distance for the warping, θ_{SUN} is the declination of the Sun relative to Saturn's equator, ρ_0 is a critical radius for the wavy magnetodisk, θ_{TILT} is the effective tilt angle of the disk axis relative to the spin axis of Saturn and $\Psi_{PS}(t, \phi, \rho)$ is the magnetic phase given by

$$\Psi_{CS}(t, \phi, \rho) = \Psi_{Mc}(t, \phi) - \psi_{PS} - \Omega_{Mc}(\rho - \rho_0)/v_{WAVE} \tag{5.3}$$

where $\Psi_{Mc}(t, \phi) = \Omega_{Mc} t + \phi$ represents the rigid co-rotation rate of the magnetic core plus the local time

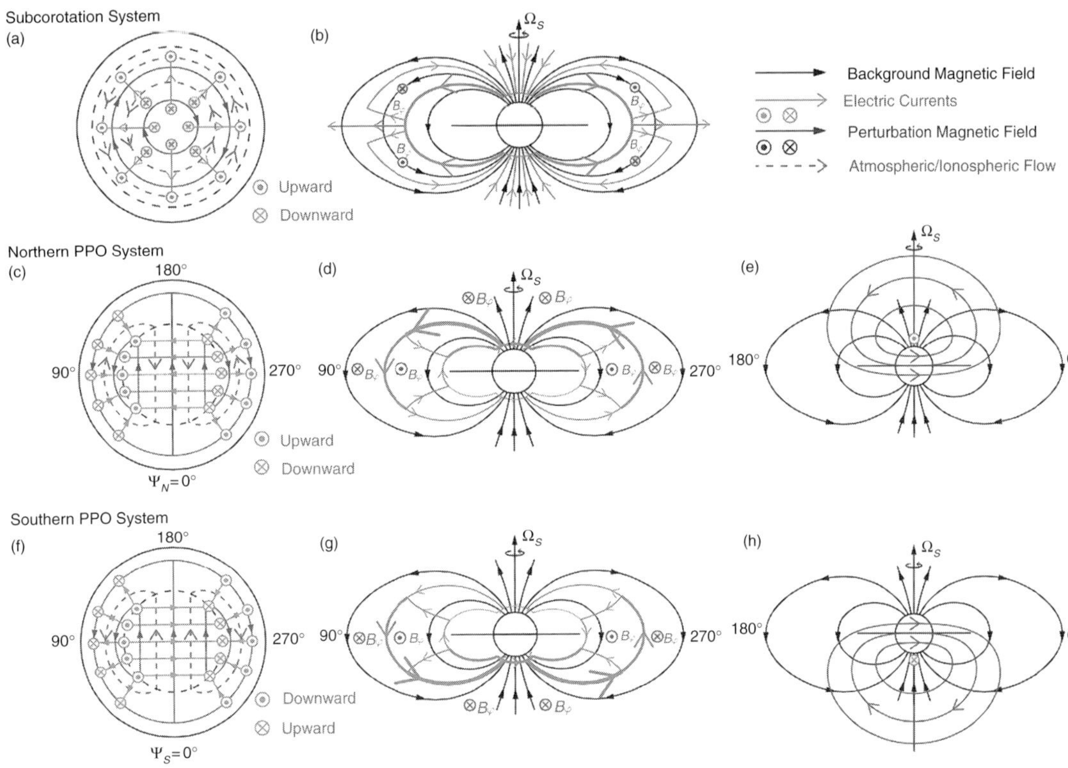

Figure 5.17 Sketches showing the form of the electric currents and perturbation magnetic fields associated with Saturn plasma subcorotation and the PPO-related current systems. Black arrowed lines indicate the background magnetic field, light grey arrowed lines and symbols the electric current system, and dark arrowed lines and symbols the associated perturbation magnetic field. Circled dots and crosses indicate vectors pointing out of and into the plane of the diagrams, respectively. The red arrowed dashed lines in the ionospheric diagrams on the left show the associated upper atmospheric and ionospheric flow, where those in the PPO-related diagrams assume that these systems are driven from the atmosphere. Panels (a) and (b) show the current system associated with magnetospheric plasma subcorotation, where panel (a) shows the northern ionosphere viewed looking down from the north and panel (b) shows a view in a magnetic meridian plane. The system is assumed axisymmetric to a first approximation. Panels (c) through (e) show the northern PPO-related system, where panel (c) shows the northern ionosphere viewed looking down from the north, panel (d) the $\Psi = 90°–270°$ meridian plane, and panel (e) the $\Psi = 0°–180°$ meridian plane. Panels (f) through (h) show the southern PPO-related system, where panel (f) shows the southern ionosphere viewed "through" the planet looking down from the north, panel (g) the $\Psi = 90°–270°$ meridian plane, and panel (h) the $\Psi = 0°–180°$ meridian plane (from Hunt et al. 2015).

phase ϕ of the spacecraft, ψ_{PS} is a "prime meridian" phase angle, Ω_{Mc} is the angular speed of the magnetic core and v_{WAVE} is the outward speed of the wave. Estimates for r_H vary from 16 to 32 R_S, for ρ_0 from 20–30 R_S, for the effective tilt angle θ_{TILT} from 2 to 5 degrees, and for the v_{WAVE} from 5 to 16 R_S/h (Arridge et al. 2011; Carbary 2013B). Figure 5.18 summarizes the wavy magnetodisk model. Within the core region ($\rho < 15\ R_S$), Birkeland current systems drive periodic magnetic perturbations, which in turn generate a rotating "cam" wave in a manner described by Southwood and Kivelson (2007, 2009). Outside this region, the cam imposes a wavy magnetodisk structure having a natural spiral shape. Periodicities in the outer region derive from the flapping motion of the magnetodisk, as suggested by Arridge et al. (2011). The wavy magnetodisk can be made compatible with dual periodicities and phase-lag structures observed in the outer magnetosphere, and can additionally generate periodicities seen at the magnetopause and bow shock (Clarke et al. 2006, 2010).

In the context of global magnetospheric dynamics, plasma mass loading and unloading can also produce a magnetospheric periodicity (Kronberg et al. 2007; Rymer et al. 2013). Plasma from an internal source such as Enceladus simply

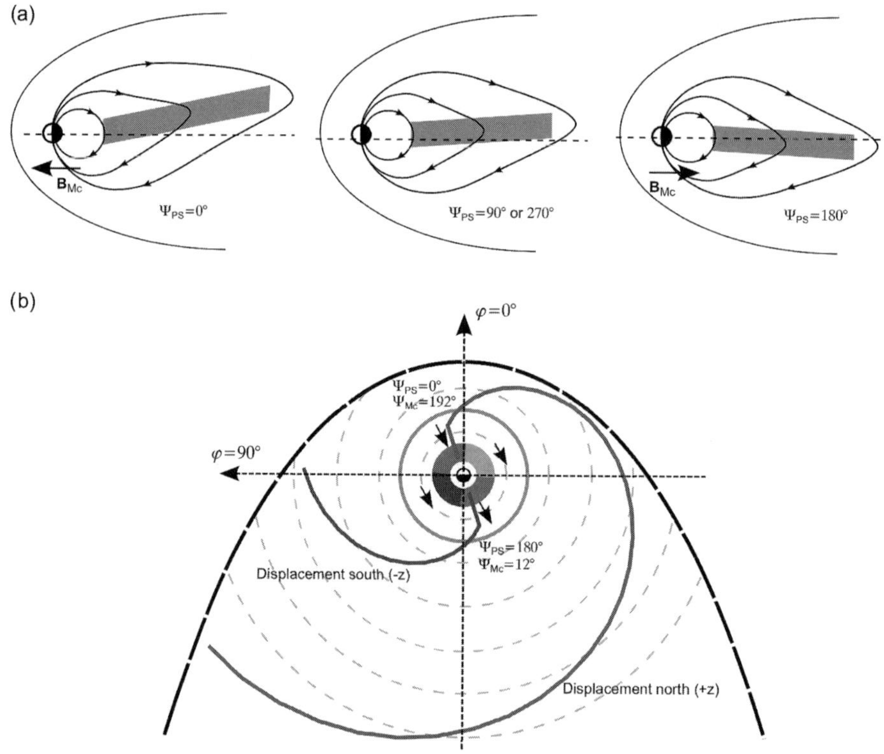

Figure 5.18 A schematic of the wavy magnetodisk and its relation to the rotation of the inner magnetic core region: (a) shows a side view (observer at dusk, Sun at left) of the magnetosdisk at various phases of its flapping motion; (b) shows a top view (observer looking down the spin axis, Sun at top) indicating the spiral wave activated by the wavy disk. A gray circle denotes the boundary of the inner "core" region where magnetic periodicities are generated by rotating Birkeland current systems. Outside the core, the cam motion of the core produces an outgoing spiral wave, which produces the periodicities in the outer magnetosphere. The outgoing wave reaches the magnetopause and generates a periodic oscillation there (from Arridge et al. 2011).

fills the magnetosphere to a critical level at which the field lines can no longer contain it, at which point they break open to release the plasma. After the field lines close back up, the cycle begins again. This model can produce a wide range of periods that may include the basic ~10.7-hour magnetospheric period, but does not attempt to address its dual nature nor its secular variations.

Finally, Saturn could very well be a differential rotator like the Sun. Visible features on the Sun (e.g. sunspots) rotate faster at low latitudes than at higher latitudes. The differential rotation has been known through observations of solar surface features since the 1600s, and until the relatively recent advent of helioseismology served as the only means for probing the rotation of the Sun (e.g. Howard 1984; Beck 1999; Paterno 2010, and references therein). Notably, all theories based on direct observations of the Sun's surface turned out to be wrong about the rotation of the solar interior when checked by the new methods of helioseismology (e.g. Paterno 2010). The prospect of differential rotation of the gas giants was introduced by Dessler (1985) soon after the Voyager flybys of Saturn, but the idea has not received much attention. According to Dessler (1985), one may expect two distinct magnetic periods. For Jupiter, these would be the wobble period of the magnetic field (the System III period) and the slightly longer period associated with oscillations in the Io torus (the System IV period) (Sandel and Dessler 1988; Brown 1995). The zonal wind speeds of Saturn's visible cloud tops definitely vary with latitude, and this variation, if transmitted upward into the ionosphere by some mechanism such as gravity waves, might cause the ionosphere to differentially rotate, which could then lead to differential rotation of the magnetosphere through the connection provided by Birkeland currents. Within the framework of the dual periodicities observed at Saturn, the differential rotation theory would imply that the northern ionosphere rotates at a slightly different rate than the southern, and that all of the rotation rates change over time scales of years.

5.14 Concluding Remarks

Saturn has multiple periodicities relating to planetary rotation, each of them with a period close to ~10.7 h. Some of these periodicities tend to split into two branches, a southern branch and a northern branch. The two branches merged a few months after 2009 equinox and have remained very similar since then. Only a few of the periodicities have been observed quasi-continuously for long periods of time (e.g. the radio, the magnetic, the particle periodicities), and these periodicities are not constant. All the periodicities are generally assumed to have a common source, but questions remain about what that source is, or whether it is in the magnetosphere, ionosphere or interior. At present, it is impossible to determine which, if any, of these periods matches the internal rotation period of Saturn, and certainly the variable periods cannot, so the length of Saturn's day remains an unknown property. None of the extant models addresses the ultimate question of the actual driver of these periodicities.

Saturn also has at least three non-rotational periodicities that are magnetospheric in nature. A "short period" of ~1 h is probably related to the transit time of Alfvén waves between northern and southern polar ionospheres or between the ionosphere and the equatorial plane, while a "long period" of ~26 days (and harmonics thereof) is related to the interaction of the magnetosphere with cyclic structures in the solar wind. A third quasi-periodicity probably arises from successive mass loading and unloading of the magnetosphere, but its period is unknown and probably variable.

Future efforts to investigate Saturn's rotational periodicities will concentrate on observations from the continuing Cassini mission to Saturn. Sometime in 2017, Cassini will enter a Grand Finale or "proximal orbits" phase when the spacecraft will fly very close to the planet – within a few thousand km of the cloud tops. This phase of the mission will provide new capabilities to measure higher-order internal moments of Saturn's mass distribution as well as any higher-order harmonics of its magnetic field. Upper limits on any longitudinal asymmetries will be refined, and these results may clarify the origins of Saturn's periodicities. In particular, Cassini will sample some of the high-latitude regions thought to contain clues about the generation of magnetic and radio periodicities. After the proximal orbits, Cassini will purposely crash into Saturn's atmosphere, leaving a wealth of data for further analysis.

Finally, the authors wish to comment on the terminology of "planetary period oscillations" (PPOs) that has become embedded in the literature. The magnetospheric periods discussed may not reveal the rotation rate of Saturn, but knowing them may enable investigators to appropriately organize certain magnetospheric phenomena. We propose that such magnetospheric periods be referred to as "magnetospheric period oscillations" or MPOs to distinguish them from the actual rotation period of the planet.

Appendix

Archives of Saturn Periodicities

The Cassini magnetospheric community has in the past made efforts to organize the discussion of Saturn's periodicities. For example, there have been specific working sessions on this topic at the Magnetospheres of the Outer Planets (MOP) Meeting in Boston and at the Magnetic Fields and Plasma Science (MAPS) meeting in Annapolis (both held in 2011). These meetings occasioned the eventual exhaustive listing of observations/models relevant to Saturn's periodicities. A collaborative/interactive webpage was opened after the MOP meeting for this purpose at http://typhon.obspm.fr/groups/saturnperiodicities/. A complete list of MOP Meetings can be found at the University of Colorado (Boulder) website: http://lasp.colorado.edu/home/mop/resources/mop-conference/. Specific details about the Cassini Mission, instruments and key data can be found at the MAPS website at the University of Michigan (Ann Arbor): http://mapsview.engin.umich.edu, although the user may need to register. The various SLS longitudes are explained and may be computed using a tool at the University of Iowa

(Iowa City) website: http://cassini.physics.uiowa.edu/sls4/. A longitude system based on magnetic field periodicities is available by contacting Gabriele Provan at gp31@leicester.ac.uk.

Acknowledgments

This research was supported by the NASA Office of Space Science under Task Order 003 of contract NAS5-97271 between NASA Goddard Space Flight Center and Johns Hopkins University. The work of TWH was supported by NASA JPL contract 1405851 with the Southwest Research Institute. The research at the University of Iowa was supported by NASA through Contract 1415150 with the Jet Propulsion Laboratory. Many of the authors also wish to acknowledge the International Space Science Institute in Bern, Switzerland, for hosting a recent meeting (October 2015) on Saturn's periodicities.

References

Acuña, M. H., J. E. P. Connerney and N. F. Ness (1983), The Z(3) zonal harmonic model of Saturn's magnetic field: Analyses and implications, *J. Geophys. Res.*, *88*, 8771.

Andrews, D. J., E. J. Bunce, S. W. H. Cowley et al. (2008), Planetary period oscillations in Saturn's magnetosphere: Phase relation of equatorial magnetic field oscillations and Saturn kilometric radiation, *J. Geophys. Res.*, *113*, A09205, doi:10.1029/ 2007JA0129376.

Andrews, D. J., A. J. Coates, S. W. H. Cowley et al. (2010a), Magnetospheric period oscillations at Saturn: Comparison of equatorial and high-latitude magnetic field periods with north and south SKR periods, *J. Geophys. Res.*, *115*, A12252.

Andrews, D. J., A. J. Coates, S. W. H. Cowley (2010b), Magnetospheric period oscillations at Saturn: Comparison of equatorial and high-latitude magnetic field periods with north and south SKR periods, *J. Geophys. Res.*, *115*, A12252, http://dx.doi.org/10.1029/2010JA015666.

Andrews, D. J., B. Cecconi, S. W. H. Cowley et al. (2011), Planetary period oscillations at Saturn: Evidence in magnetic field phase data for rotational modulation of Saturn kilometric radiation emissions, *J. Geophys. Res.*, *116*, A09206.

Andrews, D. J., S. W. H. Cowley, M. K. Dougherty et al. (2012), Planetary period oscillations in Saturn's magnetosphere: Evolution of magnetic oscillation properties from southern summer to post-equinox, *J. Geophys. Res.*, *117*, A4, A04224.

Anderson, J. D. and G. Schubert (2007), Saturn's gravitation field, internal rotation, and interior structure, *Science*, *317*, 1384–1387, doi:10.1126/science.1144835.

Arridge, C. S., K. K. Khurana, C. T. Russell et al. (2008), Warping of Saturn's magnetospheric and magnetotail current sheets, *J. Geophys. Res.*, *113*, A08217, doi:10.1029/2007JA012963.

Arridge, C. S., N. André, K. K. Khurana et al. (2011), Periodic motion of Saturn's nightside plasma sheet, *J. Geophys. Res.*, *116*, A11205, doi:10.1029/ 2011JA016827.

Badman, S. V., S W. H. Cowley, L. Lamy et al. (2008), Relationship between solar wind corotating interaction regions and the phasing and intensity of Saturn kilometric radiations bursts, *Ann. Geophys.*, *26*, 3641–3651.

Badman, S. V., A. Masters, H. Hasegawa et al. (2013), Bursty magnetic reconnection at Saturn's magnetopause, *Geophys. Res. Lett.*, *40*, 1027–1031, doi:10.1002/ grl.50199.

Baillié, K., J. E. Colwell, J. J. Lissauer et al. (2011), Waves in the Cassini UVIS stellar occultations. 2. The C ring, *Icarus*, *216*, 292–308, doi:10.1016/ j.icarus.2011.05.019.

Beck, J. G. (1999), A comparison of differential rotation measurements, *Solar Phys.*, *191*, 47–70.

Brown, M. E. (1995), Periodicities in the Io plasma torus, *J. Geophys. Res.*, *100*, 21683–21696.

Bunce, E. J., C. S. Arridge, J. T. Clarke et al. (2008), Origin of Saturn's aurora: Simultaneous observations by Cassini and the Hubble Space Telescope, *J. Geophys. Res.*, *113*, A09209, doi:10.1029/ 2008JA013257.

Burch, J. L, J. Goldstein, T. W. Hill et al. (2005), Properties of local plasma injections in Saturn's magnetosphere, *Geophys. Res. Lett.*, *32*, L14S02, doi:10.1029/ 2005GL022611.

Burns, J. A., D. P. Hamilton and M. R. Showalter (2001), Dusty rings and circumplanetary dust: Observations and simple physics, in *Interplanetary Dust*, ed. Grün, E., B. Gustafson, S. Dermott, and H. Fecthig, Springer, pp. 641–725.

Burton, M. E., M. K. Dougherty and C. T. Russell (2010), Saturn's internal planetary magnetic field. *Geophys. Res. Lett.*, *37*, L24105, http://dx.doi.org/10.1029/ 2010GL045148.

Carbary, J. F. (1980), Periodicities in the Jovian magnetosphere: Magnetodisk models after Voyager, *Geophys. Res. Lett.*, *7*(1), 29–32.

Carbary, J. F. (2013a), Longitude dependences of Saturn's ultraviolet aurora, *Geophys. Res. Lett.*, *118*, 1–5, doi:10.1002/grl.50430.

Carbary, J. F. (2013b), Wavy magnetodisk in Saturn's outer magnetosphere, *Geophys. Res. Lett.*, *40*, 1–5, doi:10.1002/grl.50994.

Carbary, J. F. (2015), Doppler effects of periodicities in Saturn's magnetosphere, *J. Geophys. Res. Space Physics*, *120*, doi:10.1002/2015JA021850.

Carbary, J. F. and S. M. Krimigis (1982), Charged particle periodicities in the Saturnian magnetosphere, *Geophys. Res. Lett.*, *9*, 1073–1076.

Carbary, J. F. and D. G. Mitchell (2013a), Periodicities in Saturn's Magnetosphere, *Rev. Geophys.*, *51*, 1–30, doi:10.1002/rog.20006.

Carbary, J. F., D. G. Mitchell and P. Brandt (2014), Local time dependences of oxygen ENA periodicities at Saturn, *J. Geophys. Res.*, *119*, doi:10.1002/jgra51230.

Carbary, J. F., D. G. Mitchell, P. Brandt et al. (2008), Periodic tilting of Saturn's plasma sheet, *Geophys. Res. Lett.*, *35*, L24101, doi:10.1029/2008GL036339.

Carbary, J. F., D. G. Mitchell, P. Brandt (2011), ENA periodicities and their phase relations to SKR emission at Saturn, *Geophys. Res. Lett.*, *38*, L16106, doi:10.1029/2011GL048418.

Carbary, J. F., D. G. Mitchell, S. M. Krimigis et al. (2007), Charged particle periodicities in Saturn's outer magnetosphere, *J. Geophys. Res.*, *112*, A06246, doi:10.1029/2007JA012351.

Carbary, J. F., D. G. Mitchell, S. M. Krimigis (2009a), Dual periodicities in energetic electrons at Saturn, *Geophys. Res. Lett.*, *36*, L20103, doi:10.1029/ 2009GL040517.

Carbary, J. F., D. G. Mitchell, S. M. Krimigis (2012), Unusually short period in electrons at Saturn, *Geophys. Res. Lett.*, *39*, L22103, doi:10.1029/2012GL054019.

Carbary, J. F., E. C. Roelof, D. G. Mitchell et al. (2009b), Solar wind periodicity in energetic electrons at Saturn, *Geophys. Res. Lett.*, *36*, L22104, doi:10.1029/GL041086.

Carbary, J. F., E. C. Roelof, D. G. Mitchell (2013b), Solar periodicity in energetic ions at Saturn, *J. Geophys. Res.*, *118*, 1–8, doi:10.1002/jgra.50282.

Chen, Y. and T. W. Hill (2008), Statistical analysis of injection/dispersion events in Saturn's inner magnetosphere, *J. Geophys. Res.*, *113*, A07215, doi:10.1029/2008JA013166.

Chenette, D. L., T. F. Conlon and J. A. Simpson (1974), Bursts of relativistic electrons from Jupiter observed in interplanetary space with the time variation of planetary rotation period, *J. Geophys. Res.*, *79*(25), 3551–3558.

Chen, Y., T. W. Hill, A. M. Rymer et al. (2010), Rate of radial transport of plasma in Saturn's inner magnetosphere, *J. Geophys. Res.*, *115*, A10211, doi:10.1029/2010JA015412.

Clarke, K.E., N. André, D. J. Andrews et al. (2006), Cassini observations of planetary-period oscillations of Saturn's magnetopause, *Geophys. Res. Lett.*, *33*, L23104, doi:10.1029/2006GL027821.

Clarke, K. E., D. J. Andrews, A. J. Coates et al. (2010), Magnetospheric period oscillations in Saturn's bow shock, *J. Geophys. Res.*, *115*, A05202, doi:10.1029/2009JA015164.

Connerney, J. E. P., M. H. Acuña and N. F. Ness (1983), Currents in Saturn's magnetosphere, *J. Geophys. Res.*, *88*, 8779.

Cowley, S. W. H., D. M. Wright, E. J. Bunce et al. (2006), Cassini observations of planetary-period magnetic field oscillations in Saturn's magnetosphere: Doppler shifts and phase motion, *Geophys. Res. Lett.*, *33*, L07104, doi:10.1029/2005GL025522.

Cowley, S. W. H. and G. Provan (2013), Saturn's magnetospheric planetary period oscillations, neutral atmospheric circulation, and thunderstorm activity: Implications, or otherwise, for physical links, *J. Geophys. Res.*, *118*, 724607261, doi:10.1002/2013JA019200.

Cowley, S. W. H. and G. Provan (2015), Planetary period oscillations in Saturn's magnetosphere: Comments on the relation between post-equinox periods determined from magnetic field and SKR emission data, *Ann. Geophys.*, accepted for publication.

Cowley, S. W. H, G. Provan and D. G. Andrews (2015), Comment on "Magnetic phase structure of Saturn's 10.7-hour oscillations" by Yates et al., *J. Geophys. Res.*, accepted for publication, doi:10.1002/2015JA021351.

Crary, F. J., J. T. Clarke, M. K. Dougherty et al. (2005), Solar wind dynamic pressure and electric field as the main factors controlling Saturn's aurora, *Nature*, *433*, 720–722, doi:10.1038/nature03333.

D'Aversa, E., G. Bellucci, P. D. Nicholson et al. (2010), The spectrum of a Saturn ring spoke from Cassini/VIMS, *Geophys. Res. Lett.*, *37*, 1, doi:10.1029/2009GL041427.

Davies, M. E., V. K. Abalakin, M. Bursa et al. (1986), Report of the IAU/IAG/COSPAR Working Group on Cartographic Coordinates and Rotational Elements of the Planets and Satellites: 1985, *Celest. Mech. 39*, 103–113.

Davis, L. and E. J. Smith (1990), A model of Saturn's magnetic field based on all available data, *J. Geophys. Res.*, *95*, 15257.

Desch, M. D. (1982), Evidence for solar wind control of Saturn radio emission, *Geophys. Res. Lett.*, *87* (A5), 4549–4554.

Desch, M. D. and M. L. Kaiser (1981), Voyager measurement of the rotation period of Saturn's magnetic field, *Geophys. Res. Lett.*, *8*, 253–256.

Desch, M. D. and H. O. Rucker (1983), The relationship between Saturn kilometric radiation and the solar wind, *J. Geophys. Res.*, *88*, 8999–9006.

Dessler, A. J. (1986), Differential rotation of the magnetic fields of the gaseous planets, *Geophys. Res. Lett.*, *12* (3), 299–302.

Dyudina, U. A., A. P. Ingersoll, S. P. Ewald et al. (2016), Saturn's aurora observed by the Cassini camera at visible wavelengths, *Icarus*, *263*, 32–43, doi:10.1016/j.icarus. 2015.05.022.

Espinosa, S. A., and M. K. Dougherty (2000), Periodic perturbations in Saturn's magnetic field, *Geophys. Res. Lett.*, *27*(17), 2785–2788, doi:10.1029/2000GL000048.

Espinosa, S. A., D. J. Southwood and M. K. Dougherty (2003a), Reanalysis of Saturn's magnetospheric field data view of spin-periodic perturbations, *J. Geophys. Res.*, *108*(A2), 1085, doi:10.1029/2001JA005083.

Espinosa, S. A., D. J. Southwood and M. K. Dougherty (2003b), How can Saturn impose its rotation period in a noncorotating magnetosphere?, *J. Geophys. Res.*, *108* (A2), 1086, doi:10.1029/2001JA005084.

Farmer, A. J. and P. Goldreich (2001), Spoke formation under moving plasma clouds, *Icarus*, *179*(2), 535–548, doi:10.1016/j.icarus.2005.07.025.

Fischer, G., D. A. Gurnett, W. S. Kurth et al. (2015), Saturn kilometric radiation periodicity after equinox, *Icarus*, *254*, 72–91, doi:10.1016/j.icarus.2015.03.014.

Fischer, G., W. S. Kurth, D. A. Gurnett et al. (2011), A giant thunderstorm on Saturn, *Nature*, *475*, 75–77, doi:10.1038/nature10205.

Franklin, F. A., G. Colombi and A. F. Cook (1982), A possible link between the rotation of Saturn and its ring structure, *Nature*, *295*, 128–130.

Galopeau, P. H. M. and A. Lecacheux (2000), Variations of Saturn's radio rotation period measured at kilometer wavelengths, *J. Geophys. Res.*, *105*(A6), 13,089–13,102.

Giampieri, G., M. K. Dougherty, E. J. Smith et al. (2006), A regular period for Saturn's magnetic field that may track its internal rotation, *Nature*, *441*, 62–64, doi:10.1038/nature04750.

Godfrey, D. A. (1988), A hexagonal feature around Saturn's north pole, *Icarus*, *76*, 335–356.

Godfrey, D. A. (1990), The rotation period of Saturn's polar hexagon, *Science*, *247*, 1206–1208.

Goertz, C. K. and G. Morfill (1983), A model for the formation of spokes in Saturn's rings, *Icarus*, *53*, 219–229.

Goldreich, P. and A. J. Farmer (2007), Spontaneous axisymmetry breaking of the external magnetic field at Saturn, *J. Geophys. Res.*, *112*, A05225, doi:10.1029/2006JA012163.

Grün, E., C. K. Goertz, G. E. Morfill et al. (1992), Statistics of Saturn's spokes, *Icarus*, *99*, 191–201.

Grün, E., G. E. Morfill, R. J. Terrile et al. (1983), The evolution of spokes in Saturn's B ring, *Icarus*, *54*, 227–252.

Gurnett, D. A., J. B. Groene, A. M. Persoon et al. (2010), The reversal of the rotational modulation rates of the north and south components of Saturn kilometric radiation near equinox, *Geophys. Res. Lett.*, *37*, L24101, doi:10.1029/2010GL045796.

Gurnett, D. A., W. S. Kurth, G. B. Hospodarsky et al. (2005), Radio and plasma wave observations at Saturn from Cassini's approach and first orbit, *Science*, *307*, 1255–1259.

Gurnett, D. A., A. Lecacheux, W. S. Kurth et al. (2009a), Discovery of a north-south asymmetry in Saturn's radio rotation period, *Geophys. Res. Lett.*, *36*, L16102, doi:10.1029/2009GL039621.

Gurnett, D. A., A. M. Persoon, J. B. Groene et al. (2009b), A north-south difference in the rotation rate of auroral hiss at Saturn: Comparison to Saturn's kilometric radio emission, *Geophys. Res. Lett.*, *36*, L21108, doi:10.1029/2009GL040774.

Hedman, M. M., J. A. Burns, M. W. Tiscareno et al. (2009), Organizing some very tenuous things: Resonant structures in Saturn's faint rings, *Icarus*, *202*, 260–279, doi:10.1016/j.icarus.2009.02.016.

Hedman, M. M. and P. D. Nicholson (2013), Kronoseismology: Using density waves in Saturn's C ring to probe the planet's interior, *Astron. J.*, *146*, 12, doi:10.1088/0004–6256/146 /1/12.

Hedman, M. M. and P. D. Nicholson (2014), More Kronoseismology with Saturn's rings, *MNRAS*, *444*, 1369–1388.

Helled, R., E. Galanti and Y. Kaspi (2015), Saturn's fast spin determined from its gravitational field and oblateness, *Nature*, *250*, 202–204, doi:10.1038/nature14278.

Helled, R., G. Schubert and J. D. Anderson (2009), Jupiter and Saturn rotation periods, *Planet. Space Sci.*, *57*, 1467–1473, doi:10.1016/j.pss.2009.07.008.

Higgins, C. A., T. D. Carr, F. Reyes et al. (1997), A redefinition of Jupiter's rotation period, *J. Geophys. Res.*, *102*, 22033–22042, doi:10.1029/97JA02090.

Hill, T. W., A. Jaggi, R. A. Wolf et al. (2014), Generation of a global longitudinal asymmetry in Saturn's magnetosphere, paper # SM23D-09 presented at Fall AGU meeting, San Francisco, CA, Dec. 16.

Hill, T. W., A. M. Rymer, J. L. Burch et al. (2005), Evidence for rotationally driven plasma transport in Saturn's magnetosphere, *Geophys. Res. Lett.*, *32*, L14S10, doi:10.1029/ 2005GL022620.

Howard, R. (1984), Solar rotation, *Ann. Rev. Astron. Astrophys.*, *22*, 131–155.

Hunt, G. J., S. W. H. Cowley, G. Provan et al. (2014), Field-aligned currents in Saturn's southern nightside magnetosphere: Sub-corotation and planetary period oscillation components, *J. Geophys. Res.*, *119*, doi:10.1002/2014JA020506.

Hunt, G. J., S. W. H. Cowley, G. Provan (2015), Field-aligned currents in Saturn's northern nightside magnetosphere: Evidence for inter-hemispheric current flow associated with planetary period oscillations, *J. Geophys. Res. Space Physics.*, *120*, doi.10.1002/2015JA021454.

Jia, X. and M. G. Kivelson (2012), Driving Saturn's magnetospheric periodicities from the upper atmosphere/ionosphere: Magnetotail response to dual sources, *J. Geophys. Res.*, *117*, A11219, doi:10.1029/2012JA018183.

Jia, X., M. G. Kivelson and T. T. Gombosi (2012), Driving Saturn's magnetosphere periodicities from the upper atmosphere/ionosphere, *J. Geophys. Res.*, *117*, A04215, doi:10.1029/ 2011JA017367.

Kane, M., D. G. Mitchell, J. F. Carbary et al. (2014), Plasma convection in the nightside magnetosphere of Saturn determined from energetic ion anisotropies, *Planet. Space Sci.*, *91*, 1–13, doi:10.1016/j.pss.2013.10.001.

Kennelly, T. J., J. S. Leisner, G. B. Hospodarsky et al. (2013), Ordering of injection evens within Saturnian SLS longitude and local time, *J. Geophys. Res.*, *118*, 1–7, doi:10.1002/jgra.50152.

Khurana, K. K. (1992), A generalized hinged-magnetodisk model of Jupiter's nightside current sheet, *J. Geophys. Res.*, *97*(A5), 6269–6276.

Khurana, K. K. and M. G. Kivelson (1989), On Jovian plasma sheet structure, *J. Geophys. Res. 94*(A9), 11791–11803.

Kidder, A., R. M. Winglee and E. M. Harnett (2009), Regulation of the centrifugal interchange cycle in Saturn's inner magnetosphere, *J. Geophys. Res.*, *114*, A02205, doi:10.1029/ 2008JA013100.

Kimura, T., L. Lamy, C. Tao et al. (2013), Long-term modulations of Saturn's auroral radio emissions by the solar wind and seasonal variations controlled by the solar ultraviolet flux, *J. Geophys. Res. Space Physics*, *118*, 7019–2035, doi:10.1002/ 2013JA018833.

Kivelson, M. G. (2014), Planetary magnetodiscs: Some unanswered questions, *Space Sci. Rev.*, doi:10.1007/ s11214-014–0046-6.

Kivelson, M. and X. Jia (2014), Control of periodic variations in Saturn's magnetosphere by compressional waves, *J. Geophys. Res.*, *119*, doi:10.1002/2014JA020258.

Kivelson, M. G., X. Jia and T. I. Gombosi (2011), A vortex in Saturn's upper atmosphere as the driver of electromagnetic periodicities at Saturn: Magnetospheric and ionospheric responses, *EPSC Abstracts*, *6*, EPSC-DPS 2011–58.

Krimigis, S. M., D. G. Mitchell, D. C. Hamilton et al. (2005), Dynamics of Saturn's magnetosphere from MIMI during Cassini's orbital insertion, *Science*, *397*, 1270–1273, doi:10.1126/science.1105978.

Kronberg, E. A., K.-H. Glassmeier, J. Woch et al. (2007), A possible intrinsic mechanism for the quasi-periodic dynamics of the Jovian magnetosphere, *J. Geophys. Res.*, *112*, A05203, doi:10.1029/2006JA011994.

Kurth, W. S., T. F. Averkamp, D. A. Gurnett et al. (2008), An update to a Saturn longitude system based on kilometric radio emissions, *J. Geophys. Res.*, *113*, A05222, doi:10.1029/2007JA012861.

Kurth, W. S., D. A. Gurnett, J. T. Clarke et al. (2005), An Earth-like correspondence between Saturn's auroral features and radio emission, *Nature*, *433*, 722–725, doi:10.1038/nature03334.

Kurth, W. S., A. Lecacheux, T. F. Averkamp, et al. (2007), A Saturnian longitude system based on a variable kilometric radiation period, *Geophys. Res. Lett.*, *34*, L02201, doi:10.1029/2006GL028336.

Lamy, L., P. Zarka, B. Cecconi et al. (2008), Saturn kilometric radiation: Average and statistical properties, *J. Geophys. Res.*, *113*, A07201, doi:10.1029/ 2007JA012900.

Lamy, L., P. Zarka, B. Cecconi et al. (2010a), Auroral kilometric radiation diurnal, semi-diurnal, and shorter-term modulations disentangled by Cassini/RPWS, *J. Geophys. Res.*, *115*, A09221, doi:10.1029/ 2010JA015434.

Lamy, L., P. Schippers, P. Zarka et al. (2010b), Properties of Saturn kilometric radiation measured within its source region, *Geophys. Res. Lett.*, *37*, L12104.

Lamy, L. (2011), Variability of southern and northern SKR periodicities, in *Planetary Radio Emissions VII*, edited by H.O. Rucker, W.S. Kurth, P. Louarn, and G. Fischer, pp. 39050, Austrian Academy of Sciences, Vienna, Austria.

Lecacheux, A., P. Galopeau and M. Aubier (1997), Revisiting Saturnian radiation with Ulysses/URAP, in *Planetary Radio Emissions IV*, ed. by H. O. Rucker, S. J. Bauer, and A. Lecacheux, pp. 313–325, Austrian Acad. of Sci. Press, Vienna.

Liu, X., T. W. Hill, R. A. Wolf et al. (2010), Numerical simulation of plasma transport in Saturn's inner magnetosphere using the Rice Convection Model, *J. Geophys. Res.*, *115*, A12252, doi:10.1029/2010JA015859.

Liu, X. and T. W. Hill (2012), Effects of finite plasma pressure on centrifugally driven convection in Saturn's inner magnetosphere, *J. Geophys. Res.*, *117*, A07216, doi:10.1029/2012JA017827.

Marley, M. S. and C. C. Porco (1993), Planetary acoustic mode seismology: Saturn's rings, *Icarus*, *106*, 508–524.

Mauk, B. H., J. Saur, D. G. Mitchell et al. (2005), Energetic particle injections in Saturn's magnetosphere, *Geophys. Res. Lett.*, *32*, L14S05, doi:10.1029/2005GL022485.

McIntyre, M. E. and T. G. Shepherd (1987), An exact local conservation theorem for finite-amplitude disturbances to non-parallel shear flows, with remarks on Hamiltonian structure and Arnol'd's stability theorem, *J. Fluid Mech*, *181*(1), 527–565.

Mitchell, D. G., P. C. Brandt, E. C. Roelof et al. (2005), Energetic ion acceleration in Saturn's magnetotail: Substorms at Saturn? *Geophys. Res. Lett.*, *32*, L20S01, doi:10.1029/2005GL022647.

Mitchell, D. G., W. S. Kurth, G. B. Hospodarsky et al. (2009), Ion conics and electron beams associated with auroral processes on Saturn, *J. Geophys. Res.*, *114*, A02212, doi:10.1029/2008JA013621.

Mitchell, D. G., J. F. Carbary, E. J. Bunce et al. (2014), Recurrent pulsations in Saturn's high latitude magnetosphere, *Icarus*, *263*, 94–100, doi:10.1016/j.icarus.2014.10.028.

Mitchell, C. J., C. C. Porco, H. L. Dones et al. (2013), The behavior of spokes in Saturn's B ring, *Icarus*, *225*, 446–474, doi:10.1016/j.icarus.2013.02.011.

Morioka, A., Y. Miyoshi, S. Kurita et al. (2013), Universal time control of AKR: Earth is a spin-modulated variable radio source, *J. Geophys. Res. Space Physics*, *118*, 1123–1131, doi:10.1002/jgra.50180.

Nichols, J. D., B. Cecconi, J. T. Clarke et al. (2010a), Variation of Saturn's UV aurora with SKR phase, *Geophys. Res. Lett.*, *37*, L15102.

Nichols, J. D., J. T. Clarke, S. W. H. Cowley et al. (2008), Oscillation of Saturn's southern auroral oval, *Geophys. Res. Lett.*, *113*, A11205, doi:10.1029/2008JA013444.

Nichols, J. D., S. W. H. Cowley and L. Lamy (2010b), Dawn-dusk oscillation of Saturn's conjugate auroral ovals, *Geophys. Res. Lett.*, *37*, 24, L24102.

Panchenko, M., M. L. Khodachenko, A. G. Kislyakov (2009), Daily variations of auroral kilometric radiation observed by STEREO, *Geophys. Res. Lett.*, *36*, L06102, doi:10.1029/ 2008GL037042.

Paranicas, C., D. G. Mitchell, E. C., Roelof et al. (2005), Periodic intensity variations in global ENA images of Saturn, *Geophys. Res. Lett.*, *32*, L21101, doi:10.1029/ 2005GL023656.

Paranicas, C., D. G. Mitchell, E. C., Roelof (2007), Energetic electrons injected into Saturn's neutral gas cloud, *Geophys. Res. Lett.*, *34*, L02109, doi:10.1029/ 2006GL028676.

Parker, E. (1958), Dynamics of the interplanetary gas and magnetic fields, *Ap. J.*, *128*, 664–676.

Paterno, L. (2010), The solar differential rotation: A historical review, *Astrophys. Space Sci.*, *328*, 269–277.

Porco, C. C. and G. E. Danielson (1982), The periodic variation of spokes in Saturn's rings, *Astron. J.*, *87*(2), 826–833.

Provan, G., D. J. Andrews, C. S. Arridge et al. (2012), Dual periodicities in planetary-period magnetic field oscillations in Saturn's tail, *J. Geophys. Res.*, *117*, A01209, doi:10.1029/2011JA017104.

Provan, G, D. J. Andrews, B. Cecconi et al. (2011), Magnetospheric period magnetic field oscillations at Saturn: Equatorial phase 'jitter' produced by superposition of southern- and northern-period oscillations, *J. Geophys. Res.*, *116*, A04225.

Provan, G., S. W. H. Cowley, J. Sandhu et al. (2013), Planetary period magnetic field oscillations in Saturn's magnetosphere: Postequinox abrupt non-monotonic transitions to northern system dominance, *J. Geophys. Res.*, *118*, 3243–3264, doi:10.1002/jgra.50186.

Provan, G., L. Lamy, S. W. H. Cowley et al. (2014), Planetary period oscillations in Saturn's magnetosphere: Comparison of magnetic oscillations and SKR modulations in the post-equinox interval, *J. Geophys. Res.*, *119*, 7380–7401, doi:10.1002/ 2014JA020011.

Provan, G., C. Tao, S. W. H. Cowley et al. (2015), Planetary period oscillations in Saturn's magnetosphere: Examining the relationship between abrupt changes in behavior and solar wind-induced magnetospheric compressions and expansions, *J. Geophys. Res. Space Physics*, *120*, doi:10.1002/2015JA021642.

Read, P. L., T. E. Dowling, and G. Schubert (2009), Saturn's rotation period from its atmospheric planetary-configuration, *Nature*, *460*, 608–610, doi:10.1038/ nature08194.

Roussos, E., N. Krupp, D. G. Mitchell et al. (2015), Quasi-periodic injections of relativistic electrons in Saturn's outer magnetosphere, *Icarus*, *263*, 101–116, doi:10.1016/ j.icarus.2015.04.017.

Rymer, A. M., D. G. Mitchell, T. W. Hill et al. (2013), Saturn's magnetospheric refresh rate, *Geophys. Res. Lett.*, *40*, 1–5, doi:10.1002/ grl.50530.

Sánchez-Lavega, A. (2005), How long is the day on Saturn?, *Science*, *307*, 1223–1224, doi:10.1126/ scence.1104956.

Sánchez-Lavega, A., J. F. Rojas, J. R. Acarreta et al. (1997), New observations and studies of Saturn's long-lived North Polar Spot, *Icarus*, *128*, 322–334.

Sánchez-Lavega, A., J. F. Rojas and P. Sada (2000), Saturn's zonal winds at cloud level, *Icarus*, *145*, 405–420, doi:10.1006/icar.2000.6449.

Sánchez-Lavega, A., T. del Rio-Gazrelurrutia, R. Hueso et al. (2014), The long-term steady motion of Saturn's hexagon and the stability of its enclosed jet stream under seasonal changes, *Geophys. Res. Lett.*, *41*, 1425–1431, doi:10.1002/2013GL059078.

Sandel, B. R. and A. J. Dessler (1988), Dual periodicity of the Jovan magnetosphere, *J. Geophys. Res.*, *93*, 5487–5504.

Seidelman, P. K., V. K. Abalakin, M. Bursa et al. (2002), Report of the IAU/IAG working group on cartographic coordinates and rotational elements of the planets and satellites: 2000, *Celest. Mech. Dyn. Astr.*, *82*, 83–110.

Seidelmann, P. K., B. A. Archinal, M. F. A'hearn et al. (2007), Report of the IAU/IAG working group on cartographic coordinates and rotational elements: 2006, *Celest. Mech. Dyn. Astr.*, *98*, 155–180, doi:10.1007/ s10569-007–9-72-y.

Shu, F. H. (1984), Waves in planetary rings, in *Planetary Rings*, ed. R. Greenberg and A. Brahic, University of Arizona Press, Tucson, AZ, pp. 513–561.

Smith, C. G. A. (2006), Periodic modulations of gas giant magnetospheres by the neutral upper atmosphere, *Ann. Geophys.*, *24*, 2709–2717, doi:10.5195/angeo-24–2709-2006.

Smith, C. G. A. (2011), A Saturnian cam current system driven by asymmetric thermospheric heating, *M.N.R.A.S.*, *410*, 2315–2328, doi:10.1111/j. 1365–2966.2010.17602.x.

Smith, C. G. A. (2014), On the nature and location of the proposed twin vortex systems in Saturn's polar upper atmosphere, *J. Geophys. Res.*, *119*, doi:10.1002/ 2014JA019934.

Southwood, D. J. and M. G. Kivelson (2007), Saturnian magnetospheric dynamics: Elucidation of a camshaft model, *J. Geophys. Res.*, *112*, A12222, doi:10.1029/ 2007JA012254.

Southwood, D. J. and M. G. Kivelson (2009), The source of Saturn's periodic radio emission, *J. Geophys. Res.*, *114*, A09201, doi:10.1029/2008JA013800.

Southwood, D. J. (2011), Direct evidence of differences in magnetic rotation rate between Saturn's northern and southern polar regions, *J. Geophys. Res.*, *116*. A01201, doi:10.1029/2010JA016070.

Southwood, D. J. and S. W. H. Cowley (2014), The origin of Saturn's magnetic periodicities: Northern and southern current systems, *J. Geophys. Res.*, *119*, doi:10.1002/ 2013JA019632.

Stevenson, J. D. (1982), Are Saturn's rings a seismograph for planetary inertial oscillations? *EOS, Trans AGU*, *63*, 1020.

Szego, K., Z. Nemeth, L. Foldy et al. (2013), Dual periodicities in the flapping of Saturn's magnetodisk, *J. Geophys. Res.*, *118*, 2883–2887, doi:10.1002/ jgra.50316.

Thomsen, M. F., D. B. Reisenfeld, D. M. Delapp et al. (2010), Survey of ion plasma parameters in Saturn's magnetosphere, *J. Geophys. Res.*, *115*, A10220, doi:10.1029/2010JA015267.

Warwick, J. W., J. B. Pearce, D. R. Evans et al. (1981), Planetary radio astronomy observations from Voyager 1 near Saturn, *Science*, *212*, 239–243, doi:10.1126/ science.212.4491.239.

Wilson, R. J., R. L. Tokar, M. G. Henderson et al. (2008), Cassini Plasma Spectrometer thermal ion measurements in Saturn's inner magnetosphere, *J. Geophys. Res.*, *113*, A12218, doi:10.1029/2008JA013486.

Wilson, R. J., R. L. Tokar, and M. G. Henderson (2009), Thermal ion flow in Saturn's inner magnetosphere measured by the Cassini plasma spectrometer: a signature of the Enceladus torus?, *Geophys. Res. Lett.*, *36*, L23104, doi:10.1029GL040225.

Winglee, R. M., A. Kidder, E. Harnett et al. (2013), Generation of periodic signatures at Saturn through Titan's interaction with the centrifugal interchange instability, *J. Geophys. Res.*, *118*, 1–17, doi:10.1002/ jgra.50379.

Yang, Y. S., R. A. Wolf, R. W. Spiro et al. (1994), Numerical simulation of torus-driven plasma transport in the Jovian magnetosphere, *J. Geophys. Res.*, *99*, 8755–8770.

Yates, J. N., D. J. Southwood and M. K. Kivelson (2015), Magnetic phase structure of Saturn's 10.7 h oscillations, *J. Geophys. Res.*, *120*, 2631–2648, doi:10.1002/ 2014JA020629.

Ye, S.-Y., D. A. Gurnett, J. B. Groene et al. (2010), Dual periodicities in the rotational modulation of Saturn narrowband emissions, *J. Geophys. Res.*, *115*, A12258, doi:10.1029/2010JA015780.

Zarka, P., L. Lamy, B. Cecconi et al. (2007), Modulation of Saturn's radio clock by solar wind speed, *Nature*, *450*, 265–267.

6

Global Configuration and Seasonal Variations of Saturn's Magnetosphere

NORBERT KRUPP, PETER KOLLMANN, DONALD G. MITCHELL, MICHELLE THOMSEN, XIANZHE JIA, ADAM MASTERS AND PHILIPPE ZARKA

Abstract

Our understanding of Saturn's magnetosphere has been drastically changed over the last decade, since the arrival of Cassini, the first spacecraft to go into orbit around the planet. The trajectory of Cassini allowed the Saturnian magnetosphere to be studied both in the equatorial plane and at high latitudes, in a wide range of radial distances and local time sectors. This chapter reviews the current picture of Saturn's global magnetospheric configuration and describes the local fields and particle properties in key regions like the radiation belts and the inner, middle and outer magnetosphere. The moon Enceladus, deep in the magnetosphere, is the major source of neutrals and charged particles in the magnetosphere, and in this chapter we describe how the particles are generated, transported and lost within the highly dynamic magnetosphere. We also describe how both particles and fields in the Saturnian magnetosphere vary with time, both on shorter timescales and with Saturn's seasons. We highlight some of the most recent findings and discoveries, including a formerly unknown electric field oriented in the noon-midnight direction. Finally, we discuss magnetospheric measurements planned for the final sequence of the Cassini mission in 2017, called the "Grand Finale," along with a list of open questions to be solved by future missions.

6.1 Introduction

The Saturnian magnetosphere is the second-largest magnetosphere in our solar system, after Jupiter's. Our core understanding of the magnetosphere stems from the flyby missions of the Pioneer 11, Voyager 1 and Voyager 2 spacecraft in the late 1970s and early 1980s, as well as extensive observations from Cassini in orbit between 2004 and 2017. Some of Saturn's magnetospheric parameters are listed in Table 6.1. Additional understanding of Saturn's magnetospheric processes comes from the detailed investigation of auroral imaging from Earth or Earth-orbit-based telescopes (see Chapter 7), as well as from simulations and other theoretical considerations. Figure 6.1 illustrates the global magnetospheric configuration of Saturn's magnetosphere in two different views. On the left, the Saturnian magnetosphere is shown embedded in the solar wind, while on the right, more details of the internal structure of the magnetosphere are emphasized. Saturn's magnetosphere is a rotationally powered system, with its major source of neutrals and plasma, the moon Enceladus, contained deep inside its magnetic environment. With more than 100 geysers on this small moon (Porco et al. 2014), a stream with a combined internal mass of around 36–1600 kg/s in particles is continuously injected in Saturn's magnetosphere. As a result, neutrals dominate the magnetospheric environment, with more than 100 times more neutrals than ions (Delamere et al. 2007; Fleshman et al. 2013), so that the equatorial neutral density is usually significantly larger than the plasma density. Saturn's magnetosphere is also unique because its rotation axis and its magnetic dipole axis are co-aligned with each other. This fundamental structure, its dynamics and the relevant plasma processes as revealed during the Cassini prime mission (2004–2008) were described by Gombosi et al. (2009), Mitchell et al. (2009c) and Mauk et al. (2009). Saturn's

Table 6.1 *Characteristic parameters of the Saturnian magnetosphere*

Parameter	Characteristic value	Reference
Semimajor axis	9.6 AU	
Orbital period	29.5 yr	
Equatorial radius	1 R_S = 60,268 km	Seidelmann et al. (2007)
Mean radius	58,232 km	Seidelmann et al. (2007)
Equatorial magnetic field	21 μT	Burton et al. (2010)
Magnetic moment	4.7 10^{18} T/m^3	Burton et al. (2010)
Typical magnetopause distance	25 R_S	Arridge et al. (2011a)
Typical bow shock distance	31 R_S	Went et al. (2011)
Maximum neutral density	104/cm^3	Section 6.3
Maximum plasma density	102/cm^3	Section 6.2
Internal mass source	36–1600 kg/s	Section 6.3
Internal plasma source	17–47% of mass	Section 6.3

magnetosphere can be subdivided into an inner magnetosphere ($L < 6$), a middle magnetosphere ($6 < L < 15$) and an outer magnetosphere ($L > 15$). The radiation belts are embedded within the inner magnetosphere, with nearly dipolar field lines. Trapped, charged ions bouncing along field lines are lost onto the surfaces of the inner icy satellites, separating the radiation belts from each other. All of Saturn's rings except the E-ring and the Phoebe-ring are located within the inner magnetosphere. The middle magnetosphere begins roughly outside the orbit of Enceladus and includes the neutral torus as well as the E-ring, which is formed from icy material thrown out by Enceladus and which extends through the entire middle magnetosphere and beyond. Also included in this region are the inner plasma torus, the ring-current region where charged particles drift in opposite directions around the planet, dependent on the sign of their charge, and the magnetodisk or plasma sheet, where the deviation from a more dipolar to a more disk-like field configuration occurs. Radially outward- and inward-directed field lines come close to each other in that disk, and stable conditions require the presence of a current sheet in the center of the

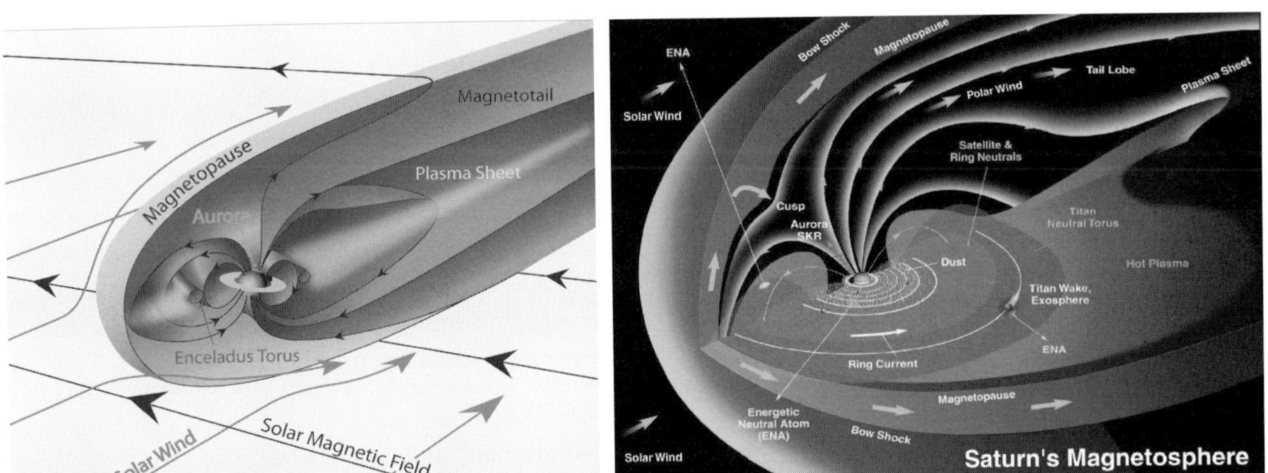

Figure 6.1 An artist's impression of the global configuration of the Saturnian magnetosphere. Left: A rotationally dominated magnetosphere, strongly affected by neutral particles ejected from Enceladus, deep within the magnetosphere. These particles produce the Enceladus Neutral Torus, the dust E-ring, and an equatorial curved plasma sheet (magnetodisk) with radially stretched magnetic field lines. At 10 AU, the distance of Saturn from the Sun, the interplanetary magnetic field and the nominal direction of the solar wind are nearly perpendicular to each other. On the nightside, the magnetosphere is significantly stretched into the elongated magnetotail (from Bagenal 2013). Right: Detailed illustration of features inside the Saturnian magnetosphere. Interaction with the solar wind leads to the formation of a bow shock and a magnetopause. The outer magnetosphere contains hot plasma (shown in red). Also shown is how the interaction between energetic charged and neutral particles can create energetic neutral atoms (ENA) and cold ions in different regions of the magnetosphere. Although a dense Titan torus was expected, Cassini has found no evidence to suggest that it exists. (A black-and-white version of this figure appears in some formats. For the color version, please refer to the plate section.)

magnetodisk. The outer magnetosphere (roughly outside Titan's orbit) is characterized by highly stretched field lines, especially in the elongated magnetotail, in which plasmoids are regularly released. Between these regions, there are many physical processes that occur on different time scales. Arridge et al. (2011a) has derived a map of Saturn's near-equatorial magnetosphere that emphasizes the physical processes in each region. Additional reviews have emphasized the implications of the dominant internal source of plasma (i.e. Enceladus) for the plasma disk structure of the magnetosphere, the transport of mass and possible loss processes (Gombosi and Ingersoll 2010; Bagenal and Delamere 2011; Thomsen et al. 2013; Achilleos et al. 2014; Krupp 2014; Jackman et al. 2014b). Bagenal and Delamere (2011) calculated radial profiles of plasma mass, pressure, thermal energy density, kinetic energy density and energy density of suprathermals. They also calculated that ≈75–630 GW of power input is needed to produce the observed plasma heating in Saturn's magnetodisk. Saturn's magnetosphere is in some aspects Earth-like and in other aspects more Jupiter-like. Saturn's appears to interact with the solar wind, driven by a process that is similar to that observed at Earth, described as the "Dungey cycle" (Dungey 1961). In addition, Saturn's equatorial surface magnetic field strength is similar to the Earth's (Gombosi et al. 2009). At the same time, the Saturnian magnetosphere is also similar in appearance to that of Jupiter because its major plasma source is the moon Enceladus, deep inside the magnetosphere, similar to the role and position of the moon Io at Jupiter. Unlike Jupiter, the neutrals within Saturn's magnetosphere are water-group ions (H_nO^+ with n = 0–3) and dust particles released from geysers near the south pole of Enceladus. The magnetospheres of Saturn and Jupiter are rotationally dominated, as the internally produced ions and neutrals spread into the magnetosphere radially, driven by the centrifugal force arising from their relatively fast rotation. The mass-loaded, frozen-in magnetic field of Saturn is stretched significantly and forms an equatorial plasma sheet outside the orbit of Enceladus. The charged particle population is (sub) co-rotating with the planet all the way to the magnetopause. Oppositely oriented magnetic field lines come close to each other, so that in the center of the plasma sheet a current sheet has to form in order to maintain stability. Continuous adding of mass to the field lines stretches them further until eventually magnetic reconnection occurs and plasmoids are released down the magnetotail. The emptied flux tubes move back planetwards and the process begins again. This is the so-called "Vasyliūnas cycle" (Vasyliūnas 1983), originally described for the Jupiter system, is equally important at Saturn. An interesting outcome of the Cassini mission is the fact that the rotation period of the system is not known, due to temporal and seasonal variations in the radio emissions formally used to determine the planet's rotation rate. This was discussed in detail in Chapter 5. In the following sections, the structure of Saturn's magnetosphere as well as a detailed description of the plasma sources in the system, loss mechanisms, transport and magnetospheric dynamics, and boundary phenomena are reviewed and discussed. In addition, the major magnetospheric science questions to be addressed during the final phase of the Cassini mission (its "Grand Finale") are summarized.

6.2 Magnetospheric Structure

In this Section, a phenomenological description of Saturn's magnetosphere is provided. Saturn's intrinsic magnetic field (Section 6.2.1) creates a fundamentally different environment than what is found at unmagnetized planets. The strong field close to the planet is able to trap MeV particles forming the radiation belts (Section 6.2.2). Plasma at lower energies dominates at larger distances, in regions that are separated into the inner, middle and outer magnetosphere and the magnetotail (Sections 6.2.3 to 6.2.4). The physical processes that shape Saturn's magnetosphere and determine its dynamics will be discussed later (Sections 6.3 to 6.5).

6.2.1 Magnetic Field Structure

Saturn's internal magnetic field is well described by an axisymmetric, spherical-harmonic model of degree 3 with a northward offset of 0.036 R_S (Gombosi et al. 2009; Burton et al. 2009, 2010). Parameters for a non-axisymmetric model rotating with the planet are summarized in Gombosi et al. (2009), and local time asymmetries were fit in Andriopoulou et al. (2012). Within $r < 6\ R_S$, the field is strongly dipolar. At $6 < r < 15\ R_S$, it is quasi-dipolar, with distortion due to the ring current.

Beyond $r > 15\ R_S$, the field describes a warped magnetodisk. This is because Saturn's magnetic axis tilts up to ±27.7° against the solar wind, depending upon Saturn's season, forcing the dayside plasma disk away from the wind (Arridge et al. 2008b; Carbary et al. 2010). The degree of warping depends on distance from the planet, local time, solar wind dynamic pressure and season. The field configuration can change significantly over the range of plasma pressures observed in the mid-tail region, affecting the location of the magnetopause and the magnetic mapping from the ionosphere (Achilleos et al. 2010). Ionospheric footpoints of the averaged field can differ by 6° latitude from those predicted by a dipole field.

The total azimuthal current peaks between 8 and 10 R_S (Kellett et al. 2010; Sergis et al. 2010), the suprathermal ion pressure dominates over the thermal plasma pressure beyond about 11 R_S and the total particle pressure dominates over the magnetic pressure beyond about 8 R_S (Sergis et al. 2009, 2010). In contrast with this, the current density varies with local time by factors of 2 or less (Kellett et al. 2011). At high latitudes, the transition from closed to open magnetic field lines can be observed (Gurnett et al. 2010). Strong field-aligned currents (FAC), probably associated with the aurora (see Chapter 7), are located just inside this boundary. Field-aligned currents that mediate the acceleration of plasma up to corotational velocities have been estimated from plasma electron data (Schippers et al. 2012), from the CAPS instrument onboard Cassini and from magnetic field perturbations (Hunt et al. 2014). FACs are modulated by the phase of the southern planetary period oscillations (PPOs). The PPO-related currents map into the slightly subcorotating equatorial main hot plasma region $L > 9–14\ R_S$. Comparison of the FACs with magnetospheric plasma velocities suggests that there are significant latitudinal variations in the ionospheric Pedersen conductivity and that the PPOs are driven from the atmosphere, rather than from the magnetosphere, downwards (Hunt et al. 2014).

6.2.2 Radiation Belts

Saturn's main ion radiation belts are located outside of the F-ring (2.3 R_S). In the energy range of several 100 keV to tens of MeV and above, they are clearly separated from each other and the rest of the magnetosphere by the major moon orbits (Janus 2.5 R_S, Mimas 3.1 R_S, Enceladus 4.0 R_S, Tethys 4.9 R_S) (Krimigis and Roelof 1983; Roussos et al. 2011). This separation is very different from the belts of other planets. It is caused by the moons, which are always close to the magnetic equator, and which sweep up charged particles they encounter in their orbits. This is so efficient that even a solar energetic particle event causing major intensity enhancements beyond the orbit of Tethys did not affect the radiation belt intensity inside of Tethys (Roussos et al. 2008).

Saturn's ion radiation belts are very stable and can be described well simply by using source, diffusion and moon losses (Cooper 1983; Gubar 2004; Kollmann et al. 2013). The energetic protons are supplied by the so-called "cosmic ray albedo neutron decay" (CRAND) process (Section 6.3). Transport via interchange or re-ionization of energetic neutral atoms (ENAs) does not contribute significantly to protons at radiation belt energies, and adding charge exchange only has a minor effect (Santos-Costa et al. 2003; Kollmann et al. 2013). However, in Krimigis et al. (2005) and the review by Gombosi et al. (2009) it was reported that another possible source population of trapped ions in the keV energy range inside the innermost D-ring exist, as concluded from ENA measurements coming out of that region after double charge exchange processes. In addition, evidence for oxygen and molecular hydrogen ions suggests the presence of a separate source (Armstrong et al. 2009) in the main ion belts. Between hundreds of keV and 10 MeV, ion pitch angle distributions in Saturn's inner magnetosphere are dominantly field aligned; above 10 MeV they tend to be peaked perpendicular to the field (Armstrong et al. 2009; Roussos et al. 2011).

The electron radiation belts in the tens of keV to tens of MeV range show, at best, weak depletions at the moon orbits and merge smoothly with the middle magnetosphere (Paranicas et al. 2010b; Kollmann et al. 2011; Tang and Summers 2012). Since the magnetic drift of electrons at Saturn is opposite to the corotational drift, electrons of hundreds of keV have a lower azimuthal velocity relative to the moons than protons, and therefore are only rarely absorbed by the moons. Instead of permanent intensity dropouts along their orbits (macrosignatures), the moons leave behind a wake (microsignature) that is refilled via diffusion (see Roussos et al. 2007 and Section 6.4). Only at

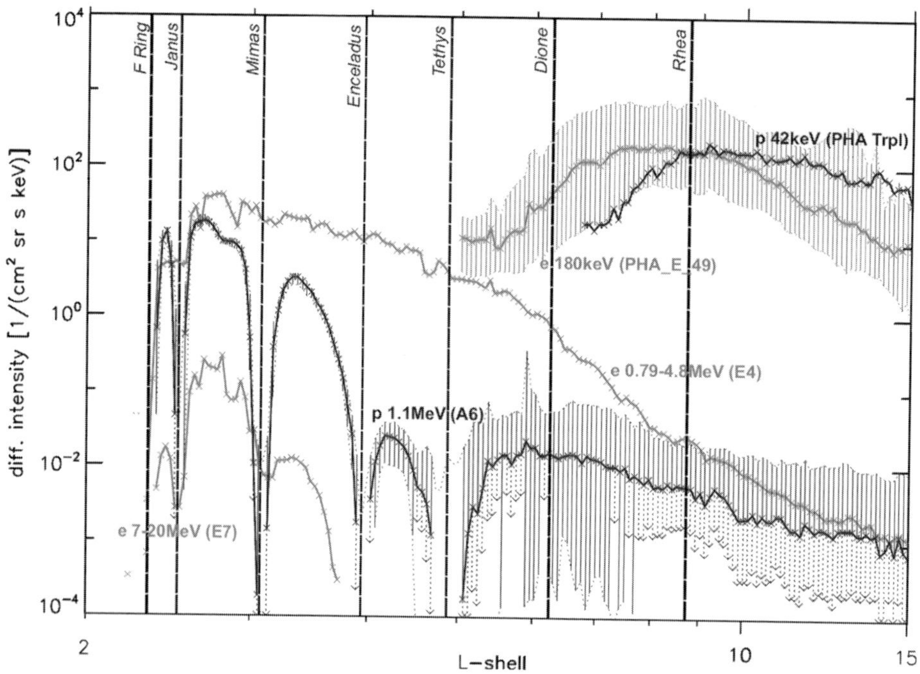

Figure 6.2 Differential intensities (particles per cm^2 per second, per steradian, and per energy in keV) of energetic protons (black, labelled with p) and electrons (gray, e) as a function of dipole L-shell in multiples of Saturn radii from the MIMI instrument (updated version from Kollmann et al. 2011). The individual channel names of the instrument are given in brackets after the labels. Intensities are median averages over the mission. The error bars shown for two curves are equivalent to the 1-sigma standard deviation, and indicate the variability of the magnetosphere. Only field-aligned protons (<30° equatorial pitch angle) within ±10° to the magnetic equator were considered. Strongly contaminated data were excluded. Labels provide geometric mean energy of the measurements taken in the corresponding instrument channels named in brackets.

energies above several MeV do the electrons encounter the moons quickly enough for a macrosignature to form (Figure 6.2).

Electrons inside of 15 R_S in the tens of keV to tens of MeV range are likely provided by transport and acceleration from larger distances (Section 6.4). Electron loss processes are discussed in Section 6.3.

6.2.3 Inner and Middle Magnetosphere

The transition from the "inner magnetosphere" ($L < 6$) to the "middle magnetosphere" ($6 < L < 15$) is characterized by Arridge et al. (2011a):

1. Distinct drop in the fluxes of radiation belt particles
2. Magnetic field starts to deviate from a dipolar configuration
3. Neutral gas density starts to fall
4. Appearance of hotter plasma (Arridge et al. 2011b)

Beyond L $\approx$ 4, the plasma densities inferred from a variety of instruments and techniques agree to within a factor of ~2–3. These include numerical moments of observed plasma ion data from the CAPS instrument (Thomsen et al. 2010, see Figure 6.3), total electron densities inferred from the upper hybrid resonance measured by the plasma wave instrument RPWS (Persoon et al. 2009, 2013), forward modeling of CAPS ion data (Sittler et al. 2006; Wilson et al. 2008; Livi et al., 2014), core electron density from RPWS/HFR and quasi-thermal noise spectroscopy (Schippers et al. 2013) and plasma density based on the RPWS/Langmuir probe (Jacobsen et al. 2009). In the inner and middle magnetosphere, the ion population up to the keV range consists of three main species, H^+, H_2^+ and W^+. W^+ dominates in the inner magnetosphere, with H^+ a factor of 20 lower. H_2^+ is only a minor constituent there, but beyond L $\approx$ 15 the light ions (H^+ and H_2^+) are comparable to or exceed the W^+.

At tens to hundreds of keV, the composition in the range $7 < L < 16$ R_S follows the same ordering as the thermal plasma, with tiny contributions from He^+ and He^{2+} (DiFabio et al. 2011). At thermal energies, the

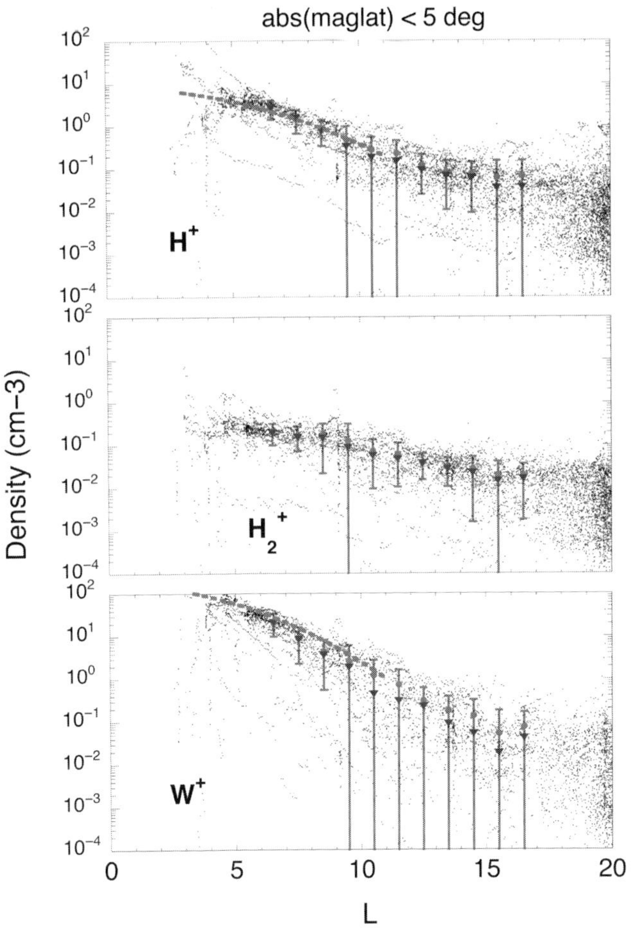

Figure 6.3 Equatorial density of hydrogen H^+, molecular hydrogen H_2^+ and water group ions W^+ as a function of L-shell (the L-shell is equal to radial distance in multiples of Saturn radii in the equatorial plane and follows the magnetic dipole field lines to higher latitudes), as derived from CAPS measurements (from Thomsen et al. 2010). Small black points show single measurements taken throughout the mission within ±5° to the magnetic equator. Large gray dots are means with their standard deviations; black triangles are medians. Dashed curves are the profiles given by Wilson et al. (2008).

major ion species can be modelled as convecting Maxwellian distributions, (e.g. Wilson et al. 2008, 2013; Livi et al. 2014), with temperature anisotropies $T_\perp/T_{||}$ = 2 or greater. Spectra have a suprathermal extension (Figure 6.7) that can be characterized above a few keV as a power law and an additional Kappa distribution above a few 10 keV (Dialynas et al. 2009). Electron spectra can be fit well with two Kappa distributions (Schippers et al. 2008; Livi et al. 2014). Spectra in the tens to hundreds of keV range can be roughly described by power laws with exponents that range from −0.5 to −3, being steepest around 13 R_S (Carbary et al. 2011). Spectra along the moon orbits between 3 and 6 R_S show a cutoff around 1 MeV (Paranicas et al. 2010b, 2014). The maximum equatorial plasma density of around 100/cm^3 is found between 4 and 5 R_S. The plasma density falls off outward of 5 R_S as a power law in L, with an exponent that depends on the species (Wilson et al. 2008; Thomsen et al. 2010; Persoon et al. 2013; Schippers et al. 2013; Livi et al. 2014) and, inward of 4 R_S, on season.

Plasma is strongly confined to the magnetic equator by the centrifugal force (Persoon et al. 2006). Heavy ions are more confined (Sittler et al. 2006; Persoon et al. 2009, 2013; Thomsen et al. 2010), resulting in a strong latitude dependence to the composition. Electrons at all measured energies (tens of eV to hundreds of keV) show a similar equatorial confinement within > 10 R_S, i.e. are mostly perpendicular to the magnetic field ("trapped" or "pancake"). They show the opposite behavior outside of this distance, i.e. are mostly bi-directional field-aligned (Carbary et al. 2011; Carbary and Rymer 2014; Clark et al. 2014). This is consistent with adiabatic transport inside 10 R_S, and an ionospheric source with field-aligned acceleration boosting the energies beyond 10 R_S, a region that maps roughly to the auroral zone. "Butterfly" distributions (with minima both in the field-aligned and perpendicular directions) are commonly seen around 10 R_S, and isotropic distributions can occur throughout the inner magnetosphere, presumably due to scattering by waves. Mission averages show an intensity maximum in suprathermal particles between 6 R_S and 10 R_S, depending on energy and local time (Figure 6.2). At > 10 keV, the intensities show considerable scatter as a function of L and energy outside of 5 R_S (see DeJong et al. 2010; Kollmann et al. 2011; and Figure 6.2). MeV electron intensities appear to vary with the solar wind stream structure and with long term changes in the solar UV intensity (Roussos et al. 2014).

Core ion temperatures for the thermal plasma constituents between 4 and 17 R_S are 10 to several 100 eV and are higher for W^+ than for H^+. The core temperature rises with distance, a dependence not consistent with adiabatic radial transport but suggesting ongoing plasma energization, perhaps through a continuing process of ion pickup to subcorotational energies.

The temperatures of the different species tend to converge toward common values beyond L ≈ 10, evidence of possible collisional sharing of thermal energy (Thomsen et al. 2010). Various studies (Livi et al. 2014; Schippers et al. 2008, 2013; Thomsen et al. 2012; Gustafsson and Wahlund 2010) find core electron temperatures in the range of 1–10 eV inside of 10 R_S, with a tendency to increase with increasing distance. Beyond 5 R_S, the temperature is higher on the nightside than the dayside (Schippers et al. 2013; Thomsen et al. 2012). The core electron temperature shows a strong increase with latitude (Gustafsson and Wahlund 2010). The total electron temperature is modified by the suprathermal (1–10s of keV) electrons delivered by interchange (see Section 6.4). Beyond 8 R_S, the hot-electron temperatures are on the order of 1 keV (Livi et al. 2014; Schippers et al. 2008). The suprathermal temperature appears to decrease inward of this. Azimuthal plasma flow has been determined by forward modeling the plasma energy/charge distributions using CAPS data (Wilson et al. 2008, 2009), CAPS spectra (Thomsen et al. 2010; Livi et al. 2014), Langmuir probe measurements (Jacobsen et al. 2009; Holmberg et al. 2012) and backtracing of dispersed injection events (Müller et al. 2010). Some of these profiles are summarized in Figure 6.4. Several studies find the plasma close to full corotation around 3 R_S, i.e. moving with the same angular velocity as Saturn's ionosphere. Subcorotation is found beyond that. There appears to be a local minimum around 5 R_S. It then recovers and decays again outward of about 7 R_S. The lowest values can approach 50% of corotation. Subcorotation has been attributed to local mass and momentum loading and outward transport. Solutions to the equation describing radial transport and interaction with a neutral gas cloud show that between $3 < L < 5$, the observed corotation lag is mainly caused by charge exchange and elastic ion-neutral collisions. Continued mass loading beyond L ≈ 5 eventually increases the mass outflux until the lag due to transport becomes dominant (Pontius and Hill 2009). Alternatively, subcorotation may be a signature of collisions between ions and dust (Sakai et al. 2013). A combination of plasma and energetic-particle observations (Sergis et al. 2010) reveals that the total plasma pressure maximizes in the inner magnetosphere, where it is dominated by the thermal plasma contribution. Beyond ≈ 12 R_S, the suprathermal pressure typically slightly exceeds that in the lower-energy component. The plasma beta exceeds 1 beyond ≈ 8 R_S, leading to the increased stretching of the outer magnetospheric magnetic structure, as discussed above. The energy transport in the magnetosphere of Saturn is discussed in detail in Bagenal and Delamere (2011).

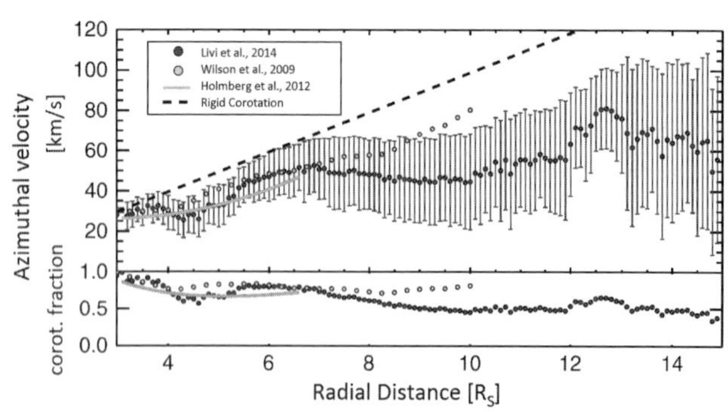

Figure 6.4 Azimuthal flow velocities of ions as a function of radial distance in the Saturnian magnetosphere (adapted from Livi et al. 2014).

6.2.4 Outer Magnetosphere and Tail Structure

The transition from the "middle magnetosphere" (≈ 6–15 R_S) to the "outer magnetosphere" (> 15 R_S) is characterized by (Arridge et al. 2011a):

1. The distortion of the field into a non-dipolar current sheet configuration
2. The disappearance of a quasi-constant cold plasma population
3. A lack of injection signatures
4. A transition to a much more variable plasma and field environment moving rapidly between lobe-type and plasma/current sheet-type regions

The issue of the spatial distribution of injection signatures will be addressed in Section 6.4.3. Comprehensive studies of the energetic-particle (Krupp et al. 2009) and plasma (Morooka et al. 2009; Thomsen et al. 2010) properties show that beyond ~10 R_S there is substantially greater variability in the particle populations compared to the inner magnetosphere. Such transitions may represent latitudinal motions across the disk-like plasma sheet, including entry into the open-field-line region of the lobes (see below). Strong bi-directional energetic electron beams are very commonly seen between 10 and 25 R_S, from open polar cap field to well into the closed-field-line region (Mitchell et al. 2009a; Krupp et al. 2009). The plasma density in the outer magnetosphere decreases with L and $| z |$. Beyond L = 15, the density is highly variable: There is a high-density sector (magnetodisk) that is 100 times denser than the low-density sector (lobe), and the transition from one to the other appears to be periodic at roughly the SKR modulation period, suggesting a longitudinally variable density model or a rotational modulation of the current-sheet latitude (Morooka et al. 2009). The light-ion densities in the outer magnetosphere are frequently comparable to or even greater than the W^+ density. Part of this is attributable to the continued centrifugal confinement of the heavier ions close to the magnetic equatorial plane, but in addition there is evidence for a significant light-ion contribution from Titan to the outer magnetospheric plasma (Thomsen et al. 2010). Plasma flow velocities out to distances of more than 30 R_S (McAndrews et al. 2009, corrected in McAndrews et al. 2014; Thomsen et al. 2010, 2013, 2014b; Kane et al. 2014) are dominantly in the corotation direction throughout most of the magnetosphere explored by Cassini, although the flow speed is below full corotation, flattening out at about 100–200 km/s beyond ~25 R_S. The dominance of the azimuthal component suggests that the ionosphere continues to exert a strong influence on the magnetospheric plasma throughout this region of the tail, but with decreasing ability to enforce corotational flow at larger radial distance. Flows in the post-dusk sector have a significant outward component, suggesting the existence of a "planetary wind" along the dusk magnetopause (Thomsen et al. 2013, 2014b; Kane et al. 2014). Through the nightside sector, the plasma flow frequently has an outward component, leading to the suggestion that beyond 20 R_S the flux tubes subcorotating through the nightside are not able to return to the dayside (McAndrews et al. 2009; Thomsen et al. 2014b; Kane et al. 2014). There is a net radial mass outflow between 18:00 and 03:00 local time (LT), estimated as ≈34 kg/s (Thomsen et al. 2014b), comparable to the estimated plasma production rate from Enceladus of 12–250 kg/s (Bagenal and Delamere 2011). The flow velocities at LT > ~3 typically show a slight inward tendency, unlike the outflows that seem to dominate across the nightside (Kane et al. 2014). These inward flows are indicative of flux tubes that have traversed the nightside and have been successfully returned to the dayside for another passage around Saturn. Figure 6.5 summarizes the current consensus regarding the global flow structure. The stretched, disk-like structure of the outer magnetosphere is more prominent on the nightside than on the dayside, where it can largely disappear under conditions of strong solar wind dynamic pressure (Arridge et al. 2008a; Szego et al. 2011). On the nightside, the flattened magnetodisk becomes the magnetotail, which extends to the furthest downtail distances probed by Cassini (≈70 R_S) and consists of a central current sheet of relatively dense plasma and relatively weak total field strength, surrounded both above and below by lobes of stronger, largely radial magnetic field lines that appear to be connected to the planet at one end and to the solar wind at the other. The tail plasma sheet densities fall off with distance at a rate similar to what is seen on the dayside inside of 20 R_S, but decline less steeply beyond 20 R_S (McAndrews et al. 2009; Thomsen et al. 2014b). The densities can vary over 1–2 orders of magnitude at a given radial distance, presumably because of strong latitudinal variation. The temperatures show little radial dependence beyond ≈20 R_S. The magnetotail is warped by the dynamic pressure of the solar wind (e.g. Arridge et al. 2008c) and has been found to flap up and down so that at any given time it is difficult to know exactly where the spacecraft is relative to the center of the current sheet. With B_r as a proxy for location relative to the current sheet in the radial range of 10–22 R_S, heavy ions are found to be more narrowly confined to the equator and to show narrow substructures. The H^+ ions typically have a lower density than that of the water-group ions at the equator, but H^+ dominates in the high-latitude layers of the plasma sheet. The H^+ density also appears to be organized by SLS3 longitude (Szego et al. 2011, 2012). Arridge et al. (2011a) developed a simple structural model of the plasma sheet based on fits to electron densities measured by CAPS/ELS, which organizes the field minima and ion density maxima of observed current-sheet crossings very well (Szego et al. 2011). The ion data suggest that the amplitude of the plasma sheet flapping is as high as 5 R_S at Titan's orbit.

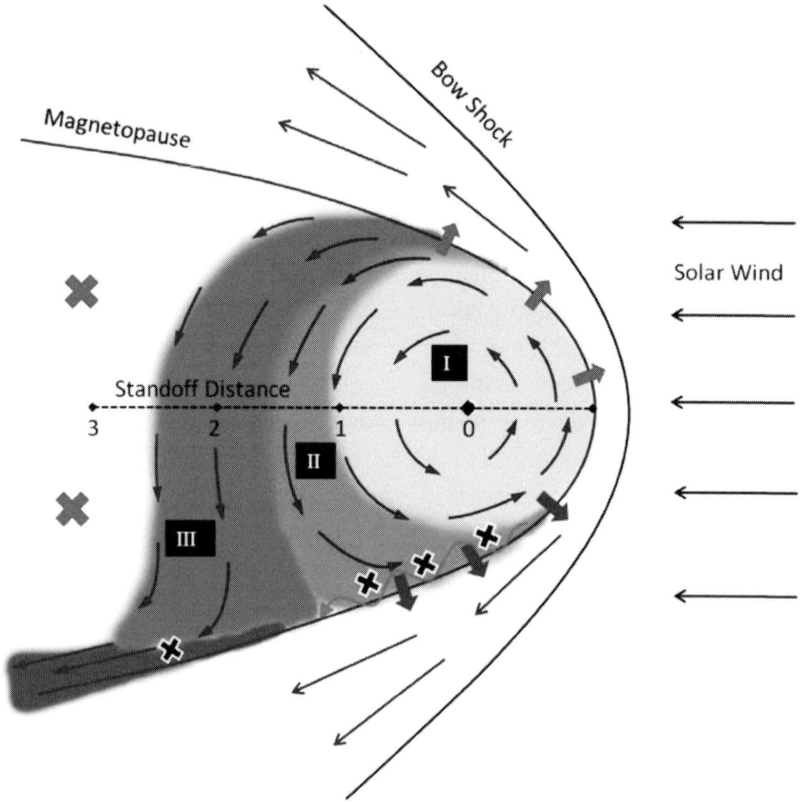

Figure 6.5 Artist's impression of Saturn's outer magnetosphere structure (from Kane et al. 2014). The arrows only indicate the direction of the assumed flows (not to scale) and the black crosses simply indicate that small-scale reconnection occurs preferentially near the dawn magnetopause. Large-scale reconnection might occur deep in the magnetotail.

6.3 Plasma Sources and Sinks

This section summarizes the relevant plasma sources at Saturn, organized by their locations. An extended summary is available in Blanc et al. (2015). Loss processes that remove plasma from the system are also discussed.

6.3.1 Enceladus and Its Neutral Torus

One of the main findings of the Cassini mission is the discovery of water plumes near the south pole of Enceladus (e.g. Dougherty et al. 2006; Porco et al. 2006; Waite et al. 2006; Hansen et al. 2006), which turns out to be the major source of neutral gas and plasma for Saturn's magnetosphere. The gas in the plumes consists of ~91% H_2O, ~3% CO_2 and 4% N_2 or CO (Waite et al. 2006). The plumes also contain ice grains composed mostly of water ice (Hillier et al. 2007). The gas production rate has been estimated to range from ~36 to ~1600 kg/s (e.g. Saur et al. 2008; Smith et al. 2010; Tenishev et al. 2010; Dong et al. 2011). An important question concerning the Enceladus plumes is the level of temporal variability of the source rate. Studies applying the same analysis technique to different flybys found changes that range from 15% to up to a factor of 10. Hedman et al. (2013) reported that the plume brightness correlates with the level of tidal stress experienced by the moon during its different orbital phases, and is strongest when the moon is furthest from Saturn.

After escaping from Enceladus, the neutral particles undergo a series of processes, including photolysis and interactions with ions, electrons and other neutrals, which not only produce daughter species of water molecules, such as OH and O, but also spread them out into a neutral torus that extends from inside Enceladus' orbit to the middle/outer magnetosphere (e.g. Jurac and Richardson 2005; Johnson 1990; Farmer 2009; Cassidy and Johnson 2010; Tadokoro et al. 2012; Smith et al. 2010; Fleshman et al. 2012, 2013). The neutral-ion ratio in the Saturnian magnetosphere is about 3 orders of magnitude higher than is the case at Jupiter (Delamere et al. 2007; Fleshman et al. 2013) and is related to the fact that the plasma flow is slower at Enceladus that at Io. As a consequence, the pickup ion temperature is lower, resulting in a net cooler ion and electron population, less efficient for ionization. Atomic H (Shemansky et al. 2009; Tseng et al. 2011), O and OH (Melin et al. 2009) and H_2O (Hartogh et al. 2011) in the neutral torus have been observed in the sub-mm and UV wavelengths.

The torus has also been observed through energetic neutral atoms (ENAs) arising from charge exchange between ions and neutrals (Dialynas et al. 2013). In situ measurements (Perry et al. 2010) reveal that the gas along Enceladus' orbit consists mostly of H_2O, but with a CO_2 density about one-third that of H_2O. These measurements are well complemented with modeling work (Cassidy and Johnson 2010). It is generally found that H_2O dominates around 4 R_S, whereas at larger distances O becomes the dominant species (Figure 6.6). A fraction of the neutral particles released from Enceladus become ionized through processes such as photoionization and electron impact ionization, thereby providing a plasma source for Saturn's magnetosphere. The photoionization lifetimes at Saturn are on the order of years, and consequently photoionization is not efficient for ion production. Instead, electron-impact ionization appears to be the main ionization process. Because the impact ionization rate depends on the electron density and temperature, which vary as a function of radial distance from the planet, the resulting ionization rate has a strong spatial dependence. Although the neutral densities peak right near Enceladus' orbit, the bulk of the electrons inside L ≈ 6 have energies below the ionization cutoff energies, so the peak of electron impact ionization rate lies outside Enceladus' orbit around 7–9 R_S (Smith et al. 2010; Tadokoro et al. 2012). The ionization rate is also species dependent (Fleshman et al. 2012). Overall about 17–47% of the neutral particles in the torus are ionized (Jurac and Richardson 2007; Cassidy and Johnson 2010; Bagenal and Delamere 2011; Fleshman et al. 2012).

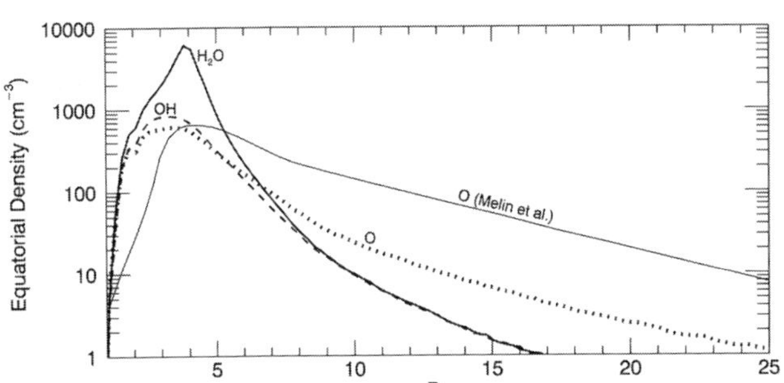

Figure 6.6 Radial profiles of the equatorial O distribution derived by Melin et al. (2009) using Cassini ultraviolet observations from the UVIS instrument (thin solid line) and the modelled distributions of various neutral species (thick solid line: H_2O; dashed line: OH; dotted line: O) from an updated model of Cassidy and Johnson (2010) constrained by the Herschel observations (Hartogh et al. 2011).

Another important interaction process is charge exchange, which results in a fast neutral and a cold ion, which is then picked up by the convection electric field. Escaping neutrals at keV energies are referred to as energetic neutral atoms (ENAs) and have been directly measured by the energetic particle spectrometer Cassini/MIMI (Mitchell et al. 2009b; Carbary and Rymer 2014). Charge-exchange thus serves as a loss process for the neutrals and a momentum-loading process for the plasma. The charge-exchange rate depends both on the densities of the participating particles and their relative speed (Lindsay and Stebbings 2005) and appears to peak right near Enceladus (Smith et al. 2010; Fleshman et al. 2012). Like the plume gas evolving into the neutral torus, the ice grains evolve into the tenuous E-ring (Nicholson et al. 1996) driven by gravity, electromagnetic forces, drag and radiation pressure (Horányi et al. 2008; Beckmann 2008). The E-ring particles also peak around 4 R_S and follow a power law size distribution (Kempf et al. 2008). The grains are eroded by sputtering by incident ions (Jurac et al. 2001b), which additionally supplies plasma and neutrals to the magnetosphere (Jurac et al. 1995).

6.3.2 Other Icy Moons

Other Saturnian icy moons of significant size within the magnetosphere are Mimas, Tethys, Dione and Rhea. Their surfaces are sources of neutrals which form an exosphere around each moon. Subsequent ionization of neutrals feeds the magnetospheric plasma. The main neutral release process for these moons is surface sputtering (Jurac et al. 2001a; Teolis et al. 2010). Below 1 keV/amu, interactions with energetic ions and electrons drive chemical reactions that produce fast products, similar to photolysis. At lower energies, the particles result from collisions (Cassidy and Johnson 2010; Johnson et al. 1989). Photolysis also contributes to releasing neutrals. The released species depend not only on the water ice bulk material but also on minor constituents that might be primordial, deposited by meteoroids, or produced by previous radiolysis. In situ measurements detected O_2^+ ions at Dione (Tokar et al. 2012). At Rhea, O_2^q (q = −1, 0, +1) and CO_2^{q+} (q = 0, 1) were found. The O_2 source rate at Rhea was estimated to be $2.2 * 10^{24}$ O_2/s = 0.12 kg/s (Teolis et al. 2010). The H_2O source rate of the icy moons is at least an order of magnitude lower than the Enceladus plumes (Jurac et al. 2001b), although they are still predicted as an important source of H_2O outside of Enceladus' orbit (Tadokoro et al. 2012).

6.3.3 Titan

Titan orbits Saturn at $\approx 20\ R_S$ and has a dense N_2 atmosphere with minor contributions of CH_4 and H_2 (Strobel 2008). UV light, photoelectrons and magnetospheric and pickup ions cause dissociation of N_2 into fast atoms and ions (Shematovich et al. 2003), some of which exceed escape velocity. Hydrodynamic escape could also release neutrals (Strobel 2008). Ions and electrons are picked up by Saturn's corotation electric field. Escaping electrons can set up an additional electric field that facilitates further ion escape (Wellbrock et al. 2012; Sittler et al. 2010). In Titan's induced magnetotail, escaping ions with masses of 1–2 (H^+, H_2^+), 16 (likely CH_4^+) and 28 (likely N_2^+) amu were detected (Sittler et al. 2010; Coates et al. 2012). The total mass loss is estimated to be on the order of 10^{24} ions/s, or 10^{-4} kg/s. Titan is a slightly dominating source of N_2^+ and N^+ beyond 4 and 10 R_S, respectively. Inward of that, Enceladus dominates by about a factor of 100 (Smith et al. 2007). Titan is also likely a major source of H_2 and H_2^+ outside of about 6 R_S (Thomsen et al. 2010; Tseng et al. 2011).

6.3.4 Main Rings

An important plasma source near Saturn is its main rings, suggested by the detection of protons, water group ions and O_2^+ above the rings (Waite et al. 2005; Tokar et al. 2005). Seasonal variation of the plasma densities suggest that the rings also supply ions to the innermost magnetosphere. A major process producing O-H compounds from water ice is photolysis (photo-sputtering, photo-desorption) by UV light (Johnson et al. 2006; Westley et al. 1995). Impact on the rings by dust and meteoroids can also produce neutrals, ions and ice grains (Morfill et al. 1983; Ip 2005). UV light, electron impact (mostly outside of ring orbits) and cosmic rays (Johnson et al. 2006; Smith et al. 2010; Elrod et al. 2012) cause ionization and dissociation of the neutrals, while interactions between neutrals and ions spread out the initial distribution away from the rings, similar to the situation at Enceladus. Photolysis rates of the main rings depend on the ring illumination angle and ring temperature (Tseng et al. 2010), both having seasonal dependence (Flandes et al. 2010), and on the UV intensity, which changes over solar cycle. Both of these patterns had a minimum around 2009 (equinox and approximate solar minimum). The densities of O_2^+ and water group ions with <10 keV between the rings and 4 R_S (Elrod et al. 2012, 2014) and of O_2^+ ranging from tens to hundreds of keV outside of 4 R_S (Christon et al. 2013, 2014) went through a minimum during this time, clearly indicating that those particles are primarily fed by the rings. The rings also contribute to the production of 10 MeV protons (Cooper 1983) via the CRAND process (discussed below).

6.3.5 Saturn's Atmosphere

Saturn's thermosphere consists mostly of H_2. Ionospheric electrons (see Chapter 8) with energies around 10 eV impacting on H_2 cause electronic excitation into fast H atoms and ions. If this happens at high enough altitudes, then the kinetic energy is not thermalized to other particles and the atoms can escape (escape velocity is 6.5 eV/nuc) (Shemansky et al. 2009; Tseng et al. 2013b). An asymmetric plume of H emission around Saturn has been observed (Shemansky et al. 2009; Melin et al. 2009). The details of the hydrogen source on Saturn are not well understood, but modeling suggests that it can contribute 20% of the H inside 5 R_S (Tseng et al. 2013b). Ionospheric ions (H^+, H_3^+) following open

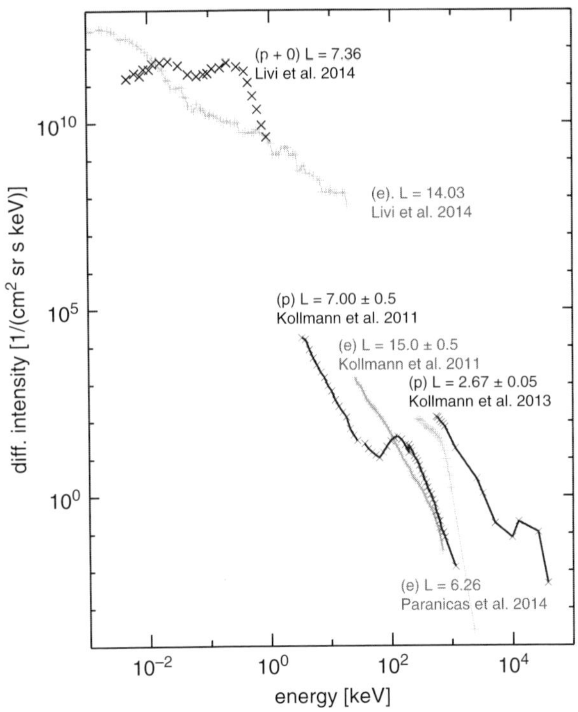

Figure 6.7 Compilation of energy spectra throughout the magnetosphere. Black × signs and (p) refer to protons, (e) and gray + signs to electrons, L to the magnetic L-shell in multiples of R_S. L = 7.36 is an example spectrum (i.e. at a given instant) of protons and oxygen by CAPS/IMS, each described well with a Maxwellian distribution. L = 14.03 is an example spectrum by CAPS/ELS. It shows that the electron distribution needs to be described by two Kappa distributions peaking around 1 eV and 1 keV. The other spectra are mission averages in L-ranges close to the equator. They show median values of the fluxes, except the L = 6.26 spectrum, which shows a logarithmic average. L = 7 combines MIMI/CHEMS and LEMMS data. The other spectra are based on MIMI/LEMMS. L = 2.67 is between Janus' and Mimas' orbits and shows a proton peak around 10 MeV from CRAND. L = 6.26 is at Dione's orbit and shows an electron cutoff around 1 MeV.

magnetic field lines may form a polar wind producing $\sim 10^{27}$ ions/s. Subsequent magnetic reconnection could bring about half of these ions back into the magnetosphere (Glocer et al. 2007). Galactic cosmic rays with tens of GeV energy impacting Saturn's atmosphere produce secondary particles, including neutrons (e.g. Glass et al. 1977). The neutrons decay to 10-MeV protons that form a peak in the energy spectrum inside the radiation belts (see Figure 6.7) (Krimigis and Armstrong 1982; Armstrong et al. 2009), a process called cosmic ray albedo neutron decay (CRAND) (Singer 1958). The power law spectrum between the peak and hundreds of keV might be caused partially by deceleration of neutrons in the atmosphere (e.g. Kollmann et al. 2013). The CRAND source is expected to change with the cosmic ray intensity over the solar cycle (Usoskin et al. 2005), consistent with a proton intensity increase between 2004 and 2011 by a factor of three (Roussos et al. 2011).

6.3.6 Solar Wind

Plasma in the solar wind can enter Saturn's magnetosphere through boundary processes (see Section 6.5) such as reconnection or the Kelvin–Helmholtz instability, thereby also providing a source of plasma to the magnetosphere. Solar wind as a plasma source for Saturn's magnetosphere is discussed in Bagenal and Delamere (2011), Delamere and Bagenal (2013) and Blanc et al. (2015). Energetic He^{2+} and He^{+}, which are tracers of the solar wind and interplanetary pickup ions, respectively, have been observed in the middle magnetosphere (DiFabio et al. 2011). Their abundances show a solar cycle dependence, decreasing towards solar minimum.

6.3.7 Loss Processes

Recombination of ions and electrons into neutrals removes plasma. Recombination depends on the energy and density of both particles and on the ion species and excitation state (Hahn 1997; Funsten et al. 1993; Richardson 1986). Recombination plays an important role inside 5 R_S (Elrod et al. 2012; Tseng et al. 2013b) but is weak outside (Richardson 1998; Sittler et al. 2008). Collisions involving ions and molecules can cause charge exchange and also chemical reactions (Tseng et al. 2013b), leading to a loss of one ion species or charge state and a source of another. Charge exchange (Section 6.3.1) into ENAs is a major loss process for energetic ions. It dominates over energy loss (see below) at hundreds of keV in the middle magnetosphere but appears insignificant for the ion radiation belts (Kollmann et al. 2011, 2013).

Ions and electrons moving through neutral gas experience energy loss (stopping, friction). Ions below a few keV lose energy mainly via collisions with the target nuclei. Ions at higher energies and electrons below a few MeV lose energy via impact ionization. Energy loss results in particle loss at a given energy for spectra that fall with energy. For rising spectra, energy loss at high energies results in a particle source at lower energies (Kollmann et al. 2013). Gyrating charged particles emit synchrotron radiation and thereby lose energy perpendicular to the magnetic field, which makes them more field aligned. This effect is important for electrons at Jupiter but negligible at Saturn (Santos-Costa et al. 2003).

Particle transport can either remove or supply plasma in a given region (Section 6.4). Moreover, radial transport exposes plasma to interactions with Saturn's moons. Except for Titan and Enceladus, the moons only weakly perturb the plasma flow, so they mainly absorb incoming plasma. Energetic ion losses can be quite significant, resulting in the clear separation of the ion belts (Section 6.2.2). Energetic electrons can escape absorption if their azimuthal drift velocity roughly equals the Keplerian velocity of the moon (Van Allen et al. 1980), but lower-energy electrons impinge on the moons from all directions and get absorbed due to their high thermal velocity (Roussos et al. 2010). Pitch angle diffusion due to wave-particle interactions can cause energetic charged particles to precipitate into Saturn's atmosphere, reducing particle intensities mainly at field-aligned pitch angles (the loss cone). Pitch angle diffusion might be important for electrons around 6 R_S, where the intensity is near the Kennel–Petschek limit (Kennel and Petschek 1966; Tang and Summers 2012; Mauk and Fox 2010).

6.4 Transport and Magnetospheric Dynamics

The magnetosphere needs to shed plasma that is continuously produced in the source region of Enceladus' extended neutral gas torus. Bagenal and Delamere (2011) gathered published information to estimate the plasma production and outflow rate, loss processes, and energy input requirements, both for Saturn and for Jupiter. Other work emphasized the importance of centrifugally driven interchange motion for outward transport of the plasma and tail processes (especially reconnection) for releasing it out to the solar wind (Thomsen et al. 2013; Mitchell et al. 2015; Louarn et al. 2014; Jackman et al. 2014a).

6.4.1 Diffusion

If Saturn's magnetic and electric fields were stable, charged particle guiding centers would drift around the planet on closed trajectories encompassing the same magnetic flux. For axisymmetric fields, this drift is on fixed L-shells (Roederer 1970; McIlwain 1966). If the fields are changing stochastically on the timescale of the drift, enclosed magnetic flux and L-shell can change. The phase space density (PSD) distribution of particles in L-shell then reorganizes in a way that can be described as diffusion in L-shell ("radial diffusion"), driven by the PSD gradient (Schulz and Lanzerotti 1974; Walt 1994). Transport by interchange is not modelled well with diffusion formalism, since it does not depend on the PSD gradient, but instead transports plasma inward and outward depending on force imbalance between the centrifugal force on mass-loaded flux tubes and the magnetic tension force on lightly mass-loaded flux tubes located radially outward of the loaded region. Diffusion is important in the proton belts (Section 6.4.1 and Kollmann et al. 2013).

The field variations leading to radial diffusion are expected to conserve the first two adiabatic invariants (Schulz and Lanzerotti 1974), which implies an increase in particle energy towards Saturn. The increase depends on pitch angle, which is the main reason why the pitch angle distribution also changes with distance from Saturn. Models assuming invariant conservation can reproduce the data quite well (e.g. Hood 1983; Lorenzato et al. 2012; Kollmann et al. 2013; Clark et al. 2014). Direct evidence for invariant conservation are the electron spectra inside of L≈6 R_S (Paranicas et al. 2010b) with a cutoff at an energy near 1 MeV that varies with radial distance, as well as the radial suprathermal proton temperature profiles (Dialynas et al. 2013). Various determinations of the diffusion coefficients have been reviewed by Van Allen (1984). Recent work was done by Roussos et al. (2007) and Kollmann et al. (2013) using Cassini measurements. Lorenzato et al. (2012) and Clark et al. (2014) assumed parameters in their models and found good fits to the data. Diffusion coefficients typically increase with radial distance and are approximated with a power law in L. Electrons in the 100 keV range and ions in the MeV range inside 5 R_S have power law exponents between 6 and 10. MeV electrons inside 15 R_S appear to have exponents around 3. These exponents can be compared with prototype calculations for recurrent magnetic compressions, variable convection electric fields (Walt 1994) or turbulent ionospheric winds (Brice and McDonough 1973; Coroniti 1974). They lead to exponents between 8 and 10, 6 and 10, and 3 (for corotating particles), respectively. Such calculations need to be redone based on observed field fluctuations (Lejosne et al. 2013).

In addition to radial diffusion, particles are also subject to pitch angle, energy and cross diffusion (the last of these mixes modes). Pitch angle diffusion can be caused by Coulomb scattering in gas and ice (Beutier et al. 1995) or by fluctuations of the electromagnetic fields on the bounce time scale. The contribution from whistler waves (see Section 6.4.7) to electron pitch angle diffusion around 6 R_S is larger than from scattering (Lorenzato et al. 2012). Other wave modes have also been investigated as contributors to diffusion (Menietti et al. 2012; Gu et al. 2013). Protons above hundreds of keV do not scatter significantly in matter (Kollmann et al. 2013) and models of proton belts yield realistic results neglecting pitch angle diffusion (Selesnick et al. 2007; Kollmann et al. 2013).

6.4.2 Interchange

Plasma production in the magnetosphere combined with Saturn's rapid rotation results in a centrifugally driven interchange instability (Sonnerup and Laird 1963; Southwood and Kivelson 1987) that appears to be the primary mechanism, at least in the middle magnetosphere, by which plasma is transported outward and shed to the solar wind. Interchange is similar to the Rayleigh–Taylor instability, where a dense liquid sits on top of a light liquid, and these liquids then change places, except that the effective centrifugal force takes the place of gravity. The primary observational signatures of interchange are brief intervals of "injected" hot plasma (see Section 6.4.3) on dipolarized flux tubes. Injections show up as intensity enhancements, since their PSD at their original location is generally higher than in the environment where they are observed. The hot plasma replaces the outward-driven cool plasma, thereby maintaining overall magnetic flux balance in the magnetosphere and also delivering electrons capable of ionizing additional neutrals.

Interchange may well supply energetic electrons in as far as 5 R_S by inward adiabatic transport. Since very few fresh injections are observed inside of this, further transport would require a different process such as radial diffusion (Paranicas et al. 2010b). Interchange injections also provide > 12 eV predominantly field-aligned electrons. DeJong et al. (2010) suggested that most of these electrons are not transported adiabatically but are of ionospheric origin, accelerated in association with the field-aligned current (FAC) that drive the injections inward (Southwood and Kivelson 1989). "Young" interchange events exhibit little gradient-and-curvature drift dispersion in the energetic injected plasma, while for "old" ones such dispersion is not only notable but enables the age of the injection to be estimated (e.g. Hill et al. 2005; Burch et al. 2005; Müller et al. 2010). Old events can wrap around the planet several times before ultimately dissipating (Paranicas et al. 2007, 2010b). A case study of one "young" interchange injection seen at ≈ 7 R_S revealed details of the plasma, field and wave properties within the injection (Rymer et al. 2009a). The injection origin, based on the pitch angles, was estimated at $9 < L < 11$ R_S. A study between 8.2 and 14.6 R_S (Thomsen et al. 2014a) concluded that some of the events probably originated beyond $L > 14$. The radial distribution of injections peaks around 8 R_S (Chen et al. 2010; Kennelly et al. 2013). The local time distribution differs between studies.

Kennelly et al. (2013), using fresh wave signatures, found occurrence peaks around 01:00 and 14:00 LT; Chen and Hill (2008) considering older, dispersed plasma signatures, found most between 04:00 and 12:00 LT; and Müller et al. (2010) using back-traced old injections, found most between 21:00 and 09:00 LT. The Kennelly et al. (2013) events near 00:00 LT are strongly ordered by SLS3, with the northern (southern) longitude system correlating best pre(post)-equinox. Inflow channels occupy only ≈ 7% of the available longitudinal space (Chen et al. 2010), implying that the inflow speed must be much greater than the outflow speed if net magnetic flux balance is to be maintained. Considering that high-energy particles drift out of the injected flux tube faster than low-energy particles,

Burch et al. (2005) estimated an inflow speed of 25 km/s. Using similar arguments, Rymer et al. (2009a) obtained 71 km/s (as corrected by Chen et al. 2010). With the average outflow speed estimated from earlier measurements of Wilson et al. (2008), inflow speeds of 2 km/s at 5 R_S up to 38 km/s at 10 R_S were derived (Chen et al. 2010). The inferred outflow speeds increase faster than linearly with radial distance and remain well below the theoretical limit of the local azimuthal corotation speed (Vasyliūnas 1994; Hill 2009). Calculating the flux tube mass content, Chen et al. (2010) estimate a total global plasma mass outflow rate $\approx$ 280 kg/s at 10 R_S, but with very large error bars. This and their estimate of the total plasma sheet mass implies a plasma residence time of only 2 days inside of 10 R_S. Global multifluid modeling has shown the global development of the instability, with fingers of cool outwelling plasma alternating with fingers of hot injected outer-magnetospheric material (Kidder et al. 2009; Winglee et al. 2013). In the nonlinear development of the instability, the inflow fingers become much narrower than the outflow sectors and have larger radial speeds (Liu et al. 2010; Liu and Hill 2012), consistent with the observations of Chen et al. (2010). Figure 6.8 shows the results of three different multifluid model runs of the so-called Rice convection model. The interplanetary magnetic field parallel to the planetary field results in fewer fingers, while higher mass loading rates result in more. In the simulations, the instability starts near 7 R_S, and when cold plasma fingers reach $\approx$ 12–14 R_S, they spread in azimuth. Pieces can break off in the afternoon/dusk region and be flung out along the dusk magnetopause. Analytical work has related the interchange instability to the development of a rotating two-cell convection pattern in the inner magnetosphere (Goldstein et al. 2014). This pattern is initiated by a ring-current pressure distribution with an m = 1 longitudinal asymmetry. That reproduces the longitudinal ion density asymmetry (Burch et al. 2009).

6.4.3 Energetic Particle Injections

Both Thomsen et al. (2013) and Mitchell et al. (2015) have noted the confusion in the literature between large-scale "injections" arising from dynamical processes in the tail, "current sheet collapse," probably associated with magnetic reconnection on plasma-laden flux tubes in the tail, analogous to substorms at the Earth (Section 6.4.6), and "injections" arising from the centrifugally driven interchange instability just discussed (Section 6.4.2). The term has been used in a phenomenological sense to refer to a (usually energy-dispersed) enhancement in the local charged-particle intensity or in remotely sensed ENAs. But it is important to try to distinguish the nature of the phenomenon giving rise to these enhancements. Mitchell et al. (2015) distinguish between the two types of events based on location: interchange occurs primarily in the middle magnetosphere, while current sheet collapse occurs primarily in the outer magnetosphere. Thomsen et al. (2013) argues that interchange events are small in scale (< 1 R_S wide, typically seen only for a few minutes), frequent (often with numerous events during a single orbit) and often of relatively limited extent in terms of energy. The large-scale particle injections are broad (frequently more than an hour wide; e.g. Paranicas et al. 2010b), typically extend to several hundred keV, occur less frequently (once every few days) and can persist for more than 20 hours (Paranicas et al. 2007). Figure 6.9 (adapted from Mitchell et al. 2015) shows a typical example of a flux tube interchange injection in the Saturnian magnetosphere. There is a distinct boundary between the inner magnetosphere, where interchange injections are typically seen, and the outer magnetosphere, where the current sheet collapse events are seen. This boundary changes from orbit to orbit, but typically resides between 12 and 15 R_S, which not coincidentally is just the region where the magnetic field transitions between nearly dipolar to magnetodisk-like. Inside this boundary, the magnetic field dominates the energetic particle motions, and energy dispersion is therefore easily modelled (Paranicas et al. 2007; Brandt et al. 2008). Outside this boundary, energy spectrograms rarely display dispersive features, probably because the field is not as well ordered and because the weak, highly curved magnetic field violates the adiabatic invariants. Both small-scale and larger-scale injections are seen in the inner magnetosphere. It is possible that the larger-scale injections are simply time-evolved versions of multiple interchange injections. It is not clear that the typical particle content of interchange channels is sufficient to account for the long-duration elevated fluxes in energetic-particle injection events. It is also possible that a current sheet

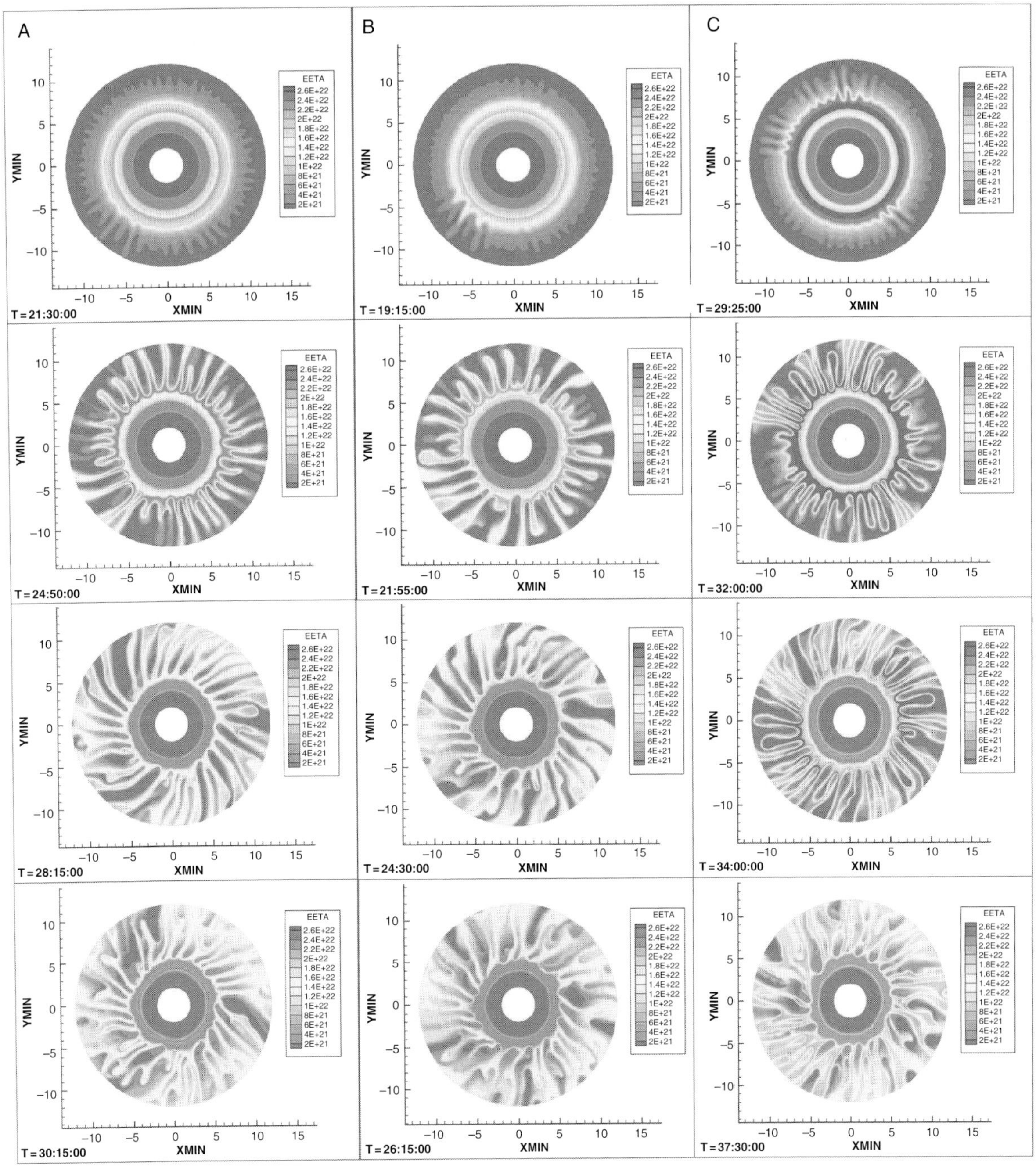

Figure 6.8 Visualization of the fluxtube interchange process in the Saturnian magnetosphere from results of the Rice convection model. Evolution over time (top to bottom) of the plasma convection in three different Rice convection model RCM simulations (A: without plasma pressure, Petersen conductance $\Sigma_p = 3S$; B: with plasma pressure observed, $\Sigma_p = 3S$; C: with plasma pressure observed, $\Sigma_p = 6S$). Grey-scale indicates the plasma ion content per unit magnetic flux in Saturn's equatorial plane inside 10 R_S (figure 3 from Liu and Hill 2012). (A black-and-white version of this figure appears in some formats. For the color version, please refer to the plate section.)

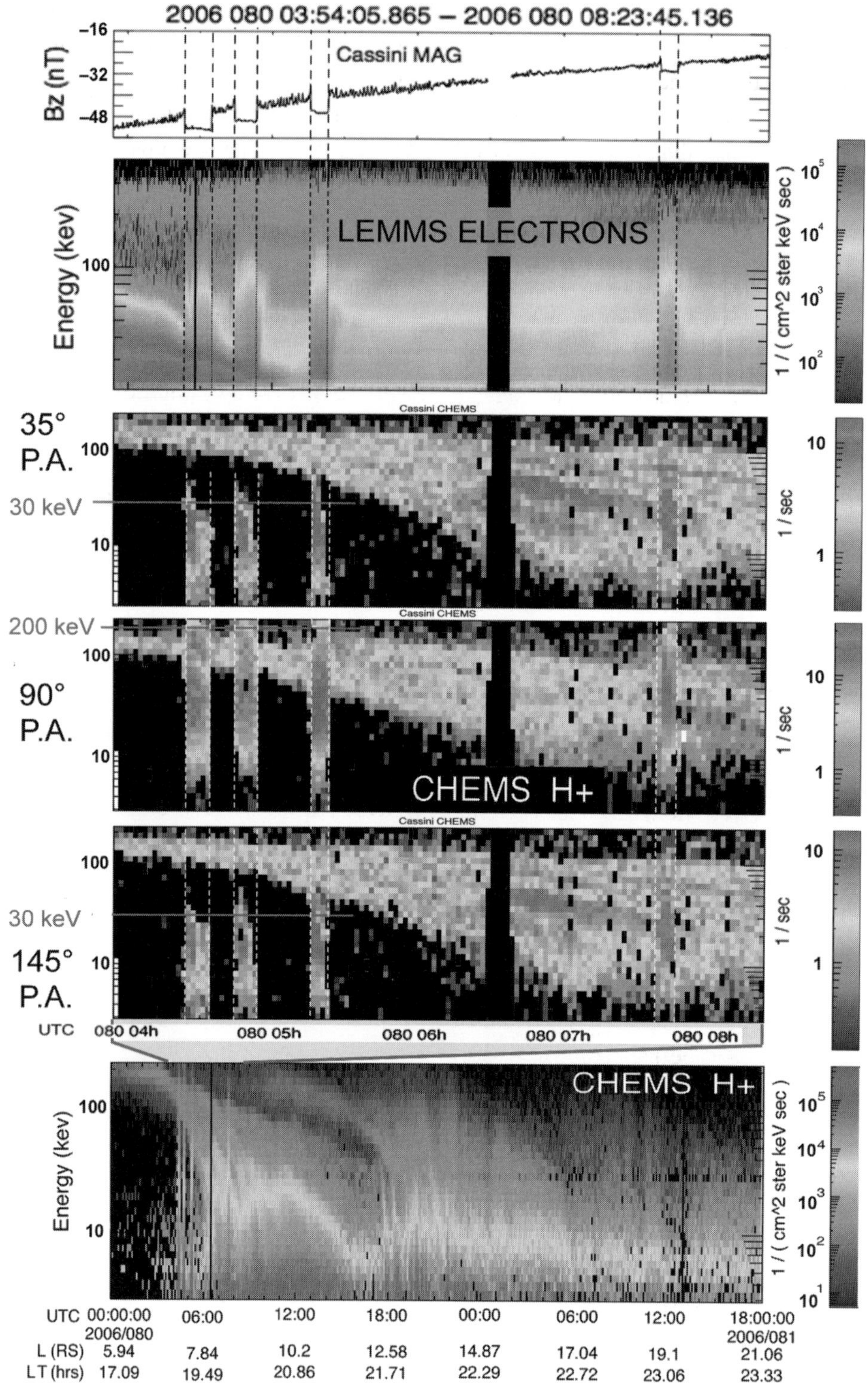

Figure 6.9 Flux tube interchange injections in Saturn's magnetosphere as seen in color-coded energetic particle intensities (lower four panels) and magnetic field (top panel) measurements onboard Cassini (from Mitchell et al. 2015). The lowest panels show data for one day (2006, day 080) when Cassini was between 6 and 21 R_S (local time, 17:00–23:30) while the other color panels show only data of hours 04:00–08:00 on that day. (A black-and-white version of this figure appears in some formats. For the color version, please refer to the plate section.)

collapse event may destabilize interchange over an unusually broad azimuthal range, giving rise to large-scale interchange events in the inner magnetosphere. Energetic particle injections have proven to be extremely useful in diagnosing the dynamics of Saturn's magnetosphere.

6.4.4 Magnetospheric Interactions with Moons

One of the most intriguing aspects of Saturn's magnetosphere is the presence of several moons, which interact with the magnetospheric fields and particles in a number of ways. First, the moons (Enceladus, in particular, but also Titan and the rings) are the sources of most of Saturn's magnetospheric plasma and some of the dust (see Section 6.3). Second, the moons absorb magnetospheric charged particles, producing both macrosignatures and microsignatures in energetic particle fluxes (see Section 6.2). Third, energetic particles impacting the moons produce weathering and charging of the surfaces as well as sputtered particles (Paranicas et al. 2012, 2014). And fourth, the electrodynamic interaction mediated by pickup of neutral gas and dust from the moons drives field-aligned currents (FAC) that result in particle energization and auroral signatures at the feet of the magnetic field lines that intersect the moons (e.g. Pryor et al. 2011). These interactions involve a rich spectrum of physical processes, and space here does not permit a full review of progress in illuminating them. Enceladus, by virtue of its striking water plumes, has perhaps the most dramatic interactions with the magnetosphere and has received the most attention, both observationally and theoretically. In addition to neutral gas, the plumes contain tiny ice grains ("dust") that become electrically charged, both positively and negatively. Negatively charged "nanograins" greatly outnumber positive ones, indicating that the charging of such grains by electron impact increases as they travel away from the source vents (Hill et al. 2012). They constitute an important source for Saturn's magnetospheric plasma and E-ring particles. Dust is detected by the plasma wave instrument onboard Cassini RPWS within ~20 Enceladus radii (R_{Enc}) from the moon in association with strong slowing of the plasma in the moon's vicinity (Farrell et al. 2010). Near Enceladus and in the E-ring, electron densities appear to be much lower than ion densities, suggesting that many electrons are being attached to the dust. The bulk of ions in this region appear to drift at a speed much less than rigid corotation, due to the interaction between the corotating plasma and charged dust grains (Morooka et al. 2011). The negatively charged dust may also lead to the so-called "anti-Hall effect," in which the Hall current is carried dominantly by water-group ions, potentially accounting for previously unexplained transverse magnetic perturbations (Simon et al. 2011a; Jia et al. 2011). The importance of negatively charged dust has been confirmed by a number of different simulations of Enceladus' interaction with Saturn's magnetosphere (Jia et al. 2012b; Omidi et al. 2012; Kriegel et al. 2009, 2011, 2014). However, 3D hybrid simulations show that even complete electron absorption by the dust decelerates the flow by only a few km/s, whereas charge exchange is the dominant mechanism that produces the observed flow stagnation at Enceladus (Omidi et al. 2012). Whistler-mode emissions (similar to terrestrial auroral hiss), field-aligned electron beams and magnetic perturbations consistent with presence of FAC are seen at Enceladus. Analysis of the whistler waves shows that the electron beams responsible for the emissions are accelerated very close to the moon (within 1–3 R_{Enc} above the surface), and the most likely acceleration mechanism is parallel electric fields associated with standing Alfvén waves (Gurnett et al. 2011; Leisner et al. 2013). The first detection of an auroral footprint associated with Enceladus in Saturn's northern ionosphere has also been reported, presumably related to field-aligned ion and electron beams seen by Cassini downstream of the moon at distances between 3 and 20 R_{Enc}. Changes in the footprint emission magnitude cannot be fully accounted for by changes in the magnetospheric conditions, and this indicates the presence of variable plume activity on the moon (Pryor et al. 2011).

At Dione (6.36 R_S), magnetic field observations (Simon et al. 2011b) and detection of O_2^+ pickup ions (Tokar et al. 2012) are consistent with a plasma interaction with a weak exosphere. Observations from the mass spectrometer INMS onboard Cassini of O_2 and CO_2 at Rhea (8.73 R_S) similarly suggest the existence of a tenuous atmosphere there (Teolis et al. 2010). At Rhea, the observed magnetic field perturbations are consistent with an Alfvén-wing type of interaction, perhaps produced by a diamagnetic current in the wake region (Simon et al. 2012). The only significant effect on the plasma at Rhea appears to be absorption by the moon and creation of a wake plasma void (Wilson et al. 2010). However, intense plasma waves in the magnetic flux tube connecting Rhea to Saturn are observed, including bursty electrostatic waves, possibly driven by a low-energy electron beam accelerated away from Rhea, and whistler-mode emissions propagating toward Rhea, generated by the loss-cone anisotropy caused by absorption of electrons at the moon's surface (Santolík et al. 2011). Although the ionospheres of Dione and Rhea are tenuous, the energetic particle signatures at both moons are not consistent with simple geometrical absorption. Rather, the wake regions are structured, with unexpected access of particles (Roussos et al. 2012; Krupp et al. 2013). Flux dropouts observed on the first flyby of Rhea were attributed to a possible particulate halo around the moon (Jones et al. 2008), but similar analysis during subsequent flybys (Roussos et al. 2012) and an optical search (Tiscareno et al. 2010) were inconsistent with that suggestion.

Although at its orbital radius of 20 R_S, Titan is occasionally exposed to the magnetosheath and solar wind, it spends most of the time within Saturn's magnetosphere (Smith and Rymer 2014; Bertucci et al. 2014). The plasma environment it encounters can be that of the plasma sheet, the lobe, the magnetosheath or "bimodal" (highly variable and mixed) (Rymer et al. 2009b; Smith and Rymer 2014). The magnetic field is similarly quite variable (Bertucci et al. 2014; Simon et al. 2010). While the field orientation is seldom steady and perpendicular to Titan's orbit, as was often assumed in pre-Cassini simulations, the observed field at Titan is still usually well described by draping of the average background field over the moon's ionosphere (Simon et al. 2013). Short-time fluctuations in the ambient field may obscure signatures of the moon's induced magnetosphere (Simon et al. 2013) and may cause "fragmentation" of Titan's magnetic lobes (Simon et al. 2014). Titan's dayside ionosphere below 1400 km altitude is produced mainly by sunlight, with little evidence of any significant effects of Saturn's magnetosphere (Luhmann et al. 2012). However, significant electron heating appears to occur on the night/wake side of Titan's ionosphere, such that the cold ionospheric plasma extends to higher altitudes there, potentially providing an important atmospheric loss region (Edberg et al. 2010). Indeed, cold plasma of ionospheric origin, both heavy and light ions, is seen streaming into Titan's tail with an estimated loss rate of 7 tons per Earth day (Coates et al. 2012; see also Section 6.3). Titan's tail regularly exhibits a split structure with two distinctly separate ionospheric outflow regions (Coates et al. 2012; Agren et al. 2007; Modolo and Chanteur 2008), perhaps a reflection of the fragmentation expected due to variability in the upstream magnetic field (Simon et al. 2014).

6.4.5 Noon-to-Midnight Electric Field

The inner moons act as particle absorbers, creating a "shadow" in the plasma population as it convects past them (see Roussos et al. 2005; Section 6.2.2). These so-called microsignatures propagate in the magnetosphere exactly as the missing particles would have done, providing a tracer of the transport. The microsignatures are often offset radially from the orbital distance of the moon (e.g. Roussos et al. 2007), revealing a radial component to the predominantly azimuthal inner-magnetospheric flows. Microsignatures are mostly found outside (inside) the moon's orbital distance on the dayside (nightside) (Roussos et al. 2007; Andriopoulou et al. 2012). This local time dependence implies a systematic dawnward component of flow in the inner magnetosphere (i.e. a global, roughly noon-to-midnight directed electric field), superimposed on the dominant azimuthal corotation (radial electric field) and magnetic gradient/curvature drifts of energetic particles (Roussos et al. 2007; Andriopoulou et al. 2012). The existence of such a flow pattern results in zero-energy drift paths that are $\approx 1\ R_S$ further from Saturn on the dayside than on the nightside (Thomsen et al. 2012). The inward (outward) flow at dusk (dawn) is confirmed by noon/midnight asymmetries in plasma ion and electron temperatures and energetic-particle phase space densities (Thomsen et al. 2012), energetic particle intensities (Kollmann et al. 2011; Paranicas et al. 2010a; Carbary and Rymer 2014), radial plasma velocities (Wilson et al. 2013) and density and azimuthal velocity (Holmberg et al. 2014). This flow might also be responsible for the systematic radial

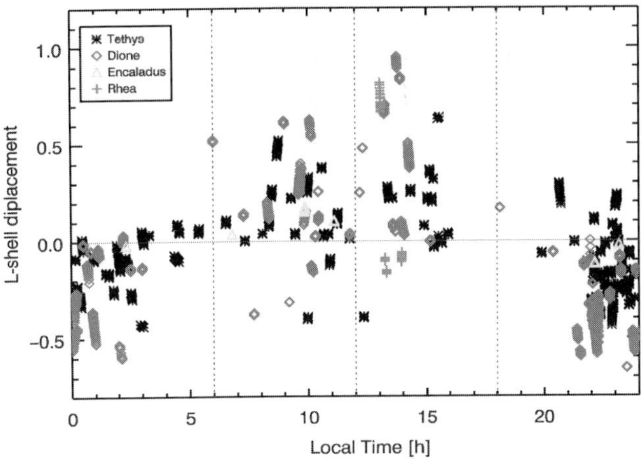

Figure 6.10 Radial displacements of observed absorption microsignatures from the expected moon's L-shell for the moons Tethys, Dione, Enceladus and Rhea (0 means no displacement, positive and negative values mean displacement radially outward or inward, respectively) as a function of local time in Saturn's magnetosphere (from Andriopoulou et al. 2012, 2014).

velocities of "blobs" of ENA emissions associated with injected high-energy ions (Carbary et al. 2008). The inferred electric field extends at least between 4 and 10 R_S, has a strength of a few 0.1 mV/m, and decreases with radial distance (Thomsen et al. 2012; Andriopoulou et al. 2014). The direction of the field varies, and is on average rotated ≈ 12°–32° east of midnight (Andriopoulou et al. 2014; Wilson et al. 2013). There is a possible long-term secular variation. The origin of this flow pattern is not known, although a number of possibilities (results of solar, ionospheric and planetary winds) have been examined (Thomsen et al. 2012). Plasma interactions with charged E-ring dust grains, which are in turn subject to radiation pressure forces, have been proposed, but this mechanism does not seem entirely consistent with the eastward rotation of the flow pattern (Holmberg et al. 2014). In addition to revealing the global convection pattern, analysis of one rather complex microsignature as a function of energy indicated the existence of a pulse of convection that could locally enhance radial transport by 3 orders of magnitude, compared to radial diffusion (Roussos et al. 2010). A numerical simulation of microsignature motion in realistic fields reproduced the observed energy dispersion profiles and revealed the limitations of previous calculations of microsignature propagation, particularly for higher electron energies (Roussos et al. 2013). The tool may be very useful for exploring temporal variations in the convection experienced by complex microsignatures (e.g. Andriopoulou et al. 2012).

6.4.6 Magnetotail Flows, Tail Reconnection, Plasmoids and Dipolarizations

The global magnetospheric flow pattern was discussed in Section 6.2. Here, we focus on deviations from this pattern associated with dynamic events occurring in the tail. Although measurements of the magnetic field orientation (e.g. Jackman et al. 2009a, 2011) and flow direction (e.g. Thomsen et al. 2013, 2014b; Kane et al. 2014) show that within the region of space sampled by Cassini there is no evidence for a quasi-steady Dungey-type reconnection line, there is considerable evidence that tail reconnection does occur, at least episodically and perhaps with restricted longitudinal extent. Clear signatures of plasmoid formation inward of 30–50 R_S in the tail have been found in both magnetic field (e.g. Jackman et al. 2007, 2009b, 2011, 2013, 2014b) and plasma data (e.g. Hill et al. 2008; Jackman et al. 2014b). Such plasmoid departures have been associated with SKR enhancements (e.g. Mitchell et al. 2005; Mitchell et al. 2009b; Jackman et al. 2009b) and ENA brightenings in the near-tail (e.g. Mitchell et al. 2005; Hill et al. 2008). The SKR and ENA enhancements are attributed to the dipolarization and sunward flow of reconnected flux returning to the inner magnetosphere (Figure 6.11). In situ signatures of these dipolarizations and return flow have also been found (Bunce et al. 2005; Russell et al. 2008; Masters et al. 2011b; Thomsen et al. 2013). Tail reconnection, plasmoid production and dipolarization have been associated with enhancements in the solar wind dynamic pressure (e.g. Bunce et al. 2005; Badman et al. 2005; Mitchell et al. 2005; Mitchell et al. 2009a; Jackman et al. 2010; Jia et al. 2012a), which are relatively infrequent. More commonly, tail reconnection events appear to occur roughly once per Saturn rotation, apparently associated with a particular longitude sector rotating into the midnight region (Mitchell et al. 2009b; Jia and Kivelson 2012). Plasmoids encountered in the tail can occur as single events or in groups over

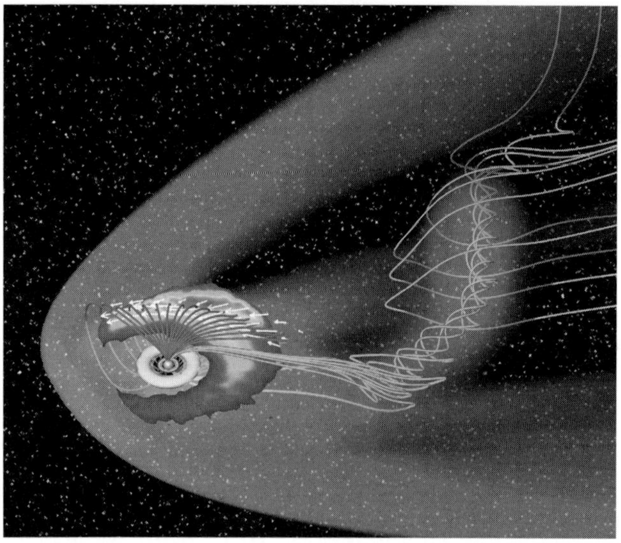

Figure 6.11 Global MHD simulation of a large-scale tail disruption event (Jia et al. 2012a). Centrifugal stresses resulting from Saturn's rapid rotation and strong mass-loading from Enceladus, lead-to-tail reconnection, releasing a large plasmoid (green spiral field lines, overlain by cyan field lines with both ends in the solar wind). The Saturn-ward portion of the reconnected magnetic field (magenta lines) snaps back toward the planet, accelerating and heating the remaining plasma. The accelerated flow is denoted by the white arrows. The color image superimposed on the simulation shows ENA fluxes observed during such an event by the Cassini/MIMI instrument (e.g. Mitchell et al. 2009b) and reveals strong plasma energization in just the region predicted by the simulation. (A black-and-white version of this figure appears in some formats. For the color version, please refer to the plate section.)

a relatively short time (Jackman et al. 2011). They exhibit either a loop-like or a flux rope-like structure (Jackman et al. 2014b). Global MHD simulations (Zieger et al. 2010; Jia et al. 2012a) also clearly show the occurrence of tail reconnection and large-scale plasmoid production. Helical plasmoids (flux ropes) are released at a rate that depends on the solar wind dynamic pressure and the mass-loading rate in the simulations (Zieger et al. 2010). There is evidence for both "Vasyliūnas-type" (Vasyliūnas 1983) and "Dungey-type" (Dungey 1961) reconnection in the simulations (Jia et al. 2012a). Plasmoid production has been viewed as a promising mechanism for shedding the plasma produced within the magnetosphere. But so far, the observed reconnection events and resultant plasmoids seem insufficient to remove most of the plasma produced (Louarn et al. 2014; Thomsen et al. 2013; Jackman et al. 2011, 2014b; Bagenal and Delamere 2011). The global numerical simulations come to the same conclusion and suggest instead that there is a significant amount of mass lost down the flanks of the magnetotail (Zieger et al. 2010; Jia et al. 2012a). On the other hand, these reconnection events do seem capable of recapturing open magnetic flux sufficiently quickly: the magnetic signatures of plasmoid passage commonly feature an extended interval of northward field following the departure of the plasmoid, which has been taken as the signature of ongoing reconnection that began with the plasmoid production and then proceeded to involve the lobe field lines ("Dungey-type"). Estimates of the closure of magnetic flux during these intervals are in reasonable agreement with that required to recapture flux opened at the magnetopause (Jackman et al. 2011, 2014b). One possibility for reconciling this inconsistency is related to the distinction between Vasyliūnas reconnection that does or does not proceed to involve Dungey lobe reconnection: the Jackman et al. (2010, 2011, 2014b) papers identified plasmoids based on the "classical" signatures in the elevation angle of the magnetic field. But those signatures were derived in the context of plasmoids that travel rapidly down the magnetotail, propelled by the tension force created as the overlying lobe field reconnects behind the plasmoid (Jia et al. 2012a). Vasyliūnas-type reconnection that does not proceed to involve the lobe results in release of a plasmoid from the corotation forces, but its velocity is not accelerated and is largely tangential to the arc it was describing before pinch-off (Jia et al. 2012a; Mitchell et al. 2015).

The magnetic signature of such a plasmoid would be entirely different and would not be identified in the Jackman et al. approach. On the other hand, the closure of open flux will only take place in cases of Dungey reconnection, which is likely to be preceded by Vasyliūnas reconnection forming a plasmoid (e.g. Thomsen et al. 2013). In this case, the plasmoid is accelerated tailward and would exhibit the sought-for magnetic signature. Thus, the plasmoids that Jackman et al. identify are just that subset that does include a Dungey reconnection phase, and it is only those that close the open flux, which is why it is found to be sufficient to satisfy the required flux closure.

The implication is that there may be many more plasmoids that carry away magnetospheric plasma than those that lead to closure of open flux. Such an alternative might be detected by the frequent occurrence of northward fields in the night-side plasma sheet.

6.4.7 Waves and Wave-Particle Interactions

Saturn's magnetosphere abounds in different plasma waves, which emerge from a wide array of wave-particle interactions. These waves are good tracers of the physical conditions and magnetospheric processes that produced them. They dissipate the free energy of their causative particle distributions and may scatter or energize other populations.

Saturn Kilometric Radiation SKR

Saturn Kilometric Radiation (SKR) is a well-known radio wave signature of auroral activity believed to be generated by the cyclotron maser instability (CMI) (Zarka 1998). Its periodic intensity variations near 10.66 hours, discovered by Voyager, were originally thought to be the rotation period of the interior of Saturn (Desch and Kaiser 1981), but long- and short-term variations of this periodicity as well as its dual nature (different for each hemisphere) raise many new questions, which were discussed at length in Chapter 5. Further discussion will be found in Chapter 7.

Resolved SKR maps (from Cassini/RPWS observations) revealed an average continuous high-latitude radio oval in the same latitude range as the UV auroral oval observed by the Hubble Space Telescope (Cecconi et al. 2009; Lamy et al. 2009). The emitted power along the oval is enhanced on the dawn side, both in UV and radio. Voyager observations suggested that the SKR source and its magnetospheric drivers were fixed in LT and modulated by the planetary rotation (≈10.88 hours) like a strobe, with a phase independent of the observer's position. Cassini's good LT coverage and multi-instrument studies (SKR/plasma/magnetic field and UV with HST) showed that SKR and UV sources actually rotate with the planet, but are more intense in the predawn-to-noon sector. Combined with the SKR anisotropic (hollow conical) beaming and with a limited LT coverage of the observations, this can mimic a strobe-like source such as was observed by Voyager (Andrews et al. 2011). During its southern periapsis pass of Revolution 89, Cassini passed in the midnight-to-dawn sector at high latitude (>65°S) and low altitude (<5 R_S) inside a region identified as an SKR source, the first auroral radio source other than Earth's explored in situ. During that period, Cassini was connected to the UV aurora at 70°–80°S. Magnetometer (+electrons and SKR) data revealed unique signatures of a contracted circumpolar open-field-line region surrounded by an unusual region of super-corotating flow containing hot keV electrons, itself surrounded by the usual region of sub-corotating flow (Bunce et al. 2010). This was interpreted as an exceptionally intense field-aligned current (FAC) system – consisting of a central region of downward current flanked by two regions of upward current – related to the occurrence of a major open flux closure event in Saturn's magnetotail (≈15 GWb), possibly triggered by solar wind compression of the magnetosphere. The spiral shape of the UV/radio auroral sources provided evidence of an ongoing reconfiguration of the magnetosphere (Bunce et al. 2010). Strong SKR emissions coincided with the high-latitude upward current region. The emission, propagating at a large angle from the magnetic field in the source region, was consistent with the CMI driven by 6–9 keV electrons (Lamy et al. 2010). Electric potential structures in the auroral acceleration region are thought to be important in forming horseshoe or shell-like electron distributions. Those are unstable to the CMI generation mechanism, hence tying SKR generation to the precipitation of electrons responsible for the UV aurorae. Loss-cone and shell distributions of electrons were indeed observed in this upward FAC regions (Schippers et al. 2011) as well as during a second encounter with the probable SKR source region (Menietti et al. 2011a). Observed levels of X and Z mode were consistent with linear wave growth computed from shell-driven CMI (Mutel et al. 2010), whereas the O mode amplitude appears to require non-linear effects (as for Earth) (Menietti et al. 2011a). In the downward FAC region of Bunce et al. (2010), electron data revealed two types of upward-accelerated electron beams: a ubiquitous broadband energetic (1–100 keV) population, with a low-energy (0.1–1 keV) population superimposed sporadically. Both populations are likely sources of very low-frequency emissions (Schippers et al. 2011). Ion cyclotron waves were also observed within the downward current region, and modeling of the observed upward electron beams successfully predicted growth of hydrogen cyclotron waves, potentially a source of auroral ion heating (Menietti et al. 2011b).

SKR emissions provide a powerful capability to monitor global magnetospheric dynamics from any position where it is visible (Kurth et al. 2005; Lamy et al. 2013; Kurth et al. 2016). SKR intensifies and expands toward lower frequencies when substorm-like activity occurs in the tail, along with auroral brightening, ENA enhancements, plasmoid production and energetic-particle injections (e.g. Mitchell et al. 2005; Mitchell et al. 2009a; Hill et al. 2008; Jackman et al. 2009b). Mitchell et al. (2009a) proposed that the accelerated electrons that emit SKR are produced as part of the FAC system that closes the partial ring current created by the localized injection of hot ions into the inner magnetosphere following a tail reconnection event. SKR emissions also increase in response to solar wind dynamic pressure enhancements (Desch 1982).

Narrowband Emissions

Cassini-RPWS detected narrowband radio emission around 5 and 20 kHz, occurring for several days after SKR intensifications, following compression of the magnetotail (Ye et al. 2009; Wang et al. 2010), similar to Jovian narrowband kilometric emissions generated at density gradients in Io's plasma torus. Emissions are mainly observed on the nightside, at mid to high latitudes. They slowly fade out while drifting toward higher frequencies after each onset. Several localization techniques point at sources at the northern and southern edges of Saturn's plasma torus at L = 4 to 10. Polarization suggests L–O mode emissions (Ye et al. 2009). Their generation has been attributed to mode conversion from electrostatic upper hybrid (Ye et al. 2009; Wang et al. 2010), or from electromagnetic Z mode at density gradients. Z mode (and weak O mode) waves can be directly produced by the CMI when $f_{UH} \sim n * f_{ce}$ where f_{UH} is the upper hybrid frequency, f_{ce} is the electron cyclotron frequency and n is an integer (Menietti et al. 2010). Emissions show

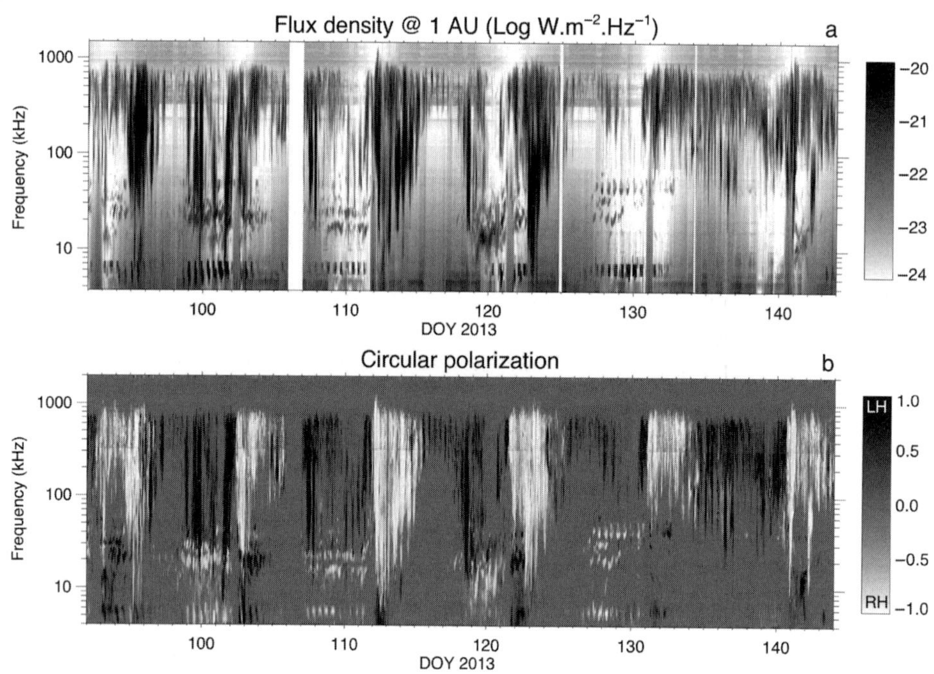

Figure 6.12 (a) Normalized flux density in the range 3.5 kHz to 16.5 MHz measured by RPWS for ≈ 2 months. SKR is the dominant emission and narrowband (NB) emissions are well visible at 5 kHz (strongest), 20–30 kHz, and even sometimes up to 40–50 kHz (harmonics of the 20–30 kHz emissions). Recurrences of NB events every 7 to 10 days illustrate the energetic events discussed by Louarn et al. (2014). Low-frequency extensions of SKR are also well visible (e.g. on DOY 95, 112, 123, . . .) and may be the signature of substorm-like activity. Long-term variations (e.g. ≈ 28-day separation between major intensifications of DOY 95 and 123) are here mostly related to solar wind fluctuations. (b) Polarization shows a cleaner picture, reveals the hemisphere of origin of SKR (LH = south, RH = north) and/or the emission mode of NB events (X/O/Z).

near-rotational modulation (~10.8 hours period), and comparisons with ENA images suggest a relation with transient hot plasma clouds – actually active flux tube interchange regions – corotating with the planet in the mid-magnetosphere (Mitchell et al. 2015).

Auroral Hiss

During high-latitude passes, RPWS observed upward-propagating, low-frequency whistler-mode emissions similar to Earth's auroral hiss. In situ CAPS data during one pass revealed intense upward-moving electron beams with energies of a few hundred eV that produce large growth rates for whistler-mode waves propagating along the resonance cone, as at Earth (Kopf et al. 2010). This auroral hiss is also well associated with energetic (40 to 500 keV) field-aligned electrons and energetic ion conics accelerated near magnetic field line footprints (Mitchell et al. 2009a). Sharp gradients in the electron density inferred from the auroral hiss suggest a plasmapause-like boundary at high latitudes that moves with the eccentric rotation of the auroral ovals (Gurnett et al. 2011).

Electron- and Ion-Cyclotron Harmonic Waves (ECH and ICH)

RPWS observed strong ECH waves associated with inward radial injections of hot plasma from Saturn's outer magnetosphere (Menietti et al. 2008; Tao et al. 2010). ECH growth is usually attributed to a loss cone in a cold (few eV) electron population. This seems unlikely at Saturn because the electron collision and bounce frequencies are quite similar, but a loss cone at higher energies (100s of eV to a few keV) inside the warm injected plasma can also drive ECH waves (Tao et al. 2010). During the crossing of a downward current region (Bunce et al. 2010; Schippers et al. 2011), CAPS detected intense upward-going electron beams and diffuse ion beams. RPWS detected associated ion cyclotron harmonic waves (Menietti et al. 2011b), possibly related to ion heating and electron acceleration.

Whistler-Mode Chorus

Chorus is an intense, structured whistler-wave emission observed near the magnetic equator. The strongest intensities (and largest bandwidths) found within a few degrees of the magnetic equator remain high up to latitudes of 30° (Menietti et al. 2012, 2013b, 2014). Frequency drift of chorus bursts at Saturn (~200 Hz) is consistent with predictions from the nonlinear wave growth theory of Omura et al. (2007) and CAPS measurements (Menietti et al. 2013a, 2014). Weaker and narrower equatorial emissions, lacking drifting tones, may be due to linear growth driven by a temperature anisotropy ($T_\perp/T_\parallel \sim 1.3$). Ion cyclotron waves may be the seed of the nonlinear growth leading to higher-frequency/fast-drifting chorus (Menietti et al. 2013b). At Jupiter, chorus intensities are sufficient to accelerate electrons by a stochastic gyroresonant process. At Saturn, the ≈ 10-times-weaker chorus intensity, narrower bandwidth (typically < 2 kHz ~ 0.6 f_{ce}) and high plasma density (larger f_{pe}/f_{ce}, with f_{pe} the electron plasma frequency) at low latitudes suggest less-efficient gyroresonant interactions, except locally within plasma injection regions (hot and depleted). This inefficiency makes chorus an unlikely acceleration mechanism for the relativistic electrons in Saturn's radiation belts, unlike what is currently understood about the Earth's (Shprits et al. 2012; Mauk and Fox 2010).

6.4.8 Temporal Variability and Seasonal Changes

Besides temporal variability at various timescales associated with the above dynamic processes, Saturn's magnetosphere also exhibits longer-term changes that may be related to season, source populations modifications, properties of the solar wind, Solar UV intensity or illumination geometry. Cassini, in orbit since mid-2004, has now covered a third of a Saturn year, from just past southern summer solstice (April 2002), through equinox (August 2009), to northern summer solstice soon (May 2017), enabling us to search for longer-term variability in magnetospheric properties.

The global configuration of the magnetosphere also changes seasonally, as the "bowl-shaped" warping of the extended plasma disk and tail are forced by the angle between the solar wind flow and Saturn's rotational axis (e.g. Arridge et al. 2008a; Sergis et al. 2011). By contrast, the temperature, pressure and number density in the center of the plasma sheet are unaffected by seasonal changes (Sergis et al. 2011). He^{2+} ions (of solar wind origin) and He^{+}(interplanetary pickup) decreased by a factor of 3 during 2009 – near solar minimum (DiFabio et al. 2011). In contrast, suprathermal H^+, O^+, O_2^+, H_2^+ ions in Saturn's near-equatorial magnetosphere ($7 < L < 16$) show no long-term trend from 2005 to 2011, indicating that the main source of magnetospheric plasma (the Enceladus gas torus) does not change appreciably on a seasonal basis, in agreement with Sergis et al. (2011), Elrod et al. (2012, 2014) and Persoon et al. (2015). On the other hand, thermal O_2^+ and W^+ plasma between the main rings and Enceladus (Elrod et al. 2012, 2014), and energetic O_2^+ and unidentified mass-28 ions outside of Enceladus (Christon et al. 2013, 2014), exhibit a clear seasonal variation (see Figure 6.13) attributed to a plasma source from the main rings that varies with changing EUV illumination (Section 6.3). The dissimilar responses of O_2^+ and mass-28 ions is attributed to the combination of a dominant ring source with a weaker secondary source, perhaps from the moons, constant through an extended interval around equinox (Christon et al. 2014). The seasonal variability of the thermal O_2^+ has been successfully modelled (Tseng et al. 2013a), suggesting strong losses of O_2^+ just outside the edge of the main rings, potentially due to the presence of small grains or the precipitation of inwardly diffusing radiation-belt particles (Elrod et al. 2014). At the beginning of the Cassini mission, several enhancements in energetic proton intensity (more than an order of magnitude above background) were observed in the magnetosphere and attributed to interplanetary energetic particle events, more common near solar maximum. The proton intensities between the moons Tethys and Dione were especially enhanced, forming a transient radiation belt (Roussos et al. 2008). Such interplanetary events recurred later in the mission, but Cassini's periapses did not reach in close enough to observe the possible resulting radiation-belt enhancements. During the "Grand Finale" (see Section 6.6) at the end of the Cassini mission in 2017 the chance to observe the radiation belt response will be quite good.

6.4.9 Global Dynamics and Drivers

Saturn's rapid rotation and strong plasma source deep inside the magnetosphere result in internally driven dynamics like interchange and tail reconnection. On the other hand, observations of the aurorae, SKR, ENAs and particles and fields have made it clear that variations in the solar wind, especially enhancements in the dynamic pressure, can directly and dramatically

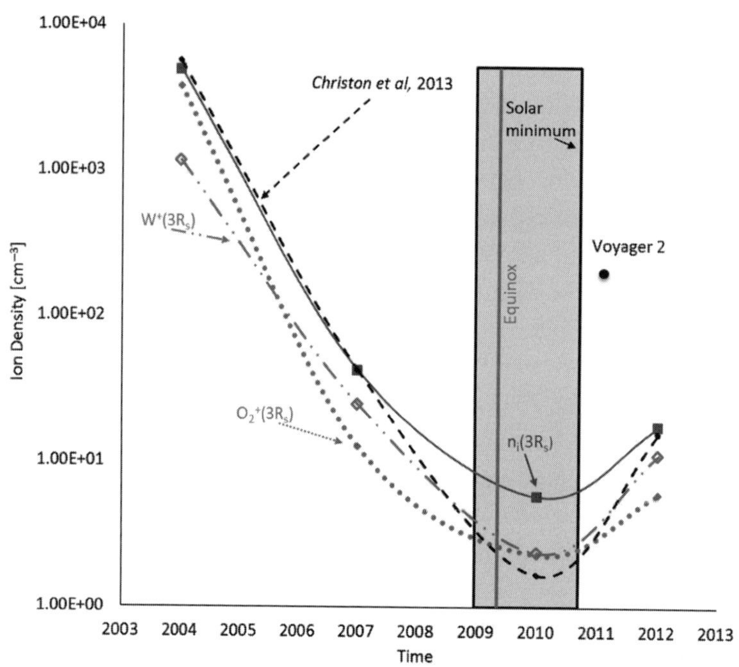

Figure 6.13 Time variation of ion density near the main rings (from Elrod et al. 2014): n_i shows total thermal ions, W^+ and O_2^+ are split by species. The line labelled as "Christon et al. 2013" is energetic O_2^+/W^+ times thermal W^+. Voyager 2 measured the total ion density in 1981, which was one year after equinox and close to solar maximum (Richardson 1986).

influence magnetospheric dynamical processes. So far, however, the anticipated signature of X-ray aurorae produced by interaction between solar wind ions and the planet's atmosphere has not been detected (Branduardi-Raymont et al. 2013).

Recent work explores the impact of magnetopause reconnection as a factor in Saturn's magnetospheric processes and finds that, except under rare circumstances, it is not expected to be a major driver (see Masters et al. 2014a; and Section 6.5). On the other hand, global multi-fluid simulations of the magnetosphere show that the interplanetary magnetic field (IMF) orientation can affect the rate at which dynamical events occur (Kidder et al. 2009). The model shows that the global development of the centrifugally driven interchange instability is more favored when the IMF is antiparallel (as opposed to parallel) to the equatorial planetary field because of changes in the bowl shape of the magnetospheric current sheet.

The contribution of magnetopause reconnection to global dynamics is also seen in 3D MHD simulations of the magnetosphere (Jia et al. 2012a; see also Figure 6.11). These simulations suggest that centrifugal acceleration of mass-loaded flux tubes stretches field lines and eventually leads to reconnection and formation of plasmoids on closed field lines, an intrinsic process (Vasyliūnas-cycle) occurring in a rapidly rotating magnetosphere with a strong internal plasma source, independent of the IMF orientation. However, when the IMF is northward and magnetopause reconnection loads open magnetic flux into the tail lobes, the tail reconnection progresses into the open field-line region, reclosing the open flux and recirculating it back into the magnetosphere. Once this Dungey-cycle reconnection can occur, it produces stronger global impacts on the magnetosphere and ionosphere. These simulations also show that the location of the tail neutral line is variable and apparently affected by solar wind conditions (Jia et al. 2012a).

Recycling of the global magnetic flux via magnetopause opening and tail reclosure has been explored with Hubble Space Telescope observations of Saturn's aurora (Badman et al. 2013, 2014). Using the poleward boundary of the main oval as a proxy for the open-closed field line boundary, the amount of open flux was found to be 10–50 GWb, with a mean of 35 GWb. While the percentage of magnetic flux opening in Saturn's magnetosphere is similar to that measured at the Earth (2–11%), the typical percentage of flux closure between consecutive observations is significantly lower (13% vs. 40–70%). Therefore, open flux is usually closed in smaller (few GWb) events at Saturn. The typical decrease of open flux of ≈5 GWb is similar to the amount of flux transported following a tail reconnection event inferred from in situ observations (e.g. Jackman et al. 2014b) and MHD simulations (Jia et al. 2012a). The exception is that large flux-closure events occur in association with strong solar wind compressions, which are known to produce particularly large tail reconnection events (e.g. Mitchell et al. 2015).

While much attention has been given to the importance of magnetic reconnection as a crucial dynamical process in Saturn's magnetosphere, Delamere and Bagenal (2013) suggest that viscous interaction between the solar wind and the magnetosphere plays an important role in shaping the magnetotail and in releasing internally produced plasma, much like the situation at a comet. Noting the slow rate of magnetic flux recycling at Jupiter (which is similarly found at Saturn by Badman et al. 2014), Delamere and Bagenal (2013) envision a rather slowly processed open-flux region, primarily in the center of the tail, with substantial "wings" of closed flux stretched out tailward on both flanks due to viscous interaction at the magnetopause. Small-scale reconnection events can occur within these wings, allowing loss of magnetospheric plasma. MHD simulations (Jia et al. 2012a) indeed show such closed-field wings along both flanks, and the duskside region of tailward flow observed (Thomsen et al. 2013, 2014b) and described as a "planetary wind" may correspond to this predicted region.

6.5 Boundary Phenomena

Data returned by Cassini has allowed significant progress in understanding the edge of Saturn's magnetosphere: the magnetopause current sheet. This boundary separates the magnetospheric environment of planetary magnetic field and plasma (mainly sourced from Enceladus) from the solar wind interplanetary magnetic field (IMF) and plasma that has to flow around the magnetospheric obstacle. Cassini observations have not only constrained the properties of the adjacent regimes, but have also revealed the detailed structure of the boundary, and the processes that can lead to transfer of mass and energy across it. A model of the location and shape of Saturn's magnetopause was constructed by Kanani et al. (2010), showing compressibility of Saturn's magnetosphere that is intermediate between that of Earth and Jupiter. Pilkington et al. (2014) have also showed that the internal "magnetodisk" flattens the magnetopause surface about the poles. Went et al. (2011) reported that there is little evidence for the Saturnian equivalent of the magnetospheric "cushion region" seen in Jupiter's outer magnetopause, and Sergis et al. (2013) and Sulaiman et al. (2014) have shown evidence for magnetospheric plasma and the influence of polar flattening on IMF draping around the magnetosphere. A planetary magnetopause region is structured. Masters et al. (2011a,b) presented evidence for Saturn's boundary layer, a mixed plasma regime just inside the current sheet. An external plasma depletion layer (PDL) was identified by Masters et al. (2014b), which results from IMF influence on the flow of solar wind plasma. Jasinski et al. (2014) reported high-latitude Cassini observations near Saturn's magnetospheric cusp, a funnel-shaped region of the magnetosphere where solar wind plasma can directly enter the system.

Two processes that operate at Saturn's magnetopause are candidates for driving magnetospheric dynamics: (1) growth of the Kelvin–Helmholtz (K–H) instability and (2) magnetic reconnection. Growth of the K–H instability at Saturn's magnetopause (first discussed by Galopeau et al. 1995) underpins the so-called "viscous" interaction between the solar wind and a magnetosphere. Large flow shears across an interface favor K–H instability, resulting in the surface waves and rolled-up vortices. Cutler et al. (2011) and Masters et al. (2012a) identified widespread evidence for K–H-driven surface waves, and Masters et al. (2010), Delamere et al. (2011, 2013) and Wilson et al. (2012) have identified evidence for K–H vortices. Grodent et al. (2011) suggested that features of Saturn's ultraviolet aurora could be caused by K–H vortices on the magnetopause, as proposed by Masters et al. (2010).

The overall picture of K–H instability growth at Saturn's magnetopause is that K–H-driven waves propagate with comparable frequency down the dawn and dusk flanks. K–H vortices also form at both dawn and dusk, but with a dusk bias, which is contrary to expectations based on the higher flow shears across the interface at dawn. Delamere et al. (2013) suggested that this is due to growth of K–H perturbations at dawn followed by propagation to dusk caused by the momentum of duskward-rotating magnetospheric plasma. Global modeling has supported this. Walker et al. (2011) reproduced K–H waves on Saturn's magnetopause in an MHD simulation, and Fukazawa et al. (2012) studied these K–H waves and the vortices that form, including the resulting FAC. Results of analytical modeling carried out by Desroche et al. (2013) imply that much of Saturn's low-latitude magnetopause is K–H unstable, particularly at dawn.

Magnetic reconnection at Saturn's magnetopause (see Figure 6.14) is the other external candidate driver of magnetospheric dynamics. Although some direct evidence for reconnection at Saturn's magnetopause has been found (e.g. McAndrews et al. 2008), the evidence from Cassini data remains scarce, e.g. the survey of Lai et al. (2012) did not lead to any identification of flux transfer events (phenomena associated with reconnection). However, we now have considerable indirect evidence for magnetopause reconnection, most notably from ion dispersion signatures measured in the cusp region reported by Jasinski et al. (2014) and from dayside auroral features studied by Radioti et al. (2011, 2013), Badman et al. (2012, 2013) and Meredith et al. (2013, 2014).

Reconnection at Saturn's magnetopause does not appear to result in a host of established, Earth-like consequences for Saturn's magnetosphere. Lai et al. (2012) were unable to resolve any "erosion" of the magnetopause location due to reconnection, and Masters et al. (2011a, 2014b) were similarly unable to identify any response of boundary layers to the orientation of the IMF.

To try to explain these Earth-Saturn differences, Masters et al. (2012b) used Cassini data to show that the ratio of plasma to magnetic pressure is higher in Saturn's magnetosheath than in Earth's, which is expected to restrict reconnection to regions of closer to anti-parallel

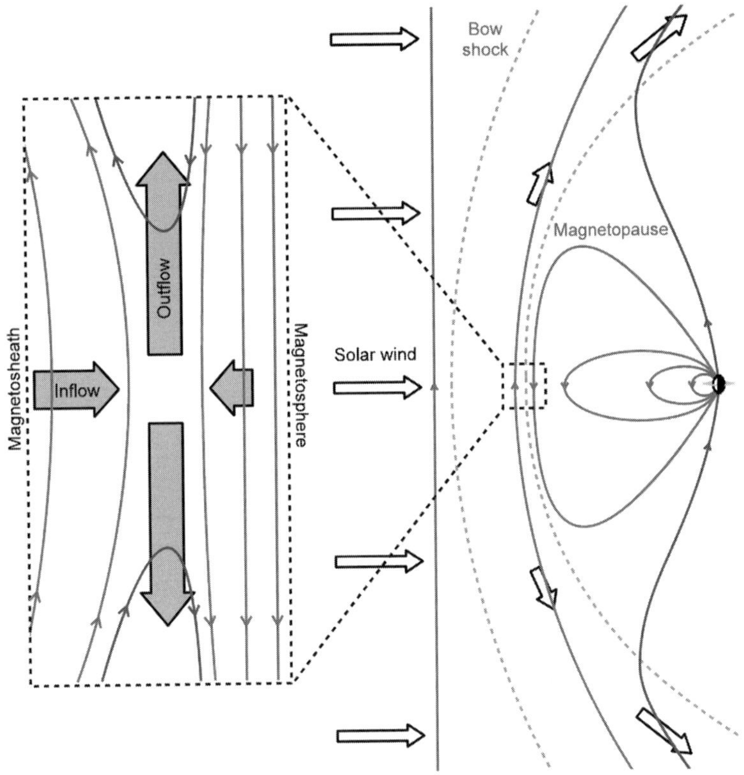

Figure 6.14 Diagram illustrating magnetic reconnection at Saturn's magnetopause for (rare) northward interplanetary magnetic field. Interplanetary (solar), planetary and reconnected (open) magnetic field lines are shown in light gray, dark gray and black, respectively (adapted from Masters et al. 2014a).

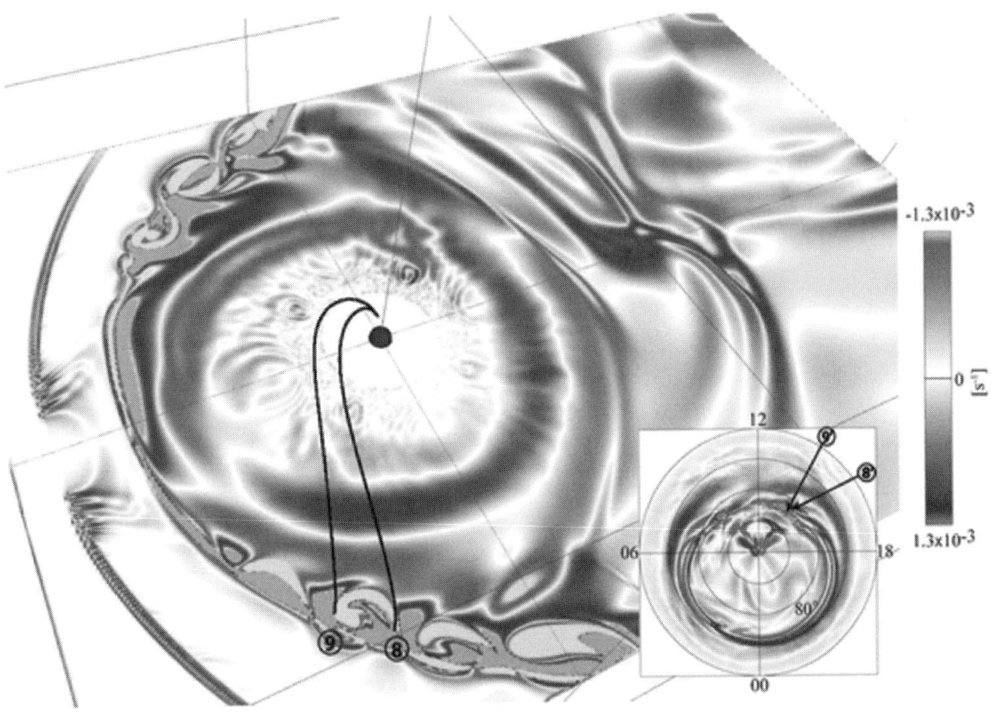

Figure 6.15 Snapshot of an MHD simulation of Saturn's magnetosphere showing K–H instability at the magnetopause. The color coding in the equatorial plane gives the difference in vorticity parallel to the magnetic field between 16:00 and 12:00 LT in the simulation (taken from Fukazawa et al. 2012). (A black-and-white version of this figure appears in some formats. For the color version, please refer to the plate section.)

magnetic fields. Desroche et al. (2013) globally applied reconnection onset conditions, also finding that conditions are less favorable for magnetopause reconnection at Saturn. Most recently, Masters et al. (2014a) placed an upper limit on the reconnection voltage applied to Saturn's magnetosphere, concluding that such voltages are likely to be lower than previously thought, and thus less capable of driving the magnetosphere.

Two major open questions concerning magnetopause boundary phenomena at Saturn are noteable:

How are growth of the Kelvin–Helmholtz instability and magnetic reconnection coupled?
What is the dominant mode of solar wind interaction with the magnetosphere: the "viscous" interaction or the "large-scale" reconnection interaction?

The combination of existing and future Cassini data sets will allow researchers to make further progress in addressing these questions.

6.6 Cassini "Grand Finale" Magnetospheric Science during the Proximal and F-Ring Orbits

The Cassini mission will end in 2016/2017 with a spectacular set of high-latitude orbits pushing the spacecraft inside the innermost D-ring and up to 67° in latitude above the poles. As illustrated in Figure 6.16, Cassini will first cross the equator along those polar orbits near the F-ring (F-ring orbits) followed by orbits where the equator is crossed inside the D-ring (proximal orbits).

These F-ring and proximal orbits are excellent for remote sensing (ENA imaging, UV, visible and IR imaging and radio science) as well as for in situ measurements of the local magnetic field and the neutral and charged particle population and dust in those regions.

Magnetospherically, this mission phase is unique because (1) Cassini will fly through a region inside the

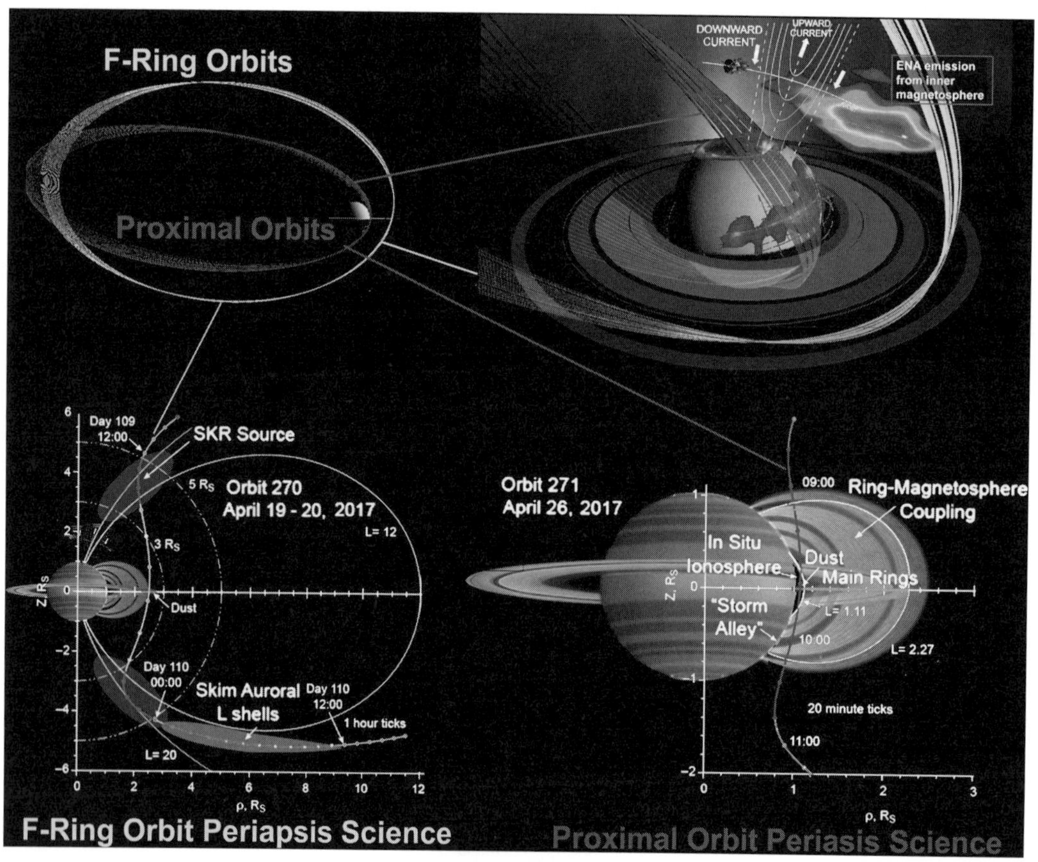

Figure 6.16 Illustration of the so-called F-ring and proximal orbits phase at the end of the Cassini mission 2016/2017, when the spacecraft will fly on high-latitude polar orbits very close to the planet. (A black-and-white version of this figure appears in some formats. For the color version, please refer to the plate section.)

D-ring where an additional trapped-particle distribution has been inferred remotely only (Krimigis et al. 2005) and (2) the spacecraft trajectory is such that it will fly along the same magnetic field lines for an extended period in time, allowing the study of field-aligned particle acceleration mechanisms and their relationship to auroral emissions, lightning, sprites and related electron acceleration. It will also probe source regions of SKR, whistler and chorus waves and identify other wave-particle interactions or ring-magnetosphere coupling processes.

Some of the most important specific questions about magnetospheric processes to be addressed in that mission phase are

1. What is the intensity, composition and energy spectrum of trapped radiation between the D-ring and Saturn's atmosphere?
2. Is CRAND the primary source of radiation belt particles in the gap?
3. What is the nature of the quasi-trapped ring current particles responsible for the ENA emissions inside the D-ring belt imaged by the neutral gas camera MIMI/INCA during Saturn orbit insertion (SOI)? What is their composition, and how durably are they trapped?
4. Are ionospheric particles (ions and/or electrons) accelerated to higher energies inside the D-ring?
5. Do sprite-like events accelerate energetic electrons along field lines that connect between the atmosphere and the rings?
6. What is the spatial distribution (local time, latitude) of ions accelerated above Saturn's auroral zones?
7. Are there wave-particle interactions associated with the inner radiation belt?
8. What are the values of the higher-order magnetic field components?
9. What is the total production of neutrals and plasma by the different satellites, and how do these sources vary with time?
10. How important is transport vs. chemistry in producing the distribution of plasma in the magnetosphere?
11. What are the plasma source contributions provided by the ionosphere and the solar wind?

6.7 Summary and Open Questions

Cassini has, for the first time, investigated the second-largest magnetosphere of our solar system for an extended period. Since 2004, it has used remote and in situ measurements along both equatorial and high-latitude orbits, and as a result, our understanding of the processes acting in the Saturnian system has been greatly extended, and in some cases our previous picture of the magnetosphere has been completely changed. In summary, Saturn's magnetosphere is a fantastic, highly dynamic plasma laboratory. The Saturnian system is sometimes more Earth-like and sometimes very similar to the Jovian system, which makes it even more complex in its nature. The local environment of the planet is dominated by the internal-active moon Enceladus, the major plasma source for the entire magnetosphere. The magnetosphere is powered by the fast rotational energy of the planet and yet the unique, nearly perfect alignment of the magnetic axis with the rotation axis makes it nearly impossible to determine the correct planetary rotation period. Plasma processes act on a whole variety of scales in time and space within the magnetosphere. Complicated transport mechanisms are responsible for the radial outward motion of the plasma originating from deep within the inner magnetosphere. In addition, processes transporting plasma inward in the magnetosphere have been identified. Measuring the interaction between both hot and cold neutrals and the charged particles in the system, as well as the interaction between magnetospheric plasma and the moons, rings and ring arcs, has significantly improved our understanding of these plasma processes. Cassini also found seasonal changes of plasma parameters in the Saturnian system and discovered a new electrical field, which is yet to be understood. Last, but not least, the unknown rotational period of Saturn has prevented us from being able to distinguish between different sources of plasma waves in different hemispheres. At the end of the mission in September 2017, Cassini will have performed 294 orbits around the planet. Even though Saturn's magnetosphere is better understood today in terms of global magnetospheric configuration, the importance of internal plasma sources, transport and dynamics in the magnetosphere, temporal and seasonal variability, and boundary phenomena of the system, the results have also opened up new questions for future investigations of the Saturnian system. Though we are not able to list a complete set of unsolved issues, several open questions are summarized below. Some of them will be addressed in Cassini's "Grand Finale," but answers to others have to wait for the next mission to Saturn:

1. How is plasma transported outward, and magnetic flux returned in the region between 12 and 20 or 25 R_S?
2. Where in the Saturnian system does Vasyliūnas-type reconnection take place and where does Dungey-type reconnection take place? Is there a region in Saturn's magnetosphere where one or the other dominates?
3. Where does the magnetosphere unload its cold, dense plasma?
4. How does the solar wind drive magnetospheric dynamics at Saturn? (Reconnection is thought to be weak at Saturn. Many of the boundary phenomena as reflected in the auroral zone are nevertheless interpreted as reconnection driven. Kelvin–Helmholtz instability is also invoked, but only at the magnetopause. Is solar wind pressure alone sufficient to explain most of the dynamics?)
5. What causes the inner magnetospheric electric field that drives the noon-midnight asymmetry in plasma temperature, density, energetic particle intensity, etc.?
6. Is the dipole component of the magnetic field truly aligned with the spin axis? Are there higher-order moments? At what period do such higher-order moments rotate?
7. What solar cycle effects can be found in Saturn's magnetosphere?
8. How important is Titan as a source of magnetospheric H_2? What is the fraction of N from Titan?
9. Is there an inner radiation belt inside the D-ring, and if so, how intense is it and what is its ion composition?
10. What is the dust population inside the D-ring?

References

Achilleos, N., N. André, X., Blanco-Cano et al. (2014), 1. Transport of Mass, Momentum and Energy in Planetary Magnetodisc Regions. *Space Sci. Rev.*, Oct.

Achilleos, N., P. Guio, C. S. Arridge et al. (2010), Influence of hot plasma pressure on the global structure of Saturn's magnetodisk. *Geophys. Res. Lett.*, **37**(Oct.), 20201.

Agren, K., J.-E. Wahlund, R. Modolo et al. (2007), On magnetospheric electron impact ionisation and dynamics in Titan's ram-side and polar ionosphere: A Cassini case study. *Ann. Geophysicae*, **25**(Nov.), 2359–2369.

Andrews, D. J., R. Cecconi, S. W. H. Cowley et al. (2011), Planetary period oscillations in Saturn's magnetosphere: Evidence in magnetic field phase data for rotational modulation of Saturn kilometric radiation emissions. *J. Geophys. Res.*, **116**(Sept.), 9206.

Andriopoulou, M., E. Roussos, N. Krupp et al. (2012), A noon-to-midnight electric field and nightside dynamics in Saturn's inner magnetosphere, using microsignature observations. *Icarus*, **220**(Aug.), 503–513.

Andriopoulou, M., E. Roussos, N. Krupp (2014), Spatial and temporal dependence of the convective electric field in Saturn's inner magnetosphere., **229**(Feb.), 57–70.

Armstrong, T. P., S. Taherion, J. Manweiler et al. (2009), Energetic ions trapped in Saturn's inner magnetosphere. *Planetary and Space Science*, **57**(Dec.), 1723–1731.

Arridge, C. S., N. André, N. Achilleos et al. (2008b), Thermal electron periodicities at $20R_S$ in Saturn's magnetosphere.*Geophys. Res. Lett.*, **35**(Aug.), 15107.

Arridge, C. S., N. André, K. K. Khurana et al. (2011b), Periodic motion of Saturn's nightside plasma sheet. *J. Geophys. Res.*, **116**(Nov.), 11205.

Arridge, C. S., N. André, H. J. McAndrews et al. (2011a), Mapping magnetospheric equatorial regions at Saturn from Cassini prime mission observations. *Space Sci. Rev.*, **164**(Dec.), 1–83.

Arridge, C. S., K. K. Khurana, C. T. Russell et al. (2008c), Warping of Saturn's magnetospheric and magnetotail current sheets. *J. Geophys. Res.*, **113**(A12), 8217.

Arridge, C. S., C. T. Russell, K. K. Khurana et al. (2008a), Saturn's magnetodisc current sheet. *J. Geophys. Res.*, **113**(Apr.), 4214.

Badman, S. V., N. Achilleos, C. S. Arridge et al. (2012), Cassini observations of ion and electron beams at Saturn and their relationship to infrared auroral arcs. *J. Geophys. Res.*, **117**(Jan.), 1211.

Badman, S. V., E. J. Bunce, J. T. Clarke et al. (2005), Open flux estimates in Saturn's magnetosphere during the January 2004 Cassini-HST campaign, and implications for reconnection rates. *J. Geophys. Res.*, **110** (Nov.), 11216.

Badman, S. V., C. M. Jackman, J. D. Nichols et al. (2014), Open flux in Saturn's magnetosphere. *Icarus*, **231** (Mar.), 137–145.

Badman, S. V., A. Masters, H. Hasegawa et al. (2013), Bursty magnetic reconnection at Saturn's magnetopause. *Geophys. Res. Lett.*, **40**(Mar.), 1027–1031.

Bagenal, F. (2013), Planetary Magnetospheres, in T. D. Oswalt, L. French and P. Kalas (eds.), *Planets, Stars and Stellar Systems. Volume 3: Solar and Stellar Planetary Systems*, Springer, Dordrecht, 251.

Bagenal, F. and P. A. Delamere (2011), Flow of mass and energy in the magnetospheres of Jupiter and Saturn. *J. Geophys. Res.*, **116**(May), 5209.

Beckmann, U. (2008), Dynamik von Staubteilchen in Saturns E-Ring. Ph.D. thesis, Ruprecht-Karls Universitt Heidelberg, Heidelberg, Germany.

Bertucci, C., D. C. Hamilton, W. S. Kurth et al. (2014), Titan interaction with the supersonic solar wind. *ArXiv e-prints*, Oct.

Beutier, T. and D. Boscher (1995), A three-dimensional analysis of the electron radiation belt by the Salammbô, *J. Geophys. Res.*, *100* (A8): 14853–14862, doi:10.1029/94JA03066.

Blanc, M., D. J. Andrews, A. J. Coates et al. (2015), Saturn plasma sources and associated transport processes. *Space Sci. Rev.*, Sept.

Brandt, P. C., C. P. Paranicas, J. F. Carbary et al. (2008), Understanding the global evolution of Saturn's ring current. *Geophys. Res. Lett.*, **35**(Sept.), 17101.

Branduardi-Raymont, G., P. G. Ford, K. C. Hansen et al. (2013), Search for Saturn's X-ray aurorae at the arrival of a solar wind shock. *J. Geophys. Res.*, **118**(May), 2145–2156.

Brice, N. M. and T. R. McDonough (1973), Jupiter's radiation belts. *Icarus*, **18**(Feb.), 206–219.

Bunce, E. J., S. W. H. Cowley, D. L. Talboys et al. (2010), Extraordinary field-aligned current signatures in Saturn's high-latitude magnetosphere: Analysis of Cassini data during revolution 89. *J. Geophys. Res.*, **115**(Oct.), 10238.

Bunce, E. J., S. W. H. Cowley, D. M. Wright et al. (2005), In situ observations of a solar wind compression-induced hot plasma injection in Saturn's tail. *Geophys. Res. Lett.*, **32**(July), 20.

Burch, J. L., A. D. DeJong, J. Goldstein et al. (2009), Periodicity in Saturn's magnetosphere: Plasma cam. *Geophys. Res. Lett.*, **36**(July), 14203.

Burch, J. L., L. Goldstein, T. W. Hill et al. (2005), Properties of local plasma injections in Saturn's magnetosphere. *Geophys. Res. Lett.*, **32**(June), L14S02.

Burton, M. E., M. K. Dougherty and C. T. Russell (2009), Model of Saturn's internal planetary magnetic field based on Cassini observations. *Planet. Space Sci.*, **57** (Dec.), 1706–1713.

Burton, M. E., M. K. Dougherty and C. T. Russell (2010), Saturn's internal planetary magnetic field. *Geophys. Res. Lett.*, **37**(Dec.), 24105.

Carbary, J. F., N. Achilleos, C. S. Arridge et al. (2010), Global configuration of Saturn's magnetic field derived from observations. *Geophys. Res. Lett.*, **37**(Nov.), 21806.

Carbary, J. F., D. G. Mitchell, P. Brandt et al. (2008), Track analysis of energetic neutral atom blobs at Saturn. *J. Geophys. Res.*, **113**(A12), 1209.

Carbary, J. F., D. G. Mitchell, C. Paranicas et al. (2011), Pitch angle distributions of energetic electrons at Saturn. *J. Geophys. Res.*, **116**(Jan.), 1216.

Carbary, J. F., C. Paranicas, D. G. Mitchell et al. (2011), Energetic electron spectra in Saturn's plasma sheet. *Journal of Geophysical Research: Space Physics*, **116** (A7), n/a–n/a.

Carbary, J. F. and A. M. Rymer (2014), Meridional maps of Saturn's thermal electrons. *J. Geophys. Res.*, **119**(Mar.), 1721–1733.

Cassidy, T. A. and R. E. Johnson (2010), Collisional spreading of Enceladus' neutral cloud. *Icarus*, **209**(Oct.), 696–703.

Cecconi, B., L. Lamy, P. Zarka et al. (2009), Goniopolarimetric study of the revolution 29 perikrone using the Cassini Radio and Plasma Wave Science instrument high-frequency radio receiver. *J. Geophys. Res.*, **114**(Mar.), 3215.

Chen, Y. and T. W. Hill (2008), Statistical analysis of injection/dispersion events in Saturn's inner magnetosphere. *J. Geophys. Res.*, **113**(A12), 7215.

Chen, Y., T. W. Hill, A. M. Rymer et al. (2010), Rate of radial transport of plasma in Saturn's inner magnetosphere. *J. Geophys. Res.*, **115**(Oct.), 10211.

Christon, S. P., D. C. Hamilton, R. D. Difabio et al. (2013), Saturn suprathermal O_2^+ and mass-28^+ molecular ions: Long-term seasonal and solar variation. *J. Geophys. Res.*, **118**(June), 3446–3463.

Christon, S. P., D. C. Hamilton, D. G. Mitchell et al. (2014), Suprathermal magnetospheric minor ions heavier than water at Saturn: Discovery of $^{28}M^+$ seasonal variations. *J. Geophys. Res.*, **119**(July), 5662–5673.

Clark, G., C. Paranicas, D. Santos-Costa et al. (2014), Evolution of electron pitch angle distributions across Saturn's middle magnetospheric region from MIMI/LEMMS. *Planet. Space Sci.*, **104**(Dec.), 18–28.

Coates, A. J., J.-E. Wahlund, K. Ågren et al. (2012), Recent Results from Titan's Ionosphere, *Space Science Rev.*, 162(1–4), 85–111.

Cooper, J. F. (1983), Nuclear cascades in Saturn's rings: Cosmic ray albedo neutron decay and origins of trapped protons in the inner magnetosphere. *Journal Geophys. Res.*, **88**(May), 3945–3954.

Coroniti, F. Ṽ. (1974), Energetic electrons in Jupiter's magnetosphere. *Astrophys. J. Suppl. Ser.*, **27**(Mar.), 261–281.

Cutler, J. C., M. K. Dougherty, E. Lucek et al. (2011), Evidence of surface wave on the dusk flank of Saturn's magnetopause possibly caused by the Kelvin–Helmholtz instability. *Journal of Geophysical Research (Space Physics)*, **116**(Oct.), 10220.

DeJong, A. D., J. L. Burch, J. Goldstein et al. (2010), Low-energy electrons in Saturn's inner magnetosphere and their role in interchange injections. *J. Geophys. Res.*, **115**(Oct.), 10229.

Delamere, P. A. and F. Bagenal (2013), Magnetotail structure of the giant magnetospheres: Implications of the viscous interaction with the solar wind. *J. Geophys. Res.*, **118**(Nov.), 7045–7053.

Delamere, P. A., F. Bagenal, V. Dols et al. (2007), Saturn's neutral torus versus Jupiter's plasma torus. *Geophys. Res. Lett.*, **34**(May), L09105.

Delamere, P. A., R. J. Wilson, S. Eriksson et al. (2013), Magnetic signatures of Kelvin–Helmholtz vortices on Saturn's magnetopause: Global survey. *J. Geophys. Res.*, **118**(Jan.), 393–404.

Delamere, P. A., R. J. Wilson and A. Masters (2011), Kelvin–Helmholtz instability at Saturn's magnetopause: Hybrid simulations. *J. Geophys. Res.*, **116** (Oct.), 10222.

Desch, M. D. (1982), Evidence for solar wind control of Saturn radio emission. *J. Geophys. Res.*, **87**(June), 4549–4554.

Desch, M. D. and M. L. Kaiser (1981), Voyager measurement of the rotation period of Saturn's magnetic field. *Geophys. Res. Lett.*, **8**(Mar.), 253–256.

Desroche, M., F. Bagenal, P. A. Delamere et al. (2013), Conditions at the magnetopause of Saturn and implications for the solar wind interaction. *J. Geophys. Res.*, **118**(June), 3087–3095.

Dialynas, K., P. C. Brandt, S. M. Krimigis et al. (2013), The extended Saturnian neutral cloud as revealed by global ENA simulations using Cassini/MIMI measurements. *J. Geophys. Res.*, **118**(June), 3027–3041.

Dialynas, K., S. M. Krimigis, D. G. Mitchell et al. (2009), Energetic ion spectral characteristics in the Saturnian magnetosphere using Cassini/MIMI measurements. *Journal of Geophysical Research (Space Physics)*, **114** (Jan.), 1212.

DiFabio, R. D., D. C. Hamilton, S. M. Krimigis et al. (2011), Long term time variations of the suprathermal ions in Saturn's magnetosphere. *Geophys. Res. Lett.*, **38**(Sept.), 18103.

Dong, Y., T. W. Hill, B. D. Teolis et al. (2011), The water vapor plumes of Enceladus. *jgr*, **116**(Oct.), 10204.

Dougherty, M. K., K. K. Khurana, F. M. Neubauer et al. (2006), Identification of a dynamic atmosphere at Enceladus with the Cassini magnetometer. *Science*, **311**(Mar.), 1406–1409.

Dungey, J. W. (1961), Interplanetary magnetic field and the auroral zones. *Physical Review Letters*, **6**(Jan.), 47–48.

Edberg, N. J. T., J.-E. Wahlund, K. Ågren et al. (2010), Electron density and temperature measurements in the cold plasma environment of Titan: Implications for atmospheric escape. *Geophys. Res. Lett.*, **37**(Oct.), 20105.

Elrod, M. K., W.-L. Tseng, R. J. Wilson et al. (2012), Seasonal variations in Saturn's plasma between the main rings and Enceladus. *J. Geophys. Res.*, **117** (Mar.), 3207.

Elrod, M. K., W.-L. Tseng, A. K. Woodson et al. (2014), Seasonal and radial trends in Saturn's thermal plasma between the main rings and Enceladus. *Icarus*, **242** (Nov.), 130–137.

Farmer, A. J. (2009), Saturn in hot water: Viscous evolution of the Enceladus torus. *Icarus*, **202**(July), 280–286.

Farrell, W. M., W. S. Kurth, R. L. Tokar et al. (2010), Modification of the plasma in the near-vicinity of Enceladus by the enveloping dust. *Geophys. Res. Lett.*, **37**(Oct.), L20202.

Flandes, A., L. Spilker, R. Morishima et al. (2010), Brightness of Saturn's rings with decreasing solar elevation. *Planet. Space Sci.*, **58**(Nov.), 1758–1765.

Fleshman, B. L., P. A. Delamere, F. Bagenal et al. (2012), The roles of charge exchange and dissociation in spreading Saturn's neutral clouds. *J. Geophys. Res.*, **117**(May), 5007.

Fleshman, B. L., P. A. Delamere, F. Bagenal (2013), A 1-D model of physical chemistry in Saturn's inner magnetosphere. *J. Geophys. Res.*, **118**(Aug.), 1567–1581.

Fukazawa, K., T. Ogino and R. J. Walker (2012), A magnetohydrodynamic simulation study of Kronian field-aligned currents and auroras. *J. Geophys. Res.*, **117** (Feb.), 2214.

Funsten, H. O., D. J. McComas and B. L. Barraclough (1993), Ultrathin foils used for low-energy neutral atom imaging of the terrestrial magnetosphere. *Optical Engineering*, **32**(Dec.), 3090–3095.

Galopeau, P. H. M., P. Zarka and D. L. Quéau (1995), Source location of Saturn's kilometric radiation: The Kelvin–Helmholtz instability hypothesis. *J. Geophys. Res.*, **100**, 26397–26410.

Glass, G., M. Jain, M. L. Evans et al. (1977), Neutron spectra at 0 from proton-proton collisions between 647 and 805 MeV. *Phys. Rev. D*, **15**(1), 36–46.

Glocer, A., T. I. Gombosi, G. Toth et al. (2007), Polar wind outflow model: Saturn results. *jgr*, **112**(Jan.), 1304.

Goldstein, J., T. W. Hill, J. H. Waite et al. (2014), Analytical model of rotating two-cell convection at Saturn. *jgr*, **119** (Mar.), 1980–1993.

Gombosi, T. I., T. P. Armstrong, C. S. Arridge et al. (2009), *Saturn's Magnetospheric Configuration*. Page 203.

Gombosi, T. I. and A. P. Ingersoll (2010), Saturn: Atmosphere, ionosphere, and magnetosphere. *Science*, **327**(Mar.), 1476.

Grodent, D., J. Gustin, J,-C. Gérard et al. (2011), Small-scale structures in Saturn's ultraviolet aurora. *J. Geophys. Res.*, **116**(Sept.), 9225.

Gu, X., R. M. Thorne, B. Ni and S.-Y. Ye (2013), Resonant diffusion of energetic electrons by narrowband Z mode waves in Saturn's inner magnetosphere, *Geophysical Res. Lett.*, **40**(Jan.), 255–261.

Gubar, Y. I. (2004), A diffusion model of radial distributions of energetic protons in Saturn's magnetosphere. *Cosmic Research*, **42**(4).

Gurnett, D. A., T. F. Averkamp, P. Schippers et al. (2011), Auroral hiss, electron beams and standing Alfvén wave currents near Saturn's moon Enceladus. *Geophys. Res. Lett.*, **38**(Mar.), 6102.

Gurnett, D. A., A. M. Persoon, A. J. Kopf et al. (2010), A plasmapause-like density boundary at high latitudes in Saturn's magnetosphere. *Geophys. Res. Lett.*, **37** (Aug.), 16806.

Gustafsson, G. and J.-E. Wahlund (2010), Electron temperatures in Saturn's plasma disc. *Planet. Space Sci.*, **58** (June), 1018–1025.

Hahn, Y. (1997), Electron-ion recombination processes: An overview. *Reports on Progress in Physics*, **60** (July), 691–759.

Hansen, C. J., L. Esposito, A. I. F. Stewart et al. (2006), Enceladus' water vapor plume. *Science*, **311**(Mar.), 1422–1425.

Hartogh, P., E. Lellouch, R. Moreno et al. (2011), Direct detection of the Enceladus water torus with Herschel. *Astron. Astrophys.*, **532**(Aug.), L2+.

Hedman, M. M., C. M. Gosmeyer, P. D. Nicholson et al. (2013), An observed correlation between plume activity and tidal stresses on Enceladus. *Nature*, **500**(Aug.), 182–184.

Hill, T. W. (2009), Speed limit for centrifugally driven convection. *AGU Fall Meeting Abstracts*, Dec., D8.

Hill, T. W., A. M. Rymer, J. L. Burch et al. (2005), Evidence for rotationally driven plasma transport in Saturn's magnetosphere. *Geophys. Res. Lett.*, **32**(June), L14S10.

Hill, T. W., M. F. Thomsen, M. G. Henderson et al. (2008), Plasmoids in Saturn's magnetotail. *J. Geophys. Res.*, **113**(A12), 1214.

Hill, T. W., M. F. Thomsen, R. L. Tokar et al. (2012), Charged nanograins in the Enceladus plume. *J. Geophys. Res.*, **117**(May), 5209.

Hillier, J. K., S. F. Green, N. McBride et al. (2007), The composition of Saturn's E ring, *Monthly Notices of the Royal Astronomical Soc.*, **377**(June), 1588–1596.

Holmberg, M. K. G., J.-E. Wahlund and M. W. Morooka (2014), Dayside/nightside asymmetry of ion densities and velocities in Saturn's inner magnetosphere. *Geophys. Res. Lett.*, **41**(June), 3717–3723.

Holmberg, M. K. G., J.-E. Wahlund, M. W. Morooka et al. (2012), Ion densities and velocities in the inner plasma torus of Saturn. *Planet. Space Sci.*, **73**(Dec.), 151–160.

Hood, L. L. (1983), Radial diffusion in Saturn's radiation belts: A modeling analysis assuming satellite and ring E absorption. *J. Geophys. Res.*, **88**(Feb.), 808–818.

Horányi, M., A. Juhász and G. E. Morfill (2008), Large-scale structure of Saturn's E-ring. *Geophys. Res. Lett.*, **35** (Feb.), 4203.

Hunt, G. J., S. W. H. Cowley, G. Provan et al. (2014), Field-aligned currents in Saturn's southern nightside magnetosphere: Subcorotation and planetary period oscillation components. *J. Geophys. Res.*, **119**(Dec.), 9847–9899.

Ip, W.-H. (2005), An update on the ring exosphere and plasma disc of Saturn. *Geophys. Res. Lett.*, **32**(July), L13204.

Jackman, C. M., N. Achilleos, S. W. H. Cowley et al. (2013), Auroral counterpart of magnetic field dipolarizations in Saturn's tail. *Planet. Space Sci.*, **82**(July), 34–42.

Jackman, C. M., C. S. Arridge, N. André et al. (2014a), Large-scale structure and dynamics of the magnetotails of Mercury, Earth, Jupiter and Saturn. *Space Sci. Rev.*, **182**(Aug.), 85–154.

Jackman, C. M., C. S. Arridge, H. J. McAndrews et al. (2009a), Northward field excursions in Saturn's magnetotail and their relationship to magnetospheric periodicities. *Geophys. Res. Lett.*, **36**(Aug.), 16101.

Jackman, C. M., C. S. Arridge, J. A. Slavin et al. (2010), In situ observations of the effect of a solar wind compression on Saturn's magnetotail. *J. Geophys. Res.*, **115** (Oct.), 10240.

Jackman, C. M., L. Lamy, M. P. Freeman et al. (2009b), On the character and distribution of lower-frequency radio emissions at Saturn and their relationship to substorm-like events. *J. Geophys. Res.*, **114**(Aug.), 8211.

Jackman, C. M., C. T. Russell, D. J. Southwood et al. (2007), Strong rapid dipolarizations in Saturn's magnetotail: In situ evidence of reconnection. *Geophys. Res. Lett.*, **34**(June), 11203.

Jackman, C. M., J. A. Slavin and J. A. Cowley (2011), Cassini observations of plasmoid structure and dynamics: Implications for the role of magnetic reconnection in magnetospheric circulation at Saturn. *J. Geophys. Res.*, **116**(A15), 10212.

Jackman, C. M., J. A. Slavin, M. G. Kivelson et al. (2014b), Saturn's dynamic magnetotail: A comprehensive magnetic field and plasma survey of plasmoids and traveling compression regions and their role in global magnetospheric dynamics. *J. Geophys. Res.*, **119**(July), 5465–5494.

Jacobsen, K. S., J.-E. Wahlund and A. Pedersen (2009), Cassini Langmuir probe measurements in the inner magnetosphere of Saturn. *Planet. Space Sci.*, **57**(Jan.), 48–52.

Jasinski, J. M., C. S. Arridge, L. Lamy et al. (2014), Cusp observation at Saturn's high-latitude magnetosphere by the Cassini spacecraft. *Geophys. Res. Lett.*, **41**(Mar.), 1382–1388.

Jia, X., K. C. Hansen, T. I. Gombosi et al. (2012a), Magnetospheric configuration and dynamics of Saturn's magnetosphere: A global MHD simulation. *J. Geophys. Res.*, **117**(May), 5225.

Jia, X. and M. G. Kivelson (2012), Driving Saturn's magnetospheric periodicities from the upper atmosphere/ionosphere: Magnetotail response to dual sources. *J. Geophys. Res.*, **117**(A16), 11219.

Jia, Y.-D., Y. J. Ma, C. T. Russell et al. (2012b), Perpendicular flow deviation in a magnetized counter-streaming plasma. *Icarus*, **218**(Apr.), 895–905.

Jia, Y.-D., C. T. Russell, K. K. Khurana et al. (2011), Cassini magnetometer observations over the Enceladus poles. *Geophys. Res. Lett.*, **38**(Oct.), 19109.

Johnson, R. E. (1990), Energetic charged-particle interactions with atmospheres and surfaces. *Physics and Chemistry in Space*, **19**.

Johnson, R. E., J. G. Luhmann, R. L. Tokar et al. (2006), Production, ionization and redistribution of O_2 in Saturn's ring atmosphere. *Icarus*, **180**(Feb.), 393–402.

Johnson, R. E., M. K. Pospieszalska, E. C. Sittler et al. (1989), The neutral cloud and heavy ion inner torus at Saturn. *Icarus*, **77**(Feb.), 311–329.

Jones, G. H., E. Roussos, N. Krupp et al. (2008), The dust halo of Saturn's largest icy moon, Rhea. *Science*.

Jurac, S., R. A. Baragiola, R. E. Johnson et al. (1995), Charging of ice grains by low-energy plasmas: Application to Saturn's E ring. *J. Geophys. Res.*, **100** (Aug.), 14821–14832.

Jurac, S., R. E. Johnson and J. D. Richardson (2001b), Saturn's E ring and production of the neutral torus. *Icarus*, **149**(Feb.), 384–396.

Jurac, S., R. E. Johnson, J. D. Richardson et al. (2001a), Satellite sputtering in Saturn's magnetosphere. *Planet. Space Sci.*, **49**(Mar.), 319–326.

Jurac, S. and J. D. Richardson (2005), A self-consistent model of plasma and neutrals at Saturn: Neutral cloud morphology. *J. Geophys. Res.*, **110**(Sept.), 9220.

Jurac, S. and J. D. Richardson (2007), Neutral cloud interaction with Saturn's main rings. *Geophys. Res. Lett.*, **34** (Apr.), 8102.

Kanani, S. J., C. S. Arridge, G. H. Jones et al. (2010), A new form of Saturn's magnetopause using a dynamic pressure balance model, based on in situ, multi-instrument Cassini measurements. *J. Geophys. Res.*, **115**(A14), 6207.

Kane, M., D. G. Mitchell, J. F. Carbary et al. (2014), Plasma convection in the nightside magnetosphere of Saturn determined from energetic ion anisotropies. *Planet. Space Sci.*, **91**(Feb.), 1–13.

Kellett, S., C. S. Arridge, E. J. Bunce et al. (2010), Nature of the ring current in Saturn's dayside magnetosphere. *J. Geophys. Res.*, **115**(Aug.), 8201.

Kellett, S., C. S. Arridge, E. J. Bunce (2011), Saturn's ring current: Local time dependence and temporal variability. *J. Geophys. Res.*, **116**(May), 5220.

Kempf, S., U. Beckmann, G. Moragas-Klostermeyer et al. (2008), The E ring in the vicinity of Enceladus. I. Spatial distribution and properties of the ring particles. *Icarus*, **193**(Feb.), 420–437.

Kennel, C. F. and H. E. Petschek (1966), Limit on stably trapped proton fluxes. *J. Geophys. Res.*, **71**(1), 1–28.

Kennelly, T. J., J. S. Leisner, G. B. Hospodarsky et al. (2013), Ordering of injection events within Saturnian SLS longitude and local time. *J. Geophys. Res.*, **118**(2), 832–838.

Kidder, A., R. M. Winglee and E. M. Harnett (2009), Regulation of the centrifugal interchange cycle in Saturn's inner magnetosphere. *J. Geophys. Res.*, **114** (Feb.), 2205.

Kollmann, P., E. Roussos, C. Paranicas et al. (2011), Energetic particle phase space densities at Saturn: Cassini observations and interpretations. *J. Geophys. Res.*, **116**(A15), A05222.

Kollmann, P., E. Roussos, C. Paranicas (2013), Processes forming and sustaining Saturn's proton radiation belts. *Icarus*, **222**(Jan.), 323–341.

Kopf, A. J., D. A. Gurnett, J. D. Menietti et al. (2010), Electron beams as the source of whistler-mode auroral hiss at Saturn. *Geophys. Res. Lett.*, **37**(May), 9102.

Kriegel, H., S. Simon, P. Meier et al. (2014), Ion densities and magnetic signatures of dust pickup at Enceladus. *J. Geophys. Res.*, **119**(Apr.), 2740–2774.

Kriegel, H., S. Simon, U. Motschmann et al. (2011), Influence of negatively charged plume grains on the structure of Enceladus' Alfvén wings: Hybrid simulations versus Cassini magnetometer data. *J. Geophys. Res.*, **116**(A15), 10223.

Kriegel, H., S. Simon, J. Müller et al. (2009), The plasma interaction of Enceladus: 3D hybrid simulations and comparison with Cassini MAG data. *Planet. Space Sci.*, **57**(Dec.), 2113–2122.

Krimigis, S. M. and T. P. Armstrong (1982), Two-component proton spectra in the inner Saturnian magnetosphere. *Geophys. Res. Lett.*, **9**(Oct.), 1143–1146.

Krimigis, S. M., D. G. Mitchell, D. C. Hamilton et al. (2005), Dynamics of Saturn's magnetosphere from MIMI during Cassini's orbital insertion. *Science*, **307**(Feb.), 1270–1273.

Krimigis, S. M. and E. C. Roelof (1983), Low-energy particle population. Chap. 4, pages 106–155 of: Dessler, A. J. (ed), *Physics of the Jovian Magnetosphere*. Cambridge Univ. Press, New York.

Krupp, N. (2014), Giant magnetospheres in our solar system: Jupiter and Saturn compared. Astron. *Astrophys.*, **22** (Sept.), 1–18.

Krupp, N., E. Roussos, H. Kriegel et al. (2013), Energetic particle measurements in the vicinity of Dione during the three Cassini encounters 2005–2011. *Icarus*, **226** (Sept.), 617–628.

Krupp, N., E. Roussos, A. Lagg et al. (2009), Energetic particles in Saturn's magnetosphere during the Cassini nominal mission (July 2004–July 2008). *Planet. Space Sci.*, **57**(Dec.), 1754–1768.

Kurth, W. S., D. A. Gurnett, J. T. Clarke et al. (2005), An Earth-like correspondence between Saturn's auroral features and radio emission. *Nature*, **433**(Feb.), 722–725.

Kurth, W. S., G. B. Hospodarsky, D. A. Gurnett et al. (2016), Saturn kilometric radiation intensities during the Saturn auroral campaign of 2013. *Icarus*, **263**(Jan.), 2–9.

Lai, H. R., H. Y. Wei, C. T. Russell et al. (2012), Reconnection at the magnetopause of Saturn: Perspective from FTE occurrence and magnetosphere size. *J. Geophys. Res.*, **117**(May), 5222.

Lamy, L., B. Cecconi, R. Prangé et al. (2009), An auroral oval at the footprint of Saturn's kilometric radio sources, colocated with the UV aurorae. *J. Geophys. Res.*, **114**(Oct.), 10212.

Lamy, L., R. Prangé, W. Pryor et al. (2013), Multispectral simultaneous diagnosis of Saturn's aurorae throughout a planetary rotation. *J. Geophys. Res.*, **118**(Aug.), 4817–4843.

Lamy, L., P. Schippers, P. Zarka et al. (2010), Properties of Saturn kilometric radiation measured within its source region. *Geophys. Res. Lett.*, **37**(June), 12104.

Leisner, J. S., G. B. Hospodarsky and G. B. Gurnett (2013), Enceladus auroral hiss observations: Implications for electron beam locations. *J. Geophys. Res.*, **118**(Jan.), 160–166.

Lejosne, S., D. Boscher, V. Maget et al. (2013), Deriving electromagnetic radial diffusion coefficients of radiation belt equatorial particles for different levels of magnetic activity based on magnetic field measurements at geostationary orbit. *J. Geophys. Res.*, **118**(June), 3147–3156.

Lindsay, B. G. and R. F. Stebbings (2005), Charge transfer cross-sections for energetic neutral atom data analysis. *J. Geophys. Res.*, **110**(Dec.), 12213.

Liu, X. and T. W. Hill (2012), Effects of finite plasma pressure on centrifugally driven convection in Saturn's inner magnetosphere. *J. Geophys. Res.*, **117**(July), 7216.

Liu, X., T. W. Hill, R. A. Wolf et al. (2010), Numerical simulation of plasma transport in Saturn's inner magnetosphere using the Rice Convection Model. *J. Geophys. Res.*, **115**(A14), 12254.

Livi, R., J. Goldstein, J. L. Burch et al. (2014), Multi-instrument analysis of plasma parameters in Saturn's equatorial, inner magnetosphere using corrections for corrections for spacecraft potential and penetrating background radiation. *J. Geophys. Res.*, **119**(May), 3683–3707.

Lorenzato, L., A. Sicard and S. Bourdarie (2012), A physical model for electron radiation belts of Saturn. *J. Geophys. Res.*, **117**(Aug.), 8214.

Louarn, P., N. Andre, C. M. Jackman et al. (2014), Magnetic reconnection and associated transient phenomena within the magnetospheres of Jupiter and Saturn. *Space Sci. Rev.*, May.

Luhmann, J. G., D. Ulusen, S. A. Ledvina et al. (2012), Investigating magnetospheric interaction effects on Titan's ionosphere with the Cassini orbiter Ion Neutral Mass Spectrometer, Langmuir Probe and magnetometer observations during targeted flybys. *Icarus*, **219**(June), 534–555.

Masters, A., N. Achilleos, J C. Cutler et al. (2012a), Surface waves on Saturn's magnetopause. *Planet. Space Sci.*, **65**(May), 109–121.

Masters, A., N. Achilleos, M. G. Kivelson et al. (2010), Cassini observations of a Kelvin–Helmholtz vortex in Saturn's outer magnetosphere. *J. Geophys. Res.*, **115** (A14), 7225.

Masters, A., J. P. Eastwood, M. Swisdak et al. (2012b), The importance of plasma β conditions for magnetic reconnection at Saturn's magnetopause. *Geophys. Res. Lett.*, **39**(Apr.), 8103.

Masters, A., M. Fujimoto, H. Hasegawa et al. (2014a), Can magnetopause reconnection drive Saturn's magnetosphere? *Geophys. Res. Lett.*, **41**(Mar.), 1862–1868.

Masters, A., D. G. Mitchell, A. J, Coates et al. (2011a), Saturn's low-latitude boundary layer: 1. Properties and variability. *J. Geophys. Res.*, **116**(June), 6210.

Masters, A., T. D. Phan, S. V. Badman et al. (2014b), The plasma depletion layer in Saturn's magnetosheath. *J. Geophys. Res.*, **119**(Jan.), 121–130.

Masters, A., M. F. Thomsen, S. V. Badman et al. (2011b), Supercorotating return flow from reconnection in Saturn's magnetotail. *Geophys. Res. Lett.*, **38**(Feb.), 3103.

Mauk, B. H. and N J. Foxz (2010), Electron radiation belts of the solar system. *J. Geophys. Res.*, **115**(A14), 12220.

Mauk, B. H., D. C. Hamilton, T. W.. Hill et al. (2009), Fundamental Plasma Processes in Saturn's Magnetosphere, in M. K. Dougherty et al. (eds.), *Saturn from Cassini–Huygens*, Springer, Page 281.

McAndrews, H. J., C. J. Owen, M. F. Thomsen et al. (2008), Evidence for reconnection at Saturn's magnetopause. *J. Geophys. Res.*, **113**(Apr.), 4210.

McAndrews, H. J., M. F. Thomsen, C. S., Arridge et al. (2009), Plasma in Saturn's nightside magnetosphere and the implications for global circulation. *Planet. Space Sci.*, **57**(Dec.), 1714–1722.

McAndrews, H. J., M. F. Thomsen, C. S., Arridge (2014), Corrigendum to "Plasma Saturn's nightside magnetosphere and the implications for global circulation" [Planet. Space Sci. 57(14-15) (2009) 1714-1722]. *Planet. Space Sci.*, **97**(July), 86–87.

McIlwain, C. E. (1966), Magnetic coordinates. *Space Sci. Rev.*, **5**(Aug.), 585–598.

Melin, H., D. E. Shemansky and X. Liu (2009), The distribution of atomic hydrogen and oxygen in the magnetosphere of Saturn. *Planet. Space Sci.*, **57** (Dec.), 1743–1753.

Menietti, J. D., G. B. Hospodarsky, Y. Y. Shprits et al. (2014), Saturn chorus latitudinal variations. *J. Geophys. Res.*, **119**(June), 4656–4667.

Menietti, J. D., Y. Katoh, G. B. Hospodarsky et al. (2013a), Frequency drift of Saturn chorus emission compared to nonlinear theory. *J. Geophys. Res.*, **118**(Mar.), 982–990.

Menietti, J. D., R. L. Mutel, P. Schippers et al. (2011a), Analysis of Saturn kilometric radiation near a source center. *J. Geophys. Res.*, **116**(Dec.), 12222.

Menietti, J. D., O. Santolik, A. M. Rymer et al. (2008), Analysis of plasma waves observed within local plasma injections seen in Saturn's magnetosphere. *J. Geophys. Res.*, **113**(May), 5213.

Menietti, J. D., P. Schippers, Y. Katoh et al. (2013b), Saturn chorus intensity variations. *J. Geophys. Res.*, **118** (Sept.), 5592–5602.

Menietti, J. D., P. Schippers, O. Santolík et al. (2011b), Ion cyclotron harmonics in the Saturn downward current auroral region. *J. Geophys. Res.*, **116**(Dec.), 12234.

Menietti, J. D., Y. Y. Shprits, R. B. Horne et al. (2012), Chorus, ECH, and Z mode emissions observed at Jupiter and Saturn and possible electron acceleration. *J. Geophys. Res.*, **117**(A16), 12214.

Menietti, J. D., P. H. Yoon, S.-Y. Ye et al. (2010), Source mechanism of Saturn narrowband emission. *Annales Geophysicae*, **28**(Apr.), 1013–1021.

Meredith, C. J., I. I. Alexeev, S. V. Badman et al. (2014), Saturn's dayside ultraviolet auroras: Evidence for morphological dependence on the direction of the upstream interplanetary magnetic field. *J. Geophys. Res.*, **119** (Mar.), 1994–2008.

Meredith, C. J., S. W. H. Cowley, K. C. Hansen et al. (2013), Simultaneous conjugate observations of small-scale structures in Saturn's dayside ultraviolet auroras: Implications for physical origins. *J. Geophys. Res.*, **118**(May), 2244–2266.

Mitchell, D. G., P. C. Brandt, J. F. Carbary et al. (2015), Injection, Interchange, and Reconnection Energetic Particle Observations in Saturn's Magnetosphere, in A. Keiling, C. M. Jackman and P. A. Delamere (eds.), *Magnetotails in the Solar System*, John Wiley, Hoboken, NJ, Page 424.

Mitchell, D. G., P. C. Brandt, E. C. Roelof et al. (2005), Energetic ion acceleration in Saturn's magnetotail: Substorms on Saturn? *Geophys. Res. Lett.*, **32**, L20S01.

Mitchell, D. G., J. F. Carbary, S. W. H. Cowley et al. (2009c), The Dynamics of Saturn's Magnetosphere, in M. K. Dougherty et al. (eds.), *Saturn from Cassini–Huygens*, Springer, Page 257

Mitchell, D. G., S. M. Krimigis, C. Paranicas et al. (2009b), Recurrent energization of plasma in the midnight-to-dawn quadrant of Saturn's magnetosphere, and its relationship to auroral UV and radio emissions. *Planet. Space Sci.*, **57**(Dec.), 1732–1742.

Mitchell, D. G., W. S. Kurth, G. B. Hospodarsky et al. (2009a), Ion conics and electron beams associated with auroral processes on Saturn. *J. Geophys. Res.*, **114**(Feb.), 2212.

Modolo, R. and G. M. Chanteur (2008), A global hybrid model for Titan's interaction with the Kronian plasma: Application to the Cassini TA flyby. *J. Geophys. Res.*, **113**(Jan.), 1317.

Morfill, G. E., H. Fechtig, E. Gruen et al. (1983), Some consequences of meteoroid impacts on Saturn's rings. *Icarus*, **55**(Sept.), 439–447.

Morooka, M. W., R. Modolo, J.-E. Wahlund et al. (2009), The electron density of Saturn's magnetosphere. *Annales Geophysicae*, **27**(7), 2971–2991.

Morooka, M. W., J.-E. Wahlund, A. I. Eriksson et al. (2011), Dusty plasma in the vicinity of Enceladus. *J. Geophys. Res.*, **116**(Dec.), 12221.

Müller, A. L., J. Saur, N. Krupp et al. (2010), Azimuthal plasma flow in the Kronian magnetosphere. *J. Geophys. Res.*, **115**(A14), 8203.

Mutel, R. L., J. D. Menietti, D. A. Gurnett et al. (2010), CMI growth rates for Saturnian kilometric radiation. *Geophys. Res. Lett.*, **37**(Oct.), 19105.

Nicholson, P. D., M. R. Showalter, L. Dones et al. (1996), Observations of Saturn's ring-plane crossings in August and November 1995. *Science*, **272**(Apr.), 509–515.

Omidi, N., R. L. Tokar, T. Averkamp et al. (2012), Flow stagnation at Enceladus: The effects of neutral gas and charged dust. *J. Geophys. Res.*, **117**(June), 6230.

Omura, Y., N. Furuya and D. Summers (2007), Relativistic turning acceleration of resonant electrons by coherent whistler mode waves in a dipole magnetic field. *J. Geophys. Res.*, **112**(June), 6236.

Paranicas, C., D. G. Mitchell, S. M. Krimigis et al. (2010a), Asymmetries in Saturn's radiation belts. *J. Geophys. Res.*, **115**(A14), 7216.

Paranicas, C., D. G. Mitchell, E. C. Roelof et al. (2007), Energetic electrons injected into Saturn's neutral gas cloud. *Geophys. Res. Lett.*, **34**(Jan.), 2109.

Paranicas, C., D. G. Mitchell, E. Roussos et al. (2010b), Transport of energetic electrons into Saturn's inner magnetosphere. *J. Geophys. Res.*, **115**(A14), 9214.

Paranicas, C., E. Roussos, R. B. Decker et al. (2014), The lens feature on the inner saturnian satellites. *Icarus*, **234**(May), 155–161.

Paranicas, C., E. Roussos, N. Krupp et al. (2012), Energetic charged particle weathering of Saturn's inner satellites. *Planet. Space Sci.*, **61**(Feb.), 60–65.

Perry, M. E., B. Teolis, H. T. Smith et al. (2010), Cassini INMS observations of neutral molecules in Saturn's E-ring. *J. Geophys. Res.*, **115**(Oct.), 10206.

Persoon, A. M., D. A. Gurnett, W. S. Kurth et al. (2006), A simple scale height model of the electron density in Saturn's plasma disk. *Geophys. Res. Lett.*, **33**(Sept.), 18106.

Persoon, A. M., D. A. Gurnett, W. S. Kurth (2015), Evidence for a seasonally dependent ring plasma in the region between Saturn's A ring and Enceladus' orbit. *J. Geophys. Res.*, **120**(Aug.), 6276–6285.

Persoon, A. M., D. A. Gurnett, J. S. Leisner et al. (2013), The plasma density distribution in the inner region of Saturn's magnetosphere. *J. Geophys. Res.*, **118**(June), 2970–2974.

Persoon, A. M., D. A. Gurnett, O. Santolik et al. (2009), A diffusive equilibrium model for the plasma density in Saturn's magnetosphere. *J. Geophys. Res.*, **114**(Apr.), 4211.

Pilkington, N. M., N. Achilleos, C. S. Arridge et al. (2014), Polar confinement of Saturn's magnetosphere revealed by in situ Cassini observations. *J. Geophys. Res.*, **119** (Apr.), 2858–2875.

Pontius, D. H. and T. W. Hill (2009), Plasma mass loading from the extended neutral gas torus of Enceladus as inferred from the observed plasma corotation lag. *Geophys. Res. Lett.*, **36**(Dec.), 23103.

Porco, C., D. DiNino and F. Nimmo (2014), How the geysers, tidal stresses, and thermal emission across the south polar terrain of Enceladus are related. *The Astrophysical Journal*, **148**(Sept.), 45.

Porco, C. C., P. Helfenstein, P. C. Thomas et al. (2006), Cassini observes the active south pole of Enceladus. *Science*, **311**(Mar.), 1393–1401.

Pryor, W. R., A. M. Rymer, D. G. Mitchell et al. (2011), The auroral footprint of Enceladus on Saturn. *Nature*, **472**(Apr.), 331–333.

Radioti, A., D. Grodent, J.-C. Gérard et al. (2011), Bifurcations of the main auroral ring at Saturn: Ionospheric signatures of consecutive reconnection events at the magnetopause. *J. Geophys. Res.*, **116** (Nov.), 11209.

Radioti, A., D. Grodent, J.-C. Gérard (2013), Auroral signatures of multiple magnetopause reconnection at Saturn. *Geophys. Res. Lett.*, **40**(Sept.), 4498–4502.

Richardson, J. D. (1986), Thermal ions at Saturn: Plasma parameters and implications. *J. Geophys. Res.*, **91**(10), 1381–1389.

Richardson, J. D. (1998), Thermal plasma and neutral gas in Saturn's magnetosphere. *Rev. of Geophys.*, **36**(Nov.), 501–524.

Roederer, J. G. (1970), *Dynamics of geomagnetically trapped radiation*, Springer.

Roussos, E., M. Andriopoulou, N. Krupp et al. (2013), Numerical simulation of energetic electron microsignature drifts at Saturn: Methods and applications. *Icarus*, **226**(Nov.), 1595–1611.

Roussos, E., G. H. Jones, N. Krupp et al. (2007), Electron microdiffusion in the Saturnian radiation belts: Cassini MIMI/LEMMS observations of energetic electron absorption by the icy moons. *J. Geophys. Res.*, **112** (A11), 6214.

Roussos, E., P. Kollmann, N. Krupp et al. (2012), Energetic electron observations of Rhea's magnetospheric interaction. *Icarus*, **221**(Sept.), 116–134.

Roussos, E., N. Krupp, T. P. Armstrong et al. (2008), Discovery of a transient radiation belt at Saturn. *Geophys. Res. Lett.*, **35**(Nov.), 22106.

Roussos, E., N. Krupp, C. P. Paranicas et al. (2010), Energetic electron microsignatures as tracers of radial flows and dynamics in Saturn's innermost magnetosphere. *J. Geophys. Res.*, **115**(A14), 3202.

Roussos, E., N. Krupp, C. P. Paranicas (2011), Long- and short-term variability of Saturn's ionic radiation belts. *J. Geophys. Res.*, **116**(A15), A02217.

Roussos, E., N. Krupp, C. P. Paranicas (2014), The variable extension of Saturn's electron radiation belts. *Planet. Space Sci.*

Roussos, E., N. Krupp, J. Woch et al. (2005), Low energy electron microsignatures at the orbit of Tethys: Cassini MIMI/LEMMS observations. *Geophys. Res. Lett.*, **32** (Dec.), 24107.

Russell, C. T., C. M. Jackman, H. Y. Wei et al. (2008), Titan's influence on Saturnian substorm occurrence. *Geophys. Res. Lett.*, **35**(June), 12105.

Rymer, A. M., B. H. Mauk, T. W. Hill et al. (2009a), Cassini evidence for rapid interchange transport at Saturn. *Planet. Space Sci.*, **57**(Dec.), 1779–1784.

Rymer, A. M., H. T. Smith, A. Wellbrock et al. (2009b), Discrete classification and electron energy spectra of Titan's varied magnetospheric environment. *Geophys. Res. Lett.*, **36**(Aug.), 15109.

Sakai, S., S. Watanabe, M. W. Morooka et al. (2013), Dust-plasma interaction through magnetosphere-ionosphere coupling in Saturn's plasma disk. *Planet. Space Sci.*, **75** (Jan.), 11–16.

Santolík, O., D. A. Gurnett, G. H. Jones et al. (2011), Intense plasma wave emissions associated with Saturn's moon Rhea. *Geophys. Res. Lett.*, **38**(Oct.), 19204.

Santos-Costa, D., M. Blanc, S. Maurice et al. (2003), Modeling the electron and proton radiation belts of Saturn. *Geophys. Res. Lett.*, **30**(20), 6–1.

Saur, J., N. Schilling, F. M. Neubauer et al. (2008), Evidence for temporal variability of Enceladus' gas jets: Modeling of Cassini observations. *Geophys. Res. Lett.*, **35**(Oct.), 20105.

Schippers, P., N. André, D. A. Gurnett et al. (2012), Identification of electron field-aligned current systems in Saturn's magnetosphere. *J. Geophys. Res.*, **117** (May), 5204.

Schippers, P., C. S. Arridge, J. D. Menietti et al. (2011), Auroral electron distributions within and close to the Saturn kilometric radiation source region. *J. Geophys. Res.*, **116**(A15), A05203.

Schippers, P., M. Blanc, N. André et al. (2008), Multi-instrument analysis of electron populations in Saturn's magnetosphere. *J. Geophys. Res.*, **113**(A12), 7208.

Schippers, P., M. Moncuquet, N. Meyer-Vernet et al. (2013), Core electron temperature and density in the innermost Saturn's magnetosphere from HF power spectra analysis on Cassini. *J. Geophys. Res.*, **118** (Nov.), 7170–7180.

Schulz, M. and L. J. Lanzerotti (1974), *Particle Diffusion in the Radiation Belts*. Physics and Chemistry in Space, Berlin: Springer.

Seidelmann, P. K., B. A. Archinal, M. F. A'Hearn et al. (2007), Report of the IAU/IAG Working Group on cartographic co-ordinates and rotational elements: 2006. *Celestial Mechanics and Dynamical Astronomy*, **98**(July), 155–180.

Selesnick, R. S., M. D. Looper and R. A. Mewaldt (2007), A theoretical model of the inner proton radiation belt, *Space Weather*, *5*, 4003.

Sergis, N., C. S. Arridge, S. M. Krimigis et al. (2011), Dynamics and seasonal variations in Saturn's magnetospheric plasma sheet, as measured by Cassini. *J. Geophys. Res.*, **116**(A15), A04203.

Sergis, N., C. M. Jackman, A. Masters et al. (2013), Particle and magnetic field properties of the Saturnian magnetosheath: Presence and upstream escape of hot magnetospheric plasma. *J. Geophys. Res.*, **118**(Apr.), 1620–1634.

Sergis, N., S. M. Krimigis, D. G. Mitchell et al. (2009), Energetic particle pressure in Saturn's magnetosphere measured with the Magnetospheric Imaging Instrument on Cassini. *J. Geophys. Res.*, **114**(Feb.), 2214.

Sergis, N., Krimigis, S. M., Roelof, E. C., et al. (2010), Particle pressure, inertial force, and ring current density profiles in the magnetosphere of Saturn, based on Cassini measurements. *Geophys. Res. Lett.*, **37**(Jan.), 2102.

Shemansky, D. E., X. Liu and X. Melin (2009), The Saturn hydrogen plume. *Planet. Space Sci.*, **57**(Dec.), 1659–1670.

Shematovich, V. I., R. E. Johnson, M. Michael et al. (2003), Nitrogen loss from Titan. *J. Geophys. Res.*, **108**(Aug.), 5087.

Shprits, Y. Y., J. D. Menietti, X. Gu et al. (2012), Gyroresonant interactions between the radiation belt electrons and whistler mode chorus waves in the radiation environments of Earth, Jupiter, and Saturn: A comparative study. *J. Geophys. Res.*, **117**(A16), 11216.

Simon, S., H. Kriegel, J. Saur et al. (2012), Analysis of Cassini magnetic field observations over the poles of Rhea. *J. Geophys. Res.*, **117**(July), 7211.

Simon, S., F. M. Neubauer, A. Wennmacher et al. (2014), Variability of Titan's induced magnetotail: Cassini magnetometer observations. *J. Geophys. Res.*, **119** (Mar.), 2024–2037.

Simon, S., J. Saur, H. Kriegel et al. (2011a), Influence of negatively charged plume grains and hemisphere coupling currents on the structure of Enceladus' Alfvén wings: Analytical modeling of Cassini magnetometer observations. *J. Geophys. Res.*, **116**(A15), 4221.

Simon, S., J. Saur, F. M. Neubauer et al. (2011b), Magnetic signatures of a tenuous atmosphere at Dione. *Geophys. Res. Lett.*, **38**(Aug.), 15102.

Simon, S., S. C. Treeck, A. Wennmacher et al. (2013), Structure of Titan's induced magnetosphere under varying background magnetic field conditions: Survey of Cassini magnetometer data from flybys TA-T85. *J. Geophys. Res.*, **118**(Apr.), 1679–1699.

Simon, S., A. Wennmacher, F, M. Neubauer et al. (2010), Dynamics of Saturn's magnetodisk near Titan's orbit: Comparison of Cassini magnetometer observations from real and virtual Titan flybys. *Planet. Space Sci.*, **58**(Oct.), 1625–1635.

Singer, S. F. (1958), Trapped albedo theory of the radiation belt. *prl*, **1**(8), 300.

Sittler, E. C., N. Andre, M. Blanc et al. (2008. Ion and neutral sources and sinks within Saturn's inner magnetosphere: Cassini results. *Planet. Space Sci.*, **56**(Jan.), 3–18.

Sittler, E. C., R. E. Hartle, R. E. Johnson et al. (2010), Saturn's magnetospheric interaction with Titan as defined by Cassini encounters T9 and T18: New results. *Planet. Space Sci.*, **58**(Feb.), 327–350.

Sittler, E. C., M. Thomsen, R. E. Johnson et al. (2006), Cassini observations of Saturn's inner plasmasphere: Saturn orbit insertion results. *Planet. Space Sci.*, **54** (Oct.), 1197–1210.

Smith, H. T., R. E. Johnson, M. E. Perry et al. (2010), Enceladus plume variability and the neutral gas densities in Saturn's magnetosphere. *J. Geophys. Res.*, **115** (Oct.), 10252.

Smith, H. T., R. E. Johnson, E. C. Sittler et al. (2007), Enceladus: The likely dominant nitrogen source in Saturn's magnetosphere. *Icarus*, **188**(June), 356–366.

Smith, H. T. and A. M. Rymer (2014), An empirical model for the plasma environment along Titan's orbit based on Cassini plasma observations. *J. Geophys. Res.*, **119** (July), 5674–5684.

Sonnerup, B. U. and M. J. Laird (1963), On magnetospheric interchange instability. *J. Geophys. Res.*, **68**.

Southwood, D. J. and M. G. Kivelson (1987), Magnetospheric interchange instability. *J. Geophys. Res.*, **92**(Jan.), 109–116.

Southwood, D. J. and M. G. Kivelson (1989), Magnetospheric interchange motions. *J. Geophys. Res.*, **94**(Jan.), 299–308.

Strobel, D. F. (2008), Titan's hydrodynamically escaping atmosphere. *Icarus*, **193**(Feb.), 588–594.

Sulaiman, A. H., A. Masters, M. K. Dougherty et al. (2014), The magnetic structure of Saturn's magnetosheath. *J. Geophys. Res.*, **119**(July), 5651–5661.

Szego, K., Z. Nemeth, G. Erdos et al. (2011), The plasma environment of Titan: The magnetodisk of Saturn near the encounters as derived from ion densities measured by the Cassini/CAPS plasma spectrometer. *J. Geophys. Res.*, **116**(Oct.), 10219.

Szego, K., Z. Nemeth, G. Erdos (2012), Location of the magnetodisk in the nightside outer magnetosphere of Saturn near equinox based on ion densities. *J. Geophys. Res.*, **117**(Sept.), 9225.

Tadokoro, H., H. Misawa, F. Tsuchiya et al. (2012), Effect of photo-dissociation on the spreading of OH and O clouds in Saturn's inner magnetosphere. *J. Geophys. Res.*, **117** (Sept.), 9226.

Tang, R. and D. Summers (2012), Energetic electron fluxes at Saturn from Cassini observations. *J. Geophys. Res.*, **117** (June), 6221.

Tao, X., R. M. Thorne, R. B. Horne et al. (2010), Excitation of electron cyclotron harmonic waves in the inner Saturn magnetosphere within local plasma injections. *J. Geophys. Res.*, **115**(Dec.), 12204.

Tenishev, V., M. R. Combi, B. D. Teolis et al. (2010), An approach to numerical simulation of the gas distribution in the atmosphere of Enceladus. *J. Geophys. Res.*, **115**(A14), 9302.

Teolis, B. D., G. H. Jones, P. F. Miles et al. (2010), Cassini Finds an oxygen-carbon dioxide atmosphere at Saturn's icy moon Rhea. **330**(Dec.), 1813.

Thomsen, M. F., C. M. Jackman, R. L. Tokar et al. (2014b), Plasma flows in Saturn's nightside magnetosphere. *J. Geophys. Res.*, **119**(June), 4521–4535.

Thomsen, M. F., D. B. Reisenfeld, D. M. Delapp et al. (2010), Survey of ion plasma parameters in Saturn's magnetosphere. *J. Geophys. Res.*, **115** (Oct.), 10220.

Thomsen, M. F., D. B. Reisenfeld, R. J. Wilson et al. (2014a), Ion composition in interchange injection events in Saturn's magnetosphere. *J. Geophys. Res.*, **119**(Dec.), 9761–9772.

Thomsen, M. F., E. Roussos, M. Andriopoulou et al. (2012), Saturn's inner magnetospheric convection pattern: Further evidence. *J. Geophys. Res.*, **117**(A16), 9208.

Thomsen, M. F., R. J. Wilson, R. L. Tokar et al. (2013), Cassini/CAPS observations of duskside tail dynamics at Saturn. *J. Geophys. Res.*, **118**, 57675781.

Tiscareno, M. S., J. A. Burns, J. N. Cuzzi et al. (2010), Cassini imaging search rules out rings around Rhea. *Geophys. Res. Lett.*, **37**(July), 14205.

Tokar, R. L., R. E. Johnson, M. F. Thomsen et al. (2005), Cassini observations of the thermal plasma in the vicinity of Saturn's main rings and the F and G rings. *Geophys. Res. Lett.*, **32**(June), 14.

Tokar, R. L., R. E. Johnson, M. F. Thomsen 2012), Detection of exospheric O_2^+ at Saturn's moon Dione. *Geophys. Res. Lett.*, **39**(Feb.), 3105.

Tseng, W.-L., W.-H. Ip, R. E. Johnson et al. (2010), The structure and time variability of the ring atmosphere and ionosphere. *Icarus*, **206**(Apr.), 382–389.

Tseng, W.-L., R. E. Johnson and M. K. Elrod (2013a), Modeling the seasonal variability of the plasma environment in Saturn's magnetosphere between main rings and Minas. *Planet. Space Sci.*, **77**(Mar.), 126–135.

Tseng, W.-L., R. E. Johnson and W.-H. Ip (2013b), The atomic hydrogen cloud in the Saturnian system. *Planet. Space Sci.*, **85**(Sept.), 164–174.

Tseng, W.-L., R. E. Johnson, M. F. Thomsen et al. (2011), Neutral H_2 and H_2^+ ions in the Saturnian magnetosphere. *J.Geophys. Res.*, **116**(Mar.), 3209.

Usoskin, I. G., K. Alanko-Huotari, G. A. Kovaltsov et al. (2005), Heliospheric modulation of cosmic rays: Monthly reconstruction for 1951–2004. *J. Geophys. Res.*, **110**(Dec.), 12108.

Van Allen, J. A. (1984), Energetic particles in the inner magnetosphere of Saturn. Pages 281–317 of: T. Gehrels and M. S. Matthews (eds), *Saturn*. University of Arizona Press, Tucson.

Van Allen, J. A., M. F. Thomsen and B. A. Randall (1980), The energetic charged particle absorption signature of Mimas. *J. Geophys. Res.*, **85**(A11), 5709–5718.

Vasyliūnas, V. M. (1994), Role of the plasma acceleration time in the dynamics of the Jovian magnetosphere. *Geophys. Res. Lett.*, **21**(Mar.), 401–404.

Vasyliūnas, V. M. (1983), Plasma Distribution and Flow, *Physics of the Jovian Magnetosphere*, Cambridge University Press, Cambridge, 395–453.

Waite, J. H., M. R. Combi, W.-H. Ip et al. (2006), Cassini ion and neutral mass spectrometer: Enceladus plume composition and structure. *Science*, **311**(Mar.), 1419–1422.

Waite, J. H., T. E. Cravens, W.-H. Ip et al. (2005), Oxygen ions observed near Saturn's A ring. **307**(Feb.), 1260–1262.

Walker, R. J., K. Fukazawa, T. Ogino et al. (2011), A simulation study of Kelvin–Helmholtz waves at Saturn's magnetopause. *J. Geophys. Res.*, **116**(Mar.), 3203.

Walt, M. (1994), *Introduction to Geomagnetically Trapped Radiation*. Cambridge University Press, Cambridge.

Wang, Z., D. A. Gurnett, G. Fischer et al. (2010), Cassini observations of narrowband radio emissions in Saturn's magnetosphere. *J. Geophys. Res.*, **115**(June), 6213.

Wellbrock, A., A. J. Coates, I. Sillanpää et al. (2012), Cassini observations of ionospheric photoelectrons at large distances from Titan: Implications for Titan's exospheric environment and magnetic tail. *J. Geophys. Res.*, **117** (Mar.), 3216.

Went, D. R., M. G. Kivelson, N. Achilleos et al. (2011), Outer magnetospheric structure: Jupiter and Saturn compared. *J. Geophys. Res.*, **116**(Apr.), 4224.

Westley, M. S., R. A. Baragiola, R. E. Johnson et al. (1995), Photodesorption from low-temperature water ice in interstellar and circumsolar grains. *Nature*, **373**(Feb.), 405–407.

Wilson, R. J., F. Bagenal, P. A. Delamere et al. (2013), Evidence from radial velocity measurements of a global electric field in Saturn's inner magnetosphere. *J. Geophys. Res.*, **118**(May), 2122–2132.

Wilson, R. J., P. A. Delamere, F. Bagenal et al. (2012), Kelvin–Helmholtz instability at Saturn's magnetopause: Cassini ion data analysis. *J. Geophys. Res.*, **117**(Mar.), 3212.

Wilson, R. J., R. L. Tokar and M. G. Henderson (2009), Thermal ion flow in Saturn's inner magnetosphere measured by the Cassini plasma spectrometer: A signature of the Enceladus torus? *Geophys. Res. Lett.*, **362**(Dec.), L23104.

Wilson, R. J., R. L. Tokar, M. G. Henderson et al. (2008), Cassini plasma spectrometer thermal ion measurements in Saturn's inner magnetosphere. *J. Geophys. Res.*, **113** (Dec.), 12218.

Wilson, R. J., R. L. Tokar, W. S. Kurth et al. (2010), Properties of the thermal ion plasma near Rhea as measured by the Cassini plasma spectrometer. *J. Geophys. Res.*, **115**(A14), A05201.

Winglee, R. M., A. Kidder, E. Harnett et al. (2013), Generation of periodic signatures at Saturn through Titan's interaction with the centrifugal interchange instability. *J. Geophys. Res.*, **118**(July), 4253–4269.

Ye, S.-Y., C. A. Gurnett, G. Fischer et al. (2009), Source locations of narrowband radio emissions detected at Saturn. *J. Geophys. Res.*, **114**(June), 6219.

Zarka, P. (1998), Auroral radio emissions at the outer planets: Observations and theories. *J. Geophys. Res.*, **103** (Sept.), 20159–20194.

Zieger, B., K. C. Hansen, T. I. Gombosi et al. (2010), Periodic plasma escape from the mass-loaded Kronian magnetosphere. *J. Geophys. Res.*, **115** (Aug.), 8208.

7

Saturn's Aurorae

TOM STALLARD, SARAH V. BADMAN, ULYANA DYUDINA, DENIS GRODENT AND LAURENT LAMY

Abstract

The aurorae of each planet are produced as a direct interaction between the upper atmosphere and magnetosphere of that planet. Energetic particles from the magnetosphere are driven into the top of the atmosphere, depositing energy there, and ultimately resulting in an electromagnetic emission. As a result, aurorae are related to conditions within the planetary magnetospheres so an understanding of the auroral emission provides a view of both the magnetospheric structure and how that magnetosphere is coupled with the underlying ionosphere. In the past, Saturn's magnetosphere, and thus its aurorae, have been seen as something of a hybrid between the solar-wind-driven interaction at Earth and the rotationally dominated system at Jupiter. However, observations across a wide wavelength range by both the Cassini spacecraft and supporting Earth-based telescopes have revealed Saturn's aurorae to be highly complex. We now recognize that Saturn's aurorae are driven by the dynamic magnetic field interactions between the atmosphere, the solar wind and plasma trapped within the magnetosphere, all strongly affected by the rapid rotation of the planet.

In this chapter, we highlight the broad variety of auroral features observed at Saturn, and discuss how these are generated by energetic particles moving within current systems that link to solar wind interactions (Section 7.2), interactions with plasma generated within the magnetosphere (Section 7.3) and with current systems that vary periodically, including those linked to weather systems within Saturn's upper atmosphere (Section 7.4). Finally, we conclude with a discussion of the major questions that remain about Saturn's aurorae, and summarize the upcoming observations that will help us answer them. We begin with a discussion of how the auroral emission is generated and the characteristics of aurorae observed at Saturn. In particular, we highlight the most recent auroral research, following on from the overview of Saturn's auroral processes presented in past reviews of the subject (for example, Kurth et al. 2009).

7.1 Generation of Auroral Emission on Saturn

Saturn's aurorae are driven by charged particles precipitating into the planet's atmosphere. These charged particles are produced by the ionization of neutral gases in two regions within the Saturnian system. First, the outgassing and sputtering of volatiles from the icy moons and rings of Saturn, in particular from the geysers of Enceladus, have resulted in a source of neutral material within the equatorial region of Saturn's magnetosphere, which, when ionized through a combination of solar photoionization and ionizing collisions within the neutral gas, results in a complex mixture of water-derived atomic and molecular ions concentrated in the equatorial plane of the magnetosphere (Young et al. 2005), along with hydrogen ions from the planet. Second, within the upper atmosphere, a combination of solar extreme ultraviolet (EUV) photoionization and auroral electron impact ionization drives the production of atomic hydrogen, molecular hydrogen and hydrocarbon ions, and triggers complex photochemical reactions leading to the production of hydrocarbon aerosols. The effects on the ionosphere caused by infalling water ions are discussed in detail within Chapter 8.

Ions and electrons within Saturn's magnetosphere are generally magnetically confined close to the equatorial plane, with densities falling off as the magnetic field lines approach the planet. Particles with sufficient energy can reach high latitudes and precipitate into the atmosphere. This process drives the production of a relatively weak form of aurorae on Saturn. Such aurorae are strongest on magnetic field lines that close through magnetospheric regions that have sources of hot plasma, as described in Chapter 6, which have enough kinetic energy to provide a constant drizzle of particles into the top of the atmosphere.

At increasing altitudes within the ionosphere, the charged particle densities drop off significantly. This means that there is a region along the magnetic field lines, between the magnetosphere and ionosphere, where the particle densities reach a relatively low minimum, but there remains an overlap between these two systems. Since there is a continuous region of electric conductivity between the ionosphere and magnetosphere, the charged particles within these two regions are directly linked by Birkeland currents, with associated Alfvén waves propagating the magnetic disturbances these currents produce along the field lines. The resultant ionospheric/magnetospheric interactions allows the flow of energy between these regions, with forces producing currents in one part of the system in turn driving responsive currents within the other part of the system.

Ultimately, the study of Saturn's aurorae provides a view of the current circuit between the ionosphere and magnetosphere that effectively measures how energy is transferred between these regions. Understanding how currents are driven between the magnetosphere and the planet requires a determination of what influences the relative motion of plasma within these regions. The plasma is controlled by two main drivers: collisions with surrounding neutral gas and the motion of charged particles relative to the surrounding magnetic field.

The magnetic field of Saturn extends out from the interior of the planet through the magnetosphere. Plasma moving relative to this magnetic field generates an electric field, this in turn produces a Lorentz force within the plasma that acts to accelerate the plasma particles back to the magnetic field velocity. However, these electric currents also act to distort the magnetic field within both the ionosphere and magnetosphere of Saturn.

Where the density of the surrounding neutral atmosphere is high enough to result in significant collisions between the ions and neutrals, the flow of ionospheric charged particles is also affected. The upper atmosphere itself is driven through friction with the underlying, denser atmosphere. As such, ions are accelerated into quasi co-rotation with the surrounding neutral thermosphere, which in turn drives a circuit of currents that closes in the surrounding magnetosphere (Ray et al. 2012, 2013). This allows the angular momentum of Saturn to be transferred into plasma within the surrounding magnetosphere. These atmospheric collisions also produce a significant amount of Joule heating within the upper atmosphere (Cowley et al. 2004; Müller-Wodarg et al. 2012).

Magnetospheric plasma is considered to be collisionless, unaffected by the numerous neutral sources within the magnetosphere, so that the magnetic field of Saturn is quasi co-rotating with the ionosphere because frozen-in field conditions apply. However, forces that act upon neutrals, such as the gravitational pull and centripetal force from Saturn, also act upon these charged particles. As plasma moves radially outwards through the magnetosphere, a process discussed in detail in Chapter 6, conservation of angular momentum requires that it slow down, producing a bend-back in the planetary magnetic field, resulting in an increasing lag behind co-rotation.

An extreme example of the distortions that can affect the localized magnetic field are regions of the ionosphere that are connected to magnetic field lines that are open to the solar wind. These open field lines are produced through reconnection between Saturn's magnetic field lines and solar magnetic field lines that were frozen into solar wind, and remain frozen-in as the solar wind travels past Saturn. The reconnected ends of the open field lines are held in the inertial frame by the solar wind as it passes Saturn. Since these field lines no longer rotate with the planet, ions within the atmosphere are forced to flow relative to the surrounding neutral atmosphere, producing strong currents within the ionosphere (Cowley et al. 2004). Since these ions experience strong Joule heating driven by the neutral atmosphere, they cannot remain static relative to the solar wind and are dragged into sub-rotating with the planet. Because the magnetic fields are frozen-in with the plasma, the magnetic fields sub-rotate with the ions so that the open field lines propagating into the tail experience a twisting force, as field lines are dragged into motion by the neutral atmosphere (Milan et al. 2005; Stallard et al. 2007).

As a result of electric circuits that close along magnetic field lines, electric currents are coupled between the ionosphere and magnetosphere. However, due to the relatively low particle densities within the boundary region between the ionosphere and magnetosphere, charged particles in this region are accelerated to carry necessary currents. This process results in the production of energetic electrons that carry the current through this boundary region. These particles then precipitate into the upper atmosphere of Saturn, typically with mean electron energies of 1–20 KeV carrying an electron energy flux of 0.1–10 mWm^{-2} (Badman et al. 2015). As these electrons undergo acceleration, they spiral in towards the planet along the magnetic field lines. Saturn's emissions were thought to be located at the boundary between closed and open field lines (Cowley et al. 2004), but the field-aligned current patterns are more complex than initial modelling work suggested (Bunce 2012). This complex sequence of currents is described in more detail in Section 7.2, and may partially be explained by the overlaying of currents associated with two current systems, one fixed in local time and the other rotating at the planetary period, as described in Section 7.4.2.

7.1.1 Generation of Radio Aurora

Particles orbit the localized magnetic field lines concentrated within the equatorial plane, perpendicular to the magnetic field of the planet. The motion of these energetic electrons around the magnetic field drives cyclotron maser instabilities that produce resonant radio emissions. The frequency with which these particles gyrate within the magnetic field is described as the electron gyro frequency, which is proportional to the local magnetic field strength. Particles within a plasma also rapidly oscillate as a result of small-scale charge-density perturbations within the plasma. The frequency of these oscillations – the plasma frequency – is independent of magnetic field strength and proportional to the number density of the plasma. Cyclotron maser instabilities (CMI) occur where the electron gyrofrequency is much larger than the plasma frequency. As a result, they occur when regions are depleted of particles (and so have a low plasma density, and thus a relatively small plasma frequency compared to the gyrofrequency) and in regions that are strongly magnetized (and so have a high gyrofrequency, dominating over the plasma frequency). Under such conditions, cyclotron maser instabilities occur when the distribution of electron flux is non-Maxwellian, with electron energies of typically a few keV (Zarka 1998). At Saturn, this effectively confines radio auroral emissions known as the Saturn Kilometric Radiation (SKR) to the low-density region above the planet, as shown in Figure 7.1. Observations have shown that these emissions are clearly auroral in nature (Kurth et al. 2005). Early modelling analysis used 20 keV electrons with a loss-cone distribution to simulate CMI-driven highly beamed emissions that fairly reproduce common features observed in SKR dynamic spectra (Lamy et al. 2008).

More recently, several studies have investigated in situ plasma and wave conditions during individual inclined Cassini orbits that passed through the SKR source region, providing a first overview of the auroral region and validating the cyclotron maser instability as the mechanism driving SKR. SKR sources were found in a region of upward field-aligned current. The observed wave emission frequency was consistent with that theoretically derived for emission perpendicular to the magnetic field and the 6–9 keV hot electron population observed in situ. These electrons displayed a shell-like distribution function able to drive a high enough wave growth rate to account for the observed SKR intensity (Lamy et al. 2010, 2011; Mutel et al. 2010; Bunce et al. 2010; Schippers et al. 2011; Menietti et al. 2011). In the nearby region of downward field-aligned currents, two types of upward accelerated electron beams were observed: a broadband energetic (1–100 keV) electron population that is observed throughout the region and a narrow-banded (0.1–1 keV) electron population that is observed sporadically,

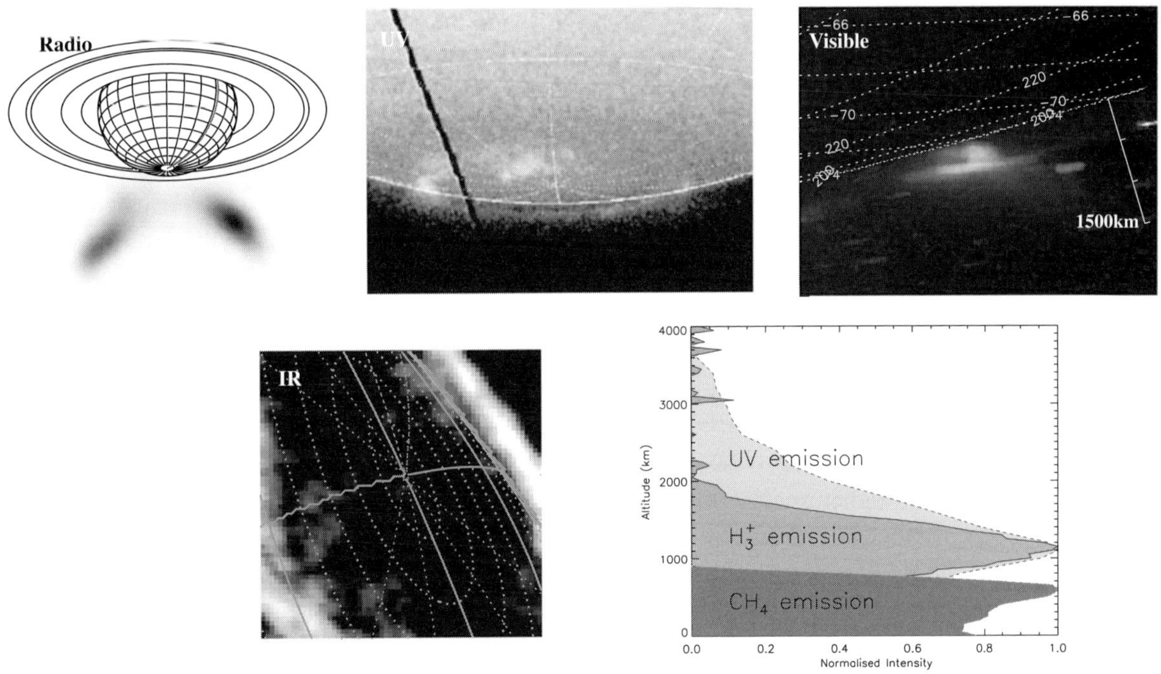

Figure 7.1 The altitudinal distribution of various aurorae above Saturn's limb. The radio aurora (adapted from Lamy et al. 2009, Figure 3) shows the measured SKR emission measured over seven consecutive hours in 17 January 2007 by Cassini/RPWS, projected onto the observation plane. The UV image (adapted from Gerard et al., 2009, Figure 1) shows an image of Saturn's UV auroral emission beyond the nightside limb of Saturn's south polar region, obtained with the HST/ACS camera. The visible image (adapted from Dyudina et al. 2016, Figure 8) shows a particularly bright aurora observed above Saturn's nightside limb by Cassini/ISS on 27 November 2010. The infrared image (adapted from Stallard et al. 2012a, Figure 1) shows bright H_3^+ emission above Saturn's southern limb, observed by Cassini/VIMS on Oct. 11, 2006. The altitudinal profile (adapted from Stallard et al. 2012a, Figure 3) compares the normalized emission derived from the H_3^+ and UV images. Also shown, for comparison, is emission from non-auroral fluorescent methane within the upper stratosphere, indicating the region below the homopause where vertical mixing still occurs. (A black-and-white version of this figure appears in some formats. For the color version, please refer to the plate section.)

producing very low-frequency whistler-mode auroral hiss (Schippers et al. 2011).

7.1.2 Atmospheric Ionization and Excitation

Highly energetic particles, accelerated down along magnetic field lines, are forced directly into the upper atmosphere of the planet. How these particles react on interacting with the neutral atmosphere varies as a function of the energy of the particle. Modelling of this process (e.g. Galand et al. 2011, and references therein) has shown that for energetic electrons, the majority of such particles' energy is lost through inelastic collisions with the neutral atmosphere, with a small minority lost through collisions with thermal electrons. Nearly all of these collisions are with molecular hydrogen (90% for 1 keV electrons), driving the excitation, ionization and dissociation of H_2. It is the combination of these processes that drive the majority of all auroral emissions within the ultraviolet, visible and infrared wavelengths at Saturn, with the remaining impacts occurring with atomic hydrogen and helium near the top of the atmosphere and with trace hydrocarbons located deeper, below the homopause.

Hydrogen Auroral Emission

About half of all inelastic collisions between precipitating energetic electrons and H_2 in Saturn's upper atmosphere result in the excitation of molecular hydrogen into upper electronic levels, dominated by the Lyman ($B^1\Sigma_g^+$) and Werner ($C^1\Pi_u$) systems, although higher-energy interactions can produce more excited

H_2 (populating the $B'^1\Sigma_u^+$, $D^1\Pi_u$, $B''^1\Sigma_u^+$, $D'^1\Pi_u$ systems):

$$e^- + H_2 \rightarrow H_2^* (B, C, B', D, B'', D') + e^-$$

The majority of these interactions are, in fact, produced by secondary electrons, produced as a result of ionization processes within the atmosphere, described in Chapter 8. This energetic molecular hydrogen quickly de-excites back to the electronic ground state, producing rovibronic (a combination of rotational, vibrational and electronic) transitions, with overall UV emission dominated by Lyman (B → X) and Werner (C → X) transitions, but higher-energy transitions (B′, D, B″, D′ → X) contributing significantly to EUV emission between 80 and 120 nm, as shown in Figure 7.2.

These de-excitations occur at energies where the internuclear distance between the two protons in H_2 is smaller than the dissociation continuum of the ground state. However, at high enough energies, the H_2 molecule is dissociated, forming two H(1s) atoms. Since the kinetic energy of these atoms is not discrete, the emitted photons produce a continuum emission. The Lyman and Werner transitions, along with these associated H_2 continua, dominate the UV aurorae from Saturn, contributing 90% of the total UV auroral emission (Gustin et al. 2012).

The pure H_2 emission spectrum is partially absorbed as the light travels out from Saturn's atmosphere through hydrocarbons beneath the homopause, dominated by methane (CH_4) absorption, but with additional absorption from ethane (C_2H_6) and acetylene (C_2H_2) (Gustin et al. 2012). The deeper into the atmosphere an auroral emission is produced, the greater the distortion of the emission spectrum produced at wavelengths where hydrocarbon absorption is strong. As such, the extent of hydrocarbon absorption provides direct information about the depth to which precipitating particles penetrate. Modelling of hydrocarbon absorption can be performed in a detailed

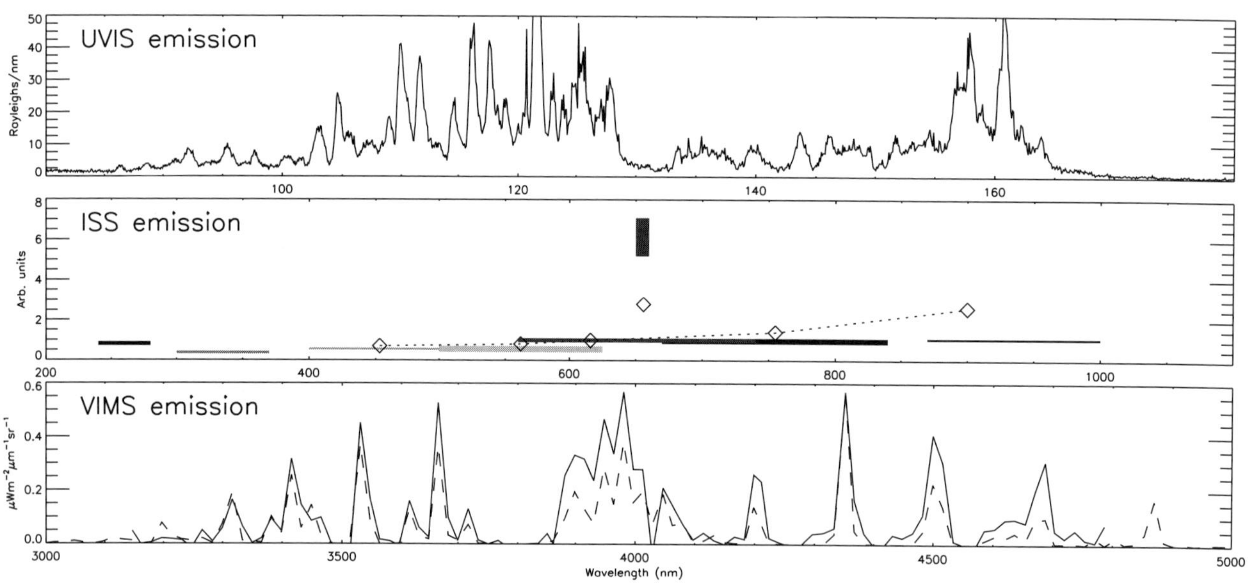

Figure 7.2 Saturn's typical UV, visible and IR auroral emission as observed by the Cassini spacecraft. In the top panel, a typical nightside UV auroral spectrum is shown, as measured by Cassini/UVIS (data provided by J. Gustin, personal communication). The UV emission comes from simultaneous observations from the FUV and EUV channels of the spectrograph. The spectrum contains very little methane absorption, typical for Saturn. Total UV emission is ~20 kR, within the upper range of values generally observed. The middle panel shows a visible spectrum measured from Cassini/ISS images taken with nine different filters (adapted from Dyudina et al. 2016, Figure 8) and a laboratory-simulated spectrum (black dotted curve; Aguilar et al. 2008). For each ISS filter, boxes show the values observed on 27 November 2010, and depict the wavelength coverage (width) and the range of intensities (box height in Rayleigh/nm) measured within the brightest auroral region, scaled to the clear filter brightness. The box showing strongest emission is the Hα (656 nm) filter. Typical near-infrared auroral spectrum observed by Cassini/VIMS (see Stallard et al. 2012b for methodology), depicting the total H_3^+ emission measured between 900 and 1500 km above the planet limb, is shown in the bottom panel (bold), together with a modeled H_3^+ spectrum at 550 K, convolved to the VIMS resolution (dashed line).

spectral analysis, but can be approximated by taking the ratio of emission seen between 155 and 162 nm (a region almost free of hydrocarbon absorption) and between 123 and 130 nm (a region strongly absorbed by hydrocarbons). These regions of emission, when unabsorbed, have an energy flux colour ratio of 1.1, but this increases when precipitating electrons are able to penetrate deep enough into the atmosphere that the resultant H_2 emission is absorbed by hydrocarbons (Gustin et al. 2013, and references therein).

The precipitation of energetic electrons also results in the electronic excitation of atomic hydrogen. Since it is the closest electronic energy states that are most easily populated, de-excitation from the H Lyman-α line within the Lyman series ($n = 2,3,4 \ldots \rightarrow n' = 1$) dominates UV atomic hydrogen emission, emitting at 121.6 nm.

Visible emission from this series is dominated by H Balmer-α emission ($n = 3 \rightarrow 2$) an emission line at 656.3 nm often referred to as Hα. Saturn's visible wavelength aurora cannot be observed from Earth because it is dominated by the underlying contribution of solar scattered light from the planetary disc. The auroral emission was detected by the Cassini/ISS at Hα, and also in several other wavelength bands, as shown in Figure 7.2 (Dyudina et al. 2016). This emission can only be observed on the dark nightside of the planet, both on the disk and above the limb (Figure 7.1). Both Cassini/ISS and laboratory simulations (Aguilar et al. 2008) show that the bright Hα (656 nm) line is clearly dominant over other hydrogen lines at visible wavelengths. The simulations model auroral emission by electron impact-induced H_2 fluorescence, with an electron energy of 20 eV. The observed Hα line in aurora is stronger than the simulations, and an additional general "red" slope of emission is observed in the background at visible wavelengths.

In addition to the production of atomic hydrogen through radiative dissociation of H_2, precipitating energetic electrons can also produce atomic hydrogen through dissociative excitation:

$$e + H_2 \rightarrow H + H + e + h\upsilon$$

Although models readily show this process occurs for any H_2 atmosphere, this process has not been studied in detail at Saturn. However, at Jupiter, this process dominates emission from atomic hydrogen within the Lyman series, with this process producing 99% of the auroral Lyman line emission on the dayside. Additional atomic hydrogen emission is produced through the direct excitation of atomic hydrogen, with additional scattering within reflected solar Lyman lines potentially obscuring the auroral emission on the dayside (Gustin et al. 2013). The processes that drive H and H_2 aurora, outlined here, are described in detail within Gustin et al. (2013).

Gérard et al. (2013) combined HST and Cassini-UVIS spectral FUV observations and found that the mean primary electron energy ranges from less than 3 to ~10 keV. Variations of the ratio between H_2 emission spectrum and the H Lyman-alpha brightness are sensitive to changes in auroral electron energies < 20 keV, as a result of the acceleration processes at work (Tao et al. 2014). Applied to measurements along Saturn's main oval, this reveals a flux-energy relationship for the precipitating electrons exhibiting two different behaviours: electrons >5 keV are found to be consistent with the Knight (1973) relationship (which describes how energetic particles can be accelerated into the atmosphere are a result of the nonlinear change in field-aligned current with field-aligned potential, which is, in turn dependent upon the density and temperature of the electron population along the magnetic field lines), as at Earth and Jupiter, while electrons <5 keV do not appear to be accelerated, and instead are precipitated by magnetospheric plasma and wave-particle interactions (Lamy et al. 2013).

Although H_2 has weak emission from transitions that emit in the visible wavelength regime, these are very weak in comparison to visible emission from atomic hydrogen. Vibrational and rovibrational transitions could, theoretically, be measured from quadrupole H_2 emission lines in the near-infrared, with such emissions seen at Jupiter and Uranus (Trafton et al. 1988, 1999). No such emissions have been detected at Saturn, though the expected emission would be difficult to observe using the broad filters on Cassini-ISS.

Auroral Thermalization

About half of all inelastic collisions between precipitating energetic electrons and Saturn's upper atmosphere result in the ionization of H_2 that is initially in the electronic ground state level, producing both H_2^+

and secondary electrons. These processes ultimately produce H_3^+ molecular ions above the homopause close to the source of ionization, under a chain of processes described in Chapter 8.

Once H_3^+ is produced, it is quickly thermalized within the atmosphere of Saturn (Tao et al. 2011), so that collisional interactions with the neutral atmosphere result in H_3^+ having the same temperature as the surrounding atmosphere. This excites both vibrational and rovibrational states of H_3^+, which then de-excite to produce a thermalized distribution of emission lines. At Saturn, where typical thermospheric temperatures range between 400–600 K, these emissions are almost entirely dominated by the H_3^+ υ_2 fundamental transitions ($\upsilon_2 \rightarrow 0$) that emit at between 3.5 and 4 μm, as shown in Figure 7.2, since ~2-μm overtone ($2\upsilon_2 \rightarrow 0$) and 3.5 to 4 μm hotband ($2\upsilon_2 \rightarrow \upsilon_2$) transitions are not well populated below ~1000 K. Since the thermosphere is directly heated by particle precipitation and by the Joule heating associated with auroral currents closing through the ionosphere (see Chapter 9), the resulting increase in H_3^+ emission caused by these raised temperatures should be considered as another aspect of Saturn's overall auroral output.

The H_3^+ density is closely related to the ionization rate, and thus will have a similar auroral morphology to both H and H_2 aurora. However, since the total emission from these molecules varies as a function of both temperature and density, significant differences can be expected between the infrared and ultraviolet/visible aurora. Typically, variations in temperature dominate the H_3^+ emission, but occur over relatively long temporal and spatial timescales, while the localized H_3^+ auroral morphology is controlled by the density of H_3^+. As a result, variations in the H, H_2 and H_3^+ aurorae should be comparable on the small scale (Melin et al. 2011), while broader regions of emission, and changes in emission on longer timescales, are more likely to differ.

At Jupiter, highly energetic heavy ions precipitate into the atmosphere, stripping electrons from the precipitating ions through charge exchange with atmospheric particles, which in turn should produce X-ray emissions (Gladstone et al. 2002). However, X-ray telescopes in orbit around Earth are not yet sensitive enough to allow such auroral X-rays to be detected at Saturn (Hui et al. 2010; Branduardi-Raymont et al. 2013).

7.1.3 Altitude Profiles

Another significant difference in the emission structures that are measured within H, H_2 and H_3^+ aurorae come from the difference in emission altitude, as shown in Figure 7.1. The altitudinal locations of both H and H_2 aurora are dependent on the energy of primary precipitating electrons, with higher-energy electrons penetrating to higher H_2 densities, and thus deeper into the atmosphere. However, although these energetic electrons will ionize H_2, beneath the homopause any H_3^+ created is quickly destroyed through charge-exchange with hydrocarbons, preventing any H_3^+ emission. In addition to this, H_3^+ cannot be generated at high altitudes, where the H_2 density is too low. At the top of Saturn's atmosphere, where atomic hydrogen dominates, significant auroral H Lyman and Balmer series emission can be generated without the H_2 ionization that ultimately produces H_3^+. As such, the altitudinal range of infrared H_3^+ auroral emission is limited to a roughly Gaussian distribution, rising from a negligible level close to the homopause (Tao et al. 2011), peaking at 1100 km above the one bar surface, and dropping away completely by 2000 km (Stallard et al. 2010). In comparison, while nightside UV auroral emission also peaks at 1100 km, the overall distribution of emission with altitude follows a broader Kappa distribution, extending emission to higher altitudes than is seen with H_3^+ emission. In addition, measurements of the extent of hydrocarbon absorption in UV auroral spectra show that in some cases, the peak H_2 auroral emission occurs at or above 610 km (Gerard et al. 2009). This later result is further supported by analysis of Saturn's auroral colour ratio and electron energy by Gérard et al. (2013), which uses Monte Carlo simulation of the electron precipitation process to show that the main emission appears to typically precipitate below 800 km, with the auroral peak as low as 640 km in some cases. The relatively deep penetration of electrons seen by Gérard et al. (2013) could indicate that the spectra studied were atypical, or might reveal a day-night difference, as the deep spectra were all observed on Saturn's dayside.

Direct imaging of the visible aurora "curtain" near Saturn's limb show emissions over a wide range of altitudes, between ~300 and 1500 km above the visible horizon (Dyudina et al. 2016). Sequences of filtered images taken by Cassini/ISS camera provided information about the visible spectrum of Saturn's aurorae. Among other filters, the aurora was detected in red, green and blue filtered images, allowing the true colour of the aurora to be determined. It was found that the auroral curtain changed from pink at a few hundred km above the horizon to purple at 1000–1500 km above the horizon (Dyudina et al. 2016), indicating a relative decrease in green light emitted at high altitudes. Colours can differ due to changes of atmospheric density, the chemical state of the element, and the energy of impacting particles, though some reddening of the aurora near the horizon may be due to atmospheric scattering of the light on the way to the observer.

The presence of aurora occurring below ~800 km suggests that this region may be important for atmosphere-ionosphere-magnetosphere coupling.

This aurora indicates that highly energetic electrons, with a mean energy of >3 keV (Gérard et al. 2013) or >10 keV (Galand et al. 2011), precipitate below the homopause. These are likely to drive the production of as-yet unobserved hydrocarbon ions to the point where these ions dominate the ionosphere below the homopause (Mueller-Wodarg et al. 2012). While this region does not contribute significantly to the total number density of the entire ionosphere, the relatively small number of ions in this region interact much more strongly with the high H_2 density in this altitude region, so that the Joule heating in this region will be comparable with that near the H_3^+ peak. As such, this region is important when considering atmosphere-ionosphere-magnetosphere coupling, despite the relatively low charged particle number densities in this region.

7.1.4 Auroral Morphology

While the infalling particles generate auroral emissions in a relatively narrow region of the upper atmosphere, the currents that drive these particles vary significantly with both latitude and longitude, so that when observed from above, a wide array of different auroral features can be observed, as outlined in Figure 7.3. These include (1) a highly fragmented main (ring of) emission; (2) cusp emission; (3) small-scale spots and arcs; (4) poleward auroral arcs; (5) bifurcations; (6) poleward auroral spots; (7) signatures of injections; (8) outer emission; (9) nightside polar arcs; (10) infilled polar cap aurora; (11) Enceladus footprint; and (12) Enceladus auroral oval.

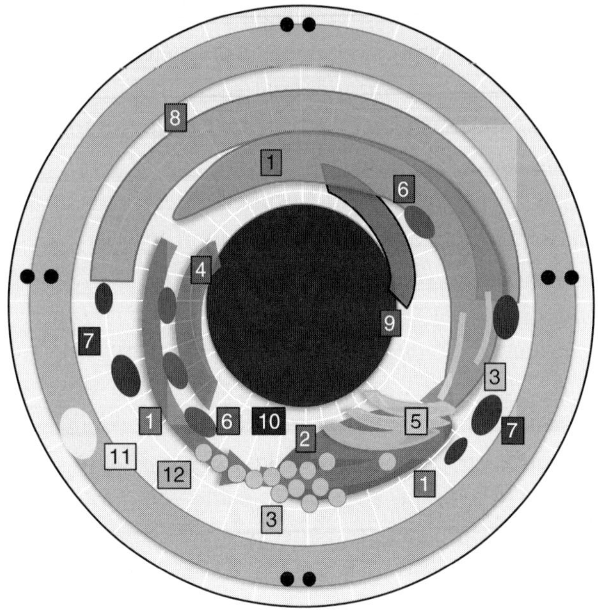

Figure 7.3 A sketch of the typical auroral components observed in Saturn's polar region (adapted from Grodent 2014, Fig. 4). The local time polar map displays parallels and meridians every 10°. These features include (1) a highly fragmented main (ring of) emission; (2) cusp emission; (3) small scale spots and arcs; (4) poleward auroral arcs; (5) bifurcations; (6) poleward auroral spots; (7) signatures of injections; (8) outer emission; (9) nightside polar arcs; (10) infilled polar cap aurora; (11) Enceladus footprint; and (12) Enceladus auroral oval. (A black-and-white version of this figure appears in some formats. For the color version, please refer to the plate section.)

This complex array of auroral features is produced by precipitating particles that are in turn driven by current systems that originate from a variety of different interactions between the ionosphere and magnetosphere. In the following sections, we will discuss each of the three major sources for these currents, detail the individual auroral features produced and relate these aurorae back to the magnetic interactions that generated them.

7.1.5 Ionospheric Electrical Conductance

Currents in a strongly magnetized ionosphere, such as Saturn's, can flow along the magnetic field (Birkeland currents), as well as perpendicular to the magnetic field (Pedersen and Hall currents). Ion and electron mobility in each direction at a particular altitude is represented by its electrical conductivity; the height-integrated conductivity is called conductance. Variations in the ionospheric electrical conductances can significantly affect a range of auroral processes, including Joule heating, plasma flows and magnetosphere-atmosphere current systems (e.g. Cowley et al. 2004, 2005). A thorough summary of ionospheric electrical conductances at Jupiter and Saturn is given in section 2.2.2 of Badman et al. (2015), including a table of Pedersen conductance estimates from the literature.

Pedersen conductances have been estimated from radio occultations at Saturn by combining observed electron density altitude profiles with modelled ion and neutral parameters. Based on Cassini radio occultations, Moore et al. (2010) estimated a range of Pedersen conductances between ~0.02 and 7.5 mho, though at non-auroral latitudes. They further demonstrated that, as conductivities are dependent on ion mass and maximize in the lower ionosphere, calculations that do not include the complete hydrocarbon ion chemistry will overestimate the associated conductances by ~0.06–0.6 mho.

Subsequent comprehensive ionospheric calculations focused on electron precipitation in Saturn's auroral region, including its effect on ionospheric electrical conductivities (Galand et al. 2011). They found that Pedersen conductances increase with the mean energy of the precipitating electrons, though only up to a value of ~10 mho near 2–3 keV. More energetic particles deposited most of their energy below the Pedersen layer, leading to a smaller Pedersen conductance. Hall conductances, meanwhile, which are generated at lower altitudes, continue to increase for electron mean energies at least up to 20 keV (reaching 11.5 mho), the maximum energy considered. Galand et al. (2011) also found conductances to be proportional to the square root of the energy flux of the incident electrons, with a slight delay in the ionospheric response to changing fluxes. This delay – approximately 10 minutes for Pedersen conductances and 4 minutes for Hall conductances – was attributed to the differences in chemical timescales associated with the altitude regions over which Pedersen and Hall conductivities peaked.

More recently, an ionosphere-magnetosphere coupling study (Ray et al. 2012) and a global upper atmosphere thermosphere-ionosphere global circulation model (Müller-Wodarg et al. 2012) have also been used to derive ionospheric electrical conductances at high latitude. Ray et al. find a range of Pedersen conductances between 3 and 18.9 mho across a range of latitudes and local times, while Müller-Wodarg et al. (2012) calculate a diurnal variation auroral Pedersen conductance of ~5–16 mho. These conductances, as with those discussed previously, depend intimately upon a few key assumptions, including precipitating particle energies and fluxes, and H_2 vibrational populations. It is therefore difficult to compare the various modelled Pedersen conductances directly. The range of values, however, seems to indicate that Saturn's ionosphere-magnetosphere coupling system is not expected to be limited by the ionospheric Pedersen conductance (Ray et al. 2012).

7.2 Solar Wind Interactions

It has long been recognized that Saturn's aurorae are strongly influenced through interactions with the solar wind. However, unlike Earth's aurorae, the current systems that drive these aurorae are also strongly modulated by the rapid rotation of the planet, resulting in significant differences in the solar wind interaction with Saturn's magnetosphere. In addition, as we shall show in later sections, there are other significant sources of aurorae at Saturn. However, as with Earth, the auroral phenomena that are driven by the Sun will occur in relation to the direction of the Sun, while aurorae driven by internal magnetospheric processes will typically occur in the rotational phase of the planet. As such, the best way to investigate the solar-wind-driven aurorae is to measure the aurorae fixed in local time, assessing changes seen in the aurora that re-occur in the same place, relative to the direction of the Sun.

Analysis of the average H_3^+ and UV auroral emission, fixed in local time, have shown that the location of the northern H_3^+ main emission (Badman et al. 2011a) is co-located with the northern UV aurora observed by the Hubble Space Telescope (HST; Nichols et al. 2009), indicating that they are driven by the same field-aligned current system. In the north, the equatorial boundary of the UV aurora was reasonably fitted by a circle of radius ~74°N, offset by 1.6° towards the nightside. Visible aurora observed in the dusk-midnight sector is also offset towards the nightside both in the southern and northern hemispheres (Dyudina et al. 2016). The co-location of the UV and IR main emission was confirmed using simultaneous observations of the H, H_2 and H_3^+ main auroral arc at 73°S in the pre-dawn sector made by Cassini UVIS and VIMS by Melin et al. (2011; shown later in the chapter in Figure 7.8). The co-location of the main arc at 73°S near midnight in UV, IR and visible was observed by the same instruments plus Cassini ISS (Melin et al. 2016).

These individual observations provide a view of the typical auroral structure, but in order to understand the mean auroral brightness, and thus the strength of the auroral currents, it is necessary to observe local-time averaged auroral intensity. Observations by HST at Saturn equinox in 2009 showed the average power emitted by the northern UV auroral region to be 17% greater than in the south (Nichols et al. 2009). This was attributed to the linear dependence of the emitted UV power on the ionospheric field strength in regions of field-aligned current, assuming equal ionospheric conductivity, where the magnetic field is known to be stronger in the northern ionosphere than the southern due to the quadrupole term of Saturn's internal field (Dougherty et al. 2005; Burton et al. 2010). Carbary (2013) conversely found the average and peak intensity of the UV northern aurora to be lower than that in the south using the UVIS data from 2007 to 2009.

In the near-infrared, the H_3^+ southern main oval was on average more intense than the northern main oval from pre-equinox observations (Badman et al. 2011b), as shown in Figure 7.4. Observations of the near-noon H_3^+ aurora in both hemispheres simultaneously using the Keck Observatory have shown that this trend continued after equinox, meaning that the effect was not simply a seasonal dependence on the ionospheric conductivity caused by solar EUV ionization (O'Donoghue et al. 2013). Instead, it was suggested that the thermospheric temperature, and hence H_3^+ total emission, could be higher in the south than the

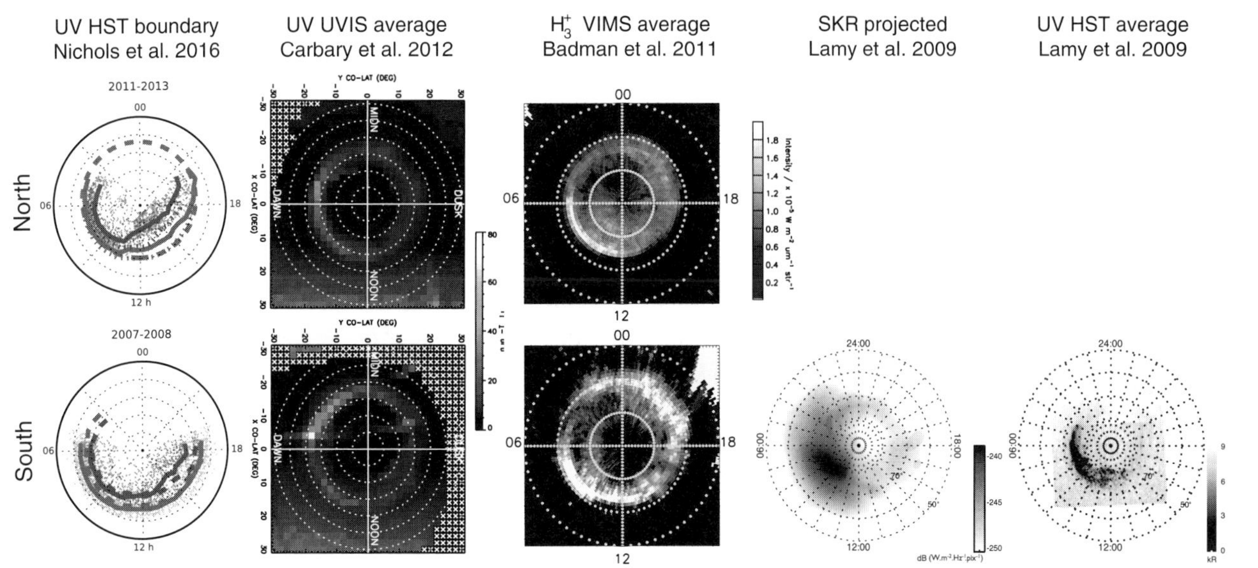

Figure 7.4 The local-time averaged aurora at various wavelengths in the northern (top row) and southern (bottom row) hemispheres, each presented with the sun at the bottom. First column: the distribution of locations for the UV auroral oval (taken from Nichols et al. 2016) from images obtained during ~60 orbits in 2007–2008 and 30 orbits in 2011–2013. Yellow and red points show the equatorward and poleward extent of the aurora, and the green and blue solid lines are the mean equatorward and poleward locations, respectively. Second column: composite bin-average maps of the FUV aurora in angular 2° × 2° planetographic co-latitude bins. Crosses indicate bins for which no data are included (adapted from Carbary et al. 2013, Figure 1). Third column: the average H_3^+ aurora created from northern winter images obtained between October 2006 and February 2009 (adapted from Badman et al. 2011b). Fourth column: the average SKR emission mapped in polar projection, recorded between 29 June 2004 and 30 March 2008, and projected down onto the planet (adapted from Lamy et al. 2009, Figure 4). Fifth column: the average UV emission combined from 383 individual images taken by HST in January 2004 and January 2007 (adapted from Lamy et al. 2009, Figure 4). (A black-and-white version of this figure appears in some formats. For the color version, please refer to the plate section.)

north because the ionospheric area subject to Joule heating from auroral currents would be larger in the south because of the weaker field strength present.

Interestingly, the high latitude auroral region inside the main oval has been observed to be more intense in the north than the south, the opposite trend to that observed for the oval emission itself (Badman et al. 2011b). This behaviour may be due to increased coupling and precipitation into the northern polar region under a preferential orientation of the interplanetary magnetic field. However, as the northern and southern observations used in this study were not simultaneous, it cannot be ruled out that a few exceptionally bright polar regions were by chance captured during northern imaging intervals but their southern counterparts were not. While the polar H_3^+ aurora is generally less intense than the main emission, the difference is less in the H_3^+ aurora than in the H and H_2 emission, where the polar aurora is typically much fainter than the main emission (Stallard et al. 2007a; Lamy et al. 2013).

Based on ground-based observations, the ion wind flows across the auroral region have shown that, on average, H_3^+ ions in the polar region sub-rotate at only one-third of Saturn's rotation rate. These values are close to the prediction of Cowley et al. (2004) that ion-neutral collisions within Saturn's atmosphere would drag ions attached to open field lines that should be fixed in the inertial frame of the Sun into only ~25% of co-rotation. More detailed analysis of the ion winds revealed that, while the majority of the polar region significantly sub-rotates, a region of co-rotation is often observed between the dawnward edge of the auroral oval and a point typically close to the center of the oval (Stallard et al. 2007a, 2012c). Milan et al. (2005) suggest that the twisting of Saturn's tail field lines would result in a "core" of field lines at the center of each of the magnetotail lobes shielded from tail reconnection by the surrounding field lines. This co-rotating region of the ionosphere could be linked to field lines embedded in the center of Saturn's magnetotail, which are shielded from the solar wind such that their rotation is controlled only by the neutral atmosphere (Stallard et al. 2007b).

The intensity of Saturn's main auroral emission displays a local time dependence, indicating an influence exerted by the external solar wind, as shown in Figure 7.4. The aurora generally takes the form of an arc on the dawnside, with more patchy forms around noon and dusk, and additional arcs occasionally extending poleward in the noon-to-dusk sector. The average location of the southern UV aurora was determined from HST images taken during 1997–2004, showing that the typical location of the aurora was an arc of latitudinal width ~2° centred on ~75°–76°S at dawn, shifting and broadening poleward around noon due to cusp signatures, and occurring at lower latitude ~74°–75°S at dusk (Badman et al. 2006). The location of the aurora, particularly the poleward boundary, varied between 67°S and 88°S overall, implying that the auroral source region in the magnetosphere is also highly spatially variable.

The inner and outer edges of the main auroral emission have been re-examined across a broad set of observations, between 2007 and 2008 in the South and between 2011 and 2013 in the North (Nichols et al. 2016). This analysis, shown in Figure 7.4, reveals that the typical shape of the main emission is not well resolved by a fitted circle, since the main emission has a wide variety of morphologies, and the region between noon and dusk is significantly poleward of the dawn-noon sector.

HST and Cassini observations between 2004 and 2007 (Lamy et al. 2009; Figure 7.4) showed that the typical source locations of both the UV aurora and SKR emissions were on the same field lines. Local time profiles of the SKR and UV auroral intensity were also shown to be similar, peaking at 08:00 LT and significantly decreasing in the pre-dusk region. Lamy et al. (2009) concluded that the radio and UV emissions were triggered by the same electron beams precipitating from the outer magnetosphere. Later analysis of simultaneous UV, IR and radio auroral emissions observed by Cassini showed co-located emissions, the intensity of which peaked on the dawnside but was further modulated by an active region rotating at the planetary-period, and the occurrence of a sub-rotating injection from the magnetotail (Lamy et al. 2013). Nichols et al. (2016) concluded that there is a good correlation between SKR power and power input to the UV aurora. A statistical analysis of the UV auroral location in the nightside region was made using pre-equinox observations provided by UVIS (Carbary

2013; Figure 7.4). This study found that the midnight sector aurora lies on average at 74°N and 73°S and that the UV aurora can be fitted by circles with offsets of 2 to 3 degrees in the direction of ~02 LT. These results are generally consistent with previous observations (Badman et al. 2006; Nichols et al. 2008), though they do show a significant brightening in southern dawn-midnight sector not seen elsewhere.

A series of crossings of the nightside auroral currents during Cassini's 2008 inclined orbits was examined (Talboys et al. 2011), comparing the location of the currents determined in the magnetic field data with the concurrent plasma measurements to determine to which region of the magnetosphere the currents mapped. This demonstrated that the upward current associated with the main auroral emission was encountered on field lines with plasma characteristics of the outer ring current in this local time sector, in rough agreement with the conclusions of Belenkaya et al. (2014). In the pre-noon region, however, an intense upward current was detected at the open-closed field line boundary, extending onto outer closed field lines (Talboys et al. 2009). This, combined with near-simultaneous imaging of the auroral oval, demonstrated that the near-noon auroral arcs are associated with field-aligned currents at and adjacent to the open-closed field line boundary (Bunce et al. 2008; Badman et al. 2012b). The upward currents associated with the auroral emission were bounded by a current directed downward into the ionosphere at higher latitudes (mapping to the open field region) and sometimes also by a downward current at lower latitudes (mapping to the ring current) (Talboys et al. 2009, 2011). The downward current regions have been shown to play an important role in the energetics of the auroral region via the detection of high-energy electrons and ions (e.g. Mitchell et al. 2009; Badman et al. 2012b).

The persistent local time asymmetry of the main auroral intensity implies that the current system driving the aurora is stronger on the dawn side than on the dusk. The auroral morphology observed by the HST has been compared with an IMF-dependent paraboloid model of the magnetospheric field (Belenkaya et al. 2014). The resultant auroral mapping and its unresponsiveness to the concurrent IMF conditions suggests that the dawn auroral arc is associated with field-aligned currents driven by pressure gradients and/or flow shears associated with co-rotation breakdown in the middle and outer ring current. The ring current is fed by injection of hot plasma from the nightside, which, because of the direction of planetary rotation, will preferentially enter on the dawnside, as has been observed in Cassini/ENA imaging (e.g. Mitchell et al. 2009). This explains the enhancement of auroral intensity on the dawnside compared to dusk. The dawnside auroral forms sometimes broaden to higher latitudes, suggesting the extension of the field-aligned currents into the outer magnetosphere region. This description is a modification of the theory proposed by Cowley et al. (2004) in which the upward current at the poleward boundary of the main aurora lies at the open-closed field line boundary, as the more recent studies have shown. A local time asymmetry arises as anti-sunward solar wind flow drags the field lines uniformly across the polar region, against the direction of planetary rotation on the dawnside and with planetary rotation on the duskside.

In addition to the persistent local time asymmetry associated with the solar wind interaction with Saturn's magnetosphere, the aurorae also display a strong response to the arrival of solar wind compressions (e.g. Clarke et al. 2005). This is primarily manifested as an intensification and expansion to higher latitudes of the UV auroral emission, as shown in Figure 7.5, accompanied by intensification and extension to lower frequencies of the SKR (e.g. Grodent et al. 2005; Clarke et al. 2009). These observations have been interpreted as the response to compression of the magnetotail, which enhances instabilities therein and instigates tail reconnection (Cowley et al. 2005). The field-aligned currents that result from the reconfiguration of the magnetotail field cause enhanced emission from the conjugate nightside ionosphere (Jia et al. 2012; Jackman et al. 2013), while the heating of plasma on the field lines contracting towards the planet intensifies the ring current in the midnight sector (Mitchell et al. 2009).

If the tail reconnection proceeds onto open field lines (Thomsen 2013), the aurorae are observed to contract to corresponding higher latitudes in the post-midnight sector of the ionosphere (Badman et al. 2016). In situ observations on one of Cassini's high-latitude passes revealed strong field-aligned currents and SKR emission at unusually high latitude and LT

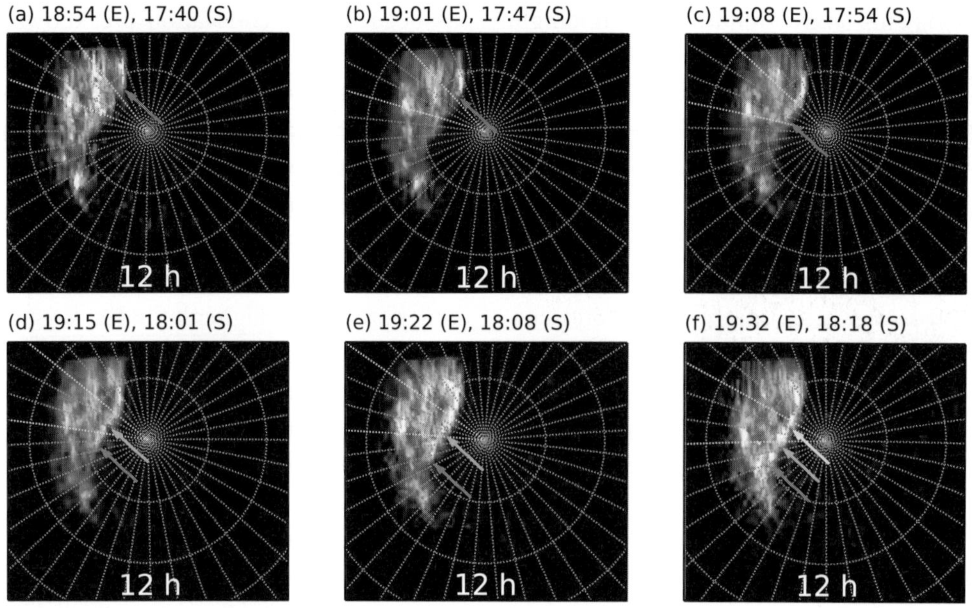

Figure 7.5 A dramatic dawn enhancement caused by compression in the solar wind. This sequence of evenly spaced HST images, taken on 5 April 2013, shows the development within an infilled dawn aurora over half an hour. The images are projected as viewed from above, with a 10° × 10° latitude-longitude grid overlain. The images are saturated at 80 kR. Arrows indicate the forward edges of three fast-propagating bursts of emission observed along the poleward edge of the aurora (from Nichols et al. 2014, Figure 1).

(79° at ~00 LT) which are suggested to occur under solar wind compression conditions (Bunce et al. 2010; Lamy et al. 2010). In general, the emitted auroral power (UV and radio) increases under solar wind compression conditions and the SKR extends to lower frequencies, consistent with sources at higher altitudes along the field line (e.g. Clarke et al. 2005; Kurth et al. 2005).

Observations of the H_3^+ aurora during the arrival of a major solar wind compression revealed the same morphological changes in the auroral structure as the UV aurora (Stallard et al. 2012d). In addition, ion wind velocities showed that this was accompanied by the loss of the open field line co-rotation region. This correlates well with the theory that this region of the ionosphere maps to an "old core" of magnetic field lines open to the solar wind, protected from reconnection due to the twisting in the magnetotail. These field lines could only be closed by a major compression, after which the polar ionosphere would begin to sub-rotate with the surrounding ionosphere. In addition, a >8-hour delay between the arrival of a major compression at Cassini and the resulting effect upon the aurora was observed, suggesting that reconnection must either occur well into the tail or that there are other processes in the chain of events that lead to the major dawn brightening seen in both these observations and previously studied UV images.

Inspection of UV auroral images taken during a solar wind compression interval in April 2013 (identified by enhanced SKR power and high energetic ion fluxes in the equatorial magnetosphere) revealed bursts of emission at the poleward edge of a broad, intense auroral feature, extending from post-midnight to dawn (Nichols et al. 2014; shown in Figure 7.5). These features propagated along the boundary at a speed of 330% of the planetary rotation rate, suggesting that they were related to the propagation of the field-aligned current associated with the onset of reconnection rather than plasma flows in the magnetosphere, which have only been observed at slower speeds (McAndrews et al. 2009). The location of the bursts suggests they are the signatures of open flux closure in Saturn's magnetotail, similar to the "poleward boundary intensifications" characterized in the terrestrial magnetosphere. In a separate event, a similar broad, intense auroral form on the dawnside was compared to the location of enhanced ENA emission from the ring current, indicative of hot plasma injected from the magnetotail.

The intense aurora has been associated with the upward current connected to the trailing region of the enhanced ring current region observed in the ENA (Nichols et al. 2014). A distinction is made, based on the relative location in both latitude and longitude of the peak ENA and auroral intensity, between this scenario and the relation of patchy, broad auroral emission located at ~70° latitude to filamentary field-aligned currents present throughout an enhanced ring current region that has previously been observed (Mitchell et al. 2009).

7.2.1 Small-Scale Features

High-resolution images of Saturn's main emission have revealed isolated features as small as 500 km across (Grodent et al. 2011; Dyudina et al. 2016; shown in Figure 7.6). The portion of main auroral emission, observed at high spatial resolution with the UVIS instrument, consists of distinct small-scale substructures. These auroral features roughly take the shape of elongated spots or narrow arcs. Up to 14 isolated spots were observed and arranged in a "bunch of grapes" configuration. Each spot is approximately shaped like an ellipse with full major axis of ~2600 km, assuming an altitude of 1100 km. Their brightness is relatively large, on the order of a few tens of kR, which translates to an emitted UV power of several gigawatts per spot. The spots are located in the dawn sector (observations were mainly restricted to the daytime hemisphere) and their latitude ranges from 75° to 80°. According to the magnetic field model of Saturn's magnetosphere described by Bunce et al. (2008), these northern latitudes map to the equatorial plane at distances between ~16 R_S and the dayside magnetopause. Their typical length scale roughly translates to ~4 R_S in the equatorial plane. Most spots appear to move at ~70% of rigid rotation. However, one isolated spot in the postnoon sector appears to move slower at ~50% of full co-rotation. This discrepancy was already observed in previous HST/STIS images (Grodent et al. 2005), where isolated auroral structures were shown to slow down from 65 to 20% of rigid rotation as they rotated from dawn to dusk.

Similar small-scale variability is seen in H_3^+ emission, and these spots and arcs have been directly associated with emission observed in the UV (Melin et al. 2011). Lamy et al. (2013) identified a UV hot spot moving at 65% of co-rotation, consistent with the simultaneous occurrence of an SKR arc. Since Lamy et al. (2008) requires field lines moving at 90% of co-rotation to correctly model SKR arcs, these results suggested that sub-rotating isolated features, colocated at different wavelengths, are common in auroral observations.

Cassini/ISS observations of the visible aurora between 70°S and 75°S revealed ~1000-km-sized clumps within nightside aurora, shown in Figure 7.6. These features typically co-rotate with Saturn, sometimes changing their speed to super- or sub-rotation by tens of percent (Dyudina et al. 2016). In sequences of images, at 30 km/pixel and 1-minute resolution, they appear as anticlockwise vortices extending poleward

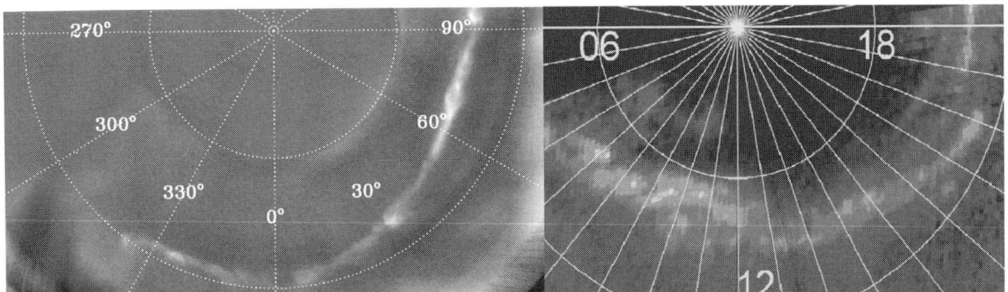

Figure 7.6 Small scale auroral features. On the left is a mosaic of projected images of the northern auroral emission within the FUV channel (111–191 nm) of Cassini/UVIS on 26 August 2008 (from Grodent et al. 2011, Figure 1). Based on observations acquired over 77 minutes, this reconstructed view is not an instantaneous snapshot, but provides instead a global view of small scale structure within the main emission with a spatial resolution close to 200 km. On the right, a Cassini/ISS mosaic of the visible night side images on 24 November 2012, in a rotating frame of Saturn (from Dyudina et al. 2016, Figure 3), also shows fine scale auroral structure in the main auroral oval.

from the main arc. When they extend poleward, they brighten and accelerate in prograde direction, consistent with shear flow with faster rotation at the polar side of the oval. The co-rotation and vorticity of these clumps is yet to be theoretically explained.

The fragmentation of the main ring of emission into small-scale spots and narrow arcs is likely the result of a complex spatial dispersion of field-aligned currents (FAC) driving energetic particles into the upper atmosphere of Saturn. Talboys et al. (2009, 2011) reported structuring of field-aligned currents, based on magnetic field perturbations observed with Cassini/MAG at high latitude near periapsis. They showed that the FAC pattern is regularly arranged in a set of upward current regions, giving rise to auroral UV emission, flanked by downward current layers unable to produce auroral emissions. Bunce et al. (2010) showed similar structures organized in reverse order, with a downward current layer bracketed by two upward FAC regions. Their size and location and the net current that they carry are fully compatible with the characteristics of the small spots and arcs observed with UVIS. Similar correspondences are found between FAC and radio emissions. Crossings of the SKR source region in 2008 resolved the auroral curtain, with measured widths between 900 and 1800 km, compared with the typical 100 km value at Earth (Lamy et al. 2013; Mutel et al. 2010). UVIS auroral observations have also identified small-scale auroral structures extending from 02 to 05 LT, indicative of a magnetotail reconnection event, perhaps associated with field-aligned currents building up in vortical flows in the magnetotail (Radioti et al. 2016).

These field-aligned currents may stem from the non-uniformity of the azimuthal plasma flow in the equatorial plane, with abrupt changes in the current layers marking strong velocity shears in the flows. As a result, the main ring of emission may not be uniform but more likely consists of substructures such as the spots and arcs that were observed with Cassini. One plausible explanation for these auroral spots could then be related to local fluctuations of the plasma flow, or vortices, induced by Kelvin–Helmholtz waves at the magnetopause and the inner edge of the boundary layer (e.g. Masters et al. 2009; Grodent et al. 2011). These vortices give rise to systems of field-aligned currents able to connect with the Kronian ionosphere. Magnetohydrodynamic simulations of Saturn's magnetosphere (e.g. Walker et al. 2011) suggest the occurrence of successive vortices with a characteristic length of several R_S, in agreement with the size and shape determined from the UV spots. However, according to Masters et al. (2010), the acceleration of electrons into the ionosphere associated with these upward field-aligned currents should produce an auroral footprint one or two orders of magnitude below the actual brightness of the spots. An alternative explanation was proposed by Meredith et al. (2013) who suggested that larger scale patches of auroral emission appearing simultaneously at the level of the main emission in both hemispheres, as observed with HST, may be similar to the very small-scale spots observed with Cassini. These features were shown to be consistent with field-aligned currents associated with ultra-low-frequency field line resonance waves propagating eastward through the equatorial plasma and driven within the magnetosphere by drift-bounce resonant interactions with trapped magnetospheric particles.

7.2.2 Polar Emissions

One significant polar auroral feature observed repeatedly at Saturn is the bifurcation of the main emission in the dusk sector, with individual arcs of emission splitting off and moving poleward as the aurora rotates, as shown in Figure 7.7. These features begin with a brightening on the main oval, close to noon, which is possibly associated with dayside reconnection opening field lines to the solar wind. The main emission splits into multiple arcs of emission as the aurora rotates towards dusk, with one end of each arc travelling poleward, while the end connected to the main oval sub-rotates. The region of open flux was observed to increase when these bifurcations appeared, and their morphology implied an overall length of the reconnection line of ~4 hours of local time and suggest that dayside reconnection at Saturn can occur at several positions on the magnetopause consecutively or simultaneously (Radioti et al. 2011, 2013b; Badman et al. 2012a, 2013).

Observations of the bifurcations of the H_3^+ main emission were directly associated with upward field-aligned currents, measured above this region by Cassini as it passed through field lines that mapped directly down into the aurora, providing the first in situ

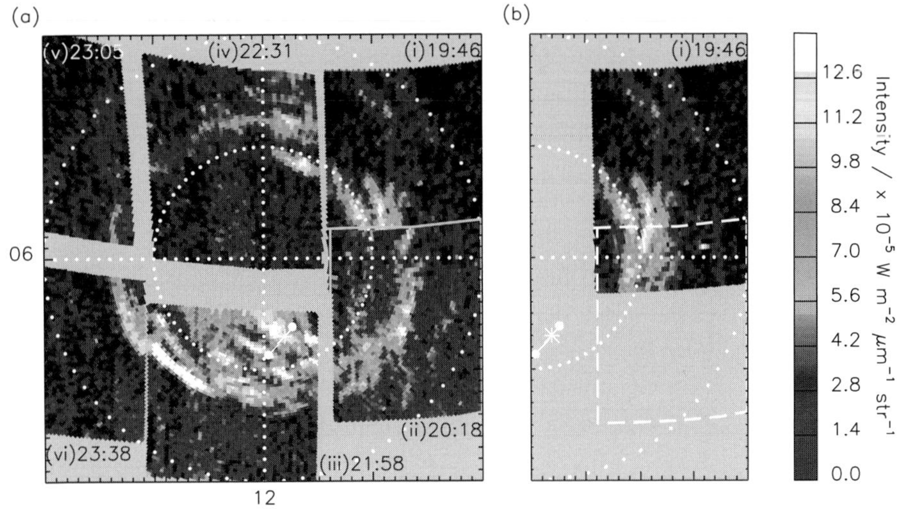

Figure 7.7 Bifurcations in Saturn's main emission. A mosaic of six Cassini/VIMS images taken 15 November 2008 between 19:46 and 23:38 UT. The main oval is clearly bifurcated close to noon, at the bottom of the mosaic. The white grid marks latitudes at intervals of 10° and the noon-midnight and dawn-dusk meridians. Also shown (right panel) are changes in the area marked by the white dashed line (from Badman et al. 2012a, Figure 2).

identification of transient reconnection events at regions magnetically conjugate to Saturn's magnetopause. This reveals the bifurcated auroral arcs as the ionospheric signature of bursts of reconnection occurring at the dayside magnetopause (Badman et al. 2012a). Observations within adjacent downward current regions showed upward bursts of 100–360 keV light ions, energetic (hundreds of keV) electrons and broadband, upward-propagating whistler waves.

The acceleration of the light ions from low altitudes was attributed to wave-particle interactions in the downward current regions (Badman et al. 2012a). The bifurcations of the main emission occur when the magnetosphere is strongly compressed with a high magnetosheath field strength, suggesting that this bursty reconnection can result in significant solar wind-driven flux transport in Saturn's outer magnetosphere (Badman et al. 2013).

Another polar auroral feature, revealed during high-inclination Cassini orbits, is a "polar arc" attached to the quasi-continuous nightside sector of the main emission that separates the region poleward of the main emission into two compartments dawnward and duskward of the arc (Radioti et al. 2014). This polar arc extends close to the dayside portion of the main emission but never reaches it. It has a brightness up to 21 kR, comparable with that of the main emission. The feature co-rotates where connected with the main emission, while its poleward portion moves slowly dawnward at a latitudinal rate of ~1° per hour. Milan et al. (2005) suggested that Earth's transpolar arcs are formed by magnetotail reconnection closing magnetospheric lobe flux, resulting in a build-up of closed flux in the tail and the formation of a transpolar arc (Fear and Milan 2012). The large differences between Saturn's and Earth's magnetospheres, such as Saturn's centrifugal effects, may explain the paucity of this feature in Saturn's aurorae, implying that the conditions for build-up of closed flux are rarely satisfied in Saturn's magnetotail.

Bright H_3^+ emission has also been observed to fill Saturn's polar cap, with broad emission across all longitudes at >82°N (Stallard et al. 2008d). Such emission is relatively rare and has never been observed in other wavelengths, suggesting it could be driven by low-energy particle precipitation or by localized heating within the thermosphere.

7.3 Effects from Magnetospheric Interactions

In addition to the aurorae generated by interactions with the solar wind, Saturn also has a significant variety of auroral features generated by plasma that then co-rotate with the magnetic field lines within the magnetosphere. This results in auroral features that typically move approximately in co-rotation with the planet, though since they are controlled by magnetospheric plasma, they are often observed to move significantly within the planetary frame. Emission on the main auroral oval and auroral hiss is also observed to vary with a 1-hour periodicity (Dyudina et al. 2016; Mitchell et al. 2016), which is likely to be associated with other 1-hour periodicities, described in more detail in Chapter 5. The processes that drive this 1-hour periodicity remain unexplained, though, given their regularity, the oscillation involved is likely to originate from within Saturn's magnetosphere or atmosphere.

7.3.1 Energetic Particle Injections

Cassini measurements provided evidence for the ubiquitous occurrence of energetic particle injections in the inner magnetosphere (e.g. Mauk et al. 2005) similar to those occurring at Jupiter, confirming that Saturn's inner magnetosphere is also rotation dominated.

These injections are likely powered by centrifugal interchange events in the inner magnetosphere in which distant magnetic flux tubes containing hot, tenuous plasma convect inward to replace flux tubes containing dense, cold plasma. The rapid radial transport preserves the particle adiabatic invariants for the gyration and bounce motions. Subsequently, the particles drift azimuthally as a result of rotational electric field drifts, magnetic gradient, and curvature drifts. According to Chen and Hill (2008), these small-scale injections/dispersion events are observed in all local times with a preference in the pre-noon quadrant between 6 and 8 R_S, although Kennelly et al. (2013) found that the post-noon and midnight sectors are favoured for injection event creation. Their lifetime is estimated to be, on average, smaller than 2 hours, and their spatial scale is < 1 R_S, with the majority of the events being under 0.4 R_S. Chen and Hill (2008) further found that the hot injected flux tubes only occupy about 5 to 10% of the full 360° of longitude, demonstrating that the interchange occurs via rapid injection of narrow channels of hot, outer magnetospheric material separated by much broader, slower outflow of cold, inner magnetospheric material.

Auroral structures equatorward of, and well separated from, the main emission have been investigated using HST (Radioti et al. 2009) and the extensive Cassini/UVIS dataset (Radioti et al. 2013a). The expected auroral characteristics produced by plasma injections were compared with observed UV emissions, and showed that pitch angle diffusion and electron scattering in the injection region could well reproduce the observed longitudinal elongation of the features and their decay in brightness. This suggests that pitch angle diffusion plays an important role in generating these UV emissions. On the other hand, field-aligned currents driven by the pressure gradient along the boundaries of the injected plasma cloud seem to be the main contribution for the ENA enhancements observed simultaneously by Cassini/INCA (Mitchell et al. 2009). Since the ENA emission co-locates with the leading part of the corresponding UV emission, field-aligned currents could contribute to this portion of the UV emission. Additionally, auroral spirals propagating from midnight to noon via dawn have been observed (Radioti et al. 2015). These may be related to large-scale magnetospheric particle injections, creating regions with strong velocity gradients and producing a spiral shape in the auroral emissions.

7.3.2 Outer Emission

A faint but discrete partial auroral ring of emission exists equatorward of Saturn's main auroral emission, shown in Figure 7.8. This auroral feature appears at both poles, mostly in the nightside sector, mapping, according to the Bunce et al. (2008) magnetic field model to ~9 R_S (Grodent et al. 2010). In the current dataset, this auroral feature consists of one or more narrow arcs (~3° of latitude) of emission usually extending equatorward of the main emission from 18LT to 06LT through midnight, although a limited number of images show the emission extending down to 09LT. The emission is not uniform in longitude, as it contains persistent patches with clear contours moving at ~70% of rigid co-rotation (Grodent et al. 2005), similar to the main emission and compatible with a magnetospheric plasma source rotating at a distance between 7 and 10 R_S from Saturn. The typical brightness of the UV outer emission, corrected for limb brightening, is on the order of a few kR, just enough to be detected with HST, but largely sufficient to appear in Cassini-UVIS observations.

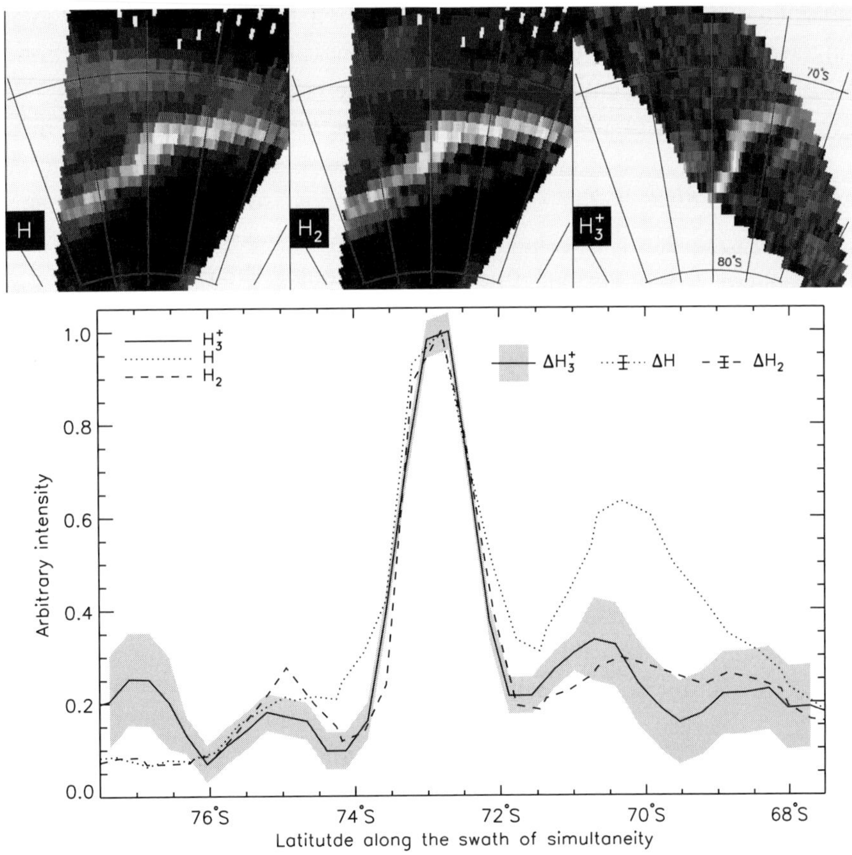

Figure 7.8 The outer aurora, at lower latitudes than the main auroral emission. Polar-projected images of the H Lyman-alpha and H_2 Lyman band emissions acquired by Cassini/UVIS, and of the H_3^+ emission acquired by Cassini VIMS are shown (top 3 panels), as contemporaneously observed on 10 September 2008. On the bottom is an intensity profile showing the co-incident H Lyman-alpha (dotted line), H_2 Lyman bands (dashed line) and H_3^+ (solid line) emission from these images. The shaded area shows the error on the H_3^+intensity, with the errors in H and H_2 indicated by respective error bars (from Melin et al. 2011, Figures 2 and 3).

Early observations suggested that the H_3^+ outer aurora is up to 25% as bright as the main auroral emission (Stallard et al. 2008d), yet more recent analysis has shown that the outer aurora is significantly brighter in H emission (122 nm) than either in the H_2 (123–166 nm) or H_3^+ (3.4–3.7 μm) (Melin et al. 2011; Figure 7.8). This suggests the energy was deposited at higher altitudes where the H_2 density is less. This observation limits the precipitating electron energy in this region to 100s of eV, as higher-energy electrons would penetrate deeper into the atmosphere (Gustin et al. 2009).

It was initially thought that pitch angle scattering of electrons into the loss cone by whistler waves would be responsible for the outer auroral emission, with a suprathermal electron population, consisting of a hot (a few keV) tenuous component and a cold (a few eV) dense component, observed with Cassini/CAPS (e.g. Schippers et al. 2008) in the nightside sector between 7 and 10 R_S, powering this process. However, more-detailed modeling has shown that an emission brightness of a couple of kR requires the injection of a few tenths of mW m^{-2}, comparable to the minimum energy flux that is available from the hot electron population of the inner plasmasphere (Schippers et al. 2008).

A more recent analysis, based on 7 years of Cassini electron plasma data, suggests an alternative mechanism for providing the power to the outer emission (Schippers et al. 2012). This study reveals the presence of layers of upward and downward field-aligned currents that appear to be part of a large-scale current system involving dayside-nightside asymmetries as well as trans-hemispheric variations. This system comprises a net upward (away from the planet) current layer, carried by hot electrons (> 400 eV) originating from equatorial regions at a distance in excess of 9 R_S and limited to the nightside sector, which may well generate the nightside outer UV auroral emission.

7.3.3 Enceladus Associated Aurora

The detection of the auroral footprint of Enceladus in Saturn's northern ionosphere (Pryor et al. 2011; Gurnett and Pryor 2011; shown in Figure 7.9), comparable with the auroral footprints of the Galilean moons Io, Europa and Ganymede at Jupiter, suggests that a universal magnetospheric mechanism connects conductive moons to their parent-magnetized planets' ionospheres. The Enceladus footprint was detected on several days with the UVIS imaging spectrograph onboard Cassini, but the spot is usually absent and/or below the UVIS detection threshold. The brightness of the best-observed spot was on the order of 1 kR, just enough to be detected with Cassini but not with the STIS and ACS UV cameras onboard HST. The dimension of the auroral spot suggests that the emission is not directly connected to Enceladus, but to a larger interaction region near the equatorial plane extending as far as 20 Enceladus radii (R_E) downstream, with a radial extent between 0 and 20 R_E, consistent with the extent of the water vapour plume resulting from Enceladus' cryovolcanic activity (e.g. Waite et al. 2006).

The brightness of Enceladus' auroral footprint was shown to vary significantly on that day, presumably in response to the time-variable cryovolcanism from Enceladus' south polar vents, suggesting that plume activity was exceptionally high at the time of this thus-far unique UVIS observation.

This footprint observation and its association with Enceladus were further supported by in situ signatures detected by Cassini, two weeks before the UV detection, in the form of azimuthal perturbation of the magnetic field, indicative of field line draping around the moon, and in magnetic-field-aligned ion and electron beams connected with Enceladus' interaction region, offset several moon radii downstream from Enceladus. The electron beam was consistently shown to carry enough power to produce the auroral footprint of Enceladus (of its interaction region) with a variable UV brightness peaking at several kR. Systematic monitoring of Enceladus' ultraviolet auroral footprint should provide evidence of the electromagnetic coupling of Enceladus with Saturn and of plume variability (Pryor et al. 2011).

Although the Enceladus footprint remains undetected in infrared observations, Stallard et al. (2008c, 2010;

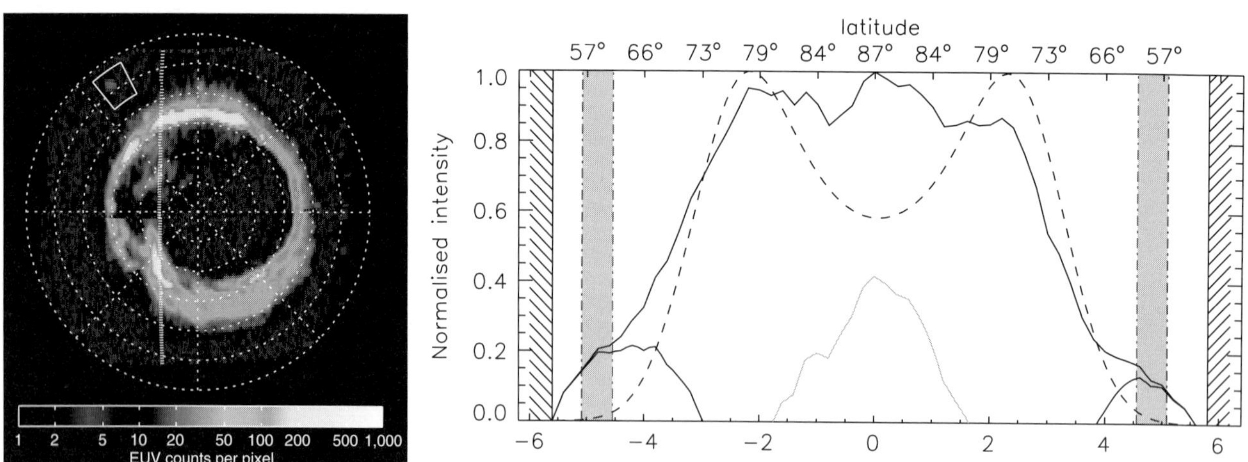

Figure 7.9 Enceladus-associated aurorae. On the left is a Cassini/UVIS image of Saturn's northern aurora, taken on 26 August 2008 (from Pryor et al. 2011, Figure 2). A white box, centred on 64.5°N and the sub-Enceladus longitude, highlights the predicted location of the Enceladus auroral footprint. On the right is a H_3^+ intensity profile (solid line), cutting east–west across Saturn's southern polar region. When compared with a simple model of the main emission (dashed line), a residual emission can be measured (bold line). The two equatorward peaks form a mid-latitude auroral oval, projecting to between 3 and 3.95 R_S (two dot-dashed gray lines, with gray infill; from Stallard et al. 2010, Figure 2).

shown in Figure 7.9) have identified and mapped an auroral oval well equatorward of the main emission to Enceladus' orbit (3.5–5 R_S). It was suggested that this oval could be associated with currents driven by co-rotation breakdown in the magnetospheric plasma close to Enceladus, since the location where observed ion winds within the ionosphere returned to co-rotation coincides with the observed emission. These currents would make this auroral feature analogous to the main auroral emission at Jupiter. Stallard et al. (2010) suggested that since no UV counterpart has been found for these emissions, they may be induced by the precipitation of electrons with low energies, and Schippers et al. (2012) identified a layer of net downward-directed thermal electrons with very low energy that could generate the secondary IR aurora. However, the detection of equatorial ionospheric features driven by interactions between Saturn's ionosphere and rings, described as "Ring-Rain" (O'Donoghue et al. 2013) shows that a strong H_3^+ brightening can occur when the ionosphere is magnetically mapped to a strong source of water group ions, a form of aurora covered in more detail in Chapter 8 of this book. As such, the H_3^+ auroral oval associated with Enceladus could instead be the strongest form of H_3^+ emission driven by water ion chemistry.

7.4 Effects from Planetary Period Oscillations

As we have seen, Saturn's auroral features can be generated at fixed local times, generated by interactions with the solar wind, or through currents produced by magnetospheric plasma sub-rotating away from the planet's magnetic field. However, in addition to these auroral features, Saturn also has aurorae that rotate much more directly with the planet, fixed to northern and southern planetary period oscillations.

In Chapter 5, these auroral features are overviewed in context with other modulations of the magnetosphere and atmosphere at the planetary rotational period. In this section, we will provide a more detailed description of auroral variability with planetary period oscillation, and then review how these are related to magnetospheric rotating current systems. We will then show the direct evidence for this modulation measured in field-aligned currents and show our current understanding of how these auroral features might be produced.

7.4.1 Rotational Modulation of Saturn's Aurorae

The apparent variation of Saturn's planetary period with time was first identified using measurements of Saturn's Kilometric Radiation (SKR). Details of how this was revealed, and how these periods relate to other planetary oscillations, are described in Chapter 5.

The SKR period was found to vary by 1% over years to weeks by successive long-term radio measurements of Ulysses in the 1990s (Galopeau and Lecacheux 2000) and Cassini after 2003 (Gurnett et al. 2005; Zarka et al. 2007). The tracking of the time-variable SKR period over years resulted in the development of extended SLS2 and SLS3 systems (Kurth et al. 2008, 2009). Cassini observations revealed that southern and northern SKR were in fact modulated at slightly different periods of ~10.8 and ~10.6 hours, respectively (Gurnett et al. 2009). The two radio periods were observed to roughly vary in anti-correlation over timescales of years from which were defined southern and northern SKR phases and SLS4 systems, until merging with each other in the year following equinox (Gurnett et al. 2010; Lamy 2011; Gurnett et al. 2011).

Cassini observations provided a major update in the interpretation of the SKR rotational modulation. As SKR sources are, similarly to atmospheric aurorae to which they are associated, found at all longitudes (Farrell et al. 2005; Cecconi et al. 2009; Lamy et al. 2009), the anisotropic beaming of radio waves makes the understanding of SKR dynamics non-trivial. Investigating the statistical locations of southern radio sources, derived directly through direction-finding methods (goniopolarimetry) from mid-2004 to mid-2010, Lamy (2011) showed that the (dominant) southern SKR modulation is an intrinsically rotating phenomenon, with the strong enhancement at dawn resulting in a strobe-like effect whenever the morning radio sources are visible, a result confirmed by an independent study of the phase relationship between SKR and magnetic oscillations over the same interval (Andrews et al. 2011). The latest simulations of SKR visibility correctly modelled the dynamic spectrum of a series of southern SKR bursts by using a rotating active region of the auroral oval, 90° wide in longitude, as observed simultaneously in the UV (Lamy et al. 2013).

The most striking evidence of a rotational modulation of H and H_2 aurorae was nonetheless provided by a series of studies based on intensive HST observations of Saturn combined with multi-instrumental measurements of Cassini. In a study of HST images collected between 2005 and 2009, the total power radiated by both southern and northern aurorae was found to vary with the corresponding hemispheric SKR phase (Nichols et al. 2010). The dawnside portion of the aurorae displayed the largest diurnal variation, varying by a factor of ~3 and peaking during SKR bursts.

By fitting the location of the southern auroral oval observed in Jan. 2007 and Feb. 2008 with circles, Nichols et al. (2008) also identified an elliptical oscillation motion of 1° to 2° with a period (~10.76h) close to that of southern SKR and magnetic field oscillations. The major oscillation axis was oriented along the pre-noon/pre-midnight direction. Looking at the phase relationship between the auroral oval motion and magnetic oscillations, Provan et al. (2009) showed that the oval is displaced in a direction opposite to the rotating equatorial perturbation field, namely sunward at SKR maxima. These results were verified by HST images of early 2009 obtained near equinox, simultaneously sampling northern and southern auroral regions (Nichols et al. 2010b). These confirmed dawn-dusk oscillations of both auroral ovals as a function of hemispheric SKR phases, with an amplitude of 1° to 2° and a displacement towards dusk lagging 90° behind the SKR bursts. Nichols et al. (2010) attributed the observed modulation of the brightness and location of UV aurorae to the rotating field-aligned currents systems described in Section 7.4.2, producing an effective tilt of the magnetic dipole (see Figure 7.10).

An independent statistical study performed with Cassini/UVIS measurements spanning the period from 2007 to early 2009 further confirmed the diurnal variation of the southern auroral oval brightness (by a factor of ~5) and location (by ~1°) in phase with the southern SKR modulation (Carbary et al. 2013). However, no comparable diurnal variation of the northern auroral oval at the northern SKR period was observed. A case study of Cassini/UVIS measurements over a two-hour interval illustrated a motion of the northern oval consistent with that of field-aligned currents tracked in situ, at the phase of the magnetospheric oscillation in the northern hemisphere (Bunce et al. 2014).

Similar studies of the dynamics of H_3^+ auroral emissions in the IR were performed using Cassini/VIMS observations. A statistical study covering the interval 2006 to 2009 revealed a clear rotational modulation of the intensity of both northern and southern aurorae at the corresponding magnetic phase systems, although not in phase with the expected position of upward currents (Badman et al. 2012a). A weak modulation of northern intensities by the southern magnetic phase system, similar to that observed for southern SKR, was also noticed.

In visible wavelengths, ~1000 km structures in the main oval nearly co-rotate with Saturn, as observed in 1-minute-resolution Cassini nightside movies (Dyudina et al. 2016). The persistence of nearly co-rotating large bright longitudinal structure in the auroral oval seen in two movies spanning 8 and 11 rotations gives an estimate on the period of 10.65h ± 0.15h for 2009 in the northern oval and 10.8h ± 0.1h for 2012 in the southern oval. These periods are broadly consistent with the SKR periods.

Finally, a multi-spectral analysis of two days of Cassini auroral observations at radio, UV and IR wavelengths illustrates that the rotational modulation of the southern auroral brightness at a given local time sector (see Figure 7.11) results from the rotation of an active region of the auroral oval (which also hosts sub-rotational features, as described in Section 7.4.2), which is significantly enhanced in the dawn sector (Lamy et al. 2013).

7.4.2 Origin of Rotating Auroral Currents

Modelling of these auroral features, along with the periodic variations within the magnetosphere, has revealed that the aurorae are generated by two independent rotating systems of field-aligned currents in the southern and northern hemispheres. Although these currents are independent of one another, each consists of a cross-polar current that is linked with the surrounding magnetosphere via field-aligned current, close to the same latitudinal region as the main auroral emission, that maps to the magnetospheric equator near radial distances of 15 R_S. This rotating current system has a layer of downward currents on one side of the pole and a layer of intense upward currents on the other. The latter drive energetic electrons into the

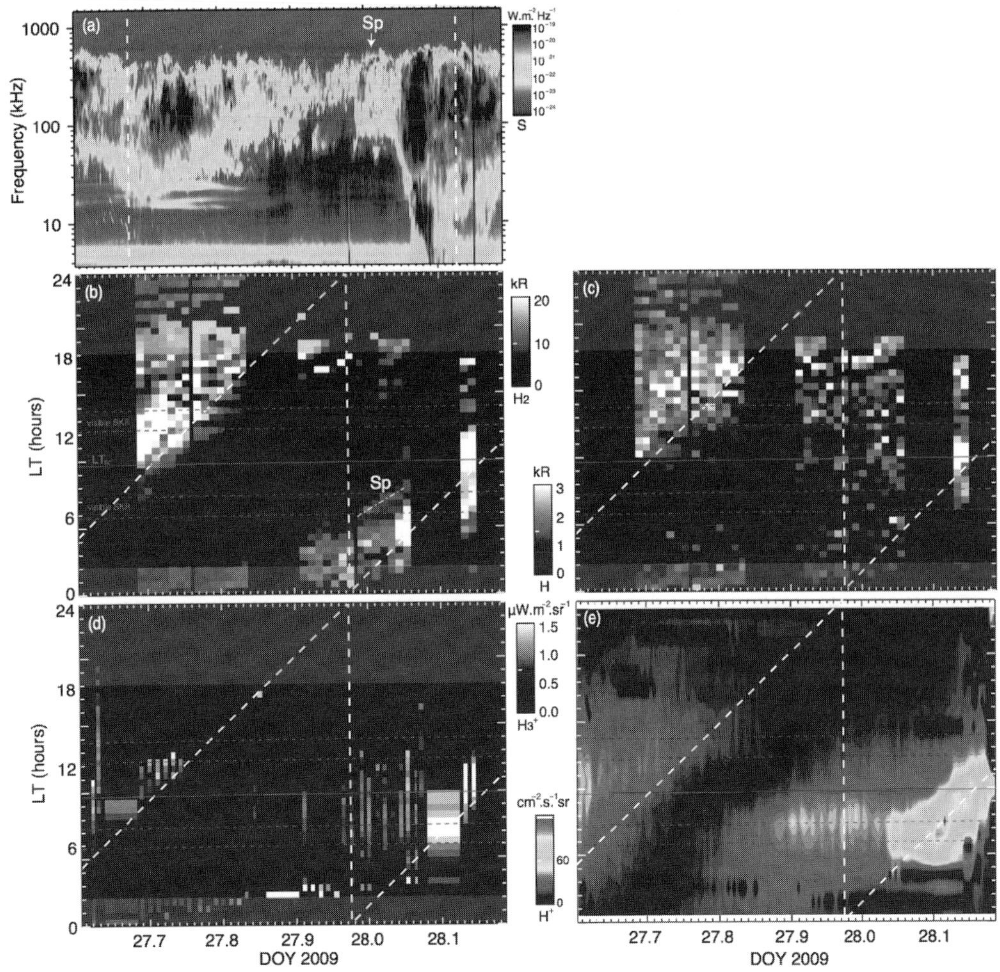

Figure 7.10 Simultaneous Cassini observations showing the clear rotational modulation of (a) the SKR emission measured by Cassini/RPWS, (b) the total H_2 auroral brightness measured by Cassini/UVIS, (c) the total H Ly-alpha brightness measured by Cassini/UVIS, (d) the total H_3^+ flux measured by Cassini/VIMS and (e) the average magnetospheric H^+ emission measured by Cassini/MIMI/INCA (from Lamy et al. 2013, Figure 2).

atmosphere, producing enhanced auroral emission, and are observed to pass through dawn during SKR maxima, therefore providing a plausible origin for radio bursts and UV/IR dawnside periodic auroral enhancements.

The hemispheric rotating field-aligned currents thus appear to trigger the rotational modulation of auroral emissions and other magnetospheric observables. A statistical study of field-aligned currents observed by Cassini as it passed repeatedly through Saturn's southern post-midnight auroral region has directly measured the currents associated with the southern planetary period oscillation (Hunt et al. 2014; shown in Figure 7.12). By measuring the auroral currents at a variety of planetary phases, both the phase-dependent and phase-independent currents were calculated. The phase-independent currents, most likely associated with solar wind interactions, show downward currents of the over-the-pole open field region, indicative of significant plasma sub-corotation, and an upward-directed current sheet co-located with the UV oval at latitudes of ~71° to 73°, mapping to the outer hot plasma region in Saturn's magnetosphere between ~11 and 16 R_S.

Overlain on this are currents that vary with the planetary period, overlapping the phase-independent currents at co-latitudes between ~17.5° and 20°. These currents are minimized at phases of 0° and 180°, produce significant downward currents at 90° and upward currents at 270°. This system of currents is exactly that predicted to flow into the ionosphere by current magnetospheric models and directly explains

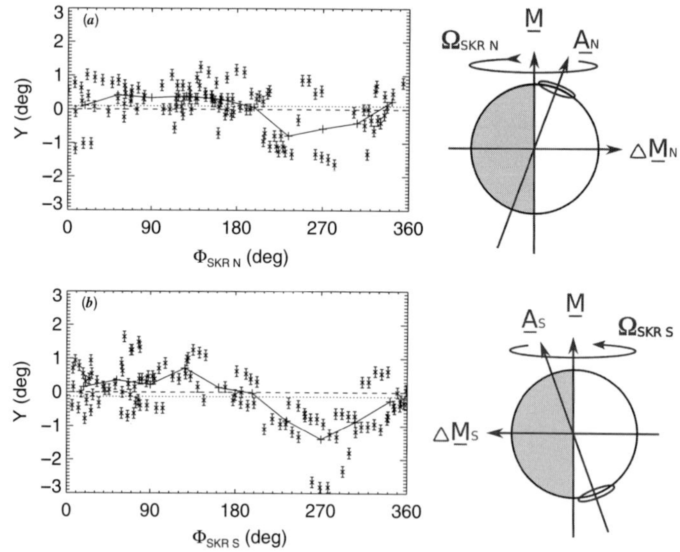

Figure 7.11 The rotational motion of the UV auroral ovals in phase with the planetary period oscillation. On the left is a schematic showing how two rotating perpendicular quasi-dipolar magnetic perturbations would affect the location of the northern (top) and southern (bottom) auroral ovals at respective SKR maxima. The vectors depicting planetary magnetic moment (M), northern and southern magnetic moments (ΔM_N, ΔM_S) and resulting northern and southern "auroral axes" (A_N and A_S) are shown. On the right are plots showing the center locations Y in degrees for the north (top) and the south (bottom) plotted versus respective SKR phases for all the images (crosses), with the mean values over 36° (solid line). The error bars indicate the 2-pixel uncertainty in the determination of the planet's center. The dashed horizontal lines indicate the location of the noon-midnight meridian at Y = 0, and the horizontal dotted lines indicate the mean locations of the center of the auroral ovals (from Nichols et al. 2010b, Figures 2 and 3).

the strong auroral brightening observed at a southern phase of 270°, as upward currents are matched by the downward acceleration of energetic electrons.

Because these two current systems close through the auroral region at similar latitudes, the overlaying of these two separate current systems should result in a complex sequence of reversing field-aligned currents, perhaps explaining the field-aligned currents observed by Cassini/MAG at high latitude near periapsis (Talboys et al. 2009, 2011).

The difficulty in driving the rotating current system from interactions with the magnetosphere strongly suggests that this system of currents is generated, instead, from thermospheric flows unconnected with magnetospheric processes, as discussed in Chapter 5. Hunt et al. (2014) show that the overall current layer co-latitude is also modulated with 1° amplitude in phase with the planetary period oscillations, reaching a maximum equatorward extent adjacent to the peak upward current at a southern phase of 270° and a maximum poleward extent adjacent to peak downward current at a southern phase of 90°. This implies that the rotating current system is driven by processes occurring within the atmosphere outwards, rather than directly from the magnetosphere inwards, though the origin of such current systems remains unexplained.

If this is the case, it would mean that Saturn is unique in having significant auroral emissions modulated by flows within the atmosphere; effectively, these would be uniquely weather-generated aurorae. However, the interaction of such thermospheric flow vortices with the surrounding ionosphere has been modelled to produce Joule heating so great that the resulting thermospheric temperatures would vastly exceed those that are observed (Smith 2014). One alternative could be that the neutral vortices occur within Saturn's upper stratosphere, around 750 km altitude, where Joule dissipation is significantly lower. The vortices would blow laterally across the Hall conductance gradient at the main auroral oval, generating field-aligned currents similar to those currently modelled at higher altitudes (Smith 2014). Unfortunately, no measurements of ionospheric, thermospheric or upper stratospheric wind systems, in phase with the planetary period oscillations, have been made at Saturn. Ultimately, such observations could reveal the source for Saturn's planetary period oscillations.

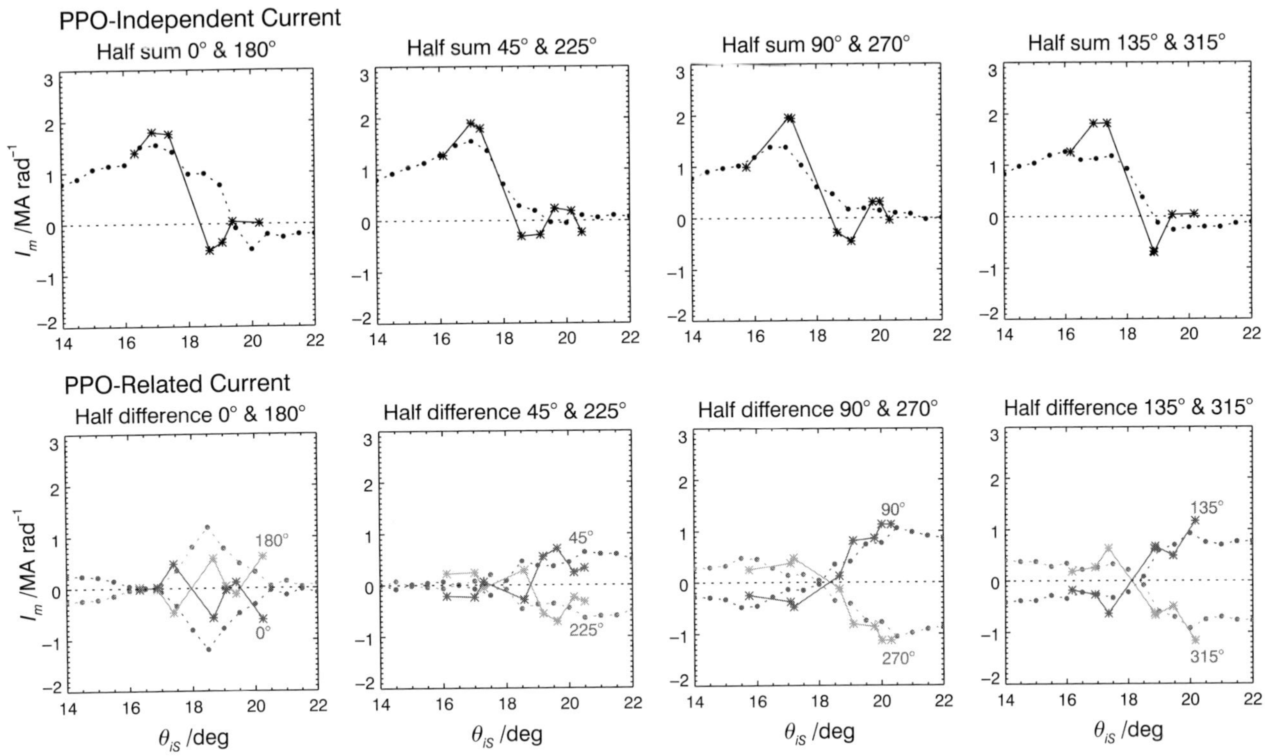

Figure 7.12 Profiles of auroral currents measured above Saturn's southern auroral region on the midnight side (from Hunt et al. 2014, Figure 14). Positive values indicate downward currents and negative currents indicate upward currents. Positive values drive electron precipitation in to the atmosphere, and thus produce aurorae. These currents are a combination of an axisymmetric current system associated with sub-rotation within the magnetosphere and a current associated with a twin-cell rotating at Saturn's variable planetary period. Combining the data in two ways reveals these individual components. The top row shows the mean currents for observations made with planetary periods that are directly out of phase; this cancels out the rotating current system, resulting in a measure of the currents independent of phase – the similarity of all four currents shows the consistency of this phase independence. The bottom row does the opposite, instead showing the differences in current for observations in planetary period anti-phase with one another; taking the difference subtracts out the axisymmetric current and leaves only the current that rotates with the planet, resulting in a measure of the planetary period dependent currents flowing into the ionosphere.

7.5 Conclusions

Our understanding of the aurorae of Saturn has greatly improved over the lifetime of Cassini, with the combination of in situ measurements above the planet, multiple wavelength observations of the planet from Cassini and the combined supporting observations from Earth producing a wealth of new information about both the aurorae and the processes that drive them. Despite this, significant questions remain about how Saturn's magnetosphere and ionosphere interact with one another. However, we expect a wealth of new data to provide essential insights into this problem in the next few years. In this section, we discuss the major outstanding questions, and outline how these might be answered.

7.5.1 Outstanding Questions about Saturn's Aurorae

The following list presents some of the most important questions about Saturn's aurorae that remain unanswered:

1. What are the origins of the current systems and acceleration processes responsible for main, polar or equatorward auroral emissions?

   a. To what extent do Kelvin–Helmholtz instabilities on the flanks of the magnetosphere drive spots of auroral emission in the planet, and what does this tell us about the magnetic interactions with that region?
   b. Is the H_3^+ emission seen at Enceladus the result of a current system formed by the breakdown in co-rotation, or is it the most dramatic example of infalling water ions, analogues to "Ring Rain"?

   c. Why is the entire H_3^+ polar cap aurora sometimes completely infilled, and is this feature unique to H_3^+ emission at Saturn?

2. Are the planetary period oscillations driven by the upper atmosphere of Saturn, and if that is the case, can we characterize and understand not only the shape and strength of this system, but ultimately what processes drive it?
3. How do sub-rotating dynamics (rotational modulation of the main auroral emission) coexist with co-rotating ones (auroral hot spots) along common magnetic flux tubes?
4. What process is driving the 1-hour periodicities seen in the visible and UV aurora, and can these periodicities be identified in radio and IR aurora?
5. How does the cyclotron maser instability mechanism operate in Saturn's auroral regions?

   a. What are the unstable electrons that are driving SKR?
   b. Is the emission saturated?
   c. What is the effect of wave propagation within the magnetospheric environment on the observed SKR polarization?

7.5.2 Cassini's Grand Finale and Beyond

To help answer these fundamental questions about Saturn's aurorae, we will need new investigations of Saturn's auroral region. Fortunately, in the next few years, the scientific community will produce a wealth of new data and analysis that will greatly improve our understanding.

The leading source of new data will come from Cassini, which will continue to orbit Saturn until 2017, providing more dedicated auroral observations that will greatly improve our understanding of the coupling between the atmosphere, ionosphere and magnetosphere that drives these aurorae. As Cassini enters the Grand Finale orbits in 2016 and 2017, it will enter high-inclination and low-altitude orbits over the auroral regions.

These orbits will provide new views of Saturn's aurora from very close to the planet, allowing Cassini to image the aurora at the highest spatial resolution at which Giant Planet aurorae have ever been observed, with resolutions of tens of km/pixel for IR and UV, and hundreds of meters/pixel in visible light (Melin et al. 2016). Cassini will fly directly through magnetospheric regions that carry the field-aligned currents that drive the aurora, allowing these fine-scale emission structures to be directly associated with the field-aligned currents that produce them. It will also provide several opportunities to fly through the SKR sources and particle acceleration regions (<~4 R_S) on the auroral field lines, allowing these regions to be measured at different local times and address the questions raised by the first SKR source crossing. Finally, observations of the auroral intensity in 2016–2017 will allow extended studies of its seasonal variation.

Cassini will continue to be supported by a significant number of auroral observations from Earth, both by the Hubble Space Telescope and numerous ground-based telescopes. These will provide essential context to high-resolution Cassini observations, and will continue to produce important scientific returns of their own.

All of the leading auroral questions described above should be addressed in some way by these ongoing observations.

1. During the final orbits of the mission, Cassini will measure the field perturbations and accelerated particles associated with the auroral field-aligned currents at low altitudes (~3 R_S) above the planet, meaning that the uncertainty in mapping from the spacecraft position to the auroral features in the ionosphere will be minimal. This will improve the association of the different structures with their auroral signatures. In addition, comparisons between Earth-based observations of auroral emissions and in situ measurements of the magnetic field will allow a better understanding of how all the auroral currents are generated. Data taken in 2016–2017 will also mean that Cassini will have observed Saturn's aurora over an entire season, from northern winter to summer. This, alongside supporting Earth-based observations, will allow the occurrence of unusual features such as the polar infilling and long-term trends in the main oval intensity, location and rotational modulation in both hemispheres to be studied.
2. Continuing Cassini observations of the magnetic fields around Saturn will improve the initial results presented in Hunt et al. (2014), providing a better understanding of the location of field-aligned currents in this region. Direct measurements of the neutral and ion winds within the auroral region, when combined with the planetary period measured by Cassini, will provide direct evidence of how planetary period currents flow through this region and will reveal the presence or lack of a weather system that could drive such currents.

3. The Grand Finale Cassini orbits will fly directly through the region of the magnetosphere where auroral field-aligned currents originate, allowing a direct sampling of the magnetospheric conditions that drive these auroral features. When combined with supporting Earth-based observations, in particular from the Hubble Space Telescope, both co-rotating and sub-rotating auroral features can be identified and their magnetospheric source understood.
4. Cassini observations of the 1-hour periodicity have shown that this feature regularly occurs simultaneously within both visible and UV emission, and so it is expected that this will also drive changes in radio and IR emission. In order to address this, several significant changes have been made to the way Cassini observers Saturn's aurora in order to allow the auroral brightness to be measured on shorter timescales across all wavelengths. These shorter-period observations will continue into the Grand Finale orbits, providing the essential data needed to better understand these 1-hour period observations.
5. During the F-ring and proximal orbits, Cassini will repeatedly fly through the region where the SKR is believed to originate. By measuring the particle energy distributions, density and variability in LT and altitude, the characteristics of the source region will be explored. Simultaneous imaging of the aurora by Cassini and Earth-based observatories will provide the global context for these single-point in situ measurements.

Even with the triumphant end of Cassini, our scientific investigations will continue, with ongoing observations from the Hubble Space Telescope and improved instrumentation on current ground-based telescopes. The launch of the James Webb Space Telescope in 2018 will allow us to observe Saturn's infrared aurora in greater detail than ever before, and with the coming 25+ meter class ground-based telescopes, the next decade will provide us with unprecedented views of Saturn's infrared aurora.

References

Aguilar, A., J. M. Ajello, R. S. Mangina et al. (2008), The electron-excited mid-ultraviolet to near-infrared spectrum of H_2: Cross sections and transition probabilities, *Astrophys. J. Supp. Ser.* 177, 388–407, DOI:10.1086/587690.

Andrews, D. J., B. Cecconi, S. W. H. Cowley et al. (2011), Planetary period oscillations in Saturn's magnetosphere: Evidence in magnetic field phase data for rotational modulation of Saturn kilometric radiation emissions, *J. Geophys. Res.*, 116, A09206, DOI:10.1029/2011JA016636.

Badman, S. V., S. W. H. Cowley, J.-C. Gérard and D. Grodent (2006), A statistical analysis of the location and width of Saturn's southern auroras, *Annales Geophysicae* 24, 3533–3545, DOI:10.5194/angeo-24–3533-2006.

Badman, S. V., N. Achilleos, K. H. Baines et al. (2011a), Location of Saturn's northern infrared aurora determined from Cassini VIMS images. *Geophys. Res. Lett.*, 38, L03102, DOI:10.1029/2010GL046193.

Badman, S. V., C. Tao, A. Grocott et al. (2011b), Cassini VIMS observations of latitudinal and hemispheric variations in Saturn's infrared auroral intensity, *Icarus* 216, 367–375. DOI:10.1016/j.icarus.2011.09.031.

Badman, S. V., D. J. Andrews, S. W. H. Cowley et al. (2012a), Rotational modulation and local time dependence of Saturn's infrared H_3^+ auroral intensity, *J. Geophys. Res.*, 117, A09228, DOI:10.1029/2012JA017990.

Badman, S. V., N. Achilleos, C. S. Arridge et al. (2012b), Cassini observations of ion and electron beams at Saturn and their relationship to infrared auroral arcs, *J. Geophys. Res.*, 117, A01211, DOI:10.1029/2011JA017222.

Badman, S. V., A. Masters, H. Hasegawa et al. (2013), Bursty magnetic reconnection at Saturn's magnetopause, *Geophys. Res. Lett.*, 40, 1027–1031, DOI:10.1002/grl.50199.

Badman, S. V., G. Branduardi-Raymont, M. Galand et al. (2015), Auroral processes at the giant planets: energy deposition, emission mechanisms, morphology and spectra, *Space Sci. Rev.*, 180, DOI:10.1007/s11214-014–0042-x.

Badman, S. V., G. Provan, E. J. Bunce et al. (2016), Saturn's auroral morphology and field-aligned currents during a solar wind compression, *Icarus* 263, 83–93, DOI:10.1016/j.icarus.2014.11.014.

Belenkaya, E. S., S. W. H. Cowley, C. J. Meredith et al. (2014), Magnetospheric magnetic field modelling for the 2011 and 2012 HST Saturn aurora campaigns: Implications for auroral source regions, *Annales Geophysicae* 32, 689–704, DOI:10.5194/angeo-32–689-2014.

Branduardi-Raymont, G.,P. G. Ford, K. C. Hansen et al. (2013), Search for Saturn's X-ray aurorae at the arrival of a solar wind shock, *J. Geophys. Res. Space Physics*, 118, DOI:10.1002/jgra.50112.

Bunce, E. J., C. S. Arridge, J. T. Clarke et al. (2008), Origin of Saturn's aurora: Simultaneous observations by Cassini and the Hubble Space Telescope, *J. Geophys. Res.*, 113, A09209, DOI:10.1029/2008JA013257.

Bunce, E. J., S. W. H. Cowley, D. L. Talboys et al. (2010), Extraordinary field-aligned current signatures in Saturn's high-latitude magnetosphere: Analysis of Cassini data during Revolution 89, *J. Geophys. Res.* 115, A10238, DOI:10.1029/2010JA015612.

Bunce, E. J. (2012), *Origins of Saturn's Auroral Emissions and Their Relationship to Large-Scale Magnetosphere*

Dynamics, Washington DC, American Geophysical Union, Geophysical Monograph Series 197, 397–410, DOI:10.1029/2011GM001191.

Bunce, E. J., D. C. Grodent, S. L. Jinks et al. (2014), Cassini nightside observations of the oscillatory motion of Saturn's northern auroral oval, *J. Geophys. Res. Space Physics*, 119, 3528–3543, DOI:10.1002/2013JA019527.

Carbary, J. F. (2013), Longitude dependences of Saturn's ultraviolet aurora, *Geophysical Res. Lett.*, 40, DOI:10.1002/grl.50430.

Cecconi, B., L. Lamy, P. Zarka et al. (2009), Goniopolarimetric study of the revolution 29 perikrone using the Cassini Radio and Plasma Wave Science instrument high-frequency radio receiver, *J. Geophys. Res.*, 114, A03215, DOI:10.1029/2008JA013830.

Chen, Y. and T. W. Hill (2008), Statistical analysis of injection/dispersion events in Saturn's inner magnetosphere, *J. Geophys. Res.*, 113, A07215, DOI:10.1029/2008JA013166.

Clarke, J. T., J.-C. Gérard, D. Grodent et al. (2005), Morphological differences between Saturn's ultraviolet aurorae and those of Earth and Jupiter, *Nature* 433, 717–719, DOI:10.1038/nature03331.

Clarke, J. T., J. Nichols and J.-C. Gérard (2009), Response of Jupiter's and Saturn's auroral activity to the solar wind, *J. Geophys. Res.*, 114, A05210, DOI:10.1029/2008JA013694.

Cowley, S. W. H., E. J. Bunce, J. M. O'Rourke (2004), A simple quantitative model of plasma flows and currents in Saturn's polar ionosphere, *J. Geophys. Res.* 109, A05212, DOI:10.1029/2003JA010375.

Cowley, S. W. H., S. V. Badman, E. J. Bunce et al. (2005), Reconnection in a rotation-dominated magnetosphere and its relation to Saturn's auroral dynamics, *J. Geophys. Res.* 110, A02201, DOI:10.1029/2004JA010796.

Desch, M. D. and M. L. Kaiser (1981), Voyager measurement of the rotation period of Saturn's magnetic field, *Geophys. Res. Lett.*, 8, 253–256.

Dyudina, U. A., A. P. Ingersoll, S. P. Ewald and D. Wellington (2016), Saturn's aurora observed by the Cassini camera at visible wavelengths, *Icarus* 263, 32–43, DOI:10.1016/j.icarus.2015.05.022

Farrell, W. M., M. D. Desch, M. L. Kaiser et al. (2005), A nightside source of Saturn's kilometric radiation: Evidence for an inner magnetosphere energy driver, *Geophys. Res. Lett.*, 32, L18107, DOI:10.1029/2005GL023449.

Fear, R. C. and S. E. Milan (2012), The IMF dependence of the local time of transpolar arcs: Implications for formation mechanism, *J. Geophys. Res.*, 117, A03213, DOI:10.1029/2011JA017209.

Galand, M., L. Moore, I. Mueller-Wodarg et al. (2011), Response of Saturn's auroral ionosphere to electron precipitation: Electron density, electron temperature, and electrical conductivity, *J. Geophys. Res.* 116, A09306, DOI:10.1029/2010JA016412.

Galopeau, P. H. M. and A. Lecacheux (2000), Variations in Saturn's radio rotation period measured at kilometer wavelengths, *J. Geophys. Res.*, 105, 13089–13101.

Gérard, J.-C., B. Bonfond, J. Gustin et al. (2009), Altitude of Saturn's aurora and its implications for the characteristic energy of precipitated electrons, Geophys. *Res. Lett.*, 36, L02202.

Gérard, J.-C., J. Gustin, W. R. Pryor et al. (2013), Remote sensing of the energy of auroral electrons in Saturn's atmosphere: Hubble and Cassini spectral observations, *Icarus* 223, 211–221, DOI:10.1016/j.icarus.2012.11.033.

Gladstone, G. R., J. H. Waite, D. Grodent et al. (2002), A pulsating auroral X-ray hot spot on Jupiter, *Nature* 415, 1000–1003.

Grodent, D., J.-C. Gérard, S. W. H. Cowley et al. (2005), Variable morphology of Saturn's southern ultraviolet aurora, *Journal of Geophysical Research* (Space Physics) 110, A07215, DOI:10.1029/2004JA010983.

Grodent, D., A. Radioti, B. Bonfond and J.-C. Gérard (2010), On the origin of Saturn's outer auroral emission, *J. Geophys. Res.*, 115, A08219, DOI:10.1029/2009JA014901.

Grodent, D., J. Gustin, J.-C. Gérard et al. (2011), Small-scale structures in Saturn's ultraviolet aurora, *J. Geophys. Res.*, 116, A09225, DOI:10.1029/2011JA016818.

Grodent, D. (2014), A brief review of ultraviolet auroral emissions on giant planets, *Space Sci. Rev.*, 187, 1-4, DOI:10.1007/s11214-014-0052-8.

Gurnett, D. A. et al. (2005), Radio and plasma wave observations at Saturn from Cassini's approach and first orbit, *Science*, 307 no. 5713, 1255–1259, DOI:10.1126/science.1105356.

Gurnett, D. A., A. M. Persoon, J. B. Groene et al. (2009), Discovery of a north-south asymmetry in Saturn's radio rotation period, *Geophys. Res. Lett.*, 36, L16102.

Gurnett, D. A., J. B. Groene, A. M. Persoon et al. (2010), The reversal of the rotational modulation rates of the north and south components of Saturn kilometric radiation near equinox, *Geophys. Res. Lett.*, 37, L24101.

Gurnett, D. A., J. B. Groene, T. F. Averkampf et al. (2011), An SLS4 Longitude System Based on a Tracking Filter Analysis of the Rotational Modulation of Saturn Kilometric Radiation, Planetary, Solar and Heliospheric Radio Emissions (PRE VII) 51–64.

Gurnett D. A. and W. R. Pryor (2011), Auroral processes associated with Saturn's moon Enceladus. In *Relationship between Auroral Phenomenology and Magnetospheric Processes*, ed. by A. Keiling, E. Donovan, F. Bagenal, T. Karlsson, Geophysical Monograph Series, American Geophysical Union, Washington, DC, pp. 305–313, DOI:10.1029/2011BK001174.

Gustin, J., J.-C. Gérard, W. Pryor et al. (2009), Characteristics of Saturn's polar atmosphere and auroral electrons derived from HST/STIS, FUSE and Cassini/UVIS spectra, *Icarus* 200, 176–187, DOI:10.1016/j.icarus.2008.11.013.

Gustin, J., B. Bonfond, D. Grodent, J.-C. Gérard (2012), Conversion from HST ACS and STIS auroral counts into brightness, precipitated power, and radiated power for H_2 giant planets, *J. Geophys. Res*, 117, A07316. DOI:10.1029/2012JA017607.

Gustin, J., J.-C. Gérard, D. Grodent et al. (2013), Effects of methane on giant planet's UV emissions and implications for the auroral characteristics, *Journal of Molecular Spectroscopy*, 291, 108–117, DOI:10.1016/j.jms.2013.03.010.

Hui, Y., T. E. Cravens, N. Ozak and D. R. Schultz (2010), What can be learned from the absence of auroral X-ray emission from Saturn?, *J. Geophys. Res.*, 115, A10239, DOI:10.1029/2010JA015639.

Hunt, G. J., S. W. H. Cowley, G. Provan et al. (2014), Field-aligned currents in Saturn's southern nightside magnetosphere: Sub-corotation and planetary period oscillation components, *J. Geophys. Res.*, 119, 9847–9899, DOI:10.1002/2014JA020506.

Jackman, C. M., N. Achilleos, S. W. H. Cowley et al. (2013), Auroral counterpart of magnetic field dipolarizations in Saturn's tail, *Planet. Space Sci.*, 82, 34–42, DOI:10.1016/j.pss.2013.03.010.

Jia, X., M. G. Kivelson and T. I. Gombosi (2012), Driving Saturn's magnetospheric periodicities from the upper atmosphere/ionosphere, *J. Geophys. Res.*, 117, A04215, DOI:10.1029/2011JA017367.

Kennelly, T. J., J. S. Leisner, G. B. Hospodarsky and D. A. Gurnett (2013), Ordering of injection events within Saturnian SLS longitude and local time, *J. Geophys. Res.*, 118, 832–838, DOI:10.1002/jgra.50152.

Knight, S. (1973), Parallel electric fields, *Planet. and Space Sci.*, 21, 741–750, DOI:10.1016/0032–0633(73)90093–7.

Kullen, A. and P. Jahnunen (2004), Relation of polar auroral arcs to magnetotail twisting and IMF rotation: A systematic MHD simulation study, *Ann. Geophys.*, 22(3), 951–970.

Kurth, W. S., D. A. Gurnett, J. T. Clarke et al. (2005), An Earth-like correspondence between Saturn's auroral features and radio emission, *Nature*, 433, 722–725, DOI:10.1038/nature03334.

Kurth, W. S., A. Lecacheux, T. F. Averkamp et al. (2007), A Saturnian longitude system based on a variable kilometric radiation period, *Geophys. Res. Lett.*, 34, L02201.

Kurth, W. S., T. F. Averkamp, D. A. Gurnett et al. (2008), An update to a Saturnian longitude system based on kilometric radio emissions, *J. Geophys. Res.*, 113, A05222.

Kurth, W. S., E. J. Bunce, J. T. Clarke et al. (2009), Auroral Processes, *Saturn from Cassini-Huygens*. 333, DOI:10.1007/978-1-4020-9217-6_12.

Lamy, L., P. Zarka, B. Cecconi et al. (2008), Saturn kilometric radiation: Average and statistical properties, *J. Geophysical Research* 113, A07201, DOI:10.1029/2007JA012900.

Lamy, L., B. Cecconi, R. Prangé et al. (2009), An auroral oval at the footprint of Saturn's kilometric radio sources, co-located with the UV aurorae, *J. Geophys. Res.*, 114, A10212, DOI:10.1029/2009JA014401.

Lamy, L., P. Schippers, P. Zarka et al. (2010), Properties of Saturn kilometric radiation measured within its source region, *Geophysical Research Letters*, 37, L12104, DOI:10.1029/2010GL043415.

Lamy, L. (2011), Variability of southern and northern SKR periodicities, in *Planetary Radio Emissions VII*, edited by H. O. Rucker et al., pp. 39–50, Austrian Acad. Sci. Press, Vienna, Austria.

Lamy, L., R. Prangé, W. Pryor et al. (2013), Multispectral simultaneous diagnosis of Saturn's aurorae throughout a planetary rotation, *J. Geophys. Res.*, 118, 4817–4843, DOI:10.1002/jgra.50404.

Masters, A., N. Achilleos, D. Bertucci et al. (2009), Surface waves on Saturn's dawn flank magnetopause driven by the Kelvin–Helmholtz instability, *Planetary and Space Science*, 57, 1769–1778, DOI:10.1016/j.pss.2009.02.010.

Masters, A., N. Achilleos, M. G. Kivelson et al. (2010), Cassini observations of a Kelvin–Helmholtz vortex in Saturn's outer magnetosphere, *J. Geophys. Res.* 115, A07225, DOI:10.1029/2010JA015351.

Mauk, B. H., J. Saur, D. G. Mitchell et al. (2005), Energetic particle injections in Saturn's magnetosphere, *Geophysical Research Letters* 32, L14S05, DOI:10.1029/2005GL022485.

McAndrews, H. J., M. F. Thomsen, C. S. Arridge et al. (2009), Plasma in Saturn's nightside magnetosphere and the implications for global circulation, *Planet. Space Sci.*, 57, 1714–1722, DOI:10.1016/j.pss.2009.03.003.

Melin, H., T. Stallard, S. Miller et al. (2011), Simultaneous Cassini VIMS and UVIS observations of Saturn's southern aurora: Comparing emissions from H, H_2 and H_3^+ at a high spatial resolution, *Geophysical Research Letters* 38, L15203, DOI:10.1029/2011GL048457.

Melin, H., S. V. Badman, T. S. Stallard et al. (2016), Simultaneous multi-scale and multi-instrument observations of Saturn's aurorae during the 2013 observing campaign, *Icarus* 263, 56–74, DOI:10.1016/j.icarus.2015.08.021.

Menietti, J. D., R. L. Mutel, P. Schippers, S.-Y. Ye, D. A. Gurnett, and L. Lamy (2011), Analysis of Saturn kilometric radiation near a source center, *J. Geophys. Res.* 116, A12222, DOI:10.1029/2011JA017056.

Meredith, C. J., S. W. H. Cowley, K. C. Hansen et al. (2013), Simultaneous conjugate observations of small-scale structures in Saturn's dayside ultraviolet auroras: Implications for physical origins, *J. Geophys. Res.* 118, 2244–2266, DOI:10.1002/jgra.50270.

Milan, S. E., B. Hubert and A. Grocott (2005), Formation and motion of a transpolar arc in response to dayside and nightside reconnection, *J. Geophys. Res.*, 110, A01212, DOI:10.1029/2004JA010835.

Mitchell, D. G., W. S. Kurth, G. B. Hospodarsky et al. (2009), Ion conics and electron beams associated with auroral processes on Saturn. *Journal Geophysical Research* 114, 2212.

Mitchell, D. G., J. F. Carbary, E. J. Bunce et al. (2016), Recurrent pulsations in Saturn's high latitude magnetosphere, *Icarus* 263, 94–100, DOI:10.1016/j.icarus.2014.10.028.

Moore, L., I. Mueller-Wodarg, M. Galand, A. Kliore and M. Mendillo (2010), Latitudinal variations in Saturn's ionosphere: Cassini measurements and model comparisons, *Journal of Geophysical Research* 115, A11317, DOI:10.1029/2010JA015692.

Müller-Wodarg, I. C. F., L. Moore and G. M. Mendillo (2012), Magnetosphere-atmosphere coupling at Saturn: 1. Response of thermosphere and ionosphere to steady state polar forcing, *Icarus* 221(2) DOI:10.1016/j.icarus.2012.08.034.

Mutel, R. L., J. D. Menietti, D. A. Gurnett et al. (2010), CMI growth rates for Saturnian kilometric radiation, *Geophys. Res. Lett.*, 37, L19105, DOI:10.1029/2010GL044940.

Nichols, J. D., J. T. Clarke, S. W. H. Cowley et al. (2008), Oscillation of Saturn's southern auroral oval, *J. Geophys. Res.*, 113, A11205, DOI:10.1029/2008JA013444.

Nichols, J. D., S. V. Badman, E. J. Bunce (2009), Saturn's equinoctial auroras, *Geophys. Res. Lett.* 36, L24102-DOI:10.1029/2009GL041491.

Nichols, J. D., S. W. H. Cowley and L. Lamy (2010), Dawn-dusk oscillation of Saturn's conjugate auroral ovals, *Geophys. Res. Lett.*, 37, L24102, DOI:10.1029/2010GL045818.

Nichols, J. D., S. V. Badman, K. H. Baines et al. (2014), Dynamic auroral storms on Saturn as observed by the Hubble Space Telescope, *Geophys. Res. Lett.*, 41, 3323–3330, DOI:10.1002/2014GL060186.

Nichols, J. D., S. V. Badman, E. J. Bunce et al. (2016), Saturn's northern auroras as observed using the Hubble Space Telescope, *Icarus* 263, 17–31, DOI:10.1016/j.icarus.2015.09.008.

O'Donoghue, J., T. S. Stallard, H. Melin et al. (2013), The domination of Saturn's low-latitude ionosphere by ring "rain," *Nature* 496, 193–195, DOI:10.1038/nature12049.

Provan, G., S. W. H. Cowley and J. D. Nichols (2009), Phase relation of oscillations near the planetary period of Saturn's auroral oval and the equatorial magnetospheric magnetic field, *J. Geophys. Res.*, 114, A04205, DOI:10.1029/2008JA013988.

Pryor, W. R., A. M. Rymer, D. G. Mitchell et al. (2011), The auroral footprint of Enceladus on Saturn, *Nature*, 472, 331–333, DOI:10.1038/nature09928.

Radioti, A., D. Grodent, X. Jia et al. (2016), A multi-scale magnetotail reconnection event at Saturn and associated flows: Cassini/UVIS observations, *Icarus*, DOI:10.1016/j.icarus.2014.12.016.

Radioti, A., D. Grodent, J.-C. Gérard et al. (2015), Auroral spirals at Saturn, *J. Geophys. Res.*, 120, DOI:10.1002/2015JA021442.

Radioti, A., D. Grodent, J.-C. Gérard et al. (2014), Saturn's elusive nightside polar arc, *Geophys. Res. Lett.*, 41, 6321–6328, DOI:10.1002/2014GL061081.

Radioti, A., E. Roussos, D. Grodent et al. (2013a), Signatures of magnetospheric injections in Saturn's aurora, *J. Geophys. Res.* 118, 1922–1933, DOI:10.1002/jgra.50161.

Radioti, A., D. Grodent, J.-C. Gérard, B. Bonfond, J. Gustin, W. Pryor, J. M. Jasinski and C. S. Arridge (2013b), Auroral signatures of multiple magnetopause reconnection at Saturn, *Geophys. Res. Lett.*, 40, 4498–4502, DOI:10.1002/grl.50889.

Radioti, A., D. Grodent, J.-C. Gérard et al. (2011), Bifurcations of the main auroral ring at Saturn: ionospheric signatures of consecutive reconnection events at the magnetopause, *J. Geophys. Res.* 116, 11209.

Radioti, A., D. Grodent, J.-C. Gérard, E. Roussos, C. Paranicas, B. Bonfond, D. G. Mitchell, N. Krupp, S. Krimigis and J. T. Clarke (2009), Transient auroral features at Saturn: Signatures of energetic particle injections in the magnetosphere, *J. Geophys. Res.*, 114, A03210, DOI:10.1029/2008JA013632.

Ray, L. C., M. Galand, L. E.., Moore and B. L. Fleshman (2012), Characterizing the limitations to the coupling between Saturn's ionosphere and middle magnetosphere, *J. Geophys.* Res 117, A07210, DOI:10.1029/2012JA017735.

Ray, L. C., M. Galand, P. A. Delamere and B. L. Fleshman (2013) Current-voltage relation for the Saturnian system, *J. Geophys. Res* 118, 3214–3222, DOI:10.1002/jgra.50330.

Sandel, B. R., D. E. Shemansky, A. L. Broadfoot et al. (1982), Extreme ultraviolet observations from the Voyager 2 encounter with Saturn, *Science*, 215, 548–553, DOI:10.1126/science.215. 4532.548.

Schippers, P., M. Blanc, N. André et al. (2008), Multi-instrument analysis of electron populations in Saturn's magnetosphere, *J. Geophys. Res.* 113, A07208, DOI:10.1029/2008JA013098.

Schippers, P., C. S. Arridge, J. D. Menietti et al. (2011), Auroral electron distributions within and close to the Saturn kilometric radiation source region, *J. Geophys. Res.* 116, A05203, DOI:10.1029/2011JA016461.

Schippers, P., N. André, D. A. Gurnett et al. (2012), Identification of electron field-aligned current systems in Saturn's magnetosphere, *J. Geophys. Res.* 117, A05204, DOI:10.1029/2011JA017352.

Smith, C. G. A. (2014), On the nature and location of the proposed twin vortex systems in Saturn's polar upper atmosphere, *J. Geophys. Res.* 119, 5964–5977, DOI:10.1002/2014JA019934.

Stallard, T. S., S. Miller, L. M. Trafton et al. (2004), Ion winds in Saturn's southern auroral/polar region, *Icarus* 167, 204–211. DOI:10.1016/j.icarus.2003.09.006.

Stallard, T., S. Miller, H. Melin et al. (2007a), Saturn's auroral/polar H_3^+ infrared emission. I. General morphology and ion velocity structure, *Icarus* 189, 1–13, DOI:10.1016/j.icarus.2006.12.027.

Stallard, T., C. Smith, S. Miller et al. (2007b), Saturn's auroral/polar H_3^+ infrared emission. II. A comparison with plasma flow models, *Icarus* 191, 678–690, DOI:10.1016/j.icarus.2007.05.016.

Stallard, T., S. Miller, H. Melin et al. (2008c), Jovian-like aurorac on Saturn, *Nature* 453, 1083–1085, DOI:10.1038/nature07077.

Stallard, T., S. Miller, M. Lystrup et al. (2008d), Complex structure within Saturn's infrared aurora, *Nature* 456, 214–217, DOI:10.1038/nature07440.

Stallard, T., H. Melin, S. W. H. Cowley et al. (2010), Location and magnetospheric mapping of Saturn's mid-latitude infrared auroral oval, *Astrophysical Journal Letters* 722, L85–L89, DOI:10.1088/2041–8205/722/1/L85.

Stallard T. S., A. Masters, S. Miller et al. (2012d), Saturn's auroral/polar H_3^+ infrared emission: The effect of solar wind compression, *J. Geophys. Res*, A12302.

Stallard, T., S. Miller and H. Melin (2012c), Clues on Ionospheric Electrodynamics from IR Aurora at Jupiter and Saturn. Auroral Phenomenology and Magnetospheric Processes: Earth and Other Planets, *Geophysical Monograph Series* 197, 215–224, DOI:10.1029/2011GM001168.

Stallard, T. S., H. Melin, S. Miller et al. (2012b), Temperature changes and energy inputs in Giant Planet atmospheres: What we are learning from H_3^+, *Phil. Trans. R. Soc. A* 370, 5213–5224, DOI:10.1098/rsta.2012.0028.

Stallard, T. S., H. Melin, S. Miller et al. (2012a), Peak emission altitude of Saturn's H_3^+ aurora, *Geophysical Research Letters* 39, L15103, DOI:10.1029/2012GL052806.

Tao, C., S. V. Badman and M. Fujimoto (2011), UV and IR auroral emission model for the outer planets: Jupiter and Saturn comparison, *Icarus* 213, 581–592, DOI:10.1016/j.icarus.2011.04.001.

Tao, C., L. Lamy, R. Prangé (2014), The brightness ratio of H Lyman-alpha H_2 bands in FUV auroral emissions: A diagnosis for the energy of precipitating electrons and associated magnetospheric acceleration processes applied to Saturn, *Geophysical Research Letters* 41, 6644–6651, DOI:10.1002/2014GL061329.

Talboys, D. L., C. S. Arridge, E. J. Bunce et al. (2009), Characterization of auroral current systems in Saturn's magnetosphere: High-latitude Cassini observations, *J. Geophys. Res.* 114, A06220, DOI:10.1029/2008JA013846.

Talboys, D. L., E. J. Bunce, S. W. H. Cowley et al. (2011), Statistical characteristics of field-aligned currents in Saturn's nightside magnetosphere, *J. Geophys. Res.* 116, A04213, DOI:10.1029/2010JA016102.

Thomsen, M. F. (2013), Saturn's magnetospheric dynamics, *Geophysical Research Letters* 40, 5337–5344, DOI:10.1002/2013GL057967.

Trafton, L., J. Carr, D. Lester and P. Harvey (1988), A possible detection of Jupiter's northern auroral S1(1) H_2 quadrupole line emission, *Icarus* 74, 351–356, DOI:10.1016/0019–1035(88)90047–4.

Trafton, L. M., S. Miller, T. R. Geballe et al. (1999), H_2 quadrupole and H_3^+ emission from Uranus: The Uranian thermosphere, ionosphere, and aurora, *Ap. J.* 524, 1059–1083, DOI:10.1086/307838.

Walker, R. J., K. Fukazawa, T. Ogino and D. Morozoff (2011), A simulation study of Kelvin–Helmholtz waves at Saturn's magnetopause, *J. Geophys. Res.*, 116, A03203, DOI:10.1029/2010JA015905.

Waite, J. H., M. R. Combi, W.-H. Ip et al. (2006), Cassini ion and neutral mass spectrometer: Enceladus plume composition and structure, *Science* 311, 1419–1422, DOI:10.1126/science.1121290.

Warwick, J. W., J. B. Pearce, D. R. Evans et al. (1981), Planetary radio astronomy observations from Voyager 1 near Saturn, *Science*, 212, 239–242.

Young, D. T., J.-J. Berthelier, M. Blanc et al. (2005), Composition and dynamics of plasma in Saturn's magnetosphere, *Science* 307, 1262–1266, DOI:10.1126/science.1106151.

Zarka, P. (1998), Auroral radio emissions at the outer planets: Observations and theories, *Journal of Geophysical Research* 103, 20159–20194, DOI:10.1029/98JE01323.

Zarka, P., L. Lamy, B. Cecconi et al. (2007), Modulation of Saturn's radio clock by solar wind, *Nature*, 450, 265–267.

8

Saturn's Ionosphere

LUKE MOORE, MARINA GALAND, ARVYDAS J. KLIORE, ANDREW F. NAGY
AND JAMES O'DONOGHUE

Abstract

This chapter summarizes our current understanding of the ionosphere of Saturn. We give an overview of Saturn ionospheric science from the Voyager era to the present, with a focus on the wealth of new data and discoveries enabled by Cassini, including a massive increase in the number of electron density altitude profiles. We discuss recent ground-based detections of the effect of "ring rain" on Saturn's ionosphere, and present possible model interpretations of the observations. Finally, we outline current model-data discrepancies and indicate how future observations can help in advancing our understanding of the various controlling physical and chemical processes.

8.1 Introduction

Saturn's upper atmosphere is typically defined to be the region above the homopause, which marks the transition between a well-mixed atmospheric region dominated by eddy diffusion below (the lower atmosphere) and a region dominated by molecular diffusion above. It can further be broken down into two coincident regions: the neutral thermosphere (Chapter 9 by Strobel et al.) and the charged ionosphere. The upper atmosphere forms the transition region between a dense neutral atmosphere below and a tenuous, charged magnetosphere above (Chapter 6 by Krupp et al.); consequently, it also mediates the exchange of particles, momentum, and energy between these two regions. External forcing on the upper atmosphere, such as by solar extreme ultraviolet (EUV) photons or energetic particles, determines the degree of ionization within the ionosphere. As magnetic fields strongly influence charged particle motions, the ionosphere tends to be more strongly ionized where magnetic field configurations favor precipitation of energetic particles into the atmosphere, namely, the auroral regions near the magnetic poles (Chapter 7 by Stallard et al.).

In this chapter we summarize our current understanding of the non-auroral ionosphere of Saturn, using ground-based and space-based measurements (Section 8.2) as well as modeling studies in order to interpret the physical processes driving Saturn's ionosphere (Section 8.3). As giant planet ionospheres are all qualitatively similar, we try to avoid reproducing material unnecessarily from previous giant planet review articles. Therefore, those reviews are still highly relevant (e.g., Atreya et al., 1984; Waite et al., 1997; Nagy and Cravens, 2002; Majeed et al., 2004; Yelle and Miller, 2004; Nagy et al., 2009; Schunk and Nagy, 2009). While we will touch upon closely related topics as appropriate, more thorough discussions of Saturn's aurorae and thermosphere can be found in Chapters 7 and 9, respectively.

8.2 Observations

There are several options for remotely observing Saturn's ionosphere, and each technique has not only particular advantages but also significant, unique limitations. To date, the vast majority of observational information regarding Saturn's ionosphere has come from spacecraft radio occultations, which yield vertical electron density structure, $N_e(h)$. Due to Sun-Saturn-Earth geometry, however, a spacecraft-to-Earth radio occultation can only be performed near the terminator and therefore can only sample the dawn or dusk ionosphere of Saturn (see Section 8.2.1). An additional method of tracking the peak ionospheric electron density, N_{MAX}, involves the detection of broadband radio emission (dubbed Saturn Electrostatic Discharge, SED) originating from powerful lightning storms in Saturn's lower atmosphere. As the storm rotates with the planet, a nearby spacecraft can use SED emission to derive the diurnal variation of N_{MAX} near the storm location. Unfortunately, this technique is reliant upon storm activity and latitude, and it does not contain any altitude information (Section 8.2.2). Finally, infrared measurements of rotational-vibrational emissions from H_3^+ near 3–4 μm yield temperature and density information for this major ion in Saturn's ionosphere. Aside from one so-far-unique observation, however, H_3^+ emission has proven to be too weak to be detected outside of Saturn's auroral regions (Section 8.2.3).

8.2.1 Radio Occultations

The technique of radio occultation, whereby Saturn's atmosphere occults the transmission of a radio signal (e.g., Lindal, 1992; Kliore et al., 2004), provides the only available remote diagnostic of electron density altitude profiles, $N_e(h)$, a basic ionospheric property. There are, at most, two opportunities for deriving atmospheric properties during a spacecraft flyby: one during the occultation ingress (or entry, often designated with "N") and one during the occultation egress (or exit, often designated with "X"). The geometry for all of the radio occultations performed at Saturn to date is such that the ingress occultations sample the dusk ionosphere while the egress occultations sample the dawn ionosphere. Radio occultation latitude, typically given in planetographic coordinates, refers to the latitude at the lowest altitude of the occultation. As the spacecraft almost never follows a trajectory in which the occultation ray path is continually above a single latitude, however, the altitude profile derived from a radio occultation actually samples a range of latitudes. Quoted radio occultation latitudes therefore usually refer either to approximate latitudes or to the latitude at the base of the occultation, near the electron density peak.

The main approach for analyzing spacecraft radio occultations is based on geometrical optics (e.g., Fjeldbo et al., 1971). In this approach, the deviation of a ray path in response to refractive index gradients in ionospheric plasma is tracked, leading ultimately to a Doppler shift in the frequency of the signal received at Earth. The time series of differences between the transmitted and received frequencies, called the frequency residuals, can then be used to derive a vertical profile of refractive index and electron number density (e.g., Withers et al., 2014). This direct approach is particularly susceptible to multipath propagation effects, wherein narrow ionospheric layers can lead to multiple, distinct signals arriving simultaneously at the receiving antenna, each with a different Doppler-shifted frequency. This effect is stronger the farther a spacecraft is away from the occulting planet. A better approach is to first use scalar diffraction theory, which transforms the data using Fourier analysis in order to mimic an occultation by a nearby spacecraft, thereby removing complications due to multipath propagation and diffraction effects (e.g., Karayel and Hinson, 1997; Hinson et al., 1998). Once this is accomplished, the geometric approach can then be used. Saturn radio occultations have so far only been analyzed by the geometrical optics technique and therefore are highly uncertain in the presence of narrow ionospheric layers, which appear to be particularly common at lower altitudes. In addition to instrumental effects, other important sources of uncertainty include the required assumption of spherical symmetry within the ionosphere, the degree of intervening plasma between the transmitting and receiving antenna and inaccurate positions or velocities. The propagation of uncertainties through all of the processing steps is non-trivial (e.g., Lipa and Tyler, 1979), but typical estimates are on the order of a few hundred electrons per cubic centimeter for Cassini radio occultations (Nagy et al., 2006; Kliore et al., 2011).

Flybys of the Saturn system by Pioneer 11 (1 September 1979), Voyager 1 (12 November 1980) and Voyager 2 (26 August 1981) yielded our first observational insights into Saturn's ionospheric electron densities (Kliore et al., 1980; Lindal et al., 1985). While there was some variation in the electron density profiles obtained by the Pioneer 11 and Voyager spacecraft, the altitudes of the electron density peaks (h_{MAX})

ranged from ~1000 to 2800 km and the maximum electron density (N_{MAX}) of five out of the six profiles was of order 10^4 cm^{-3}. The Voyager 2 ingress profile was an outlier from this trend, with narrow low-altitude layers of peak electron density near 7×10^4 cm^{-3}. Such layers appear to be relatively common in the giant planet ionospheres, as they have also been found at Jupiter (e.g., Fjeldbo et al., 1975; Yelle and Miller, 2004), Uranus (Lindal et al., 1987) and Neptune (Lindal, 1992). While there is a lack of consensus regarding the origin of these layers, a likely explanation is that vertical wind shears – such as those that might result from atmospheric gravity waves – can lead to localized electron density enhancements. This effect has been demonstrated through modeling of Saturn's upper atmosphere by Moses and Bass (2000) and Barrow and Matcheva (2013), and also of Jupiter (Matcheva and Strobel, 2001) and Neptune (Lyons, 1995).

The arrival of Cassini at Saturn in 2004 has significantly increased the number of radio occultations, allowing for a more thorough examination of possible ionospheric trends. A total of 59 Cassini radio occultations have already been obtained and analyzed, with one final occultation currently planned (Kliore et al., 2014). The first dozen Cassini radio occultations were obtained between May and September 2005 and sampled Saturn's near-equatorial region between 10°N and 10°S planetographic latitude. These profiles, shown in Figure 8.1, revealed a clear dawn-dusk asymmetry. While there is still a high degree of variability among the 12 profiles, on average the peak densities are lower and the peak altitudes are higher at dawn than at dusk (Nagy et al., 2006). This trend is also present in the complete Cassini radio occultation dataset, which includes a number of additional low-latitude profiles (Kliore et al., 2014). Such a behavior is consistent with the expectation that chemical losses during the Saturn

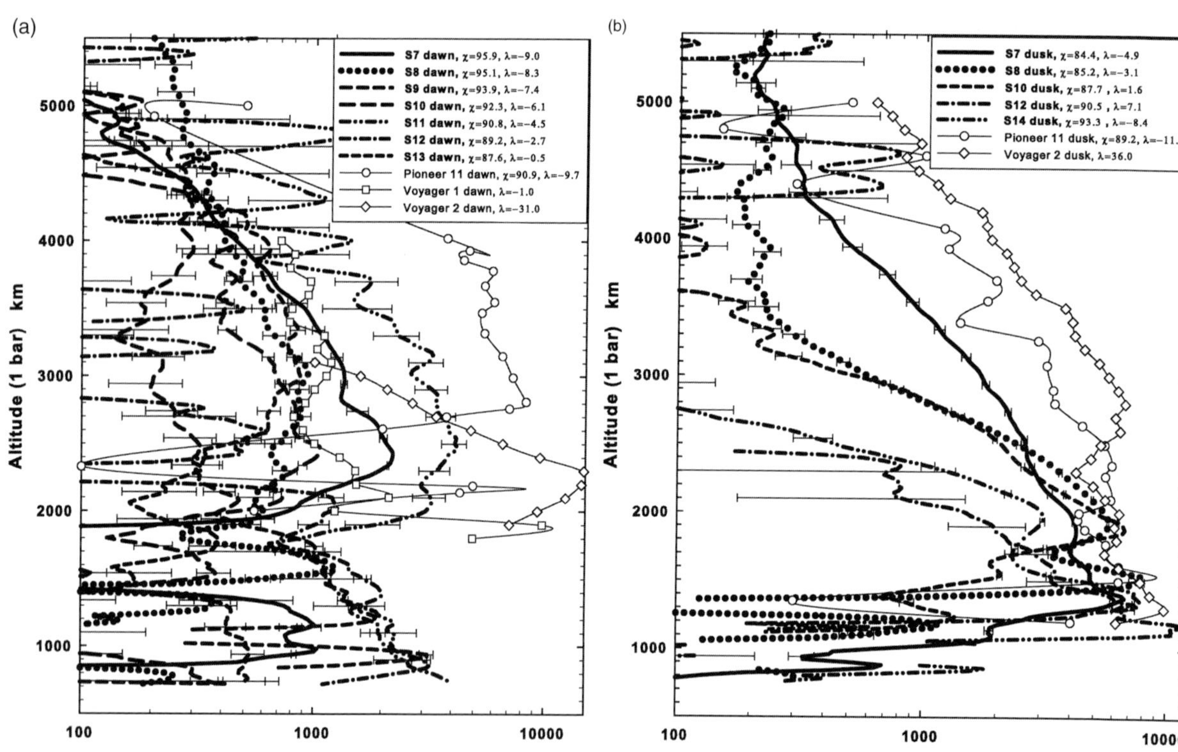

Figure 8.1 Saturn ionospheric electron density altitude profiles retrieved from radio occultations by the Cassini spacecraft at **(left)** dawn and **(right)** dusk. Also shown for comparison are Pioneer 11 and Voyager radio occultation profiles. Error bars represent the uncertainties introduced by baseline frequency fluctuations and the effects of averaging data from multiple Deep Space Network stations. All Cassini profiles are from ±10° latitude. From Nagy et al. (2006).

night would lead to a depletion of the low-altitude ionospheric electron density peak. The averages of the first dozen low-latitude Cassini profiles (five at dusk and seven at dawn) are shown in Figure 8.2.

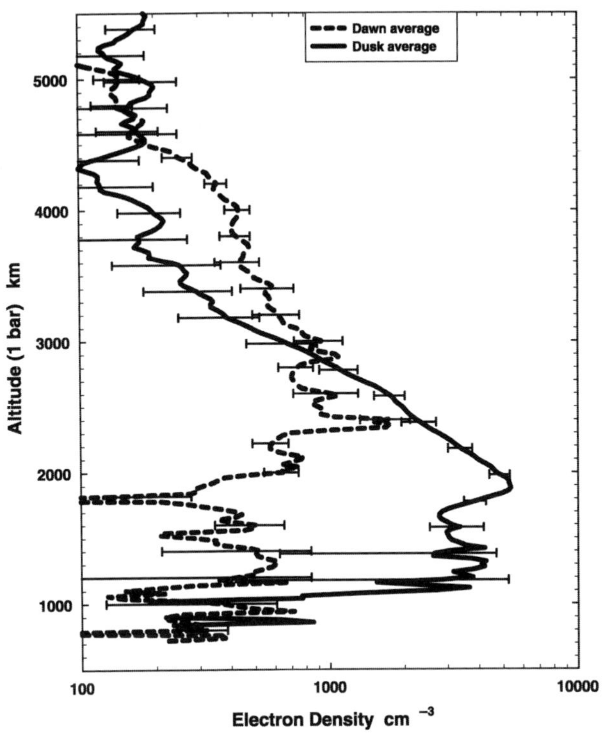

Figure 8.2 Weighted averages of the dawn and dusk electron density profiles from Figure 8.1. From Nagy et al. (2006).

Subsequent Cassini radio occultations sampled a wider range of latitudes, revealing a surprising trend in N_{MAX}. Nineteen low-, mid- and high-latitude radio occultations obtained between September 2006 and July 2008 found that peak electron densities were smallest at Saturn's equator, and increased with latitude (Kliore et al., 2009). The average sub-solar latitude during this occultation period was −8.5°, meaning that the sun primarily illuminated Saturn's low latitudes. Solar EUV photons are expected to be the primary source of ionization in Saturn's non-auroral ionosphere, and yet Cassini radio occultations revealed the minimum electron densities to be in regions of maximum insolation.

This latitudinal electron density trend was reconfirmed and bolstered with 28 additional occultations obtained between 2008 and 2013 (Kliore et al., 2014). While there are a few high-altitude occultations, none of them appear to have sampled active auroral regions (Moore et al., 2010). Figure 8.3 shows the total electron content (TEC), or column integrated electron density, for all 59 of the published Cassini radio occultations, plotted versus planetographic latitude. The trend in TEC is quite similar to that in N_{MAX}, though with considerably less scatter (Kliore et al., 2014). In both cases, there is a clear minimum in electron density at Saturn's equator, and an increase in electron density with latitude. There also appears to be a local minimum in electron density around 45°N latitude, near the region of Saturn's atmosphere that is magnetically

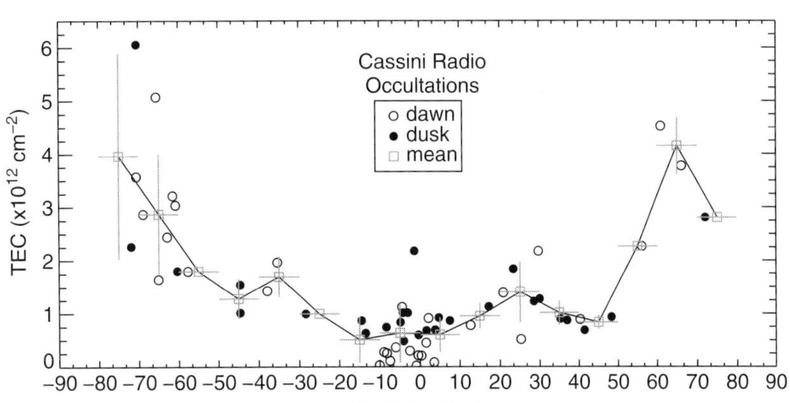

Figure 8.3 Total electron content (TEC) from all 59 published Cassini radio occultations. Dawn (exit) profiles are shown as open circles; dusk (entry) profiles are shown as filled circles. Gray squares represent the local mean over 10° latitude bins. Vertical gray bars indicate the standard deviation of the mean, while horizontal gray bars demarcate the latitude range over which the mean is calculated. Based on Table 1 of Kliore et al. (2014).

linked to the inner B ring, long predicted to be the site of an enhanced influx of water from the rings to the atmosphere (e.g., Connerney and Waite, 1984; Connerney, 1986). The introduction of oxygen, whether in the form of neutral or charged water or other oxygen-bearing molecules or charged submicrometer grains, acts to reduce the local electron density, as it converts the long-lived atomic ion H^+ into a short-lived molecular ion. While it is tempting to associate the localized minimum in electron density near 45°N with an enhanced water influx, ring-derived water influxes are expected to be stronger in the southern hemisphere due to Saturn's effectively offset magnetic dipole (Burton et al., 2010), independent of Saturn season (e.g., Northrop and Connerney, 1987; Tseng et al., 2010). No similar localized minimum is obvious in the southern hemisphere radio occultations. There is a slight hint of a minimum near 45°S, though the sampling statistics are poor in that region.

Cassini radio occultation measurements have demonstrated that Saturn's ionosphere is highly variable, with electron density altitude profiles obtained at similar latitudes from similar times often differing significantly from each other, and with frequent narrow low-altitude layers of electron density. Nevertheless, the wealth of data – at least compared with data from other giant planet ionospheres – has also allowed for identification of a number of trends. On average, peak electron densities are smaller and peak altitudes are higher at dawn than at dusk, consistent with recombination of major ions during the Saturn night. On average, the smallest electron densities are found near Saturn's equator, and electron densities increase with latitude, contrary to what would be expected from an ionosphere driven purely by solar photoionization with constant photochemical loss sources. Finally, though the statistics are poor, radio occultation observations also give some indication that there may be localized minima in electron density near 45°N planetographic latitude, which could be consistent with an influx of charged water grains from Saturn's rings to its atmosphere along magnetic field lines.

8.2.2 Saturn Electrostatic Discharges (SEDs)

During the Voyager encounters with Saturn, the Planetary Radio Astronomy (PRA) instrument detected mysterious, broadband, short-lived, impulsive radio emission (Warwick et al., 1981, 1982). These radio bursts, termed Saturn Electrostatic Discharges (SEDs), were organized in episodes lasting several hours and separated from each other by roughly 10h 10m. While there was initially some uncertainty regarding the origin of SEDs, Burns et al. (1983) suggested they were radio manifestations of atmospheric lightning storms, and Kaiser et al. (1984) demonstrated that an extended source region in the equatorial atmosphere was consistent with the observed SED recurrence pattern. Cassini's Radio and Plasma Wave Science (RPWS) instrument began detecting SEDs prior to its orbital insertion on 1 July 2004, and has since observed nine distinct storm periods, separated by SED-quiet periods of a few days to 21 months (Fischer et al., 2011b). Shortly after Cassini's arrival at Saturn, the Imaging Science Subsystem instrument detected a storm system at 35°S planetocentric latitude that correlated with the observed SED recurrence pattern (Porco et al., 2005). Dyudina et al. (2007) extended this finding by presenting three further storm systems where SED observations were correlated with the rising and setting of a visible storm on the Saturn radio horizon. Finally, lightning flashes were imaged directly by Cassini in 2009, providing a convincing demonstration that SEDs were indeed radio signatures of atmospheric storms in Saturn's lower atmosphere (Dyudina et al., 2010).

SEDs have a large frequency bandwidth, but appear as narrow, banded streaks in both Voyager PRA and Cassini RPWS dynamic spectra due to the short duration of the radio burst and the frequency sampling nature of the receivers. The number of SEDs detected in an individual storm varies significantly, from hundreds to tens of thousands (Fischer et al., 2008), with typical burst rates of a few hundred per hour (Zarka and Pedersen, 1983; Fischer et al., 2006). SED *storms* are periods of nearly continuous SED activity, modulated by *episodes* of varying SED activity. The recurrence period of the episodes within a storm represents the time between peaks of SED activity; for a single longitudinally confined storm system, therefore, this period is closely related to the local rotation rate of the atmosphere. Recurrence periods for Voyager 1 and Voyager 2 SED episodes were ~10h 10m and ~10h 00m, respectively (Evans et al., 1981; Warwick et al., 1982), and were consequently thought to originate from equatorial storm systems (Burns et al., 1983), though none were observed directly. In contrast, the majority of recurrence periods for Cassini-era SED storms are near 10h 40m (Fischer et al., 2008), implying a mid-latitude origin, as confirmed by the 35°S planetocentric latitude storm clouds and visible lightning flashes imaged by Cassini. Approximately 16 months after Saturn passed through its equinox (August 2009) toward southern winter, a giant convective storm developed at 35°N planetocentric latitude, accompanied by unprecedented levels of SED activity, with flash rates an order of magnitude higher than previously observed storms (Fischer et al., 2011a; this storm is described in detail in Chapter 13 by Sánchez-Lavega et al.). While the tendency for Saturn lightning storms to preferentially form near ±35° planetocentric latitude remains unexplained, it is important to keep in mind that SEDs appear to primarily probe either Saturn's mid-latitude (Cassini-era) or equatorial (Voyager-era) ionosphere.

SEDs originating from lightning storms deep within Saturn's atmosphere must ultimately transit the ionosphere in order to be detected by a spacecraft. Therefore, the low-frequency cutoff of each SED episode provides information about the intervening plasma densities, as only frequencies larger than the peak electron plasma frequency will pass through Saturn's ionosphere. Kaiser et al. (1984) combined the observed low-frequency cutoffs with a ray-tracing analysis in order to derive the peak electron density along the propagation path between Voyager 1 and the SED storm. Using similar assumptions, Zarka (1985) also derived peak electron densities from Voyager 1 SED measurements. (Voyager 2 data showed a decline in number and intensity of SEDs with no clear episodic behavior, meaning it could not be used for a similar analysis.) Both Kaiser et al. (1984) and Zarka (1985) found that N_{MAX} varied by more than two orders of magnitude throughout the Saturn day, with midnight electron densities below 10^3 cm^{-3} and noon densities greater than 10^5 cm^{-3}. The SED-derived dawn and dusk N_{MAX} values were of order 10^4 cm^{-3}, in rough agreement with radio occultation results. These diurnal N_{MAX} trends, along with a Cassini-era SED-derived trend, are presented in Figure 8.4.

Whereas Voyager 1 N_{MAX} trends were derived from three SED episodes, Cassini SED observations allowed for investigation of diurnal trends over different storm periods, different episodes within storm periods, and different Cassini-Saturn distances. The diurnal N_{MAX} trend shown in Figure 8.4c was based on 48 SED episodes between 2004 and 2009 when Cassini was within 14 R_S of Saturn (Figure 11 of Fischer et al., 2011b). This profile is slightly different from the

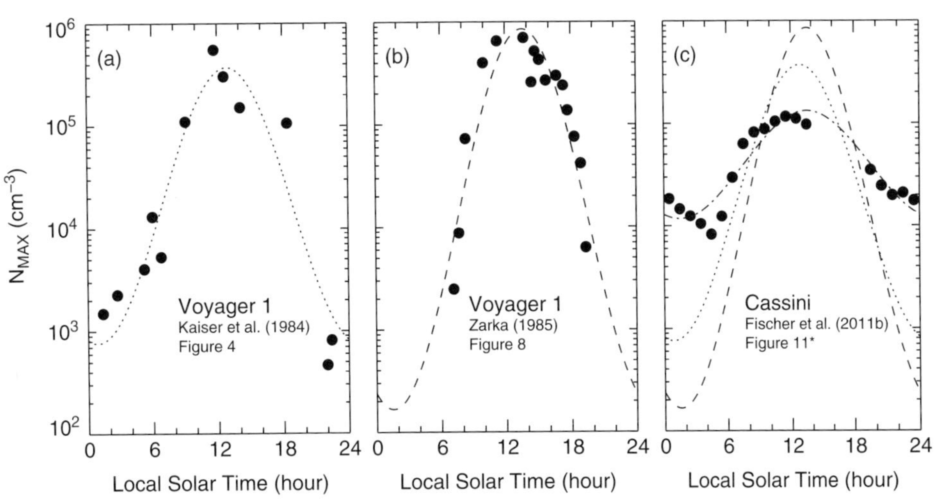

Figure 8.4 Diurnal variation in N_{MAX} derived from Voyager 1 and Cassini SED observations (filled circles), along with a least-squares fit to an equation of the form $log\ N_e = A - B\ cos(SLT - \emptyset)$, where SLT is solar local time in hours and $\emptyset$ is the phase shift (the dotted, dashed, and dot-dash curves). **(a)** Voyager 1, taken from Figure 4 of Kaiser et al. (1984); **(b)** Voyager 1, taken from Figure 8 of Zarka (1985); and **(c)** Cassini, based on Figure 11 of Fischer et al. (2011). The Voyager 1 fits (i.e., dotted and dashed curves) are repeated in (c) in order to more easily compare them with the diurnal variation derived from Cassini-era SEDs (dot-dash curve). Adapted from Figure 1 of Moore et al. (2012).

Cassini-era trend that includes all of the SED observations through 2009 (Figure 9 of Fischer et al., 2011b), most likely due to the fact that attenuation of radio waves by Saturn's ionosphere is frequency dependent, and Cassini is better able to detect weaker bursts at lower frequencies when it is nearer to Saturn. In contrast to the two-order-of-magnitude diurnal variation in N_{MAX} derived from Voyager SEDs, the Cassini SED-derived N_{MAX} varies by only one order of magnitude, between ~10^4 cm^{-3} and ~10^5 cm^{-3}, with a local minimum just before sunrise. Such a local minimum in electron density is consistent with the nighttime chemical loss due to recombination implied by radio occultations; a minimum at midnight is much more difficult to understand theoretically (e.g., Majeed and McConnell, 1996).

While peak electron densities derived from SEDs are highly complementary to the dawn/dusk electron density profiles retrieved from radio occultation measurements, there are also some unique limitations to bear in mind. First, the frequency-dependent attenuation of radio waves by Saturn's ionosphere highlights an ambiguity regarding the observed cutoff frequency. For example, Fischer et al. (2011b) find a correlation between Cassini-Saturn distance and cutoff frequency: on average Cassini observes lower cutoff frequencies (and therefore derives lower N_{MAX} values) when it is closer to Saturn. There are relatively few SED episodes with Cassini closer than 5 R_S, however, so it is not clear whether this trend continues radially inward towards Saturn. If so, it may imply that the current derived N_{MAX} values should be reduced in magnitude. It is not immediately obvious how such a correlation would affect the Voyager era results, though it is worth noting that the anomalously low frequency cutoffs occurred during Voyager's closest approach (Kaiser et al., 1984). These anomalous low-frequency cutoffs also occurred near ~300–600 kHz, where Saturn kilometric radiation, or SKR, typically dominates the frequency spectrum, a fact that can complicate SED analysis (Fischer et al., 2011b).

The second main limitation of SED observations is the uncertainty regarding the portion of the ionosphere sampled by the transiting radio waves. Over-horizon SEDs have been observed regularly by Cassini – that is, SED detections prior to the rising of their originating storm above the horizon as seen by the spacecraft (Fischer et al., 2008). These SEDs are likely a result of ducting (Zarka et al., 2006), wherein propagating radio waves are refracted by the ionosphere, and their detection emphasizes that one cannot rely on the assumption that SEDs traverse a straight line from their origin to the observer. Consequently, it is possible that SEDs sample portions of the ionosphere different from where the radio signals originate. For example, shadowing by Saturn's rings leads to patterns of depleted electron density that depend on season, and may help explain the anomalously low frequency cutoffs observed by Voyager (Burns et al., 1983; Mendillo et al., 2005). Nevertheless, despite the above limitations, SEDs provide valuable insight to Saturn's ionosphere as well as an additional observational constraint that remains to be explained: the strong diurnal variation of N_{MAX} (of 1–2 orders of magnitude). While weak SED-like radio spikes were detected by Voyager 2 during its encounters with Uranus (Zarka and Pedersen, 1986) and Neptune (Kaiser et al., 1991), no high-frequency radio emission from lightning was detected at Jupiter by any visiting spacecraft, despite whistler and optical lightning detections (Zarka et al., 2008), possibly due to ionospheric attenuation (Zarka, 1985b) or to slow lightning discharge (Farrell et al., 1999).

8.2.3 Observations of Ionospheric H_3^+

Since its initial discovery in the Jovian atmosphere (Drossart et al., 1989), H_3^+ has been an effective probe of the auroral ionospheres of Jupiter (e.g., Lystrup et al., 2008; Stallard et al., 2012b, and references therein), Saturn (e.g., Stallard et al., 2008, 2012a; O'Donoghue et al., 2014, and references therein; see also Chapter 7 by Stallard et al.), and Uranus (Melin et al., 2013). There is an abundance of strong H_3^+ rotational-vibrational emission lines available in the near-IR, as described in Chapter 7. Ionization of molecular hydrogen, the dominant constituent of Saturn's upper atmosphere, leads directly to the production of H_3^+ via the rapid ion-molecule reaction ($H_2^+ + H_2 \rightarrow H_3^+ + H$). Therefore, H_3^+, along with H^+, is expected to be a major ion in giant planet ionospheres. As H_3^+ is in quasi-local thermodynamic equilibrium with the neutral atmosphere in the collisional region of the ionosphere, it also serves as a valuable probe of upper atmospheric temperatures (e.g., Miller et al., 2000).

At Jupiter, where ionospheric densities and upper atmospheric temperatures are relatively large, H_3^+ can be observed at all latitudes. Low- and mid-latitude H_3^+ column densities at Jupiter are approximately ~$3–5\times10^{11}$ cm^{-2}, whereas auroral column densities can be more than 10^{12} cm^{-2} (Lam et al., 1997; Miller et al., 1997). Auroral column densities at Saturn have been measured to be of the same order, though variable, with reported values between ~$1–7\times10^{12}$ cm^{-2} (Melin et al., 2007). The strong dependence of H_3^+ emission on temperature, however, has inhibited searches for non-auroral H_3^+ at Saturn, due to the low equatorial temperatures described in Chapter 9 by Strobel et al. Prior to 2011, the lowest latitude of detected H_3^+ at Saturn was ~57° (planetographic south), a weak secondary auroral oval thought to be associated with the breakdown in corotation within the magnetosphere (Stallard et al., 2008, 2010). Observations made in April 2011 using the Near-InfraRed Spectrograph (NIRSPEC) instrument on the Keck telescope detected the first low- and mid-latitude H_3^+ emission at Saturn (O'Donoghue et al., 2013). The measured emission revealed significant latitudinal structure, with local extrema in one hemisphere being approximately mirrored at magnetically conjugate latitudes in the opposite hemisphere. Furthermore, those minima and maxima mapped along magnetic field lines to distinct regions of Saturn's rings, implying a direct ring-atmosphere connection. These low- and mid-latitude H_3^+ emission structures have therefore been interpreted as representing the ionospheric signatures of "ring rain," a process wherein charged water products from Saturn's rings are transported along magnetic field lines into its atmosphere (Connerney, 2013). The spectral images obtained during those observations are shown in Figure 8.5, and the H_3^+ intensities plotted as a function of planetocentric latitude are given in Figure 8.6.

While Figures 8.5 and 8.6 indicate there is some connection between Saturn's rings and its ionosphere, the nature of that connection is not definitively revealed by the measurements. The detection was too weak to derive column-integrated H_3^+ temperatures, as is commonly done for auroral H_3^+ measurements (e.g., Melin et al., 2007), and consequently there is an ambiguity behind whether the emission structure is driven primarily by temperature or density variations. Based on earlier work, an enhanced influx of ring material is expected at latitudes that map to the inner edge of Saturn's B ring, roughly 44°N and 38°S planetocentric latitude (e.g., Connerney and Waite, 1984; Connerney, 1986; see also Section 8.2.1). These regions do correspond to local maxima in H_3^+ emission (Figure 8.6). Saturn's rings are predominantly water ice bodies (e.g., Cuzzi et al., 2010, and references therein), and thus some sort of water product, such as sub-micrometer-sized grains with a high charge-to-mass ratio (Connerney, 2013), is the most likely ring material to precipitate into its atmosphere. More recent models of the ring ionosphere, prompted by the detection of O_2^+ and O^+ ions by Cassini during Saturn Orbital Insertion (Tokar et al., 2005; Waite et al., 2005), also predict a significant influx

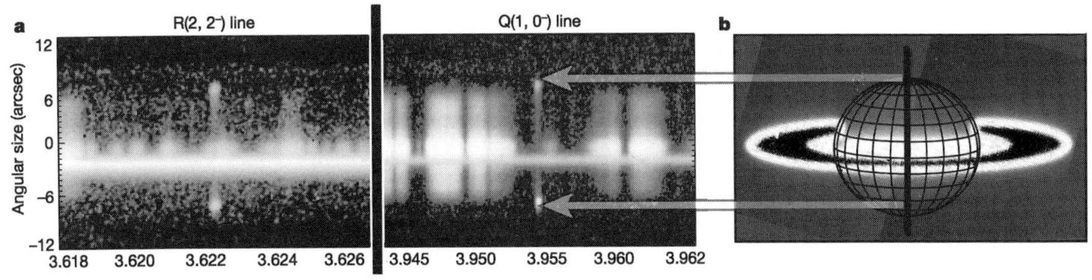

Figure 8.5 Two spectral regions are indicated in **(a)**, surrounding two bright rotational-vibrational H_3^+ emission lines: the $R(2,2^-)$ line at 3.622 μm and the $Q(1,0^-)$ line at 3.953 μm. The vertical axis corresponds to spatial position along the slit, which is oriented north-south along Saturn's rotational axis, as illustrated in **(b)**. Uniform reflection of sunlight by the rings can be seen as a bright horizontal patch of emission near −3 arcsec in (a). Remaining bright areas in the spectra are due to methane reflection. Saturn's aurorae are visible as the bright regions at either end of the $R(2,2^-)$ and $Q(1,0^-)$ lines, though emission is also clearly seen across the entire disk. From O'Donoghue et al. (2013).

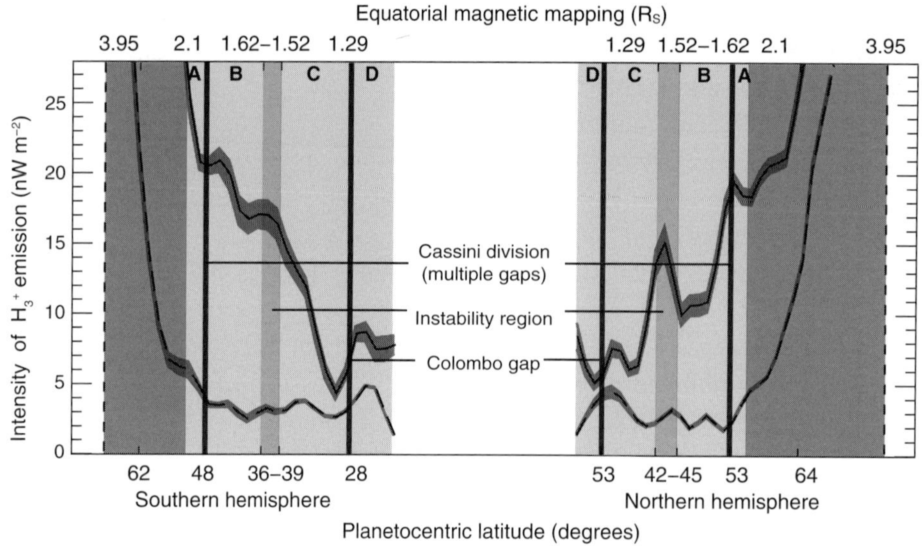

Figure 8.6 Intensity of H_3^+ infrared emission as a function of latitude along Saturn's noon meridian, based on the spectra of Figure 8.5. Planetocentric latitudes are indicated along the bottom axis, while the upper axis gives the equatorial radius each latitude element maps to along magnetic field lines. Gray shading envelopes surrounding the Q(1,0$^-$) (solid curve) and R(2,2$^-$) (dashed curve) spectral lines indicate the 1σ errors in intensity. From O'Donoghue et al. (2013).

of those ions into Saturn's atmosphere (Luhmann et al., 2006; Tseng et al., 2010). A ring-atmosphere current system may lead to localized temperature enhancements due to Joule heating (O'Donoghue et al., 2013), though preliminary calculations find the Joule heating rates to be negligible (Crary, 2014). Therefore, the observed H_3^+ emission maxima more likely represent either an enhancement in density in those regions or depletions in density in surrounding regions. An influx of water into Saturn's atmosphere has been invoked in the past in order to explain radio occultation observations, though the effect of such an influx on H_3^+ densities has been largely neglected until now. Hereafter, for simplicity, "water" is used as a catch-all term to describe ring-derived influxes into Saturn's atmosphere, as they are all expected to have the same primary effect: increased H^+ loss chemistry leading ultimately to a reduced electron density.

8.3 Models

We have outlined the available observational constraints for Saturn's ionospheric parameters in Section 8.2. We now give an overview of the basic theory used to explain these measurements, and review the past modeling studies (Section 8.3.1). A more thorough description of past ionospheric modeling studies can be found in Nagy et al. (2009), and so we touch only briefly on that history here. Contemporary model-data comparisons are highlighted in order to give context to the current state of knowledge and to emphasize outstanding model-data discrepancies (Section 8.3.2).

8.3.1 Basic Theory: Chemistry, Ionization, and Temperature

Molecular hydrogen is the dominant constituent in Saturn's upper atmosphere, with atomic hydrogen becoming important at higher altitudes. At non-auroral latitudes, the primary source of ionization is solar X-ray (0.1–10 nm) and EUV (10–110 nm) photons. As the dominant constituent, H_2 absorbs most of the incident radiation at those wavelengths, and so the vast majority of photo-produced ions are H_2^+. Maximum photoionization rates (i.e., overhead illumination for solar maximum photon fluxes) for H_2^+ and H^+ in Saturn's ionosphere are roughly 10 cm^{-3} s^{-1} and 1 cm^{-3} s^{-1}, respectively (Moore et al., 2004). Direct photoionization of methane also leads to an array of hydrocarbon ions; CH_4^+ is produced most rapidly, with a 2–3 cm^{-3} s^{-1} production rate, though these ions are produced much lower in the ionosphere, near the homopause, where their parent species are located (Kim et al., 2014). While H_2^+ is the dominant species created by photoionization, it is quickly converted to H_3^+ via the charge-exchange reaction (Theard and Huntress, 1974; see also Miller et al., 2006, and references therein):

$$H_2^+ + H_2 \rightarrow H_3^+ + H \tag{8.1}$$

Saturn's ionosphere is thus predicted to be dominated by a mix of H_3^+ and H^+ ions near the electron density peak and above, with an additional ledge of hydrocarbon ions below the peak, closer to the homopause (e.g., Majeed and McConnell, 1991; Moses and Bass, 2000; Moore et al., 2004; Kim et al., 2014). The lower ionosphere (e.g., ~2300 km and below for mid-latitudes) is expected to be in photochemical equilibrium, meaning that chemical processes dominate over plasma dynamics; for such conditions the H^+/H_3^+ number density ratio is proportional to electron density (Moore et al., 2004). Dissociative recombination between H_3^+ and electrons – the dominant loss for H_3^+ ions – is rapid relative to the Saturn day. Typical H_3^+ lifetimes are ~10–15 minutes in Saturn's ionosphere (Melin et al., 2011; Tao et al., 2011), much shorter than Saturn's ~10 hour rotation period (see Chapter 5 by Carbary et al.). Consequently, for conditions of reduced electron density, the H^+/H_3^+ ratio decreases, and the expected diurnal variation of the electron density in the main ionosphere increases (i.e., a smaller H^+/H_3^+ ratio means a larger fraction of short-lived H_3^+ ions).

Early models of Saturn's ionosphere (e.g., McElroy, 1973; Capone et al., 1977) predicted a peak electron density an order of magnitude larger than subsequent Pioneer and Voyager radio occultation measurements revealed (Kliore et al., 1980; Lindal et al., 1985). The only chemical loss included for H^+ in these early models was radiative recombination, an extremely slow process. Therefore, in order to reduce modeled electron densities to better reproduce the observed values, a method of converting H^+ to a short-lived ion was required. The two methods considered by most subsequent models are charge exchange between H^+ and vibrationally excited H_2, and charge exchange with water introduced into the atmosphere from Saturn's rings and/or icy moons.

Vibrationally Excited H_2

As was recognized early on (McElroy, 1973), the charge-exchange reaction

$$H^+ + H_2(\upsilon \geq 4) \rightarrow H_2^+ + H \tag{8.2}$$

is exothermic only when H_2 is in the 4th or higher vibrational level. While there has historically been some uncertainty regarding the 8.2 reaction rate, it has generally been assumed to be of the order 1 to 2×10^{-9} cm^3 s^{-1} (e.g., McConnell et al., 1982; Cravens, 1987). Recent extrapolation of work by the plasma fusion quantum theory community (e.g., Ichihara et al., 2000; Krstić, 2002) has been used to refine the estimated 8.2 reaction rate (at 600 K) only slightly, to 0.6 to 1.3×10^{-9} cm^3 s^{-1} (Huestis, 2008). Of far greater uncertainty is the population of non-LTE vibrationally excited H_2, where LTE refers to local thermodynamic equilibrium. (If the H_2 vibrational distribution were in LTE at 1000 K, fewer than 10^{-10} molecules would be in the $\upsilon\geq4$ state, far too low a value to significantly impact H^+ densities.) Molecular hydrogen in giant planet upper atmospheres can be vibrationally excited by electron impact, by solar and electron excitation of the H_2 Lyman and Werner bands, which then fluoresce to vibrationally excited levels in the ground state, and by dissociative recombination of H_3^+ (e.g., Waite et al., 1983; Cravens, 1987; Majeed et al., 1991). Major vibrational loss processes include de-excitation through collisions with H and H_2, reactions with H^+ (i.e., 8.2), and redistribution of vibrational quanta among molecular levels through vibrational-vibrational (V-V) collisions and in altitude through diffusion.

The first detailed model for $H_2(\upsilon)$ in the outer planets was presented by Cravens (1987) for Jupiter. Majeed et al. (1991) added solar fluorescence in their low-latitude solar input $H_2(\upsilon)$ calculations for Jupiter and Saturn, finding it to be a dominant source of vibrational excitation. Both Cravens (1987) and Majeed et al. (1991) found modest enhancements in $H_2(\upsilon\geq4)$ populations, leading to reductions in calculated electron densities that were insufficient to reproduce the observed electron density profiles. In general, calculations of $H_2(\upsilon\geq4)$ appear to fall short of the enhanced vibrational populations required to bring modeled and observed electron densities into agreement, possibly due to uncertainties in the calculations (such as rate coefficients or source mechanisms; Majeed et al., 1991), or possibly due to other processes acting to reduce H^+ densities. More recently, Hallett et al. (2005a) developed a new rotational-level hydrogen physical chemistry model, and subsequently applied it to Uranus (Hallett et al., 2005b).

Subsequent model reproductions of electron density profiles from radio occultation observations have typically modified $H_2(\upsilon \geq 4)$ populations freely or have used scaled versions of the Majeed et al. (1991) calculations (e.g., Majeed and McConnell, 1996; Moses and Bass, 2000; Moore et al., 2006). In other words, partly due to uncertainties in direct $H_2(\upsilon \geq 4)$ calculations, and partly due to attention being focused on other details of the Saturn ionosphere, contemporary models have predominantly treated the population of vibrationally excited H_2 as a free parameter. Reaction 8.2 directly reduces H^+ ion densities by converting H^+ to short-lived molecular ions, thereby indirectly reducing the net electron density and correspondingly reducing the dominant chemical loss for H_3^+ – dissociative recombination with electrons. Due to the long lifetime of H^+, 8.2 can also act as an additional source of H_2^+ (and H_3^+ via 8.1), even in regions absent of ionizing radiation or precipitating energetic particles. Therefore, as it affects all of the major chemistry, reaction 8.2 remains a major point of uncertainty for ionospheric calculations. There are no direct observational constraints of the vibrational populations of H_2 published at present.

Water in Saturn's Ionosphere

A second likely method of converting H^+ ions into short-lived molecular ions, thereby depleting the calculated electron densities, begins with an influx of water into Saturn's atmosphere. Possible external water sources include micrometeorites as well as Saturn's rings and icy satellites. This ionospheric quenching chain was first postulated by Shimizu (1980), implemented by Chen (1983), and treated comprehensively by Connerney and Waite (1984).

$$H^+ + H_2O \rightarrow H_2O^+ + H \quad (8.3)$$

$$H_2O^+ + H_2 \rightarrow H_3O^+ + H \quad (8.4)$$

$$H_3O^+ + e \rightarrow \begin{cases} H_2O + H & (8.5a) \\ H_2 + OH & (8.5b) \\ H + H + OH & (8.5c) \end{cases}$$

As noted by Connerney and Waite (1984), any OH in the system (such as that from 8.5) has a short lifetime due to its reaction with H_2, producing H_2O and H. Similarly, any ionized water products, such as O_2^+, dissociatively recombine with electrons extremely rapidly – roughly three times faster than H_3^+ in Saturn's ionosphere – leading to a chain of photochemical reactions that produce primarily OH (via $O + H_2$) and H_2O (via $OH + H_2$) in the thermosphere and lower atmosphere (e.g., Moses and Bass, 2000; Moses et al., 2000). Hence, while the exact form of exogenous influx may not always be pure H_2O, the ionospheric chemical effects are similar.

The various reaction rates for water chemistry in outer planet upper atmospheres are relatively well known (e.g., Moses and Bass, 2000; Moses et al., 2000). Of far greater uncertainty is the magnitude of water influx at Saturn, as well as its spatial and temporal distribution and variability. While a number of modeling studies have derived a range of water influxes indirectly as a means of reproducing the electron density profiles from radio occultations (e.g., Connerney and Waite, 1984; Majeed and McConnell, 1991, 1996; Moore et al., 2006, 2010), directly constraining the influxes observationally has proven more difficult. The first unambiguous detection of water in Saturn's upper atmosphere came from the Infrared Space Observatory (ISO; Feuchtgruber et al., 1997), which measured an H_2O column abundance of (0.8 to 1.7) $\times 10^{15}$ cm^{-2} and was used to derive a global water influx of $\sim 1.5 \times 10^6$ H_2O molecules cm^{-2} s^{-1} (Moses et al., 2000). Subsequent studies based on Sub-millimeter Wave Astronomy Satellite and Herschel Space Observatory measurements found global influx values within a factor of four of the Moses et al. result (Bergin et al., 2000; Hartogh et al., 2011). Despite predictions of strong latitudinal variations in water influx (e.g., Connerney, 1986), no observational confirmation of such variations has yet been published. At present, there are ambiguous detections of latitudinally varying water volume mixing ratios in the ultraviolet (e.g., a 2σ detection of 2.70×10^{16} cm^{-2} at 33°S planetocentric latitude: Prangé et al., 2006) as well as preliminary indications of larger equatorial water densities from Cassini Composite InfraRed Spectrometer (CIRS) observations (Bjoraker et al., 2010) and from further Herschel observations (Cavalié et al., 2014).

A number of different categories of theoretical studies support the preliminary observational results that favor a latitudinal variation of water influx at Saturn. The first category includes ring-modeling studies focused on the erosion of Saturn's rings (e.g., Northrop and Hill, 1982, 1983; Ip, 1983; Northrop and Connerney, 1987). One of the outcomes of such studies is the demonstration that small, negatively charged ring grains inside of a marginal stability radius of 1.525 R_S can be lost to Saturn's atmosphere along magnetic field lines. A recent related study also explores the evolution of positively charged ring grains, and finds that they are sometimes deposited in Saturn's equatorial atmosphere (Liu and Ip, 2014). A second category of studies is focused on the ring atmosphere and ionosphere, and likewise predicts a precipitation of ring ions into Saturn's atmosphere (e.g., Luhmann et al., 2006; Tseng et al., 2010). In general, all of the ring models predict an asymmetry in the particle influx due to Saturn's slightly offset magnetic dipole, with stronger influxes expected in the southern hemisphere. Finally, a third main category of modeling studies that predict a latitudinal variation of water influx at Saturn are those that track the water vapor ejected from Enceladus' plumes (Porco et al., 2006). These models estimate that approximately 10, 7, 3, and 6 percent, respectively, of the Enceladus water is lost to Saturn's atmosphere (Jurac and Richardson, 2007; Cassidy and Johnson, 2010; Hartogh et al., 2011; Fleshman et al., 2012). Latitude variations of Saturn water influx from the Enceladus models vary, though in general a stronger influx is predicted at low latitudes. Finally, temporal (and possibly seasonal) variability is also expected for both ring and Enceladus sources of water, though it is not well constrained at present.

Primary and Secondary Ionization

Ionizing radiation at the giant planets comes primarily in two forms: solar photons and energetic particles. Electrons released during ionization are usually suprathermal, meaning that they are more energetic than the ambient electron population. These suprathermal electrons – referred to as photoelectrons in the case of photoionization and secondary electrons in the case of particle impact ionization – possess enough energy to excite, dissociate, and further ionize the neutral atmosphere as well as to heat the plasma. On the one hand, photoionization production rates follow from a fairly straightforward application of the Lambert–Beer Law, at least assuming that the neutral atmospheric densities, incident solar fluxes, and photoabsorption and photoionization cross-sections are known. Conversely, a more comprehensive treatment is required in order to track the transport, energy degradation, and angular redistribution of suprathermal electrons. Two main techniques have been applied in the past: a kinetic approach, such as solution to the Boltzmann equation using a multi-stream code (Perry et al., 1999; Moore et al., 2008; Galand et al., 2009; Gustin et al., 2009), and a Monte Carlo approach (e.g., Tao et al., 2011). While ionization due to photoelectrons and secondary electrons is usually included in the case of Venus, Earth, Mars, Jupiter, and Titan (e.g., Kim et al., 1989; Kim and Fox, 1994; Schlesier and Buonsanto, 1999; Cravens et al., 2004; Fox, 2004, 2007; Fox and Yeager, 2006; Galand et al., 2006; Matta et al., 2014), it is this added complexity that has prevented secondary ionization from being treated in the majority of Saturn ionospheric models.

Higher-energy photons produce higher-energy photoelectrons, and therefore lead to more secondary ionization. In general, photoabsorption cross-sections are smallest for high-energy photons and increase nearly monotonically over their ionization range. There are certainly exceptions to this generality (e.g., methane near 4 nm), however it holds true for H_2 for photons with wavelength >1 nm – the dominant absorber in the outer planets – meaning that more energetic photons are typically absorbed lower in the atmosphere at Saturn (Moses and Bass, 2000; Galand et al., 2009).

Secondary ionization due to solar illumination, therefore, primarily increases the ion production rate in the lower ionosphere.

In the auroral regions, precipitating particles interact with the ambient atmosphere via collisions, leading to excitation, ionization and heating. About half of all inelastic collisions between precipitating energetic electrons and Saturn's upper atmosphere result in the ionization of H_2 that is initially in the electronic ground state ($X^1\Sigma_g^+$), producing both H_2^+ and secondary electrons e_s:

$$e + H_2 \rightarrow H_2^+ + e_s + e \tag{8.6}$$

Since this process does not remove the energetic electrons, the precipitating electrons and their secondary products undergo further inelastic collisions, producing additional ionization, excitation, and dissociation in the atmosphere, as well as heating of thermal, ionospheric electrons. These energetic particles lose kinetic energy with each collision until they are finally thermalized with the surrounding atmosphere. Atmospheric effects due to precipitating energetic electrons (e.g., Galand et al., 2011; Tao et al., 2011), such as ionization and heating, are discussed in more detail in Chapters 7 and 9.

There are two studies that treat secondary ionization comprehensively in Saturn's non-auroral ionosphere. The first, Galand et al. (2009), solves the Boltzmann equation for suprathermal electrons using a multistream transport model based on the solution proposed by Lummerzheim et al. (1989) for terrestrial applications. A simple parameterization of secondary ionization rates based on the Galand et al. (2009) results appears in Moore et al. (2009), accurate over a range of solar/seasonal conditions and latitudes. A number of related modeling studies include either the Galand et al. (2009) results or the Moore et al. (2009) parameterization (e.g., Moore et al., 2010; Barrow and Matcheva, 2013). The second study to calculate secondary ionization rates at Saturn directly, Kim et al. (2014), assumes that photoelectrons deposit their energy locally using a simple method described by Dalgarno and Lejeune (1971). A similar approach has been used for Jupiter (e.g., Kim and Fox, 1994). Both the Galand et al. (2009) and the Kim et al. (2014) studies are for mid-latitude, and both find that the secondary ionization production rates are roughly comparable to primary photoionization rates just below 1000 km altitude (i.e., for pressures greater than ~10^{-5} mbar). At lower altitudes (higher pressures) secondary ionization production rates are dominant by as much as an order of magnitude. The effect on calculated ion and electron densities is also altitude dependent: electron densities are increased by roughly 30 percent at the peak and by up to an order of magnitude at lower altitudes (Galand et al., 2009). The impact of secondary ionization on Saturn ionospheric electron densities is illustrated in Figure 8.7, which shows the ratio of calculated electron densities between simulations that do and do not include the extra production term. From Figure 8.7, it is clear that models of Saturn's ionosphere that do not account for secondary ionization in some form will significantly underestimate electron densities in Saturn's lower ionosphere.

Hydrocarbon Ions

Most of the preceding discussion has focused on H^+ and H_3^+, as those are the ions predicted to be dominant throughout most of Saturn's ionosphere. There is, however, an additional predicted ledge of low-altitude ionization thought to be dominated by hydrocarbon (and possibly metallic) ions, just above the homopause. Many models treat the hydrocarbon layer as an ionospheric sink, if they consider it at all, as methane readily charge-exchanges with H^+ and H_3^+, leading to a relatively short-lived molecular hydrocarbon ions (e.g., Moore et al., 2008, and related studies). Such

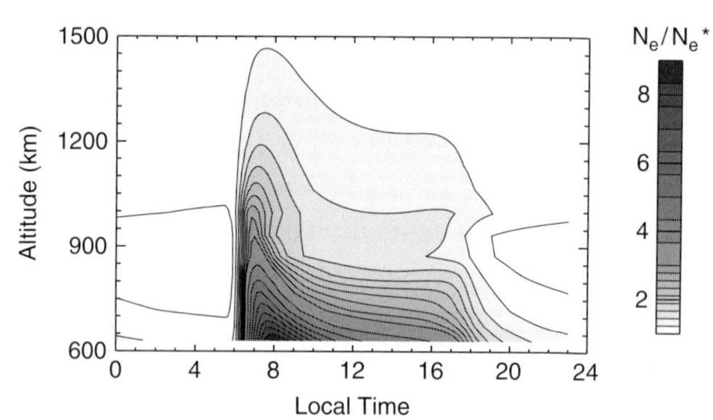

Figure 8.7 Contour plot of the electron density ratio between a simulation that includes secondary ionization (N_e) and a run that ignores secondary ionization (N_e^*). Both simulations are for solar minimum conditions at mid-latitude. From Galand et al. (2009). (A black-and-white version of this figure appears in some formats. For the color version, please refer to the plate section.)

a treatment can lead to fairly accurate electron densities within the hydrocarbon layer, at least when compared with more comprehensive models, but the resulting hydrocarbon ion composition is incorrect (e.g., Moses and Bass, 2000; Moore et al., 2008).

There are two main complications that models must address in order to treat hydrocarbon ions comprehensively. First, there are numerous hydrocarbon ions and a significantly more complex series of reactions to consider. Depending on the modeling approach, this may only hinder results by requiring a larger table of ions and reactions to be inserted, though even in that case many of the reaction rates are unknown or untested in the laboratory. The two models that treat the hydrocarbon layer at Saturn most comprehensively are Moses and Bass (2000) and Kim et al. (2014). Moses and Bass consider 109 different ion species with 845 reactions, while Kim et al. track 53 ions using 749 reactions. Note that models developed for Titan's rich high-molecular-weight hydrocarbon atmosphere include an even more complete list of reactions and ions (e.g., Waite et al., 2010; Vuitton et al., 2015). The second complication that needs to be addressed for accurate calculations of hydrocarbon ion densities is that high-resolution H_2 photoabsorption cross-sections (of the order of 10^{-4} nm) are required between ~80 nm (the H_2 ionization threshold) and 111.6 nm. Photons across this wavelength range, in which H_2 absorbs in discrete transitions – mostly in the Lyman, the Werner, and the Rydberg band systems – possess enough energy to ionize atomic hydrogen as well as methane, and H_2 photoabsorption cross-sections can differ by six orders of magnitude over very short wavelength scales. Calculations that use low-resolution cross-sections will absorb these photons higher in the atmosphere, on average, before they can ionize methane and other hydrocarbon neutrals near the homopause; such models consequently underpredict photoionization rates within the hydrocarbon layer. The only study so far to include high-resolution H_2 photoabsorption cross-sections at Saturn is Kim et al. (2014). Steady-state ion-density profiles from the Kim et al. (2014) calculations are shown in Figure 8.8, where the high-resolution (0.0001 nm) and low-resolution (0.1 nm) model densities are indicated by solid and dotted curves, respectively. Kim et al. (2014) find photoionization rates from the high-resolution model are much larger than those from the low-resolution model for some ion species. For example, at low altitude, near 830 km (0.2 μbar), Kim ct al. (2014) find the photo production rate of H^+ (from H) is enhanced by a factor of ~25. At a slightly lower altitude, the production rate of CH_4^+ is larger by a factor of ~22. Due to rapid photochemical loss rates, however, these differences are not as dramatic for the calculated ion densities. The high-resolution model ion densities (solid lines) are more moderately enhanced relative to the low-resolution model ion densities (dotted lines) in Figure 8.8: the sum of the ion densities at the hydrocarbon peak is ~3200 cm^{-3} and ~1800 cm^{-3} for the high-resolution and low-resolution models, respectively.

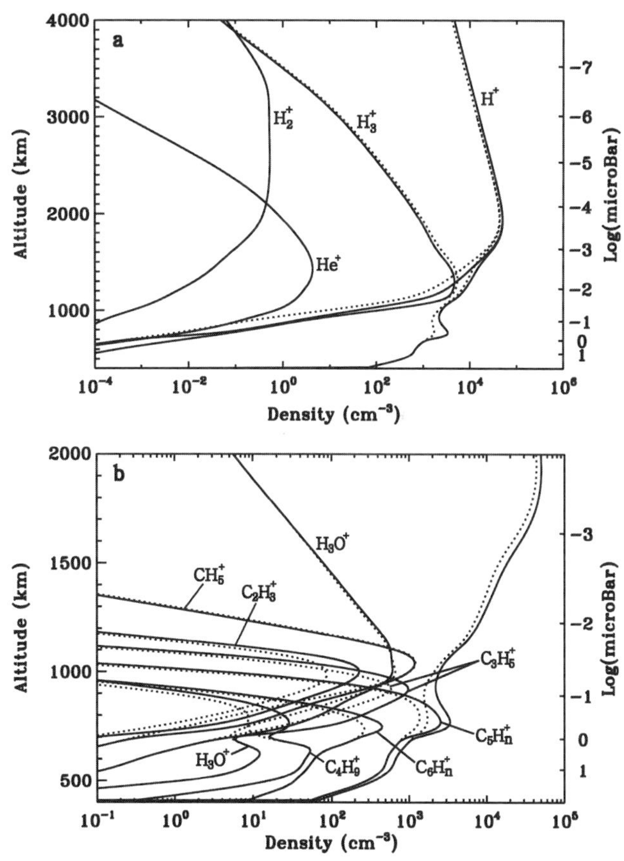

Figure 8.8 Steady-state density profiles of **(a)** (non-hydrocarbon) H- and He-bearing ions, and **(b)** major hydrocarbon and oxygen-bearing ions, as well as total electron density (thick solid line, both panels). Solid and dotted curves represent densities from the high-spectral-resolution (0.0001 nm) and low-spectral-resolution (0.1 nm) models, respectively. Note the different altitude and density scales in each panel. From Kim et al. (2014).

Plasma Temperatures

Ion, electron, and neutral temperatures are expected to deviate in the upper atmosphere of Saturn, though no *in situ* measurements have yet been made. Photoelectron (and secondary electron) interactions with the ambient plasma are likely the dominant source of heating in the non-auroral ionosphere, and therefore plasma temperature model calculations require an accurate treatment of the transport, energy degradation, and angular redistribution of suprathermal electrons. Plasma temperatures affect ionospheric model calculations primarily by altering chemical reaction and ambipolar diffusion rates.

There are two model calculations for plasma temperatures at Saturn available in the refereed literature, one at high latitudes (Glocer et al., 2007), and one at mid latitudes (Moore et al., 2008). (Two other previous studies are from Ph.D. dissertations: Waite, 1981; Frey, 1997.) Glocer et al. (2007) use a 1-D multi-fluid model to study the polar wind at Saturn, starting from below the ionospheric peak and extending to an altitude of 1 R_S, yielding densities, fluxes, and temperatures for H^+ and H_3^+. They find peak ion temperatures of roughly 2000 to 3200 K for H^+ and 1300 to 2600 K for H_3^+ – well above the main ionosphere sampled by remote auroral observations. Moore et al. (2008) self-consistently coupled a 1-D ionospheric model that solves the ion continuity, momentum, and energy equations with a suprathermal electron transport code adapted to Saturn (Galand et al., 2009). Their calculations specified a fixed mid-latitude neutral background based on results from 3-D global circulation calculations that reproduced observed thermospheric temperatures (Müller-Wodarg et al., 2006). Moore et al. (2008) found only relatively modest electron temperature enhancements during the Saturn day, calculating peak values of ~500 to 560 K (~80 to 140 K above the neutral temperature). Ion temperatures were somewhat smaller, reaching ~480 K at the topside during the day while remaining nearly equal to the neutral temperature at altitudes near and below the electron density peak. Both ions and electrons cooled to the neutral temperature within an hour after sunset. A parameterization of the thermal electron-heating rate based on primary photoionization rates was also developed (Moore et al., 2008) and then demonstrated to work for a wide variety of seasonal/solar conditions and latitudes (Moore et al., 2009).

Plasma temperatures can also be estimated from the topside scale heights of observed electron densities, though there are a number of uncertainties associated with such an estimate. For example, the ion composition has not been measured, and there may also be small altitude gradients in temperature. Both of these unknowns can lead to ambiguous results. Nonetheless, as most Saturn models predict H^+ as the dominant topside ion, especially at dawn, Nagy et al. (2006) assumed H^+ was the major topside ion and neglected possible temperature gradients in order to arrive at an estimate of 625 K based on analyzing the average low-latitude dawn radio occultation profile above 2500 km altitude. By applying the same assumptions, and by considering that dusk temperatures should be at least as large as dawn temperatures, Nagy et al. (2016) further arrive at the implication that the dusk topside might be 72 percent H_3^+, or 7.7 percent H_3O^+, or some other appropriate combination of ion fractions.

8.3.2 Model-Data Comparisons

There are five major categories of observational constraints that must be explained by models: (1) peak electron density and altitude, (2) dawn/dusk electron density asymmetry, (3) latitudinal variations in N_{MAX} and TEC, (4) diurnal variation of N_{MAX} and (5) latitudinal H_3^+ structure. While a number of these observational constraints are closely related, it is illustrative to review model-data comparisons for each separately in order to highlight the key parameters that drive each of the observables. Models typically attempt to reproduce two or more of the observables simultaneously, though this is often not possible due to the different solar, seasonal, and latitudinal conditions of the multi-instrumental observations.

Electron Density Altitude Structure

Peak electron densities in Saturn's ionosphere (N_{MAX}), based on Cassini radio occultations, range from ~0.3×10^3 to 2.6×10^4 cm^{-3} at dawn and ~3×10^3 to 1.9×10^4 at dusk. The altitude of the electron density

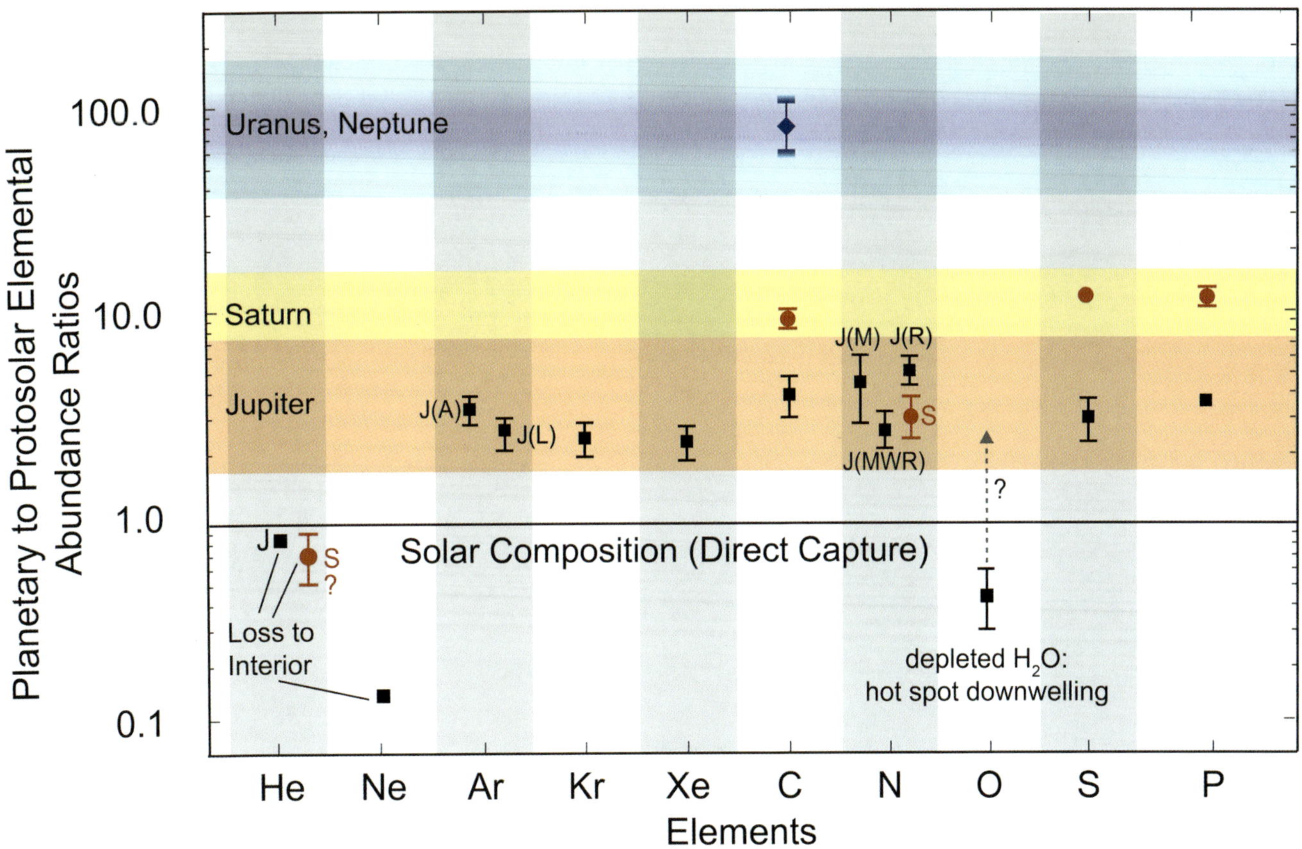

Figure 2.1

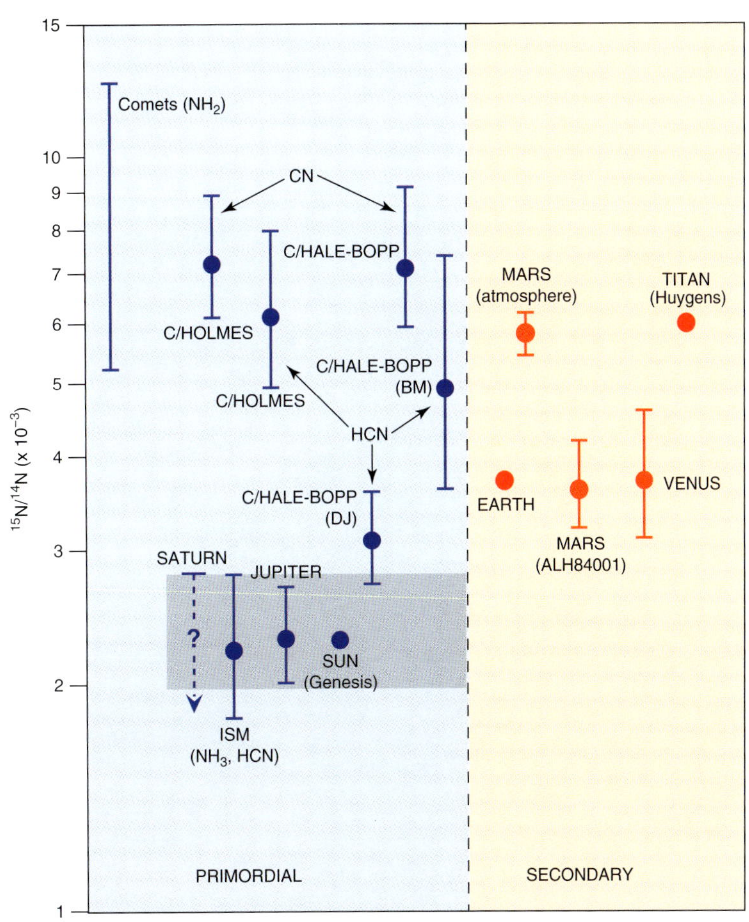

Figure 2.3

Figure 2.4

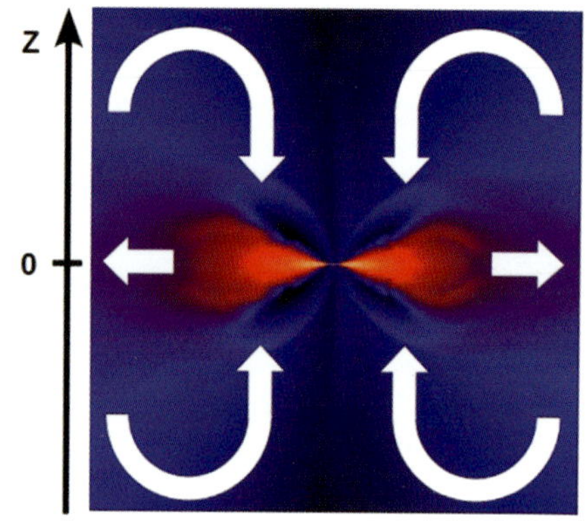

Figure 2.5

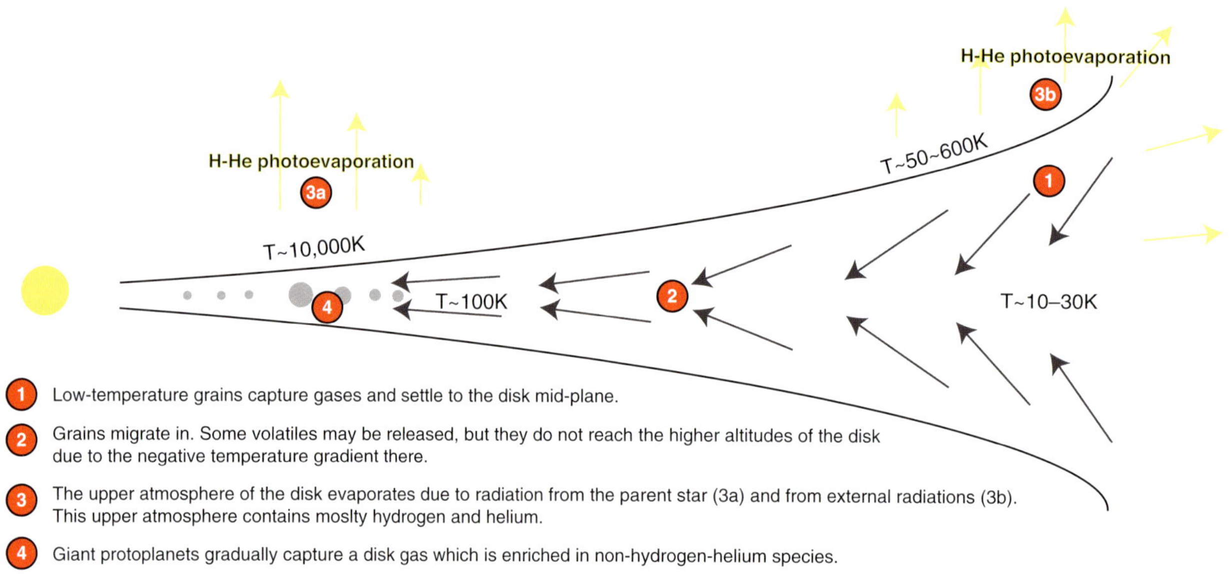

Figure 2.8

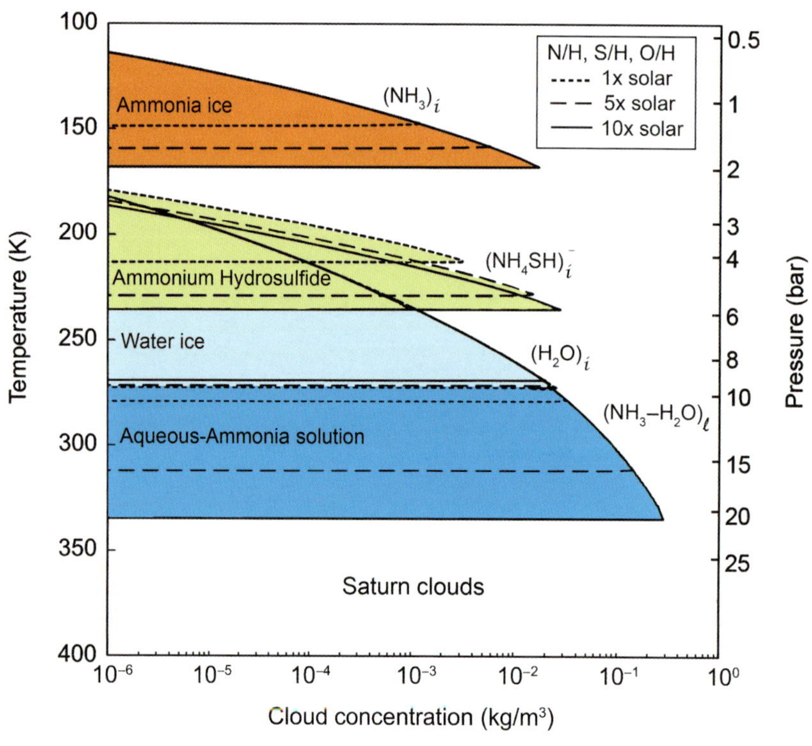

Figure 2.9

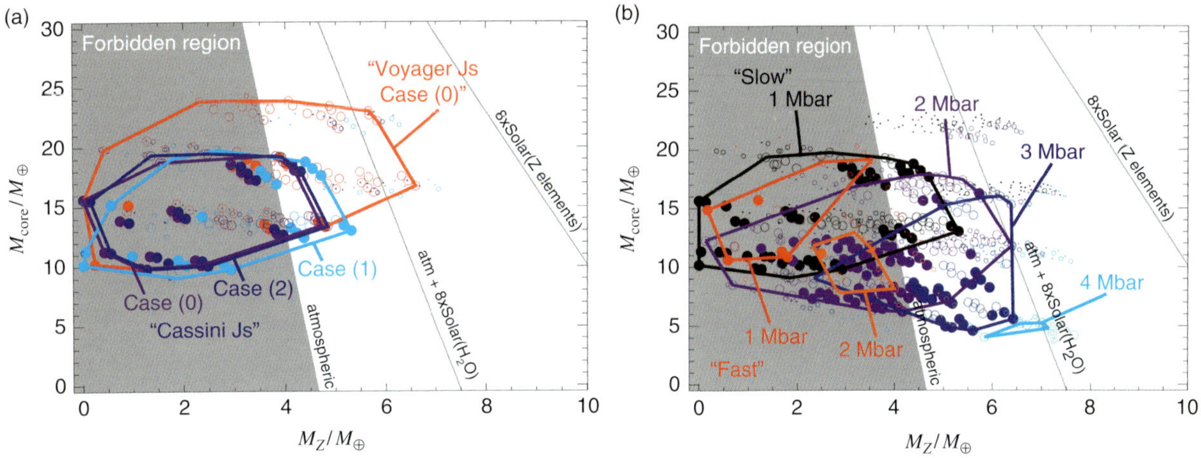

Figure 3.7

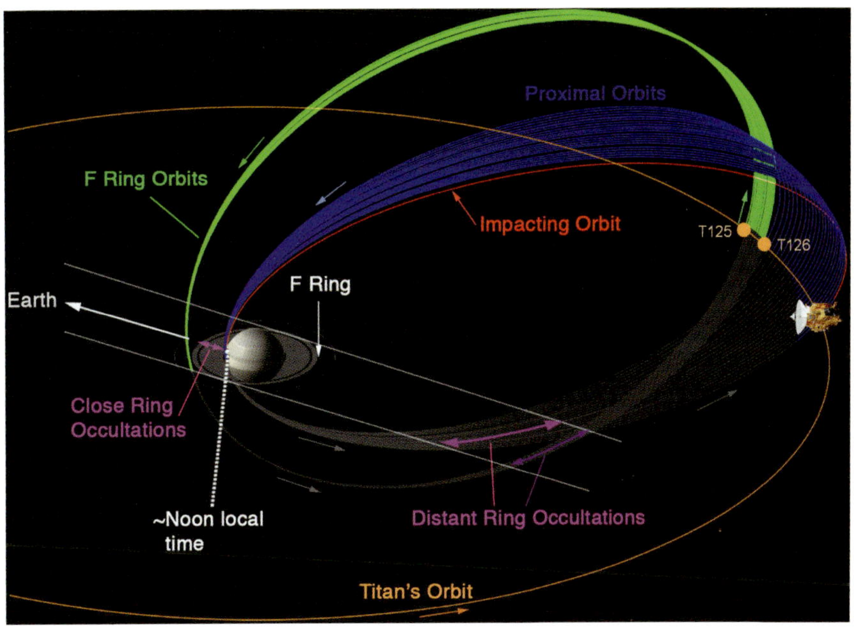

Figure 3.10

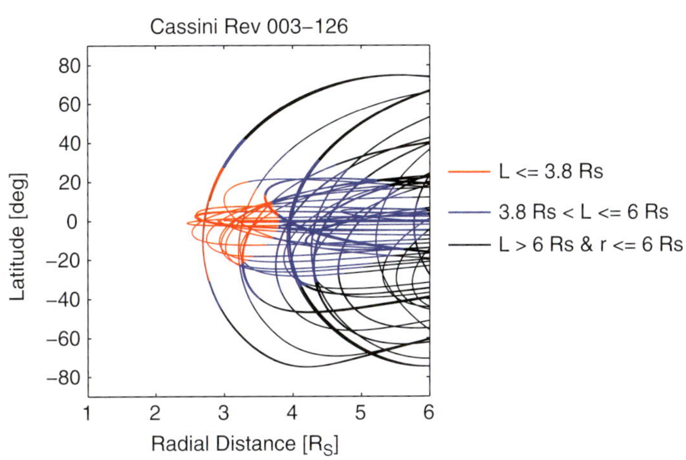

Figure 4.2

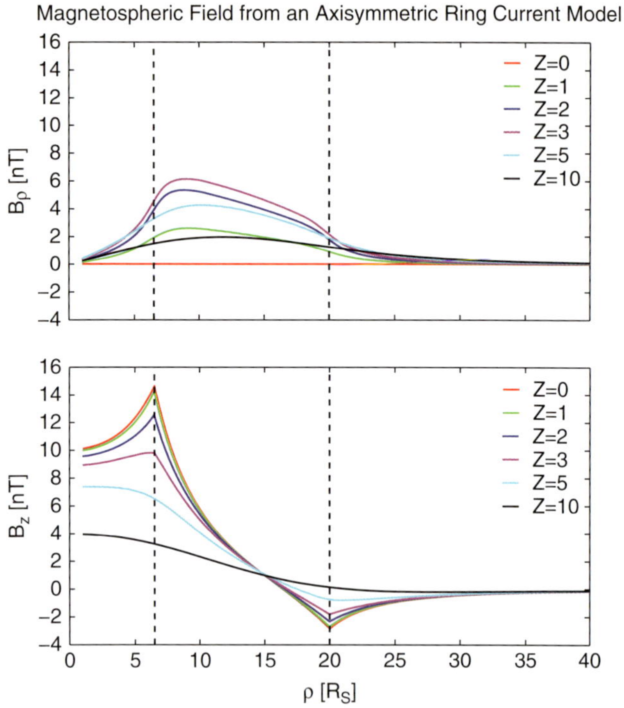

Figure 4.5

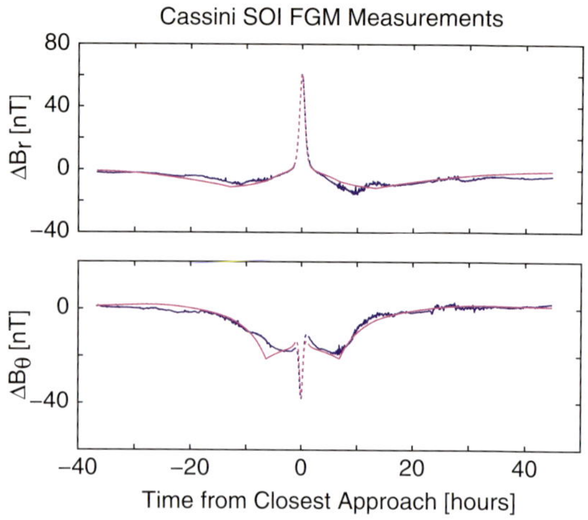

Figure 4.6

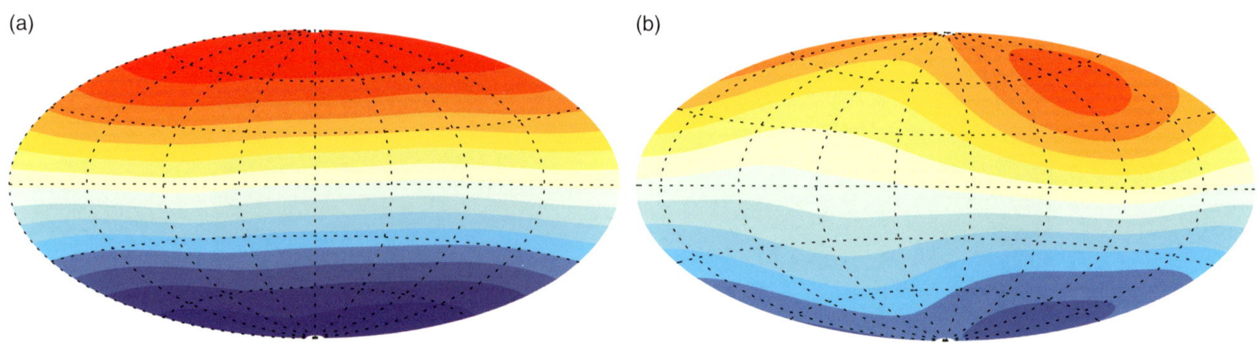

Figure 4.13

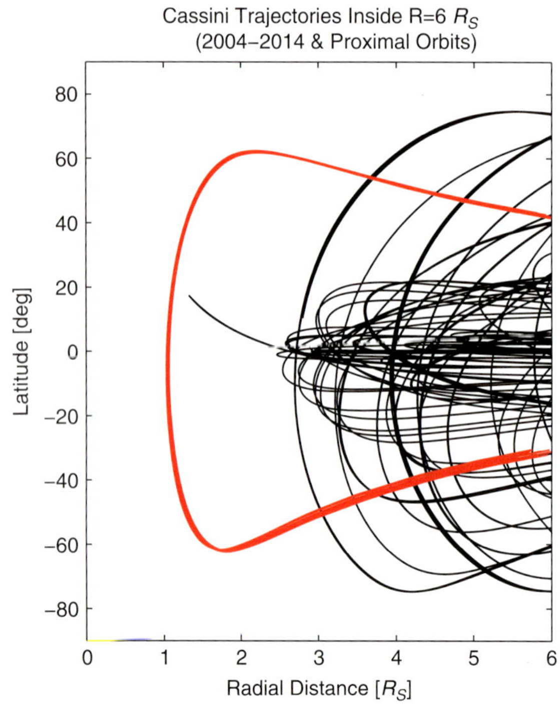

Figure 4.15

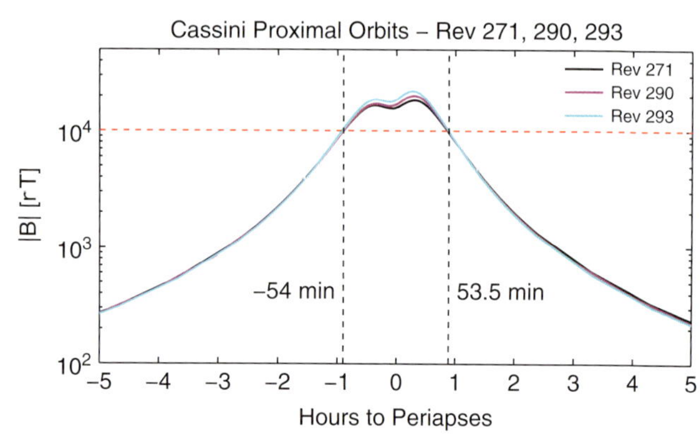

Figure 4.16

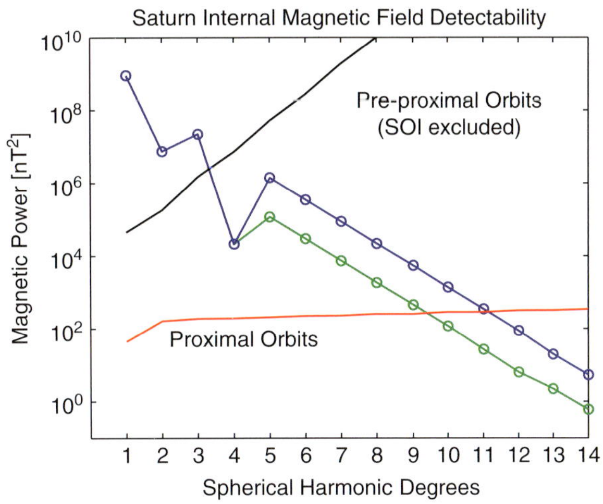

Figure 4.17

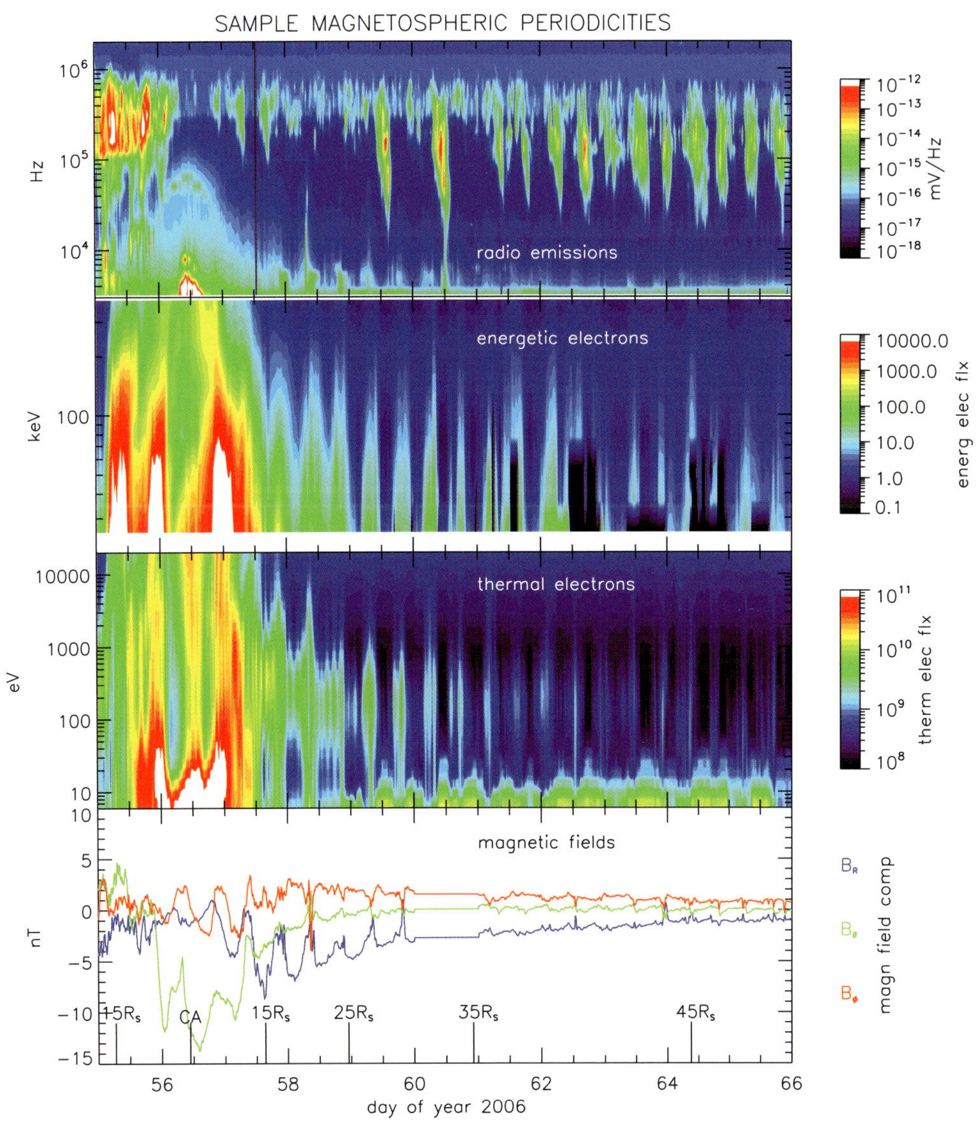

Figure 5.2

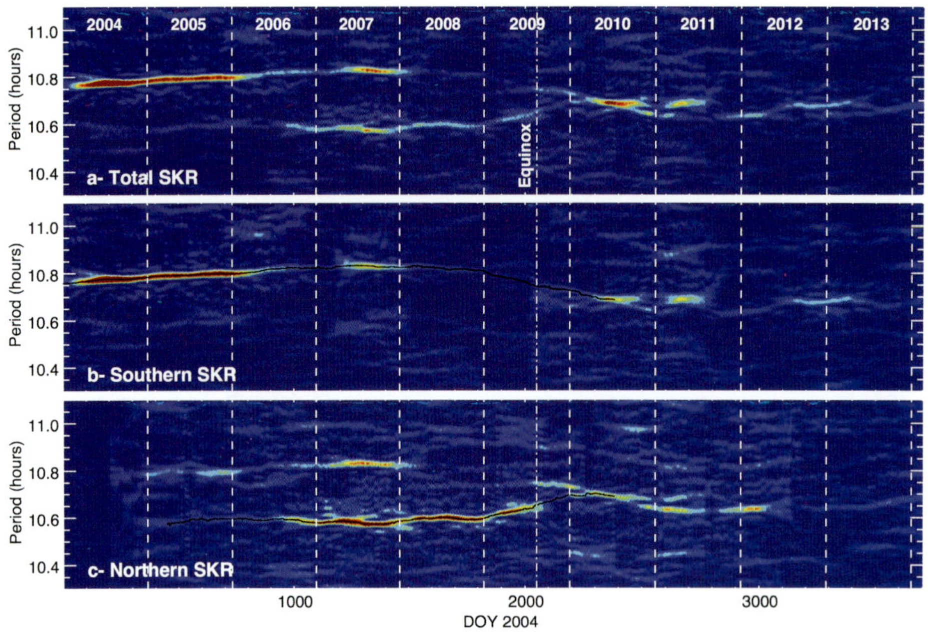

Figure 5.4

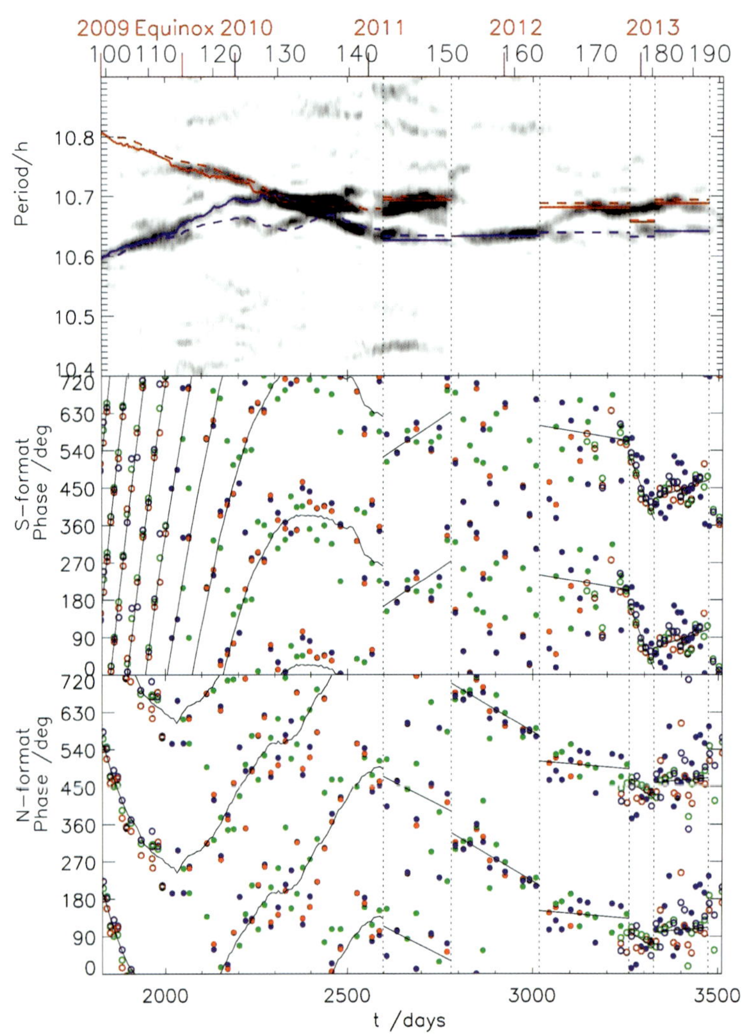

Figure 5.6

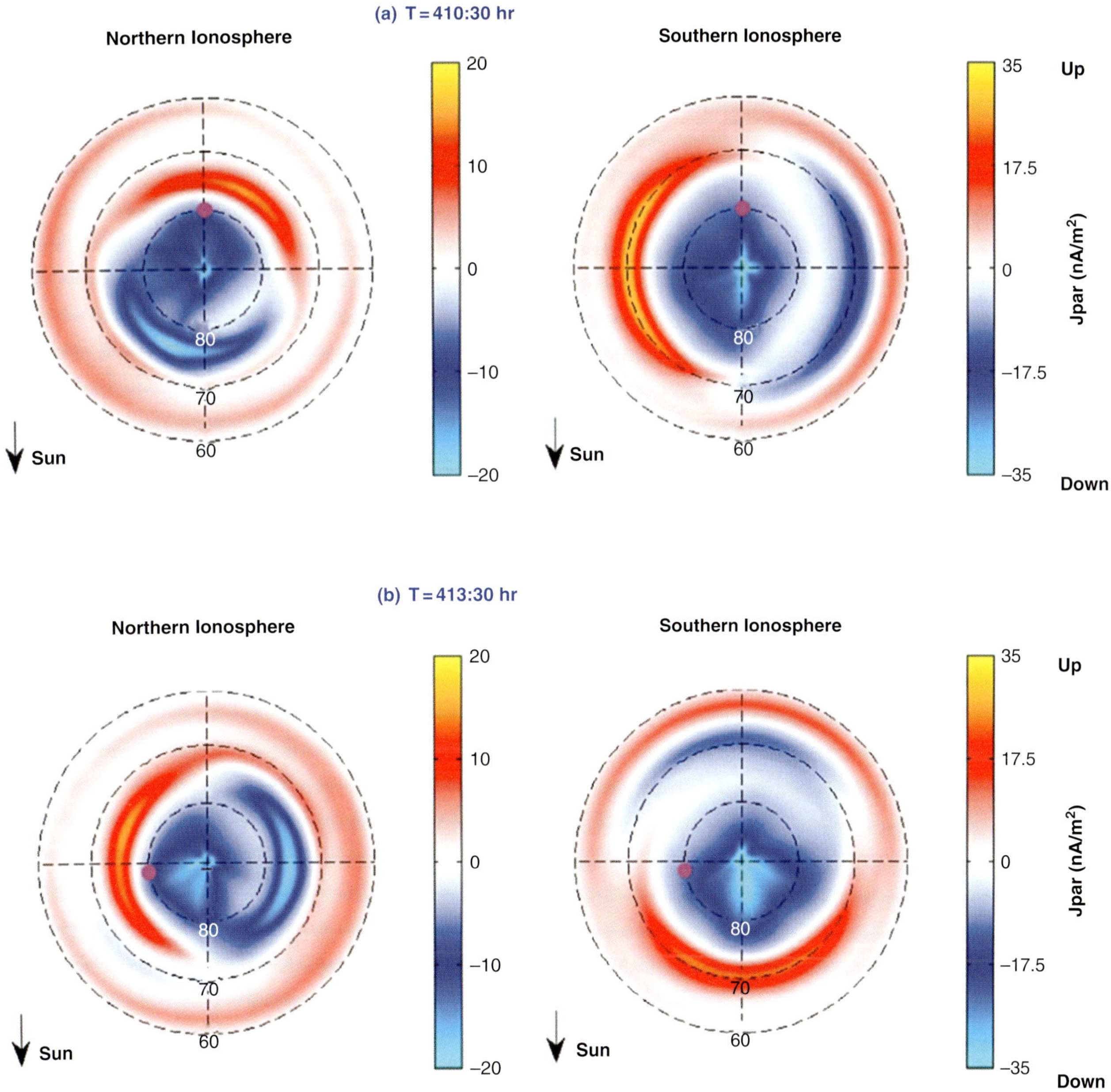
(a) T = 410:30 hr
Northern Ionosphere
Southern Ionosphere
20
10
0
–10
–20
35
17.5
0
–17.5
–35
Up
Down
Jpar (nA/m²)
80
70
60
Sun
(b) T = 413:30 hr
Northern Ionosphere
Southern Ionosphere
20
10
0
–10
–20
35
17.5
0
–17.5
–35
Up
Down
Jpar (nA/m²)
80
70
60
Sun

Figure 5.15

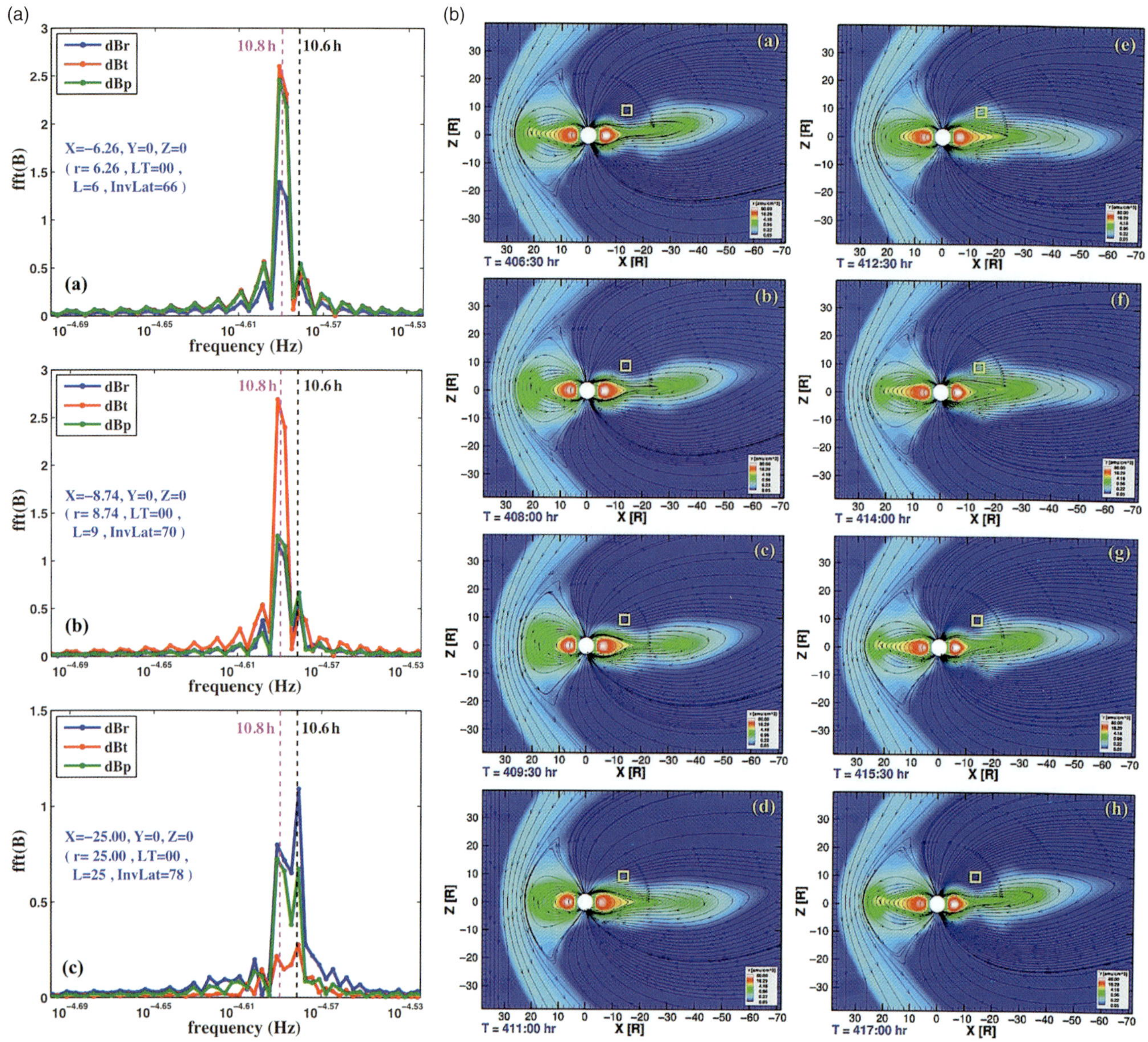

Figure 5.16

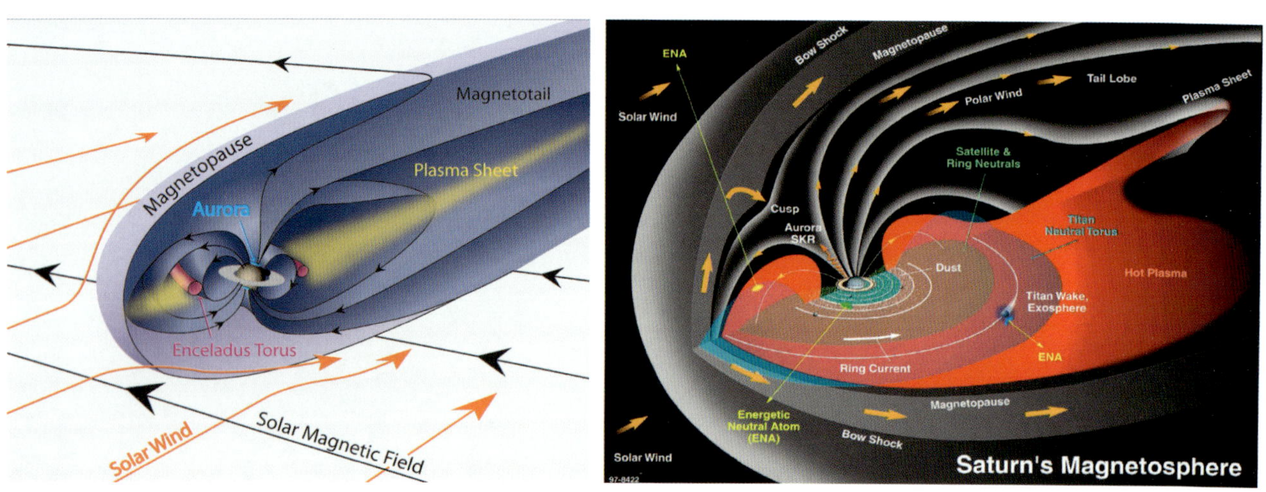

Figure 6.1

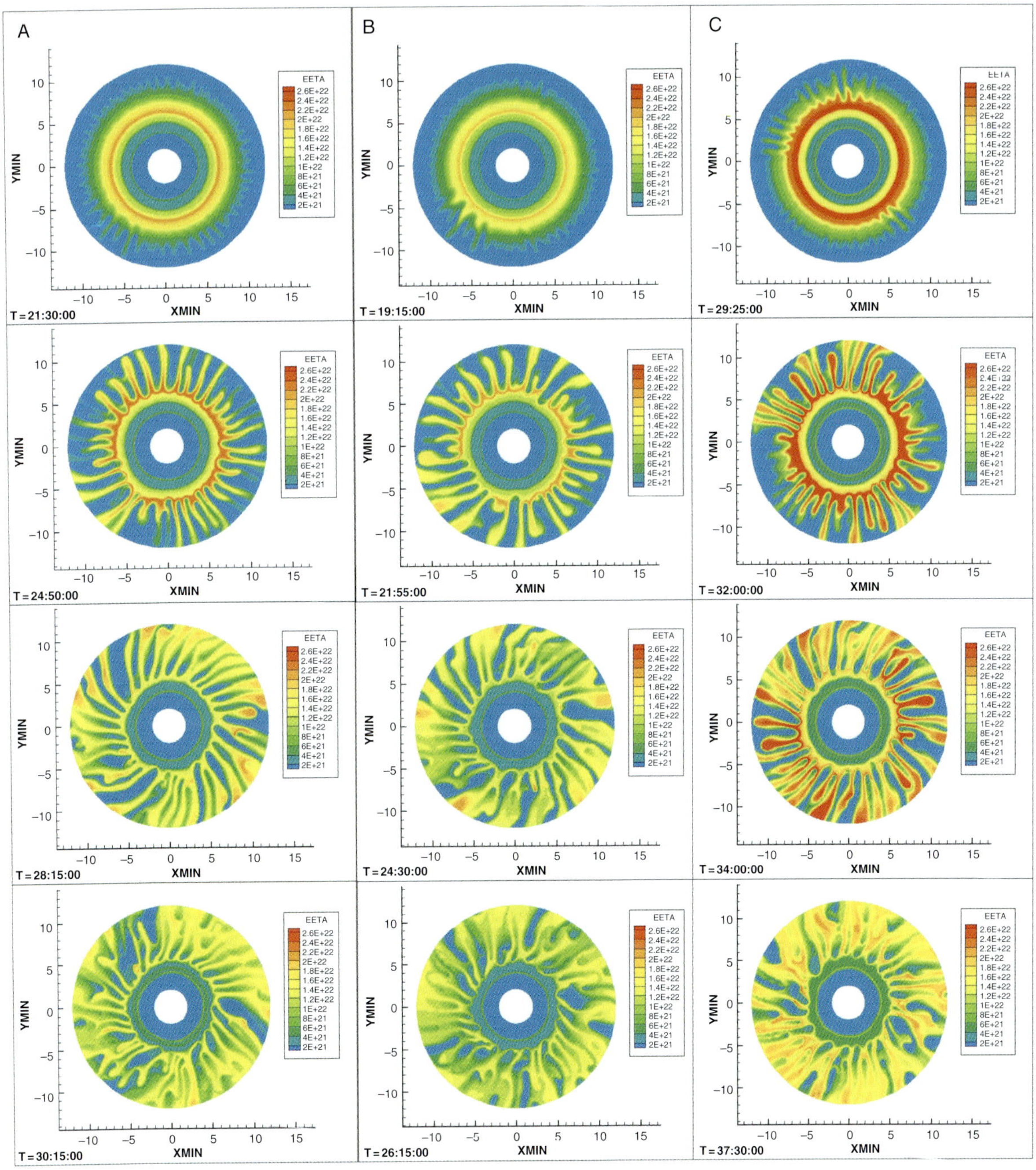

A
B
C
EETA
2.6E+22
2.4E+22
2.2E+22
2E+22
1.8E+22
1.6E+22
1.4E+22
1.2E+22
1E+22
8E+21
6E+21
4E+21
2E+21
YMIN
XMIN
T = 21:30:00
T = 19:15:00
T = 29:25:00
T = 24:50:00
T = 21:55:00
T = 32:00:00
T = 28:15:00
T = 24:30:00
T = 34:00:00
T = 30:15:00
T = 26:15:00
T = 37:30:00

Figure 6.8

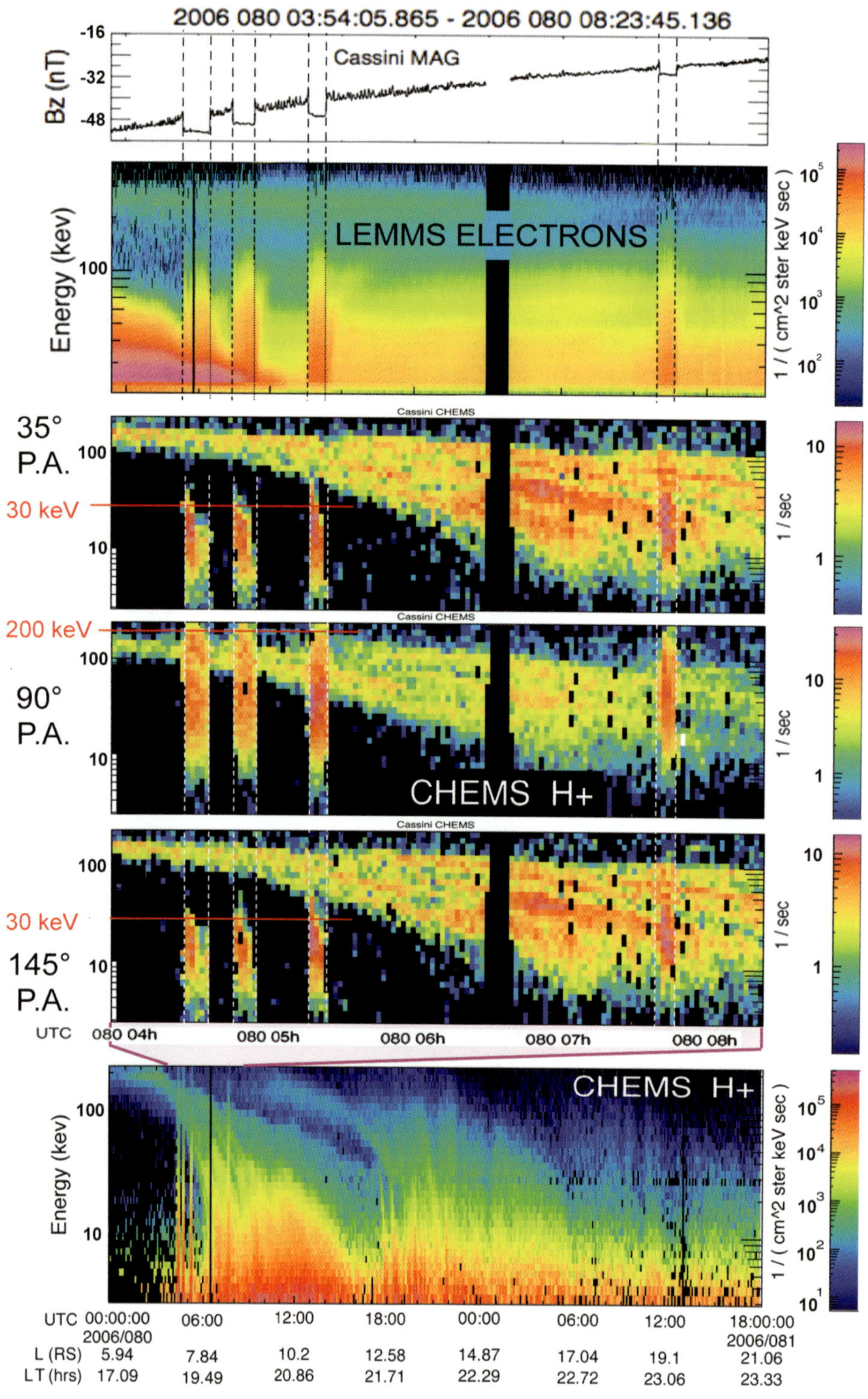
2006 080 03:54:05.865 - 2006 080 08:23:45.136
Cassini MAG
Bz (nT)
LEMMS ELECTRONS
Energy (kev)
1 / (cm^2 ster keV sec)
35° P.A.
30 keV
200 keV
90° P.A.
CHEMS H+
30 keV
145° P.A.
1 / sec
UTC 080 04h 080 05h 080 06h 080 07h 080 08h
CHEMS H+
UTC 00:00:00 2006/080 06:00 12:00 18:00 00:00 06:00 12:00 18:00:00 2006/081
L (RS) 5.94 7.84 10.2 12.58 14.87 17.04 19.1 21.06
LT (hrs) 17.09 19.49 20.86 21.71 22.29 22.72 23.06 23.33

Figure 6.9

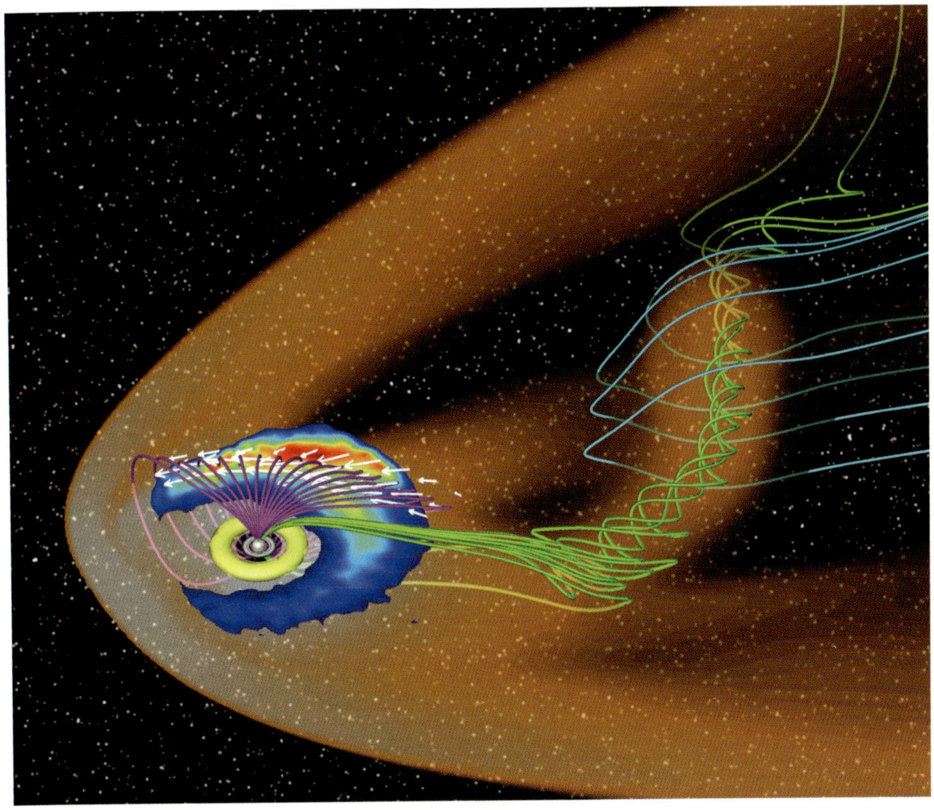

Figure 6.11

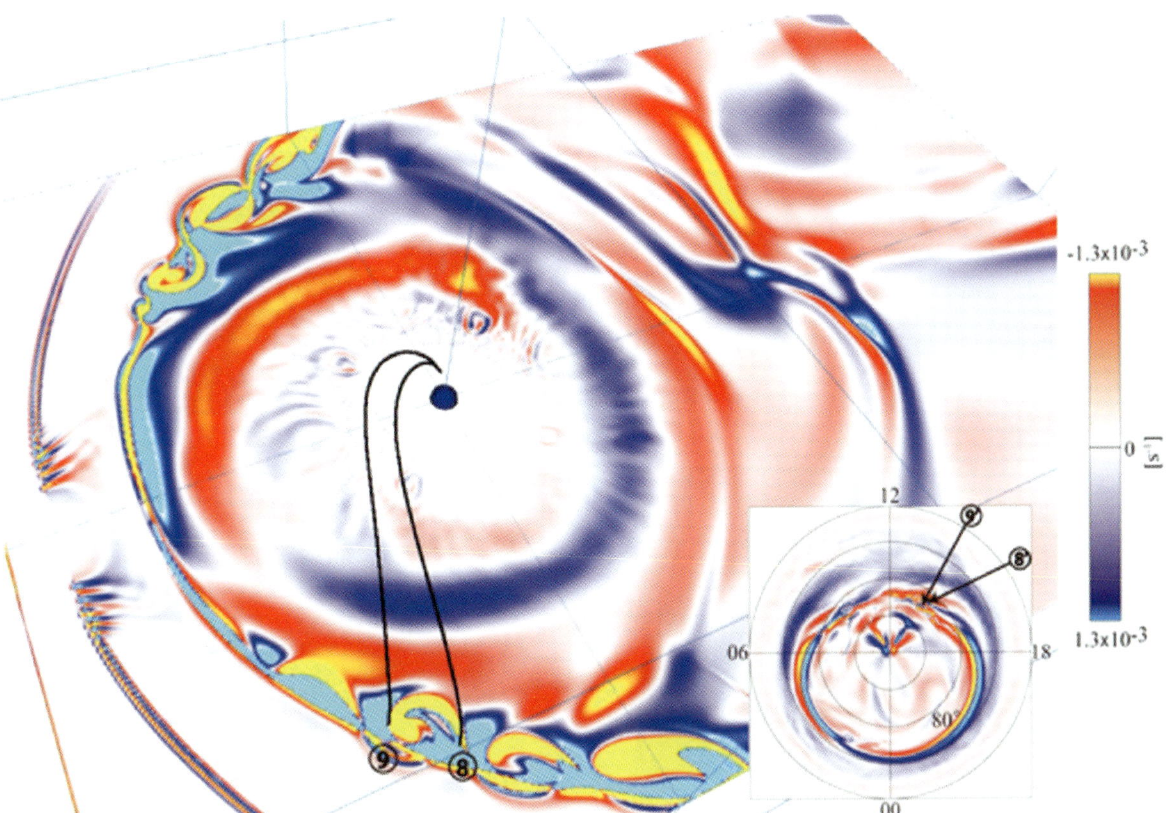

Figure 6.15

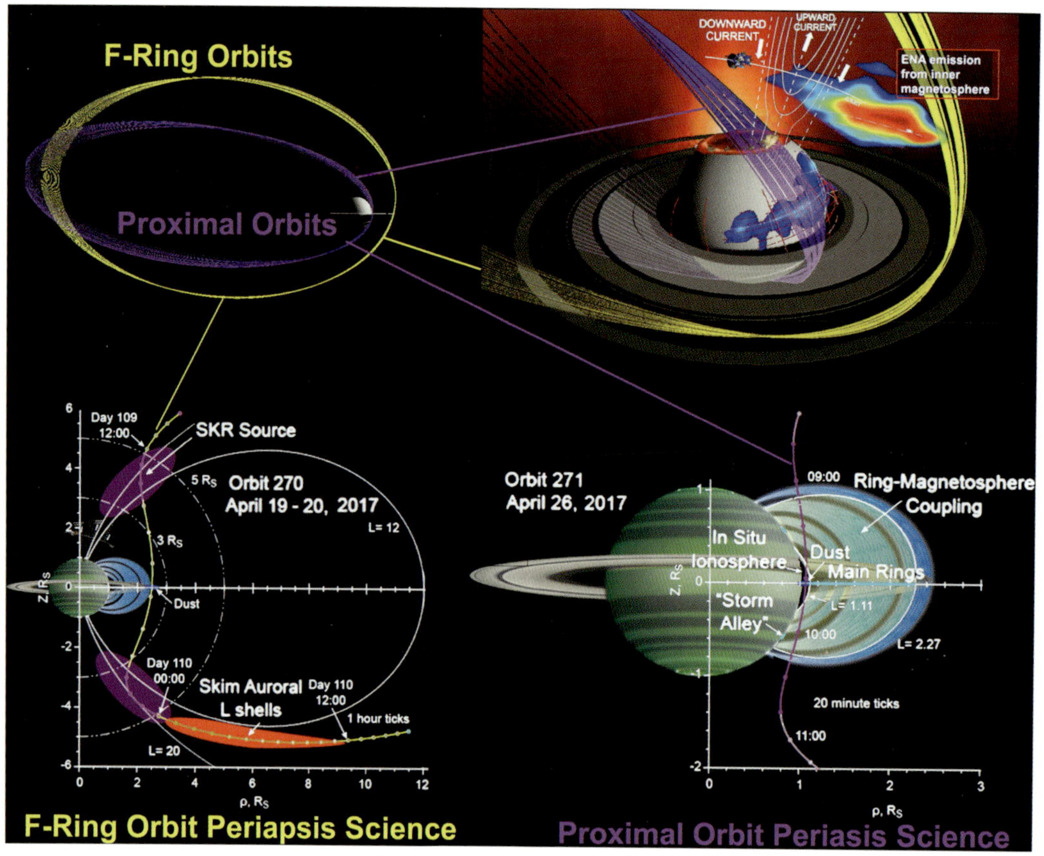

Figure 6.16

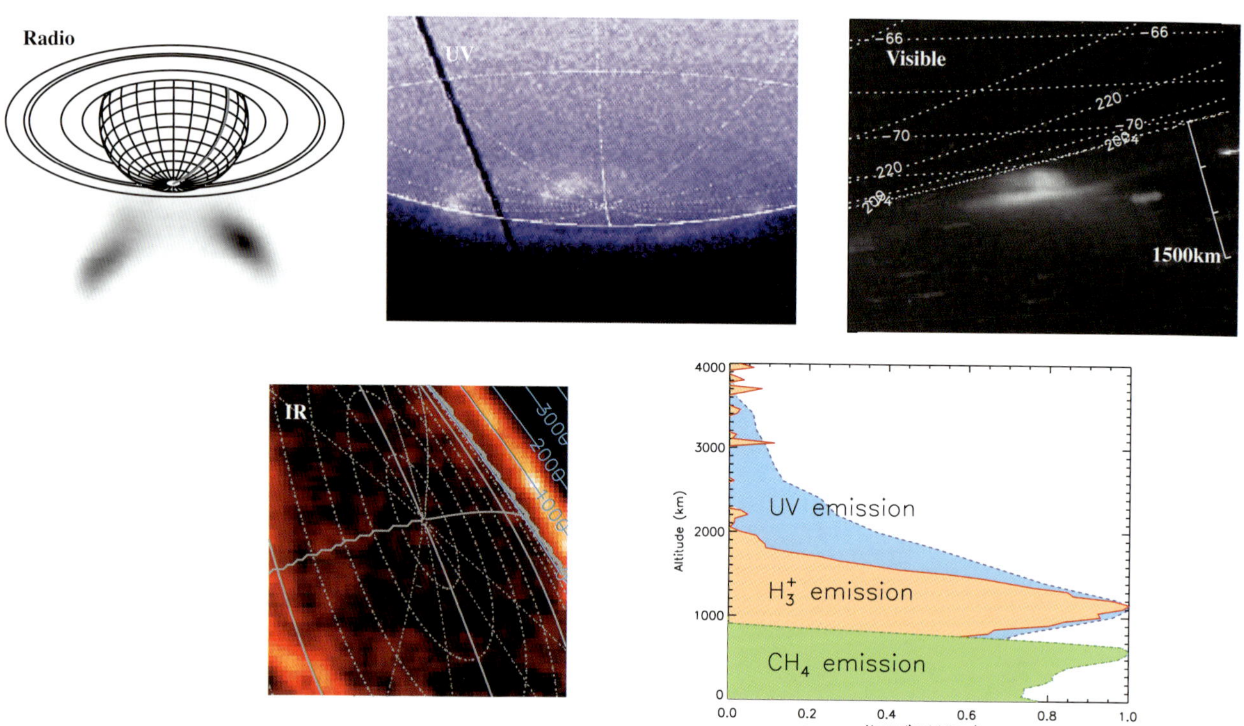

Figure 7.1

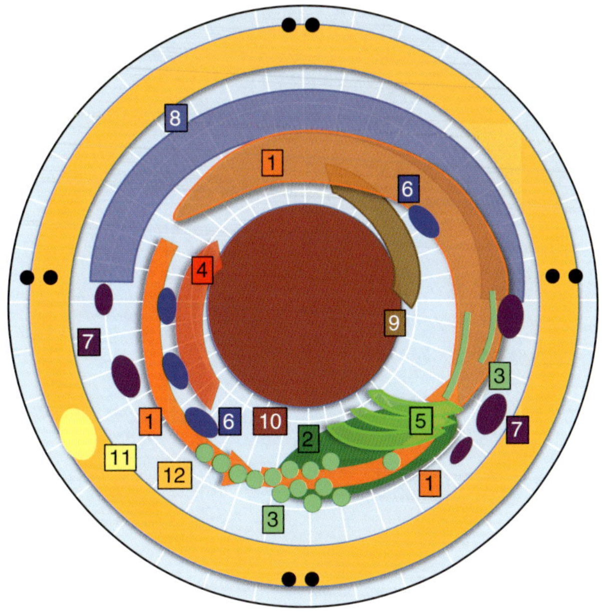

Figure 7.3

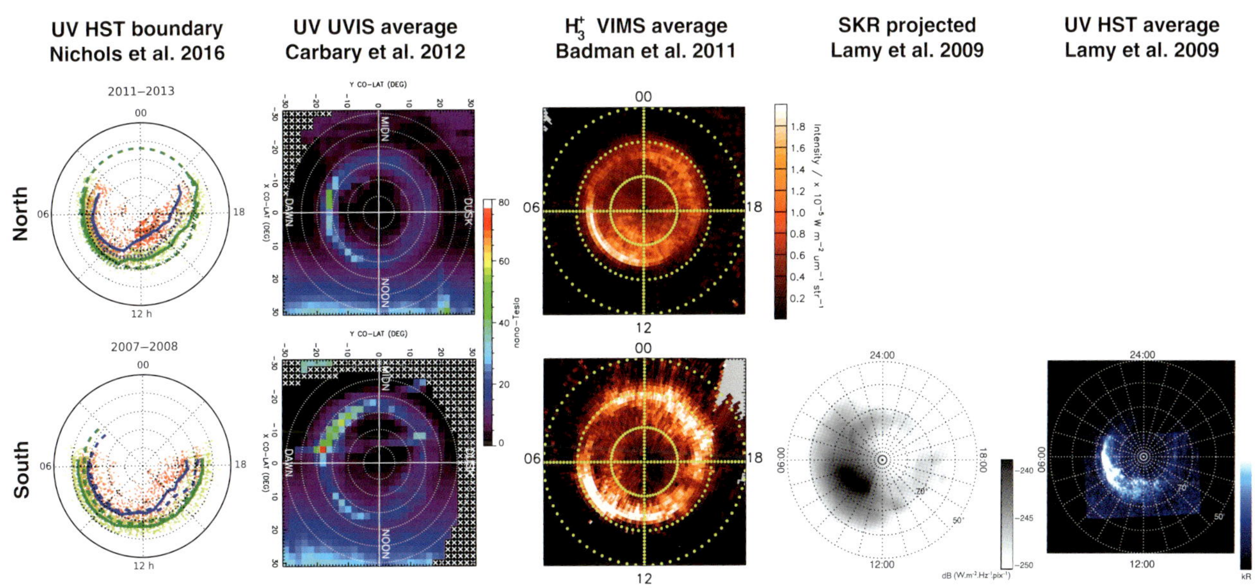

Figure 7.4

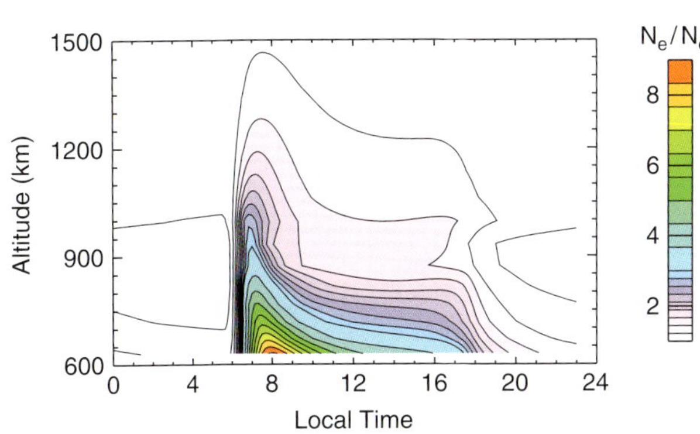

Figure 8.7

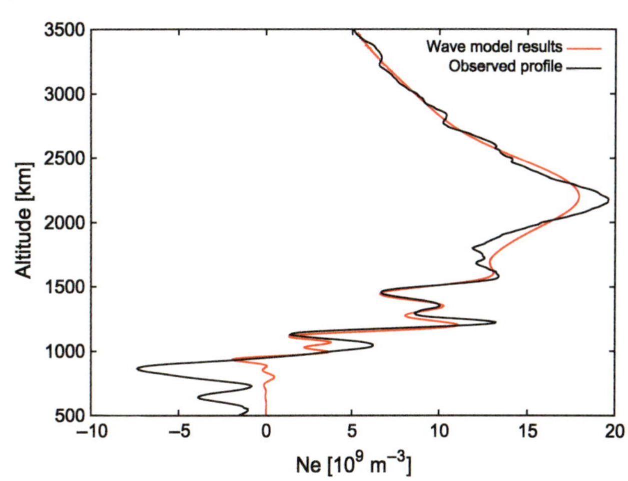

Figure 8.9

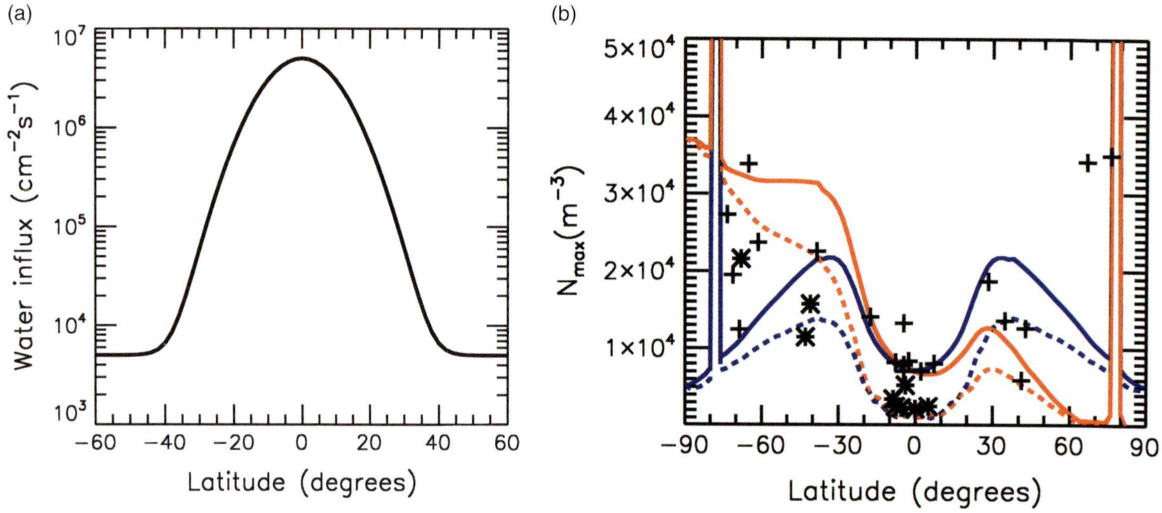

Figure 8.13

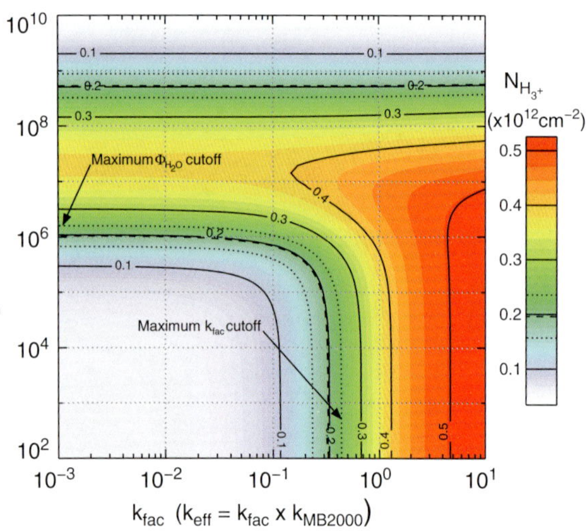

Figure 8.17

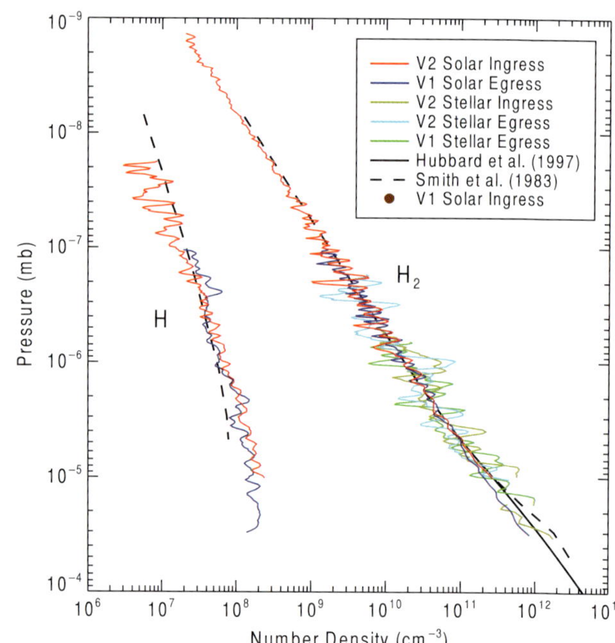

Figure 9.1

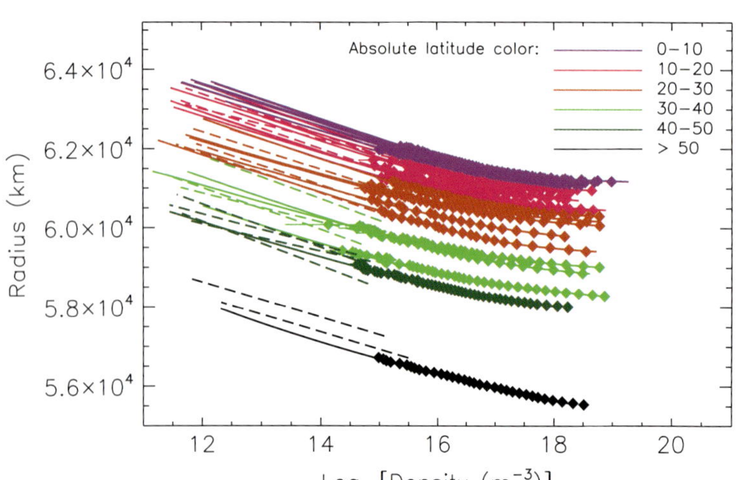

Figure 9.2

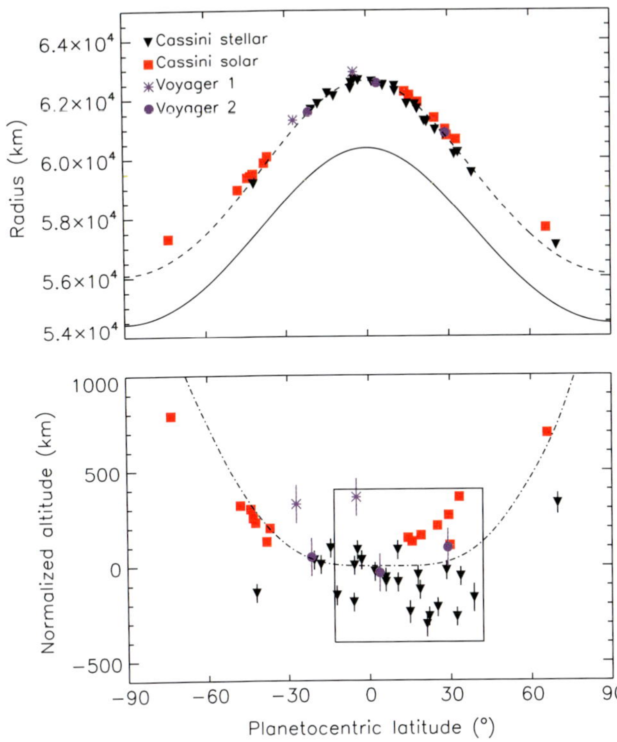

Figure 9.4

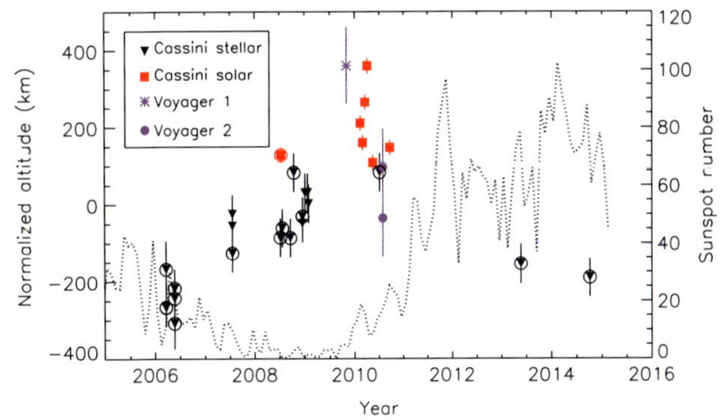

Figure 9.5

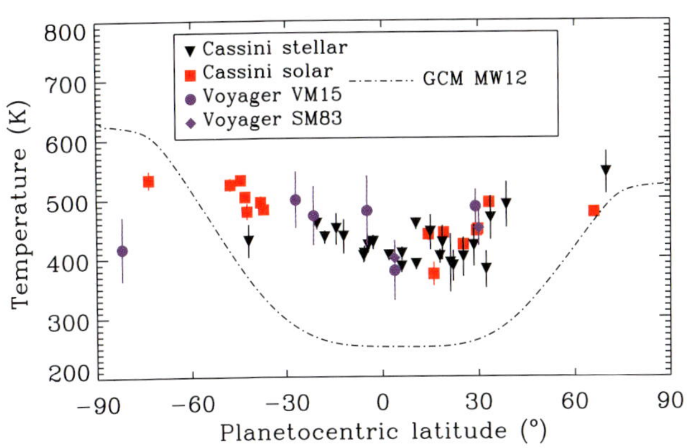

Figure 9.6

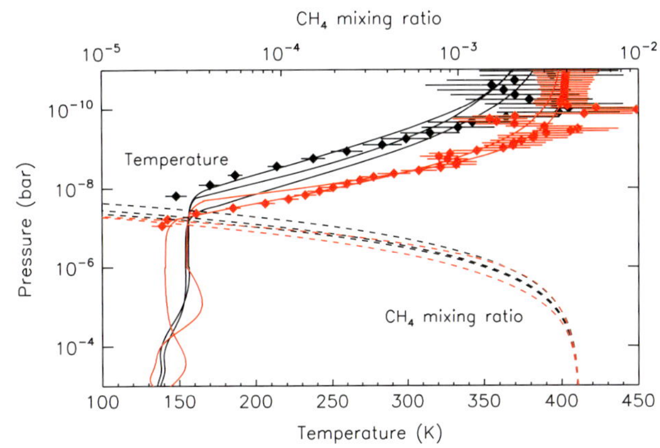

Figure 9.8

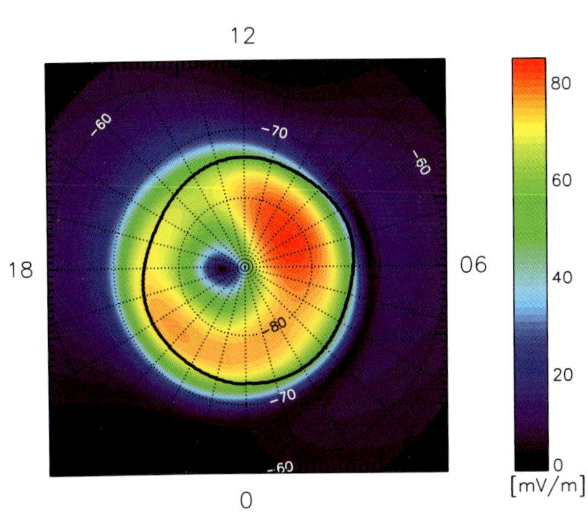

Figure 9.11

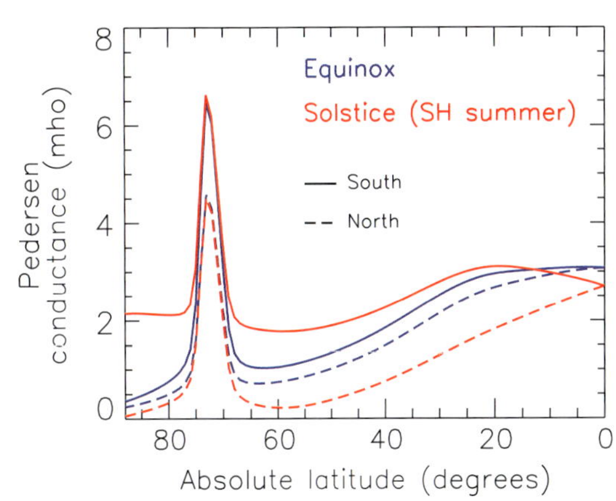

Figure 9.12

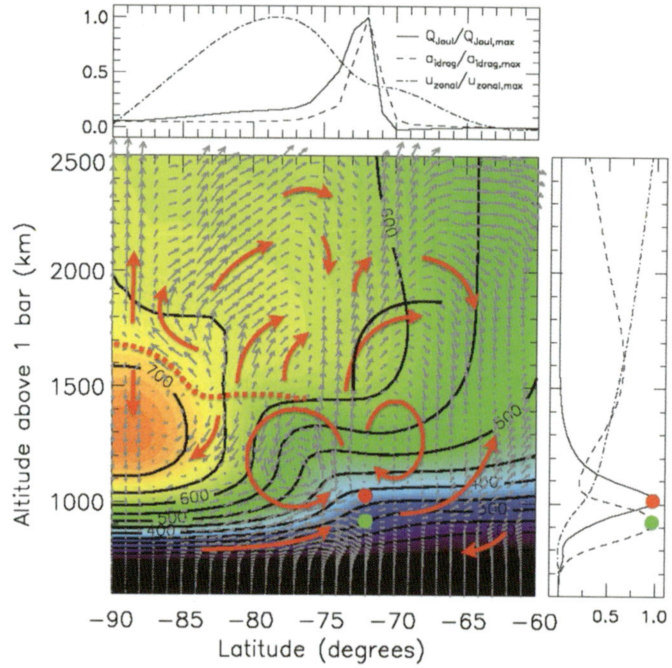

Figure 9.13

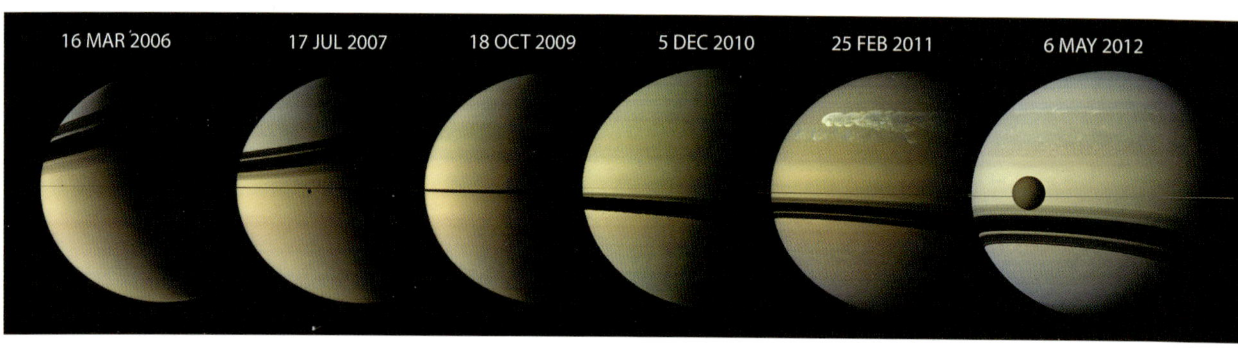

Figure 10.1

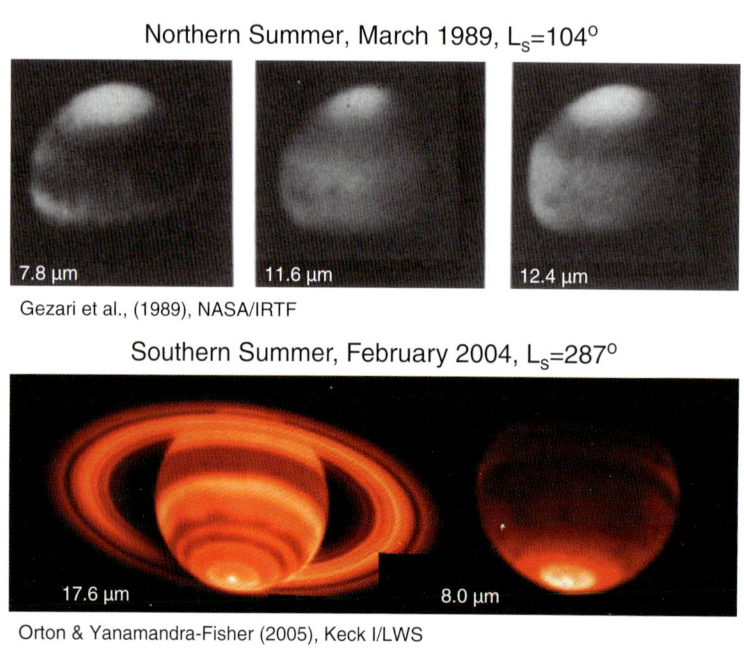

Figure 10.2

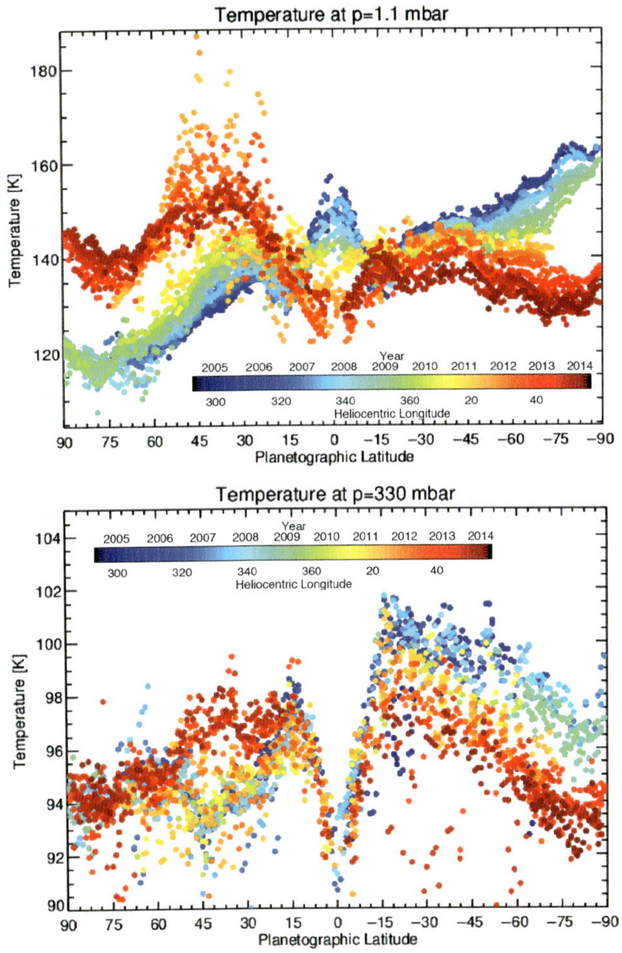

Figure 10.4

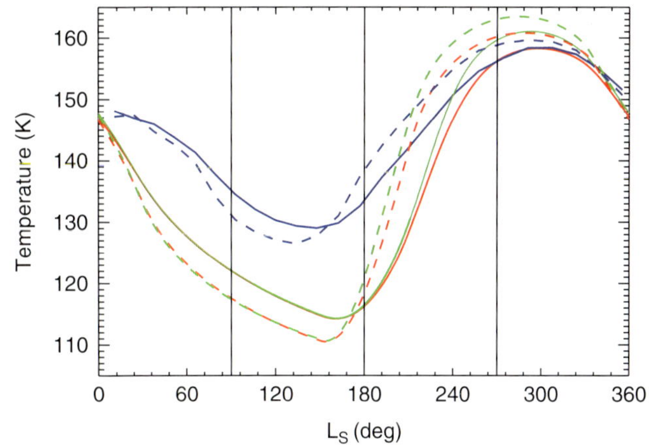

Figure 10.8

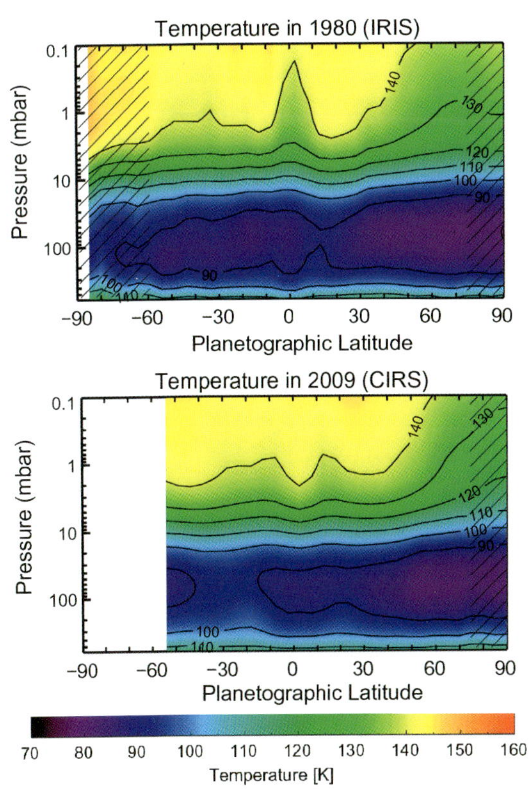

Figure 10.9

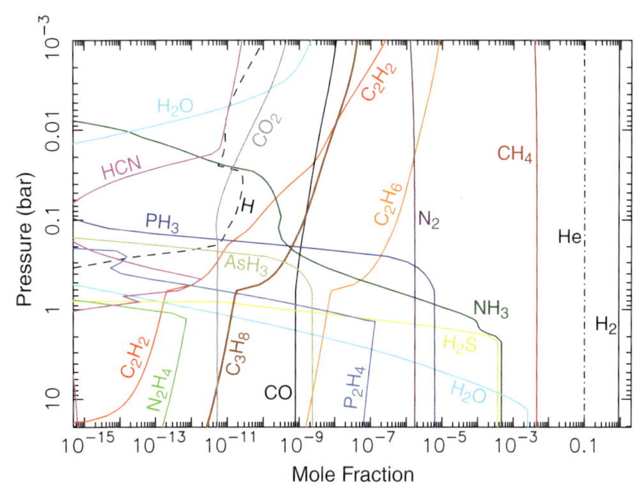

Figure 10.11

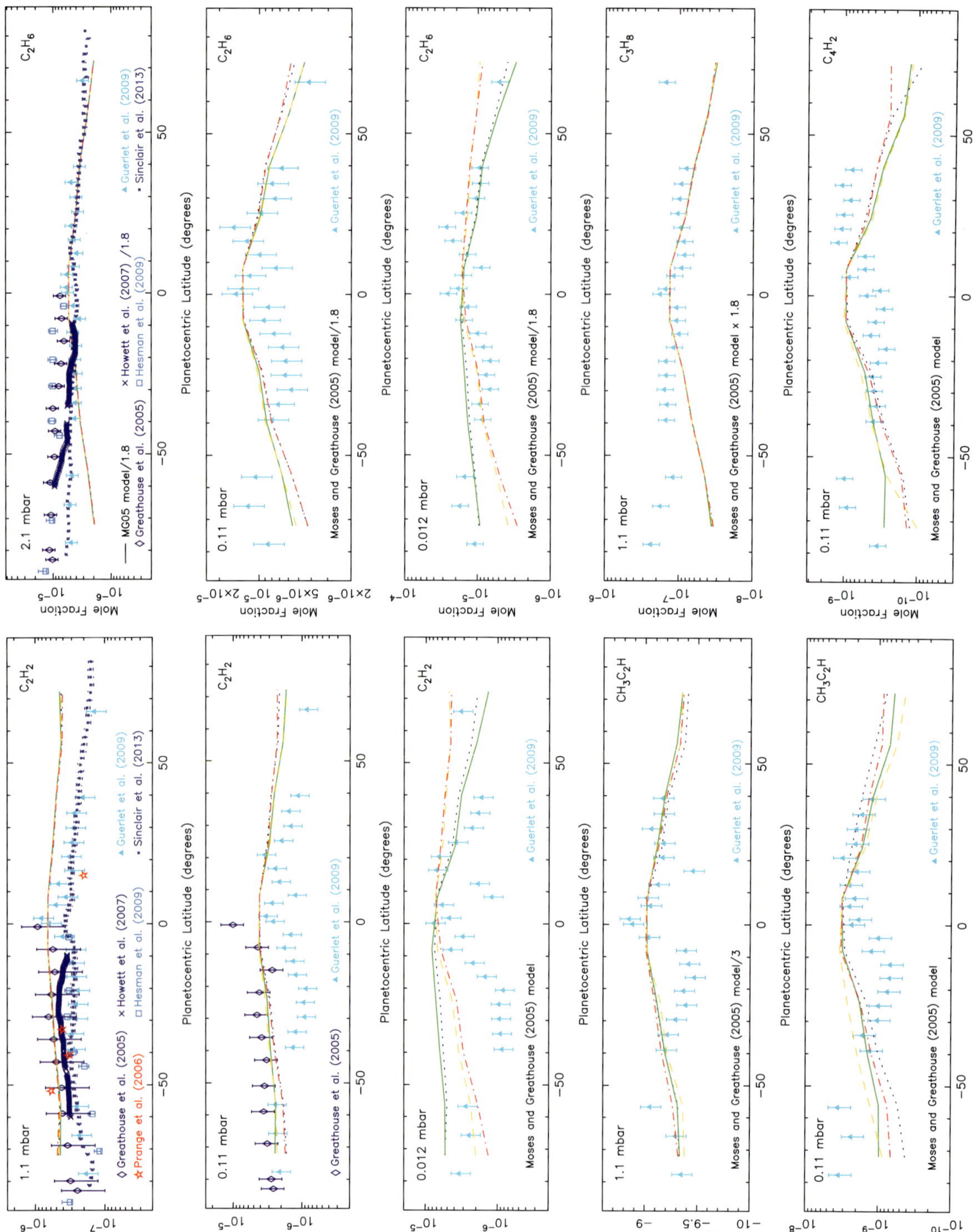

1.1 mbar
C_2H_2
Greathouse et al. (2005)
Prange et al. (2006)
Howett et al. (2007)
Hesman et al. (2009)
Guerlet et al. (2009)
Sinclair et al. (2013)
Mole Fraction
Planetocentric Latitude (degrees)
0.11 mbar
Moses and Greathouse (2005) model
0.012 mbar
CH_3C_2H
Moses and Greathouse (2005) model/3
Log_{10} (Mole Fraction)
2.1 mbar
C_2H_6
MG05 model/1.8
Howett et al. (2007) /1.8
Moses and Greathouse (2005) model/1.8
C_3H_8
Moses and Greathouse (2005) model × 1.8
C_4H_2

Figure 10.16

Figure 10.18

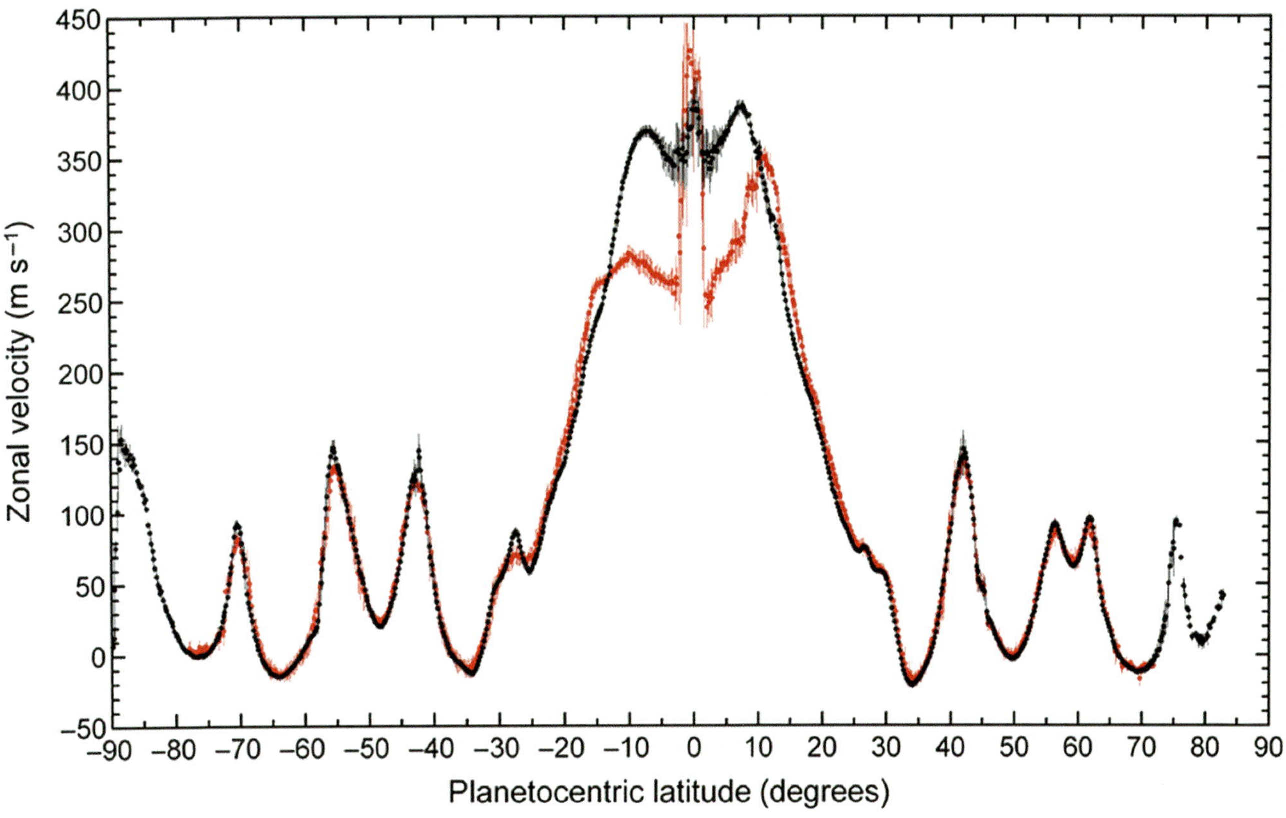

Figure 11.2

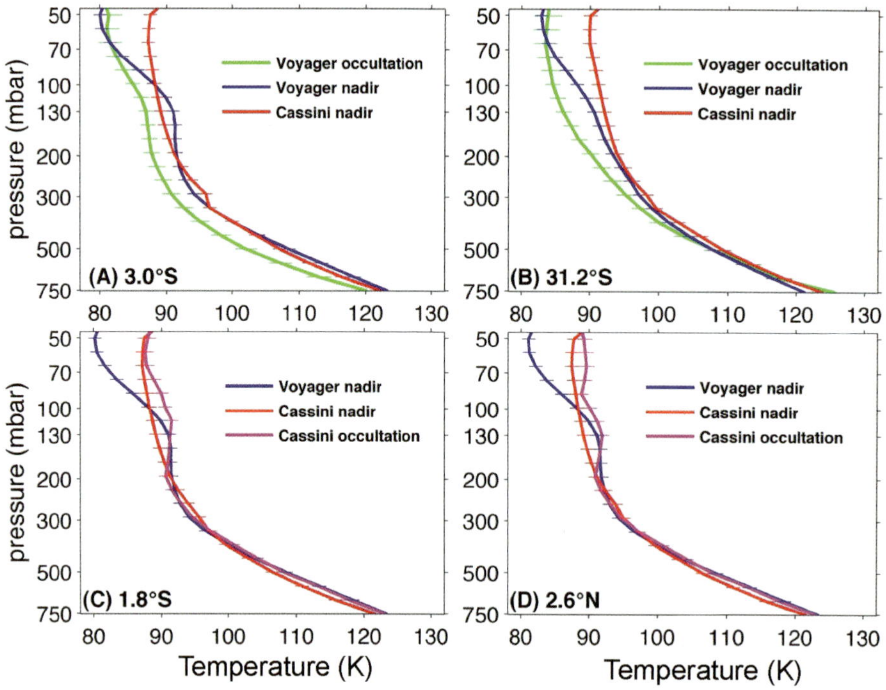

Figure 11.4

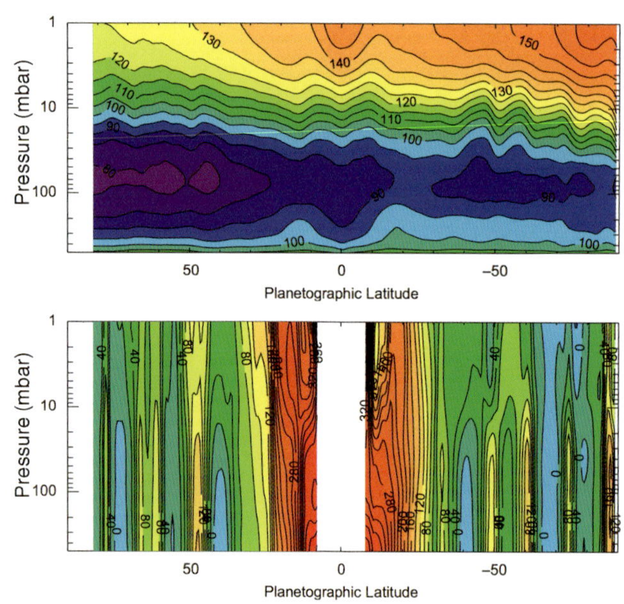

Figure 11.5

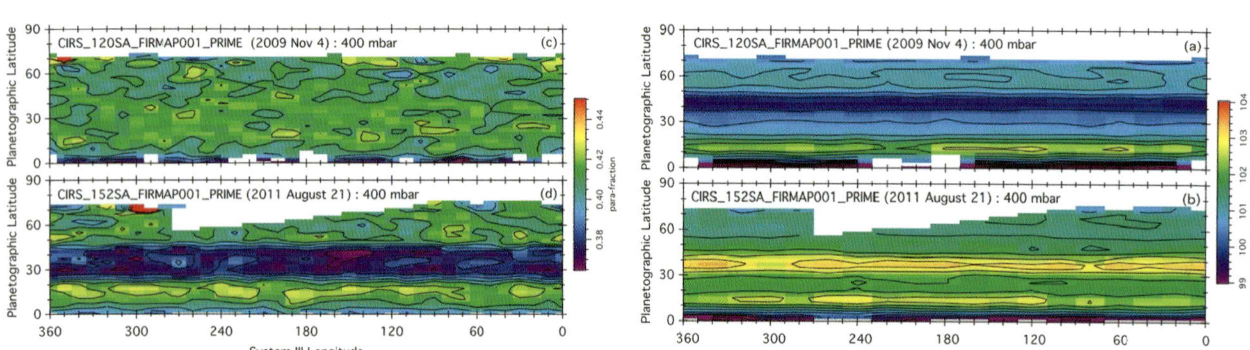

Figure 11.11

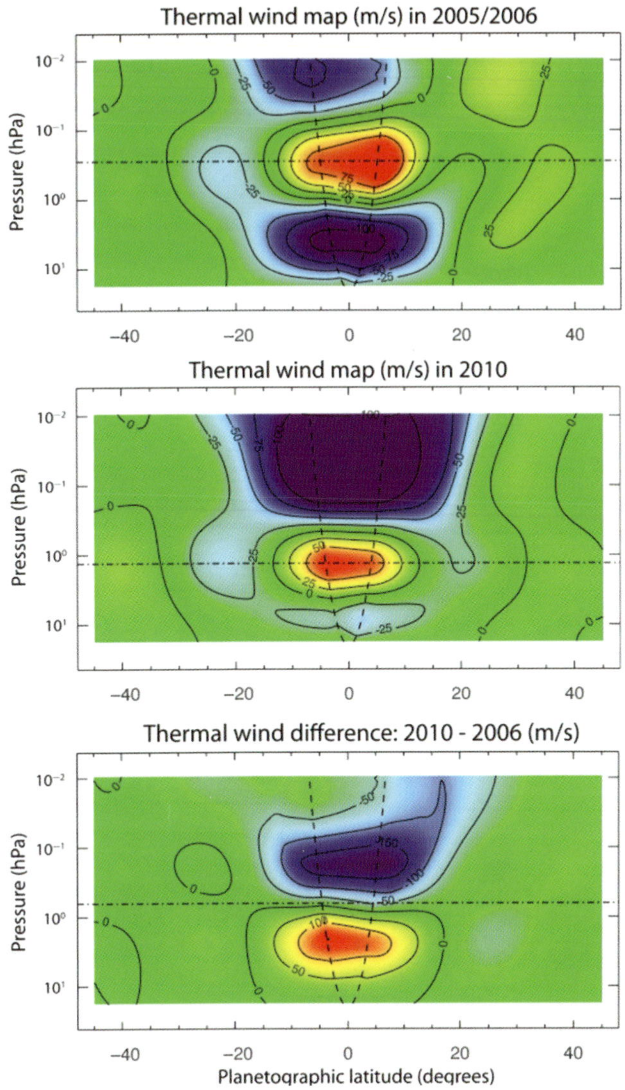

Figure 11.15

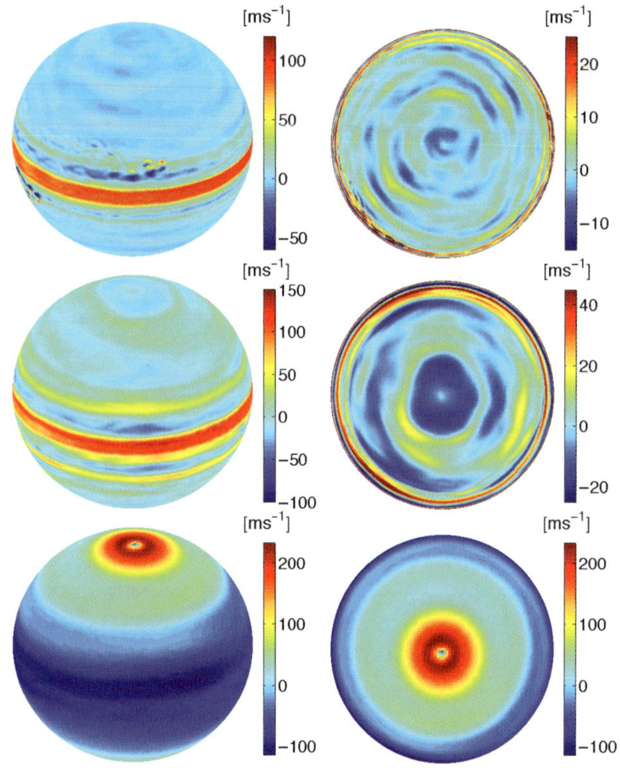

Figure 11.16

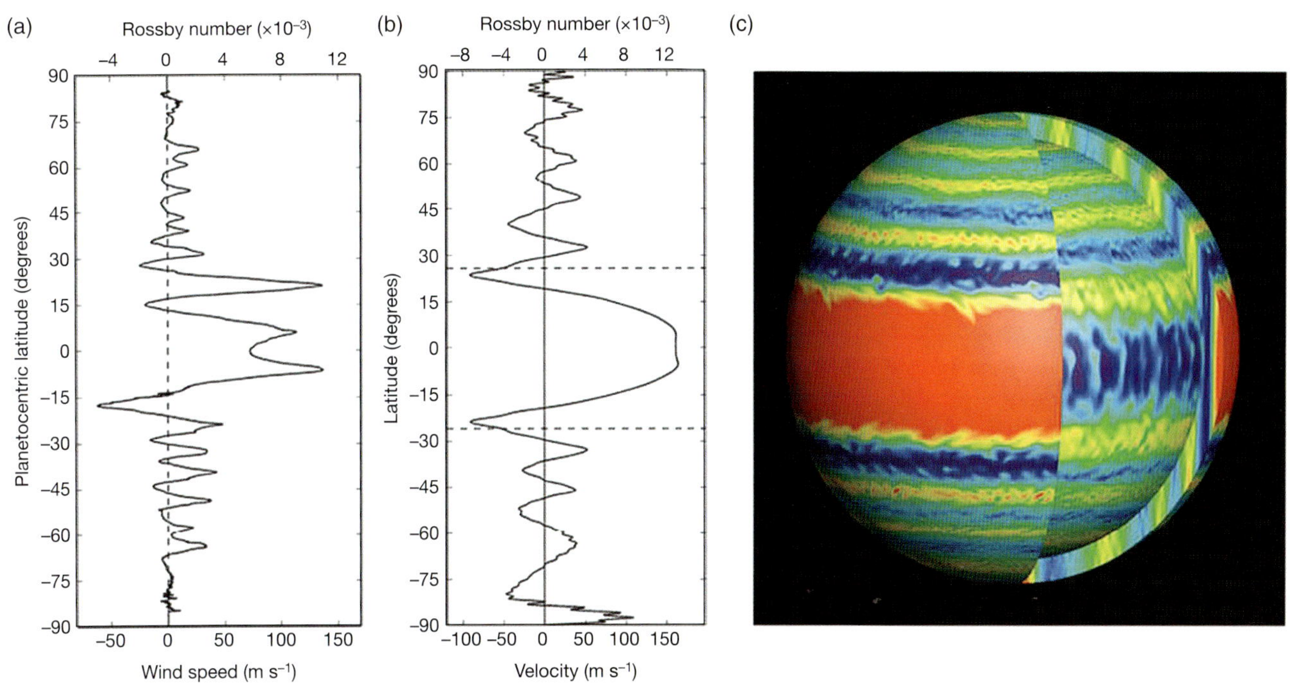

Figure 11.17

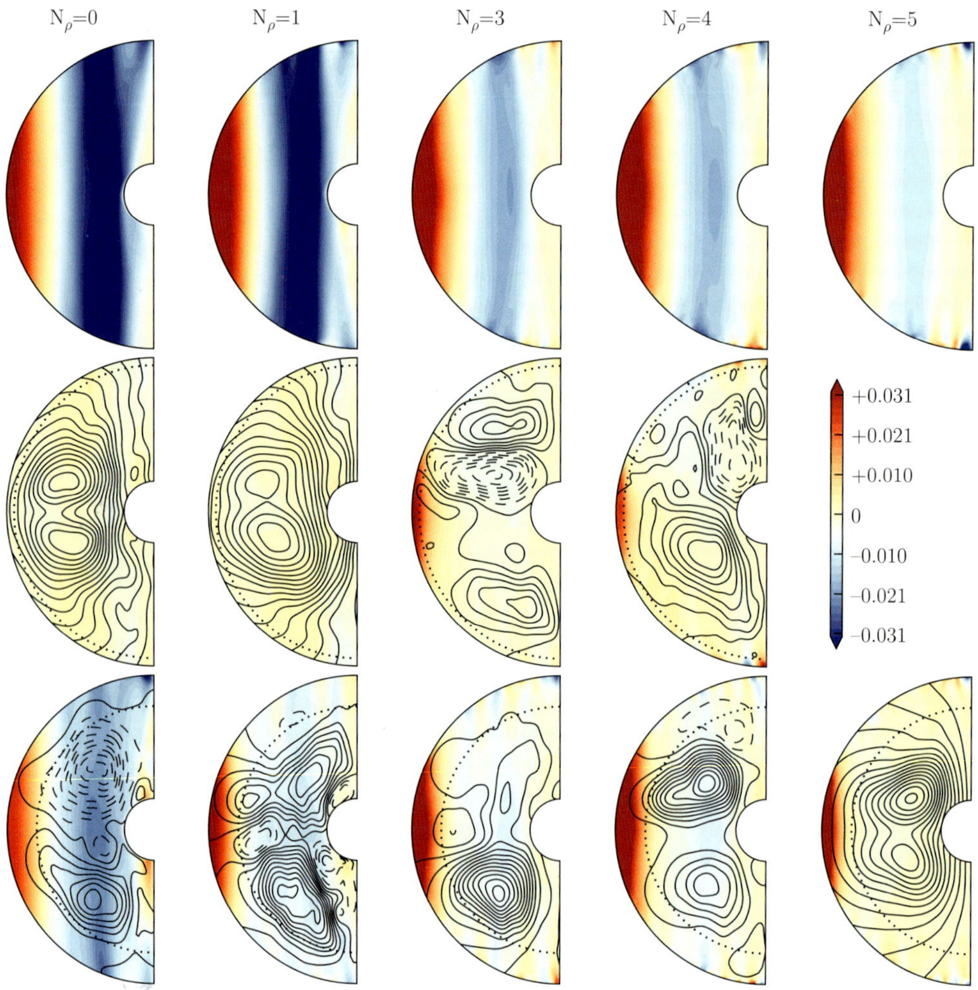

Figure 11.18

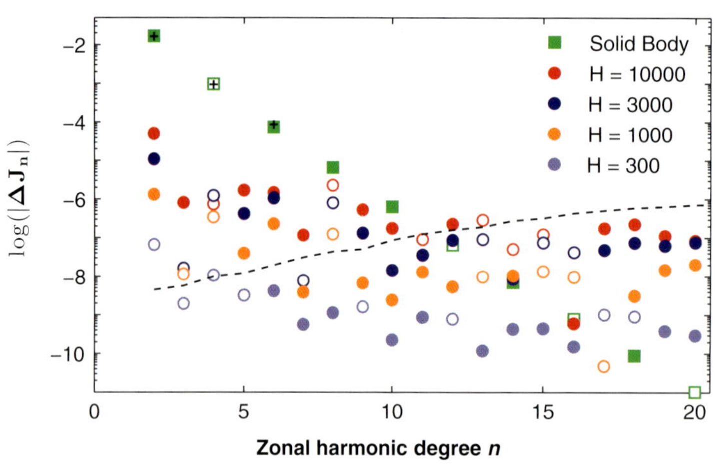

Figure 11.19

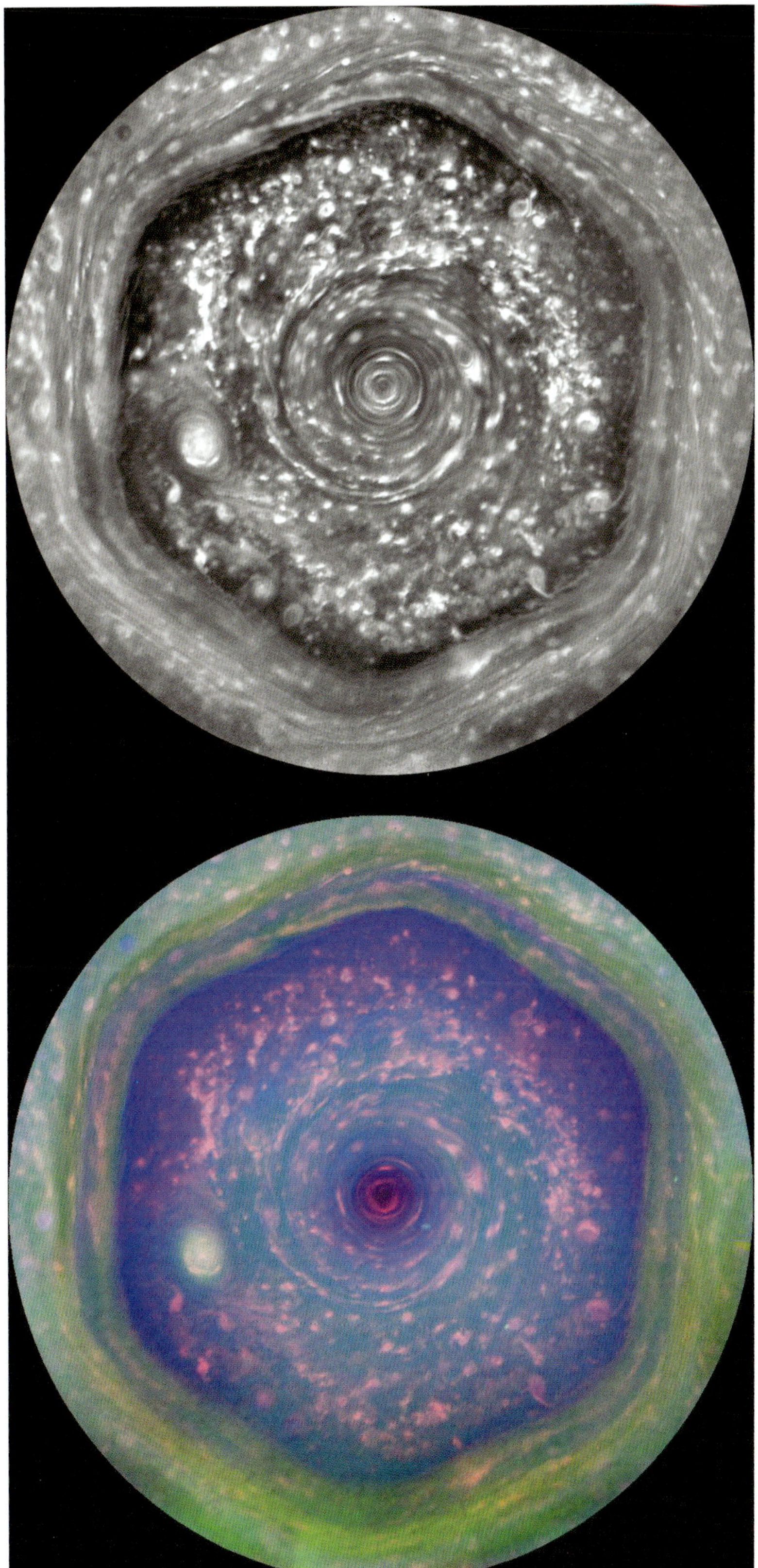

Figure 12.3

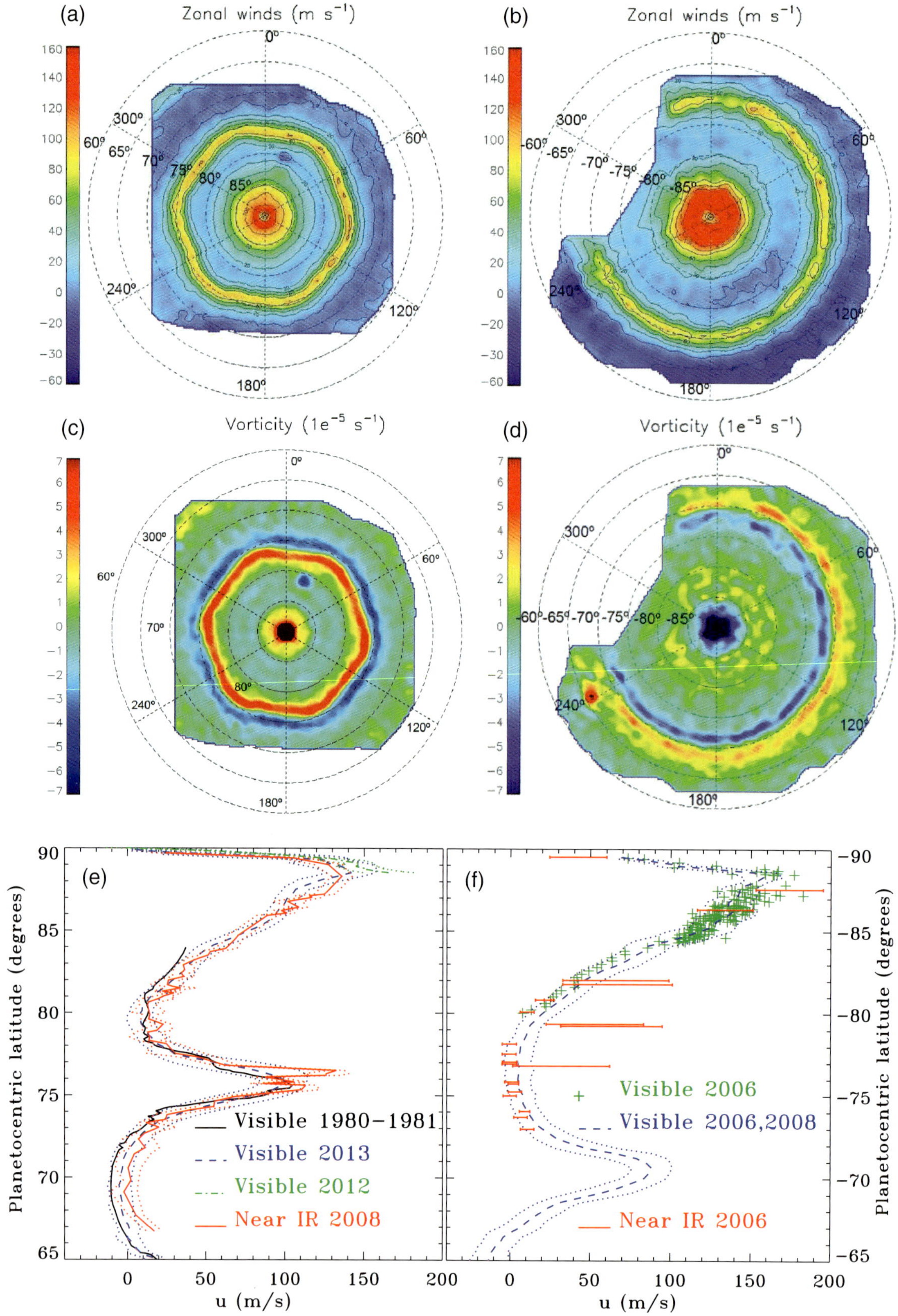
(a)
Zonal winds (m s^{-1})
(b)
Zonal winds (m s^{-1})
(c)
Vorticity (1e^{-5} s^{-1})
(d)
Vorticity (1e^{-5} s^{-1})
(e)
(f)
Planetocentric latitude (degrees)
u (m/s)
Visible 1980−1981
Visible 2013
Visible 2012
Near IR 2008
Visible 2006
Visible 2006,2008
Near IR 2006

Figure 12.5

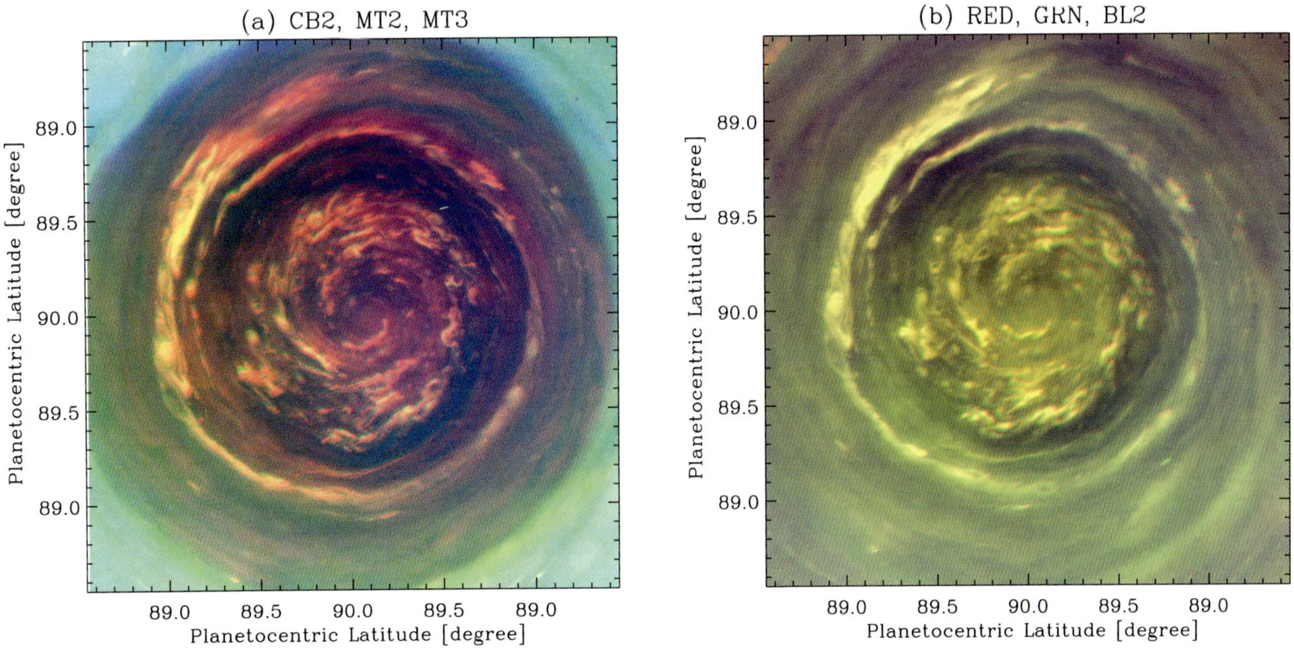

Figure 12.7

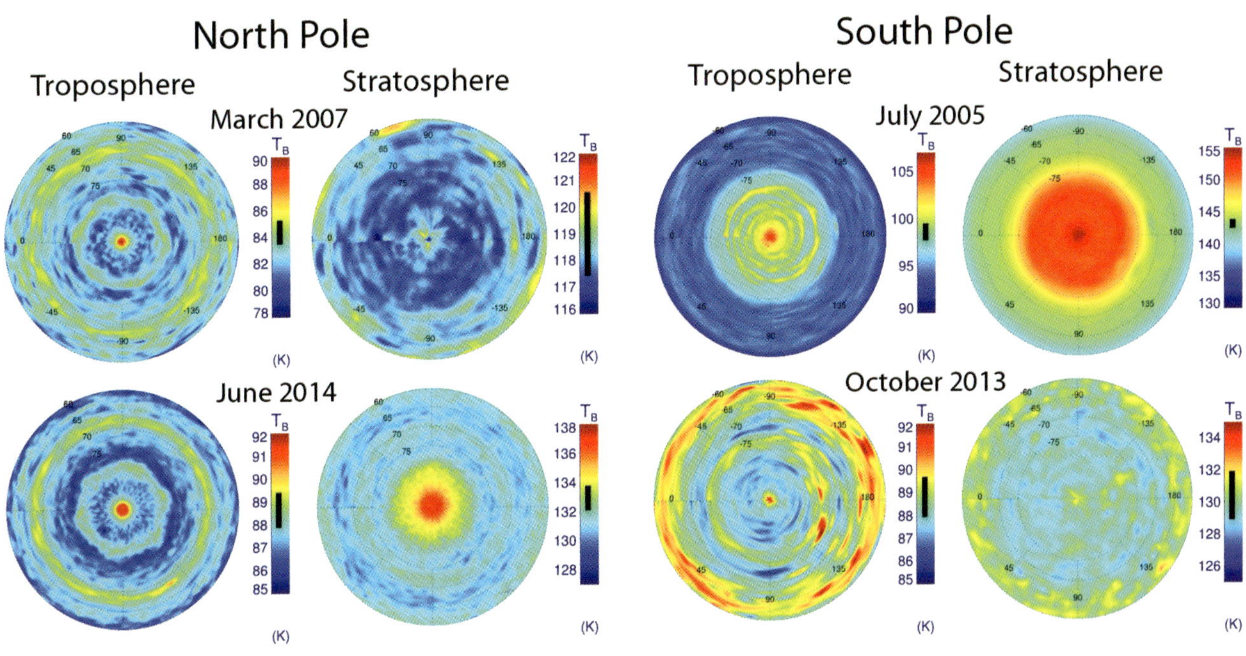

Figure 12.11

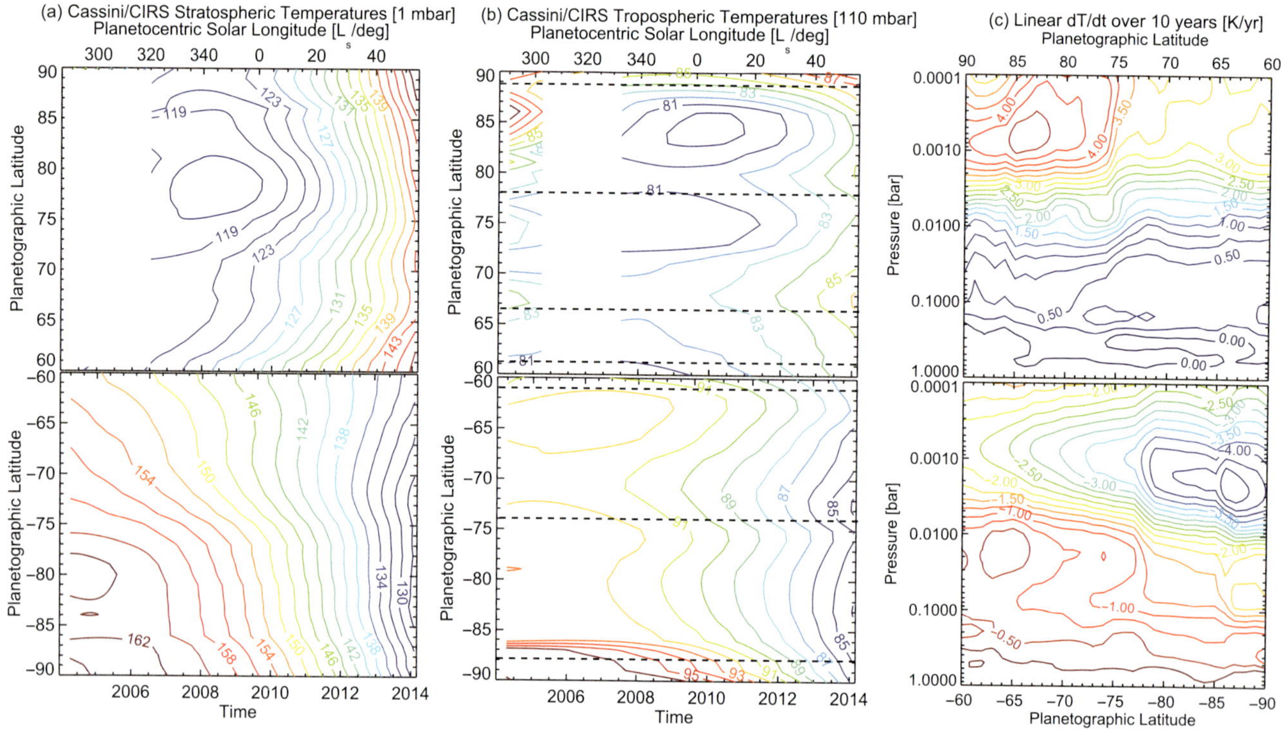

Figure 12.12

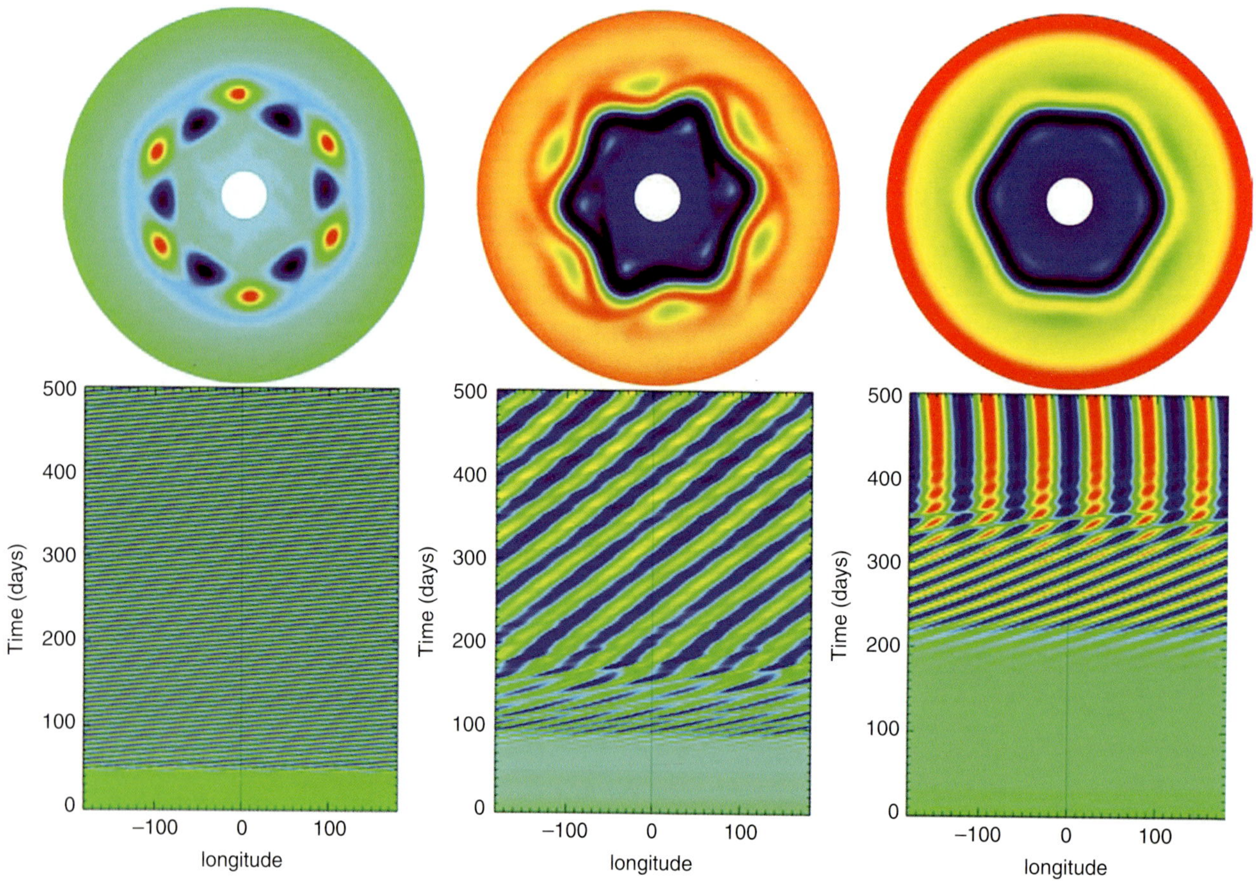

Figure 12.18

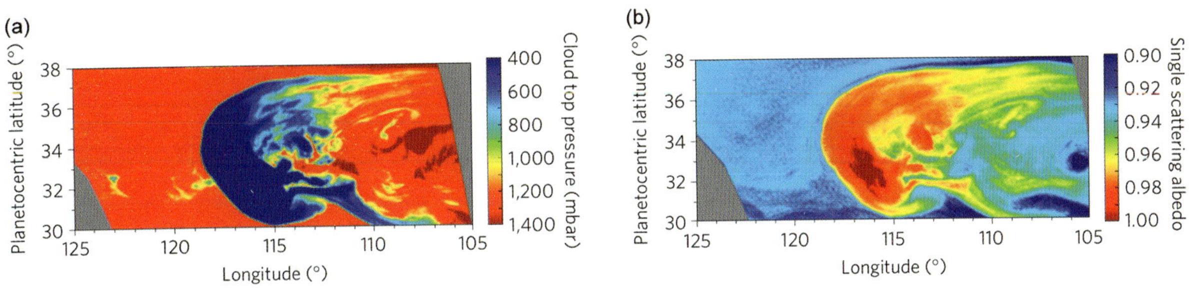

Figure 13.7

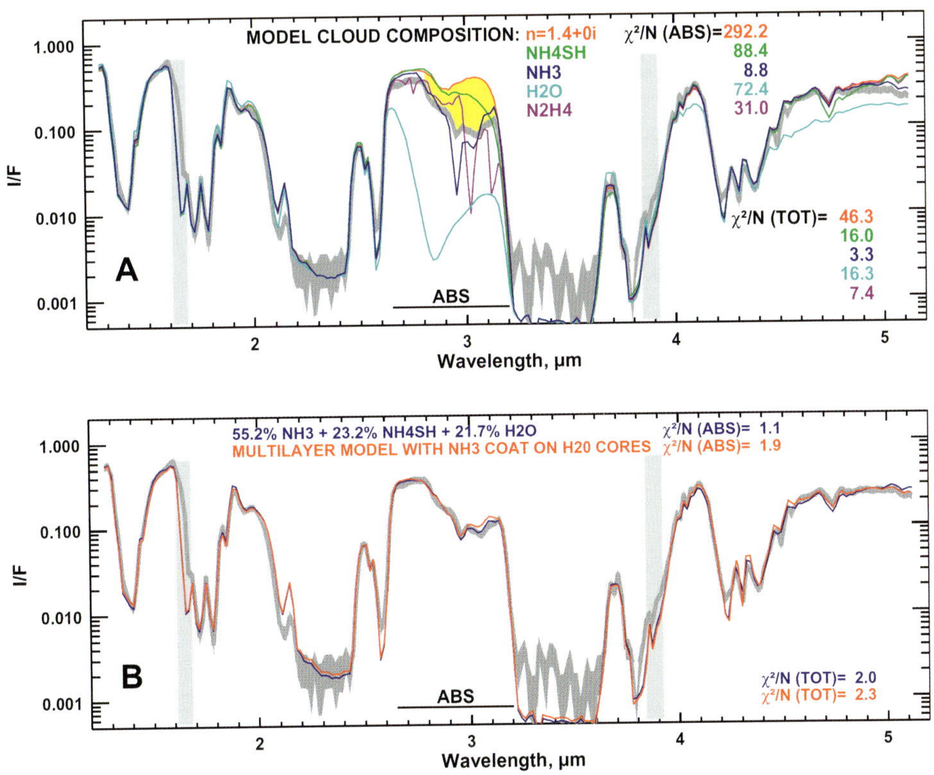

Figure 13.9

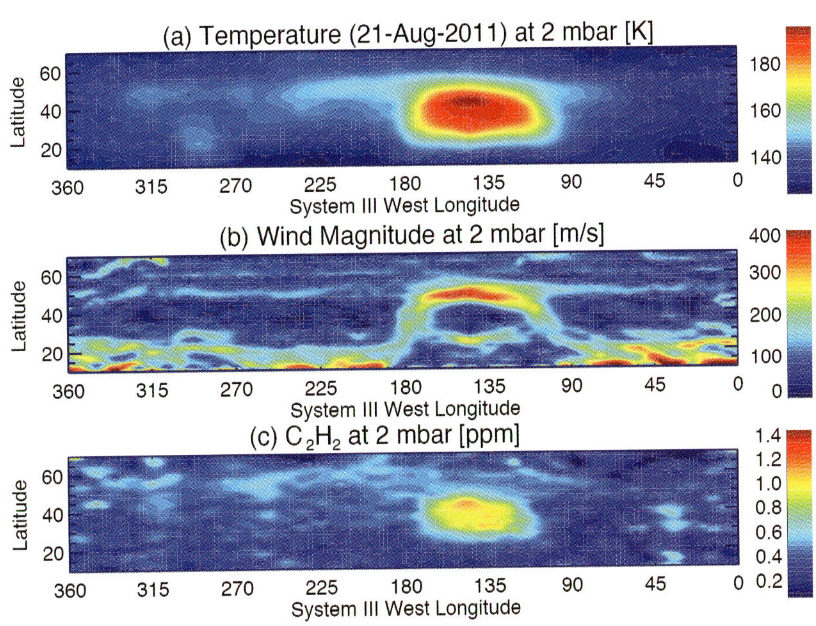

Figure 13.12

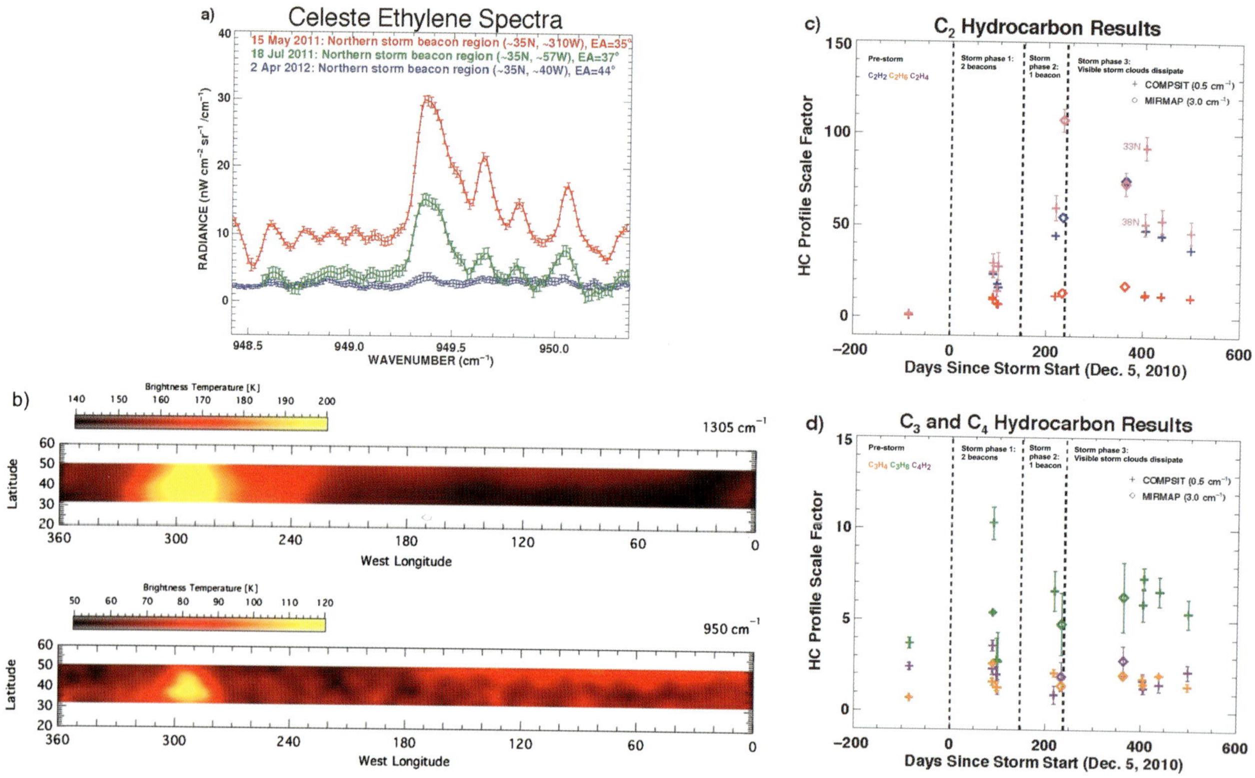

Figure 13.13

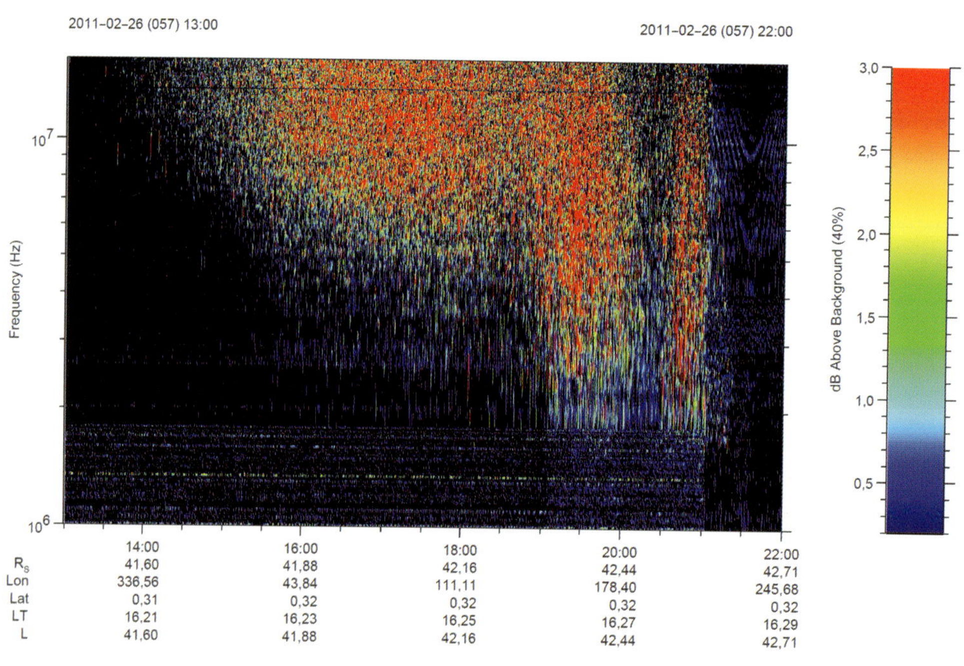

Figure 13.14

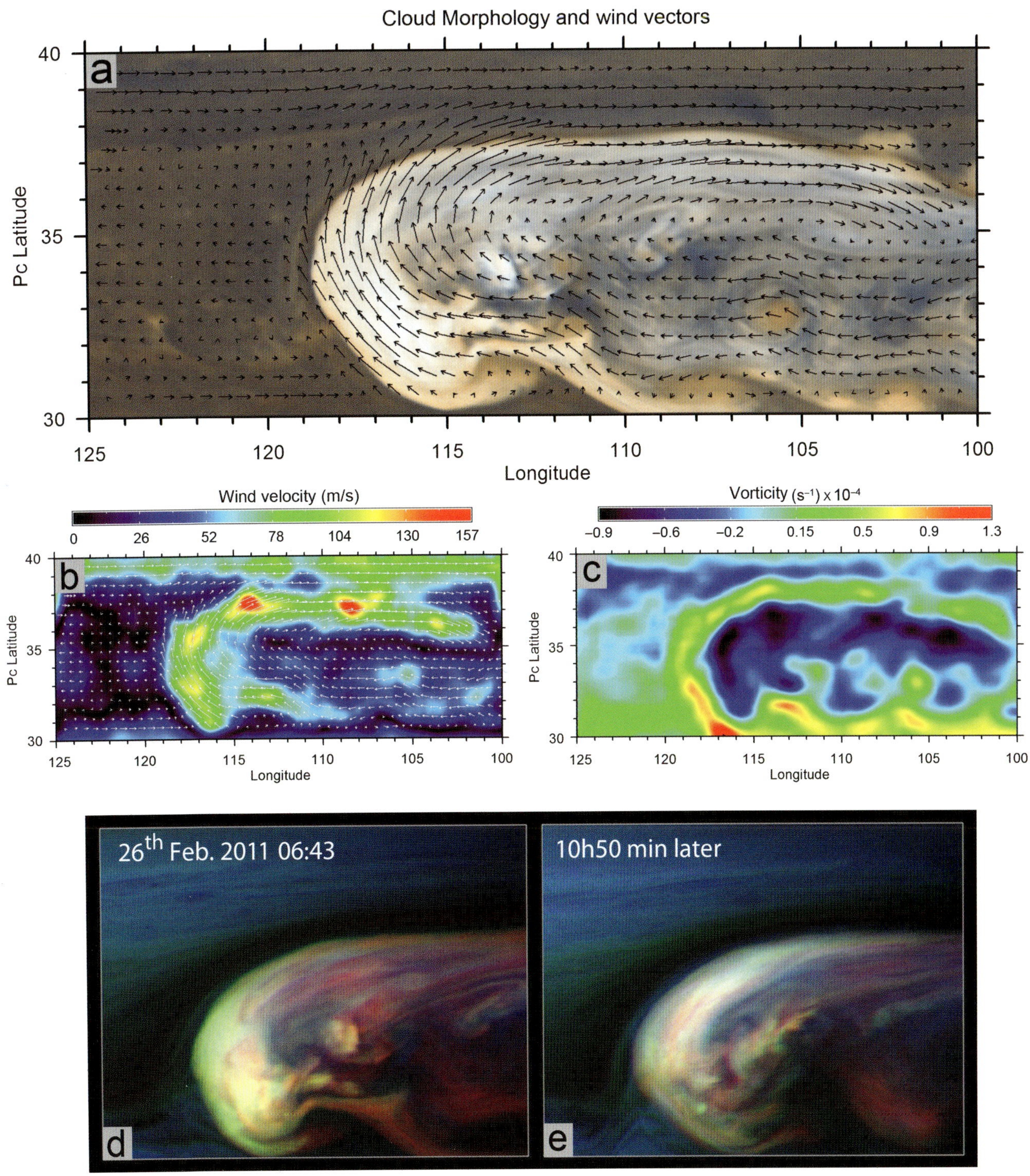

Cloud Morphology and wind vectors
a
Pc Latitude
Longitude
Wind velocity (m/s)
b
Vorticity (s⁻¹) x 10⁻⁴
c
26th Feb. 2011 06:43
10h50 min later
d
e

Figure 13.17

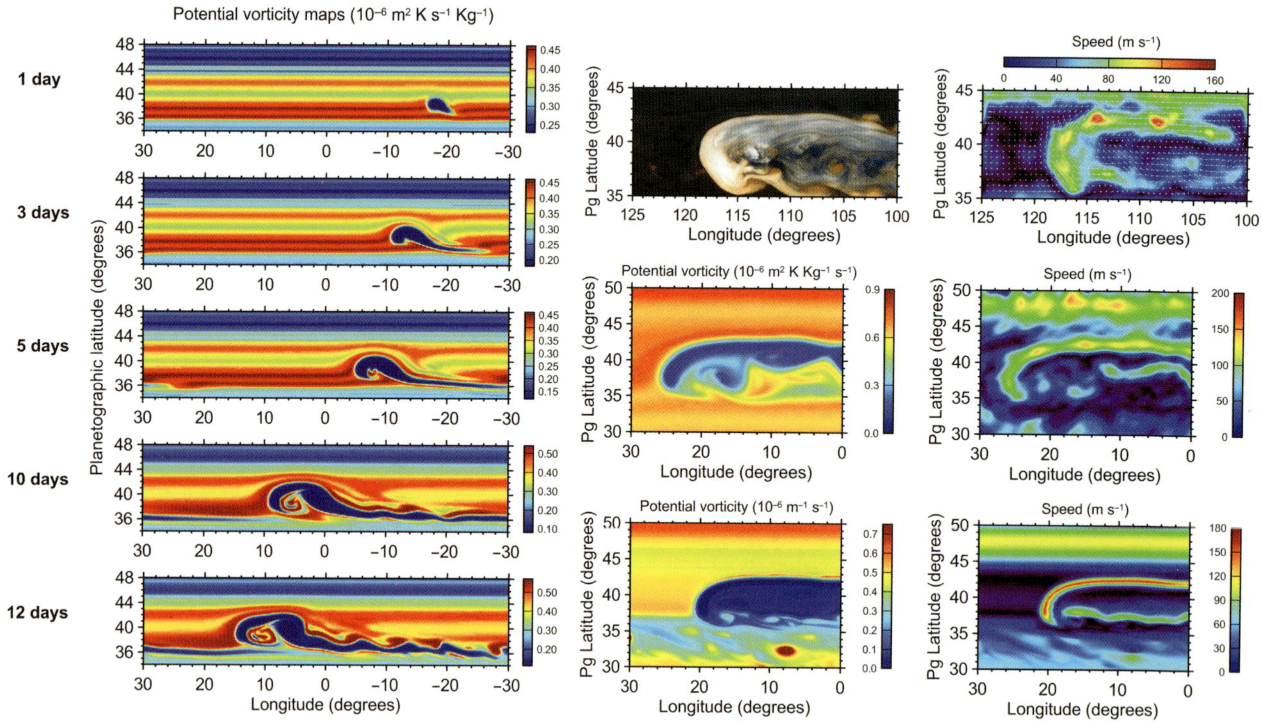

Figure 13.18

Figure 14.1

Figure 14.2

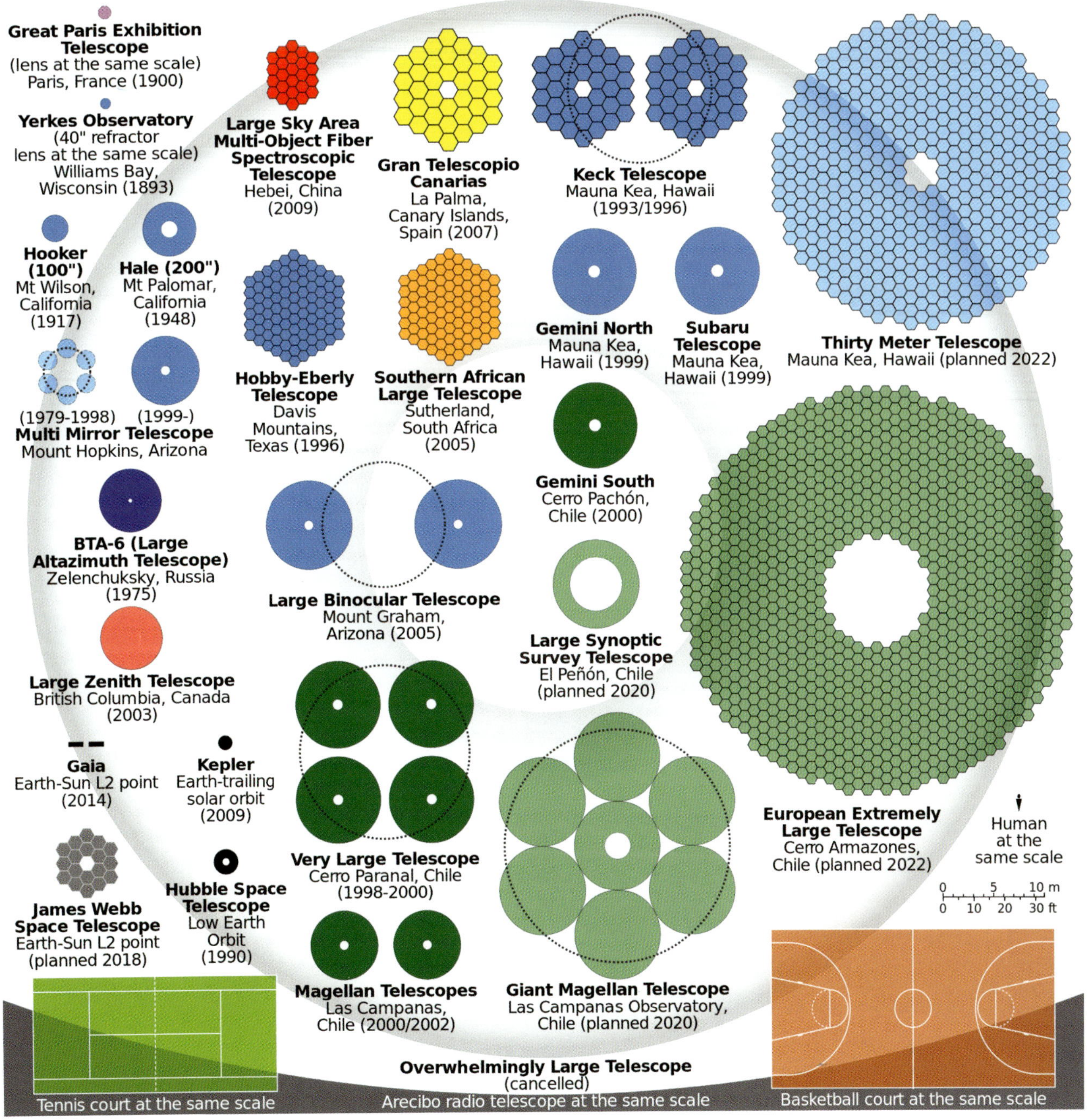
Great Paris Exhibition Telescope
(lens at the same scale)
Paris, France (1900)
Yerkes Observatory
(40" refractor
lens at the same scale)
Williams Bay, Wisconsin (1893)
Hooker (100")
Mt Wilson, California (1917)
Hale (200")
Mt Palomar, California (1948)
(1979-1998) (1999-)
Multi Mirror Telescope
Mount Hopkins, Arizona
BTA-6 (Large Altazimuth Telescope)
Zelenchuksky, Russia (1975)
Large Zenith Telescope
British Columbia, Canada (2003)
Gaia
Earth-Sun L2 point (2014)
Kepler
Earth-trailing solar orbit (2009)
James Webb Space Telescope
Earth-Sun L2 point (planned 2018)
Hubble Space Telescope
Low Earth Orbit (1990)
Large Sky Area Multi-Object Fiber Spectroscopic Telescope
Hebei, China (2009)
Gran Telescopio Canarias
La Palma, Canary Islands, Spain (2007)
Keck Telescope
Mauna Kea, Hawaii (1993/1996)
Thirty Meter Telescope
Mauna Kea, Hawaii (planned 2022)
Hobby-Eberly Telescope
Davis Mountains, Texas (1996)
Southern African Large Telescope
Sutherland, South Africa (2005)
Gemini North
Mauna Kea, Hawaii (1999)
Subaru Telescope
Mauna Kea, Hawaii (1999)
Gemini South
Cerro Pachón, Chile (2000)
Large Binocular Telescope
Mount Graham, Arizona (2005)
Large Synoptic Survey Telescope
El Peñón, Chile (planned 2020)
European Extremely Large Telescope
Cerro Armazones, Chile (planned 2022)
Human at the same scale
0 5 10 m
0 10 20 30 ft
Very Large Telescope
Cerro Paranal, Chile (1998-2000)
Magellan Telescopes
Las Campanas, Chile (2000/2002)
Giant Magellan Telescope
Las Campanas Observatory, Chile (planned 2020)
Overwhelmingly Large Telescope
(cancelled)
Tennis court at the same scale
Arecibo radio telescope at the same scale
Basketball court at the same scale

Figure 14.4

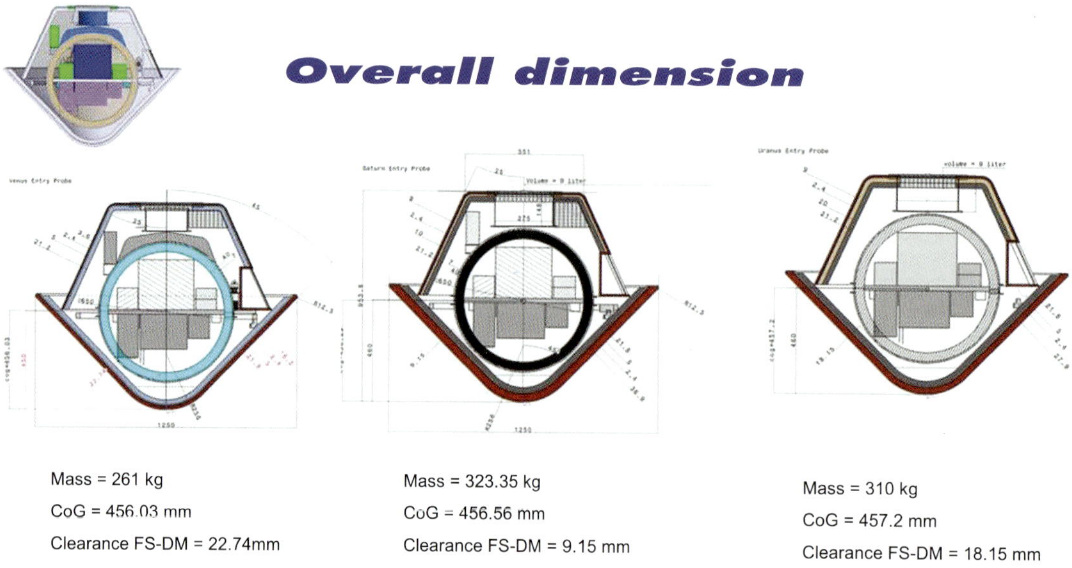

Figure 14.3

Figure 14.10

peak, h_{MAX}, has been observed at altitudes between 1100 km and 3200 km (Section 8.2.1). In general, both h_{MAX} and N_{MAX} are smallest near Saturn's equator, and increase with latitude, though there is significant scatter about this average behavior (Kliore et al., 2014).

The primary focus of early Saturn ionospheric models following the observations of the Pioneer and Voyager spacecraft was on reducing modeled electron densities to the measured ~10^4 cm^{-3} peak value. As can be seen from Figure 8.1, there is very rarely a classic Chapman-type smooth electron density profile at Saturn, and the maximum electron density is at times located in a narrow, low-altitude layer rather than what might be called the "main peak." For these reasons, models have generally focused on understanding the average trends revealed by observations rather than on reproducing exact electron density altitude profiles. Individual profiles have also been reproduced in the past, typically by varying a number of free parameters (e.g., Majeed and McConnell, 1991): the population of vibrationally excited H_2, the external water influx, and the vertical plasma drift. The first two of these parameters have been discussed already (Section 8.3.1); vertical plasma drift can arise from neutral winds (e.g., meridional winds driving plasma up or down magnetic field lines) and electric fields (e.g., **E** × **B** drifts driven by zonal electric fields; Kelley, 2009; Schunk and Nagy, 2009). Whereas the primary effect of an enhanced population of vibrationally excited H_2 or water influx is to reduce the modeled electron densities, the peak altitude is also slightly increased as these reactions are more effective at lower altitudes (e.g., Majeed and McConnell, 1991; Moses and Bass, 2000; Moore et al., 2004). Similarly, the primary effect of a vertical plasma drift is to shift h_{MAX} up or down in altitude, though N_{MAX} can also be affected if the plasma is moved into a different chemical or dynamical regime.

Though it is possible to construct a model reproduction of most of the observed electron density altitude profiles by varying unknown parameters (though likely not all of them), the derived parameters vary significantly from observation to observation. No single set of water influxes, vibrationally excited H_2 populations, and enforced vertical plasma drifts can reproduce all of the observed radio occultation electron densities simultaneously, probably an indication of spatial and/or temporal variations in these parameters.

Beyond simply comparing to N_{MAX} and h_{MAX}, modelers have also attempted to understand the narrow layers of electron density frequently observed (Figures 8.1 and 8.2). Vertically varying horizontal winds, such as might occur from atmospheric gravity waves, can cause alternating compression and extension of plasma with altitude, thus creating ionospheric layers (Kelley, 2009). Such layering is frequently observed in the terrestrial E region and the lower F region. This mechanism is especially effective for long-lived ions, such as the metallic ions introduced from the ablation of micrometeoroids. Moses and Bass (2000) were able to demonstrate the plausibility of such a layering process in Saturn's ionosphere by introducing a modest oscillatory vertical plasma drift as well as a Mg influx of 1.3×10^5 cm^{-2} s^{-1} focused in the 790 to 1290 km region. Kim et al. (2014) also suggest that such layers may result from photochemistry driven by high-resolution H_2 photoabsorption cross-sections.

Based on a wavelet analysis of 31 Cassini radio occultations (Nagy et al., 2006; Kliore et al., 2009), Matcheva and Barrow (2012) were able to detect several discrete scales of variability in Saturn's electron density profiles. Furthermore, by applying a gravity wave propagation model to Saturn's upper atmosphere, they also demonstrated that the observed features were consistent with gravity waves being present in the lower ionosphere, causing layering of the ions and electrons. In a continuation of that study, wherein a 2-D, non-linear, time-dependent model of the interaction of atmospheric gravity waves with ionospheric ions was applied to Saturn's upper atmosphere, Barrow and Matcheva (2013) were able to reproduce the structure of two Cassini radio occultations, the S08 entry and the S56 exit. Figure 8.9 presents the model-data comparison for the S56 occultation.

Extreme "bite-outs," or localized electron density depletions, frequently seen in Saturn radio occultations can also be produced from atmospheric gravity waves, as discussed above (see also Figure 6 of Kliore et al., 2009). Another possible generation mechanism for ionospheric depletions is a time-dependent water influx. Moore and Mendillo (2007) were able to reproduce the observed S7 bite-out by increasing the background water influx (of 5×10^6 cm^{-2} s^{-1}) by a factor of

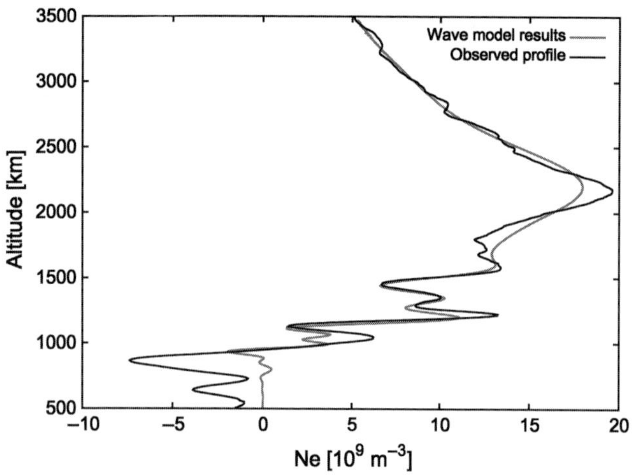

Figure 8.9 Comparison between the S56 exit observation (black line) and a model electron density profile (red line). The ionospheric model illustrates the effect of three small-amplitude gravity waves superimposed on the background electron density profile. From Barrow and Matcheva (2013). (A black-and-white version of this figure appears in some formats. For the color version, please refer to the plate section.)

50 for ~27 minutes, after which it returned to its initial value. As the resulting bulge of water density diffuses downward through the thermosphere, it undergoes charge exchange reactions with H^+, leading to a localized reduction in electron density. While a number of Cassini radio occultations contain depletions similar to the Moore and Mendillo (2007) results, the magnitude and frequency of possible water influx variations are not known at present. Therefore this possibility remains unconfirmed. Nevertheless, atmospheric gravity waves have difficulty producing ionospheric structure at high altitude due to wave dissipation lower in the thermosphere (Matcheva and Barrow, 2012), and as some electron density depletions have been observed above 2000 km (Kliore et al., 2009), an alternative mechanism for generating such structures, such as a time-variable water influx, may yet be required to reproduce some observed electron density altitude profiles.

Dawn/Dusk Electron Density Asymmetry

The first dozen radio occultations obtained by Cassini revealed a dawn/dusk asymmetry in Saturn's low-latitude ionosphere (Nagy et al., 2006). On average, the peak densities are lower and the peak altitudes higher at dawn than at dusk. A recently developed global circulation model (GCM) of Saturn's upper atmosphere, called STIM (the Saturn Thermosphere Ionosphere Model), was leveraged in order to study this new observational constraint. In comparing with the first dozen Cassini equatorial radio occultations, 1-D ionospheric calculations were performed (Moore et al., 2006) using a 3-D thermosphere that reproduced observed upper-atmospheric temperatures using a combination of auroral Joule heating and low-latitude wave heating (Müller-Wodarg et al., 2006). Moore et al. (2006) found that the average dawn and dusk equatorial electron density profiles were best reproduced by model simulations that considered a water influx at the top of the atmosphere of 5×10^6 cm^{-2} s^{-1}. Electron density comparisons are shown in Figure 8.10. This water influx was roughly a factor of three larger than the globally averaged influx derived from ISO observations, 1.5×10^6 cm^{-2} s^{-1} (Moses et al., 2000), though such a difference is likely not drastic enough to conflict with the spatial constraints evaluated by Moses et al. (2000). While the Moore et al. (2006) water influx is smaller than the favored value derived from previous comparisons with the Voyager 2 exit occultation – 2.2×10^7 cm^{-2} s^{-1} (Majeed and McConnell, 1991) – this disagreement can be explained by the fact that the Majeed and McConnell (1991) water influx was derived for a scenario in which electron loss due to reactions between protons and vibrationally excited H_2 was not considered. By including this reaction in addition to charge exchange with water, the magnitude of each required rate is reduced: Moore et al. (2006) favored a population of vibrationally excited H_2 that was 25 percent of the nominal value considered by Majeed and McConnell (1991).

The diurnal variation in peak electron density implied by the model reproductions of average dawn and dusk Cassini occultations was modest, less than a factor of 6, as illustrated in Figure 8.11. A similar set of parameters was able to reproduce the dawn and dusk N_{MAX} values measured by Cassini for a majority of the twelve equatorial occultations, with the notable exceptions being S11x, S9x, and S12 n/x (Moore et al., 2006).

The dawn/dusk asymmetry is primarily due to the presence of both atomic (H^+) and molecular (H_3^+) ions at the electron density peak. At dusk the solar source of ionization has only just shut off, so both H^+ and H_3^+ are still present and contribute to the electron density peak. During the ~5 hours of

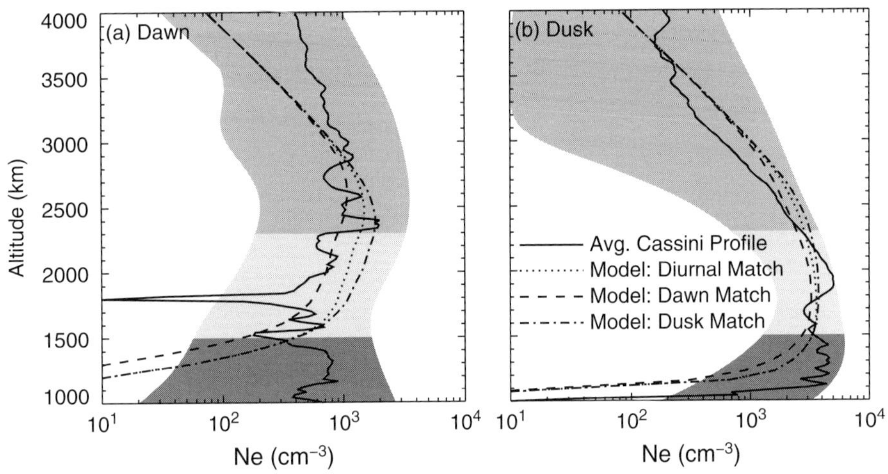

Figure 8.10 The average **(a)** dawn and **(b)** dusk Cassini electron density profiles (Nagy et al., 2006), along with model comparisons. Dotted lines represent model results that best match the observations at both dusk and dawn using a full diurnal calculation with a single set of parameters, whereas dashed and dot-dashed lines give results best matched to the average dawn or dusk profile, respectively. The width of the shaded regions corresponds to the full range of electron densities observed by Cassini, and the degree of shading represents three distinct ionospheric altitude regimes, from top to bottom: diffusive regime (dominated by H^+), photochemical regime (dominated by H^+ and H_3^+), and hydrocarbon/metallic ion regime (where these ions begin to dominate). From Moore et al. (2006).

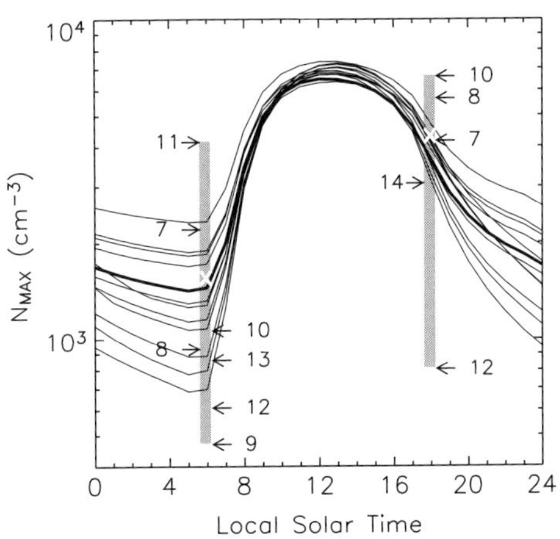

Figure 8.11 Plot of modeled local time variations of N_{MAX}. Each solid line represents the best diurnal match for one set of combinations of the two unknown chemical losses (due to vibrationally excited H_2 and water influx). Gray-shaded rectangles identify the ranges in local solar time and N_{MAX} from the Cassini radio occultations (Nagy et al., 2006). Numbers and arrows mark the individual occultation values and "x" marks the averaged dawn and dusk observations. From Moore et al. (2006).

darkness on the nightside, however, most of the H_3^+ ions have dissociatively recombined with electrons, resulting in a reduced electron density at dawn. This effect is similar to another well-known atomic and molecular ion region, the terrestrial F-layer. Modeled ion and electron altitude profiles at dawn, noon, and dusk, are presented in Figure 8.12, demonstrating the loss of H_3^+ ions during the Saturn night (Moore et al., 2008). Earlier Saturn ionospheric calculations produce the same sort of variations in H_3^+, H^+, and N_e, though for different seasonal and solar conditions (e.g., Figure 6 of Majeed and McConnell, 1996; Figure 13 of Moses and Bass, 2000).

Latitudinal Variations in N_{MAX} and TEC

Analysis of 19 additional Cassini radio occultations, mostly at mid- and high-latitudes (Kliore et al., 2009), in addition to the 12 previous equatorial occultations (Nagy et al., 2006), revealed a clear latitudinal trend in electron density: on average, electron densities were smallest at Saturn's equator and increased with latitude. This trend was recently strengthened by the addition of 28 more radio occultations across all latitudes (Figure 8.3; Kliore et al., 2014). The sub-solar point was near Saturn's equator for a majority of the occultations, and therefore the presence of a minimum in electron density in the region of peak photoionization was most likely indicative of a latitudinal dependence in the chemical loss rates.

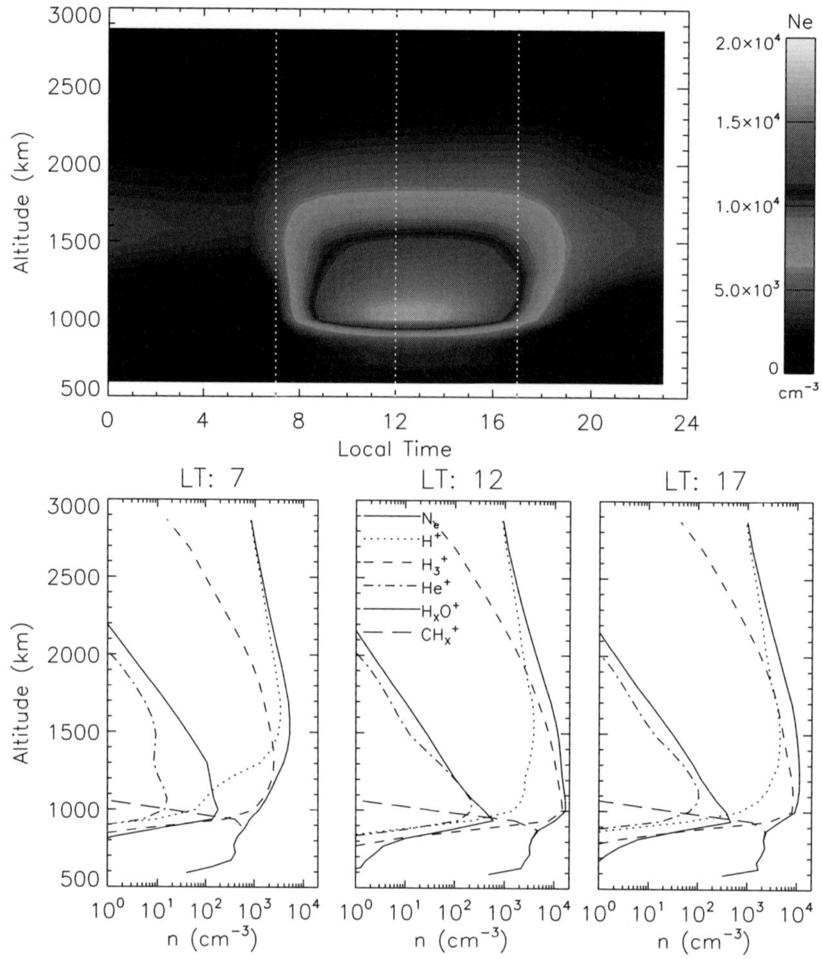

Figure 8.12 Model results for Voyager 2 conditions at 30° N latitude. Electron density contours (top) with (bottom) extracted plasma density altitude profiles near dawn, local noon, and dusk. From Moore et al. (2008).

Moore et al. (2010) explored the possibility of a latitudinally varying chemical loss using 2-D ionospheric calculations combined with a fixed 3-D thermospheric background appropriate for the average seasonal and solar conditions for the first 31 Cassini radio occultations (Kliore et al., 2009). They evaluated a Gaussian water influx centered at Saturn's equator with a peak value of $(0.1 \text{ to } 1)\times10^{7}$ cm^{-2} s^{-1} and a full-width half maximum (FWHM) between 2° and 180° (in addition to a latitudinally invariant water influx). Six populations of vibrationally excited H_2 were also considered, based on modifications to the Majeed et al. (1991) values by a factor of 0 to 2. The observed latitudinal trend in electron density was best reproduced for a Gaussian water influx centered at the equator with a FWHM of ~23.5°, illustrated in the left panel of Figure 8.13. The globally averaged water influx from such a profile is 1.1×10^{6} cm^{-2} s^{-1}, slightly less than the 1.5×10^{6} cm^{-2} s^{-1} inferred from ISO measurements (Moses et al., 2000; Moore et al., 2015). Müller-Wodarg et al. (2012) applied the Moore et al. (2010) water distribution in a subsequent study focused on magnetosphere-atmosphere coupling at high latitudes using the full 3-D STIM. Figure 8.13 illustrates the imposed water influx and the latitude variation of both modeled and observed electron densities with latitude, demonstrating that a combination of latitudinally varying water influxes and seasonal trends can explain the majority of N_{MAX} variations from radio occultation observations.

Diurnal Variation of N_{MAX}

The diurnal variation of the peak electron density, as inferred from Saturn Electrostatic Discharge (SED)

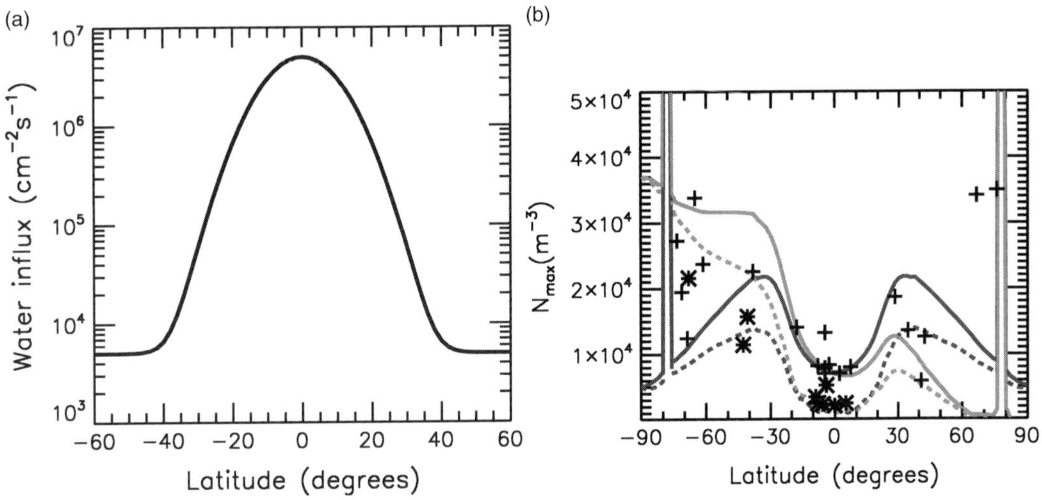

Figure 8.13 **(left)** Water influx imposed at the upper boundary of the vertical grid of STIM. A minimum base level of water influx is assumed poleward of ~40° in both hemispheres for numerical stability. **(right)** Latitudinal variation of peak electron densities in Saturn's ionosphere, as observed by Cassini, for dusk (plus symbols) and dawn (star symbols) conditions (Nagy et al., 2006; Kliore et al., 2009). Also shown are modeled peak electron densities for Saturn equinox (blue) and southern summer (red) at dusk (solid lines) and dawn (dashed lines). From Müller-Wodarg et al. (2012). (A black-and-white version of this figure appears in some formats. For the color version, please refer to the plate section.)

observations, is between 1 and 2 orders of magnitude (Figure 8.4; Kaiser et al., 1984; Zarka, 1985a; Fischer et al., 2011b). Both Voyager- and Cassini-era SED measurements find an N_{MAX} at noon of order 10^5 cm^{-3}, with dawn and dusk values approximately 10^4 cm^{-3}, in agreement with most radio occultations. Midnight electron densities derived from Voyager SEDs are ~10^3 cm^{-3}, while Cassini SEDs find a value nearer to 10^4 cm^{-3}. This disagreement may arise from complications due to Saturn Kilometric Radiation (SKR) emission, it may be related to Saturn's ring shadows leading to locally depleted electron densities or it may be attributed to something else entirely. At present, however, no model has come close to reproducing either the Voyager or the Cassini diurnal N_{MAX} trend, regardless of the nighttime electron densities.

Majeed and McConnell (1996) presented the first comprehensive model comparison with SED-derived ionospheric electron densities using a 1-D chemical diffusive model. They evaluated a wide range of combinations of vibrationally excited H_2 populations and water influxes in an attempt to reproduce the diurnal variation of N_{MAX}, as shown in Figure 8.14. Majeed and McConnell (1996) found that in the absence of these additional chemical losses the peak electron density was of order 10^5 cm^{-3} with little diurnal variation,

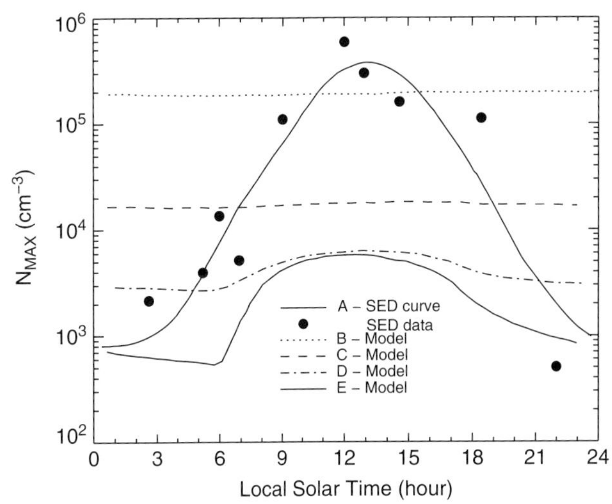

Figure 8.14 Modeled and SED-inferred diurnal variation of N_{MAX}. Curve A (solid line) shows the diurnal trend in peak electron density derived from SED measurements (filled circles). Curve B (dotted line) is the nominal model calculation for which radiative recombination is the dominant loss of H^+. Curves C (dashed line), D (dot-dash line), and E (dash-triple-dot line) are model calculations that consider enhanced chemical losses due to water influx and vibrationally excited H_2. From Majeed and McConnell (1996).

in agreement with earlier ionospheric models. By enhancing the rate of plasma recombination, the modeled diurnal variation was also enhanced, though at the cost of a reduced N_{MAX} that no longer accurately represented observed daytime electron densities.

A subsequent study by Moore et al. (2012) focused on comparisons with N_{MAX} derived from Cassini-era SEDs. Rather than explore just a limited set of realistic vibrationally excited H_2 populations and water influxes, already shown not to work by Majeed and McConnell (1996), Moore et al. (2012) also considered enhanced ion production rates. Their goal was to answer the question, "What does it take to reproduce SED-derived diurnal variation?" Figure 8.15 presents the result of 1-D model simulations that examined wide ranges of ion production and loss in order to reproduce the Cassini N_{MAX} trend. The best-fit model simulation required both drastic chemical loss and ion production in order to create and then destroy so many ions in only ~10 hours: the photoionization rate was increased by a factor of 60, the nominal population of vibrationally excited H_2 by a factor of 20, and the water influx by a factor of 540.

The primary difficulty in reproducing SED-derived diurnal electron density trends is the extremely rapid buildup of ionization in the morning hours implied by the measurements. For example, the net (i.e., production minus loss) electron production rate between dawn and noon from SED measurements is between ~9 $cm^{-3}\ s^{-1}$ (Cassini-era diurnal N_{MAX} trend) and ~30 to 79 $cm^{-3}\ s^{-1}$ (Voyager-era diurnal N_{MAX} trend), whereas the peak overhead production rate due to solar EUV is ~10 cm^{-3} s^{-1}. Therefore, an explanation of SED observations may require some sort of extreme ionization or transport process, such as due to a diurnal ionosphere-protonosphere exchange (e.g., Connerney and Waite, 1984). One alternative explanation is that SEDs may be sampling the narrow, low-altitude layers frequently seen in radio occultation electron density profiles rather than the canonical "main" ionospheric peak. Such layers are consistent with the presence of gravity waves in Saturn's lower thermosphere (Matcheva and Barrow, 2012), and can lead to narrow regions of electron density enhancements without requiring any additional sources of ionization (Barrow and Matcheva, 2013). For this explanation to hold, however, these layers must also exhibit a strong diurnal variation, must correlate with solar flux (Fischer et al., 2011b), and must be ever-present structures generated at a wide range of latitudes (Moore et al., 2012). Finally, as the atmospheric storms that give rise to SEDs tend to occur only over a limited set of specific latitudes for currently unknown reasons (primarily 35°S; Fischer et al., 2011b), any ionospheric explanation of SED-derived electron densities may also be local in nature.

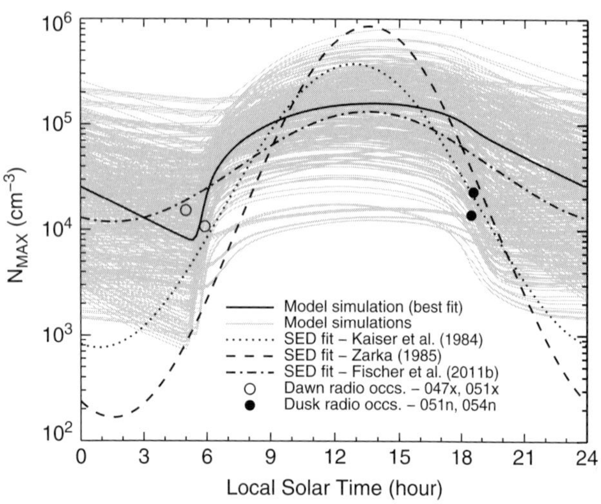

Figure 8.15 Model simulation (black solid line) that comes closest to reproducing the diurnal variation in N_{MAX} derived from Cassini SEDs (dot-dash line). Also shown are the full range of model simulations (gray solid lines), the two Voyager-era SED trends (dotted and dashed lines) and the radio occultation measurements nearest to the Cassini-era SED storm location at 35°S planetocentric latitude (filled circles for dusk, open circles for dawn). Both ion production and loss rates have been substantially increased over nominal model values. Adapted from Moore et al. (2012).

Latitudinal H_3^+ Structure: Ionospheric Signatures of Ring Rain

A new observational constraint was enabled recently following the first detection of H_3^+ emission at Saturn's non-auroral latitudes (O'Donoghue et al., 2013), which revealed magnetically conjugate extrema in opposite hemispheres that also mapped along magnetic field lines to structures in Saturn's rings. These low- and mid-latitude H_3^+ emission structures have therefore been interpreted as representing the ionospheric signatures of "ring rain," a process wherein charged water products from Saturn's rings are transported along magnetic field lines into its atmosphere (Connerney, 2013). Rather than being limited to electron density altitude profiles, like radio occultations, or to diurnal variations in peak electron density, like SED measurements, observations of ring rain signatures offer the possibility of deriving a snapshot of latitudinal variations in column-integrated H_3^+ densities and temperatures near local noon. As one of the two major ions in the upper ionosphere, such a measurement could be combined with electron density trends from radio occultation measurements in order to estimate H^+ column densities, and thereby provide a significantly improved constraint on the loss processes that control H^+ densities – vibrationally excited H_2 populations and water influx.

Emission from H_3^+ ions depends exponentially upon their temperature. Typically, the ratios between various emission lines are used to derive the column-integrated temperature of the emitting gas, which further allows a calculation of the column density (e.g., Melin et al., 2007). (Column-integrated temperature refers to the effective temperature derived for a column of gas that varies in both density and temperature along the column.) Unfortunately, due to a low signal-to-noise ratio, these parameters could not be derived from the only measured signatures of ring rain at present. Therefore, in order to make any direct model comparisons, a method of estimating the column densities is required.

In a recent study involving 1-D ionospheric calculations over a fixed 3-D background thermosphere, Moore et al. (2015) addressed this shortcoming by first estimating the neutral temperatures. They based latitudinal trends in exospheric temperature on solar and stellar EUV occultation results described in Chapter 9 by Strobel et al. As these temperatures represent conditions near the top of the atmosphere, above the altitude of peak H_3^+ ionization, an offset was then applied that accounted for the difference in neutral temperature between H_3^+ altitudes and the exobase as a function of latitude, as calculated by the 3-D STIM (Müller-Wodarg et al., 2012). These temperature estimates were then combined with the measured intensities in order to derive H_3^+ column densities from the observations, shown in Figure 8.16. The structure in Figure 8.16 is driven by the observed structure in H_3^+ emission, whereas the range of estimated column densities is dominated by temperature uncertainties.

While enhanced populations of vibrationally excited H_2 and water influxes have historically been considered as a means of reducing the modeled electron

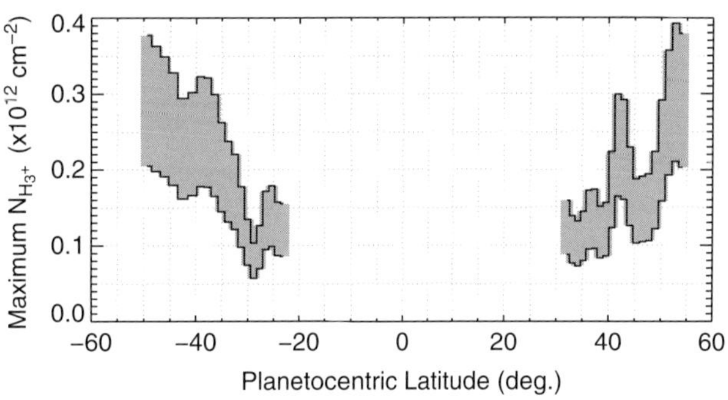

Figure 8.16 Maximum H_3^+ column densities estimated from Saturn observations of ring rain signatures (O'Donoghue et al., 2013), based on exospheric temperature measurements (Koskinen et al., 2013) and modeled temperature offsets (Müller-Wodarg et al., 2012). The range of H_3^+ column densities is driven primarily by temperature uncertainties of ~80 K. Adapted from Moore et al. (2015).

density (via destruction of the long-lived H^+ ions), there is an important secondary effect on H_3^+ densities. Specifically, because the dominant chemical loss for H_3^+ in outer planet ionospheres is due to dissociative recombination with electrons, the above loss processes also act as proxy "sources" of H_3^+ by reducing its dissociative recombination rate in correspondence with the reduced electron densities. This effect is augmented without limit by reactions between H^+ and vibrationally excited H_2, as they lead to additional production of H_2^+ (and therefore H_3^+), but it is reversed for water influxes above ~10^7 cm^{-2} s^{-1}, as H_3^+ can also be lost due to charge exchange with water products (Moore et al., 2015).

Contours of modeled H_3^+ column density as a function of vibrationally excited H_2 population (designated as k_{fac}) and water influx are presented in Figure 8.17. A dashed contour indicates the column density estimated from the ring rain measurements, and therefore represents the combinations of k_{fac} and ϕ_{H_2O} that could reproduce the observation at that latitude, 35°S planetocentric. By performing a similar series of calculations for each of the 40 latitude elements of the observations, Moore et al. (2015) were able to derive a latitudinal variation in the maximum water influxes and maximum populations of vibrationally excited H_2. They estimated the globally averaged maximum ring-derived water influx to be (1.6 to 12)×10^5 cm^{-2} s^{-1}, which represents a maximum total global influx of

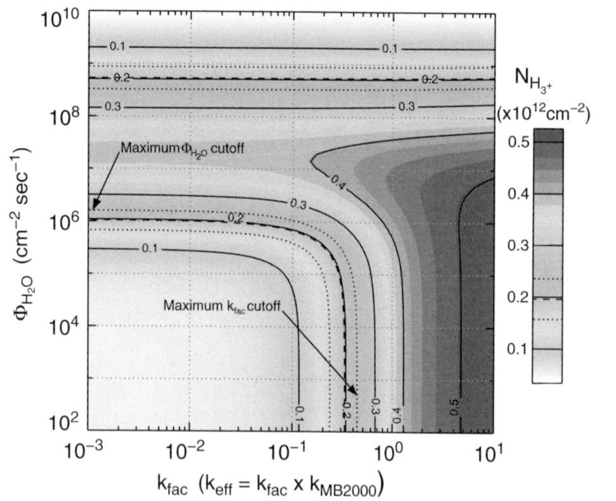

Figure 8.17 Contours of modeled H_3^+ column densities for a range of water influxes, ϕ H_2O, and populations of vibrationally excited H_2 (represented as k_{fac}). Calculation results are for 35°S planetocentric latitude at local solar noon. The dashed curves indicate the H_3^+ column density estimated from the observations (Figure 8.16), and therefore identify combinations of ϕ H_2O and k_{fac} that can reproduce the observation at that latitude. Dotted curves outline a range of H_3^+ column densities that result from accounting for a 3σ uncertainty in the observed emission intensity. From Moore et al. (2015). (A black-and-white version of this figure appears in some formats. For the color version, please refer to the plate section.)

water from Saturn's rings to its atmosphere of (1.0 to 6.8)$\times10^{26}$ s^{-1}. While there are no direct observational constraints for purely ring-derived water influxes, these upper limits compare favorably with the total oxygen influx of ~10^5 cm^{-2} s^{-1} estimated from ring atmosphere models (e.g., Tseng et al., 2010). Furthermore, the globally averaged upper limits for water influx, when combined with the assumed neutral water influx profile of Figure 8.13, are (1.3–2.3)$\times10^6$ cm^{-2} s^{-1}, fairly close to the value of ~1.5$\times10^6$ cm^{-2} s^{-1} derived by Moses et al. (2000) based on ISO observations (Feuchtgruber et al., 1997). The wide ranges in these estimates stem primarily from uncertainties in H_3^+ column integrated temperatures, and therefore future observations of the ionospheric signatures of ring rain may be able to reduce those uncertainties by determining temperatures self-consistently.

8.4 Summary

Saturn ionospheric science has advanced significantly over the past decade. Prior to the arrival of Cassini, there were six electron density profiles from radio occultations – there are now a total of 65. The diurnal variation in peak electron density derived from Voyager-era Saturn Electrostatic Discharge (SED) measurements was based on a few measurements over three SED episodes. In contrast, Cassini-era diurnal variations of electron density were derived from 48 SED episodes by 2009, and there is a corresponding wider range of additional observational parameters to explore (e.g., Cassini distance from Saturn versus derived electron density). In addition, there is the promise of a new remote observational constraint for Saturn's mid and low latitudes: H_3^+ emissions and related ring rain implications. While the conclusions that can be drawn from the current measurements of ring rain signatures are not yet definitive, future observations should allow derivations of H_3^+ and H^+ column densities, column-integrated H_3^+ temperatures, and an in-depth examination of the coupling between Saturn's rings and its atmosphere.

Ionospheric models are able to explain the gross ionospheric structure revealed by radio occultation measurements. Electron densities appear to be controlled primarily by a latitudinally varying water influx, reducing the need for enhanced populations of vibrationally excited molecular hydrogen. The observed dawn/dusk asymmetry is well explained by a mix of atomic (H^+) and molecular (H_3^+) ions forming Saturn's main ionospheric peak. Low-altitude layers are likely indicative of atmospheric gravity waves, which can lead to vertically varying plasma drifts.

The major remaining model-data discrepancy is in the diurnal variation of peak electron density. SED-derived electron densities feature large daytime values and strong diurnal variations. Models can reproduce one or the other of these constraints, but not both simultaneously, having thoroughly demonstrated that no combination of known chemistry can account for the SED-derived variations. Resolution of this discrepancy may lie in introducing additional dynamics to the modeling, such as a diurnal ionosphere-protonosphere exchange, or further investigation into the possibility that SEDs are sampling the narrow, low-altitude layers of electron density, or something else entirely.

Future observations should help refine the modeling constraints in a number of important areas. Observations of ionospheric H_3^+ emissions can provide the first ion density and temperature measurements in Saturn's ionosphere. Additional insights into ion densities resulting from *in situ* measurements are expected during Cassini's end-of-mission proximal orbits in 2017. During this mission phase, called the Cassini Grand Finale, the spacecraft will sample Saturn's upper atmosphere directly, dipping below 4000 km altitude on 22 orbits, with closest approach at low latitude near noon. With luck, Cassini may also be able to measure near-Saturn SEDs or determine the properties of the particles on magnetic field lines connecting Saturn's rings to its atmosphere. Combined with the existing observational constraints, these two new datasets – ground-based observations and Cassini Grand Finale measurements – should help to refine the remaining model-data discrepancies and to pave the way for future fully coupled models of the Saturn system by providing much-needed insight into ionospheric chemical losses due to water and to vibrationally excited H_2.

Acknowledgments

The authors wish to acknowledge the contribution of the International Space Sciences Institute (ISSI) in Bern, Switzerland, for hosting and funding the International Team on "Coordinated Numerical Modeling of the Global Jovian and Saturnian Systems," and the constructive discussions by colleagues attending the meetings. The manuscript benefitted significantly from thorough reviews by Tom Stallard and two anonymous reviewers. Support for this work at Boston University comes from NASA Grants NNX13AG21 G, NNX13AG57 G, and NNX14AG72 G. M. Galand was supported by the STFC of UK under grant ST/K001051/1 and ST/N000692/1. A. F. Nagy acknowledges support from the Cassini Mission as a member of the Radio Science Team through NASA prime Contract Task Order NMO711123.

Dedication: Arvydas J. Kliore

This chapter is dedicated to the memory of Arvydas J. Kliore, a pioneer in the field of radio science. He was involved in spacecraft radio occultations from the very first NASA planetary mission, and has shaped our knowledge of Saturn's upper atmosphere in his role as Principle Investigator of the Cassini Radio Science Subsystem. Without him, this chapter would not exist.

References

Atreya, S. K., T. M. Donahue, A. F. Nagy et al. (1984), Theory, measurements, and models of the upper atmosphere and ionosphere of Saturn, in *Saturn*, pp. 239–277, University of Arizona Press, Tucson, AZ.

Barrow, D. J. and K. I. Matcheva (2013), Modeling the effect of atmospheric gravity waves on Saturn's ionosphere, *Icarus*, 224(1), 32–42, doi:10.1016/j.icarus.2013.01.027.

Bergin, E. A., E. Lellouch, M. Harwit et al. (2000), Submillimeter wave astronomy satellite observations of Jupiter and Saturn: Detection of 557 GHz water emission from the upper atmosphere, *Astrophys. J.*, 4 (539), 147–150.

Bjoraker, G., R. K. Achterberg and D. E. Jennings (2010), Cassini/CIRS observations of water vapor in Saturn 's stratosphere, *EPSC Abstracts*, 5, 9007.

Burns, J. A., M. R. Showalter, J. N. Cuzzi et al. (1983), Saturn's electrostatic discharges: Could lightning be the cause?, *Icarus*, 54(2), 280–295, doi:10.1016/0019–1035(83)90198–7.

Burton, M. E., M. K. Dougherty and C. T. Russell (2010), Saturn's internal planetary magnetic field, *Geophys. Res. Lett.*, 37(24), doi:10.1029/2010GL045148.

Capone, L. A., R. C. Whitten, S. S. Prasad et al. (1977), The ionospheres of Saturn, Uranus, and Neptune, *Astrophys. J.*, 215, 977–983.

Cassidy, T. A. and R. E. Johnson (2010), Collisional spreading of Enceladus' neutral cloud, *Icarus*, 209(2), 696–703, doi:10.1016/j.icarus.2010.04.010.

Cavalié, T., V. Hue, P. Hartogh et al. (2014), Is Enceladus Saturn's source of water?, *EPSC Abstracts*, 9, 411.

Chen, R. H. (1983), Saturn's ionosphere: A corona of ice particles?, *Moon Planets*, 28, 37–41.

Connerney, J. (2013), Saturn's ring rain, *Nature*, 496(7444), 178–179.

Connerney, J. and J. Waite (1984), New model of Saturn's ionosphere with an influx of water from the rings, *Nature*, 312, 136–138.

Connerney, J. E. P. (1986), Magnetic connection for Saturn's rings and atmosphere, *Geophys. Res. Lett.*, 13(8), 773–776.

Crary, F. J. (2014), Saturn's other ring current, *EPSC Abstracts*, 111, 140.

Cravens, T. E. (1987), Vibrationally excited molecular hydrogen in the upper atmosphere of Jupiter, *J. Geophys. Res.*, 92(5), 11,083–11,100.

Cravens, T. E., J. Vann, J. Clark et al. (2004), The ionosphere of Titan: An updated theoretical model, *Adv. Sp. Res.*, 33(2), 212–215, doi:10.1016/j.asr.2003.02.012.

Cuzzi, J. N., J. A. Burns, S. Charnoz et al. (2010), An evolving view of Saturn's dynamic rings., *Science*, 327(5972), 1470–5, doi:10.1126/science.1179118.

Dalgarno, A. and G. Lejeune (1971), The absorption of electrons in atomic oxygen, *Planet. Space Sci.*, 19, 1653–1667.

Drossart, P., J.-P. Maillard, J. Caldwell et al. (1989), Detection of H_3^+ on Jupiter, *Nature*, 340, 539–541.

Dyudina, U., A. Ingersoll, S. Ewald et al. (2007), Lightning storms on Saturn observed by Cassini ISS and RPWS during 2004–2006, *Icarus*, 190(2), 545–555, doi:10.1016/j.icarus.2007.03.035.

Dyudina, U. A., A. P. Ingersoll, S. P. Ewald et al. (2010), Detection of visible lightning on Saturn, *Geophys. Res. Lett.*, 37(9), 1–5, doi:10.1029/2010GL043188.

Evans, D. R., J. W. Warwick, J. B. Pearce et al. (1981), Impulsive radio discharges near Saturn, *Nature*, 292 (August), 716–718.

Farrell, W. M., M. L. Kaiser and M. D. Desch (1999), A model of the lightning discharge at Jupiter, *Geophys. Res. Lett.*, 26(16), 2601–2604.

Feuchtgruber, H., E. Lellouch and T. De Graauw (1997), External supply of oxygen to the atmospheres of the giant planets, *Nature*, 389(September), 159–162.

Fischer, G., M. Desch, P. Zarka et al. (2006), Saturn lightning recorded by Cassini/RPWS in 2004, *Icarus*, 183(1), 135–152, doi:10.1016/j.icarus.2006.02.010.

Fischer, G., D. A. Gurnett, W. S. Kurth et al. (2008), Atmospheric electricity at Saturn, *Space Sci. Rev.*, 137 (1–4), 271–285, doi:10.1007/s11214-008–9370-z.

Fischer, G., D. A. Gurnett, P. Zarka et al. (2011a), Peak electron densities in Saturn's ionosphere derived from the low-frequency cutoff of Saturn lightning, *J. Geophys. Res.*, 116(A4), A04315, doi:10.1029/2010JA016187.

Fischer, G., W. S. Kurth, D. A. Gurnett et al. (2011b), A giant thunderstorm on Saturn., *Nature*, 475(7354), 75–77, doi:10.1038/nature10205.

Fjeldbo, G., A. Kliore and V. Eshleman (1971), The neutral atmosphere of Venus as studied with the Mariner V radio occultation experiments, *Astron. J.*, 76, 123.

Fjeldbo, G., A. Kliore, B. Seidel et al. (1975), The Pioneer 10 radio occultation measurements of the ionosphere of Jupiter, *Astron. Astrophys.*, 39, 91–96.

Fleshman, B. L., P. A. Delamere, F. Bagenal et al. (2012), The roles of charge exchange and dissociation in spreading Saturn's neutral clouds, *J. Geophys. Res.*, 117(E5), E05007, doi:10.1029/2011JE003996.

Fox, J. L. (2004), Response of the Martian thermosphere/ionosphere to enhanced fluxes of solar soft X rays, *J. Geophys. Res. Sp. Phys.*, 109(A11), 1–18, doi:10.1029/2004JA010380.

Fox, J. L. (2007), Near-terminator Venus ionosphere: How Chapman-esque?, *J. Geophys. Res.*, 112(E4), E04S02, doi:10.1029/2006JE002736.

Fox, J. L. and K. E. Yeager (2006), Morphology of the near-terminator Martian ionosphere: A comparison of models and data, *J. Geophys. Res.*, 111(A10), A10309, doi:10.1029/2006JA011697.

Frey, M. A. (1997), *The Polar Wind of Saturn*, University of Michigan Press, Ann Arbor, MI.

Galand, M., L. Moore, B. Charnay et al. (2009), Solar primary and secondary ionization at Saturn, *J. Geophys. Res.*, 114(A6), A06313, doi:10.1029/2008JA013981.

Galand, M., L. Moore, I. Mueller-Wodarg et al. (2011), Response of Saturn's auroral ionosphere to electron precipitation: Electron density, electron temperature, and electrical conductivity, *J. Geophys. Res.*, 116(A9), A09306, doi:10.1029/2010JA016412.

Galand, M., R. V. Yelle, A. J. Coates et al. (2006), Electron temperature of Titan's sunlit ionosphere, *Geophys. Res. Lett.*, 33(21), L21101, doi:10.1029/2006GL027488.

Glocer, A., T. I. Gombosi, G. Toth et al. (2007), Polar wind outflow model: Saturn results, *J. Geophys. Res.*, 112 (A1), A01304, doi:10.1029/2006JA011755.

Gustin, J., J.-C. Gérard, W. Pryor et al. (2009), Characteristics of Saturn's polar atmosphere and auroral electrons derived from HST/STIS, FUSE and Cassini/UVIS spectra, *Icarus*, 200(1), 176–187, doi:10.1016/j.icarus.2008.11.013.

Hallett, J. T., D. E. Shemansky and X. Liu (2005a), A rotational level hydrogen physical chemistry model for general astrophysical application, *Astrophys. J.*, 624 (1), 448–461, doi:10.1086/428935.

Hallett, J. T., D. E. Shemansky and X. Liu (2005b), Fine-structure physical chemistry modeling of Uranus H_2 X quadrupole emission, *Geophys. Res. Lett.*, 32(2), L02204, doi:10.1029/2004GL021327.

Hartogh, P., E. Lellouch, R. Moreno et al. (2011), Direct detection of the Enceladus water torus with Herschel, *Astron. Astrophys.*, 532, L2, doi:10.1051/0004–6361/201117377.

Hinson, D., J. Twicken and E. Karayel (1998), Jupiter's ionosphere: New results from Voyager 2 radio occultation measurements, *J. Geophys. Res.*, 103(A5), 9505–9520.

Huestis, D. L. (2008), Hydrogen collisions in planetary atmospheres, ionospheres, and magnetospheres, *Planet. Space Sci.*, 56(13), 1733–1743, doi:10.1016/j.pss.2008.07.012.

Ichihara, A., O. Iwamoto and R. Janev (2000), Cross sections for the reaction $H^+ + H_2$ (v=0–14) -> $H + H_2^+$ at low collision energies, *J. Phys. B At.*, 33, 4747–4758.

Ip, W.-H. (1983), On plasma transport in the vicinity of the rings of Saturn: A siphon flow mechanism, *J. Geophys. Res.*, 88(A2), 819, doi:10.1029/JA088iA02p00819.

Jurac, S. and J. D. Richardson (2007), Neutral cloud interaction with Saturn's main rings, *Geophys. Res. Lett.*, 34 (8), L08102, doi:10.1029/2007GL029567.

Kaiser, M. L., M. D. Desch and J. E. P. Connerney (1984), Saturn's ionosphere: Inferred electron densities, *J. Geophys. Res.*, 89(A4), 2371, doi:10.1029/JA089iA04p02371.

Kaiser, M. L., P. Zarka, M. D. Desch et al. (1991), Restrictions on the characteristics of Neptunian lightning, *J. Geophys. Res.*, 96, 19,043–19,047, doi:10.1029/91JA01599.

Karayel, E. and D. Hinson (1997), Sub-fresnel-scale vertical resolution in atmospheric profiles from radio occultation, *Radio Sci.*, 32(2), 411–423.

Kelley, M. C. (2009), *The Earth's Ionosphere: Plasma Physics and Electrodynamics*, 2nd ed., Academic Press, San Diego, CA.

Kim, J., A. F. Nagy, T. E. Cravens et al. (1989), Solar cycle variations of the electron densities near the ionospheric peak of Venus, *J. Geophys. Res.*, 94(89), 11,997–12,002, doi:10.1029/JA094iA09p11997.

Kim, Y. and J. Fox (1994), The chemistry of hydrocarbon ions in the Jovian ionosphere, *Icarus*, 112, 310–324.

Kim, Y. H., J. L. Fox, J. H. Black et al. (2014), Hydrocarbon ions in the lower ionosphere of Saturn, *J. Geophys. Res. Sp. Phys.*, 119, 1–12, doi:10.1002/2013JA019022.

Kliore, A. J., J. D. Anderson, J. W. Armstrong et al. (2004), Cassini radio science, *Space Sci. Rev.*, 115, 1–70.

Kliore, A. J., A. Nagy, S. Asmar et al. (2014), The ionosphere of Saturn as observed by the Cassini radio science system, *Geophys. Res. Lett.*, 41(16), 1–5, doi:10.1002/2014GL060512.Received.

Kliore, A. J., A. F. Nagy, T. E. Cravens et al. (2011), Unusual electron density profiles observed by Cassini radio occultations in Titan's ionosphere: Effects of enhanced magnetospheric electron precipitation?, *J. Geophys. Res.*, 116(A11), A11318, doi:10.1029/2011JA016694.

Kliore, A. J., A. F. Nagy, E. A. Marouf et al. (2009), Midlatitude and high-latitude electron density profiles in the ionosphere of Saturn obtained by Cassini radio occultation observations, *J. Geophys. Res.*, 114(A4), A04315, doi:10.1029/2008JA013900.

Kliore, A. J., I. R. Patel, G. F. Lindal et al. (1980), Structure of the ionosphere and atmosphere of Saturn from Pioneer 11 Saturn radio occultation, *J. Geophys. Res.*, 85(A11), 5857–5870.

Koskinen, T. T., B. R. Sandel, R. V. Yelle et al. (2013), The density and temperature structure near the exobase of Saturn from Cassini UVIS solar occultations, *Icarus*, 226(2), 1318–1330, doi:10.1016/j.icarus.2013.07.037.

Krstić, P. (2002), Inelastic processes from vibrationally excited states in slow $H^+ + H_2$ and $H + H_2^+$ collisions: Excitations and charge transfer, *Phys. Rev. A*, 66(4), 042717, doi:10.1103/PhysRevA.66.042717.

Lam, H. A., N. Achilleos, S. Miller et al. (1997), A baseline spectroscopic study of the infrared auroras of Jupiter, *Icarus*, 393, 379–393.

Lindal, G. (1992), The atmosphere of Neptune: An analysis of radio occultation data acquired with Voyager 2, *Astron. J.*, 103(3), 967–982.

Lindal, G. F., J. R. Lyons, D. N. Sweetnam et al. (1987), The atmosphere of Uranus: Results of radio occultation measurements with Voyager 2, *J. Geophys. Res.*, 92 (A13), 14987, doi:10.1029/JA092iA13p14987.

Lindal, G. F., D. N. Sweetnam and V. R. Eshleman (1985), The atmosphere of Saturn: An analysis of the Voyager radio occultation measurements, *Astron. J.*, 90(6), 1136–1146.

Lipa, B. and G. L. Tyler (1979), Statistical and computational uncertainties in atmospheric profiles from radio occultation – Mariner 10 at Venus, *Icarus*, 208, 192–208, doi:10.1016/0019–1035(79)90163–5.

Liu, C.-M. and W.-H. Ip (2014), A new pathway of Saturnian ring-ionosphere coupling via charged nanograins, *Astrophys. J.*, 786(1), 34, doi:10.1088/0004-637X/786/1/34.

Luhmann, J. G., R. E. Johnson, R. L. Tokar et al. (2006), A model of the ionosphere of Saturn's rings and its implications, *Icarus*, 181(2), 465–474, doi:10.1016/j.icarus.2005.11.022.

Lummerzheim, D., M. H. Rees and H. R. Anderson (1989), Angular dependent transport of auroral electrons in the upper atmosphere, *Planet. Space Sci.*, 37 (1), 109–129.

Lyons, J. R. (1995), Metal ions in the atmosphere of Neptune, *Science (80-.)*, 267(5198), 648–651.

Lystrup, M. B., S. Miller, N. Dello Russo et al. (2008), First vertical ion density profile in Jupiter's auroral atmosphere: Direct observations using the Keck II telescope, *Astrophys. J.*, 677(1), 790–797, doi:10.1086/529509.

Majeed, T. and J. McConnell (1991), The upper ionospheres of Jupiter and Saturn, *Planet. Space Sci.*, 39(12), 1715–1732.

Majeed, T. and J. McConnell (1996), Voyager electron density measurements on Saturn: Analysis with a time dependent ionospheric model, *J. Geophys. Res.*, 101, 7589–7598.

Majeed, T., J. McConnell and R. Yelle (1991), Vibrationally excited H_2 in the outer planets thermosphere: Fluorescence in the Lyman and Werner bands, *Planet. Space Sci.*, 39(11), 1591–1606.

Majeed, T., J. Waite, S. Bougher et al. (2004), The ionospheres–thermospheres of the giant planets, *Adv. Sp. Res.*, 33(2), 197–211, doi:10.1016/j.asr.2003.05.009.

Matcheva, K. I. and D. J. Barrow (2012), Small-scale variability in Saturn's lower ionosphere, *Icarus*, 221(2), 525–543, doi:10.1016/j.icarus.2012.08.022.

Matcheva, K. I. and D. F. Strobel (2001), Interaction of gravity waves with ionospheric plasma: Implications for Jupiter's ionosphere, *Icarus*, 152, 347–365, doi:10.1006/icar.

Matta, M., M. Galand, L. Moore et al. (2014), Numerical simulations of ion and electron temperatures in the ionosphere of Mars: Multiple ions and diurnal variations, *Icarus*, 227, 78–88, doi:10.1016/j.icarus.2013.09.006.

McConnell, J. C., J. B.. Holberg, G. R. Smith et al. (1982), A new look at the ionosphere of Jupiter in light of the UVS occultation results, *Planet. Space Sci.*, 30(2), 151–167.

McElroy, M. (1973), The ionospheres of the major planets, *Space Sci. Rev.*, 14, 460–473.

Melin, H., S. Miller, T. Stallard et al. (2007), Variability in the H_3^+ emission of Saturn: Consequences for ionisation rates and temperature, *Icarus*, 186(1), 234–241, doi:10.1016/j.icarus.2006.08.014.

Melin, H., T. S. Stallard, S. Miller et al. (2013), Post-equinoctial observations of the ionosphere of Uranus, *Icarus*, 223(2), 741–748, doi:10.1016/j.icarus.2013.01.012.

Melin, H., T. Stallard, S. Miller et al. (2011), Simultaneous Cassini VIMS and UVIS observations of Saturn's southern aurora: Comparing emissions from H, H_2 and H_3^+ at a high spatial resolution, *Geophys. Res. Lett.*, 38 (15), doi:10.1029/2011GL048457.

Mendillo, M., L. Moore, J. Clarke et al. (2005), Effects of ring shadowing on the detection of electrostatic discharges at Saturn, *Geophys. Res. Lett.*, 32, 3–7, doi:10.1029/2004GL021934.

Miller, S., N. Achilleos, G. E. Ballester et al. (2000), The role of H_3^+ in planetary atmospheres, *Philos. Trans. R. Soc. A Math. Phys. Eng. Sci.*, 358, 2485–2502.

Miller, S., N. Achilleos, G. E. Ballester et al. (1997), Mid-to-Low Latitude H_3^+ Emission from Jupiter, *Icarus*, 130, 57–67.

Miller, S., T. Stallard, C. Smith et al. (2006), H_3^+: The driver of giant planet atmospheres, *Philos. Trans. R. Soc. A*, 364(1848), 3121–3135, doi:10.1098/rsta.2006.1877.

Moore, L., G. Fischer, I. Müller-Wodarg et al. (2012), Diurnal variation of electron density in Saturn's ionosphere: Model comparisons with Saturn Electrostatic Discharge (SED) observations, *Icarus*, 221(2), 508–516, doi:10.1016/j.icarus.2012.08.010.

Moore, L., M. Galand, I. Müller-Wodarg et al. (2008), Plasma temperatures in Saturn's ionosphere, *J. Geophys. Res.*, 113(A10), A10306, doi:10.1029/2008JA013373.

Moore, L., M. Galand, I. Mueller-Wodarg et al. (2009), Response of Saturn's ionosphere to solar radiation: Testing parameterizations for thermal electron heating and secondary ionization processes, *Planet. Space Sci.*, 57(14–15), 1699–1705, doi:10.1016/j.pss.2009.05.001.

Moore, L. and M. Mendillo (2007), Are plasma depletions in Saturn's ionosphere a signature of time-dependent water input?, *Geophys. Res. Lett.*, 34, 1–5, doi:10.1029/2007GL029381.

Moore, L. E., M. Mendillo, I. C. F. Müller-Wodarg et al. (2004), Modeling of global variations and ring shadowing in Saturn's ionosphere, *Icarus*, 172(2), 503–520, doi:10.1016/j.icarus.2004.07.007.

Moore, L., I. Mueller-Wodarg, M. Galand et al. (2010), Latitudinal variations in Saturn's ionosphere: Cassini measurements and model comparisons, *J. Geophys. Res.*, 115(A11), A11317, doi:10.1029/2010JA015692.

Moore, L., A. F. Nagy, A. J. Kliore et al. (2006), Cassini radio occultations of Saturn's ionosphere: Model comparisons using a constant water flux, *Geophys. Res. Lett.*, 33(22), L22202, doi:10.1029/2006GL027375.

Moore, L., J. O'Donoghue, I. Müller-Wodarg et al. (2015), Saturn ring rain: Model estimates of water influx into Saturn's atmosphere, *Icarus*, 245, 355–366, doi:10.1016/j.icarus.2014.08.041.

Moses, J. and S. Bass (2000), The effects of external material on the chemistry and structure of Saturn's ionosphere, *J. Geophys. Res.*, 105(1999), 7013–7052.

Moses, J., B. Bézard, E. Lellouch et al. (2000), Photochemistry of Saturn's atmosphere II: Effects of an influx of external oxygen, *Icarus*, 145(1), 166–202, doi:10.1006/icar.1999.6320.

Müller-Wodarg, I. C. F., M. Mendillo, R. Yelle et al. (2006), A global circulation model of Saturn's thermosphere, *Icarus*, 180(1), 147–160, doi:10.1016/j.icarus.2005.09.002.

Müller-Wodarg, I. C. F., L. Moore, M. Galand et al. (2012), Magnetosphere–atmosphere coupling at Saturn: 1 – Response of thermosphere and ionosphere to steady state polar forcing, *Icarus*, 221(2), 481–494, doi:10.1016/j.icarus.2012.08.034.

Nagy, A. F. and T. E. Cravens (2002), Solar system ionospheres, in *Atmospheres in the Solar System: Comparative Aeronomy*, edited by M. Mendillo, A. Nagy and J. H. Waite, pp. 39–54, American Geophysical Union, Washington, DC.

Nagy, A. F., A. J. Kliore, E. Marouf et al. (2006), First results from the ionospheric radio occultations of Saturn by the Cassini spacecraft, *J. Geophys. Res.*, 111(A6), A06310, doi:10.1029/2005JA011519.

Nagy, A. F., A. J. Kliore, M. Mendillo et al. (2009), Upper atmosphere and ionosphere of Saturn, in *Saturn from Cassini-Huygens*, edited by M. K. Dougherty, L. W. Esposito and S. M. Krimigis, pp. 181–201, Springer Netherlands, Dordrecht.

Northrop, G. and J. R. Hill (1982), Stability of negatively charged dust grains in Saturn's ring plane, *J. Geophys. Res.*, 87(A8), 6045–6051.

Northrop, T. G. and J. E. P. Connerney (1987), A micrometeorite erosion model and the age of Saturn's rings, *Icarus*, 70, 124–137.

Northrop, T. G. and J. R. Hill (1983), The inner edge of Saturn's B ring, *J. Geophys. Res.*, 88(A8), 6102–6108.

O'Donoghue, J., T. S. Stallard, H. Melin et al. (2014), Conjugate observations of Saturn's northern and southern aurorae, *Icarus*, 229, 214–220, doi:10.1016/j.icarus.2013.11.009.

O'Donoghue, J., T. S. Stallard, H. Melin et al. (2013), The domination of Saturn's low-latitude ionosphere by ring "rain," *Nature*, 496(7444), 193–195, doi:10.1038/nature12049.

Perry, J., Y. Kim, J. Fox and H. Porter (1999), Chemistry of the Jovian auroralIionosphere, *J. Geophys. Res.*, 104 (E7), 16,541–16,565.

Porco, C. C., E. Baker, J. Barbara et al. (2005), Cassini imaging science: Initial results on Saturn's atmosphere, *Science (80-.)*., 307(February), 1243–1247.

Porco, C. C., P. Helfenstein, P. C. Thomas et al. (2006), Cassini observes the active south pole of Enceladus, *Science (80-.)*., 311(5766), 1393–1401, doi:10.1126/science.1123013.

Prangé, R., T. Fouchet, R. Courtin et al. (2006), Latitudinal variation of Saturn photochemistry deduced from spatially-resolved ultraviolet spectra, *Icarus*, 180(2), 379–392, doi:10.1016/j.icarus.2005.11.005.

Schlesier, A. C. and M. J. Buonsanto (1999), The Millstone Hill ionospheric model and its application to the May 26–27, 1990, ionospheric storm, *J. Geophys. Res.*, 104(A10), 22,453–22,468.

Schunk, R. W. and A. F. Nagy (2009), *Ionospheres: Physics, Plasma Physics, and Chemistry*, 2nd ed., Cambridge University Press, Cambridge.

Shimizu, M. (1980), Strong interaction between the ring system and the ionosphere of Saturn, *Moon Planets*, 22, 521–522.

Stallard, T., H. Melin, S. W. H. Cowley et al. (2010), Location and magnetospheric mapping of Saturn's mid-latitude infrared auroral oval, *Astrophys. J.*, 722(1), L85–L89, doi:10.1088/2041–8205/722/1/L85.

Stallard, T. S., H. Melin, S. Miller et al. (2012a), Peak emission altitude of Saturn's H_3^+ aurora, *Geophys. Res. Lett.*, 39(15), 1–5, doi:10.1029/2012GL052806.

Stallard, T. S., H. Melin, S. Miller et al. (2012b), Temperature changes and energy inputs in giant planet atmospheres: what we are learning from H_3^+, *Philos. Trans. A. Math. Phys. Eng. Sci.*, 370(1978), 5213–5124, doi:10.1098/rsta.2012.0028.

Stallard, T., S. Miller, H. Melin et al. (2008), Jovian-like aurorae on Saturn, *Nature*, 453(7198), 1083–1085, doi:10.1038/nature07077.

Tao, C., S. V. Badman and M. Fujimoto (2011), UV and IR auroral emission model for the outer planets: Jupiter and Saturn comparison, *Icarus*, 213(2), 581–592, doi:10.1016/j.icarus.2011.04.001.

Theard, L. P. and W. T. Huntress (1974), Ion-molecule reactions and vibrational deactivation of H_2^+ ions in mixtures of hydrogen and helium, *J. Chem. Phys.*, 60, 2840–2848, doi:10.1063/1.1681453.

Tokar, R. L., R. E. Johnson, M. F. Thomsen et al. (2005), Cassini observations of the thermal plasma in the vicinity of Saturn's main rings and the F and G rings, *Geophys. Res. Lett.*, 32(14), 1–5, doi:10.1029/2005GL022690.

Tseng, W.-L., W.-H. Ip, R. E. Johnson et al. (2010), The structure and time variability of the ring atmosphere and ionosphere, *Icarus*, 206(2), 382–389, doi:10.1016/j.icarus.2009.05.019.

Vuitton, V., O. Dutuit, M. Smith et al. (2015), Chemistry of Titan's atmosphere, *in Titan: Interior, Surface, Atmosphere, and Space Environment*, edited by I. Mueller-Wodarg, C. A. Griffith, E. Lellouch and T. E. Cravens, pp. 224–284, Cambridge University Press, Cambridge.

Waite, J. H. (1981), *The Ionosphere of Saturn*, University of Michigan Press, Ann Arbor, MI.

Waite, J. H., T. E. Cravens, J. Kozyra et al. (1983), Electron precipitation and related aeronomy of the Jovian thermosphere and ionosphere, *J. Geophys. Res.*, 88(3), 6143–6163.

Waite, J. H., T. E. Cravens, W. Ip et al. (2005), Oxygen ions observed near Saturn's A ring, *Science (80-.)*, 307 (February), 1260–1262.

Waite, J. H., W. S. Lewis, G. R. Gladstone et al. (1997), Outer planet ionospheres: A review of recent research and a look toward the future, *Adv. Sp. Res.*, 20(2), 243–252.

Waite, J. H., D. T. Young, J. H. Westlake et al. (2010), High-altitude production of Titan's aerosols, in *Titan from Cassini–Huygens*, edited by J. H. Brown, R.H. Lebreton and J.-P. Waite, pp. 201–214, Springer Netherlands, Dordrecht.

Warwick, J. W., D. R. Evans, J. H. Romig et al. (1982), Planetary radio astronomy observations from Voyager 2 near Saturn, *Science*, 215(4532), 582–587.

Warwick, J. W., J. B. Pearce, D. R. Evans et al. (1981), Planetary radio astronomy observations from Voyager 1 near Saturn, *Science*, 212(4491), 239–243, doi:10.1126/science.212.4491.239.

Withers, P., L. Moore, K. Cahoy et al. (2014), How to process radio occultation data: 1. From time series of frequency residuals to vertical profiles of atmospheric and ionospheric properties, *Planet. Space Sci.*, 101 (July 2013), 77–88.

Yelle, R. V. and S. Miller (2004), Jupiter's thermosphere and ionosphere, in *Jupiter: The Planet, Satellites and Magnetosphere*, edited by F. Bagenal, T. E. Dowling and W. B. McKinnon, pp. 185–218, Cambridge University Press, Cambridge.

Zarka, P. (1985a), Directivity of Saturn electrostatic discharges and ionospheric implications, *Icarus*, 61, 508–520.

Zarka, P. (1985b), On detection of radio bursts associated with Jovian and Saturnian lightning, *Astron. Astrophys.*, 146(L), 15–18.

Zarka, P., B. Cecconi, L. Denis et al. (2006), Physical properties and detection of Saturn's lightning radio bursts, in *Planetary Radio Emissions VI*, edited by H. O. Rucker, W. S. Kurth and G. Mann, pp. 111–122, Austrian Academy of Sciences Press, Vienna.

Zarka, P., W. Farrell, G. Fischer et al. (2008), Ground-based and space-based radio observations of planetary lightning, *Space Sci. Rev.*, 137(1–4), 257–269, doi:10.1007/s11214-008-9366-8.

Zarka, P. and B. M. Pedersen (1983), Statistical study of Saturn electrostatic discharges, *J. Geophys. Res.*, 88 (A11), 9007–9018.

Zarka, P. and M. Pedersen (1986), Radio detection of Uranian lightning by Voyager 2, *Nature*, 323, 605–608.

9

Saturn's Variable Thermosphere

DARRELL F. STROBEL, TOMMI T. KOSKINEN AND INGO MÜLLER-WODARG

Abstract

Our knowledge of Saturn's neutral thermosphere is far superior to that of the other giant planets due to Cassini Ultraviolet Imaging Spectrograph (UVIS) observations of 15 solar occultations and 26 stellar occultations analyzed to date. These measurements yield H_2 as the dominant species, with an upper limit on the H mole fraction of 5%. Inferred temperatures near the lower boundary are ~150 K, rising to an asymptotic value of ~400 K at equatorial latitudes and increasing with latitude to polar values in the range of 550–600 K. The latter is consistent with a total estimated auroral power input of ~10 TW generating Joule and energetic particle heating of ~5–6 TW that is more than an order of magnitude greater than solar EUV/FUV heating. This auroral heating would be sufficient to solve the "energy crisis" of Saturn's thermospheric heating if it can be efficiently redistributed to low latitudes. The inferred structure of the thermosphere yields poleward-directed pressure gradients on equipotential surfaces consistent with auroral heating and poleward increasing temperatures. A gradient wind balance aloft with these pressure gradients implies westward, retrograde winds ~500 m s^{-1} or Mach number ~0.3 at mid-latitudes. The occultations reveal an expansion of the thermosphere peaking at or slightly after equinox, anti-correlated with solar activity, and apparently driven by lower thermospheric heating of unknown cause. The He mole fraction remains unconstrained, as no Cassini UVIS He 58.4 nm airglow measurements have been published.

9.1 Introduction

Traditionally, the thermosphere is defined as the region characterized by a steep temperature gradient generated at its base by intense heating from absorption of short-wavelength solar ultraviolet radiation (< 170 nm) in the dissociation and/or ionization of homonuclear molecules (e.g. H_2, N_2) and the downward transport of thermal energy by heat conduction due to the absence of infrared active molecules. The base of the thermosphere is the mesopause, defined as the level where the temperature reaches a minimum and the temperature gradient vanishes due to infrared active molecules there radiating away the solar UV heating above. As with all solar system giant planets, however, solar UV heating is a minor energy source for Saturn's thermosphere.

The thermosphere is typically located above the homopause, where atmospheric species undergo gravitational diffusive separation and assume scale heights in accordance with their atomic and molecular masses rather than the atmospheric mean. Methane density profiles as determined from occultations provide relatively precise locations of the CH_4 homopause, in contrast to the much larger uncertainty found when using inferred temperature profiles to locate the base of thermosphere.

Writing an authoritative chapter on Saturn's thermosphere is challenging, for one cannot replicate the continuous vertical structure achieved by the Galileo probe at Jupiter or the Huygens probe at Titan. Over some altitude regions, temperature, composition and density profiles are retrieved as a function of pressure, whereas for others data are acquired as a function of radial distance or, in particular near the mesopause, are only marginally available. A definitive He/H_2 mixing ratio, which would enable construction of continuous atmospheric profiles at individual locations, is also lacking at Saturn, for which we have the value of 0.034 inferred from Voyager 1 infrared and radio occultation measurements, universally regarded as incorrect and at best an extreme lower bound. Instead, Conrath and Gautier (2000) inferred a ratio of ~0.13 from reanalysis of Voyager IRIS data.

The chapter is organized to establish what we know, or at least what we think we know, from density data (Section 9.2), temperature data (Section 9.3), airglow data (Section 9.4), and composition data (Section 9.5), followed by an examination of the density, pressure, and temperature structure derived from the available data and the inference of net heating rates from radial temperature profiles (Section 9.6). In Section 9.7, UVIS stellar occultation data are used to infer the location of the CH_4 homopause. Energetics and potential sources are discussed and reviewed in Section 9.8, which sets the stage for Section 9.9 on global general circulations models for the thermosphere and a discussion of the observed state of Saturn's atmosphere as compared to the models. Auroral physics and ionospheric chemistry and physics, often considered part of thermospheric physics, are dealt with in Chapters 7 and 8. We treat them here only as necessary for our purposes.

9.2 Review of Relevant Density Data

The neutral thermosphere is composed mainly of H_2 and He, with trace amounts of H and water group molecules (O, OH, H_2O). Atomic H is released by photochemical reactions below the CH_4 homopause (e.g. Moses et al. 2005) and its mixing ratio increases with altitude in the thermosphere due to molecular diffusion. Water group molecules are delivered to the upper atmosphere from Saturn's magnetosphere and rings (e.g. Connerney and Waite 1984; Feuchtgruber et al. 1997; Cassidy et al. 2010; O'Donoghue et al. 2013). Their net mixing ratio is expected to decrease with increasing pressure in the stratosphere due to condensation in the lower atmosphere and to be roughly constant with altitude in the upper atmosphere (e.g. Moses et al. 2000; Müller-Wodarg et al. 2012). The mixing ratios of CH_4 and other hydrocarbons, on the other hand, decrease rapidly with altitude above the homopause due to molecular diffusion and photolysis, and should be negligible in the thermosphere. The mixing ratio of He also decreases with altitude in the thermosphere, but it does so less rapidly than the mixing ratios of the hydrocarbons.

Observational constraints on H_2 and H can be obtained from extreme (EUV) and far-ultraviolet (FUV) occultations of bright stars and the Sun as well as airglow and auroral emissions (e.g. Broadfoot et al. 1981; Sandel et al. 1982; Smith et al. 1983; Shemansky and Ajello 1983; Yelle 1988; Gérard et al. 1995, 2009, 2013; Shemansky et al. 2009; Gustin et al. 2010). Constraints on He can be obtained from EUV airglow measurements (Sandel et al. 1982; Parkinson et al. 1998), while CH_4 near the homopause can be probed by FUV occultations (Smith et al. 1983; Shemansky and Liu 2012; Koskinen et al. 2015; Vervack and Moses 2015) as well as limb emissions and occultations in the near-IR CH_4 bands at 3.3 μm (Baines et al. 2005; Kim et al. 2012). Unfortunately, there are no direct observations of the water group molecules reported at present, even though they may play a substantial role in controlling the electron densities in the ionosphere (e.g. Moore et al. 2010; Müller-Wodarg et al. 2012; cf. Chapter 8 of the current volume by Moore et al.).

In principle, stellar and solar occultations are the most reliable sources of information about the density and temperature structure in the thermosphere. This is because the interpretation of airglow and auroral emissions depends on free parameters in complex non-LTE radiative transfer models (e.g. Liu and Dalgarno 1996; Hallett et al. 2005; Gustin et al. 2010; García-Comas et al. 2011; Adriani et al. 2011) and instrument calibration that is not always well characterized. The analysis of the occultations, on the other hand, is much simpler. The data do not need to be absolutely calibrated and the retrieved densities are simply based on the line-of-sight optical depths of the absorbers that can be obtained from the raw data with relative ease. The occultations also probe the atmosphere at a wide range of altitudes and latitudes. In this context, it may appear surprising that there are significant disagreements between different analyses of the Voyager/UVS (Broadfoot et al. 1981; Festou and Atreya 1982; Vervack and Moses 2015) and Cassini/UVIS occultations (Shemansky and Liu 2012; Koskinen et al. 2013, 2015).

The recent re-analysis of the Voyager/UVS data (Vervack and Moses 2015) and new results based on 41 solar and stellar occultations from Cassini/UVIS (Koskinen et al. 2013, 2015) have resolved this dispute.

9.2.1 Voyager/UVS Occultation Data

Voyager 1 (V1) observed two solar occultations and one stellar occultation of ι Herculis while Voyager 2 (V2) observed one solar occultation and two stellar occultations of δ Scorpii. Density and temperature profiles retrieved from all six of the Voyager/UVS occultations were published only recently by Vervack and Moses (2015). Previous results were limited to exospheric temperatures and simple forward models of the atmosphere based on only three of the occultations (Broadfoot et al. 1981; Sandel et al. 1982; Festou and Atreya 1982; Smith et al. 1983; Shemansky and Liu 2012). There are significant differences in the retrieved exospheric temperatures between these results that we discuss further in Section 9.3.2. Here, we concentrate only on the newly published density profiles of H_2 and H (Vervack and Moses 2015).

In order to better understand the results and their limitations, it is useful to develop a basic understanding of the instruments and the occultation data. The processed data consist of transmission spectra as a function of radial distance and latitude from the center of Saturn. Together, these spectra make up light curves, i.e. transmission as a function of radial distance in each wavelength band. The wavelength range of the Voyager/UVS instruments is 51–170 nm, with a spectral resolution of 1.8–3 nm for point sources. The signal-to-noise ratio (S/N) of the occultation data, however, was often so poor that in reality the data had to be binned to a much lower resolution. The observations also suffered from significant spacecraft pointing drifts and slews, interference from the rings, changes in detector gain and pixel-to-pixel sensitivity variations. All these effects significantly complicated the retrieval of density profiles from the Voyager/UVS occultations (Vervack and Moses 2015).

The density profiles of H_2 and H are retrieved by analyzing transmission at wavelengths of 51–115 nm. Hydrogen in the interstellar medium (ISM) absorbs all starlight at wavelengths shorter than 91 nm; thus the shorter wavelengths in this range are only available in the solar occultations. This is an important point because absorption at 51–80.4 nm is dominated by the ionization continuum of H_2. The absorption cross-section in the ionization continuum varies smoothly with wavelength and, even more importantly, does not depend on temperature (Samson and Haddad 1994). This is in contrast to the electronic Lyman and Werner bands of H_2 (hereafter, the LW bands) at longer wavelengths, where the cross-section is much more complicated and depends on temperature. Thus, solar occultations provide the most robust measurements of the density profiles and exospheric temperatures. The results, however, are limited to relatively high altitudes near the exobase above the 0.1 nbar level. An analysis of the LW bands is still required to probe the density and temperature profiles in the lower thermosphere and near the homopause (~0.01–0.1 μbar). The exobase is formally defined as the level in the atmosphere where the mean free path of the major species (H_2) is equal to its scale height divided by $\sqrt{2}$ and regarded as the top of the atmosphere (Strobel 2002). If the collisionless, neutral gas above the exobase is gravitationally bound, it is referred to as the exosphere.

Figure 9.1 shows the H_2 and H profiles based on the V1 and V2 data from Vervack and Moses (2015), together with results from the previous analysis of the V2 occultations by Smith et al. (1983) and ground-based observations of the occultation of the star 28 Sgr (Hubbard et al. 1997). Vervack and Moses (2015) retrieved the lowest-pressure H_2 densities (red line) from the ionization continuum in the V2 solar occultation, while the rest of the H_2 profiles below the 0.1 nbar level are based on the LW bands. They used the same canonical T-P profile to calculate the absorption cross-sections for all occultations and did not account for the change in temperature along the line of sight in this calculation. Given the relatively large uncertainties in the Voyager/UVS light curves, however, these assumptions are unlikely to result in significant systematic errors in the retrieved density profiles. As illustrated by Figure 9.1, the retrieval of the H_2 profiles is typically limited to altitudes above the 10 nbar level – a level where transmission in the LW bands vanishes. Owing to the sophisticated treatment of the instrument effects and updated retrieval techniques used by Vervack and Moses (2015), the density profiles based on the Voyager/UVS data are now

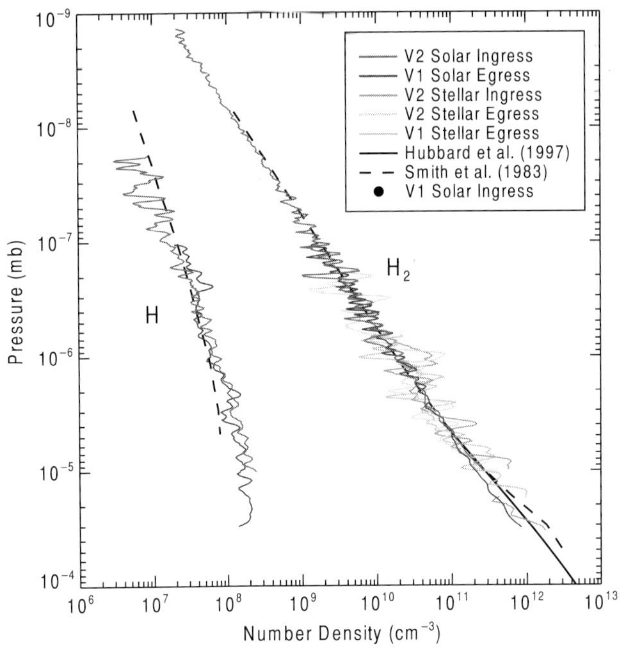

Figure 9.1 Density profiles of H_2 and H retrieved by Vervack and Moses (2015) from five low- to mid-latitude Voyager/UVS occultations. The results are compared with previous retrieval by Smith et al. (1983) and the density profile retrieved by Hubbard et al. (1997) from a ground-based stellar occultation. The figure is from Vervack and Moses (2015) and reprinted from Icarus with permission by Elsevier. (A black-and-white version of this figure appears in some formats. For the color version, please refer to the plate section.)

remarkably consistent. They also agree well with the density profile retrieved by Hubbard et al. (1997) at all relevant altitudes.

The abundance of H in Figure 9.1 is retrieved from the Lyman continuum at 80.4–91 nm in the V1 and V2 solar occultations. The analysis of the higher-resolution solar occultations from Cassini/UVIS (see Section 9.2.2 below) suggests that separating absorption by H and H_2 in this region may be more challenging than previously thought. Based on their model of the Cassini/UVIS data, Koskinen et al. (2013) found that absorption in the Lyman continuum region arises mostly from the LW bands of H_2. As a result, they were only able to retrieve upper limits on the abundance of H. These upper limits, however, are generally less than 5% below the exobase and thus consistent with the Voyager/UVS results from both Smith et al. (1983) and Vervack and Moses (2015). This means that the thermosphere of Saturn is dominated by H_2 at all altitudes below the exobase (~0.5–1 picobar).

9.2.2 Cassini UVIS Occultation Data

The Cassini/UVIS data represent a significant improvement and constitute the most extensive and reliable dataset to probe the upper atmosphere of any giant planet. Compared to the Voyager/UVS occultations, the Cassini/UVIS occultations afford higher S/N and a better point source resolution of 0.2–0.3 nm. The pointing stability of the Cassini/UVIS occultations is also excellent, thereby avoiding many of the issues that affected the Voyager observations. Finally, progress in computing, inversion techniques and retrieval algorithms have undergone substantial evolution since the Voyager era (e.g. Koskinen et al. 2011, 2013, 2015; Vervack et al., 2004; Vervack and Moses 2015). This helps to improve the accuracy of the results, and to explore the reasons for the discrepancies in past analyses.

Since the orbit insertion in 2004, Cassini/UVIS has observed more than 40 stellar and solar occultations so far. Shemansky and Liu (2012) published results based on three of the stellar occultations, and Koskinen et al. (2015) have recently analyzed 26 of the stellar occultations that were observed between 2005 and 2014 and provide useful information on the thermosphere. Koskinen et al. (2013) also analyzed 15 of the solar occultations to retrieve the density profiles and exospheric temperatures near the exobase of Saturn. The results of this study that were obtained by analyzing the ionization continuum of H_2 above the 0.1 nbar level (see Section 9.2.1) indicate that the exobase of Saturn is generally located between 2700 km and 3000 km above the 1 bar level and the composition of the thermosphere is dominated by H_2, with negligible dissociation and production of H at the probed latitudes.

Figure 9.2 shows the 41 density profiles of H_2 retrieved by Koskinen et al. (2013, 2015) from the solar and stellar occultations. The finite size of the Sun in the solar occultations means that the relationship between the observed transmission and altitude in the atmosphere is ambiguous. Thus the density profiles had to be parameterized and fitted to the data by using a forward model that accounts for the perceived size of the solar disk in the atmosphere (Koskinen et al. 2013). Stars are point sources, and a combination of direct retrieval and forward modeling was used to analyze

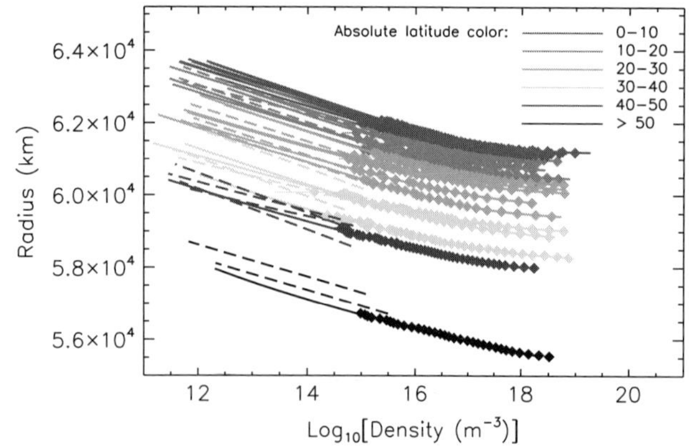

Figure 9.2 H_2 density as a function of radius (from the center of Saturn) based on the Cassini/UVIS stellar (solid lines and diamonds, Koskinen et al. 2015) and solar occultations (dashed lines, Koskinen et al. 2013). The color scale is based on latitude, which decreases from lower radial distances (black) to higher radial distances (purple). No distinction between northern and southern latitude is made. The data points (diamonds) show inverted densities, while the solid and dashed lines show forward model density profiles. The figure is based on Koskinen et al. (2015). (A black-and-white version of this figure appears in some formats. For the color version, please refer to the plate section.)

stellar occultations. In direct retrieval, each spectrum is fitted separately to obtain the column densities of the absorbers as a function of tangent altitude. The column density profiles can then be inverted to retrieve number densities (Koskinen et al. 2011, 2015). At higher altitudes ($z > 1800$ km), however, transmission is close to unity and the retrieved column density profiles are too noisy for inversion. Thus, the best-fit forward model number density profiles are shown at high altitudes in Figure 9.2. Statistical simulations show that the density profiles are accurate to within about 5 to 20%.

A comparison of the Voyager/UVS and Cassini/UVIS density profiles can be used to further test the validity of the results and detect differences in atmospheric properties between the Voyager and Cassini eras. Figure 9.3 compares four Voyager/UVS density profiles with Cassini/UVIS stellar occultation profiles at equivalent planetocentric latitudes. In general, the Voyager/UVS results agree well with the Cassini/UVIS results. Interestingly, the Cassini/UVIS densities at 21.1°N agree well with the nearly symmetric V2 stellar ingress occultation at 21.7°S, while the Cassini/UVIS densities at 28.5°N also agree well with the V2 solar egress occultation at 29°N. This comparison puts upper limits on hemispheric differences near the equator and their time evolution between the Voyager and Cassini observations. The uncertainties in the Voyager/UVS density profiles, however, are often significantly larger than the uncertainties in the Cassini/UVIS profiles and they can mask temporal and spatial variations that we discuss in the next section. As argued by Vervack and Moses (2015), these uncertainties in the density profiles have more than likely also contributed to the ambiguities in the temperature retrievals (see Section 9.3.2).

The overall density structure in Figure 9.2 illustrates the fact that surfaces of constant pressure on Saturn can be approximated as deformed ellipsoids of revolution (e.g. Zharkov and Trubitsyn 1970). Thus the density profiles reach minimum radial distances at the poles and a maximum near the equator. In order to see if the isobars in the thermosphere are similar to the lower atmosphere, we plotted the radial distances of the 0.01 nbar level (slightly below the exobase of Saturn) as a function of planetocentric latitude based on the Cassini/UVIS and Voyager/UVS occultations in Figure 9.4. In both cases, we retrieved the radial distances from the forward model density profiles (Koskinen et al. 2015; Vervack and Moses 2015). We note that the Voyager/UVS data are generally noisier and the associated uncertainty in the pressure levels is likely to be larger than the uncertainty in the Cassini/UVIS results. Thus, we assigned an uncertainty of 50 km to the Cassini/UVIS pressure levels and an uncertainty of 100 km to the Voyager/UVS occultations.

The upper panel of Figure 9.4 indicates that the occultation results are broadly consistent with the expected shape of Saturn's atmosphere. A closer inspection, however, reveals that there are differences between the shapes of the thermosphere and the lower atmosphere. The deviations of the thermosphere from the lower atmosphere are shown by the lower panel of Figure 9.4, which compares the 0.01 nbar level, based on the occultations with the 100 mbar reference model of Anderson and Schubert (2007). Here we extrapolated the reference model to 0.01 nbar by adjusting the equatorial radius of the model to match three of the equatorial stellar occultations at planetocentric latitudes of 2°N and 3°S from late 2008 and early 2009. We note that there is a lively debate on the rotation rate of Saturn (see Chapter 5 by Carbary et al.) that affects the wind-driven perturbations to the reference model.

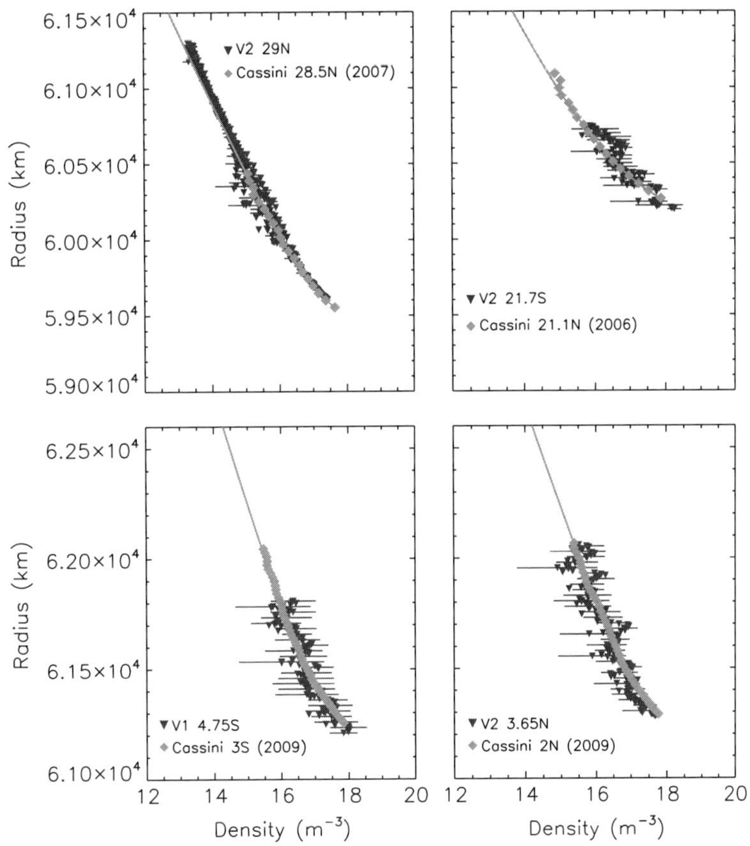

Figure 9.3 H_2 density profiles based on Cassini/UVIS stellar occultations (grey diamonds and solid grey lines, Koskinen et al. 2015) and Voyager/UVS solar (29°N) and stellar occultations (dark triangles, Vervack and Moses 2015) at roughly equivalent planetocentric latitudes. The uncertainties in the Cassini profiles are mostly too small to be visible, while the uncertainties in the Voyager profiles are larger. Note that the Cassini (2009) and Voyager V1 and V2 occultations were obtained during the equinox season.

We chose the model of Anderson and Schubert (2007) for convenience because it minimizes the wind-driven perturbations while still matching the Voyager and Cassini gravity field parameters and the observed shape of the atmosphere (Lindal et al. 1985; Jacobson et al. 2006).

The deviations of the isobars in the thermosphere from the predicted shape (hereafter, the normalized altitudes) in Figure 9.4 show two interesting trends. First, the solar occultations (purple diamonds) from Cassini/UVIS indicate that the normalized altitude along the terminator limb increases with latitude away from the equator. The solar occultations in the southern hemisphere were all obtained between late 2007 and early 2008 so that they should be relatively free of time-dependent trends. The same is true of the solar occultations in the northern hemisphere that were obtained in 2010, with only two exceptions (the high latitude data point from 2007 and one of the low latitude data points from 2008). The trend of increasing normalized altitude with latitude was noted by Koskinen et al. (2013), who explained it by arguing that the thermosphere extends to deeper pressure levels at higher latitudes. The second trend is a relatively large 600–700 km scatter of the data points at low to mid (northern) latitudes in the lower panel of Figure 9.4. To our surprise, this scatter appears not to be random. Instead, Figure 9.5 indicates that the exobase on Saturn expanded by about 500 km between 2006 and 2011, apparently followed by the onset of contraction sometime after 2011. This trend is thought to arise from changes in the energy balance near the homopause that have caused the thermosphere to warm by about 100–200 K during the same time period (Koskinen et al. 2015), probably followed by cooling during contraction.

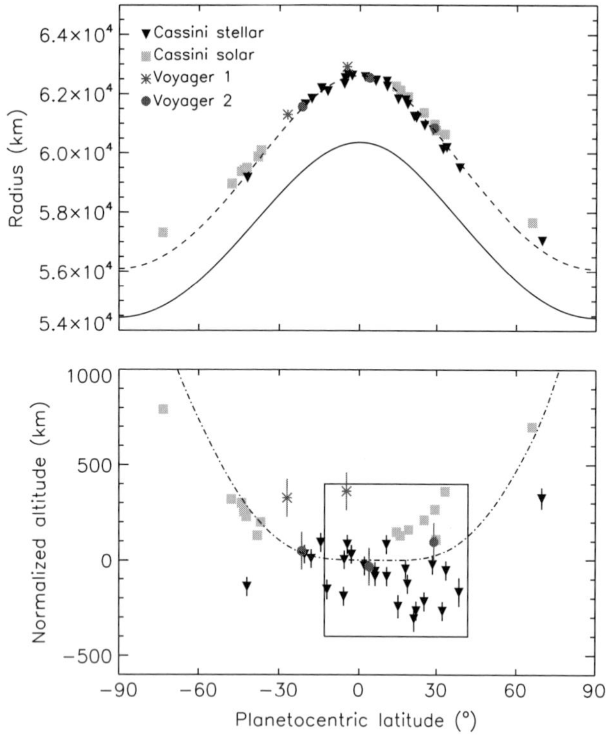

Figure 9.4 Upper panel: radial distances to the 0.01 nbar pressure level based on Cassini/UVIS solar (red squares, Koskinen et al. 2013) and stellar (black triangles, Koskinen et al. 2015) occultations as well as Voyager 1 and Voyager 2 occultations (purple stars and filled circles, respectively, Vervack and Moses 2015). The solid line is the 100 mbar reference level (Anderson and Schubert 2007) and the dashed line is an extrapolation of this reference model to 0.01 nbar that matches the equatorial occultations from 2008 to 2009. Lower panel: altitude of the data points relative to the 0.01 nbar reference level (dashed line in the upper panel). The square indicates data points included in Figure 9.5. The dashed-dotted line shows normalized altitudes predicted by the new results from the GCM of Müller-Wodarg et al. (2012) (see Section 9.8.3). The figure was taken from Koskinen et al. (2015). (A black-and-white version of this figure appears in some formats. For the color version, please refer to the plate section.)

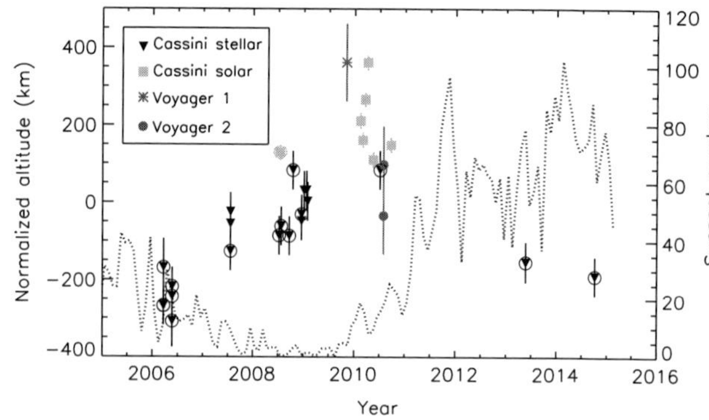

Figure 9.5 Normalized altitude above and below the 0.01 nbar reference model as a function of time, including the data points inside the square in the lower panel of Figure 9.4. The Voyager/UVS data points (based on Vervack and Moses 2015) are shown at the equivalent season and time after the equinox. The Cassini occultations that fall either into the ring shadow or probe the latitudes of the ring shadow are indicated by open circles. The sunspot number is shown by the dotted line. The figure is based on Koskinen et al. (2015). (A black-and-white version of this figure appears in some formats. For the color version, please refer to the plate section.)

The lower panel of Figure 9.4 shows that the V2 results agree well with the Cassini/UVIS results from 2008 to 2009. Within their uncertainty, they also agree with the time-dependent trend in Figure 9.5. The V1 data points in Figure 9.4, however, appear significantly more elevated than the V2 data points. This result may be compromised by the relatively large uncertainties in the V1 data, and we do not give it much significance. Overall, then, the Voyager/UVS points are consistent with the elevated state of the atmosphere that we observe in the Cassini/UVIS occultations of 2008–2009. We note that the expansion and warming of the atmosphere in the Cassini/UVIS occultations anti-correlates with solar activity between 2006 and 2010. Also, the Voyager/UVS and Cassini/UVIS observations of 2008–2009 occurred at opposite solar activity levels, indicating that the changes in the atmosphere are not driven by changes in solar activity. Instead, the Voyager and Cassini observations coincided with the same season during the northern spring. This implies that the observed changes, that are likely to arise from changes in dynamics, could be seasonal in nature.

9.3 Review of Relevant Temperature Data

9.3.1 Molecular H_3^+ Near-IR Thermal Emission

H_3^+ thermal emission has only been detected repeatedly in hotter, auroral/polar regions, where O'Donoghue et al. (2014) report average thermospheric temperatures: 527 ± 18 K in northern spring and 583 ± 13 K in southern autumn seasons, respectively (see Chapter 7 by Stallard et al.). However, on different Saturnian days, the southern aurora has exhibited a much wider range of temperatures, varying between ~400 and 600 K (Melin et al. 2007; Stallard et al. 2012; Lamy et al. 2013). In contrast to Jupiter, H_3^+ emissions from the disk were detected only recently by O'Donoghue et al. (2013) and they are of intermittent nature, as described in Chapter 8 by Moore et al. They are also not of sufficient quality to allow for the retrieval of temperatures. Thus, for low and mid latitudes one must rely on solar and stellar occultation measurements to infer temperatures in Saturn's thermosphere from the derived H_2 density scale heights. Radio occultation data yield electron density profiles from which plasma scale heights can be derived, but there is no assurance that the electron, ion and neutral gases are in thermal equilibrium so at most only the sum of electron and ion temperatures can be inferred (see Chapter 8 for plasma temperatures).

9.3.2 Inferred Temperatures from H_2 Density Profiles

In line with the H_2 density profiles (see Figures 9.2 and 9.3), the retrieval of temperatures from the occultations is limited to pressures lower than 10 nbar. As we pointed out in Section 9.2.1, there have been significant disagreements over the exospheric temperatures retrieved from the occultations. For example, Broadfoot et al. (1981) used the V1 solar egress occultation at the planetocentric latitude of 30S to derive an exospheric temperature of 850 ± 100 K. Only a year later, Sandel et al. (1982) analyzed the V2 stellar egress occultation and reported an exospheric temperature of only 400 K at the planetocentric latitude of 3.5°N. This result was contradicted by Festou and Atreya (1982), who retrieved a temperature of 800 K from the same stellar occultation.

Sandel et al. (1982) argued that the results of Broadfoot et al. (1981) may have been biased towards larger scale height and temperature by the finite size of the Sun that was not taken into account in the analysis. The angular size of the Sun at Saturn is about 1 mrad, which, depending on the distance of the spacecraft from Saturn during the occultations, can translate to an apparent diameter of the solar disk comparable to or larger than the scale height of about 150–200 km of the thermosphere. By ignoring this effect, however, we would only overestimate the temperature by about 70 K, even if the apparent diameter of the solar disk were as large as 500 km (Koskinen et al. 2013). Therefore, we consider it unlikely that ignoring the solar disk in the analysis can explain a discrepancy of 400 K in the temperatures.

Instead, Smith et al. (1983) argued that the temperature retrieved by Broadfoot et al. (1981) was too large because the latter misinterpreted the effects of an instrument gain change during the occultation and had problems in dealing with severe pointing drifts. Vervack and Moses (2015), on the other hand, found that the high-altitude light curves of the V1 solar egress occultation are actually consistent with a temperature of 800 K, but also concluded that the high-altitude results were corrupted by bad data. Thus, both Smith et al. (1983) and Vervack and Moses (2015) support the conclusion that the relatively high temperature of 800 K is erroneous.

This does not explain the disagreement between Sandel et al. (1982) and Festou and Atreya (1982) over the V2 stellar egress occultation. In our opinion, however, Smith et al. (1983) already convincingly demonstrated that an exospheric temperature of 800 K does not provide as good a fit to this occultation as a temperature of 400 K. They also derived an exospheric temperature of only 450 K at the planetocentric latitude of 29°N from the ionization continuum of H_2 in the V2 solar occultation, which provides a fundamentally more robust measurement of the temperature than the LW bands of H_2 that were used by Festou and Atreya (1982). In addition, the lower temperature of 400–500 K is supported by the recent reanalyses of the V2 stellar egress occultation (Shemansky and Liu 2012; Vervack and Moses 2015).

The lower temperature is also supported by the Cassini/UVIS occultations. To show this, Figure 9.6 compares the exospheric temperatures from Cassini/UVIS (Koskinen et al. 2013, 2015) with the Voyager/UVS results (Smith et al. 1983; Vervack and Moses 2015). The temperatures from Cassini/UVIS range from 370 K to 590 K, and the solar occultations in particular also indicate that the temperature increases by 100–150 K with latitude from the equator towards the poles. In general, the Cassini/UVIS results are in good agreement with the Voyager/UVS data, with the exception of the V1 solar ingress occultation near the south pole at the planetocentric latitude of 84°S. This occultation, however, suffered from spacecraft slewing and anomalous channel behavior, and Vervack and Moses (2015) do not place much significance on the

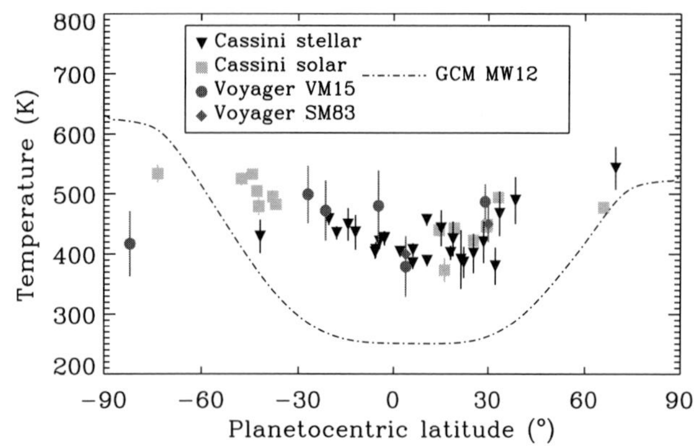

Figure 9.6 Exospheric temperatures retrieved from Cassini/UVIS solar (red squares, Koskinen et al. 2013) and stellar (black triangles, Koskinen et al. 2015) occultations, and Voyager/UVS occultations by Smith et al. (1983) (purple diamonds) and Vervack and Moses (2015) (purple circles). The dashed line shows the exospheric temperatures based on new results from the GCM of Müller-Wodarg et al. (2012) (see Section 9.8.3). The figure was taken from Koskinen et al. (2015). (A black-and-white version of this figure appears in some formats. For the color version, please refer to the plate section.)

disagreement with the Cassini/UVIS results there. In our opinion the results from Cassini/UVIS, together with the re-analysis of the Voyager/UVS data, finally settle the debate on Saturn's exospheric temperatures.

In addition to the exospheric temperatures, the occultations can be used to retrieve temperature profiles that are critical to our understanding of the energy balance in the thermosphere. In general, there are three methods that have been used to retrieve the temperature profiles in the past. First, a parameterized temperature profile can be fitted to the data by forward modeling the light curves or the retrieved column density profiles. Second, the retrieved density profiles can be integrated directly to obtain partial pressures of H_2 that can be converted to temperatures by using the ideal gas law. Third, the transmission spectra can be used to infer the rotational temperature of the H_2 molecules by analyzing the absorption bands.

Many of the past studies relied on forward modeling to estimate the temperatures (Festou and Atreya 1982; Smith et al. 1983). This approach can be dangerous, particularly if the uncertainty in the light curves and thus the density profiles is large (Vervack and Moses 2015). It is especially dangerous if the atmosphere models start from the 1 bar level because in that case the results depend on several free parameters that are not constrained by the data. Direct retrieval of temperatures has the advantage that it does not make any assumptions about the temperature profile and the uncertainties are tractable with Monte Carlo techniques (e.g. Koskinen et al. 2015). With large uncertainties in the density profiles, however, direct retrieval can introduce artificial waves to the temperature profiles and the exospheric temperature depends on the upper boundary pressure that is often not known a priori.

Both forward modeling and direct retrieval depend on the assumption of hydrostatic equilibrium. This is in general an excellent approximation in the thermosphere, but in principle it is also possible to constrain the rotational temperature of H_2 directly from the observed spectra. This approach was attempted by Shemansky and Liu (2012), who used a combination of forward modeling and spectral analysis to constrain the temperatures. They, however, concluded that the existing databases of H_2 absorption probabilities are not sufficiently extensive to reliably measure the temperatures. We note that spectral measurements of the temperature are also compromised by the insufficient wavelength resolution and S/N of the data. In addition, the absorption bands are affected by changes in temperature and level populations of H_2 along the line of sight that are not separable in the transmission spectra.

In order to reduce the associated uncertainties, Koskinen et al. (2015) used a combination of forward modeling and direct retrieval to obtain temperature and density profiles iteratively from the Cassini/UVIS occultations. For example, Figure 9.7 shows the temperature-pressure (T-P) profile based on an occultation of β Crucis from January 2009 that probes the atmosphere at the planetocentric latitude of 3°S (hereafter, ST32). In this case the exospheric temperature is 427 ± 11 K and the uncertainty along the profile ranges from a few percent to about 15%. Here the forward model profile agrees well with the direct retrieval. The uncertainty depends on the brightness of the star and altitude sampling rate of the occultations, and in this regard ST32 is one of the best datasets.

Curiously, the temperatures retrieved by Koskinen et al. (2015) disagree significantly with Shemansky and Liu (2012) for two of the three stellar occultations that were analyzed by the latter, i.e. ST32 and an occultation of δ Orionis from April 2005 that probes the atmosphere at the planetocentric latitude of 42°S (hereafter,

ST1). For ST1, Shemansky and Liu (2012) obtained an exospheric temperature of 318 ± 5 K, whereas Koskinen et al. (2015) obtained a temperature of 429 ± 28 K for the same occultation. In addition, Shemansky and Liu (2012) obtained an exospheric temperature of 612 K for ST32 (cf. Figure 9.7).

Shemansky and Liu (2012) suggested that the scale height of H_2 decreases with altitude above 1400 km (1 nbar) due to significant dissociation of H_2. We note that the stellar occultations cannot be used to directly retrieve the abundance of H, and the idea that H_2 is significantly dissociated contradicts the relatively low abundances of H below the exobase that have been retrieved from solar occultations (Koskinen et al. 2013; Vervack and Moses 2015). Furthermore, Koskinen et al. (2015) did not find evidence for the dissociation of H_2 in the light curves from the stellar occultations that are actually consistent with the scale height increasing with altitude as expected. This suggests that dissociation of H_2 is not particularly important below the exobase and the H_2 density profiles are likely to be close to diffusive equilibrium above the homopause. As a result, the retrieval of temperatures in the thermosphere should not be significantly affected by uncertainties in the composition.

Assuming that the scale height of H_2 is not a reliable measure of the temperature in the upper thermosphere, Shemansky and Liu (2012) derived the temperature for ST32 by "using a polynomic fit to the scale height" near 1400 km in altitude. We note that this method is not accurate in regions where the temperature changes with altitude, and it is typically much less accurate than forward modeling the density profiles or direct retrieval of temperatures (see above). To highlight this point, Figure 9.7 shows a comparison between the temperature profiles retrieved by Koskinen et al. (2015) and Shemansky and Liu (2012) for ST32. It is difficult to believe that a heat conducting atmosphere can support the temperature profile retrieved by Shemansky and Liu (2012) where the temperature increases by about 500 K within practically a single pressure level.

Finally, the extended analysis of the stellar occultations by Koskinen et al. (2015) allows for a more systematic exploration of the temperature structure in the thermosphere. A particularly fruitful method to probe thermal structure in the atmosphere is to combine the T-P profiles from the occultations with Cassini/CIRS data to create T-P profiles that extend from the 1 bar level to the thermosphere. For example, Figure 9.8 shows five temperature profiles based on three occultations from the spring of 2006 (hereinafter, ST5, ST10 and ST12) and two from December 2008 (hereinafter, ST30 and ST31). ST5, ST10, ST12 and ST31 probe the atmosphere near the planetographic

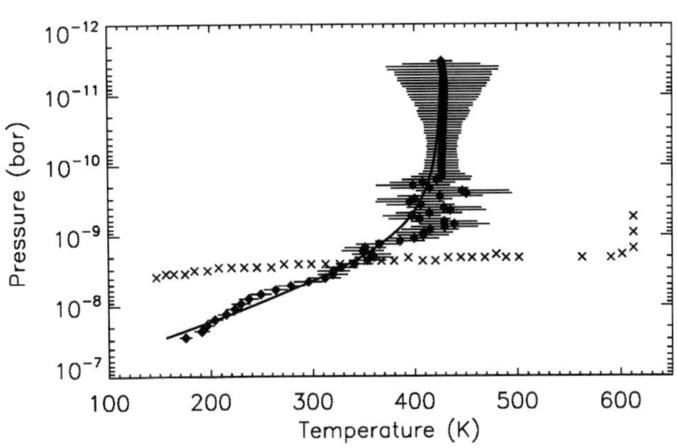

Figure 9.7 Temperature-pressure (T-P) profile based on the occultation of β Crucis from January 2009 that probes the atmosphere at the planetocentric latitude of 3°S. The diamonds and solid line show the direct retrieval and forward model profiles from Koskinen et al. (2015) while the crosses show the temperature profile from Shemansky and Liu (2012) (see text).

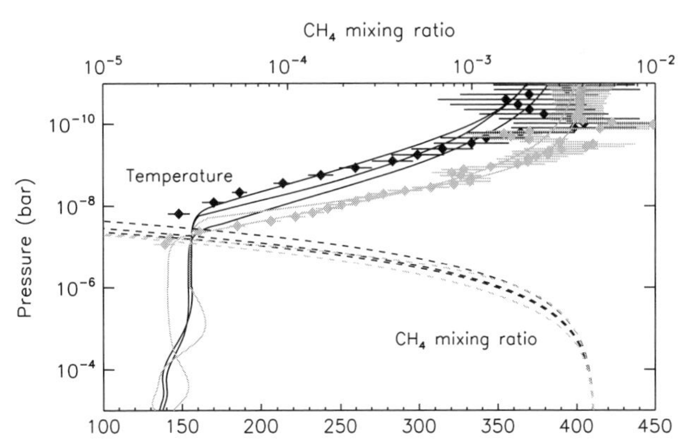

Figure 9.8 Forward model temperature profiles (solid lines) and CH_4 mixing ratio profiles (dashed lines) for spring of 2006 (black) and December 2008 (red). The occultations probe the atmosphere at planetographic latitudes of 2° to 20°N (see text). Diamonds show the direct retrieval temperatures for two of the occultations. The figure was taken from Koskinen et al. (2015). (A black-and-white version of this figure appears in some formats. For the color version, please refer to the plate section.)

latitude of 20°N, while ST30 probes the atmosphere near 2°N. These occultations were chosen because of a close coincidence between the UVIS occultations with the CIRS measurements in the spring of 2006, and for the fact that most of the data points showing the expansion of the atmosphere between 2006 and 2011 lie in this region. The figure also shows the best fit forward model mixing ratios of CH_4 for the occultations that are discussed further in Section 9.7.

The temperatures in the lower atmosphere are retrieved from CH_4 emissions in the Cassini/CIRS limb scans and the results are valid up to the 3 μbar level. As we pointed out before, the Cassini/UVIS retrievals are valid down to the 0.1–0.01 μbar level, depending on the occultations. The implied agreement between the CIRS and UVIS temperatures is relatively good and the data indicate that the location of the base of the thermosphere varies between 0.1 and 0.01 μbar. We note, however, that this region falls into a gap in coverage between the two instruments, and thus we are prevented from accurately locating the base of the thermosphere. This also introduces additional uncertainty to the hydrocarbon mixing ratio profiles that are derived from the occultations (see Section 9.7).

Interestingly, the temperature profiles from December 2008 in Figure 9.8 are generally hotter than the corresponding profiles from the spring of 2006 in the lower thermosphere (~0.1–10 nbar) while there are no detectable differences in the exospheric temperatures. The base of the thermosphere may also be at a higher pressure level in the December 2008 occultations. This supports the argument by Koskinen et al. (2015) that warming and extension of the thermosphere to deeper pressures can explain the expansion of the atmosphere and, by inference, that the contraction of the atmosphere that may have started after 2011 is accompanied by cooling of the lower thermosphere. The origin of these changes in thermal structure, however, is currently poorly understood. A more comprehensive study that uses photochemical and radiative transfer models to interpret the temperature profiles together with the hydrocarbon abundances can shed further light on these processes and may provide more detailed information on dynamics in the mesosphere and lower thermosphere.

9.4 Review of Airglow Data

In common with the H_2/He atmospheres of Jupiter, Uranus and Neptune, Saturn's airglow is dominated by H_2 electronic bands, the He 58.4 nm line, the H Lyman line series, and H_3^+ near-IR bands. Because each atmosphere has a thermosphere significantly hotter than would be predicted by solar EUV and FUV heating, other energy sources must be considered to understand the mechanisms for airglow emission. A discussion of airglow is further challenged by the long-term calibration issues in the EUV/FUV for space-borne spectrometers, the low spectral resolution of the Voyager Ultraviolet Spectrometers (UVS), and the "no resolution" of the Pioneer 10 photometer. Thus, one looks to the Hopkins Ultraviolet Telescope (HUT), Hubble Space Telescope (HST) and Cassini /UVIS for high-spectral-resolution, well-calibrated data. Without accurate absolutely calibrated data, a discussion of airglow is reduced to purely qualitative statements without any firm understanding.

9.4.1 H Lyman Alpha

In principle, the H Lyman-α (121.6 nm) dayglow on Saturn should be quite straightforward to explain. The strong solar Lyman-α line, with a line width of ~0.1 nm characteristic of line formation in a region where the temperature is ~10^{4-5} K in the solar atmosphere, is resonantly scattered by Saturn's atomic hydrogen above the CH_4 absorbing region, whose upper bound is approximately the homopause. The thermospheric temperature, ~300–600 K, governs the intrinsic planetary line width, and the H column density above the absorbing CH_4 region governs the scattering optical depth at line center, and together they determine what fraction of the solar line can be resonantly scattered out of the atmosphere. While the thermospheric scattering optical depth at line center can be very large, up to 10^5, it may be optically thin in the wings of the solar line, due to the mismatch of line widths associated with the mismatch of line formation temperatures, ~400 K vs. ~30,000 K. In addition, radiative transfer to properly compute planetary line formation and the emergent intensity from the atmosphere must include angle dependent scattering with frequency redistribution (cf. Lee and Meier 1980).

Voyager UVS and Cassini UVIS observations yield a relatively flat center-to-limb variation (Ben-Jaffel et al. 1995; Gustin et al. 2010), which would suggest optically thin emission (e.g. Ben-Jaffel et al. 2007 and references therein), even though the Saturn line is optically thick at line center. If total emission were dominated by line center photons, then a conservatively scattering atmosphere would give a center-to-limb cosine-like variation. However, as the limb is approached the optically thin wings of the Saturn line become a more important source of emission and produce a flatter center-to-limb variation.

The two Voyager UVSs measured Saturn's H Lyman-α brightnesses at ~3.3 kR (V1) and 3.0 kR (V2) (Broadfoot et al. 1981; Sandel et al. 1982). The average Lyman-α disk brightness from 29 IUE observations was 1.1 ± 0.36 kR (McGrath and Clarke 1992). This discrepancy between UVS and IUE is perplexing in light of their agreement on the Jovian Lyman-α brightness. Gustin et al. (2010) give peak limb brightness values with adjustments for solar activity for V1: 1.9, 2.5 kR; V2: 1.8 kR; to be compared with their UVIS limb scans with peak brightness of only 0.8 kR and scan averages of 0.44 kR. But in Table 3 of Shemansky et al. (2009), the UVIS non-auroral Lyman-α brightness range near the limb is stated to be higher at ~1–1.2 kR, with reference to Shemansky and Ajello (1983) that the V1 brightness was larger, ~4.9 kR, at mid-latitudes in 1980. No UVIS center of the disk nor disk averaged values have been published to facilitate a better comparison, but the prudent conclusion would be that the Voyager values need downward adjustment.

9.4.2 He 58.4 nm

Like Lyman-α, the interpretation of the He I 58.4 nm line should also be straightforward, were it not for the requirement that the He/H_2 ratio be accurately known. Planetary He absorbs solar He I 58.4 nm radiation and reemits/scatters it with a probability equal to 0.9989. In addition, knowledge of the thermospheric temperature for planetary line width, and location of the homopause for the He column density above the unit H_2 absorption optical depth are necessary for accurate interpretation of He I 58.4 nm observations and all are uncertain to various degrees.

The originally reported Voyager brightnesses were V1: 2.2 ± 0.3 and V2: 4.2 ± 0.5 R (Broadfoot et al. 1981; Sandel et al. 1982), whereas Parkinson et al. (1998) reported these measurements as disk center brightness values of V1: 3.1 ± 0.4 and V2: 4.2 ± 0.5 R, with no discussion for the increased V1 value.

Parkinson (2002) performed the most recent analysis of the Saturnian He 58.4 nm line brightness for Voyager UVS, for which some aspects were previously reported in Parkinson et al. (1998). Constrained by the Voyager IRIS He/H2 mixing ratio ~0.03 and UVS occultation data, Parkinson (2002) required an implausibly high homopause altitude and large vertical mixing of $K_{zz} > 10^9\ cm^2\ s^{-1}$, whereas if a solar He/H_2 mixing ratio ~0.13 were appropriate as Conrath and Gautier (2000) inferred from re-analysis of IRIS data, then $K_{zz} > 2 \times 10^7\ cm^2\ s^{-1}$ for V1 and $K_{zz} > 1 \times 10^8\ cm^2\ s^{-1}$ for V2, with the latter still exceedingly large. It must be kept in mind that the Voyager He 58.4 nm brightnesses might need downward adjustment.

9.4.3 H_2 Electronic Bands

The surprisingly large H_2 EUV/FUV dayglow intensities observed by Voyager for Jupiter, Saturn and Uranus generated a lively debate about excitation mechanism(s), primarily because at the time there were no rigorous calculations available on strong solar line contributions to H_2 fluorescence in the dayglow. Three principal mechanisms were advanced to explain the dayglow: (1) additional electron excitation (Shemansky 1985), (2) dynamo-plasma acceleration (Clarke et al. 1987) and (3) solar fluorescence (Yelle 1988), in addition to dayglow generated by photoelectrons (cf. Strobel et al. 1991). The "excess" dayglow was given a name "electroglow" (Broadfoot et al. 1986), yet the measured intensities exhibited a dependence on the incident solar EUV and FUV fluxes at each planet. Broadfoot et al. (1986) emphasized excitation by low-energy electrons as a necessary component of the phenomenon. However, the power requirements to energize these electrons exceeded substantially what the Sun could supply in the UV from known processes.

It was the combination of the high-resolution HUT spectra (0.3 nm) of Jupiter's dayglow (Feldman et al. 1993) and the definitive calculation performed by Liu and Dalgarno (1996), who demonstrated that solar-induced H_2 fluorescence creates a spectrum distinctly different from photoelectron impact on H_2 that explains Jupiter's dayglow. The strongest fluorescence, ~14% of the total, is due to the solar Lyman-β line at 102.572 nm (as proposed by Yelle 1988), which is coincident with the P(1) line of the H_2 Lyman 6–0 band at 102.593 nm.

Fortunately for Saturn, Cassini UVIS data has a spectral resolution of ~0.55 nm, sufficient to separate the solar fluorescence contribution from electron impact generated H_2 band emissions. Gustin et al. (2010) used UVIS limb scan data taken at low latitudes below the ring plane to derive volume emission rates for various components of the dayglow. With disk-averaged Jupiter dayglow contributions adopted from Liu and Dalgarno (1996) adjusted for solar activity and scaled to Saturn, Gustin et al. (2010) obtained 173 R for fluorescence generated dayglow and 131 R for electron impact produced dayglow; thus in the ratio of 0.57:0.43. From the UVIS limb data, Gustin et al. (2010) derived limb-averaged values of 460 and 1054 R, respectively, with a ratio of 0.3:0.7. They noted that this ratio reaches a minimum of 0.2:0.8 at a tangent altitude of 1400 km which suggests that solar fluorescence is relatively more important on disk and relatively unimportant on the limb and that electron impact becomes progressively more important at high altitudes. A detailed analysis of disk-center dayglow would be extremely enlightening to determine whether solar fluorescence plus photoelectron-generated H_2 dayglow is sufficient to explain Saturn's dayglow as it is for Jupiter's dayglow.

9.4.4 H_3^+ Thermal Emission

The H_3^+ ion plays a fundamental role as a thermospheric thermostat for the giant planets in a manner analogous to NO in the Earth's thermosphere. By near-IR thermal emissions in its ν2 band, between 3.4 and 4.1 microns (described in detail in Chapter 7 by Stallard et al.), H_3^+ regulates Saturn's thermospheric temperature. Saturn's low and mid-latitude thermosphere is colder (~400–450 K) with fewer H_3^+ ions. Thus, H_3^+ thermal emission has mostly been detected in hotter (> 500 K) auroral/polar regions.

9.5 Review of Composition

Saturn's thermosphere is mostly H_2, with an uncertain amount of He and a maximum volume mixing ratio of H atoms at the exobase of 0.05 (Koskinen et al. 2013) and proportionally decreasing with decreasing altitude, given the 2:1 ratio in H:H_2 scale heights above the homopause. We note that the UVIS occultation forward models (Section 9.7) are also more consistent with the higher He mixing ratio of 0.13 from Conrath and Gautier (2000), while the lower bound of 0.03 leads to atmospheric structure that provides a worse fit to the H_2 and CH_4 density profiles retrieved from the occultations. From the He 58.4 nm line emission analysis by Parkinson et al. (1998) and Parkinson (2002), our inferred location of Saturn's homopause from UVIS occultation data, and downward revision of the Voyager 58.4 nm brightnesses, only a He/H_2 ratio close to Jupiter's ratio of 0.157 could yield 58.4 nm intensities in the revised Voyager range.

To date there are no measurements of HD in the thermosphere, but it may be possible to detect HD with the Cassini Ion Neutral Mass Spectrometer (INMS) during the Proximal Mission when the Cassini spacecraft flies through Saturn's thermosphere. In Saturn's well-mixed lower atmosphere, there are a number of measurements of the D/H ratio in molecular hydrogen and methane, i.e. of HD and CH_3D. From the review of these measurements by Fouchet et al. (2009), one concludes that the HD/H_2 ratio is ~3.5×10^{-5} with error bars of ~ ± 50%.

With a D/H ratio (= ½ HD/H_2) in the well-mixed atmosphere, H atoms with an upper limit of 5% at the exobase, and D with the same scale of height as H_2, the atomic D mixing ratio will be in the range of 10^{-8} to 10^{-7} (Parkinson et al. 2006), and hence of limited interest in the thermosphere.

For the purposes of this chapter, the only real importance of CH_4, with a volume-mixing ratio in the lower atmosphere of 0.0047, is to locate the homopause. Chapter 10 by Fletcher et al. discusses CH_4 and its photochemistry in depth. Likewise, H_2O is another minor species in the thermosphere, which is of much greater interest for Saturn's ionosphere, in connection with a phenomenon known as "ring rain" and is discussed in depth in Chapter 8 by Moore et al. H_2O molecules are heavier than H_2 and have a large

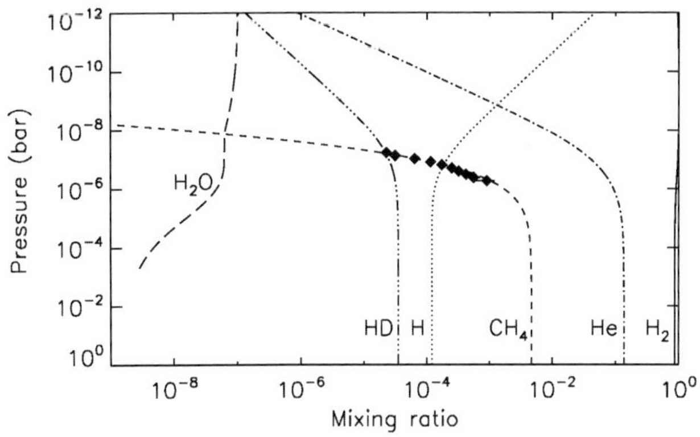

Figure 9.9 Calculated volume-mixing ratio profiles that correspond to the atmospheric structure model (temperature-pressure profile) based on the occultation of ε Eridani near the planetographic latitude of 25°N from spring of 2006 and nearly coinciding CIRS limb observations (see Figure 9.8). Diamonds show the CH_4 mixing ratios retrieved from the occultation to constrain the K_{zz} profile. The assumed mole fractions at the 1 bar level are 0.1355 for He, 4.7×10^{-3} for CH_4 and 3.5×10^{-5} for HD. For water, we assumed an influx of 10^6 cm^{-2} s^{-1} and fixed the mixing ratio to 3×10^{-9} at 0.5 mbar based on recent Herschel observations (Fletcher et al. 2012). The mixing ratio of H was set to match the upper limit of 5% in the thermosphere (Koskinen et al. 2013).

loss rate in the lower stratosphere due to chemical loss, if they survive condensation, as they diffuse downward through the atmosphere. If they diffuse at their maximum velocity, their volume mixing ratio, μ, is approximately the downward flux, φ(H_2O), multiplied by the H_2 O scale height divided by the H_2O-H_2 binary collision coefficient and in cgs units μ(H_2O) = φ(H_2O)/10^{13}, essentially the "inverse Hunten limiting diffusive flux" for heavy gases (Hunten 1973, cf. his Equation 15). Thus, for example, a flux of 1×10^6 H_2O cm^{-2} s^{-1} estimated by Müller-Wodarg et al. (2012) near the planetocentric latitude of 20°N yields a thermospheric mixing ratio of 1×10^{-7}. Figure 9.9 illustrates density profiles and volume mixing ratios representative of the above discussion.

9.6 Inferred Net Heating Rate from Radial Temperature Profile

In the thermosphere, molecular heat conduction is an important process for redistribution of thermal energy. A temperature profile yields from its gradient the heat conduction flux, $F_H = -\kappa \nabla T$, where $\kappa = 252\ T^{0.751}$ in ergs cm^{-1} s^{-1} K^{-1} (Hanley et al. 1970) for a H_2 dominated atmosphere, and from its curvature the heating/cooling rate $\nabla \cdot F_H$. Occultation data yield fundamentally the line-of-sight (los) column density. The local number density is derived by inverting the column density profile and the temperature is inferred from the retrieved density profile. The partial pressures are first obtained by integrating the equation of hydrostatic equilibrium downward, starting from an assumed temperature and thus pressure at the upper boundary of the observed density profile. The ideal gas law then yields the temperature at each altitude point based on the derived pressure and the observed densities. An alternate approach is to create a model atmosphere with the temperature lower boundary condition from CIRS and wavelength dependent light curves that match the observed UVIS light curves. A comparison of both approaches is shown in Figure 9.8 for UVIS stellar occultations obtained in 2006 and 2008, with diamonds for the data-only method and solid lines for the forward model approach.

The radial heat conduction equation, with r for radial distance, is $-\frac{1}{r^2}\frac{\partial}{\partial r}\left(r^2 \kappa \frac{\partial T}{\partial r}\right) = Q(r) - C(r)$, where Q(r) and C(r) are the heating and cooling rates, respectively, and can be due to dynamical as well as radiative processes. If one transforms the heat equation from variable r to 1/u, an analytic solution is obtained in terms of Gaussian-like functions for Q(r) and C(r) (cf. Stevens et al. 1993). Derived heating and cooling rates are most valid if their sources are spatially well separated and the temperature profile being modeled is well constrained by data over the entire profile. Referring to Figure 9.8, CIRS data constrain temperature profiles reliably up to 3 μbar and can be extrapolated to 0.01 μbar. Only in exceptional circumstances such as the December 2008 do stellar occultations yield an adequate temperature profile in this critical region.

In Figure 9.10 an illustrative solution to the above heat conduction equation is given for December 2008 stellar occultation derived temperature profiles shown in Figure 9.8. Solution of the heat conduction equation yields a net integrated heating rate of 0.072 ergs cm^{-2} s^{-1}, with peak heating at 1450 km and 0.65 nbar, while the peak cooling is inferred at 870 km and 70 nbar.

The 2008 occultation was at low latitude, 18°N, where the asymptotic temperatures are ~400 K, whereas at auroral/polar latitudes the temperature

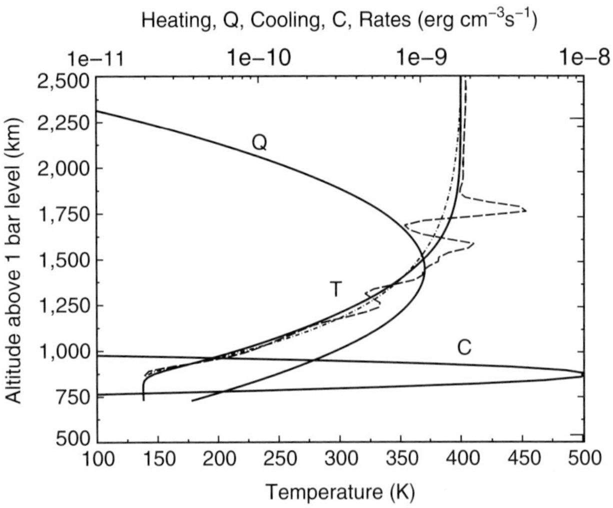

Figure 9.10 Inferred heating, Q, and cooling, C, rates from December 2008 stellar occultation derived temperature profiles illustrated in Figure 9.8 (red diamonds, there; here dashed line) and from forward model (solid red lines, there; here dash dot line). Solving the heat conduction equation with the inferred Q and C profiles yields the solid line temperature. The inferred net integrated heating rate is 0.072 ergs cm^{-2} s^{-1}, with peak heating at 1450 km and 0.65 nbar, while the peak cooling is inferred at 870 km and 70 nbar.

rises to values of 550–600 K. Thus, a solution of the heat equation at high latitudes would yield a larger heating rate. If one performed a series of solutions at discrete latitudes and then globally averaged the rates, one would find a globally average heating rate of ~0.1 erg cm^{-2} s^{-1} or ~ 5 TW total for Saturn's thermosphere, in considerable excess of what solar EUV/FUV power can deliver (~0.15–0.3 TW). Note that this is the global heating rate and the required power input is the heating rate divided by the heating efficiency, which according to Waite et al. (1983) is ~0.5 for solar UV heating and auroral energy sources. Thus the power input required is ~10 TW, for which only Joule/ion-neutral heating can supply this amount, as discussed in Section 9.8.3.

9.7 Inferred Homopause Location from CH_4 Data

Absorption by H_2 in the occultations is negligible at wavelengths higher than about 120 nm, and this allows for minor species such as CH_4, C_2H_6, C_2H_4 and C_2H_2 to be detected. The CH_4 profiles can then be used to constrain the eddy mixing coefficient K_{zz} and the location of the homopause. Unfortunately, mixing ratios are required to properly pinpoint the location of the homopause, and absorption by H_2 is saturated at the level in the atmosphere where methane densities are retrievable (i.e. roughly below the 0.01 μbar level), making it difficult to determine the mixing ratio without interpolating between regions. Combining temperature measurements in the stratosphere with the temperature profiles from the UV occultations is therefore critical for creating atmosphere models that can be used to calculate the mixing ratios of CH_4 and other hydrocarbons (cf. Figure 9.9). The wealth of observations from Cassini/CIRS makes this approach more reliable for the Cassini/UVIS data than for the Voyager/UVS occultations. The results are still subject to uncertainties, however, because CIRS and UVIS do not observe the same location at the same time, and there is a gap in the temperature coverage of the two instruments between 0.01 μbar and 3 μbar (see Section 9.3.2).

With these caveats in mind, Vervack and Moses (2015) draw two conclusions based on the Voyager/UVS occultations that can now be re-evaluated in light of the Cassini data. First, Saturn's upper atmosphere is subject to strong mixing with a relatively high altitude homopause and second, the location of the homopause may be highly variable. Based on their analysis of five Voyager/UVS occultations, Vervack and Moses (2015) found that the CH_4 profiles could only be fitted by K_{zz} profiles that increase with altitude throughout the thermosphere. Thus the homopause pressure, where by definition K_{zz} is equal to the $CH_4 - H_2$ molecular diffusion coefficient, was typically very low. The Voyager 2 solar ingress occultation near the planetocentric latitude of 29°N showed the lowest pressure homopause at 0.7 nbar with $K_{zz} = 2 \times 10^9$ cm^2 s^{-1}. We note that such high values of K_{zz} agree with the inference of strong mixing from the Voyager/UVS He 58.4 nm data (see Section 9.4.2).

Given the behavior of the K_{zz} profiles in their atmosphere models, Vervack and Moses (2015) found it more convenient to derive the pressure and K_{zz} at the level where the mixing ratio of CH_4 is 5×10^{-5} (hereafter, the CH_4 reference level), than locating the homopause, to facilitate comparison with other work and to look for variations. In the three Voyager/UVS occultations probing the southern hemisphere the CH_4 reference level was located at 0.01–0.1 μbar with $K_{zz} = 1–3 \times 10^7$ cm^2 s^{-1}. In the two occultations probing the northern hemisphere, on the other hand, the CH_4 reference level was closer to 0.01 μbar, with a higher K_{zz} of $(1–2) \times 10^8$ cm^2 s^{-1}.

Similar background atmosphere models based on the Cassini/UVIS observations have only been developed for five occultations to date. The model temperature profiles and the resulting CH_4 mixing ratios are shown in Figure 9.8. Koskinen et al. (2015) retrieved these CH_4 profiles from the FUV channel of the Cassini/UVIS instrument and created the atmosphere models. A more comprehensive analysis of all of the Cassini/UVIS occultations is in progress and it will provide highly anticipated global constraints on the variability of the homopause and associated dynamics.

Meanwhile, the results from Cassini/UVIS so far are generally more consistent than from the Voyager data, particularly because they do not confirm the peculiarly low-pressure homopause in the Voyager 2 solar ingress occultation. Four of the occultations in Figure 9.8 probe almost the same location as the Voyager 2 solar occultation near the planetographic latitude of 20°N and they indicate that the homopause pressure is 0.01–0.1 μbar with $K_{zz} = 10^6 – 10^7$ cm^2 s^{-1}. The CH_4 reference level based on these occultations, on the other hand, is located closer to the 0.1 μbar level, with similar values of K_{zz} as at the homopause. These results agree reasonably well with the Voyager/UVS results in the southern hemisphere but not in the northern hemisphere.

We note that the Cassini fits to the CH_4 profiles are in agreement with the Voyager/UVS results in that the K_{zz} profiles that are required to match the data often increase with altitude until relatively low pressures. This differs from the typical behavior of the K_{zz} profiles in many planetary atmosphere applications that are assumed to asymptote to a constant value at some point in the thermosphere. The curious behavior of the K_{zz} profiles could arise from photochemical processes that, contrary to expectations, affect the CH_4 profile, and/or waves or other dynamical processes that are not captured by the form of the K_{zz} profile assumed in the current studies.

9.8 Energetics of the Thermosphere

9.8.1 Inadequacy of Solar EUV/FUV Heating

Strobel and Smith (1973) reviewed the literature on calculations of the temperature of the Jovian thermosphere and performed new calculations for Jupiter, Saturn and Titan. For Saturn, they estimated that solar EUV/FUV heating could raise the asymptotic isothermal thermospheric temperature by only ~10 K above the mesopause temperature. As noted in Section 9.6, the inferred heating rate from thermospheric temperature profiles far exceeds what the Sun can supply at EUV/FUV wavelengths.

9.8.2 Wave Heating

The possibility of wave heating was evaluated in Strobel (2002) for the giant planets' thermospheres, based on previous detailed calculations performed by Matcheva and Strobel (1999) for gravity waves in Jupiter's thermosphere. With appropriate values for the input parameters, dynamic viscosity, μ, gravitational acceleration, g, and gas constants, cp/R, the maximum gravity wave energy flux in isothermal regions for Saturn is just $3.22 \frac{\mu g R}{c_p} = 0.13$ erg cm^{-2} s^{-1} (corrected expression from Strobel 2002), which, when coupled with the estimated heating efficiency, ~0.41, reduces the maximum integrated heating rate to ~0.055 erg cm^{-2} s^{-1}, too low by about a factor of 2, and less if wave heating were not globally distributed and continuously active. The latter conditions are extremely improbable.

Another important class of vertically propagating internal waves is Rossby waves, whose restoring force is the meridional variation of the Coriolis force and whose dynamics are based on conservation of potential vorticity. Generally, the potential vorticity of the atmosphere is dominated by planetary vorticity (f = twice the rotation rate times the sin(latitude)), with a minor contribution from the relative vorticity of the velocity field, $\nabla \times \vec{v}$. As Rossby waves propagate vertically, they must extract potential vorticity from the mean flow, q_0, in order for their wave potential vorticity, q', to grow exponentially in amplitude as $\rho_0{}^{-0.5}$, in the absence of dissipation. But the wave potential vorticity cannot exceed the basic state potential vorticity, i.e. $q' < f$, and this restricts wave amplitudes to two orders of magnitude lower than estimated by amplitude growth (Schoeberl and Lindzen 1982).

The last class of propagating waves is acoustic waves, which are generated by lightning and thunderstorms (Schubert et al. 2003). Their amplitudes and associated energy fluxes are poorly constrained. To reach the thermosphere, their horizontal phase speeds need to be supersonic relative to the local speed of sound; otherwise they will be refracted by the thermosphere's increasing index of refraction. This requires that the storms launching these waves be moving at supersonic speeds in the troposphere (>1.5 km s^{-1}) and three times the speed of the equatorial tropospheric jet.

9.8.3 Joule (Ion-Neutral) Heating Due to Magnetosphere-Ionosphere-Atmosphere Coupling

In the upper atmosphere, the presence of ionospheric plasma provides a medium which responds to the presence of magnetic and electric fields, and thereby to processes that occur in the magnetosphere. When a magnetospheric electric field maps along the magnetic field lines into the atmosphere, the ionospheric ions are accelerated and collide with the ambient neutral gas particles. This collisional interaction leads to an acceleration of the neutral gases in the direction of the zonal ion drift, generating a region of large zonal wind velocities where magnetospheric electric fields are strongest, near the auroral emission regions. The acceleration is given by $\mathbf{a} = -\mathbf{v}_{ni}(\mathbf{u}-\mathbf{v}) = (\mathbf{j} \times \mathbf{B})/\rho$, $\mathbf{v}_{ni}$ being the neutral-ion collision frequency, $\mathbf{u}$ the neutral wind vector, $\mathbf{v}$ the ion velocity vector, $\mathbf{B}$ the planetary magnetic field and ρ the mass density of the neutral atmosphere. The electrical current density perpendicular to the magnetic field in the ionosphere is $\mathbf{j} = \boldsymbol{\sigma}(\mathbf{E} + \mathbf{u} \times \mathbf{B})$, with $\boldsymbol{\sigma}$ being a tensor with components for the Pedersen conductivity and the Hall conductivity, $\mathbf{E}$ being the sum of the magnetospheric electric field mapped into the upper atmosphere and any polarization field set up by divergence of $\mathbf{j}$. Because the ionosphere is not perfectly conducting, resistive heating occurs, a process often referred to as Joule heating. The thermal heating of the atmosphere by electrical currents per unit mass can be written as $q_{Joule} = (\mathbf{j}\cdot\mathbf{E})/\rho$, where the electrical current density $\mathbf{j}$ and the electric field $\mathbf{E}$ both include the effect of neutral winds via the dynamo field term, $\mathbf{u} \times \mathbf{B}$ (Vasyliunas and Song 2005, Equation 43). We note that electrical currents also result in momentum change due to ion drag that affects the kinetic energy of the gas. Sometimes this latter effect is referred to as "ion drag heating." While the thermal heating by currents alone (without considering neutral winds) can only be a positive quantity, the ion drag heating, q_{Ion}, can also attain negative values, implying the loss of kinetic energy of the neutral atmosphere. As a result, the calculation of ion drag heating requires knowledge of the thermospheric winds.

The Saturn Thermosphere Ionosphere Model (STIM) is a General Circulation Model (GCM) which numerically solves non-linear coupled Navier–Stokes equations of energy, momentum and continuity for both neutral gas particle and ions in Saturn's thermosphere and ionosphere (Müller-Wodarg et al. 2006, 2012). The model currently relies on provision of magnetospheric electric fields and electron energy fluxes as external boundary conditions, but then calculates the magnetosphere-ionosphere-thermosphere interaction self-consistently. The model includes solar and electron impact ionization, using for the latter the parameterization of Galand et al. (2011). Once created, the ions undergo chemical reactions as described by Moore et al. (2004).

Figure 9.11 shows the magnetospheric electric field strength (color contours) as mapped into the southern polar region. Also shown are the locations of maximum field-aligned current (ring-shaped region denoted by the black line), which coincide with the regions of largest electron precipitation into Saturn's polar upper atmosphere. The values of Figure 9.11 are taken from the BATSrUS MHD model of Saturn's magnetosphere

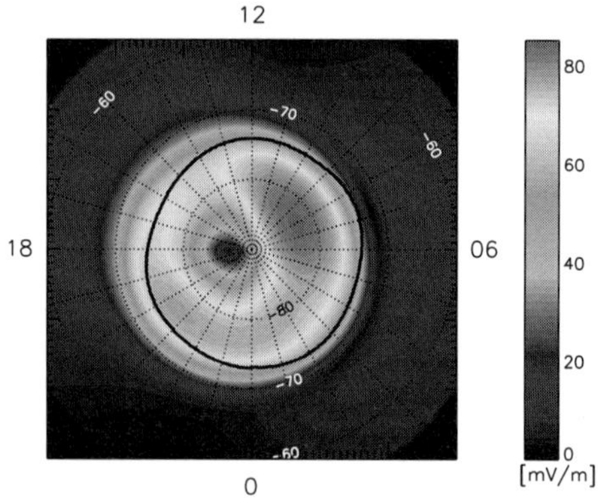

Figure 9.11 The magnetospheric electric field strength (color contours) as mapped into the southern polar region with the locations of maximum field-aligned current (black line), which coincide with the regions of largest electron precipitation into Saturn's polar upper atmosphere. The figure axes indicate local times. (A black-and-white version of this figure appears in some formats. For the color version, please refer to the plate section.)

(Jia et al. 2012) for quiet solar wind conditions. We have multiplied the original field strength by a factor of 4 in order to better reproduce observed polar temperatures and winds. The electric field is directed primarily equatorward and thereby generates a westward acceleration of the ionospheric ions.

Electron precipitation occurs along the ring-shaped region in Figure 9.11 and is local-time-dependent, not only in terms of its latitude (as seen in the figure), but also in terms of the magnitude of the electron energy flux. We apply in STIM-GCM the local time shape of electron flux consistent with that inferred from auroral observations by Lamy et al. (2009), with a maximum flux in the dawn sector near 08:00 (Müller-Wodarg et al. 2012). Near midnight, the electron flux is close to zero. In the simulation shown here we apply 10 keV electrons alone, but the model allows for implementation of other electron populations as well. We assume a longitudinally averaged auroral energy flux of 1.0 mW m^{-2}, a value based on the findings of Lamy et al. (2009). The electron impact ionization causes enhanced Pedersen and Hall conductivities in the atmosphere, which closely follow the local time changes of electron precipitation.

Figure 9.12 shows zonally averaged Pedersen conductances in Saturn's ionosphere as a function of latitude, assuming that magnetic field lines are aligned radially in the thermosphere. Since the auroral magnetospheric interaction is confined to polar latitudes, this assumption is acceptable in Saturn's almost perfect

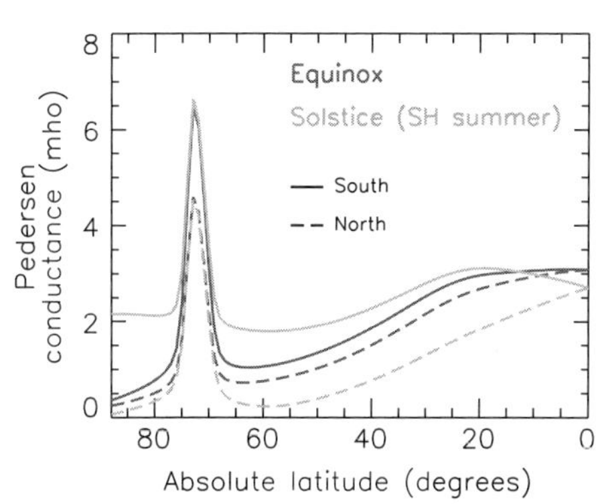

Figure 9.12 Zonally averaged Pedersen conductances in Saturn's ionosphere as a function of latitude. Solid lines denote the southern hemisphere values; dashed values are for the northern hemisphere. The blue lines are for an equinox simulation of STIM and the red lines for southern hemisphere summer conditions. (A black-and-white version of this figure appears in some formats. For the color version, please refer to the plate section.)

dipole field (see Chapter 4 by Christensen et al.). At low latitudes, the conductance results from solar radiation ionization, reaching around 2–3 mho, depending on the season (larger at equinox). These rapidly decrease towards the poles with increasing solar zenith angle. From 70° to 75° latitude, however, we see a strong enhancement to values of around 5–7 mho, which result from the 10 keV electron impact ionization. The figure shows no seasonal variation of conductances at auroral latitudes, but a hemispheric asymmetry. Southern auroral conductances may attain 7 mho, while those in the north reach around 5 mho only. This difference is a direct consequence of the magnetic field asymmetry between north and south. We assume the Saturn Pioneer Voyager (SPV) magnetic field model in our simulations (Davis and Smith, 1990).

Figure 9.13 shows neutral temperature contours and the meridional circulation wind vectors, as simulated by STIM for equinoctial conditions. The longest arrow corresponds to around 350 m s^{-1}. Also shown in the figure are two line plot panels with the normalized quantities: Joule heating rates (solid line), ion drag acceleration (dashed) and zonal wind velocities (dashed-dotted). The red dot in the temperature panel denotes the location of maximum Joule heating; the green dot denotes the region of maximum zonal ion drag. Both occur at 72°S latitude, but around 100 km apart vertically. The curves on the right panel are vertical profiles at this latitude while the curves on the top panel are latitudinal profiles at the height of peak Joule heating (solid line) and at the height of peak ion drag (dashed and dashed-dotted lines). Peaks of Joule heating and ion drag in our simulation are slightly below the region of peak H_3^+ emission (1155 ± 25 km) observed by Stallard et al. (2012). Zonal winds in the upper panel are normalized to a value of 334 m s^{-1}, and their largest values in the right panel reach 1500 m s^{-1}, the local sound speed. The peak values of zonal ion drag and Joule heating in the two panels are 0.02 m s^{-2} and 1.9 × 10^{-8} W m^{-3}, respectively.

As expected, the 72°S locations of maximum Joule heating and ion drag coincide with the region of peak electron precipitation and largest resulting conductance (see Figure 9.12). Interestingly, the largest zonal winds occur more poleward at 78°S (see top panel), showing the influence of pressure gradients and Coriolis acceleration as additional factors affecting the winds.

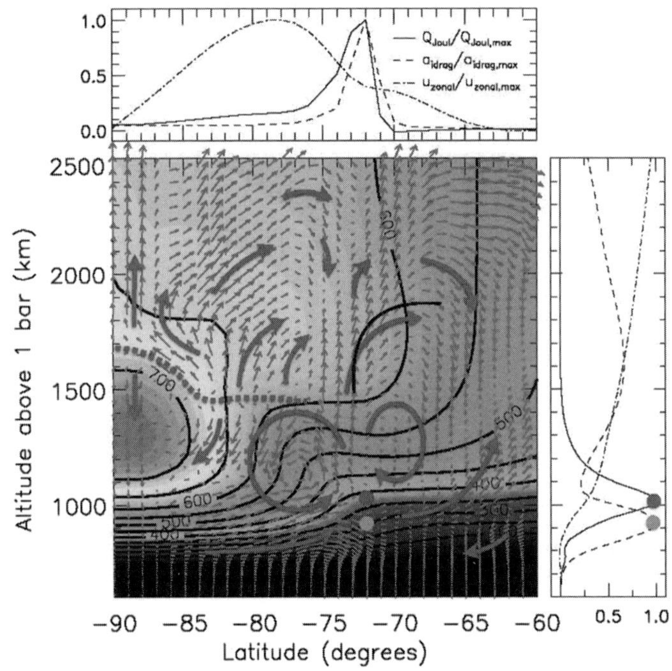

Figure 9.13 Neutral temperatures contours and meridional/vertical circulation wind vectors (grey arrows), as simulated by STIM for equinoctial conditions. The longest grey arrow corresponds to around 350 m s^{-1}. Red arrows illustrate the broad circulation pattern and the red dotted line separates regions of upward and downward vertical winds. Two line plot panels with normalized quantities show Joule heating rates (solid line, $Q_{Joule}/Q_{Joule,max}$), ion drag acceleration (dashed, $a_{idrag}/a_{idrag,max}$) and zonal wind velocities (dashed-dotted, $u_{zonal}/u_{zonal,max}$). The right panel shows vertical profiles at 72°S latitude, while the curves in the top panel are latitudinal profiles at the height of peak Joule heating (solid line) and at the height of peak ion drag (dashed and dashed-dotted lines). The red dot in the temperature panel denotes the location of maximum Joule heating shown in the line plot, also labeled there with a red dot. The green dots on the temperature panel and line plot denote the region of maximum zonal ion drag. Peak Joule heating and ion drag both occur at 72°S latitude but shifted vertically by ~100 km from one another. Zonal winds in the upper panel are normalized to a value of 334 m s^{-1}; their largest values in the right panel reach 1500 m s^{-1}. The peak values of zonal ion drag and Joule heating in the two panels are 0.02 m s^{-2} and 1.9×10^{-8} W m^{-3}, respectively. (A black-and-white version of this figure appears in some formats. For the color version, please refer to the plate section.)

A westward zonal wind will experience poleward Coriolis acceleration. While ion drag is largest in the deeper ionosphere near 900 km (green dot in main panel of Figure 9.13 and dashed line in right panel), zonal winds are essentially in a gradient wind balance aloft driven by poleward directed pressure gradients on equipotential surfaces generated by auroral heating and resulting in increasing temperatures with latitude as illustrated in Figure 9.13. Ground-based Doppler analyses of H_3^+ emissions have revealed zonal ion velocities at polar latitudes on Saturn reaching supersonic speeds of several km s^{-1} (Stallard et al. 2007), described in Chapter 7. As shown by Müller-Wodarg et al. (2012), plasma velocities for the conductances encountered in our simulations (Figure 9.12) can exceed neutral velocities by around a factor of 2, so our simulations are broadly consistent with these observations.

The temperatures, like zonal winds, are not largest in the region of strongest coupling with the magnetosphere (red and green dots) but instead peak in the polar cap region near 1300 km altitude, decrease again towards higher altitudes and reach their asymptotic values near 2000 km. Investigation of the energy equation terms in STIM (Müller-Wodarg et al. 2012) reveals this behavior to be a direct consequence of the transport terms, advection and adiabatic heating and cooling. As illustrated in Figure 9.13 (red arrows), the polar thermosphere hosts a complex circulation pattern in the meridional/altitude plane, with anti-clockwise and clockwise circulation cells, respectively, on the poleward and equatorward side of 72°S below 1500 km. The poleward flow below 1500 km transports energy away from the region of peak Joule heating, and downwelling over the polar region causes adiabatic compression and heating near 1100–1600 km. The red dotted line in Figure 9.13 indicates the boundary between upward and downward winds. At higher altitudes, the circulation is broadly from pole to equator, but broken into several clockwise circulation cells due to Coriolis forces. This cools the polar region above 1600 km adiabatically (causing the negative temperature gradient there, as seen in Figure 9.13) and transporting some of the energy equatorward (thereby causing a slight temperature increase with altitude equatorward of 80°S).

These simulations illustrate that Saturn's high latitude thermosphere and ionosphere are driven primarily by coupling to its magnetosphere. The strong westward winds, which reduce the degree of corotation of the thermosphere to only 25% near 78°S, are a signature of angular momentum transfer from the upper atmosphere into the magnetosphere. The atmospheric response to this localized coupling to the magnetosphere spreads over the entire high latitude region poleward of 60° in both hemispheres. The zonal winds rapidly decrease towards more equatorial latitudes as a result of angular momentum conservation. Joule heating in our simulations provides about 6 TW of thermal energy into Saturn's upper atmosphere (summed over both hemispheres), around 20 to 40 times the total energy deposited by solar heating. This further emphasizes the importance of magnetosphere-atmosphere coupling on Saturn and giant planets.

9.8.4 Resistive Heating and Ion Drag by Wind-Driven Electrodynamics

Outside the auroral regions, the neutral atmosphere can be affected by resistive heating and ion drag driven by electric fields that arise from the interaction of the ionospheric plasma with neutral winds, turbulence or waves. This interaction is known to be important in the E and F region of the Earth's ionosphere (e.g. Richmond et al. 1992; Richmond 1995; Richmond and Thayer 2000). It should be important also in Saturn's ionosphere, given the importance of electrodynamics in the high latitude thermosphere, and could play a role in explaining the remarkable variability in the electron density profiles retrieved from Cassini/RSS observations (Nagy et al. 2006; Kliore et al. 2009; see Chapter 8 by Moore et al.). It may also interfere with the circulation driven by auroral heating and ion drag, modify the ionospheric electric fields that are mapped down from the magnetosphere, and heat the non-auroral thermosphere.

The generation of ionospheric currents in general relies on plasma-neutral collisions that force the electrons and ions to move at different velocities across magnetic field lines, thus violating the frozen-in flux condition of ideal magnetohydrodynamics. Wind-driven electrodynamics or the ionospheric dynamo, on the other hand, is based on the generation of polarization electric fields by neutral winds that lie perpendicular to the magnetic field lines and, due to high field-aligned conductivity, remain approximately constant along magnetic field lines that traverse different layers of the atmosphere. On the Earth, for example, the dayside dynamo layer is in the E region, and the electric fields are mapped between the E and F regions. The result is an electric circuit that allows current to flow in the F region, leading to ion drag and resistive heating. At night when the E region electron density diminishes, however, the polarization electric fields that are set up in the F region significantly reduce the current density. One suggested system of thermospherically driven currents at Saturn is a polar twin-cell vortex that has been evoked to drive oscillations in the planetary period, as described in Chapter 5 by Carbary et al.

To further understand the ionospheric dynamo, it is convenient to divide Saturn's ionosphere into different "magnetization" regions M1, M2 and M3 (e.g. Koskinen et al. 2014). In the M1 region (> 10 μbar) both the electrons and ions are collisionally coupled to the neutrals and currents are generally negligible. In the M2 region (0.01–10 μbar), which is similar to the Earth's E region dynamo layer, the electron gyrofrequency is higher than the electron-neutral collision frequency, while the ions remain collisionally coupled to the neutrals. In the M3 region (< 0.01 μbar) both the ion and electron gyrofrequencies are higher than the ion/electron-neutral collision frequencies. By definition, the Hall conductivity dominates in the M2 region while the Pedersen conductivity dominates in the M3 region.

Recently, Smith (2013) suggested a new source of thermospheric heating on Jupiter based on electrodynamic coupling of the thermosphere and stratosphere that relies on wind-driven electric fields. His study shows that in principle the ionospheric dynamo can generate perpendicular electric fields on the order of 10 mV m^{-1} that result in resistive heating rates that are sufficiently large to explain the high temperatures in the Jovian thermosphere. For Saturn's equatorial thermosphere, the required peak heating rate to explain the low latitude temperature profiles retrieved from the Cassini/UVIS occultations is about 10^{-10} W m^{-3}. Given that the peak Pedersen conductivity is about 10^{-5} S m^{-1} (Moore et al. 2010), an electric field of about 3.2 mV m^{-1} in the neutral reference frame would produce the required heating rate. The maximum dynamo electric field strength can be estimated as ≈UB, where U is the wind speed, which implies winds of ~150 m s^{-1} to generate the required electric field. While not impossible, these winds are needed in the M2 region (0.01–10 μbar), where we have no data.

Whether this mechanism actually turns out to be feasible on Saturn, however, depends on a number of assumptions. For example, Smith (2013) assumed zero winds in the thermosphere. This is problematic, because the electric field in the neutral reference frame is given by $\mathbf{E_n} = \mathbf{E} + \mathbf{u} \times \mathbf{B}$, where **u** is the neutral wind and **E** is the dynamo electric field. The assumption of zero winds thus provides only a crude estimate of the current density that is likely to be an upper limit, because the strength of the dynamo field also depends sensitively on the winds in both the M2 and M3 regions (Koskinen et al. 2014). For example, if the current initiated in the M3 region cannot close in the M2 region, a polarization electric field **E** is set up that cancels out the current in the M3 region. In addition, an electric field of 3.2 mV m^{-1} at low latitudes would lead to substantial ion drag on the neutral atmosphere that affects the heating rates. Resistive heating and ion drag must therefore be modeled self-consistently by a circulation model that includes ionospheric electrodynamics.

Lastly, electrodynamic coupling and the ionospheric dynamo rely on substantial conductivities in both the M2 and M3 regions. The M2 region on Saturn lies almost entirely in the hydrocarbon ion layer where recombination rates are much faster than in the M3 region. Nevertheless, recent calculations by Kim et al. (2014) indicate that electron densities ~10^3 cm^{-3} are possible in the M2 region, but the complex ion chemistry makes the identity of major heavy ions, effective dissociation recombination rates, and calculated electron densities uncertain. While radio occultations do not rule out significant electron densities in the M2 region, this region is at the limit for retrieving reliable electron densities.

9.9 Global General Circulation Models of the Thermosphere

Apart from Doppler measurements of H_3^+ emissions at auroral latitudes, no observational evidence exists of the winds in Saturn's (or any giant planet's) mesosphere and thermosphere. We thereby rely on the use of numerical models to examine the general circulation of Saturn's mesosphere and thermosphere, for which only the STIM model has published results (Müller-Wodarg et al. 2006, 2012). For Jupiter, however, several such models have been published (Achilleos et al. 1998; Bougher et al. 2005; Tao et al. 2014). These build on a heritage of thermosphere-ionosphere models for Earth which have been adapted to simulate the giant planet environment. GCMs numerically integrate the non-linear coupled Navier–Stokes equations of momentum, energy and continuity on a global spherical grid. For giant planets, the most common vertical coordinate used is pressure, based on the hydrostatic assumption for a high-gravity environment.

The models require the inclusion of magnetosphere drivers, namely particle (mostly electron) precipitation and the magnetospheric convection electric fields. These appear as sources of ionization (alongside solar EUV), thermal energy and momentum in the codes. For Earth, detailed knowledge is available of the high-latitude electric convection fields and exact locations of electron precipitation. For giant planets, locations of electron precipitation are thought to coincide with locations of peak auroral UV emissions and can be inferred geographically, along with the electron energy fluxes, from observations and modeling (cf. Chapter 7 by Stallard et al.). The electric fields are more challenging to obtain, as no in situ measurements are available. Assuming the magnetosphere-ionosphere coupling processes via Birkeland currents, as originally proposed for Jupiter by Hill (1979), observations of the degree of corotation in the magnetosphere plasma may in principle yield estimates of electric fields. For Saturn, the model of Müller-Wodarg et al. (2012) initially assumed high latitude electric fields based on magnetospheric plasma flow patterns calculated by Cowley et al. (2004) but most recently replaced these with electric fields calculated by the BATSrUS Saturn magnetosphere MHD model (Jia et al. 2012).

The other important aspect relates to the electron precipitation. Energetic electrons from the magnetosphere are known to induce auroral emissions on Saturn in the EUV and FUV as well as the IR (Kurth et al. 2009). Electron precipitation occurs primarily along a narrow auroral oval region located between 70° and 85° latitude of width 1.5°–3.5° (Badman et al. 2006). The electrons have a mean energy of around 10 keV, but observations have identified energies ranging from 400 eV to 30 keV (Sandel et al. 1982; Gérard et al. 2004, 2009; Gustin et al. 2009). Including the effects of this electron precipitation on the ionosphere requires calculation of the collisional interaction between the electrons and atmosphere, including secondary ionization. This is best done via numerical solution of the Boltzmann equations for suprathermal electrons, which for practical reasons is done in 1-D rather than 3-D. Several such models have been proposed for Jupiter and Saturn (Grodent et al. 2001; Gustin et al. 2009; Galand et al. 2011) and STIM relies on the parameterization for Saturn which was developed by Galand et al. (2011) on the basis of full 1-D calculations. This parameterization provides vertical profiles of ionization rates for different electron populations, scaled by the background neutral densities and initial electron energy flux. Calculations have shown electron precipitation controls ionospheric plasma densities, with solar ionization in auroral regions playing only a secondary role due to the large zenith angles (Galand et al. 2011). Therefore, any realistic calculations of the high-latitude regions with GCMs require explicit inclusion of electron precipitation.

9.10 Concluding Remarks

One of the outstanding problems in the study of Saturn's atmosphere is the He/H_2 ratio as the mean molecular mass of the atmosphere is needed to correctly calculate the pressure levels as a function of radial distance. As noted the Cassini UVIS team has not reported to date any measurements of the He 58.4 nm line which could constrain this critical ratio. But this line is formed well above the He homopause shown in Figure 9.9. Thus one has to extrapolate the inferred ratio into the well mixed atmosphere with considerable uncertainty.

At the end of the Cassini Mission, known as the Grand Finale Tour, the last five orbits will be flybys through Saturn's thermosphere penetrating down to the sub-nanobar level, but above the homopause, and before a final fatal plunge into the atmosphere. The motivation is to take advantage of the onboard INMS that is capable of measuring the neutrals: H_2, He for sure, and possibly HD at these low pressures. With a spacecraft velocity ~30 km s^{-1}, the kinetic energy of H_2 molecules colliding with the spacecraft is ~8.8 eV, in excess of the 4.5 eV H_2 dissociation energy. If one of the H_2 proton nuclei is imparted more than 4.5 eV of kinetic energy in a collision with the instrument, the H_2 bond would be broken and H_2 could be undercounted relative to atomic He. The measurement of HD will be marginal if its actual mixing ratio is close to what is displayed in Figure 9.9.

But if one steps down sequentially in altitude during the last four orbits after spacecraft safety is confirmed from the first orbit, one can improve the chances of measuring HD and get better density profiles of H_2 and He. While performing the Grand Finale Tour, there will be many opportunities for UV stellar occultations to add more H_2 density profiles at a variety of latitudes to complement the more than 40 occultations discussed in Section 9.2.2. The challenge will be to translate He/H_2 and HD/H_2 density ratio profiles above the homopause into extrapolation of asymptotic values deep in the well-mixed atmosphere. For Jupiter, the latter was only achieved by dropping the Galileo Probe into the atmosphere and making measurements down to ~20 bar. In comparison to Jupiter our knowledge of the structure of Saturn's thermosphere will be far superior due to the large number solar and stellar occultations.

Far more certain is that INMS will measure the composition of Saturn's ionosphere and, hopefully, obtain clear evidence of water group ions and infer their effect on electron densities. The ion composition will provide a consistency check on the neutral composition measurements.

One of the fundamental problems in understanding the thermospheres of Saturn and Jupiter is the heating mechanism(s) that accounts for their temperatures far exceeding what solar UV radiation can generate. Auroral heating is sufficient to solve this "energy crisis" for Saturn's thermospheric heating, if it can be efficiently redistributed to low latitudes. This solution has been rejected on the basis of thermospheric GCM calculations, such as the STIM model. The net effect is considerably colder thermospheric temperatures than derived from occultations outside the polar regions. The fundamental cause of this under prediction by GCMs is that ion drag induces zonal winds in the retrograde, westward direction in the auroral regions, which when acted on by the Coriolis force generate poleward meridional winds that transport and confine auroral heating to polar latitudes rather than transport heat to the equator where it is most needed.

Almost all thermospheric GCMs addressing this problem include the assumption of hydrostatic equilibrium, which is only applicable to low Mach number flows of less than 0.3 (Kundu 1990). But the STIM model produces neutral winds up to the speed of sound in some regions and the hydrostatic assumption is no longer valid. The Coriolis force has a term in the radial direction of 2Ω u cos(lat), which cannot be ignored for flows in excess of Mach 0.5. Also, the hydrostatic approximation filters out acoustic-gravity waves which can transmit energy out of the polar regions to lower latitudes. Auroral heating is a spatial and time dependent forcing capable of generating such waves, which have horizontal group velocities close to the sound speed. Thus, auroral power pulses can be propagated to the equator in less than two Saturn days. Likewise, supersonic ions $\mathbf{E} \times \mathbf{B}$ convecting through the auroral thermosphere must generate shocks in the neutral atmosphere, which cannot be handled by hydrostatic GCMs. Thus the energy "crisis" may not be inadequate total power input but a problem in the global redistribution of power in models. Contributing to this is a further shortcoming of all GCM simulations published to-date, the neglect of dynamical coupling to regions below in the form of mean background winds and upward propagating waves, an aspect that is known to considerably affect the circulation in the Earth's lower thermosphere. Work to address this shortcoming is underway with STIM and may lead to more realistic lower thermosphere dynamics which favor polar energy redistribution towards the equator.

Finally, the thermal structure of the pressure region (~0.030–1 μbar) is not well characterized and the question of whether a mesopause is present is unanswered. A well-instrumented probe or probes could address this question as well as shed light on the He/H_2 ratio and atmospheric structure from sub-nanobar to 10+ bars.

Acknowledgments

DFS was supported by the Cassini–Huygens Mission through JPL Contract Nos. 1408487 and NASA Grant NNX10AB84 G. TTK was supported by the NASA CDAPS Grant NNX14AD51 G.

References

Achilleos, N., S. Miller, J. Tennyson et al. (1998), JIM: A time-dependent, three-dimensional model of Jupiter's thermosphere and ionosphere. *J. Geophys. Res.*, **103**, doi:10.1029/98JE00947.

Adriani, A., B. M. Dinelli, M. López-Puertas et al. (2011), Distribution of HCN in Titan's upper atmosphere from Cassini/VIMS observations at 3 μm. *Icarus*, **214**, 584–595, doi:10.1016/j.icarus.2011.04.016.

Anderson, J. D. and G. Schubert (2007), Saturn's gravitational field, internal rotation, and interior structure. *Science*, **317**, 1384–1387, doi:10.1126/science.1144835.

Badman, S. V., S. W. H. Cowley, J.-C. Gérard et al. (2006), A statistical analysis of the location and width of Saturn's southern auroras, *Ann. Geophys.*, **24**, 3533–3545.

Baines, K. H. et al. (2005), The atmospheres of Saturn and Titan in the near-infrared: First results of Cassini/VIMS. *Earth, Moon, and Planets*, **96**, 119–147, doi:10.1007/s11038-005–9058-2.

Ben-Jaffel, L. Ben, Y. J. Kim and J. Clarke (2007), The H Lyman-α emission line from the upper atmosphere of Jupiter: Parametric radiative transfer study and comparison with data. *Icarus*, **190**, 504–527, doi:10.1016/j.icarus.2007.03.013.

Ben-Jaffel, L. Ben, R. Prangé, B. R. Sandel et al. (1995), New Analysis of the Voyager UVS H Lyman-α Emission of Saturn. *Icarus*, **113**, 91–102, doi:10.1006/icar.1995.1007.

Bougher, S. W., J. H. Waite, T. Majeed et al. (2005), Jupiter Thermospheric General Circulation Model (JTGCM): Global structure and dynamics driven by auroral and Joule heating. *J. Geophys. Res.*, **110**, doi:10.1029/2003JE002230.

Broadfoot, A. L. et al. (1981), Extreme Ultraviolet Observations from Voyager 1 Encounter with Saturn, *Science*, **212**, 206–211, doi:10.1126/science.212.4491.206.

Broadfoot, A. L. (1986), Ultraviolet Spectrometer Observations of Uranus, *Science*, **233**, 74–79, doi:10.1126/science.233.4759.74.

Cassidy, T. A. and R. E. Johnson (2010), Collisional spreading of Enceladus' neutral cloud. *Icarus*, **209**, 696–703, doi:10.1016/j.icarus.2010.04.010.

Clarke, J. T., M. Hudson and Y. L. Yung (1987), The excitation of the far ultraviolet electroglow emissions on Uranus, Saturn, and Jupiter. *J. Geophys. Res.*, **92**, 15,139–115, 147, doi:10.1029/JA092iA13p15139.

Connerney, J. E. P. and J. H. Waite (1984), New model of Saturn's ionosphere with an influx of water from the rings. *Nature*, **312**, 136–138, doi:10.1038/312136a0.

Conrath, B. J. and D. Gautier (2000), Saturn Helium Abundance: A Reanalysis of Voyager Measurements. *Icarus*, **144**, 124–134, doi:10.1006/icar.1999.6265.

Cowley, S. W. H., E. J. Bunce and J. M. O'Rourke (2004), A simple quantitative model of plasma flows and currents in Saturn's polar ionosphere. *J. Geophys. Res. Space Phys.*, **109**, doi:10.1029/2003JA010375.

Curdt, W. et al. (2001), The SUMER spectral atlas of solar-disk features. *Astron. Astrophys.*, **375**, 591–613.

Davis Jr., L. and E. J. Smith (1990), A model of Saturn's magnetic field based on all available data. *J. Geophys. Res.*, **95**, 15257–15261, doi:10.1029/JA095iA09p15257.

Esposito, L. W. et al. (2004), The Cassini ultraviolet imaging spectrograph investigation. *Space Sci. Rev.*, **115**, 299–361.

Feldman, P. D., M. A. McGrath, H. W. Moos et al. (1993), The spectrum of the Jovian dayglow observed at 3 A resolution with the Hopkins Ultraviolet Telescope, *Astrophys. J.*, **406**, 279–284.

Festou, M. C. and S. K. Atreya (1982), Voyager ultraviolet stellar occultation measurements of the composition and thermal profiles of the Saturnian upper atmosphere. *Geophys. Res. Lett.*, **9**, 1147–1150, doi:10.1029/GL009i010p01147.

Feuchtgruber, H., E. Lellouch, T. de Graauw et al. (1997), External supply of oxygen to the atmospheres of the giant planets. *Nature*, **389**, 159–162, doi:10.1038/38236.

Fletcher, L. N., B. Swinyard, C. Salji et al. (2012), Submillimetre spectroscopy of Saturn's trace gases from Herschel/SPIRE. *Astron. Astrophys.*, **539**, A44, doi:10.1051/0004–6361/201118415.

Fouchet, T., J. I. Moses and B. J. Conrath (2009), Saturn: Composition and Chemistry. In *Saturn From Cassini-Huygens, eds.* M. K. Dougherty, L. W. Esposito and S. M. Krimigis, Springer, pp. 83–112, doi:10.1007/978-1-4020-9217-6.

Friedson, A. J. and J. I. Moses (2012), General circulation and transport in Saturn's upper troposphere and stratosphere. *Icarus*, **218**, 861–875, doi:10.1016/j.icarus.2012.02.004.

Galand, M., L. Moore, I. C. F. Müller-Wodarg et al. (2011), Response of Saturn's auroral ionosphere to electron precipitation: Electron density, electron temperature, and electrical conductivity. *J. Geophys. Res.*, **116**, A09306, doi:10.1029/2010JA016412.

García-Comas, M., M. López-Puertas, B. Funke et al. (2011), Analysis of Titan CH_4 3.3 μm upper atmospheric emission as measured by Cassini/VIMS. *Icarus*, **214**, 571–583, doi:10.1016/j.icarus.2011.03.020.

Gérard, J.-C., B. Bonfond, J. Gustin et al. (2009), Altitude of Saturn's aurora and its implications for the characteristic energy of precipitated electrons. *Geophys. Res. Lett.*, **36**, doi:10.1029/2008GL036554.

Gérard, J.-C., V. Dols, D. Grodent et al. (1995), Simultaneous observations of the Saturnian aurora and polar haze with the HST/FOC. *Geophys. Res. Lett.*, **22**, 2685–2688.

Gérard, J.-C., D. Grodent, J. A. Gustin et al. (2004), Characteristics of Saturn's FUV aurora observed with the Space Telescope Imaging Spectrograph. *J. Geophys. Res.*, **109**, A09207, doi:10.1029/2004JA010513.

Gérard, J.-C., J. Gustin, W. R. Pryor et al. (2013), Remote sensing of the energy of auroral electrons in Saturn's atmosphere: Hubble and Cassini spectral observations. *Icarus*, **223**, 211–221, doi:10.1016/j.icarus.2012.11.033.

Grodent, D., J. H. Waite Jr. and J.-C. Gérard (2001), A self-consistent model of Jovian auroral thermal structure, *J. Geophys. Res.*, **106**, 12,933–912, 952, doi:10.1029/2000JA900129.

Guerlet, S., T. Fouchet, B. Bézard et al. (2009), Vertical and meridional distribution of ethane, acetylene and propane in Saturn's stratosphere from CIRS/Cassini limb observations. *Icarus*, **203**, 214–232, doi:10.1016/j.icarus.2009.04.002.

Guerlet, S., T. Fouchet, B. Bézard (2011), Evolution of the equatorial oscillation in Saturn's stratosphere between 2005 and 2010 from Cassini/CIRS limb data analysis. *J. Geophys. Res. Lett.*, **38**, L09201, doi:10.1029/2011GL047192.

Gustin, J., J.-C. Gérard et al. (2013), Remote sensing of the energy of auroral electrons in Saturn's atmosphere: Hubble and Cassini spectral observations. *Icarus*, **223**, 211–221, doi:10.1016/j.icarus.2010.06.031.

Gustin, J., J.-C. Gérard, W. Pryor et al. (2009), Characteristics of Saturn's polar atmosphere and auroral electrons derived from HST/STIS, FUSE and Cassini/UVIS spectra, *Icarus*, **200**, 176–187, doi:10.1016/j.icarus.2008.11.013.

Gustin, J., I. Stewart, J.-C. Gérard et al. (2010), Characteristics of Saturn's FUV airglow from limb-viewing spectra obtained by Cassini-UVIS. *Icarus*, **210**, 270–283, doi:10.1016/j.icarus.2010.06.031.

Hallett, J. T., D. Shemansky and X. Liu (2005), A rotational-level hydrogen physical chemistry model for general astrophysical application. *Astrophys. J.*, **624**, 448–461, doi:10.1086/428935.

Hanley, H. J., R. D. McCarthy and H. Interman (1970), The viscosity and thermal conductivity of dilute hydrogen from 150 to 5000 K. *J. Res. Natl. Bur. Stds.* **A74**, 331–353.

Hill, T. W. (1979), Inertial limit on corotation. *J. Geophys. Res.*, **84**, doi:10.1029/JA084iA11p06554.

Hubbard, W. B., C. C. Porco, D. M. Hunten et al. (1997), Structure of Saturn's mesosphere from the 28 Sgr occultations. *Icarus*, **130**, 404–425, doi:10.1006/icar.1997.5839.

Hunten, D. M. (1973), The escape of light gases from planetary atmospheres. *J. Atmos. Sci.* **30**, 1481–1494.

Jacobson, R. A., P. G. Antreasian, J. J. Bordi et al. (2006), The gravity field of the Saturnian system from satellite observations and spacecraft tracking data. *Astron. J.*, **132**, 2520–2526, doi:10.1086/508812.

Jia, X., K. C. Hansen, T. I. Gombosi et al. (2012), Magnetospheric configuration and dynamics of Saturn's magnetosphere: A global MHD simulation. *J. Geophys. Res.*, **117**, doi:10.1029/2012JA017575.

Kim, S. J., C. K. Sim, D. W. Lee et al. (2012), The three-micron spectral feature of the Saturnian haze: Implications for the haze composition and formation process. *Plan. Space Sci.*, **65**, 122–129, doi:10.1016/j.pss.2012.02.013.

Kim, Y. H., J. L. Fox, J. H. Black et al. (2014), Hydrocarbon ions in the lower ionosphere of Saturn. *J. Geophys. Res. Space Phys.*, **119**, 384–395, doi:10.1002/2013JA019022.

Kliore, A. J., A. F. Nagy, E. A. Marouf et al. (2009), Midlatitude and high-latitude electron density profiles in the ionosphere of Saturn obtained by Cassini radio occultation observations. *J. Geophys. Res.*, **114**, A04315, doi:10.1029/2008JA013900.

Koskinen, T. T., B. R. Sandel, R. V. Yelle et al. (2013), The density and temperature structure near the exobase of Saturn from Cassini UVIS solar occultations. *Icarus*, **226**, 1318–1330, doi:10.1016/j.icarus.2013.07.037.

Koskinen, T. T., B. R. Sandel, R. V. Yelle (2014a), *The variability of Saturn's thermosphere from Cassini/UVIS occultations*. 46th DPS Meeting, Tucson, AZ, 511.08.

Koskinen, T. T., B. R. Sandel, R. V. Yelle (2015), Saturn's variable thermosphere from Cassini/UVIS occultations, *Icarus*, doi:10.1016/j.icarus.2015.07.008.

Koskinen, T. T., R. V. Yelle, P. Lavvas et al. (2014b), Electrodynamics on extrasolar giant planets. *Astrophys. J.*, **796**, 16, doi:10.1088/0004-637X/796/1/16.

Koskinen, T. T., R. V. Yelle, D. S. Snowden et al. (2011), The mesosphere and thermosphere of Titan revealed by Cassini/UVIS stellar occultations. *Icarus*, **216**, 507–534.

Kundu, P. K. (1990), *Fluid Mechanics*, San Diego, CA: Academic Press, 638 pp., pp. 580–582.

Kurth, W. S., E. J. Bunce, J. T. Clarke et al. (2009), Auroral processes, in: *Saturn from Cassini-Huygens*, M. Dougherty, L. W. Esposito, and S. M. Krimigis, Eds, Chapter 12, pp. 333–374, New York, NY: Springer, doi:10.1007/978-1-4020-9217-6.

Lamy, L., B. Cecconi, R. Prangé et al. (2009), An auroral oval at the footprint of Saturn's kilometric radio sources, colocated with the UV aurorae. *J. Geophys. Res.*, **114**, doi:10.1029/2009JA014401.

Lamy, L., R. Prangé, W. Pryor et al. (2013), Multispectral simultaneous diagnosis of Saturn's aurorae throughout a planetary rotation, *J. Geophys. Res.*, **118**, 4817–4843, DOI :10.1002/jgra.50404.

Lee, J. S. and R. R. Meier (1980), Angle-dependent frequency redistribution in a plane-parallel medium – External source case. *Astrophys. J.* **240**, 185–195, doi:10.1086/158222.

Lindal, G. F., D. N. Sweetnam and V. R. Eshleman (1985), The atmosphere of Saturn: An analysis of the Voyager radio occultation measurements. *Astron. J.*, **90**, 1136–1146, doi:10.1086/113820.

Liu, W. and A. Dalgarno (1996), The Ultraviolet Spectrum of the Jovian Dayglow. *Astrophys. J.*, **462**, 502–518, doi:10.1086/177168.

Matcheva, K. I. and D. F. Strobel (1999), Heating of Jupiter's thermosphere by dissipation of gravity waves due to molecular viscosity and heat conduction, *Icarus*, **140**, 328–340, doi:10.1006/icar.1999.6151.

McGrath, M. A. and J. T. Clarke (1992), H I Lyman alpha emission from Saturn (1980–1990). *J. Geophys. Res.* **97**, 13691–13703, doi:10.1029/92JA00143.

Melin, H., S. Miller, T. Stallard et al. (2007), Variability in the H_3^+ emission of Saturn: Consequences for ionisation rates and temperature. *Icarus*, **186**, 234–241, doi:10.1016/j.icarus.2006.08.014.

Melin, H., T. Stallard, S. Miller et al. (2011), Simultaneous Cassini VIMS and UVIS observations of Saturn's southern aurora: Comparing emissions from H, H_2, and H_3^+ at high spatial resolution. *Geophys. Res. Lett.*, **38**, L15203, doi:10.1029/2011GL048457.

Moore, L. E., M. Mendillo, I. C. F. Müller-Wodarg et al. (2004), Modeling of global variations and ring shadowing in Saturn's ionosphere. *Icarus*, **172**, doi:10.1016/j.icarus.2004.07.007.

Moore, L., I. C. F. Müller-Wodarg, M. Galand et al. (2010), Latitudinal variations in Saturn's ionosphere: Cassini measurements and model comparisons. *J. Geophys. Res.*, **115**, A11317, doi:10.1029/2010JA015692.

Moses, J. I. and S. F. Bass (2000), The effects of external material on the chemistry and structure of Saturn's ionosphere. *J. Geophys. Res.*, **105**, 7013–7052, doi:10.1029/1999JE001172.

Moses, J. I., T. Fouchet, B. Bézard et al. (2005), Photochemistry and diffusion in Jupiter's stratosphere: Constraints from ISO observations and comparisons with other giant planets. *J. Geophys. Res.*, **110**, E08001, doi:10.1029/2005JE002411.

Müller-Wodarg, I. C. F., M. Mendillo, R. V. Yelle et al. (2006), A global circulation model of Saturn's thermosphere. *Icarus*, **180**, 147–160, doi:10.1016/j.icarus.2005.09.002.

Müller-Wodarg, I. C. F., L. Moore, M. Galand et al. (2012), Magnetosphere-atmosphere coupling at Saturn: 1 – Response of thermosphere and ionosphere to steady state polar forcing. *Icarus*, **221**, 481–494, doi:10.1016/j.icarus.2012.08.034.

Müller-Wodarg, I. C. F., R. V. Yelle, M. J. Mendillo et al. (2003), On the global distribution of neutral gases in Titan's upper atmosphere and its effect on the thermal structure. *J. Geophys. Res.*, **108**, 1453, doi:10.1029/2003JA010054

Nagy, A. F., A. J. Kliore, E. Marouf et al. (2006), First results from the ionospheric radio occultations of Saturn by the Cassini spacecraft. *J. Geophys. Res.*, **111**, A06310, doi:10.1029/2005JA011519.

O'Donoghue, J., T. S. Stallard, H. Melin et al. (2013), The domination of Saturn's low-latitude ionosphere by ring rain. *Nature*, **496**, 193–195, doi:10.1038/nature12049.

O'Donoghue, J., T. S. Stallard, H. Melin (2014), Conjugate observations of Saturn's northern and southern H_3^+ aurorae, *Icarus*, **229**, 214–220, doi:10.1016/j.icarus.2013.11.009.

Parkinson, C. D. (2002), Photochemistry and Radiative Transfer Studies in the Atmospheres of Jupiter and Saturn. Unpublished Ph. D. thesis, York University, North York, Ontario, Canada, 193 pp.

Parkinson, C. D., E. Griffioen, J. C. McConnell et al. (1998), He 584 Å Dayglow at Saturn: A Reassessment, *Icarus*, **133**, 210–220, doi:10.1006/icar.1998.5926.

Parkinson, C. D., J. C. McConnell, L. Ben Jaffel et al. (2006), Deuterium chemistry and airglow in the jovian thermosphere, *Icarus*, **183**, 451–470, doi:10.1016/j.icarus.2005.09.02.

Read, P. L., T. E. Dowling and G. Schubert (2009), Saturn's rotation period from its atmospheric planetary wave configuration. *Nature*, **460**, 608–610.

Richmond, A. D. (1995), Modeling equatorial ionospheric electric fields. *J. Atmos. Terr. Phys.*, **57**, 1103–1115, doi:10.1016/0021–9169(94)00126–9.

Richmond, A. D., E. C. Ridley and R. G. Roble (1992), A thermosphere/ionosphere general circulation model with coupled electrodynamics. *Geophys. Res. Lett.*, **19**, 601–604, doi:10.1029/92GL00401.

Richmond, A. D. and J. P. Thayer (2000), Ionospheric electrodynamics: A tutorial. *Magnetospheric current systems, Geophysical monograph*, **118**, 131–146 (Americal Geophysical Union), doi:10.1029/GM118p0131.

Samson, J. A. R. and G. N. Haddad (1994), Total photoabsorption cross section of H_2 from 18 to 113 eV. *J. Opt. Soc. Am. B*, **11**, 277–279, doi:10.1364/JOSAB.11.000277.

Sandel, B. R. et al. (1982), Extreme Ultraviolet Observations from Voyager 2 Encounter with Saturn, *Science*, **215**, 548–553, doi:10.1126/science.215.4532.548.

Schoeberl, M. R. and R. S. Lindzen (1982), A note on the limits of Rossby wave amplitudes, *J. Atmos. Sci.*, **39**, 1171–1174.

Schubert, G., M. P. Hickey and R. L. Walterscheid (2003), Heating of Jupiter's thermosphere by the dissipation of upward propagating acoustic waves, *Icarus*, **163**, 398–413, doi:10.1016/S0019-1035(03)00078-2.

Shemansky, D. E. (1985), An explanation for the H Ly α longitudinal asymmetry in the equatorial spectrum of Jupiter: An outcrop of paradoxical energy deposition in the exosphere. *J. Geophys. Res.* **90**, 2673–2694, doi:10.1029/JA090iA03p02673.

Shemansky, D. E. and J. M. Ajello (1983), The Saturn spectrum in the EUV: Electron excited hydrogen. *J. Geophys. Res.*, **88**, 459–464, doi:10.1029/JA088iA01p00459.

Shemansky, D. E. and X. Liu (2012), Saturn upper atmosphere structure from Cassini EUV and FUV occultations. *Can. J. Phys.*, **90**, 817–831, doi:10.1139/p2012-036.

Shemansky, D. E., X. Liu and H. Melin (2009), The Saturn hydrogen plume. *Planetary Space Sci.*, **57**, 1659–1670, doi:10.1016/j.pss.2009.05.002.

Sinclair, J. A., P. G. J. Irwin, L. N. Fletcher et al. (2013), Seasonal variations of temperature, acetylene and ethane in Saturn's atmosphere from 2005 to 2010, as observed by Cassini/CIRS. *Icarus*, **225**, 257–271, doi:10.1016/j.icarus.2013.03.011.

Smith, C. G. A. (2013), Electrodynamic coupling of Jupiter's thermosphere and stratosphere: A new source of thermospheric heating. *Icarus*, **226**, 923–944, doi:10.1016/j.icarus.2013.07.001.

Smith, G. R., D. E. Shemansky, J. B. Holberg et al. (1983), Saturn's upper atmosphere from the Voyager 2 EUV solar and stellar occultations. *J. Geophys. Res.*, **88**, 8667–8678, doi:10.1029/JA088iA11p08667.

Snowden, D. S., R. V. Yelle, J. Cui et al. (2013), The thermal structure of Titan's upper atmosphere, I: Temperature profiles from Cassini INMS observations. *Icarus*, **226**, 552–582.

Stallard, T. S., H. Melin, S. Miller et al. (2012), Peak emission altitude of Saturn's H_3^+ aurora. *Geophys. Res. Lett.*, **39**, L15103, doi:10.1029/2012GL052806.

Stallard, T., S. Miller, H. Melin et al. (2007), Saturn's auroral/polar H_3^+ infrared emission, I. General morphology and ion velocity structure. *Icarus*, **189**, 1–13, doi:10.1016/j.icarus.2006.12.027

Stevens, M. H., D. F. Strobel and F. Herbert (1993), An analysis of the Voyager 2 ultraviolet spectrometer occultation data at Uranus: Inferring heat sources and model atmospheres, *Icarus*, **100**, 45–63, doi:10.1006/icar.1993.1005.

Strobel, D. F. (2002), Aeronomic Systems on Planets, Moons, and Comets. In *Atmospheres in the Solar System: Comparative Aeronomy*, eds. M. Mendillo, A. Nagy, and H. Waite. Washington, DC: American Geophysical Union, Geophysical Monograph Series, pp. 7–22.

Strobel, D. F. and G. R. Smith (1973), On the Temperature of the Jovian Thermosphere, *J. Atmos. Sci.*, **30**, 718–725.

Strobel, D. F., R. V. Yelle, D. E. Shemansky et al. (1991), The Upper Atmosphere. In *Uranus*, eds. J. Bergstrahl and M. S. Matthews. Tucson, AZ: University of Arizona Press, pp. 65–109.

Tao, C., Y. Miyoshi, N. Achilleos et al. (2014), Response of the Jovian thermosphere to variations in solar EUV flux. *J. Geophys. Res. Space Phys.*, **119**, doi:10.1002/2013JA019411.

Vasyliūnas, V. M. and P. Song (2005), Meaning of ionospheric Joule heating. *Journal of Geophysical Research*, **110**(A), doi:10.1029/2004JA010615.

Vervack Jr., R. J. and J. I. Moses (2015), Saturn's upper atmosphere during the Voyager era: Reanalysis and modeling of the UVS occultations. *Icarus*, **258**,135–163, doi:10.1016/j.icarus.2015.06.007.

Vervack Jr., R. J., B. R. Sandel and D. F. Strobel (2004), New perspectives on Titan's upper atmosphere from a reanalysis of the Voyager 1 UVS solar occultations, *Icarus*, ***170***, 91–112, doi:10.1016/j.icarus.2004.03.005

Waite, J. H., T. E. Cravens, J. Kozyra et al. (1983), Electron precipitation and related aeronomy of the Jovian thermosphere and ionosphere, *J. Geophys. Res.*, **88**, 2156–2202, doi:10.1029/JA088iA08p06143.

Yelle, R. V. (1988), H_2 emissions from the outer planets. *Geophys. Res. Lett.*, **15**, 1145–1148, doi:10.1029/GL015i010p01145.

Yelle, R. V., J. Cui and I. C. F. Müller-Wodarg (2008), Methane escape from Titan's atmosphere. *J. Geophys. Res.*, **113**, E10003.

Yelle, R. V., L. A. Young, R. J. Vervack et al. (1996), Structure of Jupiter's upper atmosphere: Predictions for Galileo. *J. Geophys. Res.*, **101**, 2149–2161.

Zhang, X., C. A. Nixon, R. L. Shia et al. (2013), Radiative forcing of the stratosphere of Jupiter, Part I: Atmospheric cooling rates from Voyager to Cassini. *Planet. Space Sci.*, **88**, 3–25.

Zharkov, V. N. and V. P. Trubitsyn (1970), Theory of the figure of rotating planets in hydrostatic equilibrium – a third approximation. *Sov. Phys. Astron.*, **13**, 981–988.

10

Saturn's Seasonally Changing Atmosphere

Thermal Structure, Composition and Aerosols

LEIGH N. FLETCHER, THOMAS K. GREATHOUSE, SANDRINE GUERLET, JULIANNE I. MOSES AND ROBERT A. WEST

Abstract

The longevity of Cassini's exploration of Saturn's atmosphere (a third of a Saturnian year) means that we have been able to track the seasonal evolution of atmospheric temperatures, chemistry and cloud opacity over almost every season, from solstice to solstice and from perihelion to aphelion. Cassini has built upon the decades-long ground-based record to observe seasonal shifts in atmospheric temperature, finding a thermal response that lags behind the seasonal insolation with a lag time that increases with depth into the atmosphere, in agreement with radiative climate models. Seasonal hemispheric contrasts are perturbed at smaller scales by atmospheric circulation, such as belt/zone dynamics, the equatorial oscillations and the polar vortices. Temperature asymmetries are largest in the middle stratosphere and become insignificant near the radiative-convective boundary. Cassini has also measured southern-summertime asymmetries in atmospheric composition, including ammonia (the key species forming the topmost clouds), phosphine and para-hydrogen (both disequilibrium species) in the upper troposphere; and hydrocarbons deriving from the UV photolysis of methane in the stratosphere (principally ethane and acetylene). These chemical asymmetries are now altering in subtle ways due to (i) the changing chemical efficiencies with temperature and insolation and (ii) vertical motions associated with large-scale overturning in response to the seasonal temperature contrasts. Similarly, hemispheric contrasts in tropospheric aerosol opacity and coloration that were identified during the earliest phases of Cassini's exploration have now reversed, suggesting an intricate link between the clouds and the temperatures. Finally, comparisons of observations between Voyager and Cassini (both observing in early northern spring, one Saturn year apart) show tantalizing suggestions of non-seasonal variability. Disentangling the competing effects of radiative balance, chemistry and dynamics in shaping the seasonal evolution of Saturn's temperatures, clouds and composition remains the key challenge for the next generation of observations and numerical simulations.

10.1 Introduction

We can achieve a greater understanding of any complex system by studying how that system evolves with time. Saturn, with its 26.7° Earth-like axial tilt, 29.5-Earth-year orbital period and orbital eccentricity of 0.057 (perihelion near northern winter solstice, aphelion near northern summer solstice), is our closest and best example of a seasonally variable giant planet atmosphere, in contrast with Jupiter (negligible seasonal influences from the 3.1° tilt), Uranus (extreme seasonal contrasts from the 98° tilt) and Neptune (slow evolution due to the 165-year period). Furthermore, Cassini has provided our best opportunity in a generation to study the seasonal evolution of a giant planet atmosphere and the influence of temporal variations in sunlight on the atmospheric temperatures,

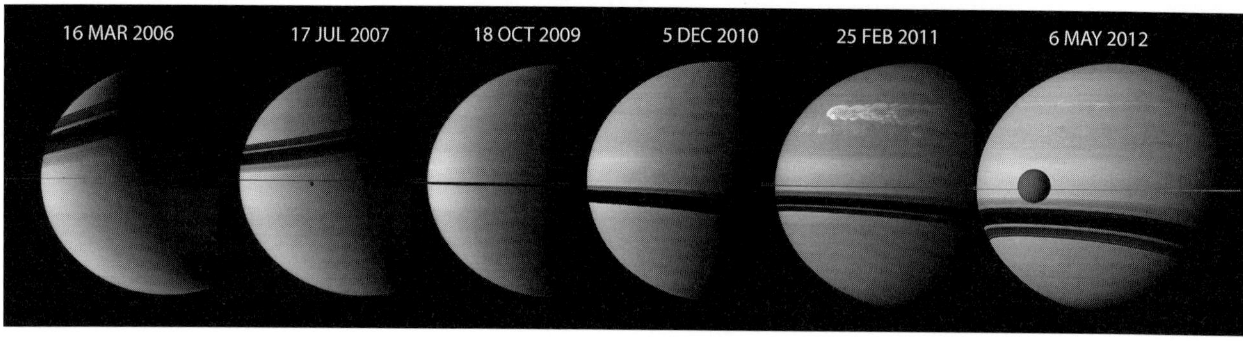

Figure 10.1 Saturn's changing insolation from 2006 to 2012, three years on either side of the northern spring equinox. The colors of Saturn's tropospheric clouds and hazes can be seen shifting as northern winter becomes northern spring. Compiled from Cassini images courtesy of NASA/JPL-Caltech. (A black-and-white version of this figure appears in some formats. For the color version, please refer to the plate section.)

clouds and chemistry. Saturn completed 13.5 orbits of the Sun between Galileo's first glimpses of the ringed planet and its "strange appendages" (July 1610) and the Cassini spacecraft's arrival at Saturn orbit in June 2004. Studies of seasonally changing atmospheric properties have only been possible for the past four decades (1.3 Saturn years, coinciding with the improved capabilities of ground-based infrared remote sensing), with the Cassini orbiter providing reconnaissance for the last third of a Saturn year (Figure 10.1). In this chapter, we review our current knowledge of Saturn's temporally variable thermal structure, composition and aerosols, from the churning tropospheric cloud decks to the middle atmosphere.

What properties might we expect to be time-variable on a giant planet? Atmospheric temperatures are governed by a delicate balance between Saturn's internal heat source and heating from the Sun. Both diabatic (balance between radiative heating and cooling) and adiabatic (atmospheric motions redistributing energy vertically and horizontally) forcings govern the spatial variations of temperature, so that we might expect warm summers and cool winters, albeit with a phase lag compared to the solstices due to the atmospheric inertia. The seasonal diabatic heating will generate hemispheric temperature gradients that are superimposed on the belt/zone structure, which then drive atmospheric transport to redistribute excess energy. These hemispheric temperature contrasts should remain in balance with Saturn's zonal wind system (via the thermal wind relation), such that vertical shears on the zonal jets could vary with time. Saturn's chemical composition may also vary with season, as variations in ultraviolet insolation drive ionization and photodissociation rates, which in turn govern the populations of hydrocarbons and hazes derived from methane photolysis in the stratosphere, and the distribution of key volatiles (e.g. NH_3) and disequilibrium species (e.g. PH_3) in the upper troposphere. Photochemically produced hazes could sediment downward to serve as cloud-condensation nuclei for condensible volatiles, which in turn could cause aerosol and cloud properties to vary with time (e.g. Figure 10.1). Finally, dynamic phenomena in the weather layer (vortices, storms, plumes and waves) respond to modifications of atmospheric stratification, so seasons could help modulate meteorological activity.

Cassini's longevity, coupled with a long baseline of ground-based observing, allows us to monitor each of these processes during a Saturn year to provide a four-dimensional understanding of Saturn's troposphere and stratosphere. Seasons are indicated by the planetocentric solar longitude (L_s), from 0° at northern spring equinox (1980, 2009), to 90° at the northern summer solstice (1987, 2017), 180° at northern autumnal equinox (1995) and 270° at the northern winter solstice (2002). Saturn's southern summers receive greater insolation than northern summers (perihelion occurs near L_s = 280°, July 2003; aphelion at L_s = 100°, April 2018). In addition to ground-based remote sensing since the mid-1970s (southern summer), Saturn's seasonal asymmetries have been observed by four visiting spacecraft: observations by Pioneer 11 and

Voyagers 1 and 2 were clustered around northern spring equinox (1979–1981); Cassini entered Saturn orbit in June 2004, two years after southern summer solstice (L_s = 293°) and aims to complete its mission at northern summer solstice (L_s = 93° in September 2017).

This chapter is organized as follows: Section 10.2 reviews investigations of Saturn's tropospheric and stratospheric temperature field, comparing them to climate models to understand the influence and variability of atmospheric circulation. In Section 10.3 we review observations and chemical modeling of the spatial distributions and variability of key atmospheric species. Section 10.4 reviews the characteristics of Saturn's clouds and hazes, focusing on their time-variable properties, before we review unanswered questions in Section 10.5. Planetographic latitudes are assumed unless otherwise stated. We confine our discussion to Saturn's seasonally variable troposphere and stratosphere; the thermosphere and ionosphere are discussed in Chapter 9.

10.2 Seasonally Evolving Thermal Structure

10.2.1 Pre-Cassini Studies

Saturn's temperature structure in the cloud-forming region and the lower troposphere is expected to follow an adiabatic gradient, with the lapse rate dominated by the heat capacity of the hydrogen-helium atmosphere but with small contributions from latent heat released by the condensation of volatile species (NH_3, NH_4SH and H_2O) and lagged conversion between the two different spin isomers (ortho- and para-H_2) of molecular hydrogen (see the review by Ingersoll et al. 1984, and Section 10.3.2). At lower pressures above the radiative-convective boundary (350–500 mbar, Fletcher et al. 2007a), the atmospheric opacity drops sufficiently to allow efficient cooling by radiation and the temperatures deviate from the adiabat, becoming more stably stratified in the upper troposphere towards the temperature minimum (the tropopause near 80 mbar). Atmospheric heating due to short-wavelength sunlight absorption by methane and aerosols causes temperatures to rise again in the stratosphere, balanced by long-wavelength cooling from methane (upper stratosphere), ethane and acetylene (middle and lower stratosphere), and the collision-induced hydrogen-helium continuum (upper troposphere and lower stratosphere). It is the seasonal dependence of the atmospheric heating (and cooling, via chemical changes) that causes Saturn's upper tropospheric and stratospheric temperatures to vary considerably with season.

Seasonal temperature variations were observed as asymmetries in emission measured at thermal infrared wavelengths, first detected in ground-based images from stratospheric ethane near 12 μm (Gillett and Orton 1975; Rieke 1975) and methane near 8 μm (Tokunaga et al. 1978) during southern summertime conditions, revealing enhanced emission from the southern pole. Tropospheric contrasts in the 17–23 μm region were still present but more muted (Caldwell et al. 1978; Tokunaga et al. 1978), consistent with a seasonal response that becomes weaker with depth. The first spacecraft measurements of Saturn's temperatures by Pioneer 11 (1979, L_s = 354°) revealed no tropospheric thermal asymmetries between 10°N and 30°S just before the northern spring equinox (Orton and Ingersoll 1980). However, inversions of Voyager 1 (1980, L_s = 8.6°) and 2 (1981, L_s = 18.2°) IRIS 14–50 μm spectra revealed tropospheric temperature asymmetries (Hanel et al. 1981, 1982; Conrath and Pirraglia 1983; Conrath et al. 1998), with the north cooler than the south at 150–200 mbar shortly after northern spring equinox. The equinoctial timing of these observations suggested a delay between increased insolation (the solar forcing) and the atmospheric response (Cess and Caldwell 1979) as a consequence of the high thermal inertia of the atmosphere. At tropospheric depths of 500–700 mbar this inertia, and hence the lagged response to the solar forcing, becomes so large that Saturn's northern and southern hemisphere temperatures do not show significant asymmetries (Conrath and Pirraglia 1983). Seasonal amplitudes are largest in the stratosphere, as we shall describe below.

Twenty-three years would pass before another spacecraft reached the Saturn system, but ground-based studies continued to provide insights into Saturn's seasons, particularly with the advent of 2D mid-infrared detector technologies and high-resolution spectroscopy (see the review by Orton et al. 2009). Saturn's north polar regions exhibited enhanced methane and ethane emission by early northern summer (observations by Gezari et al. 1989, in March 1989, L_s = 104°, Figure 10.2) just like the south pole would have in southern summer, although this enhanced emission was not readily observable in the troposphere (observations in December 1992, $L_s = 146°$ by Ollivier et al. 2000). When southern summer returned in the early 2000s, Greathouse et al. (2005) used high-resolution CH_4 emission spectroscopy to derive a stratospheric temperature asymmetry at southern summer solstice (2002, $L_s = 268°$, the south pole 10 K warmer than the equator) and an asymmetry that weakened with increasing depth; while Orton and Yanamandra-Fisher (2005) presented high resolution 7–25 μm images that revealed the southern summer hemisphere in exquisite detail (February 2004, $L_s = 287°$, Figure 10.2). Specifically, they observed a 15 K temperature contrast from the equator to the south pole at 3 mbar, a sharp temperature gradient near 70°S (the edge of the south polar warm hood) and an intense tropospheric hotspot associated with the south polar cyclone poleward of 87°S. These observations confirmed that Cassini would observe a stark asymmetry between the northern winter and southern summer hemispheres upon arrival in 2004.

10.2.2 Cassini Observations in Southern Summer

Cassini's great advantage is the ability to view both the northern and southern hemispheres near-simultaneously, whereas Saturnian winter is forever hidden to an Earth-based observer. Tropospheric and stratospheric temperatures are provided via a combination of infrared remote sensing, radio occultation and ultraviolet stellar occultations. Nadir 7–1000 μm spectroscopy with the Composite Infrared Spectrometer (CIRS, Flasar et al. 2004) measures tropospheric temperatures from the tropopause down to the topmost cloud decks (approximately 80–800 mbar) and stratospheric temperatures (from methane emission) in the 0.5–5.0 mbar range (Flasar et al. 2005; Fletcher et al. 2007a). CIRS limb observations (Fouchet et al. 2008; Guerlet et al. 2009) complement the vertical coverage of nadir observations by constraining the stratospheric temperature profile between 20 mbar and 1 μbar with a vertical resolution of 1–2 scale heights, at the expense of poorer spatial and temporal coverage. Radio-occultations of Cassini by Saturn (Schinder et al. 2011), available at a limited number of latitudes, achieve the highest vertical resolution (6–7 km, a tenth of a scale height) and constrain the temperature-pressure profile between 1500 mbar and 0.1 mbar.

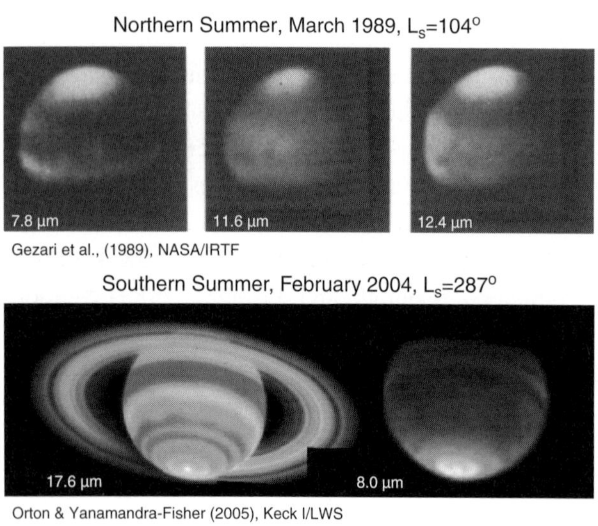

Figure 10.2 Stratospheric thermal emission imaged by ground-based facilities in March 1989 (near northern summer solstice, Gezari et al. 1989) and February 2004 (near southern summer solstice, Orton and Yanamandra-Fisher 2005), showing enhanced emission from the summer pole and similarities in the seasonal response. (A black-and-white version of this figure appears in some formats. For the color version, please refer to the plate section.)

Cassini's prime mission provided a snapshot of Saturn's hemispheric temperature asymmetries in late southern summer (2004–2008, $L_s = 293°–345°$, Flasar et al. 2005; Fletcher et al. 2007a, 2008; Guerlet et al. 2009), as shown in Figure 10.3(a). The summer pole was found to be 40 K warmer than the winter pole at 1 mbar, with the contrast decreasing with increasing pressure (Fletcher et al. 2007a). Intriguingly, the latitudinal asymmetry appeared to be smaller (≈ 24 K) at 0.1 mbar and smaller still at 0.01 mbar (Guerlet et al. 2009). The tropopause was around 10 K cooler (and slightly higher in altitude) in the winter hemisphere than the summer hemisphere (Fletcher et al. 2007a). Below the tropopause, the lapse rate increased with depth until reaching the dry adiabat at approximately 350–500 mbar (Lindal et al. 1985; Fletcher et al. 2007a), likely indicating the location of the radiative-convective boundary that separates the stably stratified upper troposphere from the convective deeper troposphere. This lapse rate change occurred at higher pressures (400–500 mbar) in the summer hemisphere than in the northern hemisphere (350–450 mbar), due to the greater penetration of solar heating in the southern hemisphere. Temperature asymmetries became negligible for $p > 500$ mbar.

Between the radiative-convective boundary and the tropopause, inversions of far-infrared spectra revealed an inflection point in the tropospheric temperature structure in the 100–300 mbar region. This perturbation was referred to as the "knee" (Fletcher et al. 2007a), suggestive of heating that was localized in altitude within Saturn's tropospheric haze. It was first noted in Voyager/IRIS retrievals at equatorial and southern latitudes (Hanel et al. 1981) and in Voyager radio occultations at 3°S and 74°S (Lindal et al. 1985). The pressure level and magnitude of this temperature perturbation varied strongly with latitude, being enhanced in southern summer and weak or absent in the northern winter hemisphere. The knee was higher and weaker over the equator, and showed local maxima at 15°N and 15°S associated with the warm equatorial belts (see Figure 10.20 at the end of this chapter). Fletcher et al. (2007a) concluded that this was a radiative effect due to solar absorption by aerosols in the upper troposphere, and explained the asymmetry in terms of both seasonal insolation and the variable distributions of tropospheric aerosols (later confirmed by radiative climate modeling by Friedson and Moses 2012 and Guerlet et al. 2014). The relationship between the "knee" and the upper tropospheric haze is discussed in Section 10.4.

Saturn's temperature distribution in Figure 10.3 reveals the influence of dynamics as well as radiative balance. The global temperature asymmetries are superimposed onto small-scale latitudinal contrasts between the cool zones (anticyclonic shear regions equatorward of prograde jets) and warmer belts (cyclonic shear regions poleward of prograde jets) (Conrath and Pirraglia 1983; Fletcher et al. 2007a). Note that the correlations between this belt/zone structure (defined in terms of the zonal jets and tropospheric temperatures) and the cloud reflectivity is not as clear-cut as for Jupiter. For example, bands of low reflectivity are often narrow and located close to the prograde jet peaks (e.g. Vasavada et al. 2006), and do not appear correlated with the thermal structure. Saturn's polar troposphere features long-lived cyclonic "hot spots" located directly at each pole irrespective of season, and the northern hexagonal jet at 77°N is related to a hexagonal warm polar belt poleward of this latitude (Fletcher et al. 2008). At lower pressures, the polar stratosphere features a warm "polar hood" in summer that is absent in winter, one of the most extreme examples of a seasonal phenomenon on Saturn (Fletcher et al. 2008). In Saturn's tropical stratosphere, contrasts between the equator and neighboring latitudes are observed to oscillate with time and altitude. This "semi-annual oscillation" in the thermal structure implies a strong vertical shear of the zonal wind at the equator and is reminiscent of the Quasi-Biennal Oscillation in the Earth's stratosphere, a dynamical phenomenon driven by wave-zonal flow interactions (Fouchet et al. 2008; Orton et al. 2008).

10.2.3 Seasonal Evolution of Temperatures

Saturn's thermal structure during southern summer was reviewed by Del Genio et al. (2009), but Cassini has since revealed how the global temperature structure has evolved with season through northern spring equinox. Fletcher et al. (2010) used CIRS observations from 2004 to 2009 ($L_s = 297° - 358°$) to determine Saturn's upper tropospheric and stratospheric temperature variability, finding (i) stratospheric warming of northern mid-latitudes by 6–10 K at 1 mbar as they emerged from ring shadow into springtime conditions; (ii) southern cooling by 4–6 K (both at mid-latitudes and within the south polar stratospheric hood poleward of 70°S) and a resulting decrease in the 40-K asymmetry between the hemispheres that was present in 2004; and (iii) a tropospheric response to the seasonal insolation shifts that seemed to be larger at the locations of the broadest retrograde jets. The "flattening" of the summertime temperature asymmetry (by Saturn's equinox, northern and southern mid-latitude 1-mbar temperatures were both in the 140–145 K range) followed the expectations of a radiative climate model (Greathouse et al. 2010, and see below), albeit perturbed by the equatorial oscillation, polar vortex dynamics and the belt/zone structure. Sinclair et al. (2013) extended this analysis to include observations in 2010 ($L_s = 15°$) and observed the continued northern warming and southern cooling, particularly intense within Saturn's south polar region (≈17 K between 2005 and 2010). At higher stratospheric altitudes, Sylvestre et al. (2015) extended the analysis of Guerlet et al. (2009) by considering limb spectroscopy in 2010–2012, finding a seasonal trend consistent with the nadir data at 1 mbar, but with smaller variations at lower pressures. For example, mid-latitude

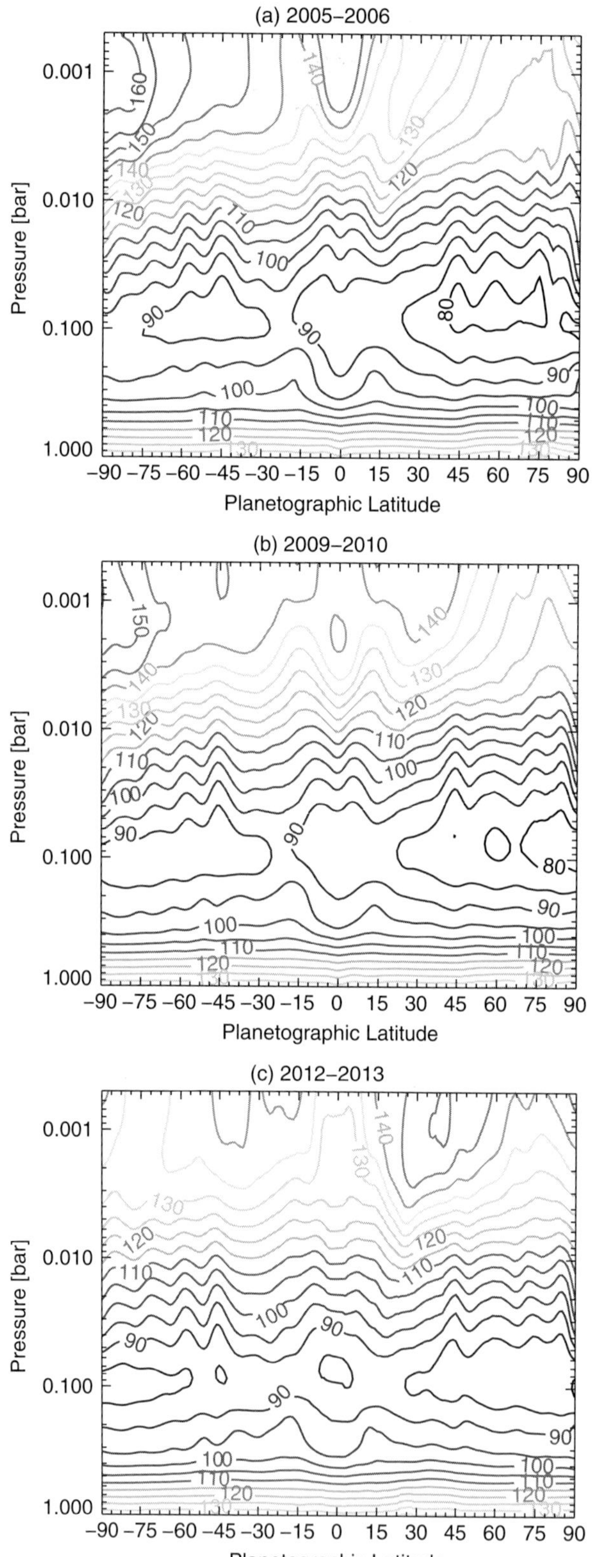

Figure 10.3 Zonal mean temperatures derived from Cassini/CIRS nadir infrared spectroscopy at three different epochs, representing a snapshot of the temperature field in (a) late southern summer (2005–2006, $L_s \approx 313°$); (b) northern spring equinox (2009–2010, $L_s \approx 5°$); and (c) early northern spring (2012–2013, $L_s \approx 41°$). Adapted and updated from Fletcher et al. (2010) and Fletcher et al. (2015).

temperatures near 0.1 mbar have remained approximately constant between 2005 and 2010.

Cassini's monitoring of the seasonal temperatures has now extended well into northern spring, and most recently Fletcher et al. (2015) extended the work of Fletcher et al. (2010) and Sinclair et al. (2013) to 2014 ($L_s = 56°$). The zonal mean temperature as a function of latitude and pressure is shown for three epochs in Figure 10.3, and all CIRS nadir retrievals are shown as a function of time in Figure 10.4. Focusing on the highest latitudes, tropospheric contrasts between the cool polar zones (80°–85° latitude) and warm polar belts (near 75°–80° latitude) have varied over the ten-year span of observations, indicating changes to the vertical shear on the zonal jets via the thermal wind equation (Figure 10.3). The warm south polar stratosphere has cooled dramatically by ≈ 5 K/yr, mirrored by warming of a similar magnitude in the north. However, while the south polar region was isolated by a strong thermal gradient near 75°S during the height of summer, no similar boundary was apparent near 75°N during spring despite rising temperatures towards the north pole at L_s = 56° (the last published data), suggesting that the northern summer vortex has yet to form (see Figure 10.5, Chapter 12 and Fletcher et al. 2015). The peak stratospheric warming in the north was occurring at lower pressures (0.5–1.0 mbar) than the peak stratospheric cooling in the south (1–3 mbar). Figure 10.5 demonstrates that north polar minima in stratospheric temperatures were detected in 2008–2010 (lagging one season, or 6–8 years, behind winter solstice); south polar maxima appear to have occurred before the start of the Cassini observations (1–2 years after summer solstice).

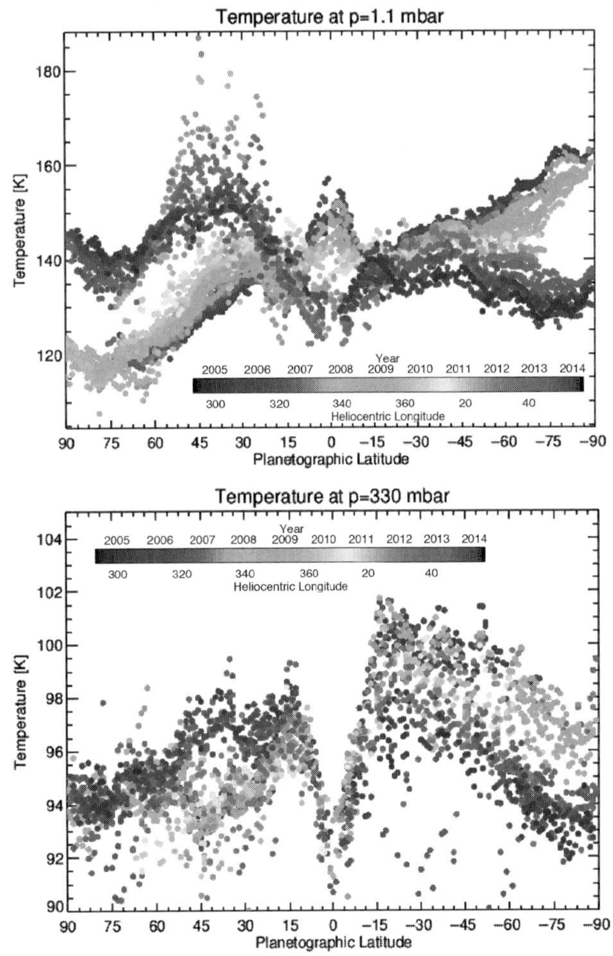

Figure 10.4 Zonal mean temperatures in the stratosphere (1 mbar) and troposphere (330 mbar) as derived from nadir Cassini/CIRS spectra over the duration of the Cassini mission. Note that the absence of high-latitude measurements between 2010 and 2012 is due to Cassini's near-equatorial orbit at that time, preventing nadir observations of the poles. Updated from Fletcher et al. (2010) and Fletcher et al. (2015). (A black-and-white version of this figure appears in some formats. For the color version, please refer to the plate section.)

10.2.4 Seasonal Climate Modeling

Seasonal climate modeling is an attempt to create models that accurately predict the temporal evolution of a planet's thermal structure as a function of altitude, latitude and longitude. Diurnal temperature variations are not expected due to Saturn's high thermal inertia, the relatively low amount of solar forcing at Saturn's distance (≈100 times less than at Earth), and Saturn's fast rotation (10 hour 39 minute long days), and current models bear this out (Greathouse et al. 2008; Guerlet et al. 2014). This fact allows modelers to reduce the complexity of the problem and focus on the temporal evolution of temperatures as a function of altitude and latitude only, calling then for detailed 2D time variable models.

The cooling of Saturn's stratosphere is dominated by the radiative emissions from C_2H_2 and C_2H_6 at pressures lower than 5 mbar, along with some cooling due to the ν_4 band of CH_4 near 8 μm. Cooling due to other hydrocarbons is estimated to account for no more than 5% of the total radiative cooling rate (Guerlet et al. 2014). Tropospheric and lower stratospheric cooling occurs via emission from the H_2-H_2 and H_2-He

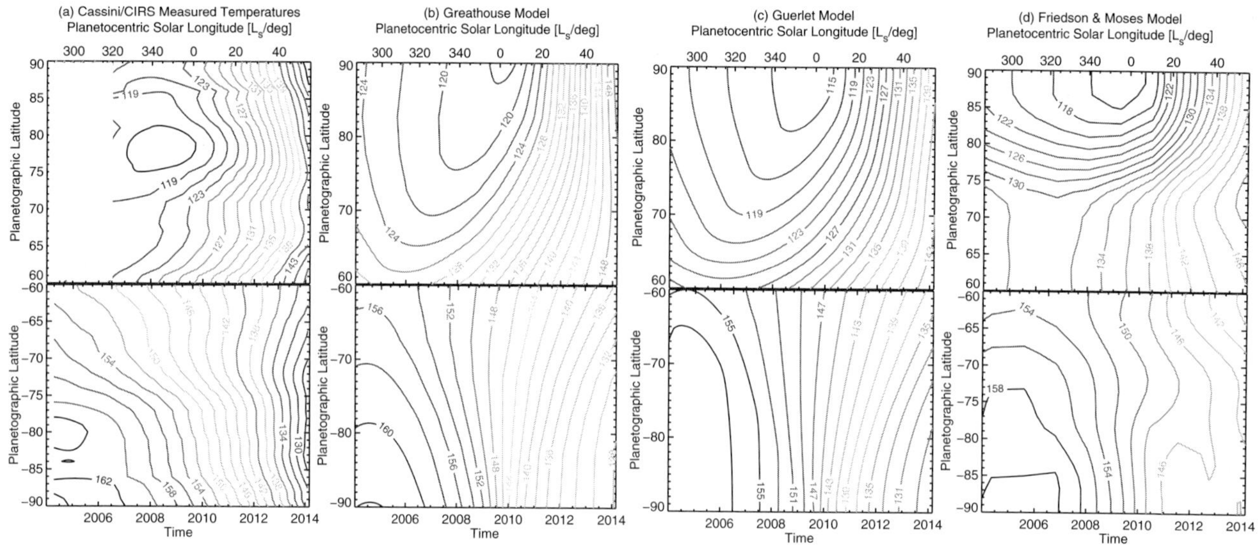

Figure 10.5 Temperatures at 1 mbar derived in the polar regions of Saturn (Fletcher et al. 2015), exposed to the most severe insolation changes over a Saturnian year. Cassini measurements are compared to the model predictions of Greathouse et al. (2010), Guerlet et al. (2014) and Friedson and Moses (2012). Although discrepancies in the absolute temperatures are evident, the timing of the polar maxima/minima in temperatures is reasonably well reproduced.

collision-induced continuum. Atmospheric heating is due to the absorption of sunlight primarily by CH_4 and aerosols. C_2H_2 and C_2H_6 are both byproducts of CH_4 photochemistry (for details see Section 10.3), which is initiated at the top of the atmosphere just below the CH_4 homopause. Since C_2H_2 and C_2H_6 are the dominant radiative coolants, and their distribution controls the extent and direction of radiant energy to space, their distribution can alter stratospheric temperatures, which in turn can induce circulation patterns that serve to redistribute the molecules. While this complicated interchange continues, ongoing photochemical processes are constantly making their own adjustments to the abundances of C_2H_2 and C_2H_6. This interconnectedness means that to produce a truly accurate seasonal climate model one needs to include accurate calculations of the absorption and emission of radiation, photochemistry, and dynamics. Any one of these would make for a complex model on its own and the combination of all three is a worthy goal. While this level of complexity is currently not achieved, we show below that much has been accomplished in all three disciplines, and the most recent incarnations of seasonal models are moving ever closer to this goal.

The first generation of seasonal climate models for Saturn were inspired by the first observations of seasonal temperature asymmetries in the 1970s and Pioneer-Voyager epoch. Radiative-convective equilibrium models in the 1970s (Caldwell 1977; Tokunaga and Cess 1977; Appleby and Hogan 1984) showed that the solar absorption by methane in the visible and near-infrared could explain the temperature inversion identified from ethane limb brightening (Gillett and Orton 1975). Early models of Saturn's stratospheric temperature response (Cess and Caldwell 1979; Carlson et al. 1980) were extended into the troposphere by Bézard et al. (1984), and indicated that an optimal fit to the measured tropospheric temperatures required an additional source of opacity (potentially from tropospheric aerosols). Bézard and Gautier (1985) incorporated a non-gray treatment of radiative transfer to construct a more sophisticated model, accounting for seasonal changes in solar forcing, ring obscuration and planetary oblateness. Though they were limited by both the lack of some lab measurements (several near-infrared bands of methane had yet to be measured in the lab, Bézard and Gautier 1985) and data on Saturn (detailed measurements of temperatures versus altitude, latitude, and time along with the variations of mixing ratio for C_2H_2 and C_2H_6 as a function of time, altitude and latitude), they were able to produce models that to first order reproduced early ground based observations and much of the data retrieved by Voyager 1 and 2, including the observation that the thermal inertia increased with depth such that there was no seasonal modulation of the deepest atmospheric temperatures. These models were one-dimensional in nature, requiring the user to model each latitude of interest and then compile the results to show how the global temperatures changed over time. Conrath et al. (1990) produced more sophisticated radiative-convective-dynamical models of Saturn's troposphere and stratosphere for a comparison to Voyager results, and Barnet et al. (1992) included the effects of ring shadowing and ring thermal emission.

The radiative climate models have evolved with time as (i) the spectroscopic parameters of opacity sources (methane, hydrocarbons and hazes) have become better constrained; (ii) the spatial distributions of the key infrared coolants (ethane and acetylene) have become better known; and (iii) the vertical distribution of clouds and hazes, which have a substantial contribution to the radiative budget, have been revealed. The richness and complexity in the Cassini seasonal dataset prompted the evolution of a new generation of radiative climate models, taking advantage of improvements in laboratory measurements and computational capabilities. The first such model (Greathouse et al. 2010, hereafter TG) was a purely radiative seasonal model of Saturn's stratosphere, with the capability of assuming any vertical, meridional, or temporal variation of the key hydrocarbon coolants. This model is one dimensional in nature, one latitude over time, requiring multiple runs to compile results from different latitudes into a 2D representation of Saturn's stratospheric seasonal evolution. However, as initially planned, this model was absorbed as a module within the Explicit Planetary Isentropic-Coordinate Global Circulation Model (EPIC GCM), allowing EPIC to accurately calculate the radiative heating and cooling rates while accounting for dynamics (Dowling et al. 2010). The second model, the Outer Planet General Circulation Model (OPGCM), is a stratospheric and tropospheric seasonal dynamical model which efficiently and accurately accounts for the radiative heating and cooling while also tracking circulation caused by the seasonal forcing in 3 dimensions (Friedson and Moses 2012, hereafter FM). Finally, the most recent seasonal model is the stratospheric and tropospheric radiative-convective model produced by Guerlet et al. (2014) (hereafter SG).

These three modern models are based largely on the same underlying assumptions, and so we compare each to Cassini measurements of the 1-mbar temperature contrasts in Figure 10.6 to illuminate interesting seasonally forced effects and localized dynamical effects. The TG model assumes the meridional and vertical mixing ratios for C_2H_2 and C_2H_6 as measured by Guerlet et al. (2009), and assumes that this distribution is fixed with time. The TG model is the coarsest (in latitude sampling) of the three models, and does not include the effects of dynamics (dashed line in Figure 10.6). The FM model uses the latitudinal average of the vertical profiles of C_2H_2 and C_2H_6 as measured by Guerlet et al. (2009) and holds these vertical profiles constant with time and latitude, even though they are accounting for dynamical redistribution of heat via diffusion and advection (dotted line in Figure 10.6). The SG model uses the average of the vertical mixing ratio profiles for C_2H_2 and C_2H_6 between 40°N and 40° S (planetocentric) as taken from Guerlet et al. (2009) (solid line in Figure 10.6). These mean vertical profiles are assumed constant with latitude and time. None of the models feature the sharp upturns in C_2H_2 and C_2H_6 abundances at high latitudes (nor the added contribution of polar stratospheric aerosols), and so they are unlikely to reproduce the radiative energy balance near the poles (Fletcher et al. 2015; Guerlet et al. 2015).

A comparison between the three models reveals how the different assumptions manifest themselves in the final predicted temperatures and how those final temperatures compare to the measurements. Given these assumptions, one would expect that the TG and SG models should return very similar results with the differences between the two being due to the different C_2H_2 and C_2H_6 distributions assumed, which is in fact what we see in Figure 10.6. Similarly, the comparison between the FM and the SG/TG models show the effects of dynamical diffusion and advection of heat on temperatures, as the radiative heating and cooling scheme in all three models are quite similar. Where the FM model (dotted) is significantly cooler/hotter than the SG model (solid), this could be related to upwelling/downwelling in the FM model causing adiabatic expansion/compression. However, this assumes identical treatment of radiative balance between the three models, and neglects other sources of heating and cooling (e.g. interactions between waves and the mean flow).

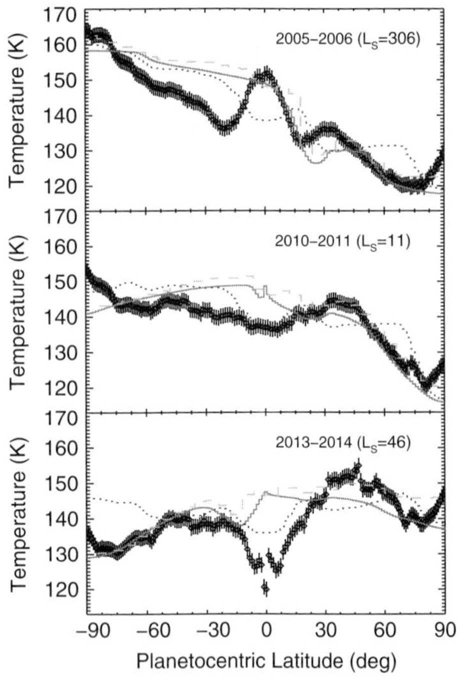

Figure 10.6 Saturn's meridional temperature variations at 1 mbar at 3 different epochs derived from nadir Cassini/CIRS observations, compared with model predictions from Guerlet et al. (2014) as the solid line, Friedson and Moses (2012) as the dotted line and Greathouse et al. (2010) as the dashed line.

As can be seen in Figure 10.6, all three models do an impressive job of reproducing the main pole to pole temperature variations as measured by Cassini/CIRS nadir observations (Fletcher et al. 2010, 2015). The strong southern summer temperature gradient seen in the top panel of Figure 10.6 (L_s = 306°) transitioning to a more equal north/south temperature profile by mid-northern spring (L_s = 46°) is well tracked by the models, with the exception of the FM model which does not cool as fast at high southern latitudes. However, several discrepancies stand out. The temperatures between ±20° planetocentric latitude are completely different compared to the three model predictions. Temperatures in this region are significantly altered by the dynamical phenomenon known as Saturn's Semi-Annual Oscillation (SSAO) (Fouchet et al. 2008; Orton et al. 2008; Guerlet et al. 2011). The TG and SG models are purely radiative and thus could not possibly match the temperature variation of this dynamical phenomenon. While the FM model could possibly reproduce such a dynamical event, they suggest that the coarseness of their vertical grid may have made it impossible to resolve the waves needed to force the SSAO. Another complication is the heating associated with Saturn's stratospheric vortex (Fletcher et al. 2012b). The nominal seasonal trend of stratospheric temperatures at mid-northern latitudes was disrupted by the production of the stratospheric vortex (known as the "beacon," see Chapter 13) in late 2010, and although this region was avoided to produce the measured temperatures in Figure 10.6, the storm-induced heating of the northern mid-latitudes is the likely cause for the mismatch with the model predictions in 2013.

One of the starkest discrepancies between the models and the data in Figure 10.6 appears at the polar latitudes. The 1-mbar temperatures measured by Cassini at the northern and southern poles are compared to the three models in Figure 10.5, to assess their capabilities for reproducing the absolute temperatures and the timing of the polar minima/maxima in temperatures. Although the timing of the south polar maximum is reproduced in the models (1–2 years after summer solstice), the south polar temperatures are elevated over the predictions of all three models, and the temperature range is larger in the data than in the model predictions. At the north pole, the models are unable to reproduce the observation that the coldest temperatures are identified between 75° and 80°N, rather than at the north pole itself, although the 6- to 8-year phase lag between the winter solstice and the coldest temperatures is consistent with the models. It is suspected that these differences are due to circulations associated with polar vortices, or the increasing importance of stratospheric aerosols or enhanced hydrocarbons in the radiative budget at high latitudes (Fletcher et al. 2015). Interestingly, the FM model stays warmer for longer at the south pole, over-predicting the southern temperatures in 2013–2014. This large temperature increase poleward of 60°S seen in the FM model is the result of extensive downwelling of gas, whereas observations show that this region has in fact cooled substantially for the duration of Cassini's observations (Figure 10.5). While the downward advection of material naturally heats the gas by adiabatic compression, one would also expect an increase in cooling due to an increase in C_2H_2 and C_2H_6 mixing ratio. If advection of material to serve as coolants (i.e. C_2H_2 and C_2H_6 concentrations, stratospheric aerosols) could be included in the FM model, it might help explain the mismatch between the data and model.

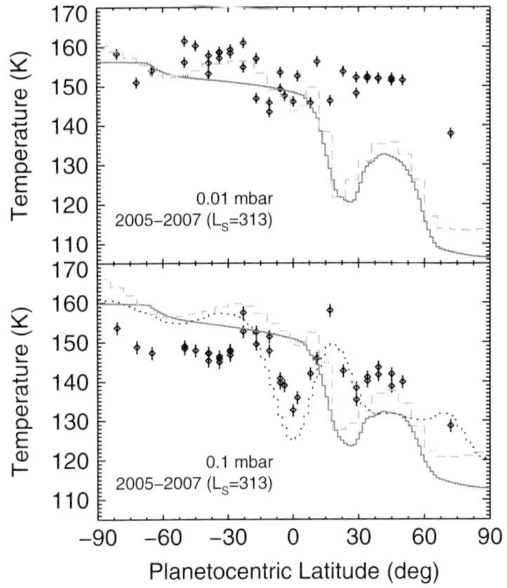

Figure 10.7 Model-data comparisons for temperatures at the 0.01 and 0.1 mbar level. The observed temperatures were retrieved using Cassini/CIRS limb observations in late southern summer Guerlet et al. (2009). Models are identified as the solid line for Guerlet et al. (2014), the dotted line for Friedson and Moses (2012) and the dashed line for Greathouse et al. (2010). Note the Friedson and Moses model is not shown at the 0.01-mbar level as it does not accurately model Saturn's atmosphere at that pressure level.

We now return to the curious fact that the seasonal response measured at higher stratospheric altitudes using CIRS limb observations (0.01–0.1 mbar) by Guerlet et al. (2009) and Sylvestre et al. (2015) appears to be smaller than that at 1 mbar. Figure 10.7 shows the comparison of the three models to the limb observations in southern summer (L_S = 313°) at 0.1 and 0.01 mbar. Although there is remarkable agreement with the temperatures in southern summer (particularly at 0.01 mbar), the measured temperatures are much higher than model predictions in the north. This could be the result of several processes: near 25°N, the effect of ring shadowing may have been overestimated, or significant mixing/advection might wash out the ring shadow effects seen in the purely radiative models. Although the FM model does not offer useful results at the 0.01-mbar level, it does at the 0.1-mbar level. There we can see that the cool region of temperatures due to the ring shadow (15°–30°N) seen in the TG and SG models is warmed significantly by subsidence in the FM model. It is likely this effect occurs at 0.01 mbar as well, and may help to explain the model-data mismatch at the higher altitudes. Away from the ring-shadowed region, additional sources of upper atmospheric heating, such as gravity wave breaking or from some interaction with the thermosphere, might be the culprit for the warmer temperatures over the rest of the northern hemisphere.

Figures 10.5, 10.6 and 10.7 demonstrate how well the models perform relative to the measured data and at specific instances in time. It is also useful to compare how the models behave over an entire Saturn year at a single latitude to learn about seasonal phase lags. In Figure 10.8 we plot the predicted temperatures of Saturn from all three models at 75°S planetocentric latitude. The solid lines represent the temperatures at 1 mbar while the dashed are from 0.1 mbar. All of the models show that the peak temperature in southern summer occurs about 30° of L_S after the southern summer solstice (L_S = 270°) at 1 mbar, but only about 15° of L_S at 0.1 mbar. This is due to the lower pressure level having less mass, and thus less thermal inertia, in addition to the fact that the atmospheric coolants are more abundant at lower pressures. The models all suggest that the maximum and minimum in temperatures are more extreme and can change more rapidly at lower pressures due to the lower thermal inertia found there, which appears to be oddly inconsistent with the limb-data analyses from Cassini. Finally, the model of FM is significantly different from those of TG and SG. The reduction of the wintertime temperatures during polar night is not surprising as the purely radiative seasonal models continually radiate to space whilst in darkness without accounting for any heat advection, but it is probable that significant diffusion and advection of heat would occur in this region. However, it is interesting to note that the models that do the best job of reproducing the measured temperatures at L_S = 46° (i.e. southern autumn, Figure 10.4) and 75°S latitude are those without dynamics.

In summary, it appears that the dominant driver of Saturn's stratospheric temperatures is solar forcing, albeit perturbed by circulation and dynamics. While purely radiative seasonal models do not reproduce the measured temperatures perfectly, they seem to do as good a job as the more complicated dynamical model of FM. Advection of the hydrocarbon abundances along with the heat may be significant, in addition to the seasonally shifting contribution of tropospheric and stratospheric aerosols and other heating sources (e.g.

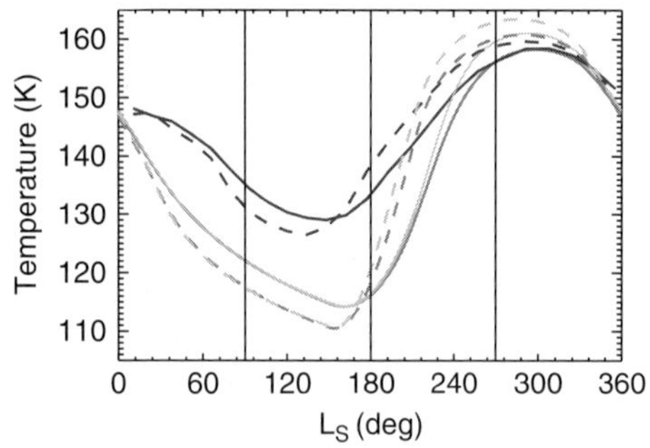

Figure 10.8 Model/model comparison at 75°S between the Guerlet et al. (2014) model in red, the Friedson and Moses (2012) model in blue and the Greathouse et al. (2010) model in green. The solid lines represent the change in temperature over time at the 1.0-mbar level while the dashed lines are for the 0.1-mbar level. The vertical black lines represent the times of Northern Summer Solstice, Northern Autumnal Equinox, and Northern Winter Solstice, from left to right respectively. (A black-and-white version of this figure appears in some formats. For the color version, please refer to the plate section.)

wave activity) not currently accounted for in the theoretical models. Hopefully, with the advent of new 3D GCMs with proper radiative forcing and higher spatial resolution (required for resolving waves and instabilities that influence the circulation), a model will be produced in the near future with a self-consistent radiative, chemical and dynamical system.

10.2.5 Implications of Seasonal Temperature Change

The seasonal shifting of hemispheric temperature asymmetries has implications for both Saturn's global energy budget and atmospheric circulation (e.g., polar atmospheric circulations, wave phenomena, tropospheric overturning, storm perturbations, Hadley circulations). Li et al. (2010) demonstrated that the shifting thermal asymmetries measured by CIRS had implications for Saturn's emitted power (average of 4.952 ± 0.035 W/m^2), which decreased by 2% from 2005 to 2009 and was found to be larger in the south (5.331 ± 0.058 W/m^2) than in the north (4.573 ± 0.014 W/m^2). The peak contribution to the outgoing emission (near 320 mbar) was found to be shallower in the northern winter hemisphere than in southern summer, potentially due to seasonal asymmetries in aerosol content in the upper troposphere, or the details of the upper tropospheric temperature structure. Thus, emitted power measurements over a full Saturn year are desirable to properly constrain the energy budget.

Meridional temperature gradients are related to the vertical shear on the zonal winds via the thermal wind equation, as the Coriolis forces on the winds should remain in geostrophic balance with the horizontal pressure gradients. The stratospheric dT/dy is indicative of positive vertical shear in the winter and negative vertical shear in the summer (Friedson and Moses 2012). Cassini's seasonal monitoring has shown that dT/dy has become increasingly positive in the mid-stratosphere over the ten years of the mission (see Figure 10.3), meaning that northern middle-atmospheric prograde jets should become more retrograde (i.e. decreasing eastward velocities) and southern prograde jets should become more prograde (increasing eastward velocities). These modifications to the stratospheric wind field, although inferred indirectly from temperatures, can have implications for the transmissivity of waves upwards from the convective troposphere (e.g. the wave transport of energy from Saturn's storm regions in 2010–2011, Fletcher et al. 2012b). Despite the large temperature fluctuations, thermal wind variability in the troposphere has been small over the duration of the Cassini mission (variations smaller than 10 m/s per scale height at the tropopause, Fletcher et al. 2016), with the largest changes in the northern flank of the equatorial jet.

Finally, although radiative climate models successfully reproduce the magnitude and scale of the observed asymmetries, they lack the dynamic perturbations that govern the temperatures on a smaller scale. When purely radiative calculations are insufficient to reproduce the observed temperature changes, we can use the thermodynamic energy relationship (Hanel et al. 2003; Holton 2004) to relate the change in the temperature field to the residual mean circulation causing net heating (subsidence) and cooling (upwelling). Conrath et al. (1990) used these techniques to show that a diffuse inter-hemispheric circulation was expected at solstice, with rising motion in the summer hemisphere and downward motion in the winter hemisphere. However, at equinox the flow consisted of two cells, with rising motion at low latitudes and subsidence poleward of $\pm 30°$ latitude, which does not appear to correspond to the equinoctial observations of Cassini. Friedson and Moses (2012) produced a more complex model to predict these inter hemispheric transports, finding a seasonally reversing Hadley-like circulation at tropical latitudes and cross-equatorial flow from the summer into the winter hemisphere twice per year. Fletcher et al. (2015) used the deviation of Saturn's measured temperatures from radiative equilibrium to show that upwelling/downwelling winds with zonally averaged vertical velocities of $|w| \approx$ 0.1 mm/s could account for the cooling/warming of the southern/northern polar stratospheres at 1 mbar, respectively. In summary, the general warming of the northern hemisphere and cooling of the southern hemisphere over the duration of the mission could be reproduced by an inter-hemispheric transport into the spring hemisphere in the stratosphere and upper troposphere, perturbed at low latitudes by a Hadley-type circulation, and at high latitudes by the formation and dissipation of polar vortices.

10.2.6 Non-Seasonal Phenomena

At low latitudes, the signature of Saturn's evolving seasons appears to be overwhelmed by the phenomenon known (perhaps incorrectly) as Saturn's Semi-Annual Oscillation (SSAO). Originally identified by Fouchet et al. (2008) and Orton et al. (2008) as temporally evolving oscillations in the vertical and latitudinal temperature structure, the downward propagation of the wave structure has been studied by Cassini infrared spectroscopy and radio occultation data (Fletcher et al. 2010; Guerlet et al. 2011; Li et al. 2011; Schinder et al. 2011). These studies revealed that the local temperature extrema observed in 2005 (an equatorial local maximum located at the 1-mbar pressure level and a local minimum located at 0.1 mbar) descended by approximately 1.3 scale heights in 4.2 years (see Figure 10.4). The downward propagation of the oscillation was consistent with it being driven by absorption of upwardly propagating waves and suggested a ~15-year period for Saturn's equatorial oscillation, as already derived from long-term ground-based observations Orton et al. (2008).

Cassini observations in early northern spring might be expected to replicate Voyager 1 and 2 observations taken one Saturnian year earlier (L_s = 8°–18°). These datasets have been compared by Li et al. (2013), Sinclair et al. (2014) and Fletcher et al. (2016), and mid-latitude temperatures were largely the same

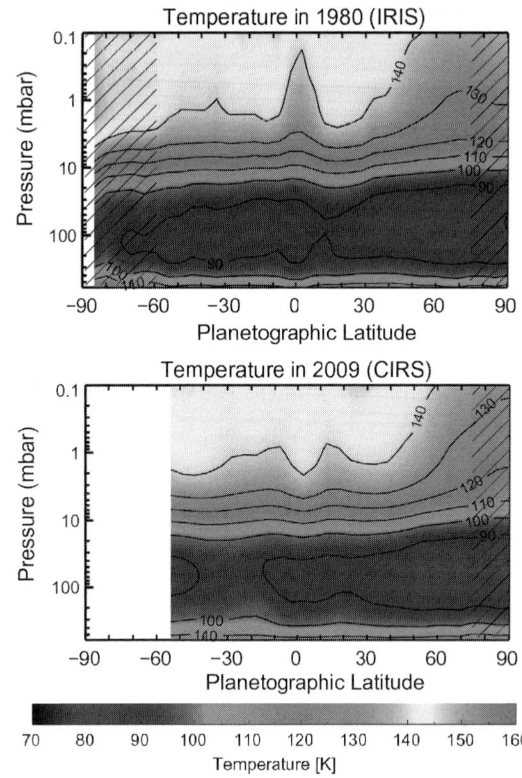

Figure 10.9 Zonal mean temperature retrievals from Voyager/IRIS (L_s = 8°) and Cassini/CIRS (L_s = 3°) as presented by Sinclair et al. (2014). Although these differ by ΔL_s = 5°, the comparison shows that the zonal mean stratospheric temperatures differ considerably at the equator between the two epochs. The upper troposphere also appears to be warmer in the southern autumn hemisphere (10°–45°S) in the Cassini measurements compared to the Voyager measurements, as discussed by Li et al. (2013), suggesting non-seasonal variability. The hashed areas are regions of low retrieval confidence. (A black-and-white version of this figure appears in some formats. For the color version, please refer to the plate section.)

between the two epochs, confirming a repeatability of the broad seasonal cycle. In the stratosphere, Sinclair et al. (2014) showed that CIRS 2009–2010 temperatures were 7.1 ±1.2 K warmer than IRIS 1980 temperatures near 2 mbar in the equatorial region (Figure 10.9), implying that Voyager and Cassini are capturing the equatorial oscillation in slightly different phases, which is inconsistent with the previously identified period of the SSAO. Indeed, this suggests that the oscillation may be quasi-periodic, as on Earth. It is possible that seasonal storm activity, such as the equatorial eruption in 1990 and the northern mid-latitude eruption of 2010 (see Chapter 13), may have disrupted this equatorial oscillation.

Evidence for non-seasonal variability was also observed in the troposphere (Li et al. 2013; Fletcher et al. 2016), potentially as a result of the wave structure impinging on the stably-stratified upper troposphere. Li et al. (2013) suggested that CIRS results were 2–4 K warmer than IRIS results at ±15° latitude near 300 mbar, and that the tropopause (50–100 mbar) is 8–10 K warmer in 2009–2010 compared to 1980. Although low-pressure ($p < 100$ mbar) changes were not confirmed by Sinclair et al. (2014) and Fletcher et al. (2016), they do confirm the changing latitudinal thermal gradients near the equator (and hence the vertical shear on zonal winds), which were rather different between the Voyager and Cassini epochs. The distribution of para-H_2 also suggested substantial differences in the tropical region between Voyager and Cassini (see Section 10.3.2 and Fletcher et al. 2016). Taking the tropospheric and stratospheric results together, it appears that Saturn's equatorial oscillation may not be strictly "semi-annual" at all, and possibly changes from year to year in response to dynamic phenomena such as storm eruptions.

10.3 Distribution of Chemical Species

10.3.1 Overview of Saturn's Atmospheric Composition

The composition and 3D distribution of gas-phase constituents in Saturn's atmosphere is controlled by the coupled influence of thermochemical equilibrium, photochemistry and other disequilibrium chemical mechanisms, global circulation and regional atmospheric dynamics, and aerosol microphysical processes. In the absence of in situ sounding of Saturn's atmosphere (see Chapter 14), we rely on remote sensing measurements (Figure 10.10) to determine the spatial distribution of these species, and on theoretical chemical models to explain *why* particular species are observed or not observed. Saturn's hydrogen-helium atmosphere contains a wealth of reduced trace species deriving from the dominant chemical elements (carbon, nitrogen, sulfur, phosphorus and oxygen), and thermochemical equilibrium in the deep hot atmosphere steers the bulk atmospheric abundances toward the dominant H_2, He, H_2O, CH_4, NH_3 and H_2S atmospheric composition at the low temperatures and pressures found in the observable upper atmosphere (e.g. Fegley and Prinn 1985). Metals, silicates and other refractory species condense out deep in the atmosphere and remain unobservable. Nitrogen, sulfur and oxygen compounds are largely hidden from the reach of remote sensing due to condensation into cloud decks (Section 10.4), with the exception of NH_3, which is volatile enough that its presence can be detected in the upper troposphere within its condensation region. Oxygen compounds can be detected in the upper atmosphere when delivered from sources external to Saturn. Temperatures at Saturn's tropopause remain too warm for methane condensation, so that this principal carbon-bearing molecule remains abundant throughout the observable upper troposphere and stratosphere.

Chemical equilibrium is not always maintained, however, as the gases at depth are advected upward into cooler regions where chemical-kinetic reactions can become more sluggish. Species mixing ratios "quench" when vertical transport time scales fall below the chemical kinetic time scales for conversion between different major forms of an element (e.g. Prinn and Barshay 1977; Lewis and Fegley 1984). Such quenching is responsible for the presence of CO, PH_3, GeH_4, and AsH_3 in Saturn's upper troposphere, whereas equilibrium arguments suggest that phosphorus, germanium and arsenic should be sequestered in condensates at depth (e.g. Fegley and Lodders 1994).

When a species is able to avoid cold-trapping by condensation, it can be transported to sufficiently high altitudes to interact with solar UV radiation, and can serve as a parent molecule for chains of chemical reactions that produce additional disequilibrium species in the upper troposphere and stratosphere (e.g. N_2H_4, P_2H_4 and complex hydrocarbons). Saturn's axial tilt and seasonally variable solar insolation as a function of latitude result in temporal and meridional variations in the photochemical production and loss rate of atmospheric species. This variable photochemistry, combined with atmospheric dynamics, controls the distribution of the atmospheric constituents in the stratosphere and upper troposphere, as shown in Figure 10.11. In this section, we review our current knowledge of Saturn's seasonally variable composition, building upon the extensive reviews by Prinn et al. (1984) and Fouchet et al. (2009), and showing how the species distributions are intricately connected with Saturn's radiative budget (Section 10.2) and aerosols

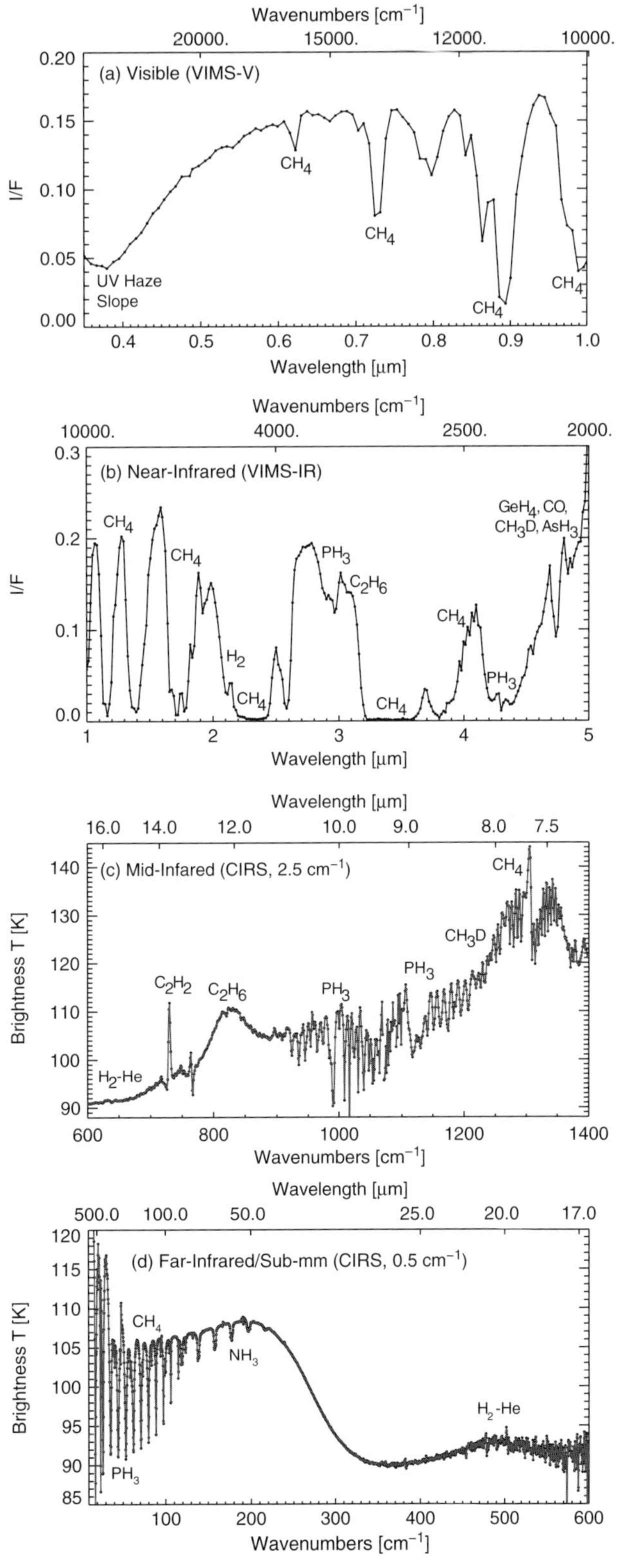

Figure 10.10 Examples of remote sensing data from the UV to the sub-millimeter as measured by Cassini/VIMS (top two panels) and Cassini/CIRS (bottom two panels), demonstrating the rich variety of absorption and emission features. Note that the VIMS panels are provided in wavelength and reflectivity units, whereas the CIRS panels are provided in wavenumber and brightness temperature units, to maximize the visibility of features. All spectra were averaged over Saturn's equatorial region, and can be expected to vary from location to location on the planet. Not shown are Cassini/UVIS observations at the shortest wavelengths (55–190 nm) and Cassini/RADAR observations at the longest wavelengths (2.2 cm).

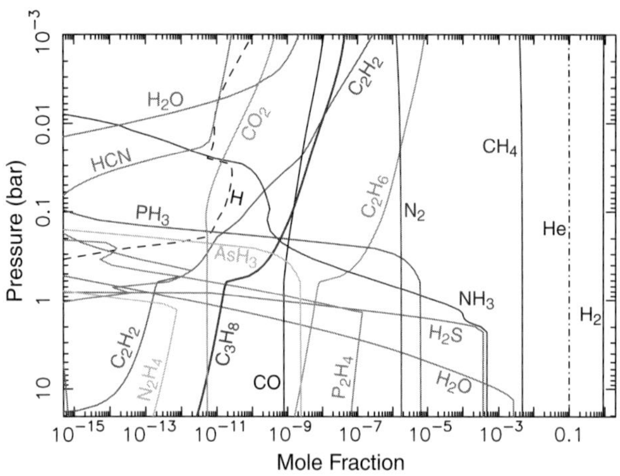

Figure 10.11 Predicted vertical distributions of several tropospheric species of interest on Saturn for 27°S planetographic latitude under summer conditions, based on the photochemical model described in Moses et al. (2010). The sharply decreasing abundance of H_2O, H_2S, NH_3, P_2H_4 and N_2H_4 with increasing altitude in the 0.1–10 bar region is due to condensation, whereas the sharp decrease in the PH_3 abundance near 0.2 bar is due to photochemistry. Molecules whose mixing ratios increase with increasing altitude have a source at higher altitudes, either from stratospheric hydrocarbon photochemistry (e.g. C_2H_6, C_2H_2, C_3H_8, HCN) or from influx of external material combined with photochemistry (e.g. H_2O, CO, CO_2). (A black-and-white version of this figure appears in some formats. For the color version, please refer to the plate section.)

(Section 10.4). We briefly describe the dominant processes and discuss the advances in our understanding of atmospheric chemistry that have occurred since the review by Fouchet et al. (2009). We confine our discussion to Saturn's troposphere and stratosphere; the thermosphere and ionosphere are covered in Chapter 9.

10.3.2 Tropospheric Composition

Condensible Volatiles

The vertical structure of Saturn's cloud-forming region is determined by thermochemical equilibrium, condensation chemistry and vertical mixing of quenched species. Thermochemical equilibrium models for Saturn have been developed by Fegley and Prinn (1985), Fegley and Lodders (1994), Lodders and Fegley (2002), Visscher and Fegley (2005) and Visscher et al. (2006). Water, the dominant equilibrium form of oxygen on Saturn, will condense in the upper troposphere to form a liquid aqueous solution cloud at depth (around 20 bar, although local temperatures, meteorology and the bulk abundances will affect these cloud condensation levels), trending to water ice at higher altitudes (e.g. Weidenschilling and Lewis, 1973). The pressure at the cloud base will depend on the unknown bulk oxygen abundance on Saturn, as well as on the details of regional atmospheric dynamics (e.g. Sugiyama et al. 2011, 2014; Palotai et al. 2014). Because some of the planet's oxygen is tied up in condensed silicates at even deeper levels, the water abundance at the base of the aqueous solution cloud represents an already depleted fraction of the bulk oxygen abundance. Visscher and Fegley (2005) determine that roughly 20% of Saturn's bulk oxygen will be tied up in these condensates if the oxygen-to-silicate-and-metal fraction in the atmosphere is in solar proportions. Indeed, signatures of Saturn's tropospheric water have only been detected near 5 μm by ISO (probing the 2- to 4-bar level above the cloud, de Graauw et al. 1997), and the mixing ratios were highly sub-solar. Cassini/VIMS spectra of the same region were unable to provide sufficient sensitivity to measure the water abundance (Fletcher et al. 2011).

Ammonia and hydrogen sulfide are the dominant equilibrium forms of nitrogen and sulfur, respectively, in Saturn's atmosphere. Some NH_3 will be dissolved in the upper-tropospheric water solution cloud. Above that level, gas-phase NH_3 and H_2S are expected to react to form a crystalline NH_4SH cloud, followed at even higher altitudes by a cloud of NH_3 ice. Briggs and Sackett (1989) used 1.3–70 cm radio observations from the VLA and Arecibo in 1980 to measure a depletion in NH_3 from 25 bar to 2 bar, suggesting that the formation of the NH_4SH cloud was the main NH_3 sink, which indirectly required H_2S to be ten times solar composition (approximately 400 ppm, following van der Tak et al. 1999, but depending on the solar sulfur abundances chosen). However, uncertainties in the spectral line shapes and data calibration in the radio region, combined with the relatively featureless spectra of the various constituents, make this result highly uncertain, and H_2S has never been directly detected. The deep formation of the H_2O and NH_4SH clouds means that Saturn's bulk oxygen and sulfur abundances therefore remain unknown, and that H_2O at least is unlikely to contribute significantly to seasonally-variable photochemistry in the upper troposphere.

Ammonia, on the other hand, is volatile enough to maintain a presence at the uppermost tropospheric altitudes (see Figure 10.11). Signatures of NH_3 were first tentatively identified at visible wavelengths by Giver and Spinrad (1966) and positively identified by Encrenaz et al. (1974), and has since been identified in the near-infrared (Larson 1980; Fink et al. 1983), far-infrared (Courtin et al. 1984) and microwave wavelengths (the 1.2-cm inversion band, de Pater and Massie 1985). Measurements at infrared wavelengths have provided globally averaged abundances at a range of altitudes: microwave and radio observations indicate mole fractions exceeding 500 ppm at $p > 3$ bar (de Pater and Massie 1985; Briggs and Sackett 1989; van der Tak et al. 1999), declining to abundances of around 100 ppm near the condensation altitude (Courtin et al. 1984; Grossman et al. 1989; Briggs and Sackett 1989; Noll and Larson 1990; de Graauw et al. 1997; Orton et al. 2000; Burgdorf et al. 2004), and becoming strongly sub-saturated above the clouds (approximately 0.1 ppm at 500 mbar, Kerola et al. 1997; Kim et al. 2006; Fletcher et al. 2009). See Orton et al. (2009) for a historical overview. These globally averaged values suggest that ammonia decreases with altitude above the clouds due to condensation (relative humidities of approximately 50%, de Graauw et al. 1997) and photochemical processes. Both of these processes depend on temperature and seasonal insolation cycles, so we might expect ammonia to be spatially and temporally variable.

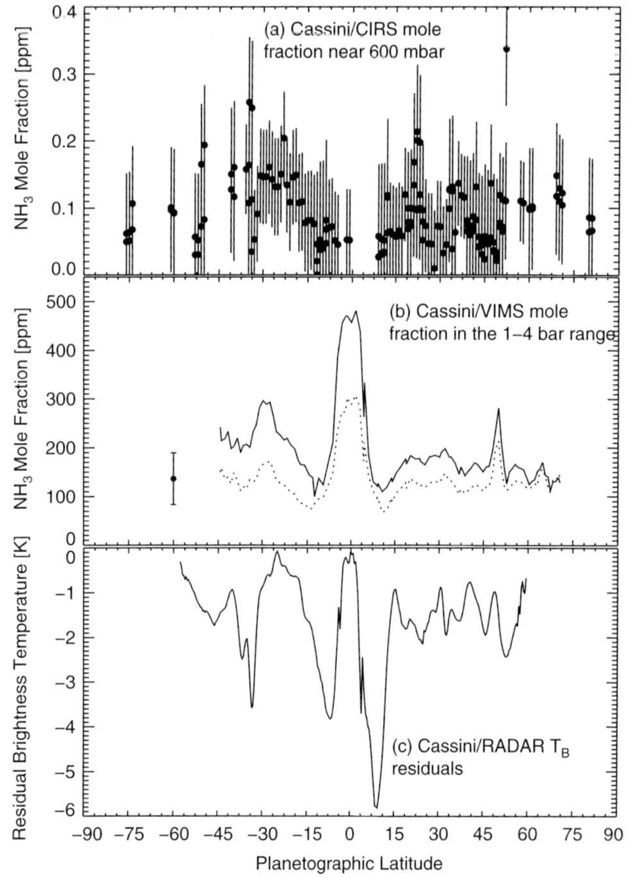

Figure 10.12 Zonal mean ammonia distributions measured (a) in the upper troposphere by Cassini/CIRS (Hurley et al. 2012) and in the cloud-forming region by (b) Cassini/VIMS (Fletcher et al. 2011) and (c) Cassini/RADAR (Laraia et al. 2013). The solid line in panel (b) represents a VIMS retrieval in the absence of aerosol scattering, the dotted line shows the effects of scattering. The isolated point with error bars at 60°S shows the typical uncertainty on the VIMS inversion. In panel (c), large negative values of the 2.2-cm brightness temperature residuals imply a low NH_3 humidity, and these regions are aligned with regions of depleted ammonia vapor in the VIMS results. The region between ±4° is from a single RADAR scan, whereas the rest is an average over five RADAR maps.

Spatially resolved distributions of NH_3 have been provided by the CIRS (Hurley et al. 2012), VIMS (Fletcher et al. 2011) and RADAR instruments (Janssen et al. 2013; Laraia et al. 2013) on the Cassini spacecraft (Figure 10.12). Hurley et al. (2012) modelled CIRS far-IR rotational NH_3 lines to determine the 600-mbar NH_3 distribution (i.e. above the clouds), finding that NH_3 was largely uncorrelated with the belt/zone structure of the thermal field, and instead showed a weak hemispheric asymmetry (with higher abundances in the southern summer hemisphere) and no evidence for peak equatorial abundances (Figure 10.12a). They concluded that condensation and photolytic processes were shaping the distribution, and that NH_3 does not trace the local circulation in this region. Fletcher et al. (2011) used VIMS observations of the 5-μm region to identify a narrow equatorial peak (within 5° of the equator) in NH_3 at $p > 1$ bar in the cloud-forming region, with peak values of 500 ppm similar to the deep abundances observed in the microwave (de Pater and Massie 1985). Elsewhere the NH_3 abundances fell to 100–200 ppm in the 1- to 4-bar region, consistent with previous studies of the condensation region, and there was no indication of a strong hemispheric asymmetry (Figure 10.12b). 2.2 cm brightness temperature maps measured by RADAR (Janssen et al.

2013; Laraia et al. 2013) were used to study NH_3 humidity near 1.5 bar, finding low brightnesses near the equator (consistent with the narrow NH_3 peak observed by VIMS) that were flanked by regions of high brightness near ±9° latitude (Figure 10.12c), and a subtle asymmetry between the northern and southern hemisphere (Janssen et al. 2013). Furthermore, Janssen et al. (2013) demonstrated a remarkable stability in the latitudinal distribution of NH_3 measured by RADAR over the 2005–2011 time span of the Cassini data, in contrast to the time-variability suggested by VLA data in the 1980s and 1990s by van der Tak et al. (1999). This raises the possibility of long-term variability in Saturn's banded structure similar to Jupiter's "global upheavals." However, during the Cassini epoch, the equatorial maximum in NH_3 observed by both VIMS and RADAR, flanked by local minima near ±9°, may be consistent with strong equatorial upwelling flanked by regions of subsidence. Laraia et al. (2013) deduced an average humidity of 70 ±15% in the cloud region, and showed that their data were consistent with a $3 - 4 \times$ solar enrichment (360–480 ppm) at depth, but depleted for $p < 2$ bar. Figure 10.12c may show a slight hemispheric asymmetry, with higher brightnesses (lower NH_3 abundances) in the northern, winter hemisphere, qualitatively consistent with the asymmetry in the upper troposphere identified by Hurley et al. (2012). The possible mechanisms creating this asymmetry in Saturn's southern summer are discussed later in this section.

Quenching and Disequilibrium

Phosphine. Thermochemical equilibrium arguments predict the absence of several species in Saturn's upper troposphere (phosphine, arsine, germane and carbon monoxide) due to sequestration in the hotter deep interior. For example, equilibrium models suggest that phosphorus would be tied up in PH_3 at great depths on Saturn, but reactions with water should convert the phosphine to P_4O_6 at temperatures below ~900 K, with condensed $NH_4H_2PO_4$ becoming the dominant phase at temperatures below ~400 K (Fegley and Prinn 1985). Nevertheless, PH_3 has been shown to have a significant effect on Saturn's spectrum, providing a clear indication that phosphorus chemical equilibrium is not achieved on Saturn. The presence of PH_3 is most likely due to rapid vertical mixing, transporting PH_3 upwards with a dynamical timescale shorter than the chemical depletion timescale, so that the observed abundances are representative of the "quenched" equilibrium conditions at deeper atmospheric levels (Prinn et al. 1984; Fegley and Lodders 1994). Furthermore, the PH_3 profile will be very sensitive to vertical winds, with downwelling winds suppressing the PH_3 abundance at the ~100- to 400-mbar level and upwelling winds enhancing it, making this species a useful tracer of tropospheric dynamics.

Before the arrival of Cassini, PH_3 studies focused on disk-integrated abundances and the vertical distribution (evidence for PH_3 photolysis). Following its first detection near 10 μm (Gillett and Forrest 1974; Bregman et al. 1975), PH_3 has been studied in the mid-IR from ground- and space-based observatories (Tokunaga et al. 1980; Courtin et al. 1984; de Graauw et al. 1997; Lellouch et al. 2001); in the 5-μm opacity window (Larson 1980; Noll and Larson 1991; de Graauw et al. 1997); the 3-μm reflected component (Larson 1980; Kerola et al. 1997; Kim and Geballe 2005; Kim et al. 2006); the sub-millimeter (Weisstein and Serabyn 1994; Davis et al. 1996; Orton et al. 2000, 2001; Burgdorf et al. 2004; Fletcher et al. 2012a) and the ultraviolet (Winkelstein et al. 1983; Edgington et al. 1997). Vertical distributions were presented by Noll and Larson (1991), de Graauw et al. (1997), Orton et al. (2001), Lellouch et al. (2001), Kim et al. (2006), and Fletcher et al. (2012a) to show that PH_3 declined with altitude above the 500- to 700-mbar level and was not present in the stratosphere, but none of these studies provided assessments of the spatial distribution, which would be required to understand seasonal contrasts and regions of powerful convective uplift.

To date, Cassini is the only facility to have provided the spatial distribution of PH_3 from Cassini/VIMS (5 μm) and Cassini/CIRS (10 μm and sub-millimeter), as shown in Figure 10.13. Although there are quantitative differences between the mole fractions derived from the two instruments, the results agree qualitatively in the upper troposphere. A key finding was the enhancement in upper tropospheric PH_3 in the equatorial region (Fletcher et al. 2007b, 2009, 2011), suggestive of rapid vertical mixing consistent with the cold equatorial temperatures (Section 10.2) and elevated hazes (Section 10.4) observed there. This PH_3-enriched zone is flanked at tropical latitudes (±23° planetographic) by

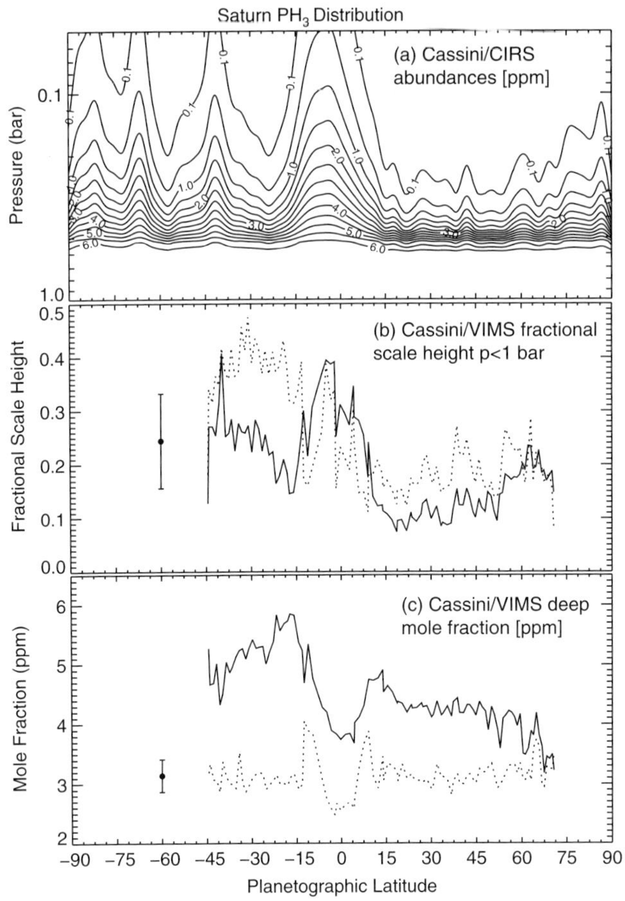

Figure 10.13 Distribution of PH_3 determined by Cassini/CIRS (Fletcher et al. 2009) and Cassini/VIMS (Fletcher et al. 2011) during southern summer, sensing different regions of the troposphere. The CIRS distribution in the upper troposphere (shown as a mole fraction contour plot for $p < 1$ bar) and VIMS-derived fractional scale height ($p < 1$ bar) are consistent with one another and show an equatorial enhancement and a north-south asymmetry in southern summertime (the fractional scale height is the scale height of PH_3 gas divided by the bulk gas scale height). The deep mole fraction derived from VIMS observations shows opposite trends, suggestive of a different circulation regime beneath the topmost clouds. Solid lines show VIMS nightside calculations without aerosol scattering, dotted lines show VIMS calculations with scattering of thermal photons. The two points with error bars at 60°S show the typical uncertainties on the VIMS inversions.

regions of depletion, consistent with a Hadley-style circulation. However, the equatorial enhancement is only identifiable in the upper troposphere, and not in the deeper adiabatic region (i.e. below the radiative-convective boundary), where the PH_3 distribution suggests depleted abundances at the equator and enhanced abundances in the neighboring belts (Figure 10.13c and Fletcher et al., 2011). CIRS does not have the altitudinal sensitivity to probe pressures greater than 1 bar, so this 'deep' distribution of PH_3 comes solely from inversions of VIMS spectra in the 5-μm window, where measurements of composition are rather degenerate with assumptions about tropospheric aerosol properties. As a result, this difference in the zonal mean PH_3 distribution between the upper and lower troposphere has yet to be explained, but could be related to a transition in the tropospheric circulation above and below the main cloud decks. In addition, PH_3 was found to be depleted within the cyclonic polar vortices identified at both the summer and winter poles (Fletcher et al. 2008). Finally, both CIRS and VIMS observations suggest a hemispheric asymmetry in the PH_3 abundance, being more abundant in the southern summer hemisphere. Whether this asymmetry is permanent or will reverse with Saturn's seasons remains to be seen (Fletcher et al., 2010) suggested that the zonal mean PH_3 distribution measured by CIRS had not changed during the 2004–2009 period. The possible seasonal nature of this contrast is discussed below.

Other Disequilibrium Species. Ground-based observations of Saturn's 5-μm opacity window in 1986–1989 provided the first detections of germane (GeH_4, Noll et al., 1988), arsine (AsH_3, Noll et al. 1989; Noll and Larson 1990; Bézard et al. 1989) and carbon monoxide (CO, Noll et al. 1986). As with PH_3, these species were not expected in the upper troposphere on the grounds of thermochemical equilibrium. Gaseous GeH_4 and AsH_3 are expected to be the dominant phases of germanium and arsenic at depth ($T > 400$ K), to be replaced by condensed Ge and As phases in the upper troposphere (e.g. Fegley and Lodders 1994), but both species are detected in the upper troposphere, and apparently decrease in abundance with altitude (de Graauw et al. 1997; Bézard et al. 1989). Cassini/VIMS data are theoretically sensitive to both species, but only AsH_3 could be mapped spatially given the low spectral resolution (Fletcher et al. 2011). As with the deep distribution of PH_3, the zonal mean AsH_3 distribution showed an equatorial minimum flanked by local maxima at ±7° and a possible north-south asymmetry (the size of which depended on the assumed scattering properties of Saturn's aerosols in the 5-μm window). The global mean abundances of AsH_3 (2.2 ±0.3 ppb with fully scattering aerosols, Fletcher et al. 2011) were consistent with ground-based determinations of 3.0 ± 1.0 ppb (Noll and Larson 1990) and $2.4^{+1.4}_{-1.2}$ ppb (Bézard et al. 1989).

Saturn's CO can originate from two sources: an external source of oxygen-bearing molecules in the upper stratosphere (detected in the sub-millimeter by Cavalié et al., 2009, 2010), and an internal source as a quenched disequilibrium species (measured at ppb levels in the upper troposphere by Noll et al. 1986; Noll and Larson 1991; Cavalié et al. 2009). These measurements are complex, as the internal and external CO distributions are difficult to separate spectrally, meaning that measurements of the intrinsic CO abundance remain highly uncertain. For the intrinsic, deep-tropospheric source of CO, recent ab initio transition-state theory rate-coefficient calculations and numerical modeling by Moses et al. (2011) and Visscher and Moses (2011) have advanced our understanding of the CH_4-CO quench process on hydrogen-rich planets. These authors find that thermal decomposition of methanol to form CH_3 + OH is the rate-limiting step in the CO → CH_4 conversion process under solar-system giant-planet conditions. The resulting thermochemical kinetics and transport models can reliably predict the resulting quenched CO abundance in the upper tropospheres of the giant planets, and possibly provide indirect constraints on the deep-water abundance and transport rates if the kinetic conversion between CH_4 and CO at high temperatures and pressures were well characterized (e.g. Prinn and Barshay 1977; Lewis and Fegley 1984; Yung et al. 1988; Fegley and Lodders 1994; Lodders and Fegley 2002; Bézard et al. 2002; Visscher and Fegley 2005; Visscher et al. 2010). The external source of CO (cometary impacts, continuous interaction with the rings and satellites) are discussed in Section 10.3.3.

Theoretical models suggest that N_2 is the most abundant quenched thermochemical species on Saturn and the other giant planets, with a mole fraction of ~1 ppm (Moses et al. 2010; see also Prinn and Fegley 1981; Lewis and Fegley 1984; Fegley and Prinn 1985; Lodders and Fegley 2002), with the rate-limiting step in the N_2 → NH_3 conversion being the reaction of H with N_2H_2 (Moses et al. 2010, 2011). Because it is a homonuclear molecule, N_2 is difficult to observe, but there are potential photochemical consequences of a large N_2 abundance in the high-altitude stratosphere and ionosphere that may be observable, particularly at UV wavelengths. Other quenched molecules such as CO_2, HCN, CH_3NH_2, H_2CO, CH_3OH, CH_2NH, and HNCO are predicted to have very small mixing ratios (Visscher et al. 2010; Moses et al. 2010), and so far only upper limits on HCN, H_2CO and CH_3OH have been reported from sub-millimeter spectroscopy (Fletcher et al. 2012a).

Tropospheric Photochemistry

Asymmetries in the zonal mean distributions of upper tropospheric PH_3 and NH_3 have been identified in Cassini/CIRS, VIMS and RADAR observations, likely related to seasonal contrasts in the efficiency of photolysis and condensation (in the case of NH_3). Because NH_3 and PH_3 are the most abundant photochemically active species in the region of Saturn's troposphere above the cloud decks, tropospheric photochemistry revolves around the coupled chemistry of these species (see the reviews of Strobel 1983, 2005; Atreya et al. 1984; West et al. 1986; Fouchet et al. 2009). Although CH_4 is more abundant than either NH_3 or PH_3, methane is only photolyzed at UV wavelengths shorter than ~145 nm, and these short-wavelength photons are absorbed by other atmospheric constituents well above the tropopause, shielding methane from photolysis in the troposphere. Moreover, NH_3 and PH_3 photolysis products do not readily react with CH_4, so methane photochemistry is more prominent in the stratosphere than the troposphere. Photolysis and photochemical destruction of H_2S may occur within and below the NH_4SH cloud deck, but little is known about the ultimate fate of the sulfur photochemical products (see Prinn and Owen 1976; Lewis and Prinn 1984, for details). The same can be said for AsH_3 and GeH_4 photochemistry (e.g. Fegley and Prinn 1985; Nava et al. 1993).

Owing to a lack of relevant rate-coefficient information for PH_3 photochemistry, in particular, our understanding of tropospheric photochemistry on Saturn (and Jupiter, which is similar) has remained fairly stagnant since the pioneering studies in the late 1970s and

1980s (Strobel 1975, 1977; Atreya et al. 1977, 1980; Kaye and Strobel 1983b,a,c, 1984). Advances since that time, which mostly revolve around the relevant kinetics and product quantum yields from PH_3 and coupled PH_3-NH_3 and NH_3-C_2H_2 photochemistry, are reviewed by Fouchet et al. (2009).

Ammonia is photolyzed by photons with wavelengths less than ~230 nm; the dominant photolysis products are NH_2 and H. The NH_2 reacts effectively with H_2 to reform NH_3, releasing another hydrogen atom in the process. Although NH_3 condenses in the upper troposphere, its vapor pressure at the tropopause is large enough that NH_3 photolysis is an important source of atomic hydrogen in the upper troposphere and lower stratosphere in the models (see Figure 10.11), affecting the photochemistry of both local PH_3 and the complex hydrocarbons that are flowing down from their production regions at higher altitudes. The NH_2 and H can also react with PH_3 to form PH_2. The reaction of NH_2 with PH_3 helps recycle the ammonia. Two PH_2 radicals can re-combine to form P_2H_4, which condenses in its production region, preventing the effective recycling of PH_3. Phosphine itself does not condense under tropospheric conditions on Saturn, but direct photolysis (to produce largely PH_2 + H) and reaction of PH_3 with NH_2 and H serve to drastically decrease the PH_3 abundance above the ammonia clouds (see Figure 10.11).

Figure 10.14 shows the details of this coupled NH_3-PH_3 photochemistry, as described by Visscher et al. (2009). The dominant product is condensed P_2H_4, which becomes a major component of the upper-tropospheric haze. Photochemistry of P_2H_4 can lead to the production of condensed elemental phosphorus as an additional major photochemical product, but the dominant pathways are more speculative (Visscher et al. 2009). Other important but less abundant photoproducts include condensed NH_2PH_2, condensed N_2H_4, and gas-phase N_2 (e.g. Kaye and Strobel 1984). The N_2 shown in Figure 10.11 is derived from thermochemical quenching and transport from the deep troposphere, with photochemical production contributing only a tiny fraction in comparison with this deep source.

NH_3 photochemistry plays only a minor role in shaping its vertical profile at and below the cloud decks. Photochemistry is responsible for removing NH_3 above the clouds (with N_2H_4 being the dominant product), but dynamics and condensation control the profile within the cloud region. Hemispheric contrasts are more strongly influenced by the thermal profile, condensation, regional dynamics (e.g. equatorial upwelling) and aerosol-microphysical processes. These competing effects have not yet been disentangled to explain the suggested weak enhancement of ammonia in Saturn's southern summer hemisphere in Section 10.3.2. Furthermore, it is not understood whether the NH_3 distribution is responding to the seasonal insolation shifts (i.e. following the thermal and aerosol changes), or remaining static with time. We might expect that northern hemisphere warming will permit the release of more NH_3 into the vapor phase during northern spring, helping to reverse the asymmetry observed during southern summer.

The PH_3 vertical profile, on the other hand, is strongly sensitive to chemistry, as well as to vertical transport and to the opacity of upper-tropospheric hazes and clouds, which help shield the PH_3 from photolysis and other photochemical losses. The model PH_3 vertical profiles drop much more sharply with altitude in the upper troposphere than the profiles derived from CIRS and VIMS retrievals (e.g. Fletcher et al. 2007b, 2009, 2011), but these differences are likely an artifact of the assumptions in the data retrievals rather than a true model-data mismatch. Indeed, Herschel/SPIRE observations (Fletcher et al. 2012a) suggested that PH_3 is not present in the lower stratosphere at the 10–20 mbar level because it is so chemically fragile. The observed hemispheric asymmetry in the upper-tropospheric PH_3 abundance (e.g. Fletcher et al. 2009, 2011) may be related to higher haze opacities of the UV-shielding aerosols produced by photochemistry in the summer hemisphere, although the global circulation system (i.e. an inter-hemispheric transport from the autumn to the spring hemisphere, see Section 10.2.5) may also play a role. Seasonal tropospheric photochemistry has not yet been investigated theoretically. Summertime insolation will promote the production of aerosols like P_2H_4, which will help shield the PH_3 and NH_3 from photolysis, allowing the PH_3 molecules to be carried to higher altitudes during southern summer and autumn, qualitatively consistent with Cassini's observations.

Unsaturated hydrocarbons like C_2H_2 are not particularly abundant in the region in which PH_3 and NH_3

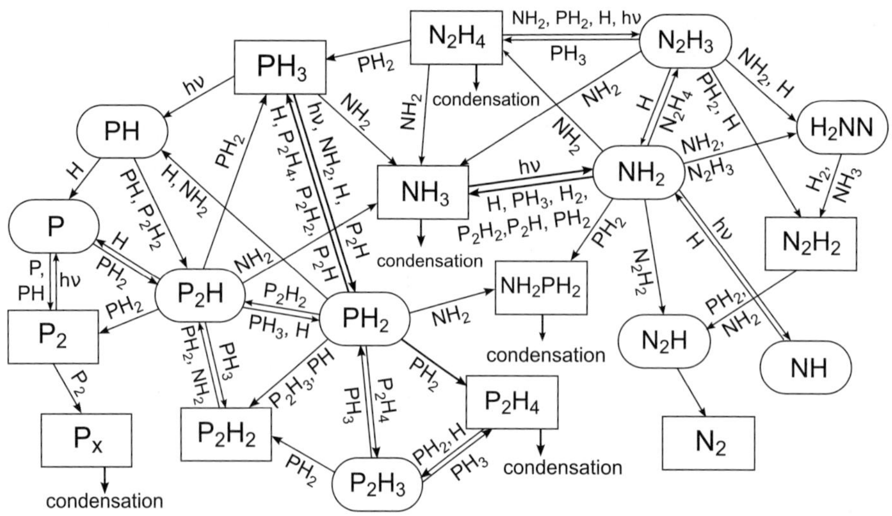

Figure 10.14 Schematic diagram illustrating the important reaction pathways for coupled PH_3 and NH_3 in Saturn's troposphere (based on Visscher et al. 2009; Kaye and Strobel 1984). The rectangular boxes represent stable molecules, rounded boxes represent radicals or less stable molecules, and $h\nu$ represents ultraviolet photolysis. The reaction network begins with the photolysis of NH_3 and PH_3 and the destruction of PH_3 from reaction with ammonia photolysis products. The dominant end products are condensed P_2H_4, P_x (elemental phosphorus), N_2H_4, NH_2PH_2 and gas-phase N_2. Ammonia is also lost via condensation.

photochemistry is active (see Figure 10.11), limiting the effectiveness of the coupled photochemistry of NH_3 and PH_3 with C_2H_x and C_3H_x species (e.g. Moses et al. 2010), which would otherwise be expected to produce HCN, acetonitrile, methylamine, ethylamine, acetaldazine, acetaldehyde hydrazone, vinylphosphine, ethylphosphine and a whole suite of other interesting organo-nitrogen and organo-phosphorus species (Kaye and Strobel 1983b,a; Ferris and Ishikawa 1988; Guillemin et al. 1995; Keane et al. 1996; Moses et al. 2010). The upper-tropospheric C_2H_2 abundance is not notably enhanced in latitude regions known to host thunderstorm activity (Hurley et al. 2012), suggesting that lightning chemistry does not play much of a role in enhancing upper-tropospheric hydrocarbons beyond their photochemically produced abundances. Based on the chemical mechanism of Moses et al. (2010), hydrogen cyanide would be the dominant product of the coupled hydrocarbon-NH_3 photochemistry on Saturn, but these models suggest that photochemically produced HCN is not abundant enough to be observable on Saturn (see Figure 10.11), and despite a tentative detection of HCN by Weisstein and Serabyn (1996), Herschel analyses have provided upper limits an order of magnitude smaller (mole fractions less than 1.6×10^{-11} if the species is well-mixed, Fletcher et al. 2012a). Small amounts of all of the aforementioned organic compounds are produced, however, and will condense, which could potentially contribute to the cloud chromophores on Saturn (see also Carlson et al. 2012). Photoprocessing of the condensed P_2H_4, elemental phosphorus, or NH_4SH is also a potential source of the yellowish colors on Saturn. The fact that the yellow-brown coloring is apparently absent in the Cassini images of the high-latitude winter hemisphere of Saturn soon after it emerges back into sunlight – thought to result from a clearing, thinning or reduction in size or depth of the aerosols in the winter hemisphere (e.g. Fletcher et al. 2007a; West et al. 2009) – further points to a photochemical product as the source of the chromophore (see Section 10.4).

Para-hydrogen

Our final tropospheric species shapes the underlying H_2-H_2 and H_2-He collision-induced continuum throughout the infrared and sub-millimeter, in addition to readily identifiable quadrupole line features near 0.62–0.64 μm and collision-induced dipole features near 0.826 μm and 2.1–2.2 μm. Despite the lack of a permanent dipole, collisions between these molecules create instantaneous dipoles that shape the continuum, and the ratio of ortho-

H_2 (corresponding to the odd rotational spin state of H_2 with parallel spins) and para-H_2 (the even spin state of H_2 with anti-parallel spins) can be deduced from the relative sizes of the broad S(0) (due to para-H_2) and S (1) (due to ortho-H_2) absorptions near 354 and 587 cm^{-1}, respectively. In "normal" hydrogen, at the high temperatures of Saturn's lower troposphere, this ratio should be 3:1 in equilibrium (a para-H_2 fraction f_p of 25%) (e.g. Massie and Hunten 1982). At the colder temperatures of the upper troposphere the equilibrium para-H_2 fraction (f_{eqm}) increases beyond 25%, as para-H_2 has the lowest rotational energy state, and we might expect the tropopause para-H_2 fraction to be in the 45–50% range, falling as we rise further into the warm stratosphere. However, parcels of air displaced upwards will retain their initial low para-H_2 fraction if the vertical mixing is faster than the chemical equilibration timescale, leading to sub-equilibrium conditions ($f_p < f_{eqm}$) (Conrath et al. 1998; Fletcher et al. 2007a). Conversely, downward displacement of cool air with high values of f_p can lead to super-equilibrium conditions ($f_p > f_{eqm}$).

The spatial distribution of the para-H_2 fraction, and its deviation from equilibrium, can therefore be used to trace both vertical mixing and the efficiency of chemical equilibration, both of which may vary with Saturn's seasons. On Jupiter, where seasonal influences are expected to be negligible, inversions of Voyager/IRIS spectra by Conrath et al. (1998) revealed a distribution of para-H_2 that was largely symmetric about the equator, with evidence for sub-equilibrium conditions at the equator and super-equilibrium at higher latitudes. Voyager observations of Saturn in early northern spring ($L_s = 8.6 - 18.2°$) revealed local sub-equilibrium conditions near 60°S, and an asymmetry with a higher f_p in the spring hemisphere (super-equilibrium conditions from 70°N to 30°S, Conrath et al. 1998). These data, most sensitive to the para-H_2 fraction between 100–400 mbar (i.e. below the tropopause but above the radiative convective boundary), were interpreted as atmospheric subsidence in the spring hemisphere.

Conversely, early measurements by Cassini during southern summer ($L_s = 297 - 316°$) revealed super-equilibrium over much of the southern summer hemisphere and sub-equilibrium in the winter hemisphere poleward of 30°N (Fletcher et al. 2007a). Much of this asymmetry was due to the gradient in f_{eqm} related to the seasonal temperature contrast, whereas f_p itself was found to be largely symmetric about the equator (e.g. left panels of Figure 10.15).

Fletcher et al. (2010, 2016) demonstrated that the super-equilibrium region was moving slowly northwards throughout the ten-year span of Cassini observations, both due to an increased f_p and warmer temperatures in the northern spring hemisphere (i.e. smaller f_{eqm}). The para-H_2 fraction in the southern hemisphere changed very little between 2004 and 2014, although the cooling temperatures as southern winter approached caused an increase in the expected f_{eqm}, thereby decreasing the discrepancy between f_p and equilibrium expectations. In summary, the stark asymmetry in $f_p - f_{eqm}$ observed during southern summer has reduced significantly (see Figure 10.15), such that the southern hemisphere is now close to equilibrium, whereas the wintertime sub-equilibrium conditions still prevail at the highest northern latitudes. As a result, the zonal mean f_p measured by Cassini in 2014 now qualitatively resembles that determined by Voyager in the last northern spring (Fletcher et al. 2016), although quantitative differences remain to be understood, particularly at Saturn's equator.

Given that the timescales for para-H_2 equilibration range from decades to centuries in the troposphere (e.g. Conrath et al. 1998; Fouchet et al. 2003), even in the presence of aerosols whose surfaces could provide paramagnetic sites to catalyze the efficient conversion of uplifted para-H_2 back to ortho-H_2, the discrepancy from a seasonally dependent equilibrium ($f_p - f_{eqm}$) may not be a good measure of tropospheric circulation. Put another way, the theoretical f_{eqm} varies instantaneously with temperatures on seasonal timescales, whereas f_p undergoes much more subtle changes with time (showing the largest changes in the northern spring hemisphere). Furthermore, Fletcher et al. (2007a) speculated that the intricate connection between aerosol catalysts and the rate of para-H_2 conversion means that equilibrium conditions would be more easily attained when the atmosphere is hazy (e.g. summertime) than when it is aerosol-free (wintertime), and that this was responsible for the asymmetry observed in $f_p - f_{eqm}$ during southern summer. In either case, it is the tropospheric temperature and/or aerosol availability that determines the magnitude of the disequilibrium, rather than large-scale overturning.

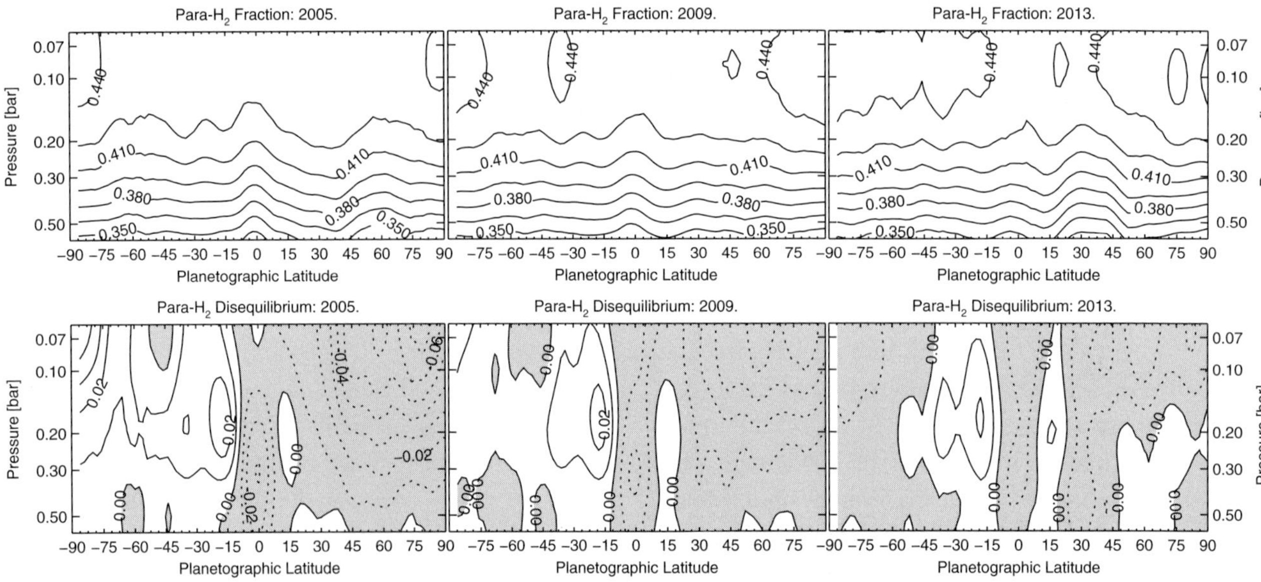

Figure 10.15 Seasonal changes in the zonal mean para-H_2 fraction measured by Cassini/CIRS using the collision-induced H_2-H_2 continuum (Fletcher et al. 2007a, 2010, 2016). The upper panels show the subtle changes in f_p between 2005 and 2013, including the rising f_p at northern high latitudes during spring. The lower panels compare the degree of disequilibrium between 2005 and 2013 (contours have a 1% spacing), showing the reversal of the strong asymmetries found in southern summer. Super-equilibrium conditions ($f_p > f_{eqm}$) are shown as solid lines; sub-equilibrium conditions ($f_p < f_{eqm}$) are shown as dotted lines and light grey shading.

Only in regions of strong localized dynamics is f_p significantly altered from equilibrium conditions (e.g. air rising at the equator, storm eruptions, and air sinking at high northern latitudes during spring). For example, the equatorial minimum in f_p (sub-equilibrium conditions) has persisted throughout the Cassini mission, consistent with the evidence of powerful equatorial upwelling in the temperature, aerosol, ammonia and phosphine fields (Fletcher et al. 2007a, 2010). The connection between the para-H_2 fraction and the tropospheric temperatures, aerosols and circulation remains a topic of active exploration.

10.3.3 Stratospheric Composition

In this section, we review spatially resolved observations and modeling of Saturn's stratospheric hydrocarbons, and their variations with time, to reveal the photochemical and dynamical processes shaping the middle atmosphere. Stratospheric photochemistry on Saturn is dominated by Lyman-alpha photolysis of methane at high altitudes, producing a multitude of hydrocarbons, many of which have been observed. Ethane (C_2H_6) was the first photochemical product of methane to be detected in Saturn's stratosphere from its ν_9 emission band at 12 μm by Gillett and Forrest (1974), confirmed shortly after by Tokunaga et al. (1975). Acetylene (C_2H_2) was detected a few years later by Moos and Clarke (1979) in the UV using data from the International Ultraviolet Explorer (IUE). A column abundance of ethane was obtained using spectra at 3 μm by Bjoraker et al. (1981). The first stratospheric mixing ratios of ethane and acetylene were derived from Voyager/IRIS spectra in the thermal infrared (Courtin et al. 1984; Sada et al. 2005). These measurements confirmed C_2H_6 and C_2H_2 as the dominant photochemical products. Constraints from the Infrared Space Observatory (ISO) measurements (de Graauw et al. 1997) were also used by Moses et al. (2000a) to demonstrate that acetylene increases with altitude, as expected for a chemical produced in the upper stratosphere and transported downward by eddy diffusion (see also Prangé et al. 2006). New hydrocarbon species were identified in the 1990s due to the increasing sensitivity of mid-infrared detectors, leading to the determination of the disk-average integrated column abundance of methlyacetylene (CH_3C_2H), diacetelene (C_4H_2) (de Graauw et al. 1997; Moses et al. 2000a), benzene (C_6H_6) (Bézard et al. 2001a) and the methyl radical (Bézard et al. 1998), all from ISO. Using ground-based high-resolution spectroscopic data from the NASA/Infrared Telescope Facility (IRTF), ethylene (C_2H_4) (Bézard et al. 2001b) and propane (C_3H_8) (Greathouse et al. 2006) were also detected.

Saturn hydrocarbon photochemistry has been recently reviewed by Fouchet et al. (2009), and there have been few significant theoretical advances since the time of that review. A reduced hydrocarbon reaction mechanism suitable for computationally expensive 3D models has been described in Dobrijevic et al. (2011), and recent improvements in photochemical models of Titan and the other giant planets are often directly applicable to Saturn (see Lavvas et al. 2011; Gans et al. 2011; Bell et al. 2011; Vuitton et al. 2012; Westlake et al. 2012; Mandt et al. 2012; Plessis et al. 2012; Moreno et al. 2012; Hébrard et al. 2013; Lara et al. 2014; Dobrijevic et al. 2014; Krasnopolsky 2014; Orton et al. 2014; Loison et al. 2015, and references therein). Our knowledge of relevant reaction rate coefficients and overall reaction pathways and products is also continually improving due to new laboratory investigations or theoretical calculations (with references too numerous to list). A recent theoretical advance is the new 2D seasonal photochemical modeling of Hue et al. (2015), which takes into account the expected seasonal variation in temperatures. Full details of methane photochemistry on Saturn and the other giant planets can be found in Atreya et al. (1984), Strobel (1983, 2005), Gladstone et al. (1996), Yung and DeMore (1999) and Moses et al. (2000a, 2005).

Methane photolysis leads to the production of CH, CH_2 (both the ground-state triplet and the excited singlet) and CH_3. The CH radicals tend to favor unsaturated hydrocarbon production, whereas the CH_3 radicals tend to recycle the methane or lead to the production of saturated hydrocarbons. The peak hydrocarbon production region at 10^{-3}–10^{-4} mbar occurs just below the methane homopause where the Lyman alpha photons are absorbed by CH_4, but there is a secondary peak in the 0.1–10 mbar region due to photosensitized destruction of CH_4 resulting from photolysis of C_2H_2 and other hydrocarbon photochemical products. This secondary production region can particularly affect the relative abundances of the hydrocarbons at the pressures at which the infrared observations are sensitive.

As we shall see, the hydrocarbon abundances vary with both altitude and latitude. One obvious cause of this variation is seasonal change: the mean daily solar insolation, atmospheric temperatures and perhaps circulation patterns on Saturn change with season, inducing a response in the vertical and meridional distributions of chemical species. Although 3D dynamical models that include chemistry have not yet been developed for Saturn or the other giant planets, Moses and Greathouse (2005) have investigated seasonal stratospheric chemistry on Saturn with a 1D time-variable photochemical model, exploring how the molecular abundances change solely due to seasonally varying solar insolation. The results of this model are compared to the measured distributions of hydrocarbons in Figure 10.16 at a variety of different pressure levels, showing the expected seasonal variability based on photochemistry alone. This figure will be referred to throughout this section. The Moses and Greenhouse model includes the effects of ring shadowing and solar-cycle variations, but neglects horizontal transport (e.g. Moses et al. 2007) or time-variable temperatures and vertical winds. The hemispheric asymmetries in the hydrocarbon abundances are more pronounced at higher altitudes where vertical diffusion time scales and chemical lifetimes are short. Our theoretical discussion of the principle production and loss mechanisms for each species will be guided by the results of this 1D model. Note, also, the recent 2D (latitude and altitude) study by Hue et al. (2015) that extends the Moses and Greathouse (2005) study by considering the effects of seasonally varying temperatures.

C_2 Hydrocarbons

Observations and models both demonstrate that C_2H_6 is the most abundant hydrocarbon photochemical product in Saturn's stratosphere. Ethane is produced predominantly by CH_3 recombination throughout the stratosphere and by sequential addition of H to unsaturated C_2H_x hydrocarbons in the lower stratosphere; ethane is lost via photolysis and reaction with C_2H produced from C_2H_2 photolysis. Production of C_2H_6 exceeds loss at most altitudes in the Saturn models, and the net C_2H_6 production is balanced by downward flow

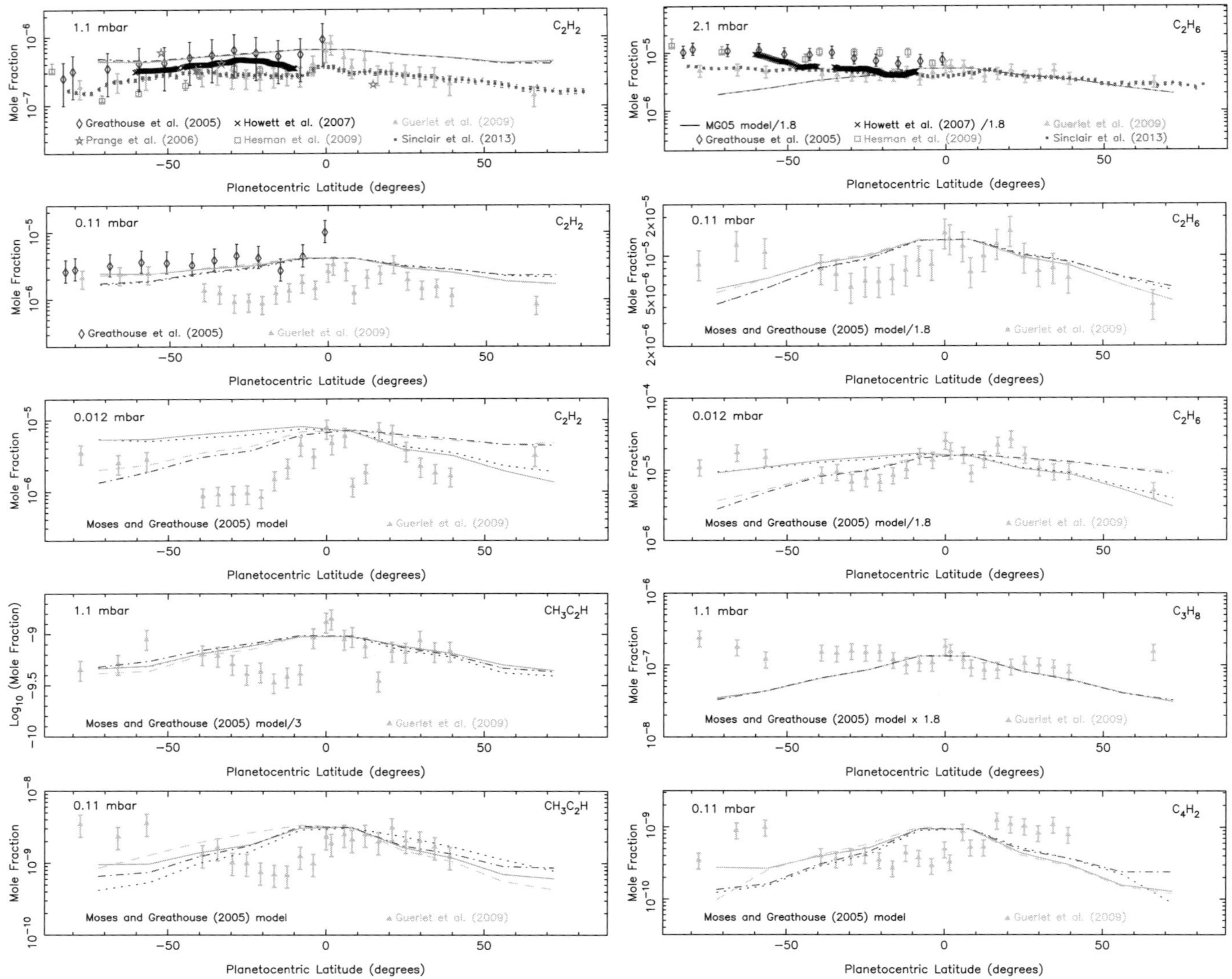

Figure 10.16 Comparing zonal mean hydrocarbon distributions measured for five species (ethane, acetylene, propane, methylacetylene and diacetylene) with the photochemical predictions of Moses and Greathouse (2005). Observers have been abbreviated as follows: Gre05 (Greathouse et al. 2005); Pra06 (Prangé et al. 2006); How07 (Howett et al. 2007); Hes09 (Hesman et al. 2009); Gue09 (Guerlet et al. 2009); and Sin13 (Sinclair et al. 2013). The model output (labelled MG05) is shown for the solstices and equinoxes to show the predicted magnitude of seasonal variability, particularly for $p < 0.1$ mbar: $L_s = 270°$ is dotted (purple), $L_s = 0°$ is solid (green), $L_s = 90°$ is dashed (orange) and $L_s = 180°$ is dot-dashed (red). The colors of the individual measurements are designed to represent the seasonal timing of the observations. The model abundances of C_2H_6, CH_3C_2H and C_3H_8 of Moses and Greathouse (2005) have been approximately scaled to match the data, as described in the figure legends. Note that C_3H_8 exhibits a similar meridional trend to C_2H_6, whereas all the other species tend to track C_2H_2. (A black-and-white version of this figure appears in some formats. For the color version, please refer to the plate section.)

in to the deeper troposphere, where the ethane will eventually be thermochemically converted back to methane. The chemical loss time scale for C_2H_6 exceeds a Saturn year at most altitudes below its peak production region, making C_2H_6 particularly stable and thus a good tracer for local dynamics and global circulation.

Ethylene is produced through the reaction of CH with CH_4, reactions of C_2H_3 with H and H_2, and through C_2H_6 photolysis. It is lost through H-atom addition to form C_2H_5 and through photolysis. Unlike C_2H_6, ethylene can be photolyzed by UV photons with wavelengths out to ~200 nm, which makes the C_2H_4 much less stable in the lower stratosphere than ethane. Its chemical lifetime is shorter than an Earth year at most altitudes, although recycling can help keep it around longer than its loss time scale would indicate. The very low levels of C_2H_4 under quiescent conditions on Saturn are consistent with photochemical models; its greatly enhanced abundance in the northern hemispheric storm beacon region (Hesman et al. 2012, and Chapter 13) is most likely the result of the high dependence of the $C_2H_3 + H_2 \rightarrow C_2H_4 + H$ reaction on temperature (e.g. Tautermann et al. 2006; Armstrong et al. 2014; Moses et al. 2014).

Acetylene is the second-most-abundant hydrocarbon photochemical product on Saturn. Its chemistry is more intricate, complicated, and non-linear, and C_2H_2 is a particularly important parent molecule for other observed hydrocarbons such as CH_3C_2H and C_4H_2. Acetylene is produced primarily from the photolysis of ethane and ethylene, although "recycling" reactions such as $C_2H + H_2 \rightarrow C_2H_2 + H$, $H + C_2H_3 \rightarrow C_2H_2 + H_2$, and $C_2H + CH_4 \rightarrow C_2H_2 + CH_3$ dominate the total column production rate of C_2H_2 in the stratosphere. Acetylene is indeed rapidly recycled throughout the stratosphere, although there is a steady leak out of the ongoing recycling reactions into C_2H_6, C_4H_2 and higher-order hydrocarbons. Acetylene has a lifetime intermediate between that of C_2H_6 and C_2H_4: its pure chemical loss time scale is of the order of a Saturn season through much of the stratosphere (~0.03–100 mbar, see Moses and Greathouse 2005), but recycling reactions give it a much longer "effective" lifetime.

In the absence of atmospheric circulation, photochemically generated species at microbar pressures (e.g. Figure 10.16) tend to have abundances that follow local insolation conditions, resulting in higher abundances in the summer hemisphere compared to the winter hemisphere (Moses and Greathouse 2005). At lower altitudes, however, vertical diffusion time scales and chemical lifetimes are much longer, and the meridional distributions track the yearly average of the mean daily solar insolation rather than the instantaneous solar fluxes. Because the yearly average insolation is largest at the equator and decreases toward the poles, the photochemical models predict that most of the hydrocarbon photochemical products should have a maximum abundance at the equator, decreasing smoothly toward both poles at pressures of 1 mbar and greater. These predictions can be tested by comparison to latitudinally resolved hydrocarbon measurements.

Ethane and Acetylene in Southern Summer. Greathouse et al. (2005) were the first to present latitudinally resolved distributions of the principle methane photolysis products, ethane and acetylene, in Saturn's southern (summer) hemisphere (Figure 10.16a). Using IRTF data acquired in 2002, they showed that acetylene decreased from equator to pole at the 1-mbar pressure level (in agreement with photochemical model predictions), whereas the opposite trend was marginally observed for ethane at 2 mbar. These results suggested that dynamical redistribution is effective in Saturn's stratosphere and operates on timescales less than ethane's chemical lifetime of ~700 years and longer than acetylene's lifetime of ~100 years (Moses and Greathouse 2005).

Since the work of Greathouse et al. (2005), the distributions of ethane and acetylene have been routinely measured by the Composite Infrared Spectrometer (CIRS) onboard Cassini, using a combination of nadir and limb sounding. Nadir and limb observations provide complementary information. On the one hand, CIRS nadir data provide an excellent spatio-temporal coverage of Saturn's atmosphere and are mainly sensitive to the abundance of ethane and acetylene, with a peak sensitivity near the ~2-mbar pressure level. Under particular conditions, such as the high temperatures encountered within the storm beacon region (see Chapter 13), other trace species can also be measured (such as ethylene, which has been measured by Hesman et al.

2012). On the other hand, CIRS data acquired in limb-viewing geometry allow the retrieval of the vertical abundance profile of ethane and acetylene over a greater pressure range (typically 5 mbar – 5 μbar), but with a much sparser meridional and temporal coverage.

Five Cassini studies have reported on the meridional distribution of ethane and acetylene during southern summertime based on Cassini/CIRS measurements (Figure 10.16a): three based on nadir data (Howett et al. 2007; Hesman et al. 2009; Sinclair et al. 2013) and two based on limb observations (Guerlet et al. 2009; Sylvestre et al. 2015). Although these studies do not necessarily agree quantitatively, they report similar trends. Howett et al. (2007) measured the southern hemisphere hydrocarbon distributions in 2004 between 10° and 70°S, showing a decrease of acetylene poleward of 30°S. They also confirmed that the ethane abundance is relatively uniform with latitude equatorward of 50° S, but suggested a rather large increase poleward of 50°S. This increase has not been reproduced by subsequent investigators, and could be ascribed to a bias with emission angle (as the higher latitude data were acquired at higher emission angles) linked to spectroscopic errors. Furthermore, all follow-up studies based on CIRS measurements rely on an updated ethane linelist by Vander Auwera et al. (2007), which has line intensities higher by approximately 30% than the values used by Howett et al. (2007). Hesman et al. (2009) then extended the CIRS retrievals towards the south pole in 2005, and combined them with low-latitude ground-based observations. They confirmed acetylene's poleward decrease but found a sharp rise in C_2H_2 right at the summer pole (87°S), within the stratospheric vortex identified by Orton and Yanamandra-Fisher (2005) and Fletcher et al. (2008). This sharp increase has also been observed for ethane (Fletcher et al. 2015), and suggests enhancement by strong subsidence within the summer polar vortex. Using limb observations acquired in 2005–2006 covering both hemispheres (80° S–45°N), Guerlet et al. (2009) confirmed that ethane and acetylene follow different meridional trends in the 1- to 5-mbar pressure range: while acetylene generally decreases from the equator towards both poles, the ethane distribution is much more homogeneous with latitude. These results are also in agreement with the study of Sinclair et al. (2013) from CIRS nadir observations covering both hemispheres.

Figures 10.16a and 10.16f compare the meridional distribution of ethane and acetylene at the 1- to 2-mbar pressure level during southern summer, as measured by various authors, and as predicted by the photochemical model of Moses and Greathouse (2005). The chemistry of C_2H_2 and C_2H_6 is highly linked, such that the C_2H_2 meridional distribution tends to track the C_2H_6 distribution in photochemical models (e.g. Moses and Greathouse 2005; Moses et al. 2007; Hue et al. 2015). The notable differences in the meridional profiles of C_2H_2 and C_2H_6 on Saturn is therefore a surprise and suggests the influence of meridional transport on long timescales, as already proposed by Greathouse et al. (2005). Guerlet et al. (2009) also report a sharp and narrow equatorial maximum in the C_2H_2 distribution at 1 mbar (and a more moderate local maximum of C_2H_6), much higher than predicted by the seasonal photochemical model. The authors interpret this strong maximum as the signature of a local subsidence associated with the equatorial oscillation. Indeed, given that C_2H_2 mixing ratio increases with height, a downwelling wind would carry C_2H_2 and enrich lower-altitude regions.

At sub-millibar pressure levels, Guerlet et al. (2009) took advantage of limb-viewing geometries from Cassini to derive the distribution of acetylene and ethane in the upper stratosphere ($p < 0.1$ mbar, Figure 10.16b,c,g,h). Contrary to the trends observed in the lower stratosphere, they find that both species follow very similar trends in the upper stratosphere. These distributions of the two hydrocarbons are asymmetric, with volume mixing ratios 2 to 6 times higher at northern mid-latitudes (with a local maximum centered at 25°N) than at southern mid-latitudes. This is opposite to what might be expected from hydrocarbon production rates alone, where production rates should have been higher in the summer southern hemisphere than in the winter hemisphere (see Figure 10.16c,h). Guerlet et al. (2009) hence suggest that a strong meridional transport from the summer to the winter hemisphere, occurring on seasonal timescales, is responsible for the observed enrichment at 25°N. This hypothesis is consistent with the predictions of the global circulation model of Friedson and Moses (2012), in which stratospheric circulation cells develop, with a descending branch at 25°N at this season.

Temporal Trends in Ethane and Acetylene. The southern summer observations by Cassini revealed asymmetries in ethane and acetylene that would be expected to shift over time, potentially reversing by northern summer solstice in 2017 (L_s = 90°). Sinclair et al. (2013) were the first to study temporal variability in the distributions of ethane and acetylene using nadir sounding at low spectral resolution (15 cm^{-1}) between 2005 and 2010 (L_s = 308° – 15°). They confirmed (i) the absence of a sharp south-poleward rise in ethane as first reported by Howett et al. (2007) from early CIRS nadir data ($L_s \approx 295°$); and (ii) the equatorial peak abundance of acetylene observed in 2005 by Guerlet et al. (2009), although this maximum appears more muted in the nadir data (see Figure 10.16a). Differences between limb and nadir results in the equatorial region could be ascribed to the temporal evolution of the equatorial oscillation, which strongly perturbs the thermal (and chemical) structure of the stratosphere on small vertical scales not fully resolved with nadir, low-spectral-resolution data.

Away from the equator, the Cassini/CIRS observations are beginning to reveal seasonal shifts in the zonal-mean ethane and acetylene distributions, as shown in Figure 10.17. At mid-latitudes, Sinclair et al. (2013) report significant enrichments in ethane (and to a lesser extent, in acetylene) at the 2-mbar pressure level in the region 20° – 65°N and depletions at 15°S, which they interpret as extended subsidence occurring in the northern hemisphere between 2005 and 2010, and upwelling at 15°S. Sinclair et al. (2013) also showed that the equatorial maximum of C_2H_2 decreases over time (while the ethane concentration remains fairly constant), possibly due to the evolution of the equatorial oscillation. At higher latitudes, Fletcher et al. (2015) use ten years of CIRS nadir data to demonstrate that the summertime south polar maximum in both C_2H_2 and C_2H_6 has been declining throughout the timespan of the observations (L_s = 293° – 54°) at 1–2 mbar, mirrored by enhancements in both species in the north polar spring (poleward of 75°N). These changes in the millibar region cannot be explained solely in terms of photochemistry, so that the authors invoke stratospheric circulation from south to north (broadly speaking, upwelling in the autumnal hemisphere and subsidence in the spring hemisphere) to explain the observed changes.

In many of these studies (Guerlet et al. 2009; Sinclair et al. 2013; Fletcher et al. 2015), the temporal behavior of C_2H_2 and C_2H_6 were found to be different (ethane variations appeared more significant than those of acetylene), despite the highly linked chemistry of the two species, and suggestive of the influence of dynamics. Acetylene is expected to behave differently from ethane in the presence of vertical winds, for example. In the ~0.1- to 10-mbar region, downwelling winds could carry more C_2H_2 from high altitudes to lower altitudes, but the increased C_2H_2 at these lower altitudes results in a non-linear chemical loss rate due to the increased photolysis (and thus C_2H and H production) in conjunction with the resulting increased rates of the non-recycling loss reactions such as $C_2H + C_2H_2 \rightarrow C_4H_2 + H$ and $H + C_2H_2 + M \rightarrow C_2H_3 + M$. Thus, the subsidence will lead to an increased fraction of the C_2H_2 photolysis products being permanently converted to other hydrocarbons rather than recycling the acetylene, resulting in less of an increase than one would expect based on subsidence alone without chemical coupling. By the same token, upwelling winds will decrease the C_2H_2 abundance and loss rate non-linearly, favoring recycling over permanent loss. Upwelling and downwelling winds can therefore lead to less significant decreases and increases in the C_2H_2 mixing ratios as compared to C_2H_6, which might help explain some of the temporal behavior observed by Sinclair et al. (2013); however, the slope of the unperturbed mixing-ratio profile also contributes to the observed magnitude of the increase or decrease during the upwelling/downwelling (e.g. Guerlet et al. 2009), so the resulting response will be complicated. Coupled 2D and 3D photochemical-dynamical models will likely be needed to fully explain the temporal and meridional variations of stratospheric hydrocarbons in the Cassini CIRS observations. No such models have been published to date.

At higher altitudes, seasonal changes in the hydrocarbon distributions have also been studied by Sylvestre et al. (2015), who analyzed CIRS limb data acquired in 2010–2012 and compared those results to the 2005–2006 observations by Guerlet et al. (2009). Contrary to Sinclair et al. (2013), Sylvestre et al. (2015) find little change in the ethane and acetylene

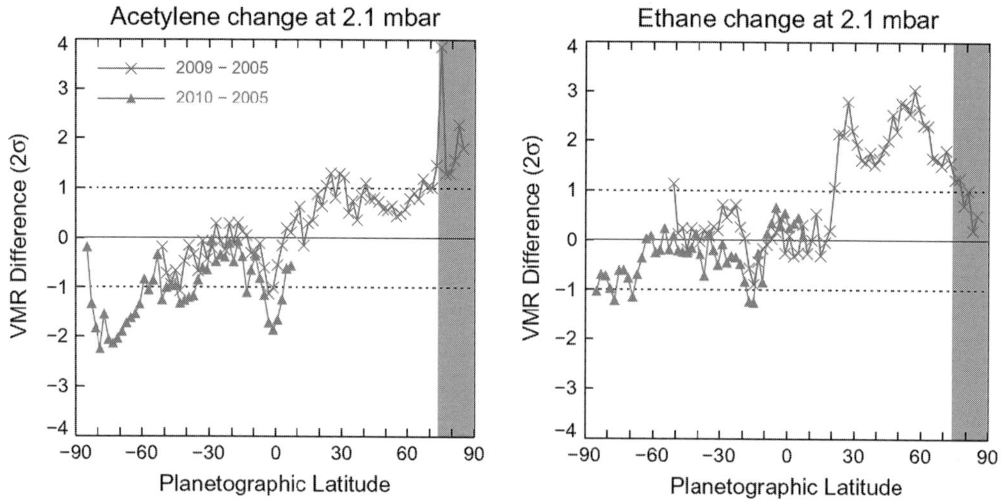

Figure 10.17 Differences in zonal mean retrievals of ethane and acetylene at 2 mbar between 2005 and 2009/2010 from Cassini/CIRS, as presented by Sinclair et al. (2013). Trends are shown in units of the uncertainty (σ) to reveal where the trends are statistically significant. In general, the results show a decline in abundance over the southern hemisphere and an increase over the northern hemisphere, suggestive of large-scale interhemispheric circulation with subsidence in northern winter/spring. Similar conclusions have been reached using temporal trends in C_2H_2 and C_2H_6 distributions at higher latitudes (Fletcher et al. 2015).

distribution in the millibar region. However, they find that the local enrichment in ethane and acetylene previously observed at 25°N in 2005–2006 (especially at $p < 0.1$ mbar) has since disappeared. If subsidence at 25° N was occurring during the northern winter, then it has stopped by northern spring, qualitatively consistent with the predictions of the seasonally reversing Hadley circulation proposed by Friedson and Moses (2012). Alternatively, the variability at 25°N could be coupled to secondary meridional circulations induced by the quasi-periodic equatorial oscillations whose stacked pattern is descending with time.

Despite quantitative differences between spatial distributions and temporal behavior identified by different authors, the general trends for ethane and acetylene are consistent: (i) C_2H_2 decreasing from equator to pole whereas C_2H_6 is largely uniform with latitude; (ii) regional enhancements of both species at the highest latitudes within polar vortices, possibly associated with auroral chemistry and dynamic entrainment; (iii) general spring hemisphere enhancements and autumn hemisphere depletions in the millibar region, suggestive of stratospheric circulation from the southern to the northern hemisphere; and (iv) complex temporal behavior in the tropics associated with vertical propagating waves and a seasonally reversing Hadley-type circulation. As ethane and acetylene are also the principal stratospheric coolants, their spatial and temporal behavior must also be incorporated in radiative climate models to better explain the thermal asymmetries observed by Cassini (see Section 10.2).

C_3 and Higher Hydrocarbons

In addition to ethane and acetylene, Cassini/CIRS has also revealed hemispheric asymmetries in the higher order hydrocarbons. Although studies of their temporal evolution are rather limited at the time of writing, we review here the main conclusions from both theory and observations.

Propane. Propane is produced in the photochemical models through termolecular reactions, which are more effective at higher pressures. Hence, the production rate for C_3H_8 peaks near ~3 mbar, where photosensitized CH_4 destruction occurs, rather than at higher altitudes where CH_4 is destroyed by Lyman alpha photolysis. Propane is lost predominantly through photolysis. Because it is shielded to some extent by C_2H_6 and C_2H_2, propane is relatively stable in the models. Figure 10.16i indicates that propane is expected to show a similar latitudinal trend to C_2H_2, with a decline from equator to pole in both hemispheres. However, the first spatial distribution of propane provided by

Greathouse et al. (2006) found that propane at the 5-mbar level was relatively uniform in the southern hemisphere (based on two measurements at 20°S and 80°S). Guerlet et al. (2009) derived propane from Cassini/CIRS limb observations between March 2005 and January 2008 ($L_s = 300°–340°$), with a peak sensitivity near 1 mbar. They pointed out that the meridional behavior of C_3H_8 closely tracks the instantaneous solar insolation (i.e. more propane in the summer hemisphere), suggesting that the chemical lifetime of C_3H_8 is shorter than the models indicate. Another possibility is that the C_3H_8 abundance is sensitive to temperature through reactions that have a higher degree of temperature dependence than is assumed in the models (Dobrijevic et al. 2011). In fact, models tend to underpredict the C_3H_8 abundance on Saturn (Greathouse et al. 2006; Guerlet et al. 2009; Dobrijevic et al. 2011). Most recently, Cassini/CIRS limb observations in 2010 (Sylvestre et al. 2015) suggest little to no change since 2005–2006.

Methylacetylene and Diacetylene Guerlet et al. (2010) continued to exploit the same Cassini/CIRS limb data ($L_s = 300°–340°$) to provide the first latitudinally resolved distributions of methylacetylene (CH_3C_2H) and diacetylene (C_4H_2), finding that mid-southern latitudes are depleted in both hydrocarbons compared to mid-northern latitudes during southern summer. The photochemical models do not predict such a hemispheric asymmetry during this season (see Figure 10.16), and Guerlet et al. (2010) suggest that the behavior is caused by upwelling at mid-southern latitudes and subsidence at mid-northern latitudes. Methylacetylene and diacetylene are important hydrocarbon photochemical products on Saturn that have much shorter chemical lifetimes than C_2H_6, C_2H_2 and C_3H_8. In the upper atmosphere where CH_4 is photolyzed by Lyman α and other short-wavelength radiation, CH insertion into C_2H_6, C_2H_2 and C_2H_4 initiates the production of C_3H_x hydrocarbons, which then can be photolyzed or react with hydrogen or other species to eventually produce CH_3C_2H (see Moses et al. 2000a, 2005 for details). At pressures greater than ~0.01 mbar, CH_3C_2H has additional sources through photolysis of C_4H_x species and reactions such as $CH_3 + C_2H_3$, which produce C_3H_x species that can again make their way to forming CH_3C_2H. In this lower-altitude region, acetylene is an important "parent" molecule for CH_3C_2H. Methylacetylene is lost through photolysis and H-atom addition. The primary (non-recycling) mechanism for C_4H_2 production is acetylene photolysis followed by $C_2H + C_2H_2 \rightarrow C_4H_2 + H$. Diacetylene formation therefore depends non-linearly on the acetylene abundance. H-atom addition to form C_4H_3 is the dominant loss reaction for C_4H_2, but the resulting C_4H_3 radical can react with H to recycle the C_4H_2. Photolysis is an effective loss process, although it, too, leads to some C_4H_2 recycling. Diacetylene is expected to condense in the lower stratosphere to contribute to a substantial fraction of the stratospheric haze burden (e.g. Moses et al. 2000b).

The meridional distribution of CH_3C_2H (Figure 10.16d,e) and C_4H_2 (Figure 10.16j) in the 0.05- to 1-mbar region appears to grossly track that of C_2H_2 (Guerlet et al. 2010), which makes sense in theory, given the short lifetimes of methylacetylene and diacetylene and given that C_2H_2 is the key "parent" molecule in the middle and lower stratosphere for both these species. The C_4H_2 abundance is typically well reproduced in 1D global-average models when the predicted C_2H_2 profiles match observations, suggesting that the C_4H_2 chemistry is well understood, but global-average photochemical models tend to notably overpredict the CH_3C_2H abundance in the 0.1- to 1-mbar region (e.g. Moses et al. 2000a, 2005; Guerlet et al. 2010; Dobrijevic et al. 2011), suggesting that the CH_3C_2H chemistry is not well described in the models. The meridional variations of CH_3C_2H and C_4H_2 appear more extreme than for C_2H_2, which is likely the result of the non-linear nature of their dependence on C_2H_2 photochemistry; however, these species also exhibit smaller-scale variations and differences between their respective distributions that are not easily explained with the chemical models. Dobrijevic et al. (2011) demonstrate that rate-coefficient uncertainties lead to a large spread in the predicted C_3H_8, CH_3C_2H and C_4H_2 abundances in the photochemical models, which could explain these model-data mismatches. Guerlet et al. (2010) suggest that seasonally variable transport is affecting the C_4H_2 and CH_3C_2H distributions.

Benzene. Benzene photochemistry is not well understood under conditions relevant to Saturn. Neutral C_6H_6 photochemical production and loss mechanisms on the giant planets are discussed by Moses et al. (2000a, 2005) and Lebonnois (2005). Recent CIRS measurements of the C_6H_6 column abundance show that benzene is at least as abundant in the polar stratosphere (80°S) compared to equatorial regions, in contradiction with photochemical model predictions including only neutral chemistry (Guerlet et al. 2015). The latter authors conclude that ion chemistry in the auroral regions plays an important role in the benzene production rates, as is the case on Jupiter (Wong et al. 2000; Friedson et al. 2002; Wong et al. 2003) and for non-auroral ion chemistry on Titan (Vuitton et al. 2009). The apparent presence of high-altitude hazes on Saturn observed during VIMS stellar occultations (Bellucci et al. 2009; Kim et al. 2012) and by CIRS (Guerlet et al. 2015) further suggests that Titan-like ion chemistry is contributing to the production of complex organics on Saturn. Benzene, like C_4H_2, is also expected to condense to form haze particles in Saturn's lower stratosphere.

In summary, Cassini has determined the spatial distribution of Saturn's stratospheric hydrocarbons (ethane, acetylene, propane, methylacetylene, diacetylene and benzene), discovering asymmetries during summertime conditions that bear little resemblance to the photochemical model predictions in the absence of circulation. Auroral chemistry is a possible contributor to increased abundances in the high-latitude regions, which should be incorporated into the radiative budget in the polar regions. Coupled 2D and 3D photochemical-dynamical models will likely be needed to fully explain the temporal and meridional variations of stratospheric hydrocarbons in the Cassini CIRS observations, but no such models have been published to date.

Oxygen Species

Finally, the reducing nature of Saturn's stratospheric composition is perturbed by a steady influx of oxygenated species from external sources, such as micrometeoroid precipitation, cometary impacts or a connection with Saturn's rings and satellites. These oxygen compounds can generate new photochemical pathways to form unexpected molecules, attenuate UV flux or provide condensation nuclei (Moses et al. 2000b). The history of ground-based observations of oxygen compounds, using rotational lines in the far-IR and submillimeter, was reviewed by Fouchet et al. (2009). Since the time of that review, investigators have continued to study disk-integrated stratospheric CO and H_2O from ground-based (Cavalié et al. 2009, 2010) and space-based (Herschel, Hartogh et al. 2011; Fletcher et al. 2012a) observatories, and Cassini/CIRS has demonstrated an absence of latitudinal CO_2 trends for $p < 10$ mbar (Abbas et al. 2013).

However, we discuss oxygen compounds briefly here because, although the chemical and diffusion timescales for CO, H_2O and CO_2 are too long to generate seasonal asymmetries, the water column abundance itself should change with season due to the altering thermal structure. Because water condenses relatively high in the stratosphere, the far-IR and sub-mm observations will be very sensitive to the altitude at which the water condenses, which is in turn controlled by the stratospheric temperatures. If the source is isotropic (interplanetary dust) or favors low latitudes (e.g. from Enceladus, Hartogh et al. 2011), we would expect to see a larger water column in the summer hemisphere and/or wherever mid-to-low stratospheric temperatures are highest. If the source is from the ring atmosphere (O'Donoghue et al. 2013; Tseng and Ip 2011), then the source itself is seasonal (greater ring atmosphere and thus greater influx when the ring has its highest opening angle at the solstices). The external oxygen supply is discussed in Chapter 9.

10.4 Clouds and Hazes

Cloud and haze particles provide yet another potential avenue to probe important atmospheric processes, with both obvious and subtle ties to the wind, temperature and chemical tracer fields. Although observation of clouds and haze is relatively straightforward, retrieval of vertical profiles, particle size, shape and composition and relationships to the other fields is not. Both types of particles can be classified generically as aerosols. We usually refer to "clouds" in the context of condensation/sublimation of volatile constituents (NH_3, NH_4SH and H_2O/NH_3) in the upper troposphere, while the term "haze" is used in association with the solid phase of photochemical products: P_2H_4 is expected to exist in the upper troposphere (although this has yet to be confirmed by observations); hydrocarbons and condensed external water in the stratosphere (Moses et al. 2000a). The two may combine: stratospheric haze sedimenting from high altitudes may serve as condensation nuclei for condensates deeper down. Conversely, ice crystals in tropospheric clouds may acquire a photochemically produced hydrocarbon coating. The reader is referred to West et al. (2009) for a comprehensive review of the literature on clouds and hazes, as a full treatment of aerosol composition and derived properties is beyond the scope of this chapter.

Common features of the various reflectivity investigations (e.g. Karkoschka and Tomasko 1992, 1993; Stam et al. 2001; Temma et al. 2005; Pérez-Hoyos et al. 2005; Karkoschka and Tomasko 2005) include (a) a stratospheric haze ($1 < p < 90$ mbar) of small radius ($r \approx 0.1 - 0.2$ μm) particles, possible originating from photochemical processes; (b) a tropospheric haze from the tropopause down to the first condensation cloud deck at 1.5–2.0 bar, possibly with aerosol-free gaps in the vertical distribution; and (c) a possible thick NH_3 cloud, although signatures of fresh NH_3 ice have so far only been detected in Saturn's 2010–2011 storm region (Sromovsky et al. 2013). Numerous explanations have been presented for the concealment of condensate signatures, including large particle sizes, non-spherical particle shapes, coating of the pure ices by photochemical products sedimenting downwards (Atreya and Wong 2005; Kalogerakis et al. 2008), or by mixing of the condensate phases to mask the spectral signatures. The polar stratospheric haze appears distinct from all other latitudes, being optically thicker and darker at UV wavelengths, with strong Rayleigh-like polarization suggestive of the importance of auroral processes in their formation. High-spatial-resolution limb images have revealed distinct haze layers at some latitudes (e.g. Rages and Barth 2012). The equatorial zone, between ±18° latitude, features consistently high clouds with perturbations from major storms (Pérez-Hoyos et al. 2005). The vertical distributions of the tropospheric aerosols have been well mapped with ground-based and Hubble data, and Cassini is beginning to add to this picture. Here we focus on the seasonal and some non-seasonal behavior of clouds and haze on the scale of the zonal jets.

10.4.1 Pre-Cassini Reflectivity Studies

Remote-sensing measurements of clouds and haze probe the upper troposphere and stratosphere, and it is in this regime where seasonal effects from atmospheric heating and photochemistry are most likely to have an effect. CCD images in near-infrared methane bands provide a measure of cloud top altitude and haze/cloud density, and it is not surprising that seasonally varying hemispheric asymmetries are apparent in ground-based and space-based images in methane absorption bands. Other diagnostics of cloud and haze vertical structure include polarization at phase angles near 90° and ultraviolet reflectivity. Polarization is sensitive to cloud altitude provided that it is dominated by Rayleigh scattering above a non-polarizing cloud. This is the case for low and middle latitudes but not so for high latitudes on Saturn and Jupiter at blue wavelengths. At high latitudes, haze particles also provide strong polarization. Ultraviolet reflectivity is diagnostic of cloud altitude, provided that it is also dominated by Rayleigh scattering from gas. Again, this seems to hold reasonably well at low latitude but not at high latitudes where UV-absorbing particles are abundant in the stratosphere. The latitudinal behavior of the highly-polarizing, UV-absorbing stratospheric aerosols is most likely a result of auroral energy deposition at high latitudes (Pryor and Hord 1991; West and Smith 1991); polar aerosols are discussed in Chapter 12.

The record of reflectivity measurements now span more than an entire seasonal cycle on Saturn. The Pioneer 11 flyby occurred just prior to northern spring equinox ($L_s = 354°$); the Voyager 1 and 2 encounters were a little after northern spring equinox ($L_s = 8.6°$–$18.2°$, respectively). Near the northern spring equinox in the early 1980s, methane-band imagery from 1979 (West et al. 1982), polarization imagery from Pioneer 11 in 1979 and Voyager 2 photopolarimeter scans in 1981 (Lane et al. 1982) all showed hemispheric contrasts at mid-latitudes consistent with deeper clouds and/or a smaller column density of haze in the southern mid-latitudes during early northern spring. Tomasko and Doose (1984) summarized those results using simple cloud/haze models showing effective cloud top pressures at 40°N and 40°S to be near 300 and 460 mbar, respectively. Viewing geometry and spatial resolution limited the northern analysis to latitudes less than 50°–60°N, while at southern latitudes the methane-band imagery indicated the effective cloud top pressure rising poleward of 40°S to 320 mbar at 70°S. The observed hemispheric difference was initially attributed to higher static stability and suppressed convection at southern latitudes after an extended period of solar heating during southern summer.

The equinoctial snapshot provided by Pioneer and Voyager was supplemented by ground- and space-based observations prior to Cassini's observations, most notably using the Hubble Space Telescope. During northern spring and summer, Karkoschka and Tomasko (1992) studied ground-based images from 1986 to 1989; and Karkoschka and Tomasko (1993) presented some of the earliest Hubble imaging in 1991 ($L_s = 68°–130°$). Stam et al. (2001) presented ground-based images of Saturn near the northern autumn equinox in 1995 ($L_s = 176°$), able to view both northern and southern hemispheres simultaneously. A seasonal asymmetry was observed, with both the tropospheric and stratospheric optical thicknesses appearing larger over northern mid-latitudes (approaching autumn) than over southern mid-latitudes (approaching spring), interpreted by Stam et al. (2001) as a thicker tropospheric cloud deck in the autumn hemisphere.

As the northern hemisphere receded from view and Saturn approached southern summer solstice, images from the Hubble Space Telescope (HST) provided a data set of significant value to the study of seasonal and non-seasonal behavior of Saturn's haze and clouds. Pérez-Hoyos et al. (2005, 2006) used HST data to study the equator and southern hemisphere between 1994 and 2003 ($L_s = 158°–286°$); and Karkoschka and Tomasko (2005) presented analyses of 134 images of Saturn taken by the Hubble Space Telescope between 1991 and 2004 ($L_s = 130°–289°$), and performed a principal-component analysis of many latitudes on Saturn. Four statistically meaningful principal components emerged. The first principal variation is a strong mid-latitude variation of the aerosol optical depth in the upper troposphere. This structure shifts with Saturn's seasons, but the structure on small scales of latitude stays constant. This is what is most apparent in a casual comparison of images taken in different seasons. The second principal variation is a variable optical depth of stratospheric aerosols. The optical depth is large at the poles and small at mid- and low latitudes, with a steep gradient in between. This structure remains essentially constant in time. The third principal variation is a variation in the tropospheric aerosol size, which has only shallow gradients with latitude, but large seasonal variations. Aerosols are largest in the summer and smallest in the winter, broadly consistent with the 1980s-equinox observation of a haze free southern autumn hemisphere. The fourth principal variation is a feature of the tropospheric aerosols with irregular latitudinal structure and fast variability, on the time scale of months.

10.4.2 Cassini's Observations of Seasonal Aerosol Changes

The seasonal asymmetry in tropospheric aerosols was therefore well established prior to Cassini's arrival just after southern summer solstice. The tropospheric haze optical thickness was expected to be the largest and most extended in the summer hemisphere, and smallest in the winter hemisphere, with the transition occurring at some time near to the equinox as previously observed by Voyager ($L_s = 8°–18°$) and Pioneer ($L_s = 354°$). The Cassini spacecraft arrived at Saturn at an earlier seasonal phase ($L_s \approx 290°$, southern summer), and has now provided the opportunity to track these changing hemispheric asymmetries. Indeed, Cassini observations pre-equinox revealed a hemispheric asymmetry that was opposite to those seen by Voyager post-equinox. As shown in Figure 10.18, the high northern latitudes showed a vibrant blue color in 2004. The interpretation (Edgington et al. 2012) is that Rayleigh scattering by gas molecules is responsible, and that the colored haze material is suppressed in the northern (winter) latitudes relative to southern (summer) latitudes. Subsequent Cassini ISS images have shown that the blue color persisted into 2008 but by 2009 (near equinox) the blue color had dissipated at northern latitudes (see the images in Figure 10.1). At the current epoch (2014) the southern high latitudes are beginning to show a blue color as they recede into winter conditions. These observations are consistent with the idea that the blue color indicates reduced production of haze throughout the winter season.

These inferences from the Cassini/ISS visible data are consistent with infrared observations from Cassini/VIMS, particularly at 5 μm where the dearth of hydrogen and methane opacity permits the escape of radiation from relatively deep (3–6 bar) levels, and cloud opacity serves to attenuate this 5 μm flux. As first

Figure 10.18 This true-color image was obtained in 2004 by the Cassini ISS camera on approach to Saturn. It illustrates the strong hemispheric color difference observed during southern summer. Figure 10.1 shows how Saturn's colors evolved with time during the Cassini mission. Courtesy NASA/JPL-Caltech. (A black-and-white version of this figure appears in some formats. For the color version, please refer to the plate section.)

Figure 10.19 Saturn at 5 microns wavelength, showing cloud features in silhouette against the bright background of Saturn's own thermal emission (Baines et al. 2006; West et al. 2009). In addition to fine-scale banding and numerous dynamic features, note the hemispheric differences in brightness and contrast, caused by a thicker tropospheric haze in the southern hemisphere. Courtesy NASA/JPL-Caltech.

reported by Baines et al. (2006), Figure 10.19 shows that Saturn's northern hemisphere was brighter than the southern summer hemisphere, with southern hemisphere contrasts muted due to the relatively higher aerosol opacity overlying the contrast-producing clouds. Fletcher et al. (2011) performed a quantitative analysis of Cassini/VIMS cubes from April 2006 (L_s = 317°), finding opacity in two regimes: a compact cloud deck centered in the 2.5- to 2.8-bar region, symmetric between the two hemispheres with small-scale opacity variations responsible for the numerous light/dark axisymmetric lanes; and secondly a hemispherically asymmetric population of aerosols at $p < 1.4$ bar which was ≈2.0× more opaque in the southern summer hemisphere. The upper tropospheric haze asymmetry is shown in Figure 10.20b, compared to the CIRS-derived 400-mbar temperatures (panel c) and the observed contrast in 5 μm brightness temperature (panel a). The vertical structure of this upper-level "haze" could not be constrained by the nightside VIMS observations, but is likely to be the same material responsible for the homogenous haze observed in CCD methane-band imaging. The deep cloud was at higher pressures than the predicted condensation altitude for NH_3 (1.8 bar for a 5× enrichment of heavy elements, Atreya et al. 1999), but at lower pressures than the predicted levels for NH_4SH condensation (5.7 bar), so its composition could not be identified unambiguously. Unfortunately, there are no studies of the expected reversing of the 5 μm cloud asymmetry available in the literature today.

Roman et al. (2013) conducted a quantitative study of Cassini/ISS images of the southern summer hemisphere between 2004 and 2007 (L_s = 296°–333°), reproducing the data with a stratospheric haze merging into a tropospheric haze that sits within the convectively stable region of the upper troposphere (i.e. above the R-C boundary discussed in Section 10.2). The tropospheric haze was found to reach the greatest heights (40 ±20 mbar) at the equator, but to sit deeper (140 ±20 mbar) at southern mid-latitudes. Figure 10.20 compares the southern hemisphere optical depths per bar at 619 nm derived by Roman et al. (2013) with the haze opacities at 5 μm derived from Fletcher et al. (2011). We have normalized to the maximum haze

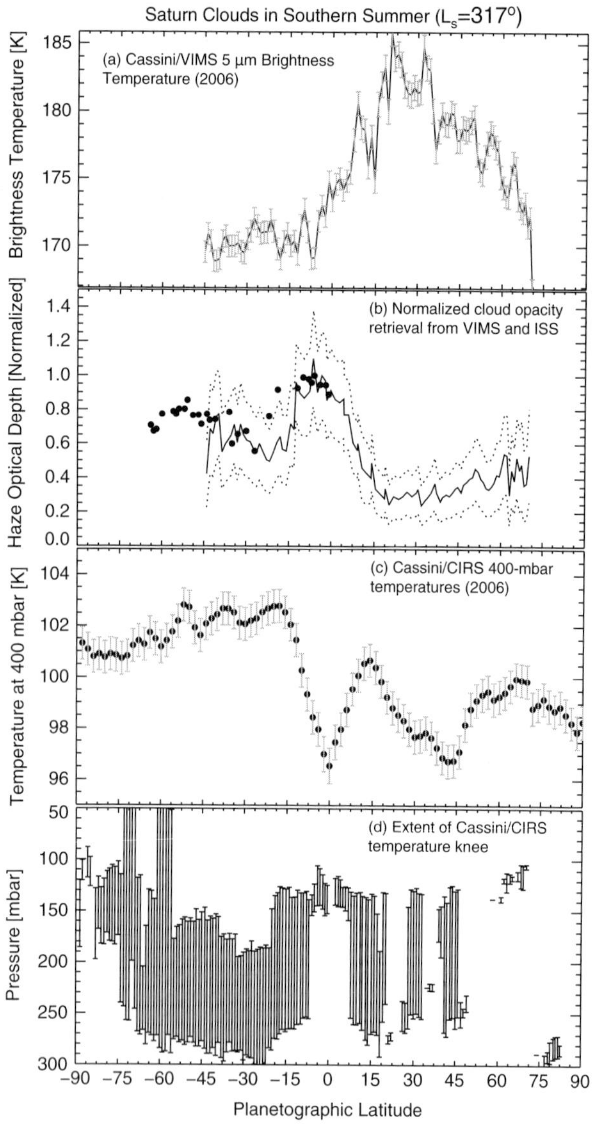

Figure 10.20 Comparison of the observed 5 μm asymmetry in brightness temperature in southern summer in panel a (2006, L_s = 317°) with upper tropospheric haze opacities derived from Cassini/ISS for the southern hemisphere (black circles, Roman et al. 2013) and Cassini/VIMS for both hemispheres (solid line with dotted error range, Fletcher et al. 2011). These are found to be largely correlated with the upper tropospheric thermal structure (400 mbar temperatures from Cassini/CIRS) in panel c. Furthermore, the asymmetry correlates with the extent of the temperature "knee" observed by CIRS in the upper troposphere (panel d), and believed to be caused by localized heating in the tropospheric haze layer (Fletcher et al. 2007a).

opacity at the equator to allow intercomparison, highlighting the north-south asymmetry and the comparison with the CIRS-derived temperature structure. The haze location derived by Roman et al. (2013) is consistent with the localized aerosol heating observed in Saturn's thermal field by Fletcher et al. (2007a) (Figure 10.20d), and Roman et al. (2013) suggest that the tropospheric haze is correlated with Saturn's mid-summer temperatures (Figure 10.20c). At higher pressures, the authors find discrete cloud structures in the 1- to 2-bar range, which may or may not be the same as the 2.5- to 2.8-bar cloud inferred from VIMS. Quantitative ISS studies have yet to extend into the northern winter hemisphere, and resolving the apparent discrepancy in cloud vertical structure with VIMS is a source of ongoing activity.

Finally, Sromovsky et al. (2013) analyzed VIMS reflectivity observations in the vicinity of the northern storm, and found that the main haze (of particles 1 μm in radius or smaller) in Saturn's northern hemisphere in 2011 (away from the storm-perturbed regions) was located between 111–178 mbar (top) and 577–844 mbar (bottom), depending on the latitude, with a deep, compact and opaque cloud near 2.6–3.2 bar. This is broadly consistent with the 5-μm thermal emission studies. They confirmed that this haze contained no spectroscopically identifiable features of pure condensates in the VIMS spectral range, and no signatures of hydrazine (N_2H_4), although diphosphine cannot be definitively ruled out. But, as for ISS, the latitudinal and seasonal dependence of Saturn's reflectivity has not yet been investigated.

In summary, the historical record of Saturn's aerosol distributions has shown that seasonal insolation changes induce hemispheric asymmetries in the tropospheric (and potentially stratospheric) hazes, with higher opacity in the summer hemisphere and low opacity (and blue color) in the winter hemisphere (Figure 10.20). The haze sits in the region approximately between the tropopause and the radiative-convective boundary, above the main convective region. Cassini images are showing that the asymmetry is reversing, along with the upper tropospheric temperatures, although quantitative studies of VIMS and ISS reflectivity have only been published for single epochs.

10.5 Conclusions and Outstanding Questions

The longevity and broad wavelength coverage of the Cassini mission, coupled with the decades-long record of ground-based observations, have revealed intricate connections between Saturn's atmospheric temperatures, chemistry and aerosol formation mechanisms. Environmental conditions in the stably stratified upper troposphere (approximately situated above the radiative-convective boundary at 400–500 mbar) and stratosphere have been observed to vary over time in response to the shifting levels of solar energy deposition and the efficiency of radiative cooling to space. Atmospheric temperatures have tracked the seasonal insolation changes, albeit with a phase lag that increases with depth into the atmosphere; the upper tropospheric haze has changed in optical thickness, causing differences in the coloration of reflected sunlight in the summer and winter hemispheres; and the zonal mean distributions of both tropospheric and stratospheric gaseous constituents exhibit hemispheric asymmetries that may be subtly shifting with time. The atmospheric soup of gases and aerosols in turn affects the radiative properties of the atmosphere (i.e. the rates of heating and cooling), which further influences the seasonal temperature shifts that we observe. Atmospheric circulation and localized dynamics can redistribute energy and material from place to place, implying that thermal and chemical perturbations are superimposed onto the large-scale seasonal asymmetries and sometimes (in the case of equatorial uplift and vertical waves; or within the polar vortices) can dominate the observed spatiotemporal trends.

Disentangling all of these competing effects is the key challenge for the next generation of modeling activities, towards a complete simulation of the seasonal behavior of Saturn's cloud-forming weather layer, upper troposphere and stratosphere. Historically, models have been developed in isolation to explain one subset of the larger problem – for example, radiative climate simulations (with or without convective adjustment and advection of heat via circulation) have been used to understand the magnitude of the seasonal temperature changes; 1D photochemical modeling with parameterized vertical mixing (and an absence of horizontal mixing and circulation) demonstrate how stratospheric hydrocarbons vary with time and location; and equilibrium cloud condensation models predict where key condensates should be forming in Saturn's troposphere. However, each of these models should be intricately linked, as gaseous and aerosol distributions influence the temperatures (and vice versa), which influences chemical and cloud microphysics time scales, as well as dynamic redistribution of heat. The latest generation of numerical simulations are moving in this direction – for example, Friedson and Moses (2012) combined radiative modeling with atmospheric circulation; Hue et al. (2015) connects the photochemically predicted hydrocarbon distributions with seasonal temperature changes in the context of the model of Greathouse et al. (2010); and Spiga et al. (2014) is aiming to incorporate the radiative model of Guerlet et al. (2014) into a full general circulation model. However, some key pieces of the puzzle (the radiative effect of poorly understood and seasonally variable aerosols; the influence of atmospheric circulation) continue to elude the community and remain the subject of ongoing theoretical development.

To date, Cassini has monitored Saturn's complex atmosphere for only a third of a Saturnian year. The upper tropospheric and stratospheric temperature fields have been measured from late northern winter through to late northern spring, which happens to overlap with Voyager observations just after the previous northern spring equinox, three decades ago. Surprisingly, the atmospheric temperatures measured by Cassini and Voyager *at the same point in the seasonal cycle* are not identical (Li et al. 2013; Sinclair et al. 2014; Fletcher et al. 2016), suggesting that Saturn might experience non-seasonal variability in its thermal field and circulation. Another mystery is why the stratospheric seasonal response at very low pressures ($p < 0.1$ mbar) appears to be more muted than that at 1 mbar, counter to the expectations of radiative climate models in the absence of stratospheric circulation. Atmospheric temperatures will continue to be monitored in the thermal-infrared using Cassini until northern summer solstice, and afterwards with ground-based and space-based observations (e.g. JWST), albeit restricted to the Earth-facing summer hemisphere. Long-term consistent datasets are essential to confirm whether Saturn truly does undergo non-seasonal variability.

Compared to the study of Saturn's temperature field, measurements of the distribution of gaseous and aerosol species are less mature. Cassini has identified hemispheric asymmetries in tropospheric species (para-H_2, the disequilibrium species PH_3 and the condensible volatile NH_3, Section 10.3.2), tropospheric haze opacities and cloud coloration (Section 10.4) and photochemically produced stratospheric hydrocarbon species (ethane, acetylene, propane, diacetylene and methylacetylene, see Section 10.3.3). It remains to be seen whether the timescales for the processes generating these asymmetries match or exceed a Saturnian year. If the timescales are short, then we may expect to see some reversal of the asymmetries as northern summer solstice approaches. If the timescales are long (potentially related to the southern-summer timing of perihelion and the northern-summer timing of aphelion) then we might expect the asymmetries to be quasi-permanent features of Saturn's atmosphere. Temporal studies of these distributions have only been published for the two principle hydrocarbons (ethane and acetylene) and the tropospheric para-H_2 fraction, showing slow and subtle changes to their distributions that may be more influenced by atmospheric circulation than by seasonally variable production and loss (e.g. a seasonally reversing Hadley cell at the equator; inter-hemispheric transport from the autumn to the spring hemisphere; and strong subsidence over the north polar region that has recently emerged into spring sunlight). In the near future, new data from Cassini will hopefully determine the magnitude of seasonal shifts in the tropospheric and stratospheric haze distributions, in addition to the higher-order hydrocarbons and the tropospheric species. By completing Cassini's observational characterization of Saturn's seasonal atmosphere through to northern summer solstice, we hope to inform and guide the development of the next generation of numerical simulations to establish Saturn as the paradigm for seasonal change on a giant planet.

Acknowledgments

The authors wish to thank A. Laraia, J. Hurley, J. Sinclair, J. Friedson and M. Roman for sharing the results of their studies. The manuscript benefited from thorough reviews by K. Baines, R. Achterberg, F. M. Flasar and G. Bjoraker. Fletcher was supported by a Royal Society Research Fellowship at the University of Oxford and University of Leicester.

References

Abbas, M. M., A. LeClair, E. Woodard et al. (2013), Distribution of CO_2 in Saturn's atmosphere from Cassini/CIRS infrared observations. *Astrophys. J.*, 776 (Oct.), 73.

Appleby, J. F. and J. S. Hogan (1984), Radiative-convective equilibrium models of Jupiter and Saturn. *Icarus*, 59 (Sept.), 336–366.

Armstrong, E. S., J. I. Moses, L. N. Fletcher et al. (2014) (Nov.), The chemistry of ethene in the storm beacon region on Saturn. Page submitted of: *AAS/Division for Planetary Sciences Meeting Abstracts*. AAS/Division for Planetary Sciences Meeting Abstracts, vol. 46.

Atreya, S. K., T. M. Donahue and W. R. Kuhn (1977), The distribution of ammonia and its photochemical products on Jupiter. *Icarus*, 31, 348–355.

Atreya, S. K., T. M. Donahue, A. F. Nagy et al. (1984), Theory, measurements, and models of the upper atmosphere and ionosphere of Saturn. Pages 239–277 of: T. Gehrels, and M. S. Matthews (eds), *Saturn*. Tucson: University of Arizona Press.

Atreya, S. K., W. R. Kuhn and T. M. Donahue (1980), Saturn: Tropospheric ammonia and nitrogen. *Geophys. Res. Lett.*, 7, 474–476.

Atreya, S. K. and A.-S. Wong (2005), Coupled clouds and chemistry of the giant planets: A case for multiprobes. *Space Sci. Rev.*, 116 (Jan.), 121–136.

Atreya, S. K., M. H. Wong, T. C. Owen et al. (1999), A comparison of the atmospheres of Jupiter and Saturn: deep atmospheric composition, cloud structure, vertical mixing, and origin. *Plan. & Space Sci.*, 47, 1243–1262.

Baines, K. H., T. W. Momary, M. Roos-Serote et al. (2006), North vs. south on Saturn: Discovery of a pronounced hemispherical asymmetry in Saturn's 5-micron emission and associated deep cloud structure by Cassini/VIMS. *B.A.A.S.*, 38, 488.

Barnet, C. D., R. F. Beebe and B. J. Conrath (1992), A seasonal radiative-dynamic model of Saturn's troposphere. *Icarus*, 98, 94–107.

Bell, J. M.,J. Westlake and J. H. Waite, Jr. (2011), Simulating the time-dependent response of Titan's upper atmosphere to periods of magnetospheric forcing. *Geophys. Res. Lett.*, 38, L06202.

Bellucci, A., B. Sicardy, P. Drossart et al. (2009), Titan solar occultation observed by Cassini/VIMS: Gas absorption and constraints on aerosol composition. *Icarus*, 201, 198–216.

Bézard, B., P. Drossart, T. Encrenaz et al. (2001a), Benzene on the Giant Planets. *Icarus*, 154 (Dec.), 492–500.

Bézard, B., P. Drossart, E. Lellouch et al. (1989), Detection of arsine in Saturn. *Astrophys. J.*, 346, 509–513.

Bézard, B., H. Feuchtgruber, J. I. Moses et al. (1998), Detection of methyl radicals (CH-3) on Saturn. *Astron. Astrophys.*, 334, L41–L44.

Bézard, B. and D. Gautier (1985), A seasonal climate model of the atmospheres of the giant planets at the Voyager encounter time: I. Saturn's stratosphere. *Icarus*, 61, 296–310.

Bézard, B., D. Gautier and B. Conrath (1984), A seasonal model of the Saturnian upper troposphere Comparison with Voyager infrared measurements. *Icarus*, 60, 274–288.

Bézard, B., E. Lellouch, D. Strobel et al. (2002), Carbon monoxide on Jupiter: Evidence for both internal and external sources. *Icarus*, 159, 95–111.

Bézard, B., J. I. Moses, J. Lacy et al. (2001b) (Nov.), Detection of Ethylene (C2H4) on Jupiter and Saturn in Non-Auroral Regions. Pages 1079 ff. of: *Bulletin of the American Astronomical Society.*

Bjoraker, G. L., H. P. Larson and U. Fink (1981), A study of ethane on Saturn in the 3 micron region. *Astrophys. J.*, 248, 856–862.

Bregman, J. D., D. F. Lester and D. M. Rank (1975), Observation of the nu-squared band of PH_3 in the atmosphere of Saturn. *Astrophys. J.*, 202.

Briggs, F. H. and P. D. Sackett (1989), Radio observations of Saturn as a probe of its atmosphere and cloud structure. *Icarus*, 80, 77–103.

Burgdorf, M. J., G. S. Orton, T. Encrenaz et al. (2004), Far-infrared spectroscopy of the giant planets: measurements of ammonia and phosphine at Jupiter and Saturn and the continuum of Neptune. *Advances in Space Research*, 34, 2247–2250.

Caldwell, J. (1977), The atmosphere of Saturn: An infrared perspective. *Icarus*, 30(Mar.), 493–510.

Caldwell, J., F. C. Gillett, I. G. Nolt et al. (1978), Spatially resolved infrared observations of Saturn: I. Equatorial limb scans at 20 microns. *Icarus*, 35 (Sept.), 308–312.

Carlson, B. E., J. Caldwell and R. D. Cess (1980), A model of Saturn's seasonal stratosphere at the time of the Voyager encounters. *Journal of Atmospheric Sciences*, 37 (Aug.), 1883–1885.

Carlson, R. W., K. H. Baines, W. S. Anderson et al. (2012), (Oct.), Chromophores from Photolyzed Ammonia Reacting with Acetylene: Application to Jupiter's Great Red Spot. Page #205.01 of: *AAS/Division for Planetary Sciences Meeting Abstracts*. AAS/Division for Planetary Sciences Meeting Abstracts, vol. 44.

Cavalié, T., F. Billebaud, M. Dobrijevic et al. (2009), First observation of CO at 345 GHz in the atmosphere of Saturn with the JCMT: New constraints on its origin. *Icarus*, 203 (Oct.), 531–540.

Cavalié, T., P. Hartogh, F. Billebaud et al. (2010), A cometary origin for CO in the stratosphere of Saturn? *Astron. Astrophys*, 510 (Feb.), A88.

Cess, R. D. and J. Caldwell (1979), A Saturnian stratospheric seasonal climate model. *Icarus*, 38, 349–357.

Conrath, B. J., P. J. Gierasch and S. S. Leroy (1990), Temperature and circulation in the stratosphere of the outer planets. *Icarus*, 83, 255–281.

Conrath, B. J., P. J. Gierasch and E. A. Ustinov (1998), Thermal structure and para hydrogen fraction on the outer planets from Voyager IRIS measurements. *Icarus*, 135, 501–517.

Conrath, B. J. and J. A. Pirraglia (1983), Thermal structure of Saturn from Voyager infrared measurements: Implications for atmospheric dynamics. *Icarus*, 53, 286–292.

Courtin, R., D. Gautier, A. Marten et al. (1984), The composition of Saturn's atmosphere at northern temperate latitudes from Voyager IRIS spectra – NH_3, PH_3, C_2H_2, C_2H_6, CH_3D, CH_4, and the Saturnian D/H isotopic ratio. *Astrophys. J.*, 287, 899–916.

Davis, G. R., M. J. Griffin, D. A. Naylor et al. (1996), ISO LWS measurement of the far-infrared spectrum of Saturn. *Astron. Astrophys.*, 315, L393–L396.

de Graauw, T., H. Feuchtgruber, B. Bézard et al. (1997), First results of ISO-SWS observations of Saturn: detection of CO_2, CH_3C_2 H, C_4H_2 and tropospheric H_2O. *Astron. Astrophys.*, 321, L13–L16.

de Pater, I. and S. T. Massie (1985), Models of the millimeter-centimeter spectra of the giant planets. *Icarus*, 62, 143–171.

Del Genio, A. D., R. K. Achterberg, K. H. Baines et al. (2009), *Saturn Atmospheric Structure and Dynamics, In: Saturn from Cassini-Huygens*. Springer. Chap. 6, pages 113–159.

Dobrijevic, M., T. Cavalié and F. Billebaud (2011), A methodology to construct a reduced chemical scheme for 2D-3D photochemical models: Application to Saturn. *Icarus*, 214, 275–285.

Dobrijevic, M., E. Hébrard, J. C. Loison et al. (2014), Coupling of oxygen, nitrogen, and hydrocarbon species in the photochemistry of Titan's atmosphere. *Icarus*, 228, 324–346.

Dowling, T. E., T. K. Greathouse, M. G. Sussman et al. (2010) (Oct.), New Radiative Transfer Capability in the EPIC Atmospheric Model with Application to Saturn and Uranus. Page 1021 of: *AAS/Division for Planetary Sciences Meeting Abstracts #42*. Bulletin of the American Astronomical Society, vol. 42.

Edgington, S. G., S. K. Atreya, L. M. Trafton et al. (1997), Phosphine mixing ratios and eddy mixing coefficients in the troposphere of Saturn. *Bulletin of the American Astronomical Society*, 29, 992.

Edgington, S. G., S. K. Atreya, E. H. Wilson et al. (2012), Photochemistry in Saturn's Ring Shadowed Atmosphere: Production Rates of Key Atmospheric Molecules and Preliminary Analysis of Observations. *AGU Fall Meeting Abstracts*, Dec., B1946.

Encrenaz, T., T. Owen and J. H. Woodman (1974), The abundance of ammonia on Jupiter, Saturn and Titan. *Astron. Astrophys.*, 37 (Dec.), 49–55.

Fegley, Jr., B. and K. Lodders (1994), Chemical models of the deep atmospheres of Jupiter and Saturn. *Icarus*, 110 (July), 117–154.

Fegley, B. and R. G. Prinn (1985), Equilibrium and nonequilibrium chemistry of Saturn's atmosphere: Implications for the observability of PH_3, N_2, CO, and GeH_4. *Astrophys. J.*, 299, 1067–1078.

Ferris, J. P. and Y. Ishikawa (1988), Formation of HCN and acetylene oligomers by photolysis of ammonia in the presence of acetylene: Applications to the atmospheric chemistry of Jupiter. *J. Am. Chem. Soc.*, 110, 4306–4312.

Fink, U., H. P. Larson, G. L. Bjoraker et al. (1983), The NH_3 spectrum in Saturn's 5 micron window. *ApJ*, 268 (May), 880–888.

Flasar, F. M., R. K. Achterberg, B. J. Conrath et al. (2005), Temperatures, winds, and composition in the Saturnian system. *Science*, 307, 1247–1251.

Flasar, F. M., V. G. Kunde, M. M. Abbas et al. (2004), Exploring the Saturn system in the thermal infrared: The composite infrared spectrometer. *Space Science Reviews*, 115, 169–297.

Fletcher, L. N., R. K. Achterberg, T. K. Greathouse et al. (2010), Seasonal change on Saturn from Cassini/CIRS observations, 2004–2009. *Icarus*, 208 (July), 337–352.

Fletcher, L. N., K. H. Baines, T. W. Momary et al. (2011), Saturn's tropospheric composition and clouds from Cassini/VIMS 4.6–5.1 μm nightside spectroscopy. *Icarus*, 214 (Aug.), 510–533.

Fletcher, L. N., B. E. Hesman, R. K. Achterberg et al. (2012b), The origin and evolution of Saturn's 2011–2012 stratospheric vortex. *Icarus*, 221 (Nov.), 560–586.

Fletcher, L. N., P. G. J. Irwin, R. K. Achterberg et al. (2016), Seasonal variability of Saturn's tropospheric temperatures, winds and para-H_2 from Cassini Far-IR Spectroscopy. *Icarus*, 264, 137–159.

Fletcher, L. N., P. G. J. Irwin, G. S. Orton et al. (2008), Temperature and composition of Saturn's polar hot spots and hexagon. *Science*, 319 (Jan.), 79–82.

Fletcher, L. N., P. G. J. Irwin, J. A. Sinclair et al. (2015), Seasonal evolution of Saturn's polar temperatures and composition. *Icarus*, 131–153.

Fletcher, L. N., P. G. J. Irwin, N. A. Teanby et al. (2007a), Characterising Saturn's vertical temperature structure from Cassini/CIRS. *Icarus*, 189, 457–478.

Fletcher, L. N., P. G. J. Irwin, N. A. Teanby (2007b), The meridional phosphine distribution in Saturn's upper troposphere from Cassini/CIRS observations. *Icarus*, 188 (May), 72–88.

Fletcher, L. N., P. G. J. Irwin, N. A. Teanby (2009), Phosphine on Jupiter and Saturn from Cassini/CIRS. *Icarus*, 202 (Aug.), 543–564.

Fletcher, L. N., B. Swinyard, C. Salji et al. (2012a), Submillimetre spectroscopy of Saturn's trace gases from Herschel/SPIRE. *Astron. Astrophys*., 539 (Mar.), A44.

Fouchet, T., S. Guerlet, D. F. Strobel et al. (2008), An equatorial oscillation in Saturn's middle atmosphere. *Nature*, 453(7192), 200–202.

Fouchet, T., E. Lellouch and H. Feuchtgruber (2003), The hydrogen ortho-to-para ratio in the stratospheres of the giant planets. *Icarus*, 161, 127–143.

Fouchet, T., J. I. Moses and B. J. Conrath (2009), Saturn: Composition and Chemistry, In: *Saturn from Cassini-Huygens*, Springer, Chap. 5, pages 83 ff.

Friedson, A. J. and J. I. Moses (2012), General circulation and transport in Saturn's upper troposphere and stratosphere. *Icarus*, 218 (Apr.), 861–875.

Friedson, A. J., A.-S. Wong and Y. L. Yung (2002), Models for polar haze formation in Jupiter's stratosphere. *Icarus*, 158, 389–400.

Gans, B., S. Boyé-Péronne, M. Broquier et al. (2011), Photolysis of methane revisited at 121.6 nm and at 118.2 nm: quantum yields of the primary products, measured by mass spectrometry. *Physical Chemistry Chemical Physics*, 13, 8140.

Gezari, D. Y., M. J. Mumma, F. Espenak et al. (1989), New features in Saturn's atmosphere revealed by high-resolution thermal infrared images. *Nature*, 342, 777–780.

Gillett, F. C. and W. J. Forrest (1974), The 7.5- to 13.5-MICRON spectrum of Saturn. *Astrophys. J.l*, 187 (Jan.), L37.

Gillett, F. C. and G. S. Orton (1975), Center-to-limb observations of Saturn in the thermal infrared. *Astrophys. J.*, 195 (Jan.), L47–L49.

Giver, L. P. and H. Spinrad (1966), Molecular Hydrogen Features in the Spectra of Saturn and Uranus. *Icarus*, 5, 586–589.

Gladstone, G. R., M. Allen and Y. L. Yung (1996), Hydrocarbon photochemistry in the upper atmosphere of Jupiter. *Icarus*, 119, 1–52.

Greathouse, T. K., J. H. Lacy, B. Bézard et al. (2005), Meridional variations of temperature, C_2H_2 and C_2H_6 abundances in Saturn's stratosphere at southern summer solstice. *Icarus*, 177, 18–31.

Greathouse, T. K., J. H. Lacy, B. Bézard (2006), The first detection of propane on Saturn. *Icarus*, 181 (Mar.), 266–271.

Greathouse, T., J. Moses, L. Fletcher et al. (2010) (May), Seasonal Temperature Variations in Saturn's Stratosphere: Radiative Seasonal Model vs. Observations. Page 4806 of: *EGU General Assembly Conference Abstracts*. EGU General Assembly Conference Abstracts, vol. 12.

Greathouse, T. K., S. Strong, J. Moses et al. (2008), A General Radiative Seasonal Climate Model Applied to Saturn, Uranus, and Neptune. *AGU Fall Meeting Abstracts*, Dec., P21B–06.

Grossman, A. W., D. O. Muhleman and G. L. Berge (1989), High-resolution microwave images of Saturn. *Science*, 245, 1211–1215.

Guerlet, S., T. Fouchet, B. Bézard et al. (2009), Vertical and meridional distribution of ethane, acetylene and propane in Saturn's stratosphere from CIRS/Cassini limb observations. *Icarus*, 203, 214–232.

Guerlet, S., T. Fouchet, B. Bézard (2010), Meridional distribution of CH_3C_2 H and C_4H_2 in Saturn's stratosphere from CIRS/Cassini limb and nadir observations. *Icarus*, 209 (Oct.), 682–695.

Guerlet, S., T. Fouchet, B. Bézard (2011), Evolution of the equatorial oscillation in Saturn's stratosphere between 2005 and 2010 from Cassini/CIRS limb data analysis. *Geophysical Research Letters*, 38 (May), 9201.

Guerlet, S., T. Fouchet, S. Vinatier et al. (2015), Stratospheric benzene and hydrocarbon aerosols detected in Saturn's auroral regions. *Astronomy and Astrophysics*, 580 (Aug.), A89.

Guerlet, S., A. Spiga, M. Sylvestre et al. (2014), Global climate modeling of Saturn's atmosphere. Part I: Evaluation of the radiative transfer model. *Icarus*, 238 (Aug.), 110–124.

Guillemin, J.-C., T. Janati and L. Lassalle (1995), Photolysis of phosphine in the presence of acetylene and propyne, gas mixtures of planetary interest. *Advances in Space Research*, 16, 85–92.

Hanel, R., B. Conrath, F. M. Flasar et al. (1981), Infrared observations of the Saturnian system from Voyager 1. *Science*, 212, 192–200.

Hanel, R., B. Conrath, F. M. Flasar (1982), Infrared observations of the Saturnian system from Voyager 2. *Science*, 215, 544–548.

Hanel, R. A., B. J. Conrath, D. E. Jennings et al. (2003), *Exploration of the Solar System by Infrared Remote Sensing: Second Edition*. Exploration of the Solar System by Infrared Remote Sensing, by R. A. Hanel and B. J. Conrath and D. E. Jennings and R. E. Samuelson. Cambridge: Cambridge University Press, April 2003.

Hartogh, P., E. Lellouch, R. Moreno et al. (2011), Direct detection of the Enceladus water torus with Herschel. *Astron. Astrophys*, 532 (Aug.), L2.

Hébrard, E., M. Dobrijevic, J. C. Loison et al. (2013), Photochemistry of C_3H_p hydrocarbons in Titan's stratosphere revisited. *Astron. Astrophys.*, 552, A132.

Hesman, B. E., G. L. Bjoraker, P. V. Sada et al. (2012), Elusive Ethylene Detected in Saturn's Northern Storm Region. *Astrophys. J.*, 760 (Nov.), 24.

Hesman, B. E., D. E. Jennings, P. V. Sada et al. (2009), Saturn's latitudinal C_2H_2 and C_2H_6 abundance profiles from Cassini/CIRS and ground-based observations. *Icarus*, 202 (July), 249–259.

Holton, J. R. (2004), *An Introduction to Dynamic Meteorology*. Academic Press.

Howett, C. J. A., P. G. J. Irwin, N. A. Teanby et al. (2007), Meridional variations in stratospheric acetylene and ethane in the Saturnian atmosphere as determined from Cassini/CIRS measurements. *Icarus*, 190(2), 556–572.

Hue, V., T. Cavalié, M. Dobrijevic et al. (2015), 2D photochemical modeling of Saturn's stratosphere. Part I: Seasonal variation of atmospheric composition without meridional transport. *Icarus*, 257 (Sept.), 163–184.

Hurley, J., L. N. Fletcher, P. G. J. Irwin et al. (2012), Latitudinal variation of upper tropospheric NH_3 on Saturn derived from Cassini/CIRS far-infrared measurements. *Plan. & Space Sci.*, 73 (Dec.), 347–363.

Ingersoll, A. P., R. F. Beebe, B. J. Conrath et al. (1984), *Structure and dynamics of Saturn's atmosphere*. Saturn. Pages 195–238.

Janssen, M. A., A. P. Ingersoll, M. D. Allison et al. (2013), Saturn's thermal emission at 2.2-cm wavelength as imaged by the Cassini RADAR radiometer. *Icarus*, 226 (Sept.), 522–535.

Kalogerakis, K. S., J. Marschall, A. U. Oza et al. (2008), The coating hypothesis for ammonia ice particles in Jupiter: Labora-tory experiments and optical modeling. *Icarus*, 196 (July), 202–215.

Karkoschka, E. and M. Tomasko (2005), Saturn's vertical and latitudinal cloud structure 1991 – 2004 from HST imaging in 30 filters. *Icarus*, 179, 195–221.

Karkoschka, E. and M. G. Tomasko (1992), Saturn's upper troposphere 1986–1989. *Icarus*, 97, 161–181.

Karkoschka, E. and M. G. Tomasko (1993), Saturn's upper atmospheric hazes observed by the Hubble Space Telescope. *Icarus*, 106, 428–441.

Kaye, J. A. and D. F. Strobel (1983a), Formation and photochemistry of methylamine in Jupiter's atmosphere. *Icarus*, 55, 399–419.

Kaye, J. A. and D. F. Strobel(1983b), HCN formation on Jupiter: The coupled photochemistry of ammonia and acetylene. *Icarus*, 54, 417–433.

Kaye, J. A. and D. F. Strobel (1983c), Phosphine photochemistry in Saturn's atmosphere. *Geophys. Res. Lett.*, 10, 957–960.

Kaye, J. A. and D. F. Strobel (1984), Phosphine photochemistry in the atmosphere of Saturn. *Icarus*, 59 (Sept.), 314–335.

Keane, T. C., F. Yuan and J. P. Ferris (1996), Potential Jupiter atmospheric constituents: Candidates for the mass spectrometer in the Galileo atmospheric probe. *Icarus*, 122, 205–207.

Kerola, D. X., H. P. Larson and M. G. Tomasko (1997), Analysis of the near-IR spectrum of Saturn: A comprehensive radiative transfer model of its middle and upper troposphere. *Icarus*, 127, 190–212.

Kim, J. H., S. J. Kim, T. R. Geballe et al. (2006), High-resolution spectroscopy of Saturn at 3 microns: CH_4, CH_3D, C_2H_2, C_2H_6, PH_3, clouds, and haze. *Icarus*, 185 (Dec.), 476–486.

Kim, S. J. and T. R. Geballe (2005), The 2.9–4.2 micron spectrum of Saturn: Clouds and CH_4, PH_3, and NH_3. *Icarus*, 179, 449–458.

Kim, S. J., C. K. Sim, D. W. Lee et al. (2012), The three-micron spectral feature of the Saturnian haze: Implications for the haze composition and formation process. *Planet. Space Sci.*, 65, 122–129.

Krasnopolsky, V. A. (2014), Chemical composition of Titan's atmosphere and ionosphere: Observations and the photochemical model. *Icarus*, 236, 83–91.

Lane, A. L., C. W. Hord, R. A. West et al. (1982), Photopolarimetry from Voyager 2: Preliminary results on Saturn, Titan, and the rings. *Science*, 215(Jan.), 537–543.

Lara, L. M., E. Lellouch, M. González et al. (2014), A time-dependent photochemical model for Titan's atmosphere and the origin of H_2O. *Astron. Astrophys.*, 566, A143.

Laraia, A. L., A. P. Ingersoll, M. A. Janssen et al. (2013), Analysis of Saturn's thermal emission at 2.2-cm wavelength: Spatial distribution of ammonia vapor. *Icarus*, 226(Sept.), 641–654.

Larson, H. P. (1980), Infrared spectroscopic observations of the outer planets, their satellites, and the asteroids. *Annual Review of Astronomy and Astrophysics*, 18, 43–75.

Lavvas, P., M. Galand, R. V. Yelle et al. (2011), Energy deposition and primary chemical products in Titan's upper atmosphere. *Icarus*, 213, 233–251.

Lebonnois, S. (2005), Benzene and aerosol production in Titan and Jupiter's atmospheres: a sensitivity study. *Planet. Space Sci.*, 53, 486–497.

Lellouch, E., B. Bézard, T. Fouchet et al. (2001), The deuterium abundance in Jupiter and Saturn from ISO-SWS observations. *Astron. Astrophys.*, 370, 610–622.

Lewis, J. S. and M. B. Fegley, Jr. (1984), Vertical distribution of disequilibrium species in Jupiter's troposphere. *Space Sci. Rev.*, 39(Oct.), 163–192.

Lewis, J. S. and R. G. Prinn (1984), *Planets and Their Atmospheres: Origin and Evolution*. Orlando: Academic Press.

Li, L., R. K. Achterberg, B. J. Conrath et al. (2013), Strong temporal variation over one Saturnian year: From Voyager to Cassini. *Scientific Reports*, 3 (Aug.).

Li, L., B. J. Conrath, P. J. Gierasch et al. (2010), Saturn's emitted power. *Journal of Geophysical Research (Planets)*, 115 (Nov.), 11002.

Li, L., X. Jiang, A. P. Ingersoll et al. (2011), Equatorial winds on Saturn and the stratospheric oscillation. *Nature Geoscience*, 4 (Nov.), 750–752.

Lindal, G. F., D. N. Sweetnam and V. R. Eshleman (1985), The atmosphere of Saturn: An analysis of the Voyager radio occultation measurements. *Astron. J.*, 90, 1136–1146.

Lodders, K. and B. Fegley (2002), Atmospheric chemistry in giant planets, brown dwarfs, and low-mass dwarf stars: I. Carbon, nitrogen, and oxygen. *Icarus*, 155 (Feb.), 393–424.

Loison, J. C., E. Hébrard, M. Dobrijevic et al. (2015), The neutral photochemistry of nitriles, amines and imines in the atmosphere of Titan. *Icarus*, 247 (Feb.), 218–247.

Mandt, K. E., D. A. Gell, M. Perry et al. (2012), Ion densities and composition of Titan's upper atmosphere derived from the Cassini Ion Neutral Mass Spectrometer: Analysis methods and comparison of measured ion densities to photochemical model simulations. *J. Geophys. Res.*, 117, E10006.

Massie, S. T. and D. M. Hunten (1982), Conversion of para and ortho hydrogen in the Jovian planets. *Icarus*, 49, 213–226.

Moos, H. W. and J. T. Clarke (1979), Detection of acetylene in the Saturnian atmosphere, using the IUE satellite. *Astrophys. J.*, 229, L107.

Moreno, R., E. Lellouch, L. M. Lara et al. (2012), The abundance, vertical distribution and origin of H_2O in Titan's atmosphere: Herschel observations and photochemical modelling. *Icarus*, 221, 753–767.

Moses, J. I., E. S. Armstrong, L. N. Fletcher et al. (2014) (Nov.), Evolution of stratospheric chemistry in the Saturn storm beacon. Page submitted of: *AAS/Division for Planetary Sciences Meeting Abstracts*. AAS/Division for Planetary Sciences Meeting Abstracts, vol. 46.

Moses, J. I. and T. K. Greathouse (2005), Latitudinal and seasonal models of stratospheric photochemistry on Saturn: Comparison with infrared data from IRTF/TEXES. *Journal of Geophysical Research (Planets)*, 110 (Sept.), E09007.

Moses, J. I., B. Bézard, E. Lellouch et al. (2000a), Photochemistry of Saturn's atmosphere. I. Hydrocarbon chemistry and comparisons with ISO observations. *Icarus*, 143, 244–298.

Moses, J. I., T. Fouchet, B. Bézard et al. (2005), Photochemistry and diffusion in Jupiter's stratosphere: Constraints from ISO observations and comparisons with other giant planets. *Journal of Geophysical Research (Planets)*, 110 (Aug.), E08001.

Moses, J. I., E. Lellouch, B. Bézard et al. (2000b), Photochemistry of Saturn's atmosphere. II. Effects of an influx of external oxygen. *Icarus*, 145 (May), 166–202.

Moses, J. I., M.-C. Liang, Y. L. Yung et al. (2007) (Mar.), Two-Dimensional Photochemical Modeling of Hydrocarbon Abundances on Saturn. Page 2196 of: *Lunar and Planetary Science Conference*. Lunar and Planetary Science Conference, vol. 38.

Moses, J. I., C. Visscher, J. J. Fortney et al. (2011), Disequilibrium carbon, oxygen, and nitrogen chemistry in the atmospheres of HD 189733b and HD 209458b. *Astrophys. J.*, 737 (Aug.), 15.

Moses, J. I., C. Visscher, T. C. Keane et al. (2010), On the abundance of non-cometary HCN on Jupiter. *Faraday Discussions*, 147, 103–136.

Nava, D. F., W. A. Payne, G. Marston et al. (1993), The reaction of atomic hydrogen with germane: Temperature dependence of the rate constant and implications for germane photochemistry in the atmospheres of Jupiter and Saturn. *J. Geophys. Res.*, 98, 5531–5537.

Noll, K. S., T. R. Geballe and R. F. Knacke (1989), Arsine in Saturn and Jupiter. *Astrophys. J.*, 338, L71–L74.

Noll, K. S., R. F. Knacke, T. R. Geballe et al. (1986), Detection of carbon monoxide in Saturn. *ApJ Letters*, 309 (Oct.), L91–L94.

Noll, K. S., R. F. Knacke, T. R. Geballe (1988), Evidence for germane in Saturn. *Icarus*, 75, 409–422.

Noll, K. S. and H. P. Larson (1990), The spectrum of Saturn from 1990–2230 cm^{-1}: abundances of AsH_3, CH_3D, CO, GeH_4, and PH_3. *Icarus*, 89, 168–189.

Noll, K. S. and H. P. Larson (1991), The spectrum of Saturn from 1990 to 2230 cm^{-1}: Abundances of AsH_3, CH_3D, CO, GeH_4, NH_3, and PH_3. *Icarus*, 89 (Jan.), 168–189.

O'Donoghue, J., T. S. Stallard, H. Melin et al. (2013), The domination of Saturn's low-latitude ionosphere by ring 'rain'. *Nature*, 496 (Apr.), 193–195.

Ollivier, J. L., F. Billebaud, P. Drossart et al. (2000), Seasonal effects in the thermal structure of Saturn's stratosphere from infrared imaging at 10 microns. *Astron. Astrophys.*, 356, 347–356.

Orton, G. S., K. H. Baines, D. Cruikshank et al. (2009v *Review of Knowledge Prior to the Cassini-Huygens Mission and Concurrent Research, In: Saturn from Cassini-Huygens*. Springer. Chap. 2, pages 9–54.

Orton, G. S. and A. P. Ingersoll (1980), Saturn's atmospheric temperature structure and heat budget. *Journal of Geophysical Research*, 85 (Nov.), 5871–5881.

Orton, G. S., J. I. Moses, L. N. Fletcher et al. (2014), Mid-infrared spectroscopy of Uranus from the Spitzer infrared spectrometer: 2. Determination of the mean composition of the upper troposphere and stratosphere. *Icarus*, 243 (Nov.), 471–493.

Orton, G. S., E. Serabyn and Y. T. Lee (2000), Vertical distribution of PH_3 in Saturn from observations of its 1–0 and 3–2 rotational lines. *Icarus*, 146, 48–59.

Orton, G. S., E. Serabyn and Y. T. Lee (2001), Erratum, Volume 146, Number 1, pages 48–59 (2000), in the article Vertical distribution of PH_3 in Saturn from observations of its 1–0 and 3–2 rotational lines, *Icarus*, 149, 489–490.

Orton, G. S. and P. A. Yanamandra-Fisher (2005), Saturn's temperature field from high-resolution middle-infrared imaging. *Science*, 307, 696–698.

Orton, G. S., P. A. Yanamandra-Fisher, B. M. Fisher et al. (2008), Semi-annual oscillations in Saturn's low-latitude stratospheric temperatures. *Nature*, 453 (May), 196–198.

Palotai, C., T. E. Dowling and L. N. Fletcher (2014), 3D Modeling of interactions between Jupiter's ammonia clouds and large anticyclones. *Icarus*, 232, 141–156.

Pérez-Hoyos, S., A. Sánchez-Lavega and R. G. French (2006), Short-term changes in the belt/zone structure of Saturn's Southern Hemisphere (1996–2004). *Astron. Astrophys.*, 460 (Dec.), 641–645.

Pérez-Hoyos, S., A. Sánchez-Lavega, R. G. French et al. (2005), Saturn's cloud structure and temporal evolution from ten years of Hubble Space Telescope images (1994–2003). *Icarus*, 176, 155–174.

Plessis, S., N. Carrasco, M. Dobrijevic et al. (2012), Production of neutral species in Titan's ionosphere through dissociative recombination of ions. *Icarus*, 219, 254–266.

Prangé, R., T. Fouchet, R. Courtin et al. (2006), Latitudinal variation of Saturn photochemistry deduced from spatially-resolved ultraviolet spectra. *Icarus*, 180, 379–392.

Prinn, R. G. and S. S. Barshay (1977), Carbon monoxide on Jupiter and implications for atmospheric convection. *Science*, 198, 1031–1034.

Prinn, R. G. and B. Fegley, Jr. (1981), Kinetic inhibition of CO and N_2 reduction in circumplanetary nebulae – Implications for satellite composition. *Astrophys. J*, 249 (Oct.), 308–317.

Prinn, R. G., H. P. Larson, J. J. Caldwell et al. (1984), Composition and chemistry of Saturn's atmosphere. *Saturn*. Tucson: University of Arizona Press, pages 88–149.

Prinn, R. G. and T. Owen (1976), Chemistry and spectroscopy of the Jovian atmosphere. Pages 319–371 of: T. Gehrels, (ed.), *Jupiter*. Tucson: University of Arizona Press.

Pryor, W. R. and C. W. Hord (1991), A study of photopolarimeter system UV absorption data on Jupiter, Saturn, Uranus, and Neptune: Implications for auroral haze formation. *Icarus*, 91, 161–172.

Rages, K. A. and E. L. Barth (2012) (Oct.), Saturn Limb Hazes as Seen from Cassini. Page #500.05 of: *AAS/Division for Planetary Sciences Meeting Abstracts*. AAS/Division for Planetary Sciences Meeting Abstracts, vol. 44.

Rieke, G. H. (1975), The thermal radiation of Saturn and its rings. *Icarus*, 26 (Sept.), 37–44.

Roman, M. T., D. Banfield and P. J. Gierasch (2013), Saturn's cloud structure inferred from Cassini ISS. *Icarus*, 225 (July), 93–110.

Sada, P. V., G. L. Bjoraker, D. E. Jennings et al. (2005), Observations of C_2H_6 and C_2H_2 in the stratosphere of Saturn. *Icarus*, 173, 499–507.

Schinder, P. J., F. M. Flasar, E. A. Marouf et al. (2011), Saturn's equatorial oscillation: Evidence of descending thermal structure from Cassini radio occultations. *Geophys. Res. Lett.*, 38 (Apr.), 8205.

Sinclair, J. A., P. G. J. Irwin, L. N. Fletcher et al. (2013), Seasonal variations of temperature, acetylene and ethane in Saturn's atmosphere from 2005 to 2010, as observed by Cassini-CIRS. *Icarus*, 225 (July), 257–271.

Sinclair, J. A., P. G. J. Irwin, L. N. Fletcher (2014), From Voyager-IRIS to Cassini-CIRS: Interannual variability in Saturn's stratosphere? *Icarus*, 233 (May), 281–292.

Spiga, A., S. Guerlet, M. Indurain et al. (2014) (Nov.), An exploration of Saturn's stratospheric dynamics through Global Climate Modeling. Page #508.09 of: *AAS/Division for Planetary Sciences Meeting Abstracts*. AAS/Division for Planetary Sciences Meeting Abstracts, vol. 46.

Sromovsky, L. A., K. H. Baines and P. M. Fry (2013), Saturn's Great Storm of 2010–2011: Evidence for ammonia and water ices from analysis of VIMS spectra. *Icarus*, 226 (Sept.), 402–418.

Stam, D. M., D. Banfield, P. J. Gierasch et al. (2001), Near-IR Spectrophotometry of Saturnian Aerosols-Meridional and Vertical Distribution. *Icarus*, 152, 407–422.

Strobel, D. F. (1975), Aeronomy of the major planets: Photochemistry of ammonia and hydrocarbons. *Reviews of Geophysics and Space Physics*, 13, 372–382.

Strobel, D. F. (1977), NH_3 and PH_3 photochemistry in the Jovian atmosphere. *Astrophys. J. Lett.*, 214, L97–L99.

Strobel, D. F. (1983), Photochemistry of the reducing atmospheres of Jupiter, Saturn, and Titan. *International Reviews in Physical Chemistry*, 3, 145–176.

Strobel, D. F. (2005), Photochemistry in outer solar system atmospheres. *Space Science Reviews*, 116, 155–170.

Sugiyama, K., K. Nakajima, M. Odaka et al. (2011), Intermittent cumulonimbus activity breaking the three-layer cloud structure of Jupiter. *Geophys. Res. Lett.*, 38, L13201.

Sugiyama, K., K. Nakajima, M. Odaka (2014), Numerical simulations of Jupiter's moist convection layer: Structure and dynamics in statistically steady states. *Icarus*, 229 (Feb.), 71–91.

Sylvestre, M., S. Guerlet, T. Fouchet et al. (2015), Seasonal changes in Saturn's stratosphere inferred from Cassini/CIRS limb observations. *Icarus*, 258 (Sept.), 224–238.

Tautermann, C. S., B. Wellenzohn and D. C. Clary (2006), Rates of the reaction $C_2H_3 + H_2 \rightarrow C_2H_4 + H$. *Molecular Physics*, 104, 151–158.

Temma, T., N. J. Chanover, A. A. Simon-Miller et al. (2005), Vertical structure modeling of Saturn's equatorial region using high spectral resolution imaging. *Icarus*, 175, 464–489.

Tokunaga, A. T., J. Caldwell, F. C. Gillett et al. (1978), Spatially resolved infrared observations of Saturn: II. The temperature enhancement at the South Pole of Saturn. *Icarus*, 36 (Nov.), 216–222.

Tokunaga, A. and R. D. Cess (1977), A model for the temperature inversion within the atmosphere of Saturn. *Icarus*, 32 (Nov.), 321–327.

Tokunaga, A. T., H. L. Dinerstein, D. F. Lester et al. (1980), The phosphine abundance on Saturn derived from new 10-micrometer spectra. *Icarus*, 42, 79–85.

Tokunaga, A., R. F. Knacke and T. Owen (1975), The detection of ethane on Saturn. *Astrophys. J.*, 197, L77.

Tomasko, M. G. and L. R. Doose (1984), Polarimetry and photometry of Saturn from Pioneer 11 Observations and constraints on the distribution and properties of cloud and aerosol particles. *Icarus*, 58, 1–34.

Tseng, W.-L. and W.-H. Ip (2011), An assessment and test of Enceladus as an important source of Saturn's ring atmosphere and ionosphere. *Icarus*, 212 (Mar.), 294–299.

van der Tak, F., I. de Pater, A. Silva et al. (1999), Time Variability in the Radio Brightness Distribution of Saturn. *Icarus*, 142 (Nov.), 125–147.

Vander Auwera, J., N. Moazzen-Ahmadi and J.-M. Flaud (2007), Toward an Accurate Database for the 12 μm Region of the Ethane Spectrum. *Astrophysical Journal*, 662 (June), 750–757.

Vasavada, A. R., S. M. Hörst, M. R. Kennedy et al. (2006), Cassini imaging of Saturn: Southern hemisphere winds and vortices. *Journal of Geophysical Research (Planets)*, 111(E10), 5004.

Visscher, C. and B. J. Fegley (2005), Chemical Constraints on the Water and Total Oxygen Abundances in the Deep Atmosphere of Saturn. *Astrophys. J.*, 623, 1221–1227.

Visscher, C., K. Lodders and B. Fegley, Jr. (2006), Atmospheric Chemistry in Giant Planets, Brown Dwarfs, and Low-Mass Dwarf Stars. II. Sulfur and Phosphorus. *Astrophys. J. Letters*, 648 (Sept.), 1181–1195.

Visscher, C. and J. I. Moses (2011), Quenching of carbon monoxide and methane in the atmospheres of cool brown dwarfs and hot Jupiters. *Astrophys. J.*, 738, 72.

Visscher, C., J. I. Moses and S. A. Saslow (2010), The deep water abundance on Jupiter: New constraints from thermochemical kinetics and diffusion modeling. *Icarus*, 209, 602–615.

Visscher, C., A. D. Sperier, J. I. Moses et al. (2009) (Mar.), Phosphine and ammonia photochemistry in Jupiter's troposphere. Page 1201 of: *Lunar and Planetary Science Conference*. Lunar and Planetary Science Conference, vol. 40.

Vuitton, V., R. V. Yelle and P. Lavvas (2009), Composition and chemistry of Titan's thermosphere and ionosphere. *Royal Society of London Philosophical Transactions Series A*, 367, 729–741.

Vuitton, V., R. V. Yelle, P. Lavvas et al. (2012), Rapid association reactions at low pressure: Impact on the formation of hydrocarbons on Titan. *Astrophys. J.*, 744, 11.

Weidenschilling, S. J. and J. S. Lewis (1973), Atmospheric and cloud structures of the jovian planets. *Icarus*, 20, 465–476.

Weisstein, E. W. and E. Serabyn (1994), Detection of the 267 GHz J = 1–0 rotational transition of PH_3 in Saturn with a new fourier transfer spectrometer. *Icarus*, 109, 367–381.

Weisstein, E. W. and E. Serabyn (1996), Submillimeter line search in Jupiter and Saturn. *Icarus*, 123, 23–36.

West, R. A., K. H. Baines, E. Karkoschka et al. (2009), *Clouds and Aerosols in Saturn's Atmosphere, In: Saturn from Cassini-Huygens*. Springer. Chap. 7, pages 161–179.

West, R. A. and P. H. Smith (1991), Evidence for aggregate particles in the atmospheres of Titan and Jupiter. *Icarus*, 90 (Apr.), 330–333.

West, R. A., D. F. Strobel and M. G. Tomasko (1986), Clouds, aerosols, and photochemistry in the Jovian atmosphere. *Icarus*, 65, 161–217.

West, R. A., M. G. Tomasko, B. A. Smith et al. (1982), Spatially resolved methane band photometry of Saturn: I. Absolute reflectivity and center-to-limb variations in the 6190-, 7250-, and 8900-A bands. *Icarus*, 51, 51–64.

Westlake, J. H., J. H. Waite, Jr., K. E. Mandt et al. (2012), Titan's ionospheric composition and structure: Photochemical modeling of Cassini INMS data. *J. Geophys. Res.*, 117, E01003.

Winkelstein, P., J. Caldwell, S. J. Kim et al. (1983), A determination of the composition of the Saturnian stratosphere using the IUE. *Icarus*, 54, 309–318.

Wong, A.-S., A. Y. T. Lee, Y. L. Yung et al. (2000), Jupiter: Aerosol chemistry in the polar atmosphere. *Astrophys. J. Lett.*, 534, L215–L217.

Wong, A.-S., Y. L. Yung and A. J. Friedson (2003), Benzene and haze formation in the polar atmosphere of Jupiter. *Geophys. Res. Lett.*, 30, 1447.

Yung, Y. L. and W. B. DeMore (1999), *Photochemistry of Planetary Atmospheres*. Oxford: Oxford University Press.

Yung, Y. L., W. A. Drew, J. P. Pinto et al. (1988), Estimation of the reaction rate for the formation of CH_3O from H + H_2CO: Implications for chemistry in the solar system. *Icarus*, 73, 516–526.

11

The Global Atmospheric Circulation of Saturn

ADAM P. SHOWMAN, ANDREW P. INGERSOLL, RICHARD ACHTERBERG AND YOHAI KASPI

Over the past decade, the Cassini spacecraft has provided an unprecedented observational record of the atmosphere of Saturn, which in many ways now surpasses Jupiter as the best-observed giant planet. These observations, along with data from the Voyager spacecraft and Earth-based telescopes, demonstrate that Saturn, like Jupiter, has an atmospheric circulation dominated by zonal (east-west) jet streams, including a broad, fast eastward equatorial jet and numerous weaker jets at higher latitudes. Imaging from Voyager, Cassini and ground-based telescopes also document a wide range of tropospheric features, including vortices, waves, turbulence and moist convective storms. At large scales, the clouds, ammonia gas and other chemical tracers exhibit a zonally banded pattern whose relationships to the zonal jets remain poorly understood. Infrared observations constrain the stratospheric thermal structure and allow the derivation of stratospheric temperatures; these exhibit not only the expected seasonal changes, but also a wealth of variations that are likely dynamical in origin and highlight dynamical coupling between the stratosphere and the underlying troposphere. In parallel to these observational developments, significant advances in theory and modeling have occurred over the past decade, especially regarding the dynamics of zonal jets, and we survey these new developments in the context of both Jupiter and Saturn. Highly idealized two-dimensional models illuminate the dynamics that give rise to zonal jets in rapidly rotating atmospheres stirred by convection or other processes, while more realistic three-dimensional models of the atmosphere and interior are starting to identify the particular conditions under which Jupiter- and Saturn-like flows – including the fast equatorial superrotation, multiple jets at higher latitudes, storms and vortices – can occur. Future data analysis and models have the potential to greatly increase our understanding over the next decade.

11.1 Introduction

The dynamics of Jupiter and Saturn have long been the subject of fascination. Although Galileo trained the telescope on both objects starting in 1610, the pace of atmospheric discoveries over subsequent centuries was far slower for Saturn than for Jupiter because of the fact that Saturn is dimmer, smaller in Earth's sky, and more muted in its cloud-albedo contrasts. The first description of Jupiter's zonal banding occurred in 1630. Cassini and others discovered short- and long-lived spots, the equatorial current, and other atmospheric features on Jupiter starting in the latter decades of the 1600s, and ever-more-refined observations have continued episodically since then (e.g. Rogers 1995). In contrast, early observations of Saturn could not make out features on the planet's surface, instead emphasizing the discovery of Saturn's moons and the nature of its rings (van Helden 1984). Cassini reported an equatorial belt on Saturn in 1676, and hints of non-zonal atmospheric features appeared a century later in the 1790s, but it took until the latter decades of the 1800s before spots could reliably be observed on Saturn (van Helden 1984). This disparity in pace of discovery between Jupiter and Saturn continued through the twentieth century, and even over the last few decades the relative difficulty of observing Saturn relative to Jupiter means that Saturn has received much less attention than its sister planet. Yet the Cassini mission, in orbit around Saturn since 2004, has revolutionized the quality and quantity of observational constraints, which now rival or exceed those available for Jupiter. This for the first time places Saturn at the forefront of understanding giant planets as a class.

The atmospheric circulations on Saturn and Jupiter are dominated by numerous fast east-west (zonal) jet streams, which interact with a wealth of vortices, waves, turbulent filamentary structures, convective storms and other features in poorly understood ways. The jet profiles exhibit qualitative similarities – on each planet, there exists a broad, fast prograde (eastward) equatorial jet and numerous narrower, weaker jets at higher latitudes. Nevertheless, the zonal jets are considerably broader, faster and fewer in number on Saturn than Jupiter – peak wind speeds reach $\sim$500 m s^{-1} on the former but never exceed 200 m s^{-1} on the latter. Both planets exhibit cloud banding that is modulated by the zonal jets and, at large scales, is far more zonally symmetric than the cloud pattern on Earth. Yet the detailed relationship of the cloud-band structure to the zonal jet structure differs considerably between Jupiter and Saturn. Moreover, despite the existence of many dozens of vortices on both planets, Saturn lacks prominent large vortices like Jupiter's Great Red Spot and White Oval. These differences remain poorly understood. As case studies in similarities and differences, comparative investigations of the two planets therefore have great potential for insights.

Motivations for studying the atmospheric dynamics of Jupiter and Saturn are several. Our understanding of Earth's circulation is well developed (see, e.g. Vallis 2006), but necessarily confined to the specific conditions relevant to Earth. The giant planets illustrate how an atmosphere's circulation operates when there is no planetary surface, when the incident solar energy flux is weak ($\sim$2 W m^{-2} on Saturn versus 240 W m^{-2} for Earth), when the energy flux transported through the interior is comparable to the energy flux absorbed from the Sun, and when the background gas comprises hydrogen rather than high-molecular-weight species like N_2 or O_2 (which, among other things, promotes larger atmospheric scale heights on Jupiter and Saturn than on Earth and implies that, opposite to the situation on Earth, moist air is denser than dry air at a given temperature and pressure). Moreover, as they lack many of the complexities of the Earth system, the giant planets serve as ideal laboratories for testing and understanding fundamental issues in geophysical fluid dynamics (GFD) such as the dynamics of zonal jets, vortices and stratified turbulence. The giant planets are dynamic and exhibit meteorological and climatic changes on timescales ranging from minutes to centuries. Most of the phenomena that are richly characterized in the observational record remain poorly understood; continued observations, theory and modeling will be necessary to understand even some of the zeroth-order aspects of the circulation. Finally, Saturn and Jupiter can serve as a prototype for helping to understand the behavior of the numerous brown dwarfs and giant planets that are being discovered outside the solar system.

The primary goal of this chapter is to survey observations and theory of Saturn's atmospheric dynamics near the close of the Cassini mission. Our scope is Saturn's global atmospheric circulation, emphasizing the thermal and dynamical structure of the large-scale zonal jets, vortices, turbulence and interior structure. We summarize the current dynamical state of Saturn – emphasizing observations but including dynamical interpretation and analysis where appropriate – beginning with the troposphere (Section 11.2), followed by the stratosphere (Section 11.3). In Section 11.4, we summarize Saturn's dynamics, beginning with basic balance arguments for jet structure, and proceeding to a summary of models for the formation of the zonal jets. Anticipating the Grand Finale stage of the Cassini Mission, we end with a brief discussion of constraints provided by gravity data on the deep jet structure. Our review follows the prior reviews of Ingersoll et al. (1984) and Del Genio et al. (2009), who summarized the state of knowledge after the Voyager flyby and nominal Cassini missions, respectively. It also complements other reviews of Jupiter and the giant planets generally, including Ingersoll (1990), Marcus (1993), Dowling (1995a), Ingersoll et al. (2004) and Vasavada and Showman (2005). Reviews of Saturn's poles, stratospheric thermal structure and convective storms can be found in Chapter 12 by Sayanagi et al., Chapter 10 by Fletcher et al. and Chapter 13 by Sanchez-Lavega et al. in this volume.

11.2 Observations of the Troposphere

On Saturn, the winds blow primarily east-west and are organized into alternating jet streams, each peaking at its particular latitude. At large scales, the cloud structure is zonally banded – to a much stronger degree than on Earth – but also contains numerous vortices, waves, storms, and various cloud features at small scales (Figure 11.1). Viewed from a non-rotating reference frame, cloud features near the equator exhibit periods of about 10h 10 min. At latitudes of 35°–40° in each hemisphere, the recurrence period is about 10h 40 min, and at higher latitudes the periods vary within this range up to the pole.

These observations indicate that rotation plays a crucial role in the atmospheric dynamics of Saturn. The importance of rotation can be characterized using the Rossby number, which is the ratio of the advection force to the Coriolis force in the horizontal momentum equation. The advection force per mass has characteristic magnitude U^2/L, and the horizontal Coriolis force per mass has characteristic magnitude fU, where U is the characteristic wind speed, L is the characteristic horizontal length scale of the circulation (e.g. the jet width), $f = 2\Omega \sin \phi$ is the Coriolis parameter, Ω is the planetary rotation rate (2π over the rotation period), and ϕ is latitude. Thus, the Rossby number is given by $Ro = U/fL$. The cloud measurements described above suggest a characteristic rotation period of order 10.5 hours, with zonal speed differences between the cloud bands

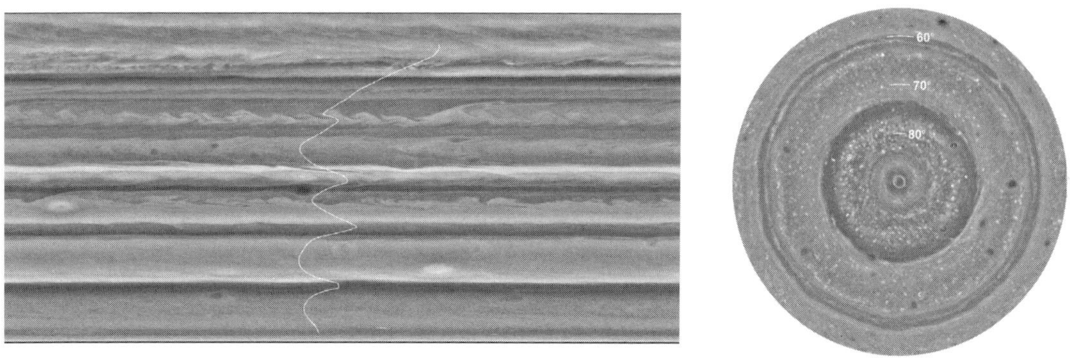

Figure 11.1 Cassini images of Saturn's southern hemisphere. *Left:* A cylindrical projection mosaic comprised of Cassini ISS images using the MT2 filter, centered in a methane absorption band at 727 nm wavelength and showing the equator to ~85°S latitude and 180°–360°W longitude. The image is overlain with the zonal-wind profile obtained from cloud tracking (plotted such that 0 m s^{-1} in System III is at the center of the panel). *Right:* Polar stereographic projection of Cassini ISS images in the CB2 filter, which is centered at 750 nm, in the continuum between methane absorption bands. From Vasavada et al. (2006).

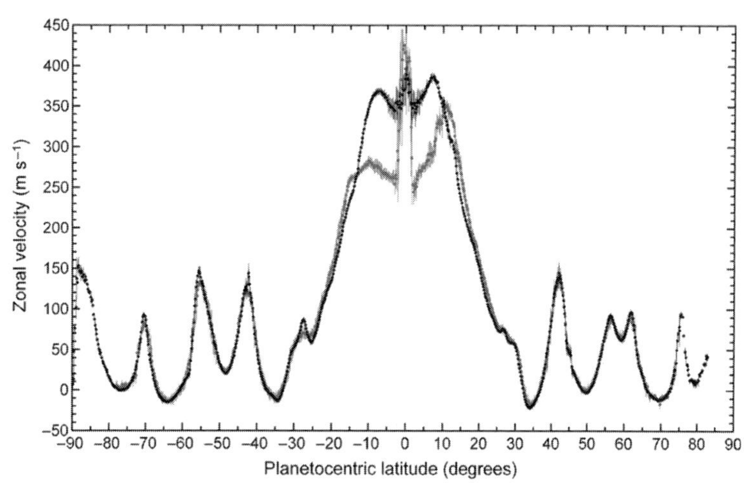

Figure 11.2 Mean zonal wind profile for the 2004–2009 time period. The black curve is from clear filters that sense clouds in the 350–700 mbar pressure range. The red curve is from methane band filters that sense clouds in the 60–250 mbar pressure range. From García-Melendo et al. (2011). (A black-and-white version of this figure appears in some formats. For the color version, please refer to the plate section.)

of typically ~100 m s^{-1} (Figures 11.1 and 11.2). Inserting appropriate values of $f \approx 2 \times 10^{-4}$ s^{-1}, $U \approx$ 100 m s^{-1}, and $L \approx 10^7$ m (relevant to Saturn's zonal jets), we obtain $Ro \approx 0.05$. Thus, on Saturn – as well as on Jupiter, Uranus, Neptune and Earth's atmosphere and oceans – the large-scale circulation exhibits small Rossby number. This estimate implies that, for the large-scale flows on the giant planets, advective forces are weak compared to the Coriolis force. Since the pressure-gradient force is the only other significant force in the horizontal momentum equation, the smallness of Ro implies that the dominant balance in the horizontal momentum equation is between the Coriolis force and the pressure-gradient force – called geostrophic balance (see, e.g. Holton and Hakim 2013, chapter 2).

Although the observed differential rotation is evidence of wind, we do not know the precise wind speeds relative to the planetary interior – simply because Saturn's interior rotation rate is not well determined. The other giant planets have internally generated magnetic fields that are tilted with respect to the rotation axis, and those fields are presumably locked to the electrically conducting fluid interiors, at least on time scales of years to decades, so the daily wobble of the field gives us a good estimate of the internal rotation period. Saturn has an internal field as well, but it is not tilted. Thus far, the instruments on Cassini and other spacecraft have been unable to detect a wobble, and have therefore been unable to provide an exact period from which to measure the winds.

Voyager flew past Saturn in 1980 and measured magnetic fields, radio emissions, and modulations of the charged particle distribution around the planet that pointed to a period that was close to the longest periods determined by tracking clouds in the atmosphere. Accordingly, the System III rotation period of 10h 39 min 24 s was chosen as the reference frame from which to measure the longitudes and drift rates of phenomena on Saturn (Desch and Kaiser 1981). This longitude system is the official standard, even though the electromagnetic phenomena on which it was based have not kept pace but have wandered in ways that are uncharacteristic of the planet's interior. In recent years, several authors have attempted to obtain improved estimates of the interior rotation period. Anderson and Schubert (2007) and Read et al. (2009b) used theoretical arguments combined with the observed wind field to suggest rotation periods that are 5–6 minutes shorter than the System III period, and Helled et al. (2015) used the measured low-order gravitational coefficients and the shape of the planet to suggest a rotation period of 10h 32 min 45 s ±46 s. If these estimates are correct, then the winds measured relative to the interior would be shifted westward (by a latitude-dependent value) relative to those plotted in Figure 11.2.

On an oblate planet like Saturn, where the rotational flattening $(1 - R_p/R_e)$ is 9.8%, where R_p and R_e are the polar and equatorial radii, there are two ways to define the latitude. Planetocentric latitude ϕ_{pc} is the angle of a line from the center of the planet relative to the equatorial plane. Planetographic latitude ϕ_{pg} is the angle of the local vertical relative to the equatorial plane. If we approximate the planet's shape, e.g. on a constant-pressure surface, as an ellipse rotated about its short axis, then the relation between ϕ_{pg} and ϕ_{pc} is $\tan\phi_{pg} = (R_e/R_p)^2 \tan\phi_{pc}$. Thus $|\phi_{pg}|$ is always greater than $|\phi_{pc}|$ except at the poles and equator, where they are equal. Both kinds of latitude are used in the literature and this review.

11.2.1 Zonal Velocity

Figure 11.2, from García-Melendo et al. (2011), shows the zonal mean zonal wind as a function of planetocentric latitude, measured in the System III reference frame. The curves are time averages from 2004 to 2009, but at high and mid latitudes, the variations from year to year are generally less than the uncertainty of the measurement, which falls in the range 5 to 9 m s^{-1}, depending on the latitude. Departures from the zonal mean – the eddies – do not show up on this figure. They have been averaged out, but at least on a large scale the eddy winds are small.

The two curves in Figure 11.2 refer to different altitudes. In both cases, the winds are obtained by tracking small clouds in images separated by one Saturn rotation period. The black curve uses images in the CB2 and CB3 filters, centered at 752 and 959 nm, where the main sources of opacity are the clouds themselves. Generally, this band-pass senses clouds at the 350–700 mbar pressure range, but clouds outside this range may show up at certain times and places. The red curve uses images in the MT2 and MT3 filters, centered at 727 and 890 nm, where methane, one of the well-mixed gases in the atmosphere, limits the depth one can see. Thus the red curve shows the motion of clouds and haze at the 60–250 mbar pressure range, depending on latitude. This range spans the tropopause – the temperature minimum where the pressure scale height (the *e*-folding scale) is of order 30 km, so the clouds contributing to the red curve are ~40 km higher than those contributing to the black curve.

One notable feature of the zonal wind profile is the high speed in the equatorial band – at least 360 m s^{-1} faster than the minimum speeds at higher latitudes. These differential speeds are two times larger than those on Jupiter and are 7–8 times larger than the difference between the easterlies and westerlies on Earth. Nevertheless, they are comparable to the speed differential on Uranus, and less than that on Neptune. These differences are not well understood.

The four high-latitude eastward jets, separated by four high-latitude westward jets (eastward jet minima) in each hemisphere are another notable feature. Both in terms of their speeds and their numbers, these jets are more like Jupiter's jets, although the latter are somewhat more numerous and somewhat less speedy. Away from the equator, the winds are remarkably steady in time and remarkably constant in altitude. There is a general tendency for the profiles at the zonal jet minima to be more rounded than the profiles at the zonal jet maxima, which are sharper. The zonal wind must go to zero at the pole, and Figure 11.2 hints at the extremely rapid drop in winds speed within one degree of the south pole. This is the south polar vortex, like the eye of a hurricane, with 150 m s^{-1} winds circling around the pole at −89° latitude (Dyudina et al. 2008; O'Neill et al. 2015, 2016). The circulation direction is cyclonic, meaning that it is in the same direction that the planet is rotating as seen looking down from above. In the southern hemisphere, a cyclonic vortex is clockwise. A cyclonic vortex exists at the north pole as well (Antuñano et al. 2015), although Figure 11.2 doesn't show it because the north pole was in darkness when the data were taken. Hurricanes on Earth are also cyclonic, although they form in the subtropics and drift around, unlike the polar vortices on Saturn (Dyudina et al. 2008). See Chapter 12 of this volume by Sayanagi et al. on polar phenomena.

The final notable feature is the vertical variation near the equator. From 10° to 25° in each hemisphere, the wind is slightly stronger at the higher altitude. From 2° to 10°, the high-altitude winds are weaker, and within the narrow jet inside ±2°, the high-altitude winds are stronger. The equatorial winds are also variable in time. García-Melendo et al. (2011) point out that the equatorial winds were 450 m s^{-1} at the time of the Voyager encounters in 1980 and 1981 (Sanchez-Lavega et al. 2000), and ranged up to 420 m s^{-1} as measured by Hubble in 1990 (Barnet et al. 1992). García-Melendo et al. (2011) conclude that there was a real slowdown of the wind speed at the equator. Nevertheless, Choi et al. (2009) showed by tracking features in 5 μm images from Cassini VIMS – which sense as deep as ~2 bars – that the near-equatorial winds reach speeds of ~450 m s^{-1}, similar to the Voyager wind measurements. This suggests that any real slowdown in the winds may have been confined to the lower-pressure levels of the upper troposphere (significantly above the level where VIMS 5 μm images sense). Moreover, a stratospheric oscillation in temperature has been observed from Earth-based telescopes since 1980 (Orton et al. 2008) and implies an oscillation in the winds via the thermal-wind equation (see Section 11.3), which may have some connection to the oscillation in equatorial jet speeds seen in Figure 11.2.

11.2.2 Clouds and Temperatures

The clouds that we see in the clear filters CB2 and CB3 are thought to be crystals of ammonia. From spectroscopic observations, we know that ammonia vapor is close to saturation at these levels, so it is likely that ammonia is condensing. Solid particles do not give as clear a spectroscopic signature as the vapor, so it is difficult to say what other substances are present in the cloud particles. What condenses below the ammonia cloud base depends on composition and temperature, and both are uncertain. One must rely on models based on plausible assumptions, which we now describe.

Figure 11.3, from Atreya (2006), is a cartoon showing a possible cloud structure for Saturn for three different assumptions about the mixing ratios of condensable gases relative to hydrogen and helium, which are the major constituents. One takes the ratios O/H, N/H, S/H, and Ar/H on the Sun, determined

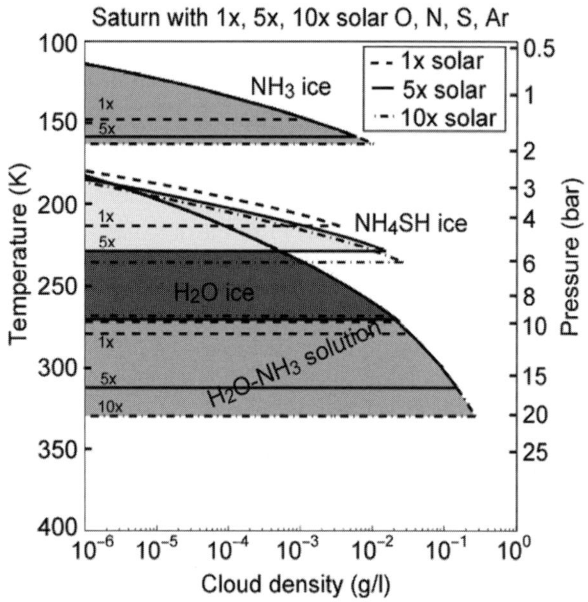

Figure 11.3 Cloud structure for 1-, 5- and 10-fold enrichment of H_2O, NH_3 and H_2S relative to solar composition. An enrichment factor of 1 means taking the solar elemental ratios of O/H, N/H and S/H and allowing the mixture to cool to planetary temperatures. Higher enrichment factors are applied uniformly to all elements relative to hydrogen and helium. Even with an enrichment factor of 10, the number of H_2 + He molecules in the gas is still more than 98%. From Atreya (2006), calculated using updated solar composition data from Grevesse et al. (2005).

spectroscopically, and multiplies (enriches) them all by a single factor, either 1, 5 or 10. Then one allows the reactive elements to combine with hydrogen to form H_2O, NH_3 and H_2S. The Ar and He stay in atomic form, and the remaining hydrogen becomes H_2. Even with tenfold enrichment, the minor constituents make up less than 2% of the molecules in the gas. One assumes the atmosphere is convecting up to the tops of the ammonia clouds, which means temperature T and pressure p follow an adiabat – a moist adiabat in this case to take into account the latent heat released when the vapors condense. The particular adiabat is chosen to match the observed T and p at the top of the clouds, which is about as deep as we can see with remote sensing instruments. In the figure the clouds are labeled and color-coded by their composition, and T, p values are given along the sides. For a rising parcel, the less volatile substance, water, condenses out first – at higher temperatures, and the more volatile substance, ammonia, condenses out last – at lower temperatures. The location of cloud base is fairly certain provided the composition is known. The cloud density is much less certain, since it assumes that none of the condensate falls out of the cloud and none is carried upward from where it condenses. Despite these uncertainties, the three-cloud structure is generally accepted. Enrichment factors up to 10 times the solar abundances are supported by the C/H ratio implied by the methane abundance (Fletcher et al. 2009, 2012). Methane, which doesn't condense at Saturn atmospheric temperatures, is easier to measure above the clouds by remote sensing and is therefore a good measure for the planet as a whole.

Temperature profiles obtained from radio occultations and infrared spectra are shown in Figure 11.4. They show that the temperature is close to an adiabat below the 300–400 mbar level, presumably indicating that convection is occurring, but becomes significantly stratified at higher levels (Li et al. 2013). The two techniques agree reasonably well, and can only sense to levels near ~1 bar, below which we have few observational constraints. The temperatures are fairly steady deeper than the ~300-mbar level, but exhibit greater spatial and temporal variations in the stratosphere.

11.2.3 Inferences from Dynamical Balance

Given the observations of zonal winds at cloud level and temperatures in the overlying layers, basic dynamical arguments can be used to *infer* the zonal winds above the cloud deck. On planets that rotate rapidly, with small Rossby number, there exists a dynamical link between winds and temperatures. Specifically, combining geostrophic balance, hydrostatic equilibrium, and the ideal-gas law leads to the thermal-wind equation for a shallow atmosphere (e.g. Holton and Hakim 2013, p. 82). In the meridional direction, this reads:

$$\frac{\partial u}{\partial \log p} = \frac{R}{f}\left(\frac{\partial T}{\partial y}\right)_p, \qquad (11.1)$$

where u is the zonal wind, p is pressure, R is the specific gas constant, $f = 2\Omega \sin\phi$ is the Coriolis parameter, Ω is the planetary rotation rate, ϕ is latitude, T is temperature, y is northward distance and the derivative on the right-hand side is taken at constant pressure. This equation can only be applied away from the equator since

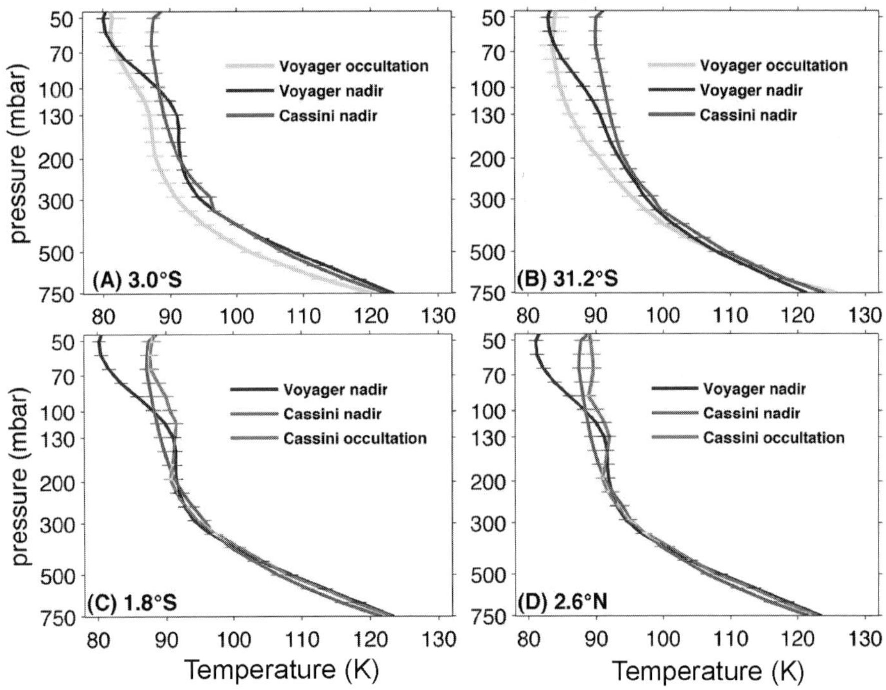

Figure 11.4 Temperature profiles (abscissa, in K) versus pressure, measured by Voyager and Cassini by two different methods. The nadir observations use upwelling thermal radiation to infer temperature as a function of pressure. The occultations use the spacecraft's radio signal passing tangentially through the atmosphere to infer density, from which temperature and pressure are derived. The profiles are close to adiabatic below the 300-mbar level. From Li et al. (2013). (A black-and-white version of this figure appears in some formats. For the color version, please refer to the plate section.)

geostrophy breaks down at the equator. The equation implies that, away from the equator, vertical gradients of the zonal wind are proportional to meridional temperature gradients. Given the known winds at the cloud level from cloud tracking (Figure 11.2), and given observations of temperature and its meridional gradient in the layers above the clouds, we can therefore integrate Equation (11.1) to estimate the zonal winds above the clouds.

The top panel of Figure 11.5, from Read et al. (2009a), shows temperatures derived from Composite Infrared Spectrometer (CIRS) observations from Cassini (Fletcher et al. 2007, 2008), with an interpolation between 6 and 35 mbar where the Cassini CIRS instrument has no spectral sensitivity. The axes are latitude and log-pressure, with pressure increasing downwards. The contours show that the temperature increases upward above the 80-mbar level. The lower panel of Figure 11.5 depicts the zonal-mean zonal wind inferred for Saturn's upper troposphere and lower stratosphere by integrating the thermal-wind Equation (11.1). The integration adopts as a lower boundary condition the zonal-mean zonal winds obtained from visually tracking clouds in the 350–700 mbar pressure range. At pressures greater than several mbar, the temperatures are correlated with the winds such that the equatorward flanks of eastward jets – the anticyclonic regions – are colder than the surrounding regions (on an isobar), and the poleward flanks of eastward jets – the cyclonic regions – are warmer than the surrounding regions. Given the thermal-wind equation, this correlation implies that the change of zonal wind with altitude is opposite to the direction of the wind – the winds are decaying with altitude. Figure 11.5 indicates that they decay to an average speed of about 40 m s^{-1}, which is the average speed of the stratosphere at the 1- to 2-mbar level.

This result is not new, but it has not been fully explained. The IRIS instrument on Voyager observed the same correlation of thermal gradients with the zonal jets during the encounters with Jupiter (e.g. Conrath and Pirraglia 1983; Gierasch et al. 1986). The problem is that there is no obvious radiative or thermodynamic heat source that would produce temperature gradients that correlate with the jets. The IRIS team therefore proposed a mechanical origin. They postulated that, in the upper troposphere above the clouds, the net zonal

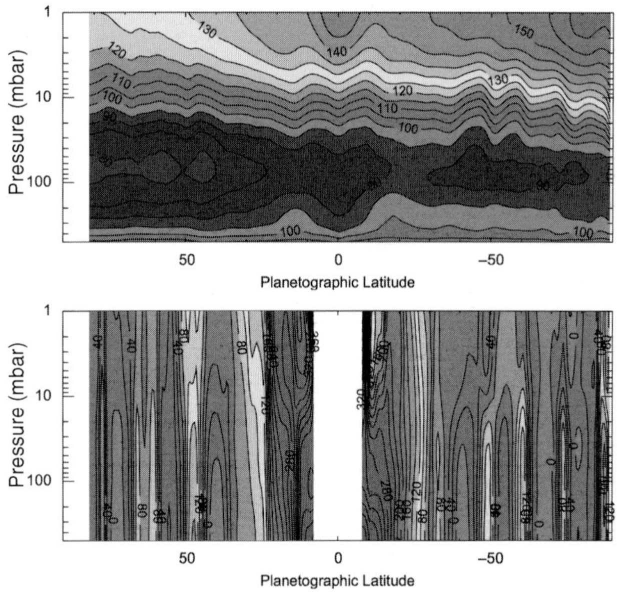

Figure 11.5 Top: Saturn's zonal-mean temperature versus latitude and pressure from Cassini CIRS nadir-viewing observations taken in 2004–2006, during Saturn's northern spring. Analysis is from Fletcher et al. (2007). Note the temperature minimum around 80 mbar and the increase in temperatures toward the south pole at all levels above the 6-mbar level. The instrument is not sensitive to the regions from 6 to 35 mbar and the temperatures have been interpolated across this pressure range. Bottom: Zonal-mean zonal winds (in Voyager System III) versus latitude and pressure obtained by integrating the thermal-wind equation using the temperature structure in the top panel, and assuming the zonal winds at 500 mbar correspond to the cloud-tracked wind profile. Warm colors represent eastward wind, and cool colors represent weaker and/or westward wind. From Read et al. (2009a). (A black-and-white version of this figure appears in some formats. For the color version, please refer to the plate section.)

acceleration due to eddies acts as a drag force in the *opposite* direction from the zonal jets. Such a zonal eddy force would – in statistical steady state – be balanced by the Coriolis force acting on the meridional wind. In other words, the postulated eddy accelerations would induce a meridional circulation.

Specifically, this force balance predicts that, for the Coriolis acceleration to balance the eddy acceleration, the resulting meridional circulation would be *poleward* across eastward jets and *equatorward* across westward jets.[1] This configuration implies that the meridional flow converges on the poleward flank of eastward jets, leading to downward motion.[2] Because the upper troposphere is stably stratified (specific entropy increases with altitude), the downwelling advects high-entropy air downward from above, leading to warmer temperature on constant-pressure surfaces. Similarly, the meridional flow diverges on the equatorward flank of eastward jets, leading to upward motion, which advects low-entropy air from below and produces cooler temperatures on constant-pressure surfaces. In other words, this scenario explains the temperature anomalies observed by Voyager and Cassini, and in thermal-wind balance, the resulting zonal jets must decay with altitude. Note that the poleward flank of eastward jets exhibit cyclonic vorticity and the equatorward flanks of eastward jets exhibit anticyclonic vorticity. Based mostly on differences in Jupiter's cloud colors and morphology, the amateur astronomers called the cyclonic and anticyclonic regions belts and zones, respectively. With a westward eddy acceleration on eastward zonal jets and an eastward eddy acceleration on westward zonal jets, the belts are regions of downwelling and the zones are regions of upwelling. This picture holds above the clouds, where observations are best, but it says nothing about how the jets are driven at deeper levels. The situation is likely very different within and below the clouds where the jets are driven, a situation that we address in the next several subsections.

11.2.4 Chemical Tracers

Although one cannot measure upwelling and downwelling directly – the velocities are too small – one can detect these motions using chemical tracers. If a substance is removed from the air at high altitudes, either by condensation or by chemical reactions, then finding it above that altitude means that it was brought there by an updraft. Conversely, if a substance is

[1] For poleward motion, the Coriolis force induces an eastward acceleration, whereas for equatorward motion, it induces a westward acceleration. If there were an eastward jet with a westward eddy-induced acceleration, the Coriolis force must be eastward in order to balance the eddy acceleration. This implies a poleward meridional flow. Likewise, if there were a westward jet with an eastward eddy acceleration, the Coriolis force must be westward to achieve steady state. This implies equatorward meridional flow.

[2] In principle, mass conservation requires that a meridional convergence be balanced by a vertical divergence of the flow. Thus, one could theoretically imagine that a meridional convergence could be balanced either by upward motion above the convergence level, or by downward motion below the convergence level (or a combination). Because of the strong stable stratification in the stratosphere, the magnitude of the upward flow above the convergence level should be limited, and thus we expect a significant degree of downward motion below the convergence level. Analogous arguments lead to the conclusion that upward motion should occur below locations of meridional flow divergence.

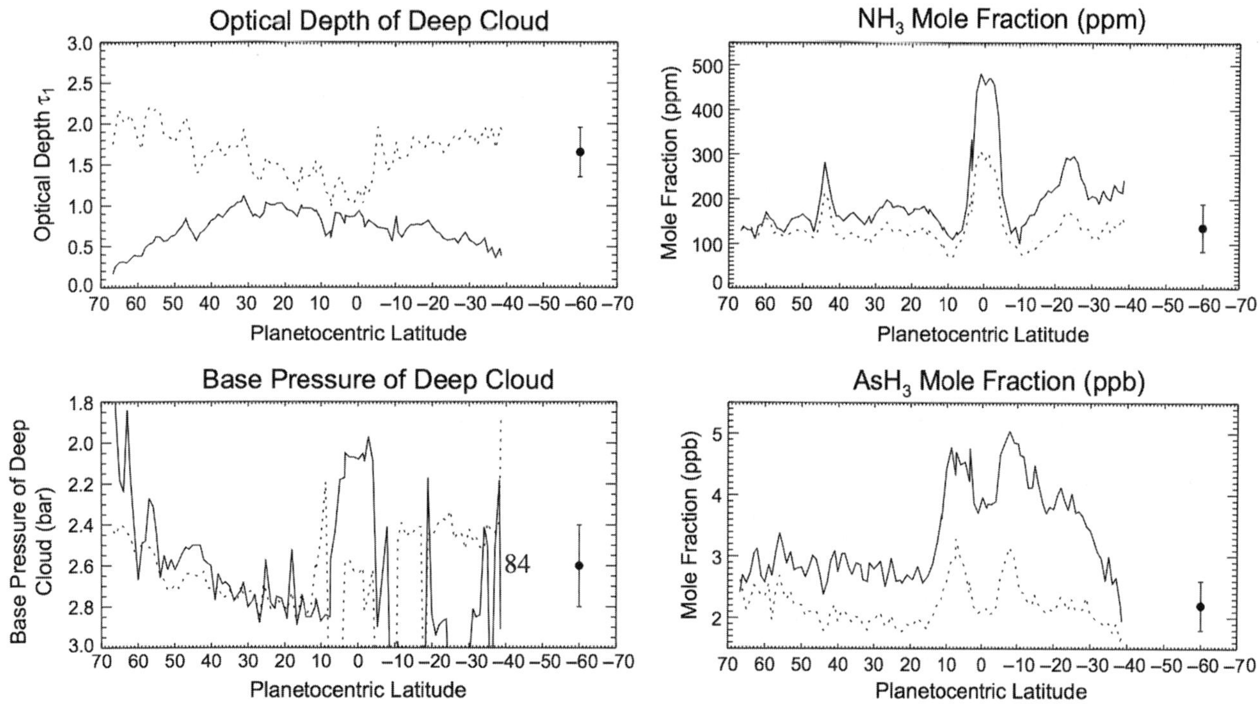

Figure 11.6 Tracers of vertical motion. The graphs are derived from 5-μm spectra obtained by the Cassini VIMS instrument, which senses thermal emission from the 1–3 bar pressure range. The solid and dotted curves employ two different assumptions about cloud opacity. Although the optical depth of the deep cloud is relatively constant, the cloud base pressure is lower (cloud height is greater), which is suggestive of equatorial upwelling. The higher NH_3 mole fraction out to latitudes of $\pm 8°$ is suggestive of upwelling, and the troughs at $\pm(8–20)°$ are suggestive of downwelling. The AsH_3 mole fraction shows a distribution that is almost opposite to that of NH_3, which is not fully understood. One possible explanation is that the pattern of upwelling and downwelling reverses at some level and the two gases are reflecting that reversal. From Fletcher et al. (2011).

severely depleted below the altitude where the chemical reactions are occurring, then the depleted air was probably brought there by a downdraft. If one knows the rate of the chemical reaction, one can often estimate the speed of the updraft or downdraft.

Figure 11.6, from Fletcher et al. (2011), shows some of the tracers of vertical motion. The data are from the Cassini VIMS instrument and are from spectra in the 4.6–5.1 μm atmospheric window. A window is where the gases of the atmosphere are especially transparent, and photons in this spectral range are sampling the 1- to 3-bar pressure region. Clouds are the main source of opacity, and the solid and dotted curves are from two different models of the clouds. Ammonia (upper right graph) is generally abundant at low altitude and is depleted by precipitation at high altitude. Abundant ammonia usually means it was brought there by an updraft. Thus, the updrafts have high ammonia and the downdrafts have low ammonia. Thus, the spike at $|\phi| < 8°$, where ϕ is the latitude, is a sign of upwelling at the equator. Similarly, the troughs at $8° < |\phi| < 20°$ are a sign of downwelling. The broad downwelling in the cyclonic regions on either side of the equator is consistent with the expected convergence of meridional flow into a region with an eastward jet on the equatorward side and a westward jet on the poleward side.

Figure 11.7, from Janssen et al. (2013), tells a similar story. The data show thermal emission at 2.2-cm wavelength, in the microwave region, and were taken by the Cassini radar instrument acting as a radiometer. Ammonia vapor is the main source of opacity at this wavelength – clouds are unimportant – so the bright areas are places of reduced ammonia abundance, which allows radiation from the warmer, deeper levels to escape into space. The depleted bands within $\pm 10°$ of the equator are evidence of downwelling. The black band on the equator is an image of Saturn's rings, which are colder than the planet. The dotted line gives the time of ring plane crossing, and the dashed line gives the time of the closest approach of the

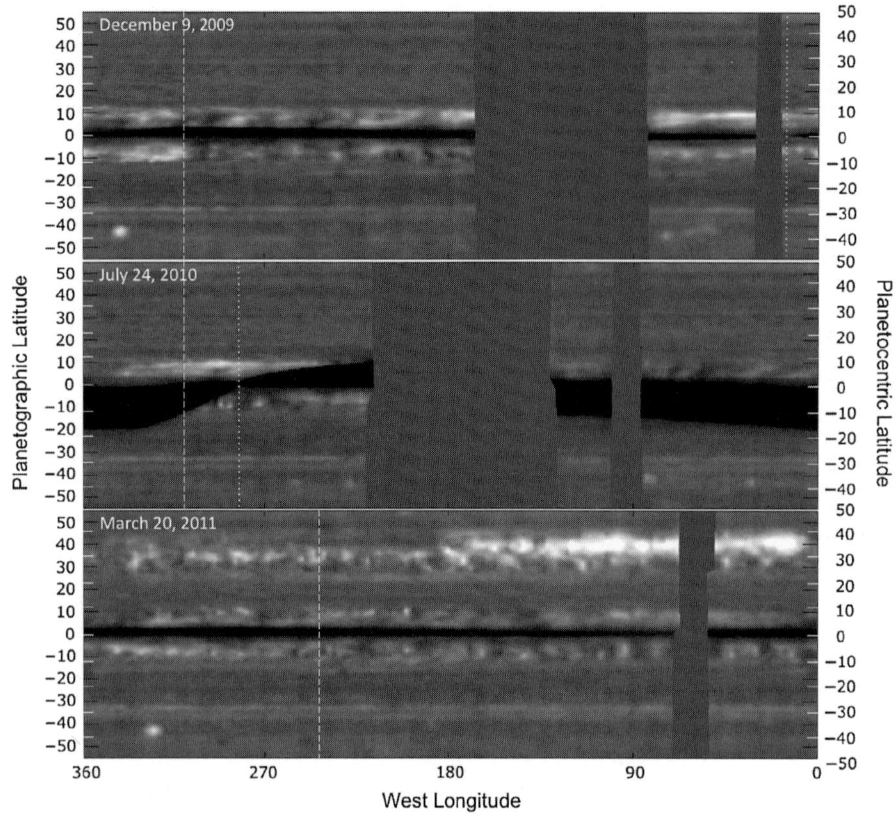

Figure 11.7 Thermal emission at 2.2-cm wavelength obtained by the Cassini radar instrument operating as a radiometer. The black band at the equator is the rings, which are low emitters because they are cold. Gaseous NH_3 blocks the radiation from warmer, deeper levels and emits at colder levels, so regions that are depleted in NH_3 appear bright. Latitudes out to ±10° appear to be depleted in NH_3, which implies downwelling. Although difficult to see due to blockage by the rings, the equator itself is dark, implying upwelling. The great northern storm is visible in the March 20, 2011 map at 30°–40° planetographic latitude and 0°–180° west longitude. A smaller storm is visible at −45° latitude and 330° west longitude. Both storms show NH_3 depletion. From Janssen et al. (2013).

spacecraft to the planet. As the spacecraft moved in and out of the ring plane, which it did on July 24, 2010, the rings covered a part of the opposite hemisphere, affording a brief look at the equator. The equator itself is not so bright, and is consistent with ammonia abundance near the saturation value, as it is over most of the planet outside the subtropical (±10°) band. Within that band, the ammonia must be depleted down to 1.5 bars to fit the high brightness temperatures (Laraia et al. 2013).

The distribution of NH_3 on Saturn resembles the distribution of H_2O on Earth. On both planets, there is rising in the tropics and sinking in the subtropics, although on Earth the subtropics extend further out, to ±30°. On Earth the circulation is called the Hadley cell, and the band of rising motion at the equator is called the Intertropical Convergence Zone (ITCZ), a feature of which are the high cumulus clouds associated with deep convection. The clouds of Saturn show a similar pattern (lower left of Figure 11.6) in which the base pressure of the deep cloud is less, indicating displacement to higher altitude. However, the Hadley cell analogy may be misleading, because Earth has equatorial easterlies (westward winds) and Saturn has equatorial westerlies (eastward winds). The eddy sources and the configuration of zonal accelerations they induce may therefore differ significantly between the two planets.

Gases other than ammonia tell a somewhat different story, and the differences are not fully understood. Figure 11.6 shows that the latitude distribution of arsine, AsH_3 is almost the opposite of the NH_3 distribution. Arsine has a dip at the equator and broad peaks on either side of the equator out to ±20°. One of the cloud models shows an extension of the peak to a latitude of −30°, but the other model does not. The distribution of phosphine, PH_3 (not shown), is a lot like that of arsine. Degeneracies between the retrievals of the cloud properties and the abundances of trace species like AsH_3 can occur (e.g. Giles et al. 2017), so the AsH_3 results shown in Figure 11.6 should perhaps be viewed as tentative until more analysis has been performed; nevertheless, the different cloud models explored by Fletcher et al. (2011) all pointed toward an equatorial dip in AsH_3 relative to the surrounding latitudes, suggesting that it should be taken seriously. Both arsine and phosphine are in chemical equilibrium at much deeper levels than those probed in the 4.6–5.1 μm data, and they are destroyed by photolysis at high altitudes. That they don't show the same patterns as NH_3 has interesting implications.

Fletcher et al. (2011) discuss the possibility of stacked meridional circulation cells, the top one with rising air at the equator, as in Earth's Hadley circulation, and the bottom one – the reverse cell – with sinking at the equator. The NH_3 cloud base is around 1.5 bars, as shown in Figure 11.3, so its mixing ratio should be large up to that level in an up draft. Above that level it loses ammonia by condensation. A downdraft could carry ammonia-depleted air down below the 1.5-bar level until it eventually mixes with air from the planet's interior. Thus, the NH_3 distribution shown in Figure 11.6 is consistent with an Earth-like Hadley cell extending down at least to the ~2-bar level. The AsH_3 distribution is consistent with a reverse Hadley cell since its chemical transformations are taking place at much deeper levels. The base pressure of the deep cloud, shown in the lower left panel of Figure 11.6, is about 2.0 bars at the equator and 2.8 bars on either side of the equator. This could be interpreted in two ways. If the cloud particles are NH_4SH or H_2O, and have been carried up from deeper levels, then the higher altitude (lower base pressure) at the equator would signify an updraft. However, if the cloud base pressure were entirely dependent on the abundance of a condensing gas, then the lower base pressure would signify a lower abundance of the condensing gas and hence a downdraft. The latter would be consistent with a reverse cell, with sinking motion at the equator. Note however that the upper cell is consistent with a meridional circulation that is driven by an eddy acceleration acting in the opposite direction as the zonal winds in the upper troposphere. The lower meridional cell would seem to require forces that accelerate the zonal winds, and such forces do exist, as we now describe.

11.2.5 Eddies and the Momentum Budget

The above analysis takes the zonal winds at cloud top (Figure 11.2) as given. The eddy-induced acceleration opposing the jet direction was postulated to account for the wind's decay with height above the cloud-top level. In contrast, observations demonstrate that at the cloud level, eddies transport momentum that acts to drive the jets. Here we describe these observations and their interpretation.

By themselves, the zonal-mean of the eddy winds average out to zero, but the eddies can have a net effect when the northward and eastward components act together. Consider a latitude band with an eastward jet to the north and a westward jet to the south. The eddy winds, with components u' and v', induce motion of air parcels north and south in the space between the two jets. (Here, as before, u is the zonal wind, and v is the meridional wind; primes denote deviations from the zonal average.) But if the parcels going north have more eastward momentum than the parcels going south, there will be a net transfer of momentum to the north. This would add to the eastward momentum of the eastward jet and it would subtract it from the westward jet, speeding up each jet in its respective direction.

Such a hypothesis is testable if one has good measurements of the eddy velocities. Parcels moving north ($v' > 0$) would also be moving east ($u' > 0$), and parcels moving south ($v' < 0$) would also be moving west ($u' < 0$). In both cases the parcel trajectories would be tilted along a northeast-southwest line, and the measured eddy winds would have $\overline{u'v'} > 0$, where the overbar denotes the zonal mean. The eddy-momentum flux is $\rho\overline{u'v'}$, where ρ is the density. The eddy-momentum flux is a stress, and it has units of force per unit area. A positive tilt ($\overline{u'v'} > 0$) implies a northward transport of eastward momentum by the eddies. A tilt the other way ($\overline{u'v'} < 0$) would represent a negative eddy-momentum flux–a northward transport of westward momentum by the eddies. In either case, if $\overline{u'v'}$ and $\partial u/\partial y$ have the same sign, where u is the mean zonal

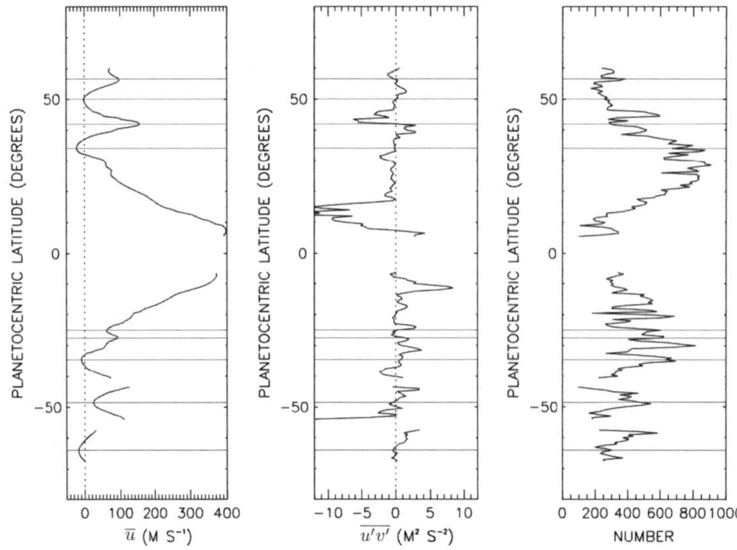

Figure 11.8 Eddy momentum transport and zonal winds. The left panel shows the mean zonal wind profile $\bar{u}$ obtained by tracking individual cloud features. The right panel shows the number of measured wind vectors in each 1° bin of latitude. The middle panel shows the covariance $\overline{u'v'}$ between the eastward eddy wind u' and the northward eddy wind v', where an eddy is defined as the residual after the mean zonal wind has been subtracted off. Multiplied by the density, this covariance is the eddy momentum transport – the northward transport of eastward momentum by the eddies. This transport tends to have the same sign as $\partial\bar{u}/\partial y$, where y is the northward coordinate, which indicates that the eddies are putting energy into the mean zonal wind and not the reverse. From Del Genio and Barbara (2012).

wind and y is the northward coordinate, then the eddies are adding momentum to the zonal jets.

Figure 11.8, from Del Genio and Barbara (2012), shows the sign of the correlation. At almost all latitudes, $u'v'$ and $\partial u/\partial y$ have the same sign. Both quantities are positive at latitudes 50°–57°N, 33°–42°N, 10°–25°S, 28°–35°S, and 45°–48°S. Both quantities are negative at latitudes 42°–50°N, 10°–33°N, 35°–40°S, and 48°–52°S. The data are noisy because the eddy winds are weak – only a few m s^{-1}, but the data clearly indicate that Saturn's eddies are acting as a force that accelerates the zonal jets. Both Voyager and Cassini observed similar behavior at Jupiter (Beebe et al. 1980; Ingersoll et al. 1981; Salyk et al. 2006).

This sounds like negative viscosity, and indeed that term was used to describe such phenomena, which are observed not only in Earth's atmosphere but also the oceans, the Sun and laboratory experiments (Starr 1968). However, using an eddy viscosity to relate a local stress like $\rho u'v'$ to a local rate of strain like $\partial u/\partial y$ is often a meaningless exercise (Phillips 1969). Energy transfer from smaller to larger scales does not violate any thermodynamic principle, and an eddy-momentum transfer that generates a shear flow – zonal jets – can arise naturally through the interaction of turbulence with the planetary rotation. Since the eddies are putting energy into the zonal flow, they must have their own source of energy. That could come from below, as internal heat powers convection currents that rise into the cloud layer. Or it could come from the sides, as lateral temperature gradients release their potential energy into a longitudinally varying wave in a process called baroclinic instability (e.g. Vallis 2006; Holton and Hakim 2013).

This eddy force at cloud level, which acts to accelerate the zonal jets, is opposite in sign to that postulated in the upper troposphere that tends to decelerate the jets, and it has the opposite effect on the meridional overturning. This reversal in sign of the eddy acceleration with height could lead to stacked meridional circulation cells, the eddy-momentum forces that oppose the jets driving the upper cell and the eddy-momentum forces that accelerate the jets driving the lower cell. Neither of these meridional cells however, is like the Earth's Hadley circulation (see Vallis 2006 or Schneider 2006 for reviews). On Earth, the eddy accelerations in the subtropics are westward, whereas the eddy accelerations occurring within Saturn's equatorial jet are eastward, leading to eddies driving a reverse meridional cell on Saturn. The direct circulation cell on Saturn, which seems to start at ~2 bars and extend into the stratosphere (Figure 11.5), is driven by an eddy acceleration opposing the zonal jets, which is opposite to the eddy momentum force at deeper levels.

The eddy-momentum flux driving the jets at cloud level has been observed directly on Jupiter and Saturn (Beebe et al. 1980; Ingersoll et al. 1981; Salyk et al. 2006; Del Genio and Barbara 2012). In contrast, there is no direct observational confirmation of the eddy acceleration above the clouds that is postulated to oppose the jets. The term "eddy" includes all motions that remain after the zonal mean has been subtracted off – waves, vortices, convective plumes and any other non-axisymmetric component of the motion. Evidence for the eddy force opposing the jets comes from the decay of the zonal winds with height as inferred from temperature gradients (Figure 11.5). Evidence for a reverse meridional circulation cell comes from the AsH_3 and PH_3 distributions, but further evidence comes from the distribution of lightning, as we discuss below.

11.2.6 Lightning and Moist Convection

Lightning is evidence of moist convection. The violent updrafts in a thunderstorm are a crucial element for electrical charge separation, which occurs when large ice particles fall through upwelling air that contains small liquid droplets (Uman 2001; MacGorman and Rust 1998). The charging mechanism is not well understood, but three phases of water – solid, liquid and vapor – in the cloud at once seem to be necessary. On giant planets, with their three-tiered structure (Figure 11.3), it is not immediately clear which cloud produces the lightning (Yair et al. 2008). For a solar composition atmosphere, the mole fractions of H_2O, NH_3 and H_2S, in units of 10^{-4} are 9.7, 1.3 and 0.29, respectively (these values depend on the abundance of helium, which is poorly known; here we have used the latest He/H_2 estimate from Sromovsky et al. 2016). For Saturn, enrichment by a factor of 10 is likely (Fletcher et al. 2009; Fletcher et al. 2012), and uniform enrichment would raise all these numbers by the same factor. Thus, on the basis of abundance alone, water is the condensable gas most likely to produce lightning. Also, water is the only gas where the temperatures and abundances allow three phases to exist at the same level in the atmosphere. For Saturn, the triple point of water occurs at a pressure of about 10 bars (Figure 11.3). This number depends on the temperature profile, but is relatively independent of the enrichment factor. Increasing the latter mostly affects the depth of the liquid cloud, extending it down to $p = 20$ bars and $T = 330$ K for 10 times solar abundances.

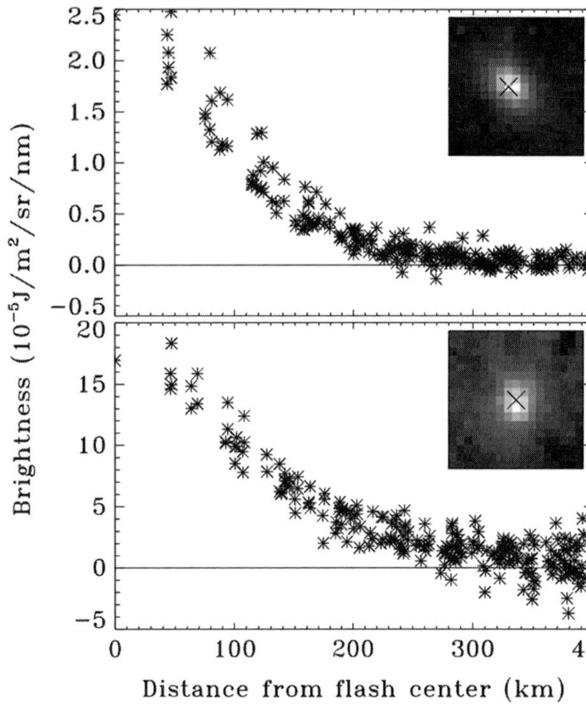

Figure 11.9 Spatial distribution of brightness of two lightning flashes vs. distance from the flash center. The distributions would appear circular if viewed from directly overhead. The half width at half maximum, defined as the radius where the brightness is half of the central brightness, is about 100 km. From these data, scattering models indicate that the vertical distance from the lightning source to the top of the clouds where the photons are detected is in the range 125–250 km. The data are consistent with a lightning source in the water cloud (Figure 11.3). From Dyudina et al. (2013).

Figure 11.9, from Dyudina et al. (2013), shows the brightness distribution of a typical lightning flash as seen from the Cassini spacecraft. Every pixel in the vicinity of the flash is plotted as a function of its horizontal distance from the flash center. This removes the effects of foreshortening and shows that the flashes are roughly circular. The half width at half maximum (HWHM) of the brightness distribution is about 100 km, and that is related to the depth of the lightning relative to the tops of the clouds – the level where the photons emerge. That depth is approximately equal to 1.5 times the HWHM. Cloud parameters – both their height and their scattering properties – create a large uncertainty. Dyudina et al. (2013) estimate that the tops are probably NH_3 or NH_4SH clouds at depths exceeding 1.2 bars, in which case the lighting is in the water cloud.

Nighttime imaging of Jupiter by the Galileo spacecraft showed that lightning is concentrated in the belts (Little et al. 1999; Gierasch et al. 2000), and that was a surprise. The traditional view, which was based on Voyager observations of the upper troposphere (Gierasch et al. 1986), was that the belts are regions of descent and the zones are regions of ascent, for several reasons: the zones exhibit a thick, bright, relatively uniform cloud deck, whereas the belts exhibit patchier, generally thinner clouds. The ammonia mixing ratios at and above the clouds are small in the belts and large in the zones, signifying descent and ascent, respectively. The belts are warm, signifying high-entropy air brought down from above, and the zones are relatively cold. Lighting therefore seemed unlikely in the belts, since moist convection requires moist air brought in from below. This led Ingersoll et al. (2000) to postulate that perhaps the belts have upwelling at the base of the cloud and downwelling at the tops. This amounted to a pair of meridional circulation cells turning in opposite directions, one on top of the other. Showman and de Pater (2005) showed that this stacked-cell scenario also best explains the deep ammonia abundances from 1 to 5 bars, which indicates that both belts and zones are depleted in ammonia relative to the presumed deeper atmosphere. According to this new picture, the traditional view was not wrong, but it was based only on the properties of the upper circulation cell. Lightning and ammonia had provided a view to deeper levels, at least on Jupiter. The data for Saturn are less clear, partly because lightning is such a rare event on that planet.

Figure 11.10, from Dyudina et al. (2007), shows lightning intensity over a 2-year period from 2004 to 2006 as recorded by the Radio and Plasma Wave Science (RPWS) instrument on Cassini. The RPWS is a radio receiver and spectrometer, and it is "on" all the time. The figure shows that there was lighting activity on days 200–270 in 2004, then a 445-day gap with no lightning, followed by activity on days 350–385 in 2005. Only one storm was active during each of the two periods, and it might have been the same storm, as indicated by the drift rate in longitude shown on the bottom part of the figure. The storm was located at −35° planetocentric latitude, at the center of the westward jet shown in Figure 11.2. This latitude band became known as "storm alley" during the first few years of the Cassini mission. The storm was a multi-armed complex, 2000 km in diameter, which spawned smaller spots at a rate of one every 1–2 days. These drifted off to the west relative to the central complex. The smaller spots began with high, thick clouds that dissipated in the first day, leaving a dark spot that consisted either of a hole in the clouds or else a deep cloud with a low reflectivity (Dyudina et al. 2007). The small bright spot in the lower left corner of Figure 11.7, the image taken on March 20, 2011, is also at −35° latitude. Its brightness at 2.2-cm wavelength indicates that it is a region of low abundance of ammonia vapor, which is consistent with a hole in the clouds. The latitude band at the center of the westward jet at −35° seems to have been active over the 10+ years of the Cassini orbital mission.

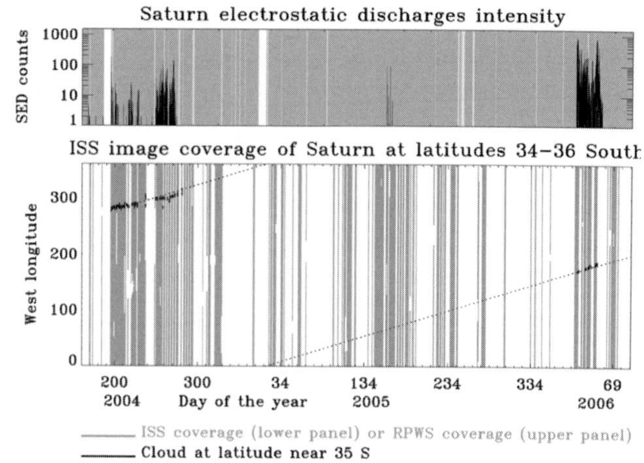

Figure 11.10 Lightning activity during the first two years of the Cassini orbital mission. The top panel shows signals detected by the RPWS instrument, which is on all the time and is sensitive to lightning all over the planet. The lower panel shows when storm clouds were seen by the Cassini imaging system. Both panels show that there was no lightning for 445 days from the end of 2004 to the beginning of 2006. For comparison, there are ~2000 lightning storms at any given moment on Earth. From Dyudina et al. (2007).

During the Cassini mission, the northern hemisphere did not have a lightning storm until 5 December 2010, when the RPWS suddenly started detecting lightning signals and the imaging system detected a new cloud at

35° latitude. As in the southern hemisphere, the latitude of the storm coincided with the center of a westward jet (Figure 11.2). However, the northern storm was a different beast from those in storm alley. Its head remained the most active part, but by late January its tail had spread around the planet to the east and was encountering the head from the west (Fischer et al. 2011; Sánchez-Lavega et al. 2011). The tail was depleted in ammonia, as seen in the March 20, 2011 map at 2.2 cm (Figure 11.7). The head spawned a series of large anticyclonic vortices, 10,000 km in north-south dimension, which also drifted to the east, but more slowly than the debris in the tail. When the first of these – also the largest – encountered the head in June 2011, the head broke up and essentially ceased to exist (Sánchez-Lavega et al. 2012; Sayanagi et al. 2013). Chapter 13 by Sanchez-Lavega et al. in this book is devoted to the giant storm, and we will only discuss a few features that are especially relevant to this dynamics chapter.

Figure 11.11, from Achterberg et al. (2014), shows changes from 2009 to 2011, before and after the storm, using infrared data from the Cassini CIRS instrument. Figures 11.11a and 11.11b show that the storm warmed the latitude band from 30° to 45°. The warming also showed up as an increase in the emitted power at that latitude (Li et al. 2015). Figures 11.11c and 11.11d show evidence of gas having been brought up from the level of the water cloud, as inferred from the relative abundance of the two states of the H_2 molecule. The observed warming made $\partial T/\partial y$ more negative on the north side of the disturbed band and more positive on the south side. The thermal-wind equation then says there should have been an increase in $\partial \bar{u}/\partial z$ on the north side and a decrease on the south side. If we are seeing the tops of the clouds, and if $\bar{u}$ is constant underneath, we should see an increase in $\bar{u}$ to the north and a decrease to the south, and that is what was observed (Sayanagi et al. 2013).

Lightning on Jupiter was seen mostly in the belts, which are the regions of cyclonic shear (Little et al. 1999; Gierasch et al. 2000). Lightning on Saturn was seen near the centers of the westward jets, which have cyclonic shear on the equatorward side and anticyclonic shear on the poleward side. Thus, the two planets appear to be different. However, Dyudina et al. (2013) noted that within the giant storm there were small regions of cyclonic shear, and they were where the lightning occurred. The giant storm consisted of an anticyclonic head with a chain of large anticyclones stretching off to the east, each one rolling in a clockwise direction. The narrow regions where they came close together were cyclonic. A cyclonic region has low pressure at the center, with the inward pressure gradient force balancing the outward Coriolis force. In a terrestrial hurricane, this leads to convergent flow in the boundary layer, where friction at the ocean surface slows the wind and weakens the Coriolis force,

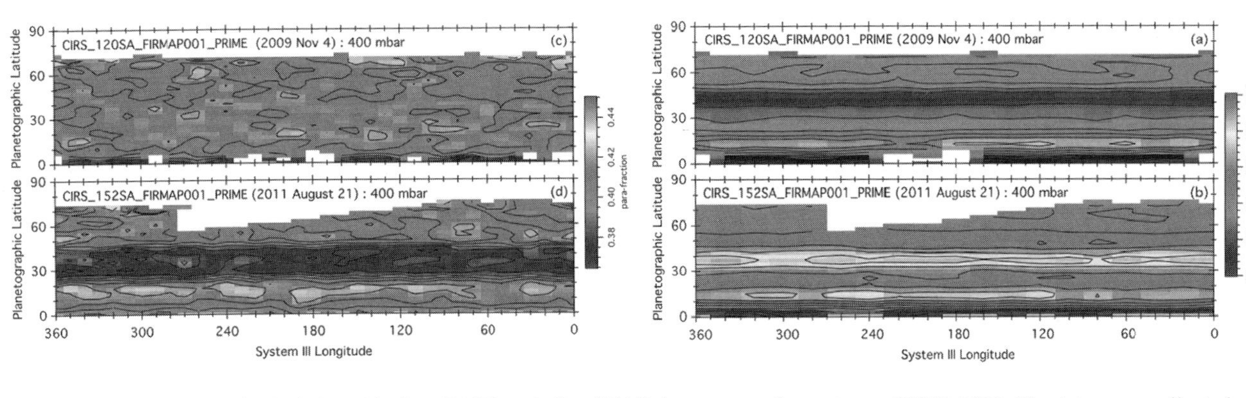

Figure 11.11 Saturn in the infrared before (2009) and after (2011) the great northern storm of 2010–2011. The data were collected by the Cassini CIRS instrument and are sensitive to the ~300-mbar level. The right panels show that the storm warmed the latitude band from 30° to 45° by ~3 K. The left panels, illustrating the hydrogen para fraction, show that the warming was due to air rising from depths near the water cloud (Figure 11.3). The equilibrium ratio of the two states of the H_2 molecule, ortho and para, depends on temperature. At depth, the expected para fraction is smaller, whereas at the cloud tops, it is larger. The conversion between the two states occurs on relatively long timescales. Thus, the sudden emergence of a zonal band of air with low para fraction-as seen in panel (d) suggests that this air was transported from depth, but that there has not yet been sufficient time for the para fraction of this air to relax into the larger para fraction associated with equilibrium at the cold temperatures of the cloud tops. From Achterberg et al. (2014). (A black-and-white version of this figure appears in some formats. For the color version, please refer to the plate section.)

leading to an unbalanced pressure gradient force acting inward. The inward flow picks up moisture from the ocean surface, leading to intense moist convection around the center. One might think that the same process is happening on the giant planets, since moist convection occurs in the cyclonic regions. But how this might work without an ocean, or a physical surface against which friction can occur, is unclear. The deep atmosphere below the clouds would have to provide a frictional force on the cloud layer that leads to convergence. If the flow below the clouds were sufficiently turbulent, with large vertical stresses, it might do the job. But that theory needs to be worked out.

11.2.7 Stability of the Zonal Jets

The cloud-top zonal jets on Jupiter and Saturn are remarkably constant in time. Their positions and wind speeds haven't changed much over 100 years, although small differences between Voyager and Cassini were observed, especially at the equator (Peek 1958; Porco et al. 2003; Vasavada et al. 2006; Li et al. 2013; García-Melendo et al. 2011). This remarkable constancy motivates a consideration of the extent to which the jets are dynamically stable. The definition of a stable flow is one in which infinitesimal perturbations cannot grow in amplitude. There exists a class of theorems that provide information on jet stability for the idealized case of zonally symmetric jets, with no forcing or damping, and several of these have been evaluated for Saturn.

Most stability theorems are cast in terms of the potential vorticity (PV), which, for the case of a single-layer, shallow-water flow is

$$PV = \frac{\varsigma + f}{h}, \tag{11.2}$$

where $\zeta = \mathrm{k} \cdot (\nabla \times v)$ is the relative vorticity, v is the fluid velocity, and k is the vertical unit vector. In this shallow-water form, h is the depth of the fluid. Under frictionless, adiabiatic conditions, the PV is a materially conserved quantity (e.g. Vallis 2006). The inverse dependence on h reflects the fact that stretching along the rotation axis of the fluid parcel (thereby increasing h) causes a contraction toward the parcel's rotation axis and a decrease in the moment of inertia, causing the parcel to spin faster and $\zeta + f$ to increase. The PV of the parcel is conserved because the numerator and denominator change together. Note that, for a stratified, three-dimensional atmosphere, the PV can also be written in the form (11.2) if one uses the component of vorticity that is perpendicular to surfaces of constant specific entropy, and if one treats h as a differential mass per unit area between surfaces of constant specific entropy.

The simplest stability theorem, which dates back to Rayleigh, with modification for rotating planets, is the Charney–Stern criterion (Vallis 2006). It says that a zonal flow is stable – infinitesimal disturbances cannot grow – if PV varies monotonically in the cross-stream direction. If we ignore the stretching term by treating h as a constant, we obtain the barotropic stability criterion, which states that a steady zonal flow $\bar{u}(y)$ is stable, provided

$$\frac{d(\bar{\zeta} + f)}{dy} = \beta - \bar{u}_{yy} \geq 0, \tag{11.3}$$

where $\beta = df/dy = 2\Omega \cos\phi / a$ and $d\bar{\zeta}/dy = -\bar{u}_{yy}$. Here, β is the planetary vorticity gradient, ϕ is latitude and a is the radius of the planet. Note that β is always positive but it goes to zero at the poles. The subscript y is a derivative, and $\bar{u}_{yy}$ is the curvature of the zonal velocity profile, which is positive at the centers of the westward jets (Figure 11.2). If this curvature is less than β at every latitude, the flow is stable provided we can ignore the stretching term. Jupiter's zonal jets violate the barotropic stability criterion (Ingersoll et al. 1981; Limaye 1986; Li et al. 2004). Here we describe an application of stability criteria to Saturn, starting with the barotropic stability criterion and moving on to the more general Ertel and quasi-geostrophic versions.

Figure 11.12, from Read et al. (2009a), shows the zonal-mean velocity $\bar{u}(y)$, the vorticity $\bar{\zeta} = -\bar{u}_y$, and the curvature $u_{yy} = -\zeta_y$ in the left, center, and right panels, respectively. The smooth curve in the right panel is β, which varies with latitude as $\cos\phi$. One can see that the curvature exceeds β at the latitudes of the westward jets, indicating a violation of the barotropic stability criterion. The violation is more evident near the poles, i.e. the difference $\beta - \bar{u}_{yy}$ is more negative there, partly because the jets have more curvature there, but also because β is approaching zero.

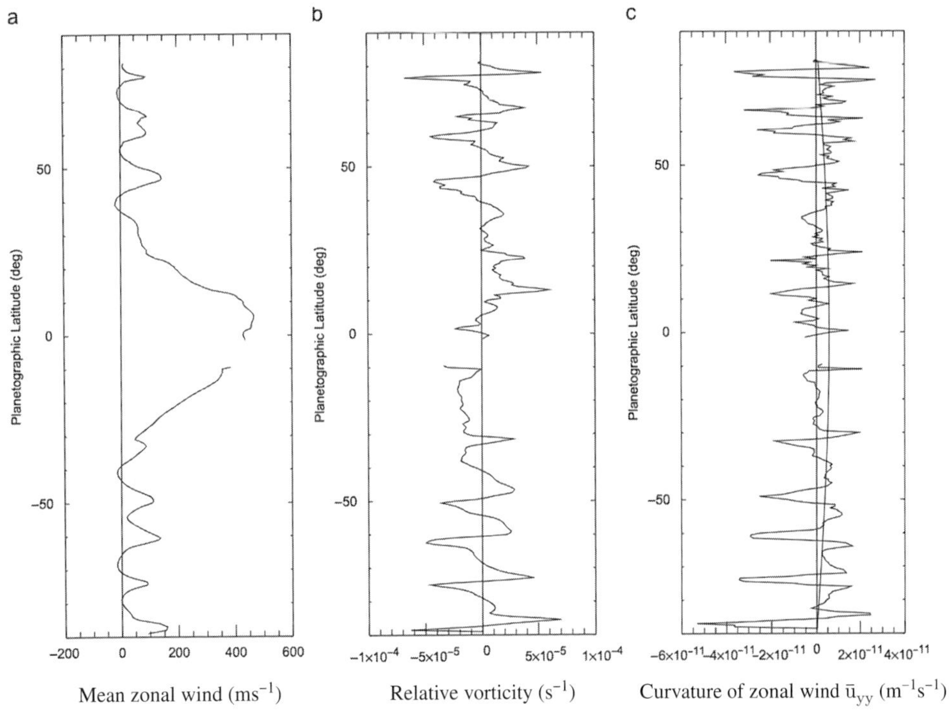

Figure 11.12 Stability of Saturn's zonal jets. The barotropic stability criterion states that a zonal flow is stable provided the curvature $\bar{u}_{yy}$ of the zonal velocity profile does not exceed the planetary vorticity gradient β. Stable means that disturbances to the profile cannot grow, but violation of the criterion does not mean that they will grow. The left panel shows the zonal velocity profile $\bar{u}$. The middle panel shows the relative vorticity $-\bar{u}_y$ associated with the zonal flow, where the subscript means differentiation of $\bar{u}$ with respect to the northward coordinate y. The right panel shows the curvature $\bar{u}_{yy}$ (jagged curve) and β (smooth curve, given by $2\Omega \cos \phi/a$), along with the zero-curvature line for reference (straight vertical line). At the latitudes where the curvature is positive, it clearly exceeds β, indicating that the barotropic stability criterion is violated. The barotropic stability criterion belongs to a more general class of stability theorems that use potential vorticity (PV) as the diagnostic quantity. The rigorous form uses Ertel's PV, but the results are the same, at least for the upper troposphere where the velocity profile is measured. From Read et al. (2009a).

Read et al. (2009a) also use temperature profiles derived from infrared observations by the Cassini CIRS instrument to estimate the magnitude of the stretching term. They calculate Ertel PV and quasi-geostrophic PV, including the stretching term, as functions of latitude and show that the stretching term has little effect, at least in the upper troposphere and lower stratosphere where the analysis was done. At these altitudes, the atmosphere is so stably stratified that the stretching is small. Although Ertel PV is the right quantity to use, the barotropic stability criterion gives the same result, that the centers of the westward jets are where the stability criterion is violated. One might get a different result if one could measure Ertel PV within and below the clouds, but that is not possible with the remote sensing data that we have.

Violation of the barotropic stability criterion or any of the related stability criteria does not mean that the flow is unstable. Further, instability does not mean that the flow cannot exist. The hexagon that surrounds the pole at mean planetocentric latitude of 76° marks the path of a meandering jet that violates the barotropic stability criterion (Antuñano et al. 2015). The jet may be unstable, but the disturbance has grown into a steady finite-amplitude wave. Presumably the wave stopped growing due to non-linear effects that are not included in the linear stability analysis. Ingersoll and Pollard (1982) describe a stability criterion that invokes interior flow, in which the zonal winds are aligned on cylinders concentric with the planets axis of rotation. Dowling (1995a,b) has argued that Jupiter's jets could be stable according to Arnol'd's second stability criterion, which states that the flow will be stable to nonlinear perturbations if there is a reference frame in which the reversals of the PV gradient coincide with reversals in the sign of velocity. Read et al. (2009b) found that the upper tropospheric and stratospheric winds appear to be close to neutral stability according to this criterion; see Dowling (2014) for a detailed discussion. The reference period they derived is ~5 minutes shorter than the System III reference period, and they propose that the period derived from Arnol'd second stability criterion is the rotation period of Saturn's interior.

11.3 Observations of the Stratosphere

In the absence of visible tracers, the primary source of data constraining the winds and circulation in Saturn's stratosphere is from observations in the thermal infrared. Observations in the collision-induced hydrogen continuum longward of 16 μm provide information on the temperatures in the upper troposphere between roughly 50 to 500 mbar. Observations in the v_4 band of methane near 7.7 μm provide information on temperatures in the middle and upper stratosphere, between 0.5 and 5 mbar for typical observations, extending to about 0.01 mbar for observations with very high spectral resolution, or limb viewing observations from Cassini. However, these thermal observations have vertical resolution of a pressure scale height, and are thus not sensitive to wave structure on small vertical scales.[3] Temperature profiles with vertical resolution of a few kilometers can be obtained through radio occultation observations, in which the radio signal from a spacecraft is observed as the spacecraft passes behind a planet at seen from Earth, and from stellar occultations. Radio occultations are generally sensitive to the lower stratosphere and upper troposphere between about 0.1 mbar and 1 bar, while stellar occultations generally sense the upper stratosphere or higher.

The first spatially resolved thermal observations of Saturn were made in the v_9 band of ethane at 12 μm (Gillett and Orton 1975; Rieke 1975) during southern summer, and showed emission increasing from north to south, peaking at the south pole. Subsequent observations in the v_4 methane band at 8 μm (Tokunaga et al. 1978; Sinton et al. 1980) showed similar variations. Since methane is expected to be meridionally mixed, these observations indicated that the emission variations are caused by a north-to-south temperature gradient, with stratospheric temperatures warmest at the south (summer) pole.

The first spacecraft observation in the thermal infrared was made by the Pioneer 11 infrared radiometer (IRR) in 1979, just prior to northern spring equinox. IRR observed in two bands at 20 μm and 45 μm, sensitive to the upper troposphere, between 10°N and 30°S (Ingersoll et al. 1980; Orton and Ingersoll 1980), finding the temperatures within 10° of the equator ~2.5 K colder than higher latitudes. Soon afterward, the Infrared Interferometer Spectrometer (IRIS) on Voyagers 1 (1980) and 2 (1981) mapped Saturn at 5 to 45 μm. Retrievals of upper tropospheric temperatures (Hanel et al. 1981, 1982; Conrath and Pirraglia 1983) showed no equator to pole temperature gradient in the southern hemisphere, and north polar temperatures roughly 5 K colder than the equator in the upper troposphere at pressures less than about 300 mbar. In addition to the large-scale gradient, the upper tropospheric temperatures showed variations of around 2 K, with the gradient anticorrelated with the zonal winds (Conrath and Pirraglia 1983). A limited set of temperature profiles, covering roughly between 1 bar and 1 mbar, were also obtained by the radio occultation experiments on Pioneer 11 (Kliore et al. 1980) and Voyagers 1 and 2 (Lindal et al. 1985). These profiles showed a broad tropopause with the temperature minimum at about 50 mbar, with mid-stratospheric temperatures of ~140 K at midlatitudes and ~120 K near the equator. The near equatorial profiles also showed small-scale vertical oscillations that were interpreted as vertically propagating waves.

Subsequent ground-based observations taken during northern summer showed enhanced emission at the north pole in the stratospheric methane and ethane bands (Gezari et al. 1989), but not at wavelengths sensitive to the troposphere (Ollivier et al. 2000). These observations, when compared to earlier observations, were interpreted as evidence for seasonal evolution of the temperatures as predicted by radiative models (Cess and Caldwell 1979; Bezard and Gautier 1985), with warm temperatures at the summer pole and

[3] Atmospheric waves generated in the troposphere can propagate into the stratosphere where they can exert a strong effect on the dynamics. Broadly, such waves fall into two classes (neglecting acoustic waves which tend to be unimportant for the meteorology). Gravity (or buoyancy) waves arise by virtue of buoyancy forces associated with vertical motion of air parcels in a stably stratified medium; they are the internal atmosphere equivalent of the waves one observes on the surface of the ocean. Rossby waves arise from a restoring force associated with meridional motion of air parcels in the presence of a latitudinal gradient of the planetary vorticity (i.e., the Coriolis parameter f). Rossby waves tend to have larger scales and longer periods than gravity waves. Mixed wave modes also exist, as well as wave types specific to the equatorial regions, such as Kelvin wave modes. All of these waves can influence the mean flow when they break or are absorbed. See Holton and Hakim (2013) for introductions to these waves.

the strength of the radiative response weaker at higher pressures where the radiative time constants are longer.

Just prior to the arrival of Cassini at Saturn, Greathouse et al. (2005) observed Saturn at high spectral resolution in 2002 near southern summer solstice, using the methane ν_4 band to retrieve southern hemisphere temperatures between 0.01 and 10 mbar. They found a general equator to pole temperature gradient, with the south pole 10 K warmer than the equator. In early 2004, Orton and Yanamandra-Fisher (2005) used the Keck telescope to image Saturn in the CH_4 and H_2 bands, allowing the retrieval of temperatures at 100 mbar and 3 mbar at 3000 km spatial resolution. They also found a 10 K temperature increase between the equator and the summer pole in the stratosphere, with a sharp ~5 K increase between 69°S and 74°S planetocentric latitude.

The arrival of Cassini at Saturn in mid-2004, near southern midsummer, began a period of quasi-continuous monitoring of Saturn's thermal emission by the Cassini Composite Infrared Spectrometer (CIRS) which is planned to continue until northern summer solstice in 2017. Flasar et al. (2005) showed temperature retrievals of the southern hemisphere from data taken during the approach of Cassini to Saturn. They found a 15 K temperature gradient between the equator and south pole at 1 mbar, decreasing with increasing pressure. This temperature gradient is larger than the 5 K expected from radiative models for the season, which Flasar et al. (2005) suggested may be caused by adiabatic heating from subsidence at the pole. The first global temperature cross-sections were made by Fletcher et al. (2007, 2008), using CIRS data to retrieve zonal-mean temperatures between 0.5 and 5 mbar in the stratosphere and 50 to 800 mbar in the troposphere; an updated version of their results is shown in Figure 11.5. At the 1-mbar level, they found a 30 K temperature difference between the warm southern summer pole and the cold northern winter pole; the temperature gradient is nearly monotonic from pole to pole except for a local temperature maximum at the equator, and a roughly 3 K warming between 80°N and the north pole. At higher pressures, the global temperature gradient becomes weaker, and small-scale variations correlated with the zonal winds appear. CIRS limb viewing observations allowed Fouchet et al. (2008) and Guerlet et al. (2009) to extend temperature retrievals up to the 0.01 mbar pressure level. Surprisingly, they found that the large-scale pole-to-pole temperature gradient was weaker at 0.1 mbar and 0.01 mbar than at 1 mbar, despite the thermal timescales becoming shorter at lower pressures. They also found that the temperature profile oscillates with altitude at equatorial and low latitudes, with equatorial temperature alternately warmer and colder than temperature at ±15°. Long-term monitoring from ground-based observations by Orton et al. (2008) showed that the difference between equatorial and low-latitude temperatures also oscillates in time, with a period near 15 Earth years. This equatorial oscillation will be discussed in more detail in Section 11.3.3.

Cassini CIRS has continued to monitor Saturn's thermal structure through equinox and into northern spring (Fletcher et al. 2010; Fletcher et al. 2015; Guerlet et al. 2011; Sinclair et al. 2013; Sylvestre et al. 2015). Outside of the equatorial region, the seasonal temperature variations are broadly consistent with models (Fletcher et al. 2010; Friedson and Moses 2012; Guerlet et al. 2014), with a warming of the northern hemisphere and cooling of the southern hemisphere. The seasonal variations are discussed in detail in Chapter 10 by Fletcher et al. in this volume.

11.3.1 Zonal Mean Winds

As described in Section 11.2, the meridional temperature gradients observed in the upper troposphere are positively correlated with the measured cloud-top zonal winds at many latitudes. Using the thermal-wind Equation (11.1), this result implies that the zonal winds observed at cloud level decay with altitude into the stratosphere, over a vertical scale of about six pressure scale heights. Furthermore, because the eastward and westward wind gradients are of similar magnitude, while the cloud top winds are predominantly eastward, the decay of the jets is to a finite eastward velocity and not zero, as measured in the System III reference frame. Further evidence for net eastward stratospheric winds was provided by the analysis of the 1989 occultation of 28 Sgr by Nicholson et al. (1995), who found that winds of around 40 m s^{-1} were needed near the 2.5 mbar level between 25°N and 70°N to fit the observed timings of the central flash. As can be seen in Figure 11.5, many – though not all – of the off-equatorial jets decay in speed with decreasing pressure. Figure 11.5 also makes plain that the middle stratosphere winds are generally eastward as measured System III, except at high southern latitudes. As mentioned previously, however, the System III reference frame may not represent the true interior rotation rate, and other possible interior rotation rates would imply different values for the overall stratospheric winds (in particular, several proposals for the interior rotation rate are faster than the System III rotation rate, which would imply stratospheric winds weaker–that is, more westward – than shown in Figure 11.5). That said, there is no dynamical reason to expect that the mean zonal velocity in the troposphere or stratosphere is the same as in the deep interior.

Observations of seasonal changes in the zonal mean stratospheric temperatures into early northern spring (Fletcher et al. 2010; Sinclair et al. 2013; Fletcher et al. 2015) show a weakening of the equator-to-pole temperature gradients, with the hemispheric average stratospheric $\partial T/\partial y$ becoming less negative. This implies corresponding changes in the stratospheric zonal winds, with northern hemispheric winds becoming more westward and southern hemispheric winds more eastward, as can be seen in the results of Fletcher et al. (2015), who used the thermal wind approximation to estimate the winds at 100, 1 and 0.5 mbar poleward of 60° from 2005 to 2013.

11.3.2 Meridional Circulation

The zonal-mean circulation in the stratosphere can be inferred from the observed temperature field using the thermodynamic energy equation

$$\frac{\partial \overline{T}}{\partial t} + \overline{v}\frac{\partial \overline{T}}{\partial y} + \overline{w}\left(\frac{\partial \overline{T}}{\partial z} + \frac{g}{c_p}\right) = \frac{\overline{q}}{c_p}, \qquad (11.4)$$

where v and w are the meridional and vertical velocities, c_p is the specific heat, and q is the specific radiative heating/cooling rate. Due to the strong stratification in Saturn's stratosphere, the vertical advection term is much larger than the meridional advection term, and so the meridional advection is often ignored when estimating the vertical velocity.[4] Observations of stratospheric temperature allow estimates of the time derivative term and the factor in parentheses. Given estimates of the radiative heating/cooling q/c_p from radiative-transfer models and observations, Equation (11.4) can be used to estimate the vertical velocity. The velocities v, w in (11.4) are the so-called "residual mean velocities" (see e.g. Andrews et al. 1987, section 3.5) which include eddy fluxes as well as the advective transport. In the Earth's stratosphere, the residual mean circulation is a good approximation to the net transport of conserved constituents (Dunkerton 1978).

Flasar et al. (2005) estimated the vertical velocities needed to produce the observed warm temperatures at the south pole by balancing the vertical advection with the time derivative of the temperature, assuming temperature variations on the order of the observed equator-to-pole gradient on a seasonal timescale, finding $w \sim 0.1$ mm s^{-1}. Fletcher et al. (2015) used the thermodynamic energy equation to estimate the vertical velocity at latitudes poleward of 60° averaged over the first ten years of the Cassini mission (2004–2014), using temperatures from Cassini CIRS and net radiative heating from the model of Guerlet et al. (2014). They also found velocities of the order of 0.1 mm s^{-1} near 1 mbar, becoming weaker at higher pressures, with rising motion in the southern hemisphere and subsidence in the north (Figure 11.13). At this speed, it would take an air parcel 15 years to rise 50 km, which is about one scale height in Saturn's stratosphere.

Information on the stratospheric meridional circulation can also be obtained from the observed distribution of disequilibrium chemical species. The most useful species are ethane (C_2H_6) and acetylene (C_2H_2), which have photochemical timescales in excess of a Saturn year in the middle and lower stratosphere, with ethane having a somewhat longer timescale. Photochemical models predict that in the absence of meridional transport, ethane and acetylene will have a meridional distribution that follows the yearly mean solar insolation at pressures of 1 mbar and greater, with the abundance a maximum at the equator and

[4] The meridional advection term scales as $\overline{v}\,\Delta T$merid/L, where ΔTmerid is the characteristic meridional temperature contrast, which occurs over a meridional scale L. Considering a vertically isothermal temperature profile for simplicity, the vertical advection term scales simply as $\overline{w}\; g/c_p$. The continuity equation implies that $\overline{v}\,/L \sim \overline{w}\,/D$, where D is the characteristic vertical scale of the circulation. The ratio of the meridional to the vertical advection term is therefore $c_p\Delta T$merid/(gH). We insert numbers for the global-scale seasonal meridional circulation, for which ΔTmerid $\sim$ 20 K (see Figure 11.5). Noting that the circulation is coherent over many scale heights vertically (Figure 11.13), and that the scale height is $H \sim 50$ km at these temperatures, we adopt $D \sim 300$ km. Using $c_p \approx 1.3 \times 10^4$ J kg^{-1} K^{-1} and $g \approx 10$ m s^{-2} shows that the ratio of horizontal to vertical advection terms is $\sim$0.1.

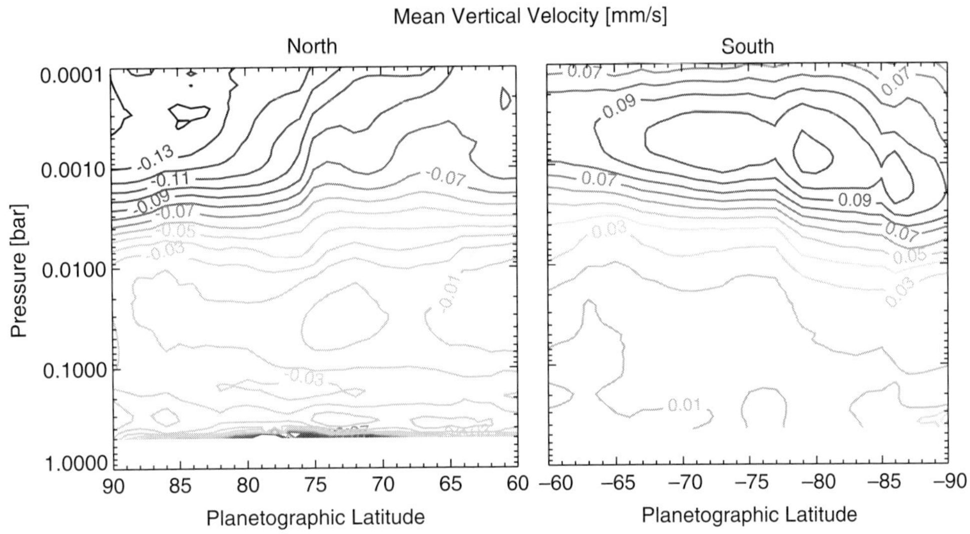

Figure 11.13 Zonal-mean vertical velocities (in mm s^{-1}) versus latitude and pressure obtained diagnostically from Equation (11.4) using the observed temperature and its seasonal variation, combined with estimates of radiative heating and cooling. From Fletcher et al. (2015).

decreasing toward the poles; at lower pressures the chemical timescales become shorter and the expected profiles follow the seasonally varying insolation (Moses and Greathouse 2005; Fouchet et al. 2009, see also Chapter 10 by Fletcher et al. in this volume). Observations of ethane and acetylene from ground-based (Greathouse et al. 2005) and Cassini CIRS data (Howett et al. 2007; Guerlet et al. 2009; Hesman et al. 2009; Sinclair et al. 2013; Sylvestre et al. 2015) found the 1- to 2-mbar acetylene abundance decreased from the equator towards the poles as predicted by photochemical models, while the ethane abundance was approximately constant with latitude. These observations have been interpreted as evidence that meridional mixing occurs on a timescale intermediate between the photochemical timescale of acetylene and ethane (Greathouse et al. 2005; Guerlet et al. 2009; Hesman et al. 2009). However, Moses et al. (2007) have pointed out that the ethane and acetylene photochemistry are strongly coupled, such that the meridional profile of acetylene should track that of ethane even in the presence of transport. Furthermore, CIRS observations by Sinclair et al. (2013) and Fletcher et al. (2015) show an increase in the 2-mbar abundances of ethane and acetylene in the northern hemisphere and decrease in the southern hemisphere, which they interpret as evidence of hemispheric transport on seasonal timescales. Thus, the global meridional distribution of ethane and acetylene is poorly understood.

In addition to the large-scale variations, acetylene and ethane both show a localized enhancement at the south pole at millibar pressures (Hesman et al. 2009; Sinclair et al. 2013; Fletcher et al. 2015). As the acetylene and ethane abundances increase with altitude, this is consistent with downward transport at the south pole indicated by temperature data. Using CIRS limb data, Guerlet et al. (2009) found a local maximum in the acetylene and ethane abundance at 25°N at 0.1 and 0.01 mbar, which they attribute to subsidence in the downward branch of a meridional circulation. This interpretation is consistent with the general circulation model of Friedson and Moses (2012) which produces a seasonally varying low-latitude circulation with rising motion in the summer hemisphere and subsidence in the winter hemisphere. The Equatorial Semi-Annual Oscillation (see Section 11.3.3) may also contribute to the subsidence at this latitude (Fouchet et al. 2008).

The earliest, pre-Cassini, models of the meridional circulation (Conrath and Pirraglia 1983; Conrath et al. 1990) assumed a circulation that is mechanically forced in the troposphere, modeled by imposing the observed zonal winds at the lower boundary, and assumed zonal momentum balance between the Coriolis acceleration of the meridional wind and damping of the zonal wind by eddies. Parameterizing the zonal acceleration due to the eddies as $-\overline{u}/\tau_f$, where τ_f is the eddy-damping timescale, this balance can be written $f\overline{v} = \overline{u}/\tau_f$. Conrath et al. (1990) also included radiative forcing from solar heating and radiative cooling. This model produces upward motion and cold temperatures on the equatorward side of eastward jets, and downward motion and warm temperatures on the poleward side, as well as the observed decay of the zonal jet with altitude. The resulting upper tropospheric temperature perturbations are consistent with the Voyager IRIS observations when the frictional and radiative timescales are comparable, $\tau_f \sim \tau_r$.

A more detailed model of the stratospheric circulation was produced by Friedson and Moses (2012), using a fully 3D general circulation model adapted to Saturn. Their model also uses a mechanical forcing at the lower boundary to produce the observed tropospheric zonal winds, along with random thermal perturbations in the troposphere to simulate generation of waves by convection, and includes realistic radiative forcing. Their model produced global scale temperature gradients generally consistent with observations, though temperatures were ~5 K too warm in the summer stratosphere. The modeled meridional transport was dominated at low latitudes by a seasonally reversing Hadley circulation, with broad upwelling at low latitudes and strong subsidence near 25° in the winter hemisphere, consistent with the circulation inferred from the ethane and acetylene measurements of Guerlet et al. (2009). At mid-latitudes they found 1-mbar eddy mixing timescales of just over 100 years, slightly longer than the photochemical timescale for acetylene and shorter than for ethane, consistent with the observations. Note, however, that the model was unable to reproduce the Semi-Annual Oscillation (see next section), which compromised the model's ability to capture the observed variations in temperature and constituents at low latitudes associated with the oscillation.

11.3.3 Equatorial Semi-Annual Oscillation

A common feature of equatorial stratospheres in rapidly rotating atmospheres is the presence of quasi-periodic oscillations, in both time and altitude, of the zonally averaged temperature and wind fields. The best studied of these is the Quasi-Biennial Oscillation (QBO) in the Earth's lower stratosphere, in which alternating layers of eastward and westward zonal mean winds, with associated warm and cold temperatures, slowly descend with a variable periodicity averaging approximately 28 months (Baldwin et al. 2001). A similar oscillation, with a more regular semi-annual period, is also seen in the upper stratosphere and mesosphere. An approximately four-year periodicity in Jupiter's equatorial temperatures was found in an analysis of ground-based data by Orton et al. (1991). Leovy et al. (1991) proposed that this Quasi-Quadrennial Oscillation (QQO) was analogous to the terrestrial QBO, and calculation of the stratospheric winds from CIRS temperature measurements by Flasar et al. (2004) showed the presence of a vertical oscillation in the zonal winds.

Using 24 years of ground-based data, Orton et al. (2008) found an oscillation of Saturn's equatorial brightness temperatures, in both 7.8 μm methane emission and 12.2 μm ethane emission, with a period of roughly 15 Earth years, which has been labeled the Saturn Semi-Annual Oscillation (SSAO). Using CIRS limb observations from 2005 to 2006, Fouchet et al. (2008) retrieved temperatures between 20 and 0.003 mbar and found equatorial temperature oscillations similar to those seen in the QBO and QQO (Figure 11.14), confined within ~10° of the equator and oscillating in height with a wavelength of several scale heights. These equatorial perturbations are flanked by temperature perturbations of opposite sign from ~10° to 20° latitude. Calculation of the zonal winds from the thermal wind equation reveals vertically alternating eastward and westward jets (Figure 11.15); the presence of a stratospheric jet was also inferred from CIRS nadir viewing data (Li et al. 2008). CIRS limb observations from 2010 (Guerlet et al. 2011) showed that the pattern of winds and temperatures had descended in altitude over the intervening five years, as is seen in the QBO. The descent of the temperature field was also observed in equatorial temperature profiles from Cassini radio occultations (Schinder et al. 2011), and cloud tracking measurements from Cassini imaging have shown variations in the velocity of the equatorial jet at the tropopause (Li et al. 2011). Comparisons of Cassini CIRS temperatures from 2009 with a reanalysis of Voyager IRIS thermal data from one Saturn year earlier (Sinclair et al. 2013) found that temperatures at 1–2 mbar were

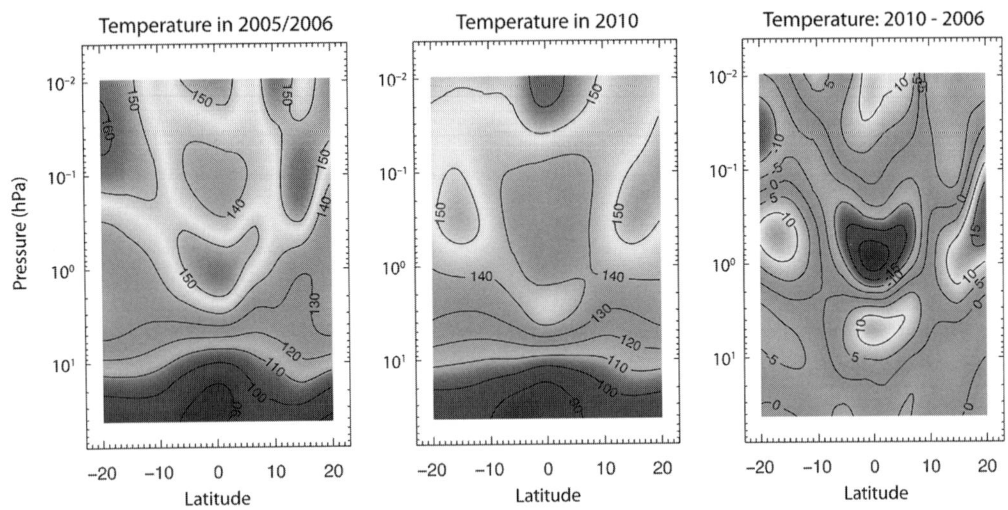

Figure 11.14 Low-latitude stratospheric temperatures on Saturn (in K) determined from Cassini/CIRS observations in 2005–2006 (left) and 2010 (middle). The difference is shown in the right panel. Note the existence of a low-latitude oscillatory structure whose phase shifted downward between 2005–2006 and 2010. From Guerlet et al. (2011).

~7 K warmer in 2010 than 1980 at the equator, and ~4 K cooler at ±10° latitude, indicating the that period of the SSAO is not exactly one-half of a Saturnian year.

The terrestrial QBO is driven by the interaction of a spectrum of vertically propagating waves – of both eastward and westward phase velocities – with the zonal wind. Absorption of the waves by the zonal flow, through radiative or frictional damping, produces a momentum transfer between the wave and zonal flow that accelerates the flow toward the zonal phase velocity of the wave (Lindzen and Holton 1968). The momentum transfer is most effective as the wave approaches a critical level where the zonal phase velocity of the wave matches the zonal wind velocity and the vertical wavelength and vertical velocity go to zero. In the presence of a jet with vertical wind shear, this results in the waves being absorbed at altitudes slightly below where the wind speed matches the wave phase velocity – and below the altitude where the jet speed peaks. This causes the pattern of zonal winds to slowly migrate downward over time (note that, as contours of zonal wind are not material surfaces, this does not imply downward transport of air parcels themselves). As the jet descends, its lower boundary becomes sharper and as the jet descends to the troposphere it dissipates, allowing the waves to now propagate to higher altitudes, where they accelerate a new jet. Equatorial confinement of the oscillation can be explained by two possible mechanisms. First, the waves driving the oscillation may be equatorially confined. Secondly, at latitudes away from the equator, Coriolis forces become important, and the wave induced momentum convergences may be partially balanced by Coriolis forces on the meridional circulation induced by the waves, instead of by the momentum changes to the winds. Full details on the terrestrial QBO can be found in the review by Baldwin et al. (2001). The wave modes responsible for the observed oscillations are poorly constrained, even on the Earth, although Kelvin and eastward-propagating gravity waves are most likely for contributing the eastward jet accelerations, and mixed Rossby-gravity waves and westward-propagating gravity waves are most likely for contributing the westward jet accelerations.

In case of Saturn's SAO, the amplitude and period of the oscillation likely contain information about the types, spectra and amplitudes of the wave modes that drive it, and detailed dynamical modeling may thus provide information about the wave properties in Saturn's stratosphere. Analogy with the Earth would suggest that such waves might arise from convective forcing in the troposphere, so in principle the properties of the SSAO might lead to improved understanding of the extent to which tropospheric convection on Saturn couples to the overlying stratified atmosphere. Moreover, although primarily an equatorial phenomenon, Earth's QBO exerts a global influence, and this is also likely to be the case on Saturn, though detailed observations and modeling will be necessary to quantify this possibility.

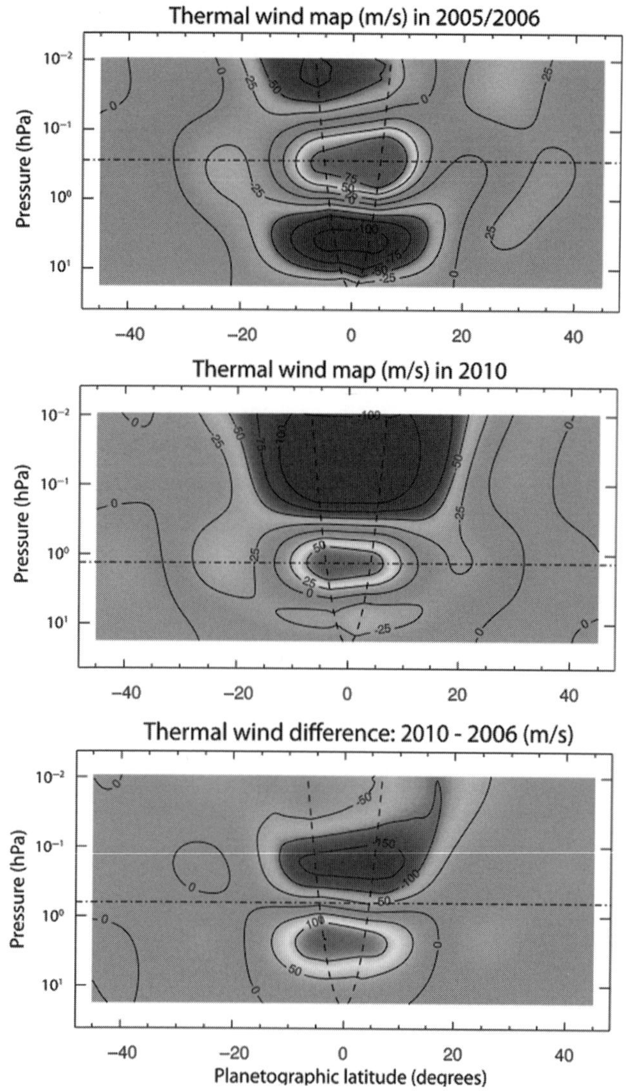

Figure 11.15 Low-latitude, stratospheric zonal winds on Saturn calculated based on the observed meridional temperature gradients using the thermal-wind Equation (11.1). The equation is integrated from the 20-mbar level and so the plotted winds are differences relative to those at 20 mbar. The thermal-wind equation breaks down near the equator, and the winds equatorward of the dashed lines are interpolated on isobars from regions outside the dashed lines. Note the existence of an oscillatory structure in the magnitude of the zonal wind with height, which has shifted downward in 2010 relative to its phase in 2005–2006. The difference is plotted in the bottom row. From Guerlet et al. (2011). (A black-and-white version of this figure appears in some formats. For the color version, please refer to the plate section.)

11.3.4 Stratospheric Response to the 2010 Storm

One of the most surprising consequences of the large convective storm, known as a Great White Spot (GWS) that began in December 2010 (see Chapter 13 by Sanchez-Lavega et al.), was the observation of large changes in the thermal structure of the stratosphere at northern mid-latitudes. Observations taken in January 2011 by Cassini CIRS and ground-based telescopes (Fletcher et al. 2011) showed a cool region in the stratosphere above the anticyclonic vortex associated with the storm, flanked by two warmer regions nicknamed the "beacons," 16 K warmer than the cold region at roughly 0.5 mbar, much larger than any zonal contrasts previously seen in thermal observations of Saturn's stratosphere. Subsequent observations from CIRS and ground-based telescopes (Fletcher et al. 2012) showed the beacons moving westward at differing rates, with one of the beacons remaining located above the convective plume associated with the storm, while increasing in temperature. At the end of April 2011, the two beacons merged, resulting in a single warm oval, with a longitudinal extent of 70°, a latitudinal extent of 30° and a peak temperature of 220 K at 2 mbar, 80 K warmer than the background atmosphere and the warmest stratospheric temperatures ever observed on Saturn. This new warm oval moved westward at a velocity intermediate between the two original beacons, and was centered about 1.5 scale heights lower in the atmosphere. The temperature of the beacon dropped by 20 K over the next two months, after which the temperature slowly declined by 0.11 ±0.01 K day^{-1} and the width of the warm spot shrank by 0.16 ±0.01 deg day^{-1}. Fletcher et al. (2012) used the thermal-wind equation to calculate the wind fields for the initial beacons in March 2011, and for the merged beacon in August 2011. The winds reveal that the hot spots are coherent anticyclonic vortices, with tangential winds of 70–140 m s^{-1} for the pre-merger vortices, and ~200 m s^{-1} for the final merged vortex.

The beacons were associated with perturbations of chemistry as well as temperature, providing additional insight into the dynamics. The abundances of ethylene (C_2H_4) and acetylene increased significantly in the hot beacon core (e.g. at pressures of order ~1 mbar) relative to pre-storm conditions, and ethane apparently also increased in abundance, though more modestly (Fletcher et al. 2012; Hesman et al. 2012; Moses et al. 2015). One-dimensional photochemical models show that the temperature increase alone, while modifying the chemistry, is insufficient to explain the observed increases (Cavalié et al. 2015; Moses et al. 2015). Therefore, dynamics likely played a role.

In particular, all three species increase with altitude at pressures of ~1–10 mbar, so a likely explanation is that subsidence occurred inside the beacon, advecting the greater high-altitude abundances of these species downward into the beacon core region. The compressive temperature increase associated with such descent would also naturally explain the high beacon temperatures. Nevertheless, the descent velocity required to explain the chemical abundances in photochemical models is surprisingly large – Moses et al. (2015) suggest that ~10 cm s^{-1} is required. It is not clear whether this is consistent with the magnitude of descent needed to explain the temperature increase, nor how it would be generated.

Models of moist convection on Saturn indicate that the convective plumes will not penetrate above the upper troposhere, 150 mbar or so (Hueso and Sánchez-Lavega 2004), and modeling of Cassini imaging data of the plume indicated that the cloud tops were at 400 mbar or deeper. It is thus unlikely that the stratospheric perturbations were produced directly by the convective plumes associated with the GWS (Garcia-Melendo et al. 2013). Sayanagi and Showman (2007) showed that tropospheric storms can cause significant wave generation, and that such waves can propagate into the stratosphere where they exert a significant dynamical effect when they break or become absorbed. Fletcher et al. (2012) noted that the storm was located at a latitude where the zonal winds have been suggested to be barotropically unstable (Achterberg and Flasar 1996; Read et al. 2009a), and where quasi-stationary Rossby waves may potentially propagate upward from the troposphere to the stratsophere. They therefore suggested that the strong perturbations in the stratosphere were the result of waves, either gravity or Rossby waves, generated by the convection impinging on the statically stable tropopause, propagating into the stratosphere. This conjecture still needs to be tested with additional numerical modeling.

11.4 Dynamics of the Zonal Jets

11.4.1 Jet Structure

As described in Section 11.2, Saturn rotates rapidly, and the large-scale dynamics is in approximate geostrophic balance. Although we lack detailed observations of the deep interior, dynamical balance arguments can be used to constrain the structure of the jets there. In particular, for a zonal flow at low Rossby number, it can be shown that, to good approximation (Showman et al. 2010)

$$2\Omega\frac{\partial u}{\partial z_*}\approx\frac{g}{\rho}\left(\frac{\partial \rho}{\partial y}\right)_p, \tag{11.5}$$

where z_* is the coordinate parallel to the rotation axis, g is gravity and ρ is density. This is a generalized thermal-wind equation that is valid in the deep molecular interior even in the presence of non-hydrostatic motions. If the density is constant on isobars – that is, if the fluid is barotropic – then Equation (11.5) implies that the zonal wind is constant on surfaces parallel to the rotation axis, i.e.,

$$\frac{\partial u}{\partial z_*}\approx 0. \tag{11.6}$$

This result is analogous to the standard Taylor–Proudman theorem (Pedlosky 1987), except that it does not make any assumption of constant density. It implies that, for a barotropic flow, the zonal wind is constant along lines parallel to the rotation axis, and is valid even if the mean density varies by orders of magnitude from the atmosphere to the deep interior, as it does on Saturn. On the other hand, if density varies on isobars – that is, if the fluid is baroclinic – then Equation (11.5) implies that the zonal wind varies along surfaces parallel to the rotation axis.

These results have been used to argue for several endpoint scenarios regarding the interior wind structure. Convective mixing is normally thought to homogenize the interior entropy, which would suggest a nearly barotropic interior in which variations of density on isobars are small. Convective plumes of course involve thermal perturbations, but simple mixing-length estimates suggest that at Saturn's heat flux these convective density variations induce only slight deviations from a barotropic state (Showman et al. 2010). Based on arguments of this type, Busse (1976) suggested that the observed zonal winds on Jupiter and Saturn extend downward into the interior on surfaces parallel to the rotation axis following Equation (11.6); the observed zonal winds would then be the surface manifestation of concentric, differentially rotating cylinders of fluid centered about – and parallel to – the rotation axis. On the other hand, other authors have suggested that perhaps the jets on Jupiter and Saturn extend only a few scale heights into the interior (e.g. Ingersoll and Cuzzi 1969; Stone 1973; Ingersoll 1976). In this case, Equation (11.5) demands the existence of latitudinal thermal variations, and would imply that, just below the cloud deck, the cyclonic bands are cooler (denser) than the anticyclonic bands. The required temperature differences, $\sim$5–10 K, are modest, and could plausibly be supplied by variations in latent or radiative heating between latitude bands. Intermediate scenarios are also possible, with horizontal temperature gradients and vertical variation of the jets in the cloud layer superposed on deep, nearly barotropic jets in the interior (Vasavada and Showman 2005). Liu et al. (2013, 2014) suggested yet another scenario based on the assumption that the interior entropy is only homogenized in the direction of z_*, while allowing for entropy variations in the direction toward/away from the rotation axis. Via Equation (11.5), this scenario also implies significant thermal wind shear and a gradual decay of the jets with depth.

It has long been recognized that the above arguments may break down in the metallic hydrogen interior at pressures $\geq 10^6$ bars. There, the high electrical conductivity allows sufficiently strong Lorentz forces to alter the forces balances, causing a breakdown of geostrophy. It has been suggested that the Lorentz forces will act to brake the zonal flows, leading to weak winds in the metallic region (e.g. Busse 1976, 2002). The transition from molecular to metallic hydrogen is gradual (Nellis 2000), and several authors have pointed out that significant magnetic effects on the flow can occur even in the extended semiconducting region between the metallic interior and the overlying molecular (electrically insulating) envelope (Kirk and Stevenson 1987; Liu et al. 2008). In particular, Liu et al. (2008) argued that the observed zonal winds cannot penetrate deeper than 96 and 86% of the radius on Jupiter and Saturn, respectively, because otherwise the Ohmic dissipation would exceed the planets' observed luminosities. If such Lorentz force braking occurs at the base of the molecular region and leads to weak winds there, and if the overlying jets are nearly barotropic, then Equation (11.6) would suggest that the winds may be weak throughout much of the molecular interior – and not just in the metallic and semiconducting regions (Liu et al. 2008). Nevertheless, the models to date have been largely kinematic, and more work is needed to determine self-consistent solutions to the full magnetohydrodynamic problem (Glatzmaier 2008). The main region that can escape such coupling is at low latitudes – there, it is possible for Taylor columns to extend throughout the planet without ever encountering electrically conducting layers. Interestingly, the width of this region is similar to the observed width of the equatorial jets on Jupiter and Saturn. An attractive scenario is therefore that the equatorial jets penetrate throughout the planet along surfaces parallel to the rotation axis (approximately following Equation (11.6) in the deep atmosphere), while the higher-latitude jets truncate at some as-yet poorly determined depth. Juno and Cassini will provide observational constraints on this question in the next year.

11.4.2 Models of Jet Formation

Thorough reviews of models for jet formation on Jupiter and Saturn were provided by Vasavada and Showman (2005) and Del Genio et al. (2009); here, we summarize only the highlights, emphasizing developments within the last ten years. Two classes of model have been introduced to explore the zonal jets on Jupiter and Saturn. In one approach, which we dub the "shallow-forcing" scenario, it is assumed that baroclinic instabilities, moist convection, and other processes within the cloud layer are responsible for driving the jets. This scenario has mostly been explored with thin-shell, one-layer and multi-layer atmospheric models from the terrestrial atmospheric dynamics community. In another approach, which we dub the "deep-forcing" scenario, it is assumed that convection throughout the interior leads to differential rotation in the molecular envelope that manifests as the zonal jets at the surface. This scenario has primarily been explored with Boussinesq and anelastic models of convection in spherical shells that are derived from the geodynamo and stellar convection communities. The distinction between the approaches is artificial, but because of the different communities involved, and the difficulty of constructing a truly coupled atmosphere-interior circulation model, the distinction has persisted over the $\sim$40-year history of the field. Nevertheless, new attempts are being made to bridge this gap, an area that will see major progress in the next decade.

It is important to emphasize that the issue of whether the jets exhibit deep or shallow structure is distinct from whether the jet driving occurs primarily in the atmosphere versus the interior (Vasavada and Showman 2005; Showman et al. 2006). Because of the nonlocal nature of the atmospheric response to eddy driving, zonal jets that extend deeply into the interior can result from eddy accelerations that are primarily confined to the atmosphere (Showman et al. 2006; Lian and Showman 2008; for theory, see Haynes et al. 1991). Likewise, under some conditions, convection throughout the interior can produce jets that may exhibit confinement near the outer margin of the planet (Kaspi et al. 2009).

Models for the atmospheric jet formation generally assume that the zonal jets result from the interaction of turbulence with the β effect that is associated with latitudinal variations in the Coriolis parameter. Early work emphasized two-dimensional, horizontally non-divergent models and one-layer shallow-water models. More recently, several three-dimensional models have been performed.

One-layer models are intended to represent an appropriate vertical mean of the atmospheric flow. They are motivated by the observation that the atmospheric winds on Jupiter and Saturn are approximately horizontal – with generally small horizontal divergence – and were originally intended to capture the flow in a presumed shallow weather layer below the clouds. More broadly, however, they serve as an ideal process model to explore the dynamics of zonal jets and vortices in the simplest possible context, thereby shedding insights into dynamical mechanisms that may also occur (but be harder to identify) in more realistic systems. In one-layer models, the effect of convection must be parameterized, typically by introducing small-scale vorticity sources and sinks that fluctuate in time. Damping is commonly represented using a frictional drag.

Most work to date has been performed with the two-dimensional, horizontally non-divergent model. This system captures the interaction of turbulence with the β effect, but neglects buoyancy, gravity waves, and any effects associated with a finite Rossby deformation radius (an assumption that is not strictly valid on Jupiter and Saturn, where the deformation radius is comparable to or smaller than the meridional jet width). When forced with small-scale turbulence under conditions of fast rotation and relatively weak frictional drag, these models generally produce multiple zonal jets whose meridional widths scale with the Rhines scale $(U/\beta)^{1/2}$, where U is a characteristic wind speed (e.g. Williams 1978; Nozawa and Yoden 1997; Huang and Robinson 1998; Sukoriansky et al. 2007; and many others).[5] Interestingly, these models lack an unusually strong equatorial jet as occurs on Jupiter and Saturn – instead, the equatorial jet tends to resemble the higher-latitude jets in speed and structure, and it often is not centered precisely around the equator. These models also tend not to produce large, long-lived coherent vortices that coexist with the jets, such as Jupiter's Great Red Spot or smaller but analogous vortices on Saturn. Both of these discrepancies likely result from the lack of a finite deformation radius in the 2D non-divergent model. Despite these failures, the ease of analyzing this model has led to crucial insights into the workings of inverse energy cascades (Sukoriansky et al. 2007) and the physical mechanisms for jet formation (e.g. Dritschel and McIntyre 2008).

The one-layer shallow-water model represents a more realistic system because it includes the effect of buoyancy (via a horizontally variable vertical layer thickness), gravity waves, and a finite deformation radius, albeit still in a context that does not capture detailed vertical structure. In the context of giant planets, the model captures the behavior of an active atmospheric weather layer that overlies an abyssal layer (representing the deep planetary interior) whose winds are specified, usually to be zero. The related, one-layer quasi-geostrophic (QG) model provides a simplification by restricting the flow to be nearly geostrophic, which filters gravity waves while retaining the effect of a finite deformation radius. Showman (2007) and Scott and Polvani (2007) presented the first forced turbulence calculations of jet formation in the shallow water system, and Li et al. (2006) performed a similar study in the QG system. These authors showed that multiple zonal jets can result from small-

[5] In Rhines' original theory, this corresponds to an eddy velocity associated with the turbulence, but in other contexts other velocity scales, such as the zonal-mean zonal wind can also be appropriate (see, e.g. Scott and Dritschel 2012).

scale forcing, and that in many cases, the jet widths are comparable to the Rhines scale $(U/\beta)^{1/2}$. Interestingly, the deformation radius influences jet formation, and when it is sufficiently small, can suppress the formation of jets entirely, leading to a flow dominated by vortices – an effect first described in the QG system (Okuno and Masuda 2003; Smith 2004) before being extended to the shallow-water system (Showman 2007; Scott and Polvani 2007). Thomson and McIntyre (2016) showed in a QG model that the presence of specified, zonally symmetric jets in the abyssal layer can help to increase the straightness of the weather-layer jets, causing the flow to more closely follow latitude circles. Using a simple parameterization of convective forcing, their model also naturally produces a preference for convection in belts rather than zones, as observed on Jupiter. In some cases, especially when jets in the abyssal layer exist, the weather-layer jets in one-layer shallow-water and QG models can violate the Charney–Stern stability theorem, in agreement with Jupiter and Saturn (Showman 2007; Scott and Polvani 2007; Thomson and McIntyre 2016).

Under conditions appropriate to Jupiter and Saturn – namely, small Rossby number and a deformation radius a few percent of the planetary radius – shallow-water models typically produce a broad, fast westward equatorial jet (Cho and Polvani 1996; Iacono et al. 1999; Showman 2007; Scott and Polvani 2007). This equatorial flow intensification could be considered a step forward from the 2D non-divergent model, since the equatorial jets of Jupiter and Saturn are faster and wider than the jets at higher latitudes. However, although Uranus and Neptune exhibit westward equatorial flow, this result represents a major failing for Jupiter and Saturn, where the equatorial flow is eastward. Nevertheless, Scott and Polvani (2008) showed that under certain conditions, shallow-water models can generate eastward equatorial jets resembling those on Jupiter and Saturn. They speculate that the key property allowing emergence of eastward (rather than westward) equatorial flow is the usage of radiative rather than frictional damping. However, this is inconsistent with the findings of Showman (2007), who adopted radiative damping and yet always obtained westward equatorial jets. Most likely, specific types of both forcing and damping are necessary. A similar "thermal shallow water" model presented by Warneford and Dellar (2014) exhibits equatorial superrotation at short radiative time constant, but subrotation at long radiative time constant, although the value of the radiative time constant at the transition between these regimes is too short to explain the transition from the Jupiter/Saturn regime to the Uranus/Neptune regime. The dynamical mechanisms controlling the equatorial jet properties in all these shallow-water-type models remain poorly understood and deserve further study.

Over the past 15 years, several three-dimensional models have been published showing how zonal jets can develop in the atmosphere. Following on earlier work by Williams (2003), Lian and Showman (2008) showed that baroclinic instabilities in the weather layer (induced by meridional temperature differences associated with the latitudinal gradients in radiative heating) can generate multiple zonal jets. These jets do not remain confined to the cloud layer where they are driven, but rather can extend deep into the interior as long as friction there is weak. The model produces statistical distributions of eddy momentum fluxes that match observations on Jupiter and Saturn, as well as jets that remain stable while violating the Charney–Stern stability criterion. Both Williams (2003) and Lian and Showman (2008) showed that sharp latitudinal temperature gradients near the equator can lead to equatorial superrotation. Schneider and Liu (2009) presented a shallow 3D model (truncated at 3 bars), which, as in Williams (2003) and Lian and Showman (2008), produces multiple mid- to high-latitude jets in response to baroclinic instabilities associated with latitudinal temperature gradients. Moreover, they introduced a simple convective parameterization which allows the emergence of Jupiter-like equatorial superrotation in response to equatorial convection in the model.

Several 3D models now exist that explain the overall features of the circulation on all four giant planets, including the transition from equatorial superrotation on Jupiter and Saturn to equatorial subrotation on Uranus and Neptune. Lian and Showman (2010) incorporated a hydrological cycle that captures the advection, condensation, rain out, and latent heating associated with water vapor, providing the first test of the long-standing hypothesis that such latent heating is crucial in driving the circulation (e.g. Barcilon and Gierasch 1970; Ingersoll et al. 2000). Their results are

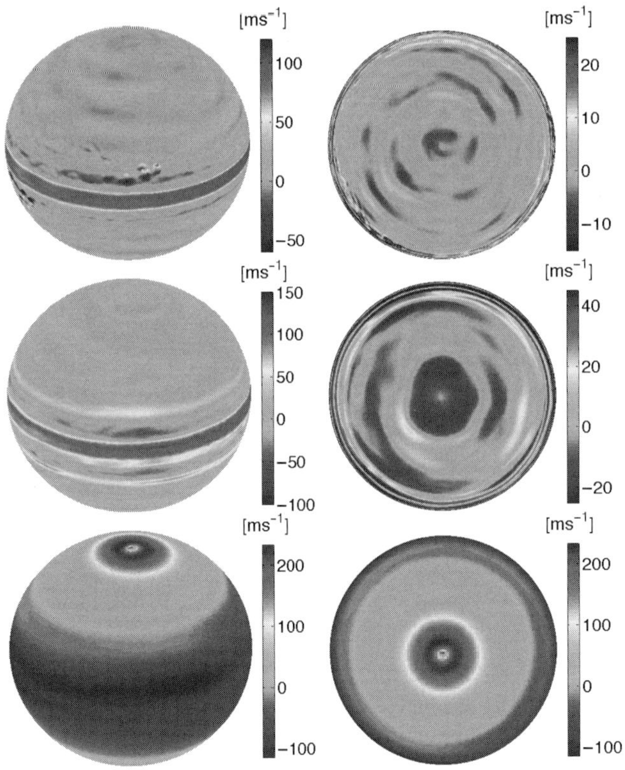

Figure 11.16 3D atmosphere simulations demonstrating that latent-heat release associated with condensation of water can lead to zonal jet patterns qualitatively resembling all four giant planets. Top, middle, and bottom rows are three models representing Jupiter, Saturn and Uranus/Neptune, which are assumed to have water abundances of 3, 5, and 30 times solar, respectively. Color scale represents zonal wind in m s^{-1}. Left column shows oblique view and right column shows view looking down over the north pole. The Jupiter and Saturn cases develop equatorial superrotation and many off-equatorial jets at higher latitudes, while the Uranus/Neptune model exhibits a three-jet pattern comprising broad equatorial subrotation and high-latitude eastward jets. From Lian and Showman (2010). (A black-and-white version of this figure appears in some formats. For the color version, please refer to the plate section.)

shown in Figure 11.16. In addition to explaining the observed transition from superrotation to subrotation, as well as mid-to-high latitude jets qualitatively similar to those observed on the giant planets, their model also produces local "storm"-like features that crudely resemble convective thunderstorm events observed on Jupiter and Saturn. In some cases, their model exhibits large storm events, which bear resemblance to Saturn's recent Great White Spot of 2010–2011. Liu and Schneider (2010) extended the work of Schneider and Liu (2009) to all four planets, likewise showing a transition from equatorial superrotation on Jupiter and Saturn to equatorial subrotation on Uranus and Neptune, as well as other jet properties similar to those on the giant planets. All of these models represent a major step forward. Nevertheless, the equatorial jet under Saturn conditions is insufficiently fast and broad, and more work is warranted to determine whether equatorial jets better resembling Saturn can be produced in this class of model.

We now turn to discuss models of jet formation via convection in the interior. Busse (1976) first envisioned that the convection in the interior could organize into the form of Taylor columns aligned with the rotation axis, and that the Reynolds stresses associated with this convection would drive jets that would manifest as concentric, differentially rotating cylinders centered on the rotation axis. The observed zonal jets would then represent the outcropping of this differential rotation at the cloud level. Early investigations of this hypothesis involved analytical calculations, laboratory studies and numerical simulations in the linear and weakly nonlinear regime. These studies generally confirmed the columnar nature of the convection at low amplitude, and showed how the spherical planetary shape would promote the emergence of equatorial superrotation (see Busse 1994 and Busse 2002 for reviews). Only in recent years, however, have computational resources become sufficient to investigate the dynamics in the more strongly nonlinear regime relevant to the giant planets.

The first such models adopted the Boussinesq approximation wherein the background density is assumed constant (i.e. the continuity equation is $\nabla \cdot v = 0$). This assumption is not valid for the giant planets, but the dynamics are nevertheless rich and provide significant insights. Generally, these models explore convection in a spherical shell with boundaries that are free-slip in horizontal velocity, and that are maintained at constant temperature (with the interior boundary being hotter than the outer boundary, allowing convection to occur). The earliest models explored relatively thick shells, with an inner-to-outer radius ratio less than 0.7. These simulations showed that when the convection is sufficiently vigorous and friction sufficiently weak (i.e. the Rayleigh number sufficiently high and Ekman number sufficiently low), the convection can produce zonal jets whose speeds are significantly faster than the convective velocities (Aurnou and Olson 2001; Christensen 2001, 2002).

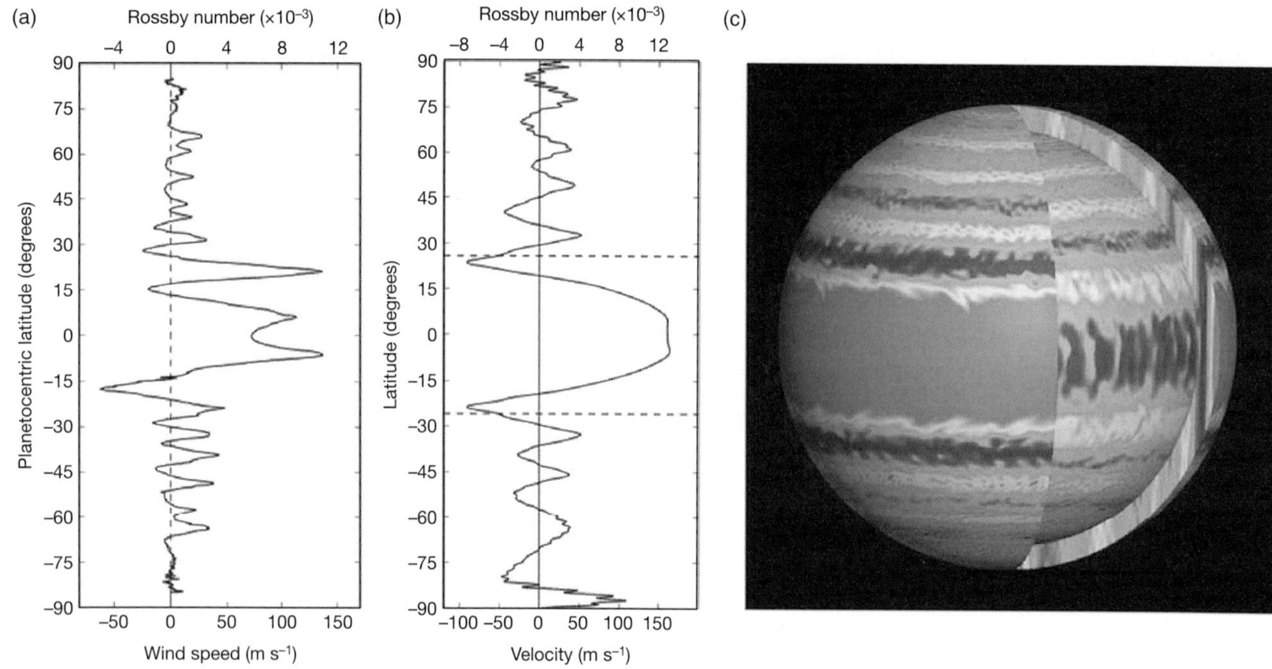

Figure 11.17 The zonal flow on the outer and inner surface of a 3D convection simulation with an aspect ratio of 0.9, showing the eastward (red) and westward (blue) zonal velocities. The Boussinesq equations were solved, in which the background (reference) density is independent of radius. Free-slip boundary conditions are used on both the inner and outer boundaries. Equatorial superrotation, and numerous higher-latitude eastward and westward jets, develop. Adapted from Heimpel et al. (2005). (A black-and-white version of this figure appears in some formats. For the color version, please refer to the plate section.)

The jets take the form of a broad equatorial superrotating jet and, in some cases, a small number of weaker jets at high latitudes. Despite the existence of the superrotation, the zonal jet patterns in these models do not resemble that on Jupiter and Saturn. Subsequent studies considered thinner shells, with an inner-to-outer radius ratio of $\sim$0.9, and when appropriately tuned, these models are able to produce superrotation comparable to that of Jupiter and Saturn, along with multiple high-latitude jets (Heimpel et al. 2005; Heimpel and Aurnou 2007; Aurnou et al. 2008; see Figure 11.17). An issue is that, in all of these models, the meridional width of the superrotation is controlled by the depth of the inner boundary, and yet such a boundary is artificial in the context of a giant planet (which lack any such impermeable boundary in their molecular/metallic envelopes).

In Jupiter and Saturn, the density varies by a factor of $\sim 10^4$ from the cloud deck to the deep interior, and recently several models have started to include such radial density variations via the anelastic approximation, which allows the background density to vary radially, while (like the Boussinesq system) filtering sound waves, which is a reasonable approximation for the giant planet interiors. The first such models were two-dimensional, considering convection in the equatorial plane and investigating the emergence of differential rotation (Evonuk and Glatzmaier 2004, 2006, 2007; Glatzmaier et al. 2009). These models demonstrated a new mechanism for generating equatorial superrotation that does not exist in the Boussinesq system; as convective plumes rise and sink, the compressibility alters their vorticity in such a way as to promote Reynolds stresses that cause superrotation (Glatzmaier et al. 2009).

More recently, several 3D anelastic convection models have been developed (Kaspi 2008; Kaspi et al. 2009; Jones and Kuzanyan 2009; Jones et al. 2011; Gastine and Wicht 2012; and others). Like the Boussinesq models, these anelastic simulations generally exhibit equatorial superrotation and (in some cases) several higher-latitude zonal jets. If the spherical shell is thin, then the shell thickness controls the meridional width of the equatorial jet – just as in Boussinesq simulations. However, if the shell thickness becomes sufficiently deep, then the equatorial jet width tends to become invariant to the location of the bottom boundary, which is an improvement over the situation in Boussinesq models. Nevertheless, in this situation, the superrotation in the simulations tends to be too broad and the higher latitude jets fewer in number compared to Jupiter and Saturn. In the models of Kaspi (2008) and Kaspi et al. (2009), the jets exhibit significant shear, becoming weaker with depth; in contrast, some other models exhibit weaker shear, with a more barotropic structure (Jones and Kuzanyan 2009; Gastine and Wicht 2012; Cai and Chan 2012; Gastine et al. 2013; Chan and Mayr 2013). The difference likely result from the different treatment of the fluid's thermodynamic properties, including the radial variation of density and especially the entropy expansion coefficient, in these different models.

A concern with these models is that the simulated parameter regime is far from that of the giant planets; the simulated heat fluxes and viscosities are both too large by many orders of magnitude. Even within the simulated range, the jet speeds vary strongly with Rayleigh number and Ekman number. Therefore, even though it is possible to choose values of the Rayleigh number and Ekman number that yield realistic jet speeds, it is unknown whether such models would produce realistic jet speeds at the values of Rayleigh and Ekman number relevant to Jupiter and Saturn. Unfortunately, computational resources will be insufficient for directly simulating the planetary regime for the foreseeable future. Therefore, answering this question requires the development of scaling laws that can be extrapolated from the simulated regime to the planetary regime. Christensen (2002) made a first attempt at this, suggesting a scaling law that could bridge the gap based on an empirical

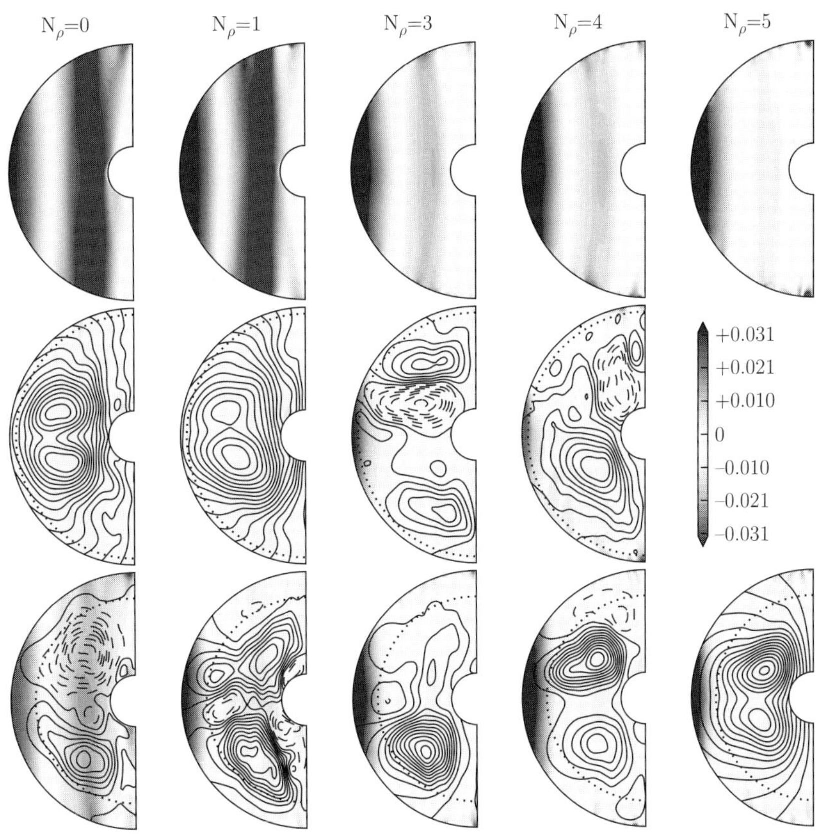

Figure 11.18 Anelastic 3D simulations of convection in a giant planet interior including coupling to the magnetic field, from Duarte et al. (2013). Each image shows the zonal-mean structure of a distinct simulation; the color scale shows the zonal-mean zonal wind (expressed as a Rossby number) and the contours show poloidal magnetic field. N_ρ denotes the number of density scale heights spanned from the outer to the inner boundary; the left column ($N_\rho = 0$) denotes Boussinesq simulations, while the right column ($N_\rho = 5$) presents simulations with five density scale heights (inner density 148 times the outer density). The top row represents hydrodynamic models (no MHD effects). The middle and bottom rows represent MHD simulations, with electrically insulating outer regions and electrically conducting inner regions. The transition occurs at a fractional radius of 0.95 and 0.8 for simulations in the middle and bottom rows, respectively. Robust equatorial superrotation can be seen in all models, which is confined to the electrically insulating region when MHD is used. The inclusion of compressibility and magnetic effects suppress the formation of zonal jets at mid-to-high latitudes. (A black-and-white version of this figure appears in some formats. For the color version, please refer to the plate section.)

assessment of his Boussinesq simulations. Showman et al. (2011) used anelastic simulations to show that, within the simulated range of Rayleigh and Ekman numbers, two distinct regimes exist, leading to distinct dependences of jet speeds on convective heat flux and viscosity. Christensen (2002) and Showman et al. (2011) both suggested an additional regime may exist wherein jet speeds become independent of viscosity when the viscosity is sufficiently small. More recent work, however, challenges the existence of a regime independent of viscosity and thermal diffusivity (Gastine et al. 2013, 2014). Additional work is warranted.

All of the above models neglect any coupling to the magnetic field. Recently, however, several models have included the effect of an electrically conducting interior and its coupling to the dynamics, allowing a joint investigation of zonal jet formation and dynamo generation of a magnetic field. This is a challenging problem because the electrical conductivity varies by many orders of magnitude with radius, a situation different from that encountered in the geodynamo problem. This situation was first investigated in Boussinesq models (Heimpel and Gómez Pérez 2011) and more recently in anelastic models (Duarte et al. 2013; Yadav et al. 2013; Jones 2014) that allow both density and electrical conductivity to vary strongly with radius. These models show that the equatorial jet penetrates to a depth where the Lorentz force becomes comparable to the Reynolds stress associated with the convection (which effectively means the jet is confined to the electrically insulating outer envelope), as illustrated for example in Figure 11.18. Jets poleward of the equatorial superrotation tend to be suppressed, because their bottoms penetrate into the electrically conducting region where Lorentz forces can act to brake the flows – an effect that extends throughout the molecular envelope due to the nearly barotropic nature of these jets (cf. Equation 11.6). This suggests the possibility that the equatorial jet results from interior convection, but the higher latitude jets result from atmospheric processes including baroclinic instability and moist convection (Vasavada and Showman 2005).

11.4.3 Detection of Deep Dynamics by Gravity Measurements

Information to date on Saturn's deep winds has been indirect. This is likely to change in 2017, as towards the culmination of its 13-year-long survey of the Saturnian system, the Cassini spacecraft will shift into a highly inclined orbit with a periapse between the planet and its rings, an orbit ideally suited to measuring the small-scale structure of Saturn's gravitational field. During this phase, known as the Cassini Grand Finale, the spacecraft will complete 22 orbits, 6 of which will be dedicated to gravity science. This will be the final maneuver of Cassini before it descends into the planet, terminating the mission. These gravity measurements will allow the determination of Saturn's gravity field to much higher accuracy than exists today (Jacobson et al. 2006). The gravitational field is commonly represented using a spherical harmonic expansion (Hubbard 1984), and so far only the lowest zonal harmonics J_2, J_4 and J_6 have been measured. These reflect the long-wavelength gravitational perturbations associated primarily with the planet's rotational bulge, and provide essentially no information on interior flows. Cassini's proximal orbits will allow the measurement of the gravity field at least up to J_{10}, including the possibility of measuring the odd gravity harmonics for the first time (Kaspi 2013).

If the strong cloud-level winds extend sufficiently deep into the planet's interior, then the density perturbations associated with the jets will induce gravitational perturbations that are measurable. Because the equatorial bulge is an extremely broad-scale (long-wavelength) feature, it affects primarily the lowest harmonics, and contributes only a much smaller signal at the higher harmonics. In contrast, the meridional scale associated with the zonal jets is smaller, and thus – if the jets extend sufficiently deep – they will induce substantial gravity signal at small meridional scales, corresponding to higher harmonics. This was first noted by Hubbard (1999), who used potential theory to show that if differential rotation on Jupiter penetrates the depth of the planet, then the high-order gravity moments will be dominated by the dynamical contribution rather than the contribution from rotational flattening. This study inspired the Juno mission gravity experiment of Jupiter, and in turn motivated the Cassini Grand Finale gravity experiment on Saturn. This potential-theory approach has recently been further developed using more accurate concentric Maclaurin-based interior models (Hubbard 2012; Kong et al. 2012; Hubbard 2013; Hubbard et al. 2014; Kaspi et al. 2016).

A difficulty with these potential-theory approaches, however, is that they are limited to fully barotropic systems satisfying the Taylor–Proudman theorem (Equation 11.6), where the flows are constant along cylinders. Therefore, this approach cannot explore the possible consequences for the gravity field if the flows penetrate only partway into the interior and die off below some critical depth.

A second approach that allows an evaluation of winds varying on cylinders is to use thermal-wind balance models (Kaspi et al. 2010; Kaspi 2013; Kaspi et al. 2013; Liu et al. 2013). Given some assumed structure of the zonal winds, the thermal-wind relation (Equation 11.5) can be used to calculate the dynamically induced density gradients in the interior. These density perturbations in turn cause perturbations to the gravity field. The dynamical contributions to the gravity harmonics are defined as (Kaspi et al. 2010)

$$\Delta J_n^{\mathrm{dyn}} = -(Ma^n)^{-1} \int P_n \rho' r^n d^3 r, \qquad (11.7)$$

where ρ' is the deviation of the density caused by dynamics (i.e. the deviation from some static density distribution, ρ_{static}, that would exist in the absence of dynamics), P_n is the nth Legendre polynomial, M is the planetary mass and a is the mean planetary radius. Kaspi et al. (2016) showed that in the barotropic limit the thermal-wind and potential-theory approaches give very similar results. However, the thermal-wind models have typically been limited to spherical symmetry, resulting in inability to calculate the static (solid-body) gravity spectrum and neglecting the effect of the oblateness of the planet on the dynamic gravity moments. Zhang et al. (2015) suggested that such effects, namely the self-gravity due to the dynamical variation itself $g(r, \theta)$, can be important, and Cao and Stevenson (2015) suggested that the effect of the oblateness on the background mean state can be important for the gravity moments. However, solving self-consistently for the full oblate system, Galanti et al. (2017) have shown that these oblateness effects give a very small contribution to the gravity moments, meaning that the spherical thermal wind captures well the leading order balance, and therefore can be used for calculating the gravity moments.

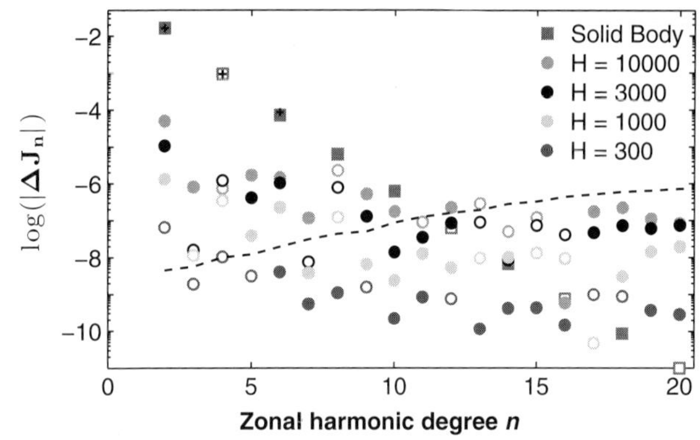

Figure 11.19 The gravity harmonics of Saturn corresponding to the static, rotationally flattened contribution (squares) and the contribution from dynamics (circles) calculated using the thermal-wind model of Kaspi (2013). The dynamical contributions to the gravity harmonics (ΔJ_n) are shown for four different assumptions about the characteristic length scale, H, over which the jets decay with depth into the interior: H = 300 km (purple), H = 1000 km (orange), H = 3000 km (blue) and H = 10000 km (maroon). Filled (open) symbols indicate positive (negative) zonal harmonics. Black plus signs show the observed values of J_n, and the dashed line is the expected detectability limit of the Cassini proximal orbits. The static values (Hubbard 1999) have only even components (odd harmonics are identically zero). From Kaspi (2013). (A black-and-white version of this figure appears in some formats. For the color version, please refer to the plate section.)

Figure 11.19 shows the gravity field resulting from a model using the thermal-wind approach, where the gravity harmonics are shown as a function of the depth that the cloud-level winds extend beneath the cloud-level. It shows that beyond n = 10 the gravity signal induced by the wind structure becomes larger than the solid-body (static) gravity harmonics.

Because Jupiter and Saturn are gaseous, aside from the cloud-level winds there is no apparent asymmetry between the northern and southern hemispheres. Therefore, the gravitational moments resulting from the shape and vertical structure of the planets have identically zero odd moments (Hubbard 1984, 1999). Unlike the even gravity moments that have a contribution both from the static density distribution and the dynamics, the odd moments are caused purely by dynamics. Thus, any odd signal detected (J_3, J_5, J_7, ...) will be a sign of a dynamical contribution to the gravity. Indeed, the cloud-level winds are not precisely symmetrical around the equator, and if these hemispheric differences extend into the interior, they could produce measurable odd harmonics (Figure 11.19)

(Kaspi 2013). Using also the thermal-wind approach, Liu et al. (2013) calculated the penetration depth of the winds on Jupiter with the additional assumption that the entropy gradient in the direction of the spin axis must be zero. This requirement sets the penetration depth of the winds, and they have also found that such a wind structure (which extends deep throughout the molecular envelope) should be detectable by Juno and Cassini (Liu et al. 2013, 2014).

The main challenge of interpreting the gravity data will be to invert the gravity measurements into meaningful wind fields. Until recently all studies have been only in the direction of forward modeling; thus, given a hypothetical wind structure (based on the observed surface winds and some assumption regarding the penetration depth) the gravity moments are calculated via the effect of the winds on the density structure (e.g. Hubbard 1999; Kaspi et al. 2010; Liu et al. 2013; Kong et al. 2012). However, in order to analyze the gravity field that will be detected by Juno and Cassini the inverse problem needs to be solved – calculating the zonal wind profile given the gravity field. This causes a difficulty since a gravity field is not necessarily invertible and a given gravity field might not have a unique corresponding wind structure. Galanti and Kaspi (2016, 2017) propose to address this using an adjoint-based inverse method that will allow the investigation of the giant planet dynamics using the observed measured gravity field. This method has been used extensively in the study of oceanic and atmospheric fluid dynamics (e.g. Tziperman and Thacker 1989; Moore et al. 2011). This method can be applied to any forward relation between the gravity and the wind structure (e.g. thermal-wind, potential-theory, etc.), and to 3D wind structure which give the full 3D gravity fields including contributions from longitudinal variations in the wind structure and meridional winds (Parisi et al. 2016).

The coincidental timing of the Cassini proximal orbits with the Juno gravity experiment will provide an opportunity to simultaneously probe the deep dynamics of both Saturn and Jupiter. Comparing the atmospheric dynamics on the two planets shows that the equatorial jet is much stronger and wider on Saturn, extending to latitude 30°, with fewer high-latitude jets – but the jets are much stronger on Saturn reaching about 400 m s^{-1}, depending on the reference frame chosen (see Section 11.2). The fact that the jets are stronger on Saturn means that the induced gravity anomalies will be generally larger as well. However, as Saturn is more oblate than Jupiter (Jupiter has an oblateness of 6.5%), the static gravity harmonics due to the non-dynamic density distribution will likewise be larger than on Jupiter. These two factors imply that the ratio between the static and the dynamic gravity harmonics are roughly similar on both planets (Kaspi 2013).

Despite the fact that no spacecraft will be visiting Uranus and Neptune in the near future, the strong and broad wind structure and smaller mass on these planets allows the use of these same techniques to infer the depth of the winds using gravity moments that have already been measured. In particular, the speed and meridional breadth of the jets on Uranus and Neptune allow the possibility that – unlike the case of Jupiter and Saturn – even the low harmonics such as J_4 can be significantly affected by dynamics. Using the gravity data then available, Hubbard et al. (1991) first showed that the fast winds must be confined to the outermost few percent of these planets' masses. Exploiting recent updates on the gravity harmonic J_4 and its uncertainty (Jacobson 2007, 2009), Kaspi et al. (2013) were able to refine this constraint considerably, showing that the observed flows must be constrained to the outermost 1000 km of the planetary radius – corresponding to no more than 0.15 and 0.2% of the mass of Uranus and Neptune, respectively. Thus, the fast jets on Uranus and Neptune are relatively shallow.

In summary, within the next few years, using the fact that if deep flows exist on giant planets they will induce a measurable gravity signal, we will finally address one of the longest-standing debates in planetary science regarding the depth of the observed zonal flows on all four giant planets (Vasavada and Showman 2005). Understanding if the flows are deep or shallow will allow tests between the theories of jet formation and energetics as discussed in Sections 11.4.1–11.4.2. The upcoming few years therefore provide an exciting opportunity in the study of the atmospheric dynamics of Saturn and the other giant planets.

11.5 Conclusions

Saturn's atmosphere exhibits a wealth of interacting dynamical processes that present an interesting contrast with Jupiter. The dynamics are dominated by fast zonal jets whose meridional width, latitudinal positions, and speeds have remained relatively constant over many decades. Saturn's clouds are organized into zonal bands that are modulated by (and perhaps affect) the zonal jets, but the relationship of the cloud bands to the jets themselves differs from that on Jupiter and remains poorly understood. Observations indicate that eddies transport momentum upgradient into the cores of the zonal jets at cloud level, thereby maintaining the jets. This points toward the importance of cloud level processes in jet dynamics, but the exact processes that power these jet-pumping eddies – baroclinic instability, moist convection, or interaction of dry convection with the stratified atmosphere – remain poorly understood. Lightning and storm activity – including small storms as well as the Great White Spot of 2010–2011 – provide constraints on the thermal structure and vertical motion below the clouds. There also exist a variety of vortices, waves, and other local features at cloud level, with lifetimes of months to decades, whose relationship to the jet dynamics remains enigmatic. Notable features include the string of pearls (Sayanagi et al. 2014), the Ribbon, the polar hexagon (Baines et al. 2009) and a significant population of small vortices that tend to cluster predominantly at the latitudes of the westward jets (e.g. Vasavada et al. 2006; Choi et al. 2009). Interestingly, at and above the cloud level, the jets exhibit reversals in the meridional gradient of potential vorticity, and the relative stability of the jets despite these reversals has important implications for the flow below the clouds.

A variety of observations place constraints on the circulation in the upper troposphere and stratosphere, which has important analogies to the stratospheric circulation on the other giant planets as well as Earth. Saturn's obliquity of 27° causes seasonal temperature changes in the stratosphere, but there also exist a wealth of temperature variations at a variety of length and timescales that are likely dynamical in origin. The temperature patterns above the clouds suggest, through thermal-wind balance, that most of the zonal jets decay with altitude in the upper troposphere. These temperature patterns, along with the ammonia abundance near the clouds, are best explained by a meridional circulation comprising ascent in the anticyclonic regions, producing the cold temperatures and high ammonia abundances, and descent in the cyclonic regions, producing the warm temperatures and low ammonia abundances. Interestingly, such a circulation is thermally indirect and is likely driven by absorption of waves propagating from below. This is directly analogous to the wave-driven circulations that exist in Earth's stratosphere and that have been inferred in the stratospheres of Jupiter, Uranus and Neptune. The source of the waves and the specific details of this circulation in the Saturn context remain poorly understood. Additionally, stratospheric thermal perturbations have been observed at low latitudes that suggest an oscillation of equatorial winds and temperatures with a period close to 15 years, which seems to be analogous to the terrestrial Quasi-Biennial Oscillation, which is driven by a population of upward propagating waves generated by convection and other processes in the troposphere.

Dynamical models do not yet come close to explaining this wealth of phenomena; nevertheless, significant progress in understanding the dynamics of zonal jets and local features has been made over the past $\sim$15 years. Over the past decade, highly idealized models have greatly improved our understanding of the processes that cause the flow to self-organize into zonal jets. More realistic three-dimensional models of the weather-layer flow are now capable of explaining the equatorial superrotation and multiple high-latitude jets on Jupiter and Saturn, although the superrotation in these models tends to be narrower and slower than that on Saturn. Models of convection throughout the planetary interior now include the multi-order-of-magnitude variation of density with radius and are likewise starting to incorporate the coupling to magnetic fields, including the transition from electrically insulating in the molecular envelope to electrically conducting in the deeper interior. These models are now elucidating how the compressibility and coupling to magnetic fields influences the zonal jets and other aspects of the dynamics. These models show that convection in the interior can naturally produce a Saturn-like equatorial jet, but the coupling to magnetic effects tends to suppress any higher latitude jets. An attractive hybrid scenario, which combines the strengths of the shallow and deep models, might suggest that convection in the molecular envelope outside some tangent cylinder causes the broad equatorial jet, but that the mid-to-high-latitude jets result from atmospheric processes such as baroclinic instabilities or moist convection in the cloud layer (Vasavada and Showman 2005).

Future progress in modeling and theory will occur on several fronts. Investigations of atmospheric, weather-layer processes and of deep, convective dynamics are usually conducted separately, using distinct classes of dynamical model that solve distinct equation sets and use distinct numerical codes. The distinction between "shallow" and "deep" processes is artificial, however, and in reality atmospheric and interior processes may interact in a variety of ways not included in current models. A new generation of GCMs that spans both atmosphere and interior processes is needed to explore this coupling, and this will be a major area of progress over the next decade. Global atmospheric models of jet formation do not yet include clouds, and yet clouds represent a significant fraction of the observational database for both Jupiter and Saturn, as well as influencing the dynamics through their mass loading, interaction with radiation, and latent heating and cooling due to their formation and sublimation. Thus, significant progress is possible by including clouds in models, both in widening the range of observations against which the models can be tested, and by illuminating the extent to which clouds influence the atmospheric circulation. Inclusion of realistic radiative transfer and chemistry into stratospheric GCMs will be a growth area, and will help shed light on the implications of tracers such as ethane and acetylene for the dynamics. The nature of the eddies or waves that apparently cause a deceleration of the upper tropospheric and lower stratospheric zonal jets, and that cause the SSAO, are poorly understood. These and many related topics concerning the zonal jets are amenable to continued idealized theory and modeling.

Future progress will also depend crucially on improved observations. Due to the Cassini mission, Saturn is perhaps now the best-observed giant planet – outpacing Jupiter – but many of the Cassini observations have yet to be analyzed, and significant progress is possible with the existing data. For all four giant planets, improvements over the past decade in ground-based astronomical facilities have led to a renaissance in observations of cloud features and upper tropospheric temperature structure from the ground – by professional and amateur astronomers alike. These ground-based observations will continue to play a crucial role in filling in the temporal gaps between higher-quality, but episodic, coverage from spacecraft missions. Moreover, over the next several years, the Juno mission to Jupiter and the Cassini Grand Finale at Saturn will provide gravity measurements that promise to constrain the depth to which the zonal jets extend below the visible cloud decks. Close-in microwave and IR sounding of Jupiter by Juno, as well as close-in sensing of Saturn with Cassini's suite of instruments during the proximal orbits, may produce qualitatively new constraints on the cloud-level dynamics of both planets. The Juno microwave radiometer will measure the global water abundance and the latitudinal variation of water and ammonia below cloud base. Future missions to the giant planets include the NASA Europa Multiple Flyby Mission and the European Jupiter Icy Moon Explorer (JUICE) mission, which will obtain observations of Jupiter in addition to their prime targets of icy satellites starting around 2030. Prospects for giant-planet missions beyond the 2030s are currently nebulous, but there is strong scientific merit – and interest – in missions to Uranus and/or Neptune, as well as an entry probe mission to Saturn. This activity will not only revolutionize our understanding of the grandest planets in our solar system, but will also provide a foundation for understanding the many brown dwarfs and extrasolar giant planets that are being discovered and characterized light years away.

Acknowledgments

We thank M. Flasar and an anonymous reviewer for helpful comments. Adam P. Showman acknowledges support from the NASA Planetary Atmosphere program. Yohai Kaspi acknowledges support from the Israeli Ministry of Science and the Minerva Foundation.

References

Achterberg, R. K. and F. M. Flasar (1996), Planetary-scale thermal waves in Saturn's upper troposphere. *Icarus*, **119**, 350–369.

Achterberg, R. K., P. J. Gierasch, B. J. Conrath, L. N. Fletcher, B. E. Hesman, G. L. Bjoraker and F. M. Flasar (2014), Changes to Saturn's zonal-mean tropospheric thermal structure after the 2010–2011 northern hemisphere storm. *Astrophys. J.*, **786**, 92.

Anderson, J. D., and G. Schubert (2007), Saturn's gravitational field, internal rotation, and interior structure. *Science*, **317**, 1384–1387.

Andrews, D. G., J. R. Holton, and C. B. Leovy (1987), *Middle Atmosphere Dynamics*. New York: Academic Press.

Antuñano, A., T. Río-Gaztelurrutia, A. Sánchez-Lavega, and R. Hueso (2015), Dynamics of Saturn's polar regions. *Journal of Geophysical Research (Planets)*, **120**, 155–176.

Atreya, S. K. (2006), Saturn probes: Why, where, how? *Proceedings: International Planetary Probe Workshop, IPPW-4, Pasadena, California*.

Aurnou, J., M. Heimpel, L. Allen, E. King and J. Wicht (2008), Convective heat transfer and the pattern of thermal emission on the gas giants. *Geophysical Journal International*, **173**, 793–801.

Aurnou, J. M. and P. L. Olson (2001), Strong zonal winds from thermal convection in a rotating spherical shell. *Geophys. Res. Lett.*, **28**, 2557–2560.

Baines, K. H., T. W. Momary, L. N. Fletcher, A. P. Showman, M. Roos-Serote, R. H. Brown, B. J. Buratti, R. N. Clark and P. D. Nicholson (2009), Saturn's north polar cyclone and hexagon at depth revealed by Cassini/VIMS. *Planet. Space Sci.*, **57**, 1671–1681.

Baldwin, M. P. et al. (2001), The quasi-biennial oscillation. *Reviews of Geophysics*, **39**, 179–230.

Barcilon, A. and P. Gierasch (1970), A moist, Hadley cell model for Jupiter's cloud bands. *Journal of Atmospheric Sciences*, **27**, 550–560.

Barnet, C. D., J. A. Westphal, R. F. Beebe and L. F. Huber (1992), Hubble Space Telescope observations of the 1990 equatorial disturbance on Saturn: Zonal winds and central meridian albedos. *Icarus*, **100**, 499–511.

Beebe, R. F., A. P. Ingersoll, G. E. Hunt, J. L. Mitchell and J.-P. Muller (1980), Measurements of wind vectors, eddy momentum transports, and energy conversions in Jupiter's atmosphere from Voyager 1 images. *Geophys. Res. Lett.*, **7**, 1–4.

Bezard, B. and D. Gautier (1985), A seasonal climate model of the atmospheres of the giant planets at the Voyager encounter time: I. Saturn's stratosphere. *Icarus*, **61**, 296–310.

Busse, F. H. (1976), A simple model of convection in the Jovian atmosphere. *Icarus*, **29**, 255–260.

Busse, F. H. (1994), Convection driven zonal flows and vortices in the major planets. *Chaos*, **4**, 123–134.

Busse, F. H. (2002), Convective flows in rapidly rotating spheres and their dynamo action. *Physics of Fluids*, **14**, 1301–1314.

Cai, T. and K. L. Chan (2012), Three-dimensional numerical simulation of convection in giant planets: Effects of solid core size. *Planet. Space. Sci*, **71**, 125–130.

Cao, H. and D. J. Stevenson (2015), Gravity and zonal flows of giant planets: From the Euler equation to the thermal wind equation. *ArXiv e-prints*.

Cavalié, T., M. Dobrijevic, L. N. Fletcher, J.-C. Loison, K. M. Hickson, V. Hue and P. Hartogh (2015), Photochemical response to the variation of temperature in the 2011–2012 stratospheric vortex of Saturn. *Astron. Astrophys.*, **580**, A55.

Cess, R. D. and J. Caldwell (1979), A Saturnian stratospheric seasonal climate model. *Icarus*, **38**, 349–357.

Chan, K. L. and H. G. Mayr (2013), Numerical simulation of convectively generated vortices: Application to the Jovian planets. *Earth Planet. Sci. Lett.*, **371**, 212–219.

Cho, J. Y.-K. and L. M. Polvani (1996), The emergence of jets and vortices in freely evolving, shallow-water turbulence on a sphere. *Physics of Fluids*, **8**, 1531–1552.

Choi, D. S., A. P. Showman and R. H. Brown (2009), Cloud features and zonal wind measurements of Saturn's atmosphere as observed by Cassini/VIMS. *Journal of Geophysical Research (Planets)*, **114**, E04,007.

Christensen, U. R. (2001), Zonal flow driven by deep convection in the major planets. *Geophys. Res. Lett.*, **28**, 2553–2556.

Christensen, U. R. (2002), Zonal flow driven by strongly super-critical convection in rotating spherical shells. *Journal of Fluid Mechanics*, **470**, 115–133.

Conrath, B. J., P. J. Gierasch and S. S. Leroy (1990), Temperature and circulation in the stratosphere of the outer planets. *Icarus*, **83**, 255–281.

Conrath, B. J. and J. A. Pirraglia (1983), Thermal structure of Saturn from Voyager infrared measurements: Implications for atmospheric dynamics. *Icarus*, **53**, 286–292.

Del Genio, A. D., R. K. Achterberg, K. H. Baines, F. M. Flasar, P. L. Read, A. Sánchez-Lavega and A. P. Showman (2009), Saturn Atmospheric Structure and Dynamics. *Saturn from Cassini-Huygens*, eds. M. K. Dougherty, L. W. Esposito and S. M. Krimigis, Springer, pp. 113–159.

Del Genio, A. D. and J. M. Barbara (2012), Constraints on Saturn's tropospheric general circulation from Cassini ISS images. *Icarus*, **219**, 689–700.

Desch, M. D. and M. L. Kaiser (1981), Voyager measurement of the rotation period of Saturn's magnetic field. *Geophys. Res. Lett.*, **8**, 253–256.

Dowling, T. E. (1995a), Dynamics of jovian atmospheres. *Annual Review of Fluid Mechanics*, **27**, 293–334.

Dowling, T. E. (1995b), Estimate of Jupiter's deep zonal-wind profile from Shoemaker-Levy 9 data and Arnol'd's second stability criterion. *Icarus*, **117**, 439–442.

Dowling, T. E. (2014), Saturn's longitude: Rise of the second branch of shear-stability theory and fall of the first. *International Journal of Modern Physics D*, **23**, 1430006.

Dritschel, D. G. and M. E. McIntyre (2008), Multiple jets as PV staircases: The Phillips effect and the resilience of eddy-transport barriers. *Journal of Atmospheric Sciences*, **65**, 855–874.

Duarte, L. D. V., T. Gastine and J. Wicht (2013), Anelastic dynamo models with variable electrical conductivity: An application to gas giants. *Phys. Earth Planet. Int.*, **222**, 22–34.

Dunkerton, T. (1978), On the mean meridional mass motions of the stratosphere and mesosphere. *J. Atmos. Sci.*, **35**, 2325–2333.

Dyudina, U. A., A. P. Ingersoll, S. P. Ewald, C. C. Porco, G. Fischer and Y. Yair (2013), Saturn's visible lightning, its radio emissions, and the structure of the 2009–2011 lightning storms. *Icarus*, **226**, 1020–1037.

Dyudina, U. A. et al. (2007), Lightning storms on Saturn observed by Cassini ISS and RPWS during 2004–2006. *Icarus*, **190**, 545–555.

Dyudina, U. A. (2008), Dynamics of Saturn's South Polar Vortex. *Science*, **319**, 1801.

Evonuk, M. and G. A. Glatzmaier (2004), 2D studies of various approximations used for modeling convection in giant planets. *Geophysical and Astrophysical Fluid Dynamics*, **98**, 241–255.

Evonuk, M. and G. A. Glatzmaier (2006), A 2D study of the effects of the size of a solid core on the equatorial flow in giant planets. *Icarus*, **181**, 458–464.

Evonuk, M. and G. A. Glatzmaier (2007), The effects of rotation rate on deep convection in giant planets with small solid cores. *Planet. Space Sci.*, **55**, 407–412.

Fischer, G. et al. (2011), A giant thunderstorm on Saturn. *Nature*, **475**, 75–77.

Flasar, F. M. et al. (2004), An intense stratospheric jet on Jupiter. *Nature*, **427**, 132–135.

Flasar, F. M. (2005), Temperatures, winds, and composition in the Saturnian system. *Science*, **307**, 1247–1251.

Fletcher, L. N., K. H. Baines, T. W. Momary, A. P. Showman, P. G. J. Irwin, G. S. Orton, M. Roos-Serote and C. Merlet (2011), Saturn's tropospheric composition and clouds from Cassini/VIMS 4.6–5.1 μm nightside spectroscopy. *Icarus*, **214**, 510–533.

Fletcher, L. N., G. S. Orton, N. A. Teanby, P. G. J. Irwin and G. L. Bjoraker (2009), Methane and its isotopologues on Saturn from Cassini/CIRS observations. *Icarus*, **199**, 351–367.

Fletcher, L. N. et al. (2007), Characterising Saturn's vertical temperature structure from Cassini/CIRS. *Icarus*, **189**, 457–478.

Fletcher, L. N. (2008), Temperature and composition of Saturn's polar hot spots and hexagon. *Science*, **319**, 79–81.

Fletcher, L. N. (2010), Seasonal change on Saturn from Cassini/CIRS observations, 2004–2009. *Icarus*, **208**, 337–352.

Fletcher, L. N. (2011), Thermal structure and dynamics of Saturn's northern springtime disturbance. *Science*, **332**, 1413–1417.

Fletcher, L. N. (2012), The origin and evolution of Saturn's 2011–2012 stratospheric vortex. *Icarus*, **221**, 560–586.

Fletcher, L. N. (2015), Seasonal evolution of Saturn's polar temperatures and composition. *Icarus*, **250**, 131–153.

Fouchet, T., S. Guerlet, D. F. Strobel, A. A. Simon-Miller, B. Bézard and F. M. Flasar (2008), An equatorial oscillation in Saturn's middle atmosphere. *Nature*, **453**, 200–202.

Fouchet, T., J. I. Moses and B. J. Conrath (2009), Saturn: Composition and chemistry. *Saturn from Cassini-Huygens*, eds. M. K. Dougherty, L. W. Esposito and S. M. Krimigis, Springer, pp. 83–112.

Friedson, A. J. and J. I. Moses (2012), General circulation and transport in Saturn's upper troposphere and stratosphere. *Icarus*, **218**, 861–875.

Galanti, E. and Y. Kaspi (2016), An adjoint-based method for the inversion of the Juno and Cassini gravity measurements into wind fields. *Astrophys. J.*, **820**, 91.

Galanti, E. and Y. Kaspi (2017), Decoupling Jupiter's deep and atmospheric flows using the upcoming Juno gravity measurements and a dynamical inverse model. *Icarus*, **286**, 46–55.

Galanti, E., Y. Kaspi and E. Tziperman (2017), A full, self-consistent treatment of thermal wind balance on fluid planets. *J. Fluid Mech.*, **810**.

Garcia-Melendo, E., R. Hueso, A. Sánchez-Lavega, J. Legarreta, T. del Rio-Gaztelurrutia, S. Pérez-Hoyos and J. F. Sanz-Requena (2013), Atmospheric dynamics of saturn's 2010 giant storm. *Nature Geoscience*, **6**, 525–529.

García-Melendo, E., S. Pérez-Hoyos, A. Sánchez-Lavega and R. Hueso (2011), Saturn's zonal wind profile in 2004–2009 from Cassini ISS images and its long-term variability. *Icarus*, **215**, 62–74.

Gastine, T. and J. Wicht (2012), Effects of compressibility on driving zonal flow in gas giants. *Icarus*, **219**, 428–442.

Gastine, T., J. Wicht and J. M. Aurnou (2013), Zonal flow regimes in rotating anelastic spherical shells: An application to giant planets. *Icarus*, **225**, 156–172.

Gastine, T., J. Wicht, L. D. V. Duarte, M. Heimpel and A. Becker (2014), Explaining Jupiter's magnetic field and equatorial jet dynamics. *Geophys. Res. Lett.*, **41**, 5410–5419.

Gezari, D. Y., M. J. Mumma, F. Espenak, D. Deming, G. Bjoraker, L. Woods and W. Folz (1989), New features in Saturn's atmosphere revealed by high-resolution thermal infrared images. *Nature*, **342**, 777–780.

Gierasch, P. J., J. A. Magalhaes and B. J. Conrath (1986), Zonal mean properties of Jupiter's upper troposphere from Voyager infrared observations. *Icarus*, **67**, 456–483.

Gierasch, P. J. et al. (2000), Observation of moist convection in Jupiter's atmosphere. *Nature*, **403**, 628–630.

Giles, R. S., L. N. Fletcher and P. G. J. Irwin (2017), Latitudinal variability in Jupiter's tropospheric disequilibrium species: GeH_4, AsH_3, and PH_3. *Icarus*, in press.

Gillett, F. C. and G. S. Orton (1975), Center-to-limb observations of Saturn in the thermal infrared. *Astrophys. J. Lett.*, **195**, L47–L49.

Glatzmaier, G. A. (2008), A note on "Constraints on deep-seated zonal winds inside Jupiter and Saturn." *Icarus*, **196**, 665–666.

Glatzmaier, G. A., M. Evonuk and T. M. Rogers (2009), Differential rotation in giant planets maintained by density-stratified turbulent convection. *Geophys. Astrophy. Fluid Dyn.*, **103**, 31–51.

Greathouse, T. K., J. H. Lacy, B. Bézard, J. I. Moses, C. A. Griffith and M. J. Richter (2005), Meridional variations of temperature, C_2H_2 and C_2H_6 abundances in Saturn's stratosphere at southern summer solstice. *Icarus*, **177**, 18–31.

Grevesse, N., M. Asplund, and J. Sauval (2005), The new solar composition. In *Element Stratification in Stars: 40 Years of Atomic Diffusion*, eds. G. Alecian, O. Richard and S. Vauclair, pp. 21–30, EAS Publications Series.

Guerlet, S., T. Fouchet, B. Bézard, F. M. Flasar and A. A. Simon-Miller (2011), Evolution of the equatorial oscillation in Saturn's stratosphere between 2005 and 2010 from Cassini/CIRS limb data analysis. *Geophys. Res. Lett.*, **38**, 9201.

Guerlet, S., T. Fouchet, B. Bézard, A. A. Simon-Miller and F. Michael Flasar (2009), Vertical and meridional distribution of ethane, acetylene and propane in Saturn's stratosphere from CIRS/Cassini limb observations. *Icarus*, **203**, 214–232.

Guerlet, S. et al. (2014), Global climate modeling of Saturn's atmosphere. Part I: Evaluation of the radiative transfer model. *Icarus*, **238**, 110–124.

Hanel, R. et al. (1981), Infrared observations of the Saturnian system from Voyager 1. *Science*, **212**, 192–200.

Hanel, R. (1982), Infrared observations of the Saturnian system from Voyager 2. *Science*, **215**, 544–548.

Haynes, P. H., M. E. McIntyre, T. G. Shepherd, C. J. Marks and K. P. Shine (1991), On the "Downward Control" of extratropical diabatic circulations by eddy-induced mean zonal forces. *Journal of Atmospheric Sciences*, **48**, 651–680.

Heimpel, M. and J. Aurnou (2007), Turbulent convection in rapidly rotating spherical shells: A model for equatorial and high latitude jets on Jupiter and Saturn. *Icarus*, **187**, 540–557.

Heimpel, M., J. Aurnou and J. Wicht (2005), Simulation of equatorial and high-latitude jets on Jupiter in a deep convection model. *Nature*, **438**, 193–196.

Heimpel, M. and N. Gómez Pérez (2011), On the relationship between zonal jets and dynamo action in giant planets. *Geophys. Res. Lett.*, **38**, L14,201.

Helled, R., E. Galanti and Y. Kaspi (2015), Saturn's fast spin determined from its gravitational field and oblateness. *Nature*, **520**, 202–204.

Hesman, B. E. et al. (2009), Saturn's latitudinal C_2H_2 and C_2 H_6 abundance profiles from Cassini/CIRS and ground-based observations. *Icarus*, **202**, 249–259.

Hesman, B. E. (2012), Elusive ethylene detected in Saturn's northern storm Region. *Astrophys. J.*, **760**, 24.

Holton, J. R. and G. J. Hakim (2013), *An Introduction to Dynamic Meteorology, 5th Ed.* San Diego, CA: Academic Press.

Howett, C. J. A., P. G. J. Irwin, N. A. Teanby, A. Simon-Miller, S. B. Calcutt, L. N. Fletcher and R. de Kok (2007), Meridional variations in stratospheric acetylene and ethane in the southern hemisphere of the Saturnian atmosphere as determined from Cassini/CIRS measurements. *Icarus*, **190**, 556–572.

Huang, H.-P. and W. A. Robinson (1998), Two-dimensional turbulence and persistent zonal jets in a global barotropic model. *Journal of Atmospheric Sciences*, **55**, 611–632.

Hubbard, W. B. (1984), *Planetary Interiors*. New York, NY: Van Nostrand Reinhold Co., 1984, 343 p.

Hubbard, W. B. (1999), Gravitational signature of Jupiter's deep zonal flows. *Icarus*, **137**, 357–359.

Hubbard, W. B. (2012), High-precision Maclaurin-based models of rotating liquid planets. *Astrophys. J. Lett.*, **756**, L15.

Hubbard, W. B. (2013), Conventric Maclaurian spheroid models of rotating liquid planets. *Astrophys. J.*, **768**(1).

Hubbard, W. B., W. J. Nellis, A. C. Mitchell, N. C. Holmes, P. C. McCandless and S. S. Limaye (1991), Interior structure of Neptune: Comparison with Uranus. *Science*, **253**, 648–651.

Hubbard, W. B., G. Schubert, D. Kong and K. Zhang (2014), On the convergence of the theory of figures. *Icarus*, **242**, 138–141.

Hueso, R. and A. Sánchez-Lavega (2004), A three-dimensional model of moist convection for the giant planets II: Saturn's water and ammonia moist convective storms. *Icarus*, **172**, 255–271.

Iacono, R., M. V. Struglia and C. Ronchi (1999), Spontaneous formation of equatorial jets in freely decaying shallow water turbulence. *Physics of Fluids*, **11**, 1272–1274.

Ingersoll, A. P. (1976), The atmosphere of Jupiter. *Space Sci. Rev.*, **18**, 603–639.

Ingersoll, A. P. (1990), Atmospheric dynamics of the outer planets. *Science*, **248**, 308–315.

Ingersoll, A. P., R. F. Beebe, B. J. Conrath and G. E. Hunt (1984), Structure and dynamics of Saturn's atmosphere. In *Saturn*, eds. T. Gehrels and M. S. Matthews, pp. 195–238, Tucson, AZ: University of Arizona Press.

Ingersoll, A. P., R. F. Beebe, J. L. Mitchell, G. W. Garneau, G. M. Yagi and J.-P. Muller (1981), Interaction of eddies and mean zonal flow on Jupiter as inferred from Voyager 1 and 2 images. *J. Geophys. Res.*, **86**, 8733–8743.

Ingersoll, A. P. and J. N. Cuzzi (1969), Dynamics of Jupiter's cloud bands. *Journal of Atmospheric Sciences*, **26**, 981–985.

Ingersoll, A. P., T. E. Dowling, P. J. Gierasch, G. S. Orton, P. L. Read, A. Sánchez-Lavega, A. P. Showman, A. A. Simon-Miller and A. R. Vasavada (2004), Dynamics of Jupiter's Atmosphere, pp. 105–128 in *Jupiter. The Planet, Satellites and Magnetosphere.* Cambridge: Cambridge University Press.

Ingersoll, A. P., P. J. Gierasch, D. Banfield, A. R. Vasavada and A3 Galileo Imaging Team (2000), Moist convection as an energy source for the large-scale motions in Jupiter's atmosphere. *Nature*, **403**, 630–632.

Ingersoll, A. P., G. S. Orton, G. Munch, G. Neugebauer and S. C. Chase (1980), Pioneer Saturn infrared radiometer: Preliminary results. *Science*, **207**, 439–443.

Ingersoll, A. P. and D. Pollard (1982), Motion in the interiors and atmospheres of Jupiter and Saturn: Scale analysis, anelastic equations, barotropic stability criterion. *Icarus*, **52**, 62–80.

Jacobson, R. A. (2007), The gravity field of the Uranian system and the orbits of the Uranian satellites and rings. *AAS/Division for Planetary Sciences Meeting Abstracts #39*, vol. 38, pp. 453–453.

Jacobson, R. A. (2009), The orbits of the Neptunian satellites and the orientation of the pole of Neptune. *Astrophys. J.*, **137**, 4322–4329.

Jacobson, R. A. et al. (2006), The gravity field of the Saturnian system from satellite observations and spacecraft tracking data. *Astrophys. J.*, **132**, 2520–2526.

Janssen, M. A., et al. (2013), Saturn's thermal emission at 2.2-cm wavelength as imaged by the Cassini RADAR radiometer. *Icarus*, **226**, 522–535.

Jones, C. A. (2014), A dynamo model of Jupiter's magnetic field. *Icarus*, **241**, 148–159.

Jones, C. A., P. Boronski, A. S. Brun, G. A. Glatzmaier, T. Gastine, M. S. Miesch and J. Wicht (2011), Anelastic convection-driven dynamo benchmarks. *Icarus*, **216**, 120–135.

Jones, C. A. and K. M. Kuzanyan (2009), Compressible convection in the deep atmospheres of giant planets. *Icarus*, **204**, 227–238.

Kaspi, Y. (2008), Turbulent convection in an anelastic rotating sphere: A model for the circulation on the giant planets, Ph.D. thesis, Massachusetts Institute of Technology.

Kaspi, Y. (2013), Inferring the depth of the zonal jets on Jupiter and Saturn from odd gravity harmonics. *Geophys. Res. Lett.*, **40**, 676–680.

Kaspi, Y., J. E. Davighi, E. Galanti and W. B. Hubbard (2016), The gravitational signature of internal flows in giant planets: Comparing the thermal wind approach with barotropic potential-surface methods. *Icarus*, **276**, 170–181.

Kaspi, Y., G. R. Flierl and A. P. Showman (2009), The deep wind structure of the giant planets: Results from an anelastic general circulation model. *Icarus*, **202**, 525–542.

Kaspi, Y., W. B. Hubbard, A. P. Showman and G. R. Flierl (2010), Gravitational signature of Jupiter's internal dynamics. *Geophys. Res. Lett.*, **37**, L01,204.

Kaspi, Y., A. P. Showman, W. B. Hubbard, O. Aharonson and R. Helled (2013), Atmospheric confinement of jet streams on Uranus and Neptune. *Nature*, **497**, 344–347.

Kirk, R. L. and D. J. Stevenson (1987), Hydromagnetic constraints on deep zonal flow in the giant planets. *Astrophys. J.*, **316**, 836–846.

Kliore, A. J., I. R. Patel, G. F. Lindal, D. N. Sweetnam, H. B. Hotz, J. H. Waite and T. McDonough (1980), Structure of the ionosphere and atmosphere of Saturn from Pioneer 11 Saturn radio occultation. *J. Geophys. Res.*, **85**, 5857–5870.

Kong, D., K. Zhang and G. Schubert (2012), On the variation of zonal gravity coefficients of a giant planet caused by its deep zonal flows. *Astrophys. J.*, **748**.

Laraia, A. L., A. P. Ingersoll, M. A. Janssen, S. Gulkis, F. Oyafuso and M. Allison (2013), Analysis of Saturn's thermal emission at 2.2 cm wavelength: Spatial distribution of ammonia vapor. *Icarus*, **226**, 641–654.

Leovy, C. B., A. J. Friedson and G. S. Orton (1991), The quasiquadrennial oscillation of Jupiter's equatorial stratosphere. *Nature*, **354**, 380–382.

Li, L., A. P. Ingersoll and X. Huang (2006), Interaction of moist convection with zonal jets on Jupiter and Saturn. *Icarus*, **180**, 113–123.

Li, L., A. P. Ingersoll, A. R. Vasavada, C. C. Porco, A. D. del Genio and S. P. Ewald (2004), Life cycles of spots on Jupiter from Cassini images. *Icarus*, **172**, 9–23.

Li, L. et al. (2008), Strong jet and a new thermal wave in Saturn's equatorial stratosphere. *Geophys. Res. Lett.*, **35**, 23,208.

Li, L. (2011), Equatorial winds on Saturn and the stratospheric oscillation. *Nature Geoscience*, **4**, 750–752.

Li, L. (2013), Strong temporal variation over one Saturnian year: From Voyager to Cassini. *Scientific Reports*, **3**, 2410.

Li, L. (2015), Saturn's giant storm and global radiant energy. *Geophys. Res. Lett.*, **42**, 2144–2148.

Lian, Y. and A. P. Showman (2008), Deep jets on gas-giant planets. *Icarus*, **194**, 597–615.

Lian, Y. and A. P. Showman (2010), Generation of equatorial jets by large-scale latent heating on the giant planets. *Icarus*, **207**, 373–393.

Limaye, S. S. (1986), Jupiter: New estimates of the mean zonal flow at the cloud level. *Icarus*, **65**, 335–352.

Lindal, G., D. Sweetnam and V. Eshleman (1985), The atmosphere of Saturn: An analysis of the voyager radio occultation measurements. *The Astronomical Journal*, **90**, 1136–1146.

Lindzen, R. S., and J. R. Holton (1968), A theory of the quasi-biennial oscillation. *Journal of Atmospheric Sciences*, **25**, 1095–1107.

Little, B., C. D. Anger, A. P. Ingersoll, A. R. Vasavada, D. A. Senske, H. H. Breneman, W. J. Borucki and The Galileo SSI Team (1999), Galileo images of lightning on Jupiter. *Icarus*, **142**, 306–323.

Liu, J., P. M. Goldreich and D. J. Stevenson (2008), Constraints on deep-seated zonal winds inside Jupiter and Saturn. *Icarus*, **196**, 653–664.

Liu, J. and T. Schneider (2010), Mechanisms of jet formation on the giant planets. *Journal of Atmospheric Sciences*, **67**, 3652–3672.

Liu, J., T. Schneider and L. N. Fletcher (2014), Constraining the depth of Saturn's zonal winds by measuring thermal and gravitational signals. *Icarus*, **239**, 260–272.

Liu, J., T. Schneider and Y. Kaspi (2013), Predictions of thermal and gravitational signals of Jupiter's deep zonal winds. *Icarus*, **224**, 114–125.

MacGorman, D. R. and D. Rust (1998), *The Electrical Nature of Storms*. Oxford: Oxford University Press.

Marcus, P. S. (1993), Jupiter's Great Red Spot and other vortices. *Annu. Rev. Astron. Astrophys.*, **31**, 523–573.

Moore, A. M., H. G. Arango, G. Broquet, B. S. Powell, A. T. Weaver and J. Zavala-Garay (2011),

The regional ocean modeling system (ROMS) 4-dimensional variational data assimilation systems. Part I: System overview and formulation. *Prog. Oceanogr.*, **91**, 34–49.

Moses, J. and T. Greathouse (2005), Latitudinal and seasonal models of stratospheric photochemistry on Saturn: Comparison with infrared data from IRTF/TEXES. *Journal of Geophysical Research*, **110**(E9), E09,007.

Moses, J. I., E. S. Armstrong, L. N. Fletcher, A. J. Friedson, P. G. J. Irwin, J. A. Sinclair and B. E. Hesman (2015), Evolution of stratospheric chemistry in the Saturn storm beacon region. *Icarus*, **261**, 149–168.

Moses, J. I., M.-C. Liang, Y. L. Yung and R.-L. Shia (2007), Two-Dimensional Photochemical Modeling of Hydrocarbon Abundances on Saturn. *Lunar and Planetary Science Conference*, vol. 38 of *Lunar and Planetary Inst. Technical Report*, p. 2196.

Nellis, W. J. (2000), Metallization of fluid hydrogen at 140 GPa (1.4 Mbar): Implications for Jupiter. *Planet. Space Sci.*, **48**, 671–677.

Nicholson, P. D., C. A. McGhee and R. G. French (1995), Saturn's central flash from the 3 July 1989 occultation of 28 SGR. *Icarus*, **113**, 57–83.

Nozawa, T. and S. Yoden (1997), Formation of zonal band structure in forced two-dimensional turbulence on a rotating sphere. *Physics of Fluids*, **9**, 2081–2093.

Okuno, A. and A. Masuda (2003), Effect of horizontal divergence on the geostrophic turbulence on a beta-plane: Suppression of the Rhines effect. *Physics of Fluids*, **15**, 56–65.

Ollivier, J. L., F. Billebaud, P. Drossart, M. Dobrijévic, M. Roos-Serote, T. August-Bernex and I. Vauglin (2000), Seasonal effects in the thermal structure of Saturn's stratosphere from infrared imaging at 10 microns. *Astronomy and Astrophysics*, **356**, 347–356.

O'Neill, M. E., K. A. Emanuel and G. R. Flierl (2015), Polar vortex formation in giant-planet atmospheres due to moist convection. *Nature Geoscience*, **8**, 523–526.

O'Neill, M. E., K. A. Emanuel and G. R. Flierl (2016), Weak jets and strong cyclones: Shallow-water modeling of giant planet polar caps. *Journal of Atmospheric Sciences*, **73**, 1841–1855.

Orton, G. S. and A. P. Ingersoll (1980), Saturn's atmospheric temperature structure and heat budget. *J. Geophys. Res.*, **85**, 5871–5881.

Orton, G. S. and P. A. Yanamandra-Fisher (2005), Saturn's temperature field from high-resolution middle-infrared imaging. *Science*, **307**, 696–698.

Orton, G. S. et al. (1991), Thermal maps of Jupiter: Spatial organization and time dependence of stratospheric temperatures, 1980 to 1990. *Science*, **252**, 537–542.

Orton, G. S. (2008), Semi-annual oscillations in Saturn's low-latitude stratospheric temperatures. *Nature*, **453**, 196–199.

Parisi, M., E. Galanti, S. Finocchiaro, L. Iess and Y. Kaspi (2016), Probing the depth of Jupiter's Great Red Spot with the Juno gravity experiment. *Icarus*, **267**, 232–242.

Pedlosky, J. (1987), *Geophysical Fluid Dynamics, 2nd Ed.* New York, NY: Springer-Verlag.

Peek, B. M. (1958), *The Planet Jupiter*. London: Faber and Faber.

Phillips, O. M. (1969), Turbulent flow. *Science*, **163**, 1441–1442.

Porco, C. C. et al. (2003), Cassini imaging of Jupiter's atmosphere, satellites, and rings. *Science*, **299**, 1541–1547.

Read, P. L., B. J. Conrath, L. N. Fletcher, P. J. Gierasch, A. A. Simon-Miller and L. C. Zuchowski (2009a), Mapping potential vorticity dynamics on Saturn: Zonal mean circulation from Cassini and Voyager data. *Planet. Space Sci.*, **57**, 1682–1698.

Read, P. L., T. E. Dowling and G. Schubert (2009b), Saturn's rotation period from its atmospheric planetary-wave configuration. *Nature*, **460**, 608–610.

Rieke, G. H. (1975), The thermal radiation of Saturn and its rings. *Icarus*, **26**, 37–44.

Rogers, J. H. (1995), *The Giant Planet Jupiter*. New York, NY: Cambridge University Press.

Salyk, C., A. P. Ingersoll, J. Lorre, A. Vasavada and A. D. Del Genio (2006), Interaction between eddies and mean flow in Jupiter's atmosphere: Analysis of Cassini imaging data. *Icarus*, **185**, 430–442.

Sanchez-Lavega, A., J. F. Rojas and P. V. Sada (2000), Saturn's zonal winds at cloud level. *Icarus*, **147**, 405–420.

Sánchez-Lavega, A. et al. (2011), Deep winds beneath Saturn's upper clouds from a seasonal long-lived planetary-scale storm. *Nature*, **475**, 71–74.

Sánchez-Lavega, A. (2012), Ground-based observations of the long-term evolution and death of Saturn's 2010 Great White Spot. *Icarus*, **220**, 561–576.

Sayanagi, K. M., U. A. Dyudina, S. P. Ewald, G. Fischer, A. P. Ingersoll, W. S. Kurth, G. D. Muro, C. C. Porco and R. A. West (2013), Dynamics of Saturn's great storm of 2010–2011 from Cassini ISS and RPWS. *Icarus*, **223**, 460–478.

Sayanagi, K. M., U. A. Dyudina, S. P. Ewald, G. D. Muro and A. P. Ingersoll (2014), Cassini ISS observation of Saturn's string of pearls. *Icarus*, **229**, 170–180.

Sayanagi, K. M. and A. P. Showman (2007), Effects of a large convective storm on Saturn's equatorial jet. *Icarus*, **187**, 520–539.

Schinder, P. J. et al. (2011), Saturn's equatorial oscillation: Evidence of descending thermal structure from Cassini radio occultations. *Geophys. Res. Lett.*, **38**, 8205.

Schneider, T. (2006), The general circulation of the atmosphere. *Annual Review of Earth and Planetary Sciences*, **34**, 655–688.

Schneider, T. and J. Liu (2009), Formation of jets and equatorial superrotation on Jupiter. *J. Atmos. Sci.*, **66**, 579–601.

Scott, R. K. and D. G. Dritschel (2012), The structure of zonal jets in geostrophic turbulence. *J. Fluid Mech.*, **711**, 576–598.

Scott, R. K. and L. Polvani (2007), Forced-dissipative shallow water turbulence on the sphere and the atmospheric circulation of the giant planets. *J. Atmos. Sci*, **64**, 3158–3176.

Scott, R. K. and L. Polvani (2008), Equatorial superrotation in shallow atmospheres. *Geophys. Res. Lett.*, **35**, L24, 202.

Showman, A. P. (2007), Numerical simulations of forced shallow-water turbulence: Effects of moist convection on the large-scale circulation of Jupiter and Saturn. *J. Atmos. Sci.*, **64**, 3132–3157.

Showman, A. P., J. Y.-K. Cho and K. Menou (2010), Atmospheric circulation of exoplanets. In *Exoplanets*, ed. S. Seager, pp. 471–516, Tucson, AZ: University of Arizona Press.

Showman, A. P. and I. de Pater (2005), Dynamical implications of Jupiter's tropospheric ammonia abundance. *Icarus*, **174**, 192–204.

Showman, A. P., P. J. Gierasch and Y. Lian (2006), Deep zonal winds can result from shallow driving in a giant-planet atmosphere. *Icarus*, **182**, 513–526.

Showman, A. P., Y. Kaspi and G. R. Flierl (2011), Scaling laws for convection and jet speeds in the giant planets. *Icarus*, **211**, 1258–1273.

Sinclair, J. A., P. G. J. Irwin, L. N. Fletcher, J. I. Moses, T. K. Greathouse, A. J. Friedson, B. Hesman, J. Hurley and C. Merlet (2013), Seasonal variations of temperature, acetylene and ethane in Saturn's atmosphere from 2005 to 2010, as observed by Cassini-CIRS. *Icarus*, **225**, 257–271.

Sinton, W. M., W. W. Macy, J. Good and G. S. Orton (1980), Infrared scans of Saturn. *Icarus*, **42**, 251–256.

Smith, K. S. (2004), A local model for planetary atmospheres forced by small-scale convection. *J. Atmos. Sciences*, **61**, 1420–1433.

Sromovsky, L. A., K. H. Baines, P. M. Fry and T. W. Momary (2016), Cloud clearing in the wake of Saturn's Great Storm of 2010–2011 and suggested new constraints on Saturn's He/H_2 ratio. *Icarus*, **276**, 141–162.

Starr, V. P. (1968), *Physics of Negative Viscosity Phenomena*. New York, NY: McGraw-Hill.

Stone, P. H. (1973), The dynamics of the atmospheres of the major planets (article published in the Space Science Reviews special issue on "Outer Solar System Exploration: An Overview," ed. by J. E. Long and D. G. Rea. *Space Sci. Rev.*, **14**, 444–459.

Sukoriansky, S., N. Dikovskaya and B. Galperin (2007), On the "arrest" of inverse energy cascade and the Rhines scale. *J. Atmos. Sci.*, **64**, 3312–3327.

Sylvestre, M., S. Guerlet, T. Fouchet, A. Spiga, F. M. Flasar, B. Hesman and G. L. Bjoraker (2015), Seasonal changes in Saturn's stratosphere inferred from Cassini/CIRS limb observations. *Icarus*, **258**, 224–238.

Thomson, S. I. and M. E. McIntyre (2016), Jupiter's unearthly jets: A new turbulent model exhibiting statistical steadiness without large-scale dissipation. *Journal of Atmospheric Sciences*, **73**, 1119–1141.

Tokunaga, A. T., J. Caldwell, F. C. Gillett and I. G. Nolt (1978), Spatially resolved infrared observations of Saturn. II: The temperature enhancement at the south pole of Saturn. *Icarus*, **36**, 216–222.

Tziperman, E. and W. C. Thacker (1989), An optimal-control/adjoint-equations approach to studying the oceanic general circulation. *J. Phys. Oceanogr.*, **19**, 1471–1485.

Uman, M. A. (2001), *The Lightning Discharge*. Mineola, NY: Dover Books.

Vallis, G. K. (2006), *Atmospheric and Oceanic Fluid Dynamics: Fundamentals and Large-Scale Circulation*. Cambridge: Cambridge University Press.

van Helden, A. (1984), Saturn through the telescope: A brief historical survey. In *Saturn*, eds. T. Gehrels and M.S. Matthews, pp. 23–43, Tucson, AZ: University of Arizona Press.

Vasavada, A. R., S. M. Hörst, M. R. Kennedy, A. P. Ingersoll, C. C. Porco, A. D. Del Genio and R. A. West (2006), Cassini imaging of Saturn: Southern-hemisphere winds and vortices. *J. Geophys. Res.*, **111**, E05004(E10), 1–13.

Vasavada, A. R. and A. P. Showman (2005), Jovian atmospheric dynamics: An update after Galileo and Cassini. *Reports of Progress in Physics*, **68**, 1935–1996.

Warneford, E. S. and P. J. Dellar (2014), Thermal shallow water models of geostrophic turbulence in Jovian atmospheres. *Physics of Fluids*, **26**(1), 016,603.

Williams, G. P. (1978), Planetary circulations. I: Barotropic representation of Jovian and terrestrial turbulence. *Journal of Atmospheric Sciences*, **35**, 1399–1426.

Williams, G. P. (2003), Jovian dynamics. Part III: Multiple, migrating, and equatorial jets. *Journal of Atmospheric Sciences*, **60**, 1270–1296.

Yadav, R. K., T. Gastine and U. R. Christensen (2013), Scaling laws in spherical shell dynamos with free-slip boundaries. *Icarus*, **225**, 185–193.

Yair, Y., G. Fischer, F. Simões, N. Renno and P. Zarka (2008), Updated review of planetary atmospheric electricity. *Space Sci. Rev.*, **137**, 29–49.

Zhang, K., D. Kong and G. Schubert (2015), Thermal-gravitational wind equation for the wind-induced gravitational signature of giant gaseous planets: Mathematical derivation, numerical method and illustrative solutions. *Astrophys. J.*, **806**, 270–279.

12

Saturn's Polar Atmosphere

KUNIO M. SAYANAGI, KEVIN H. BAINES, ULYANA DYUDINA, LEIGH N. FLETCHER, AGUSTÍN SÁNCHEZ-LAVEGA AND ROBERT A. WEST

Abstract

This chapter reviews the state of our knowledge about Saturn's polar atmosphere that has been revealed through Earth- and space-based observation as well as theoretical and numerical modeling. In particular, the Cassini mission to Saturn, which has been in orbit around the ringed planet since 2004, has revolutionized our understanding of the planet. The current review updates a previous review by Del Genio et al. (2009), written after Cassini's primary mission phase that ended in 2008, by focusing on the north polar region of Saturn and comparing it to the southern high latitudes. Two prominent features in the northern high latitudes are the northern hexagon and the north polar vortex; we extensively review observational and theoretical investigations to date of both features. We also review the seasonal evolution of the polar regions using the observational data accumulated during the Cassini mission since 2004 (shortly after the northern winter solstice in 2002), through the equinox in 2009, and approaching the next solstice in 2017. We conclude the current review by listing unanswered questions and describing the observations of the polar regions planned for the Grand Finale phase of the Cassini mission between 2016 and 2017.

12.1 Introduction: Views of Saturn's Poles from the Cassini Orbiter

The Cassini mission has explored the Saturnian system in ways that are possible only from an orbiting spacecraft. Cassini is the fourth spacecraft to visit Saturn after Pioneer 11 and Voyagers 1 and 2, and the first spacecraft to orbit the planet. Cassini has been in orbit around Saturn since 2004, has accumulated a decade of near-continuous observational data of the planet as of this writing and will capture almost half a Saturn-year by the planned end of the mission in 2017.

Among the new data returned from Cassini, its observations of Saturn's polar regions in particular have opened a new era in atmospheric studies of the giant planets. The data returned from the Cassini mission represent a significant advance in two major ways. The first is the long temporal coverage, spanning nearly two solstices, that has captured diverse atmospheric phenomena including seasonal changes. In comparison, each of the previous flyby missions captured data for only several months, which essentially represent temporal snapshots in the context of planetary climatology on a planet for which the orbital period is 30 Earth years. Second, Cassini's dynamic orbital trajectory has allowed the spacecraft to tour the Saturnian space between the equatorial plane to high above the poles, enabling observations of every latitude of Saturn from a wide range of illumination and observation angles. From those various observational geometries, infrared observations coupled with stellar occultation and radio occultation measurements can also provide a crucial vertical dimension to our understanding of the polar regions, extending hundreds of kilometers above the visible cloud deck and tropospheric hazes. These advantages have enabled detailed studies of both poles of Saturn, revealing surprising new results in the nature of polar winds, vortex features, aerosol properties and thermal structure, as we detail in this review.

12.1.1 Context of the Current Review

This review of Saturn's poles integrates the most recent Cassini data, presenting multi-faceted results on cloud morphology, thermal structure, aerosol properties and atmospheric dynamics. As such, this review builds on and extends previous work based on the Voyager flyby missions (e.g. Gehrels and Matthews 1984; in particular, Tomasko et al. 1984 and Ingersoll et al. 1984), and on the first 5 years of the Cassini mission (Dougherty et al. 2009; specifically, the reviews by Del Genio et al. 2009 and West et al. 2009). Reviews of Jupiter's atmosphere also provide important points of comparison for the current review. The book by Gehrels (1976) contains reviews of Jupiter's polar atmosphere using the Pioneer 11 Infrared Radiometer data (Ingersoll et al. 1976; Orton and Ingersoll 1976). Another review book on Jupiter by Bagenal et al. (2004) treats the state of the knowledge after the Galileo mission (West et al. 2004; Ingersoll et al. 2004; Moses et al. 2004). In addition, Dowling (1995) and Vasavada and Showman (2005) provide in-depth reviews of processes related to jet stream formation and maintenance in giant planet atmospheres, and form the basis of our review of giant planet polar atmospheric dynamics. A textbook by Sánchez-Lavega (2011) also provides a thorough introductory treatment of physical and chemical processes relevant to planetary atmospheres.

The Cassini observations of Saturn's poles represent our most in-depth polar dataset for a giant planet. The picture revealed for Saturn's poles provides a key point of comparison for Jupiter, Uranus and Neptune at a critical time when knowledge of the polar environments of these planets is rapidly growing.

From Earth, Jupiter's poles are not visible because the planet orbits in the ecliptic plane and its rotation axis has virtually no tilt; consequently, the poles can be viewed adequately only by spacecraft flying over the non-equatorial region of the planet. To date, only Pioneer 11 and Ulysses have departed the ecliptic significantly, and only the former carried an imaging instrument. In 2016, NASA's Juno spacecraft will enter orbits that pass directly over Jupiter's north and south poles, exploring these environments in detail for the first time.

Uranus, with its rotation axis tilted by 98°, offers a unique opportunity to observe its poles from Earth. In the images of Uranus returned by Voyager 2, only a small number of discrete features were seen at latitudes poleward of 45°S latitude (Allison et al. 1991; West et al. 1991); however, modern digital image processing analysis by Karkoschka (2015) revealed many more features of Uranus near the south pole for the first time. Hubble Space Telescope (HST) and ground-based observations leading up to Uranus' equinox were similarly free of discrete cloud features in the southern high latitudes (e.g. Sromovsky et al. 2012a). In contrast, in the northern hemisphere, a large number of discrete features have been observed north of 45°N in ground-based images (Sromovsky et al. 2009, 2012a,b). Continued studies of Uranus' hemispheric asymmetry, polar winds and the north polar clouds are providing new insights into this planet's polar circulation (e.g. Sromovsky et al. 2015).

Neptune's polar environments may be most analogous to Saturn's in terms of seasonal effects because their obliquities are similar; Neptune's obliquity is 28.32°. Indeed, several observations of Neptune's south pole have resulted in comparisons with Saturn (e.g. Hammel et al. 2007; Orton et al. 2007). In particular, Neptune's hotspot near the south pole, revealed by large ground-based telescopes armed with adaptive optics (e.g. Luszcz-Cook et al. 2010; Orton et al. 2012; Fletcher et al. 2014; de Pater et al. 2014), appears to be similar to those on Saturn (e.g. Fletcher et al. 2008, and this chapter).

As our knowledge of the poles of Jupiter, Uranus and Neptune continues to grow, the rich Saturn dataset returned by Cassini provides the framework for understanding the polar environments of all four giant planets.

12.1.2 Summary of Cassini Polar Datasets

In the current review, the primary sources of data are the remote-sensing instruments onboard the Cassini spacecraft that cover the wavelengths between the ultraviolet and the thermal infrared. Here, we provide brief descriptions of the instruments that have played key roles in the studies presented in this review.

There are two infrared instruments that have played key roles. First, the Composite Infrared Spectrometer (CIRS; Flasar et al. 2004) is a Fourier transform spectrometer that provides spatially resolved spectral data at thermal infrared wavelengths between 7 μm and 1 mm; for the purpose of observing Saturn's atmosphere, the data returned by the instrument allow determinations of atmospheric thermal structure as well as chemical composition. The other infrared instrument, the Visible and Infrared Mapping Spectrometer (VIMS; Brown et al. 2004), is also an imaging spectrometer and covers the wavelength range between 0.3 and 5.1 μm. In particular, images captured in its 5-μm channel reveal clouds silhouetted against the background thermal emissions; the cloud features that appear in the VIMS 5-μm channel likely reside at the depth of several bars (Baines et al. 2005). The 5-μm images' spatial resolutions are often high enough to track some cloud features to determine the wind fields (e.g. Choi et al. 2009; Baines et al. 2009). High-resolution images in wavelengths between near-infrared and near-ultraviolet are captured using the Imaging Science Subsystem (ISS; Porco et al. 2004). ISS consists of two framing cameras with Charge-Coupled Device (CCD) detectors: the Narrow-Angle Camera (NAC) and the Wide-Angle Camera (WAC). The field of view (FOV) of WAC is ten times wider than that of NAC. Both cameras have multiple spectral filters; several of them have been used extensively in polar atmospheric measurements. In this review, we refer to these filters with their filter IDs as follows. The continuum-band filters with central wavelengths of 750 nm (filter ID: CB2) and 935 nm (CB3) sense in the spectral windows between methane absorption bands, and show the top of the deep tropospheric cloud deck. The methane absorption band filters with central wavelengths of 727 nm (MT2) and 889 nm (MT3) are used to image the upper tropospheric haze layer that is believed to reside around the 100–150 mbar level. The broadband color filters with central wavelengths of 420 nm (VIO), 460 nm (BL1), 567 nm (GRN) and 648 nm (RED) are used to capture the colors of the clouds and hazes, which are indicative of compositional information. Each of these filters can be combined with a polarizing filter to analyze the properties of cloud and haze particles. At ultraviolet wavelengths, the Ultraviolet Imaging Spectrograph (UVIS; Esposito et al. 2004) captures spectrally resolved images in wavelengths between 56 and 190 nm that can be used to study the stratospheric aerosol properties and photochemical processes.

ISS and UVIS were able to image the south pole until the arrival of south polar night, beginning with the 2009 equinox. As Saturn approached the equinox, the northern hemisphere gradually emerged from the winter polar night and the extensive shadows cast by the rings. ISS and UVIS did not start returning images of the north pole until late 2012. As CIRS and VIMS are sensitive to thermal emissions of Saturn, they can resolve features in the winter polar night when the region is not illuminated by sunlight. The north pole has been illuminated since the 2009 equinox; however, the Cassini spacecraft was then in the equatorial plane where it had a poor view of the poles. The first post-equinoctial data of the poles were returned in late 2012 when the spacecraft raised its orbital inclination.

12.1.3 Scope and Organization of the Current Review

This chapter first provides an overview of recent observations of the north polar region of Saturn, devoting sections to two of the major features that have received extensive observational coverage, namely, the northern hexagon (in Section 12.2) and the north polar vortex (in Section 12.3). We compare the north polar region to its southern counterpart in Section 12.4, updating the previous review by Del Genio et al. (2009). We then focus on the physical processes that operate in the polar regions. Section 12.5 discusses the thermal structure, seasonal forcing and chemical processes in the polar stratosphere. The properties of the polar clouds and hazes are described in Section 12.6. A review of polar atmospheric dynamics theories and modeling are presented in Section 12.7. Finally, Section 12.8 concludes the chapter with a summary of unanswered questions, and an outlook of the data expected to be returned by Cassini during the final years of the mission.

12.2 Northern Hexagon

12.2.1 Observations of the Hexagon

The hexagonal pattern in the cloud morphology that encircles the north pole of Saturn is a distinct feature that is not found on any of the other solar system planets; in this chapter, we refer to the feature simply as the hexagon. The feature was discovered by Godfrey (1988) in images taken at visible wavelengths during the Voyager 2 flyby of Saturn in 1981 (Figure 12.1, left panel). The hexagon went unnoticed in these images for years after they were captured because the two Voyager spacecraft had poor views of the poles due to their near-equatorial trajectories; the hexagon was discovered when the images were mosaicked and mapped in polar projections by Godfrey (1988).

After the Voyagers, the hexagon was not detected again until after Saturn's north pole came into a favorable view from Earth around the solstice in 1987. The existence of the hexagon was confirmed in the images captured by the HST in 1991 (Caldwell et al. 1993); HST had been launched in 1990. The feature was also imaged by Sánchez-Lavega et al. (1993) using ground-based telescopes armed with the CCD imaging technology. Sánchez-Lavega et al. (1997) tracked the feature until the 1995 solstice, when the pole began tilting away from the Earth, obscuring it from view. After the early 1990s, the north polar region was not observed again until the Cassini era. The orbiter arrived at Saturn in 2004; however, it did not have a good view of the poles until Cassini's first high-inclination mission phase between late 2006 and early 2007. During this period, the northern high latitudes were still shrouded in polar winter night, so it was the VIMS instrument that first confirmed the hexagon's presence using its 5-μm channel on October 29, 2006 (publicly released as PIA09188). The hexagonal shape was also recorded in the temperature map constructed using CIRS data captured on 2 March 2007 (Figure 12.2, panel C) (Fletcher et al. 2008). In late 2008, sunlight started shining on the hexagon at 75°N planetocentric (PC) latitude, or 78°N planetographic (PG) latitude, which allowed visible-wavelength imaging using the ISS cameras; the first full view of the hexagon was revealed in a polar mosaic captured on 3 January 2009 (Figure 12.1, right panel) during another set of high-inclination orbits. The north pole has been illuminated by the sun since the equinox in August 2009; however, the next view of the hexagon had to wait until late 2012 because Cassini stayed in the equatorial plane during the intervening period. The first post-equinox set of images that showed the entire hexagon region was captured on 10 December 2012 (Figure 12.3).

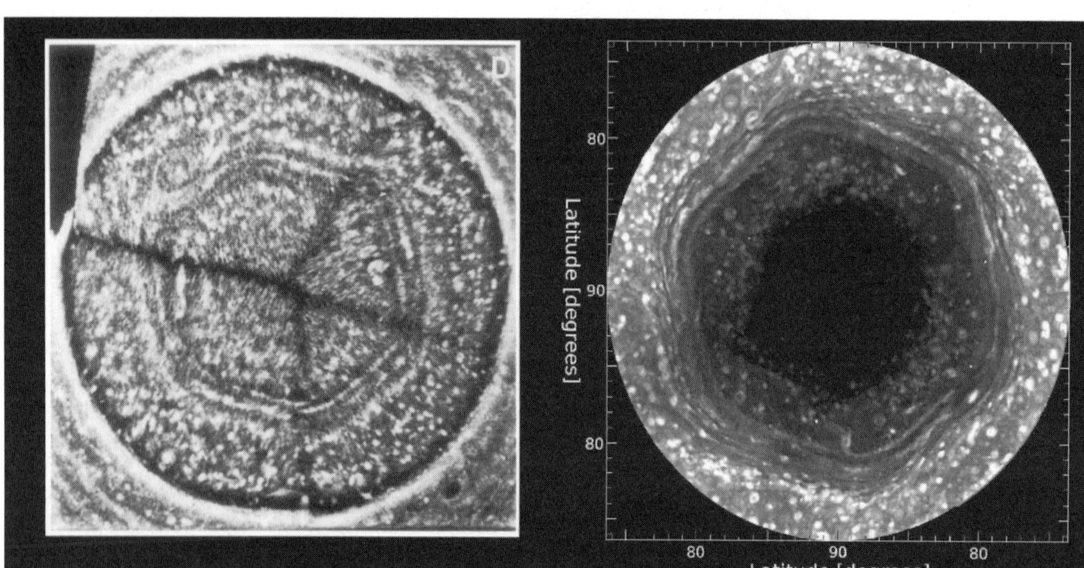

Figure 12.1 Polar-projected mosaics of Saturn's north pole. The left map is a mosaic of Voyager 2 images captured in August 1981 using the green filter (Reprinted from Godfrey 1988 with permission from Elsevier). The right map was generated using images captured by Cassini ISS camera using the CB2 filter on January 3, 2009. Originally released by NASA as PIA11682.

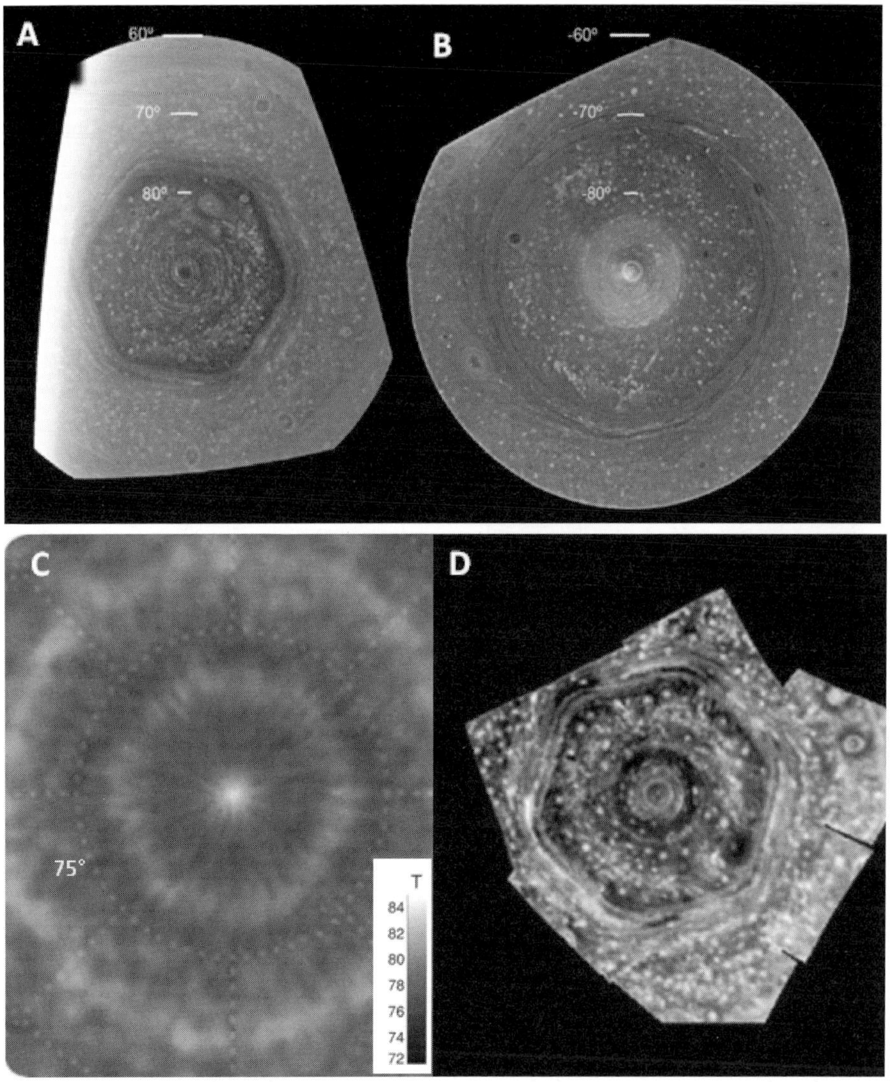

Figure 12.2 Polar-projected maps of Saturn's poles seen in three different Cassini imaging instruments. Panels A and B present the morphology of north and south polar regions, respectively, seen by the ISS camera using the CB2 (750 nm) filter (reprinted from Antuñano et al. 2015 with permission from John Wiley and Sons). Panel C shows the temperature at the 100-mbar level retrieved from the CIRS spectra near 600 cm^{-1} that is sensitive to the atmospheric temperatures near the tropopause (from Fletcher et al. 2008) (reprinted with permission from AAAS). Panel D shows the VIMS 5-μm thermal radiation map (reprinted from Baines et al. 2009 with permission from Elsevier); the contrast is inverted in this panel such that high radiance is black and low radiance is bright, rendering clouds white, as they appear in the ISS images of Panels A and B.

In summary, the hexagon was discovered in the Voyager images captured in 1980–1981 and confirmed using the HST and ground-based telescopes in 1991–1995. After the Cassini–Huygens spacecraft arrived at Saturn, it was seen in 2006–2007 using infrared instruments while the feature was still in winter polar darkness. Since sunlight began to illuminate the feature in early 2008, the hexagon has been observed every year by the ISS cameras. Since 2012, the hexagon has also been clearly discernible in images captured by amateur astronomers armed with modest-sized telescopes (Sánchez-Lavega et al. 2014; cf. Chapter 14). Saturn's polar latitudes experience a strong seasonal forcing (Pérez-Hoyos and Sánchez-Lavega 2006); even though there are large gaps in the observational records of the northern high latitudes of Saturn, these observations clearly demonstrate that the hexagon is a long-lived structure that persists for longer than a Saturnian year.

The hexagon is a unique feature that has been found only around the north pole of Saturn, although some of its morphological characteristics are shared by features

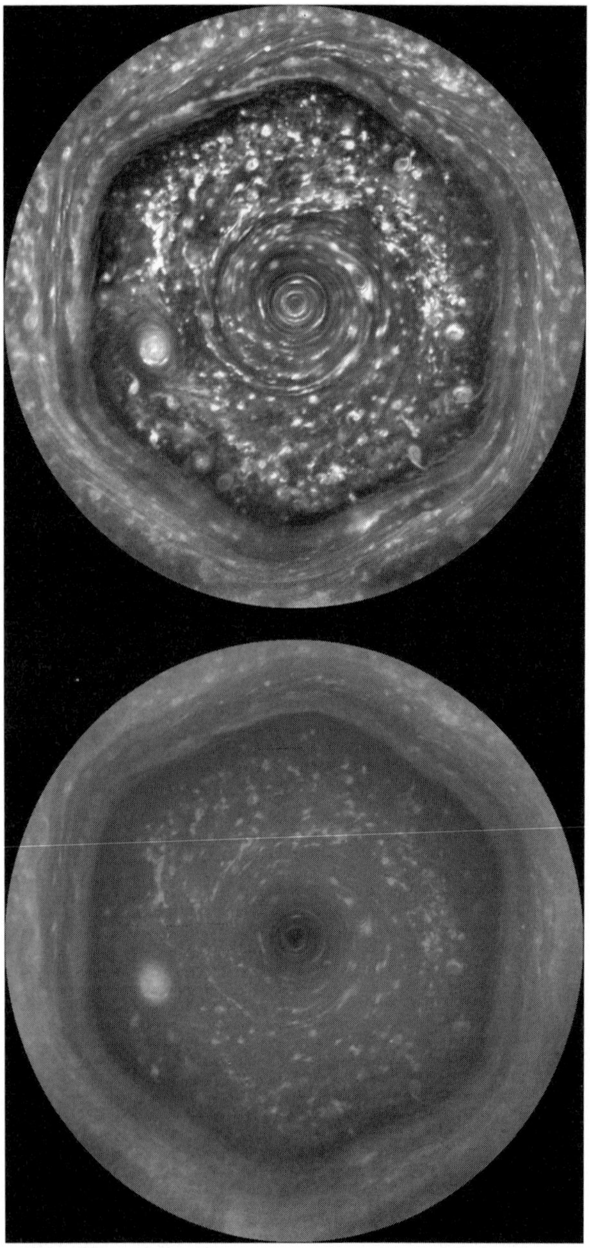

Figure 12.3 Polar-projected mosaic of Saturn's north pole captured by Cassini ISS camera on 10 December 2012. The images were obtained at a distance of approximately 533,304 km from Saturn. The smallest resolved features at the latitude of the hexagon have a horizontal scale of approximately 30 km. The top panel is CB2 view. The bottom panel is a color composite in which the red channel is assigned to CB2, green is assigned to MT2 and blue is the sum of BL1 and VIO. Originally released by NASA as PIA17652. (A black-and-white version of this figure appears in some formats. For the color version, please refer to the plate section.)

elsewhere in our solar system. For example, terrestrial hurricanes sometimes develop a polygonal pattern inside the eye (Lewis and Hawkins 1982; Schubert et al. 1999; Kossin and Schubert 2001). The polar vortex of Venus exhibits transient polygonal characteristics (Limaye et al, 2009; Piccioni et al. 2007). Earth's mid-latitude jet stream in the winter hemisphere often develops a stationary meandering pattern (e.g. Held et al. 2002). On Saturn, the eastward jet at 55.2°N PC (60.5°S PG) has been seen to develop a transient wavy appearance (Vasavada et al. 2006). On Jupiter, almost every jet harbors waves; for example, the southern hemisphere jets at 49.3°S, 58.4°S and 64.1°S PC latitudes (53.1°S, 61.7°S and 67.0°S PG latitudes, respectively) sometimes exhibit wavy cloud morphologies (Sánchez-Lavega et al. 1998). In the northern hemisphere of Jupiter, the westward jet at 27°N PC (30°N PG) (Cosentino et al. 2015) and the eastward jet at 62°N PC (65°N PG) (Orton et al. 2003; Barrado-Izagirre et al. 2008) latitudes also develop transient wavy morphologies. Although these features exhibit morphological similarities to Saturn's northern hexagon, the mechanisms that generate these morphological characteristics may not be the same.

12.2.2 Cloud Morphology

The noticeable characteristic that distinguishes the hexagon is the six-sided shape of the feature that encircles the north pole at around 75°N PC (78°N PG) latitude. The hexagonal shape is approximately 30,000 km across, centered on the north pole. This striking geometric shape is a pattern in the atmospheric reflectivity.

In a latitude-longitude cylindrical projection, the hexagon's outline is very close to a simple sinusoidal function with zonal wavenumber-6 that meanders between 75.3°N and 76.3°N PC (78.0°N and 78.9°N PG) latitude. This sinusoidal amplitude is such that the vertices are formed by the latitudinal minima in the sinusoid (i.e. the point of maximum northward curvature) and the flat sides correspond to the latitudinal maxima (the point of maximum southward curvature) (Antuñano et al. 2015). The numerical models of Morales-Juberìas et al. (2015) showed that the sinusoidal amplitude can be tuned by the vertical wind shear, as discussed further in Section 12.7.3 of this chapter. Note that the hexagonal morphology becomes noticeable only when the planet is viewed from a viewpoint above the north pole, multiple Saturn radii away; an in situ, traditional meteorological measurement would first reveal a remarkably monochromatic wavenumber-6 atmospheric structure, but a hexagonal geometry does not become apparent until it is mapped in a polar orthographic projection.

The hexagonal shape stands out especially in the near-infrared continuum filters CB2 (Figures 12.4a and 12.2A) and CB3 (Figure 12.4b) and the methane absorption filters MT2 and MT3 (Figures 12.4c and 12.4d). In those filters, the interior (i.e. the region poleward of the hexagon) appeared significantly darker than the exterior, with a sharp, well-defined boundary. Long-term observations of the hexagon over 5 Earth years show that the vertices of this boundary maintain a mean latitude of 74.7°N PC (77.4°N PG) (Sánchez-Lavega et al. 2014).

In the CB2 and CB3 view, the interior of the hexagon is filled with vortices of various sizes from the limit of image resolution to several thousand kilometers. In MT2 and MT3, most of those vortices are not visible, except for the large anticyclone, which appears as a bright area around the 5 o'clock position of Figures 12.4c and 12.4d; this large anticyclone has been seen since the arrival of Cassini at Saturn, but it was not seen in Voyager images captured in 1980–1981. (Godfrey 1988 found a large anticyclone *outside* of the hexagon.) In MT2 and MT3, the albedo darkening toward the pole happens in three steps; the outer two steps form two concentric hexagons that surround the hexagonal jet. The inner and outer edges of the hexagon are separated by approximately 2.8° of latitude (Sánchez-Lavega et al. 2014). The last step towards the pole is associated with the north polar vortex, to be discussed further in Section 12.3.

In the visible wavelengths, the brightness contrast gradually reverses with wavelength from red to violet, such that the interior of the hexagon becomes brighter relative to the exterior. In the RED filter (Figure 12.4e), the appearance is similar to the continuum filter; the interior of the hexagon is darker than the exterior. In GRN, the brightness of the interior and exterior are approximately equal, and the hexagon presents a muted appearance (Figure 12.4f). The interior of the hexagon is brighter than the exterior in BL1 and VIO (Figures 12.4g and 12.4f). At even shorter wavelengths in ultraviolet, the interior of the hexagon outline is darker than the surrounding, as shown in Figure 12.17. In BL1 and VIO, the hexagon shape consists of two parallel tracks of dark bands; these tracks roughly follow the inner and outer edges apparent in the MT2 and MT3 images. These double tracks are also apparent in the VIMS 5-μm images (Figure 12.2D), which sense the opacity structure of the cloud layer predominantly residing at around 1.5–4 bar.

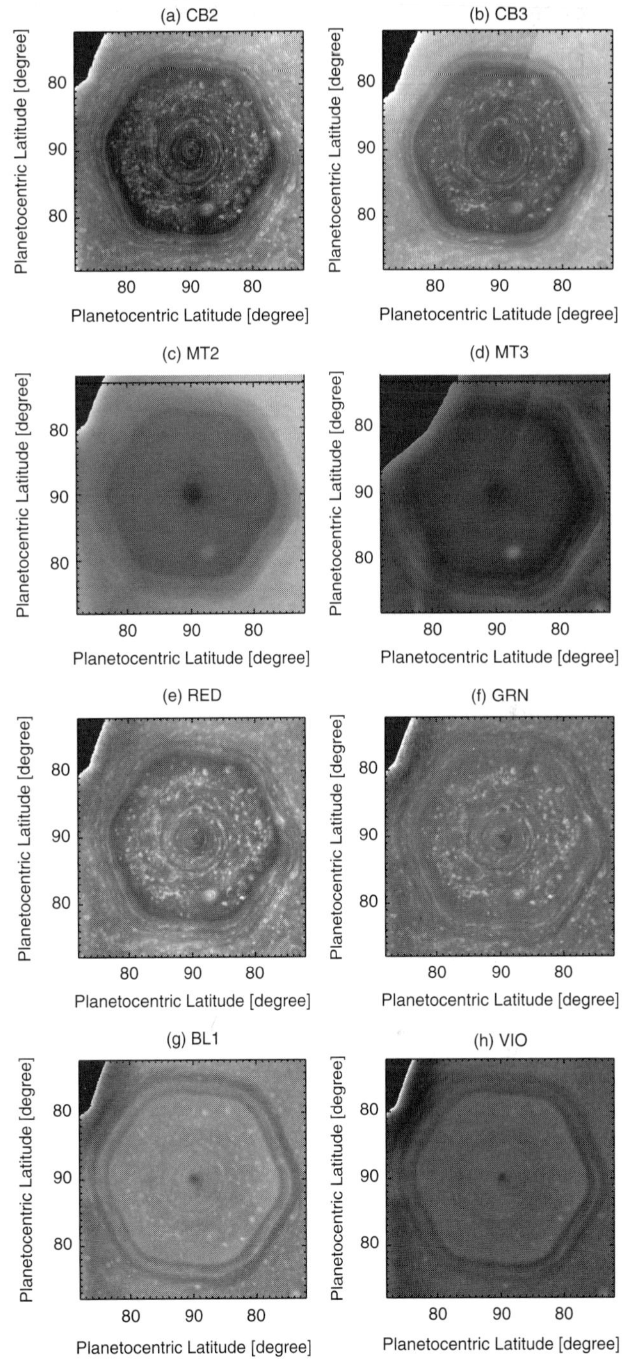

Figure 12.4 Cassini ISS WAC images of northern high latitudes in polar orthographic projection (Sayanagi et al. 2015). The filter used to make each of the maps is denoted in the panel header (see main text for their central wavelengths).

A long-term tracking from 1980 to 2014 of the hexagon's vertices in the System III longitude shows the hexagon to be nearly stationary, propagating westward with a zonal velocity of 0.036 m s^{-1}, or a longitudinal propagation rate of $\langle\omega\rangle = -0.0129 \pm 0.0020°$ (negative number denotes a westward motion, and the brackets $\langle\ \rangle$ denote averaging in time). In the inertial reference frame, this translates to an absolute angular velocity of $\langle\Omega\rangle = 810.7810°$ day $^{-1}$ or a rotation period of $\langle\tau\rangle$ = 10 h 39 min 23.01 ±0.01 s (Sánchez-Lavega et al. 2014). Note that the System III rotation rate is defined as the angular rotation rate of Ω_{III} = 810.7939024° day $^{-1}$ by Seidelmann et al. (2007) and Archinal et al. (2011), based on the primary periodicity in the Saturn Kilometric Radiation (SKR) during the Voyager flybys (Desch and Kaiser 1981); however, the SKR period has been found to be variable in subsequent measurements (e.g. Galopeau and Lecacheux 2000; Fischer et al. 2014). As such, there is currently no evidence that Saturn's deep interior rotates at the System III rate derived from the Voyager-era SKR period. As of this writing, System III continues to be used as the standard reference frame; however, this convention may change in the future.

In the Voyager images, a large anticyclone that measured 11,000 km in the east-west dimension was seen lined up outside one of the hexagon edges (Godfrey 1988) – the vortex is identified as the North Polar Spot (NPS) in the literature. The NPS was re-observed by the HST (Caldwell et al. 1993) and ground-based images between 1990 and 1995 (Sánchez-Lavega et al. 1993, 1997). In 1980–1981 and 1990–1995, the NPS drifted at a longitudinal rate of $\langle\omega\rangle = 0.035°$ day^{-1} (i.e. eastward), and it was suggested to be dynamically coupled to the hexagon. Allison et al. (1990) proposed that the hexagon was a Rossby wave excited by the large vortex impinging on the jet stream; however, the NPS has not been present since Cassini arrived at Saturn in 2004, indicating that the hexagon can exist without the NPS. Sánchez-Lavega et al. (2014) interpret the hexagon as a stationary three-dimensional unforced Rossby wave trapped in the jet, which acts as a waveguide. The trapped Rossby wave manifests as a meandering in the jet as demonstrated by Morales-Juberías et al. (2015), as discussed further in Section 12.7.

Around the south pole of Saturn, there is no hexagon or other permanent polygon in the southernmost zonal jet at 70.4 ±0.1°S PC latitude (73.4°S PG) with a peak speed of $\bar{u}$ = 87 ±12 m s^{-1}, which can be considered the southern counterpart of the hexagon jet at 75°N PC (78°N PG), as documented using HST images (Sánchez-Lavega et al. 2002) and Cassini ISS images (Antuñano et al. 2015). The southern jet does not follow a path that oscillates in latitude; instead, it follows the latitude circle (Figure 12.5, panels E and F). Both jets have similar width and a maximum relative vorticity $d\bar{u}/dy$ = 5 $\times10^{-5}$ s^{-1}, which is about 10^{-1} of the planetary vorticity (the Coriolis parameter at Saturn's ~75° latitude is $f \approx 3.2\times10^{-4}$ s^{-1}). Although the southernmost jet does not harbor a polygonal wave, a transient polygonal wave was noted in the eastward jet at 55.2°S PC (60.5°S PG) by Vasavada et al. 2006.

12.2.3 Observed Dynamical Characteristics

Voyager images revealed that small clouds were moving alongside the outline of the hexagon, showing that the hexagon is associated with a strong eastward jet (Godfrey 1988). Cassini ISS and VIMS images confirmed the Voyager observation (Baines et al. 2009; Sánchez-Lavega et al. 2014; Antuñano et al. 2015). The zonal wind structure is shown in Figure 12.5.

The jet is centered at 75.5°N PC (78.1°N PG) latitude and has a mean peak velocity $\bar{u} = 104 \pm 15$ m s^{-1} with some features reaching 120 m s^{-1} and a meridional width (defined by the Gaussian FWHM) of L_{jet} = 2,900 km (see Figure 12.5).

The hexagon wind field shows a prominent wavenumber-6 structure in the meridional component of the wind with a magnitude of $v \approx 28$ m s^{-1} – this structure arises from the meridionally meandering path of the hexagon jet.

The hexagon is also detected in the thermal infrared in a wavelength sensitive to the upper tropospheric temperatures in the 100–800 mbar altitude; however, it is not apparent in the wavelength range of 7.4–8 μm sensitive to the stratosphere at the 1–6 mbar altitude level (Fletcher et al. 2008; Figures 12.2C and 12.11, lower left panel). The temperature maps show that the hexagon is manifested as a warm band at 77°N PC (79°N PG), with a cold band outside at 73°N PC (76°N PG), coincident with the 5 μm bright inner band (cloud clearing, cyclonic vorticity) and dark band (cloud covered, anticyclonic vorticity) on VIMS

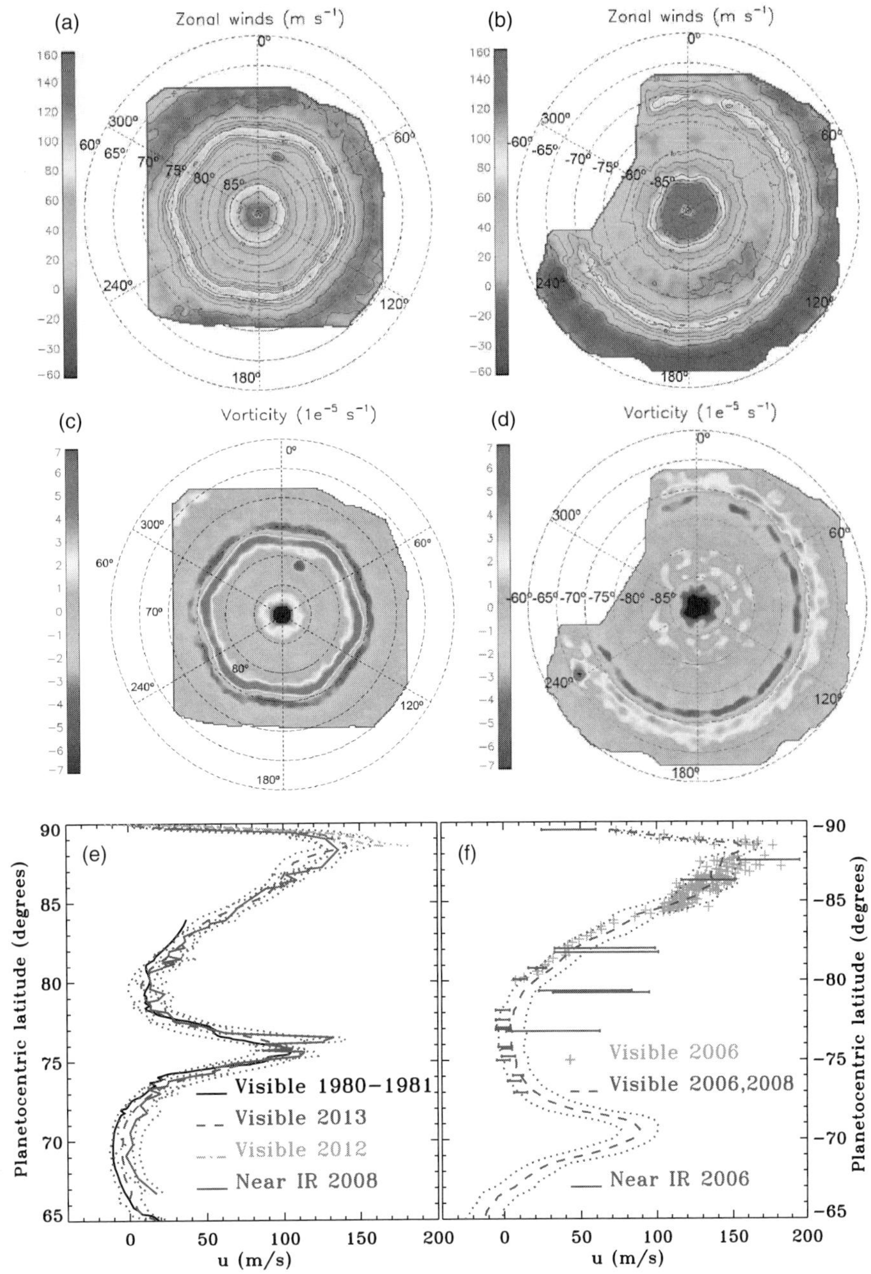

Figure 12.5 **Panels A–D:** Maps of zonal wind speeds and vorticity. The panels on the left show measurements in the north polar region, and those on the right are for the south. Panels A and B show the zonal winds, and panels C and D are vorticity measured from Cassini ISS images by Antuñano et al. (2015). **Panels E and F:** Black solid line shows zonal winds measured by Voyager cameras during flybys in 1980–1981 (Godfrey 1988). Blue dashed lines with dotted lines denoting error bars show winds measured from Cassini ISS images in 2006, 2008 and 2013 (Antuñano et al. 2015). The green dot-dashed line with dotted lines denoting error bars in panel E shows measurements from Cassini ISS in 2012 (Sayanagi et al. 2017). Green + symbols in panel F are measurements from Cassini ISS in 2006 (Dyudina et al. 2009). Red lines are from Cassini VIMS 5-micron images in 2008 (Baines et al. 2009) and 2006 (Dyudina et al. 2009). (A black-and-white version of this figure appears in some formats. For the color version, please refer to the plate section.)

images, respectively (Baines et al. 2009). This temperature-cloud opacity relationship has been explained as due to upwelling on the equator side and subsidence on the poleward side of the hexagon (Fletcher et al. 2008).

From the observed characteristics, Sánchez-Lavega et al. (2014) deduced the following dynamical parameters for the hexagon. The notation and the derivations follow those in Chapter 8 of Sánchez-Lavega

(2011). The zonal, meridional and vertical wavenumbers k, l, and m, respectively, are defined as

$$k = \frac{2\pi}{L_x}; l = \frac{2\pi}{L_y}; m = \frac{2\pi}{L_z}, \tag{12.1}$$

where L_x, L_y and L_z are characteristic zonal, meridional and vertical length scales, respectively. We take the zonal wavelength of the hexagon as the zonal length scale L_x and the jet's width as the meridional length scale L_y. We take the measured value of the phase speed for c_x. The values of these dynamical parameters are summarized in Table 12.1. Substituting the dynamical parameters into the three-dimensional Rossby wave dispersion relationship,

$$c_x - \overline{u} = -\frac{\beta - \mathrm{d}^2\overline{u}/\mathrm{d}y^2}{k^2 + l^2 + (f_0{}^2/N_\mathrm{B}{}^2)m^2 + 1/4L_\mathrm{D}{}^2} \tag{12.2}$$

reveals that, for the wave to propagate, the vertical wavenumber must be imaginary, $m^2 < 0$. In the equation, $\bar{u}$ is the zonal mean wind profile, $\mathrm{d}^2\bar{u}/\mathrm{d}y^2$ is the curvature of the zonal wind at the center of the jet, y is the north-south distance and $L_\mathrm{D} = N_\mathrm{B}H/f$ is the deformation radius, with H as the atmospheric scale height. β is the gradient of the planetary vorticity; at the latitude of the hexagon, $\beta = \mathrm{d}f/\mathrm{d}y = 2\Omega_{\mathrm{III}} \cos\varphi_0 / R_\mathrm{S} = 1.5\times10^{-12}$ m^{-1} s^{-1} (R_S = 55,250 km is the radius of the local surface curvature on Saturn at the jet's latitude). The upper limit for the Brunt-Väisälä frequency N_B is estimated by Lindal et al. (1985) to be $N_\mathrm{B} \lesssim 10^{-2}$ s^{-1}; Sánchez-Lavega et al. (2014) used $N_\mathrm{B} = 8.3\times10^{-3}$ s^{-1}. This suggests that the Rossby wave is vertically trapped within the region of positive static stability in Saturn's atmosphere, which is in agreement with a previous analysis by Allison et al. (1990). The depth of the layer that is trapping the Rossby wave is unknown – the layer could extend well below the water condensation level at ~20-bar; however, such interpretation would contradict the results of Morales-Juberìas et al. (2015), as discussed in Section 12.7.3. The lower limit to the vertical wavenumber m (or upper limit to the vertical length scale L_z) can be estimated by deriving the condition for which the vertical wavenumber is real, $m^2 > 0$ (i.e. the wave is vertically propagating with a real wavelength L_z). The resulting condition demands a meridional length scale of $L_y >$ 15,700 km,

Table 12.1 *Hexagon atmospheric dynamics parameters determined by Sánchez-Lavega et al. (2014)*

Notation	Parameters	Value
L_x	Zonal length scale	14,500 km
L_y	Meridional length scale	5,785 km
f_0	Coriolis parameter at hexagon jet center	3.17×10^{-4} s^{-1}
β	Coriolis gradient at hexagon jet center	1.5×10^{-12} m^{-1} s^{-1}
c_x	Zonal wave phase speed	−0.036 m s^{-1}
$\overline{u}$	Mean hexagon jet speed	120 m s^{-1}
$\mathrm{d}^2\overline{u}/\mathrm{d}y^2$	Zonal wind curvature	$-6 - 8\times10^{-11}$ m^{-1} s^{-1}
N_B	Brunt–Väisälä frequency	8.3×10^{-3} s^{-1}
L_D	Deformation radius	~1000 km

which is not consistent with the observed characteristics of the hexagon; a jet centered at 75°N PC (78°N PG) with a meridional width of L_y = 15,700 km would overtake the pole to the north and the neighboring jet to the south. As noted earlier, Saturn's internal rotation rate remains uncertain; however, note that above interpretation is not affected by the uncertainties in the rotation rate because $c_x - \overline{u}$ is invariant to the rotation rate except for a minor dependence on f_0 and β. Since c_x and $\overline{u}$ are both observed in a common reference frame of System III, shifting them to another common reference frame with a different rotation rate does not substantially affect $c_x - \overline{u}$. Furthermore, note that the potential vorticity is invariant to the planetary rotation; thus, we do not expect that a small uncertainty in the planet's rotation rate would significantly affect atmospheric dynamics.

The eastward hexagon jet centered at 75°N PC (78°N PG) is in geostrophic balance, indicated by the Rossby number $R_\mathrm{o} = u/f_0L$ ~0.13. The jet violates the Rayleigh–Kuo stability criterion, which is a necessary but not sufficient condition that states that a flow is stable if the gradient of the absolute vorticity $\beta - \mathrm{d}^2\bar{u}/\mathrm{d}y^2$ does not change sign anywhere in the flow; the condition is applicable to a geostrophic barotropic zonal jet (Kuo 1949). The zonal wind curvature is negative at the peak of the jet, $\mathrm{d}^2\bar{u}/\mathrm{d}y^2 \sim -7 \times 10^{-11}$ m^{-1} s^{-1} < 0, so the absolute vorticity gradient is positive $\beta - \mathrm{d}^2\bar{u}/\mathrm{d}y^2 > 0$. The zonal wind curvature has maxima to the south of the jet at 74°N PC (77°N PG) $d^2\bar{u}/dy^2 \sim 2 \times 10^{-11}\ m^{-1}\ s^{-1}$ and to the north at 78°N PC (80°N PG) $d^2\bar{u}/dy^2 \sim 4 \times 10^{-11}\ m^{-1}\ s^{-1}$, both of which can overcome the local planetary vorticity gradient $\beta \sim 1.5 \times 10^{-12}\ m^{-1}\ s^{-1}$ (Antuñano et al. 2015). Thus $\beta - d^2\bar{u}/dy^2$ changes sign in the flanks of the hexagon jet and violates the Rayleigh–Kuo criterion. The jets on Saturn also violate the Charney–Stern criterion (a jet is stable when the gradient of potential vorticity does not change sign; Charney and Stern 1962; Read et al. 2009a). Saturnian jets may be neutrally stable according to Arnol'd's Second Criterion (Read et al. 2009a,b),

$$\frac{\overline{u} - c_x}{\partial\overline{q}/\partial y} \geq L_D^2, \qquad (12.3)$$

where c_x is the fastest propagating wave's phase speed and $\overline{q}$ is the zonal mean potential vorticity. The criterion states that, for the wind to be stable, the magnitude of the wind speed must be greater than the phase speed of the fastest upstream-propagating wave by $L_D{}^2\partial\overline{q}/\partial y$ (Arnol'd 1966; Dowling 2014). Dowling (1995) and Vallis (2006) discuss these stability criteria. The depth of the jet structure and the stability of the zonal jets on Saturn are related and still need to be explained (to be discussed further in Section 12.7). These violations of stability criteria are not unique to Saturn's atmospheric jets; the forcing and maintenance mechanisms of giant planet atmospheric zonal jets is a continuing topic of research, as also discussed in Chapter 11.

12.3 North Polar Vortex

12.3.1 Cloud Morphology

The north polar vortex (NPV) is an intense cyclonic vortex revealed in the Cassini era that resides inside of the hexagon and is centered on the north pole. The presence of the NPV was detected in CIRS data by Fletcher et al. (2008). The cloud morphology of the NPV was first documented by Baines et al. (2009) using VIMS 5-μm images captured in 2008 (which can be seen at the center of Figure 12.2D) before the 2009 equinox while the north pole was still under the winter polar darkness. The VIMS observation by Baines et al. (2009) also detected cloud motions and provided a direct confirmation that there is an intense cyclonic vortex over the north pole of Saturn (Figure 12.5E). The cyclonic cloud motions have also been shown later by cloud tracking in Cassini ISS visible observations after the Sun illuminated the North Pole following the 2009 equinox (Antuñano et al. 2015; Sayanagi et al. 2017). The ISS observations also show that there is a large spiraling cloud structure centered on the north pole, as can be seen in Figure 12.4. The spiral was cyclonic (i.e. a spiraling arm traces a line that turns clockwise as it moves away from the pole), which is consistent with the thermal hotspot revealed by Fletcher et al. (2008). The south polar vortex also has cyclonic vorticity (Sánchez-Lavega et al. 2006; Dyudina et al. 2008, 2009).

The morphological context that surrounds the polar vortex is presented in Figure 12.4. The high-resolution cloud and wind structure of the north polar vortex are captured in ISS visible images (Antuñano et al. 2015; Sayanagi et al. 2017), which yield higher spatial resolution than the infrared VIMS images. The ISS instrument also returns images in multiple wavelength bands using different filters. Images captured in CB2 and CB3 show a large spiraling cloud morphology (Figures 12.4a and 12.4b). The NPV's spiraling appearance in RED (Figure 12.4e) and GRN (Figure 12.4f) are similar to that in the CB2. In BL1 and VIO (Figures 12.4g and 12.4h), the spiraling structure is barely discernible, and poleward of about 89.5°N PC (89.6°N PG) latitude is dark, appearing as a circular hole. In the methane-absorption bands MT2 and MT3 (Figures 12.4c and 12.4d), the contrast is dominated by the dark circular hole centered on the pole; the rim of the dark hole follows the 88.5°N PC (88.8°N PG) latitude circle.

High-resolution morphology of the center of the polar vortex is presented in Figure 12.6. The maps show a region within 1.5° of latitude around the north pole. The morphology in CB2 (Figure 12.6a), RED (Figure 12.6d) and GRN (Figure 12.6e) are very similar. The BL2 view (Figure 12.4f) exhibits the same dark center as in the context view (Figure 12.4g). In BL2, the hole lacks a clear outline and its reflectivity gradually darkens toward the center of the hole. In MT2, the circular dark center (possibly a hole in the upper tropospheric haze) is apparent; this feature can also be seen in the context image (Figure 12.4c); inside the hole,

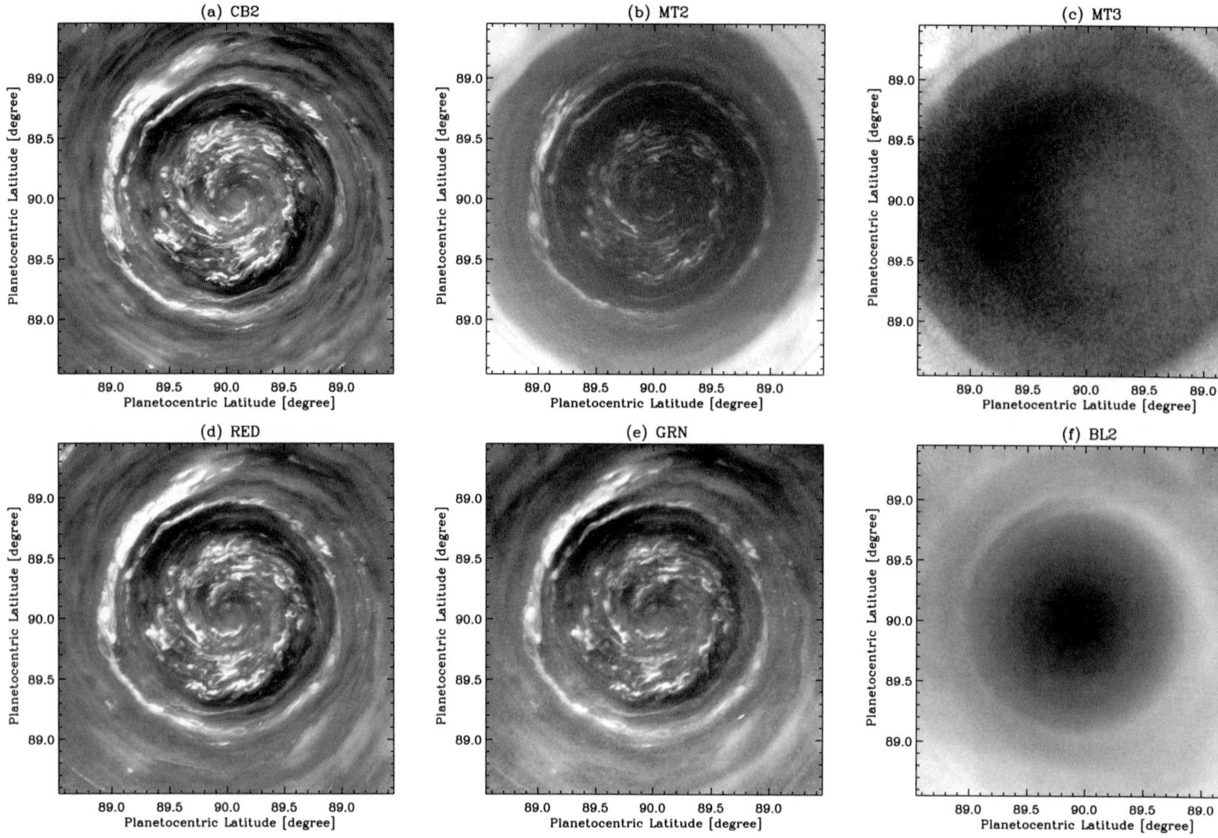

Figure 12.6 Cassini ISS NAC images of the northern high latitudes in polar orthographic projection. The filter used to make each of the map is denoted in the panel header. Reprinted from Sayanagi et al. (2017) with permission from Elsevier.

only the brightest of the features that are visible in CB2 (Figure 12.6a) are apparent. Figure 12.6c shows the view in MT3; the circular hole with a well-defined outline that roughly follows the 88.5°N PC (88.8°PG) latitude is the same feature as that seen in MT2 (Figure 12.6b); however, no cloud feature is visible inside of the dark region, indicating a lack of light-scattering aerosols above the upper troposphere. Figure 12.7 shows the near-infrared false color and enhanced visible-color maps of the NPV.

12.3.2 North Polar Wind Structure

The cyclonic sign of the NPV's relative vorticity was predicted by the CIRS measurement of the temperature field (Fletcher et al. 2008); however, the region's vorticity was not directly measured until Baines et al. (2009) used VIMS images to analyze the motion of clouds to deduce the zonal mean wind and vorticity structure in the northern high latitudes. Higher-resolution wind measurements have been obtained using ISS images; Antuñano et al. (2015) analyzed the dynamical context surrounding the NPV (Figure 12.5E), while Sayanagi et al. (2017) analyzed the high-resolution wind structure at the center of the NPV between the north pole and 88°N PC (Figure 12.8).

To the north of the hexagon jet at 75°N PC (78°N PG), the zonal wind profile has a minimum at around 80° N PC (82°N PG) latitude (Baines et al. 2009; Antuñano et al. 2015). From the minimum, the zonal mean wind speed monotonically increases poleward until reaching the rim of the "eye" of the NPV such that the profile has a uniform absolute vorticity (i.e. $\omega_a = constant$ where the absolute vorticity is $\omega_a = f + \zeta$, ζ is the relative vorticity, $f = 2\Omega \sin\phi$ is the Coriolis parameter, Ω is the planetary rotation rate and ϕ is the PG latitude). The VIMS measurements by Baines et al. (2009) detected a peak wind speed of 136.0 ± 6.5 m s^{-1} at 88.3°N PC (88.6°N PG) latitude, which is near the rim of the NPV's eye.

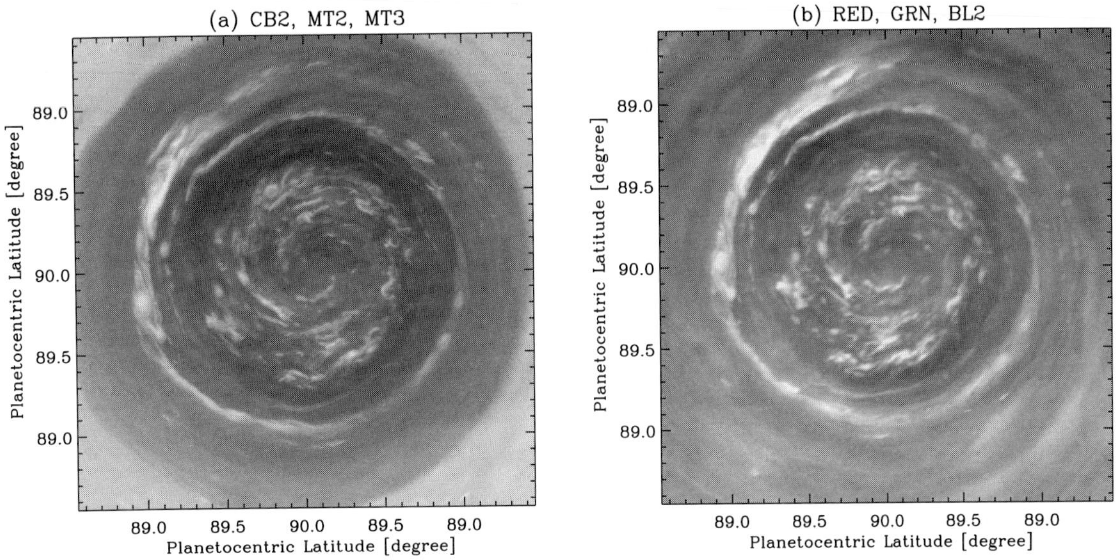

Figure 12.7 Color composites of northern high latitudes in polar orthographic projection. The red, green and blue channels of the composites are assigned to the filters identified in each of the panel headers. Reprinted from Sayanagi et al. (2017) with permission from Elsevier. (A black-and-white version of this figure appears in some formats. For the color version, please refer to the plate section.)

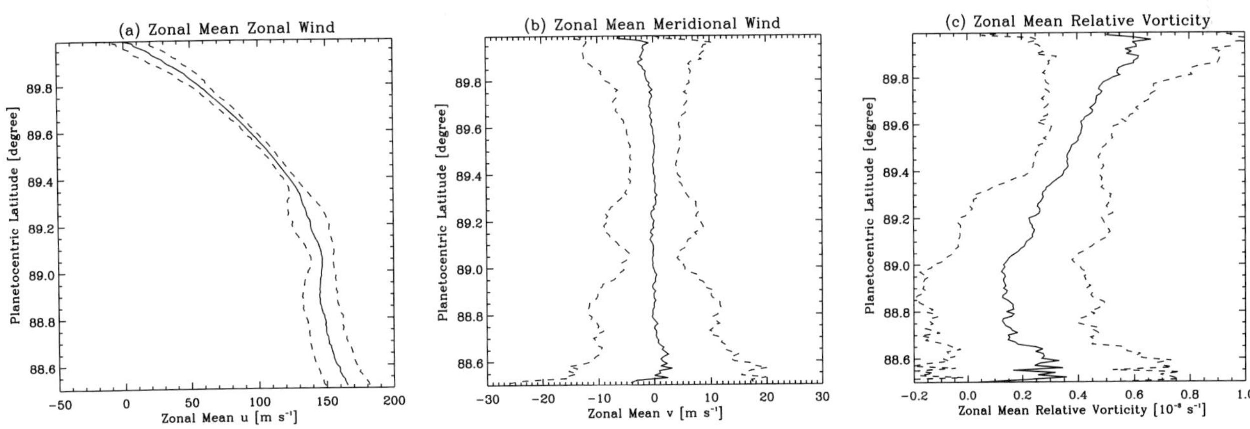

Figure 12.8 The zonal wind structure at the center of the NPV measured using Cassini ISS NAC images. (a) Zonal mean of eastward wind component. (b) Zonal mean of northward wind component. (c) Zonal mean of the relative vorticity. The dashed lines in each panel denote zonal mean plus and minus the zonal standard deviation. Reprinted from Sayanagi et al. (2017), with permission from Elsevier.

The analysis by Antuñano et al. (2015) using ISS images returned a similar result (Figure 12.5).

Sayanagi et al. (2017) used ISS images to analyze the wind structure at the center of the NPV poleward of 88.5°N PC (88.5°N PG) latitude. The zonal mean and standard deviation of the eastward wind component u are presented in Figure 12.8a. From the north pole, the zonal wind speed steadily increases equatorward to the end of the measurement at 88.5°N PC latitude. The zonal mean of the meridional wind component v, shown in Figure 12.8b, is essentially zero throughout the domain of measurement. Figure 12.8c presents the result of the zonal relative vorticity around the north pole, showing that the cyclonic relative vorticity peaks at the pole at $\sim 6 \times 10^{-4}$ s^{-1}. This value is slightly greater than that measured by Baines et al. (2009), $\sim 5 \times 10^{-4}$ s^{-1}, which is not surprising, because the ISS measurements give higher spatial resolution and the peak speed at the pole is most likely not resolved in VIMS measurements. The relative vorticity analysis by Antuñano et al. (2015) gives a value of $(2.5 \pm 0.1) \times 10^{-4}$ s^{-1} at 89.8°N PC (89.0°N PG), which is less than half of that measured by

Sayanagi et al. (2017); this difference is also likely due to the lower-resolution data and spatial averaging employed by Antuñano et al. (2015).

We note a confusion in the literature regarding the vorticity in the polar regions of Saturn. Dyudina et al. (2009) used an unconventional definition of vorticity that reversed the sign (defined by the first equation in their section 4), resulting in a positive sign of vorticity for the cyclonic south polar vortex. Baines et al. (2009) employed the same convention as Dyudina et al. (2009) to define vorticity, resulting in a negative sign of vorticity for the cyclonic north polar vortex. Antuñano et al. (2015) employs a conventional definition of relative vorticity (their equation 4); however, their calculation contains a sign error, resulting in a negative sign of relative vorticity for the cyclonic north polar vortex. In this review, we compare the measurements by these authors using the conventional definition of the relative vorticity, i.e. a cyclonic vortex has a positive relative vorticity in the northern hemisphere, and negative in the southern hemisphere.

12.4 Comparison of the Northern High Latitudes to the South Polar Region

In this section, we compare the cloud morphology of the north polar vortex to that of the south polar vortex. The morphology and dynamics of Saturn's south pole has been documented by Sánchez-Lavega et al. (2006), Dyudina et al. (2008, 2009), Antuñano et al. (2015) and Sayanagi et al. (2017). The existence of a cyclonic vortex at the south pole was first indicated by a hotspot found in earth-based observations by Orton and Yanamandra-Fisher (2005) and later in a CIRS-derived temperature map at the 100-mbar level by Fletcher et al. (2008). Cassini imaging observations (Vasavada et al. 2006; Sánchez-Lavega et al. 2006) revealed cyclonic rotation around the spot in 2005.

Figure 12.2 panels A and B (from Antuñano et al. 2015) compare the large-scale structure of north and south polar regions poleward from 60° north and south latitude. The polar vortex at the south pole appears brighter, but this is an artifact of the mosaicking and overcorrection at the latitudes where the illumination incident angle is nearly 90°. Without limb-darkening correction, the south polar vortex has a contrast similar to that of the northern vortex, as can be seen in

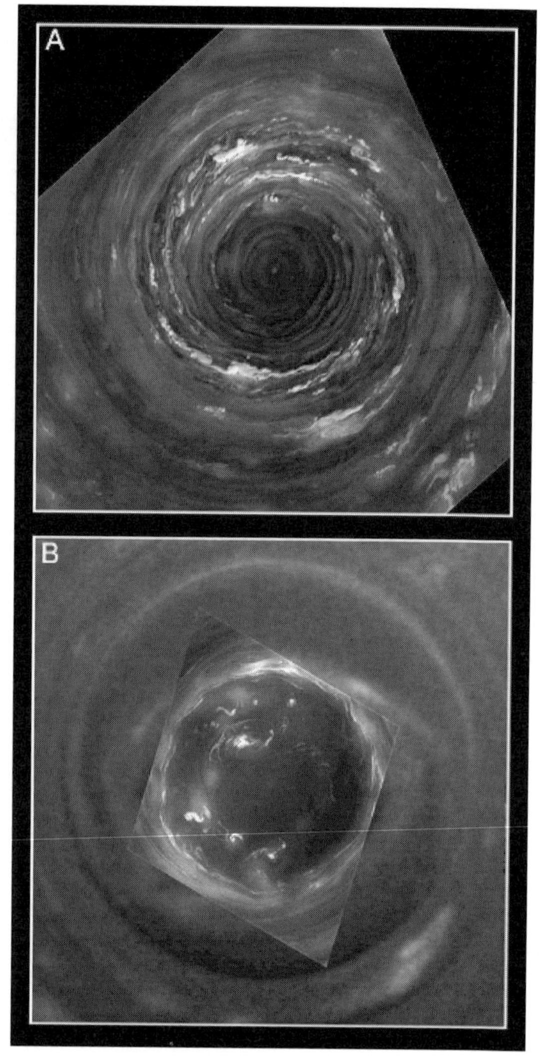

Figure 12.9 (a) Polar projection of the north polar region from 87.5° to 90° latitude on 14 June 2013, using an image with CB2 filter and spatial resolution at the pole of 5.3 km/pixel. (b) Polar projection of the south polar region from 87.5° to 90° on 14 July 2008. Panel b is a composite of two different images with different resolutions. The image on the front is a NAC image with CB2 filter at 2.4 km/pixel, while the one at the back is a WAC image with a CB2 filter at 29.5 km/pixel. Reprinted from Antuñano et al. (2015), with permission from John Wiley and Sons.

Figure 12.9, which shows a high-resolution polar projection of both poles of Saturn at the same scale.

Figure 12.10 presents polar-projected maps of the southern high-latitude region using various filters. In CB2 and CB3 images (Figures 12.10a and 12.10b), the SPV exhibit cyclonically spiraling morphology

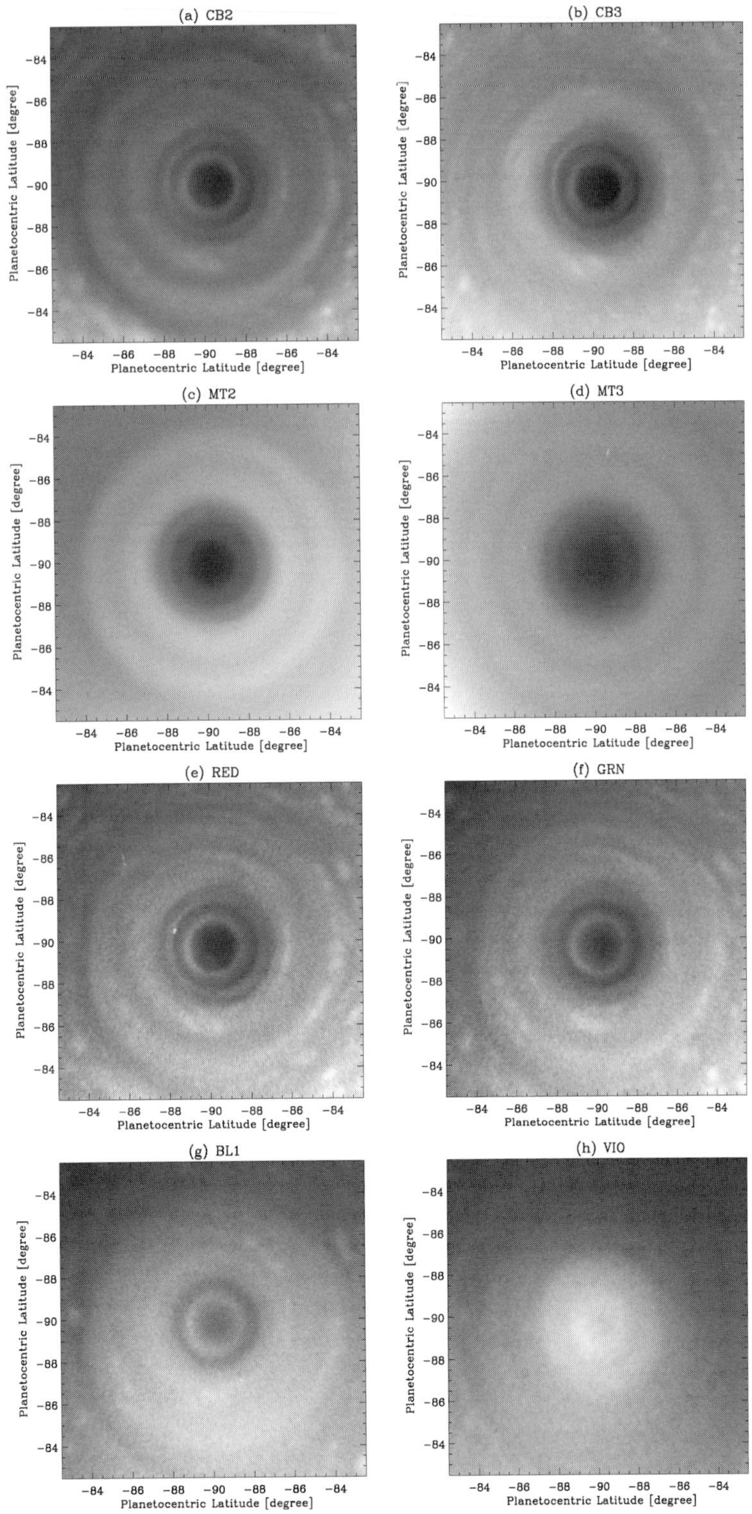

Figure 12.10 Same as Figure 12.4, except that south polar data captured in 2007 are shown here. Reprinted from Sayanagi et al. (2017) with permission from Elsevier.

centered on the pole and extending down to around 83° S PC (84°S PG) latitude. The filamentary clouds in the images demonstrate wind shear in the polar vortices. The cloud texture seems richer and more complex around the north pole, which contains more fine scale structures. The difference is primarily due to the difference in the high-altitude haze, although effects of the observation geometry are also not negligible. The eastward wind speed peaks at the edge of the "eye" of the cyclonic vortex in both the NPV and the SPV. For both NPV (see Figure 12.4) and SPV (Figure 12.9), the edge of the eye appears as rings of bright clouds. The peak wind speeds at 88°– 89° north and south latitudes reach about 150 m s^{-1}, as shown in Figure 12.5 panels E and F and Figure 12.8a.

The double-wall structure surrounding the "eye" of the south polar vortex, originally reported by Dyudina et al. (2008), is apparent when viewed in CB2 (Figure 12.10a), CB3 (Figure 12.10b), RED (Figure 12.10e), GRN (Figure 12.10f) and BL1 (Figure 12.10g). In those filters, the interior of the eye appears dark. Outside of the eye, a concentric albedo structure dominates the morphology, the interior of which appears substantially darker than the outside. A few diffuse, discrete features can also be seen inside the eye. The interior of the eye also appears dark in MT2 (Figure 12.10c) and MT3 (Figure 12.10d). While the interior of the eye of the south polar vortex appears darker than the surroundings in all above filters, the contrast is reversed in VIO (Figure 12.10h), in which a bright hood covers the south polar vortex eye region poleward of 87°S PC (88°S PG) latitude; the polar hood is apparent only in the VIO filter because it is sensitive to particulate scattering in the stratosphere. Radiative transfer analyses are needed to deduce the altitudes of these cloud and haze layers and to infer the properties of the aerosol particles that make up those clouds and hazes; such analyses are yet to be done.

A cyclonic vortex with a compact region of enhanced thermal infrared emissions centered on each of the poles has been a persistent feature of the polar tropospheres for the full ten-year span of the observations in cloud-tracking wind measurements as well as thermal fields, despite seasonal shifts in the absolute temperatures (Figure 12.11). The warm air is isolated from lower latitudes (cool tropospheric polar zones

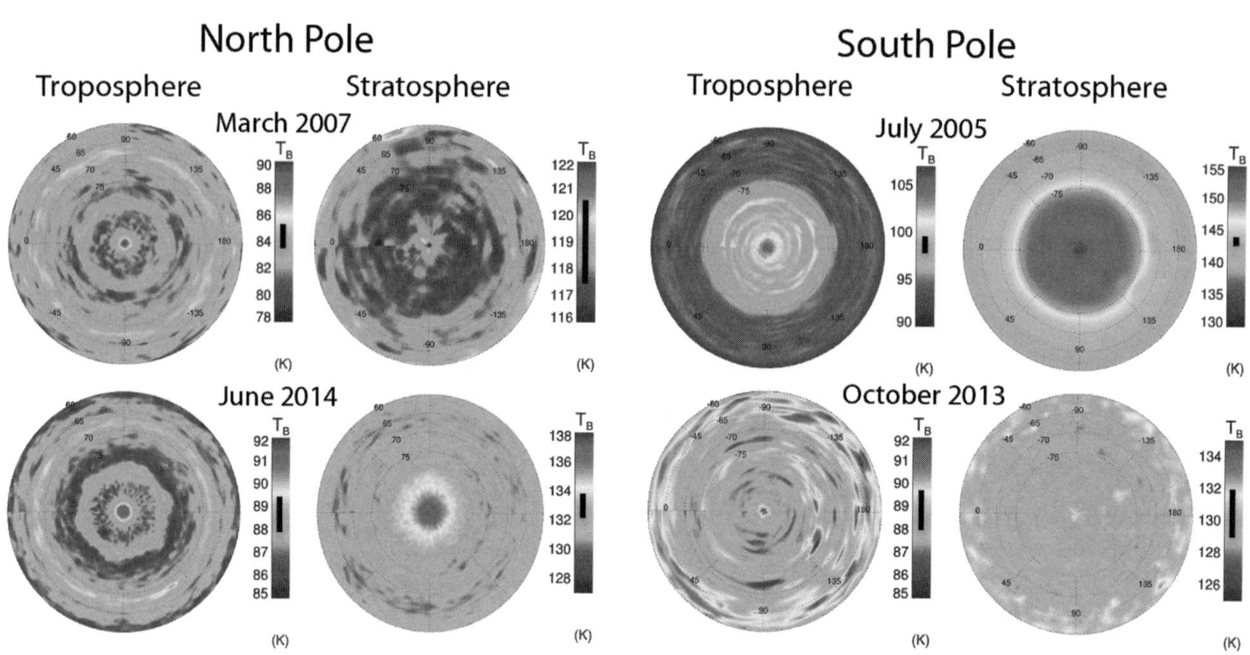

Figure 12.11 Brightness temperature maps of the troposphere and stratosphere constructed from Cassini/CIRS spectroscopy. Tropospheric maps use the H_2-He collision-induced continuum between 600 and 620 cm^{-1} to sense the 150-mbar level; stratospheric maps use methane emission between 1290 and 1310 cm^{-1} to sense the 1-mbar level. Black lines in each scale bar indicate the uncertainty on the measurements. The top row shows the polar regions in northern winter, and the bottom row shows the polar regions in northern spring. Reprinted from Fletcher et al. (2015), with permission from Elsevier. (A black-and-white version of this figure appears in some formats. For the color version, please refer to the plate section.)

near 84°PC (85°PG) latitude of both poles) by peripheral jets at 88.6°N PC (88.9°N PG) and 87.5°S PC (88.0°S PG), respectively. Warm tropospheric belts exist equatorward of the polar zones, and poleward of the eastward jets at 75.4°N PC (78.0°N PG; i.e. the hexagon jet) and 70.5°S PC (73.9°S PG) latitudes. The visibility of the south polar belt has increased as the upper tropospheric temperature gradients have changed, and the warm north polar belt has maintained its hexagonal wave structure throughout the mission (Fletcher et al. 2015). Because the northern jet's peak at 75.4°N PC (78.0°N PG) is located at a higher latitude than the southern jet at 70.5°S PC (73.9°S PG), as can be seen in Figures 12.5E and 12.5F, the morphologies of the northern and southern polar belts are fundamentally different – the northern belt sits closer to the pole than the south, confined by the eastward jets.

12.5 Polar Stratosphere

This section discusses the thermal structure and hydrocarbon distribution in the polar regions as well as their response to seasonal radiative forcing. Atmospheric temperatures are determined by a complex interplay between radiative energy balance and atmospheric circulation, which is in turn determined by the contributions of composition and aerosols to radiative heating and cooling (see Chapter 10). By deriving the thermal, chemical and aerosol distributions within and surrounding the polar vortices, we are able to gain insights into the circulations shaping these unique environments on Saturn from the upper troposphere into the stratosphere. Vertical motions are inferred from both the temperature field (via the thermodynamic energy relation) and the use of gaseous species as quasi-conservative tracers (via the continuity equation). Horizontal motions are determined from latitudinal temperature gradients via the thermal wind equation, assuming that zonal winds are in geostrophic balance.

12.5.1 Pre-Cassini Observations

Prior to Cassini's arrival in 2004 (planetocentric solar longitude of L_S = 293°), our knowledge of the polar stratosphere was restricted to the Earth-facing summer hemisphere. Orton and Yanamandra-Fisher (2005) discovered a region of elevated south polar stratospheric methane emission in Keck observations in February 2004 (L_S = 287°, just after southern summer solstice), poleward of approximately 65°S PC (70° S PG) and sensitive to the 0.5- to 20-mbar pressure range. This mirrored a similar region of enhanced emission observed in March 1989 in the northern hemisphere using the Infrared Telescope Facility (L_S = 104.5°, just after northern summer solstice, Gezari et al. 1989), and suggested the presence of a seasonally varying stratospheric vortex poleward of approximately 65°PC (70°PG) latitude in both hemispheres. The ground-based record has insufficient temporal sampling to constrain the timescales for the formation and dissipation of these summertime vortices, presenting a unique opportunity for Cassini.

In addition to the large region of enhanced methane emission poleward of ~65°S PC (~70°S PG) latitude, Orton and Yanamandra-Fisher (2005) also detected a warm, compact cyclonic vortex in the summer stratosphere and upper troposphere, centered on the south pole with a radius of 2°–3° latitude. This warm cyclone was spatially coincident with the hurricane-like eyewalls identified in the tropospheric cloud decks in visible light (Sánchez-Lavega et al. 2006; Dyudina et al. 2008, 2009; Baines et al. 2009; Antuñano et al. 2015; Sayanagi et al. 2017), suggesting that the circulation responsible for the cloud-deck vortex was penetrating through the stably stratified upper troposphere and high into the stratosphere over the summer pole. Later measurements by Cassini/CIRS (Fletcher et al. 2008) measured 1-mbar temperatures in excess of 165 K within the compact south polar cyclone.

12.5.2 Thermal Structure of the North Polar Vortex

Cassini's arrival at Saturn in 2004 allowed characterization of the thermal structure of the north pole, which had been hidden from Earth-based observers since the mid-1990s. Fletcher et al. (2008) utilized Cassini/CIRS observations to compare the summer and winter poles using data acquired between 2005 and 2007 (L_S = 300°–340°), and discovered a compact spot of enhanced thermal emission in the troposphere of the northern winter pole, mirroring that observed at the south pole. As mentioned in Section 12.3, the north polar hotspot was later shown to be coincident with the north polar vortex by the motion of the clouds observed in 5 μm by Baines et al. (2009). However, unlike the summertime south polar cyclone, there was no evidence of the compact vortex penetrating high into the wintertime northern polar stratosphere (see North Pole in March 2007 in Figure 12.11). As discussed in Section 12.2, the north polar belt near 70°–78°N PC (75°–80°N PG) exhibited the sinusoidal manifestation of the hexagon in the upper troposphere that did not penetrate into the overlying stratosphere. However, contrary to the expectations of radiative climate models that predict the coldest stratospheric temperatures to occur right at the north pole (see Chapter 10), Fletcher et al. (2008) demonstrated that the high northern latitudes were subtly warmer than lower latitudes, presumably as a result of adiabatic heating caused by large-scale atmospheric subsidence.

As of this writing, the temperature and composition of the northern and southern polar regions (upper troposphere and stratosphere) have been tracked by Cassini from northern winter (L_s = 293°) through to northern spring (L_s = 56°), using deviations from radiative-climate and chemical expectations to infer the general circulation of Saturn's polar stratosphere (Fletcher et al. 2015).

12.5.3 Seasonal Variation

Above the tropopause, radiative climate models that balance radiative heating via methane and aerosol absorption with radiative cooling via hydrocarbons (Bézard and Gautier 1985; Greathouse et al. 2010; Friedson and Moses 2012; Guerlet et al. 2014) predict the rate of change of temperatures $\mathrm{d}T/\mathrm{d}t$ to vary monotonically with latitude such that the coldest winter temperatures would be expected right at the poles. Given that the atmosphere has a substantial thermal inertia that increases with depth, these coolest stratospheric temperatures lag behind the winter solstice by approximately a season. Indeed, Cassini detected the coldest northern temperatures between 2008 and 2010, 6 to 10 years after the winter solstice (Fletcher et al. 2015; Figure 12.12). Peak stratospheric temperatures in the southern hemisphere occurred 1 to 2 years after southern summer solstice, consistent with radiative climate predictions. However, the *location* of the coolest northern temperatures was at 72°–78°N PC (75°–80°N PG), with the stratosphere warming at higher latitudes. Indeed, the northern stratosphere within 15° latitude of the pole has been warming throughout the Cassini observations, by approximately 5 K year^{-1} at 0.5–1.0 mbar. This has been mirrored at the south pole by a cooling within 15° of the pole at a similar rate, but with the peak cooling occurring at slightly higher pressures of 1–3 mbar. A crucial observational difference is that a strong north-south temperature gradient at the 1-mbar level occurred at 72°S PC (75°S PG) during southern summer, giving the appearance of an isolated stratospheric vortex, whereas no such gradients are yet visible in the northern stratosphere as of the time of writing (Spring 2015). Based on the observations of Gezari et al. (1989) shortly after the previous northern summer solstice, we can expect that such a sharp boundary will form near 72°N PC (75° N PG) between now and the end of the Cassini mission (L_s = 93°), signaling the onset of the summertime vortex. The 72°S PC (75°S PG) boundary vanished after the autumnal equinox as the southern vortex cooled by more than 35 K, implying the dissipation of the southern summer stratospheric vortex.

12.5.4 Implications for Zonal Wind Vertical Shear

The shifting latitudinal temperature gradients in the upper troposphere and stratosphere discussed in Section 12.5.3 have implications for the shear on the zonal jets measured at the tropospheric cloud level. For example, Figure 12.12 column (a) shows that the jet at 70.5°S PC (73.9°S PG) (see Figure 12.5) exists in a region of temporally evolving horizontal temperature gradient; the thermal wind equation (see, e.g. Holton 2004) can be used to calculate the vertical shear and extrapolate the vertical wind profile from the tropospheric wind measurements like those shown in Figure 12.5. The change in the horizontal temperature gradient indicates that the stratospheric wind speed at 1 mbar was westward at the start of the Cassini mission, but weakened and became eastward between 2004 and 2014. In comparison, the horizontal temperature gradient around the jet at 75.4°N PC (78.0°N PG) has not evolved substantially in that time period. The overlying stratospheric trends promote positive vertical shear in the winter hemisphere and negative vertical shear in the

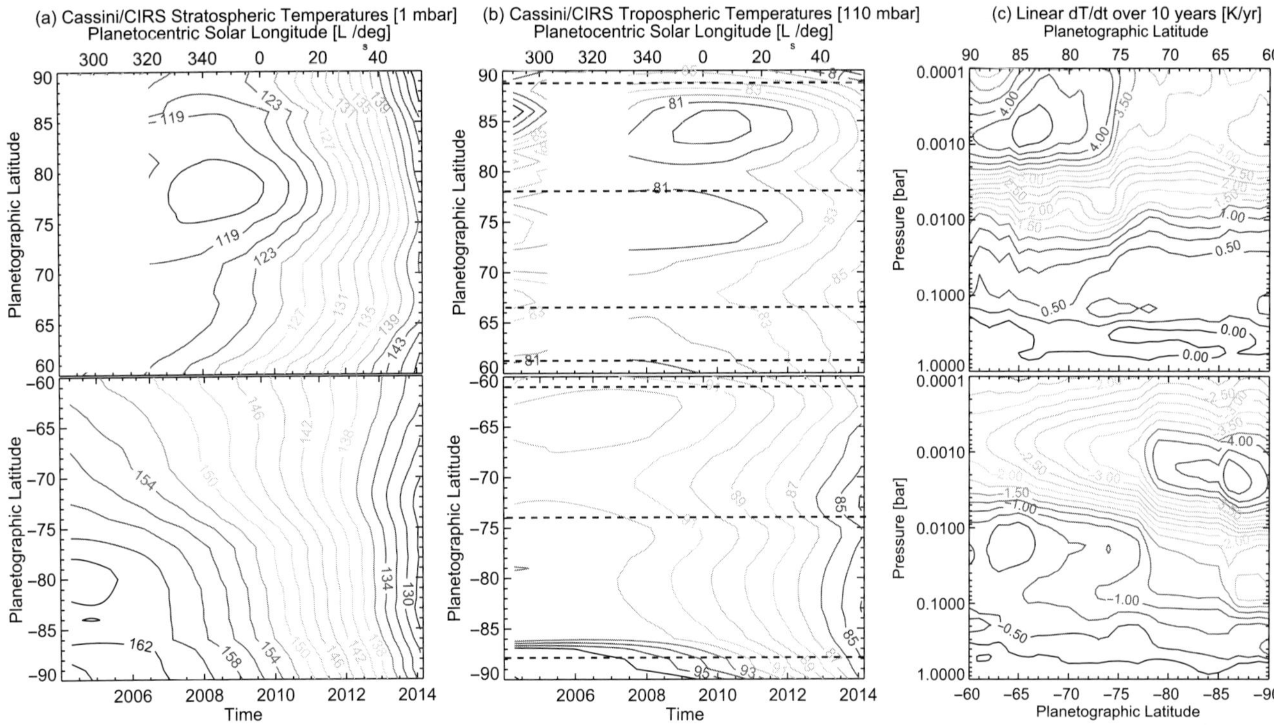

Figure 12.12 Measured temperature changes over the ten years of the Cassini mission for the stratosphere (1 mbar, column a) and troposphere (110 mbar, column b), reconstructed from Cassini/CIRS measurements. In column b, the horizontal dashed lines indicate the peaks of the eastward tropospheric jets to show the correspondence with the temperature field. Column c provides a simple way of visualizing the temperature changes dT/dt (K/yr), assuming a linear change. Although this is evidently not the case in columns a and b, it serves to demonstrate that the most intense heating and cooling is occurring within ±15° of the poles. Reprinted from Fletcher et al. (2015), with permission from Elsevier. (A black-and-white version of this figure appears in some formats. For the color version, please refer to the plate section.)

summer hemisphere (Friedson and Moses 2012). Similar oscillations in the stratospheric jet velocities were observed for *some* high-latitude jets in the radiative dynamical model of Friedson and Moses (2012), but not for all. However, calculating the vertical shear by integrating the thermal wind equation is extremely sensitive to uncertainties the horizontal temperature gradient, and the inferred vertical shear remains to be tested via direct measurements; remote-sensing wind measurements at the 1-mbar level is difficult due to the absence of observable cloud tracers in the stratosphere. Changes in jet velocities in the upper troposphere are expected to be smaller than 10 m s^{-1}, and are likely to be undetectable in cloud-tracking observations in the main cloud deck. The implications of the thermal wind equation are that the strong negative dT/dy at 72°S PC (75°S PG) in southern summer is related to a westward jet encircling the region of warm emission at the south pole. No similar westward jet is yet present at 72°N PC (75°N PG).

12.5.5 Effects on the Hydrocarbon Distribution

In both hemispheres of Saturn, the polar-most zonal jets may act as a transport barrier that isolates the unique chemistry and aerosols of the polar regions from the lower and mid-latitudes. Over much of Saturn's stratosphere, neutral photochemistry dominates the sources and sinks of the soup of hydrocarbon species produced by the UV photolysis of methane (e.g. Moses and Greathouse 2005). These hydrocarbon products sediment downwards to ultimately be recycled in the troposphere, but some can condense to form high-altitude stratospheric haze layers that typically are optically thin to remote sensing, but can be seen in limb views from Cassini (e.g. Rages and Barth 2012). At the poles, however, energy inputs associated with auroral energy and heating, in addition to ion chemistry, have the potential to shift the atmospheric chemistry away from the expectations of neutral photochemistry. Numerical simulations of ion chemistry have yet to fully explain observed aerosol properties (Friedson et al. 2002; Wong et al. 2000, 2003), so we still rely on photochemistry to inform our expectations for the polar composition, which suggests that the 1-mbar abundances of ethane and acetylene should not be varying significantly with time (although lower bar pressures will see relatively rapid increases in hydrocarbon production when sunlight returns to the high latitudes in the spring, peaking at solstice and then declining through autumn). Data from Guerlet et al. (2015) on thermal infrared detection of polar aerosols provide observational evidence that the auroral chemistry leads to the production of these aerosols.

Cassini has captured snapshots of the zonal mean hydrocarbon distributions during southern summertime conditions (see Chapter 10), finding considerable asymmetries between northern and southern hemispheres. Measurements at high latitudes have been sparse. Sinclair et al. (2013) identified a minimum in the 2-mbar acetylene (C_2H_2) distribution near 78°S PC (80°S PG), followed by a sharp rise at higher latitudes within the summer stratospheric vortex. Similar, shallower latitudinal gradients were also observed in ethane towards the south pole and acetylene towards the north pole. Higher-order hydrocarbons might also be expected to be enhanced at high polar latitudes, such as the recent detection of benzene (potentially related to ion chemistry) at Saturn's south pole (Guerlet et al. 2015). Fletcher et al. (2015) report on the rate of change of the C_2H_6 and C_2H_2 abundance at both poles over ten years of Cassini observations (Figure 12.13). Hydrocarbon emissions were barely discernible at the north pole during winter because of the low temperatures, and although the abundances of both C_2H_6 and C_2H_2 have been steadily rising poleward of 72°N PC (75°N PG) during spring (0.45 ppm year^{-1} for ethane and 0.03 ppm year^{-1} for acetylene), there is no peak in the hydrocarbon abundances within 5° of the north pole as there appeared to be in the south. Poleward of 72°S PC (75°S PG), the abundances have been steadily falling, by 0.35 ppm year^{-1} for ethane and 0.03 ppm year^{-1} for acetylene. As these rates far exceed the expectations of diffusive photochemistry models, we invoke subsidence of hydrocarbon-rich air at the north pole, and upwelling of hydrocarbon-poor air at the south pole, to explain these observed changes to the hydrocarbon distributions.

12.5.6 Vertical Motion

In a planetary stratosphere, the vertical component of the general circulation is controlled by radiative heating and cooling. Even though the speed of the vertical motion is too slow for direct measurement, it can be estimated from the deviation of the temperature from the radiative equilibrium. Thus, the calculation of radiation energy balance is the first step in understanding the stratospheric dynamics. The key radiative coolants in Saturn's stratosphere are ethane and acetylene; however, the current generation of radiative-dynamical models (i) lack the spatial resolution to deal with the hydrocarbon excesses within the polar regions (and the enhanced cooling they may produce), (ii) hold the hydrocarbons fixed with time rather than allowing them to vary seasonally and (iii) do not permit the advection of these hydrocarbons as quasi-conserved tracers. Furthermore, polar stratospheric aerosols are expected to provide an additional term to the radiative budget (Fletcher et al. 2015; Guerlet et al. 2015), enhancing the rate of summertime heating and wintertime cooling. To address these issues, ongoing efforts focus on more accurate modeling of the radiative balance in Saturn's stratosphere.

Although the current models are incomplete, deviations from expected radiative equilibrium can be assessed in terms of vertical motions causing adiabatic heating and cooling (e.g. Conrath et al. 1990). In this sense, the warming north polar temperatures, which increased at their fastest rates poleward of 72°N PC (75°N PG), could be related to subsiding air masses at vertical velocities of $w \approx -0.1$ mm s^{-1} (Fletcher et al. 2015). Similarly, the cooling of the southern polar vortex could be related to upwelling at $w \approx +0.1$ mm s^{-1}. These estimates, derived from a modified form of the heat equation (Conrath et al. 1990), are only approximate, given the large uncertainties in the absolute values of the radiative equilibrium temperature. Nevertheless, vertical motions of similar magnitude can explain the observed variability of the hydrocarbon distributions, and suggest that springtime subsidence and autumnal upwelling are indeed occurring, and that these are related to the formation/dissipation of the northern/southern summertime stratospheric vortices.

This stratospheric circulation, potentially confined to latitudes poleward of ±72° PC (75°PG) in both

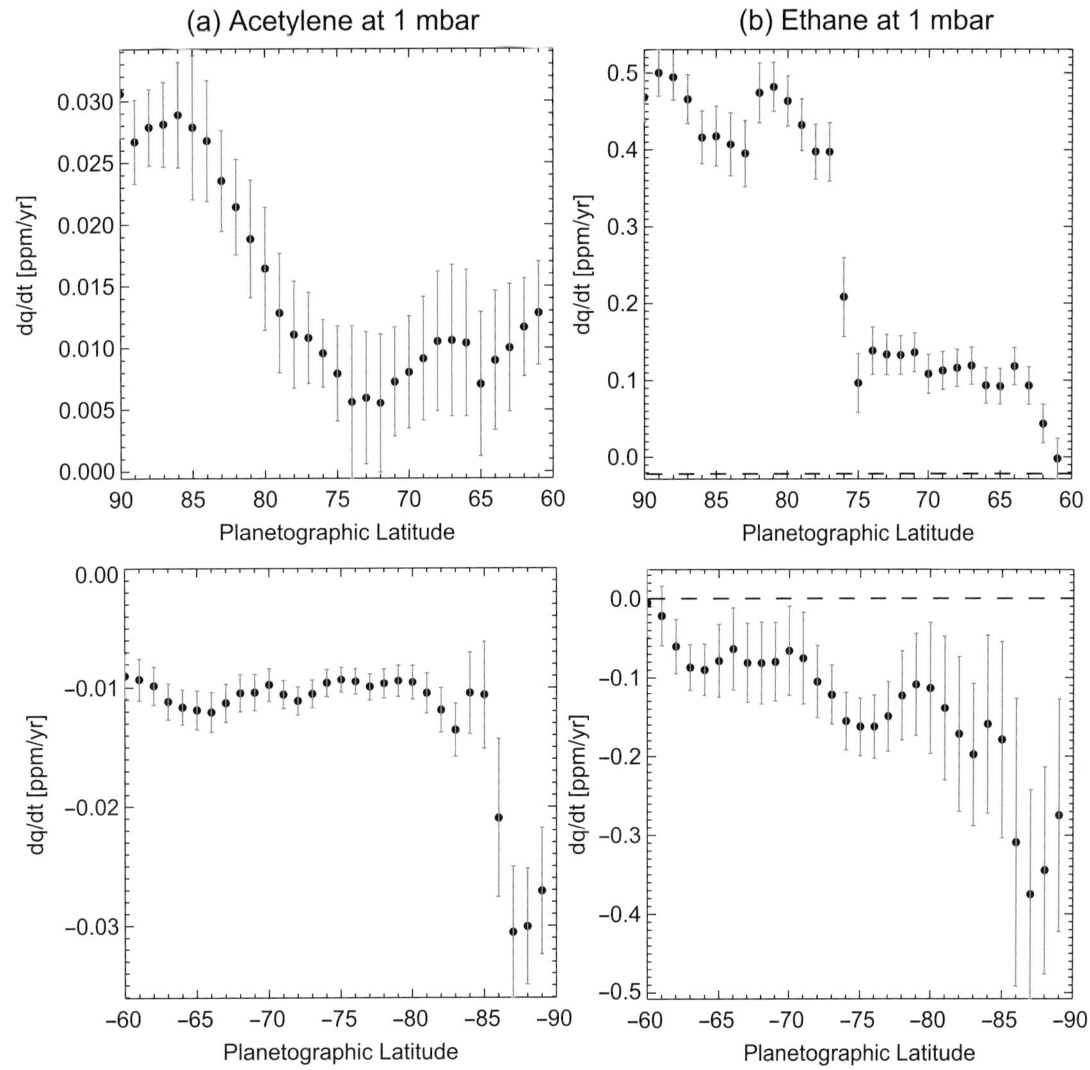

Figure 12.13 Rate of change of mole fraction (q) with time (ppm/year) at the 1-mbar level as measured by Cassini/CIRS spectroscopy, assuming a linear dependence over the ten-year span of the Cassini mission. Acetylene and ethane respond to the seasonal changes in insolation in subtly different ways, but are consistent with subsidence enhancing hydrocarbons over the north polar region, and upwelling depleting hydrocarbons over the south polar region. Reprinted from Fletcher et al. (2015), with permission from Elsevier.

hemispheres, may have consequences for the distributions of clouds, hazes and chemicals throughout the polar regions. For example, stratospheric polar hazes are known to change character within the polar regions (Sánchez-Lavega et al. 2006), being composed of smaller (0.1 μm radius at high latitudes, compared to 0.2 μm at low latitudes, Pérez-Hoyos et al. 2005), more UV-absorbent (Karkoschka and Tomasko 1993; also see Section 12.6 and Figure 12.17) aerosols and with an increased optical depth (Karkoschka and Tomasko, 2005). In the upper troposphere, the disequilibrium species PH_3 is known to be suppressed within the core of the two matching polar cyclones, potentially due to atmospheric subsidence within the centers of the two vortices (Fletcher et al. 2008). The connection between the observed gaseous and aerosol properties and the residual mean circulation of the stratosphere remains unclear.

In summary, stratospheric and upper-tropospheric circulation can be inferred from measurements of the temporal variability of the thermal structure and the distribution of chemicals (notably stratospheric C_2H_2 and C_2H_6) as quasi-conserved tracers. Since Cassini's arrival at Saturn, a region within 15° to 20° (14,000 to

19,000 km) of the south pole, accompanied by bright infrared emission, has been cooling with time (compare the July 2005 and October 2013 maps in Figure 12.11) with declining hydrocarbon abundances, whereas the northern springtime pole has been steadily warming (compare the March 2007 and June 2014 panels in Figure 12.11) with rising hydrocarbon abundances. At the time of writing ($L_s \approx 60°$), the north polar stratosphere lacks a strong meridional temperature gradient near 72°N PC (75°N PG) that would signify the establishment of a summertime vortex, which we expect to occur before northern summer solstice in 2017. In addition to the seasonally variable stratospheres, both poles feature warm cyclonic vortices in their upper tropospheres, which extend well into the stratosphere during summertime conditions.

12.5.7 Long-Term Prediction

Finally, the success of radiative-climate modeling in reproducing the general thermal trends allows us to make some predictions. Northern summer solstice will be quickly followed by peak 1-mbar temperatures at the north pole (1–2 years after solstice, around 2018–2019) and the presence of a warm northern stratospheric vortex, but given that northern summer solstice occurs near aphelion, the magnitude of the temperatures will be smaller than those observed during southern summer (near perihelion). The south pole is currently descending into winter, but we would expect the coolest 1-mbar stratospheric temperatures to occur near to the next equinox (i.e. ring-plane crossing) in 2025. By then, the subsidence at the north pole may cease, the subsidence at the south pole may start, the northern summer vortex would dissipate, and we would start to observe renewed heating of the south pole at 1 mbar for the creation of a new, seasonal, south polar stratospheric vortex.

12.6 Saturn's Polar Aerosols and Clouds

Polar latitudes are of particular interest for clouds and aerosols, as they are for atmospheric dynamics and chemistry. Both Jupiter and Saturn show enhanced ultraviolet absorption at high latitudes and there is strong evidence that energy deposition from the aurora acting on methane contributes to the production of UV absorbers and aerosols in the auroral deposition region (high atmosphere). These effects are more subtle on Saturn than on Jupiter because the auroral energetics are weaker on Saturn. For an introduction to the relevant chemistry, see papers by Friedson et al. (2002) and Wong et al. (2000, 2003). At the same time, images at mid-visible to near-infrared continuum wavelengths show abundant small-scale convective clouds and vortices in the deeper (pressure of ~2 bars or more) atmosphere. The challenges are to sort out the various contributions from this vast stratigraphy, to understand how they influence data at different wavelengths, polarization states, viewing geometry, and time, and then to properly attribute the observed features to atmospheric processes. This effort is still ongoing.

Evidence for a population of fractal aggregate particles in the polar stratosphere comes from a combination of images in near-infrared methane bands, images in the near-UV, and polarization images from the Cassini ISS (Imaging Science Subsystem) cameras. In these images, it is clear that the light scattered by the stratospheric haze particles deviate significantly from the scattering phase function predicted by the Mie theory that assumes spherical particles. The polarization signatures suggestive of aspherical aerosols for the southern and northern high latitudes are evident in Figure 12.14 and Figure 12.15. However, to date, few analyses have taken advantage of these polarization signatures, and quantitative analysis of light scattering by these polar clouds and hazes is yet to be done – studies to date have focused on lower latitudes, assuming spherical particles (e.g. Roman et al. 2013).

In Figure 12.14, taken on Cassini's approach to Saturn, the degree of linear polarization increases at the high latitudes and the angle of polarization at those latitudes is 90° (perpendicular to the scattering plane; West et al. 2009). These attributes are consistent with scattering by gas or by very small (relative to visible wavelengths) particles or aggregates of small particles. The polarization signature is more interesting at northern high latitudes where high polarization is confined to the region inside the hexagon.

In Figure 12.15, the material inside the hexagon appears dark in the images captured using the polarizer oriented with principal axis in the scattering plane (West et al. 2015). The different appearance in the two polarizer orientations is a signature of

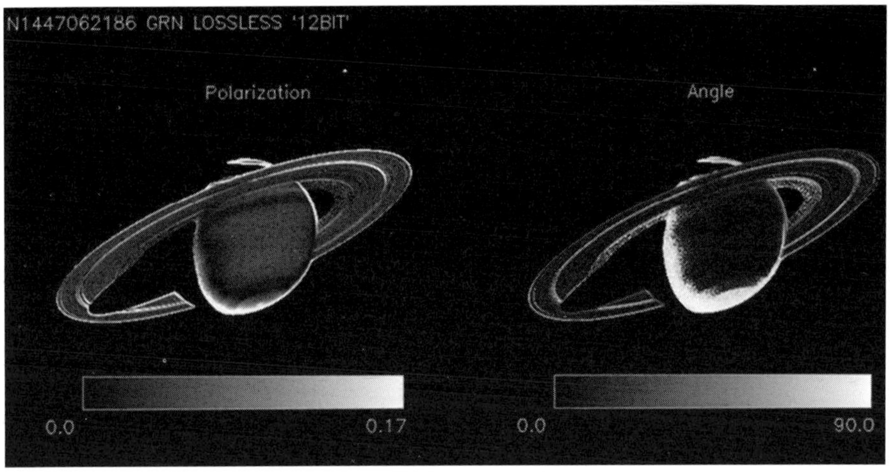

Figure 12.14 Polar polarization measurements. The degree of linear polarization (left) and angle of polarization (right) with respect to the scattering plane captured in 2003 by the Cassini ISS NAC using the GRN filter are shown, depicting enhanced positive polarization (angle of the electric vector near 90° with respect to the scattering plane) at high southern latitudes. Reprinted from West et al. (2009), with permission of Springer Nature.

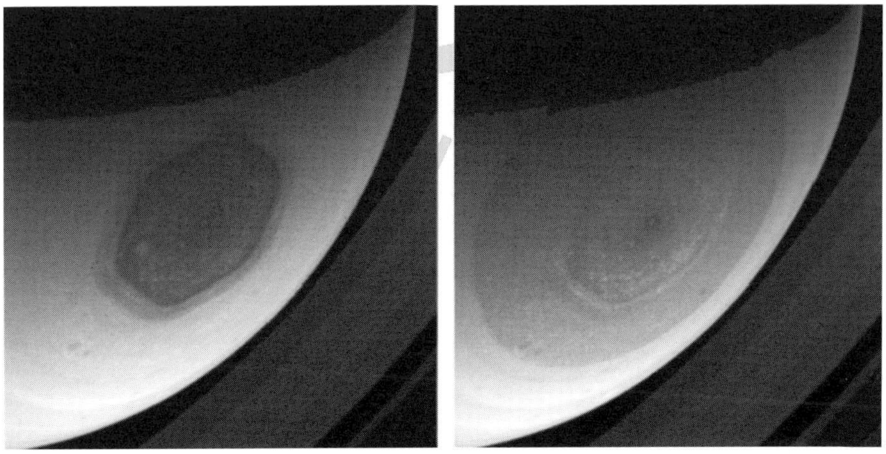

Figure 12.15 Cassini ISS images of Saturn taken 90 seconds apart in the 935-nm filter, coupled with polarizers with the electric vector parallel (left) and perpendicular (right) to the scattering plane. These images capture Saturn's northern polar region, including the hexagon and polar vortex. Structural differences inside and outside the hexagon are apparent. Figure from West et al. (2015).

Rayleigh-type linear polarization. Some of this may be due to a larger optical path through molecular hydrogen and helium in this region (due to the absence of haze particles). However, methane-filter images at high phase angle show enhanced scattering from an aerosol layer at high latitudes (Figure 12.16). The combination of high polarization and strong forward scattering is also seen in the polar regions of Jupiter (West and Smith 1991) and points to a population of fractal aggregate particles in the stratosphere (probably near the 10-mbar pressure level). The small monomers, or primary particles, that combine to form the aggregate are responsible for the high polarization, while the large effective size of the aggregate is responsible for strong forward scattering. The details concerning particle optical depths, monomer size and average number of monomers in the aggregate remain to be worked out.

The latitudinal distribution of UV-absorbing haze provides another clue to the processes that control aerosol formation and distribution. UV-absorbing haze in the north is concentrated within the hexagon, as is the region of high polarization shown in

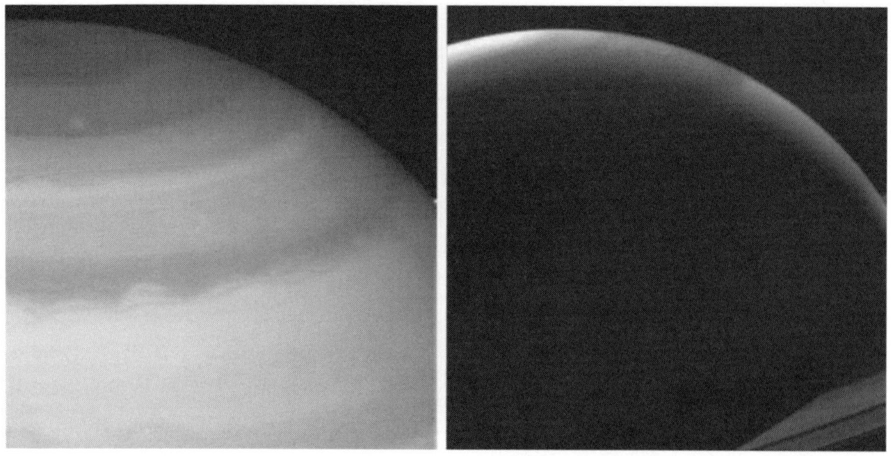

Figure 12.16 These images in the Cassini ISS MT2 filter (727 nm methane band) obtained at low (2.6°) and high (151.3°) phase angles show the forward-scattering property of polar stratospheric aerosols. Saturn's spin axis is vertical in the left image and tipped to the left by 16° in the right image. In both images, the north pole is near the upper left corner. Figure from West (2014).

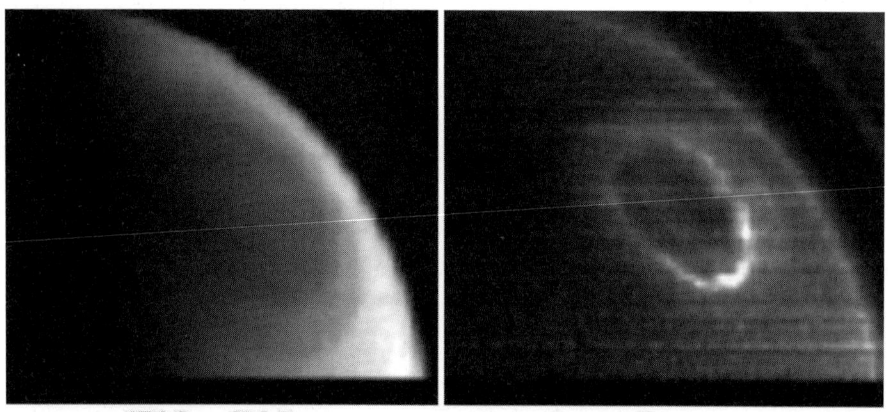

Figure 12.17 UV imagery of Saturn's north polar region. The image on the left was constructed from a scan across the northern high latitude of Saturn by the Cassini UVIS instrument. It is the sum of spectral channels spanning the range 171.2–191.0 nm (reflected sunlight). On the right is the image in the hydrogen Ly-alpha line (121.6 nm) showing the auroral oval near 75°N latitude (near the latitude of the hexagon). The images were from the same spectral cubes and are shown at the same scale. The UVIS slit moved across the horizontal direction; the streaks in the Ly-alpha image are from uncorrected residual at field errors. Figure from West (2014).

Figure 12.15. The region inside the hexagon is one of strong near-UV haze absorption, as Figure 12.17 shows (West 2014).

Acetylene gas also absorbs in the 171–191 nm region, but the acetylene signature in the UV is nearly uniform over the image in Figure 12.17 and does not contribute to the contrasts (West 2014). As discussed in Section 12.5, CIRS data reveal that acetylene concentration is increasing in the northern high latitudes in the infrared (Figure 12.13); the UV signature is yet to be analyzed in conjunction with the CIRS data.

The confinement of the high polarization and UV absorption to the region inside the hexagon suggests that auroral processes lead to hydrocarbon production, which ultimately leads to aerosol production in the high stratosphere in the form of fractal aggregate particles. Furthermore, the jet that forms the hexagon appears to provide a confining mechanism whereby particles produced within the hexagon are trapped. This mechanism is familiar in the terrestrial winter polar stratosphere, where ozone is trapped within the winter polar vortex. Much of the Saturn data collected by the UVIS instrument remains to be analyzed, which could return results on stratospheric haze particle properties, photochemical processes that produce the particles and dynamic transport processes that redistribute the particles.

12.7 Polar Atmospheric Dynamics

In this section, we review recent theoretical and modeling developments relevant to understanding giant planet polar atmospheric dynamics. As points of comparison, we use Jupiter and Saturn as the prototypical examples of the Jovian-class planets (as opposed to the ice-giant class) that have been studied in detail. For Saturn specifically, we discuss the progress towards understanding the formation and maintenance of the hexagon and the polar vortices.

12.7.1 Statement of the Problem: Explaining the Three Dynamical Regimes on Jupiter and Saturn

Spacecraft observations of Jupiter and Saturn since the early 1970s have identified three distinct dynamical regimes in the cloud-top winds. In the equatorial region, a fast, broad jet stream blows eastward, where no vortices are found. In the mid-latitudes of both Jupiter and Saturn, many vortices exist between the numerous jet streams that alternate in wind direction between eastward and westward – Jupiter and Saturn have ~30 and ~15 jet streams, respectively (here, we define each of the eastward and westward peaks in the zonal wind profile as a *jet*; nevertheless, such a definition inevitably carries a level of arbitrariness as exemplified by the opening sentence of Rhines (1994): "It is not easy to give a precise definition of a jet of fluid motion, but most of us know one when we see it").

Closer to the poles, vortices become increasingly prevalent with latitude; however, Jupiter and Saturn critically differ in their atmospheric dynamics around the poles. On Jupiter, both the visible appearances and the dynamical characteristics change dramatically poleward of ~60° latitude compared to the lower latitudes. While the low to mid-latitudes are characterized by alternating dark and bright bands whose boundaries loosely correspond to the peaks of the zonal jets, the high-latitude regions lack clear banding and are instead marked by countless small vortices. Cassini observations of the northern high latitudes of Jupiter reveal a wind flow dominated by numerous vortices ranging in size from the limit of image resolution to thousands of kilometers (Porco et al. 2003). The motions of those high-latitude vortices are zonally organized, as can be seen in the NASA public release movie PIA03452, and cloud tracking wind measurements such as that by Porco et al. (2003) are sensitive to vortex drifts. However, like the Great Red Spot and the White Ovals, those drifting vortices do not move at the background prevailing wind speed. We identify this vortex-filled dynamical regime as *polar turbulence*. Saturn, on the other hand, maintains zonally organized cloud bands up to the poles and lacks Jupiter-like polar turbulence. As discussed in the previous sections, zonal structure of Saturn culminates in each of the hemispheres with a hurricane-like cyclonic vortex residing precisely at the poles. As of this writing, Jupiter's poles are yet to be observed with sufficient resolution to answer whether the planet harbors polar vortices – our knowledge of Jupiter's poles is expected to be significantly updated by the upcoming Juno mission, which will go into orbit around Jupiter in 2016.

A comprehensive gas-giant planetary atmospheric dynamics model must explain these three dynamical wind regimes: (1) the vortex-free equatorial region with a fast, broad eastward jet, (2) the middle latitudes where stable vortices are embedded in the numerous robust zonal jets that violate the barotropic stability criterion and (3) the vortex-dominated flow in the high latitudes. Such a model should also explain the difference between Jupiter and Saturn (and, eventually, the difference between the Jovian and Ice-Giant classes of planets). In this review, we focus on our current understanding of the mechanism that separates the mixed jet-vortex flow regime to the polar turbulence, and the processes that differentiates the polar atmospheric dynamics of Jupiter and Saturn.

12.7.2 Transition from Mid-latitude Jets to Polar Turbulence

A remarkable common property of the mid-latitude jet streams of Jupiter and Saturn is that they are extremely steady even though they violate the barotropic stability criterion at some latitudes. The spatial steadiness is signified by the observation that the longitudinal wind speed variation at a given latitude is much smaller than the variations of the zonal mean zonal wind in latitude (Limaye 1986; Sánchez-Lavega et al. 2000). Comparison of the zonal mean wind profiles from the Voyager measurements and the Cassini measurements illustrates the spatial steadiness of the zonal jets in time, in which only minor changes in the locations and the speeds of those jets are found for both Jupiter (between 1979 and 2000; Porco et al. 2003) and Saturn (between 1980–1981 and Cassini measurements since 2004; Porco et al. 2005; García-Melendo et al. 2011).

These stable mid-latitude jets are shaped by the effects of fast planetary rotation on the atmospheric turbulence. In giant planet atmospheres, the sources of turbulent motions such as thunderstorms and baroclinic instabilities can contain spatial scales smaller than the size of jets and vortices. In quasi-two-dimensional large-scale flows in stratified atmospheres, these small seeds of turbulence self-organize to form larger structures. The tendencies for the kinetic energy contained in small spatial features to be transferred to larger structures in two-dimensional fluid flows is often called the upscale (or inverse) cascade (early theoretical works include Kraichnan 1967, 1971, Charney 1971 and Salmon 1980; also see textbooks such as Salmon 1998 and Vallis 2006). When this process occurs on a fast-rotating planet such as Jupiter or Saturn, the resulting structures become elongated in east-west directions until they form zonal jets. This topic, the generation of zonal flow through geostrophic turbulence, has been extensively reviewed by Rhines (1979), Rhines (1994), Vasavada and Showman (2005) and Galperin et al. (2006); in this review, we focus on the transition in the outcome of the upscale energy transfer from a jet-dominated regime to a vortex-dominated flow.

The effects of planetary rotation on the upscale energy transfer were first explained in numerical experiments by Rhines (1975). Using a barotropic model, Rhines showed that the upscale redistribution of the kinetic energy in the spectral space results in structures elongated in the zonal directions with the north-south characteristic width

$$L_\beta \approx (U/\beta)^{1/2}, \qquad (12.4)$$

Where $\beta = df/dy > 0$, y is northward distance, $f = 2\Omega\sin(\phi)$, Ω is planetary rotation rate and ϕ is latitude. L_β is called the Rhines length (note that the original form by Rhines is $L_\beta = \sqrt{2U/\beta}$, while the form presented in Equation 12.4 has become more common in recent years). Recent studies further show that this upscale energy transfer is mediated by waves that transfer energy non-locally in the spectral space, i.e. the energy does not cascade between adjacent spectral modes (Sukoriansky et al. 2007; Srinivasan and Young 2012). Two-dimensional and quasi-two-dimensional geophysical turbulence remain an active area of investigation through numerical experiments (e.g. Yoden and Yamada 1993; Chekhlov et al. 1996; Nozawa and Yoden 1997; Huang and Robinson 1998; Yoden et al. 1999; Huang et al. 2001; Danilov and Gurarie 2004; Danilov and Gryanik 2004; Sukoriansky et al. 2007; Sayanagi et al. 2008), laboratory experiments (e.g. Read et al. 2007; Espa et al. 2010; Slavin and Afanasyev 2012; Di Nitto et al. 2013; Smith et al. 2014; Zhang and Afanasyev 2014; Read et al. 2015), and observational analyses (e.g. Galperin et al. 2001, 2004; Barrado-Izagirre et al. 2009; Choi and Showman 2011; Galperin et al. 2014).

If the turbulent kinetic energy redistribution is indeed responsible for the generation of the mid-latitude zonal jets on Jupiter and Saturn, the transition from the mid-latitude flow regime to the polar turbulence must involve suppression of the effect. Using quasi-geostrophic (QG) models, Okuno and Masuda (2003) and Smith (2004) showed that the kinetic energy transfer can be suppressed when the Rossby deformation radius is small, as is the case in high-latitude regions of fast-rotating planets. For a single-layer fluid flow (e.g. shallow-water model), the Rossby deformation radius is $L_D = \sqrt{gh/f}$, where g and h are the surface gravity acceleration and depth of flow, respectively. The barotropic model used in the original derivation by Rhines (1975) has $L_D \approx \infty$. When the deformation radius is finite, Okuno and Masuda (2003) showed that the Rhines length takes the form

$$L'_\beta \approx \left(\frac{1}{L_\beta^2} - \frac{1}{L_D^2} \right)^{-1/2} \qquad (12.5)$$

When $L_D \ll L_\beta$, Equation 12.5 indicates that L'_β becomes imaginary, and the kinetic energy transfer is suppressed (Okuno and Masuda 2003; Smith 2004).

Theiss (2004) proposed that the boundary between the mid-latitude jet regime and the polar-turbulence region is marked by $L_D \approx L_\beta$. Theiss (2006) further investigated the effect of the background flow on the value of effective β, essentially taking account of the change in the Rossby wave frequency due to the zonal flow. Theiss (2006) and Penny et al. (2010) showed that the distributions of vortices on Jupiter and Saturn are consistent with the suppression of the upscale kinetic energy transfer by the background zonal flow. The difference in the high-latitude atmospheric dynamics between vortex-filled Jupiter and zonally organized Saturn could be explained by the relative scales between the deformation radius L_D and the Rhines length L'_β on each of the planets – L_D is indeed generally smaller on Jupiter than on Saturn in low and mid-latitudes (Read et al. 2006, 2009a); however, L_D and L'_β remain to be estimated for high-latitude regions of either planet.

Studies using shallow-water (SW) models show that the suppression of the upscale kinetic energy transfer under small L_D also occurs when the ageostrophic effects are fully included. Simulations of freely evolving (Cho and Polvani 1996; Iacono et al. 1999; Kitamura and Ishioka 2007) and forced (Showman 2007; Scott and Polvani 2007) turbulence showed that the flows in such systems form multiple zonal jets in low latitudes accompanied by vortex-dominated regions in the high latitudes. Note that, under horizontally uniform stratification, L_D decreases with latitude as f increases. In these 1-layer simulations, a clear critical latitude divides the jets and the vortex-dominated regions. However, these shallow-water models do not explain two aspects of the zonal jets on Jupiter and Saturn. First, under small L_D like those of Jupiter and Saturn, their equatorial jet becomes westward (Showman 2007; Scott and Polvani 2007; Kitamura and Ishioka 2007), which is the opposite of Jupiter or Saturn. Second, they do not reproduce the mid-latitude flow regime in which small vortices and stable zonal jets coexist.

A mixed jet-vortex flow regime like that of Jovian mid-latitudes can be reproduced using three-dimensional multi-layer numerical models that solve the primitive equations. Sayanagi et al. (2008) demonstrated that, in freely evolving simulations of geostrophic turbulence on a beta-plane, the outcome of the turbulent self-organization transitioned from the mixed jet-vortex regime to the vortex-dominated regime as the deformation radius L_D was decreased. Their experiment can be seen as a three-dimensional counterpart to the SW simulations by Cho and Polvani (1996) and Iacono et al. (1999). To date, studies of the effects of forced three-dimensional turbulence on giant planet atmospheres mostly focus on jet formation through baroclinic instabilities, and few studies focus on high latitudes. Nevertheless, a gradual increase in the vortical activities towards the poles can be discerned in the results by Williams (2003) and Liu and Schneider (2010). Unlike the 1-layer SW models, these many-layer, three-dimensional models can have multiple baroclinic deformation radii that become relevant in their dynamics (e.g. Achterberg and Ingersoll 1989). Sayanagi et al. (2008) speculated that a mixed jet-vortex regime occurs when the Rhines length is smaller than some, but not all, deformation radii; in such a scenario, the upscale energy transfer would produce vortices at those deformation radii smaller than the Rhines length and co-exist with jets. Much work remains to be done before we gain a full understanding of the effect of three-dimensionality on the jet formation and the upscale energy transfer.

It is worth noting that some recent studies show that the upscale energy transfer may not be solely responsible for the generation of the zonal jets (for example, Sukoriansky et al. 2007; Scott and Tissier 2012; Verhoeven and Stellmach 2014). However, we limit the scope of our current discussion to the transition of flow regime from the mixed jet-vortex regime to the polar turbulence through the suppression of the upscale energy transfer.

12.7.3 Models of the Hexagon

As discussed in Section 12.2, the hexagon was first discovered by Godfrey (1988) in images captured by Voyager in 1980–1981. The key characteristics of the feature that must be explained are (1) that its hexagonal outline follows the path of a fast eastward jet stream that blows at around 75°N PC latitude (Godfrey 1988; Baines et al. 2009; Sánchez-Lavega et al. 2014; Antuñano et al. 2015); (2) the feature does not accompany large vortices (evident in the measurements by Antuñano et al. 2015); (3) the hexagonal pattern's longitudinal drift is extremely slow in the System III reference frame (0.8 ±1.1 m s^{-1} as measured by Godfrey 1988 using Voyager data, and 0.036 ±0.004 m s^{-1} by Sánchez-Lavega et al. 2014 using a combination of Cassini and ground-based data); and (4) the potential vorticity gradient across the hexagon jet is substantially steeper than the surrounding region (Read et al. 2009a), i.e. it forms a *PV front*.

A substantial body of literature about meandering geophysical jets accompanied by PV fronts can be found in studies of Earth's meandering ocean currents (e.g. Allen et al. 1991; Flierl 1999) and mid-latitude atmospheric jets (e.g. Polavarapu and Peltier 1993; Swanson and Pierrehumbert 1994). A jet stream with a sharp PV front can develop finite amplitude instability (i.e. an instability in which amplitude growth is damped by nonlinear effects); Hart (1979) is an early review on the topic that introduced the idea. To the best of our knowledge, Flierl et al. (1987) conducted the first numerical experiments to demonstrate that an unstable jet can develop instabilities, which subsequent amplitude growth halts nonlinearly in some cases and finds a stable configuration. Using a barotropic model, they tested the stability of various initial flows that contain single-mode perturbations and examined the instability growth. Their results demonstrate that, depending on the initial flow, the flow can equilibrate in either (1) a vortex street, (2) a stably meandering jet with a monochromatic wavelength or (3) a jet with chaotic unsteady meandering. Here, a vortex street is a fluid flow structure that has a chain of vortices with alternating signs of vorticity (e.g. Kàrmàn 1921). Many numerical models and laboratory experiments have examined each of these outcomes as summarized by Sayanagi et al. (2010) in the context of Saturn's ribbon jet at 41°N PC (47°N PG) latitude, which exhibits chaotic meandering (Sromovsky et al. 1983; Godfrey and Moore 1986). The hexagon jet is a case of a steadily meandering zonal jet with a monochromatic wavelength; the geometry of the meandering appears polygonal when viewed from above, due to its high latitude.

Numerical and laboratory experiments to date have demonstrated that a vortex street and a meandering jet can both have polygonal flow path even though their vorticity structures are different. In a vortex street, each of the vortices has a local peak of relative vorticity surrounded by closed streamlines. In comparison, a meandering jet does not accompany closed streamline vortices, and the streamlines are instead roughly parallel to the jet's path. The emergence of a polygonal structure associated with a meandering, barotropically unstable zonal jet was first shown in a laboratory experiment by Sommeria et al. (1989), in which they produced a pentagonal flow in a rotating annular tank that accompanies a sharp potential vorticity front. Solomon et al. (1993) and Marcus and Lee (1998) performed laboratory and numerical experiments to demonstrate that shear instability can equilibrate to form a vortex street. In particular, Marcus and Lee (1998) showed that, when the vortex street consisted of 6 pairs of interlocking cyclones and anticyclones around the jet stream, the path of the jet resembled the hexagon; however, unlike these vortex street flows, the Saturnian jet at 75°N PC latitude lacks strong vortices adjacent to the jet.

Barbosa Aguiar et al. (2010) applied these previous results to parameter regimes relevant to the hexagon through linear stability analysis and laboratory experiments. They studied a flow with parameters relevant to the 75°N hexagon jet ($L_D = N_B H / f_0$, where N_B is the Brunt-Väaisäalää frequency and H is the atmospheric scale height) and showed that, when Rossby deformation radius $L_D = \infty$, the maximum growth rate occurs for a zonal wavenumber $k_{max} = 13$. However, when L_D = 2500 km, the most unstable modes become $k_{max} \sim$ 5–6 and the instability growth rate also decreases, meaning that the flow is more stable. For $L_D \leqq$ 2000 km, the large-scale instability and the growth rate increases monotonically with the wavenumber. Barbosa Aguiar et al. (2010) applied the result of this analysis to conduct fluid dynamic laboratory experiments in a rotating tank (similar setup to those by Sommeria et al. (1989), Solomon et al. (1993) and Marcus and Lee (1998), but with a different forcing mechanism) to simulate the hexagon formation mechanism to show that the jet equilibrates into a wavenumber-6 mode, and demonstrated that barotropic instabilities lead to a meandering jet that forms a hexagon. One aspect of the experimental result from Barbosa Aguiar et al. (2010) that is not consistent with the observation is the presence of large vortices that sandwich the meandering jet – such vortices are absent in the observed wind profile (Antuñano et al. 2015). Morales-Juberìas et al. (2011, 2015) discuss that the results from Barbosa Aguiar et al. (2010) resemble a vortex street.

Morales-Juberìas et al. (2011, 2015) studied the instabilities arising in an unstable jet encircling the north pole of Saturn using nonlinear three-dimensional atmospheric dynamics models. Their first study showed that, when the jet is deep-seated (i.e. when assumed to have a negligible vertical shear below the observed cloud top), the jet relaxed into a vortex street. They demonstrated that the dominant wavenumber of the final relaxed state depends on the width of the initial jet by tuning the wavenumber between 3 and 8. However, all their outcomes were vortex streets, and the feature propagated in longitude much faster than the observed behavior of the Saturnian hexagon. Their second study demonstrated that, when the initial jets were shallow (i.e. their wind speed decreased with depth below the observed cloud top such that it approached zero near the cloud condensation level at 10 bar), the instabilities relaxed the jets into a steady, near-stationary meandering state accompanied by a PV front (Figure 12.18), thus reproducing all of the four key characteristics observed in the hexagon noted earlier. Morales-Juberìas et al. (2011) also showed that a vortex street model cannot explain the slow propagation rate of the observed hexagon. The strong vertical shear of the jet presumably makes the instabilities in the model from Morales-Juberìas et al. (2015) more baroclinic rather than barotropic, which would make the dynamics more like Saturn's ribbon wave (Godfrey and Moore 1986; Sayanagi et al. 2010); however, Morales-Juberìas et al. (2015) do not analyze the energetics of the flow and the details remain unclear.

Results from Morales-Juberìas et al. (2015) suggest that Saturn's 75°N jet must be a shallow atmospheric structure. Numerical simulations from Liu and Schneider (2010), which also assumed shallow jet structures, also produced meandering polygonal high-latitude jets for

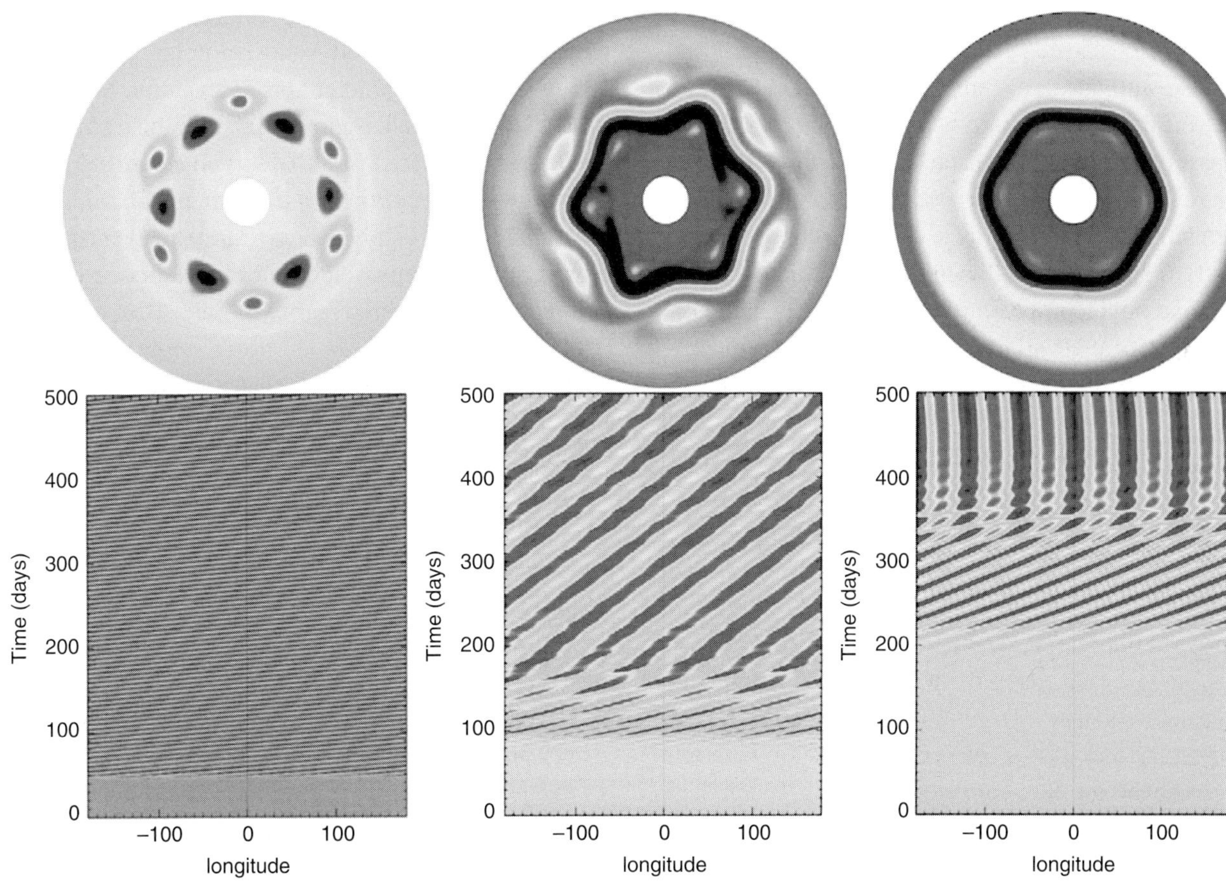

Figure 12.18 Morales-Juberìas et al. (2015) showed that, depending on the vertical wind structure, a barotropically unstable jet stream around the pole can relax into a vortex street (left), a meandering jet with a star-shaped path (middle) and a meandering jet with a polygonal path (right). The top panels show maps of potential vorticity at the 1-bar level, in which the value increases in order of red-green-blue. The bottom panels are Hovmöoller diagrams that depict the propagation of the hexagonal disturbance in time, which shows that the polygonal flow pattern is nearly stationary. (A black-and-white version of this figure appears in some formats. For the color version, please refer to the plate section.)

Saturn. Such a zonal jet with high vertical wind shear must accompany a horizontal temperature gradient through the thermal wind balance, but temperature measurements from these relevant depths below the cloud tops do not yet exist – the results by Morales-Juberias et al. (2015) are within the uncertainties of the observed temperature structure of Fletcher et al. (2008). Measurements of Saturn's high-order gravitational moments (Liu et al. 2014) may be able to test this hypothesis that the 75°N jet is shallow. This shallow-jet scenario is not consistent with a deep-rooted jet advocated by Sánchez-Lavega et al. (2014); this disagreement could be resolved by the gravity measurements, which are sensitive to depth of jets during Cassini's Grand Finale orbits. Morales-Juberìas et al. (2015) also showed that an identical jet centered at 70°N does not develop meandering, suggesting that the slightly equatorward center latitude of the southern jet at 70°S PC is responsible for the lack of polygonal meandering in the south.

12.7.4 Models of Polar Vortices

The polar vortices of Saturn show many similarities to other vortices elsewhere in the solar system. Table 12.2 compares the dynamical characteristics of the Saturnian polar vortices to a few other notable vortices. In the table, Saturnian polar vortices (Saturn PVs) are compared with terrestrial hurricanes (Anthes 1982; Emanuel 2003), Arctic and Antarctic polar vortices on Earth (Earth PV) (Nash et al. 1996), Jupiter's Great Red Spot (GRS) (Simon-Miller et al. 2002; Choi et al. 2007; Simon et al. 2014), south polar vortex on Venus (Venus PV) (Piccioni et al. 2007; Limaye et al. 2009; Patsaeva et al. 2015), and the warm spot at the south pole of Neptune (Orton et al. 2012; Sromovsky et al. 2001). The "radius" refers to the major axis of the area enclosed within the maximum winds (Emanuel 2003 defines the radius of a cyclone as the radius where the wind speed becomes indistinguishable from the background wind; however, we use the radius of the maximum wind here because, for polar vortices, vortex wind speeds and background zonal winds are impossible to distinguish). For the Saturnian PVs and terrestrial hurricanes, the radius of the eye is used. The temperature anomaly refers to the emission temperature at the cloud top relative to its surrounding, which may have different sign higher or lower in the atmosphere. The "unknown" eyewall clouds may be blocked from view by higher clouds. The vorticity ratio in the table ζ/f is effectively the Rossby number R_o.

The table reveals that the Saturnian polar vortices, terrestrial hurricanes and Venusian polar vortices share the warm-cored, cyclonic properties. In Table 12.2, Jupiter's Great Red Spot is classified as anticyclonic and cold-cored based on its observed cloud-top properties; however, its structure at depth remains unknown (note that the circulations in terrestrial hurricanes transition vertically from cyclonic near the surface to anticyclonic aloft; Frank 1977a).

Table 12.2 *Comparison of the Saturnian Polar Vortices (Saturn PVs) with other vortices in the solar system (reprinted from Dyudina et al. 2009, with permission by Elsevier).* [(1)]*This assumes that the eastward zonal wind peak at ~72°S in the measurement by Sromovsky et al. (2001) is the polar-most zonal wind peak, even though the wind structure poleward remains unresolved.* [(2)]*Luszcz-Cook et al. (2010) showed that Neptune's south polar vortex was split into two in July 2007.*

Parameters	Saturn PVs	Hurricane	Venus PVs	Earth PV	GRS	Neptune pole
Radius (km)	2000	10–100	~5000	~2500	7000–20000	~7500[(1)]
Radius (% planetary radius)	3	0.15–1.5	~85	~40	10–30	~30
Lifetime	>11 years	weeks	years	seasonal	>150 years	unknown
Max wind speed (m s^{-1})	150	85	35	50	120–190	250[(1)]
Temperature anomaly	warm	warm	warm	cold	cold	warm
Vorticity	cyclonic	cyclonic	cyclonic	cyclonic	anticyclonic	cyclonic
Eyewall clouds	yes	yes	unknown	no	unknown	unknown
Fixed to pole	yes	no	yes	yes	no	yes[(2)]
Convective clouds on periphery	yes	yes	no	no	yes	unknown
Convective clouds on periphery	no	yes	no	yes	no	no
$\lvert\zeta/f\rvert \sim R_o$	1	100	100	1	0.2	0.25[(1)]

In contrast to the warm-cored Saturnian polar vortices, Earth's polar vortices are cold cored near the surface (Holton 2004). The cold surface temperature occurs because radiative effects are more dominant in the troposphere and stratosphere than are horizontal transport and mixing of heat. Consequently, in winter, the pole cools radiatively, and a stronger vertical shear develops in the eastward winds of the circumpolar vortex to balance the horizontal temperature gradient via the thermal wind equation. However, higher up in the mesosphere, radiative effects are smaller and the adiabatic heating associated with subsidence over the winter pole becomes dominant, and temperatures become relatively warmer than those at lower latitudes. As a result, the polar vortex decays with altitude in the mesosphere. Indeed, the mesopause (~90 km) is warmest over the winter pole of Earth. (The situation on Titan is similar, except that it is the stratopause that is warm over the winter pole, and meridional contrasts actually decrease at higher altitudes.)

Unlike Earth's polar vortices, Saturn's polar vortices probably do not have the vertical dipolar structure described in the previous paragraph. Saturn's internal heat source tends to reduce lateral variations in entropy at depth (e.g. Allison 2000), and radiative damping times in the lower troposphere are too large to form a cold core over the poles in winter. Given that warm cores are observed at both poles in the upper troposphere (150 mbar), apparently independent of season (Figure 12.11; Fletcher et al. 2008, 2015), it is clear that the terrestrial polar analog does not apply. An analogy to hurricanes would offer a way around this problem, in which cyclonic vorticity develops via low-level convergence and uplift, with maximum ascent in the eyewall. In terrestrial hurricanes, the swirling winds are at maximum near the surface and decrease with altitude, and the warm temperatures within the eye, implied by the thermal wind equation, are maintained by a weaker descending flow in the eye (e.g. Frank 1977a,b).

The morphological similarity between the Saturnian polar vortices and the terrestrial tropical cyclones (such as hurricanes) are especially notable. First, both examples have a clear hole in the clouds ("eye") at the center that allows space-based remote-sensing instruments to probe deeper into the atmosphere (down to the surface, for the case of Earth). Second, both types of vortices have cyclonic vorticity and swirling cloud patterns that reflect the wind shear surrounding the central eye. However, a problem in drawing an analogy between Saturnian polar vortices and terrestrial hurricanes is that the dominant forcing necessary for generation of the latter is the latent heat release from the warm ocean surface underlying the vortex (Anthes 1982; Houze 1993; Emanuel 2003), whereas there is no analogous process on Saturn.

A mechanism that has been proposed for the formation of the Saturnian polar vortices involves the tendency for cyclonic vortices to drift poleward called the β-drift. The β-drift is caused by the difference in the potential vorticity of a vortex with its surrounding; a vortex drifts toward a direction where the environmental potential vorticity matches more closely with that of the vortex (an analogy could be drawn with motion caused by buoyancy force, in which a parcel of fluid rises or sinks toward a direction where the environmental density matches more closely with that of the parcel). The β-drift is the main mechanism that drives terrestrial tropical cyclones poleward, and has been studied extensively (e.g. Reznik 1992; Nycander 1993; Sutyrin et al. 1994; Sutyrin and Morel 1997; Montgomery et al. 1999; Sai-Lap Lam and Dritschel 2001). Elsewhere in the solar system, the β-drift mechanism is responsible for the equatorward drift of the anticyclonic dark spots on Uranus (Hammel et al. 2009; de Pater et al. 2011) and Neptune (Lebeau and Dowling 1998; Stratman et al. 2001).

The accumulation of cyclonic vorticity can be seen in the outcomes of freely evolving and forced turbulence simulations of geostrophic turbulence (Cho and Polvani 1996; Scott and Polvani 2007; Scott 2010), though the formation of polar vortices was not the focus of these studies. Taking hints from these studies, Scott (2011) systematically tested the formation of polar vortices through the β-drift mechanism by running freely evolving simulations of polar turbulence using a one-layer quasi-geostrophic shallow-water model. When the flow is initialized with a random distribution of small cyclones and anticyclones, the cyclones indeed β-drifted poleward as expected and merged at the pole to form an intense polar cyclone. O'Neill et al. (2015, 2016) further tested the idea with simulations of forced turbulence using a two-layer shallow-water model (i.e. not quasi-geostrophic) to show that the mechanism works in generating a polar vortex when forcing ageostrophic effects, and some three-dimensional effects are included.

O'Neill et al. (2015, 2016) also showed that a strong polar cyclone emerges only when the deformation radius L_D is relatively large (~1/20 of the planetary radius) and cumulus convection is energetic enough. When L_D is reduced to ~1/40 of the planetary radius (a Jupiter-like regime), a polar vortex does not form, but instead, weak jets emerge like those in the high-latitude regions of Jupiter. When L_D is held at ~1/20 of the planetary radius and cumulus convective energy flux is reduced, a transient polar vortex emerged similar to that observed on Neptune by Luszcz-Cook et al. (2010). Results from O'Neill et al. (2015, 2016) suggest that differences in the polar atmospheric dynamics between Jupiter, Saturn and Neptune are controlled by the deformation radius and cumulus convective energy flux; based on this result, they predict that Jupiter's poles have weak jets and lack a Saturn-like polar vortex. This prediction of reduced zonal organization for Jupiter compared to Saturn is consistent with the prediction that the region affected by polar turbulence (described by a region of $L_D \ll L_\beta$ following Equation 12.5) should expand when L_D is reduced.

On Earth, tropical cyclones require a warm sea surface in order to form, but their intensification follows a process similar to that tested by Scott (2011). When a terrestrial tropical cyclone becomes intense enough, it creates a local peak in the potential vorticity field such that it generates a local *effective* β-effect linked to the local gradient of potential vorticity. The process involving local effective-β is analogous to that of the planetary-β, which is the consequence of the gradient in the planetary vorticity, and the results are similar; when cumulus convection generates eddies inside of a terrestrial tropical cyclone, it causes an inward flux of cyclonic vorticity, and intensifies the cyclone (Montgomery and Enagonio 1998). Another implication of the locally modified effective β is the propagation of vortex Rossby waves. As a cyclonic vortex has a potential vorticity local maximum at its center, Rossby waves propagate azimuthally around the center of vortex and transport the azimuthal momentum component. Montgomery and Kallenbach (1997) proposed the link between the excitation and absorption of the vortex Rossby waves to the intensification of hurricanes, which was further studied by Montgomery et al. (1999) and Möller and Montgomery (2000). The intensification of terrestrial hurricanes and subsequent formation of an eye has also been linked to the vortex Rossby waves (Wang 2002). Also, similar to the pumping of zonal jets on Jupiter (e.g. Salyk et al. 2006) and Saturn (Del Genio et al. 2007), eddy momentum flux divergence is proposed as an important diagnostic quantity in tropical cyclone intensification (Nolan and Farrell 1999). The vortex Rossby waves and eddy momentum flux divergence should also play a role in the Saturnian polar vortices' intensification and eye formation, but these effects are yet to be studied for Saturn.

12.8 Discussion and Unanswered Questions

Over the past decade, the Cassini Orbiter has discovered a wealth of information about Saturn's poles, including the following.

1. The existence of large cyclonic vortices centered over both poles, covering ~10° (9,500 km) and ~15° (14,000 km) of latitude in radius for the north and south polar vortices, respectively. (Here, a vortex radius is defined by the region poleward of the polar-most minimum in the eastward wind.)
2. These polar cyclonic vortices extend over 100 km in depth, as revealed (i) in the classic latitudinal vortex structure observed in the tropospheric zonal winds down to at least the 2-bar level and (ii) in the latitudinal thermal temperature structure throughout the upper stratosphere, up to at least the 1-mbar level.
3. The polar vortices have morphological and dynamical similarities to terrestrial hurricanes.
4. Tropospheric winds in both polar vortices exceed 120 m s^{-1} within 1° of the pole, decreasing steeply to calm winds at the poles. Maximum winds of 136 m s^{-1} and 174 m s^{-1} occur near 88.3°N PC (88.6°PG) and 87.5°S PC (88.0°S PG), respectively.
5. While seasonal changes in stratospheric temperatures are apparent in polar regions, the classical latitudinal thermal structure of cyclonic polar vortices has persisted over each pole for one-third of a Saturn year since their discovery by Cassini.
6. High-speed jets border the polar vortices in both hemispheres: the Hexagon centered at 75°N PC (78°N PG) with a maximum wind speed of ~130 m s^{-1}, and a southern jet at 70°S PC (73°PG) with a maximum wind speed of ~87 m s^{-1}.
7. The north polar hexagon has now been consistently observed for over 35 years, from its discovery in Voyager images acquired in 1980 through Cassini and

high-resolution ground-based images acquired today, in the Spring of 2015, including many acquired with newly developed techniques by amateur astronomers (cf. Chapter 14).
8. The hexagon is a meandering jet that is not accompanied by large accompanying vortices.

Despite these revelations, fundamental questions remain, including the following.

A. Why does Saturn display intense polar vortices, complete with "hurricane eyes" on both poles? Does Jupiter, its fellow gas giant, lack such vortices, as predicted by O'Neill et al. (2015, 2016)?
B. What sustains the hexagonal and other high-speed jets?
C. How deeply do the hexagon and the polar vortices extend below the visible cloud top?
D. How do seasonal effects affect the cloud morphologies and meridional circulation associated with the polar vortices and the hexagon?
E. What are the depth of cloud layers, each exhibiting different morphology, sensed at different wavelengths?
F. What is the role of hydrocarbon photochemical products, including aerosols, on the radiative energy balance?
G. What are the properties (e.g. altitude, density, radius, composition) of the haze aerosol particles, which may be fractal aggregates of many particles?
H. How does the ion chemistry associated with aurorae work, and how do they affect haze particle production?

12.8.1 Outlook: End of Cassini Mission

To help answer such questions, the Cassini Orbiter will provide some of its most detailed and comprehensive imagery and spectra yet during and just prior to the Grand Finale Phase in late 2016 and 2017. In November 2016, while flying a series of inclined orbits that begin in February 2016, Cassini encounters Titan, reducing its closest approach distance from ~279,000 km above the clouds of Saturn to ~158,000 km. This provides a 24-hour period for clear, sunlit viewing of the north polar region. During the orbiter's inbound trajectory, its sub-spacecraft latitude will first increase from 45°N at an altitude of ~809,000 km to a peak latitude of ~60°N (PC); the periapsis altitude will be ~290,000 km above 55°N latitude, from where the north polar region will be severely foreshortened. Encounters on 18 and 27 November 2016 from similar viewing geometries allow detailed investigations of polar winds and structure by the ISS Camera, the VIMS visual-near-infrared spectral mapper and the CIRS thermal spectral mapper. During the orbiter's outbound trajectory after each periapsis pass, over a 10-hour period, the spacecraft will maintain favorable observing geometries of the southern high latitudes when its sub-spacecraft latitude first varies from ~55°S to ~60°S, and then equatorward to ~45°S from altitudes of ~205,000 to ~450,000 km, enabling VIMS and CIRS instruments to capture thermal maps of the south polar region under nighttime conditions.

On 19 November 2016, another Titan encounter lowers the closest approach altitude distance to ~90,000 km, setting up the first phase of the Grand Finale mission. For twenty orbits, the spacecraft will maintain highly eccentric orbits that pass over the north polar region; as the spacecraft approaches the periapsis, its altitude decreases from ~600,000 km to ~165,000 km over a ~15.2-hour period as the sub-spacecraft latitude varies from 55°N to 45°N. The south polar region is then viewed from distances ranging from ~126,000 to ~300,000 km over ~7.4 hours. Presently, Cassini is scheduled to observe Saturn during four of these twenty orbits with periapses on 16 January, 14 February, 7 March and 29 March 2017.

A final Titan pass on 22 April 2017 starts the second and final stage of the Grand Finale mission phase, sending the Cassini orbiter to a periapsis altitude of less than 3000 km over the Saturnian clouds. For the next 22 encounters, until the final plunge of the spacecraft into Saturn on 15 September 2017, the Cassini Orbiter obtains the highest resolution images of Saturn ever obtained. The orbiter will observe the north pole while its sub-spacecraft latitude varies from 55°N to 45°N from altitudes of ~270,000 to ~42,000 km over ~4.5 hours. The southern high latitudes will be observed while the sub-spacecraft latitude varies from 55°S to 45°S from an altitude of ~26,000 to ~160,000 km over ~2.3 hours. Cassini is currently scheduled to observe the poles extensively during four of the twenty orbits, on 26 April, 29 June, 1 August and 20 August 2017. From within 50,000 km, the VIMS instrument will produce spectral maps of the hexagon and north polar regions with an individual pixel scale of less than 25 km at the cloud top. CIRS will obtain thermal maps with better than 13-

km sampling, and ISS will produce visual images with just 300-meter pixel resolution. However, due to the swift movement of the spacecraft (exceeding 34 km s^{-1}, or 122,000 km $hour^{-1}$), only a few such high-resolution images are expected to be captured during each pass. Nevertheless, virtually all imagery taken within the 270,000 km start of each polar pass will be of higher spatial resolution than heretofore obtained during the mission.

The final phase of the Cassini mission will address many of the unanswered questions by extending the temporal observational coverage that stretches between almost two solstices, and by obtaining unprecedentedly high spatial resolution data enabled by the low-altitude orbits during the final phase of the mission. We expect that the extended seasonal coverage will address those questions related to seasonal forcing on the thermal structure and aerosol distribution, and confirm the predicted formation of the warm north polar stratospheric vortex, mirroring that observed over the south pole at the start of the mission. The questions on the vertical structure of the zonal jets may be answered through the analysis of the gravity field during the Grand Finale orbits. Higher resolution images to be returned during the final phase of the mission will also help reveal the detailed structure of the hexagon and the polar vortices. Other questions on the details of light-scattering properties of aerosol particles, auroral ion chemistry and radiative energy balance will require further theoretical and laboratory analyses. Cassini will perform the Grand Finale polar observations while the Juno mission will make similar observations of Jupiter; together, these missions are expected to significantly update our knowledge of polar atmospheric science of Jupiter and Saturn.

In addition, many images captured by VIMS, UVIS and ISS have not been analyzed, which could be used to reveal, for example, quantitative information on the haze and cloud particles' altitude, density, radius and composition. Continued analysis of the enormous dataset collected by the Cassini mission must be part of the effort to address many of these unanswered questions for many years to come. The Cassini mission's decade-long tour of the Saturnian system has rewritten our understanding of various processes that shape Saturn's polar atmosphere. The remaining years of the mission promise to return further data that will add to our understanding of the ringed planet.

Acknowledgments

Part of this work was performed by the Jet Propulsion Laboratory and was supported by the Cassini project. K. M. Sayanagi was supported in part by grants from NASA (Outer Planet Research Program NNX12AR38 G, Planetary Atmospheres Program NNX14AK07 G, and Cassini Data Analysis Program NNX15AD33 G) and NSF (Astronomy and Astrophysics Program 1212216). L. N. Fletcher was supported by a Royal Society Research Fellowship at the University of Leicester. Sánchez-Lavega was supported by the Spanish MICIIN AYA2012-36666 with FEDER funds, Grupos Gobierno Vasco IT765-013, and UPV/EHU UFI11/55.

References

Achterberg, R. K. and A. P. Ingersoll (1989), A normal-mode approach to Jovian atmospheric dynamic. *Journal of Atmospheric Sciences*, **46**(Aug.), 2448–2462.

Allen, J. S., L. J. Walstad and P. A. Newberger (1991), Dynamics of the coastal transition zone jet. 2. Nonlinear finite amplitude behavior. *Journal of Geophysical Research*, **96**(Aug.), 14995–15016.

Allison, M. (2000), A similarity model for the windy Jovian thermocline. *Planetary and Space Science*, **48**(June), 753–774.

Allison, M., R. F. Beebe, B. J. Conrath et al. (1991), Uranus atmospheric dynamics and circulation. Pages 253–295 in: J. T. Bergstralh, E. D. Miner and M. S. Matthews (eds.), *Uranus*. Tucson, AZ: University of Arizona Press.

Allison, M., D. A. Godfrey and R. F. Beebe (1990), A wave dynamical interpretation of Saturn's polar hexagon. *Science*, **247**(Mar.), 1061–1063.

Anthes, R. A. (1982), *Tropical Cyclones. Their Evolution, Structure and Effects*. Boston, MA: American Meteorological Society.

Antuñano, A., T. Río-Gaztelurrutia, A. Sánchez-Lavega et al. (2015), Dynamics of Saturn's polar regions. *Journal of Geophysical Research (Planets)*, **120**(Feb.), 155–176.

Archinal, B. A., M. F. A'Hearn, E. Bowell et al. (2011), Report of the IAU Working Group on Cartographic Coordinates and Rotational Elements: 2009. *Celestial Mechanics and Dynamical Astronomy*, **109**(Feb.), 101–135.

Arnol'd, V. I. (1966), On an a priori estimate in the theory of hydro-dynamic stability. *Izv Vyssh. Uchebn. Zaved. Matematika*, **54(5)**, 3–5. English translation: *American Mathematical Society Translation Series* 2 79: 267–269 (1969).

Bagenal, F., T. E. Dowling and W. B. McKinnon (2004), *Jupiter: The Planet, Satellites and Magnetosphere*. Orlando, FL: Academic Press.

Baines, K. H., P. Drossart, T. W. Momary et al. (2005), The atmospheres of Saturn and Titan in the near-infrared first results of Cassini/VIMS. *Earth Moon and Planets*, **96**(June), 119–147.

Baines, K. H., T. W. Momary, L. N. Fletcher et al. (2009), Saturn's north polar cyclone and hexagon at depth revealed by Cassini/VIMS. *Planetary and Space Science*, **57**(Dec.), 1671–1681.

Barbosa Aguiar, A. C., P. L. Read, R. D. Wordsworth et al. (2010), A laboratory model of Saturn's North Polar Hexagon. *Icarus*, **206**(Apr.), 755–763.

Barrado-Izagirre, N., S. Pérez-Hoyos and A. Sánchez-Lavega (2009), Brightness power spectral distribution and waves in Jupiter's upper cloud and hazes. *Icarus*, **202**(July), 181–196.

Barrado-Izagirre, N., A. Sánchez-Lavega, S. Pérez-Hoyos et al. (2008), Jupiter's polar clouds and waves from Cassini and HST images: 1993–2006. *Icarus*, **194** (Mar.), 173–185.

Bézard, B. and D. Gautier (1985), A seasonal climate model of the atmospheres of the giant planets at the Voyager encounter time. I: Saturn's stratosphere. *Icarus*, **61**, 296–310.

Brown, R. H., K. H. Baines, G. Bellucci et al. (2004), The Cassini Visual And Infrared Mapping Spectrometer (VIMS) investigation. *Space Science Reviews*, **115**(Dec.), 111–168.

Caldwell, J., X.-M. Hua, B. Turgeon et al. (1993), The drift of Saturn's north polar SPOT observed by the Hubble Space Telescope. *Science*, **260**(Apr.), 326–329.

Charney, J. G. (1971), Geostrophic turbulence. *Journal of Atmospheric Sciences*, **28**, 1087–1095.

Charney, J. G. and M. E. Stern (1962), On the stability of internal baroclinic jets in a rotating atmosphere. *Journal of Atmospheric Sciences*, **19**(Mar.), 159–172.

Chekhlov, A., S. A. Orszag, S. Sukoriansky et al. (1996), The effect of small-scale forcing on large-scale structures in two-dimensional flows. *Phys. D*, **98**(2–4), 321–334.

Cho, J. Y.-K. and L. M. Polvani (1996), The emergence of jets and vortices in freely evolving, shallow-water turbulence on a sphere. *Physics of Fluids*, **8**(June), 1531–1552.

Choi, D. S., D. Banfield, P. Gierasch et al. (2007), Velocity and vorticity measurements of Jupiter's Great Red Spot using automated cloud feature tracking. *Icarus*, **188** (May), 35–46.

Choi, D. S. and A. P. Showman (2011), Power spectral analysis of Jupiter's clouds and kinetic energy from Cassini. *Icarus*, **216**(Dec.), 597–609.

Choi, D. S., A. P. Showman and R. H. Brown (2009), Cloud features and zonal wind measurements of Saturn's atmosphere as observed by Cassini/VIMS. *J. Geophy. Res.*, **114**(E13), E4007.

Conrath, B. J., P. J. Gierasch and S. S. Leroy (1990), Temperature and circulation in the stratosphere of the outer planets. *Icarus*, **83**(Feb.), 255–281.

Cosentino, R. G., A. Simon, R. Morales-Juberias et al. (2015), Observations and numerical modeling of the Jovian ribbon. *Astrophysical Journal Letters*, **810** (Sept.), L10.

Danilov, S. and V. M. Gryanik (2004), Barotropic beta-plane turbulence in a regime with strong zonal jets Revisited. *Journal of Atmospheric Sciences*, **61**(Sept.), 2283–2295.

Danilov, S. and D. Gurarie (2004), Scaling, spectra and zonal jets in beta-plane turbulence. *Physics of Fluids*, **16** (July), 2592–2603.

de Pater, I., L. N. Fletcher, S. Luszcz-Cook et al. (2014), Neptune's global circulation deduced from multi-wavelength observations. *Icarus*, **237**(July), 211–238.

de Pater, I., L. A. Sromovsky, H. B. Hammel et al. (2011), Post-equinox observations of Uranus: Berg's evolution, vertical structure, and track towards the equator. *Icarus*, **215**(Sept.), 332–345.

Del Genio, A. D., R. K. Achterberg, K. H. Baines et al. (2009), Saturn atmospheric structure and dynamics. Page 113 of: M. K. Dougherty, L. W. Esposito and S. M. Krimigis (eds.), *Saturn from Cassini-Huygens*. Springer.

Del Genio, A. D., J. M. Barbara, J. Ferrier et al. (2007), Saturn eddy momentum fluxes and convection: First estimates from Cassini images. *Icarus*, **189**(Aug.), 479–492.

Desch, M. D. and M. L. Kaiser (1981), Voyager measurement of the rotation period of Saturn's magnetic field. *Geophysical Research Letters*, **8**(Mar.), 253–256.

Di Nitto, G., S. Espa and A. Cenedese (2013), Simulating zonation in geophysical flows by laboratory experiments. *Physics of Fluids*, **25**(8), 086602–086602.

Dougherty, M. K., L. W. Esposito and S. M. Krimigis (2009), *Saturn from Cassini-Huygens*. Springer.

Dowling, T. E. (1995), Dynamics of Jovian atmospheres. *Annual Review of Fluid Mechanics*, **27**, 293–334.

Dowling, T. E. (2014), Saturn's longitude: Rise of the second branch of shear-stability theory and fall of the first. *International Journal of Modern Physics D*, **23**(Feb.), 30006.

Dyudina, U. A., A. P. Ingersoll, S. P. Ewald et al. (2008), Dynamics of Saturn's south polar vortex. *Science*, **319** (Mar.), 1801.

Dyudina, U. A., A. P. Ingersoll, S. P. Ewald (2009), Saturn's south polar vortex compared to other large vortices in the Solar System. *Icarus*, **202**(July), 240–248.

Emanuel, K. (2003), Tropical cyclones. *Annual Review of Earth and Planetary Sciences*, **31**, 75–104.

Espa, S., G. Di Nitto and A. Cenedese (2010), The emergence of zonal jets in forced rotating shallow water turbulence: A laboratory study. *EPL (Europhysics Letters)*, **92**(Nov.), 34006.

Esposito, L. W., C. A. Barth, J. E. Colwell et al. (2004), The Cassini Ultraviolet Imaging Spectrograph investigation. *Space Science Reviews*, **115**(Dec.), 299–361.

Fischer, G., S.-Y. Ye, J. B. Groene et al. (2014), A possible influence of the Great White Spot on Saturn kilometric radiation periodicity. *Annales Geophysicae*, **32**(Dec.), 1463–1476.

Flasar, F. M., V. G. Kunde, M. M. Abbas et al. (2004), Exploring the Saturn system in the thermal infrared:

The composite infrared spectrometer. *Space Science Reviews*, **115**(Dec.), 169–297.

Fletcher, L. N., I. de Pater, G. S. Orton et al. (2014), Neptune at summer solstice: Zonal mean temperatures from ground-based observations, 2003–2007. *Icarus*, **231** (Mar.), 146–167.

Fletcher, L. N., P. G. J. Irwin, G. S. Orton et al. (2008), Temperature and composition of Saturn's polar hot spots and hexagon. *Science*, **319**(Jan.), 79–81.

Fletcher, L. N., P. G. J. Irwin, J. A. Sinclair et al. (2015), Seasonal evolution of Saturn's polar temperatures and composition. *Icarus*, **250**(Apr.), 131–153.

Flierl, G. (1999), Thin jet and contour dynamics models of Gulf Stream meandering. *Dynamics of Atmospheres and Oceans*, **29**(July), 189–215.

Flierl, G. R., P. Malanotte-Rizzoli and N. J. Zabusky (1987), Nonlinear waves and coherent vortex structures in barotropic β-plane jets. *Journal of Physical Oceanography*, **17**(Sept.), 1408–1438.

Frank, W. M. (1977a), The structure and energetics of the tropical cyclone I. Storm structure. *Monthly Weather Review*, **105**, 1119.

Frank, W. M. (1977b), The structure and energetics of the tropical cyclone II. Dynamics and energetics. *Monthly Weather Review*, **105**, 1136.

Friedson, A. J. and J. I. Moses (2012), General circulation and transport in Saturn's upper troposphere and stratosphere. *Icarus*, **218**(Apr.), 861–875.

Friedson, A. J., A.-S. Wong and Y. L. Yung (2002), Models for polar haze formation in Jupiter's stratosphere. *Icarus*, **158**(Aug.), 389–400.

Galopeau, P. H. M. and A. Lecacheux (2000), Variations of Saturn's radio rotation period measured at kilometer wavelengths. *Journal of Geophysical Research*, **105** (June), 13089–13102.

Galperin, B., H. Nakano, H.-P. Huang et al. (2004), The ubiquitous zonal jets in the atmospheres of giant planets and Earth's oceans. *Geophysical Research Letters*, **31**(July), L13303.

Galperin, B., S. Sukoriansky, N. Dikovskaya et al. (2006), Anisotropic turbulence and zonal jets in rotating flows with β-effect. *Nonlinear Processes in Geophysics*, **13** (Apr.), 83–98.

Galperin, B., S. Sukoriansky and H.-P. Huang (2001), Universal n^5 spectrum of zonal flows on giant planets. *Physics of Fluids*, **13**(June), 1545–1548.

Galperin, B., R. M. B. Young, S. Sukoriansky et al. (2014), Cassini observations reveal a regime of zonostrophic macro-turbulence on Jupiter. *Icarus*, **229**(Feb.), 295–320.

Garcia-Melendo, E., S. Pérez-Hoyos, A. Sánchez-Lavega et al. (2011), Saturn's zonal wind profile in 2004–2009 from Cassini ISS images and its long-term variability. *Icarus*, **215**(Sept.), 62–74.

Gehrels, T. (ed.) (1976), *Jupiter: Studies of the Interior, Atmosphere, Magnetosphere and Satellites*. Tucson, AZ: University of Arizona Press.

Gehrels, T. and M. S. Matthews (eds.) (1984), *Saturn*. Tucson, AZ: University of Arizona Press.

Gezari, D. Y., M. J. Mumma, F. Espenak et al. (1989), New features in Saturn's atmosphere revealed by high-resolution thermal infrared images. *Nature*, **342**, 777–780.

Godfrey, D. A. (1988), A hexagonal feature around Saturn's North Pole. *Icarus*, **76**(Nov.), 335–356.

Godfrey, D. A. and V. Moore (1986), The Saturnian ribbon feature: A baroclinically unstable model. *Icarus*, **68** (Nov.), 313–343.

Greathouse, T., J. Moses, L. Fletcher et al. (2010) (May), Seasonal temperature variations in Saturn's stratosphere: Radiative seasonal model vs. observations. Page 4806 of: *EGU General Assembly Conference Abstracts*. EGU General Assembly Conference Abstracts, vol. 12.

Guerlet, S., T. Fouchet, S. Vinatier et al. (2015), Stratospheric benzene and hydrocarbon aerosols detected in Saturn's auroral regions. *Astronomy and Astrophysics*, **580**(Aug.), A89.

Guerlet, S., A. Spiga, M. Sylvestre et al. (2014), Global climate modeling of Saturn's atmosphere. Part I: Evaluation of the radiative transfer model. *Icarus*, **238** (Aug.), 110–124.

Hammel, H. B., L. A. Sromovsky, P. M. Fry et al. (2009), The Dark Spot in the atmosphere of Uranus in 2006: Discovery, description, and dynamical simulations. *Icarus*, **201**(May), 257–271.

Hammel, H. B., M. L. Sitko, D. K. Lynch et al. (2007), Distribution of Ethane and Methane Emission on Neptune. *Astronomical Journal*, **134**(Aug.), 637–641.

Hart, J. E. (1979), Finite amplitude baroclinic instability. *Annual Review of Fluid Mechanics*, **11**, 147–172.

Held, I. M., M. Ting and H. Wang (2002), Northern winter stationary waves: Theory and modeling. *Journal of Climate*, **15**(Aug.), 2125–2144.

Holton, J. R. 2004), *An Introduction to Dynamic Meteorology*. New York, NY: Academic Press.

Houze, R. A. (1993), *Cloud Dynamics*. San Diego, CA: Academic Press.

Huang, H.-P., B. Galperin and S. Sukoriansky (2001), Anisotropic spectra in two-dimensional turbulence on the surface of a rotating sphere. *Physics of Fluids*, **13** (Jan.), 225–240.

Huang, H.-P. and W. A. Robinson (1998), Two-dimensional turbulence and persistent zonal jets in a global barotropic model. *Journal of Atmospheric Sciences*, **55**(Feb.), 611–632.

Iacono, R., M. V. Struglia and C. Ronchi (1999), Spontaneous formation of equatorial jets in freely decaying shallow water turbulence. *Physics of Fluids*, **11**(May), 1272–1274.

Ingersoll, A. P., R. F. Beebe, B. J. Conrath et al. (1984), Structure and dynamics of Saturn's atmosphere. Pages 195–238 in: T. Gehrels, M. S. Matthews (eds.), *Saturn*. Tucson, AZ: University of Arizona Press.

Ingersoll, A. P., T. E. Dowling, P. J. Gierasch et al. (2004), Dynamics of Jupiter's atmosphere. Pages 105–128 in: *Jupiter: The Planet, Satellites and Magnetosphere*. Cambridge: Cambridge University Press.

Ingersoll, A. P., G. Muench, G. Neugebauer et al. (1976), Results of the infrared radiometer experiment on Pioneers 10 and 11. Pages 197–205 of: T. Gehrels (ed.), *Jupiter: Studies of the Interior, Atmosphere, Magnetosphere and Satellites*. Tucson, AZ: University of Arizona Press.

Karkoschka, E. (2015), Uranus' southern circulation revealed by Voyager 2: Unique characteristics. *Icarus*, **250**(Apr.), 294–307.

Karkoschka, E. and M. Tomasko (2005), Saturn's vertical and latitudinal cloud structure 1991 2004 from HST imaging in 30 filters. *Icarus*, **179**(Dec.), 195–221.

Karkoschka, E. and M. G. Tomasko (1993), Saturn's upper atmospheric hazes observed by the Hubble Space Telescope. *Icarus*, **106**, 428–441.

Kàrmàn, Th. V. (1921), Über laminare und turbulente Reibung. *Z. angew. Math. Mech.*, **1**, 233–252.

Kitamura, Y. and K. Ishioka (2007), Equatorial jets in decaying shallow-water turbulence on a rotating sphere. *Journal of Atmospheric Sciences*, **64**, 3340.

Kossin, J. P. and W. H. Schubert (2001), Mesovortices, polygonal flow patterns, and rapid pressure falls in hurricane-like vortices. *Journal of Atmospheric Sciences*, **58**(Aug.), 2196–2209.

Kraichnan, R. H. (1967), Inertial ranges in two-dimensional turbulence. *Physics of Fluids*, **10**(July), 1417–1423.

Kraichnan, R. H. (1971), Inertial-range transfer in two- and three-dimensional turbulence. *Journal of Fluid Mechanics*, **47**, 525–535.

Kuo, H.-L. (1949), Dynamic instability of two-dimensional non-divergent flow in a barotropic atmosphere. *Journal of Atmospheric Sciences*, **6**(Apr.), 105–122.

Lebeau, R. P. and T. E. Dowling (1998), EPIC simulations of time-dependent, three-dimensional vortices with application to Neptune's Great Dark SPOT. *Icarus*, **132** (Apr.), 239–265.

Lewis, B. M. and H. F. Hawkins (1982), Polygonal eye walls and rainbands in hurricanes. *Bulletin of the American Meteorological Society*, **63**(Nov.), 1294–1301.

Limaye, S. S. (1986), Jupiter: New estimates of the mean zonal flow at the cloud level. *Icarus*, **65**(Mar.), 335–352.

Limaye, S. S., J. P. Kossin, C. Rozoff et al. (2009), Vortex circulation on Venus: Dynamical similarities with terrestrial hurricanes. *Geophysical Reesarch Letters*, **36** (Feb.), 4204.

Lindal, G. F., D. N. Sweetnam and V. R. Eshleman (1985), The atmosphere of Saturn: An analysis of the Voyager radio occultation measurements. *Astronomical Journal*, **90**(June), 1136–1146.

Liu, J. and T. Schneider (2010), Mechanisms of jet formation on the giant planets. *Journal of Atmospheric Sciences*, **67**(Nov.), 3652–3672.

Liu, J., T. Schneider and L. N. Fletcher (2014), Constraining the depth of Saturn's zonal winds by measuring thermal and gravitational signals. *Icarus*, **239**(Sept.), 260–272.

Luszcz-Cook, S. H., I. de Pater, M. Adamkovics et al. (2010), Seeing double at Neptune's south pole. *Icarus*, **208**(Aug.), 938–944.

Marcus, P. S. and C. Lee (1998), A model for eastward and westward jets in laboratory experiments and planetary atmospheres. *Physics of Fluids*, **10**(June), 1474–1489.

Moller, J. D. and M. T. Montgomery (2000), Tropical cyclone evolution via potential vorticity anomalies in a three-dimensional balance model. *Journal of Atmospheric Sciences*, **57**(Oct.), 3366–3387.

Montgomery, M. T. and J. Enagonio (1998), Tropical cyclogenesis via convectively forced vortex Rossby waves in a three-dimensional quasigeostrophic model. *Journal of Atmospheric Sciences*, **55**(Oct.), 3176–3207.

Montgomery, M. T. and R. J. Kallenbach (1997), A theory for vortex Rossby-waves and its application to spiral bands and intensity changes in hurricanes. *Quarterly Journal of the Royal Meteorological Society*, **123**(Jan.), 435–465.

Montgomery, M. T., J. D. Moller and C. T. Nicklas (1999), Linear and nonlinear vortex motion in an asymmetric balance shallow water model. *Journal of Atmospheric Sciences*, **56**(Mar.), 749–768.

Morales-Juberìas, R., K. M. Sayanagi, T. E. Dowling et al. (2011), Emergence of polar-jet polygons from jet instabilities in a Saturn model. *Icarus*, **211**(Feb.), 1284–1293.

Morales-Juberìas, R., K. M. Sayanagi, A. A. Simon et al. (2015), Meandering shallow atmospheric jet as a model of Saturn's north-polar hexagon. *Astrophysical Journal Letters*, **806**(June), L18.

Moses, J. I., T. Fouchet, R. V. Yelle et al. (2004), The stratosphere of Jupiter. Pages 129–157 of: F. Bagenal, T. E. Dowling and W. B. McKinnon (eds.), *Jupiter. The Planet, Satellites and Magnetosphere*. Cambridge: Cambridge University Press.

Moses, J. I. and T. K. Greathouse (2005), Latitudinal and seasonal models of stratospheric photochemistry on Saturn: Comparison with infrared data from IRTF/TEXES. *Journal of Geophysical Research (Planets)*, **110**(E9), 9007.

Nash, E. R., P. A. Newman, J. E. Rosenfield et al. (1996), An objective determination of the polar vortex using Ertel's potential vorticity. *Journal of Geophysical Research*, **101**(Apr.), 9471–9478.

Nolan, D. S. and B. F. Farrell (1999), The intensification of two-dimensional swirling flows by stochastic asymmetric forcing. *Journal of Atmospheric Sciences*, **56** (Dec.), 3937–3962.

Nozawa, T. and S. Yoden (1997), Formation of zonal band structure in forced two-dimensional turbulence on a rotating sphere. *Physics of Fluids*, **9**(July), 2081–2093.

Nycander, J. (1993), The difference between monopole vortices in planetary flows and laboratory experiments. *Journal of Fluid Mechanics*, **254**, 561–577.

Okuno, A. and A. Masuda (2003), Effect of horizontal divergence on the geostrophic turbulence on a beta-plane: Suppression of the Rhines effect. *Physics of Fluids*, **15** (Jan.), 56–65.

O'Neill, M. E., K. A. Emanuel and G. R. Flierl (2015), Polar vortex formation in giant-planet atmospheres due to moist convection. *Nature Geoscience*, **8**(July), 523–526.

O'Neill, M. E., K. A. Emanuel and G. R. Flierl (2016), Weak jets and strong cyclones: Shallow-water modeling of giant planet polar caps. *Journal of Atmospheric Sciences*, **73**(Apr.), 1841–1855.

Orton, G., B. Conrath, R. Achterberg et al. (2003) (Apr.), The exploration of Jupiter's Arctic Polar Vortex by NASA IRTF and Cassini CIRS observations. Pages 6819–+ of: *EGS -AGU – EUG Joint Assembly*.

Orton, G. S., T. Encrenaz, C. Leyrat et al. (2007), Evidence for methane escape and strong seasonal and dynamical perturbations of Neptune's atmospheric temperatures. *Astronomy and Astrophysics*, **473**(Oct.), L5–L8.

Orton, G. S., L. N. Fletcher, J. Liu et al. (2012), Recovery and characterization of Neptune's near-polar stratospheric hot spot. *Planetary and Space Science*, **61**(Feb.), 161–167.

Orton, G. S. and A. P. Ingersoll (1976), Pioneer 10 and 11 and ground-based infrared data on Jupiter: The thermal structure and He-H2 ratio. Pages 206–215 of: T. Gehrels (ed.), *Jupiter: Studies of the Interior, Atmosphere, Magnetosphere and Satellites*. Tucson, AZ: University of Arizona Press.

Orton, G. S. and P. A. Yanamandra-Fisher (2005), Saturn's temperature field from high-resolution middle-infrared imaging. *Science*, **307**(Feb.), 696–698.

Patsaeva, M. V., I. V. Khatuntsev, D. V. Patsaev et al. (2015), The relationship between mesoscale circulation and cloud morphology at the upper cloud level of Venus from VMC/Venus Express. *Planetary and Space Science*, **113**(Aug.), 100–108.

Penny, A. B., A. P. Showman and D. S. Choi (2010), Suppression of the Rhines effect and the location of vortices on Saturn. *Journal of Geophysical Research (Planets)*, **115**(Feb.), 2001.

Pérez-Hoyos, S. and A. Sánchez-Lavega (2006), On the vertical wind shear of Saturn's equatorial jet at cloud level. *Icarus*, **180**(Jan.), 161–175.

Pérez-Hoyos, S., A. Sánchez-Lavega, R. G. French et al. (2005), Saturn's cloud structure and temporal evolution from ten years of Hubble Space Telescope images (1994 – 2003). *Icarus*, **176**(July), 155–174.

Piccioni, G., P. Drossart, A. Sánchez-Lavega et al. (2007), South-polar features on Venus similar to those near the north pole. *Nature*, **450**(Nov.), 637–640.

Polavarapu, S. M. and W. R. Peltier (1993), The structure and nonlinear evolution of synoptic-scale cyclones. Part II: Wave-mean flow interaction and asymptotic equilibration. *Journal of Atmospheric Sciences*, **50** (Sept.), 3164–3184.

Porco, C. C., E. Baker, J. Barbara et al. (2005), Cassini imaging science: Initial results on Saturn's atmosphere. *Science*, **307**(Feb.), 1243–1247.

Porco, C. C., R. A. West, A. McEwen et al. (2003), Cassini imaging of Jupiter's atmosphere, satellites, and rings. *Science*, **299**(Mar.), 1541–1547.

Porco, C. C., R. A. West, S. Squyres et al. (2004), Cassini imaging science: Instrument characteristics and anticipated scientific investigations at Saturn. *Space Science Reviews*, **115**(Dec.), 363–497.

Rages, K. A. and E. L. Barth (2012) (Oct.), Saturn limb hazes as seen from Cassini. Page #500.05 of: *AAS/Division for Planetary Sciences Meeting Abstracts*. AAS/Division for Planetary Sciences Meeting Abstracts, vol. 44.

Read, P. L., B. J. Conrath, L. N. Fletcher et al. (2009a), Mapping potential vorticity dynamics on Saturn: Zonal mean circulation from Cassini and Voyager data. *Planetary and Space Science*, **57**(Dec.), 1682–1698.

Read, P. L., T. E. Dowling and G. Schubert (2009b), Saturn's rotation period from its atmospheric planetary-wave configuration. *Nature*, **460**(July), 608–610.

Read, P. L., P. J. Gierasch, B. J. Conrath et al. (2006), Mapping potential-vorticity dynamics on Jupiter. I: Zonal-mean circulation from Cassini and Voyager 1 data. *Quarterly Journal of the Royal Meteorological Society*, **132**(July), 1577–1603.

Read, P. L., T. N. L. Jacoby, P. H. T. Rogberg et al. (2015), An experimental study of multiple zonal jet formation in rotating, thermally driven convective flows on a topographic beta-plane. *Physics of Fluids*, **27**(8), 085111.

Read, P. L., Y. H. Yamazaki, S. R. Lewis et al. (2007), Dynamics of convectively driven banded jets in the laboratory. *Journal of Atmospheric Sciences*, **64**, 4031–+.

Reznik, G. M. (1992), Dynamics of singular vortices on a beta-plane. *Journal of Fluid Mechanics*, **240**(July), 405–432.

Rhines, P. B. (1975), Waves and turbulence on a beta-plane. *Journal of Fluid Mechanics*, **69**, 417–443.

Rhines, P. B. (1979), Geostrophic turbulence. *Annual Review of Fluid Mechanics*, **11**, 401–441.

Rhines, P. B. (1994), Jets. *Chaos*, **4**(June), 313–339.

Roman, M. T., D. Banfield and P. J. Gierasch (2013), Saturn's cloud structure inferred from Cassini ISS. *Icarus*, **225** (Apr.), 93–110.

Sai-Lap Lam, J. and D. G. Dritschel (2001), On the beta-drift of an initially circular vortex patch. *Journal of Fluid Mechanics*, **436**(June), 107–129.

Salmon, R. (1980), Baroclinic instability and geostrophic turbulence. *Geophysical and Astrophysical Fluid Dynamics*, **15**, 167–211.

Salmon, R. (1998), *Geophysical Fluid Dynamics*. Oxford: Oxford University Press.

Salyk, C., A. P. Ingersoll, J. Lorre et al. (2006), Interaction between eddies and mean flow in Jupiter's atmosphere: Analysis of Cassini imaging data. *Icarus*, **185**(Dec.), 430–442.

Sánchez-Lavega, A. (2011), *An Introduction to Planetary Atmospheres* Boca Raton, FL: Taylor and Francis.

Sánchez-Lavega, A., R. Hueso, S. Pérez-Hoyos et al. (2006), A strong vortex in Saturn's South Pole. *Icarus*, **184** (Oct.), 524–531.

Sánchez-Lavega, A., R. Hueso and J. Ramon Acarreta (1998), A system of circumpolar waves in Jupiter's stratosphere. *Geophysical Research Letters*, **25**, 4043–4046.

Sánchez-Lavega, A., J. Lecacheux, F. Colas et al. (1993), Ground-based observations of Saturn's north polar SPOT and hexagon. *Science*, **260**(Apr.), 329–332.

Sánchez-Lavega, A., S. Pérez-Hoyos, J. R. Acarreta et al. (2002), No hexagonal wave around Saturn's southern pole. *Icarus*, **160**(Nov.), 216–219.

Sánchez-Lavega, A., T. Rìo-Gaztelurrutia, R. Hueso et al. (2014), The long-term steady motion of Saturn's hexagon and the stability of its enclosed jet stream under seasonal changes. *Geophysical Research Letters*, **41** (Mar.), 1425–1431.

Sánchez-Lavega, A., J. F. Rojas, J. R. Acarreta et al. (1997), New observations and studies of Saturn's long-lived north polar SPOT. *Icarus*, **128**(Aug.), 322–334.

Sánchez-Lavega, A., J. F. Rojas and P. V. Sada (2000), Saturn's zonal winds at cloud level. *Icarus*, **147**(Oct.), 405–420.

Sayanagi, K. M., J. J. Blalock, U. A. Dyudina et al. (2017), Cassini ISS observation of Saturn's north polar vortex and comparison to the south polar vortex. *Icarus*, **285** (Mar.), 68–82.

Sayanagi, K. M., R. Morales-Juberìas, J. J. Blalock et al. (2015) (Nov.), Effect of the 77 degree N Jet on Saturn's Hexagon Cloud Morphology. Page 311.18 of: *AAS/Division for Planetary Sciences Meeting Abstracts*. AAS/Division for Planetary Sciences Meeting Abstracts, vol. 47.

Sayanagi, K. M., R. Morales-Juberìas and A. P. Ingersoll (2010), Saturn's northern hemisphere ribbon: Simulations and comparison with the meandering Gulf Stream. *Journal of Atmospheric Sciences*, **67** (Aug.), 2658–2678.

Sayanagi, K. M., A. P. Showman and T. E. Dowling (2008), The emergence of multiple robust zonal jets from freely evolving, three-dimensional stratified geostrophic turbulence with applications to Jupiter. *Journal of Atmospheric Sciences*, **65**, 1947–1962.

Schubert, W. H., M. T. Montgomery, R. K. Taft et al. (1999), Polygonal eyewalls, asymmetric eye contraction, and potential vorticity mixing in hurricanes. *Journal of Atmospheric Sciences*, **56**(May), 1197–1223.

Scott, R. K. (2010), The structure of zonal jets in shallow water turbulence on the sphere. Pages 243–252 of: D. G. Dritschel (ed.), *Proceedings of the IUTAM symposium on Rotating Stratified Turbulence*.

Scott, R. K. (2011), Polar accumulation of cyclonic vorticity. *Geophysical and Astrophysical Fluid Dynamics*, **105** (Aug.), 409–420.

Scott, R. K. and L. M. Polvani (2007), Forced-dissipative shallow-water turbulence on the sphere and the atmospheric circulation of the giant planets. *Journal of Atmospheric Sciences*, **64**, 3158–3176.

Scott, R. K. and A.-S. Tissier (2012. The generation of zonal jets by large-scale mixing. *Physics of Fluids*, **24**(12), 126601.

Seidelmann, P. K., B. A. Archinal, M. F. A'Hearn et al. (2007), Report of the IAU/IAG Working Group on cartographic co-ordinates and rotational elements: 2006. *Celestial Mechanics and Dynamical Astronomy*, **98**(July), 155–180.

Showman, A. P. (2007), Numerical simulations of forced shallow-water turbulence: Effects of moist convection on the large-scale circulation of Jupiter and Saturn. *Journal of Atmospheric Sciences*, **64**(Sept.), 3132–3157.

Simon, A. A., M. H. Wong, J. H. Rogers et al. (2014), Dramatic change in Jupiter's Great Red Spot from spacecraft observations. *Astrophysical Journal Letters*, **797**(Dec.), L31.

Simon-Miller, A. A., P. J. Gierasch, R. F. Beebe et al. (2002), New observational results concerning Jupiter's Great Red Spot. *Icarus*, **158**(July), 249–266.

Sinclair, J. A., P. G. J. Irwin, L. N. Fletcher et al. (2013), Seasonal variations of temperature, acetylene and ethane in Saturn's atmosphere from 2005 to 2010, as observed by Cassini-CIRS. *Icarus*, **225**(July), 257–271.

Slavin, A. G. and Y. D. Afanasyev (2012), Multiple zonal jets on the polar beta plane. *Physics of Fluids*, **24**(1), 016603–016603.

Smith, C. A., K. G. Speer and R. W. Griffiths (2014), Multiple zonal jets in a differentially heated rotating annulus*,+. *Journal of Physical Oceanography*, **44** (Sept.), 2273–2291.

Smith, K. S. (2004), A local model for planetary atmospheres forced by small-scale convection. *Journal of Atmospheric Sciences*, **61**(June), 1420–1433.

Solomon, T. H., W. J. Holloway and H. L. Swinney (1993), Shear flow instabilities and Rossby waves in barotropic flow in a rotating annulus. *Physics of Fluids A*, **5**(Aug.), 1971–1982.

Sommeria, J., S. D. Meyers and H. L. Swinney (1989), Laboratory model of a planetary eastward jet. *Nature*, **337**(Jan.), 58–61.

Srinivasan, K. and W. R. Young (2012), Zonostrophic Instability. *Journal of Atmospheric Sciences*, **69**(May), 1633–1656.

Sromovsky, L. A., I. de Pater, P. M. Fry et al. (2015), High S/N Keck and Gemini AO imaging of Uranus during 2012–2014: New cloud patterns, increasing activity, and improved wind measurements. *Icarus*, **258**(Sept.), 192–223.

Sromovsky, L. A., P. M. Fry, T. E. Dowling et al. (2001), Coordinated 1996 HST and IRTF imaging of Neptune and Triton. III. Neptune's atmospheric circulation and cloud structure. *Icarus*, **149**(Feb.), 459–488.

Sromovsky, L. A., P. M. Fry, H. B. Hammel et al. (2009), Uranus at equinox: Cloud morphology and dynamics. *Icarus*, **203**(Sept.), 265–286.

Sromovsky, L. A., P. M. Fry, H. B. Hammel (2012b), Post-equinox dynamics and polar cloud structure on Uranus. *Icarus*, **220**(Aug.), 694–712.

Sromovsky, L. A., H. B. Hammel, I. de Pater et al. (2012a), Episodic bright and dark spots on Uranus. *Icarus*, **220** (July), 6–22.

Sromovsky, L. A., H. E. Revercomb, R. J. Krauss et al. (1983), Voyager 2 observations of Saturn's northern mid-latitude cloud features: Morphology, motions, and evolution. *Journal of Geophysical Research*, **88**(Nov.), 8650–8666.

Stratman, P. W., A. P. Showman, T. E. Dowling et al. (2001), EPIC simulations of bright companions to Neptune's Great Dark Spots. *Icarus*, **151**(June), 275–285.

Sukoriansky, S., N. Dikovskaya and B. Galperin (2007), On the arrest of inverse energy cascade and the Rhines scale. *Journal of Atmospheric Sciences*, **64**, 3312–3327.

Sutyrin, G. G., J. S. Hesthaven, J. P. Lynov et al. (1994), Dynamical properties of vortical structures on the beta-plane. *Journal of Fluid Mechanics*, **268**, 103–131.

Sutyrin, G. G. and Y. G. Morel (1997), Intense vortex motion in a stratified fluid on the beta-plane: An analytical theory and its validation. *Journal of Fluid Mechanics*, **336**(Apr.), 203–220.

Swanson, K. and R. T. Pierrehumbert (1994), Nonlinear wave packet evolution on a baroclinicaily unstable jet. *Journal of Atmospheric Sciences*, **51**(Feb.), 384–396.

Theiss, J. (2004), Equatorward energy cascade, critical latitude, and the predominance of cyclonic vortices in geostrophic turbulence. *Journal of Physical Oceanography*, **34**, 1663–1678.

Theiss, J. (2006), A generalized Rhines effect and storms on Jupiter. *Geophysical Research Letters*, **33**(Apr.), 8809.

Tomasko, M. G., R. A. West, G. S. Orton et al. (1984), *Clouds and aerosols in Saturn's atmosphere*. Pages 150–194 in: T. Gehrels and M. S. Matthews (eds.), *Saturn*. Tucson, AZ: University of Arizona Press.

Vallis, M. E. (2006), *Atmospheric and Oceanic Fluid Dynamics: Fundamentals and Large-scale Circulation*. Cambridge: Cambridge University Press.

Vasavada, A. R., S. M. Hörst, M. R. Kennedy et al. (2006), Cassini imaging of Saturn: Southern hemisphere winds and vortices. *Journal of Geophysical Research (Planets)*, **111**(E10), E05004.

Vasavada, A. R. and A. P. Showman (2005), Jovian atmospheric dynamics: An update after Galileo and Cassini. *Reports of Progress in Physics*, **68**(Aug.), 1935–1996.

Verhoeven, J. and S. Stellmach (2014), The compressional beta effect: A source of zonal winds in planets? *Icarus*, **237**(July), 143–158.

Wang, Y. (2002), Vortex Rossby waves in a numerically simulated tropical cyclone. Part II: The role in tropical cyclone structure and intensity changes. *Journal of Atmospheric Sciences*, **59**(Apr.), 1239–1262.

West, R. A. (2014), The UV hazes of Staturn. In: Asia-Oceania GeoSciences Meeting, Sapporo, Japan. Asia-Oceania GeoSciences Meeting, Sapporo, Japan.

West, R. A., K. H. Baines, A. J. Friedson et al. (2004), Jovian clouds and haze. Pages 79–104 of: F. Bagenal, T. E. Dowling and W. B. McKinnon (eds.), *Jupiter. The Planet, Satellites and Magnetosphere*. Cambridge: Cambridge University Press.

West, R. A., K. H. Baines, E. Karkoschka et al. (2009), Clouds and aerosols in Saturn's atmosphere. Page 161 of: M. K. Dougherty, L. W. Esposito and S. M. Krimigis (eds.), *Saturn from Cassini-Huygens*. Springer.

West, R. A., K. H. Baines and J. B. Pollack (1991), Clouds and aerosols in the Uranian atmosphere. Pages 296–326 of: J. T. Bergstralh, E. D. Miner and M. S. Matthews (eds.), *Uranus*. Tucson, AZ: University of Arizona Press.

West, R. A. and P. H. Smith (1991), Evidence for aggregate particles in the atmospheres of Titan and Jupiter. *Icarus*, **90**(Apr.), 330–333.

West, R. A., P. A. Yanamandra-Fisher and V. Korokhin (2015), Gas giant planets, Saturn's rings, and Titan. In: A.-C. Levasseur-Regourd (ed.), *Polarimetry of Stars and Planetary Systems*. Cambridge: Cambridge University Press.

Williams, G. P. (2003), Jovian dynamics. Part III: Multiple, migrating, and equatorial jets. *Journal of Atmospheric Sciences*, **60**(May), 1270–1296.

Wong, A.-S., A. Y. T. Lee, Y. L. Yung et al. (2000), Jupiter: Aerosol chemistry in the polar atmosphere. *Astrophysical Journal Letters*, **534**(May), L215–L217.

Wong, A.-S., Y. L. Yung and A. J. Friedson (2003), Benzene and haze formation in the polar atmosphere of Jupiter. *Geophysical Research Letters*, **30**(Apr.), 1447.

Yoden, S., K. Ishioka, Y.-Y. Hayashi et al. (1999), A further experiment on two-dimensional decaying turbulence on a rotating sphere. *Nuovo Cimento C Geophysics Space Physics C*, **22**(Nov.), 803–812.

Yoden, S. and M. Yamada (1993), A Numerical Experiment on two-dimensional decaying turbulence on a rotating sphere. *Journal of Atmospheric Sciences*, **50**(Feb.), 631–644.

Zhang, Y. and Y. D. Afanasyev (2014), Beta-plane turbulence: Experiments with altimetry. *Physics of Fluids*, **26** (2), 026602.

13

The Great Saturn Storm of 2010–2011

AGUSTÍN SÁNCHEZ-LAVEGA, GEORG FISCHER, LEIGH N. FLETCHER, ENRIQUE GARCÍA-MELENDO, BRIGETTE HESMAN, SANTIAGO PÉREZ-HOYOS, KUNIO M. SAYANAGI AND LAWRENCE A. SROMOVSKY

Abstract

In December 2010, a major storm erupted in Saturn's northern hemisphere near 37° planetographic latitude. This rather surprising event, occurring at an unexpected latitude and time, is the sixth "Great White Spot" (GWS) storm observed over the last century and a half. Such GWS events are extraordinary, planetary-scale atmospheric phenomena that dramatically change the typically bland appearance of the planet. Occurring while the Cassini mission was on orbit at Saturn, the Great Storm of 2010–2011 was well suited for intense scrutiny by the suite of sophisticated instruments onboard the Cassini spacecraft as well by modern instrumentation on ground-based telescopes and onboard the Hubble Space Telescope. This GWS erupted on 5 December close to the peak of a westward jet and generated a major dynamical disturbance that affected the whole latitude band from 25° to 48°N. At the upper cloud level, following the rapid growth of the bright outbreak spot, a blunt aerodynamic-shaped head formed due to interaction of the spot with the westward zonal jet, with the winds reaching velocities of 160 m s^{-1} along the periphery of the arc. Eastward of the head, the disturbance progressed in the following months forming a turbulent wake or tail with growing vortices, one of them a major enduring anticyclone (called AV) with a size of ~11,000 km. Lightning events were prominent and detected as outbursts and flashes at the head and along the disturbance at both optical and radio wavelengths. The activity of the head ceased after about seven months when AV reached it, leaving the cloud structure and ambient winds perturbed. The tops of the optically dense clouds of the head reached the 300-mbar altitude level (~50 km below tropopause), where a mixture of ices was detected, including (1) a component of water ice lofted over 200 km altitude from its 10-bar condensation level, (2) ammonia ice as the predominant component and (3) a component that might be ammonium hydrogen sulfide ice. The energetics of the frequency and power of lightning, as well as the estimated power generated by the latent heat released in the water-based convection to create the observed dynamical three-dimensional flows, both indicate that the power released for much of the 7-month lifetime of the storm (~10^{17} Watts) was a significant fraction of Saturn's total radiated power (~2.2 10^{17} W). A post-storm depletion of ammonia vapour was also measured in the upper troposphere. The effects of the storm propagated into the stratosphere, forming two warm air masses at the ~0.5- to 5-mbar pressure level altitude that later merged into a so-called "beacon" because of its 80 K temperature excess relative to its surroundings. Related to the stratospheric disturbance, hydrocarbon composition excesses were found, in particular for ethylene (C_2H_4), in the high stratosphere at the ~0.1- to 0.5-mbar altitude level. Numerical models of the storm dynamics explain the major observed features that essentially result from two processes: (1) a huge and sustained, moist, convective storm at the water clouds (altitude level 10–12 bar, or ~250–275 km below the tropopause) and (2) the interaction of the updraft columns with the ambient winds that generates the turbulent wake consisting of vortices and waves. Model simulations of the GWS require a low vertical shear of the zonal winds and low static stability across the weather layer where the disturbance develops. Its upward propagation into the stratosphere involves Rossby waves and their breaking and energy deposition to form the beacon and induce chemical changes.

The decades-long interval between storms is probably related to the insolation cycle and the long radiative time constant of Saturn's atmosphere, and several theories for temporarily storing energy have been proposed.

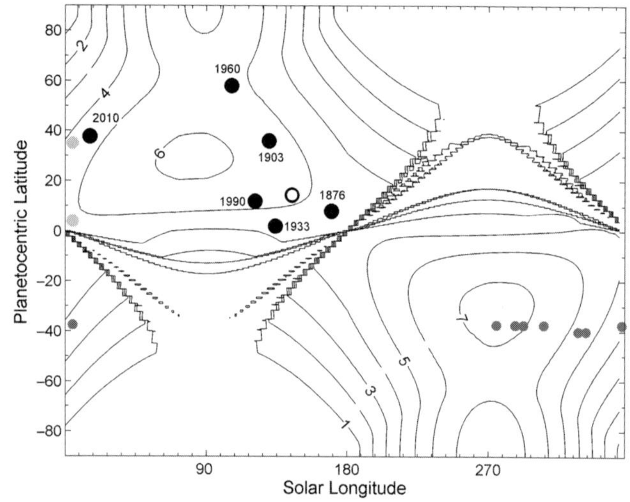

Figure 13.1 Seasonal insolation cycle at the top of Saturn's atmosphere and location in latitude and solar longitude of the major planetary scale storms (Great White Spots, GWS) and other small regular convective events: (1) lines: heat flux in W m^{-2}; (2) solid dots: GWS (years indicated), and the empty circle is the 1994 case; (3) small grey dots: small convective events (2002–2010); (4) light grey dots: small convective events (1981–1982, Voyager 1 and 2). See Pérez-Hoyos and Sánchez-Lavega (2006).

13.1 Introduction

The Great Saturn Storm phenomenon that took place in 2010–2011 is the most recent example of a type of huge and rare planet-encircling disturbance that occurs on this planet, a phenomenon nicknamed as a Great White Spots (GWS) event (Sánchez-Lavega 1982). The 2010 GWS was the sixth recorded event in the last 140 years (Sánchez-Lavega 1994; Sánchez-Lavega et al. 2011). The GWS phenomenon is easily recognizable because of the prominent brightness of the outbreak spot at visible wavelengths in a usually bland disk, and because of its size and rapid growth and expansion in longitudinal direction (Sánchez-Lavega 1994). These disturbances have appeared in intervals of approximately 30 years at different latitudes and always in the northern hemisphere: at the equator (years 1876, 1933 and 1990), at mid-latitudes (1903, 2010) and at a subpolar latitude (1960) (Figure 13.1). Because of the 26.7° tilt of the rotation axis of Saturn relative to its orbital plane, the North-South hemispheric visibility of the planet from the Earth changes along Saturn's 29.457-year orbital cycle. This visibility is further reduced due to the shadow of the rings on the globe, and thus full hemispheric visibility occurs during short periods of about 1 to 2 years coincident with near-simultaneous ring-plane crossings by the Sun and Earth-based observer. This means that the GWS statistics and perceived periodicity can be biased by such sampling effects; nevertheless, it is well established that all the recorded events have occurred in the Northern hemisphere. Thus, any conclusion of their possible seasonal origin must be regarded with caution.

Saturn's GWS represent a singular phenomenon in planetary atmospheres, similar in some aspects to the other planetary-scale disturbances that occur in Jupiter's South Equatorial Belt (SEB Disturbances, Sánchez-Lavega and Gomez 1996a) and temperate latitudes (NTB Disturbance Sánchez-Lavega et al. 2008). The GWSs represent a class of meteorological phenomena that can potentially reveal new insights into the vigorous processes of moist convection in a hydrogen-dominated atmosphere, very different from that of the Earth. As well, they allow the study of dynamic phenomena associated with the interaction of storms with a surrounding atmosphere dominated by zonal winds and its propagation along a whole latitude band, eventually encircling the planet. Finally, they give clues on the poorly known structure of Saturn's atmosphere below the upper hazes and clouds, in the weather layer where thermochemical models predict active cloud formation.

In this chapter we review the observations and theoretical studies so far performed on these giant storms. We review the GWS phenomena, including the historical cases, but we mainly focus on the well-studied event of 2010–2011 on which were trained the full suite of Cassini instruments as well as numerous advanced ground-based telescopic and Hubble Space Telescope instruments.

13.2 Review of the Historical "Great White Spot" Phenomenon

Until the adoption of CCD technology into ground-based telescopic observations during the mid-1980s, the detection of discrete atmospheric features in Saturn's atmosphere was a very difficult task due to their meagre contrast in the hazy atmosphere. Sánchez-Lavega (1982) summarized the small number of "spots" tracked on the planet from the time of W. Herschel (1793–1794) until the Voyager 1 and 2 encounters in 1980–1981. Of all these features, the most conspicuous ones were those characterized by their rapid appearance, high brightness, growth and expansion to a planetary-scale disturbance, and by their several-month lifetimes, becoming known as GWS events. To date there have been six GWSs observed, as summarized in Table 13.1. The data forthcoming from observations of these global-scale phenomena were scarce until the 1990 GWS. For the four previous cases (1876, 1903, 1933, 1960), details were captured mainly in drawings showing their temporal evolution, except for the best case studied, i.e. that of 1933, for which there is a photographic series obtained at various times and wavelengths (Sánchez-Lavega 1982; Sánchez-Lavega and Battaner 1987). For the great equatorial event of 1990, CCD images were obtained from ground-based telescopes, and the Hubble Space Telescope acquired several good imaging series, making it the best-studied case until the 2010–2011 event (Sánchez-Lavega et al. 1991; Barnet et al. 1992; Beebe et al. 1992; Westphal et al. 1992). Figure 13.2 shows a mosaic of the appearance of the GWS for these five historical cases.

In essence, the GWS phenomenon involves three stages of development (Sánchez-Lavega 1994): (1) the outbreak of a small but very bright "spot" that represents the onset phase; (2) the rapid growth and horizontal expansion of this "spot" into the full GWS feature, with divergence in the forming ammonia cloudtops near the ~1-bar level, reaching ~10,000–20,000 km in size in about ten days; and (3) the planetary-scale mature disturbance phase, when the convective source, remaining active for months, interacts with the wind-sheared atmosphere. The dynamics of the mature disturbance strongly depends on the latitude of the outbreak and the meridional wind profile in its surroundings.

The source spot is brighter than its surrounding clouds over the entire visible spectrum, from ~380 nm^{-1} μm, as observed in spectral observations of the events of 1933, 1990 and 2010 (Sánchez-Lavega and Battaner 1987; Sánchez-Lavega et al. 1991; Westphal et al. 1992; Acarreta and Sánchez-Lavega 1999; Sánchez-Lavega et al. 2011; Sanz-Requena et al. 2012; Section 13.4). The high brightness and spectral behavior is consistent with high cloud tops formed by fresh, mildly polluted ices (Section 13.4). For these three cases (1933, 1990, 2010), it has been possible to measure the expanding velocity of the initial spot $V \sim$ 30–50 m s^{-1} and its area growing rate with maximum values $dA/dt(max)$ ~200 km^2 s^{-1}. Both facts suggest that the source is driven by vigorous moist convection starting at the deep water clouds at level ~10 bars, or ~250 km below the visible cloud tops (Sánchez-Lavega and Battaner 1987; Hueso and Sánchez-Lavega 2004) (see Section 13.7). This hypothesis has been recently supported by the observation of lightning events in the optical and radio wavelengths in the 2010–2011 GWS (Fischer et al. 2011a; Section 13.6) and the detection of water ice at the clouds tops (Sromovsky et al. 2013; Section 13.4).

The evolution from the localized eruption to the mature stage that constitutes the planetary-scale disturbance is dictated essentially by the structure of the ambient zonal wind at the outbreak latitude (Figure 13.3). The GWS evolution has been studied in some detail for the 1990 case using ground-based telescopes and the Hubble Space Telescope (Sánchez-Lavega et al. 1991; Beebe et al. 1992; Barnet et al. 1992) and in great detail for the 2010 case using a variety of ground-based telescopes and instruments onboard the Cassini spacecraft (Fletcher et al. 2011;

Table 13.1 *Great White Spots' basic properties*

Event	First detection	Orbital longitude	Latitude:planetographic (planetocentric)	Velocity ($m\ s^{-1}$)	Affected latitude band	Ambient vorticity (s^{-1})	Planetary vorticity (s^{-1})
GWS 1876	1876.9	170°	8°N ± 3° (6.5°)	396	~0° – 20°N	-4×10^{-6}	4.6×10^{-5}
GWS 1903	1903.5	130°	36°N ± 2° (30.6°)	19	~30° – 45°N	-7×10^{-6}	1.8×10^{-4}
GWS 1933	1933.7	134°	2°N ± 3° (1.6°)	400	~–5° – 20°N	2.0×10^{-5}	1.1×10^{-5}
GWS 1960	1960.25	106°	58°N ± 1° (52.5°)	4	48°N – 60°N [78°N]*	10^{-5}	2.8×10^{-4}
GWS 1990	1990.9	121°	12°N ± 1° (9.8°)[5°N ± 2°]† (4.1°)	365.0 [402.0]†	15°S – 25°N	-4×10^{-5} [2.0×10^{-6}]†	6×10^{-5}
GWS 2010	2010.93	16°	38°N (32.4°)[41°N ± 1°]‡ (35.2°)	–27.8	25°N – 48°N	3×10^{-6}	1.9×10^{-4}

Notes: Data refer to the center of the main spot (usually known as the GWS storm head) that is assumed to be the convective source. The latitude of a point on an equipotential surface of an oblate planet is defined by the angle between a line through the point and the equatorial plane. It is planetocentric if the line intersects the planet's center. It is planetographic if the line is normal to the equipotential surface. For the ambient vorticity (meridional shear of the zonal wind), the positive sign means anticyclonic and the negative cyclonic. See Figure 13.2. The planetary vorticity is given by the Coriolis parameter f. * The 1960 GWS clouds expanded meridionally up to ~80°N after a month of its outbreak. † The onset of the 1990 storm took place at 12°N, but after two weeks the stormhead center migrated towards the equator. ‡ The onset of the 2010 GWS took place at a westward jet but the head center migrated towards the north.

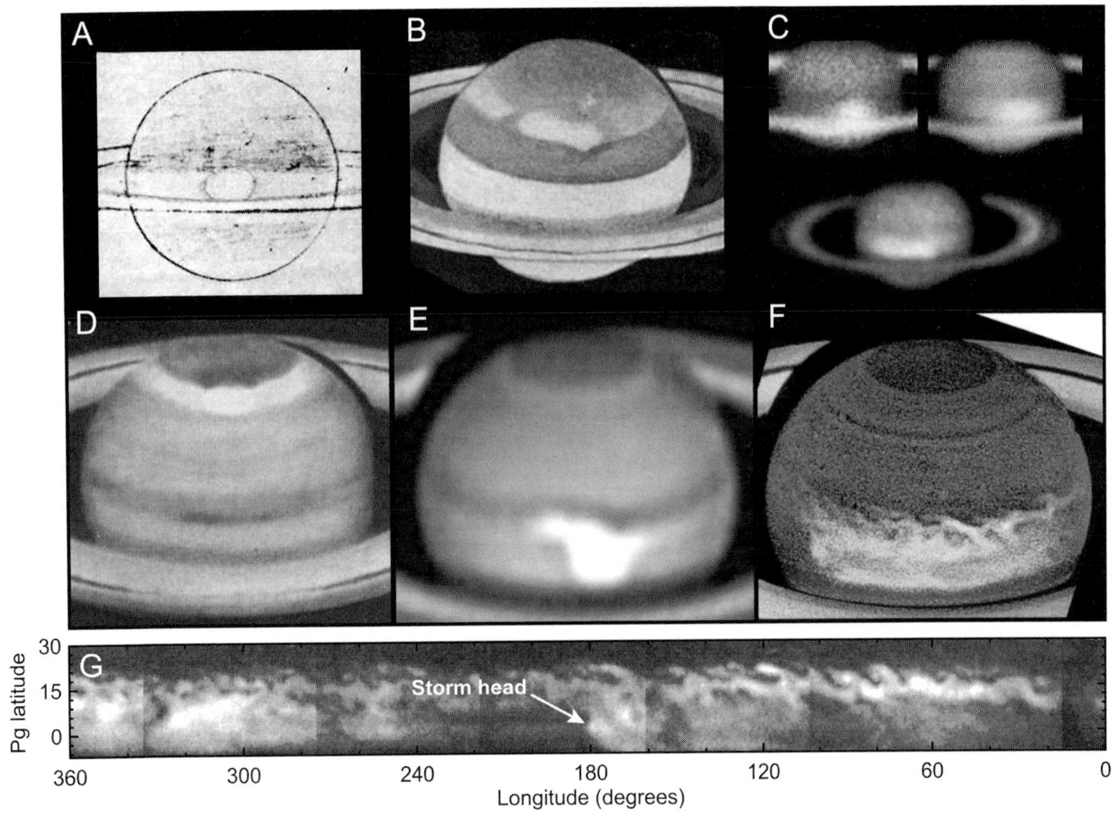

Figure 13.2 Mosaic of historical Great White Spots: (A) GWS 1876 (10 December), drawing by T. W. Webb published in *Sky and Telescope* (January 1974); (B) GWS 1903 (16 July), drawing by W. F. Denning published in *Splendour of the Heavens* (1925); (C) GWS 1933, upper left photograph from Lick Observatory (7 August, blue filter), upper right photograph from Lowell Observatory (9 August, yellow filter), lower photograph from Lowell Observatory (5 September, yellow filter); (D) GWS 1960 (27 April), drawing by A. Dollfus (1963); (E) GWS 1990 (2 October), CCD image from Pic-du-Midi Observatory (Sánchez-Lavega et al. 1991); (F) GWS 1990 (20 November), image taken with the Hubble Space Telescope WFPC; (G) GWS 1990 (November 17), composite image over 360° longitude taken with the Hubble Space Telescope WFPC at 547 nm (Westphal et al. 1992). Panels D through G are adapted from García-Melendo and Sánchez-Lavega (2017), with permission from Elsevier.

Fisher et al. 2011; Sánchez-Lavega et al. 2011, 2012; Sayanagi et al. 2013; Sromovsky et al. 2014; Section 13.3). These studies show that the latitudinal band occupied by the disturbance (Table 13.1) depends on the meridional size reached by the main spot during its growing stage and on the zonal wind meridional profile. The propagation of the disturbance along the latitude circles depends on the zonal wind velocity at each latitude, but the details of the disturbance's cloud morphology depends, according to the models, on the thermal structure, vertical wind shear and Coriolis force (see Section 13.7). For example, the mid-latitude 2010 event took place close to the peak on a westward jet and the disturbance propagated eastward, forming a tail that encircled the planet in about 2 months (Sánchez-Lavega et al. 2012; Sayanagi et al. 2013). The 1990 equatorial event took place in a sheared region of the strong eastward equatorial jet, and the disturbance propagated eastward (northern branch) and westward (southern branch) from the source, covering the whole latitude band in just one month (Sánchez-Lavega, 1994).

The convective head of the 2010–2011 GWS persisted for 7 months, until it was destroyed when the large anticyclone vortex (AV) of the disturbance arrived at its position (Sánchez-Lavega et al. 2012; Sayanagi et al. 2013; García-Melendo et al. 2013). The underlying reason for the durability of the convective source for such a long period of time is still a mystery, although such persistence has also been observed in Jupiter's similar NTBD (North Temperate Belt Disturbance) events (Sánchez-Lavega et al. 2008).

The 1990 GWS was a huge event that disturbed the entire equatorial region with extended spot

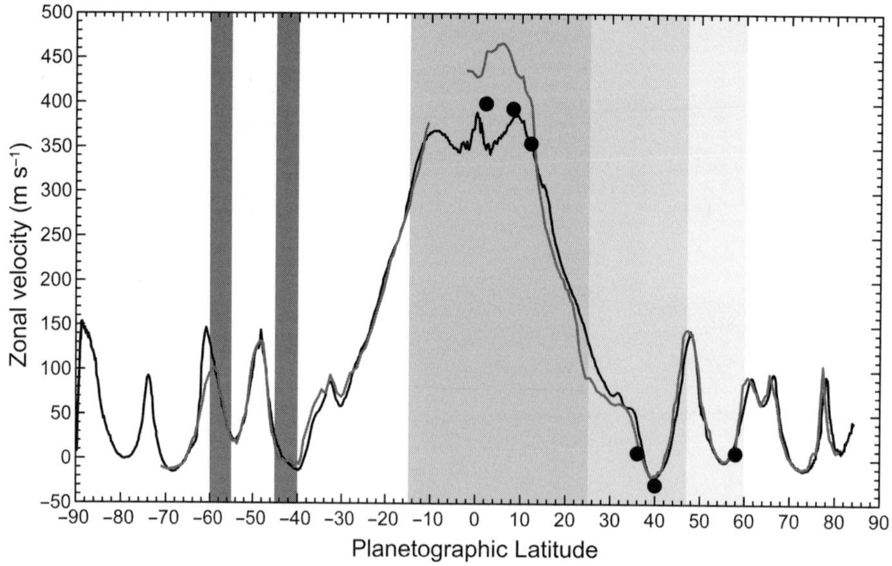

Figure 13.3 Saturn's zonal wind meridional profile at Voyager times (solid grey line, 1980–1981) and Cassini epoch (solid dark line, 2005–2011). The dots mark the GWS outbreak latitudes and the vertical bands, the affected latitudes (see Table 13.1). The narrow dark grey bands in the southern latitudes are those where small convective events have been observed.

activity into mid-1991 (Sánchez-Lavega et al. 1993). Moreover, it re-activated and formed a large spot at 9.4°N that reached a size of ~27,000 km (zonal length) during 1994–1995 (Sánchez-Lavega et al. 1996b). This activity coupled to vertical wind shear seems to be responsible for the change in the velocity of the equatorial jet between 1996 and 2002 (Sánchez-Lavega et al. 2003, 2004). The 2010 disturbance also produced a change in the profile of the westward jet where the storm emerged (Sayanagi et al. 2013; see Section 13.3). These changes are consistent with adjustment to maintain thermal wind balance following a heating event at the latitude of the storm. The mechanisms responsible for this adjustment are not fully identified, although shallow layer models point toward wave activity and momentum injection and dissipation (see Section 13.7).

13.3 The 2010–2011 Great Storm in the Cassini Epoch

The Great Storm of 2010–2011 provided the first-ever (and, possibly, once-in-a-lifetime) opportunity to observe a GWS phenomenon from the vantage point of a spacecraft in orbit around Saturn. Intervals of 27–30 years occurred between the five previous GWS events, with the last event happening in 1990, so the next event would have been expected in 2017–2020. The early occurrence of the 2010–2011 storm and the availability of the Cassini spacecraft was quite fortunate, as the planned end of the mission, when the spacecraft plunges into the atmosphere and is vaporized, is in September 2017. It remains to be seen whether the "2020 storm" will still occur, or whether the 2010–2011 storm disrupted the apparent 30-year cycle of the GWS storms.

In this section, we review the storm's temporal evolution, horizontal structure (including the cloud morphology) and wind structure using the data captured by multiple instruments onboard Cassini as well as Earth-based observations. Among the Cassini instruments, the Imaging Science Subsystem (ISS; Porco et al. 2004) returned high-resolution cloud images (Fischer et al. 2011; Garcia-Melendo et al. 2013; Sayanagi et al. 2013, 2014; Dyudina et al. 2013); the Visible and Infrared Mapping Spectrometer (VIMS; Brown et al. 2004) returned spatially resolved spectra that allow mapping of cloud composition (Sromovsky et al. 2013; Baines et al, 2011); the Composite Infrared Spectrometer (CIRS; Flasar et al. 2004) gathered stratospheric thermal structure and trace gas concentration (Fletcher et al. 2012; Hesman et al. 2012; Hurley et al. 2012), and the radar instrument (in the passive radiometer mode; Elachi et al. 2004) mapped the distribution of ammonia vapour (Janssen et al. 2013; Laraia et al. 2013). The Radio and Plasma Wave Science (RPWS; Gurnett et al. 2004) instrument can constrain the longitudes of lightning discharges through the timing of radio signals (Fischer et al. 2011; Sayanagi et al. 2013). Key observations were also made from Earth. In particular, ground-based observations using relatively small telescopes (many of which were operated by amateur astronomers) captured the development of the large-scale disturbance (Sánchez-Lavega et al. 2011, 2012; Sanz-Raquena et al. 2012; cf. Chapter 14).

13.3.1 String of Pearls: A Precursor to the 2010–2011 Storm?

The 2010–2011 storm erupted within an unusual feature known as the String of Pearls (SoP), detected in VIMS and ISS images (VIMS: Choi et al. 2009; Baines et al. 2010, 2011; ISS: Sayanagi et al. 2014). Figure 13.4 presents images of the SoP captured by the ISS (Sayanagi et al. 2014) and VIMS (Baines et al. 2011). In VIMS 5-micron images, the SoP appeared as a string of bright spots spanning some 90° of longitude, which were interpreted to be holes in the cloud deck from which thermal radiation escapes from depth. In the ISS images, the SoP appeared as dark holes, which is consistent with the VIMS view. In the ISS images, the SoP were best viewed in the CB2 continuum filter (750-nm wavelength), and it was also visible in the visible bands (RGB) and CB3 filter (950 nm), albeit with lower contrast; they are completely indecipherable in MT2 and MT3, suggesting that the structure does not extend above the tropopause. Sayanagi et al. (2014) showed that the individual *pearls* have cyclonic vorticity, and the feature's longitudinal drift can be fitted by a constant speed of -22.3 ± 0.2 m s^{-1} (i.e. −2.26° of longitude per Earth day in System III; negative denotes westward) between September 2006

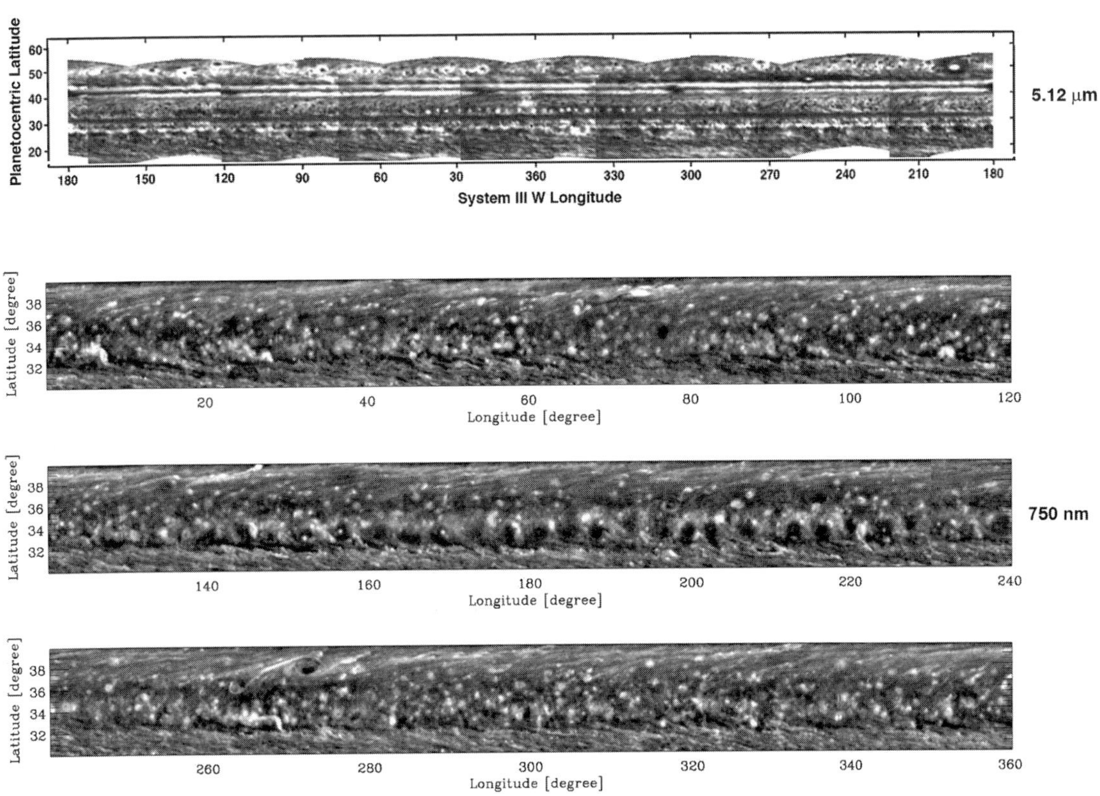

Figure 13.4 Top panel: cylindrical map (equirectangular projection) of the northern mid-latitude region. Observations at VIMS 5.1 μm channel, which uses the internal heat of Saturn as the primary lighting source, show deep features near the 3-bar level, including the thermally bright "String of Pearls" between around 40°W and 310°W longitudes. Observation was obtained 13 January 2008 (Baines et al. 2011). Lower three panels: similar view of the String of Pearls region in ISS CB2 filter (750 nm). The 23 dark spots that are aligned at 33.2°N latitude comprise the String of Pearls. The three panels show different longitudinal sections. The observation was made on 29 March 2008 (from Sayanagi et al. 2014, with permission from Elsevier).

and July 2010. On 5 December 2010, the storm erupted at 114.1°E longitude and 32.6°N planetocentric latitude (see Sánchez-Lavega 2011, for latitude definitions on an oblate planet). On 5 December, the SoPs occupied longitudes between 62°E and 161°E. Thus, the storm's initial latitude-longitude coordinates coincided with those of the SoPs. The SoPs was last detected in an ISS image on 23 December 2010 (Sayanagi et al. 2014) before evidently being covered by higher-altitude storm clouds or being physically disrupted by dynamics associated with the storm. By June 2011, the storm engulfed the entire latitude zone, which subsequently became remarkably clear of clouds for several years through this time of this writing (2014), as indicated by the remarkably intense, prolonged 5-μm brightness of the region that still exceeds any other region of the planet (Momary and Baines 2014). With such a clear, cloud-free atmosphere, it is not surprising that the SoPs has not been seen since the storm.

13.3.2 Temporal Evolution of the 2010–2011 Storm

The storm erupted on December 5, 2010, as serendipitously detected in an ISS image (Sayanagi et al. 2013) and observed by ground-based amateur astronomers (Sánchez-Lavega et al. 2011). The Cassini RPWS instrument also detected an onset of electrostatic bursts that would continue until August 2011 (Fischer et al. 2011a; Sayanagi et al. 2013). On 5 December, the storm spanned 1920 ±130 km east-west and 1300 ±130 km north-south, covering an area of approximately 1.5 × 10^6 km. Outside the storm, the surroundings appeared undisturbed from the relatively placid appearance it had maintained since emerging from the ring shadows around 2006 (Sayanagi et al. 2013).

The temporal evolution of the cloud morphology in the storm region is illustrated in the Cassini/ISS imagery shown in Figure 13.5. The red, green and blue colour channels of the mosaic are assigned, respectively, to images acquired in a continuum band (CB2), a moderate methane absorption band (MT2), and a strong methane absorption band (MT3). Because methane is well-mixed, features that appear bright in the MT2 and MT3 absorption bands must reside at relatively high altitudes (Tomasko and Doose 1984). Consequently, in this colour scheme, the altitude of detected features generally increases in order of red-green-blue, and white indicates that light is scattered in all three wavelength bands by high-altitude aerosols.

At the westernmost point of the storm (Figure 13.5), there is a bright cloud identified as the storm's convective *storm head*, where, from VIMS spectra, Sromovsky et al. (2013) detected water and ammonia ice lofted over 50–250 km in altitude, the first detection of water ice ever observed on Saturn. Bright clouds billow to the east of the head out to the large anticyclonic vortex (AV). To the east of the AV, clouds acquire a turbulent appearance without well-defined edges, identified as the storm's *tail*. The storm's cumulus convection appeared to be confined to the area between the head and AV. Here, the most reflective storm clouds appeared (Sánchez-Lavega et al. 2012; Sayanagi et al. 2013), within which lightning flashes were observed in ISS imagery (Dyudina et al. 2013) and electrostatic radio bursts were detected by RPWS (Fischer et al. 2011a; Sayanagi et al. 2013). This region is identified as the *body* of the storm.

The AV formed between 5 and 24 December 2010. Initially, its core was centred at 76°E longitude and 33° N planetocentric latitude and appeared roughly circular, with a diameter of approximately 4250 km. It appeared particularly bright in the MT3 strong methane absorption band (blue in Figures 13.5 and 13.6), indicating an unusually high altitude, as verified in a detailed multi-wavelength analysis of ISS images in CB2, MT2 and MT3 filters (Sayanagi et al. 2013). By January 11, 2011, the AV measured 11,000 km by 12,000 km in the north-south and east-west dimensions, a size that rivalled Jupiter's Oval BA (Simon-Miller et al. 2006).

Between the onset of the storm on 5 December 2010 and the head's last sighting on 14 June 2011, the *body* of the storm between the head and AV grew longitudinally due to the difference in their drift speeds, with the storm's head drifting westward at 2.79 ±0.08° of longitude per Earth day (26.9 ±0.8 m s^{-1}) (Sánchez-Lavega et al. 2011, 2012; Sayanagi et al. 2013) versus 0.85 longitude per day (8.4 m s^{-1}) for AV. As the space between the head and the AV widened at ~2.0° per day, the area covered by the storm's body changed from 1.75 × 10^8 km^2 on 24 December 2010 to 2.82 × 10^8 km^2 on 2 January 2011, and to 4.62 × 10^8 km^2 on 11 January 2011 (Sayanagi et al. 2013), corresponding to an approximately constant expansion rate of

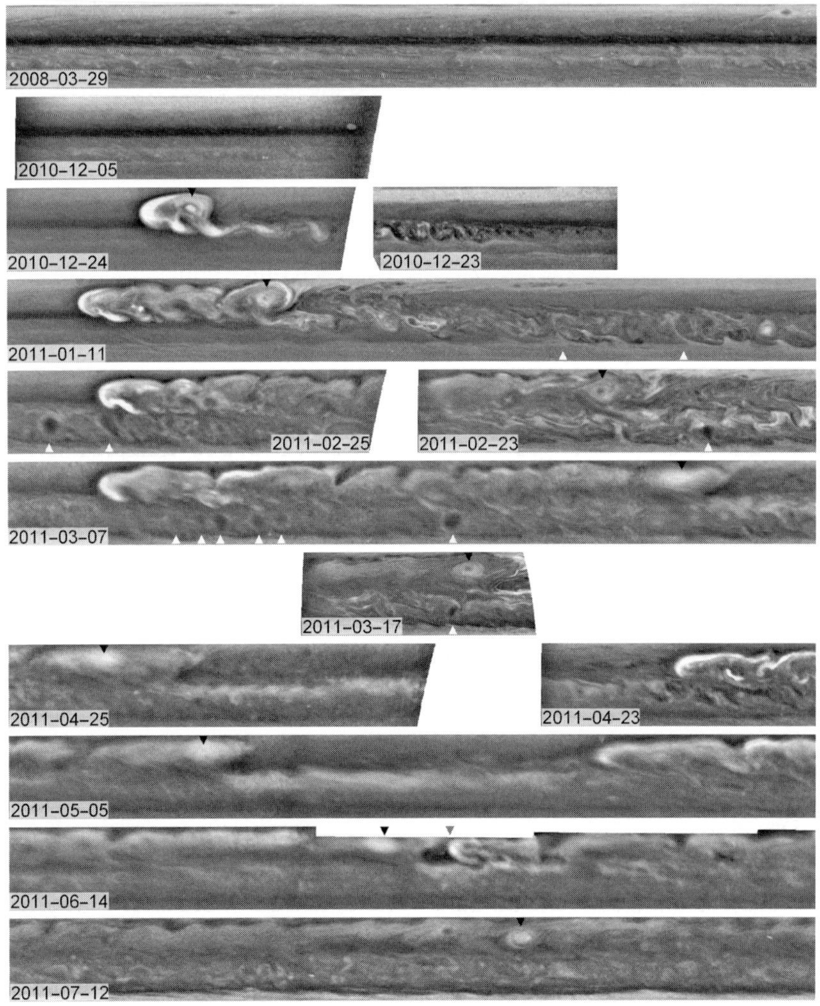

Figure 13.5 Cloud morphology before, during and after the storm. The panel IDs denote the image acquisition dates. Each panel is a latitude-longitude projected mosaic and shows 20.5°–38.9°N planetocentric latitude. The width of the entire figure represents 200° of longitude. Mosaics are cropped to highlight the spacing between the head (marked with grey triangle) and the AV (black triangle). Small triangles denote the longitudes of the dark ovals. Cassini ISS camera's CB2 (750 nm), MT2 (727 nm), and MT3 (889 nm) filters were used to compose the mosaic images. (From Sayanagi et al. 2013, with permission from Elsevier.)

140 km^2s^{-1}. Sánchez-Lavega et al. (2011) measured a peak expansion rate of 212 km^2 s^{-1} at the cloud top height of 150 mbar during 14–16 December. As the body of the storm grew, its boundary developed wave-like undulating patterns (Sánchez-Lavega et al. 2011, 2012). The Cassini RADAR radiometer examined this region and found that the body of the storm appeared unusually bright in 2.2-cm wavelength thermal emission, indicating, rather surprisingly, that the region was sub-saturated in ammonia vapour (Janssen et al. 2013; Laraia et al. 2013). Li and Ingersoll (2015) show, in an axisymmetric numerical model, that this comes about due to uplift and precipitation of ammonia followed by a dry downdraft during the geostrophic adjustment process that follows a moist convective event. Dyudina et al. (2013) revealed that lighting flashes occurred in the body of the storm away from the head, suggesting that moist convection is occurring in the body as well. One potential explanation is that the thick clouds obscured lightning in the most convective parts of the clouds from view. In late June 2011, six months after its formation, the head of the storm caught up with the AV after the vortex's position fell 360° of longitude behind (to the east) of the head due to the AV's slower westward

drift (Sánchez-Lavega et al. 2012; Sayanagi et al 2013). Figure 13.5 panel 2011-06-14 shows the head and the AV 10 days before their positions were predicted to intersect on June 24. Ground-based observation by Sánchez-Lavega et al. (2012) showed that the head and the AV made contact around 15 June, and the bright cloud materials of the head first disintegrated into three parts before disappearing between 14 June and 19 June. On 12 July (Figure 13.5 panel 2011-07-12), the Cassini ISS imaged the storm for the first time after the predicted collision between the head and the AV; after the collision, the head was no longer visible, while the AV was still present.

After the head-AV collision and the ensuing end of the convective activity, a dark, featureless region appeared in the storm-affected planetocentric latitudes between 31°N and 38°N (Sayanagi et al. 2013). The region appeared dark in CB2 and visible continuum wavelengths and had a relatively bright appearance in MT2 and MT3, consistent with the absence of clouds (Sayanagi et al. 2013). Momary and Baines (2014) showed that the region was intensely bright in 5-μm thermal emission of light escaping from the deep atmosphere; indeed, it is the brightest 5-μm region observed thus far on Saturn, thus indicating a clear, almost cloudless "desert" of air in the troposphere down to several bars, consistent with the ~2-bar-level desiccation of ammonia vapour observed by the Cassini radiometer (Janssen et al. 2013; Laraia et al. 2013). Thus, what was perhaps one of the most powerful storms ever witnessed anywhere in the Solar System left in its wake a dry, cloudless, almost barren sky that has now lasted for several years.

13.3.3 The Effects of the Storm on the Wind Field

The storm had a significant impact on the local wind field as well as the zonal mean wind speed. Dyudina et al. (2013) and Garcia-Melendo et al. (2013) found that the head of the storm had a large-scale anticyclonic circulation. In addition, Garcia-Melendo et al. (2013) found a diverging motion in the head, suggesting that the cloud-top was an anvil cloud spreading out of a cumulus convective core. Dyudina et al. (2013) also showed each crest in the undulations on the north side of the storm body is associated with an anticyclonic vortex; the large anticyclone that defined the eastern edge of the storm body was the first of the series of anticyclones that line the storm every 10° to 15° in longitude.

Garcia-Melendo et al (2013) showed that peak wind speeds in the anticyclonic circulation around the head of the storm reached 160 m s^{-1} with a relative vorticity reaching -9×10^{-5} s^{-1}. Dyudina et al. (2013) reported that each of the anticyclones in the body of the storm had a peak speed reaching 100 m s^{-1} and a relative vorticity of -5×10^{-5} s^{-1}. Sayanagi et al (2013) also measured the wind structure of the easternmost anticyclone and showed that the peak wind speed exceeded 120 m s^{-1} and the relative vorticity reached -6×10^{-5} s^{-1}.

Sayanagi et al (2013) showed that the zonal mean wind speeds in the region affected by the storm were different before and after the storm (Figure 13.6). They showed that the zonal wind profiles before the storm (on 7 May 2008 and 7 September 2010) and in the early phase of the storm (on 11 January 2011) were essentially the same. After the end of the convective phase (on ~5 August 2011), the wind speed changed. The planetocentric latitudes to the south of 34°N exhibited a deceleration of the zonal wind profile by ~30 m s^{-1}, while the profile to the north indicated an increase by ~35 m s^{-1}. Sayanagi et al. (2013) pointed out that these changes are consistent with warming at the central latitude of the storm and thermal wind balance to the north and south. Interestingly, the zonal wind measurements during the Voyager flybys in 1980–1981 given by Sánchez-Lavega et al. (2000) indicate wind speeds between these pre- and post-storm measurements.

Together with the thermal wind equation, Fletcher et al. (2011) used Cassini CIRS temperature measurements to estimate the change in the vertical shear between September 2010 and January 2011 – that is, just prior to and after the advent of the storm. They detected wind speed changes greater than 40 m s^{-1} above the 50-mbar level. In the deeper 1-bar to 100-mbar region, Fletcher et al. detected little change, consistent with the Sayanagi (2013) cloud-track measurements noted above for early storm measurements at comparable altitudes. Later in the storm development, Achterberg et al. (2014) and Fletcher et al. (2012) presented CIRS data that showed that, at the storm latitude, the tropospheric temperature increased between January and August of 2011. Achterberg

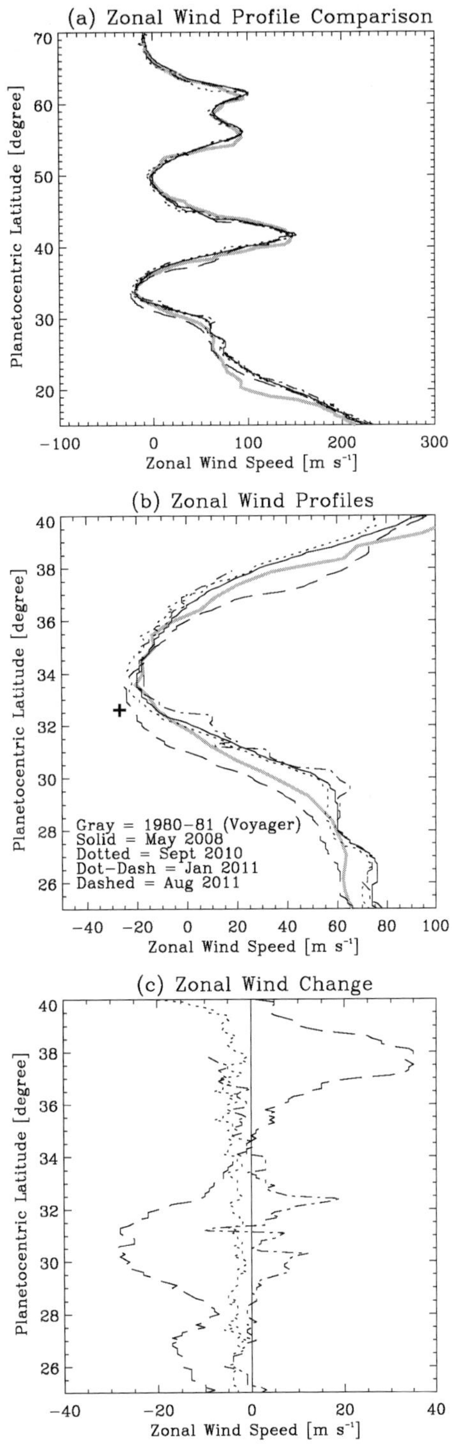

Figure 13.6 Comparison of the zonal wind measurements before, during and after the great storm of 2010–2011. All images analysed here were captured using the CB2 filter. Panel (a) shows the maximum correlation wind profiles for 7 May 2008 (solid), 7 September 2010 (dotted), 11 January 2011 (dash-dot) and 5 August 2011 (dashed). The gray line is the zonal mean wind measurement by Sánchez-Lavega et al. (2000). Panel (b) is the same except that it enlarges the latitudes affected by the storm to highlight the wind speed change. The plus (+) symbol marks the latitude and propagation speed of the storm's head. Panel (c) shows the wind change compared to the 7 May 2008 measurement, illustrating that the measurement after the storm on 5 August 2011 shows a faster wind to the north of the storm and a slower wind to the south. (From Sayanagi et al. 2013, with permission from Elsevier.)

et al. measurements indicate dT/dy > 0 equatorward of the storm and dT/dy < 0 poleward of the storm (with T being temperature and y being the northward distance). The temperature gradient suggests that the wind speed above the 1-bar level decreased to the south of the storm and increased to the north. Thus, the CIRS vertical shear measurements are consistent with the interpretation by Sayanagi et al. (2013) that the change in the cloud-tracked winds post-July 2011 represents a real wind speed change.

13.4 Vertical Cloud Structure and Composition

The Great Storm of 2010–2011 was the most well-studied meteorological event ever observed on Saturn, enabling detailed state-of-the-art analyses and modelling of cloud, temperature, compositional and dynamic structure. This event was truly four-dimensional in nature, with its morphology and thermal characteristics changing widely over latitude, longitude, depth and time. Here we first explore the vertical structure and composition of the aerosols and clouds, constrained from visual and near-infrared images, maps and spectra. The temperature structure and dynamics are covered in Sections 13.5 and 13.7, while results on lightning emissions are covered in Section 13.6.

13.4.1 Models of the Cloud Structure from Visible Wavelength Imaging

As previously discussed, the storm was first detected in visible light using small-aperture, ground-based telescopes (Fischer et al. 2011a; Sánchez-Lavega et al. 2011) and profusely imaged at high resolution by the ISS camera onboard the Cassini spacecraft in subsequent months (Sánchez-Lavega et al. 2012; Sayanagi et al. 2013). Most of the dynamical simulations of the storm (e.g. García-Melendo et al. 2013) were therefore based on the visual aspect of the storm and the evolution of the cloud field as imaged in wavelengths not far from those visible by the human eye. For this reason, even though longer wavelengths are more sensitive at higher and lower atmospheric levels and provide complete information on the distribution and nature of particles inside the storm (e.g. Sromovsky et al. 2014), some information is also required on how the visible light is scattered, mostly by the uppermost particles raised by the convective event. This provides altitude levels for the location of cloud tracers and thus the cloud-tracked winds and possible wind shears, which can then be investigated and compared against dynamical simulations. Also, historically storms have been viewed almost exclusively in the visible realm; therefore, any comprehensive comparison of the vertical cloud structure between past storms involves the analysis of images and spectra taken at wavelengths below 1 μm.

The initial report on the Great Storm of 2010–2011 by Sánchez-Lavega et al. (2011) included the estimation of the cloud top pressure and other parameters of the area disturbed by the storm. This work used images taken at a variety of filters including wide- (U, B, V, R) and narrow-band filters (624, 727, 883, 889, 914 and 954 nm), all of the latter related to red and near-infrared methane absorption bands. The center-to-limb variation of the absolute reflectivity showed the GWS to be embedded inside the tropospheric haze. The main cloud deck was below the levels sounded by some filters,

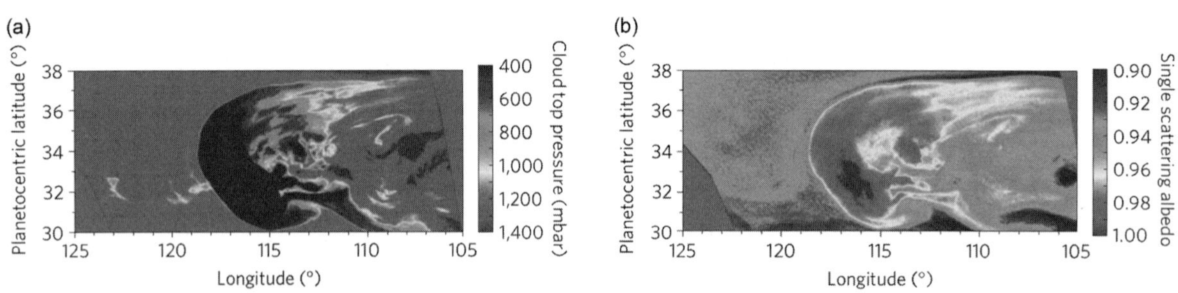

Figure 13.7 Detailed cloud properties modeling of the region of the head of the GWS in February 2011, based on Cassini ISS data. (a) Cloud top pressures ranging from 1.4 bar at the undisturbed regions to 400 mbar at the front of the disturbance. (b) Single scattering albedo at blue wavelengths (BL1) showing a dramatic increase of particle reflectivity at the compression wave formed in the front of the storm. Adapted from García-Melendo et al. (2013). (A black-and-white version of this figure appears in some formats. For the color version, please refer to the plate section.)

such as the deep methane band at 889 nm, which showed the disturbed area as darker than its surrounding. This implied that the haze above the storm itself had largely dissipated in some way. With such a haze-free atmosphere that only the storm clouds contributed to the observed reflectivity, which showed the storm particulates to be substantially more reflective at all wavelengths, including the normally absorbing blue wavelengths. This suggested for the first time that the material comprising the main cloud deck in the disturbed area was fresh, less blue-absorbing than the common hazes and clouds found in Saturn (Pérez-Hoyos et al. 2005; Karkochka and Tomasko 2005; West et al. 2009).

A similar approach was taken by Sanz-Requena et al. (2012) with observations taken one month after the outbreak of the storm. Wavelength coverage was from near-ultraviolet to the near-infrared, including methane bands of various strengths. As in Sánchez-Lavega et al. (2011), they found that the GWS cloud tops reached the altitude level 300^{-100}_{+200} mbar higher than the ~1400-mbar ammonia condensation level. Similarly García-Melendo et al. (2013) performed radiative transfer models on high-resolution images taken in February 2011 by Cassini ISS instrument in blue (BL1) and near-infrared (MT2, CB2, MT3) filters, including two methane bands and an intermediate continuum (Figure 13.7). They also found cloud tops at 400 mbar. The second-most important change produced by the GWS was an overall increase in the single-scattering albedo of the particles, with values close to 1 even at blue wavelengths. These results were analysed and discussed in the context of the dynamical modelling by García-Melendo et al. (2013) and found to be in agreement with their numerical models of the 2010 event.

In short, there are three main characteristics of the 2010 GWS that could be retrieved from imaging at wavelengths below 1 μm: (1) the GWS can be modelled as elevated thick clouds extending up to levels between 300 and 400 mbar; (2) the particles injected into the atmosphere are substantially more reflective at all wavelengths than those commonly found at the same levels in Saturn's atmosphere; and (3) the levels above the thick clouds are not substantially enriched in particles, so they appear darker than their surroundings in the 889 methane absorption band, except perhaps at the front of the disturbance. Such results are in relatively good agreement with the analysis of the 1990 GWS by Acarreta and Sánchez-Lavega (1999) based on the imaging by Sánchez-Lavega et al. (1994) and Barnet et al. (1992), which placed cloud tops at 300 mbar.

There is still much work to be done regarding the analysis of the visible data below 1 μm. Firstly, the evolution of the cloud field showed a complex evolution in the mature phase of the event (Sánchez-Lavega et al. 2012; Sayanagi et al. 2013), and there are currently no works describing the corresponding evolution of the vertical cloud structure above the 1- to 2-bar level. Secondly, the properties of the small particles potentially located above the storm itself could be best sounded using visual wavelengths. Indeed, the large coverage of phase angles provided by the Cassini spacecraft in 2011 can be used to explore the size or phase function of high-altitude particles. Finally, since the blue absorption of tropospheric particles remains as one of the most elusive problems in the atmospheres of Jupiter and Saturn, how the storm affected its colour could also provide a substantial piece of information on this issue.

13.4.2 Composition of Cloud Particles from VIMS Spectral Images

13.4.2.1 VIMS Observations of the Storm

The VIMS instrument (Miller et al. 1996; Brown et al. 2004) covers a spectral range of 0.35–5.1 μm, with an effective pixel size of 0.5 milliradians on a side and near-IR spectral resolution of approximately 15 nm. Samples of the earliest VIMS observations of the Great Storm, acquired on 24 February 2011, are shown in Figure 13.8 as remapped equirectangular projections at four key near-IR wavelengths. The most dramatic spectral characteristic of the storm is its low reflectivity at 3.05 μm (panel B), something not previously seen anywhere else on Saturn (Sromovsky et al. 2014). Outside the storm region, the I/F at 1.887 μm (panel A) is a good match to the I/F at 3.05 μm, but in the stormhead the I/F at 3.05 μm is only about 40% of the outside value. At 4.081 μm, the storm is nearly twice as reflective as non-storm clouds, indicating the presence of larger particles in

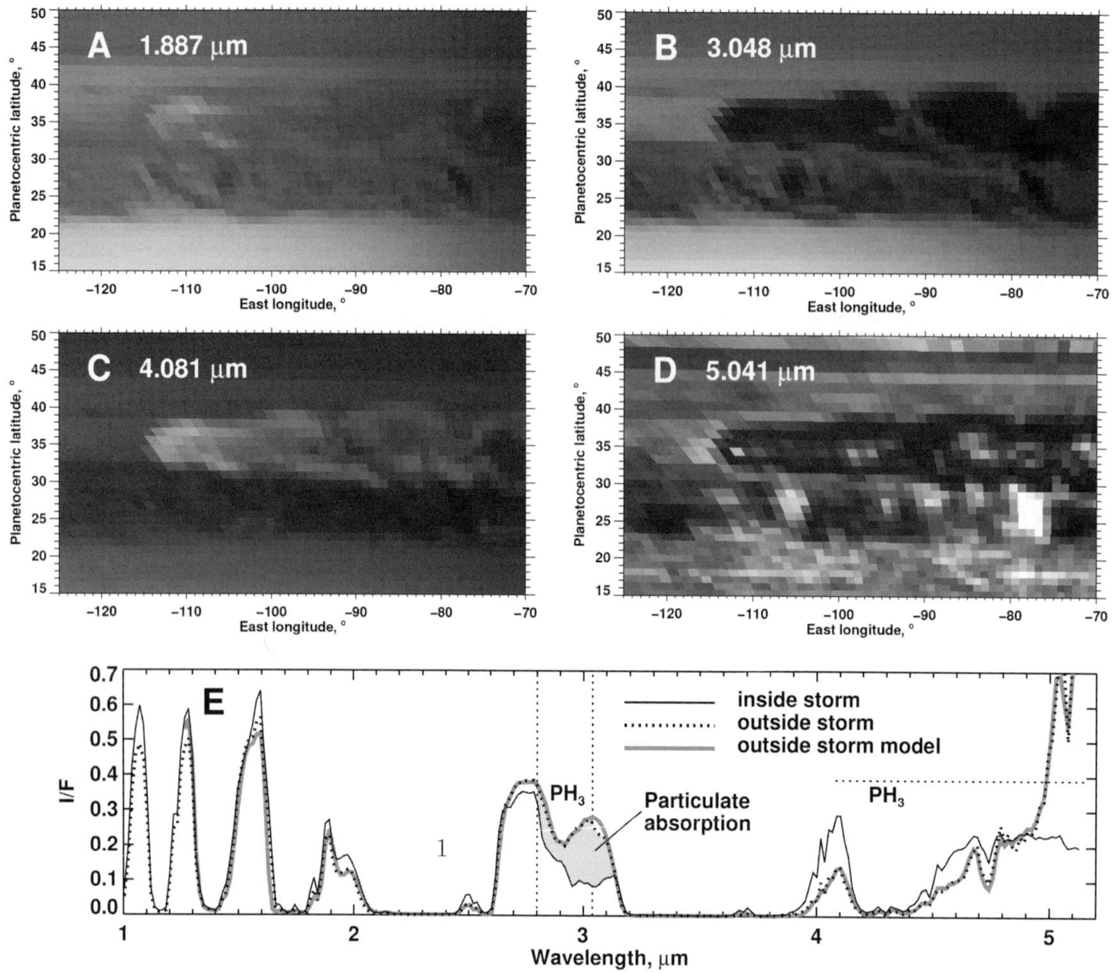

Figure 13.8 (A–C) VIMS spectral images of the storm in February 2011, sampled at wavelengths indicated within each panel, remapped to an equirectangular projection, and linearly stretched so that white corresponds to a reflectivity (I/F) of 0.4, except for D, in which white corresponds to I/F = 1. The parts of the storm that are bright at 4.081 μm are much darker at 3.048 μm than the region ahead of the storm. 1.887 μm and 4.081 μm are pseudo-continuum wavelengths, subject to little gas absorption, and thus show reflective properties of clouds without overlying atmospheric absorption. The strong reflectivity at 4.081 μm indicates that the cloud particles are relatively large (1 μm or more). The strong absorption in the gray region near 3.05 μm is due to particles in the storm cloud and not atmospheric gases. Adapted from figure 2 of Sromovsky et al. (2013), with permission from Elsevier.

the storm region. At 5.041 μm (panel D), the thermal emission of Saturn's deeper layers (near 3 bars) contributes to the observed *I*/*F*. The dark regions of the storm head at this wavelength indicate the presence of optically thick cloud layers. The spectral characteristic of the 3-μm absorbing region is illustrated in Figure 13.8E, where a spectrum from inside the storm head (solid line) is compared with a spectrum from the region just west of the storm head (dotted line). Absorption by PH_3 affects the 2.8–3.03 μm region, but the spectral difference seen in the 3.03–3.14 μm region is almost entirely due to particulate absorption. Spectra obtained upstream display absorption between 2.8 and 3 μm that is due to phosphine, but show little absorption at 3.05 μm. The upstream clouds are also more transparent at thermal wavelengths, evident from their large apparent *I*/*F* values near 5 μm. The storm particles are brighter than the upstream cloud particles in most window regions, slightly more absorbing at 2.7 μm, and dramatically more absorbing at 3.048 μm.

13.4.2.2 Vertical Structure Models

Near-IR spectra outside the storm regions can be very well fit by relatively simple cloud structures similar to those used successfully to fit spectra in the CCD range from 0.3 to 1 μm (Muñoz et al. 2004; Karkoschka and Tomasko 2005; Pérez-Hoyos et al. 2005). These structures contain a vertically diffuse stratospheric haze and an optically thick and physically thick main cloud layer, which requires visible and UV absorption, but can fit near-IR spectra with conservative cloud particles. Sromovsky et al. (2013) used a similar structure to obtain the excellent fit shown by the model spectrum (grey curve) in Figure 13.8E. For this fit the main haze layer contained Mie particles of real index 1.4 and effective radius 0.6 μm, extended from 120 mbar to 700 mbar, and had an optical depth of 6.8 at 2 μm. An optically thick layer with an effective pressure near 3 bars is also needed to control thermal emission.

13.4.2.3 Single-Component Models of 3-μm Absorption

Sromovsky et al. (2013) first tried to fit the storm spectra using pure substances in a single middle cloud layer of spherical particles, then adjusted the optical depth, particle size and top and bottom pressures to optimize fits for the region from 1.268 μm to 5.15 μm, while ignoring the region from 2.65 μm to 3.2 μm where particulate absorption is most prominent. That allowed the vertical structure of the aerosols to be primarily constrained by methane gas absorption and not much affected by particulate absorption. They then compared the model spectra with measured spectra in the 2.65 to 3.2 μm region to determine which provided the best compositional match. They considered five possible cloud materials: (1) a conservative substance with refractive index n = 1.4 + 0i, (2) NH_3, (3) NH_4SH, (4) N_2H_4 and (5) water ice. The fits for the five pure substances are displayed in Figure 13.11. Ammonia provided by far the best match to the measured spectrum of the storm head, yielding a χ^2 value in the absorbing region that is 3.5 times better (smaller) than that of the next best fit. However, this best spectral fit among these simplistic single-component models still contains substantial differences from the VIMS measured spectrum, especially the sharp ammonia ice absorption feature near 2.96 μm, but also the excess brightness near 3.1 μm.

13.4.2.4 Multi-Component Models of 3-μm Absorption

Sromovsky et al. (2013) next considered linear combinations of pure spectra, which can be thought of as spatially heterogeneous mixtures of different cloud types, perhaps as different particles visible in spaces between clouds of other particles. The best linear combination, shown by the blue curve in Figure 13.9B, used a mix of 55 ±4% NH_3, 22 ±2% water ice and 23 ±2% NH_4SH, where percentages represent fractional area coverage. This improved the fit in the absorbing region to $\chi^2/N(ABS) = 1.1$, a factor of 8 better than provided by pure NH_3 (see Figure 13.9 caption). A big improvement was also obtained using the conservative (n = 1.4) material in place of NH_4SH, with a mix of 55 ±3% NH_3, 22 ±3% water ice and 23 ±2% n = 1.4, although the improvement was only by a factor of five in this case. The best linear combination without water ice produced $\chi^2/N(ABS) = 5.2$, which is a factor of 5 worse than produced by the best combination with water ice. The water ice component reduced the excess *I/F* gradient between 3 μm and 3.15 μm and the excess *I/F* otherwise present between 2.65 μm and 2.85 μm. The third component (either n = 1.4 or NH_4SH) served to partially fill in the deep minimum produced by pure NH_3 at 2.96 μm. Their linear combination analysis clearly favours NH_3, water ice and NH_4SH, in that order.

They also considered more physically plausible, locally horizontally homogeneous cloud models, for which their best fit to VIMS spectra contained a deep water cloud, perhaps beginning at 10 bars, but certainly containing significant opacity above the 1- to 2-bar range, a main cloud layer consisting of two sub-layers, one of low optical depth ($\tau \approx 1$) above a lower sub-layer of significant optical depth ($\tau \approx 9$). Among two options for the main layer that produced good fits, the most plausible was one with particles of n = 1.4 in the top sub-layer and NH_3 in the lower sub-layer as a coating on H_2O cores, with a core radius that is 40% of the total radius (1.5 μm) and a core volume of 6%. A slightly better fit was obtained using smaller particles with some long-wave absorption in the upper main cloud layer and somewhat larger NH_3-coated water ice particles in the lower main sub-layer. Such a model is qualitatively consistent with deep convection from the water layer, as suggested by Hueso and

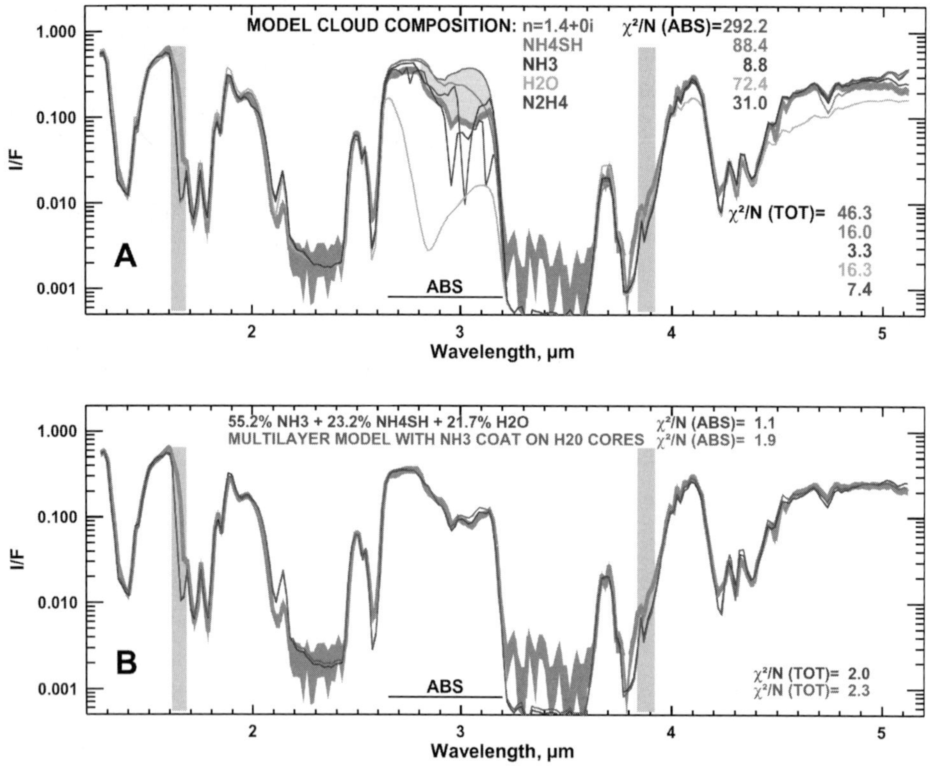

Figure 13.9 (A) A storm head spectrum, and fits to that spectrum using a main cloud layer consisting of either conservative particles (red), pure NH_4SH particles (green), pure NH_3 particles (blue), or pure H_2O particles (cyan), or pure N_2H_4 particles (magenta). The vertical gray bars indicate regions excluded from the analysis due to calibration uncertainties near order sorting filter joints. Relative fit quality for each entire pure VIMS spectrum (excluding the two regions indicated by vertical gray bars) is indicated by the χ^2/N (TOT) values, and for the 2.65 µm–3.2 µm region of particulate absorption by χ^2/N (ABS) values, where there are 33 points in the ABS region and 223 in total. The yellow region between the conservative particle spectrum (red) and the VIMS measured spectrum is due to cloud particle absorption. (B) As in A except that the blue model spectrum is a linear combination of pure spectra, with fractions adjusted to minimize χ^2 in the particle absorbing (ABS) region, and the red spectrum is a multi-layer horizontally homogeneous model spectrum with the main upper tropospheric haze layer containing an upper sub-layer of conservative particles and a lower sub-layer of ammonia-coated water ice particles. Figure from Sromovsky et al. (2013), with permission from Elsevier. (A black-and-white version of this figure appears in some formats. For the color version, please refer to the plate section.)

Sánchez-Lavega (2004). Both heterogeneous and homogeneous models strongly favoured NH_3 to be the most prominent substance producing 3-µm absorption, with water ice providing an important contribution in both cases.

The relative roles of NH_4SH and the n = 1.4 material were less clearly defined. It is also plausible that storm cloud particles might contain three materials: a core of water ice, a shell of NH_4SH, and a final coat of NH_3. Whether such particles would lead to better spectral fits remains to be determined. Also somewhat unclear is how much these relative contributions might be changed by revising the VIMS calibration (Clarke et al. 2012).

A firm conclusion of the VIMS Near-IR study of Sromovsky et al. (2013) is that frozen water is indeed present in the convective storm head of the Great Storm of 2010–2011. The presence of water ice lofted 200 km above its freezing level near 10 bar proves that such Saturnian storms are much more powerful, and extend over much more extensive depths, than their terrestrial counterparts.

13.5 Temperature Structure

Optical studies of the storm provided important insights into the physical processes at work within Saturn's cloud-forming regions, but the presence of Cassini in the Saturn system, combined with two decades of technology development in the mid-infrared, meant that Saturn's 2010 storm was the first to be studied in detail at thermal infrared wavelengths. Thermal infrared studies provide insights into the environmental conditions (temperature structure and gaseous composition) both within and surrounding the storm system. By reconstructing the temperature field in three dimensions and monitoring its evolution through time, we gained powerful new insights into the effects of a tropospheric storm system on the overlying stratosphere, including the birth of an enormous hot anticyclonic vortex (referred to as the "beacon") high above the storm region. This enormous stratospheric vortex persisted long after the tropospheric convective activity had abated and was last observed in 2013, although observations in 2013–2014 still showed residual wave activity in Saturn's stratosphere. Thermal contrasts driven by the storm were first detected by Fletcher et al. (2011), and their temporal evolution has been reported in both the troposphere (Achterberg et al. 2014) and the stratosphere (Fletcher et al. 2012; Hesman et al. 2012) using a synthesis of data from the Cassini Composite/Infrared Spectrometer (CIRS) and a host of ground-based observatories (notably the Very Large Telescope (VLT) and Infrared Telescope Facility). The stratospheric response to the tropospheric storm has resulted in the discovery of the greatest atmospheric temperature change ever seen due to a storm (Fletcher et al. 2012; Hesman et al. 2012), in addition to completely unexpected changes in species abundances (Hesman et al. 2012; Bjoraker et al. 2012; Hesman et al. 2013). This section summarizes what has been discovered about Saturnian storms (and therefore giant planet convective plumes in general) from assessments of the thermal structure.

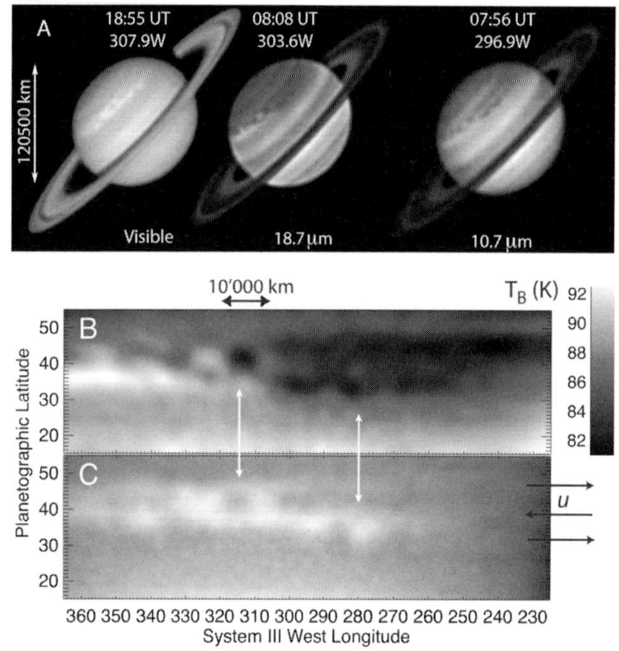

Figure 13.10 Montage showing VLT thermal infrared images of Saturn's tropospheric storm temperatures at 18.7 and 10.7 μm, compared to a visible-light image by T. Barry, all acquired on 19 January 2011. The bottom panels B and C show the direct correspondence between the brightness temperatures (sensitive to approximately 200 mbar) and the cloud albedo variations, highlighting the warm temperatures over the western storm head, the cold-core anticyclone and the two extended cool tails to the east. The latitudinal peaks of the prograde and retrograde jets are also indicated by arrows. Adapted from Fletcher et al. (2011); reprinted with permission from AAAS.

13.5.1 Tropospheric Temperatures

Remote sensing in the mid- and far-infrared (5–1000 μm) is typically sensitive to altitudes above the topmost condensate clouds but below the tropopause (approximately 80–700 mbar), in addition to sensitivity in the mid-stratosphere (0.1–10 mbar from nadir observations) from hydrocarbon emissions (primarily methane, ethane and acetylene). Fletcher et al. (2011) presented the first 7–25 μm images of the storm region on 19 January 2011 using the VLT/VISIR instrument 45 days after the initial eruption of cloud material. Troposphere-sensitive images (Figure 13.10) were centred on the cold core of a newly formed anticyclonic vortex embedded in the westward jet near 39°N (planetographic), which appeared dark bluish in amateur imaging surrounded by the collar of whiter cloud material. The cold oval near 41°N had a core temperature of 79 ±2 K at 100 mbar, consistent with upwelling and adiabatic cooling within an anticyclone of approximate dimensions of 4000 × 5500 km. To the west of the oval, the bright white storm head extended 60° longitude away from the anticyclone, and was characterized by warmer tropospheric temperatures (5 ±2 K warmer than the atmosphere east of the storm anticyclone in the 70–300 mbar region) surrounding discrete cold spots. The warmest regions appeared restricted to the southern edge (35–38°N) of the storm band, possibly co-located with the southern band of bright white clouds.

To the east of the anticyclone, the VLT images in Figure 13.10 show that the flow bifurcated into two cool branches (Fletcher et al. 2011): a northern branch at 42°–48°N and a southern branch between 31° and 38°N, separated by a fainter warm sector. These two branches were characterized by undulating white clouds and contrasts between warm and cooler spots suggestive of atmospheric thermal waves remaining in the wake of the westward-moving storm head. Although VLT images provided only a snapshot of the thermal field on a few days during the storm's lifetime, they were capable of resolving the small-scale longitudinal structures (i.e. small plumes) that would be hidden to Cassini's spectrometer (CIRS) because the signal-to-noise level on individual CIRS spectra precludes the generation of high-resolution tropospheric maps. Nevertheless, Cassini/CIRS spectra are essential to provide the spectral information necessary to retrieve the vertical temperature structure.

Achterberg et al. (2014) utilized 20–200 μm spectra of Saturn's smooth H_2-He collision-induced continuum to determine zonal mean temperatures and para-H_2 distribution, monitoring these parameters from 2009 to 2012 to understand the tropospheric evolution. During the storm, they discovered that the zonal mean temperatures for $p > 300$ mbar increased by 3 K, accompanied by a fall in the para-H_2 fraction that is consistent with the idea of upwelling associated with the convective storm. As Saturn's winds are in approximate geostrophic balance, Achterberg et al. (2014) use the rising temperatures to show that the post-storm winds become more eastward with height, poleward of the storm band, and more westward with height, equatorward of the storm band, by 5 m s^{-1} per scale height. This is in approximate agreement with the wind speeds determined by Sayanagi et al. (2013) from cloud tracer measurements.

No complete picture of the storm-induced tropospheric heating has yet emerged. Vertical advection of air heated by the latent heat from water condensation could have warmed the storm head during the active phase (Fischer et al. 2011a; Achterberg et al. 2014), particularly if restricted to the southern edge of the storm band as shown in Figure 13.10B. Alternatively, a thermally indirect circulation forced by waves and eddies could result in localized updrafts cooling due to adiabatic expansion in the stably stratified upper troposphere (the colder spots in Figure 13.10B), surrounded by a larger area of subsidence and adiabatic heating by compression. Indeed, microwave observations in the post-storm phase suggest NH_3 depletion consistent with atmospheric subsidence over wide areas of the storm remnant (Janssen et al. 2013; Laraia et al. 2013). Such small-scale uplift, balanced by subsidence over a wider area, is a typical hallmark of a moist-convective process, and we note that plumes associated with the revival of Jupiter's South Equatorial Belt also tend to be cold cored (Giles et al. 2013) and followed by large-scale warming. It remains unclear whether the rise in temperature in the storm band was due to moist convective heating on an updraft or adiabatic warming on a downdraft; whether it flipped from the former to the latter at some stage during the storm's evolution, and whether warming by latent heat might dominate the deep troposphere, whereas warming by large-scale subsidence dominates the upper troposphere. It was shown that the storm increased the emitted power in a wide latitudinal band (20°–55°N) with a maximum change of 9.2 ±0.1% around 45°N from 2010 to 2011 (Li et al. 2015). Tropospheric temperatures appeared to still be rising in early 2012 (Achterberg et al. 2014) and IRTF imaging in 2013 (Fletcher et al. 2014) showed the continued presence of the warm tropospheric band near 40°N, consistent with subsidence in the band causing desiccation and depleting the atmosphere of aerosols remaining from the convective storm (the band is extremely bright when viewed at 5 μm, Orton et al. 2013).

13.5.2 Stratospheric Temperatures

Perhaps the greatest surprise revealed by the thermal infrared studies was the intense perturbation created in Saturn's stratosphere, completely invisible to observers using reflected sunlight. Emissions from stratospheric hydrocarbons (namely methane, ethane and acetylene) permit the retrieval of the stratospheric thermal structure, under the assumption that methane remains well mixed despite the powerful dynamics associated with the storm. North-south and east-west retrievals of 0.5 to 5 mbar temperatures from Cassini suggested a substantial stratospheric cooling over a wide area approximately co-located with the tropospheric anticyclone (Fletcher et al. 2011). To the east and west, this cool region was flanked by two warm stratospheric "beacons" near 40°N, so-called because of their intensity when viewed in the infrared, and some 16 ±3 K warmer than the cold centre in January 2011. As we shall describe below, the evolution and merger of these stratospheric beacons into a single large anticyclone dominated the appearance of Saturn's northern stratosphere for some time after the tropospheric storm had ended.

The origin of the stratospheric beacon continues to be a source of debate. Although some degree of convective overshooting to the 60-mbar level might have been expected from moist convection models (e.g. Hueso and Sánchez-Lavega 2004), vertical mass motion from the storm genesis region should not penetrate high into the stratosphere. Saturn's atmosphere is so stably stratified in the tropopause region that updrafts would quickly become negatively buoyant, preventing further penetration into the stratosphere. Furthermore, even if the updrafts could be maintained by some mechanical forcing, the adiabatic cooling from expansion would be enormous, yielding regions at 1-mbar that were tens of Kelvin cooler than their surroundings, which was not observed. Thus, wave propagation, generated by the strong mechanical forcing associated with the tropospheric disturbance, is a plausible explanation for the origins of the warm beacons. However, this is only part of the story – we initially see diffuse anticyclonic vortices instead of periodic temperature structures that we might expect from vertically propagating waves, suggesting a markedly nonlinear phenomenon. Furthermore, we might have expected these two initial beacons to cool and fade from view, but the subsequent evolution was completely unprecedented.

Fletcher et al. (2012) studied the evolution of the two warm regions from January 2011 to March 2012, dividing the evolutionary period into three phases as shown in Figure 13.11. This figure has been extended to early 2013, identifying a fourth phase in the vortex evolution where it was moving with a higher velocity. Phase 1 (January to May 2011) saw the formation and intensification of the two warm regions (B1 and B2) with peak temperatures near 0.5 mbar. Both moved westward with time, with B1 residing directly above the tropospheric storm head and moving with the same velocity (2.73° ±0.1° day^{-1}), and B2 moving more slowly but in the same direction (0.6° ±0.1° day^{-1}). Over time, the storm head beacon caught up with the second, smaller warm region. These two warm air masses behaved like coherent vortices and merged when they encountered one another in late April 2011, forming a single large vortex (B0) with peak temperatures of 221.6 ±1.4 K at 2 mbar (80 K warmer than the quiescent stratosphere) measured on 5 May 2011, immediately after the merger at the start of Phase 2 (May to July 2011). This single vortex became disassociated from the tropospheric storm head and moved with an independent (slower) westward velocity (1.6 ±0.2° day^{-1}) that was intermediate between the initial velocities of B1 and B2. Furthermore, the peak temperatures in the vertical

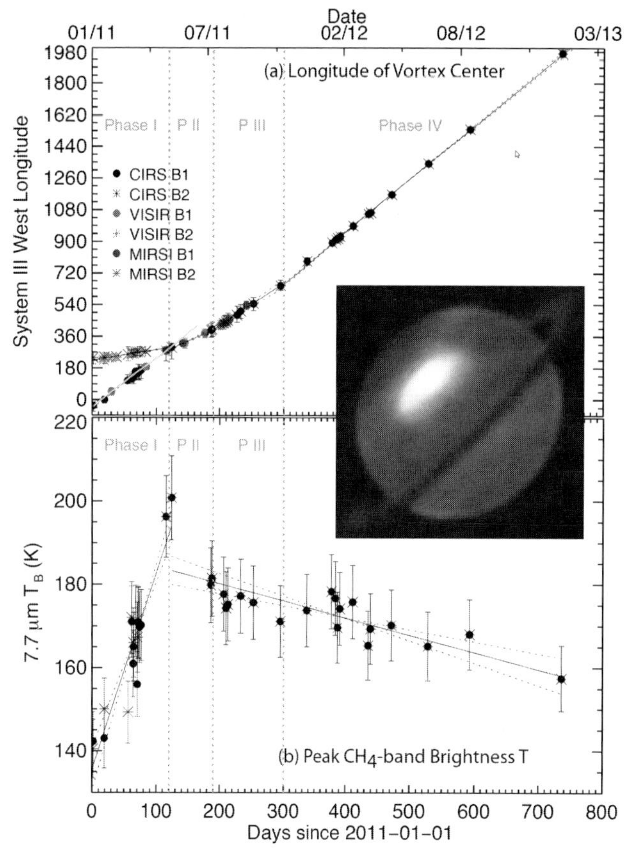

Figure 13.11 Evolution of the stratospheric vortex over two years as measured by Cassini/CIRS, updated from the analysis of Fletcher et al. (2012, with permission from Elsevier), which stopped at day 500. The upper panel shows the System III west longitude of the vortex centre, the lower panel shows the peak 7.7-µm brightness temperatures (i.e. the centre of the vortex). Different shaped points represent the data source – either Cassini (black solid dot), the VLT (light grey solid dot) or IRTF (grey solid dot). Linear regression lines are shown as solid lines with dashed uncertainties. The light grey line in the top panel is the longitude of the storm head as measured by Sayanagi et al. (2013). The inset VLT image shows the stratospheric beacon at 13.1 µm on 20 July 2011.

profile were found to be 1.4 scale heights deeper (approximately 162 km) post-merger in Phase 2, with the sudden subsidence presumably contributing to the rapid rise in temperature within the vortex. At its largest, B0 was 65° in longitudinal extent (50,000 km at 39°) and extended from 20° to 60°N latitude. The final decay phases (Phases 3 and 4) after July 2011 (i.e. after the tropospheric storm had abated) were characterized by slow cooling (0.11 ±0.01 K day^{-1}) and longitudinal shrinkage. During this phase, B0 accelerated from a constant velocity of 2.40° ±0.29° day^{-1} at the end of 2011 (−25.0 ±3.0 m s^{-1} at 40°N during Phase 3) to 3.01° ±0.05° day^{-1} throughout 2012–2013 (−31.3 ±0.5 m s^{-1} during Phase 4), meaning that it completed a circuit of the planet once every 120 days and was actually moving faster than the underlying retrograde jet in the troposphere. Guerlet et al. (2014) demonstrated that processes in addition to radiative cooling must be causing the temperature decay within the vortex, and that pure radiative decay would theoretically imply that the warm anomaly could persist until 2016. However, at the observed rate of cooling and shrinkage, Fletcher et al. (2012) predicted that the temperatures would return to their quiescent values by January 2013. CIRS observations have since suggested that the distinct vortex was absent in late 2013, having been replaced by substantial longitudinal wave activity that continues to create thermal contrasts within the 25° to 50°N region of the stratosphere as of the time of writing.

The 80–100 K stratospheric thermal contrast measured during Phase 2 was unprecedented, and led to unique chemical and dynamical consequences within the vortex itself. In the vertical direction, temperatures appeared to drop towards the quiescent stratospheric temperatures for P < 2 mbar, although CIRS nadir observations lack sensitivity to P < 0.5 mbar and limb observations were not able to sample the vortex itself. Calculations of the static stability at lower pressures (above the peak temperatures at 2 mbar) suggest that the atmosphere remained sub-adiabatic and therefore stable to convection. Assuming no horizontal motion at the tropopause and that the zonal and meridional temperature contrasts are in approximate geostrophic balance with the winds, integration of the thermal wind shear relation predicts that the hot air mass would be isolated by a peripheral clockwise wind of 200–400 m s^{-1} at 2 mbar (Figure 13.12). As upwelling and divergent tropospheric plumes tend to inject anticyclonic vorticity, this suggested that B0 was a large

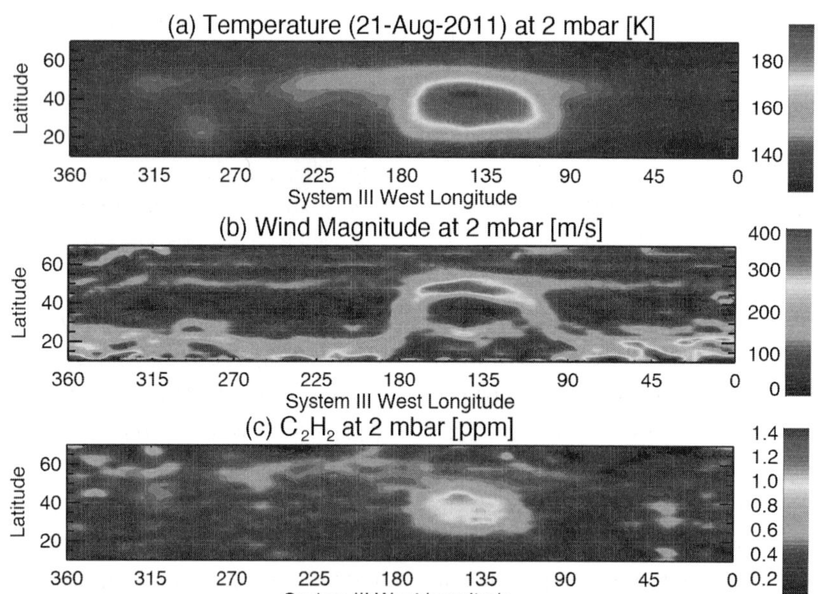

Figure 13.12 Retrieved stratospheric temperatures, wind magnitudes and acetylene abundance in the stratospheric vortex at 2 mbar derived from Cassini/CIRS low-spectral resolution mapping on 21 August 2011. Latitudes are planetographic and motion around the vortex is in a clockwise direction. Adapted from Fletcher et al. (2012), with permission from Elsevier. (A black-and-white version of this figure appears in some formats. For the color version, please refer to the plate section.)

stratospheric anticyclone only visible in the thermal infrared (Fletcher et al. 2012). Although the origin of the beacons was likely related to non-linear phenomena associated with waves rising from the tropospheric storms, the mature anticyclone can be likened to terrestrial anticyclones, which are regions of high pressure associated with subsiding, dry air. Extending this analogy to Saturn, friction at the lower boundary of the beacon (i.e. the tropopause) could cause outflow that is balanced by subsidence within the anticyclone to generate the warm temperatures by adiabatic compression. Indeed, after the merger, the warmest beacon temperatures were at a higher pressure (2 mbar) and the beacon became visible at wavelengths of 17–20 μm, suggesting subsidence and a warming of the tropopause region.

Stratospheric anticyclones are common features of the Earth's stratosphere, most apparent in northern winter and southern spring (Harvey and Hitchman 1996; Harvey et al. 2004), and possibly associated with the upward extension of planetary waves from tropospheric sources providing the source of energy (e.g. Charney and Drazin 1961). On Earth, these stratospheric anticyclones are regions of strong gradients of potential vorticity and trace species (e.g. Manney et al. 1995). On Saturn, the peripheral collar apparently isolated a region of air enhanced in stratospheric hydrocarbons (Figure 13.12), most notably acetylene (C_2H_2) and ethylene (C_2H_4), and a smaller enhancement in ethane (C_2H_6) (Fletcher et al. 2012; Hesman et al. 2012). These enhancements are most likely the result of the downward shift from 0.5 to 2 mbar (1.4 scale heights) post-merger, as stratospheric diffusive photochemistry models (Moses et al. 2015; Cavalié et al. 2015) have so far been unable to reproduce the chemical enhancements observed by Cassini without invoking downwelling winds. There was no evidence for effects of the stratospheric vortex in Saturn's tropospheric clouds and hazes as measured by Cassini/ISS and Cassini/VIMS (Fletcher et al. 2012). However, the evaporative effect on stratospheric hazes and water ice (condensed from externally supplied oxygen species) has not yet been investigated, but these may provide an additional gaseous source for species found enhanced within B0. Finally, B0 left in its wake a new warm stratospheric band centred at 48°–50°N and 0.3–0.8 mbar. Thermal wind balance therefore suggests increasingly eastward flow with altitude at latitudes poleward of 50°N and increasingly westward flow with altitude equatorward of 50°N.

Although B0 has now met its demise, mysteries surrounding its origin remain unresolved. Convective plumes impinging on the tropopause act as strong sources of waves (gravity waves, planetary waves) which propagate in 3D away from the source. The stratospheric transmissivity to these waves depends on the details of the temperature and wind structure near the tropopause (and therefore the refractive index), and simple calculations by Fletcher et al. (2012) suggest that upward propagation of Rossby waves was permitted in the springtime epoch of this storm. Nonlinear effects and wave breaking may have led to the formation of the original vortices near 0.5 mbar, although this phenomenon has not yet been reproduced by any numerical simulation of Saturn's stratosphere. Indeed, the formation and evolution of Saturn's stratospheric beacon should be a key challenge for the next generation of coupled radiative-dynamical-chemistry models. Key observational constraints would come from the presence (or absence) of stratospheric vortices spawned by similar convective storms on Saturn and the other giant planets.

13.5.3 Stratospheric Molecular Composition

In this section reviewing stratospheric molecular composition, we continue to refer to the four phases of evolution of Saturn's stratospheric anticyclone: Phase 1 when the two initial beacons were present and growing; Phase 2 when the beacons had merged; and Phases 3 and 4 during the decay phase of the mature beacon. The high stratospheric temperatures led to enhanced emission from both major hydrocarbon species (ethane C_2H_6 and acetylene C_2H_2; Fletcher et al. 2012; Hesman et al. 2012) and minor hydrocarbon species (C_2H_4, C_3H_4, C_3H_8, C_4H_2; Bjoraker et al. 2012). For many of the hydrocarbons, the enhanced emission cannot be solely attributed to the rise in atmospheric temperature (Bjoraker et al. 2012). Neither are they solely the result of the shifting photochemical reaction rates associated with the warmer temperatures (Cavalie et al. 2015; Moses et al. 2015). Indeed, the elevated abundances of the hydrocarbons can be used as diagnostic tracers of the subsiding motion within the stratospheric anticyclone, moving photochemically produced molecules downwards (e.g. Fletcher et al. 2012; Hesman et al. 2012; Moses et al. 2015).

The unique high temperatures, chemistry and dynamics of the stratospheric beacon resulted in the detection of ethylene (C_2H_4) by Cassini's Composite Infrared Spectrometer (CIRS) and the high-resolution ground-based spectrometer Celeste deployed on the McMath–Pierce telescope on Kitt Peak, Arizona. CIRS was also used to identify other rare hydrocarbon species such as diacetylene (C_4H_2) and methylacetylene (C_3H_4) (storm Phases 1, 2 and 3). Saturn's ethylene emission has been difficult to detect on Saturn (Encrenaz et al. 1975; Bezard et al. 2001). Indeed, prior to the 2010 storm, ethylene was not detected by CIRS. However, stellar occultation data acquired by Cassini's Ultraviolet Imaging Spectrograph (UVIS) yielded a vertical profile of C_2H_4 under pre-storm conditions between 0.1 μbar and 0.5 mbar at 15.2°N (Shemansky and Liu 2012). This UVIS profile then provides a profile to compare the CIRS storm results at 0.5 mbar against. The May 2011 CIRS (Figure 13.15b) and Celeste (Figure 13.13a) spectra were used to determine a C_2H_4 mole fraction of $(5.9 \pm 4.5) \times 10^{-7}$ at 0.5 mbar (CIRS) and $(2.7 \pm 0.45) \times 10^{-6}$ at 0.1 mbar (Celeste) (storm Phase 2). These values are both two orders of magnitude higher than the amount measured by UVIS under pre-storm conditions. It is also much higher than predicted by photochemical models, indicating that perhaps another production mechanism is required or a loss mechanism is being inhibited (Hesman et al. 2012). Because ethylene leads directly to the formation of other, more-prominent

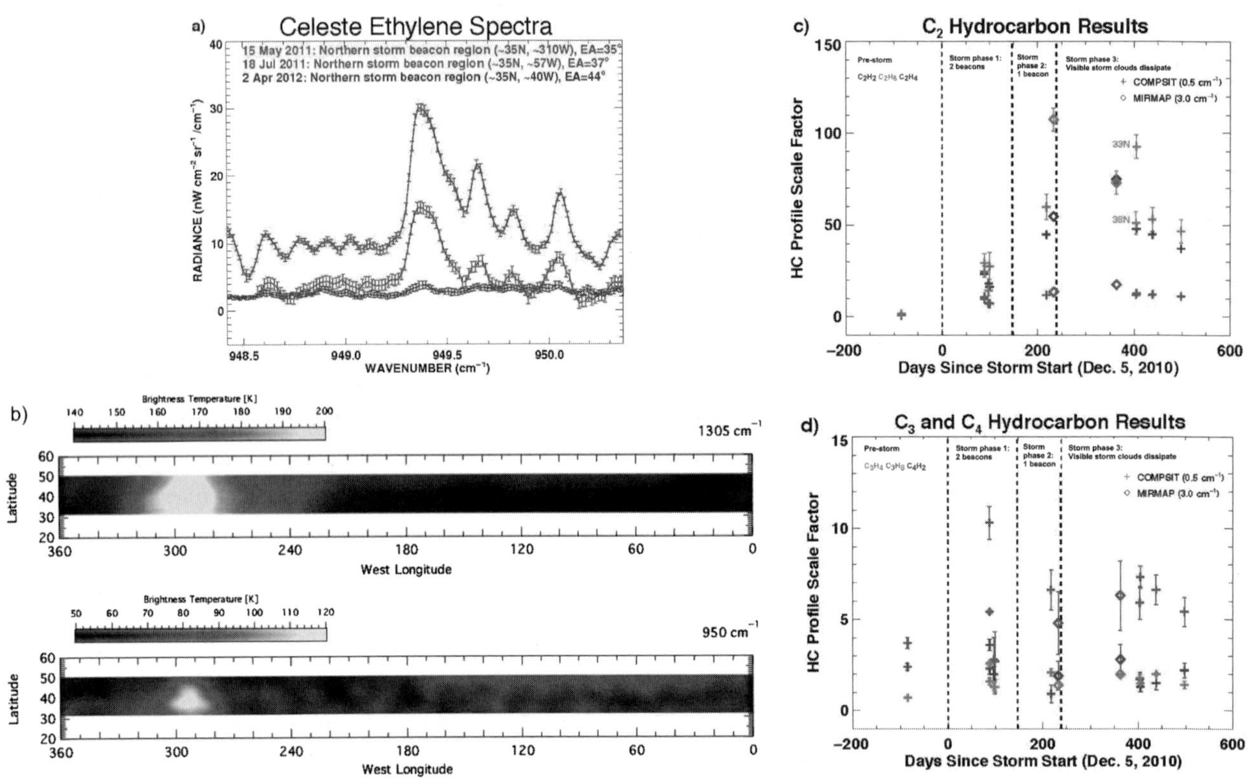

Figure 13.13 (a) (top-left): Celeste spectra of ethylene (C_2H_4) measured on 3 occasions in 2011 and 2012. The red curve (May 2011) and green (July 2011) curves were measured in phase 2 and the blue (April 2012) curve was measured in phase 3. (b) (bottom-left): CIRS maps of methane and ethylene measured in May 2011 (phase 2). The maps show that the two beacons of phase 1 have been merged into one warm bright beacon in the methane band. In the ethylene region the map shows that ethylene emission exhibits latitudinal structure (Hesman et al. 2012). (c) (top-right): The C2 hydrocarbon temporal response in the pre-storm atmosphere and all phases of the storm. These abundances were measured by scaling photochemical profiles to the CIRS spectra. The scale factor on the HC profiles is shown on the y-axis (Hesman et al. 2013). (d) (bottom-right): The C3 and C4 hydrocarbon scale factors for pre-storm and all phases of the storm. These abundances were measured by scaling photochemical profiles of these species to the CIRS spectra (Hesman et al. 2013). (A black-and-white version of this figure appears in some formats. For the color version, please refer to the plate section.)

hydrocarbons such as ethane, understanding how it is created and destroyed is important to understanding the photochemistry in giant planet atmospheres.

Full radiative transfer analyses are required to disentangle the effects of temperatures and abundances, and these have been performed for multiple hydrocarbon species by Hesman et al. (2012, 2013), Fletcher et al. (2012) and Moses et al. (2015). The results from Hesman et al. (2013) are shown in Figure 13.13c and 13.13d. Scale factors were applied to profiles produced from photochemical models of Hesman et al. (2012). These results confirm that the C_2 hydrocarbons are significantly enhanced in Saturn's beacons compared with pre-storm values. C_2H_4 exhibits the greatest enhancement followed by C_2H_2. For the January 2012 dataset, there is latitudinal structure within the beacon, with larger abundances for C_2H_4 at 33°N than at 38°N. And finally, the C_3 and C_4 hydrocarbon abundances exhibit only minor enhancement in Saturn's beacons, with emission that can be mostly explained by the rise in temperature.

Moses et al. (2015) and Cavalie et al. (2015) attempt to explain these hydrocarbon observations via photochemical modeling. Moses et al. find that the temperature rise would leave ethane, acetylene and propane relatively unaffected, whereas we would expect an ethylene increase due to the elevated temperatures. Nevertheless, these photochemistry models still underpredict the observed abundances by factors of 2 to 7 in the beacon core in May 2011, shortly after merger. Downwelling winds of order 10 cm s^{-1} near 0.1 mbar, in addition to the aforementioned chemistry, are required to explain the elevated abundances in the C_2 hydrocarbons (Moses et al. 2015), although we caution that these 1-dimensional (1D) winds are likely a gross simplification of the 3-dimensional (3D) dynamics at work within this anticyclone. With these winds in place, the model shows that we would expect a factor of 2.2 increase in C_4H_2 (primarily driven by evaporation of ices), whereas the effects on C_3H_8 and C_3H_4 would be relatively minor, consistent with the preliminary findings of Hesman et al. (2013). Evaporation of water ice and C_6H_6 aerosols could also lead to increases in the gas phase abundances of water and benzene within the beacon.

In the coming years, further analysis is needed to compare the hydrocarbon enhancements predicted from chemistry and dynamical models (including 3D simulations of the vortex dynamics) to the reality provided by Cassini and Earth-based studies of Saturn's hydrocarbon emissions. Photochemical models should be updated with the time-dependent temperatures, species abundances and rate mechanisms. In addition, other sources of these species, such as aerosol vaporization, should be examined in conjunction with photochemical models. It is only by doing these studies that the compositional changes in Saturn's Great Storm of 2010–2011 will be understood.

13.6 Lightning Events

The observation of Saturn lightning events has had a long history. The first indication of the existence of lightning in Saturn's atmosphere was obtained in November 1980 by the radio instrument onboard Voyager 1, which recorded strong and impulsive radio signals (Warwick et al. 1981), referred to as Saturn Electrostatic Discharges (SEDs), They were recorded again during the Voyager 2 flyby in August 1981, and, since 2004, by the RPWS (Radio and Plasma Wave Science) instrument (Gurnett et al. 2004) onboard the Cassini orbiter (Fischer et al. 2006). The multi-year presence of Cassini in Saturn's orbit has enabled the first observations of lightning during a GWS event (Fischer et al. 2011a), confirming – along with the water-laden cloud particle evidence mentioned in Section 13.4.2.4 – that GWS events are giant thunderstorms with strong vertical convection (e.g. Sánchez-Lavega 1994).

13.6.1 Properties of Saturn Electrostatic Discharges

The SEDs are short-duration (~100 ms), impulsive radio bursts with a large frequency bandwidth. In order to penetrate Saturn's ionosphere and reach the spacecraft, their frequency has to be greater than the ionospheric cut-off frequency, the latter being of the order of a few hundred kHz to a few MHz on the nightside and dayside, respectively (Kaiser et al. 1984; Fischer et al. 2011b). In the frequency range of SEDs, the RPWS instrument acts as a sweeping receiver that detects the radio burst at whatever frequency the receiver happens to be tuned to at the time of the flash. This leads to the typical occurrence of SEDs as

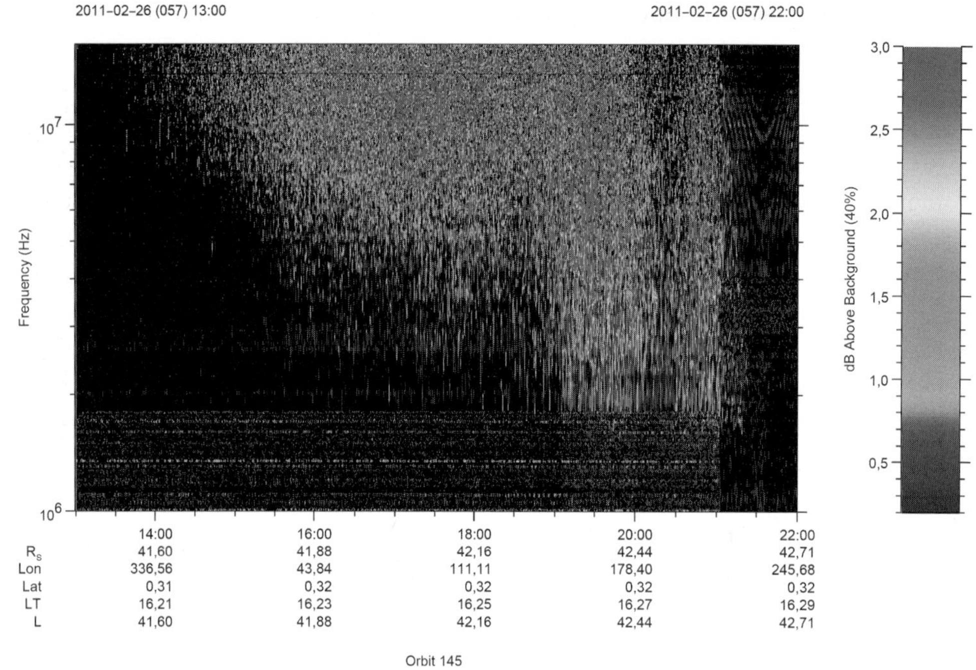

Figure 13.14 Dynamic spectrogram of one SED episode during the GWS on 26 February 2011. The color bar on the right side indicates the radio wave intensity in dB above the galactic background, which is displayed as a function of time over 9 hours and frequency in logarithmic scale from 1 to 16 MHz. Besides spacecraft event time (SCET), the following Cassini orbital parameters are indicated at the abscissa: Cassini distance to Saturn's center in Saturn radii (RS), sub-spacecraft western longitude (Lon), Cassini planetocentric latitude (Lat), and its local time (LT in hours). The decrease of SED rate around 20:30 is due to the disappearance of the storm's head at the eastern limb of Saturn. Later, the over-horizon effect allows observing SEDs also beyond the visible horizon. (A black-and-white version of this figure appears in some formats. For the color version, please refer to the plate section.)

short vertical bursts in the time-frequency spectrum of Figure 13.14, despite the fact that short bursts as SEDs are intrinsically broadband emissions. Figure 13.14 shows that single vertical bursts can only be discerned around the edges of the SED episode, whereas in its centre and especially around 19:30 SCET the flash rate is so high that the entirety of SED bursts appear as a patchy emission with no isolated single bursts. Here the observed peak flash rate is > 10 s^{-1}, a rate that surpasses the typical flash rates of smaller SED storms by about 2 orders of magnitude. Another remarkable feature of the SED episode in Figure 13.16, recorded during the GWS on 26 February 2011, is its long duration of almost 8 hours (13:20 to 21:20 SCET). The frequency range of SEDs in Figure 13.16 ranges from ~16 MHz (highest frequency of the RPWS instrument) to a lower cut-off frequency near 1 to 2 MHz. SEDs can be used as a natural tool to study Saturn's ionosphere. The measurement of the lower cut-off frequency combined with the knowledge of the SED source position is the only method which can reveal peak electron densities of Saturn's ionosphere at all local times (Fischer et al. 2011b).

Before we discuss the SED properties observed during the 2010–2011 GWS in more detail, we highlight the main properties of regular (smaller) SED storms as they were observed from 2004 to 2010 by the Cassini spacecraft (blue dots in Figure 13.1). The typical flash rate of regular SED storms observed by Cassini is a few SEDs per minute (Fischer et al. 2008). Most of the 10 regular SED storms observed from 2004 until 2010 lasted for several weeks to months, and the longest one raged in 2009 for almost the entire year. However, there were also long intervals of time with no SED activity, e.g. from February 2006 until November 2007. The SED storms were correlated to cloud features of about 2000 km in diameter located at the so-called storm alley at planetocentric latitude of 35°S (Fischer et al. 2011c; Baines et al. 2009). The first link between SEDs and optical storm cloud observations was found by combined observations of the Cassini cameras and the RPWS instrument (Porco et al. 2005). They revealed consistent longitudes and longitudinal drift rates of the cloud feature in comparison to the SED occurrence, and the white storm clouds were also found to be brighter when the SED rates were high (Dyudina et al. 2007; Fischer et al. 2007a). A white cloud captured by Cassini/VIMS in February, 2008 was also observed to be at the longitude where SED lightning was being contemporaneously observed (Baines et al. 2009). The prominent 2000-km sized storm clouds turned out to be easily detectable by amateur astronomers on Earth (Fischer et al. 2011c), who for almost a decade have provided valuable Saturn storm cloud observations on the website of the Planetary Virtual Observatory (Hueso et al. 2010).

SEDs usually occur in episodes lasting for about 6–7 hours, which is somewhat longer than half a Saturn rotation (~5 h 20 min or over 180° in sub-spacecraft longitude). This is due to the fact that most SED storms consist of a single storm cell from which SEDs are observed when the storm is on the side of the planet facing the spacecraft (the orbital period of Cassini is always much longer than a Saturn rotation). A radio wave propagation effect also allowed SEDs to be detected "over the horizon" by up to 45° in longitude when the SED source rose or set at the nightside horizon (Zarka et al. 2006). Consequently, an SED episode lasting for 8 hours (or over ~270° in sub-spacecraft longitude) as in Figure 13.15 must be caused by at least two or several longitudinally separated storm cells.

SEDs have a spectral radio source power of the order of 10 to 100 W/Hz, which is about 10,000 times stronger than the radio signals of terrestrial lightning (in the frequency range of a few MHz) (Fischer et al. 2006; Zarka et al. 2006). Taking an SED duration of 0.1 s and an SED frequency range of 10 GHz (similar to terrestrial lightning) one would arrive at a value of 10^9–10^{10} J for the radio energy of one SED, which is similar to their optical energy (Dyudina et al. 2013). This is only a rough estimate, since we have no information on the SED frequency range and do not know how the radio intensity varies with frequency. The high SED intensity (flux ~100 Jansky at Earth) and Cassini's reports of ongoing lightning activity enabled the first SED detection by the large ground-based radio telescope UTR-2 in the Ukraine (Zakharenko et al. 2012; Konovalenko et al. 2013).

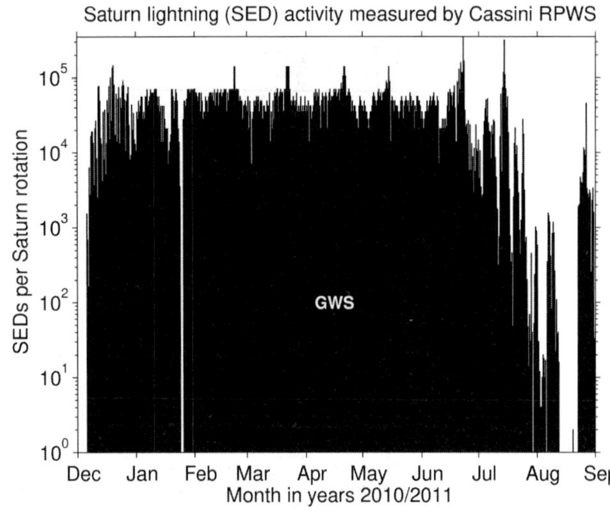

Figure 13.15 SED pixels per Saturn rotation as a function of time during the GWS. The numbers of observed SED pixels (single time-frequency measurements) are plotted as black bars in a logarithmic scale. The small gap around the end of January 2011 is due to a data gap when RPWS was turned off for a short time.

Below ~2 MHz, SEDs are highly polarized emissions with a high degree of circular polarization. Fischer et al. (2007b) found SEDs to be right-hand polarized when emitted from the southern hemisphere, and they predicted that SEDs would be left-hand polarized when emitted from the northern hemisphere, a difference that is related to the direction of the magnetic field vector in the two hemispheres relative to the radio wave propagation vector. This prediction was confirmed by polarization measurements of SEDs from the 2010–2011 GWS, which was the first SED storm located in the northern hemisphere during the Cassini era (Fischer et al. 2011a).

13.6.2 Radio and Optical Observations of Lightning During the GWS

Figure 13.15 shows the number of detected SEDs as a function of time over nine months. Precisely speaking, it is the number of time-frequency events ("SED pixels") that is displayed in the figure for the following reason. Single SEDs normally extend over a few frequency channels, but high flash rates of 5–10 SEDs per second can lead to a temporal superposition of SEDs (Fischer et al. 2011a). As a consequence, single

SEDs can no longer be resolved, because the most common receiver sweep rate is just 28 frequency channels per second (35.2 ms per channel). It can be seen in Figure 13.15 that the number of SED pixels per Saturn rotation quickly (within 1–2 weeks) rose to values of 10^4 to 10^5, and remained at that level until mid-June 2011. This level is about an order of magnitude higher than the rate of regular (smaller) SED storms that typically emit 10^3 to 10^4 SEDs per Saturn rotation. Since the intensity of SEDs decreases with distance squared, the SED rate also depends on the distance of Cassini from Saturn. SEDs from regular SED storms and the 2010–2011 GWS have about the same spectral radio wave intensity; instead, only the SED rates are significantly different for these two classes of Saturn lightning storms (few SEDs per minute for regular storms and few SEDs per second for GWS). Flash rates of terrestrial thunderstorms are proportional to the horizontal area of the charge region (Larsen and Stansbury 1974), suggesting that the charge region with convective activity was at least ten times larger for the GWS than for the regular, smaller Saturn lightning storms. The total number of SED pixels observed by the Cassini RPWS instrument during the great storm of 2010–2011 was almost 30 million.

Consistent with optical observations, the first SEDs of the GWS event were recorded on 5 December 2010, and the last ones on 28 August 2011. Figure 13.15 shows a large decrease of the SED rate around 20 June 2011. This is most likely the consequence of the collision of the storm's head with the large anticyclonic vortex (Section 13.3). The anticyclonic vortex in the tail had drifted westward with a rate of ~2° day^{-1} slower than the storm's head, leading to the head-vortex collision as described in Section 13.3. Figure 13.16 shows the storm's head on the left side and the anticyclonic vortex in the tail on the right side, separated by ~120° in longitude at the end of February 2011. After the first large decrease in lightning activity on 20 June, the storm regained some strength and showed intermittent strong SED activity around 1 July, 10 July, and 20 July. For the rest of the time until the end of August, the SED activity was comparable to regular, smaller storms, with almost no activity during mid-August. Some additional small SED activity not shown in Figure 13.15 was present at the end of September/early October 2011 and at the end of December 2011. However, it is not clear where the source of this SED activity was located: either there was a small storm cell in the remaining visible clouds of the GWS at 35°N or in some other clouds at 50°N.

During the GWS the Cassini cameras made the first optical observation of Saturn lightning flashes on the day side (Dyudina et al. 2013). This was quite surprising, since it was even difficult to detect Saturn lightning flashes on the night side. Except at equinox, the light reflected from Saturn's rings causes the night side of Saturn to be brighter than Earth under a full moon, making it impossible to distinguish clouds that are illuminated from below by lightning from clouds that are illuminated from above by ring shine. Due to this, the first visible detection of lightning on Saturn was made on 17 August 2009, just a few days after Saturn northern vernal equinox (Dyudina et al. 2010),

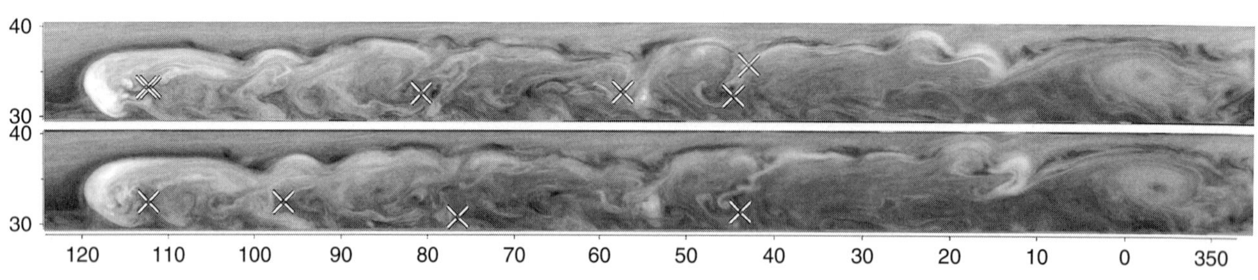

Figure 13.16 High-spatial-resolution views of the 2010–2011 GWS by the Cassini cameras on 26 February 2011. One can clearly see the head of the storm (between 110° and 120° west longitude) and the large anticyclonic vortex (between 350° to 10° longitude). The flashes detected with the blue filter are indicated as white crosses. The lower panel shows the storm ~11 h after the view in the upper panel. Both panels go over ~10° in latitude and ~140° in western longitude. Adapted from Dyudina et al. (2013), with permission from Elsevier.

when the rings were edge-on to the Sun. The Cassini camera detected flash-illuminated cloud tops with diameters around 200 km on Saturn's night side at ~36°S latitude, accompanied by lightning-related radio emissions recorded by the Cassini RPWS instrument. The diameters of the bright spots suggested that the lightning is taking place 125–250 km below the cloud tops, which is above the base of the liquid H_2O-NH_3 (water-ammonia) cloud and most likely in the water ice cloud (Dyudina et al. 2010).

The lightning flashes during the GWS event were detected on the dayside in blue light only, leaving the lightning spectrum still unknown (Dyudina et al., 2013). The flashes could be detected on the dayside due to the high flash rate and by the technique of image subtraction. Since sunlit clouds do not change much within a few minutes, repeated images of the same area were used for background subtraction to reveal lightning. It turned out that the optical flashes occurred in the diagonal gaps between large anticyclones, which can be seen in Figure 13.16. The head of the storm periodically spawned large anticyclones, which drifted off to the east with a longitudinal spacing of 10°-15°. The flashes marked by white crosses in Figure 13.16 occur between them in cyclonic gaps, where the atmosphere looked clear down to the level of deep clouds (Dyudina et al., 2013). The flashes in Figure 13.16 are distributed over a longitude range of ~70° and the western edge of the head at ~120° longitude is ~75° away from easternmost flashes at ~45°. The first SEDs in Figure 13.16 are detected at 315° sub-spacecraft western longitude, which is consistent with an SED source at ~45° rising on the western dayside horizon. Due to Saturn's eastward rotation the easternmost SED sources are detected first before the storm's head rises on the horizon leading to an increase in SED activity. The longitudinal range of the optical flashes in Figure 13.16 is thus consistent with the extended sub-spacecraft longitudinal range (~270°) over which SEDs were detected in late February 2011.

The optical energies of single Saturn lightning flashes were found to be typically ~ 10^9–10^{10} J, which is comparable to the total energy of a terrestrial flash (Dyudina et al., 2010; 2013). Since optical energies of flashes are just 0.1% of their total energy (Borucki and McKay, 1987), the total energy of an average Saturnian lightning flash should be in the range of 10^{12}–10^{13} J. This energy is comparable to the energy of a terrestrial *superbolt*. The vast amount of lightning energy goes into heating the channel and creating a shock wave. Fischer et al. (2011a) also estimated the total power of the GWS from the volume of the thundercloud and the latent heat release through condensation of water and arrived at a value of 10^{17} W, which is comparable to Saturn's total radiated power (~5 W/m²; corresponding to ~ 2.2 10^{17} W over Saturn's surface area of 4.3 10^{16} m²). Dyudina et al. (2013) arrived at the same value via the following calculation. An average SED rate of a few flashes per second (peak rate >10 s^{-1}) for the GWS, with a total energy of 10^{12}–10^{13} J per flash, leads to a power of ~10^{13} W released by lightning. The energetic efficiency of convective updrafts in producing lightning flashes is unknown for Saturn, but for Earth and Jupiter it is estimated to be around 10^{-4} (Borucki et al. 1982). Hence, a flash power of ~10^{13} W for the SEDs of the GWS should be related to a total storm power of ~10^{17} W, similar to the Fischer et al. (2011a) result. This suggests that storms such as the 2010-2011 GWS are important players in the energy budget of Saturn's atmosphere. Furthermore, Fischer et al. (2014) suggested that the great storm of 2010–2011 might even have influenced the periodicity of Saturn kilometric radiation, an auroral radio emission which is modulated by Saturn's rotation. The GWS was most likely a source of intense gravity waves, similar to terrestrial thunderstorms (Pierce and Coroniti 1966) that may have caused a global change in Saturn's thermospheric winds via energy and momentum deposition. This supports the theory that Saturn's magnetospheric periodicities are driven by the planet's upper atmosphere.

13.7. Storm Dynamics

As noted in Section 13.2, the dynamics of a GWS can be divided in three steps that involve the (1) outbreak, (2) rapid growth to a large spot in interaction with the local wind and (3) the zonal expansion to the planetary-scale disturbance. Here, we review our current understanding of the dynamics involved in each of these steps, as gleaned in particular from state-of-the-art computer modelling compared to the wealth of observations provided by Cassini and ground-based observers.

13.7.1 The Convective Source

It is generally accepted that the outbreak of a GWS is due to moist convection in Saturn's cloudy weather layer (Sánchez-Lavega and Battaner 1987; Hueso and Sánchez-Lavega 2004). Thermochemical models predict the existence of three cloud types (NH_3-ice, NH_4SH-ice, and H_2O-ice on top of an H_2O-NH_3 aqueous solution) that extend approximately between 1 and 10 bar depending on the abundances of N, O and S (Atreya and Wong 2005; West et al. 2009). The convective hypothesis involves fast upward motions and is supported by the observed initial rapid evolution, fast-expanding velocities with related horizontal divergence at cloud tops, high spot brightness (dense cloud containing fresh ices), lightning events and spectroscopic ice detection at the cloud tops (Section 13.4).

One-dimensional early parcel models of moist convection in Saturn showed that water located deep at 10–12 bar is the most effective fuel to heat the parcels through latent release and feed the updrafts, supported by an additional vertical driving given by ammonia at higher levels as the parcel ascends (Sánchez-Lavega and Battaner 1987). Since water and ammonia have molecular weight greater than the ambient hydrogen atmosphere (by a factor ~7–8), the ascent is favoured when the parcel conditions are close to saturation with low entrainment of dry, lighter environmental air into the parcel. Triggering moist convection under such conditions requires an initial heating of the parcel or a vertical initial velocity (see Hueso and Sánchez-Lavega 2004). The 1D models showed that for a highly enriched water atmosphere (5–10 times the solar oxygen abundance), the parcels are able to ascend the ~260 km vertical distance from the water clouds to the tropopause located at the ~100 mbar, reaching peak velocities up to ~150 ms^{-1}. This can be seen to first order calculating the "Convective Available Potential Energy" (CAPE) given by Sánchez-Lavega (2011),

$$\mathrm{CAPE} = \int_{z_1}^{z_2} g\frac{\Delta T}{T}dz = \frac{w_{\max}^2}{2}, \qquad (13.1)$$

and then the maximum vertical velocity is given $w_{\max} = \sqrt{2\mathrm{CAPE}} = \sqrt{2g\frac{\Delta T}{T}h} \sim 150$ ms^{-1} for $g = 10$ ms^{-1}, $T = 250$ K, $h = 260$ km, and assuming a temperature difference between the ascending parcel and the environment of just 1 K (Hueso and Sánchez-Lavega 2004). When the updrafts reach the tropopause or the upper troposphere, overshooting could occur and mass flow conservation produces the horizontal expansion at the top and divergence in the velocity field. Vertical velocities in the diverging altitude can be simply computed by $w \leq h \cdot$ divV where $\mathrm{div}_h V = (1/A)(dA/dt) \approx 1 - 5 \times 10^{-6}$ s^{-1} (A is the area), as observed for the GWS 1933, 1990, 2010, yielding $w \leq 2.5$ ms^{-1}.

Three-dimensional water and ammonia moist convective models were developed by Hueso and Sánchez-Lavega (2004). The model uses the anelastic approach and parameterized microphysics (Hueso and Sánchez-Lavega 2001) in a mesoscale domain of few hundred kilometres (i.e. at the thermal plume scale) to study the trigger conditions and storm development. The most energetic thermal plumes are those resulting from water convection, which reaches the 150 mbar altitude level, developing peak vertical velocities of 150 ms^{-1}. When the atmosphere approaches instability, a very small temperature perturbation of $\Delta T = 0.15$ K is enough to initiate the onset. Because of the large path, the Coriolis force plays a major role in the thermal plume morphology.

Figure 13.17 shows a high-resolution image of the 2010 GWS storm head and the measured wind field (García-Melendo et al. 2013). The convective source is the brightest area, spanning a meridional width ~5,000 km and a zonal length of ~2,500 km. Individual cumulus-like convective spots ~200 km in size and cumulus clusters can be observed. The interaction of the storm head with the westward jet where it resides generated a bowshock-like front with an arc-shaped morphology that approximately follows the zonal wind profile. However, the motions along this arc are intense, with top velocities reaching 160 ms^{-1} in an anticyclone open flow in gradient wind balance (García-Melendo et al. 2013). The radial pressure gradient was 2–5 times that of classical Earth mid-latitude baroclinic anticyclones and ~2 times that of Jupiter's Great Red Spot, the largest anticyclone in the solar system. Numerical models that substitute the convective storm by a source injecting mass vertically

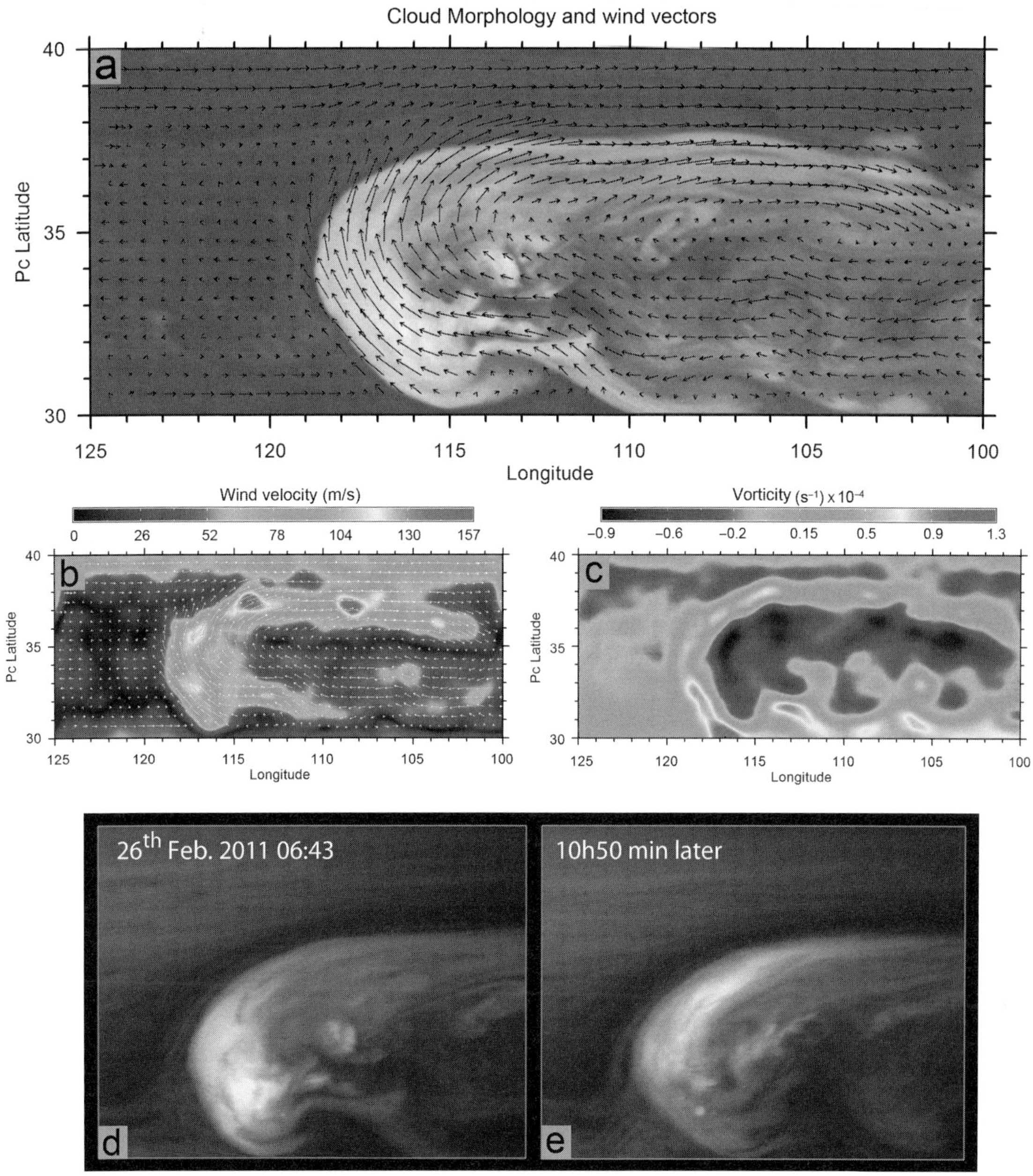

Figure 13.17 The head of the Great Storm GWS 2010–11 from the analysis of Cassini ISS images on 26 February 2011. (a) Wind velocity field vector superimposed to an image of the head of the storm; (b) mean wind velocity map; (c) vorticity map; (d) and (e) images of the head separated by 10.5 h showing rapid changes of the cloud tops. Adapted from García-Melendo et al. (2013). (A black-and-white version of this figure appears in some formats. For the color version, please refer to the plate section.)

are able to reproduce the observed morphology and dynamics under appropriate conditions in the weather layer. This requires that the static stability of the surrounding atmosphere measured in terms of the Brunt–Väisälä frequency (Sánchez-Lavega 2011) should be low with $N \sim 1.7 \times 10^{-3}$ s^{-1} to keep ambient conditions close to vertical instability. Importantly, to keep the disturbance morphology as observed, the source must inject mass continuously, which means that moist convection must be active for a long period. This is consistent with the continuously observed lightning activity (Section 13.6). At the same time, the zonal winds must extend vertically across the weather layer (from ammonia to water clouds) but with a low wind shear $\partial u/\partial z \approx 5$ ms$^{-1}/H$ or lower, with $H \sim 40$–100 km being the scale height.

13.7.2 The Planetary-Scale Disturbance

High-resolution data obtained by the Cassini spacecraft on the Great Storm of 2010–2011 and the intensive ground-based coverage allowed scientists, for the first time, to collect a unique wealth of data on the onset and evolution dynamics of a giant storm on Saturn (Sánchez-Lavega et al. 2011; Sánchez-Lavega et al. 2012; Sayanagi et al. 2013). The observed dynamical and morphological details provided a unique opportunity to perform numerical experiments to constrain Saturn's atmospheric structure in terms of fundamental variables such as vertical wind shear and vertical thermal structure. The 2010 event was simulated using the 3D Explicit Planetary-Isentropic Coordinate (EPIC) Atmospheric Model (Dowling et al. 1998), and shallow-water (SW) models (García-Melendo and Sánchez-Lavega 2016). By using these models, it was found that the most important morphological and dynamical features of the giant storm could be reproduced remarkably accurately, as we now discuss.

13.7.2.1 EPIC Models of the Great Storm

The atmosphere of Saturn can be represented as a fluid under rotation, whose motions are governed by classical geophysical fluid dynamics equations (Pedlosky 1979; Sánchez-Lavega 2011): the Navier–Stokes equation, the continuity equation, the state equation and the first law of thermodynamics. Simplified versions of these equations valid under certain conditions met by the Great Storm allow numerical models to be calculated significantly faster without leaving out the essential physics. The EPIC model assumes that air parcel movements are adiabatic, or in other words, the potential temperature obeys $d\theta/dt = 0$ (an air parcel conserves its potential temperature). As a consequence, the Ertel potential vorticity (*PV*)

$$PV = \frac{(\vec{\omega} + 2\vec{\Omega})}{\rho} \bullet \nabla\theta \qquad (13.2)$$

is also a conserved quantity where $\vec{\omega} = \nabla \times \vec{u}$ is the parcel vorticity vector ($\vec{u}$ is the parcel 3D velocity on the planet), $\vec{\Omega}$ is the rotation vector whose modulus is the planet's rotation angular velocity and whose direction follows the planet's rotation axis in the south-north direction, and ρ is fluid's density. Since potential vorticity is conserved by air parcels, it can be used as a passive tracer to represent atmospheric movements. In addition, EPIC uses isentropic coordinates where the potential temperature substitutes for the natural vertical coordinate z. Under this choice, the vertical domain can be divided in layers of constant θ where movements are purely horizontal ($\dot{\theta} = d\theta/dt \equiv 0$). The adiabatic approximation has proven to be very useful to model meteorological features in the giant planets (e.g. LeBeau and Dowling 1998; Garcia-Melendo et al. 2007; Legarreta and Sánchez-Lavega 2008; Sayanagi et al. 2010).

The model needs a set of free parameters to specify the atmospheric structure. The most important are the zonal wind profile *U(y)*, the vertical wind shear *U(z)* and the vertical static stability *N(z)*. The simulated vertical domain extends from the stratosphere at 10 mbar, down to the water condensation level at ~10 bar where the storm originated (Sánchez-Lavega and Battaner 1987; Hueso and Sánchez-Lavega 2004).

There is precise information on *U*(*y*) at the visible cloud deck (~500 mbar) in the latitude band where the 2010 GWS appeared before the storm's onset. Since orbital insertion in 2004, Cassini has monitored the global zonal circulation at different wavelengths, in particular establishing its stability during the time preceding the storm's outbreak between 2004 and 2011 (Porco et al. 2005; Vasavada et al. 2005; García-Melendo et al. 2011). The vertical thermal profile is also well known above the visible cloud deck down to ~1 bar. It was retrieved by the radio occultation experiments performed during the Voyager mission flybys (Lindal et al. 1985) down to ~1 bar, and later corrected for the updated helium molar fraction of 0.13 (Conrath and Gautier 2000), which was not well determined at the time of the Voyager experiments. Vertical wind shear is mostly unknown under the visible cloud deck. Assuming that the 2010 GWS source is at the 10-bar water-condensation level and its observed velocity is rooted at this depth, then the winds must strengthen with depth (Sánchez-Lavega et al. 2011; Sánchez-Lavega et al. 2012), or at least maintain its intensity. Therefore, Sánchez-Lavega et al. (2011) 3D simulations tested constant and slightly increasing winds with depth.

The vertical thermal profile is also unknown at deep levels under the visible cloud deck. The corrected thermal profile is somewhat unstable for p > 400 mbar (Brünt–Väisälä frequency $N^2 < 0$), but due to the presence of convective activity and vortices, stable and unstable conditions can coexist (Sánchez-Lavega et al. 2003, 2004; Porco et al. 2005). EPIC does not support convective motion; therefore, the simulations assume that the atmosphere is statically stable. This is a realistic assumption for atmospheric levels above the tropopause, and a reasonable approximation for deeper levels, if low static stability values are used, taking in mind that static stability may oscillate around a neutral value and likely approaches it as we get closer to the 10-bar level where $N^2 \approx 0$. Vertical thermal profiles were tested in the range $0.03 \times 10^{-4}\ \mathrm{s}^{-2} \leq N^2 \leq 0.7 \times 10^{-4}\ \mathrm{s}^{-2}$ for the deepest levels (Sánchez-Lavega et al. 2011).

13.7.2.2 *The Shallow-Water Model*

The relevant fluid equations can be simplified further by assuming that atmospheric movements take place in a relatively thin sheet of fluid compared to the much larger horizontal scale of movements, and that the atmospheric density is spatially constant within each pressure level. The first assumption is a very reasonable one if we consider that there is a ~260 km excursion from the water condensation level at ~10 bar to the tropopause (Hueso and Sánchez-Lavega 2004), much smaller than the ~10^4 to 10^5 km horizontal scale covered by the giant storm. Assuming a lack of vertical stratification may seem to give a poor approximation, but the model is still able to capture much of the essential physics, and in many situations the atmosphere behaves as an incompressible fluid when movements are horizontal. There are several consequences under these assumptions: thermodynamics and dynamics become uncoupled, the system becomes hydrostatic, and movements become two dimensional. Carrying out these approximations, we obtain the Shallow-Water (SW) equations which form a much more simplified version of the Navier–Stokes equations (Vallis 2006):

$$\frac{d\vec{u}}{dt} + f\vec{k} \times \vec{u} = -g\nabla h \qquad (13.3)$$

$$\frac{\partial h}{\partial t} = -\nabla \cdot (h\vec{u}) + S(\vec{r}, t) \qquad (13.4)$$

Equations (13.3) and (13.4) are the respective momentum and continuity equations, $\vec{u} = (u, v)$ is the horizontal velocity field, with u and v the respective zonal and meridional velocities, d/dt is the temporal derivative, $f = 2\Omega \sin\varphi$ is the Coriolis parameter (where φ is the latitude), h is the model's total layer thickness, $\vec{r}$ is the vector position, g is the local gravity, S is the source of mass which models the storm, and $\vec{k}$ is the unit vector in the vertical direction. SW models have been previously used in the study of dynamics on giant planets in both our solar system and other solar systems (Vasavada and Showman 2005). In all cases, the upper, lower-density layer represents the weather layer under study, and a bottom, denser layer mimics the deep interior of the planet. We performed 1-layer and 2-layer SW models to simulate the response of Saturn's atmosphere to the massive storm onset. Topography, after applying geostrophic equilibrium, may introduce the zonal wind profile, but a flat bottom may work equally well if the effect of $U(y)$ is numerically included during the computation of the prognostic variables in the form $U(y) + u$. In this way, $U(y)$ is maintained fixed, but it will let u evolve freely under the influence of zonal winds. Rayleigh damping was also introduced to prevent long-term turbulence.

13.7.2.3 *The Storm Source*

A fundamental observed aspect of the Great Storm of 2010–2011 is its strong, continuous activity over at least a 7-month period. Dynamically, the convective activity within the storm head appeared to cease when it met the long-lived vortex (Sánchez-Lavega at al. 2012). However, SED lightning detections indicate that there was additional convective activity within the storm head or nearby for about 2 months longer (Sayanagi et al. 2013), perhaps indicating strong dynamics at depth under the cloud tops. It is therefore fundamental for modellers to simulate long-lived convective storm systems in numerical experiments in order to adequately reproduce the response of Saturn's atmosphere.

The EPIC modelling allows an injection of mass and energy in the form of heat in all or part of the vertical domain. Since SW models are totally decoupled from the thermodynamic processes, the storm's action on the atmosphere is modelled as a combined effect of mass and energy perturbations in the form of a perturbed surface elevation (García-Melendo et al. 2013). For both models, a continuously active Gaussian pulse is used for the perturbation injection function *S*. In the SW model, a bump with a Gaussian shape is injected in the surface of the upper layer only (representing Saturn's weather layer) at the storm location, where the storm energy and mass are deposited, interacting with the background zonal winds. In the EPIC model, we introduce a spatial Gaussian heat distribution around the injection point. Following observations, storm's horizontal size is limited to a circular patch about 1° to 2° in radius, with a spatial Gaussian decay. We moved the simulated perturbation at the zonal velocity of ~28 m s^{-1}, the same one as the real storm. It must be noted that both the SW and EPIC models are approximations to the real atmosphere; the perturbation parameters are introduced ad hoc and tuned until an adequate response of the model atmosphere is obtained, which then yields clues on the storm's dynamics. As a consequence, it is not possible to derive absolute realistic values for the amount of energy and mass injected by the storm. In order to infer realistic values, modellers need to know something about the source of the storm and implement it in a more realistic model which includes convection, cloud formation, latent heat release due to phase changes of condensable species, etc.

13.7.2.4 Simulation Results

Both the EPIC and SW models yielded relevant information about the structure of Saturn's atmosphere and the dynamics of the 2010 perturbation (Sánchez-Lavega et al. 2011; García-Melendo and Sánchez-Lavega 2015). One important result was the inability to reproduce the observed morphology of the storm unless a specific combination of parameters initiated the calculations. For instance, they found that inadequate latitude injection within a narrow range of ±2° latitude always led to a morphology pattern totally different from that observed. In the case of the SW model, good results were obtained only when the magnitude of the Rossby deformation radius L_R was within certain limits (300 km ≤ L_R ≤ 1000 km). For large L_R, the model turned out to be unstable with the production of large vortices (Showman 2007). In the end, Sánchez-Lavega et al. (2011) and Garcia-Melendo and Sánchez-Lavega (2016) found they could reproduce the dynamics and morphology quite well, giving them confidence in the simulation results. In the 3D EPIC models, they obtained a PV field that was best able to simulate the storm with almost constant or slightly increasing winds below the cloud deck and low values of static stability, $N^2 = 0.03 \times 10^{-4}$ s^{-2}, especially during its onset (Figure 13.18). Such an atmospheric structure is also consistent with previous simulation results (García-Melendo et al. 2007; del Rio-Gaztelurrutia et al. 2007).

The SW and 3D numerical experiments of Sánchez-Lavega et al. (2011) and Garcia-Melendo and Sánchez-Lavega (2016) revealed the dynamic mechanism in the storm's head that drove the development of most of the remainder of the storm, such as the evolution of the tail and the north and south branches. According to the simulations, the head of the storm was adequately reproduced only when strong relative anticyclonic vorticity, interacting with the prevailing zonal winds, was injected into the atmosphere after the addition of mass or heat (see Dyudina et al. 2013). In turn, this indicates that convection lifted enormous amounts of both heat and mass. This is fully consistent with the observations (Sánchez-Lavega et al. 2011). Neither EPIC nor SW models can simulate convection, but geostrophic adjustment can be reproduced in both models. The effect is dramatic: as well produced in the 3D and SW models (Figure 13.2), when the perturbation moves at a speed faster that the local winds (−29 m s^{-1} for the 2010 GWS), in a background with strong meridional wind shear of the zonal wind profile, a bow front at the storm's head appears with intense anticyclonic circulation, ~160 ms^{-1} along its periphery (García-Melendo et al. 2013), constituting the tail up to the observed, long-lived anticyclonic vortex (Dyudina et al. 2013). According to these model analyses, anticyclonic circulation is also responsible for shedding the long-lived anticyclone, during the onset of the storm, which drifted to the end of the storm's tail (Sánchez-Lavega et al. 2012; Sayanagi et al. 2013). To survive as a compact anticyclone, it translated to the north to rest entirely within the anticyclonic flank of the ~50°N jet, at ~41°N.

Morphologically important details are also reproduced by numerical experiments (Figures 13.18

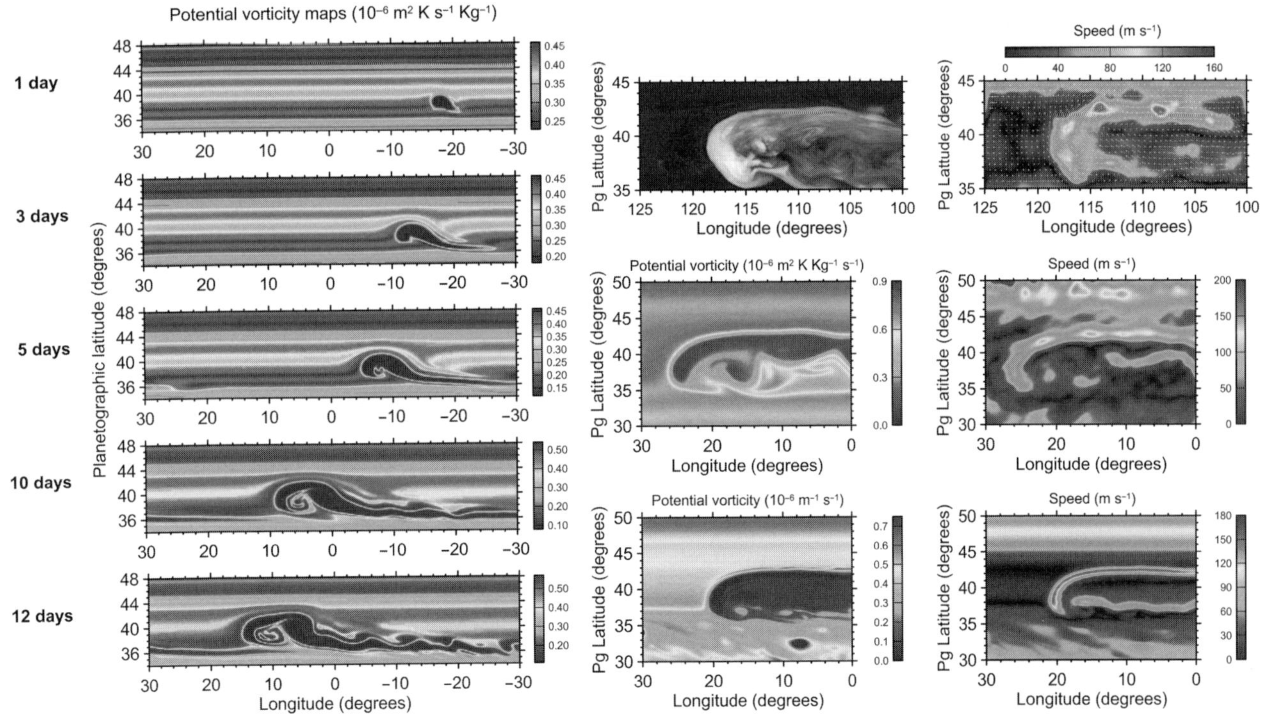

Figure 13.18 Simulation results of the GWS for the EPIC and SW models. Left column panels: evolution of the first 12 days (top to bottom) of the PV field according to EPIC simulations. This result was obtained for the lowest values of N^2 for the deepest layers and zonal winds constant or near constant and reproduces the observations quite well (compare with Figure 13.5; see also Sánchez-Lavega et al. 2011). Right: potential vorticity and velocity maps according to EPIC (upper pair) and SW (lower pair) model simulations. (A black-and-white version of this figure appears in some formats. For the color version, please refer to the plate section.)

and 13.19). The frontal arc-shaped structure forms at the beginning of the storm, reproducing very well the observed initial stages of the storm (compare Figures 13.5 and 13.17 with Figures 13.18 and 13.19). One interesting aspect captured by the numerical models is the quick expansion of part of the storm material towards the south, forming a turbulent south branch with chains of small vortices. Rapid expansion towards the south is due to the background relative vorticity where the storm was embedded. The 2010 event erupted near the peak of the westward jet centred at planetographic latitude ~40°N (Sánchez-Lavega et al. 2012), just in between the cyclonic north flank of the equatorial jet and the anticyclonic north part of the ~48°N eastward jet. It is well known that vortices survive in regions with the same relative vorticity, and they are rapidly sheared apart when the ambient vorticity is the opposite (Ingersoll et al. 1984). Simulations show that the particular point where the 2010 GWS erupted is responsible for its quick dispersion on its south side. The highly anticyclonic material injected by the storm was in part quickly sheared and dispersed southwards of ~38°N, forming wavy patterns which turned into a turbulent region. Simulations show that the chain of small vortices are little anticyclones formed at the 34°N to 30°N planetographic latitude interval (Sánchez-Lavega et al. 2012), a region where the undisturbed zonal wind profile presented two small inflexions. These vortices disappeared later, perhaps due to the alteration of the zonal profile in this region at the visible cloud top level (~500 mbar), which smoothed out the zonal wind field (Figure 13.6) (Sayanagi et al. 2013). On the other hand, the material injected by the storm reaching the anticyclonic side of the profile was accelerated to reinforce the anticyclonic cells and accelerate the material on the edge of the northern part of the storm head.

Sayanagi et al. (2013) observed that the local zonal wind speed around the 2010 storm eruption latitude had changed by as much as ~40 m s^{-1} (Figure 13.6). The authors proposed two possible mechanisms for such a change: relative vorticity injection due to

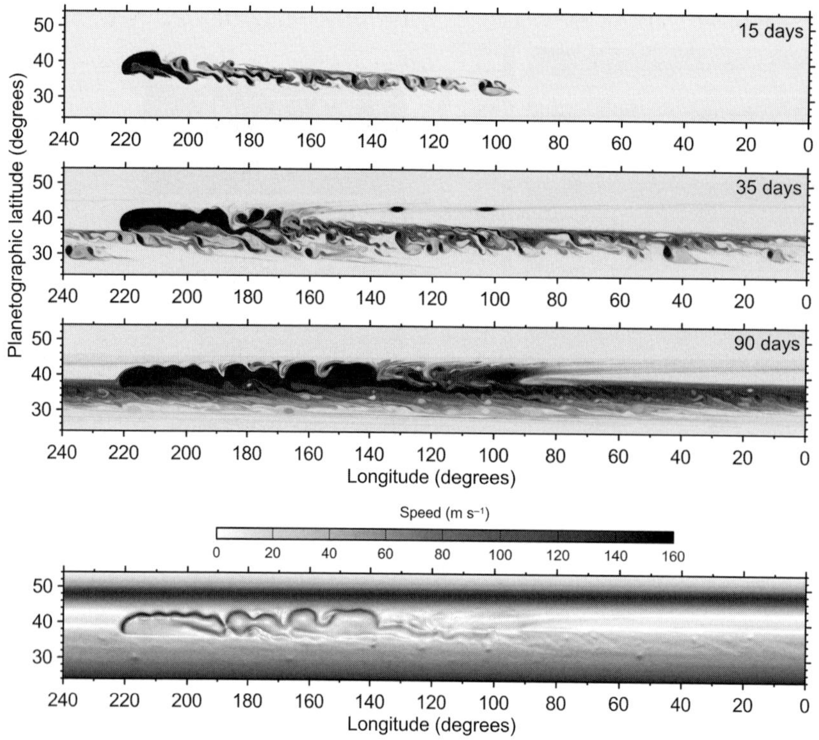

Figure 13.19 Global evolution of the storm during the first 90 days according to SW models. The three upper bottom panels represent the concentration of a passive digital "dye" introduced in the simulation to reproduce the growth and expansion of the storm. Tracer distribution is indistinguishable from the resulting PV fields. Some of the main morphological aspects of the storm are well reproduced, such as the turbulent expansion of the south branch, including chains of small vortices, the formation of the northern branch, and of a long-lived vortex at the end of the tail. The bottom panel represents the velocity field showing that the main structure of the tail is formed by anticyclonic cells at latitude 40°–43°N generated by the storm. Simulations were performed with a horizontal resolution of ~200 km and LR ≊ 350 km. Based on García-Melendo and Sánchez-Lavega (2016).

convective cloud divergence under rotational conditions, and a change in the latitudinal gradient of temperature due to latent heat release, which would change vertical wind shear and therefore the observed cloud top velocities due to the thermal wind effect. Relative vorticity injection is well reproduced in SW models. Figure 13.20 shows how the zonal wind profile is affected by the injection of vorticity according to simulations. The change is similar to that measured in real images (compare Figures 13.6 and 13.20), confirming that relative vorticity injection may, in part, alter zonal wind speed.

13.7.3 The Recurrence Interval

Several authors have addressed the intermittency between convective events on Jupiter and Saturn. Guillot (1995) was the first to point out that hydrogen gas saturated with water or ammonia vapour will have a density minimum as a function of temperature, provided the mixing ratios of the vapours are high enough. For instance, a parcel saturated with water vapour and a mixing ratio greater than 1% will first become less dense as it cools. This is because the heavier water vapour precipitates out, which overwhelms the normal thermal contraction due to cooling. A mixing ratio of 1% is about 10 times the abundance in a solar composition atmosphere. Such tenfold enrichment is suspected for Saturn but not for Jupiter, based largely on methane and phosphine, two gases whose abundances have been inferred spectroscopically.

Li and Ingersoll (2015) applied this mass-loading effect to Saturn. After a giant storm, which is treated as a giant convective event, the troposphere is warm and saturated. As it cools, it becomes less dense, and a stable interface develops at cloud base. The cooling takes decades due to the long radiative time constant of Saturn's troposphere – it has high mass per unit area

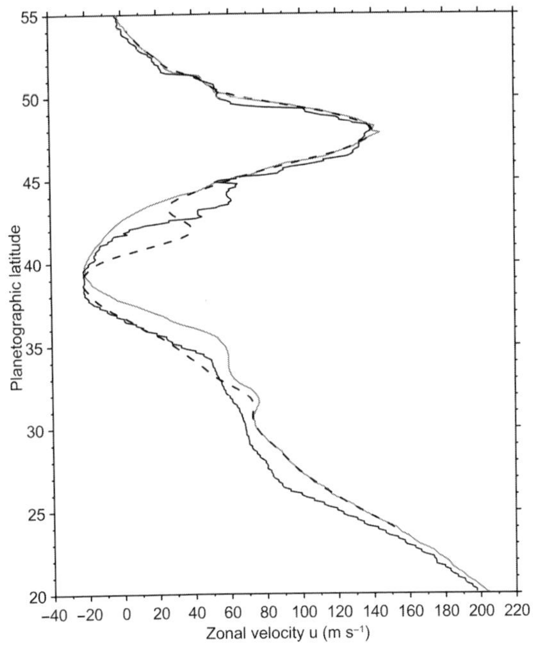

Figure 13.20 Zonal wind alterations around the latitude of the 2010 GWS onset. The grey line represents the observed zonal winds before the storm outbreak. The solid line is the new zonal wind profile measured in February 2012, six months after the storm demise, showing an altered profile. The dashed line is the result of shallow-water simulations after six months of storm evolution (starting from the grey profile), showing the winds affected by the anticyclonic vorticity injected by the storm. Based on García-Melendo and Sánchez-Lavega (2016).

and it radiates to space at a low temperature. When the troposphere has unloaded most of its water vapour, further cooling makes the troposphere denser, which erodes the stable interface until another giant convective event occurs. Li and Ingersoll model the convection and the slow cooling using a 1D radiative-convective equilibrium model, and they model the geostrophic adjustment following a convective event using a 2D axisymmetric dynamical model. The latter shows how the atmosphere can be sub-saturated with respect to ammonia after a giant storm (Janssen et al. 2013; Laraia et al. 2013).

Sugiyama et al. (2011, 2014) have a 2D cloud-resolving model that explicitly represents the convective motion and the microphysics of the three cloud components H_2O, NH_3 and NH_4SH. In their model, stable layers associated with condensation develop and act as dynamical barriers to convection. As cooling proceeds, the moist convective layer becomes potentially unstable and is triggered by the updrafts from below that are forced by neighbouring downdrafts, which themselves are forced by the cooling effect of falling raindrops as they evaporate. Sugiyama et al. find that the period of intermittency is roughly equal to the time obtained by dividing the mean temperature increase, which is caused by active cumulonimbus development, by the body cooling rate. In essence, this means that the radiative time constant is setting the long time interval between giant storms. Although the mechanisms are different, the models of Sugiyama et al. and Li and Ingersoll agree on this important point.

Alternatively, the long recurrence interval between GWSs, of the order of the Saturn year, could be controlled by the seasonal insolation heating and cooling cycle that suffers the upper cloud and hazes (see Figure 13.1). Heating rates in Saturn's upper atmosphere that have been calculated by Pérez-Hoyos and Sánchez-Lavega (2006) show that the solar flux is deposited above ~200 mbar. It remains to be seen if the solar insolation cycle affects the deep cloudy layers to trigger convection on them.

13.8 Conclusions

As the most recent and best-studied example of a GWS, the Great Storm of 2010–2011 represents a unique and major dynamical phenomenon in planetary atmospheres. Figure 13.21 summarizes schematically our ideas of the structure and major dynamical drivers of the 2010–2011 GWS. As a class of meteorological events that can be used to elucidate comparative planetology, the GWS can be considered as a testing bench of the state and structure of the atmospheres of the giant planets or, more generally, of hydrogen-dominated atmospheres. Their importance transcends the mere study of huge, quasi-periodic dynamical disturbances on Saturn. They involve a large spectrum of dynamical phenomena: moist convection, a variety of waves, vortex generation, turbulence etc., at all the scales of motion, from mesoscale to planetary scale. Their study yields information about the thermal structure, composition (for example oxygen and water), compound generation and their transport, and microphysics (particle formation) from the troposphere to the stratosphere. They also involve the study of electrical

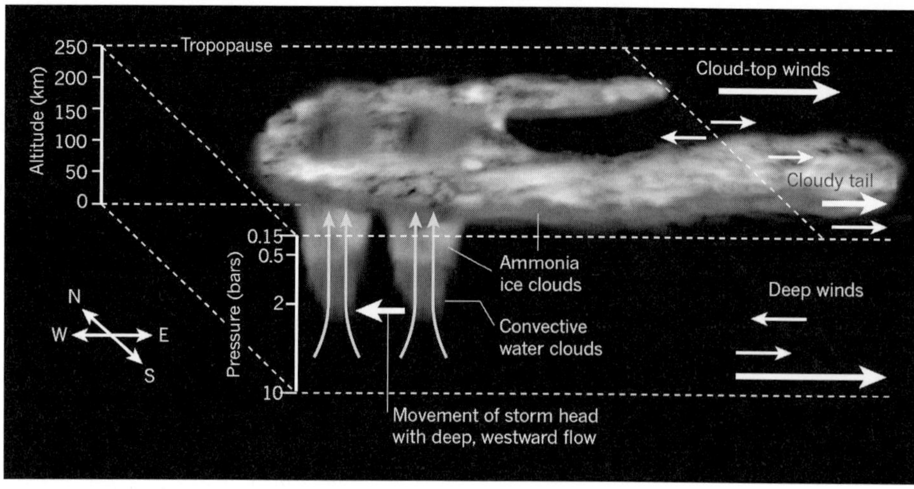

Figure 13.21 This cartoon summarizes the authors' views of the three-dimensional structure of the Great Saturn Storm of 2010–2011. From P. L. Read (2011).

phenomena (lightning) in the clouds of the giant planets. They provide insights into the nature and vertical structure of the zonal jets in giant planets and how they respond to the outbreak of huge disturbances. Finally, such energetic storms – in the case of the 2010–2011 Great Storm, emitting a significant fraction of a planet's total radiated power for many months – could also play an important role in the planet's global energy balance.

The 2010–2011 event has been a breakthrough in the understanding of the GWS, but there are still a large number of major unknowns about the phenomenon. How, exactly, periodic are they, and is this potential periodicity a key to explaining their power? Do they occur in the southern hemisphere? If not, then what is the reason for their concentration in the northern hemisphere, taking into account the hemispheric symmetry in the insolation cycle? Can they outbreak at any latitude? How are they triggered? And how is their immense power maintained over many months? As well, a plethora of physical-chemical phenomena produced by GWS events in the troposphere and stratosphere remain to be explained, including the formation of powerful, high-temperature "beacons" in the high stratosphere. Particularly important are the equatorial events, for which we have relatively sparse information. A major difficulty in answering these questions through the acquisition of new data is the rarity of the phenomenon, historically occurring about every three decades. Therefore, it is important to exploit to the maximum all the information that Cassini has obtained and is still obtaining on Saturn, and together with it, to develop more advanced models than those thus far employed, to fully explain the phenomenon and its global implications on the structure and dynamics of Saturn's atmosphere.

Acknowledgements

A. Sánchez-Lavega, E. Garcia-Melendo and S. Pérez-Hoyos are supported by the Spanish project AYA2012-36666 and AYA2015-65041-P with FEDER support, Grupos Gobierno Vasco IT-765–13 and by Universidad del País Vasco UPV/EHU through program UFI11/55. L. Sromovsky is supported for this work by NASA through its Outer Planets Research Program under grant NNX11AM58. G. L. Fletcher was supported by a Royal Society Research Fellowship at the University of Leicester. G. Fischer was supported by the Austrian Science Fund FWF through project P24325-N16. B. Hesman was supported by the NASA Cassini/CIRS project, by the NASA Planetary Astronomy Program (PAST) under grant number NNX11AJ47 G, and by the NASA Cassini Data Analysis and Participating Scientists Program (CDAPS) under grant number NNX12AC24. G. K. M. Sayanagi was supported in part by grants from NASA (Outer Planet Research Program NNX12AR38 G, Planetary Atmospheres Program NNX14AK07 G, and Cassini Data Analysis Program NNX15AD33 G) and NSF (Astronomy & Astrophysics Program 1212216).

References

Acarreta, J. R. and A. Sánchez-Lavega (1999), Vertical cloud structure in Saturn's 1990 Equatorial storm. *Icarus*, 137, 24–33. doi:10.1006/icar.1998.6034.

Achterberg, R. K., P. J. Gierasch, B. J. Conrath et al. (2014), Changes to Saturn's zonal mean tropospheric thermal structure after the 2010–2011 northern hemisphere storm. *Astrophysical Journal*, 786, 92, pp 8. doi:10.1088/0004-637X/786/2/92.

Atreya, S. K. and A.-S. Wong (2005), Coupled clouds and chemistry of the giant planets: A case for multiprobes. *Space Science Reviews*, 116, 121–136. doi:10.1007/s11214-005-1951-5.

Baines, K. H., M. L. Delitsky, T. Momary et al. (2009), Storm clouds on Saturn: Lightning-induced chemistry and associated materials consistent with Cassini/VIMS spectra. *Planetary and Space Sci.* 57, 1650–1658. doi:10:1016/j.pss.2009.06.025

Baines, K., T. Momary, L. Fletcher, A. Showman, R. Brown, B. Buratti, R. Clark, P. Nicholson, C. Go and A. Wesley (2011), Saturn's enigmatic "String of Pearls" and northern storm of 2010–2011: Manifestations of a common dynamical mechanism. EPSC abstracts, Vol 6., EPSC-DPS2011-1658, 2011.

Baines, K. H., T. W. Momary, L. N., Fletcher et al. (2010), Saturn's "String of Pearls" after five years: Still there, moving backwards faster in the Voyager system. *Bulletin of the American Astronomical Society*, 42, 1039.

Barnet, C. D., J. A. Westphal, R. F. Beebe and L. F. Huber (1992), Hubble Space Telescope observations of the 1990 equatorial disturbance on Saturn: Zonal winds and central meridian albedos. *Icarus*, 100, 499–511. doi:10.1016/0019–1035(92)90113-L.

Beebe R. F., C. Barnet, P. V. Sada and A. S. Murrell (1992), The onset and growth of the 1990 equatorial disturbance on Saturn, *Icarus*, 95, 163–172. doi:10.1016/0019–1035(92)90035–6.

Bézard, B., J. I. Moses, J. Lacy et al. (2001), Detection of ethylene (C_2H_4) on Jupiter and Saturn in non-auroral regions. *Bulletin of the American Astronomical Society*, 33, 1079.

Bjoraker, G., B. E. Hesman, R. K. Achterberg and P. N. Romani (2012), The evolution of hydrocarbons in Saturn's northern storm region. *Bulletin of the American Astronomical Society*, 44, #403.05.

Brown, R. H., K. H. Baines, G. Bellucci et al. (2004), The Cassini Visual and Infrared Mapping Spectrometer (VIMS) investigation. *Space Science Reviews* 115, 111–168. doi:10.1007/s11214-004–1453-x.

Borucki, W. J., A. Bar-Nun, F. L. Scarf, A. F. Look and G. E. Hunt (1982), Lightning activity on Jupiter, *Icarus*, 52, 492–502. doi:10.1016/0019–1035(82)90009–4.

Borucki, W. J. and C. P. McKay (1987), Optical efficiencies of lightning in planetary atmospheres, *Nature*, 328, 509–510. doi:10.1038/328509a0.

Cavalié, T., M. Dobrijévic, L. N. Fletcher et al. (2015), The photochemical response to the variation of temperature in Saturn's 2011–2012 stratospheric vortex. *Astronomy & Sstrophysics*, 580, A55.

Charney, J. G. and P. G. Drazin (1961), Propagation of planetary-scale disturbances from the lower into the upper atmosphere. *Journal of Geophysical Research*, 66, 83–109. doi:10.1029/JZ066i001p00083.

Choi, D. S., A. P. Showman and R. H. Brown (2009), Cloud features and zonal wind measurements of Saturn's atmosphere as observed by Cassini/VIMS. *Journal of Geophysical Research*, 114, E04007. doi:10.1029/2008JE003254.

Clark, R. N. and 13 colleagues (2012), The surface composition of Iapetus: Mapping results from Cassini VIMS. *Icarus*, 218, 831–860.

Conrath, B. J. and D. Gautier (2000), Saturn helium abundance: A reanalysis of Voyager measurements, *Icarus*, 144, 124–134. doi:10.1006/icar.1999.6265.

Dollfus, A. (1963), Mouvements dans l'atmosphère de Saturne en 1960. Observations coordonnées par l'Union Astronomique Internationale. *Icarus*, 2, 109–114. doi:10.1016/0019–1035(63)90010–1.

del Río-Gaztelurrutia, T., J. Legarreta, R. Hueso, S. Pérez-Hoyos and A. Sánchez-lavega (2010), A long-lived cyclone in Saturn's atmosphere: Observations and models, *Icarus*, 209, 665–681. doi:10.1016/j.icarus.2010.04.002

Dowling, T. E., A. S. Fischer, P. J. Gierasch, J. Harrington, R. P. Lebeau and C. M. Santori (1998), The Explicit Planetary Isentropic-Coordinate (EPIC) atmospheric model. *Icarus*, 132, 221–238. doi:10.1006/icar.1998.5917.

Dyudina, U. A., A. P. Ingersoll, S. P. Ewald et al. (2007), Lightning storms on Saturn observed by Cassini ISS and RPWS during 2004–2006. *Icarus*, 190, 545–555. doi:10.1016/j.icarus.2007.03.035.

Dyudina, U. A., A. P. Ingersoll, S. P. Ewald (2010), Detection of visible lightning on Saturn. *Geophysical Research Letters*, 37, L09205. doi:10.1029/2010GL043188.

Dyudina, U. A., A. P. Ingersoll, S. P. Ewald, C. C. Porco, G. Fischer and Y. Yair (2013), Saturn's visible lightning, its radio emissions, and the structure of the 2009–2011 lightning storms. *Icarus*, 226, 1020–1037. doi:10.106/j.Icarus.2013.07.013.

Elachi C., M. D. Allison, L. Borgarelli et al. (2004), Radar: The Cassini Titan Radar Mapper. *Space Science Reviews*, 115, 71–110. doi:10.1007/s11214-004–1438-9.

Encrenaz, T., M. Combes, Y. Zeau, L. Vapillon and J. Berezne (1975), A tentative identification of C_2H_4 in the spectrum of Saturn. *Astronomy & Astrophysics*, 42, 355–356.

Fischer, G., M. D. Desch, P. Zarka et al. (2006), Saturn lightning recorded by Cassini/RPWS in 2004. *Icarus*, 183, 135–152, doi:10.106/j.icarus.2006.02.010.

Fischer, G., U. A. Dyudina, W. S. Kurth et al. (2011c), Overview of Saturn lightning observations. Pages 135–144 in: H. O. Rucker, W. S. Kurth, P. Louarn and G. Fischer (eds.), *Planetary Radio Emissions VII*. Vienna: Austrian Academy of Sciences Press.

Fischer, G., D. A. Gurnett, W. S. Kurth et al. (2008), Atmospheric electricity at Saturn, *Space Science Review*, 137, 271–285, doi:10.1007/s11214-008–9370-z.

Fischer, G., D. A. Gurnett, A. Lecacheux, W. Macher and W. S. Kurth (2007b), Polarization measurements of Saturn electrostatic discharges with Cassini/RPWS below a frequency of 2 MHz, *Journal of Geophysical Research*, 112, A12308, doi:10.1029/2007JA012592.

Fischer, G., D. A. Gurnett, P. Zarka, L. Moore and U. A. Dyudina (2011b), Peak electron densities in Saturn's ionosphere derived from the low-frequency cutoff of Saturn lightning. *Journal of Geophysical Research*, 116, A04315, doi:10.1029/2010JA016187.

Fischer, G., W. S. Kurth, U. A. Dyudina et al. (2007a), Analysis of a giant lightning storm. *Icarus*, 190, 528–544, doi:10.1016/j.icarus.2007.04.002.

Fischer, G., W. S. Kurth, D. A. Gurnett et al. (2011a), A giant thunderstorm on Saturn. *Nature*, 475, 75–77. doi:10.1038/nature10205.

Fischer, G., S.-Y. Ye, J. B. Groene et al. (2014), A possible influence of the Great White Spot on Saturn kilometric radiation periodicity. *Annales Geophysicae*, 32, 1463–1476. doi:10.5194/angeo-32–1463-2014.

Flasar F. M., V. G. Kunde, M. M. Abbas et al. (2004), Exploring the Saturn system in the thermal infrared: The composite infrared spectrometer. *Space Science Reviews*, 115, 169–297. doi:10.1007/s11214-004–1454-9.

Fletcher, L. N., T. K. Greathouse, G. W. Orton et al. (2014), The origin of nitrogen on Jupiter and Saturn from the $^{15}N/^{14}N$ ratio. *Icarus*, 238, 170–190. doi:10.1016/j.icarus.2014.05.007.

Fletcher, L. N., B. E. Hesman, R. K. Achterberg et al. (2012), The origin and evolution of Saturn's 2011–2012 stratospheric vortex. *Icarus*, 221, 560–586. doi:10.1016/j.icarus.2012.08.024.

Fletcher, L. N., B. E. Hesman, P. G. Irwin et al. (2011), Thermal structure and dynamics of Saturn's northern springtime disturbance. *Science*, 332, 1413–1417. doi:10.1126/science.1204774.

García-Melendo, E., R. Hueso, A. Sánchez-Lavega et al. (2013), Atmospheric dynamics of Saturn's 2010 giant storm, *Nature Geoscience*, 6, 525–529. doi:10.1038/NGEO1860.

García-Melendo, E., S. Pérez-Hoyos, A. Sánchez-Lavega and R. Hueso (2011), Saturn's zonal wind profile in 2004–2009 from Cassini ISS images and its long-term variability. *Icarus*, 215, 62–74. doi:10.1016/j.icarus.2011.07.005.

García-Melendo, E. and A. Sánchez-Lavega (2017), Shallow water simulations of Saturn's giant storms at different latitudes. *Icarus*, 286, 241–260. doi:10.1016/j.icarus.2016.10.006.

García-Melendo, E., A. Sánchez-Lavega and R. Hueso (2007), Numerical models of Saturn's long-lived anticyclones. *Icarus*, 191, 665–677. doi:10.1016/j.icarus.2007.05.02

Guerlet, S., A. Spiga, M. Sylvestre et al. (2014), Global climate modeling of Saturn's atmosphere. Part I: Evaluation of the radiative transfer model. *Icarus*, 238, 110–124. doi:10.1016/j.icarus.2014.05.010.

Guillot, T. (1995), Condensation of methane, ammonia, and water and the inhibition of convection in giant planets. *Science*, 269, 1697–1699. doi:10.1126/science.7569896.

Gurnett, D. A., W. S. Kurth, D. L. Kirchner et al. (2004), The Cassini radio and plasma wave science investigation. *Space Science Reviews*, 114, 395–463. doi:10.1007/s11214-004–1434-0.

Harvey, V. L. and M. H. Hitchman (1996), A climatology of the Aleutian high. *Journal of Atmospheric Sciences*, 53, 2088–2102.

Harvey, V. L., R. B. Pierce, M. H. Hitchman, C. E. Randall and T. D. Fairlie (2004), On the distribution of ozone in stratospheric anticyclones. *Journal of Geophysical Research (Atmospheres)*, 109, 24308. doi:10.1029/2004JD004992.

Hesman, B. E., G. L. Bjoraker, R. K. Achterberg et al. (2013), The Evolution of Hydrocarbon Compounds in Saturn's Stratosphere During the 2010 Northern Storm. *AGU Fall Meeting Abstracts, Dec. 2013*, **C2205**.

Hesman, B. E., G. L. Bjoraker, P. V. Sada et al. (2012), Elusive ethylene detected in Saturn's northern storm region. *The Astrophysical Journal*, 760, 24–30. doi:10.1088/0004-637X/760/1/24.

Hueso, R., J. Legarreta, S. Pérez-Hoyos, J. F. Rojas, A. Sánchez-Lavega and A. Morgado (2010), The international outer planets watch atmospheres node database of giant-planet images. *Planetary and Space Science*, 58, 1152–1159. doi:10.1016/j.pss.2010.04.006.

Hueso R., and A. Sánchez-Lavega (2001), A three-dimensional model of moist convection for the giant planets: The Jupiter case. *Icarus*, 151, 257–274. doi:10.1006/icar.2000.6606.

Hueso R., and A. Sánchez-Lavega (2004), A three-dimensional model of moist convection for the giant planets II: Saturn's water and ammonia moist convective storms. *Icarus*, 172, 255–271. doi:10.1016/j.icarus.2004.06.010.

Hurley, J., P. G. J. Irwin, L. N. Fletcher et al. (2012), Observations of upper tropospheric acetylene on Saturn: No apparent correlation with 2000 km-sized thunderstorms. *Planetary and Space Science*, 65, 21–37. doi:10.1016/j.pss.2011.12.026.

Ingersoll, A. P., R. F. Beebe, B. J. Conrath and G. E. Hunt (1984), Structure and dynamics of Saturn's atmosphere. In: T. Gehrels and M. S. Matthews (eds.), *Saturn*. Tucson, AZ: University of Arizona Press, 195–238.

Janssen, M. A., A. P. Ingersoll, M. D. Allison et al. (2013), Saturn's thermal emission at 2.2-cm wavelength as imaged by the Cassini RADAR radiometer. *Icarus*, 226, 522–535. doi:10.1016/j.icarus.2013.06.008.

Kaiser, M. L., M. D. Desch and J. E. P. Connerney (1984), Saturn's ionosphere: Inferred electron densities. *Journal of Geophysical Research*, 89, A4, 2371–2376. doi:10.1029/JA089iA04p02371.

Karkoschka, E. and M. G. Tomasko (1993), Saturn's upper atmospheric hazes observed by the Hubble Space Telescope. *Icarus*, 106, 428–441. doi:10.1006/icar.1993.1183.

Karkoschka, E. and M. G. Tomasko (2005), Saturn's vertical and latitudinal cloud structure 1991–2004 from HST imaging in 30 filters. *Icarus* 179, 195–221. doi:10.1016/j.icarus.2005.05.016.

Konovalenko, A. A., N. N. Kalinichenko, H. O. Rucker et al. (2013), Earliest recorded ground-based decameter wavelength observations of Saturn's lightning during the giant E-storm detected by Cassini spacecraft in early 2006, *Icarus*, 224, 14–23. doi:10.1016/j.icarus.2012.07.024.

Laraia, A. L., A. P. Ingersoll, M. A. Janssen, S. Gulkis, F. Oyafuso and M. Allison (2013), Analysis of Saturn's thermal emission at 2.2-cm wavelength: Spatial distribution of ammonia vapor. *Icarus*, 226, 641–654. DOI:10.1016/j.icarus.2013.06.017.

Larsen, H. R. and E. J. Stansbury (1974), Association of lightning flashes with precipitation cores extending to height 7 km. *Journal of Atmospheric and Terrestrial Physics*, 36, 1547–1533.

Lebeau, R. P. and T. E. Dowling (1998), EPIC simulations of time-dependent, three-dimensional vortices with application to Neptune's great dark spot. *Icarus*, 132, 239–265. doi:10.1006/icar.1998.5918.

Legarreta, J. And A. Sánchez-Lavega (2008), Vertical structure of Jupiter's troposphere from nonlinear simulations of long-lived vortices, *Icarus*, 196, 184–201. doi:10.1016/j.icarus.2008.02.018.

Li, C. and A. P. Ingersoll (2015), Moist convection in hydrogen atmospheres and the frequency of Saturn's giant storms. *Nat. Geoscience*, 8, 398–403. doi:10.1038/ngeo2405

Li L., X. Jiang, H. J. Trammell, Y. Pan, J. Hernandez, B. J. Conrath, P. J. Gierasch, R. K. Achterberg, C. A. Nixon, F. M. Flasar, S. Pérez-Hoyos, R. A. West, K. H. Baines and B. Knowles (2015), Saturn's giant storm and global radiant energy. *Geophys. Res. Lett.*, 42, 2144–2148. doi:10.1002/2015GL063763.

Lindal, G. F., D. N. Sweetnam and V. R. Eshleman (1985), The atmosphere of Saturn: An analysis of the Voyager radio occultation measurements. *Astronomical Journal*, 90, 1136–114. doi:10.1086/113820.

Manney, G. L., L. Froidevaux, J. W. Waters et al. (1995), Formation of low-ozone pockets in the middle stratospheric anticyclone during winter. *Journal of Geophysical Research-Atmospheres*, 100, 13939–13950. doi:10.1029/95JD00372.

Miller, E. A., G. Klein, D. W. Juergens et al. (1996), The visual and infrared mapping spectrometer for Cassini. In: L. Horn (ed.), Society of Photo-Optical Instrumentation Engineers (SPIE) Conference Series, 2803, 206–220.

Momary, T. W. and K. H. Baines (2014), The anticyclonic eye of the storm. Evolution of Saturn's Great Storm region and associated anticyclone as seen by Cassini/VIMS. Abstract 422.11. Proceedings of the 46th Annual Meeting of the Division of Planetary Sciences.

Moses, J. I., E. S. Armstrong, L. N. Fletcher, P. G. J. Irwin, B. E. Hesman and P. N. Romani (2015), Evolution of stratospheric chemistry in the Saturn storm beacon. *Icarus* 261, 149–168

Muñoz, O., F. Moreno, A. Molina, D. Grodent, J. C. Gérard and V. Dols (2004), Study of the vertical structure of Saturn's atmosphere using HST/WFPC2 images. *Icarus*, 169, 413–428. doi:10.1016/j.icarus.2003.12.018.

Orton, G. S., L. N. Fletcher, T. Fouchet et al. (2013), Ground-based observations of the aftermath of the 2010–2011 Great Northern Springtime Storm in Saturn (Invited), *AGU Fall Meeting Abstracts*, Dec., A6.

Pedlosky, J. (1979), *Geophysical Fluid Dynamics*, Springer-Verlag, New York, pp. 624.

Pérez-Hoyos, S. and A. Sánchez-Lavega (2006), Solar flux in Saturn's atmosphere: maximum penetration and heating rates in the aerosol and cloud layers, *Icarus*, 180, 368–378.

Pérez-Hoyos, S., A. Sánchez-Lavega, R. G. French and J. F. Rojas (2005), Saturn's cloud structure and temporal evolution from ten years of Hubble Space Telescope images (1994–2003). *Icarus*, 176, 155–174. doi:10.1016/j.icarus.2005.01.014

Pierce, A. D. and S. C. Coroniti (1966), A mechanism for the generation of acoustic-gravity waves during thunderstorm formation. *Nature*, 210, 1209–1210. doi:10.1038/2101209a0.

Porco, C. C., E. Baker, J. Barbara et al. (2005), Cassini Imaging Science: Initial results on Saturn's atmosphere, *Science*, 307, 1243–1247. doi:10.1126/science.1107691.

Porco, C. C., R. A. West, S. Squyres et al. (2004), Cassini Imaging Science: Instrument characteristics and anticipated scientific investigations at Saturn, *Space Sci. Rev.*, 115, 363–497.

Sánchez-Lavega A. (1982), Motions in Saturn's atmosphere: Observations before Voyager encounters. *Icarus*, 49, 1–16. doi:10.1016/0019–1035(82)90052–5.

Sánchez-Lavega A. (1994), Saturn's Great White Spots, *Chaos*, 4, 341–353. doi:10.1063/1.166012.

Sánchez-Lavega A. (2011) *An Introduction to Planetary Atmospheres*, Boca Raton, FL: Taylor-Francis, CRC Press, pp. 629.

Sánchez-Lavega, A. and E. Battaner (1987), The nature of Saturn's Great White Spots. *Astronomy & Astrophysics*, 185, 315–326.

Sánchez Lavega A., F. Colas, J. Lecacheux, P. Laques, I. Miyazaki and D. Parker (1991), The Great White Spot and disturbances in Saturn's equatorial atmosphere during 1990, *Nature*, 353, 397–401. doi:10.1038/353397a0.

Sánchez-Lavega, A., T. del Río-Gaztelurrutia, M. Delcroix et al. (2012), Ground-based observations of the long-term evolution and death of Saturn's 2010 Great White Spot. *Icarus*, 220, 561–576, doi:10.1016/j.icarus.2012.05.033.

Sánchez-Lavega A., T. del Río-Gaztelurrutia, R. Hueso et al. (2011) Deep winds beneath Saturn's upper clouds from a seasonal long-lived planetary-scale storm. *Nature*, 475, 71–74. doi:10.1038/nature10203.

Sánchez-Lavega, A. and J. M. Gómez (1996a), The south equatorial belt of Jupiter. I: Its life cycle. *Icarus*, 121, 1–17. doi:10.1006/icar.1996.0067.

Sánchez-Lavega, A., R. Hueso, S. Pérez-Hoyos, J. F. Rojas and R. G. French (2004) Saturn's cloud morphology and zonal winds before the Cassini encounter. *Icarus*, 170, 519–523. doi:10.1016/j.icarus.2004.05.002.

Sánchez Lavega A., J. Lecacheux, F. Colas and P. Laques (1993), Temporal behavior of cloud morphologies and motions in Saturn's atmosphere. *Journal of Geophysical Research*, 98, 18857–18872. doi:10.1029/93JE01777.

Sánchez Lavega A., J. Lecacheux, F. Colas and P. Laques (1994), Photometry of Saturn's 1990 equatorial disturbance. *Icarus*, 108, 158–168. doi:10.1006/icar.1994.1048.

Sánchez Lavega A., J. Lecacheux, J. M. Gómez et al. (1996b), Large-scale storms in Saturn's atmosphere during 1994, *Science*, 271, 631–634. doi:10.1126/science.271.5249.631.

Sánchez-Lavega A., G. S. Orton, R. Hueso et al. (2008), Depth of a strong Jovian jet from a planetary-scale disturbance driven by storms, *Nature*, 451, 437–440. doi:10.1038/nature06533.

Sánchez-Lavega, A., S. Pérez-Hoyos, J. F. Rojas, R. Hueso and R. G. French (2003), A strong decrease in Saturn's equatorial jet at cloud level. *Nature*, 423, 623–625. doi:10.1038/nature01653.

Sánchez-Lavega A., J. F. Rojas and P. V. Sada (2000), Saturn's zonal winds at cloud level, *Icarus*, 147, 405–420.

Sanz-Requena, J. F., S. Pérez-Hoyos, A. Sánchez-Lavega et al. (2012), Cloud structure of Saturn's 2010 storm from ground-based imaging. *Icarus*, 219, 142–149. doi:10.1016/j.icarus.2012.02.023.

Sayanagi, K. M., U. A. Dyudina, S. P. Ewald et al. (2013), Dynamics of Saturn's great storm of 2010–2011 from Cassini ISS and RPWS. *Icarus*, 223, 460–478. doi:10.1016/j.icarus.2012.12.013.

Sayanagi, K. S., U. A. Dyudina, S. P. Ewald, G. D. Muro and A. P. Ingersoll (2014), Cassini ISS observation of Saturn's String of Pearls. *Icarus*, 229, 170–180. doi:10.1016/j.icarus.2013.10.032.

Sayanagi, K. M., R. Morales-Juberias and A. P. Ingersoll (2010), Saturn's northern hemisphere ribbon: Simulations and comparison with the meandering gulf stream. *Journal of the Atmospheric Sciences*, 67, 2658–2678. doi:10.1175/2010JAS3315.1.

Shemansky, D. and X. Liu (2012), Saturn upper atmospheric structure from Cassini EUV and FUV occultations. *Canadian Journal of Physics*, 90, 817–831. doi:10.1139/p2012-036.

Showman, A. (2007) Numerical simulations of forced shallow-water turbulence: Effects of moist convection on the large-scale circulation of Jupiter and Saturn. *Journal of the Atmospheric Sciences*, 64, 3132–3157. doi:10.1175/JAS4007.1.

Simon-Miller, A. A., N. J. Chanover, G. S. Orton, M. Sussman, I. G. Tsavaris and E. Karkoschka (2006), Jupiter's White Oval turns red. *Icarus*, 185, 558–562. doi:10.1016/j.icarus.2006.08.002.

Sromovsky, L. A., K. H. Baines and P. Fry (2013), Saturn's Great Storm of 2010–2011: Evidence for ammonia and water ices from analysis of VIMS spectra. *Icarus*, 226, 402–418. doi:10.1016/j.icarus.2013.05.043.

Sromovsky, L. A., K. H. Baines and P. M. Fry (2014), Vertical structure of Saturn lightning storms and storm-related dark ovals. *American Astronomical Society*, #46, #511.09.

Sugiyama, K., K. Nakajima, M. Odaka, M. Ishiwatari, K. Kuramoto, Y. Morikawa, S. Nishizawa, Y. O. Takahashi and Y.-Y. Hayashi (2011), Intermittent cumulonimbus activity breaking the three-layer cloud structure of Jupiter. *Geophys. Res. Lett.*, 38, L13201; doi:10.1029/2011GL047878.Unlinked

Sugiyama, K., K. Nakajima, M. Odaka, K. Kuramoto and Y.-Y. Hayashi (2014), Numerical simulations of Jupiter's moist convection layer: Structure and dynamics in statistically steady states, *Icarus*, 229, 71–91. doi:10.1016/j.icarus.2013.10.016.

Tomasko, M. G. and L. R. Doose (1984), Polarimetry and photometry of Saturn from Pioneer 11 Observations and constraints on the distribution and properties of cloud and aerosol particles, *Icarus*, 58, 1–34. doi:10.1016/0019-1035(84)90096-4.

Vallis, G. K. (2006), *Atmospheric and Ocean Fluid Dynamics*, Cambridge: Cambridge University Press, 745.

Vasavada, A. R., S. M. Hörst, M. R. Kennedy et al. (2005), Cassini imaging of Saturn: Southern hemisphere winds and vortices. *Journal of Geophysical Research*. 111, E05004. doi:10.1029/2005JE002563.

Vasavada, A. R. and A. P. Showman (2005), Jovian atmospheric dynamics: an update after Galileo and Cassini, *Reports on Progress in Physics*, 68, 1935–1996. doi:10.1088/0034-4885/68/8/R06.

Warwick, J. W., J. B. Pearce, D. R. Evans et al. (1981), Planetary radio astronomy observations from Voyager 1 near Saturn. *Science*, 212, 239–243. doi:10.1126/science.212.4491.239.

West, R. A., K. H. Baines, E. Karkoschka and A. Sánchez-Lavega (2009), Clouds and aerosols in Saturn's atmosphere. Pages 113–159 in: M. K. Dougherty, L. W. Esposito and S. M. Krimigis (eds.), *Saturn from Cassini–Huygens*. New York, NY: Springer.

Westphal, J. A., W. A. Baum, A. P. Ingersoll et al. (1992), Hubble Space Telescope observations of the 1990 equatorial disturbance on Saturn: Images, albedos and limb darkening. *Icarus*, 100, 485–498. doi:10.1016/0019-1035(92)90112-K.

Zakharenko, V., C. Mylostna, A. Konovalenko et al. (2012), Ground-based and spacecraft observations of lightning activity on Saturn. *Planetary and Space Science*, 61, 53–59. doi:10.1016/j.pss.2011.07.021.

Zarka, P., B. Cecconi, L. Denis et al. (2006), Physical properties and detection of Saturn's lightning radio bursts. Pages 111–122 in: *Planetary Radio Emissions VI*, H. O. Rucker, W. S. Kurth and G. Mann (eds.), Vienna: Austrian Academy of Sciences Press.

14

The Future Exploration of Saturn

KEVIN H. BAINES, SUSHIL K. ATREYA, FRANK CRARY, SCOTT G. EDGINGTON, THOMAS K. GREATHOUSE, HENRIK MELIN, OLIVIER MOUSIS, GLENN S. ORTON, THOMAS R. SPILKER AND ANTHONY WESLEY

Abstract

Despite the lack of another Flagship-class mission such as Cassini–Huygens, prospects for the future exploration of Saturn are nevertheless encouraging. Both NASA and the European Space Agency (ESA) are exploring the possibilities of focused interplanetary missions (1) to drop one or more in situ atmospheric entry probes into Saturn and (2) to explore the satellites Titan and Enceladus, which would provide opportunities for both in situ investigations of Saturn's magnetosphere and detailed remote-sensing observations of Saturn's atmosphere. Additionally, a new generation of powerful Earth-based and near-Earth telescopes with advanced instrumentation spanning the ultraviolet to the far-infrared promise to provide systematic observations of Saturn's seasonally changing composition and thermal structure, cloud structures and wind fields. Finally, new advances in amateur telescopic observations brought on largely by the availability of low-cost, powerful computers, low-noise, large-format cameras, and attendant sophisticated software promise to provide regular, longterm observations of Saturn in remarkable detail.

14.1 Introduction

Since the first planetary flyby in 1962, planetary exploration via spacecraft has proven time and again to be the most effective means to expand our knowledge of the Solar System. The Cassini/Huygens mission is arguably one of the most successful of such missions, achieving a remarkable record of discoveries about the entire Saturn system, including its icy satellites, the large atmosphere-enshrouded moon Titan, the planet's surprisingly intricate ring system and the planet's complex magnetosphere, atmosphere and interior. Far from being a small (500 km diameter) geologically dead moon, Enceladus proved to be exceptionally active, erupting with numerous geysers that spew liquid water vapor and ice grains into space – some of which falls back to form nearly pure white snowfields and some of which escapes to form a distinctive ring around Saturn (e.g. Spencer et al. 2009). Titan revealed itself to be an eerily Earthlike satellite covered with dozens of liquid hydrocarbon lakes and landscapes carved by river channels, both of which are linked to a complex methane-based meteorology complete with transient storm clouds that rain down from a nitrogen-based, smoggy sky some ten times thicker than that of Earth (e.g. Aharonson et al. 2014; Griffith et al. 2014). Similarly, Saturn's resplendent system of planet-girdling rings has revealed itself as host to a surprising number of organized and evolving features such as spiral waves – some of which may have been generated by slight gravitational perturbations within Saturn's interior (e.g. Chapter 3 by Fortney et al.; Chapter 5 by Carbary et al.) – and mysterious elongated clumps known as "propellers" that hint at processes analogous to the formation of solar systems around our Sun and other stars (e.g. Colwell et al., 2009).

Saturn itself has revealed that its typically serene, unperturbed appearance masks an underlying reality. At the poles, flickering aurorae constantly change shape and intensity, perhaps in response to variations in the solar wind brought on by solar storms, possibly augmented by the effects of Enceladus's geysers and ring rain (cf. Chapter 7 by Stallard et al.; Chapter 8 by Moore et al.). The north pole is capped by a mysterious, long-lived hexagonal feature bordering a polar vortex (cf. Chapter 12 by Sayanagi et al.). Underneath Saturn's upper skin of planet-girdling hazes, a surprising array of dynamical processes roil at depth. As detailed in Chapter 13 by Sánchez-Lavega et al., periodically – every three decades or so – a surprisingly powerful storm erupts, whose spreading clouds encircle the planet in a ~10°-wide latitudinal band centered on the storm. Indeed, the major storm of 2010–2011 well witnessed by Cassini and Earth-based observers has proved to be the most powerful convective storm system observed thus far on any planet, lofting materials over 150 km in altitude (e.g. Fletcher et al. 2011; Sanz-Requena et al. 2012; Sromovsky et al. 2013; Li and Ingersoll 2015).

The end of the Cassini-Huygens Mission in September 2017 concludes a remarkable thirteen-year era of exploration and discovery of Saturn, its moons, ring system and magnetosphere. For the foreseeable future, no comparable comprehensive orbital reconnaissance of the Saturn system is planned by any of the world's space agencies. Thus, it is unclear when the next concerted effort to explore the entire Saturn system via interplanetary spacecraft will be undertaken. In this chapter we detail the major science objectives for the future exploration of Saturn and the prime observational techniques that will likely be used over the next several decades.

14.2 Outstanding Science Issues

Here, we first review the science goals of the Grand Finale phase which concludes the Cassini Mission. Next we discuss scientific priorities for future Saturn research as defined by the Planetary Science Decadal Survey (PSDS) Committee, established by the National Research Council, in the PSDS 2011 Visions and Voyages report (NRC 2011). This document is of particular importance as it has been adopted by NASA's Science Mission Directorate as the primary guiding document for the Directorate's strategic planning of planetary exploration for the current decadal period of 2013–2023. In our discussion, we provide additional details on the major science priorities, especially regarding the need to make measurements diagnostic of the planet's origin and evolution.

14.2.1 Priority Science Goals of the Cassini Grand Finale Mission

In orbit since 1 July 2004, the Cassini Orbiter began its final phase of science investigations in December, 2016. Entitled the Grand Finale – as it concludes with a 34-km s^{-1} plunge into Saturn on 15 September 2017 – the Cassini Orbiter will spend its final five months circling the planet in a highly elliptical, highly inclined orbit, diving 22 times inside the innermost rings to fly swiftly (> 36 km s^{-1}) past the planet just ~2000 km above its equatorial cloud tops (Figure 14.1). These novel orbits are achieved by a gravitational kick from Titan in April 2017 that jumps the spacecraft's periapsis (i.e. the closest point to Saturn's center) from just outside Saturn's main rings, some 87,000 km above the cloud tops, into the clear gap between the innermost D ring and the upper atmosphere. The unique observational geometry allows Cassini to gather unprecedented close-up views of the planet and its rings, while also obtaining in situ samples of gases, dust, plasma and

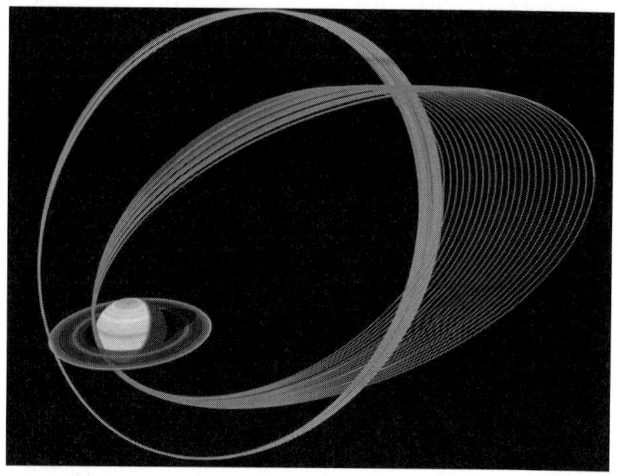

Figure 14.1 The final ten months of the Cassini–Huygens Mission to Saturn: F Ring and Grand Finale Orbit phase. Twenty orbits just outside the F Ring (grey) transition to twenty-two Grand Finale Orbits (blue) between Saturn and its innermost rings and the planet, when the orbital periapsis drops to ~3,000 km above Saturn's cloud tops, vs ~200,000 km during previous closest approaches. The last half-orbit (orange) plunges Cassini into the atmosphere of Saturn (image courtesy of NASA/JPL-Caltech). (A black-and-white version of this figure appears in some formats. For the color version, please refer to the plate section.)

magnetic fields. Altogether, the unique data collected during the Grand Finale phase addresses key issues about Saturn, its rings and its innermost plasma environment, similar to those being addressed by the Juno mission at Jupiter, thus promising new synergistic insights into how each planet formed, evolved and works today.

Salient Grand Finale science measurement objectives, as are only obtainable by such close-up passes, include (1) the first measurements of the higher-order gravitational and magnetic field moments to constrain Saturn's interior structure and possibly determine its currently unknown internal rotation rate; (2) in situ measurements of Saturn's ionosphere, innermost radiation belts, dust environment and auroral acceleration region; (3) accurate measurement of main ring mass, currently uncertain by about an order of magnitude; (4) high-spatial-resolution studies of the main rings, including novel probing by radar; and (5) high-spatial-resolution Saturn atmospheric observations and in situ sampling of outflowing ions. With an orbital periapse near local solar noon and slightly south of the ring plane, Cassini will be well placed to obtain optimized gravity measurements and high-resolution imaging and occultation measurements of the main rings. Cassini will approach the periapse over the northern hemisphere for inbound observations of the sunlit rings and Saturn's north polar regions, including the long-lived polar hexagon feature (see Chapter 12). Outbound trajectories will provide excellent views of southern aurora and the south pole in winter, as well as the unlit side of the rings.

14.2.1.1 Interior and Magnetic and Gravitational Fields

During the Grand Finale, Cassini's magnetometer will measure the magnetic field as close as 1.03 Saturn radii from Saturn's center to derive high-order coefficients to degree-9 and possibly degree-11, which then will determine the depth of Saturn's metallic hydrogen core, important for understanding the dynamo mechanism in the deep interior. Furthermore, as discussed in Section 14.2.4, these close-up measurements are Cassini's best chance for characterizing any asymmetry in the magnetic field, which will not only help to characterize the planet's dynamo but may finally determine the bulk rotation rate of the planet, perhaps to better than ±0.1%. Existing data show some statistical evidence of asymmetries at the 20 nT-R_S^3 level (Burton and Dougherty 2014), implying a small ~0.05° tilt, thus making direct measurements challenging.

The low-order harmonics of the gravity field give information about the mass of the core and possible internal layering, while the high-order harmonics are diagnostic of the winds in the deep interior. Radio science data from a number of dedicated gravity passes will determine the zonal gravity harmonics up to degree-12 (J_{12}) with an error of less than 2×10^{-7} – at least two orders of magnitude better than current values for J_{10}. This will provide a determination of the vertical extent of the observed upper-level winds and may also reveal clues about the planet's rotation rate. In addition, Cassini may be able to measure the relativistic Lense–Thirring frame-dragging effect (Helled 2011), which would determine the planet's angular momentum and further constrain the rotation rate.

14.2.1.2 Atmosphere and Exosphere In Situ Sampling

For the first time, Cassini's ion and neutral mass spectrometer will sample in situ the neutral and ion composition and density of Saturn's ionosphere and thermosphere down to approximately the ~1000 km level above the clouds during the last five orbits of the mission. Composed mostly of molecular and atomic hydrogen and helium along with trace amounts of hydrocarbons and oxygen-bearing molecules, the directly measured inventory of both atomic and molecular species will provide new insights into the energetics and dynamics of the thermosphere, as well as provide a valuable constraint on the ever-elusive abundance of helium in the deep interior. Ionospheric structure will be investigated with in situ measurements of electron density and temperature by a Langmuir probe.

Flying interior to the D ring, Cassini will characterize the enhanced radiation environment that exists in the gap between the D ring and the upper atmosphere, a region never explored in situ. Cassini's dust analyzer will determine the composition and abundance of the so-called "ring rain" material that precipitates into the planet from the rings along magnetic field lines (cf. Chapter 8). Finally, these orbits will also provide a unique opportunity for studying the properties of lightning by searching for their characteristic whistler radio waves.

Cassini's Grand Finale phase thus amounts to a brand-new mission that is quite distinct in science and measurement objectives assigned heretofore to the Cassini Mission. Overall, Cassini sets the stage for future missions, leaving an incredibly rich legacy of discoveries that have changed our views of the Saturn system, how solar systems form from proto-planetary disks, and the potential for life elsewhere in our universe.

14.2.2 The 2011 Decadal Survey: Saturn Science and Exploration Goals

To establish priorities in NASA's planetary exploration program for the decade of 2013 to 2023, the Committee on the Planetary Science Decadal Survey (hereafter, the Decadal Committee), led by the National Research Council, was formed in 2009. Over 75 scientists and engineers were involved as committee members; several hundred other scientists participated in developing and submitting 199 white papers for committee consideration.

In their final report (NRC 2011), the Decadal Committee recommended that a Saturn Probe mission be included in the list of five candidate missions for the next New Frontiers competition, formally known as New Frontiers Mission 4. Key mission science objectives, similar to those promoted by ESA's Cosmic Vision Program (cf. Section 14.3.2.2), are to (1) determine the noble gas abundances and isotopic ratios of hydrogen, carbon, nitrogen and oxygen in Saturn's atmosphere, and to (2) determine the atmospheric structure including dynamics at the probe site.

The Decadal Committee found that the Saturn probe concept directly addresses two of its major themes (NRC 2011): (1) building new worlds and (2) understanding the workings of solar systems. In particular, the mission would directly address three of the committee's ten priority questions. Under the "building" theme, the mission would address the key questions of "How did the giant planets and their satellite systems accrete?" and "Is there evidence that they migrated to new orbital positions?" Under the "workings" theme, the mission addresses two priority questions: (1) "How do the giant planets serve as laboratories to understand the Earth, the solar system and extrasolar planetary systems?" and (2) "How have the myriad chemical and physical processes that shaped the solar system operated, interacted and evolved over time?"

14.2.2.1 Building New Worlds: Clues from Elemental Abundances

Perhaps the highest priority question for Saturn pertains to its origin and overall history: *How did Saturn form and evolve?* A prime mystery linking Saturn to the primordial past is why the planet emits more than twice the energy it absorbs from the Sun, indicating a large source of internal heat associated with its origin and evolution over the æons.

The observed heavy element enrichment in Jupiter (via Galileo probe; Niemann et al. 1998) and Saturn (carbon from Cassini spectral observations of methane; Fletcher et al. 2009a) and the frequency and characteristics of exoplanets provide overwhelming evidence in favor of the core accretion model for Saturn's formation (for a detailed discussion, see Chapter 2 by Atreya et al.). It is generally accepted that Saturn's core formed from heavy elements (mass > ^{4}He), originally delivered in water ice. Although the composition of Saturn's core cannot be studied directly, being 40,000–50,000 km below Saturn's visible clouds, key core-forming heavy elements are released from the core during accretionary heating and mixed to the upper troposphere, where they can be measured. A comparison of the abundances of Saturn's heavy elements and their isotopes with those in the Sun and the other gas giant planet, Jupiter, would then provide the set of essential constraints needed to develop accurate models of the formation and evolution of Saturn and the other giant planets.

As discussed in Chapter 2 by Atreya et al., carbon – as determined from remote-sensing measurements of well-mixed methane gas in Saturn's visible atmosphere – is the only heavy element in Saturn whose abundance is determined precisely, with a C:H ratio nine times solar (Table 2.1). Due to the inability of remote sensing instruments to probe condensable volatiles such as ammonia, hydrogen sulfide and water below their respective clouds, reliable measurements of Saturn's bulk N:H, S:H and O:H are lacking. Similarly, the P:H ratio is determined precisely from PH_3 in the upper troposphere, but due to the possible loss and fractionation of PH_3 in convective upwelling from its thermochemical equilibrium level at ~1000 bar to the upper atmosphere, such P:H upper tropospheric measurements may not be indicative of the value in the deep atmosphere. As well, the noble gases cannot be measured by remote sensing. For distinguishing between Saturn formation models, it is essential to determine the abundances of all of the noble gases and their isotopic ratios, together with the abundances of C, N, S and O, and the isotopic ratios, D:H in H_2, ^{13}C:^{12}C in CH_4, ^{15}N:^{14}N in NH_3, ^{34}S:^{32}S in H_2S and ^{18}O:^{16}O in H_2O (cf. Chapter 2).

Entry probes at relatively shallow depths (~10 bar) can measure the elemental abundances and isotopes of the nobles, and N and S. However, the accurate measurement of the oxygen abundance derived from its primary reservoir, condensable water, requires that probe measurements be made at depths well below the 10–20-bar water condensation level, that is, at 50–100 bar (cf. Figure 2.9 and associated discussion in Chapter 2). However, water in Saturn's deep atmosphere can also be measured by passive microwave radiometry from Saturn orbit, as at Jupiter by the Juno mission. Including this technique, entry probes sampling down to the 10-bar level would be sufficient.

Direct in situ measurement of Saturn's helium abundance yields additional constraints on the models of Saturn's formation, interior and the heat budget. As discussed in Chapter 2, Saturn's He:H ratio is highly uncertain, with a current best estimate of 0.58–0.84 of the solar ratio (Table 2.1). The depletion of helium relative to the Sun measured in the upper troposphere of Jupiter is attributed to the condensation of helium gas into liquid helium at the 1–2 megabar level in the planet's interior. If the current He:H estimate at Saturn is confirmed, it would imply that the condensation of helium also occurs in Saturn's interior, as currently predicted by interior models (see Chapter 3 by Fortney et al. and references therein). The condensation of helium and subsequent differentiation of helium droplets from hydrogen in the planet's interior would release both latent heat and gravitational potential energy, which contribute to the planet's thermal evolution, anomalously high temperature, and energy balance (see Chapter 2).

14.2.2.2 Building New Worlds: Understanding Saturn's Rotation Rate

As discussed in Chapter 5 by Carbary et al., a surprisingly large uncertainty currently surrounds Saturn's rotation rate. If attempts to measure Saturn's bulk rate of rotation from magnetic field measurements during Cassini's Grand Finale (cf. Section 14.2.1.1) fail, the answer may ultimately be determined through Earth-based observations of Saturn's rings. Near equinox, Earth crosses Saturn's ring plane. Nicholson et al. (1999) noted that the observed timing of these crossings can be used to determine the direction of Saturn's spin axis and the precession rate of the pole, which yields constraints on the planet's angular momentum. Given constraints on the planet's moment of inertia from Cassini's close-up gravity measurements during the Grand Finale, ring-plane-crossing observations should, eventually, accurately determine the rotational period. With only two sets of observations possible per Saturn year (6–7 per century) the required data may take many decades to acquire, but this approach may finally definitively determine Saturn's rotation rate.

14.2.2.3 Understanding the Workings of Solar Systems: Saturn's Atmospheric Variability, Seasonal and Otherwise

Saturn displays a wide range of atmospheric phenomena that vary over a large variety of time scales. These phenomena include changes in temperature, composition and cloud properties (cf. Chapter 10 by Fletcher et al.). Long-term changes that take place on the order of thirty years are likely largely driven by the seasonal variation of sunlight, modified by the nonlinear dynamical and thermodynamic response of the atmosphere. Shorter-term changes may be driven by seasonal forcing or could be completely uncorrelated with seasonal changes. Cassini and ground-based observations have identified seasonal trends in temperature (Chapter 10) as well as events that seem at first glance isolated in time, e.g. the Great Storm of 2010–2011 (cf. Fischer et al. 2011; Chapter 13 by Sánchez-Lavega et al.). It is clear that the atmosphere is driven by more long-term influences than merely seasonal changes of incident sunlight. For example, dynamical features, such as the long-lived (> 1 Saturn year) polar hexagon (cf. Chapter 12 by Sayanagi et al.) and the "string of pearls"

(Chapter 13), along with episodic (~30 years) great storms appear to be confined to the northern hemisphere (cf. Figure 14.6 and Chapter 13), while "storm alley" (cf. Figure 14.5 and Chapter 11 by Showman et al.) seems to be confined to the southern hemisphere near 35° south latitude. What, exactly, are the timescales – seasonal or non-seasonal – of such hemispherically asymmetric features and what are their drivers? Zonal thermal waves also appear to differ significantly in the northern and southern hemispheres during the ~40% of a Saturn year observed by Cassini, but this is too short a time to clearly establish the seasonality of atmospheric phenomena. Therefore, it is left to future ground-based and space-based observations to identify and characterize the full range of phenomena that are distinctly the result of seasonal changes of insolation. As detailed in Section 14.4.1, characterizing these changes will include combinations of observations at all wavelengths and at high-spatial and high-spectral resolution, by both state-of-the-art ground and space-based platforms.

14.2.2.4 Understanding the Workings of Solar Systems: Saturn's Magnetospheric Interactions and Aurorae

Significant gaps in our understanding of Saturn's auroral processes exist (cf. Chapter 7 by Stallard et al). The variable rotating phenomenon, known as the planetary period oscillation is a major outstanding issue. Found at thermospheric/stratospheric altitudes by Cassini, they occur at different rates in the northern and southern hemispheres. From Earth, this phenomenon had been previously positively identified in HST ultraviolet auroral observations (Nichols et al. 2010). Auroral radio emissions are observable by spacecraft located many AU from Saturn (Lecacheux and Aubier 1997), but not from the Earth's surface, as these emissions are below the ionospheric plasma cutoff frequency of ~5 Mhz. Apart from equinox, Earth-based platforms see only a single pole. This renders the characterization of the two periodicities impossible after Cassini's 2017 end until at least the 2024 equinox. However, if the ultimate source of the periodicities rests within the atmosphere (e.g. Smith 2006; Jia and Kivelson 2012; Southwood and Cowley 2014), then the thermospheric or upper stratospheric vortices may be observable to ground-based telescopes equipped with high-spectral-resolution infrared spectrographs and adaptive optics (AO).

Voyager found that the global upper atmosphere is much hotter than predicted by current models (e.g. Müller-Wodarg et al. 2012). Although auroral processes are capable of injecting significant energy into the uppermost reaches of the atmosphere, it remains unclear how this energy could be redistributed to lower regions. The Cassini mission has provided a number of important constraints on ionospheric and thermospheric phenomena via, for example, multiple UV stellar occultations (Koskinen et al. 2013) and Cassini/VIMS H_3^+-derived temperature measurements (Stallard et al. 2012). Similar ground-based observations of H_3^+ emissions as well as ring-rain (O'Donoghue et al. 2013) will continue post Cassini, providing new insights into how energy is injected at low latitudes, how this interaction evolves as the ionosphere changes with season, and thus overall how the mechanisms that govern the energy balance of Saturn's upper atmosphere operate.

14.3 Future Saturn Missions

The remarkable record of surprising discoveries from the Cassini-Huygens mission has given rise to a plethora of new science issues as well as reinforced the quest to obtain long-sought-after measurements, such as the planet's unknown composition at depth. Thus, the scientific community and chief governing space agencies, NASA and ESA, have been spurred to investigate possible future missions to the Saturn system. NASA studies have included both large ~$2 billion Flagship missions and more modest (~$1 billion) New Frontiers Missions for potential flights in the 2020s. In Europe, several missions have been proposed to ESA's Cosmic Visions opportunities for flight in the 2020s and 2030s. All such missions must contend with the relatively complex nature of travel to and operations at Saturn, as we now explore.

14.3.1 Issues Common to Saturn Missions

Prime issues dictating the cost and complexity of Saturn missions are (1) the time-of-flight and energy needed to rendezvous with a planet 9.5 AU from the Sun, (2) the data rate achievable from that distance and

(3) the means to maintain operational systems, including the generation of adequate electric power. Perhaps the most challenging issue is how to reliably fly to Saturn within viable constraints of cost and travel time, which fundamentally depends on spacecraft mass. Beyond the mass of the science payload and associated reliable systems for operations, for orbiter missions the additional mass of the fuel required for velocity adjustments to visit other Saturn system targets such as moons and rings can amount to a significant fraction of the total vehicle mass.

Travel to Saturn involves a change in gravitational potential energy of approximately 8×10^8 J per kg of transferred mass, roughly the chemical energy available from 100 1-kg sticks of dynamite. This energy has been typically supplied by some combination of propulsion (chemical rocket engines, electric propulsion engines, etc.) and gravity assist by intervening planets (e.g. Zander 1964; Doody 2004). Future missions might use more exotic forms of propulsion such as solar sails.

The availability of multiple means for accelerating spacecraft results in a wide variety of Earth-to-Saturn trajectories of varying complexity. Simpler trajectories rely on the launch vehicle and post-launch active propulsion for most of the transfer energy. Some of these are similar to the classical Hohmann transfer (Hohmann 1925), an energy-efficient transit that utilizes an eccentric orbit to connect a pair of concentric circular, coplanar orbits at two tangent points. However, the orbits of Earth and Saturn are neither exactly circular nor coplanar; thus, trajectories that accommodate these "quasi-Hohmann" transfers require somewhat more energy. All of these direct trajectories require at least a six-year cruise and a relatively large amount of energy, resulting in large launch vehicle costs. No Saturn mission has used a direct Earth-to-Saturn transfer, testimony to the effectiveness of using gravity assists (GA) instead of rocket engines to provide a significant fraction of the required energy.

Jupiter's large mass renders it an efficient GA engine for reaching Saturn. Pioneers 10 and 11 and both Voyagers used Jupiter gravity assists (JGA) to reach Saturn and beyond, saving about 25% of the direct Earth-to-Saturn transfer energy. This also decreased the trip time with respect to the direct trajectory: Voyager 1 traveled to Saturn in slightly less than 3.2 years. The main disadvantage of JGAs is the long, ~20-year interval between the openings of relatively short ~2-year usable windows. The Voyagers, launched in 1977, utilized the 1979–1980 Jupiter window, while the Cassini-Huygens mission, launched in 1997, used the 1999–2000 window. With no missions making use of the 2018–2020 window, 2038 is the next opportunity for an efficient JGA push to Saturn.

Significant launch energy can be saved through the use of additional planetary GA pushes and/or the use of solar electric propulsion (Lam et al. 2009). One fruitful strategy, as employed by both Galileo and Cassini, is to conduct multiple encounters of terrestrial planets to reach Jupiter. In particular, Cassini used two GAs at Venus and one at Earth, saving more than 80 of those 100 sticks of dynamite per kg of transferred mass.

The use of terrestrial planets for GA opens up more Saturn launch opportunities. Many such trajectories use combinations of Venus, Earth and even Mars GAs, and some are ballistic, meaning they do not need pre-planned propulsive maneuvers. Another technique uses an Earth GA (EGA) without a flyby of Venus or Mars, but instead requires a propulsive deep space maneuver (DSM) before the EGA, hence the designation "ΔVEGA," where "ΔV" denotes a change in velocity provided by the DSM (Doody 2004). This technique offers more flexible schedules since opportunities occur every Earth-Saturn synodic period, ~1.034 years, but they require significantly larger propulsion when Saturn is far out of Earth's ecliptic plane at arrival. However, solar electric propulsion may render this a viable option.

14.3.2 Saturn Mission Studies

Armed with such trajectory strategies, NASA and ESA have explored a number of potential post-Cassini Saturn missions, focusing on (1) key aspects of unachieved Saturn measurements – such as the abundances of noble gases and their isotopes and the composition of volatiles such as ammonia and water at depth – and (2) answering fundamental new questions arising from the Cassini discoveries. Both NASA and ESA studies have focused on using a descent probe for direct sampling of atmospheric constituents and measurements of winds, particularly in the unexplored region below the upper tropospheric cloud layers.

Figure 14.2 The ESA-proposed M5 Hera Saturn entry probe descending under parachute through the Saturn skies. Having safely delivered the probe through atmospheric entry, the conical heat shield is seen falling away, enabling atmospheric sampling to begin. From Mousis et al. (2016). Artwork by T. Balint. (A black-and-white version of this figure appears in some formats. For the color version, please refer to the plate section.)

Figure 14.2, from the ESA 2016 Hera M5 proposal (Mousis et al. 2016; Section 14.3.2.2) depicts the descent of such an entry probe on parachute just after release of its conic heatshield.

14.3.2.1 NASA Saturn Mission Studies

The Decadal Committee developed and evaluated the viability of two Saturn system mission concepts, with a key objective of keeping total mission cost below ~\$2 billion in 2012 dollars. These were (1) a Titan-oriented mission concept that built on the NASA 2009 Flagship-class "Titan Saturn System Mission" (TSSM) mission study (NASA/ESA 2009; Reh et al. 2009), and (2) a New Frontiers class entry probe mission to explore, in situ, Saturn's atmosphere.

14.3.2.1.1 Titan Saturn System Mission (TSSM)

The 2009 Flagship-class TSSM concept included (1) a Titan orbiter that would visit Enceladus en route to entering Titan orbit, and (2) two probe elements for exploring Titan in situ – a Montgolfière hot-air balloon and a lander that would set down on a hydrocarbon lake. Utilizing remote-sensing and mass spectrometers on all three platforms, the mission would spectrally map and sample the gases and aerosols of both Titan and Enceladus, as well as study the magnetosphere and atmosphere of Saturn, including in-depth scrutiny of Saturn's storms and other meteorological features. However, the Decadal Committee estimated that TSSM cost exceeded the \$2 billion budget, and thus judged TSSM non-viable for the 2013–2022 period.

14.3.2.1.2 The New Frontiers Class Saturn Probe Mission

The Decadal Committee established the mission science objectives for a New Frontiers class Saturn probe (SP) mission, as outlined in Section 14.2.2, and studied a number of SP mission concepts to determine their viability. For these studies, only the most important objectives – known as Tier 1 – were addressed.

The resulting SP mission concept resembles a simpler version of the Jupiter Galileo orbiter/probe mission. The flight system consists of a probe and a flyby carrier-relay spacecraft (CRSC) that would both deliver the probe to an entry trajectory and relay the probe data to Earth. The probe would enter Saturn's atmosphere at a velocity of 26–30 km s^{-1}, significantly less than the 47.4 km s^{-1} experienced by the Galileo probe. After the entry heating and deceleration period, the probe's aeroshell would be jettisoned near the 0.1-bar pressure level, allowing the probe to commence in situ science experiments and data transmission. The descent module would descend to the 5-bar level some 55 minutes later and 210 km below where transmissions began. (Descending to the 10-bar level is preferred, but is challenging, as noted below.) Throughout the descent, the CRSC's antenna would continuously point at the entry site, recording and relaying to Earth about 2 Mb of data, some 3 times that returned by the Galileo probe.

The main conclusions of this SP mission study were

1. Such a mission can fit into the New Frontiers Program resource constraints.
2. Solar power is a feasible option.

3. The cost difference between solar and nuclear power systems is relatively small.
4. A mid-sized launcher can deliver a flight system to Saturn without a Jupiter GA in an acceptable time frame.
5. No new technologies are needed.

Conclusion 3 assumes that the Advanced Stirling Radioisotope Generator (ASRG) would be available at the NASA-stated cost. However, NASA has since cancelled the ASRG development program. The alternative nuclear electric source, the Multi-Mission Radioisotope Thermoelectric Generator (MMRTG), is more massive and expensive, so that trade must be revisited. Conclusion 4 stems from the finding that the flight system and trajectory easily fit the relatively small Atlas 401. However, a launch at a time when Saturn arrival is far out of Earth's ecliptic plane might require the next larger launcher. Conclusion 5 stems from the ongoing development of NASA's Heat-shield for Extreme Entry Environment Technology (HEEET) task for use at Saturn. NASA plans for this technology to be available at Technology Readiness Level (TRL) 6, appropriate for competed mission proposals, by late 2017.

The study also produced a noteworthy technical result concerning descent profiles. Due to Saturn's larger atmospheric scale height and weaker gravitational acceleration, a descent under a single Galileo-like parachute for the entire desired 10-bar descent exceeds the duration of the 60- to 90-minute data-relay window provided by the CRSC overflight. However, a reduction in parachute size for faster descents results in an inadequate vertical sampling rate. Possible options include (1) deployment of a parachute sized for the slower descent through the upper levels, then at a pre-determined pressure level reefing or releasing the parachute to continue the remaining descent in free-fall, or (2) following the Huygens–Titan probe design, deployment of a smaller parachute after the first is released.

The Saturn probe study suggests that such a mission is a feasible candidate for NASA's New Frontiers Program (NFP). Indeed, NASA has added a Saturn probe mission to NFP's list of desired missions for the 2016 Announcement of Opportunity, with Step 1 proposals due by May, 2017.

14.3.2.2 ESA Saturn Probe Mission Studies

Several Saturn probe concepts utilizing both simple carriers and more complex orbiters have been submitted by European scientists and engineers to calls by the ESA Cosmic Vision Program. In recent years, three configurations of varying complexities have been considered (Coustenis et al. 2014; Mousis et al. 2014):

- Configuration 1 (Probe and Carrier). The least costly option, this concept specifies that the probe transmit its data via a direct-to-Earth (DTE) radio-frequency link. The carrier follows the probe into the atmosphere, perhaps performing pre-entry science observations which it would also transmit DTE prior to its atmospheric plunge.
- Configuration 2 (Probe and Carrier/Relay). This concept is similar to the New Frontiers Class concept discussed in Section 14.3.2.1.2, where the carrier relays probe data to Earth. The carrier would release the probe several weeks prior to entry, after which the carrier trajectory would be deflected for overflight phasing of the probe descent location to conduct both probe data relay and approach and flyby science.
- Configuration 3 (Probe and Orbiter). This configuration would be similar to the Flagship Class Galileo Orbiter/Probe mission. As with Configuration 2, the carrier would release the probe several months prior to probe entry and then subsequently deflect to prepare for overflight phasing of the probe descent location. After probe relay, the orbiter would go into Saturn orbit to conduct orbital science.

In all three configurations, solar panels would be used to provide electrical power for the carrier/orbiter, with batteries included for high power demand periods, such as during probe entry. Nuclear power would be considered for the carrier/orbiter only if solar power technology were not feasible. All configurations would allow the carrier/orbiter to perform several months of approach science. Configuration 1 allows in situ pre-entry science by the carrier. Configuration 2 would possibly provide flyby science, and would allow redundant retransmissions of the probe data. With its orbital capabilities, Configuration 3 would be both the most scientifically interesting and most expensive option. As detailed below, the Large-Mission-class Kronos Mission Concept developed in 2007 expanded on Configurations 2 and 3, calling for two probes. As discussed in Section 14.3.2.2.2, the 2015 Medium-Mission-class Hera concept called for a single probe and a relay carrier, adhering closely to Configuration 2.

14.3.2.2.1. The Kronos Mission Concept for ESA Cosmic Vision 2015–2025 The Kronos (Greek name for Saturn) mission (Marty et al. 2009) was submitted to the ESA Cosmic Vision call (ESA 2007) for large missions (900 million euros). The mission's aim was to perform (i) in situ measurements of the chemical and isotopic composition of Saturn's atmosphere via two ESA probes delivered by a single NASA carrier spacecraft and (ii) remote sensing from space by carrier spacecraft instruments to better understand the formation of Saturn and the origin of its atmosphere. The probes were designed to perform in situ measurements of Saturn's atmosphere down to a depth of about 10 bar. Compared to a carrier flyby option, an orbiter configuration would have enabled a more global investigation of Saturn's atmospheric H_2O and NH_3, the planet's gravity and internal magnetic fields, the ring system and the inner magnetosphere. However, the mission would have required additional propellant tanks and a modified power system to deliver higher power.

Because Kronos was proposed as an international collaboration between ESA and NASA, both US and European solutions were considered for the launch vehicle. The total mass of the two-probe flyby-based architecture was estimated at less than 3000 kg, implying that the launch vehicle for the mission could be an Atlas V-551 or equivalent. Chemical and/or solar electric propulsion systems were envisaged for the spacecraft, leading to trajectories with flight times ranging from ~6 to 17 years, with <12 year trajectories much preferred. For carrier power, Low Intensity Low Temperature (LILT) solar panels inherited from the Juno and Bepi Colombo missions were considered. To provide the least risky mission architecture possible, the probes were expected to use batteries and a DTE communication strategy, and to arrive at or near the sub-Earth point on Saturn. For ring science, the Kronos proposal also suggested that the carrier could drop off a fleet of two or three very small, identical probes directly into the rings, with each probe carrying a single efficient camera to acquire high-resolution ring images.

The proposed Kronos Saturn probes used significant heritage from the Galileo probe, including the Thermal Protection System (TPS) and aeroshell design. The Huygens Titan probe provided technological heritage for the operation of the descent modules deployed from the two Kronos atmospheric probes.

The two key Kronos probe instruments were a mass spectrometer (Saturn Probe Gas Analysis System; SPGAS) and an Atmospheric Structure Instrument (ASI). The proposed SPGAS quadrupole mass spectrometer was inherited from the successful Galileo Jupiter and Cassini/Huygens probe composition/isotope investigations (Niemann et al. 1998, 2005). Similar to the Cassini/Huygens ASI experiment (Fulchignoni et al. 2002), the Kronos ASI consisted of three primary sensor packages: (1) a three-axis accelerometer, (2) a pressure profile instrument and (3) temperature sensors. In addition, in the upper atmosphere, deceleration measurements provided the atmospheric density, pressure and temperature profiles prior to pressure/temperature sensor deployment.

Other envisaged probe instruments included a nephelometer to determine the aerosol number density, and the sizes, shapes and index of refraction of aerosols, and a Doppler Wind Experiment (DWE) – based on the successful Galileo and Huygens designs (e.g. Atkinson et al. 1998) – to measure the vertical profiles of the zonal (east-west) winds along the probe descent paths.

For the carrier, the payload science objectives were close to those of the Juno mission. In particular, key objectives were to secure the deep, global abundances of H_2O and NH_3, as well as their distributions over all latitudes, using microwave radiometry.

14.3.2.2.2 Recent ESA Probe Concepts: The Hera Mission

In 2010, ESA assessed Planetary Entry Probes (PEP) for Venus, Saturn, Uranus and Neptune (ESA 2010) for their feasibility to investigate these planets down to 100 bar (target)/30 bar (threshold). In this study, the designs of the four probes were similar in general layout, mass and payload (see Figure 14.3). In the case of Saturn, solutions were found allowing descent from ~1 to 100 bar in ~90 minutes.

Capitalizing on these results, in January 2015 a proposed mission entitled Hera – after the daughter of Kronos in Greek mythology – was submitted to the ESA call for medium-class missions (M4). This latest proposal followed the Configuration 2 concept discussed above, with a flyby telecom carrier and a single probe. Given the reduced cost of the M4 mission, this concept was less ambitious than the PEP Saturn concept, reaching a depth of just 10 bar, sufficient to measure the noble gases,

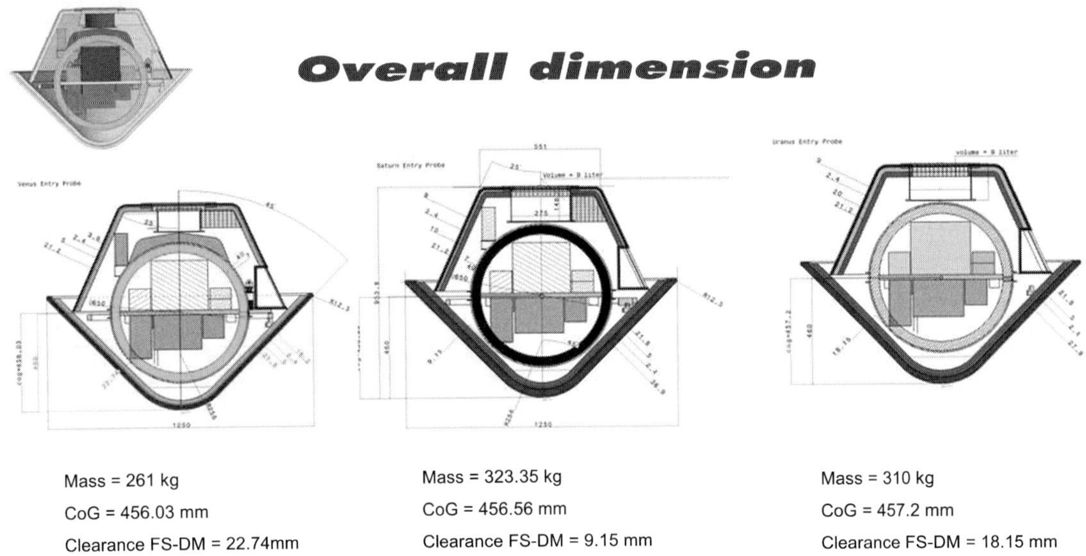

Figure 14.3 Typical dimensions of planetary entry probes investigated by ESA, for the exploration of Venus (left), Saturn (middle) and Uranus (right). CoG: Probe *Center Of Gravity*, Measured from tip of conical front shield (FS); Clearance FS-DM: Minimum space between descent module (DM) and FS (adapted from the PEP Assessment Study Presented at ESTEC on 30 June 2010). (A black-and-white version of this figure appears in some formats. For the color version, please refer to the plate section.)

including helium. As with Kronos, the science payload included an ASI, a sensitive mass spectrometer, a nephelometer, a DWE and a net flux radiometer (NFR). The carrier would have carried a science camera to obtain high-resolution images of Saturn and rings during the flyby. Following the Kronos concept, the Hera carrier would have used LILT solar panels while the probe would use batteries. The mission would have been an international collaboration with NASA who would have supplied the TPS and LILT, and perhaps other systems and instruments.

ESA is currently discussing alternative mission designs. These include a more ambitious concept than Hera to fly a carrier with two probes into the Saturn system, possibly as a large-class mission with budget of ~1 billion euros. One probe would explore Saturn's atmosphere while the other would investigate the composition of Titan's atmosphere and/or that of the geysers of Enceladus. An alternative possibility would be to send a carrier containing several similarly equipped probes to Saturn, Uranus and Neptune that would enable comparisons between the gas and icy giants. However, such a mission would require radioisotope thermoelectric generators, a technology that is not yet sufficiently mature within ESA.

14.3.3 Science Opportunities from Ancillary Saturn Missions

Other opportunities for Saturn science could arise from spacecraft targeted to Saturn's moons Enceladus and Titan, missions supported by the 2016 New Frontiers Announcement of Opportunity. If selected, a Saturn system mission would likely be launched by the mid-2020s. On such a mission, remote sensing cameras and spectrometers, magnetometers and plasma-sensing instruments could provide remarkable new observations of Saturn for several weeks/months on approach to these moons, perhaps for much longer if these missions allow for either satellite or Saturn orbit.

Valuable Saturn science could also be achieved by any appropriately equipped flyby spacecraft that use a Saturn gravity assist. This includes the Decadal Committee's Flagship class Uranus Orbiter and Probe (UOP) mission (NRC 2011) and possible future missions to Neptune and Pluto. Such a mission could provide months of useful flyby observations of Saturn. For Uranus missions, appropriate Saturn GA launch windows of 4–5 years duration occur every 45 years. Voyager 2 utilized the last opportunity when it flew past Saturn in August 1981, encountering Uranus less than 4.5 years later in January 1986. The center of the next opportunity for a Saturn gravity assist occurs in 2026, improbably near to mount a new Uranus mission. The subsequent opportunity for a Saturn kick to the ice giant occurs in 2071.

14.4 Future Observations by Ground-Based and Near-Earth Observatories

With no plans to fund missions to the Saturn System by any of the world's space agencies, ground-based and near-Earth space observatories will likely provide the only certain observations for decades to come. Premier facilities include the large, multi-meter aperture telescopes presently operational and under development in South America and Hawaii, the NASA Hubble Space Telescope, expected to be operational until 2020, and the HST-follow-on facility – the James Webb Space Telescope (JWST) that is currently slated for launch in October, 2018. The JWST is the most complex and most sensitive space-based visual-infrared astronomical telescope developed to date – encompassing a 6.5-meter aperture telescope operating at temperatures less than 50 K – and will be placed at the second Lagrange point (L_2) in a solar orbit some 1.5 million km outside of the Earth, fixed on the Sun-Earth-JWST line, thus ensuring the telescope's high sensitivity by shielding it from both the visible and thermal radiation produced by the Sun, Earth, and Moon, all of which will be on the same side of the spacecraft.

Observations from JWST and large ground-based facilities promise to provide fundamental new information on key Saturn processes, especially as pertains to the planet's response to temporally varying solar inputs. The daily average solar flux at any Saturn latitude varies due to (1) the ~11-year solar cycle that modulates the solar ultraviolet flux and (2) the seasonal variability over Saturn's 29.5-year orbital period produced by both Saturn's orbital eccentricity of 0.054 (resulting in a solar insolation variation of 24% between perihelion and aphelion) and the planet's obliquity of 26.7° degrees (vs 23.4° for Earth and 3.1° for Jupiter). As well, observations from much smaller and less complex ground-based telescopes – particularly those provided by amateur astronomers – promise to provide key information, particularly on the onset and early development of large storms and cometary impacts, as has proven valuable during the past decade (cf. Section 14.5).

14.4.1 Outstanding Science Issues Addressable by Ground-Based and Near-Earth Observatories

14.4.1.1 Long-Term Variability of the Stratosphere and Upper Troposphere

The long-term variability of phenomena in Saturn's stratosphere and upper troposphere typically manifest themselves in meridional variations of temperatures, chemical make-up and winds over latitude (cf. Chapter 10 by Fletcher et al.), as well as more locally in the formation and dissipation of storms and vortices, such as the Great Storm of 2010–2011 (cf. Chapter 13 by Sánchez-Lavega et al.). All such phenomena ultimately provide information on two fundamental issues: (1) "What is the association of the phenomena with seasonal and solar-cycle-induced changes of insolation?", and (2) "what extent is the atmosphere inherently hemispherically asymmetric in its properties?" Resolution of these issues requires observations well past the Cassini time frame.

For over a decade, Cassini has been observing the seasonal variability of Saturn's haze/cloud reflectivity, chemical make-up and temperature. In particular, Cassini/CIRS has been measuring the rate of change of temperature structure and composition over latitude, showing in the short term (several years) favorable comparisons with radiative climate models (Fletcher et al. 2010; and Chapter 10). However, over the longer, decadal term, amounting to about one-third of a Saturn year, these measurements have proven somewhat inconsistent with theory (e.g. Guerlet et al. 2014). Only data gathered throughout Saturn's orbit will provide sufficient coverage to determine the primary sources of heating, including otherwise difficult-to-model radiative absorptions from aerosols. This is particularly the case if the relevant particulates themselves form from seasonally dependent chemistry or are products of phase changes involving seasonally dependent temperature differences.

Other seasonal effects are expected, but have not been documented sufficiently, including the formation of polar vortices, such as the broad, warm polar vortex discovered in an already formed state in the south (Orton and Yanamandra-Fisher 2005) and in the north (Fletcher et al. 2008) that are likely to be the combined result of seasonal warming with boundaries that are governed by dynamical interactions. Any differences between the evolution of the north and south pole vortices could provide clues to fundamental differences in the internal structure between the hemispheres. The low-latitude oscillation of temperatures discovered in limb sensing by Cassini/CIRS (Fouchet et al., 2008) appeared to be the manifestation of a semi-annual oscillation (SAO; Orton et al. 2008). However, more recent observations noted deviations from that behavior (Sinclair et al. 2014). It remains to be seen whether these departures happen to be the result of the Great Storm of 2010–2011 (cf. Sánchez-Lavega et al. 2012 and Chapter 13) and represent a minor perturbation or whether they are a major phase change of an otherwise stable long-term cycle.

The storm itself and similar precursors (e.g. Sánchez-Lavega et al., 1991) represent unknown time-dependent phenomena whose origins and forcing are still largely uncertain but nevertheless have been the subject of intriguing hypotheses that can be tested over time. In particular, pseudo-cyclical release of latent heat energy from atmospheric water at depth near 20 bar has been recently proposed as the source of power for such eruptions (Li and Ingersoll 2015; cf. Chapter 13), with major eruptions predicted at "storm alley" latitudes (e.g. near the equator, or at mid-latitudes) about every 60 Earth years. As well, the nature and variability of key storm characteristics such as the strength and durability of long-term perturbations of cloud properties or the formation, strength and lifetime of a stratospheric vortex (Fletcher et al., 2012), both so well displayed by the Great Storm of 2010–2011, need to be explored to determine whether such phenomena are similar in all such storms or whether they are different for storms of different strengths.

14.4.1.2 The Variability of Saturn's Auroral Activity with the Solar and Seasonal Cycles

Saturn's aurorae are known to vary partly as a function of solar wind dynamic pressure (Clarke et al. 2009). Therefore, as the solar wind changes over the 11-year solar cycle, Saturn's auroral activity should vary as well. By the end of its 13-year orbital reconnaissance of Saturn in September, 2017, the Cassini Orbiter will have covered an entire solar cycle, obtaining an observational record of myriad solar-induced processes, including auroral processes. However, mixed with the solar cycle is Saturn's 29-year seasonal cycle, which continually changes the geometry of the magnetosphere with respect to the solar wind, likely resulting in a seasonal dependency of auroral activity. This convolution of the length of each season with the period of the solar cycle produces a complex interplay of dependencies that can only be untangled by long-term monitoring of auroral activity. Ground-based infrared observations of H_3^+ emission and space-based ultraviolet observations of H and H_2 emission over several decades would be key components of such a monitoring campaign.

14.4.1.3 Understanding Ring Rain

As detailed in Chapter 8 by Moore et al., the influx of ring-associated water into Saturn was inferred from the presence of dark bands on the planet observed by the Voyager 2 visible camera (Connerney 1986). These dark bands magnetically mapped to features in the rings, indicating that ionized molecules of water from the rings move along magnetic field lines into the planet, chemically altering the makeup of the planet's atmosphere. Using a medium resolution near-infrared spectrograph on the Keck telescope, O'Donoghue et al. (2013) observed this mechanism on a global scale within the ionosphere, revealing an intricate pattern of H_3^+ emissions across the entire northern hemisphere, with a nearly identical companion pattern present at southern magnetic latitudes. Since the magnetic field lines at low latitudes intersect the ring-plane, this strongly suggested that the ionosphere is attenuated by the influx of charged water from the rings. This precipitating water can result in either the net destruction or net production of H_3^+, depending on the flux. This can be investigated with sensitive ground-based telescopes – such as the Keck – during the Cassini Grand Finale in 2017, correlating the planet's H_3^+ pattern with the field lines connecting the rings and the atmosphere as observed in situ by the spacecraft flying just several thousand km above the cloud tops over near-equatorial latitudes. Such data should provide new insights into both (1) the atmospheric chemistry induced by ring particles precipitating onto the planet and (2) changes to the radiative balance of the upper atmosphere generated by these precipitating-particle-induced constituents. Post-Cassini, continued ground-based observations will provide information on the temporal variability of ring rain due to solar cycle variability, particularly as potentially caused by large solar storms.

14.4.2 Observations Obtainable over the Next Three Decades from Near-Earth and Ground-Based Observatories

14.4.2.1 The Prospects from Space Observatories

The 2.4-m Hubble Space Telescope is expected to operate until at least 2020 (www.space.com/29206-how-will-hubble-space-telescope-die.html), providing approximately two years of simultaneous operations with JWST. Until the end of Hubble, Saturn and the other outer planets are slated to be observed with WFPC3 annually under HST's Outer Planet Atmospheres Legacy (OPAL) program (www.stsci.edu/hst/phase2-public/13937.pro). With a spatial resolution of better than 0.1″, the globe of Saturn will be covered over two full planetary rotations each year with multi-filtered images between 220 and 1000 nm, providing an annual determination of large-scale changes in Saturn's mean zonal flow that complement previously observed major changes in the zonal wind (Sánchez-Lavega et al. 2003). Individual General Observer (GO) observations will be possible as well with the WPC3 visible camera and other instruments. Accessing spectral regions unavailable to ground-based observatories, the Space Telescope Imaging Spectrograph (STIS) operates in the near-ultraviolet from 165 to 310 nm, the Advanced Camera for Surveys (ACS) operates over the UV-near-IR from 115 to 1100 nm and the Cosmic Origins Spectrograph (COS) operates in the UV-visible from 115 to 320 nm.

However, at best, even the extended HST operations through 2020 supplement Cassini's on-orbit observations for just three years past the end of the Cassini mission, corresponding to only 10% of a Saturn year. Post-HST, there are currently no plans for UV-visible observations from a funded space-based observatory, although a planetary-dedicated space telescope was proposed to the 2013–2022 Planetary Decadal Survey (Wong et al. 2009). As well, the Large Ultraviolet/Optical/Infrared Surveyor (LUVIOR) 8–16-m-diameter space telescope is currently one of four space-based concepts under study by the National Research Council for the 2020 Astrophysics Decadal Survey (Thronson et al. 2016).

At longer wavelengths, JWST will achieve unprecedentedly sharp Saturn images from near-Earth space when it begins operations in 2018. JWST instrumentation includes the NIRCam near-infrared camera that acquires filtered images from 1.4 to 4.8 μm and includes highly diagnostic spectral regions blocked to ground-based telescopes by telluric absorption, such as the 2.7–3.0-μm region that effectively characterizes ammonia- and water-based condensate hazes and clouds (e.g. Sromovsky et al. 2013). The NIRSpec 1–5 μm spectrometer has a choice of slit widths of 0.1 or 3.3 arcsec and offers an integral field unit (IFU) option with both imaging and spectroscopic capabilities covering a 3×3 arcsec^2 area. Resolving powers of ~1000 or 2700 can be selected (where the resolving power is defined as the observed wavelength, λ, divided by the width of the resolution element, i.e. $\lambda/\Delta\lambda$). At longer wavelengths, MIRI can image Saturn in filters centered between 5.6 and 15 μm without saturation. However, because the planet is so bright, images must be acquired with a 7×7 arcsec sub-array field-of-view (FOV), considerably smaller than Saturn's 20-arcsec diameter. Thus, nine pointings are required to image the entire disk with a 3×3 mosaic. For spectroscopy, similar to NIRSpec, an IFU concurrent imaging/spectroscopy mode targets specific planetary regions, such as the north polar vortex and hexagon, equatorial convective cloud regions and localized cyclones and anti-cyclones. Spatial coverage ranges from 3×4 arcsec^2 at 5 μm to 7.6×7.6 arcsec^2 at 28 μm. Further details of JWST's observational capabilities are presented in Norwood et al. (2016).

In the early years of the twenty-first century, ESA's X-ray Multi-Mirror Mission, known as XMM-Newton, and NASA's Chandra X-ray Observatory observed X-rays from both Saturn and Jupiter. As reviewed by Badman et al. (2015), distinct differences exist in the character of X-rays emitted by these planets. For Jupiter, X-ray emission is associated with auroral emission, with soft X-rays (<2 keV) attributed to charge-exchange processes and hard X-rays (>2 keV) to electron bremsstrahlung origins. By analogy with Jupiter, Saturn should show soft X-ray emission generated from charge-exchange or bremsstrahlung processes, but thus far no X-ray emissions have been observed apart from solar X-rays reflected by Saturn's atmosphere. A far more sensitive X-ray instrument concept has been developed for the 2028-launched Advanced Telescope for High-Energy Astrophysics (ATHENA), which will use its X–ray Integral Field Unit (X-IFU; Barret et al. 2016) to conduct a thorough search for Saturn's auroral emissions.

14.4.2.2 The Next 30 Years: Possible Imaging and Reflection Spectroscopy Observations by Large, Ground-Based Observatories

Although space-based missions offer incredible sensitivity and access to spectral regions inaccessible to ground-based telescopes, nevertheless the short lifetime of such missions – typically less than Saturn's 30-Earth-year orbit – necessitates the use of long-lived ground observatories for the consistent monitoring of Saturn's long-term atmospheric evolution. Currently, there are more than a dozen extremely capable medium-to-large diameter telescopes throughout the world (Figure 14.4). In addition to these extant facilities, the US and ESA are developing three more massive, very large-aperture telescopes: the Thirty Meter Telescope (TMT, 30 m), the Giant Magellan Telescope (GMT, 24.5 m) and the European-Extremely Large Telescope (E-ELT, 39 m). With its baselined adaptive optics systems, the TMT promises to achieve diffraction-limited spatial resolution longward of 1 μm, providing 0.007-arcsec imaging at 1 μm (Figure 14.5) with comparable performances expected from the other new facilities. This is sufficient to image Saturn at better than 100 km resolution, superior to all but the best near-infrared images produced by Cassini/VIMS.

These three new exceptionally powerful telescopes will be equipped with state-of-the-art instruments. The TMT, with first light now delayed beyond the 2022 planned opening due to legal disputes over construction on sacred ground at Mauna Kea, Hawaii, plans to support a number of advanced instruments including (1) the Wide Field Optical Spectrometer (0.3–1.0 μm) with imaging and spectroscopy covering a 40-square-arcmin field-of-view (Pazder et al. 2006), (2) the Infrared Imaging Spectrometer (0.8–2.5 μm) that achieves diffraction-limited imaging integral field spectroscopy (Larkin et al. 2016) and (3) the Infrared Multi-Object spectrometer (0.8–2.5 μm) that acquires near-diffraction-limited imaging and spectroscopy over a 2-arcmin diameter field-of-view (Eikenberry et al. 2006). The GMT, with first light planned for 2020, will offer (1) the Visible Echelle Spectrograph (0.35–0.95 μm) with spectral resolving power from 25,000 to 120,000, well suited for precision radial velocity observations (Szentgyorgyi et al. 2014), (2) the Visible Multi-Object Spectrograph (0.35–1.1 μm) with 9 × 9 arc-minute field of view obtained at moderate-spectral resolution (DePoy et al. 2012), (3) the Near-IR IFU and AO imager (0.9–2.5 μm) with 8- to 50-milliarcsec spatial scales feeding an R = 5000 spectrograph (McGregor et al. 2012) and (4) the GMT Near-IR spectrometer (1.2–5 μm) echelle spectrograph that will deliver high R = 50,000–100,000 spectra over five (JHKLM) atmospheric windows in a single observation (Lee et al. 2010).

The E-ELT, with first light planned in the early 2020s, plans as well to have two first-light instruments. MICADO, the Multi-AO Imaging Camera for Deep Observations (Davies et al. 2016), has two imaging modes: (1) a 20-arcsec spatial coverage mode with fine pixel sampling and (2) a 1-arcmin field-of-view mode that maintains diffraction-limited imaging in H- and K-bands. HARMONI, the High Angular Resolution Monolithic Optical and Near-Infrared field spectrograph (Thatte et al. 2014) achieves spectral resolving powers of 500 to 20,000 between 0.47 and 2.45 μm over field of views of 6.42 × 9.12, 3.04 × 4.28, 1.52 × 2.14 and 0.61 × 0.86 arcsec^2. Additionally, a third instrument under construction is METIS, the Mid-Infrared E-ELT Imager and Spectrograph (Brandl et al., 2014). Over 3 to 19 μm, METIS will offer imaging, coronography and medium-resolution spectroscopy over the entire spectral region and high-resolution integral field spectroscopy over 3–5 μm.

For existing 8-to-10-meter-class facilities, improved instrumentation along with improvements in adaptive optics correction techniques promise to make these telescopes even more productive and more sensitive. One such instrument, MUSE, the Multi Unit Spectroscopic Explorer (Bacon et al. 2006), has recently come online at the Very Large Telescope (VLT) in Chile. MUSE spans 0.465–0.93 μm and offers either wide-field (1 × 1 arcmin^2, 0.3–0.4-arcsec spatial

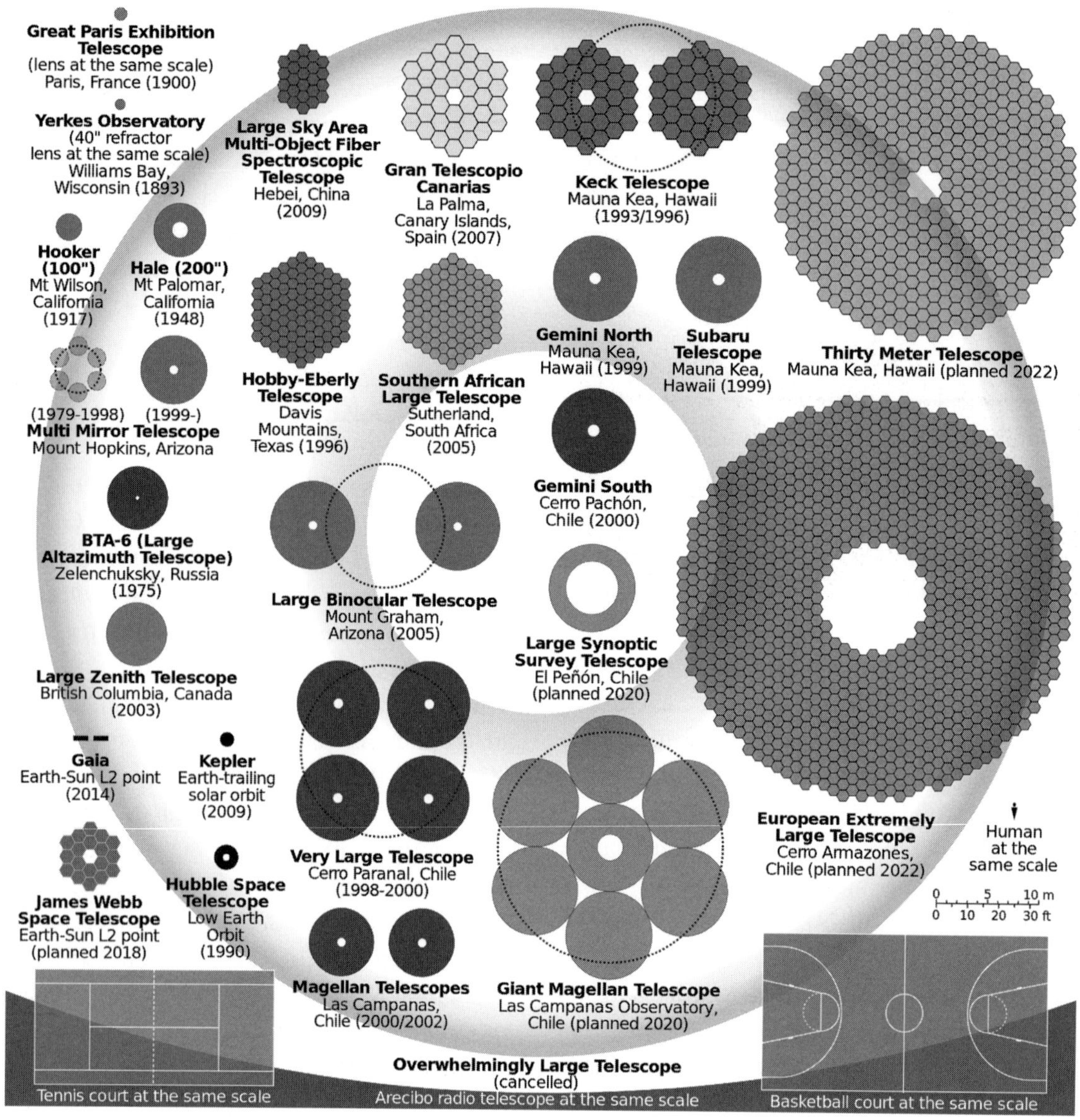

Figure 14.4 Size comparison of the world's largest telescopes. © CMG Lee / http://commons.wikimedia.org/wiki/file:comparison_optical_telescope_primary_mirrors.svg/ CC-BY-SA-3.0 (A black-and-white version of this figure appears in some formats. For the color version, please refer to the plate section.)

resolution) or narrow-field (7.5 × 7.5 arcsec^2, 0.03–0.05-arcsec resolution) integral field spectroscopy at 2000–4000 resolving power. This and other concurrent imaging/spectroscopic instruments are making spectral and spatially resolved observations even more efficient, producing more useful information with less time and overhead.

Although these premier facilities can achieve a vast amount of new Saturn science, the high oversubscription rate from all fields of astronomy will likely significantly restrict their use for studies of Solar System planets. For visible and near-infrared observations, an excellent approach to optimize the time for such in-demand facilities is to use more plentiful, smaller-scale 1- to 3-meter telescopes. Indeed, the ability to perform high-level adaptive optics corrections via laser guide stars is breathing new life into these facilities. One such system, Robo AO (Baranec et al. 2013), can achieve near-diffraction-limited performance in the visible. For example, it is currently possible to achieve 0.1″ spatial resolution in the visible with the Palomar 1.5-meter telescope (Baranec et al. 2014), achieving ~600 km spatial resolution on Saturn at opposition.

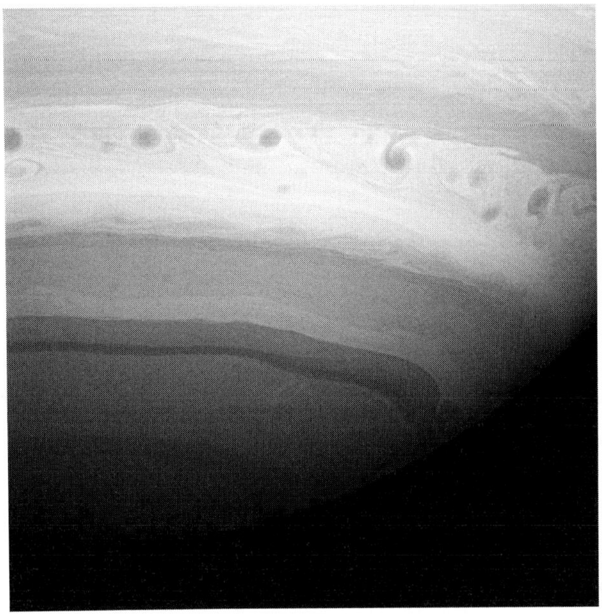

Figure 14.5 Cassini wide-angle camera image of the southern hemisphere Storm Alley in 2008, with a spatial sampling of 48 km/pixel at 728 nm (96 km resolution), comparable to the resolution the TMT will achieve at 1 μm (image courtesy of NASA/JPL-Caltech).

14.4.2.3 The Next 30 Years: Possible Thermal Observations from Large, Ground-Based Observatories

Thermal infrared imaging and high-spectral resolution observations of Saturn can provide detailed insights into the structure and evolution of the atmosphere (e.g. Orton et al. 2008). The mid-infrared allows for the tracking of the temporal evolution of temperatures, photochemical by-products, disequilibrium molecules and clouds. Currently, a number of ground-based instruments are producing valuable mid-infrared observations of Saturn and the other giant planets, including a refurbished MIRSI (Mid-IR Spectrometer and Imager), CELESTE (a high-resolution 5- to 25-μm echelle spectrometer) and HIPWAC (the Heterodyne Instrument for Planetary Wind and Composition) at the 3-m NASA IRTF. TEXES, the Texas Echelon cross-dispersed Echelle Spectrograph, operating at both the IRTF (Greathouse et al. 2005, 2006) and Gemini North, offers both long-slit spectral-scan mapping of Saturn at R = 4,000 and 15,000 as well as high-spectral-resolution scan mapping at R = 50,000–100,000, depending on wavelength. VISIR at ESA's 8-meter VLT is a mid-infrared imager and spectrograph that achieves spectral resolving powers of 150–30,000. Additionally, COMICS, the Cooled Mid-Infrared Camera and Spectrometer mounted on the 8-m Subaru telescope on Mauna Kea in Hawaii, is capable of imaging and spectral measurements with resolving powers of 250–8500 (e.g. Fletcher et al. 2009b). Although no thermal instruments are currently under development for the large 30-meter-aperture telescopes discussed in Section 14.4.2.2, several such instruments are currently planned to be built as second-generation instruments, such as MICHI for the TMT (Packham et al. 2012) and METIS for the E-ELT (Brandl et al. 2014). These could be used, for example, to study the evolution of the stratospheric polar vortex (see Chapter 12 by Sayanagi et al.) in great detail, achieving spatial resolutions of 0.12 arcsec, or ~800 km on Saturn, at 14 μm and 0.24 arcsec, or ~1600 km on Saturn, at 28 μm, for a 30-meter, diffraction-limited telescope.

Submillimeter through millimeter observations of Saturn at high spectral and spatial resolution can be achieved with the Atacama Large Millimeter Array (ALMA). Lellouch (2008), outlining possible very high spectral- and spatial-resolution spectroscopy observations of planetary atmospheres, noted the potential for resolving the 3-dimensional distribution of several species on Saturn and the other giant planets. Determining the vertical profile and meridional distribution of CO in Saturn would provide important clues to the origin of external water and therefor oxygen precipitating into the planet. As well, determining the D:H ratio of high-altitude H_2O will also enable the distinction between its origin from a distance via comets or locally from the ring system. The distribution of stratospheric winds can be mapped in particular at equatorial latitudes, possibly providing evidence of waves perhaps associated with the semi-annual oscillation (Fouchet et al. 2008; Orton et al. 2008).

Spatially resolved microwave measurements can also be made using the Karl G. Jansky Very Large Array (VLA) to measure the distribution of ammonia gas deep in Saturn's troposphere, similar to the mapping of Jupiter by de Pater et al. (2016). Unlike Jupiter, Saturn does not emit synchrotron radiation that could be confused with thermal emission from the neutral atmosphere. Thus, in principal, one could potentially explore the NH_3 distribution deeper than the ~8-bar level of the Jupiter VLA measurements.

Straddling the line between orbital and ground-based observing platforms, SOFIA, the Stratospheric Observatory for Infrared Astronomy (Temi et al. 2014), is a 2.5-m telescope mounted within a 747 aircraft which flies at ~45,000 feet to make observations above most of the Earth's water vapor. EXES, a near twin to the ground-based TEXES instrument, can be used to measure trace Saturn constituents such as H_2O and CH_3, that are currently impossible to measure from the ground due to their low abundance and telluric interference (DeWitt et al. 2014).

14.4.2.4 The Next 30 Years: Auroral Observations

After Cassini, Earth-based observations will once again become the primary means to observe Saturn's aurorae. For nearly three decades, the Earth-orbiting Hubble Space Telescope (HST) has been acquiring crucial ultraviolet observations of Saturn's aurorae. Unfortunately, such space-based observations will end with the demise of HST. JWST will have exceptional spatial resolution and spectral sensitivity in the near-infrared to observe highly diagnostic H_3^+ auroral emissions.

Ground-based infrared telescopes have been observing Saturn auroral H_3^+ emissions since the early nineties (Geballe et al. 1993). Instrumentation on existing facilities include imagers and high-spectral-resolution spectrographs that have provided observations highly complementary to Cassini's measurements. At Mauna Kea Observatory, these include the ISHELL infrared spectrometer and the SpeX medium-resolution near-infrared spectrometer at the NASA Infrared Telescope Facility (IRTF), the NIRSPEC high-spectral-resolution spectrograph at the Keck II telescope and the GNIRS near-infrared spectrometer at the Gemini North telescope. In Cerro Paranal, Chile, the CRIRES high-resolution near-infrared spectrograph is available on the Very Large Telescope. As well, the development of new or updated instrumentation, such as the updated CRIRES on VLT, will provide significant improvements in the quality of auroral observations.

Auroral observations will benefit as well from the development of the three extremely large and adaptive-optics-equipped telescopes described in Section 14.4.2.2, i.e. the TMT, the E-ELT and the GMT. Each of these telescopes will collect over ten times more light than the largest telescope available today. This, in combination with their Adaptive Optics systems, will enable very detailed, high-cadence views of auroral processes. Of particular importance is the detailed characterization of mid-to-low latitude "Ring Rain" (see Chapter 8 by Moore et al.), which utilizes H_3^+ emissions that are very weak compared to the main auroral emissions, requiring the sensitivity of these new instruments.

14.5 Ground-based Amateur Astronomy: Prospective Contributions

The army of experienced amateur astronomers spread across the world is a key asset for the future exploration of Venus. Utilizing commercially available modern yet inexpensive telescopes, computers and software, these volunteer observers regularly produce planetary images that rival those achieved by professional astronomers in spatial clarity and detail, enabling them to periodically make profound discoveries. Indeed, amateur astronomers discovered two of the most dramatic recent phenomena on Jupiter and Saturn: debris clouds produced by an asteroidal collision with Jupiter in 2009 (Sánchez-Lavega et al. 2010) and the breakout of Saturn's Great Storm of 2010–2011, one of the most powerful storms ever witnessed (Sánchez-Lavega et al. 2011; cf. Chapter 13 by Sánchez-Lavega et al. and Figure 14.6).

The major technical advance responsible for such progress has been the development of Video Astronomy. By recording many thousands of short-exposure video frames and then combining only those that show the clearest detail as obtained in moments of the best seeing, amateur astronomers have effectively created a "poor man's adaptive optics" technique. This method uses sophisticated, low-noise video cameras incorporating very sensitive CMOS or CCD sensors, combined with modern, high-quality optics, filters and software to produce high-resolution images that push the level of detail close to the theoretical aperture-determined limit on nights of good seeing, effectively removing much of the detrimental blurring caused by the Earth's atmosphere.

14.5.1 The Development of Video Astronomy

Prior to the advent of affordable, high-sensitivity video cameras suitable for planetary imaging, amateur

Figure 14.6 Amateur astronomer views of the evolution of the 2010/11Great Storm. A: 14 December 2010, B: 22 December 2010, C: 30 December 2010, D: 2 March 2011 (images © Copyright 2010–2011 by Anthony Wesley).

astronomers typically captured single images with ~0.5-sec exposures, which regularly resulted in blurred images caused by atmospheric turbulence and image motion from local wind-induced tracking errors. In addition, the slow several-second-per-image downloading speed then available limited the number of individual images that could be captured in a single session.

The acquisition of clear, highly resolved images was also hampered by focus drift. In practice, for an F/4 instrument, focus errors as small as 20 microns result in visible image degradation in good seeing. In the 1990s, a large fraction of available observing time was required to obtain and maintain the necessary focus due primarily to (1) flexure in the equipment during tracking, and (2) temperature changes that caused instrument expansion/contraction and thus changes in the optical path.

The late 1990s saw the development of a new breed of low-light-sensitive video cameras for use in internet video conferencing. With transfer rates of 1.1 Mbit/s, corresponding to about 5 frames per second, these video cameras were adopted by the amateur community for relatively rapid planetary imaging. The popular Philips ToUcam camera combined a Sony color CCD (ICX098 BQ) and a lightweight case that could easily be adapted for the modest telescopes in use by amateur astronomers. This sensitive camera spanned 640 × 480 pixels, allowing a usable image scale of ~6 pixels per arc second when operated on a 250-mm-diameter telescope, yielding images of Saturn spanning 70–100 pixels across at the equator. To double the spatial sampling/resolution, a grey-scale version of the ICX098BL detector was used to replace the color RGB Bayer matrix color-sensor (wherein multiple pixels are used to cover the red, green and blue colors necessary to produce a color image; Bayer 1976). A discrete-filter wheel was then typically added to the optical system to obtain images in various colors, including those somewhat beyond the normal visible range.

The advantages of the modern Video Astronomy technique, still used today with much more capable cameras (see Section 14.5.2) include

- Continuous viewing of the target, enabling more accurate, time-efficient and regular focus and camera gain and shutter checks;
- Real-time monitoring of the seeing with improved decision-making on when to record;
- Capturing of thousands of images in a video sequence spanning several minutes, together with efficient automatic editing to find, preserve and combine the best frames.

During the late 1990s, a new breed of free software became available to analyze and reduce the video stream to a final processed image that was greatly superior to anything previously achievable. In particular, the Registax and AutoStakkert software, created by Cor Berrevorts and Emil Kraaikamp, respectively, both of The Netherlands, allowed (1) the re-ordering of the available video frames from best to worst image quality as determined by a user-chosen metric and (2) registration of the chosen best images to a common frame registration using image morphing and inter-frame multipoint alignment techniques to correct for spatial distortions in the individual frames. Combining results obtained in individual color channels enabled the creation of detailed color images.

14.5.2 Recent Advances in Camera Technology

Significant progress has recently occurred in enhanced sensor capabilities, in more sophisticated onboard processing capabilities, and in miniaturization. Today the most powerful commercially available cameras are no more than 50 mm on a side, weigh less than 0.1 kg, and are capable of transmitting up to 5 Gbit, or 100 frames, per second.

The 2010s have seen the maturation of CMOS sensor technology, thus enabling observations at exceptionally low light levels. Currently in early 2017, the 2064 × 1544 pixel Sony IMX252 is perhaps the most advanced sensor, incorporating significant improvements in noise floor reduction and in increased video readout and transmission speed, thus allowing a reduction in shutter speed. This assists in "freezing" the seeing and reducing image distortion in individual frames while maintaining a usable signal-to-noise ratio. For Saturn, a rate of 50 frames/sec is currently achievable, resulting in two-minute sequences of 6000 raw frames per color channel that can be combined into a final image.

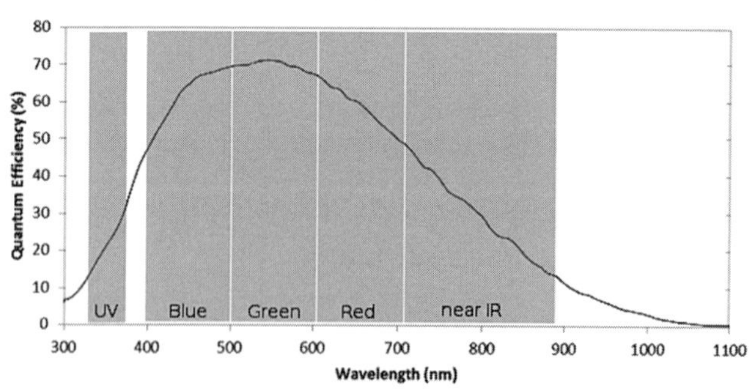

Figure 14.7 Typical response curve for sensors used in modern video astronomy, with the wavelength ranges of typical color filters depicted. Overall, the response is 10% or greater between 330 nm and 900 nm.

The current (early 2017) video astronomy cameras are based on CCD or CMOS silicon sensors with response curves similar to that shown in Figure 14.7. High-quality filters are available in both the 31.7 mm and 51 mm sizes from several vendors, including Astronomik of Germany and Astrodon in the United States.

14.5.3 Contributions of Amateurs to Saturn Science During the Cassini Mission

The eruption of the 2010/2011 Great Storm on Saturn (cf. Figure 14.6) doubled the number of Saturn observations sent to the Association of Lunar and Planetary Observers (ALPO), with 1489 images contributed in 2010/2011. Amateurs followed the storm's formation and evolution, from the first discovery images captured by amateur astronomers Sadegh Ghomizadeh and Teruaki Kumamori on 8 and 9 December 2010 until the storm eventually disappeared during the summer of 2011. It is notable that when the storm erupted, Cassini was not routinely observing the planet, leaving Ghomizadeh and Kumamori to capture the first images of one of the most powerful storms ever observed on any planet.

Continuing to the present day (2017), a compact dark storm remnant has remained visible in the latitudinal region occupied by the storm, as shown in panels B and C of Figure 13.10 in Chapter 13 by Sánchez-Lavega et al. Figure 14.8 shows the location and movement of this feature, revealed by Cassini to be an anticyclonic vortex (cf. Chapter 13), over five years from 2011 to 2016, as predominantly supplied by the amateur community. As well, as shown in Figure 14.9, amateur observers have captured transient dark spots in the rings (so-called "spokes") in good seeing.

In late 2012, as northern summer increasingly tilted the pole toward the Sun and Earth, and aided by an apparent color change, the North Polar Hexagon became visible to amateur astronomers for the first time. The previously uniform dark blue coloration of the north polar region began to change (Image A in Figure 14.10), with the exterior "collar" to the hexagon undergoing rapid changes in visible light. By March 2013 this external collar had changed to a light green,

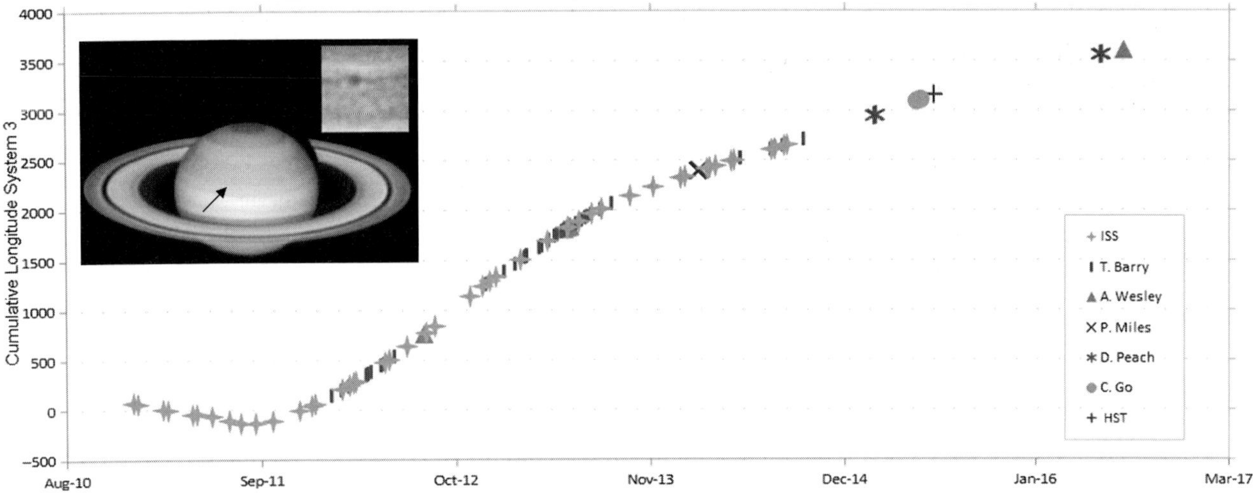

Figure 14.8 Observations of the longitudinal drift of a long-lived compact anti-cyclone on Saturn, 2011–2016, as obtained by five ground-based amateur observers (Trevor Barry, Chris Go, Phil Miles, Damian Peach and Anthony Wesley), the Hubble Space Telescope (HST), and Cassini/ISS (adapted from Sayanagi et al. 2013). Inset shows the feature as observed 18 March 2013 in a red filter image (cf. Figure 14.7) obtained by Anthony Wesley. HST point based on observations made with the NASA/ESA Hubble Space Telescope, obtained from the data archive at the Space Telescope Science Institute. STScI is operated by the Association of Universities for Research in Astronomy, Inc. under NASA contract NAS 5–26555. Graph Credit: Trevor Barry.

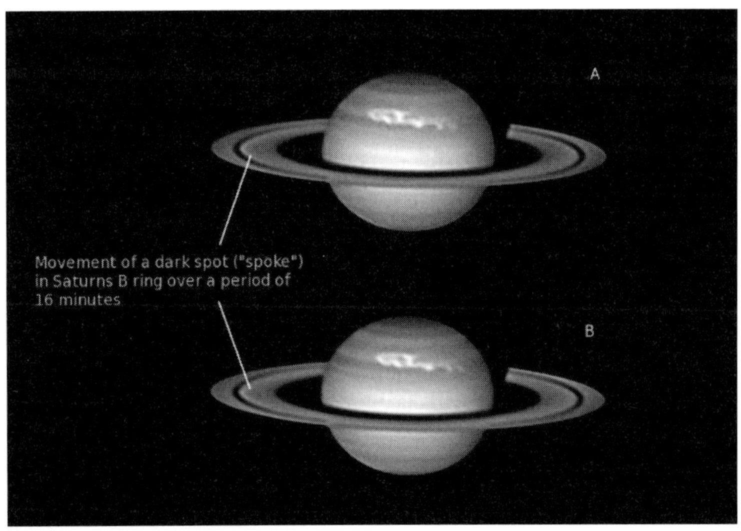

Figure 14.9 Amateur images showing a dark spot ("spoke") in Saturn's B ring. A: 30 December 2010, 1833UTC, B: 1849UTC. Images © Copyright 2010 by Anthony Wesley.

which contrasted with the continuing dark blue interior to render the edges of the hexagon readily visible to amateurs (Image B in Figure 14.10). A short time later – June 2013 – the color of this external collar had changed to yellow (Image C in Figure 14.10), and by April 2014 the color was a distinct red (Image D in Figure 14.10). Throughout, the dark blue interior of the hexagon remained unchanged.

14.5.4 After Cassini: Prospects for 2018 and Beyond

With the demise of the Cassini Orbiter in September, 2017, it will largely be up to the amateur astronomical community to continue monitoring Saturn for the onset of significant phenomena such as great storms, seasonal color changes, and dark ring spokes. In 2018, a

Figure 14.10 Saturn's North Polar Hexagon and color changes 2012–2014. A: March 2012, B: March 2013, C: June 2013, D: April 2014. All images © Copyright 2012–2014 by Anthony Wesley. (A black-and-white version of this figure appears in some formats. For the color version, please refer to the plate section.)

novel astronomical alignment involving Saturn as well as two other bright, photogenic planets occurs as Saturn, Jupiter and Mars appear close together during the perihelic Mars opposition, likely producing enhanced interest in observing Saturn and helping to build an additional army of Saturn observers for many years to come.

14.6 Conclusion

Despite the lack of a funded Cassini–Huygens-style Flagship mission, there is a solid basis for expecting that the exploration of Saturn will continue, with major discoveries being made via both (near) earth-based remote-sensing techniques and small probes that may sample the planet as early as the next decade. In particular, the forthcoming completion of large-aperture, state-of-the-art astronomical facilities on several continents and in near-Earth space (i.e. the James Webb Space Telescope slated for launch in 2018) will undoubtedly provide invaluable new observations on the planet's seasonal and dynamical variability and the workings of its aurorae and major storm systems. The growing army of amateur astronomers, using their own state-of-the-art telescopic equipment teamed with modern digital imaging analysis techniques, will provide an invaluable record of visual images of Saturn of near Hubble Space Telescope quality in spatial clarity, if not in photometric quality, thus alerting professional astronomers to novel new phenomena. Thus, the prospects look bright for effectively addressing many of the prime outstanding issues discussed earlier in this chapter as well as those promulgated by the other thirteen chapters of this book as we witness the continued scientific exploration of the Saturn in the twenty-first century.

References

Aharonson, O., A. G. Hayes, P. O. Hayne et al. (2014), Titan's surface geology. Pages 63–101 in: I. Müller-Wodarg, C. A. Griffith, E. Lellouch and T. E. Cravens (eds.), *Titan: Interior, Surface, Atmosphere, and Space Environment.* Cambridge: Cambridge University Press, Pp.

Atkinson, D. H., J. B. Pollack, A. Seiff (1998), The Galileo probe Doppler Wind Experiment: Measurement of the deep zonal winds on Jupiter. *J. Geophys. Res.* 103, 22911–22928.

Bacon, R., S.-V. Bauer, R. Böhm et al. (2006), Probing unexplored territories with MUSE: A second-generation instrument for VLT. *The Messenger* 124, 5–10.

Badman, S., G. Branduardi-Raymont, M. Galand et al. (2016), Auroral processes at the giant planets: Energy, deposition, emission mechanisms, morphology and spectra. *Space Sci. Rev.* 185, 99–179. doi : 10.1007/s/11214–014-0042–x.

Baranec, C., R. Riddle, N. M. Law et al. (2013), Bringing the visible universe into focus with Robo-AO. *J. Visualized Experiments* 72, e50021.

Baranec, C., R. Riddle, N. M. Law et al. (2014), High-efficiency autonomous laser adaptive optics. *Astrophys. J. Letts*. 790, L8.

Barret D., T. Lam, J.-W. Den Herder et al. (2016), The Athena X-ray Integral Field Unit (X-IFU). Proc. SPIE 9905 Space Telescopes and Instrumentation 2016: Ultraviolet to Gamma Ray. doi 10.111712.2232432. arXiv: 1608.08105v1 (astro-ph.IM).

Bayer, B. E. (1976), Eastman Kodak Co. US patent 3971065. http:/worldwide.spacenet.com/textdoc?DB=EPOC&IDX=US3971065

Brandl, B. R., M. Feldt, A. Glasse et al. (2014), METIS: The mid-infrared E-ELT imager and spectrograph. *Society of Photo-Optical Instrumentation Engineers (SPIE) Conference Series*, Vol. 9147, pp. 21.

Burton, M. and M. Dougherty (2014), Saturn's internal magnetic field: Expectations for Cassini's upcoming proximal orbits, Fall AGU meeting, 2014.

Clarke, J. T., J. Nichols, J.-C. Gérard et al. (2009), Response of Jupiter's and Saturn's auroral activity to the solar wind. *J. Geophys. Res. (Space Physics)*, 114, A05210, doi:10.1029/2008JA013694.

Colwell, J. E., P. D. Nicholson, M. S. Tiscareno et al. (2009), The structure of Saturn's rings. Pages 375–412 in: M. K. Dougherty, L. W. Esposito and S. M. Krimigis (eds.), *Saturn from Cassini-Huygens*. Springer.

Connerney, J. E. P. (1986), Magnetic connection for Saturn's rings and atmosphere. *Geophys. Res. Lett*., 13, 773–776, doi:10.1029/GL013i008p00773.

Coustenis, A., J.-P. Lebreton, O. Mousis et al. (2014), Possible concepts for an in situ Saturn probe mission. *Lunar and Planetary Science Conference* 45, 1244.

Davies, R., J. Schubert, M. Hartl et al. (2016), MICADO: First light imager for the E-ELT. *Society of Photo-Optical Instrumentation Engineers (SPIE) Conference Series*, Vol. 9908.

de Pater, I., R. J. Sault, B. Butler et al. (2016), Peering through Jupiter's clouds with radio spectral imaging. *Science* 352, 1198–1201.

DePoy, D. L., R. Allen, R. Barkhouser et al. (2012), GMACS: A wide field, multi-object, moderate-resolution, optical spectrograph for the Giant Magellan Telescope. *Ground-based and Airborne Instrumentation for Astronomy* IV, Vol. 8446.

DeWitt, C. N., M. J. Richter, K. R. Kulas et al. (2014), First light with the EXES instrument on SOFIA. *AAS/Division for Planetary Sciences Meeting Abstracts*, Vol. 46.

Doody, D. (2004), Basics of Space Flight, Sep. 2004; available at http://www.jpl.nasa.gov

Eikenberry, S., D. Andersen, R. Guzman et al. (2006), IRMOS: The near-infrared multi-object spectrograph for the TMT. *Society of Photo-Optical Instrumentation Engineers (SPIE) Conference Series*, Vol. 6269.

ESA. (2007), Cosmic Vision Call http://sci.esa.int/cosmic-vision/

ESA. (2010), Internal Study, http://sci.esa.int/sre-fp/47568-pep-assessment-study-internal-final-presentation

Fischer, G., W. S. Kurth, D. A. Gurnett et al. (2011), A giant thunderstorm on Saturn. *Nature*. 475, 75–77.

Fletcher, L. N., R. K. Achterberg, T. K. Greathouse et al. (2010), Seasonal change on Saturn from Cassini/CIRS observations, 2004–2009. *Icarus* 208, 337–352.

Fletcher, L. N., B. E. Hesman, R. K. Achterberg et al. (2012), The origin and evolution of Saturn's 2011–2012 stratospheric vortex. *Icarus* 221, 560–586.

Fletcher, L. N., B. E. Hesman, P. G. J. Irwin et al. (2011), Thermal structure and dynamics of Saturn's northern springtime disturbance. *Science* 332, 1413–1417.

Fletcher. L. N., P. G. J. Irwin, G. O. Orton et al. (2008), Temperature and composition of Saturn's polar hot spots and hexagon. *Science* 319, 79–821.

Fletcher, L. N., G. S. Orton, N. A. Teanby et al. (2009a), Methane and its isotopologues on Saturn from Cassini/CIRS observations. *Icarus* 199, 351–367.

Fletcher, L. N., G. S. Orton, P. Yanamandra-Fisher et al. (2009b), Retrievals of atmospheric variables on the gas giants from ground-based mid-infrared imaging. *Icarus* 200, 154–175.

Fouchet, T., S. Guerlet, D. F. Strobel et al. (2008), An equatorial oscillation in Saturn's middle-atmosphere. 2008. *Nature* 453, 200–202.

Fulchignoni, M., F. Ferri, F. Angrilli et al. (2002), The characterization of Titan's atmospheric physical properties by the Huygens Atmospheric Structure Instrument (HASI). *Space Sci. Rev.*104, 395–431.

Geballe, T. R., M. Jagod, T. Oka (1993), Detection of H_3^+ infrared emission lines in Saturn. *Astrophys. J. Lett.* 408, L109–L112, doi:10.1086/186843.

Greathouse, T., J. Lacy, B. Bézard et al. (2006), The first detection of propane on Saturn. *Icarus* 181, 266–271.

Greathouse, T. K., H. G. Roe, M. J. Richter (2005), Changes in the temperature of Saturn's stratosphere from 2002 to 2004 and direct evidence of a mesopause. 36th Annual Lunar and Planetary Science Conference, League City, Texas.

Griffith, C. A., S. Rafkin, P. Rannou et al. (2014), Storms, cloud, and weather. Pages 190–223 in: I. Müller-Wodarg, C. A. Griffith, E. Lellouch and T. E. Cravens (eds.), *Titan: Interior, Surface, Atmosphere, and Space Environment*. Cambridge: Cambridge University Press.

Guerlet, S., A. Spiga, M. Sylvestre et al. (2014), Global climate modeling of Saturn's atmosphere. Part I: Evaluation of the radiative transfer model. *Icarus* 238, 110–124.

Helled, R. (2011), Constraining Saturn's core properties by a measurement of its moment of inertia: Implications to the *Cassini* Solstice Mission, *Ap.J.Lett.* 735, L16.

Hohmann, W. (1925), *Die Erreichbarkeit der Himmelskörper (The Accessibility of Celestial Bodies)*, München: Verlag Oldenbourg.

Jia, X., M. G. Kivelson (2012), Driving Saturn's magnetospheric periodicities from the upper atmosphere/ionosphere: Magnetotail response to dual sources. *J. Geophys. Res.*, *117*, A11219, doi:10.1029/2012JA018183.

Koskinen, T. T., B. R. Sandel, R. V. Yelle et al. (2013), The density and temperature structure near the exobase of Saturn from Cassini UVIS solar occultations. *Icarus* 226, 1318–1330, doi:10.1016/j.icarus.2013.07.037.

Lam, T., D. Landau, N. Strange (2009), Broad search solar electric propulsion trajectories to Saturn with gravity assists, AAS Paper 09–355, Proceedings of the AAS/AIAA Astrodynamics Specialist Conference, August 2009, Pittsburgh, PA.

Larkin, J. E., A. M. Moore, S. A. Wright et al. (2016), The Infrared Imaging Spectrograph (IRIS) for TMT: Instrument overview. *Society of Photo-Optical Instrumentation Engineers (SPIE) Conference Series*, Vol. 9908.

Lecacheux, A. and M. Aubier (1997), Re-visiting Saturnian kilometric radiation with Ulysses/URAP. Page 313 in: H. O. Rucker, S. J. Bauer and A. Lecacheux (eds.), *Planetary Radio Emissions IV.*

Lee, S., I.-S. Yuk, H. Lee et al. (2010), GMTNIRS (Giant Magellan Telescope near-infrared spectrograph): Design concept. *Ground-based and Airborne Instrumentation for Astronomy III*, Vol. 7735.

Lellouch, E. (2008), Planetary atmospheres with ALMA. *Astrophys. Space Sci.* 313, 175–181.

Li, C. and A. P. Ingersoll (2015), Moist convection in hydrogen atmospheres and the frequency of Saturn's giant storms. *Nature GeoSci*. DOI:10.1038/NGE02405.

Marty, B., T. Guillot, A. Coustenis et al. (2009), Kronos: Exploring the depths of Saturn with probes and remote sensing through an international mission. *Exp. Astron.* 23, 947–976.

McGregor, P. J., G. J. Bloxham, R. Boz et al. (2012), GMT Integral-Field Spectrograph (GMTIFS) conceptual design. *Ground-based and Airborne Instrumentation for Astronomy IV*, Vol. 8446.

Mousis, O., D. H. Atkinson, M. Blanc et al. (2016), Hera Saturn Entry Probe Mission. A proposal in response to the ESA call for a medium-size mission opportunity in ESA's science programme for launch in 2029–2030 (M5). Submitted to ESA.

Mousis, O., L. N. Fletcher, J.-P. Lebreton et al. (2014), Scientific rationale of Saturn's in situ exploration. *Planet. and Space Sci.*, 104, 29–47.

Müller-Wodarg, I. C. F., L. Moore, M. Galand et al. (2012), Magnetosphere-atmosphere coupling at Saturn. 1: Response of thermosphere and ionosphere to steady state polar forcing. *Icarus* 221, 481–494, doi:10.1016/j.icarus.2012.08.034.

NASA, ESA (2009), *Titan Saturn System Mission (TSSM) NASA/ESA Joint Summary Report*, January 19, 2009.

National Research Council (NRC), Committee on the Planetary Science Decadal Survey (2011), *Vision and Voyages for Planetary Science in the Decade 2013–2022.* Washington, DC: The National Academies Press, http://www.nap.edu/catalog/13117.html

Nichols, J. D., B. Cecconi, J. T. Clarke et al. (2010), Variation of Saturn's UV aurora with SKR phase. *Geophys. Res. Lett.* 37, L15102, doi:10.1029/2010GL044057.

Nicholson, P. D., R. G. Drench, A. S. Bosh (1999), Ring plane crossings and Saturn's pole precession. *Bull. Am. Astron. Soc.* 31, 44.01.

Niemann, H. B., S. K. Atreya, S. J. Bauer et al. (2005), The abundances of constituents of Titan's atmosphere from the GCMS instrument on the Huygens probe. *Nature* 438, 779–784.

Niemann, H. B., S. K. Atreya, G. R. Carignan et al. (1998), The composition of the Jovian atmosphere as determined by the Galileo probe mass spectrometer. *J. Geophys. Res.* 103, 22831–22846.

Norwood, J., J. Moses, L. N. Fletcher et al., (2016), Giant planet observations with the James Webb Space Telescope. doi:10.1088/1538–3873/128/959/018005.

O'Donoghue, J., T. S. Stallard, H. Melin et al. (2013), The domination of Saturn's low- latitude ionosphere by ring "rain." *Nature* 496, 193–195, doi:10.1038/nature12049.

Orton, G. S. and P. A. Yanamandra-Fisher (2005), Saturn's temperature field from high- resolution middle-infrared imaging. *Science* 307, 696–698.

Orton, G. S., P. A. Yanamandra-Fisher, M. Brendan et al. (2008), Semi-annual oscillations in Saturn's low-latitude stratospheric temperatures. *Nature* 453, 196–199.

Packham, C., M. Honda, M. Richter et al. (2012), Key science drivers for MICHI: A mid- IR instrument concept for the TMT. Society of Photo-Optical Instrumentation Engineers (SPIE) Conference Series, Vol. 8446. doi:10.1117/12.924996.

Pazder, J. S., S. Roberts, R. Abraham et al. (2006), WFOS: A wide field optical spectrograph for the Thirty Meter Telescope. *Society of Photo-Optical Instrumentation Engineers (SPIE) Conference Series*, Vol. 6269.

Reh, K. et al. (2009), Titan Saturn System Mission Study Final Report. JPL D–48148.

Sánchez-Lavega, A., F. Colas, J. Lecacheux et al. (1991), The great white spot and disturbances in Saturn's equatorial atmospheres during 1990. *Nature* 353, 397–401.

Sánchez-Lavega, A., T. del Río-Gaztelurrutia, M. Delcroix et al. (2012), Ground-based observations of the long-term evolution and death of Saturn's 2010 Great White Spot. *Icarus* 220, 561–576.

Sánchez-Lavega A., T. del Río-Gaztelurrutia, R. Hueso et al. (2011), Deep winds beneath Saturn's upper clouds from a seasonal long-lived planetary-scale storm. *Nature* 475, 71–74, doi:10.1038/nature10203.

Sánchez-Lavega, A., S. Pérez-Hoyos, J. F. Rjoas et al. (2003), A strong decrease in Saturn's equatorial jet at cloud level. *Nature* 423, 623–625.

Sánchez-Lavega, A., A. Wesley, G. Orton et al. (2010), The impact of a large object with Jupiter in July 2009. *Astrophys. J.* 715, L155–L159.

Sanz-Requena, J. F., S. Pérez-Hoyos, A. Sánchez-Lavega et al. (2012), Cloud structure of Saturn's 2010 storm from ground-based visual imaging. *Icarus* 219, 142–149.

Sayanagi, K. M., U. A. Dyudina, S. P. Ewald et al. (2013), Dynamics of Saturn's great storm of 2010–2011 from Cassini ISS and RPWS. *Icarus* 223, 460–478.

Sinclair, J. A., P. G. J. Irwin, L. N. Fletcher et al. (2014), From Voyager-IRIS to Cassini-CIRS: Interannual variability in Saturn's stratosphere? *Icarus* 233, 281–292.

Smith, C. G. A. (2006), Periodic modulation of gas giant magnetospheres by the neutral upper atmosphere. *Annales Geophysicae* 24, 2709–2717, doi:10.5194/angeo-24-2709–2006.

Southwood, D. J. and S. W. H. Cowley (2014), The origin of Saturn's magnetic periodicities: Northern and southern current systems, *J. Geophys. Res.* (Space Physics) 119, 1563–1571, doi:10.1002/2013JA019632.

Spencer, J. R., A. C. Barr, L. W. Esposito et al. (2009), Enceladius: An active cryovolcanic satellite. Pp. 683–724 in: M. K. Dougherty, L. W. Esposito and S. M. Krimigis (eds), *Saturn from Cassini-Huygens*. Springer.

Sromovsky, L. A., K. H. Baines and P. M. Fry (2013), Saturn's Great Storm of 2010-2011: Evidence for ammonia and water ices from analysis of VIMS spectra. *Icarus* 226, 402–418.

Stallard, T. S., H. Melin, S. Miller et al. (2012), Temperature changes and energy inputs in giant planet atmospheres: what we are learning from H_3^+. *Royal Society of London Philosophical Transactions Series A*, 370, 5213–5224, doi:10.1098/rsta.2012.0028.

Szentgyorgyi, A., J. Bean, B. Bigelow et al. (2014), A preliminary design for the GMT- Consortium Large Earth Finder (G-CLEF). *Ground-based and Airborne Instrumentation for Astronomy V*, Vol. 9147.

Temi, P., P. M. Marcum, E. Young et al. (2014), The SOFIA observatory at the start of routine science operations: Mission capabilities and performance. *The Astrophysical Journal Supplement Series* 212, 24.

Thatte, N. A., F. Clarke, I. Bryson et al. (2014), HARMONI: The first light integral field spectrograph for the E-ELT. *Ground-based and Airborne Instrumentation for Astronomy V*, Vol. 9147.

Thronson, H., A. Mandell, R. S. Polidan et al. (2016), Special section guest editorial: Future large-aperture ultraviolet/optical/infrared space observatory. *J. Astron. Telesc. Instrum. Syst.* 2(4), 041201. doi:10.1117/1JATIS.2.4.041201.

Wong, M. H., M. Ádámkovics, S. K. Atreya et al. (2009), A dedicated space observatory for time-domain solar system science. White Paper submitted to the 2009–2011 Planetary Science Decadal Survey, online at www8.nationalacademies.org/ssbsurvey/DetailFileDisplay.aspx?id=186&parm_type=PSDS (see pp. 208–209 of the 2013–2022 Planetary Decadal Survey).

Zander, F. A. (1964), Problems of flight by jet propulsion: interplanetary flights, original publication 1925. *NASA Technical Translation F-147*.

Index

ab initio, calculations, 48–49, 57, 65, 270
ab initio, simulations, 61, 79
abundances, elemental, 6–9, 15, 18, 27, 30–32, 34–35, 46, 58
abundances, molecular, 31
accretionary heating, 23
acetylene (C_2H_2), 7, 170, 251, 253, 259, 274, 277, 282, 288, 314, 318, 330, 356–357, 360, 393–394, 397, 399
adiabatic assumption, 52
adiabatic compression, 242, 259–260, 397
adiabatic invariants, 138, 140, 182
adiabatic temperature gradient, 52–53, 59, 61, 70, 80, 82, 242, 252–254, 300, 313, 354, 356, 367, 393, 395, 406
adiabatic transport, 131, 139
adjoint based inverse method, 328
Advanced Telescope for High-Energy Astrophysics (ATHENA), 431
advection, 394
aerosols, 252–253, 255, 258–261, 264, 270–271, 273–274, 282–285, 287, 348, 353, 355–356, 358, 369
Alfvén waves, 111, 119, 143, 167
alkali chlorides, 32
amateur astronomers, 302, 330, 341, 369, 383–384, 401, 428, 434–436, 438
ammonia (NH_3), 7, 23–24, 31–32, 251, 266–267, 271, 274, 295, 299, 303–305, 308–309, 329–330, 377, 379, 383, 385, 389, 391–392, 404, 410–411
ammonia ice, 377, 384, 391
ammonium hydrosulfide (NH_4SH), 11–12, 32, 253, 266, 270, 272, 282, 285, 305, 308, 377, 391–392, 404, 411
amorphous ice, 26–27
amorphous matrix, 26
angular momentum, 21–22, 167, 243, 419, 421
anticyclone, 309, 343–344, 364, 367, 377, 381, 386, 393–395, 397, 399, 403–404, 408–409, 430, 437
aphelion, 251–252, 288, 358, 428
Arnol'd's second stability criterion, 311–312, 347
Arsine (AsH_3), 7, 12, 264, 269–270, 303–305, 307
aspect ratio, 21
asteroid belt, 21, 26
Atacama Large Millimeter Array (ALMA), 433
atmospheric circulation, 251, 253, 262, 277, 287–288, 295–296, 330, 353
atmospheric composition, 28, 35, 45, 251, 264
atmospheric dynamics, 30, 32, 53–55, 264, 266, 296–297, 320, 328, 338–339, 346, 358, 361, 363, 368
aurora, 106, 111, 116, 129, 152, 155, 166–167, 172–178, 180–182, 184, 186, 189–191, 196–197, 203, 208, 225, 230, 243, 245–246, 282, 417, 419, 422, 429, 434, 438
  atmospherically generated, 185, 188
  atomic hydrogen, 166, 169, 171–172, 204, 209, 234, 271, 419
  Enceladus, 173, 184
  energetic particle injections, 182
  generation of, 173
  H_2, 169–172, 174–175, 186–187
  H_3+, 172, 174–176, 178, 180, 183, 185–186, 190, 434
  hemispheric differences, 183, 186
  hydrocarbon, 166, 169, 172–173
  hydrogen, 175, 186
  infrared, 170, 172, 176, 185, 187, 190–191
  ionization, 166
  Knight relationship, 171
  local time effects, 174–175, 177, 183, 185
  magnetospheric interactions, 181, 185, 188, 190–191, 241
  morphology, 172–173, 177
  one-hour period, 182, 190–191
  outer emission, 182–183
  outstanding questions, 189–190
  planetary period, 185
  planetary period currents, 185
  polar emissions, 176, 180–182
  precipitating particles, 174
  SKR pulsations, 147, 150, 168, 176, 185–186
  small scale features, 179–180
  solar wind interactions, 176, 178, 181
  vertical profile, 169
  visible, 111, 169–172, 174, 179, 190–191
  X-ray, 151, 172
auroral curtain, 172–173, 180
auroral hiss, 101, 116, 143, 149, 169, 182
auroral precipitation, 106, 174
auroral radio emission, 54, 111, 147, 168, 176, 187, 190, 403, 422
axial tilt, 251, 264

baroclinic, 319, 363, 365, 404
baroclinic instability, 306, 320, 322, 326, 329–330, 362–363
barotropic stability criterion, 310–311, 361
beacon, 100, 260, 277, 318–319, 377–378, 393–395, 397–399, 412
benzene (C_6H_6), 7, 275, 281–282, 356, 399
Big Bang, 8, 14, 16
bow shock, 115, 117, 127, 404
Brunt-Väisälä frequency, 62, 346, 364, 405

C/O ratio, 7, 18, 20, 31, 33–35
C_2 hydrocarbons, 275, 399
canonical interior, 30
carbon dioxide (CO_2), 23, 26, 31, 35, 134, 143, 266, 270, 282
carbon monoxide (CO), 7, 24, 269
Cassini
  cameras, 111, 402, 433
  Grand Finale Mission, 35, 44, 64–65, 69, 71, 78, 109, 119, 126, 128, 150, 154–155, 190–191, 219, 245, 296, 326, 330, 366, 370, 418, 420–421, 429
  instruments, 102, 104, 119, 147, 298, 330, 341, 379, 382
  orbiter, 35, 252, 337, 368–369, 399, 418, 429, 437
  primary mission, 126, 254, 337
  proximal orbits, 7, 11, 35, 92–93, 119, 154, 168, 177, 181, 191, 219, 326–327, 330

spacecraft, 7, 46, 72, 102, 166, 170, 252, 267, 284, 295, 307, 326, 337–339, 341, 377, 379, 382, 388–389, 400, 406
CB2 filter, 297–299, 339–343, 347–348, 350, 352, 383–387, 389
CB3 filter, 298–299, 339, 343, 347, 350, 352, 383
CELESTE, 398, 433
center-to-limb variation, 235, 388
centrifugal interchange instability (CII), 113–114, 139–140, 182
Chandra X-ray Observatory, 430
charge exchange, 129, 132, 135, 137, 143, 172, 205, 212, 217
charged particles, 97, 100, 104, 112, 115, 126–127, 129, 138, 143, 155, 166–168
Charney-Stern stability theorem, 310, 322, 347
chondrites, 6
chondrules, 6
circumplanetary disk, 20–22
clathrates, 26–28
climate model, 251, 253, 255, 257–259, 262, 280, 287, 354, 358, 428
climatology, 337
cloud composition, 32, 251, 370, 382, 388–389, 391, 393, 399, 421
cloud structure, 286, 297, 299–300, 347, 377, 388–390, 417
clouds, 11–12, 16, 32–33, 54, 108, 149, 201, 251–253, 259, 267, 269, 271, 282–285, 295, 297–301, 303–304, 307–311, 321, 329, 338–339, 341, 344, 348, 352, 354, 357–358, 366–367, 369–370, 377–379, 384–386, 389–391, 393–394, 397, 401–404, 412, 417–420, 430, 433–434
  tracking, 54, 108, 297–298, 301, 306, 316, 347, 352, 355, 361
collar, 393, 397, 436–437
comet, 16, 19, 24, 151, 270, 282, 428, 433
Comet 67P/Churyumov-Gerasimenko, 24
Comet Hartley 2, 24
COMICS, 433
Composite-Infrared Spectrometer (CIRS), 8, 10, 12, 207, 233–234, 237–238, 254–257, 260–265, 267–271, 274, 277–282, 285–286, 301–302, 309, 311, 313–314, 318, 339–340, 352–353, 355, 357, 369, 382, 393–395, 398, 412, 428–429
composition, clouds/aerosol/haze species, 283, 352
composition, abundance, 35, 356, 419
composition, gas species, 7, 287
condensible, 14, 32, 36, 252, 266, 287
continuity equation, 323, 353, 406–407
convection, 53, 61, 69–70, 75, 80–82, 84–86, 89–91, 114, 135, 139–140, 145, 244, 284, 295, 300, 304, 306, 316–317, 319–324, 326, 329, 368, 377, 384, 391, 396, 399, 404, 408, 411
  double-diffusive, 65, 81, 90
  moist, 12, 307–308, 310, 319–320, 326, 329–330, 378–379, 385, 395, 404–405, 411
Convective Available Potential Energy (CAPE), 404
convective head, 381
convective source, 379–381, 404
core accretion, 5–7, 20, 33, 35, 59, 61
Coriolis force, 82–83, 87, 239, 243, 246, 262, 297–298, 302, 309, 317, 381, 404
corotation, 132–133, 136, 140, 143–144, 146, 180, 182, 203, 243–244
Cosmic Ray Albedo Neutron Decay (CRAND), 129, 136, 155
Cowling's theorem, 70, 78
CRIRES, 434
critical mass threshold, 5
crystalline ice, 26–27
cumulus, 304, 368, 384, 386, 404
current density, 114, 129, 240, 243–244
cyclotron maser instability, 147, 168, 190

D/H ratio, 16, 23–24, 236, 433
decimal exponent (dex), 9–10, 13–14
deformation radius, 321–322, 346, 362–363, 368
deformations, anelastic, 25
density profile, 50, 62, 110, 197, 199, 202, 205–206, 209, 211–213, 216, 218, 224, 226–229, 231–233, 236–237, 243, 245
deuterium, 16
diabatic heating, 252
diacetylene (C_4H_2), 276, 281–282, 288, 398
differential rotation, 47, 55–56, 58, 69, 71, 82, 85–87, 90–91, 118–119, 298, 320, 323–324, 326
diffusion, 9, 13, 49, 83–84, 88, 129, 138, 145, 205, 259, 261, 275, 277, 282
  ambipolar, 210
  eddy, 196, 274
  molecular, 196, 225, 238
diphosphine (P_2H_4), 286
dipole axis, 2, 70, 74, 86, 89, 126
dipole moment, 47, 72, 78, 86
dipole reversals, 84, 88
dipole tilt, 46, 72, 76, 78, 88–91, 93
dipole, axial, 47, 71, 79, 84, 87–88
dipole, equatorial, 86–89
disequilibrium species, 12, 14, 251–252, 264, 269–270, 287, 357
disk instability, 6, 20
disk mid-plane, 28
dispersion relationship, 346
divergence, 240, 321, 368, 379, 404, 410
Doppler shift, 46, 64, 103, 197
doppler spectroscopy, 28, 31
downdraft, 303, 305, 385, 394, 411
downwelling, 242, 259–260, 263, 268, 278–279, 302–304, 308, 397, 399
drift speed, 138, 143, 383–384
Dungey cycle, 128
dynamic pressure, 129, 145, 148, 150, 429
dynamical instability, 21, 113
dynamo, 47, 69–71, 77–80, 82–92, 114, 243–244, 326, 419

Earth, comparison to, 3, 6, 16, 19, 28, 33, 47, 70, 72, 77, 84–85, 88, 114, 128, 149, 151–152, 155, 171, 180–181, 190, 251, 257, 263, 296–299, 304, 317, 342, 367, 378, 403, 417, 423, 428
eccentric orbit, 107, 369, 423
eddy mixing coefficient, 238
eddy momentum flux, 322, 368
Ekman number, 83, 89, 323, 325–326
electroglow, 235
electron impact ionization, 135, 166, 240–241
electron impact produced H_2 dayglow, 235–236
electrostatic bursts, 384
ellipsoids of revolution, 228
Enceladus, 2, 23–24, 73–74, 114, 117, 126–129, 133–134, 136, 138, 143, 145–146, 150–151, 155, 166, 173, 184–185, 189, 207, 282, 417–418, 424, 427
energetic neutral atoms (ENA), 105–107, 127, 129, 145, 148–149, 155, 178–179, 182
energy balance, 8, 45, 229, 232, 412, 421–422
energy budget, 45, 262, 403
enrichment, factors, 8, 13–14, 28, 57, 300, 307
enrichment, heavy elements, 2, 5–7, 12–13, 15, 18, 27, 45, 49, 57, 285, 420
envelope opacity, 6
equation of state, 8, 44, 47, 49, 55–56, 58
equatorial oscillation, 251, 255, 263–264, 278–280, 313
equatorial region, 10, 77, 166, 184, 198, 263, 265, 268, 279, 282, 313, 361, 381
equinox, 74, 101, 103–104, 119, 136, 149–150, 175, 185–186, 201, 215, 224, 229–230, 241, 252–253, 255–256, 262, 276, 283–284, 287, 312–313, 337–340, 347, 354, 358, 402, 421–422
Ertel potential vorticity (EPV), 406
ethane (C_2H_6), 7, 170, 251, 253, 258, 274–275, 277–280, 312, 314–316, 318, 330, 356–357, 393–394, 397
ethylene (C_2H_4), 275, 277, 318, 377, 397–399
European Space Agency, 417
European-Extremely Large Telescope, E-ELT, 431
evolution, 5, 16, 20–21, 24–26, 29–30, 35–36, 44, 47, 59–61, 69, 81, 102, 207, 227–228, 251, 257, 259, 279–280, 379, 382, 384, 388–389, 393–395, 397, 404, 406, 409–411, 429, 433, 435–436
EXES, 434
exobase, 217, 226–229, 233, 236
exoplanets, 1, 5, 7, 12, 15, 23, 28–36, 49–50, 420

exosphere, 4, 135, 143, 226, 419
exospheric temperature, 217, 226–227, 231–234
Explicit Planetary Isentropic Coordinate (EPIC), 259, 406–409
extrasolar giant planets, 2, 5, 28, 32, 35, 330
extreme ultraviolet (EUV), 166, 196

field aligned currents (FAC), 85, 114, 129, 138–139, 147, 180
fluxgate magnetometer (FGM), 71–72, 76
f-mode oscillations, 63

Galileo Probe Mass Spectrometer (GPMS), 9, 11, 13, 16–18
gas accretion, 6, 21, 23, 59
Gauss coefficients, 74–77, 85
Genesis, 16–18
geostrophic balance, 262, 298, 300, 319, 346, 353, 394, 396
germane (GeH_4), 7, 12, 264, 268–270
Giant Magellan Telescope (GMT), 431
GNIRS, 434
gradient wind balance, 224, 242, 404
grain distribution, 6, 28
gravitational diffusive separation, 224
gravitational field, 46, 54–55, 58–59, 107, 109, 326, 419
gravitational instability, 5–6, 33, 35
gravity, 20, 46, 55, 64, 83, 139, 244, 296, 319, 326–328, 330, 366, 407, 419, 421, 426–427
gravity acceleration, 362
gravity anomalies, 328
gravity assist, 423, 427–428
gravity experiments, 52, 65, 326, 328
gravity field, 2, 4, 35, 44–47, 51, 63–65, 72, 109–111, 229, 326–328, 370, 419
gravity field measurements, 64–65
gravity harmonics, 326–328, 419
gravity mode, 63
gravity moments, 326–328
gravity passes, 419
gravity potential, 51
gravity signal, 326–328
gravity waves, 62, 118, 198, 211–212, 216, 218, 239, 246, 261, 312, 317, 319, 321, 397, 403
Great Storm of 2010–2011, 377, 382, 387–389, 392, 399, 402–403, 405–407, 411–412, 421, 428–429, 434–436
Great White Spot (GWS), 318, 323, 329, 377–379, 381

H Lyman alpha, 171, 183, 234–235, 274–275, 280–281
H_2 Lyman and Werner bands, 205
H_2 solar fluorescence, 205, 235–236
H_2-He collision-induced continuum, 253, 258, 272, 352, 394
H_3^+ near-IR thermal emission, 33, 230, 236, 286, 313, 353, 386, 390
Hadley circulation, 262, 280, 305–306, 316
Hall conductivity, 240, 243
Hall current, 143, 173
haze, 3, 32–33, 252–253, 259, 268, 271, 281–286, 299, 339, 352, 355, 357–361, 369–370, 378, 389, 391, 397, 411, 418, 428, 430
hazy gas planets, 33
HD 189733b, 31–32
HD 209458b, 7, 31–32
He I 58.4 nm, 224, 234–236, 238, 245
heat conduction, 224, 237–238
heating rate, 83, 204, 210, 225, 237–239, 241–242, 244, 411
heavy elements, 2, 5–8, 12, 14–15, 18, 20, 27–28, 35, 46–47, 49–50, 55–59, 61–62, 79, 285, 420
helical fluid motion, 82
helical plasmoids, 146
helicity, 90–91
helium (He), 7–8, 11, 13, 16–17, 28, 35, 44–46, 48–49, 52–53, 55–57, 59–60, 62, 65, 70–71, 79–82, 86, 169, 253, 264, 299–300, 307, 359, 406, 419, 421, 427
helium depletion, 8, 13, 45, 59–60, 62, 421
helium magnetometer, 71, 76
helium rain, 8, 13, 49, 55, 59–61, 65, 70, 80–81, 86, 421
helium segregation/demixing, 53, 57, 59–60, 69–70, 79–81
helium-hydrogen immiscibility layer, 8
Hera Mission concept, 426
Herschel, 206–207, 237, 271–272, 282, 379
Heterodyne Instrument for Planetary Wind And Composition (HIPWAC), 433
hexagonal jet, 255, 343–344, 346–348, 353, 360, 364, 369
hexagonal pattern/shape, 340, 342–343, 353, 363, 365
high-resolution spectra, 33
h_{MAX}, 197, 211
homopause, 11, 170, 172–173, 196, 204, 208–209, 224–226, 229, 233–236, 238–239, 245, 258, 275
  effect on aurora, 169, 172–173
hot disk, 24
hot Jupiters, 29–35
Hubble Space Telescope (HST), 147, 151, 174, 190–191, 234, 284, 340, 377, 379, 381, 428, 430, 434, 437–438
hydrazine (N_2H_4), 264, 266, 271–272, 286, 391–392
hydrocarbon haze, 33, 252, 259
hydrocarbons, 31, 170–171, 225, 238, 251–252, 257, 260, 264, 271–272, 274–275, 277–282, 287–288, 354, 356–357, 394, 397, 399, 419
hydrogen, 7–9, 13, 17, 23–24, 28, 44, 47–49, 52–53, 56, 59–60, 70, 79–80, 85, 131, 136, 148, 166, 169, 171, 206, 226, 236, 270–271, 273, 281, 296, 299–300, 312, 360, 378, 404, 410–411, 420–421
hydrogen cyanide (HCN), 272
hydrogen equation of state, 47, 56–57
hydrogen isotope ratio, 16
hydrogen sulfide (H_2S), 7, 32, 266, 420
hydro-geochemistry, 23
hydrological cycle, 322
hydrostatic equilibrium, 50–51, 232, 237, 246, 300

Iapetus, 6, 25
iceline, 26
icy planetesimals, 24, 26–27
Imaging Science Subsystem (ISS), 107, 172, 200, 284–286, 339–340, 342–344, 358–360, 382–386, 388–389, 397, 405, 437
impulsive radio bursts, 187, 200, 384, 399
infrared, i, 1–2, 11, 17, 30–31, 33, 70, 80, 169, 171, 175, 184, 197, 204, 224–225, 252–255, 258–259, 267, 272, 274–275, 283, 287, 295, 300, 309, 311–312, 337–339, 341, 344, 348, 352, 356, 358, 360, 388–389, 392–394, 397, 417, 422, 428–434
Infrared Space Observatory (ISO), 16, 206, 212, 214, 218, 266, 274
Infrared Telescope Facility (IRTF), 16, 275, 277, 353, 393–395, 433–434
interchange, 101, 104, 113, 129, 132, 138–139, 139–140, 149–150, 182, 258
interchange motion, 112, 138
internal structure, 5, 20, 25, 30, 35, 49, 51, 53–55, 58–59, 61–62, 70, 79–80, 85, 93, 110, 126, 429
International Ultraviolet Explorer (IUE), 235, 274
interplanetary dust, 282
interplanetary magnetic field (IMF), 127, 140, 151–153, 176
Ion and Neutral Mass Spectrometer (INMS), 11, 24, 143, 236, 245, 419
ion drag heating, 240
Ion Neutral Camera (INCA), 105
ionization, 114, 129, 135, 135–136, 138, 166, 169–172, 175, 196, 199, 202, 204, 207–209, 212, 216–217, 224, 226–227, 231, 241, 244–245, 252
ionosphere, 3, 111, 114–119, 143, 151, 167–168, 172–178, 184, 188–189, 196–198, 200–205, 207–208, 210–212, 217, 225, 240, 242–243, 245, 253, 270, 422, 429
  "magnetization" regions, 115, 129, 167, 173, 243
  auroral structure, 116, 166–167, 173, 185, 190, 202
  Dione and Rhea, 144
  ion winds, 114, 155, 167–168, 173, 180, 185, 187, 196
  Kronian, 180
  Titan's, 144
ionosphere-protonosphere exchange, 216, 218
ionospheric, 97–98, 114–115, 117, 129, 131, 136, 139, 144, 155, 167, 174–176, 181, 185, 188, 196–198, 202–204, 206, 208, 210–214, 216–219, 225, 240–241, 243, 245, 399, 419, 422
ionospheric electric current, 173–175, 243
ionospheric electron density, 3, 136, 197–199, 208, 215
ionospheric models, 207, 210–212, 215, 218, 244

isentrope, 47–48, 53
ISHELL, 434
isotopes, 3, 7, 16, 36, 420–421, 423
isotopes, fractionation, 16

James Webb Space Telescope (JWST), 3, 287, 428, 430, 434, 438
jetstream, 338, 342, 344, 361, 363–365
Joule heating, 167–168, 172–174, 176, 188, 204, 212, 240–243
Juno microwave radiometer (MWR), 10–13, 330
Juno mission, 4, 6, 15, 35–36, 46, 52, 59, 65, 86, 93, 320, 326, 328, 330, 338, 361, 370, 419, 421, 426
Jupiter, i, 2–8, 10–16, 19–24, 26–36, 44–49, 51–55, 59–63, 70, 72, 74, 79–86, 89–90, 93, 99–100, 106, 108, 114, 116, 118, 126, 128, 135, 138, 149, 151–152, 166, 171–172, 174, 182, 184–185, 198, 202–203, 205, 207–208, 224, 230, 234–236, 239, 243–245, 251, 255, 268, 270, 273, 282–283, 295–296, 298–299, 301–302, 306–311, 316, 319–324, 326–330, 338, 342, 358–359, 361–363, 366, 368–370, 378, 381, 384, 389, 394, 403–404, 410, 419–421, 423–426, 428, 430, 433–434, 438

Keck Observatory, 175, 203, 313, 353, 429, 434
Kelvin waves, 312, 317
Kelvin-Helmholtz instability (KHI), 112, 137, 152, 154, 156
  aurora produced, 180, 189
keogram, 111–112
Kronos Mission Concept, 425

laboratory experiment, 306, 362, 364
Lambert-Beer Law, 207
lapse rate, 253–254
latent heating, 53, 253, 300, 322, 330, 367, 377, 394, 403, 408, 410, 421, 429
layer thickness, 88–89, 321, 407
Lense-Thirring effect, 419
lightning (Saturn), 3, 12, 15, 108, 155, 197, 200–201, 307–309, 377, 384–385, 388, 399, 401–405, 407, 420
local thermodynamic equilibrium (LTE), 205
longitude (West), 297, 304, 383–384, 393, 395, 400, 402–403
Lorentz force, 83, 85, 167, 320, 326
Low Energy Magnetospheric Measurement System (LEMMS), 106
L-shell, 73, 130–131, 137–138, 145
luminosity, 44, 48, 52, 55, 59–62, 70, 80–81, 90
Lyman continuum, 227
Lyman system, 169–172, 183, 205, 209, 226, 234, 236

macrosignature, 129–130, 143
magnetic field, 1–2, 35, 44, 46–47, 52, 54–55, 61, 65, 69–79, 82–86, 89–91, 93, 97–105, 107, 110–112, 115, 117–118, 120, 129–131, 138, 140–147, 151–156, 166–168, 173, 175, 177, 179–180, 182, 184–186, 190, 196, 240–241, 298, 326, 329, 401, 419, 421, 426
magnetic field lines, 2–3, 73, 101, 106, 111, 127–129, 133, 137, 143, 149, 153, 155, 167–169, 171, 178, 181, 200, 203–204, 207, 211, 216, 219, 240–241, 243, 419, 429
Magnetic Fields and Plasma Science (MAPS), i, 119
magnetic flux, 69–70, 77, 79, 90–92, 138–139, 144, 146, 151, 156, 182, 190
magnetic pressure, 129, 152
magnetic Reynolds number, 80, 82–85, 87, 89–90
magnetodisk, 74, 97–98, 116–118, 127–129, 133, 140, 152
magnetohydrodynamic (MHD), 113–114, 146, 146, 151–153, 180, 240, 243, 245, 325
Magnetometer (MAG), 46, 69, 71–73, 76, 78, 92–93, 102, 104–105, 180, 188, 419
magnetopause, 97, 112, 115, 117–118, 127–129, 133, 140, 146, 151–153, 156, 179–181
magnetosphere, 2, 4, 73–75, 97–98, 102–108, 112–119, 126–156, 166–168, 173–174, 176–182, 185–191, 196, 203, 225, 240, 243–245, 417–418, 424, 426, 429
Magnetosphere-Ionosphere (MI) coupling, 116, 168, 173–174, 189, 214, 240, 242–244
magnetospheric electric field, 156, 240
Magnetospheric Imaging Instrument (MIMI), i, 106, 130, 135, 146, 155, 187
Magnetospheric Period Oscillation (MPO), 119
magnetospheric periodicity, 98–99, 101–102, 104, 106–108, 114–115, 117–118, 120, 403
magnetotail, 2, 98, 105, 115, 127–128, 132–133, 136, 138, 145–151, 176–178, 180–181
meridional circulation, 22, 241, 280, 302, 305–308, 314–315, 317, 329, 369
mesopause, 224, 239, 246, 367
metallic hydrogen (H^+), 8, 20, 48–49, 53, 56, 61–62, 69–70, 77, 79–81, 84, 86, 130–131, 187, 204–205, 208, 212, 217–218, 320, 419
metallicities, 6
metals, 52, 264
methane (CH_4), 7, 12, 23–24, 31, 36, 46, 136, 169–170, 203–204, 207–209, 224–225, 234, 236–239, 251–254, 257–259, 264, 270, 274–275, 277, 280–281, 283–284, 297, 299–300, 312–313, 316, 339, 343, 347, 352–355, 358, 360, 384, 388–389, 391, 393–394, 398, 410, 420
methyl radical (CH_3), 7, 275, 277, 281, 434
methylacetylene (CH_3C_2H), 275–277, 281–282, 288, 398
MHz, 10–11, 148, 399–401, 422
MICHI, 433
microsignature, 129, 143–145
microwave, 11, 36, 64, 267, 303, 330, 394, 421, 426, 433
Mid-Infrared E-ELT Imager and Spectrograph (METIS), 431, 433
Mid-IR Spectrometer and Imager (MIRSI), 433
migration, 7, 21, 24–26, 34–35
  type I, 21
  type II, 21
model, 8, 11, 13–14, 17–18, 20, 26–27, 30, 32, 35–36, 45, 51, 53, 55–56, 62, 65, 69–71, 74–76, 79–80, 82–87, 89–90, 93, 97, 99, 109, 113–116, 119, 128, 133, 138, 140, 151, 171, 177, 184, 203–211, 215, 218–219, 225, 232, 237–241, 244–246, 257–262, 264, 268, 270–272, 275, 279–281, 287, 295, 299, 303–304, 316, 320–326, 329, 342, 356, 363–365, 377, 379, 382, 385, 389, 391–392, 397, 399, 404, 406–412, 420–422
  Anderson and Schubert, 229
  anelastic, 320, 326, 404
  atmospheric dynamics, 365
  Aurnou, 90, 324
  barotropic, 362, 364
  Boussinesq, 85, 320, 324, 326
  CA, 33
  Cassidy and Johnson, 135, 135
  Cassini 3, 76–77
  Cassini 5, 47, 76–78
  clathrate hydrate, 15, 28
  core accretion, 2, 5–6, 14, 20, 34, 420
  Couette, 90–91
  Davis and Smith, 241
  disk accretion, 27
  disk instability, 2, 5, 7
  Dobrijevic, 275, 281
  Dodson-Robinson, 6, 20
  Enceladus, 207
  equilibrium cloud condensation, 32
  Fletcher, 305, 313
  forward, 130, 132, 226–228, 232–234, 236–238, 328
  French, 49, 80, 82, 85
  Friedson and Moses, 259–261, 278, 315–316, 355
  gas-giant planetary atmospheric dynamics, 361
  Gastine, 85–86, 326
  Grand Tack, 21
  Greathouse, 259
  Gudkova and Zharkov, 57
  Guerlet, 259–261, 287, 314
  Guillot, 56–58
  Hayashi, 5, 26
  isentropic interior, 47–48, 55, 58, 60–61
  jet formation, 320–321, 323, 330
  Jupiter, 59
  Jupiter's origin, 44
  Koskinen, 227
  Li and Ingersoll, 411
  magnetospheric, 113, 118–119, 179, 187, 240
  Moses and Greathouse, 278
  Moses and Greenhouse, 275

model (cont.)
Müller-Wodarg, 244
Nettelmann, 18, 48, 52, 56–58, 61, 79–80
Nice, 21, 59
periodicity, 113
photochemical, 275, 277–278, 280–282, 314–315, 318–319, 398–399
planetesimal, 26–28
quasi-geostrophic (QG), 310–311, 321–322, 362, 367
quasi-homogeneous Saturn, 56
radiative, 312–313
radiative seasonal, 259
radiative transfer, 33, 225, 234, 258, 330, 389, 399
radiative-convective dynamical, 259
radiative-convective equilibrium, 258, 356, 411
radiative-dynamical, 356
radiative-transfer, 314
satellite formation, 24–25
Saturn dynamo, 71, 83, 86–87, 89–90
Saturn formation, 421
Saturn homogeneous, 61, 71, 81
Saturn inhomogeneous, 61–62
Saturn's global circulation, 3, 212, 259
Saturn's interior, 2, 8, 17–18, 20, 30, 44–47, 50, 55, 58–60, 65
Saturn's magnetic field, 74
Saturn's origin, 2, 5–7, 12, 15, 20, 26, 34–35, 59
Saumon and Guillot, 47, 49, 56–57
Schneider and Liu, 322
shallow water, 321–322, 362–363, 367, 406–407, 411
Stevenson's, 71
Sugiyama, 411
thermal-wind balance, 327
thermochemical equilibrium, 266
thermosphere-ionosphere global circulation, 174
Vervack and Moses, 238
wavy magnetodisk, 117
molecular hydrogen aurora, 2, 85, 129, 166, 169–170, 202, 204–205, 218, 253
momentum equation, 297–298
MT2 filter, 297–298, 339, 342–343, 347–348, 352, 360, 383–386, 389
MT3 filter, 298, 339, 343, 347–348, 352, 383–386, 389
Multi Unit Spectroscopic Explorer (MUSE), 431

Navier-Stokes equation, 83, 240, 406–407
Near-InfraRed Spectrgraph Instrument (NIRSPEC), 203, 430, 434
nebular parameters, 6
neon, 8, 13, 45, 82
Neptune, 7, 14, 46, 49, 59, 99, 198, 202, 234, 251, 298, 322–323, 328–330, 338, 366–368, 426–427
Neptune, comparisons to, 13, 299
neutral particles, 2, 127, 134–135
neutral torus, 127, 134–135
New Frontiers, 420, 422, 424–425, 427
NH_3 ice cloud, 266, 283, 379, 384, 391, 403–405, 411
NH_3-H_2O (aqueous-ammonia) solution cloud, 11–12
NH_3-PH_3 photochemistry, 271–272
NH_4SH ice cloud, 11, 266, 305, 308, 377, 392, 404, 411
nitrogen (N_2), 11, 15–16, 19, 23–26, 134, 136, 224, 264, 266, 270–271, 296, 420
N_{MAX}, 197–199, 201–202, 210–216
noble gases, 3, 13–16, 27–29, 34–36, 65, 420–421, 423, 426
non-spherical particle, 283
non-zonal magnetic field, 70–71, 76–78, 85–90, 92
north polar hexagon, 3, 97, 108, 311, 329, 337, 339–347, 354, 358–361, 363–365, 368–370, 418–419, 421, 430, 436–437
North Polar Spot (NPS), 344
North Temperate Belt Disturbance (NTBD), 381
nucleated instability, 5

occultations, 110, 197, 199, 212–213, 224–234, 236, 238–239, 244–246, 301
radio, 174, 196–200, 202, 206, 210–216, 218–219, 244, 254–255, 300, 312, 316
solar, 3, 224–227, 229–233, 245
stellar, 3, 109–110, 224–229, 232–233, 237, 245, 254, 282, 312, 422
octupole (magnetic), 47, 69–70, 76, 79, 93
ohmic dissipation, 82, 85, 320
Oort cloud, 24
optical depth, 234–235, 284–285, 303, 357, 359, 391
optical depth unity, 52
optical energy, 401
ortho-H_2, 253, 273
Outer Planet General Circulation Model (OPGCM), 259
oxygen species, 282, 397

pancake distribution, 131
para-H_2, 251, 253, 264, 272–274, 287–288, 394
paramagnetic conversion, 273
particle drift, 104, 127, 129, 138–139, 139, 144, 180, 182, 211, 218
particle injections, 113, 140, 148, 182
pebble accretion, 21
Pedersen conductivity, 129, 174, 240, 243–244
perihelion, 251–252, 288, 358, 428
periodicities, 2, 55, 97–98, 101–102, 104, 106–107, 111, 113–117, 119–120, 182, 422
auroral, 106, 190
charged particle, 104–106
non-rotational, 111, 119
radio, 2, 98–99, 119, 190
ring, 106–107, 111
solar wind, 112–113
periodogram, 102, 105–107, 112, 115
phase diagram of H-He mixtures, 45, 48, 60–61
phase separation, 45, 48, 53, 59–62, 65, 80–82
phase space density (PSD) distribution, 138
phosphine (PH_3), 7, 12–14, 251–252, 264, 266, 268–272, 287, 305, 357, 390, 410, 420
photoabsorption cross sections, 207, 209, 211
photochemistry, 16, 211, 236, 258, 264, 266, 270–272, 274–275, 279, 281, 283, 315, 355–356, 397, 399
photoelectrons, 136, 207–208, 235
photoevaporation, 28–29
photoionization, 135, 166, 200, 204–205, 207–210, 213, 216
photolysis, 23, 134, 136, 225, 251–252, 268, 270–272, 274–275, 277, 279–281, 305, 355
photosphere, 13
Pioneer Mission, 46–47, 55, 69–72, 75–76, 79, 88, 100, 126, 197–198, 205, 211, 241, 252–253, 258, 283–284, 312, 337–338, 423
pitch angle, 129–130, 138–139, 183
pitch angle diffusion, 138–139, 182
planetary period oscillations, 97, 99, 102–104, 116–117, 119, 129, 185, 188, 190
auroral rotational modulation, 185
SKR, 104, 112, 116
twin-cell vortex, 243
Planetary Science Decadal Survey (PSDS), 418, 420, 430
planetocentric solar longitude, 252, 353
planetographic latitude, 198–200, 234, 237, 239, 253, 266, 298, 304, 377, 409
plasma sheet, 97, 105, 116, 127–128, 132–133, 140, 144, 147, 149
Plasma Spectrometer (CAPS), 129, 135, 183
plasmoids, 128, 145, 151
plumes, 23–24, 134–135, 143, 207, 252, 307, 319, 324, 393–394, 396–397, 404
polar cyclone, 254, 353–354, 357, 367–368
polar hood, 255, 352
polar vortex, 3, 255, 278, 299, 337, 339, 342–344, 347, 350, 352–353, 356, 359–360, 366–368, 418, 428, 430, 433
polarization, 100–101, 148, 190, 240, 243–244, 283, 358–360, 401
pole
north, 35, 257, 260, 299, 312–313, 323, 339–342, 347–349, 352–354, 356, 358, 360, 365, 369, 418
south, 134, 222, 231, 253–254, 260, 278, 299, 302, 312–315, 338–339, 344, 350, 353–356, 358, 366, 370, 419, 429
position vector, 407

potential temperature, 406
potential vorticity, 54, 72, 108, 239, 310–311, 329, 346–347, 364–365, 367–368, 397, 406, 409
power spectrum, 69–70, 77, 84, 86, 90–93, 104, 108
Prandtl number, 83, 86
Prandtl number (magnetic), 83, 89–90
primordial solar nebula, 5, 8, 16
probe, 7, 36, 45, 202, 246, 423–427, 438
  entry, 8, 11, 13, 33, 36, 65, 421, 424–425, 427
  Galileo, 6, 8, 11–13, 15–16, 26–28, 36, 45, 60, 224, 245, 420, 424–426
  Huygens, 224, 425–426
  RPWS/Langmuir, 130, 132, 419
  Saturn, 420, 424–426
  seismological, 109
  spectrum, 30
propane (C_3H_8), 266, 275–276, 280–282, 288, 397, 399
proto-atmosphere, 24
protoplanetary disk, 5, 20–21, 28, 33–34
protoplanets, 6, 29

quadrupole, 171, 426
quadrupole (magnetic), 47, 69–70, 75, 78–79, 93, 175, 272
Quasi-Biennial Oscillation (QBO), 255, 316–317, 329

RADAR, 11, 265, 267, 303–304, 385
Radau-Darwin approximation, 50–51
radial velocity (RV), 28, 431
radiated power, 377, 403, 412
radiation belts, 2, 126–130, 137, 149–150, 155–156
radiative
  accelerations, 13
  cooling, 12, 237, 252, 257, 259, 287, 314, 316, 353–354, 356, 396
  energy balance, 251, 255, 259, 353, 356, 369–370, 429
  heating, 52, 237, 252, 259, 287, 301, 314, 316, 320, 322, 353–354, 356
  timescales, 316, 322, 378, 410–411
radiative-convective boundary, 251, 253–255, 269, 273, 286–287
Radio and Plasma Wave Science (RPWS), 130, 143, 147–149, 169, 187, 200, 308, 383–384, 399–403
radio emissions, 3, 97, 99, 101, 111–112, 128, 168, 180, 298, 403
Radio Science Subsystem (IRIS-RSS), 8, 219, 225, 235, 243, 253, 255, 263–264, 273–274, 301, 312, 316
Rayleigh damping, 407
Rayleigh number, 83–84, 323, 325
Rayleigh scattering, 32, 283–284
Rayleigh-Kuo stability criterion, 346–347
Rayleigh-Taylor instability, 139
recombination, 137, 200, 202, 215, 244, 275
  dissociative, 205–206, 217, 244
  radiative, 205, 215
reconnection, 112, 128, 134, 137–138, 140, 145, 148, 150–154, 156, 167, 176–178, 180–181
recurrence interval, 411
reflectivity (I/F), 255, 265, 283, 286, 308, 347, 388–391, 428
refractive index, 197, 391, 397
refractory compounds, 32
relative vorticity, 239, 310–311, 344, 348–350, 364, 386, 409–410
resistive heating, 240, 243–244
resonances, 63, 107, 109–110, 130, 180
  cone, 149
  Jupiter-Saturn, 21
  Kozai, 34
Rhines scale, 321–322
Ribbon, 329, 364–365
Rice Convection Model, 114, 140
ring current, 73–76, 127–128, 148, 155, 177–179
ring rain, 189, 196, 203, 216–218, 236, 418–419, 429–430, 434
ring seismology, 45, 63–64, 109
ring shadow, 215, 230, 255, 259, 261, 275, 339, 378, 384
Rings of Saturn, i, 1–5, 20, 23, 25, 35, 59, 61–63, 97, 106–107, 109–110, 113, 127, 136, 143, 150, 155, 166, 185, 200, 202–203, 205–207, 216, 218–219, 225–226, 270, 282, 295, 303–304, 326, 352, 402–403, 417–419, 421, 423, 426–427, 429, 436
Rossby deformation radius, 321, 362, 364, 408
Rossby number, 297–298, 300, 319, 322, 325, 346, 366
Rossby wave, 239, 312, 319, 344, 346, 363, 368, 378, 397
rotation rate, 1–2, 4, 35, 44, 46, 50, 53, 55, 57–58, 69–70, 72–73, 77–79, 82–83, 86, 90, 97, 101, 104, 107–111, 119, 128, 176, 178, 200, 228, 239, 297–298, 300, 314, 344, 346, 348, 362, 419, 421
rotation vector, 406
runaway accretion, 21, 33

Saturn Electrostatic Discharges (SEDs), 3, 108, 200–202, 215–216, 218–219, 399–403
Saturn kilometric radiation (SKR), 54, 70, 75, 98–102, 104–107, 109–113, 115–116, 133, 145, 147–148, 155, 168, 175–180, 185–187, 190–191, 202, 215, 344, 403
Saturn Longitude System (SLS), 100–101, 106, 114, 119, 139, 185, 298
Saturn Orbital Insertion (SOI), 76–77, 92
Saturn Probe Mission (ESA), 417, 422–423, 425–427
Saturn Probe Mission (NASA), 417, 422–423, 425
Saturn seasonal variability, 155, 428
Saturn Thermosphere Ionosphere Model (STIM), 212, 240, 244–246
Saturn's aurorae, 2, 166, 173, 185, 189
Saturn's composition, 58, 420
Saturn's cooling history, 60, 65
Saturn's core, 7, 20, 35, 49, 57–58, 63, 80, 420
Saturn's evolution, 1–2, 5, 7, 20, 23, 26, 35, 44, 46, 52, 60–61, 65, 259, 418, 420–421, 431
Saturn's interior, 2–4, 7–8, 17, 20, 25, 44–45, 48–49, 52–53, 56, 60–61, 63–65, 69, 77, 79–81, 91, 97, 107–109, 111, 147, 298, 312, 320, 417, 419, 421
Saturn's ionosphere, 3, 75, 97, 102, 132, 185, 190, 196–197, 199–206, 208, 210–211, 215, 219, 236, 241, 243, 245, 266, 399–400, 419
Saturn's magnetic field, 2, 35, 46, 59, 61–62, 64, 69–73, 75–79, 81, 83, 90, 93, 119, 128, 142, 167, 419
Saturn's north polar region, 253, 288, 337, 339–341, 345, 350, 357, 360, 369, 419, 436
Saturn's origin, 5, 25, 119, 418, 420
Saturn's south polar region, 169, 255, 257, 341, 350, 357, 369
Saturn's thermosphere, 3, 136, 167, 172, 181, 196, 206, 212, 216, 224–229, 231–234, 236–245, 253, 261, 266, 419
scale height, 80, 82, 85, 210, 224, 226, 231, 233, 236–237, 254, 262–263, 269, 296, 299, 312–314, 316, 318, 320, 325, 364, 394, 396–397, 405, 425
seasonal evolution, 251, 255, 259, 312, 337
secondary electrons, 170–171, 207–208
secular change, 70, 77
secular variation, 69, 71, 79, 84–88, 90, 118, 145
sedimentation, 22, 57, 61, 252, 282–283, 355
seismology, 62–63, 65
Semi-Annual Oscillation (SSAO), 255, 260, 263, 315–317, 330, 429, 433
serpentinization, 23–24
shallow water (equations), 321, 407
silicates, 18, 20, 32–33, 264, 266
solar composition, 13, 15, 34, 79, 266, 300, 307, 410
solar nebula, 8, 15, 20, 24, 27, 44
solar wind, 14, 19, 74, 98, 100, 112–113, 119, 126–129, 131, 133, 137, 139, 144, 146–152, 154–156, 166–168, 174, 176–178, 180–181, 185, 187, 241, 418, 429
solstice, 149, 251–254, 257, 260–262, 276, 279, 282, 284, 287–288, 313, 337, 340, 353–354, 356, 358, 370
South Equatorial Belt Disturbance (SEBD), 378, 394
spacecraft event time (SCET), 400
spherical harmonics, 47
spiral wake, 21
spokes (in rings), 106–108, 436–437
stability (of zonal jets), 310–311, 329
stable layer, 69, 81, 86–91, 411
static stability, 62, 284, 346, 378, 396, 405–408
stellar equatorial plane, 34
stoichiometric hydrates, 26–27

storm body, 386
storm cell, 401–402
storm tail, 377, 381, 384, 393, 402, 408, 410
stratosphere, 3, 7, 12, 33, 188, 225, 237–238, 243, 251–253, 255, 257, 259, 262–264, 266, 268, 270–271, 273–275, 277–279, 281–283, 287, 295–296, 300–301, 307, 311–314, 316–319, 329, 339, 344, 352–360, 367–368, 377–378, 393–397, 406, 411–412, 428
stratospheric chemistry, 275
stratospheric circulation, 278–280, 287, 316, 329, 356
Stratospheric Observatory For Infrared Astronomy (SOFIA), 434
stratospheric vortex, 260, 278, 353–354, 356, 358, 370, 393, 395–397, 429
String of Pearls (SoP), 329, 383, 421
subcorotation, 117, 131–132, 187
subsidence, 261–262, 268, 273, 278–281, 288, 313–316, 319, 345, 354, 356–358, 367, 394, 396–397
superbolt, 403
suprathermal electrons, 128, 131–132, 207–208, 210, 245
synchrotron radiation, 138, 433

Taylor-Proudman theorem, 85, 319, 327
temperature asymmetry, 254–255
temperature gradient, 28, 33, 52–53, 60–61, 65, 210, 224, 243, 252, 254, 260, 262, 301, 306–307, 312–314, 316, 318, 320, 322, 353–355, 358, 366–367, 388
temperature inversion, 33, 258
temperature knee, 255, 286
temperature structure, 30, 36, 53, 80, 225, 233, 253, 255, 262–263, 286, 302, 330, 366, 368, 388, 392–395, 428
Texas Echelon cross Echelle Spectrograph (TEXES), 16, 433–434
thermal inertia, 253, 257, 259, 261, 354
thermal wind, 86, 90, 262, 314, 320, 327, 366, 382, 386, 396–397, 410
thermal wind equation, 257, 262, 316, 353–355, 367, 386
thermal wind relation, 252
thermochemical equilibrium, 264, 266, 268–269, 421
thermodynamic energy equation, 314
thermo-resistive instability, 33
thermosphere
  temperature, 172, 175, 188, 210, 230, 234–236, 239, 246
  winds, 240, 403
thermospheric composition
  CH_4, 225
  H, 225, 227, 236–237
  H_2O, 225, 236
  H_3, 225
  HD, 236
  He, 236
Thirty Meter Telescope (TMT), 431
thunderstorm, 239, 307, 323, 362, 399, 402–403
tidal dissipation, 25, 30
tidally circularized, 30
Titan, 16, 19, 23–25, 74, 114, 128, 133, 136, 138, 143–144, 156, 207, 209, 224, 239, 275, 282, 367, 369, 417–418, 424, 427
Titan Saturn System Mission (TSSM), 424
total electron content (TEC), 199
trapped particle distribution, 155
troilite (FeS), 15
tropopause, 253–255, 262, 264, 270–271, 273, 283, 286, 299, 312, 316, 319, 341, 354, 377–378, 383, 393, 395–397, 404, 407
troposphere, 1, 3, 7–8, 11–15, 32, 35–36, 240, 252–255, 257–259, 262–264, 266–273, 277, 282–287, 295–296, 299, 301–302, 305–306, 308, 311–317, 319, 329, 348, 352–355, 357–358, 367, 377, 386, 393–394, 396, 404, 410–412, 420–421, 428, 433
tropospheric chemistry, 8, 255, 258, 261, 266, 268–272, 283–284, 287–288
tropospheric haze, 252, 255, 271, 283–288, 337, 339, 347, 388, 392, 397

Ultraviolet Imaging Spectrograph (UVIS), 135, 170–171, 175–176, 179–180, 182–184, 186–187, 224–225, 227–232, 234–239, 244–245, 265, 339, 360, 370, 398
Ulysses, i, 54, 100, 107, 185, 338
updraft, 302–303, 305, 307, 378, 394–395, 403–404, 411
upper stratosphere auroral interaction, 169
upwelling, 259, 262–263, 268, 271, 274, 279, 281, 301–304, 307–308, 316, 345, 356–357, 393–394, 396, 421
Uranus, 13, 46, 49, 59, 61, 99, 198, 202, 206, 234–235, 251, 298, 322–323, 328–330, 338, 367, 426–428
Uranus, comparisons to, 13–14, 171, 299

Vasyliūnas cycle, 128
vertical instability, 405
vertical shear, 252, 255, 257, 262, 264, 354–355, 365, 367, 378, 386, 388
vertical velocity, 314, 317, 404
Very Large Array (VLA), 11, 266, 268, 433
Very Large Telescope (VLT), 393–395, 431, 434
video astronomy, 434–436
viscosity, 21–22, 53, 82–83, 90–91, 239, 306, 326
viscous disk, 34
VISIR instrument, 393, 433
Visual and Infrared Mapping Spectrometer (VIMS), 11, 169–170, 175, 181, 183, 186–187, 265–271, 282, 284–286, 299, 303, 339–341, 343–345, 347–349, 369–370, 382–384, 389–392, 397, 401, 422, 431
vortex, 75, 114–116, 257, 318, 337, 344, 350, 352–354, 358, 361, 364–365, 367–368, 381, 384–386, 393, 395–396, 402, 407–408, 410–411, 436
vorticity, 153, 180, 302, 310–311, 321, 324, 344–348, 350, 364, 366–368, 383, 396, 405, 408–411
  absolute, 346, 348
Voyager (1,2), i, 8, 23, 45–47, 50, 54–56, 58, 60, 69–72, 75, 79, 88, 100, 102, 104, 107–108, 110, 112, 118, 126, 147, 150, 196–198, 200–202, 205, 211–212, 215–216, 218, 225–232, 234–236, 238–239, 251, 253, 255, 258–259, 263–264, 273–274, 283–284, 287, 295–296, 298–299, 301–302, 306, 308, 310, 312, 316, 340–341, 343–345, 362–364, 368, 378–379, 382, 386, 399, 406, 422–423, 427, 429

WASP-17b, 30–31
WASP-43b, 7, 31–32
water ice, 3, 7, 23–27, 34, 134, 136, 136, 203, 266, 377, 384, 391–392, 397, 399, 403, 420
water vapor, 7, 12, 23, 27, 33, 114, 184, 207, 322, 410–411, 417, 434
water-rock reaction, 23
wave heating, 212, 239
wide field camera-3 (WFC3), 31
wind field, 262, 298, 316, 318, 328, 339, 344, 386, 404, 409, 417
wind shear, 198, 317, 342, 352, 366–367, 381–382, 388, 405–406, 408, 410
windex cloud, 32

xenon (Xe), 13, 15–16, 28, 32, 35
XXM-Newton, 430

zero-torque migration, 21
zonal flow, 65, 85, 88–89, 91, 109, 255, 306, 310–311, 317, 319–320, 324, 328, 430
zonal magnetic field, 71, 74, 76, 87–88, 91, 93
zonal wind, 108, 110, 118, 240–243, 246, 252, 255, 262, 264, 297–302, 305–307, 311–320, 323, 325, 327–328, 344–349, 353–354, 361, 366, 368, 378–379, 381–382, 386–387, 404–409, 411, 430